Low-Power Electronics Design

Computer Engineering Series

Series Editor: Vojin Oklobdzija

Low Power Electronics Design
Edited by Christian Piguet

Digital Image Sequence Processing, Compression, and Analysis
Edited by Todd R. Reed

Coding and Signal Processing for Magnetic Recording Systems
Edited by Bane Vasic and Erozan Kurtas

Low-Power Electronics Design

Edited by
Christian Piguet
CSEM&LAP-EPFL
Switzerland

CRC PRESS
Boca Raton London New York Washington, D.C.

Library of Congress Cataloging-in-Publication Data

Low-power electronics design / edited by Christian Piguet.
p. cm. — (Computer engineering ; 1)
ISBN 0-8493-1941-2
1. Low voltage integrated circuits—Design and construction. 2. Low voltage systems—Design and construction. I. Piguet, Christian. II. Title. III. Series: Computer engineering (CRC Press); 1

TK7874.66.L65 2004
621.381—dc22 2004045729

Direct all inquiries to CRC Press LLC, 2000 N.W. Corporate Blvd., Boca Raton, Florida 33431.

Visit the CRC Press Web site at www.crcpress.com

International Standard Book Number 0-8493-1941-2
Library of Congress Card Number 2004045729
Printed in the United States of America 1 2 3 4 5 6 7 8 9 0
Printed on acid-free paper

Preface

Purpose and Background

The goal of this book is to cover all the main aspects of low-power design of integrated circuits (ICs) in deep submicron technologies. Today, the power consumption of ICs is considered as one of the most important problems for high-performance chips as well as for portable devices. For the latter, it is due to the limited cell battery lifetime, while it is the chip cooling for the first case. As a result, for any chip design, power consumption should be taken into account very seriously.

Before 1993–1994, only speed and silicon area were important in the design of integrated circuits, and power consumption was not an issue. Just after, it was recognized that power consumption has to be taken into account as a main design parameter. Many papers and books were written to describe all the first design methodologies to save power limited to circuit design. Today, however, we have to cope with many new problems implied by very deep submicron technologies, systems on chips, embedded software, and the future of microelectronics. This book not only covers the aspects of all low-power designs, but it will also include new topics mainly related to future designs (i.e., nanotechnologies, optical chips, systems on chips, embedded software, and energy sources).

The design of more than 1 billion transistor chips, down to 0.10 μm and below, supplied at less than half a volt but working at some GHz, is a very challenging task, certainly considered as an impossible task only a few years ago. The design of such chips seem miraculous, but, as pointed out by Thomas Edison:

"Genius is one percent inspiration and ninety-nine percent perspiration."

Thomas Edison, *LIFE Magazine,* 1932

Organization

Many "low-power" techniques have been proposed during the last 10 years. Today, the power consumption of integrated circuits and of systems on chips is considered one of the most important problems for high-performance chips as well as for portable devices. Written by authors who are specialists in their respective fields, this book covers most of the low-power design techniques used today along with their results, as well as an interesting look into the future, including the serious problems due to the new deep submicron or nanometric technologies. As listed in the table of contents, the book contains seven sections:

1. Metal-oxide semiconductor (MOS) devices and technologies
2. Logic circuits
3. Processors and memories
4. Systems on chips
5. Embedded software
6. Computer-aided design (CAD) tools
7. Energy sources

In this way, many different domains and disciplines (i.e., technologies, circuit design, processors, complex circuits, software, CAD tools, and energy sources and management) are covered to give the reader a complete picture of what impacts power consumption.

This book also describes what many specialists think about the future, by presenting, for instance, nanotechnologies, optical circuits, ad hoc networks, e-textiles, and human powered sources of energy — all techniques that are promising but not really used today. The key benefits for the reader will not only be this complete picture of today's methods for reducing power, with some spectacular results, but also a serious look into the future to be fully aware of advances in chip design 10 or 15 years from now.

Locating Your Topic

Several avenues are available to access the desired information. A complete table of contents is presented at the front of the book. Each chapter is also preceded with an individual table of contents. Each contributed chapter contains comprehensive references including books, journal and magazine articles, and sometimes Web sites.

Acknowledgments

The value of this book is completely based on the many excellent contributions of experts. I am very grateful to them because they spent so much time writing excellent texts without any compensation. Their sole motivation was to provide readers and students who are interested in low-power design with excellent contributions about this topic. I sincerely thank all these authors. I am indebted to Professor Vojin G. Oklobjzija for asking me to edit this book and trusting me with this project. I sincerely thank Nora Konopka, Jamie Sigal, and Marsha Hecht of CRC Press for their excellent work in putting all this material in its present form.

Christian Piguet
Editor-in-Chief
CSEM SA
Jaquet-Droz 1
2000 Neuchâtel, Switzerland
Christian.piguet@csem.ch

Editor-in-Chief

Christian Piguet was born in Nyon, Switzerland, on January 18, 1951. He received M.S. and Ph.D. degrees in electrical engineering from the Ecole Polytechnique Fédérale de Lausanne, Switzerland, in 1974 and 1981, respectively.

Dr. Piguet joined the Centre Electronique Horloger S.A., Neuchâtel, Switzerland, in 1974. He worked on CMOS digital integrated circuits for the watch industry, on low-power embedded microprocessors, as well as on computer-aided design (CAD) tools based on a gate matrix approach. He is now head of the ultra-low-power sector at the Centre Suisse d'Electronique et de Microtechnique (CSEM) S.A., Neuchâtel, Switzerland. He is presently involved in the design and management of low-power and high-speed integrated circuits in CMOS technology. His main interests include the design of very low-power microprocessors and Digital Signal Processors (DSPs), low-power standard cell libraries, gated clock and low-power techniques, as well as asynchronous design.

He is a professor at the Ecole Polytechnique Fédérale Lausanne (EPFL), Switzerland. He also lectures on VLSI and microprocessor design at the University of Neuchâtel, Switzerland, and in the Advanced Learning and Research Institute (ALaRI) masters program at the University of Lugano, Switzerland. He is also a lecturer for many postgraduate courses in low-power design.

Christian Piguet holds about 32 patents in digital design, microprocessors, and watch systems. He has authored and coauthored more than 150 articles and chapters in various technical journals and books on low-power digital design. He has served as reviewer for many technical journals. He also served as guest editor for the July 1996 issue of *Journal of Solid-State Circuits* (JSSC). He is a member of the steering and program committees for numerous conferences and has served as program chairman of PATMOS '95 in Oldenburg, Germany, cochairman at FTFC '99 in Paris, Chairman of the ACiD 2001 Workshop in Neuchâtel, Switzerland, co-chair of VLSI-SOC 2001 in Montpellier, and co-chair of ISLPED 2002 in Monterey. He was chairman of the PATMOS executive committee during 2002. At DATE 2004, he was the low-power topic chair.

Contributors

Amit Agarwal
Purdue University
West Lafayette, Indiana

Amara Amara
ISEP
Paris, France

Claude Arm
CSEM
Neuchâtel, Switzerland

Daniel Auvergne
LIRMM, University of Montpellier
Montpellier, France

Nadine Azémard
LIRMM, University of Montpellier
Montpellier, France

Marc Belleville
CEA-LETI
Grenoble, France

Luca Benini
University of Bologna
Bologna, Italy

Davide Bertozzi
University of Bologna
Bologna, Italy

Didier Bloch
CEA-LETI-DIHS
Grenoble, France

Youcef Bouchebaba
University of Nantes
Nantes, France

Aimen Bouechhima
TIMA Laboratory
Grenoble, France

Erik Brockmeyer
IMEC
Leuven, Belgium

Francky Catthoor
IMEC
Leuven, Belgium
and
Katholiek University
Leuven, Belgium

Wander Cesario
TIMA Laboratory
Grenoble, France

Yuen Hui Chee
University of California-Berkeley
Berkeley, California

Lawrence T. Clark
Arizona State University
Tempe, Arizona

Fabien Coelho
Ecole des Mines
Paris, France

Stefan Cserveny
CSEM
Neuchâtel, Switzerland

Raphaël David
ENSSAT/University of Rennes
Lannion, France

Vivek De
Intel Labs
Santa Clara, California

Peter Dytrych
Philips Digital Systems Laboratories
Leuven, Belgium

Olivier Faynot
CEA-LETI
Grenoble, France

Antoni Ferré
UPC
Barcelona, Spain

Laurent Fesquet
TIMA Laboratory
Grenoble, France

Joan Figueras
UPC
Barcelona, Spain

Joao Fragoso
TIMA Laboratory
Grenoble, France

Jerry Frenkil
Sequence Design
Santa Clara, California

Frédéric Gaffiot
Ecole Centrale de Lyon
Lyon, France

Simone Gambini
Universita di Pisa
Pisa, Italy

Lovic Gauthier
Fleets
Fukuoka, Japan

Catherine H. Gebotys
University of Waterloo
Waterloo, Ontario, Canada

Cedric Ghez
IMEC
Leuven, Belgium

Dimitris Gizopoulos
University of Piraeus
Piraeus, Greece

Gert Goossens
Target Compilers Technologies
Leuven, Belgium

Domenik Helms
OFFIS
Oldenburg, Germany

Ed Huijbregts
Magma Design Automation
Eindhoven, The Netherlands

Koji Inoue
Fukuoka University
Fukuoka, Japan

Ahmed A. Jerraya
TIMA Laboratory
Grenoble, France

Pradeep K. Khosla
Carnegie Mellon University
Pittsburgh, Pennsylvania

Chris H. Kim
Purdue University
West Lafayette, Indiana

Ulrich Kremer
Rutgers University
Piscataway, New Jersey

Lars Kruse
Magma Design Automation
Eindhoven, The Netherlands

Chidamber Kulkarni
University of California-Berkeley
Berkeley, California

Dirk Lanneer
Philips Digital Systems Laboratories
Leuven, Belgium

Dake Liu
Department of Electrical Engineering
Linköping University
Linköping, Sweden

Richard Lu
University of California-Berkeley
Berkeley, California

Mark Lundstrom
Purdue University
West Lafayette, Indiana

Alberto Macii
Politecnico di Torino
Torino, Italy

Enrico Macii
Politecnico di Torino
Torino, Italy

Morteza Maleki
University of Southern California
Los Angeles, California

Diana Marculescu
Carnegie Mellon University
Pittsburgh, Pennsylvania

Radu Marculescu
Carnegie Mellon University
Pittsburgh, Pennsylvania

Jean-Marc Masgonty
CSEM
Neuchâtel, Switzerland

Philippe Maurine
LIRMM, University of Montpellier
Montpellier, France

Tycho van Meeuwen
IMEC
Leuven, Belgium

Renu Mehra
Synopsys Inc.
Mountain View, California

Giovanni De Micheli
Stanford University
Stanford, California

Miguel Miranda
IMEC
Leuven, Belgium

Vasily G. Moshnyaga
Fukuoka University
Fukuoka, Japan

Wolfgang Nebel
Oldenburg University
Oldenburg, Germany

Ian O'Connor
Ecole Centrale de Lyon
Lyon, France

Vojin G. Oklobdzija
University of California-Davis
Davis, California

Thierry J.-F. Omnès
Philips Semiconductors
Eindhoven, The Netherlands

Brian P. Otis
University of California-Berkeley
Berkeley, California

Barry Pangrle
Synopsys Inc.
Santa Clara, California

Joseph A. Paradiso
Massachusetts Institute of Technology
Cambridge, Massachusetts

Massoud Pedram
University of Southern California
Los Angeles, California

Pierre-David Pfister
CSEM
Neuchâtel, Switzerland

Christian Piguet
CSEM & LAP-EPFL
Neuchâtel, Switzerland

Sébastien Pillement
ENSSAT/University of Rennes
Lannion, France

Nathan M. Pletcher
University of California-Berkeley
Berkeley, California

Massimo Poncino
Universita di Verona
Verona, Italy

Jan M. Rabaey
University of California-Berkeley
Berkeley, California

Flavio Rampogna
CSEM
Neuchâtel, Switzerland

Marc Renaudin
TIMA Laboratory
Grenoble, France

Kaushik Roy
Purdue University
West Lafayette, Indiana

Philippe Royannez
Texas Instruments
Villeneuve Loubet, France

Mohammed Es Sahlienne
TIMA Laboratory
Grenoble, France

Eric Seelen
Magma Design Automation
Eindhoven, The Netherlands

Olivier Sentieys
ENSSAT/University of Rennes
Lannion, France

Kamel Slimani
TIMA Laboratory
Grenoble, France

Dimitrios Soudris
Democritus University of Thrace
Xanthi, Greece

Phillip Stanley-Marbell
Carnegie Mellon University
Pittsburgh, Pennsylvania

Thad E. Starner
Georgia Institute of Technology
Atlanta, Georgia

Philip Strenski
IBM Watson Research Center
Yorktown Heights, New York

Christer Svensson
Linköping University
Linköping, Sweden

Lars Svensson
Chalmers University
Göteborg, Sweden

Eric Tell
Linköping University
Linköping, Sweden

Arnaud Tisserand
INRIA LIP Arénaire
Lyon, France

Harry Veendrick
Philips Research Laboratories
Eindhoven, The Netherlands

Ingrid Verbauwhede
University of California–Los Angeles
Los Angeles, California

Eric A. Vittoz
CSEM
Neuchâtel, Switzerland

Patrick Volet
CSEM
Neuchâtel, Switzerland

Jing Wang
Purdue University
West Lafayette, Indiana

Terry Tao Ye
Stanford University
Stanford, California

Sungjoo Yoo
TIMA Laboratory
Grenoble, France

Jiren Yuan
Lund University
Lund, Sweden

Yervant Zorian
VirageLogic
Fremont, California

Victor Zyuban
IBM Watson Research Center
Yorktown Heights, New York

Contents

I Technologies and Devices

II Low-Power Circuits

V Embedded Software

VI CAD Tools for Low Power

VII Battery Cells, Sources of Energy, and Chip Cooling

I

Technologies and Devices

1

History of Low-Power Electronics

Christian Piguet
CSEM & LAP-EPFL

1.1 Introduction

Power consumption awareness began worldwide around 1990–1992. Before that, only niche markets required low-power integrated circuits (ICs). Today, every circuit has to face the power consumption issue, for both portable devices aiming at longer battery life and high-end circuits avoiding cooling packages and reliability issues that are too complex.

It was not anticipated that microprocessors would consume 100 watts today and perhaps 300 watts in 2016, as predicted by the International Technology Roadmap for Semiconductors (ITRS). Looking at a 10-year prediction proposed in 1986 (Figure 1.1), the frequency and throughput were accurate as well as the number of transistors based on Moore's law; however, no prediction was made for power consumption, while a simple calculation would yield 40 watts. If 40 watts had been predicted in 1986, it is conceivable that the awareness about the increase of microprocessor power would have been much better.

Roughly speaking, power was not an issue during the development of microelectronics from the invention of early computers in 1940 and of the transistor in 1947 to the early 1990s; however, many ideas proposed during this period for improving electronic circuits have been rediscovered and reused in the last 10 years, focusing on power consumption reduction [1]. This chapter begins with a brief history of early computers [2,3,4] to continue by the invention of the transistor and of the IC. Besides the mainstream microelectronics evolution, some aspects of low-power consumer applications are also described before the dramatic power increase during 1990–1992.

This first chapter is far from being an exhaustive history of low-power electronics. It contains some flashes on some well-known or ignored events and provides some considerations or interpretations

0-8493-1941-2/05/$0.00+$1.50

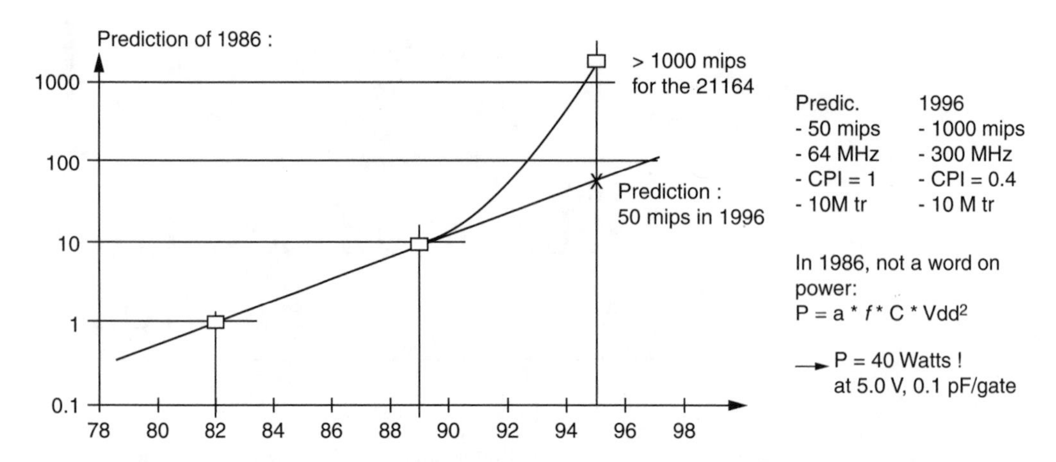

FIGURE 1.1 Predictions.

regarding low power. The history of techniques is not of general interest today, even if more and more companies try to tell their own histories within the context of the "good old days."

1.2 Early Computers

The first computer architecture, which we owe to Charles Babbage (1791–1871), was a mechanical analytical engine [5]. Although this machine was never completed, the estimated power required would certainly be very high. The first electronic computers or calculating machines were designed with vacuum tubes to significantly improve the speed over electromechanical machines.

1.2.1 Power Consumption of Early Computers

The ENIAC (1944) is generally considered to be the first electronic computer. It was programmed manually by using wires and connections between the execution units; therefore, it was very fast, achieving 100 kHz. Designed by Mauchly (1907–1980) and Eckert (1919–1995), it required 18,000 vacuum tubes and weighed 20 tons, and it was more of a huge calculator than a computer. The power consumption was 150,000 watts.

Such huge power consumption was not the highest achieved for a computer: the Whirlwind, designed by IBM in 1952 for the Semi-Automatic Ground Environment (SAGE) network (75,000 tubes, 275 tons), consumed 750,000 watts.

The introduction of transistors in the design of computers, although not really aiming at power reduction, nevertheless achieved a significant decrease of their power consumption. A transistor consumes roughly 1000 times less than a vacuum tube. Among the first transistorized computers, the TX0 designed by the Lincoln Laboratory in 1957 was an 18-bit machine containing 3500 transistors and consuming 1000 watts; however, huge mainframe computers still consumed a very large amount of power. For instance, the IBM 360 Model 91, announced in 1964, consumed a significant fraction of 1 MW [9]. The 12-bit PDP 8 minicomputer from Digital, designed in 1965, consumed 780 watts.

1.2.2 Reused Concepts for Low Power

Many concepts and ideas have been introduced for the design of early computers. Some of these ideas were rediscovered recently and used to reduce the power consumption of systems on chip (SoC). For instance, instruction formats of the early computers [6] were based on one-word instructions that could be read in one step or one clock cycle. This is much more energy efficient than the multi-byte instruction formats so common in Complex Instruction Set Computers (CISC) microprocessors. Multi-byte instruc-

FIGURE 1.2 Baby Computer of Manchester University, the world's first stored-program computer, running for the first time on June 21, 1948 (From the University of Manchester, Manchester, U.K. With permission.).

tions require several memory fetches as well as program counter updates that consume much power. It is quite interesting that the first computers were all designed with Reduced Instruction Set Computers (RISC)-like instruction sets.

Another "low-power" feature is the Harvard architecture, which comes from the name of the Harvard Mark I, designed in 1939 by Howard Aiken (1900–1973). This architecture is well-known today for providing two separate data and instruction memories, contrary to the "Von Neumann architecture" that contains only one memory (or a unified cache memory) for both instructions and data [7,8]. It results in a high sequencing of instruction execution (and a large number of clocks per instruction [CPI]), as successive instruction and operand fetches have to be performed. Today, it is well-known that this higher sequencing significantly increases power consumption.

The same applies for bit–serial architectures used for the first computers (e.g., EDVAC, Ferranti Mark I), due to the use of serial delay line memories. Consequently, many clock cycles were necessary to execute a single instruction [10]. A few years later, however, bit–parallel architectures (e.g., Von Neumann's IAS) featured a much simpler control unit and a reduced sequencing, although the execution unit was more complex. Such an observation is also valid for low power; a higher sequencing always results in larger power consumption.

Pipelined computers were introduced in the 1960s. For instance, the well-known IBM 360 Model 91, announced in 1964, was the first 360-pipelined computer with 20 stages. For scientific code, the number of clocks per instruction was about 1 (CPI = 1). Such a low CPI is very beneficial for reducing power consumption (see Section 1.5). Superscalar and parallel architectures were also introduced early in the history of computers, for instance, the 1964 CDC 6600, with 10 parallel execution units and the Illiac IV with 64 parallel processors. Parallelism is also beneficial for power consumption reduction (see Section 1.5).

1.3 Transistors and Integrated Circuits

The history of low-power electronics really starts with the invention of the bipolar transistor in late 1947. Compared to a vacuum tube, which consumes several watts, the transistor, with a range in the tens of milliwatts, is really a low-power device.

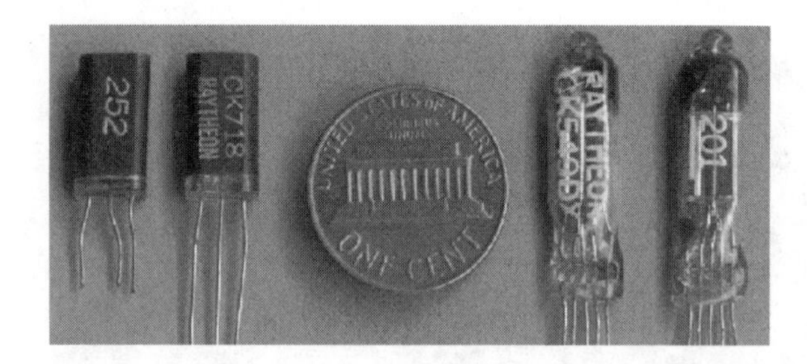

FIGURE 1.3 On the right are submini tubes used in a Zenith Royal hearing aid. The 201 date code represents week 1, 1952. On the left are examples of CK718 junction germanium transistors produced by Raytheon and used in the Zenith Royal "T" hearing aid, with 252 representing week 52, 1952. In less than 1 year, transistors had replaced the dominant vacuum tube technology in hearing aids. (From the Transistor Museum, Jack Ward, curator. With permission. [35])

1.3.1 Invention of the Transistor

When Bell Telephone Laboratories announced the invention [34] of the transistor on June 30, 1948, the general press was almost indifferent. The *New York Times* only published four short paragraphs on a back page of the paper. The inventors of the transistor were William Shockley, John Bardeen, and Walter H. Brattein. In 1956, they received the Nobel Prize for Physics for the invention of the transistor on December 23, 1947. The first working transistor was a germanium point-contact transistor; it was difficult to produce, not very reliable, and consisted of two wires pressed onto a small block of germanium. William Shockley proposed a junction or bipolar transistor as early as January 1948.

Even technical people were reluctant to recognize the benefits of the transistor. Its direct competitor was the vacuum tube, a strongly established commercial product; but transistors had very long lives, were small, and required no filament current. Despite all these advantages, however, Bell Labs decided to license it freely, and publicized it extensively in seminars and papers [14]. Imagine a free license for the transistor. This means that nobody really understood what this device was.

Since the beginning of the century, more studies have been devoted to solid-state physics, metals, and semiconductors. In 1929, a patent on a metal-oxide semiconductor (MOS)-like transistor was issued to Julius Lillienfeld (i.e., insulated material such as glass coated with a metal film having unidirectional conductivity); however, this device was impossible to fabricate with the available materials, and the world was in the midst of the Great Depression. In 1935, a patent was issued to Oscar Heil for a field-effect triode, although he was not able to explain how it worked. It is ironic that the concept of field-effect transistors, so marvelously simple, provides practical implementations after the invention of the far more complex bipolar transistor [14]. The latter was just much simpler to fabricate.

The first market pull came from telephone switching equipment and military computers, but also from hearing aids and portable radios, for which miniaturization and low power were a must. Compared to vacuum tubes, the power could be reduced by a factor of 1000; therefore, the introduction of the transistor was really a significant first step to lower power devices. Sonotone announced the first transistorized hearing aid in February 1953; it contained five transistors, but still required a pair of miniaturized tubes for the input and driver stages. Figure 1.3 and Figure 1.5 show such transistors and mini tubes. It is likely that power consumption was already an issue for a product like a hearing aid.

The second transistorized product to be introduced was a portable four-transistor radio. This is why people simply call "transistors" commercial transistorized radios, for instance, those produced by RCA and Sony. It was really the first consumer market for transistors. Figure 1.4 shows such a "radio" transistor.

The first bipolar silicon transistors were introduced in 1954 by Texas Instruments. One million transistors were produced in 1953, 3.5 million in 1955, and 29 million in 1957. Companies producing these transistors were Raytheon, Western Electric, RCA, Philco, General Electric, Texas Instruments (TI), and Fairchild. The average price was about $4 per transistor.

FIGURE 1.4 RCA introduced the 2N109 in 1955 (Germanium PNP Alloy Junction). It was an affordable and reliable germanium audio transistor used in many transistorized radios. In 1956, the 2N109 cost a little over $2 and had dropped to approximately $1 by the early 1960s (From the Transistor Museum, Jack Ward, curator. With permission. [35]).

FIGURE 1.5 CK722 is one of the well-known transistors from the 1950s and 1960s. Raytheon introduced this device in early 1953. The CK722 was the first mass-produced germanium alloy junction transistor. Raytheon was the major manufacturer of hearing aid transistors, and those units that were not quite "good enough" for the demanding hearing aid market (i.e., not enough gain or too noisy) were sold to hobbyists as the CK722 (From the Transistor Museum, Jack Ward, curator. With permission. [35]).

Finally, the significance of the invention and the introduction of the transistor became larger year after year as more transistors and integrated circuits were embedded in equipment and devices.

1.3.2 Invention of the IC

The second major step in low-power electronics history was the invention of the IC because on-chip interconnects consume much less power than off-chip connections. The IC was invented in 1958 [15] by Jack Kilby of TI, who won the 2000 Nobel Prize, and Robert Noyce, of Fairchild, and then co-founder of Intel. In October 1958, Kilby began the design of a flip-flop on a monolithic germanium chip. The device was completed in early 1959 and was revealed "as the most significant development by Texas Instruments since … the commercial silicon transistor." The announcement was widely reported in the press, but engineers were skeptical about the devices regarding optimization and yield [14]. By February 1960, Noyce, at Fairchild, announced that his company was manufacturing resistor–transistor logic (RTL) as an integrated circuit family named Micrologic. According to a recent book about this story [15], we completely ignore these famous people today: "Do you have a PC? "Yes, I do." "Do you know who Jack Kilby is?" "No, I don't."

Even before the introduction of metal oxide semiconductor (MOS) and complementary metal oxide semiconductor (CMOS) technologies, several low-power principles were understood in the 1960s [28], such as the reduction of the supply voltage, the use of analog circuits replacing digital ones, and the design of fast and parallel circuits allowing a supply voltage reduction just satisfying to the speed constraints. A few years later, these ideas would be applied to CMOS design.

1.3.3 MOS Transistors

In 1958, the first field-effect transistor became operational. Its creator, Stanislas Teszner, working in France, called it "Tecnitron." Even before Teszner, many studies about the possibilities of such a device

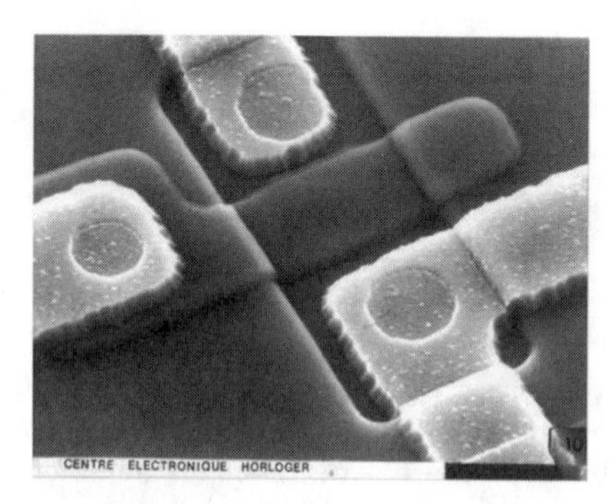

FIGURE 1.6 MOS transistor (1972).

were under way in the U.S. In 1959, RCA was already working on fluid-effect transistors (FETs) to implement logic circuits; but Dr. John T. Wallmark of the RCA Laboratories, although he had a patent, never achieved success. Two years later, Paul Weimer, another researcher from RCA, succeeded in obtaining a working FET. In 1962, RCA was fabricating multipurpose logic blocks comprising 16 MOSFETs on a single chip. By 1963, RCA had fabricated large arrays of several hundred MOS devices; however, these devices were extremely sensitive to static charge and oxide defects, and they were slower. In mid-1965, only two companies were producing MOS ICs: General Microelectronics and General Instruments. The other companies were simply waiting. Fairchild offered a 64-bit random-access memory (RAM) MOS memory in 1967, but even the first electronic watch IC was designed with bipolar transistors in 1967 [16].

Nevertheless, the move toward MOS technology was on its way. It was mainly for packing more transistors on a single chip and not for power consumption considerations. Intel developed the first microprocessor in 1971 and the first MOS EPROM memories in 1972. This first microprocessor was the famous 4004 in P-channel silicon-gate technology (Figure 1.7). The clock was 750 kHz, with an average CPI of 10 (10 clocks per instruction on average). With 0.075 MIPS (million instructions per second) performance, it consumed 0.3 watts, resulting in 0.25 MIPS /watt. (Today, 8-bit microcontrollers reach several tens of thousands of MIPS/watt.)

N-channel technology was faster than P-channel due to mobility, however. The first VLSI textbooks, such as the famous Mead/Conway [17], presented only N-MOS circuits, which were considered a dominant technology that would be used for a very long time. Figure 1.6 is an MOS transistor from 1972 in 6 μm technology. CMOS technology was, at that time, only used for special consumer markets, such as electronic watches in the early 1970s [16]; however, 10 years later, the move toward CMOS was achieved due to heat dissipation, electromigration, and reliability problems. In low-power electronics history, the introduction of the CMOS technology was the third major step after the invention of the transistor and the IC. Today, CMOS technology is clearly the dominant technology, but the CMOS power increase in the 1990s had simply followed the bipolar transistor power increase with a 10-year time shift [30].

1.3.4 Early Microprocessors

After the 4004, the technology pace could be measured by the new microprocessors introduced year after year (Figure 1.7). In 1974, the N-channel technology was chosen by Intel to produce the 8080, which was 10 times faster than the software-compatible 8008, designed by Masatoshi Shima, who later designed the Z80. The Intel 8080 was roughly equivalent to the mainframes of the 1950s. The CMOS technology was used for the RCA 1802. In 1974, Texas introduced its 4-bit TMS 1000. Motorola kept a low profile and introduced its first 8-bit microprocessor in 1974, the MC 6800. Rockwell proposed the 8-bit PPS-8 and 4-bit PPS-4 microprocessors. National Semiconductor presented a 16-bit machine called PACE, while Signetics proposed its 2650 8-bit microprocessor. By Fall 1975, almost 40 different microprocessors crowded the market. The addition time was 10.8 μsec for the 4004, 20 μsec for the 8008, 2 μsec for the 8080, 1.3 μsec for the 8085, 0.375 μsec for the 8086, 0.25 μsec for the 80296, and 0.125 μsec for the

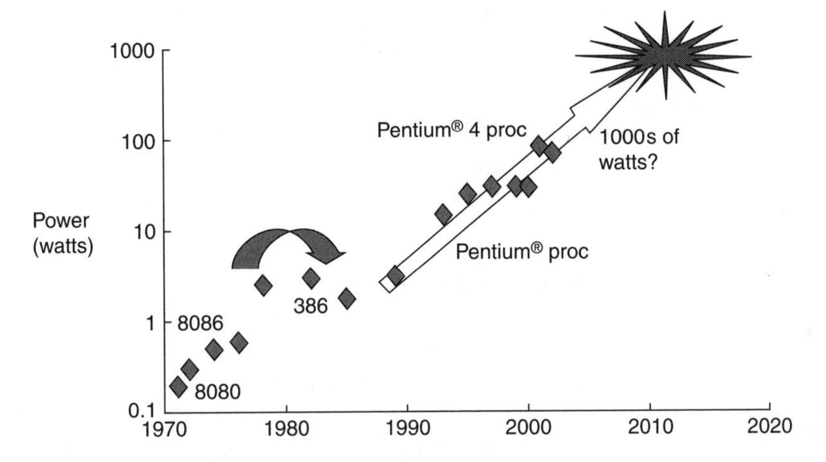

FIGURE 1.7 Power dissipation of Intel microprocessors (Courtesy of Intel).

TABLE 1.1 First Intel Microprocessors [18]

Year	μP	Technology	Nb of MOS	Address
1971	4,004	P-MOS 8 μm	2,300	4K
1971	8,008	P-MOS 8 μm	3,500	16K
1974	8,080	N-MOS 6 μm	5,000	64K
1976	8,085	N-MOS 4 μm	6,000	64K
1978	8,086	N-MOS 3 μm	29,000	1M
1982	80,286	N-MOS 2.3 μm	130,000	16M
1985	80,386	CMOS 2 μm	275,000	4G

TABLE 1.2 First Motorola Microprocessors [19]

Year	μP	Nb of MOS	Technology	Frequency
1974	6,800	5,000	N-MOS 6 μm	2 MHz
1979	68,000	68,000	N-MOS 4 μm	8 MHz
1984	68,020	200,000	CMOS 2 μm	16 MHz
1987	68,030	275,000	CMOS 1.3 μm	20 MHz
1989	68,040	2,000,000	CMOS 0.8 μm	25 MHz

80386. Table 1.1 presents some data about the first Intel microprocessors. It is interesting to note that CMOS was introduced in Intel microprocessors in 1985. Table 1.2 is a similar presentation for Motorola microprocessors, with a shift to CMOS in 1984.

1.3.5 RISC Machines

In 1981, two opposite approaches to designing future computers were possible [20]:

1. Continue the mainstream trend to design increasingly complex CISC machines.
2. Take the opposite direction and build simpler processors (i.e., RISC machines).

The alternative to complexity was obviously simplicity: less instructions, instructions executed in one clock cycle, load/store architectures, and hardware control units [20]. The 1975 IBM 801 is the first RISC machine, but the RISC concept was made popular by the RISC I and RISC II architectures from the University of California-Berkeley in 1980. The concept of RISC machines, however, rediscovered from Harvard Mark I and EDVAC early computers, is very beneficial for reducing power consumption.

The dramatic increase in the number of transistors per chip, as well as architectural advances, including the use of RISC ideas, pipelining, and caches, have produced an improvement in the performance of microprocessors at a rate of 1.5 to 2 times per year between 1980 and 1990 [21]. At that time, however, power consumption was still not an issue, the supply voltage was 5.0 volts, and no one predicted any change, as confirmed by Moore's law [22].

> "It's a 5 volt world, and to change to 1.5 volt would mean that the whole world would have to change!"
>
> **Gordon Moore**

1.4 Low-Power Consumer Electronics

Until 1990, power consumption was not an issue for a huge majority of ICs. Only a few niche applications had to take care of power, such as electronic watches, hearing aids, pacemakers, pocket calculators, pagers, and some battery-less applications. It was already for "portable" products, but only wristwatches, hearing aids, and pacemakers were, at that time, considered "portable." The history of each of these niche products would be quite interesting, but, for the most part, no documents are available. Recently, a book was published about the first electronic watch designed in Switzerland [23].

1.4.1 First Electronic Wristwatch

The Horological Electronics Center or Centre Electronique Horloger (CEH) developed the first electronic watch, a Swiss quartz watch named Beta [16, 23]. Such research was performed at that time without any public support, but pushed by some visionary people. Fortunately, this research was quite successful, producing the first quartz electronic watch. Commercialization was quite difficult, however, due to the structure of the Swiss watch industry. The impact of this research went far beyond the watch industry because it opened the way to very low-power ICs and microprocessors developed by CEH and, from 1984, by CSEM.

In the late 1950s, the Swiss watch industry was very successful in the production of mechanical watches. The president of the Swiss Horological Federation, Gérard Bauer, a visionary and powerful president, however, was under the impression that electronics could be a source of trouble for the watch industry. He was not an engineer, and he had been Swiss ambassador in Paris, so he was not able to support his contention with technical evidence, but his vision was that electronics or microelectronics would be a very dynamic discipline, producing many inventions and new products. He succeeded in convincing the Swiss watch industry to create a new laboratory known as CEH. CEH was officially created in Neuchâtel, Switzerland, on January 30, 1962, with the mission "to develop an electronic wristwatch with at least one advantage over existing watches." It was clear that few watchmakers believed that electronic watches were a major threat. It was decided that the CEH scientific results should be kept secret to prevent one company from using them before the other companies.

A silicon process was clearly required. Kurt Hübner, coming from Shockley Semiconductors (inventor of the transistor), was hired to set up a microelectronic technology. In 1963, the bipolar technology was more reliable than the MOS technology, and the first CEH circuits were bipolar circuits (Figure 1.8). It was the first silicon process installed in Switzerland. The goal was to have 1.0 to 1.3 volts and a 10-μA power consumption for the complete circuit. CMOS technology was chosen by CEH only a few years later — 10 μm in 1964 and 6 μm in 1966 — largely before the main semiconductor companies, which only switched to CMOS around 1984 (Table 1.1 and Table 1.2).

Under the direction of Max Forrer, several design projects were also defined, such as a 8.2-kHz quartz resonator (considered as a very risky project) quartz oscillators, frequency dividers in bipolar technology (consuming a few microwatts at 1.3 volts), and a vibrating motor at 256 Hz. So, from 8192 Hz, only five frequency dividers by 2 were necessary.

The first electronic watch was presented to the CEH board members in August 1967. Because all the research projects had been carried out secretly, it was a major shock for the Swiss watchmakers. A seminar

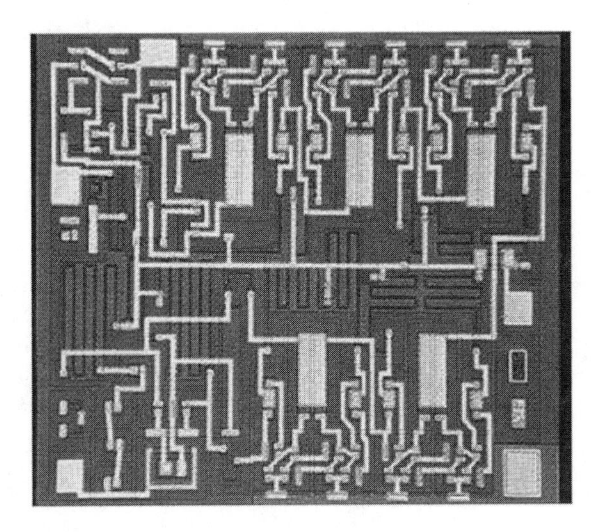

FIGURE 1.8 The bipolar circuit of the Beta 21.

was organized for CEH shareholders in December 1967, and this was the beginning of CEH's very strong reputation in low power.

The Beta wristwatch was able to reach a 1-year autonomy with the chosen battery. There were several bipolar chips: a 8192-Hz quartz oscillator and other circuits consisting of five frequency dividers to provide a 256-Hz signal used for the mechanical motor and the hands. The technology was a 10 μm bipolar process. Each divider by 2 consumed approximately 1 μA. The frequency divider was a flip-flop designed by Eric Vittoz with four NPN transistors and some integrated resistors and capacitors. Two frequency dividers (12 elements) were integrated on the same die of 2.1 mm^2. The motor control circuit was also a digital circuit, producing the right motor pulses. The total chip power consumption was 15 to 30 μA at 1.3 volts, and the chosen battery cell was supposed to deliver 18 μA during 1 year; therefore, the lifetime of 1 year was satisfied. Figure 1.9 is the Beta 2, with the electromechanical part on the left of the photograph and the printed circuit with the IC and the quartz crystal on the right.

FIGURE 1.9 The first Beta electronic watch.

Ten Beta electronic watches were presented to the "Observatoire de Neuchâtel" in December 1967. The result was 10 CEH watches at the first 10 places with only a few tenths of a second offset per day, followed by four other quartz watches from Seiko, Japan. This competition was suspended the next year. The improvement was a factor of 10 in precision compared with the mechanical wristwatches presented during the prior year.

1.4.2 Electronic Watches in Japan [24]

In the development of a quartz wristwatch, Seiko was close to CEH. Seiko began to look at quartz timekeeping in 1958 with the development of a quartz crystal clock. In 1959, Seiko was developing a quartz watch. Obviously, they had to reduce the size of a quartz-based clock to that of a quartz wristwatch. The result of this project was the world's first analog quartz watch to reach the market, the Seiko 35SQ Astron, introduced at Christmas 1969. The next Seiko, model 36SQC, was introduced in 1970; it was the first quartz watch to use a CMOS chip. Seiko and other Japanese watch companies, such as Citizen and Casio, quickly and successfully switched to electronics. As a result, Japan took the lead in worldwide watch production in 1978.

1.4.3 Electronic Watches in the U.S. [14]

The first electronic watch from the U.S. was announced in 1970 and introduced to the market by Hamilton in Fall 1971: a digital model called Pulsar. A button was necessary to display the time using light-emitting diodes (LEDs). The company bought the chip from RCA, which was the first U.S. company to produce CMOS chips.

Nevertheless, other U.S. companies were thinking of developing CMOS watch chips. For instance, Motorola offered the first integrated electronic watch kit to manufacturers in early 1972. The CMOS circuit, the quartz crystal, and a miniature microwatt motor were offered for only $15, thus beginning a move toward very cheap electronic watches.

In 1974, National Semiconductor introduced six watches. American Microsystems came out with a digital watch module that was smaller and consumed less power. By February 1975, about 40 companies were offering electronic watches with digital displays. Industry experts believed that solid-state digital watches would soon occupy a large part of the market, if not all of it.

In 1976, TI introduced its plastic-cased, five-function LED watch that sold for $19.95. As a result, National Semiconductor reduced its watch prices. Six months later, watchmakers were predicting a $9.95 digital watch would be on the market by Christmas. As liquid-crystal display (LCD) prices dropped, TI introduced a watch with liquid-crystal hands in August 1976.

By 1977, the price of digital watches had fallen from more than $100 to less than $10, in just two years. Profits evaporated. As with calculators, in 1977 only three real survivors remained: again, two Japanese competitors, Casio and Seiko, and TI [25]. Twenty years later, Intel chairman Gordon Moore was still wearing his ancient Microna watch ("My $30 million watch," he called it) to remind him of that lesson.

The U.S. watch market rise and fall was largely due to a brilliant but dangerous strategy by TI. By so dramatically reducing the TI electronic watch prices, they succeeded in eliminating all their U.S. competitors. Ultimately, however, TI found itself with watch prices that had been driven so low that the company did not make any profit. Worse than that, Japan's Seiko and Casio, with relatively low labor costs at that time, were fierce competitors. These firms soon did to TI what TI had done to its American competitors [25].

1.5 The Dramatic Increase in Power

Until 1990, CMOS integrated circuits, showing only moderate power consumption, were not designed with a serious interest in saving power [27, 28]. In 1992, however, the first very fast microprocessor, Alpha, which ran at 200 MHz, consumed about 30 watts [26]. It was a big shock for the semiconductor industry to see that power consumption was higher than expected. The second issue was the market growth of portable devices beyond the classical wristwatches, pacemakers, and pocket calculators (i.e., personal digital assistants (PDA), cellular phones, global positioning system (GPS) receivers, and palmtop computers and notebooks). New types of applications, such as ad hoc networks, did require extremely low power consumption for each network node. The reduction of power consumption was also in line with the global awareness of environmental issues.

Due to the exponential microprocessor frequency increase in the early 1990s, the increase of power dissipation, if it did not result in panic, was suddenly a major issue. A third low-power constraint was suddenly added to the well-known speed and silicon area constraints in the design of integrated circuits. It was also a major issue for chip cooling techniques [30]. In the early 1990s, therefore, engineers and managers were eager to learn how to reduce power consumption. Consequently, any course or conference labeled "low-power" was sure to be successful. Low-power postgraduate courses, such as Mead Education courses [29], attracted many participants in the early 1990s, perhaps more in the U.S. than in Europe.

1.5.1 Low-Power Workshops

The first "Low-Power" conferences and workshops were organized around 1993. Before the famous ISLPED (International Symposium on Low Power Electronics and Systems), some U.S. workshops were organized, such as the 1993 Low-Power Electronics Conference in Arizona and the 1994 Workshop on Low-Power Design in Napa, CA. These workshops were merged to create the first ISLPED conference that was held in 1995 at Dana Point, and is now regularly organized each year. The conference was held for the first time in Europe in 2000 (Rapallo, Italy), and was held for the first time in Asia (Seoul, Korea) in 2003. In 2001 and 2002, ISLPED was held at Huntington Beach, CA, and Monterey, CA, respectively.

Interestingly enough, it was in Europe that the first low-power workshops appeared, such as the PATMOS (Power and Timing Modeling Optimization and Simulation) project, followed by PATMOS workshops organized in 1993 in Montpellier, France, and in 1994 in Barcelona, Spain, with some attendees from the U.S. The PATMOS conference was originally a European project about timing and power modeling (1990–1993); however, it was decided to continue the organization of annual meetings on timing with more information on low-power issues. The PATMOS conference was then organized in 1995 at Oldenburg, Germany, Bologna, Italy, Louvain la Neuve, Belgium, Lyngby, Denmark, Kos Island, Greece, Göttingen, Germany, Yverdon, Switzerland, and Sevilla, Spain. It was organized in Torino, Italy, in 2003, in Santorini Island, Greece, in 2004, and will be in Leuven, Belgium, in 2005.

In 1999, the IEEE Alessandro Volta Memorial Workshop on Low-Power Design (VOLTA '99) was organized in Como, Italy. It was dedicated to low power as well as to recall that Volta invented the electric battery 200 years earlier in that town. A French-speaking conference, called Faible Tension Faible Consommation (FTFC), is also organized in Paris every 2 years since 1997, and the fourth edition was organized in May 2003.

One-day low-power workshops were also organized by Dimes Delft University, Netherlands, within the framework of the ESD-LPD (Electronic System Design — Low-Power Design) from 1997 to 2001. These 1-day workshops with invited speakers were usually held on the day before or after PATMOS, VOLTA, and ISLPED conferences. Some aspects of the projects of this European Low-Power Initiative are described in three books [31,32,33].

1.5.2 Low-Power Design Techniques

Many of the techniques described in the first low-power conferences were not really new ideas or concepts, but often the reuse of old techniques for achieving low power, such as asynchronous, pipelined or parallel machines (see Section 1.2), state assignment of finite state machines, reduced swing, and transistor sizing.

Pipelining and parallelism were proposed [11,13] to reduce power consumption by increasing the throughput of logic blocks and processors to reduce frequency and supply voltage. Pipelining is used today to reduce power consumption, as illustrated in Figure 1.10. A pipelined execution unit presents a shorter

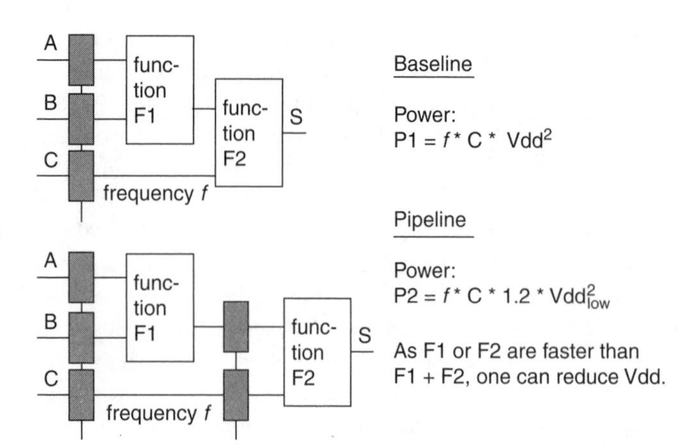

FIGURE 1.10 Pipeline for low power.

stage delay than a nonpipelined execution unit. It is therefore possible to work at the same operating frequency while reducing the supply voltage [11,13]. A lower Vdd helps to save a lot of dynamic power.

Parallelism has also been proposed [11] to lower frequency and supply voltage while maintaining the same throughput. Examples are parallel datapaths, memories (Figure 1.11), and shift registers (Figure 1.12). Figure 1.11 presents, for instance, interleaved memories accessed in an overlapped fashion at the

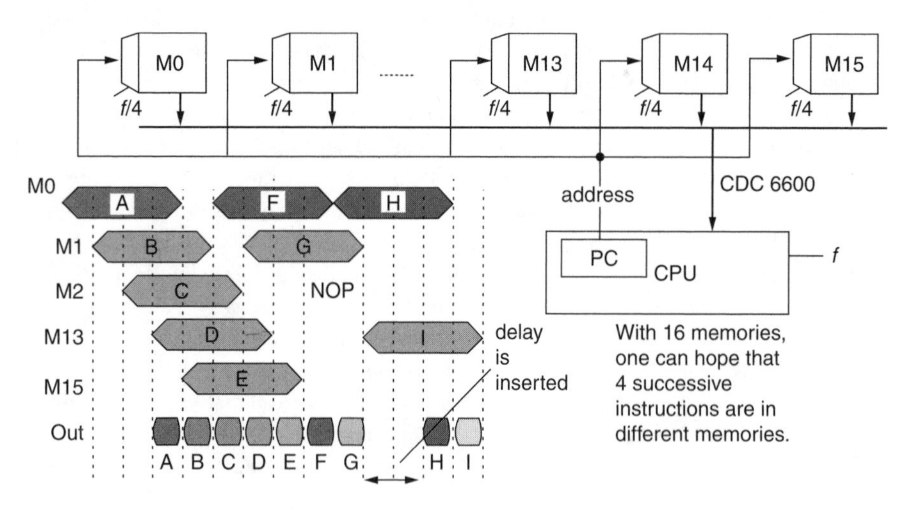

FIGURE 1.11 Interleaved memories.

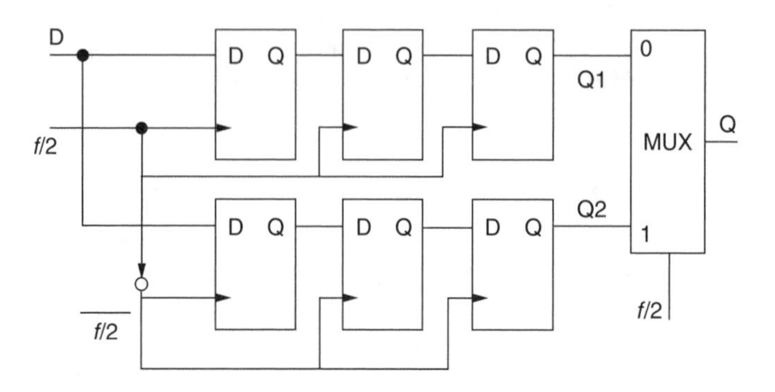

FIGURE 1.12 Parallel shift register.

frequency f/N. Each of the N memories (size 1/N) provides one instruction executed at the frequency f. The supply voltage of the memory modules can therefore be reduced because the access time is N times larger; however, this architecture comes from the early CDC 6600 computer [6].

The same idea has been proposed for parallel shift registers, as presented in Figure 1.12 [13]. The input is successively provided to the upper or to the lower half shift register at a reduced frequency, while the output multiplexer restores the output at the frequency f. The total number of D-flip-flops is the same as in the nonparallelized shift register. Parallelized shift registers with the same throughput present a reduced power consumption according to $P = f/2*C*Vdd^2$. Furthermore, as flip-flops are working at a reduced frequency (f/2), the supply voltage can be reduced to save dynamic power. This idea was first proposed for bubble memories with major and minor loops [12].

Asynchronous or self-timed architectures have been proposed to reduce power consumption by removing the clock tree known to be a large consumer. Today, some asynchronous microprocessors are available, such as the Amulet, Titac, MiniMIPS, ASPRO, or Philips 8051. John Von Neumann proposed this idea for the first time for the Institute of Advanced Studies (IAS) computer designed at Princeton University. The IAS execution units worked at their own speed, and each unit had to send a completion signal to indicate when it had finished.

Some new but obvious ideas were also proposed, such as gated clock and activity reduction, which are applied today to a majority of ICs. Adiabatic logic was also proposed, which was an old idea rediscovered and adapted to modern technologies. Although supply voltage reduction had already been proposed in the 1960s [28], in the early 1990s, it was considered a major shift in the way circuits had to be designed. The trend to very low supply voltages down 0.2 volt is still a major issue. New schemes proposed include dynamic voltage scaling (i.e., supply voltage variation depending on the application load).

In 2003, the major themes in low-power conferences deal with the increasing complexity of SoCs in very deep submicron technologies. This complexity has to be considered at all design levels because most of the power can be saved at the highest levels. At the system level, which is strongly application dependent, designers have to consider many design parameters, such as partition, activity, number of executed steps, simplicity, data representation, locality, cache memories, and distributed or centralized memories. Furthermore, many new dramatic issues result from the use of very deep submicron technologies, such as leakage, variable V_T, very low supply voltages, interconnect delays, networks on chip, cross talk, and soft errors, and require very innovative design techniques. Today, leakage in standby and active modes is certainly a main issue in very deep submicron technologies, pioneered by some authors in Japan [36,37] and now studied all over the world.

1.6 Conclusion

The design of nearly 1 billion transistor chips, down to 0.10 μm and below, and supplied at less than 1 volt but working at several GHz, is a very challenging task. It was certainly considered an impossible task a few years ago.

The microelectronics revolution is fascinating: the transistor was invented only 55 years ago, and today we are using 130 and 90 nanometers technologies. For 2016, the 2001 SIA Roadmap predicts a 0.022-μm CMOS process (probably silicon-on-insulator (SOI)) with 16 billion transistors for high-performance chips, with 0.4 volts, 288 watts, and 28 GHz as the local frequency. As a result, more transistors exist in the world today (10^{17}) than ants (10^{16}).

What is the future of microelectronics? Are we close to the end of this marvelous story? Does the future belong to nanotechnologies that could completely replace microelectronics (although 0.015 μm transistors are 15 nanometers long)? Nanodevices have been constructed, and they are capable of switching a current or single electrons with a ratio between the on/off current of 1000 to 1 million. Such elements could be promising because their size of a few nanometers and their extremely low power consumption are very attractive. Carbon nanotubes, quantum dots, single-electron devices, or molecular switches are the most promising nanodevices. For instance, depending on its diameter, a carbon nanotube is a semiconductor device (otherwise, it is a conductor, and is not usable as a switch). If one of 10 nanotubes is a semiconductor, however, how do we select and interconnect the semiconducting ones to provide a useful logic function?

Quantum dots are based on the Coulomb blockade effect, and electrons are moved one by one from dot to dot. They have been constructed atom by atom by atomic force microscopes. Due to noise, it is better to construct cellular automata with several dots, and to define a given state of the automata as the logic "0" and another state as "1." Majority gates have been demonstrated as well as AND/OR gates. The main problem is still how to interconnect these gates to provide useful functions. Furthermore, it is hard to construct a complete chip atom by atom with several billion elements.

Design methods could be completely different from today because nanodevices could be constructed randomly, without any predefined schematic or layout; however, a useful function could emerge from this huge number of nanodevices, or some auto-organization could occur. It is somewhat similar to natural selection, for which only the useful functions will survive, but it will be hard to design a predefined and very complex function like a Pentium microprocessor.

Most likely, microelectronics will be used until about 2020. Nanoelectronics will probably not replace microelectronics because the two technologies will coexist with possibly different applications.

References

[1] C. Piguet, Are early computer architectures a source of ideas for low-power? Invited paper, Volta '99, Como, Italy, March 4–5, 1999.

[2] C. Piguet, Histoire des ordinateurs. Invited paper at FTFC '99, Paris, May 26–28, 1999, pp. 7–16.

[3] M. R. Williams, *History of Computing Technology, 2nd Ed.*, IEEE Computer Society Press, Los Alamitos, CA, 1997.

[4] H. D. Huskey, and V. R. Huskey, Chronology of computing devices, *IEEE Trans. on Computers,* Vol. C-35, No. 12, December 1976, pp. 1190–1199.

[5] C. Piguet, Babbage, l'inventeur de l'ordinateur. Invited paper at FTFC '01, Paris, May 30–31, June 1, 2001.

[6] J. P. Hayes, *Computer Architecture and Organization,* McGraw-Hill Book Company, New York, 1978.

[7] W. Aspray, *John von Neumann and the Origins of Modern Computing,* MIT Press, Cambridge, MA, 1990.

[8] H. H. Goldstine, *The Computer from Pascal to von Neumann*, Princeton University Press, Princeton, NJ, 1972.

[9] M. J. Flynn, *Computer engineering 30 years after the IBM Model 91, IEEE Computer,* Vol. 31, No. 4, April 1998, pp. 27–31.

[10] R. F. Krick and A. Dollas, The evolution of instruction sequencing, *IEEE Computer,* Vol. 24, No. 4, April 1991, pp. 5–15.

[11] A. P. Chandrakasan, S. Sheng, and R. W. Brodersen, Low-power CMOS digital design *IEEE J. of Solid-State Circuits,* Vol. 27, No. 4, April 1992, pp. 473–484.

[12] D. Rutland, *Why Computers Are Computers,* Wren Publishing, 1995.

[13] W. Nebel and J. Mermet, eds., Low-Power Design in Deep Submicron Electronics, NATO ASI Series, E 337, 1997, Kluwer Academic Publishers, Dordrecht, Chapters 4.2 and 9.1.

[14] Editors of Electronics, *An Age of Innovation, The World of Electronics 1930–2000,* McGraw-Hill, New York, 1981.

[15] R. Reid, *The CHIP: How Two Americans Invented the Microchip and Launched a Revolution, 2nd ed.*, Random House Trade Paperbacks, New York, 2001.

[16] C. Piguet, The First Quartz Electronic Watch. Invited talk at PATMOS, Sevilla, Spain, September 11–13, 2002, pp. 1–15.

[17] C. Mead and L. Conway, *Introduction to VLSI Systems,* Addison-Wesley, Reading MA, 1980.

[18] G. J. Myers et al., Microprocessors technology trends, *Proc. IEEE,* Vol. 74, No. 12, December 1986, pp. 1605–1622.

[19] *IEEE MICRO Issue,* December 1996.

[20] R. Bernhard, ed., More hardware means less software, *IEEE Spectrum,* December 1981, pp. 30–37.
[21] J. L. Henessy and N. P. Jouppi, Computer technology and architecture: an evolving interaction, *Computer,* September 1991, pp. 18–29.
[22] C. Freeman, *The Economics of Industrial Innovation,* Penguin Books, 1974.
[23] M. Forrer et al., *L'aventure de la montre à quartz, Mutation technologique initiée par le Centre Electronique Horloger, Neuchâtel,* O. Attinger, ed., Neuchâtel, Switzerland, 2002.
[24] Smithsonian Museum, http://www.si.edu/lemelson/Quartz/index.html.
[25] M. S. Malone, *The Microprocessor. A Biography,* Springer-Verlag, New York, December, 1995.
[26] D. W. Dobberpuhl et al., A 200 MHz 64b dual issue CMOS microprocessor, *IEEE JSSC,* Vol. 27, No. 11, November 1992, pp. 1555–1567.
[27] J. D. Meindl, A history of low power electronics: how it began and where it's headed," *Proc. 1997 Int. Symp. On Low Power Electronics and Design, ISLPED '97,* August 18–20, pp. 149–151.
[28] J. D. Meindl, Low-*Power Microelectronics: Retrospect and Prospect, Proc. IEEE,* Vol. 83, No. 4, April 1995, pp. 619–635.
[29] Mead Education, http://www.mead.ch.
[30] A. Kaveh, The history of power dissipation, *Electronics Cooling,* January 2000, Vol. 6, No. 1.
[31] F. Catthoor, ed., *Unified Low-Power Design Flow for Data-Dominated Multi-Media and Telecom Applications,"* Kluwer Academic Publishers, Dordrecht, 2000.
[32] J. Sparso, S. Furber, eds., *Principles of Asynchronous Circuit Design, A Systems Perspective,* Kluwer Academic Publishers, Dordrecht, 2001.
[33] D. Soudris, C. Piguet, and C. Goutis, eds., *Designing CMOS Circuits for Low Power,* Kluwer Academic Publishers, Dordrecht, 2002.
[34] W. F. Brinkman, D. E. Haggan, and W. W. Troutman, *A history of the invention of the transistor and where it will lead us," IEEE JSSC,* Vol. 32, No. 12, December 1997, pp. 1858–1865.
[35] Transistor Museum, http://semiconductormuseum.com/Museum_Index.htm.
[36] T. Sakurai, Perspectives on power-aware electronics, plenary talk 1.2, *Proc. ISSCC 2003,* San Francisco, CA, Feb. 9–13, pp. 26–29.
[37] K. Itoh, *VLSI Memory Chip Design, Springer Series in Advanced Microelectronics,* Springer-Verlag, New York, 2001.

2

Evolution of Deep Submicron Bulk and SOI Technologies

Marc Belleville
Olivier Faynot
CEA-LETI

2.1 Introduction

Metal-oxide semiconductor (MOS) transistor behavior has already been demonstrated at the research level, down to a 6-nm gate length on fully depleted silicon on insulator (SOI) [1]. As complementary metal oxide semiconductor (CMOS) technologies continue to shrink, however, new physical phenomena are becoming increasingly important in the device behavior, setting up new challenges especially for low-power design. Some authors are even suggesting that power consumption will set the limits of scaling on an application dependent way [2]. Compared to traditional CMOS bulk technologies, SOI technologies are foreseen as alternative technologies that could lead to a better trade-off between active and leakage power. After a brief overview of the various scenarios proposed by the International Technology Roadmap for Semiconductors (ITRS) [3], the four main causes of limitations are discussed:

1. Voltage limits and subthreshold leakage
2. Tunneling currents

0-8493-1941-2/05/$0.00+$1.50

3. Statistical dispersions
4. Poly depletion and quantum effects

In addition, for each of those limitations, design challenges and proposed solutions are briefly presented, for bulk and SOI technologies. Finally, new innovative transistor architectures and technologies are described, and their relevance regarding the previous problems discussed.

2.2 Overview of ITRS Roadmap

2.2.1 Major Evolutions

Moving a design from an old technology to a newer one, with smaller design rules, has always been, up to now, an interesting way to lower the power consumption. Indeed, the overall parasitic capacitances (i.e., gates and interconnects) are decreased, the available active current per device is higher, and, consequently, the same performance can be achieved with a lower supply voltage. Moving to a new technology generation, however, induces a scale down of the power supply voltage (Vdd), the threshold voltage (VT), and the gate oxide thickness (T_{ox}). Beginning with the 0.18-μ technologies, it appeared that building a transistor with a good active current (I_{on}) and a low leakage current (I_{off}) was becoming more difficult. Therefore, two families of transistors were introduced: high-speed transistors and low-leakage transistors. The threshold voltages of the two families are tuned differently, thanks to a different channel doping. When moving to more advanced technologies, those two families are not sufficient anymore, regarding technological constraints. The ITRS introduces three main groups of transistors:

1. High performance (HP)
2. Low operating power(LOP)
3. Low standby power (LSTP)

At this stage, the channel doping is not only different, but also the gate oxide thickness.

2.2.2 Bulk CMOS Technologies

Table 2.1 summarizes the main parameters required for the next generations of bulk metal-oxide semiconductor field-effect transistor (MOSFET) devices, in case of HP, LOP, and LSTP technology options. The HP technology uses the shortest gate lengths in order to achieve the higher drive current. A higher leakage current is also allowed in the technology. For the LOP technology, the main target is to reduce the operating power of the circuit. Compared to the HP technology, the LOP one uses a longer physical gate length, a thicker gate oxide in order to achieve a leakage current hundred of times lower, for a given node.

The main purpose of LSTP technology is to achieve transistors with a very low leakage current (roughly five orders of magnitude smaller than the HP technology). To satisfy this criteria, gate length and gate oxide scalings are relaxed, compared to both HP and LOP technologies. In addition, threshold voltage values must be significantly increased to lower the leakage current. As discussed in the following sections, many key issues have no available solution today.

How to shrink the gate length and achieve good performances
How to shrink the gate oxide thickness and match the leakage current targets
How to reduce the supply voltage, while keeping operational circuits and low leakage current

2.2.3 SOI Technologies

For several years, SOI technologies have been developed to improve the performance of bulk technologies. The main difference between bulk and SOI substrates is the buried oxide layer located below the active silicon layer (i.e., layer where the MOSFET devices are processed). Therefore, each transistor can be electrically isolated from the others. Depending on the silicon thickness used for the SOI wafer, the transistor can operate in partially depleted or fully depleted modes [4]. When the SOI film is thick enough

TABLE 2.1 Main Device Characteristics for HP, LOP, and LSTP Technologies, Based on ITRS 2002 Update [3]

Year	2001	2002	2003	2004	2005	2006	2007	2010	2013	2016
					LOP					
Physical L_G (nm)	90	75	65	53	45	37	32	22	16	11
Physical EOT (nm)	2.0–2.4	1.8–2.2	1.6–2.0	1.4–1.8	1.2–1.6	1.1–1.5	1.0–1.4	0.8–1.2	0.7–1.1	0.6–1
V_{DD} (V)	1.2	1.2	1.1	1.1	1	1	0.9	0.8	0.7	0.6
I_{ON} (μA/μm)	600	600	600	600	600	600	700	700	800	900
I_{OFF} (pA/μm)	100	100	100	300	300	300	700	1000	3000	10000
					LSTP					
Physical L_G (nm)	100	90	75	65	53	45	37	28	20	16
Physical EOT (nm)	2.4–2.8	2.2–2.6	2.0–2.4	1.8–2.2	1.6–2	1.4–1.8	1.2–1.6	0.9–1.3	0.8–1.2	0.7–1.1
V_{DD} (V)	1.2	1.2	1.2	1.2	1.2	1.2	1.1	1	0.9	0.9
I_{ON} (μA/μm)	300	300	400	400	400	400	500	500	600	700
I_{OFF} (pA/μm)	1	1	1	1	1	1	1	3	7	10
					HP					
Physical L_G (nm)	65	53	45	37	32	28	25	18	13	9
Physical EOT (nm)	1.3–1.6	1.2–1.5	1.1–1.6	0.9–1.4	0.8–1.3	0.7–1.2	0.6–1.1	0.5–0.8	0.4–0.6	0.4–0.5
V_{DD} (V)	1.2	1.1	1	1	0.9	0.9	0.7	0.6	0.5	0.4
I_{ON} (μA/μm)	900	900	900	900	900	900	900	1200	1500	1500
I_{OFF} (pA/μm)	10,000	30,000	70,000	100,000	300,000	700,000	1E + 06	3E + 06	7E + 06	1E + 07

Note: L_G = gate length; EOT = equivalent oxide thickness.

(thicker than the depletion region), a neutral floating body region exists below the channel, inducing the partially depleted electrical behavior. The floating body effects increase the speed of the circuits. When the SOI film thickness is thinner than the depletion region, the entire film is depleted, inducing the fully depleted electrical behavior. In this case, the floating body effects are suppressed, and an ideal subthreshold swing of 60 mV/decade is theoretically achievable thanks to the constant depletion charge. This is a strong advantage compared to bulk devices. To achieve such ideal performances, ultra-thin silicon layers have to be used (with a ratio of 3 to 5 between the gate length and the SOI film thickness), which induce many technological issues, such as implantation-induced amorphization layer, thin SOI layer uniformity control, and silicon epitaxy growth.

SOI technologies are providing a complete set of transistors (HP, LOP, LSTP) just like bulk technologies. In the next sections, the advantages of SOI technologies compared with bulk technologies will be discussed.

2.3 Transistors Saturation and Subthreshold Currents

2.3.1 Subthreshold Leakage and Voltage Limits

The subthreshold current of a transistor is typically described by the following equation:

$$I_{OFF} \approx a \frac{1}{L_{EFF}} \exp\left(\frac{q(V_G - V_T)}{kT}\right) \tag{2.1}$$

where a is a constant, L_{EFF} is the effective gate length, V_G is the gate voltage, V_T is the threshold voltage, and kT/q is the thermal voltage.

In a typical scaling scenario, the electric fields are kept constant in the device by shrinking all the voltages and dimensions by the same factor. All doping levels are increased by the same scaling factor. As I_{off} increases exponentially when V_T decreases, however, static power consumption sets a lower limit to the scaling down of threshold voltages of the transistors. As the dynamic performance is directly related

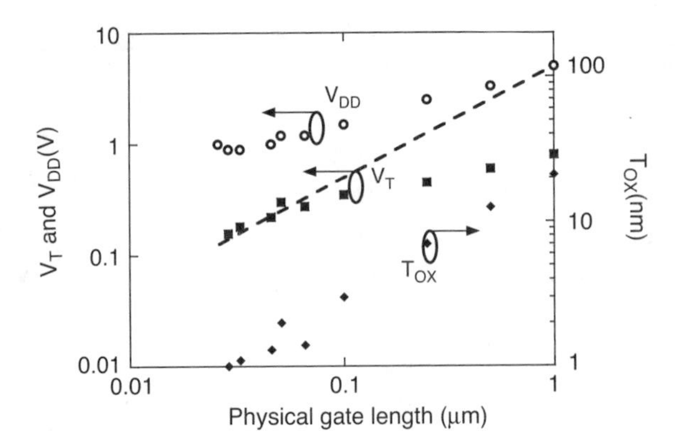

FIGURE 2.1 Supply voltage, threshold voltage, and gate oxide evolutions vs. gate length.

to the V_{dd}/V_T ratio, the power supply voltage also does not scale down easily. Consequently, in the ITRS roadmap scenario [3] (Figure 2.1) supply voltages do not shrink as rapidly as device dimensions. This results in a higher electric field in the device that has to be handled at the device level. Another consequence is the lower benefit granted to the dynamic power consumption which is proportional to Vdd^2. This is mainly because it is getting impossible to have simultaneously good active and leakage currents that several sets of transistors are required in advanced technologies.

To minimize the active power consumption, trading extra transistors against lower clock frequencies has been proposed by many authors and is well used today. The strong increase of the static power in respect to the active power in the upcoming technologies could lead to an opposite scenario: the number of transistors will have to be as small as possible regarding the targeted performance.

2.3.2 SOI Benefits

A significant active power reduction can be achieved by using SOI devices. Indeed, SOI devices are well-known for achieving the same performances than bulk devices, but with a lower power supply. This is achieved thanks to (a) lower parasitic capacitances (thick buried oxide for electrical isolation instead of junctions) and (b) lower threshold voltage in dynamic mode in case of partially depleted SOI devices. Active power reduction up to 50% can then be achieved with SOI [5, 6].

Another interesting advantage of single and multiple gates fully depleted SOI is their capability to achieve a nearly ideal (i.e., meaning 60 mV/decade at 300 K) subthreshold swing, compared to other devices. For a given OFF current, the fully depleted devices should achieve a higher drive current (due to a smaller threshold voltage) than its bulk counterpart.

To achieve subthreshold swing lower than 60 mV/decade (at 300 K), a new device structure is proposed [7]. According to the authors, 10 mV/decade can be achieved, which makes this type of device an excellent candidate for low leakage technology. The drawback is that it is a new kind of transistor, for which all the technology and design expertise have to be rebuilt.

2.3.3 Bulk CMOS Design Solutions for Subthreshold Leakage

For logical operators that can be stopped for a while, various techniques are used or foreseen to minimize standby power consumption. Those techniques can be used with the different kinds of CMOS transistors (HP, LOP, LSTP).

Triple well, or equivalent insulating technologies, are allowing an individual biasing of each independent pwell and nwell. Therefore, it is possible to tune the N-channel MOSFET (Nmos) and P-channel MOSFET (Pmos) substrate potentials to the required activity: a positive substrate potential (for an Nmos) will lower the threshold voltage of the transistors, therefore increasing its dynamic characteristics; on the contrary, a negative substrate voltage will increase the threshold voltage, consequently minimizing the subthreshold leakage. This technique, sometimes called variable threshold CMOS (VTCMOS) [8], requires efficient DC-to-DC converters. For a given technology, there is an optimum in reverse body bias, as the improvement in subthreshold leakage is compensated by an increase in source/drain to body junction leakage. Unfortunately, this technique is getting less effective with technologies scaling down [9].

With high-VT and low-VT transistors simultaneously available, other techniques are proposed: using low-VT transistors only in critical paths, or in multi-threshold CMOS (MTCMOS), introducing high-VT power switches to limit leakage current in standby mode [10]. A further level of optimization introduces multiple V_{dd} in a design [11].

More advanced concepts are now proposed to help minimize this subthreshold leakage. For instance, Abdollahi [12] is setting up, during sleep mode, the logical internal states so that the total leakage current of the circuit is minimized.

2.3.4 SOI CMOS Design Solutions for Subthreshold Leakage

Each SOI transistor has its own individual substrate usually called the "body." In fully depleted SOI technologies, the body potential roughly follows the source potential. No special design techniques are foreseen regarding subthreshold leakage in fully depleted SOI technologies; the advantage of this technology will be directly related to the better subthreshold swing of the devices. Regarding partially depleted SOI technologies, the body cannot be directly accessed in floating body SOI transistors. Only when using body-contacted SOI transistors, a body electrical node is available; the drawback is a much larger layout and because of that, their use is reserved to seldom situations when a floating body implementation is not possible.

Circuits based only on body-contacted transistors can use all the subthreshold leakage reduction mechanisms that are proposed for CMOS Bulk technologies. In MTCMOS SOI implementations, mixing floating body transistors for the logic and body contacted transistors for the standby current control can lead to a very efficient solution: driving independently the body of this transistor allows a low VT (and high current) in active mode, and a high VT in standby mode [13].

Regarding floating body transistors, substrate biasing, such as in triple-well technologies, is not possible. Other mechanisms have to be used to control the body potential and consequently the threshold voltage of the transistors. One very interesting feature of the partially depleted SOI transistors is the dynamic threshold modulation: the gate-to-body and drain-to-body capacitive couplings dynamically control body potential. Regarding subthreshold leakage, one key point is the ratio between those two capacitances. Figure 2.2 presents the current flowing through an inverter for three different values of this ratio. When the drain to body capacitance is lower than the gate to body capacitance, the leakage current of the inverter is, for a period after the input transition, lower than the DC leakage value. Then the body potential of the Nmos transistor comes back to a higher value, through the influence of DC phenomenon (impact ionization, gate tunneling current), and the leakage current rises back to the DC point. This low leakage state depends on the technology, but is frequently in the 100 microseconds–1 millisecond range.

Consequently, leakage control in SOI has to be preferably considered in a dynamic way and not in a static way. A first approach, for a block in standby mode, is to refresh the leakage current of the logic gates to lower values simply by generating one transition in the largest possible number of gates. Scan path, already inserted for test purpose can be used for this generation process.

Another dynamic approach to SOI standby was proposed by Morishita et al. [14] using a different way to control the body potential of SOI transistors. In this technique, the body potential is lowered by pulling out accumulated carriers from the floating body region with the forward biased current of the body source diode; consequently, the threshold voltage is increased. The extraction of the carriers occurs when the source line voltage is pulled down to a negative voltage; the source line is then pulled back up to 0 V. This technique requires additional switching transistors to pull up and down the power and ground lines; it also requires on chip DC-to-DC converters to generate the two extra power supply voltages. Notice however that as this occurs during standby mode, those power supplies have only to handle the leakage currents and the power line transitions. Refresh control is again required. Considering

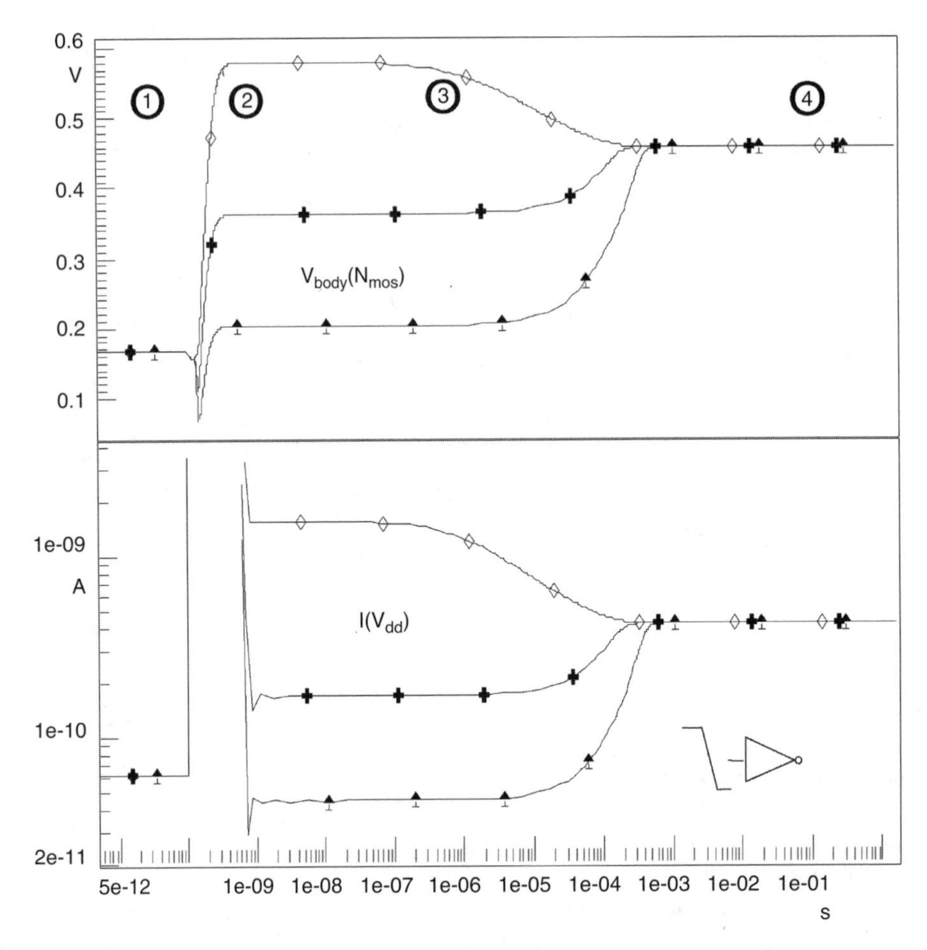

FIGURE 2.2 Power supply current of an inverter and body potential of the Nmos, before and after an input transition, for three different gate-to-body and drain-to-body capacitive ratios: 1 corresponds to the Pmos DC leakage; 2 to the input transition; 3 to the posttransition leakage evolution; and 4 to the Nmos DC leakage.

circuit architecture and layout, using this technique in memories like static random access memories (SRAMs) and standard-cell blocks is promising. The only constraint is that functions with independent shutdown modes cannot share the same rows.

2.4 Gate and Other Tunnel Currents

2.4.1 Tunneling Effects

Quantum mechanical tunneling of carriers through the energy barriers becomes more important as the dimensions of the transistors are scaled down to the nanometer range. Three main forms of leakage appear: the tunneling through the gate oxide, the band-to-band tunneling between body and drain and the direct drain to source tunneling current. The band-to-band tunneling current between the body and the drain of the transistor is strongly dependent on the electric field in the junction. The increased doping level of the body region, required by the scaling laws, emphasizes this tunneling current. Because direct band-to-band tunneling current depends on conduction-band states lining up with valence-band states, a forward bias of the body (with $V_{BS}>V_{DS}$) should limit this tunneling current. Nevertheless, this forward bias will induce a drastic increase of the direct junction leakage, which will not be acceptable. The easiest way to limit this tunneling current is to use intrinsic body devices with metal gate.

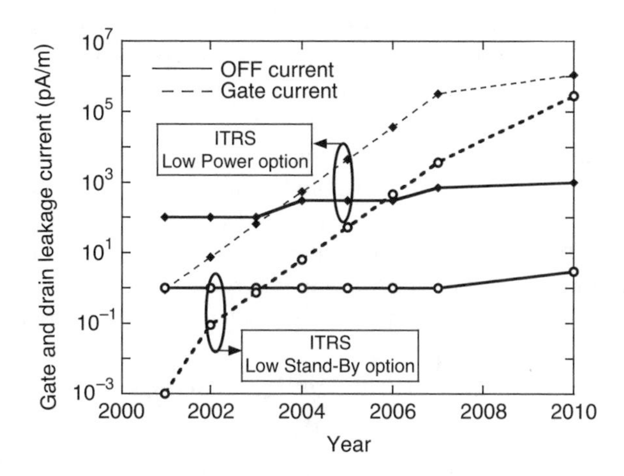

FIGURE 2.3 Evolution of drain and gate leakage current for LOP and LSTP technologies.

The second tunneling current is the direct source to drain tunneling, through the channel barrier. Recent analyses have shown that it becomes problematic for gate length smaller than 10 nm [15].

2.4.2 Gate Current

The third source of leakage current is the tunneling current through the gate oxide. The interface between silicon and silicon dioxide is still considered perfect in terms of abruptness and in terms of electrical properties. Silicon dioxide has carried us this far without limitations by extrinsic factors such as defect density, surface roughness or large-scale thickness, and uniformity control. Interface trap and fixed charge densities are so low that it corresponds to less than one surface defect in 10^5 surface silicon atoms [16]. This has made silicon dioxide become the only insulator used for MOS transistors. Based on the scaling laws, the gate oxide thickness must be few percent of the device gate length. This law implies that oxides thinner than 1.0–1.5nm must be employed for 35-nm gate lengths. These thicknesses comprise only a few layers of atoms and approach the fundamental limits. With such thicknesses, direct tunneling currents become very large, as shown in Figure 2.3 (data based on [17]). The basic equation describing this phenomenon is:

$$J_n \propto e^{\frac{-8\pi\sqrt{2m_{ox}}\Phi_B^{3/2}\left[1-\left(1-\frac{V_{ox}}{\Phi_B}\right)^{3/2}\right]}{3hqV_{ox}/T_{ox}}} \tag{2.2}$$

where Φ_B is the tunneling barrier height, m_{ox} is the oxide effective mass, T_{ox} is the gate oxide thickness, and V_{ox} is the voltage applied to the oxide [18].

This current is an exponential function of the gate oxide thickness and the applied voltage. As gate oxide decreases rapidly with the new upcoming technologies, the gate leakage current will become larger than the required leakage current of the transistor after 2004. Direct-tunneling current depends on a combination of tunneling probability and on the number of tunneling carriers. Due to the higher oxide tunneling barrier for holes (4.5 eV) than for electrons (3.1 eV) and due to the heavier effective mass of holes, the hole tunneling current is roughly one order of magnitude less than the electron tunneling current.

2.4.3 Design Issues and Possible Solutions

Several authors suggest that the upper limit of acceptable gate leakage current is in the range of 1–10 A/cm2 or even 100 A/cm2 [2] in very high performance chips. These currents become too large to accommodate the standby power requirements of integrated circuit (IC) applications. For low-power applications, the minimum oxide thickness is thicker (by 0.3–0.4nm) than for high performance. This thicker oxide reduces by two orders of magnitude the gate leakage current (for the same node).

When compared with subthreshold leakage for which several circuit design techniques are proposed to minimize it, circuit designers are facing with gate leakage a new problem with only limited solutions foreseen. A first possibility is to lower the power supply voltage in standby mode, taking advantage of the exponential relationship described in Equation 2.2. Another proposal [19] takes advantage of the difference between Pmos and Nmos gate leakage (one order of magnitude less for Pmos). For instance, by changing in MTCMOS logic Nmos gating to Pmos gating, leakage reduction between 41 and 60% are achieved [19]. This can be an interesting solution at some point; nevertheless, as gate leakage will increase exponentially from one generation to the next, solutions has to be found at the device level.

2.4.4 High-K Materials and Other Device Options

Many options are available to overcome the gate leakage problem. The first approach is to replace the silicon dioxide by a higher permittivity gate insulator. In such a case, due to the higher K, we can achieve a small electrical thickness with a thicker material. Up to now, the only successful insulator used is a silicon oxide/nitride composite. Higher-K insulators are currently studied (HfO_2, HfSiON), but even the most advanced works are still research works [20]. Nevertheless, as the dielectric constant of those binary insulators increases, their bandgap tends to decrease, as illustrated in Campbell et al. [21]. Because of the linear dependence on insulator thickness, and of the square root dependence on the barrier height shown in Equation 2.2, a large bandgap (hence a larger barrier height) is desirable when the aim is to reduce the gate tunneling current; however, other constraints exist on high-K materials. The two main constraints are: increased short channel effects and mobility degradation.

Due to its higher K value, the physical thickness of the insulator is larger than the one of silicon dioxide, and the ratio between their thicknesses is equal to the ratio between their permitivities. This thicker physical thickness increases the drain penetration under the gate, increasing drastically the short channel effects. In reference [22], it is shown that a permitivity ratio larger than 20 degrades significantly the scaling of the transistor. Therefore, the upper limit for the high-K permittivity and thickness is driven by the control of short channel effects, while the lower limit is driven by the tunneling current through the insulator [2].

The second drawback of using high-K material is related to the mobility degradation observed on MOSFET devices. Based on literature [23], from 10 to 40% of mobility degradation has been evaluated with high-K dielectric. Lots of analyses and modeling are under progress to explain this degradation. The high-K quality, the interface states density, as well as the fixed charges in the thin oxide localized between the silicon and the high K can be considered to be responsible for this degradation: the interface quality is not as good as the one obtained with silicon dioxide. Many experiments are in progress to explain and reduce this mobility degradation, which is one of the main problems to be solved for a use of high K in the semiconductor industry.

The second approach to limit the gate tunneling current is to stop the scaling of the gate oxide. Unfortunately, the use of a thicker gate oxide reduces the control of the gate on the channel conduction, leading to higher short channel effects and DIBL (drain induced barrier lowering) effects. Furthermore, the subthreshold swing expression is proportional to $1.0 + C_{si}/C_{ox}$, where C_{si} is the depletion layer capacitance, and C_{ox} is the gate capacitance. As C_{si} is proportional to $N^{1/2}$ (where N is the doping level), and C_{ox} proportional to $1/T_{ox}$, the gate oxide thickness is the most effective way to control the subthreshold leakage current. Thicker gate oxide will induce higher subthreshold swing value, and other parameters need to be scaled, in order to compensate this thicker gate oxide. Increased body doping and/or highly doped pockets can be used to optimize the device without reducing the gate oxide thickness, but the increase of the body doping reduces simultaneously the carrier mobility and the depletion depth of the channel region (i.e, reduces short channel effects). Forward biasing the body-to-source diode can also achieve this depletion depth reduction. The main drawback is an increase of the diode leakage currents and a degradation of the subthreshold swing of the device (and hence of I_{off}).

The third approach is the use of novel architectures, such as Double gate, FinFET, Triple gate, or Gate All Around, which are well known to improve the scaling of short transistors, compared to planar CMOS. These new types of architecture will be commented in the last section. With multiple gate devices, the vertical control is ensured by 2, 3, or 4 gates, enabling the gate control to be better than planar CMOS for a given gate oxide thickness. We could thus imagine using a thicker gate oxide in order to achieve the same short channel effects.

2.5 Statistical Dispersion of Transistor Electrical Parameters

2.5.1 Dopant Fluctuation

As the device dimensions shrink, the doping in the channel must be increased, in order to achieve acceptable short channel effects and VT, compatible with standard CMOS technologies. One of the main sources of variation in MOSFETs at the limit of scaling is randomness in the exact location of dopant atoms. Ion implantation leads to randomness at the atomic scale in the form of spatial fluctuations in the local doping concentration. This leads to dramatic variations of the transistor parameters. The number of dopant atoms is decreasing with scaling, as illustrated in Figure 2.4. For gate lengths smaller than 50 nm, the total number of dopant atoms is less than 100 in the device depletion region. As the standard deviation of fluctuations is equal to the square root of the number of dopant atoms, the $\pm 3\sigma$ boundaries shown in Figure 2.4 becomes extremely large when the gate length is scaled down to 30 nm.

Many publications have investigated the impact of the dopant fluctuations on the parameters of the transistors. Most of the publications use stochastically placed dopants in three-dimensional (3D) simulations [24–26]. Figure 2.5 illustrates an example of the discrete distribution of dopant atoms in the channel region. The fluctuation of the number of atom dopants induces a large variation of the parameters of the device (leakage current, subthreshold swing, etc.).

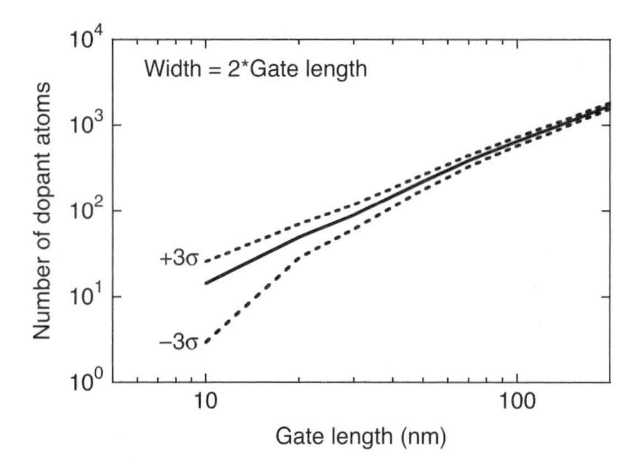

FIGURE 2.4 Variation of the number of dopant atoms and the associated fluctuations (in dashed lines) vs. gate length for the polysilicon gate.

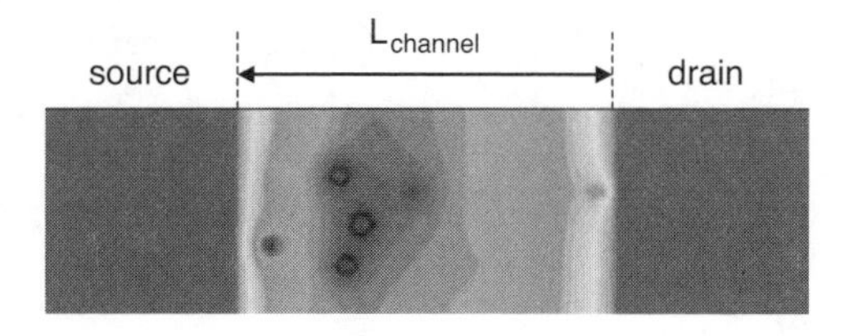

FIGURE 2.5 2D simulation of the discrete distribution of the dopant atoms in the channel region of a MOSFET transistor.

TABLE 2.2 Variation of the Number of Dopant Atoms and Its Associated Fluctuation Standard Deviation as a Function of Doping Level for a Transistor with L/W = 20 nm/40 nm

Doping level (at/cm^3)	10^{19}	10^{18}	10^{17}	10^{16}	10^{15}
N = Number of dopant atoms	93	28	8	2	0.7
$\sigma = \sqrt{N}$	9	5	2.8	1.5	0.8

It is obvious that such a device will be useless from a circuit point of view. Two solutions can be investigated to overcome this problem. First, the dopants can be moved in the body, back away from the surface. This will lead to retrograde channel doping profiles, for which the VT uncertainty is significantly reduced [26]. The second solution is the elimination of the dopants, combined with the use of metal gates. In such case, VT will be set by the gate work function, and not by the dopants. This trend is confirmed by the calculations presented in Table 2.2.

It is interesting to note that when metal gates are used with intrinsic devices, the number of dopant atoms can be smaller than one.

2.5.2 Design Issues and Possible Solutions

When considering digital circuit behavior, the major parameter to consider is the induced threshold voltage variation. An approximate expression for the delay variation of a logic gate is given as [27]:

$$\sigma Tpd \approx \frac{\sigma VT}{(Vdd - VT)^{2.5}} \tag{2.3}$$

When moving to advanced technologies, VT does not scale down as fast as Vdd; this is especially true for low-power devices. Consequently, the impact of the statistical VT variation on the gate delays will be very large. Eisele et al. [28] studied the impact of local delay variations due to VT variations on path delays in low-voltage circuits. For a given nominal VT, the relative delay variation increases with reduced logic depth, reduced supply voltages, and smaller device dimensions. In synchronous digital design, keeping a good yield consequently implies to increase the ratio "nominal path delay over clock period." One consequence is that the extra speed (or the lower active power) brought by a new technology will be partly lost in compensating for delay variations. Furthermore, all sensitive signals, like clock trees, will require large enough devices to minimize the statistical variations. Kishor et al. [27] compared the robustness of various CMOS logic design families regarding threshold voltage statistical variations; overall, static CMOS logic performs the best. It has also been demonstrated that the stability of memory cells is affected [29]. Self-Timed Logic, thanks to its inherent robustness and lower sensitivity to such variations could be a solution to overcome those limitations.

Regarding subthreshold leakage, dopant fluctuation induces variations on the threshold voltage, and on the subthreshold swing. As those variations are strongly non linear, the average standby current of a set of transistors will be higher than the sum of all the nominal standby currents. Figure 2.6 illustrates the impact of VT fluctuation on the static power consumption of a large number of inverters. The VT estimations are based on a doping level stable over the generations; as, when scaling down, doping levels are usually increased, the effective degradation will be worse. Another additive degrading factor will be the subthreshold swing fluctuation.

2.6 Physical and Electrical Gate Oxide Thickness

2.6.1 Poly Depletion

Polysilicon has been the widely used material for the gate electrode for more than 25 years. Its main advantage is that it can be doped n-type or p-type, leading to suitable work function for PMOS and

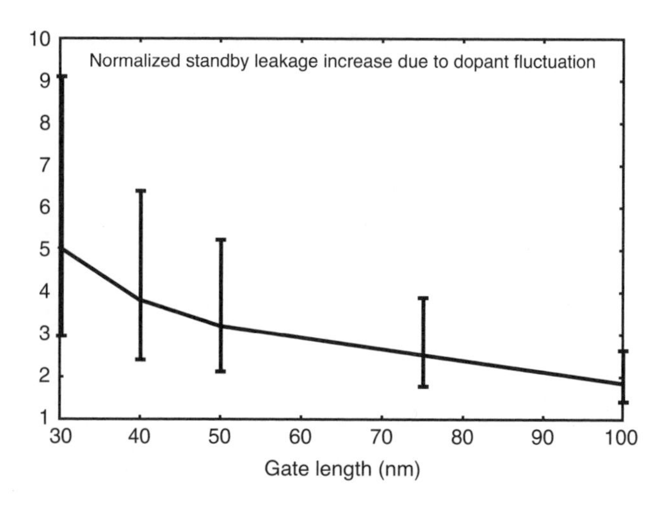

FIGURE 2.6 Normalized standby leakage increase for a set of 100 inverters (LOP), due to the impact of dopant fluctuation on VT (data from [30]).

NMOS devices. The polysilicon depletion effect occurs when the device is biased toward inversion. The applied voltage begins to deplete the highly doped polysilicon region near its interface with the gate oxide. The result of this effect is an apparent increase of the oxide thickness by a few angstroms.

As gate oxide thickness is reduced, the poly depletion effect becomes more severe. To prevent those effects, the increase of the active doping level near the gate oxide interface is required, but this increase is limited by two factors. The first factor is the maximum activation level that is not as good as the one of single crystal (maybe mainly due to the grain boundaries of polysilicon). The second one is the dopant outdiffusion from the polysilicon to channel through the gate insulator. This second effect is enhanced by the use of thinner and thinner gate dielectric, that makes the optimization of the polysilicon doping profile very difficult. This is particularly true with P+ gates because boron diffuses rapidly through SiO_2. The obvious solution to overcome this problem is the use a metal gate instead of polysilicon. In such a case, the depletion is completely suppressed. Another important advantage of metal is related to the resistivity of the gate, which is significantly reduced compared to silicided polysilicon. This can be a strong benefit, especially for radio frequency (RF) applications. As the work function of metal is close to midgap, the use of very low doping levels in the channel region is required, and many efforts must be made on work function engineering to achieve multiple VT for NMOS and PMOS.

2.6.2 Quantum Effects

Quantum effects occur in the silicon because of carrier quantum confinement in a potential field. The carrier concentration is low at the interface, and the peak of carrier concentration moves deeper into the silicon, by a few angstroms [31] (Figure 2.7). If d is the effective distance of carriers below the interface, and ε_{SI} the silicon dielectric constant, the main contribution of quantum effects to the gate capacitance is the addition of an extra capacitance (ε_{SI}/d) in series with the gate oxide capacitance. This results in a thicker effective gate oxide thickness. For 1-nm oxide thickness, the increase of the electrical thickness can reach a few tenths of a percent.

2.6.3 Circuit Dynamic Performances

Combining poly depletion and quantum effects leads to an effective gate oxide electrical thickness larger; consequently, the saturation current will be smaller. For a given I_{off}, the circuit dynamic characteristics will be degraded. For a chain of unloaded inverters in 25-nm technology, the delay is expected to be slightly degraded (~5%), mainly because the input capacitance of the next stage also decreases; for heavily loaded inverters, delay degradation approaches those of the saturation currents (10 to 20%) [32].

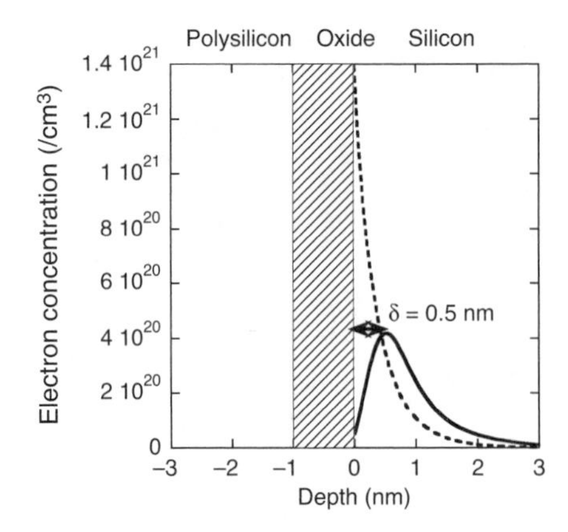

FIGURE 2.7 Quantum simulation of the inversion charge distribution with classical model (dashed line) and with quantum effects (solid line) for 1-nm gate oxide.

2.7 Innovative Transistor Architectures

This section presents innovative solutions for transistor architectures, in an attempt to solve most of the problems listed in the previous section.

2.7.1 Strained Silicon

To overcome the reduction of carrier mobility encountered in advanced devices (mainly due to the increasing doping level), two main types of strained material are under development. Strained SiGe substrates are the first option for the improvement of the mobility. By increasing the Ge concentration in the $SiGe_x$ alloy, the effective mass of holes is significantly reduced, leading to a strong improvement of the hole mobility in the case of long channel devices [33]. The drawback of such substrate is the decrease of the electron mobility. On short-channel PMOS, no significant improvement has been demonstrated, as illustrated on Figure 2.8 [34].

The second type of investigated material is strained silicon on relaxed SiGe. In such a case, both electron and hole mobilities can be slightly improved. As for the strained SiGe case, the gain in mobility decreases

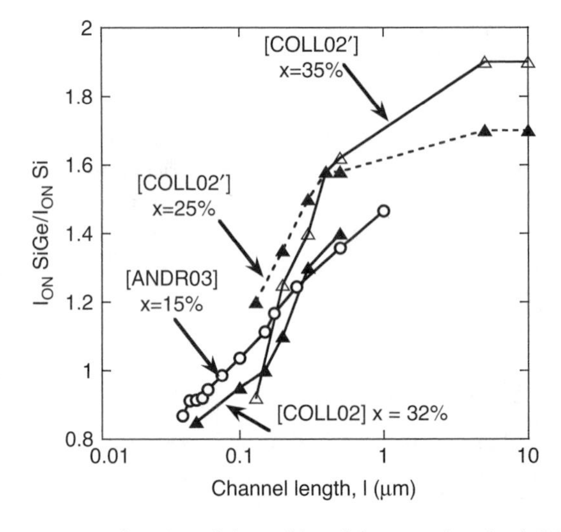

FIGURE 2.8 PMOS current gain as a function of channel length in case of strained SiGe substrates.

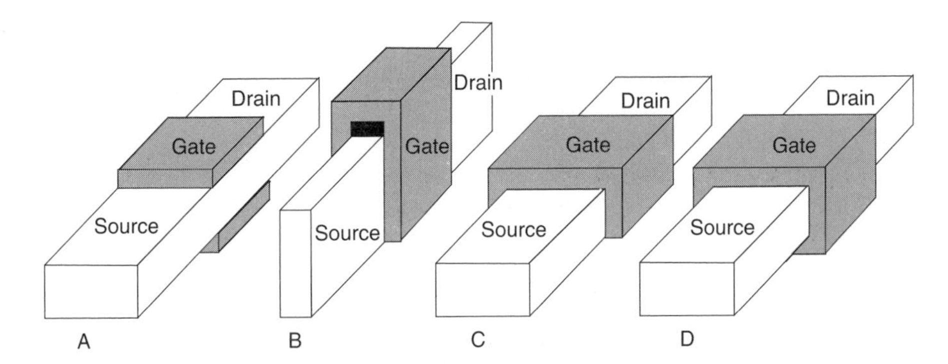

FIGURE 2.9 Scheme of multiple gate SOI devices. Planar double gate (a) FinFET, (b) triple gate, (c) quadruple gate, or (d) gate all around.

as the gate length is reduced. A 20–35% current gain has been obtained on NMOS for 70-nm gate length [35], but no significant gain has been published on short PMOS transistors. Only Intel has recently incorporated such material for its new 90-nm node.

Based on those two families of strained material, many other combinations of SiGe alloys on more or less strained layers are under analysis.Up to now, however, no real product has been fabricated with those materials.

2.7.2 Multiple Gate Devices

The use of SOI material makes the fabrication of multiple gate devices easier. Different kinds of transistors are developed: double gate [36], triple gate [37], FinFET [38] or gate all around [39]. Those architectures are summarized on Figure 2.9. Two, three, or four gates are used to control the channel regions, leading to an increase of the electric field induced by the gate. With multiple gates, the transverse electric field (i.e., gate-to-channel electric field) is reinforced compared to lateral electric field (i.e., source-to-drain electric field). The main purpose of these architectures is to:

- Improve the scaling of the transistors (i.e., by limiting DIBL and short channel effects for a given drive current) for gate length below 50 nm
- Achieve levels of performance that are as good as classical planar CMOS devices.

Planar Double gate devices [36] are the architecture that is the closest to classical planar devices. The channel conduction is achieved in the same <100> crystal orientation, compared to planar CMOS devices. The two channels (the top and the bottom ones) are controlled by two gates. In the ideal case, the current level and the gate capacitance are doubled, compared to classical devices. The main drawback of this device is the process [36] that has to be used to fabricate such a transistor: the bonding of SOI and bulk wafers, and the alignment of both top and bottom gates makes its manufacturability difficult to achieve with the equipment used today. Non-planar double gate (i.e., FinFET) and triple gate have significant easier processes compared to the previous transistor. The technology used is close to planar devices, and this makes these types of transistors very good candidates for the next generations. The main drawback is that most of conduction is ensured in vertical gates, for which the mobility and the interface quality are not well-known. Many publications are now focused on those aspects [41] to achieve drive currents that are as good as classical planar devices. The gate all around structure is the ideal case for short channel effect control: four gates control the body region potential. The gate all around process is comparable to the planar double gate process, in terms of complexity: the oxide below the channel region must be etched and the deposited gate must have the same length all around the transistor to ensure the ideal performances of the transistor (i.e., current level and gate capacitance multiplied by 4). If this is not the case (i.e., bottom and sidewall gates larger than the top one), the multiplication factor of the capacitance will be more than 4, while that of the current level will be less than 4.

With the improved performances achieved by those multiple gate SOI devices, we can imagine:

1. Having a lower I_{OFF} for a given VT criterion of a bulk transistor, possibly thanks to its steeper subthreshold swing
2. Having a lower gate tunneling current
3. Having smaller performance variations, with the combination of intrinsic devices and metal gate material

2.8 Conclusion

This chapter presented an overview of the problems generated by the scaling of MOSFETs transistors, outlining their impacts on circuit design. For each limitation, the investigated solutions have been presented. High-K materials are proposed to reduce the gate leakage current, metal gate is used to suppress the polysilicon gate depletion, and SOI technologies with single or multiple gate transistors offer opportunities for further scaling down of the transistor dimensions. Most of the proposed solutions for the transistor performance enhancement are new processes (high K, metal gate, SOI technologies). It is also clear that technology alone will not be able to solve all the problems foreseen in the coming technologies. Innovative circuit design techniques will be required to team with advanced devices in order to make circuits with acceptable dynamic performances and power consumption. Statistical dispersions and DC leakage currents, combined with low-voltage design, will be two challenges for circuit designers in the coming years.

References

[1] B. Doris et al. Extreme scaling with ultra-thin Si channel MOSFETs, *IEDM '02, Tech. Dig.*, pp. 267–270, 2002.

[2] D. J. Frank et al. Device scaling limits of SiMOSFETs and their application dependencies, *Proc. IEEE,* Vol. 89, No. 3, March 2001.

[3] The International Technology Roadmap for semiconductors, 2001 and 2002 update, http://public.itrs.net.

[4] C. Raynaud, SOI Process Integration, Short Course, *IEEE SOI Conf.,* Durango, Colorado, October 2001.

[5] L. E. Thon et al. 250–600 MHz 12b digital filters in 0.8–0.25 μm bulk and SOI CMOS technologies, *Int. Symp. on Low-Power Electronics,* pp. 89–92, 12–14 August 1996.

[6] M. Itoh et al. Fully depleted SIMOX SOI process technology for low-power digital and RF device, Silicon on Insulator Technology and Devices, *Proc. 10th Int. Symp. Electrochemical Society,* Washington, D.C., March 25–29, 2001, pp. 331—336.

[7] K. Gopalakrishnan et al. I-MOS: a novel semiconductor device with a subthreshold slope lower than kT/q, *IEDM Tech. Dig.*, pp. 289–292, 2002.

[8] T. Kuroda et al. "Variable threshold-voltage CMOS technology, *IEICE Trans. Electron.*, Vol. E83-C, No. 11, November 2000.

[9] A. Keshavarzi et al. Technology scaling behavior of optimum reverse body bias for standby leakage power reduction in CMOS IC's, *ISLPED '99,* San Diego, California, August 16–17, 1999, pp. 252–254.

[10] S. Mutoh et al. 1-V high-speed digital circuit technology with 0.5 μm multi-threshold CMOS, *ASIC Conf. and Exhibit,* September 27–October 1, 1993, pp. 186–189.

[11] T. Kuroda et al. Optimization and control of VDD and V_{th} for low-power, high-speed CMOS design, *ICCAD '02,* November 2002, pp. 28—34.

[12] A. Abdollahi et al. Runtime mechanisms for leakage current reduction in CMOS VLSI circuits, *ISPLED '02,* August 12–14, Monterey, CA, pp. 213—218.

[13] T. Douseki et al. A 0.5-V SIMOX–MTCMOS circuit with 200 ps logic gate, *Solid-State Circuits Conf., 1996. Digest of Technical Papers,* pp. 84–85, February 1996.

[14] F. Morishita et al. Dynamic floating body control SOI CMOS for power-managed multimedia ULSIs, *IEICE Trans. Electron.,* Vol. E84-C, No. 2, pp. 253–259, February 2001.
[15] Y. Naveh et al. Modeling of 10-nm scale ballistic MOSFETs, *IEEE Electron. Device Lett.,* 21, 242–244, 2000.
[16] J. D. Plummer et al. Material and process limits in silicon VLSI technology, *Proc. IEEE,* Vol. 89, No. 3, March 2001.
[17] S. H. Lo et al. Quantum mechanical modeling of electron tummeling current from the inversion layer of ultra-thin-oxide nMOSFETs, *IEEE Electron. Device Lett.,* 18, 209–211, 1997.
[18] W. C. Lee et al. Modeling gate and substrate currents due to conduction- and valence-band electron and hole tunneling, *VLSI Symp., 2000,* pp. 198–199.
[19] F. Hamzaoglu et al. Circuit-level techniques to control gate leakage for sub-100nm CMOS, *ISPLED '02,* August 12–14, Monterey, CA, pp. 60–63.
[20] A. L. P. Rotondaro et al. Advanced CMOS transistors with a novel HfSiON Gate dielectric, *VLSI Symp. 2002, Tech. Dig.,* p. 148.
[21] S. A. Campbell et al. MOSFET transistors fabricated with high permitivity Ti02 dielectrics, *IEEE Trans. on Electron. Devices,* Vol. 44, p. 104, 1997.
[22] D. J. Frank et al. Generalized scale length for two-dimensional effects in SOI MOSFETs, *IEEE Electron. Device Lett.,* Vol. 19, pp. 385–387, October 1998.
[23] B. Guillaumot et al. 75-nm damascene metal gate and high K integration for advanced CMOS devices, *IEDM '02, Tech. Dig.,* pp. 355–358, 2002.
[24] H. S. Wong et al., Three-dimensional atomistic simulation of discrete microscopic random dopant distributions effects in nanometer-scale MOSFETs, *Microelectron. Reliability,* 38, 1447–1456, 1998.
[25] H. S. Wong et al. discrete random dopant distribution effects in nanometer-scale MOSFETs, *Microelectron. Reliability,* 38, 1447–1456, 1998.
[26] D. J. Frank et al. Monte Carlo modeling of threshold variation due to dopant fluctuations, *Symp. on VLSI Technol., Dig. of Tech. Papers,* pp. 1169–170, 1999.
[27] M. Kishor et al. Threshold voltage and power supply tolerance of CMOS logic design families, *Symp. on Defect and Fault Tolerance in VLSI Systems,* October 25–27, 2000, Yamanashi, Japan, pp. 349–357.
[28] M. Eisele et al. The impact of intra-die device parameter variations on path delays and on the design for yield of low voltage digital circuits, *IEEE Trans. on VLSI Systems,* Vol. 5, No. 4, pp. 360–368, December 1997.
[29] D. Burnett et al. Implications of fundamental threshold voltage variations for high-density SRAM and logic circuits, *Symp. on VLSI Technol., 1994,* pp. 15–16.
[30] A. Asenov, Efficient 3D "atomistic" simulation technique for studying of random dopant induced threshold voltage lowering and fluctuations in decanano MOSFETs, *6th Int. Workshop on Computational Electronics, 1998,* pp. 263–266.
[31] J. D. Plummer et al. Material and process limits in silicon VLSI technology, *Proc. IEEE,* Vol. 89, No. 3, pp. 240–258, 2001.
[32] Y. Taur et al. 25-nm CMOS design considerations, *Int. Electron. Devices Meeting, IEDM '98,* December 6–9, 1998, pp. 789–792.
[33] M. V. Fischetti et al. Band structure, deformation potentials, and varrier mobility in strained Si, Ge, and SiGe alloys, *J. Appl. Phys.,* Vol. 80, No. 4, pp. 2234–2252, 1996.
[34] F. Andrieu et al. "SiGe channel p-MOSFETs scaling-down, to be published at *ESSDERC '03,* Estoril, Portugal, September 16–18, 2003, pp. 267–270.
[35] K. Rim et al. Strained Si MOSFETs for high-performance CMOS technology, *Symp. on VLSI Technol.* pp. 59–60, 2001.
[36]] H. S. P. Wong et al. Self-Aligned (Top and Bottom) Double Gate MOSFET with a 25nm Thick Silicon Channel, *IEDM Tech. Dig.,* 1997, pp. 427–430.
[37] J. T. Park et al. Pi-gate SOI MOSFET, *IEEE Electron. Dev. Lett.,* Vol. 22, No. 8, pp. 405–408, 2000.
[38] D. Hisamoto et al. FinFET- A self-Aligned Double-Gate MOSFET Scalable to 20nm, *IEEE Trans. Electron Dev.,* Vol. 47, No. 12, pp. 2320–2325, 2000.

[39] J. P. Colinge et al. Silicon-on-Insulator Gate-All-Around device, *IEDM Tech. Dig.*, pp. 595-598, 1990.

[40] J. H. Lee et al. Super self-aligned double-gate (SSDG) MOSFETs utilizing oxidation rate difference and selective epitaxy, *IEDM Tech. Dig.*, 1999.

[41] Y. K. Choi et al. FinFET process refinements for improved mobility and gate work function engineering, *IEDM Tech. Dig.*, pp. 259–262, 2002.

3

Leakage in CMOS Nanometric Technologies

Antoni Ferré
Joan Figueras
UPC

3.1 Introduction

This chapter is devoted to the characterization of the different sources of leakage current that appear in complementary metal-oxide semiconductor (CMOS) devices; physical origins of I_{LEAK} are outlined and data on how voltage and temperature affect the various components of I_{LEAK} are provided. In addition, the impact of technology scaling is presented. This characterization is used in order to predict the circuit level impact of I_{LEAK} of a CMOS circuit. The characterization of leakage components in nanometric technologies is an important issue: Whereas old technology, long channel transistors had basically one leakage mechanism, the well-known reverse-biased pn junction leakage, deep submicron, and nanometric transistors may have different leakage sources depending on the fabrication process.

3.2 I_{LEAK} Components of MOSFET Devices

Different physical phenomena contribute to the leakage currents causing the static consumption when one or more transistors in the V_{DD} to *GND* paths are in OFF-state. These currents are listed next, and are separated into five classes according to the physical origin of the current:

1. Tunneling currents of electrons across the thin gate oxide between the gate and the substrate I_G [1, 6], due to the high electric field in the gate oxide. The responsible mechanism in nanometric devices is direct tunneling through the oxide bands.
2. Subthreshold conduction producing leakage currents I_{SUBTH}, which flow from the drain to source [1, 2]. When the MOSFET has a gate voltage below the threshold voltage, the device surface is in weak inversion or depletion. When gate-to-source voltage V_{GS} is applied, even below the device

0-8493-1941-2/05/$0.00+$1.50

threshold voltage, sufficient charge carriers are on the surface region that can still create a significant current flow [3].

3. Gate-induced drain leakage I_{GIDL} currents flowing from the drain to the substrate. These currents are due to the tunneling of electrons from the valence to conduction band in the transition zone of the drain-substrate junction below the gate-to-drain overlap region where a high electric field exists [4, 5].
4. Reverse-biased pn junctions in the circuit. The leakage currents I_D of reverse-biased pn junctions are due to various mechanisms such as diffusion and thermal generation in the depletion region of the junctions [7, 8]. In nanometric technologies, junction-tunneling current due to bulk band-to-band tunneling (BTBT) current I_{BTBT} may appear [5, 9].
5. Bulk punchthrough current I_P from the source to the drain due to lateral bipolar transistor formed by the source (emitter), the bulk (base), and the drain (collector) [3]. If the drain voltage is large enough to deplete the neutral base region, a direct current I_p flows between the source and drain.

In general, CMOS technologies have one dominant OFF-state leakage mechanism and may have some secondary OFF-state leakage mechanisms. These mechanisms have evolved due to technological changes in MOSFET fabrication. The evolution is illustrated in Figure 3.1 for an OFF-state NMOS transistor.

Old technologies using long channel transistors, approximately defined as those above 0.7–1 μm channel length, had normally reverse-biased pn junction leakage current of the drain–substrate (well)

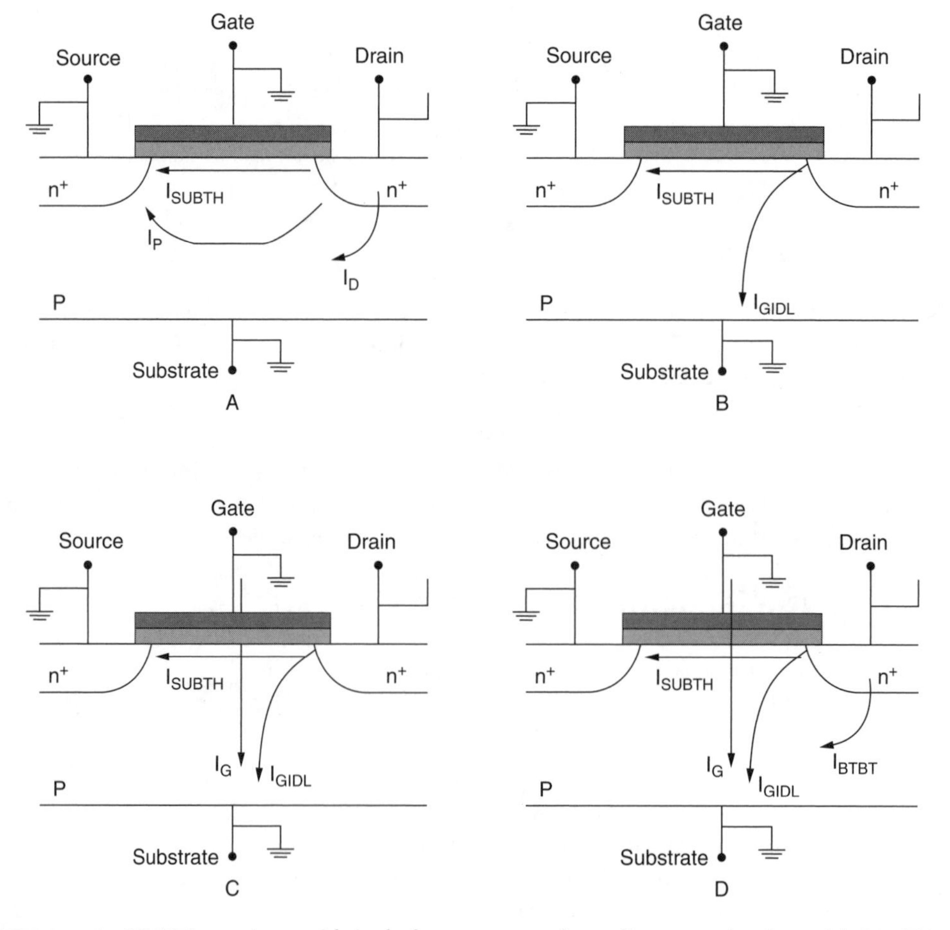

FIGURE 3.1 An NMOS transistor with its leakage currents depending on technology: (a) $L \geq 500$ nm; (b) 500 nm $\geq L \geq$ 100 nm; (c) 100 nm $\geq L \geq$ 50 nm; (d) 50 nm $\geq L$.

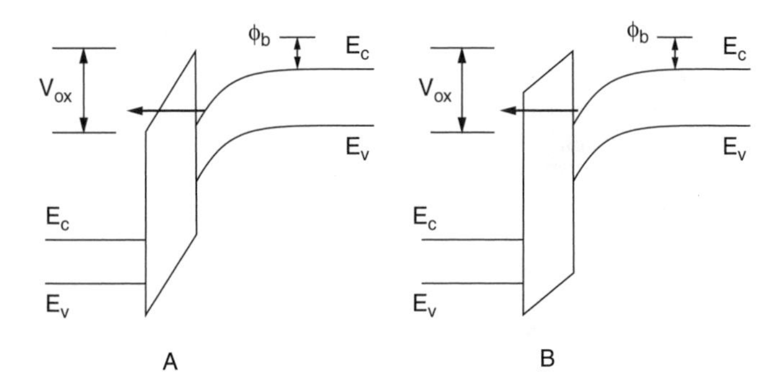

FIGURE 3.2 Gate Tunneling Mechanisms in MOSFETs: (a) Fowler–Nordheim Tunneling; (b) Direct Tunneling.

and substrate–well pn junctions as the dominant mechanism. The contribution from subthreshold leakage currents, the secondary mechanism, was usually negligible [10]. As the technology reached the 0.5 μm, the dominant mechanism changed to subthreshold leakage current [11, 12]. The punchthrough current was present also as secondary mechanism in some technologies. It is usually negligible in present technologies. This component is well controlled by raising the impurity concentration in the bulk channel region [3].

For submicron technologies below 0.5 μm, the dominant mechanism is the subthreshold leakage current. As a secondary leakage mechanism, reverse-biased pn junction leakage and gate-induced drain leakage current have been reported [10, 13].

For nanometric technologies, below 100 nm, the decrease in the gate oxide thickness needed to achieve a high current drive capability and to reduce the short-channel effects causes the magnification of nonideal effects such as gate tunneling currents. For ultrathin gate oxides, the direct tunneling increases and becomes one of the dominant leakage current mechanisms.

For sub-50-nm MOSFETs, the body-to-drain junction tunneling current is expected to become one of the dominant mechanisms due to high doping concentration [1].

Let us review the main characteristics of these currents and quantify the evolution when scaling down the technology.

3.2.1 Gate Tunneling Currents

For nanometric technologies, tunneling currents become a major issue. These currents are also greatly enhanced when scaling down the technology. Gate direct-tunneling current is produced by the quantum mechanical wave function of a charged carrier through the gate oxide potential barrier into the gate, which depends not only on the device structure, but also on its bias conditions [23–26].

The high electric field in the gate oxide may cause tunneling currents through the gate by means of two mechanisms: direct tunneling or Fowler–Nordheim tunneling through the oxide bands as illustrated in Figure 3.2. For the voltages and structures of modern MOSFETs, direct tunneling is the dominant component. Fowler–Nordheim tunneling typically appears when the oxide layer is thicker than 6 nm, and the applied field is higher than the electric field found at present day technologies.

The gate-tunneling current from the Si inversion layer to the poly-Si gate has been traditionally computed using an independent electron approximation and an elastic tunneling process. Because the exact form of the electronic tunneling barrier is not generally known, the potential barrier was commonly assumed as triangular for potentials higher than the Si/SiO_2 barrier voltage $\phi_b = 3.2V$. Whenever oxide voltage is lower than 3.2 V the electron-tunneling barrier changes from being triangular to trapezoidal. The silicon surface is strongly inverted or strongly accumulated, and the surface electric field on the silicon side and the potential barrier on the SiO2 side confine electrons at the Si/SiO2 interface. In these cases, the contribution to the leakage due to direct tunneling is given by [27]:

$$J_G = J_0 \cdot E_{ox}^2 \cdot e^{-k \cdot tox} \tag{3.1}$$

where J_0 is a technology dependent parameter adjusted to match experimental data, E_{ox} is the oxide electric field, and t_{ox} is the gate oxide thickness. The imaginary part of the wave vector k when a V_G voltage is applied is given by [27]:

$$k = \frac{2k_0}{3}\frac{\phi_b}{V_G}\left[1-\left[1-\min\left(1,\frac{V_G}{\phi_b}\right)\right]\right] \tag{3.2}$$

where V_G is a technological parameter. These expressions show the high sensitivity of direct tunneling on oxide thickness and power supply voltage. In Figure 3.3, the tunneling current density is plotted as a function of gate bias for several oxide thicknesses.

Temperature variations have a low impact on gate tunneling. The gate leakage current depends on temperature through the energy ground level, which leads to a reduction of the effective barrier height ϕ_b. The temperature also affects the mean-free path of electrons in the oxide conduction band. Consequently, when the ambient temperature is increased, the gate tunneling increases very slightly [28].

Note that the tunneling leakage in current SiO_2 dielectrics will be dominating in NMOS devices because PMOS has a higher barrier for hole tunneling and, therefore, a lower leakage current. With high-k dielectrics, such as Si_3N_4 dielectric, however, the gate current is higher in p^+ poly-Si PMOS than in n^+ poly-Si NMOS, and the scaling limit due to excessive tunneling leakage current will be first reached for PMOS.

In MOSFETs having ultrathin gate oxide thicknesses (1.4–2.4 nm), a direct current of electrons from n^+ poly-Si to underlying n-type drain extension in off-state n-channel contributes also to the gate leakage current [29]. This effect was reported experimentally by Henson *et al.* [30] and Yang et al. [31]. They found that the gate current weakly depends on channel length. The off-state bias configuration $V_{GS} = 0$ V and $V_{DS} = V_{DD}$ exhibited a non-negligible gate current but not as significant as the case when $V_{GS} = V_{DD}$ and $V_{DS} = 0$. This means that the tunneling current is localized in the edge region for the off-state condition. Figure 3.4 illustrates these various gate-tunneling components in a scaled NMOS; the gate-to-channel current I_{go}, and the direct tunneling current appearing between the source drain extension (SDE) and the gate overlap, usually called the edge direct tunneling (EDT) currents (I_{gso} and I_{gdo}) and become dominant in front of the gate-to-channel current I_{go}. In long-channel devices, EDT currents are less important than because the gate overlap length is small compared to the channel length. In very short channel devices, the portion of the gate overlap compared to the total gate length increases.

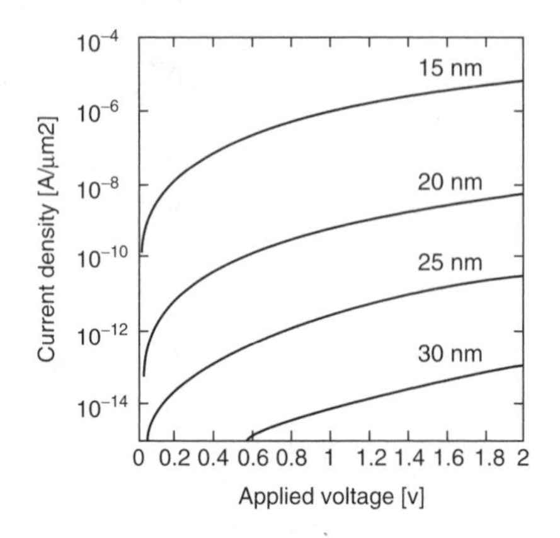

FIGURE 3.3 Gate-tunneling current as a function of applied voltage and oxide thickness.

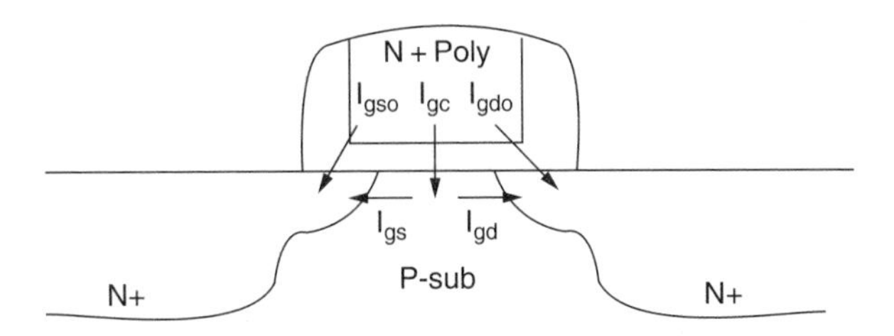

FIGURE 3.4 Gate-tunneling current components.

Different models have been proposed for EDT currents. Yu *et al.* [32] presented a model where gate direct tunneling currents are described using voltage-dependent current sources as a function of the terminal voltages. The partitioning of channel gate current is modeled by using variable resistances in each part of the channel. The channel currents of each region are obtained by adjusting the BSIM3-model parameters to fit the current-voltage curves obtained from simulation at device level.

Lee and Hu [33] proposed an accurate dielectric leakage model for metal-oxide semiconductor (MOS) capacitors based on modeling the electron conduction band (ECB), electron valence band (EVB), and hole valence band (HVB) currents. A physical source-drain current partition model is introduced using these contributions. This model has been implemented into the BSIM4 transistor model [34].

We propose a simpler but sufficiently accurate model. The use of two antiparallel current sources with an exponential (diode-like) dependence connected between the poly gate and each edge of the channel allows the modeling of the current flowing in both directions (see Figure 3.5). Currents flowing through D_{GS} and D_{SG} account for $I_{gso} + I_{gs}$, while currents flowing through D_{GD} and D_{DG} account for $I_{gdo} + I_{gd}$. The voltage drop by poly-Si gate is also taken into account by introducing a series resistance (R_{poly} in Figure 3.5). A similar model is used for PMOS with the only difference that the current driven by the antiparallel current sources is smaller than for NMOSs according to experimental data for transistors using SiO_2 as dielectric. Similar models can be derived for high-k dielectrics. The parameters of each current source are adjusted to match as close as possible the direct tunneling currents.

Comparison performed between the values obtained from expression and the simplified model shows an error lower than 2.5% through all the input gate voltage range $[0,V_{DD}]$ for $V_{DD} = 1.5$V. Figure 3.6 and Figure 3.7 show the simulation results of this model for a device with $L = 70$ nm at different V_{DS} values. The observed "dips" in the measured gate tunneling currents in these Figures are due to the combined

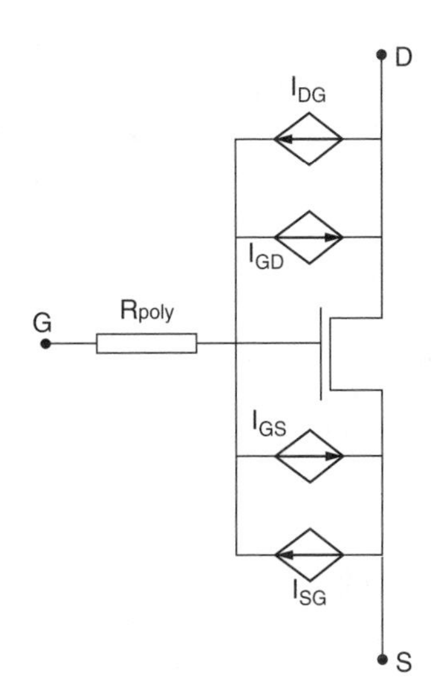

FIGURE 3.5 Gate-tunneling current model using antiparallel current sources.

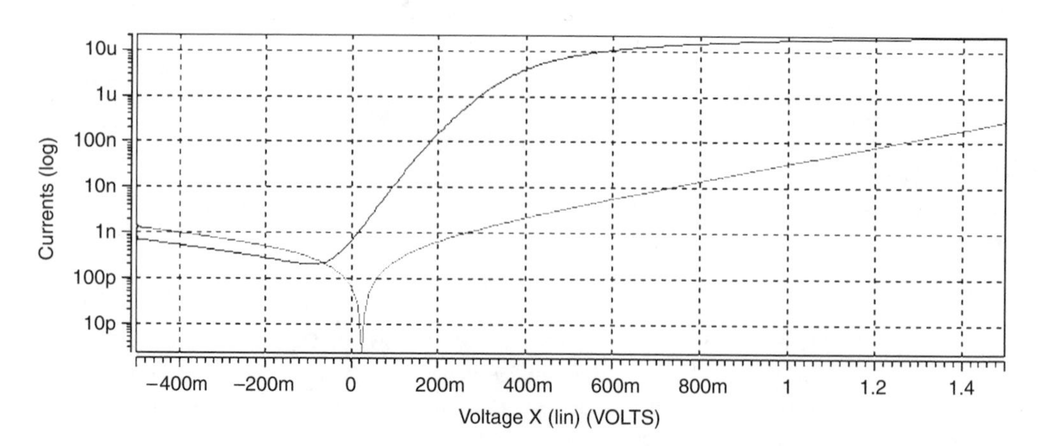

FIGURE 3.6 Leakage Current Simulation using MOSFET model including antiparallel current sources. W = 140 nm, L = 70 nm, V_{DS} = 0.05 V.

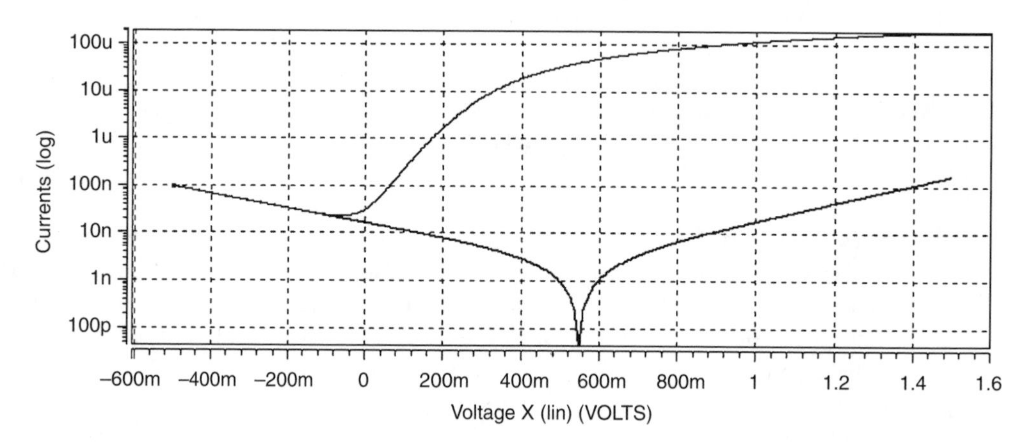

FIGURE 3.7 Leakage Current Simulation using MOSFET model including antiparallel current sources. W = 140 nm, L = 70 nm, V_{DS} = 1.5 V.

effect of source-to-channel tunneling and drain-to-gate tunneling currents in opposite directions. This effect shows the importance of the EDT currents, especially for small transistors.

This model is used later (see Section 3.4) to estimate the leakage of CMOS circuits.

3.2.2 Subthreshold Leakage Currents

When gate voltage is lower than the threshold voltage and there is a voltage applied between drain and source of a MOS transistor, a diffusion current appears due to the different carrier concentrations at the inversion layer in source and drain terminals. This current depends exponentially on gate-to-source voltage V_{GS} and drain-to-source voltage V_{DS} through the carrier concentrations. For an NMOS transistor, the subthreshold current is given by [14]:

$$I_{SUBTH} = \mu_N C_{ox} \frac{W_N}{L_N} V_t^2 \exp\left[\frac{V_{GS} - V_{TH}}{nV_t}\right]\left[1 - \exp\left[-\frac{V_{DS}}{V_t}\right]\right] \tag{3.3}$$

where μ_N is the electron carrier mobility, C_{ox} is the gate capacitance per unit area, W_N is the channel width, L_N is the channel length, V_t is the thermal voltage, and V_{TH} is the threshold voltage. The inverse slope of the subthreshold current n is given by [14]:

$$n = 1 + \frac{C_D}{C_{ox}} \tag{3.4}$$

where C_D is the depletion channel region capacitance per unit area. The subthreshold parameter n is related to the subthreshold swing S — the gate voltage change needed to raise the subthreshold current by one decade:

$$S = \ln 10 \cdot n \cdot V_t \tag{3.5}$$

At room temperature, the minimum theoretical value for S is about 60 mV/dec ($n = 1$). For present CMOS technologies, S takes values in the range of 80-90 mV/dec.

The threshold voltage for an NMOS long channel transistor with substrate bias V_{BS} is expressed by [14]:

$$V_{TH} = V_{TH0} + k_1\sqrt{\phi_S - V_{BS}} - k_2 V_{BS} \tag{3.6}$$

where V_{TH0} is the long-channel threshold voltage without substrate-bias, ϕ_S is the surface potential. The body effect and the nonuniform doping effect are modeled using the parameters k_1 and k_2.

In short-channel transistors, V_{TH} is further modified as a function of the channel length and the drain-to-source bias: these are the so-called short-channel and drain-induced barrier lowering (DIBL) effects [15].

Figure 3.8 illustrates the origin of short-channel effect. In a long-channel device, the depth of the depletion region in source and drain regions is relatively unimportant. As the channel length is reduced, however, these depletion regions occupy more space of the channel region. The depletion regions near the source and drain edges are shared with the channel. This effect produces a reduction of the threshold voltage when decreasing channel length and, therefore, increases subthreshold current.

The short-channel effect may be modeled following the Phillips model by reducing the effective threshold voltage as a function of the effective channel length L_{eff} [16]:

$$\Delta V_{TH}^{SCE}(L_{eff}) = \frac{u_{L1}}{L_{eff}} - \frac{u_{L2}}{L_{eff}^2} \tag{3.7}$$

where u_{L1} and u_{L2} are technology dependent parameters.

The drain-induced barrier lowering (DIBL) effect consists of lowering the energy barrier between the source and the channel. This causes excess injection of charge barriers into the channel and gives rise to an increased subthreshold current. Figure 3.9 presents qualitatively the band diagram at the interface of the channel, for short- and long-channel transistors. At the interface, the channel consists of three regions: the source-channel junction, the middle region, and the drain-channel junction. For the long-channel

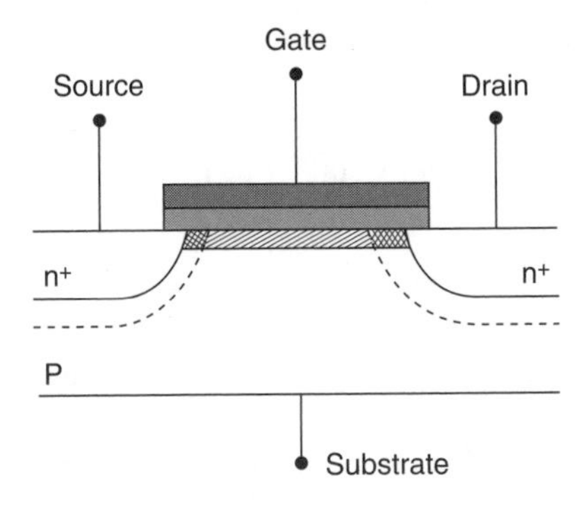

FIGURE 3.8 Physical origin of short-channel effect.

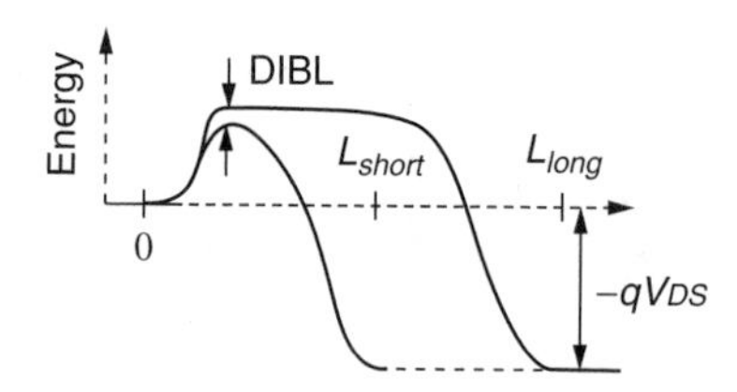

FIGURE 3.9 Physical origin of the DIBL effect.

transistor, the energy bands in the central part of the channel can be taken to be approximately constant because the voltage almost drops at the drain-channel junction. As channel length is reduced, however, this situation is no longer true. Consequently, a reduction in the interface energy barrier occurs at the source-channel junction where the maximum of the barrier is reached. This is the so-called DIBL effect [15]. The DIBL effect may be modeled, following the Phillips model, by additionally reducing the effective threshold voltage as a function of the drain to source voltage V_{DS} in the following amount [16]:

$$\Delta V_{TH}^{DIBL}(L_{eff}, V_{DS}) = \frac{S_L}{L_{eff}^2} \cdot (\phi_S + V_{SB})^{1/2} \cdot V_{DS} \tag{3.8}$$

where S_L is a constant for a given technology, ϕ_S is the surface potential, and V_{SB} is the source-to-bulk voltage.

The leakage current I_{LEAK} also depends on the power supply voltage V_{DD} through the dependence on the drain-to-source voltage V_{DS}. Two main factors are responsible for this:

1. Carrier concentration at the drain. This factor refers to the term

$$\left[1 - \exp(-V_{DS}/V_t)\right]$$

in Equation 3.3. For

$$V_{DS} \geq 4 \cdot V_t$$

this effect becomes negligible.
2. DIBL effect. For reduced channel lengths, the leakage current I_{LEAK} depends exponentially on V_{DS}. If the channel length increases, the DIBL effect reduces. For large enough channel lengths, I_{LEAK} will be almost independent on V_{DS}.

Another important issue is the variation on the subthreshold current due to the temperature. MOS transistor characteristics are strongly dependent on temperature. One of the main parameters responsible for this is the effective mobility, which is known to decrease with temperature [17]:

$$\mu(T) = \mu(T_r)\left(\frac{T}{T_r}\right)^{-\kappa 1} \tag{3.9}$$

where T is the absolute temperature [in Kelvin], T_r is room absolute temperature [in degrees Kelvin], and κ_1 is a constant technology-dependent. This value varies usually from 1.2–2.0 [17]. Other temperature-dependent parameters are the surface potential ϕ_S (through the variation of the intrinsic concentration) and the flat-band voltage V_{FB} (trough the variation in the work-function ϕ_{MS}). These effects are manifested in the value of the threshold voltage, V_{TH}, as an almost straight-line decrease with temperature [17]:

$$V_{TH}(T) = V_{TH}(T_r) - \kappa_2(T - T_r) \tag{3.10}$$

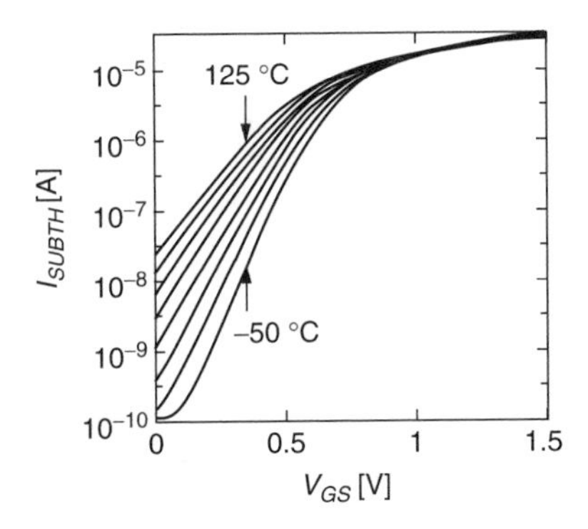

FIGURE 3.10 Subthreshold leakage current in an NMOS (L = 70 nm) depending on temperature.

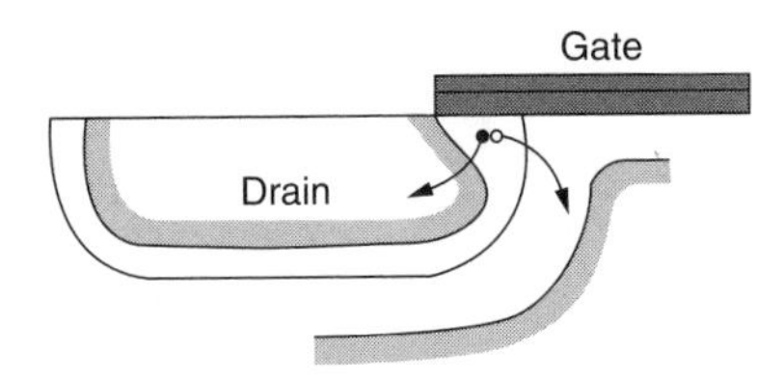

FIGURE 3.11 Physical origin of GIDL current.

where κ_2 is usually between 0.5 and 3 mV/K [17], with larger values in this range corresponding to heavier doped substrates, thicker oxides, and higher values of V_{BS}.

Thus, a temperature increase tends to increase the drain current (exponentially in the subthreshold region) through the threshold voltage variation and to decrease it through the mobility variation. At the subthreshold region, the decrease of the threshold voltage dominates. Therefore, increasing the temperatures produces an exponential increase in subthreshold current.

The temperature also affects the slope of the leakage current curves through the thermal voltage. Figure 3.10 plots the variation of leakage current of a 70-nm NMOS transistor as found by HSPICE simulation.

3.2.3 Gate-Induced Drain Leakage Currents

In some nanometric technologies, gate-induced drain leakage (GIDL) current I_{GIDL} may appear, usually for relatively high power supply voltages [2, 11]. I_{GIDL} current of a NMOS transistor flows from the drain to the substrate. This is caused by the effects of the high electric field region under the gate in the region of the drain overlap as illustrated in Figure 3.11. In this region, pair creation can occur.

Several possible mechanisms contribute to this current, which are presented in Figure 3.12. These include thermal emission, trap-assisted tunneling, and band-to-band tunneling [2, 18, 19]. It is the BTBT that has the metal-oxide semiconductor fluid-effect transistor (MOSFET) relevance at the voltages and structures of modern devices. This current is due to the direct tunneling of electrons from the valence to conduction band in the gate-to-drain overlap region where a high normal electric field E_n exists. This current may be further enhanced because the generated carriers are accelerated by the longitudinal electric field E_l in the drain-substrate junction, and this causes impact ionization [20].

The expression to estimate this leakage component as a function of the longitudinal E_l and normal E_n components of the electric field in the gate drain overlap area is [20]:

$$I_{GIDL} = A_{b1} \cdot W \cdot E_n \cdot e^{\frac{-B_{b2}}{E_n}} \cdot E_l \cdot e^{\frac{-B_{b2}}{E_l}} \tag{3.11}$$

where A_{b1}, B_{b1}, and B_{b2} are technology dependent parameters. W is the width of the device. In Figure 3.13, the contributions of subthreshold and GIDL currents in a 70-nm NMOS with V_{DD} = 1.5 V are

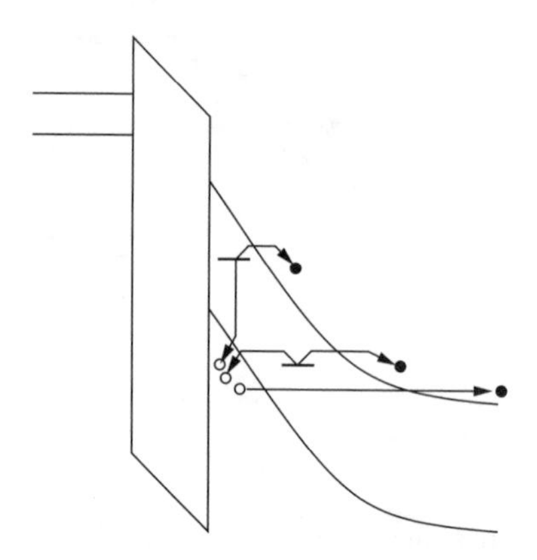

FIGURE 3.12 Tunneling mechanisms in GIDL: BTBT and trap-assisted tunneling.

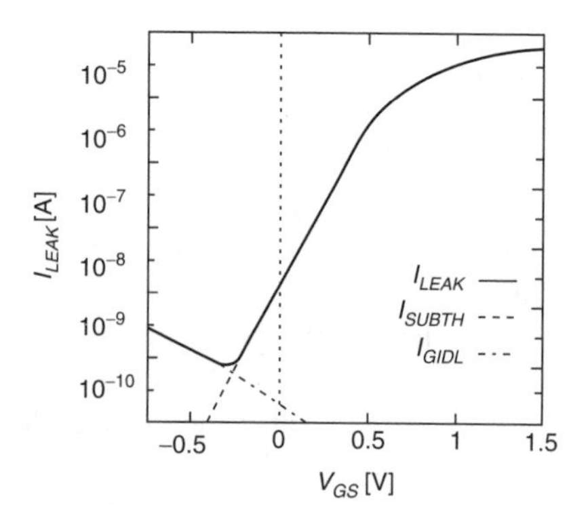

FIGURE 3.13 Subthreshold and GIDL leakage currents in an NMOS (L = 70 nm).

plotted shown as well as the total leakage current I_{LEAK}. Notice the significant effect of the I_{GIDL} leakage as the gate voltage in an NMOS transistor is brought to high negative levels.

The dependence on power supply voltage also depends on power supply. The increase in V_{DD} implies an increase of the normal electric field and, therefore, an exponential increase of I_{GIDL}.

GIDL currents may be especially important in buried channel devices. Comparison of GIDL in surface channel PMOS devices (p^+ poly gate) and buried channel PMOS devices (n^+ poly gate) have shown that buried channel PMOS has higher GIDL than the equivalent surface device for a given supply voltage [21]. This is due to the large flat band voltage between the n^+ poly gate and the p^+ junction in the overlap region of the drain. In this case, the GIDL current may become the dominant component of leakage current as illustrated in [22]. GIDL may be also a limiting factor when applying leakage reduction techniques such as body bias control (BBC) [66].

3.2.4 Junction Leakage Currents

In micrometric technologies, the leakage currents of reverse-biased pn junctions I_D are due to two mechanisms: diffusion of carriers and thermal generation currents in the depletion region of the silicided junctions [7]. Generation and diffusion currents depend strongly on temperature, through the intrinsic carrier concentration n_i. For low to moderate temperatures, I_D is dominated by generation mechanisms and increases with temperature at a rate proportional to n_i. At high temperatures, I_D is determined primarily by diffusion mechanisms and increases more rapidly at a rate proportional to n_i^2 [35]. For present day technologies, reverse-biased pn junction leakage current is lower than subthreshold leakage current and can be neglected.

For sub-50-nm technologies with highly doped pn junctions, the narrow depletion region produce tunneling currents I_{BTBT} [1, 67]. In addition, if V_{DD} increases and approaches the junction breakdown voltage, avalanche current appears from impact ionization in the depletion region [1, 27]. An important contribution to the total leakage in these devices is observed in [67].

3.2.5 Punchthrough Currents

In CMOS circuits, parasitic lateral bipolar transistors formed by the source (emitter), the bulk or the well (base) and the drain (collector) of MOS transistors are formed. If the drain voltage is large enough to deplete the neutral base region, the potential barrier height between the source and the channel region is lowered not only by the gate bias but also by the drain bias. Therefore, a punchthrough current I_P flows between the source and drain. The punchthrough current causes a large leakage current not controllable by the gate bias voltage. This component is well controlled by raising the impurity concentration in the bulk channel region [3].

3.3 Scaling

For the last four decades, silicon technology has been progressively reducing the channel length of MOSFETs from 25 μm at 5–10 V supply voltage to nanometric lengths and power supplies below 1 V in current production technologies. To maintain the transistor performance at lower voltages, the oxide thickness has also been reduced from 100 to a few nanometers in accordance with Dennard constant-field scaling law. [36–44].

As discussed in the previous section, as technology evolves and channel length becomes nanometric, total leakage current increases. This fact is mainly due to: (a) the lowering of threshold [1], which increases the subthreshold current, (b) the increased short-channel effects when reducing the channel length [45], which also increases the subthreshold current, and (c) the reduction of oxide thickness, which increases the gate tunneling current.

3.3.1 Scaling of V_{TH} and its Impact on Subthreshold Current

To avoid reliability degradation effects such as hot-carrier injection or oxide breakdown due to the high electrical fields, reduction of the power supply voltage is required. Reduction of the supply voltage has a negative effect on circuit performance: propagation delays may increase as the supply voltage decreases. To reduce these undesirable effects, the threshold voltage should be reduced. The impact of threshold voltage reduction on subthreshold current and, consequently, on leakage current consumption is illustrated in Figure 3.14. I_{TH} is defined to be the drain current when the gate voltage is equal to the threshold voltage. If the threshold voltage is reduced from V_{TH} to V'_{TH}, the OFF-state leakage current I_{OFF}, defined as the drain current when the gate voltage is zero, increases exponentially. Because S is about 80–100 mV/dec at room temperature for present technologies, a reduction of 80–100 mV in V_{TH} implies an increase of I_{OFF} by an order of magnitude.

3.3.2 Short-Channel Effects

Short-channel effects are another main factor responsible of the increase of leakage current when downsizing the technology, as we have seen before. These undesirable effects may be reduced by scaling the gate length, the source-drain junction depth, the bulk doping concentrations and the gate oxide thickness. All these quantities must be scaled together. The use of shallow drain/source junctions reduces the charge shared by the junction and the channel; however, these ultrashallow junctions increase the resistance of the device and, therefore, degrades its performance. To improve the performance, the use of elevated source and drain structures has been introduced [46].

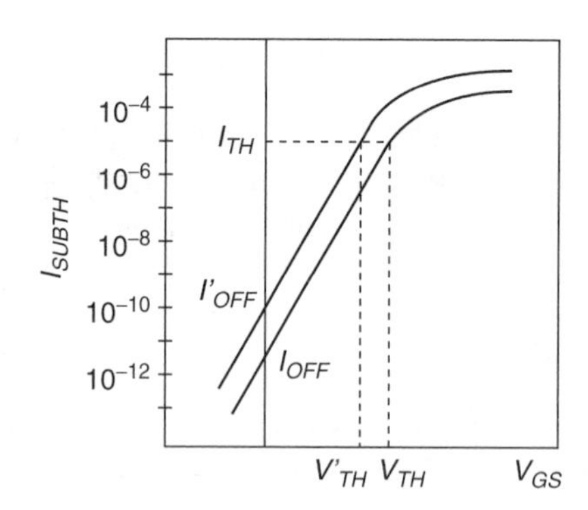

FIGURE 3.14 Impact of scaling on subthreshold leakage current.

In CMOS technologies, new doping profiles have been introduced to reduce short-channel and DIBL effect. By counter-doping the channel surface, threshold voltages can be fixed while still obtaining acceptable V_{TH} roll-off characteristics [47]. In addition, the locally high doping concentration in the channel near source-drain junctions (also known as *pocket* or *halo* implants) improves the short-channel effects [48]; however, many trade-offs exist between the improvement of short-channel immunity and the other device electrical performance that needs further research. These doping techniques may enhance I_{GIDL} and BTBT in general.

Scaled technologies have thin oxides and high channel doping concentrations. If the power supply voltage is kept at a high value, high normal electric fields may appear and, therefore, GIDL currents, even using LDD [22]. In addition, Guo *et al.* [19] have shown that the difference of GIDL between single-diffused drain (SD) devices and diffused drain (DD) devices is reduced when forward substrate bias is applied. Applying a band-trap-band tunneling model, they found the lateral electric field E_l due to the V_{BD} voltage and the ratio of lateral field to total field (E_l/E_{ox}) as the two key factors responsible for the tunneling barrier lowering and enhancement of I_{GIDL} [49].

3.3.3 Gate-Tunneling Currents

The exponential increase of direct-tunneling current should be considered due to the reduction of the oxide thickness in order to better control the inversion layer by the gate should be considered. Previous results show the high sensitivity of direct tunneling on oxide thickness and power supply voltage. Furthermore, as mentioned in previous section, scaled gate oxide thickness approaches the direct-tunneling regime, the EDT of electron from n^+ poly-Si to underlying n-type drain dominates the gate leakage. This phenomenon is more pronounced for thinner oxide thicknesses. At 1-V operation, the direct-tunneling current remains high for a gate oxide thickness of about 2 or 2.5 nm [2]. Every 0.2-nm reduction between 2 and 1 nm implies an increase of an order of magnitude of gate-tunneling current. For devices with 2-, 1.5-, and 1-nm oxide depths, the direct-tunneling current would be around 5 pA/μm, 2 nA/μm, and 50 μA/μm, respectively. A direct-tunneling current of 5 pA/μm is acceptable for high performance circuits, and 2 nA/μm is approaching the upper limit of OFF-state subthreshold leakage current (I_{OFF}); however, 50 μA/μm is certainly not acceptable.

Considering operation at high temperatures (105°C), the tunneling current is found to be ten times lower than the subthreshold leakage, since the latter increases rapidly with temperature while the former does not. Therefore, it is projected that gate oxide can be scaled to 1.5–2.0 nm before running into such a limit. Below these limits, the gate tunneling current quickly becomes problematic. When the gate becomes thinner, extremely larger direct tunneling hinders the formation of the inversion layer, and the drain current will not increase. Therefore, the limit of gate oxide thickness would be between 1.2–1.5 nm. This sets a limit for bulk CMOS scaling [45]. A solution is the use of high dielectric insulators [6, 50–53, 67]. If a good high-k dielectric insulator is developed and the direct tunneling suppressed, the gate oxide thickness may be reduced to 0.7–0.5 nm [45]. In conclusion, Figure 3.15 plots data on the

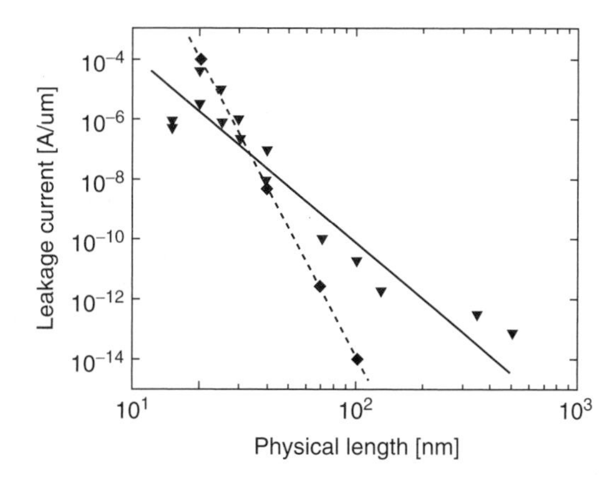

FIGURE 3.15 Trends in leakage current components as technology scales down based on published data: gate current (♦), subthreshold current (▼).

increasing leakage (subthreshold leakage current and gate tunneling current) of deep submicron transistors from different sources [10, 57, 65] and the trends computed using the expressions presented in previous sections. Notice, however, that prediction on future leakage current values, which is always based on the Semiconductor Industry Association (SIA) Roadmap [56], may be inaccurate because equal geometry transistors may have different current characteristics.

For future sub-40-nm technologies, with highly doped pn junctions, the narrow depletion region may produce worse problems due to the already mentioned BTBT currents [1, 27]. The very thin depletion depths needed in future CMOS require very high doping concentration, perhaps into the 5×10^{18} cm^{-3} range for sub-40-nm MOSFETs. At these doping levels, junction-tunneling current may appear [1, 39, 67]. Further, the combination of these effects with other effects associated to controllability and reliability of the MOSFET devices at these very small dimensions are becoming an issue. For instance, lithography variation and doping fluctuation in channel regions affecting V_{TH} control [54, 55]. To avoid these problems, novel three-dimensional (3D) double-gate transistor structures are emerging [39, 50].

3.4 Circuit Level

This section addresses the estimation and computation of leakage current in basic CMOS gates. CMOS circuits are built by series-parallel combination networks of MOS transistors. This implies that I_{LEAK} is also dependent on the input vector and circuit state: as the input vector applied to the circuit changes, the configuration of the transistors also changes. For circuits driving only subthreshold leakage currents, this dependence has been studied extensively [59–61], including the well-known stack effect. When gate currents are taken into account, the estimation of the leakage current is complicated by the state dependence of both the gate and subthreshold currents. I_{SUBTH} through OFF transistors and I_G through both ON and OFF transistors combine at internal nodes. In general, these currents are interdependent and must be analyzed simultaneously. This topic is currently an active area of research [62–64].

The models presented in Section 3.2 have been used to analyze the behavior and compute the leakage current of basic CMOS gates. At cell level, the current consumption depending on the input vector for each kind of cell in the circuit is obtained using SPICE simulation of the circuit with BSIM3 or BSIM4 (without I_G) transistor models and the added antiparallel current sources with an exponential dependence. The model is also very useful in order to highlight the current sharing between gates. Other models such as Lee et al. [63] or Mukhopadhyay et al. [64] produce similar results.

For instance, let us consider several inverters in series as illustrated in Figure 3.16. Two possible situations are shown: input V_G low (second inverter in the chain) and input V_G high (second inverter in the chain). All tunneling paths are also illustrated. In this case, for inverters with high V_G, the subthreshold current of NMOS transistor combines with the gate current of the previous transistor, while for inverters

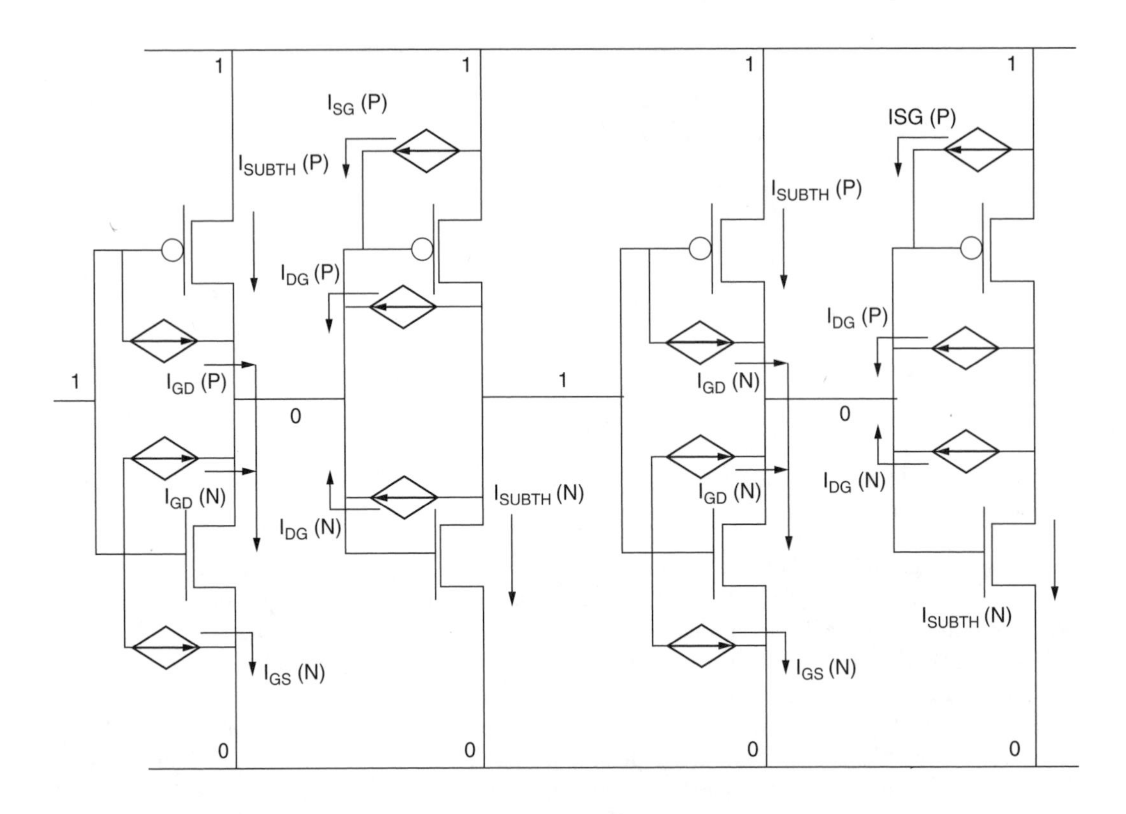

FIGURE 3.16 Inverter chain showing all gate tunneling and subthreshold leakage current using the MOSFET model including antiparallel contributing current sources.

with high V_G the subthreshold current of PMOS transistor combines with the gate current of the next transistor. In this case, gate and subthreshold currents can be computed independently.

Let us analyze the behavior of the NAND cell (Figure 3.17). The analysis of the NOR cell is similar. The input 11 produces an output equal to 0, and the leakage current is simply the sum of the subthreshold leakage current of the PMOS transistors and the SDE-to-gate tunneling current of PMOS and NMOS transistors. The leakage current flows from V_{DD} to GND internally through the NAND gate.

The inputs 00-01-10 produce an output of 1. In these cases, the internal path to GND is blocked by one or two transistors. Let us examine these cases in more detail:

1. AB = 10. Transistor N1 (connected to ground) is ON while transistor N2 (connected to VDD through the PMOS) is OFF. In this case, tunneling currents and subthreshold currents may be computed separately again. Notice that the current is shared between the NAND gate and the driving gate.
2. AB = 01. Transistor N1 (connected to ground) is OFF, while transistor N2 (connected to V_{DD} through the PMOS) is ON. In this case, the drain of the N1 transistor is held at V_{DD}–V_{TH}. Therefore, the voltage applied to the current source I_{GS} (N2) is one order of magnitude smaller than the previous case while the I_{DG} (N1) has an applied voltage equal to V_{DD}–V_{TH} and, although smaller than the previous case, cannot be neglected.
3. AB = 00. Both transistors are OFF. In this case, the subthreshold current exhibits the stack effect and the internal nodes has a voltage in the range of $\eta \cdot V_{DD}$ = 100–200mV (η models the DIBL effect). The tunneling currents I_{DG} (N1) and I_{SG} (N2) are one order of magnitude smaller than case a) and can be neglected.

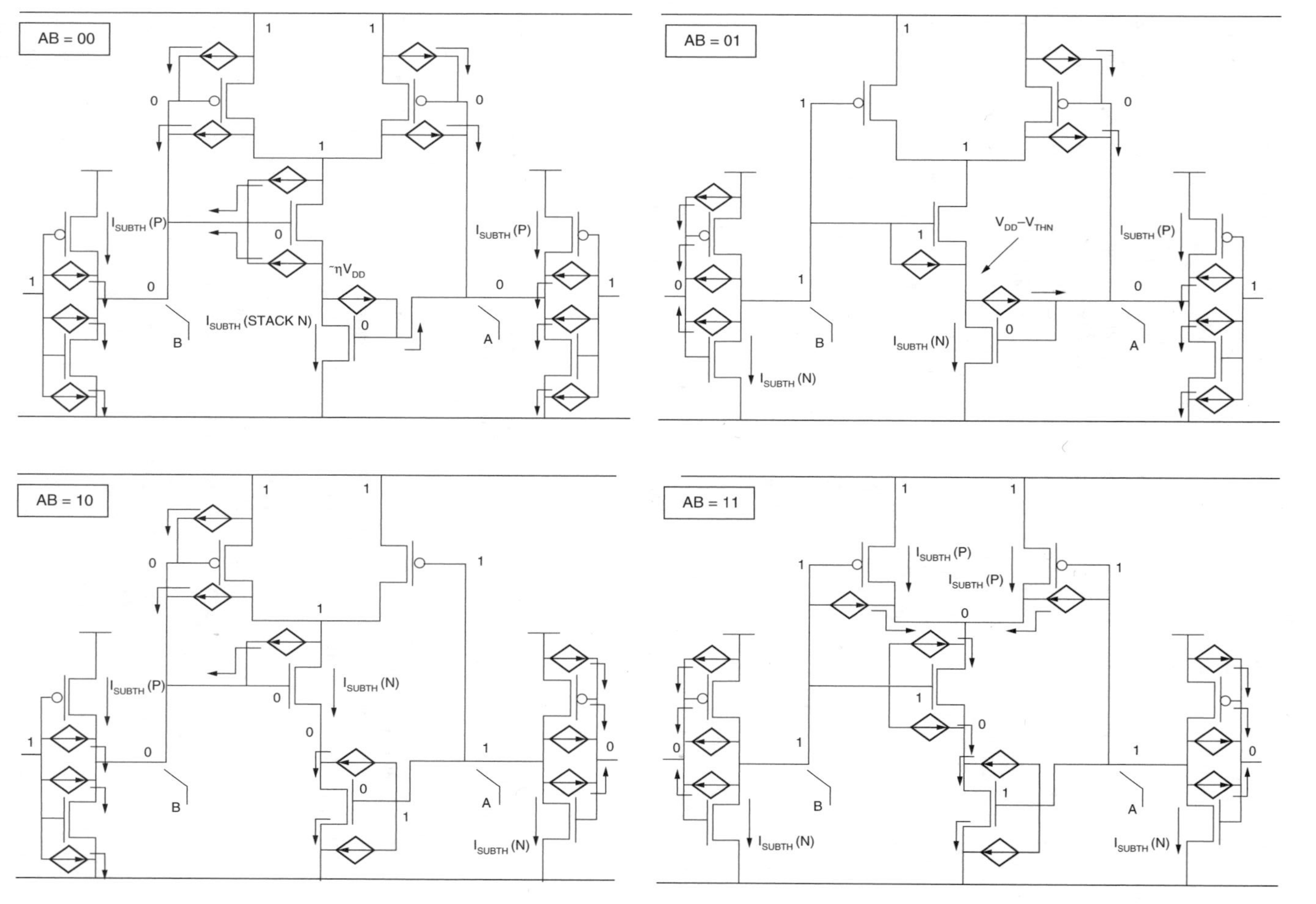

FIGURE 3.17 Two-input NAND showing all gate tunneling and subthreshold leakage current using the MOSFET model including antiparallel current sources.

In three input NAND gates, the different cases are very similar to these except the case 010. In this case, well discussed in Lee et al. [63], the current flowing from the SDE of the MOSFET in the middle of the stack increases the internal voltages and produces a reduction of the subthreshold leakage current. Therefore, the gate tunneling current therefore effectively displaces the subthreshold current, leaving the total leakage current relatively unchanged.

From these examples, it is clear that the EDT currents can degrade the performance of CMOS gates by increasing the OFF-state driving current. Therefore, an analysis of the leakage current and its bounds including gate-tunneling current are mandatory to avoid undesirable and unexpected functioning of the circuit.

For circuits with subthreshold leakage current, different methods to estimate bounds on total leakage consumption depending on the state have been presented [61]. In general, for very large circuits, it was found that the structural dependence between cells, and therefore, between the leakage current of these cells, decreased with the distance (in terms of levels) between the gates. Then, for large enough circuits, leakage current distributions were found to be Gaussian or nearly Gaussian. In other words, for large enough circuits with subthreshold leakage current as dominant leakage, some parts may be set independently in a certain logic state and control the leakage consumption [61].

When gate current is taken into account, some new issues appear. The first difference between the state dependence of I_{SUBTH} and I_G is that the value of I_{SUBTH} depends of the number of stacked transistors, while I_G depends strongly on the position of the OFF transistors in the stack [63]. Second, the current flowing in actual circuits with complex gates and the structural dependence between cells have not been deeply investigated.

These issues should be addressed to find maximum or minimum I_{LEAK} and the vectors producing these extremes. Because only in small circuits and in some special cases, it is possible to find exactly the maximum and minimum I_{LEAK}, for large circuits, the use of heuristics in order to find input vectors producing near maximum and minimum leakage currents is requested. The feasibility of using previously developed methodologies should be investigated and new methods should be introduced.

3.5 Conclusions

As CMOS technologies are scaled to nanometric ranges, the power consumption caused by leakage currents is becoming a significant part of the global power consumption. This fact has motivated a growing interest from technologists to very large scale integration (VLSI) designers to the leakage mechanisms influencing the different leakage currents and its impact on the future of CMOS.

This chapter has attempted to provide the necessary physical concepts to understand the causes of leakage and the main technology factors used to quantify the degree of leakage. The impact of scaling on the different components is expected to help in predicting the future evolution of leakage power at device and circuit levels.

The different leakage components of MOS transistors — gate tunneling, subthreshold conduction, GIDL, junction leakage, and punchthrough leakage — have been analyzed. For each component, the dependence on technology parameters, power supply, and temperature has been quantified to assess its importance as technologies evolve.

The impact of gate tunneling currents and its future trends has been studied. It is interesting to note that current available data show that for technologies with thin gate dielectrics below 40 nm for SiO_2 the contribution of the tunneling currents may become dominant over the subthreshold leakage.

The estimation of leakage at circuit level is of prime importance for VLSI designers to explore adequate solutions in the design space. The estimation of the leakage power in nanometric CMOS is a challenging problem due to the fact that gate tunneling provides new consumption paths involving driver and load gates. A model to help in estimating these currents has been proposed and used to estimate the leakage in simple circuits. Current research efforts in this area were reported.

References

[1] H-S. P. Wong *et al.* Nanoscale CMOS. *Proc. IEEE*, Vol. 87, April 1999.
[2] S. M. Sze, Ed. *Modern Semiconductor Device Physics.* John Wiley & Sons, New York, 1998.
[3] S. M. Sze, Ed. *High-Speed Semiconductor Devices.* John Wiley & Sons, New York, 1990.
[4] K-F. You and C-Y. Wu. A new quasi-2-D model for hot-carrier band-to-band tunneling current. IEEE *Trans. Electron. Devices*, Vol. 46, June 1999.
[5] M-J. Chen *et al.* Back-Gate Bias Enhanced Band-to-Band Tunneling Leakage in Scaled MOSFETs. *IEEE Electron. Device Lett.*, Vol. 19, April 1998.
[6] C.T. Liu. Circuit requirement and integration challenges of thin gate dielectrics for ultra small MOSFETs. In *IEDM Tech. Dig.*, pp. 747–750, 1998.
[7] H-D. Lee and J-M. Hwang. Accurate extraction of reverse leakage current components of shallow silicided p^+-n junction for quarter- and sub-quarter-micron MOSFETs. *IEEE Trans. Electron. Devices*, Vol. 45, August 1998.
[8] Y. Murakami and T. Shingyouji. Separation and analysis of diffusion and generation components of pn junction leakage current in various silicon wafers. *J. Applied Physics*, Vol. 75, April 1994.
[9] Y. Taur *et al.* CMOS scaling into the nanometer regime. *Proc. IEEE*, Vol. 85, April 1997.
[10] A. Keshavarzi, K. Roy, and C. F. Hawkins. Intrinsic IDDQ: origins, reduction, and applications in deep sub-um low power CMOS ICs. *Proc. Int. Test Conf. (ITC)*, pp. 167–176, 1997.
[11] A. Keshavarzi, K. Roy, and C. F. Hawkins. Intrinsic leakage in deep submicron CMOS ICs. Measurement-based test solutions. *IEEE Trans. VLSI Syst.*, Vol. 8, December 2000.
[12] D. Josephson, M. Storey, and D. Dixon. Microprocessor IDDQ testing: a case study. *IEEE Design & Test of Computers*, Vol. 12, Summer 1995.
[13] P. C. Maxwell and J. R. Rearick. A simulation-based method for estimating defect-free IDDQ. *IEEE Int. Workshop on IDDQ Testing, Digest of Papers*, pp. 80–84, 1997.
[14] G. Massobrio and P. Antognetti. *Semiconductor Device Modeling with SPICE.* McGraw-Hill, New York, 1993.
[15] T. A. Fjeldly and M. Shur. Threshold voltage modeling and the subthreshold regime of operation of short-channel MOSFETs. *IEEE Trans. Electron. Devices*, Vol. 40, January 1993.
[16] R. Velghe, D. Klaassen, and F. Klaassen. MOS Model 9, Level 902. Technical report, also available at http://www.semiconductors.philips.com/Philips_Models/Retrieval date:2004.
[17] Y. P. Tsividis. *Operating and Modeling of the MOS Transistor.* McGraw-Hill, New York, 1999.
[18] M. Rosar, B. Leroy, and G. Schweeger. A new model for the description of gate voltage and temperature dependence of gate-induced drain leakage (GIDL) in the low electric field region. *IEEE Trans. Electron. Devices*, Vol. 47, January 2000.
[19] J-Y. Guo *et al.* A three-terminal band-trap-band tunneling models for drain engineering and substrate bias effect on GIDL in MOSFET. *IEEE Trans. Electron. Devices*, Vol. 45, July 1998.
[20] M. Tanizawa *et al.* A complete substrate current model including band-to-band tunneling current for circuit simulation. *IEEE Trans. Computer-Aided Design*, Vol. 12, November 1993.
[21] N. Lindert *et al.* Comparison of GIDL in p^+-poly PMOS and n^+-poly PMOS devices. *IEEE Electron. Device Lett.*, Vol. 17, June 1996.
[22] R. Ghodsi, S. Sharifzadeh, and J. Majjiga. Gate-induced drain leakage in buried-channel PMOS — a limiting factor in developement of low-cost, high-performance 3.3-V, 0.25-mm technology. *IEEE Electron. Device Lett.*, Vol. 19, September 1998.
[23] M. Stadele, B. R. Tuttle, and K. Hess. Tunneling through ultrathin sio2 gate oxides from microscopic models. *J. Applied Physics*, Vol. 89, January 2001.
[24] S. T. Ma and J. R. Brews. Comparison of deep-submicrometer conventional and retrograde n-MOSFETS. *IEEE Trans. Electron. Devices*, Vol. 47, August 2000.
[25] B. Majkusiak and M. H. Badri. Semiconductor thickness and back-gate voltage effects on the gate tunnel current in the MOS/SOI system with an ultrathin oxide. *IEEE Trans. Electron. Devices*, Vol. 47, December 2000.

[26] S.-H. Lo, D. A. Buchanan, and Y. Taur. Modeling and characterization of quantization, poly-Si depletion, and direct tunneling effects in MOSFETs with ultrathin oxides. *IBM J. Research Dev.*, Vol. 43, No. 3, 1999.

[27] C. A. Mead. Scaling of MOS technology to submicrometer feature sizes. *Analog Integrated Circuits-Signal Processing,* Vol. 6, No. 1, 1994.

[28] L. Larcher, A. Paccagnella, and G. Ghidini. Gate current in ultrathin MOS capacitors: a new model of tunnel current. *IEEE Trans. Electron. Devices,* Vol. 48, February 2001.

[29] K. N. Yang *et al.* Characterization and modeling of edge direct tunneling (EDT) leakage in ultrathin gate oxide MOSFETs. *IEEE Trans. on Electron. Devices,* Vol. 48, June 2001.

[30] W. K. Henson *et al.* Analysis of leakage currents and impact on OFF-state power consumption for CMOS technology in the 100-nm regime. *IEEE Trans. Electron. Devices,* Vol. 47, July 2000.

[31] N. Yang, W. K. Henson, and J. J. Wortman. A comparative study of gate direct tunneling and drain leakage currents in n-MOSFETs with sub-2-nm gate oxides. *IEEE Trans. Electron. Devices,* Vol. 47, August 2000.

[32] Z. Yu *et al.* Impact of gate direct tunneling current on circuit performance: a simulation study. *IEEE Trans. Electron. Devices,* Vol. 48, December 2001.

[33] W-C. Lee and C. Hu. Modeling CMOS tunneling currents through ultrathin gate oxide due to conduction- and valence-band electron and hole tunneling. *IEEE Trans. Electron. Devices,* Vol. 48, July 2001.

[34] K. M. Cao *et al.* BSIM4 gate leakage model including source-drain partition. *IEDM Tech. Dig.,* pp. 815–818, 2000.

[35] G. W. Neudeck. *The PN Junction Diode, 2nd ed.,* John Wiley & Sons, New York, 1988.

[36] M. T. Bohr. Nanotechnology goals and challenges for electronic applications. *IEEE Trans. Nanotechnology,* Vol. 1, March 2002.

[37] Y-S. Lin *et al.* Leakage scaling in deep submicron CMOS for SoC. *IEEE Trans. Electron. Devices,* Vol. 49, June 2002.

[38] Y-S. Lin, *et al.* On the SiO2-based gate dielectric scaling limit for low-standby power applications in the context of a 0.13-μm CMOS logic technology. *IEEE Trans. Electron. Devices,* Vol. 49, March 2002.

[39] D. J. Frank *et al.* Device scaling limits for Si MOSFETs and their application dependencies. *Proc. of the IEEE,* Vol. 89, March 2001.

[40] A. O. Adan and K. Higashi. OFF-state leakage current mechanisms in BulkSi and SOI MOSFETs and their impact on CMOS ULSIs standby current. *IEEE Trans. Electron. Devices,* Vol. 48, September 2001.

[41] K. A. Bowman et al. A circuit-level perspective of the optimum gate oxide thickness. *IEEE Trans. Electron. Devices,* Vol. 48, August 2001.

[42] R. D. Isaac. The future of CMOS technology. *IBM J. Research Dev.,* Vol. 44, No. 3, 2000.

[43] A. J. Bhavnagarwala et al. A minimum total power methodology for projecting limits on CMOS GSI. *IEEE Trans. VLSI Syst.,* Vol. 8, May 2000.

[44] F. Assad *et al.* On the performance limits for Si MOSFETs: a theoretical study. *IEEE Trans. Electron. Devices,* Vol. 47, January 2000.

[45] H. Iwai. CMOS technology — year 2010 and beyond. *IEEE J. Solid-State Circuits,* Vol. 34, March 1999.

[46] J. J. Sun *et al.* The effect of the elevated source/drain doping profile on performance and reliability of deep submicron MOSFETs. *IEEE Trans. Electron. Devices,* Vol. 44, June 1997.

[47] D. Hisamoto *et al.* A low-resistance self-aligned T-shaped gate for high-performance sub-0.1-mm CMOS. *IEEE Trans. Electron. Devices,* Vol. 44, June 1997.

[48] B. Yu *et al.* Short-channel effect improved by lateral channel-engineering in deep-submicronmeter MOSFETs. *IEEE Trans. Electron. Devices,* Vol. 44, April 1997.

[49] S-C. Lin *et al.* A closed-form back-gate-bias related inverse narrow-channel effect model for deep-submicron VLSI CMOS devices using shallow trench isolation. *IEEE Trans. Electron. Devices*, Vol. 47, April 2000.

[50] B. Doyle *et al.* Transistor elements for 30nm physical gate lengths and beyond. *Intel Technol. J.*, Vol. 06, June 2002.

[51] S. Mudanai *et al.* Modeling of direct tunneling current through gate dielectric stacks. *IEEE Trans. Electron. Devices*, Vol. 47, October 2000.

[52] I. C. Kizilyalli *et al.* MOS transistors with stacked SiO_2-Ta_2O_5-SiO_2 gate dielectrics for giga-scale integration of CMOS technologies. IEEE Electron Device Letters, Vol. 19, November 1998.

[53] Y-C. Yeo *et al.* MOSFET gate leakage modeling and selection guide for alternative gate dielectrics based on leakage considerations. *IEEE Trans. Electron. Devices*, Vol. 50, April 2003.

[54] M. Koh *et al.* Limit of gate oxide thickness scaling in MOSFETs due to apparent threshold voltage fluctuation induced by tunnel leakage current. *IEEE Trans. Electron. Devices*, Vol. 48, February 2001.

[55] K. Takeuchi, R. Koh, and T. Mogami. A study of the threshold voltage variation for ultra-small bulk and SOI CMOS. *IEEE Trans. Electron. Devices*, Vol. 48, September 2001.

[56] National Technology Roadmap for Semiconductors 2002. Available at http://public.itrs.net/, 2002.

[57] B. Davari, R.H. Dennard, and G. G. Shahidi, CMOS scaling for high performance and low power-the next ten years. Proc. of the IEEE, Vol. 83, April 1995.

[58] K. Roy, S. Mukhopadhyay, and H. Mahmoodi-Meimand. Leakage current mechanisms and leakage reduction techniques in deep-submicrometer CMOS circuits. *Proc. of the IEEE*, Vol. 91, February 2003.

[59] R. X. Gu and M. I. Elmasry, Power dissipation analysis and optimization of deep submicron CMOS digital circuits. *IEEE J. Solid-State Circuits*, Vol. 31, May 1996.

[60] R. M. C. Johnson, D. Somasekhar, and K. Roy, Models and algorithms for bounds on leakage in CMOS circuits. *IEEE Trans. Computer-Aided Design*, Vol. 18, June 1999.

[61] A. Ferré and J. Figueras. Leakage power bounds in CMOS digital technologies. *IEEE Trans. Computer-Aided Design*, Vol. 21, June 2002.

[62] A. Ferré and J. Figueras. Leakage power analysis considering gate tunneling currents. DEE - UPC Internal Report, No. 03/06, April 2003.

[63] D. Lee *et al.* Analysis and minimization techniques for total leakage considering gate oxide leakage. *Proc. of the DAC*, pp. 175–180, 2003.

[64] S. Mukhopadhyay, A. Raychowdhury, and K. Roy. Accurate estimation of total leakage current in scaled CMOS logic circuits based on compact current modeling. *Proc. of the DAC*, pp. 169–174, 2003.

[65] T. Mak, Leakages and its implication to test, Private Communication, June 2002.

[66] Y-F. Tsai *et al.* Implications of technology scaling on leakage reduction techniques, *Proc. of the DAC*, pp. 187–190, 2003.

[67] Brian Doyle *et al.* Transistor elements for 30-nm physical gate lengths and beyond, *Intel Technol. J.*, May 2002. Available online.

4

Microelectronics, Nanoelectronics, and the Future of Electronics

Jing Wang
Mark Lundstrom
Purdue University

4.1 Introduction

Silicon technology continues to progress rapidly, with current generation technologies having physical gate lengths well below 100 nm. At the same time, remarkable advances in nonsilicon nano- and molecular technologies are occurring. It is time to think seriously about the role that nanoelectronics and nontraditional technologies could play in future electronic systems. Moore's law describes device scaling-down in integrated circuits, which has led an unprecedented growth of the semiconductor industry. At the same time, it also carried device researchers into the nano world. Well-established concepts from mesoscopic physics [1] are now entering the working knowledge of device physicists and engineers as silicon transistors enter the nanoscale [2]. At the micrometer scale, transistors were well described by drift-diffusion equations, but now people are beginning to use a new language to describe nanoscale transistors. In addition, several interesting new devices that may have important applications are also being developed [3–6].

Nanoelectronics can play an important role in future electronic systems, if the design community is engaged to exploit the opportunities that nanoelectronics offers. Therefore, we appreciate this opportunity to give an overview of the current developments of nanoscale transistors. The chapter begins by defining nanotechnology and discussing how a metal-oxide-semiconductor field-effect transistor (MOSFET) performs in the nanometer regime (Section 4.2), then examines the ultimate scaling limit and

0-8493-1941-2/05/$0.00+$1.50

practical limits of the silicon MOSFET (Section 4.3 and Section 4.4). After that, several new types of field-effect transistors (FETs) are introduced, which may become the substitutes for the silicon MOSFET (Section 4.5) and other nanotransistors beyond the FET (Section 4.6). Several important issues in the research of nanoelectronics are also discussed (Section 4.7). To be concise, we do not include the detailed mathematical formalism of the device theory, but the references are listed to help the reader who has particular interests find the sources.

4.2 The Silicon MOSFET as a Nanoelectronic Device

4.2.1 What Is Nanotechnology?

Nanotechnology has been defined as work at the 1–100-nm length scale to produce structures, devices, and systems that have novel properties because of their nanoscale dimensions [7]. Some insist that two dimensions lie in the 1–100 nm regime, which would rule out traditional technologies such as thin films. A key part of the definition is that new phenomena occur (caused, for example, by the dominance of interfaces and quantum mechanical effects), and that these new phenomena may be exploited to improve the performance of materials, devices, and systems. Nanotechnologies also involve the manipulation and control of matter at the nanoscale. Semiconductor technology does much of this with a "top-down" approach that lithographically imposes a pattern, and then etches away bulk material to create a nanostructure. Some argue that self-assembly is an essential component of nanotechnology. The hope is that nanostructures can be self-assembled from the "bottom up," molecule by molecule. We argue that current-day silicon technology meets the definition of nanoelectronics, that future silicon technologies will meet it even better, and that nontraditional technologies could play an important role in future electronic systems by complementing the capabilities of nanoscale silicon technology, rather than by attempting to replace it.

4.2.2 Silicon MOSFETs in the Nanometer Regime

The International Technology Roadmap for Semiconductors (ITRS) [8] calls for 9-nm physical gate lengths for integrated circuit (IC) transistors in 2016. At the same time, major IC manufacturers have reported transistors with 10-nm (or shorter) gate lengths on IEDM 2002 [9,10], which demonstrate the promise of pushing IC technology to the 10-nm regime.

To scale silicon transistors down to the 10-nm scale, new device structures are needed to suppress the short channel effects [11]. Figure 4.1 is a schematic illustration of a fully depleted, double-gate (DG) MOSFET, a device that offers good prospects for scaling silicon transistors to their limits [12]. Other approaches (e.g., the FinFET [9] and the tri-gate MOSFET [13]) are also being explored. At a 9-nm gate length, acceptable short-channel effects require a fully depleted silicon body thickness of 3 nm or less, and an equivalent gate oxide thickness of less than 1 nm. At such dimensions, the properties of the silicon material will be affected by quantum confinement (e.g., the bandgap will increase), and device properties will be influenced by quantum transport.

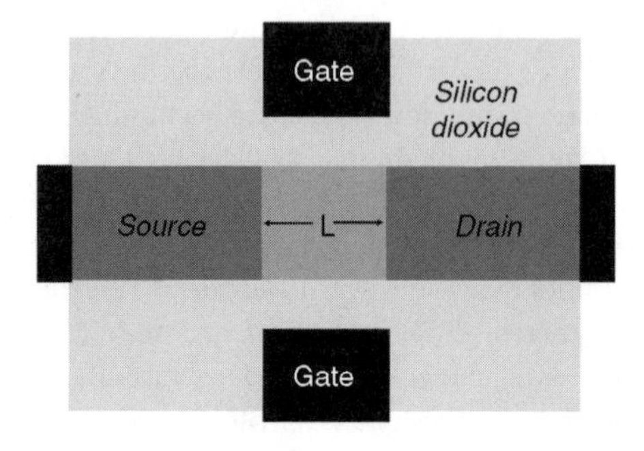

FIGURE 4.1 The double-gate MOSFET structure.

Traditional device equations are based on the drift-diffusion theory [11], which assumes that the device scale is much larger than the electron wavelength (~ 8nm at room temperature) and the electron mean-free-path (the average distance an electron travels between two collisions, ~ 10nm for electrons in the inversion layer). The first assumption allows us to treat electrons as classical particles with zero size, and the second one justifies the "local transport" property (the electron velocity at a position is solely determined by the local electric field and mobility). Unfortunately, at the nanoscale, neither of these assumptions is well satisfied. As a result, to capture the new physical effects that occur at the nanoscale, the old device theory must be modified or even completely replaced by a new quantum transport theory.

Four important phenomena need to be properly treated in the modeling of nanotransistors:

1. Quantum confinement
2. Gate tunneling
3. Quasi-ballistic transport
4. Source-to-drain (S/D) tunneling

The first two effects occur in the confinement direction (normal to the gate electrode(s)) of the MOSFET. As silicon technology entered the sub-100-nm regime (the corresponding oxide thickness < 3nm), those effects became significant and began to affect the MOSFET threshold voltage and leakage currents in the "OFF-state." Extensive work has been done to explore the physics of the first two effects, and numerous device models have been developed to capture them in device and circuit simulations [14–17]. (Here, we will not give the details of those models. Readers with particular interests should refer to the related references.) In contrast to the quantum confinement and gate tunneling, quasi-ballistic transport and source-to-drain tunneling begin to significantly affect the device performance of the silicon MOSFET when the gate length scales down to 10 nm or less [18]. Therefore, the exploration of these mesoscopic transport effects is important for the description of silicon MOSFETs at their scaling limit, as well as the understanding of device physics of other nanoscale devices (to be discussed later). In the following paragraph, a simple description of the ballistic/quasi-ballistic transport [19–23] is presented, which gives us the upper performance limit of nanoscale transistors. Section 4.3 discusses the source-to-drain tunneling in silicon MOSFETs at the scaling limit.

In a conventional MOSFET, the channel length is much longer than the electron mean-free-path, so an electron will experience numerous collisions during its travel from the source to the drain. Nevertheless, when the channel length shrinks to less than the mean-free-path, an electron may go through the channel with no or little scattering, which is called ballistic/quasi-ballistic transport. According to the quasi-ballistic transport theory [2,22,23], the current under low drain bias can be written as (assuming nondegenerate statistics),

$$I_{DS} = \frac{\lambda}{L+\lambda} W Q_i(0) \frac{\upsilon_T}{2k_BT} V_{DS} \tag{4.1}$$

where λ is the electron mean-free-path, L is the channel length of the MOSFET, $Q_i(0)$ is the sheet electron density at the beginning of the channel, $V_T \sqrt{2k_BT/\pi m^*}$ is the unidirectional thermal velocity of nondegenerate electrons, and other symbols have their common meanings. For a long channel device, $L >> \lambda$, so Equation 4.1 becomes

$$I_{DS} = W Q_i(0) \frac{\upsilon_T \lambda}{2k_BT} \frac{V_{DS}}{L} \tag{4.2}$$

Because the mobility for nondegenerate electrons can be defined as $\mu_0 = \upsilon_T \lambda / (2k_BT)$ [21], Equation 4.2 is simply the well-known classical device equation based on the drift-diffusion theory [11]. When $L << \lambda$, the current approaches its upper (ballistic) limit,

$$I_{DS} = I_{ballistic} = WQ_i(0)\frac{\upsilon_T}{2k_BT}V_{DS} \tag{4.3}$$

The point is that the conventional device equations are an approximation valid when $L >> \lambda$. As MOSFET channel lengths approach the nanoscale, the classical MOSFET equations must be modified to capture quasi-ballistic transport. It is important for both device simulation and the development of circuit models for nanotransistors.

4.3 Ultimate Limits of the Silicon MOSFET

As the gate lengths of Si MOSFETs continue to shrink, the two-dimensional (2D) electrostatics become increasingly important, which causes the well-known short-channel effects (SCEs). At the same time, for the MOSFET with a gate length < 10nm, the quantum mechanical tunneling from source to drain may also be significant. It will degrade the subthreshold slope and increase the leakage current in the OFF-state. According to our previous work [18], the ultimate scaling limit of Si MOSFETs is determined by both the semiclassical SCEs (i.e., DIBL, V_T roll-off) and the S/D tunneling.

In Wang and Lundstrom [18], S/D tunneling has been extensively examined using the nonequilibrium Green's function (NEGF) approach [24], a general and rigorous quantum model for nanoscale transistors. (The 2D quantum simulator for double-gate Si MOSFETs, nanoMOS-2.5, is available at http://nano-hub.purdue.edu.) The main conclusions are summarized next:

1. For the well-designed devices (with very thin silicon body and oxide layers that provide good electrostatics), S/D tunneling sets an ultimate scaling limit that is well below 10 nm.
2. S/D tunneling dominates OFF-current in the devices at scaling limit, and it may play an important role in the ON-state of ballistic devices.
3. Due to S/D tunneling, the sub-threshold slope saturates at low temperature (see Figure 4.2). Therefore, the leakage current in the OFF-state may still be high even at low temperature.

We also found that for a double-gate MOSFET with a 1-nm-thick silicon body and 0.6-nm(equivalent)-thick oxide layers, S/D tunneling sets a scaling limit of L = 5 nm if we require that *the subthreshold swing*

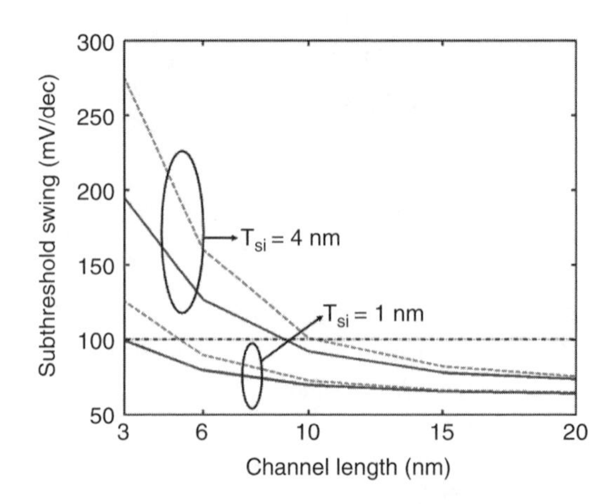

FIGURE 4.2 The subthreshold swing vs. temperature. The simulated device structure is a double-gate MOSFET with 0.6-nm(equivalent)-thick oxide layers. Two silicon body thicknesses (1 nm and 4 nm) are adopted in this simulation. The solid curves are for the semiclassical Boltzmann simulation (without S/D tunneling), while the dashed curves are for the quantum NEGF simulation (with S/D tunneling). (Obtained from J. Wang and M. Lundstrom, Does source-to-drain tunneling limit the ultimate scaling of MOSFETs? *IEEE Int. Electron. Devices Meeting (IEDM), Tech. Dig.*, pp. 707–710, San Francisco, CA, Dec. 2002. With permission.)

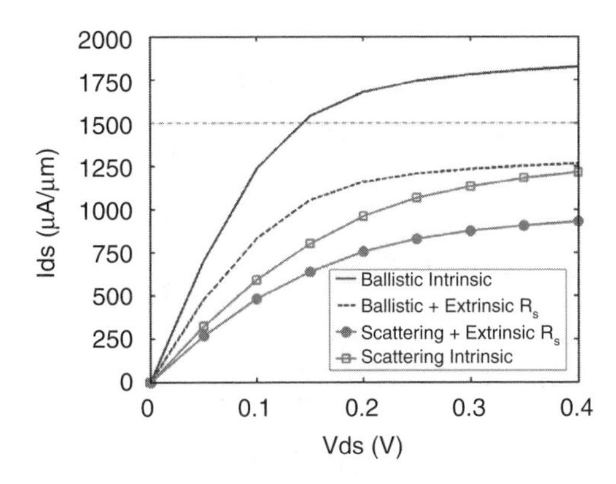

FIGURE 4.3 The top curve is the intrinsic ballistic current I^i_{ball}, and the dashed curve is I^e_{ball}, so the top two curves represent the ballistic intrinsic device. The bottom two are for intrinsic device with scattering. The curve with square markers represents I^i_{scatt}, and the fourth curve is I^e_{scatt}. (Obtained from S. Hasan, J. Wang, and M. Lundstrom, Device design and manufacturing issues for 10nm-scale MOSFETs: a computational study, *Solid State Electronics,* 48, 6, 867–875, 2004. With permission.)

is smaller than 100 mV/dec and the ON-OFF current ratio is larger than 100. Obviously, we could have different criteria to determine the ultimate scaling limit of a MOSFET. Likharev [25] proposed a criterion that the voltage gain of a CMOS inverter is larger than one, and used it to find a scaling limit of L = 2 nm. So two very important questions arise: How is the scaling limit of a MOSFET determined? What is the *worst* performance of a transistor that can be accepted by a very large scale integration (VLSI) circuit designer to build an IC chip? Clear answers to these questions require cooperation between device researchers and circuit engineers.

4.4 Practical Limits of the Silicon MOSFET

Section 4.3 discussed the ultimate scaling limit of a MOSFET. In practice, some technical issues (e.g., the source/drain series resistances, process variations, and power dissipation) may greatly affect device performance and set practical limits for MOSFETs.

In Hasan et al. [26], a computational study of the end-of-roadmap (L_G = 9nm) MOSFETs (high-performance) was presented. It was found that:

1. With a double-gate structure and a 3-nm-thick silicon body, the 10-nm-scale MOSFET can be realized but the ON-current is ~ 40% below the ITRS prediction. S/D series resistance and low gate overdrive ($V_{GS}-V_T$) were identified as limiting factors for the ON-current (see Figure 4.3 for details).
2. Process variations will seriously affect the device performance for the 10-nm-scale MOSFET. For example, a single monolayer (~ 0.3nm) variation in the silicon body thickness will cause more than 50% variation in the OFF-current (see Figure 4.4).

In summary, as the silicon MOSFET approaches its scaling limit, maintaining drive current at low supply voltages (~ 0.5V) will be very difficult, and device parasitics will be much more important than for current technology. Devices will be extremely sensitive to manufacturing variations. New design techniques will be needed to make use of devices with low drive current, high leakage, and large process variations.

4.5 Beyond the Silicon MOSFET

This section discusses several new types of FETs that are being explored by device physicists and engineers. Those devices could become either substitutes for the silicon MOSFET or complementary circuit elements that might be implemented into silicon IC circuits to improve their performance in the nanometer regime.

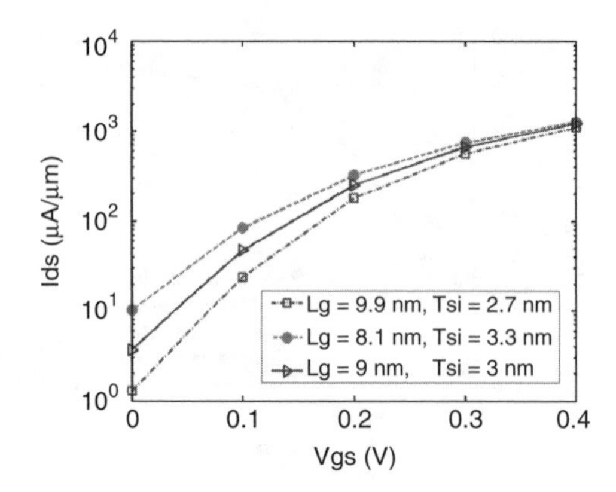

FIGURE 4.4 Intrinsic transfer characteristics of three different transistors. The top curve represents the worst transistor in terms of SCE, with L_G 10% smaller and t_{si} 10% larger, the middle one is the nominal device, and the bottom one represents the best device, with L_G 10% larger and t_{si} 10% smaller. (Obtained from S. Hasan, J. Wang, and M. Lundstrom, Device design and manufacturing issues for 10nm-scale MOSFETs: a computational study, *Solid State Electronics,* 48, 6, 867–875, 2004. With permission.)

4.5.1 Carbon Nanotube Transistors

One can think of a carbon nanotube as a 2D sheet of graphene (in which carbon atoms in a hexagonal lattice are bonded to three nearest neighbors as illustrated in Figure 4.5) that is rolled up into a tube. Depending on how the sheet is rolled up to produce a tube (in a "zigzag" pattern, "armchair," or in between (chiral), the nanotube can be either metallic or semiconducting). For semiconducting tubes, the bandgap is inversely proportional to the nanotube diameter. A diameter of 1 nm (a typical value) gives a bandgap of about 0.8 eV.

The interest in carbon nanotubes arises from the unique material properties they display. The one-dimensional (1D) energy band structure suppresses scattering, so ballistic transport can be achieved over relatively long distances. The thermal conductivity is exceptional, even higher than diamond, and nanotubes display excellent resistance to electromigration. These properties make nanotubes interesting for interconnects and heat removal in gigascale systems. Semiconducting nanotubes also display excellent transport properties, and the absence of dangling bonds may make it easier to incorporate high-K gate dielectrics into carbon nanotube field effect transistors (CNTFETs). Because the valence and conduction bands are mirror images of each other, n-type and p-type transistors should display essentially identical

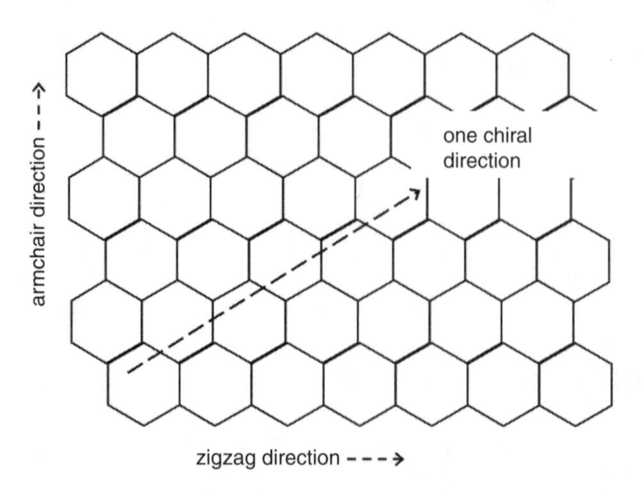

FIGURE 4.5 A 2D sheet of graphene showing the roll-up directions for different nanotubes.

characteristics, a significant advantage for complementary metal-oxide semiconductor (CMOS) circuits. Initially, CNTFETs suffered from high series resistance and low gate capacitance. Improved contacts are being developed, and new structures employ high-K gate dielectrics. The ITRS calls for an ON-current of 750 μA/μm (0.75 μA/nm) for PMOS transistors in 2016, which will be very difficult to meet by the silicon material at the low supply voltages needed (V_{DD} ~ 0.5 V). Experimental CNTFETs have already achieved over 7 μA/nm at 0.9 V [27].

It is clear that carbon nanotubes have great promise, but what are the challenges? The growth of CNTs with well-defined electronic properties is a critical issue. Growth from a catalytic seed can be used to control the CNT diameter, but it is more difficult to control the CNTs chirality (i.e., how it is rolled up). For applications in terascale systems, we will need to grow at least 10^{12} CNTs — all semiconducting with well-controlled diameters. Device structures and process flows are still primitive. One approach is to produce planar FETs with arrays of CNTs to provide sufficient current for conventional digital applications [28]. This approach aims to replace the silicon CMOS transistor with a higher-performance device. Another approach would be to explore the use of single nanotube electronics in dense locally interconnected architectures that could complement silicon CMOS. As CNT materials and device work proceeds, work at the system design level is needed to identify the most promising opportunities.

4.5.2 Organic Molecular Transistors

The organic molecular transistor is another possibility for post-CMOS devices. Figure 4.6 shows a schematic structure of the molecular FET. Compared with silicon MOSFETs and other nanotransistors, molecular FETs might have advantages on both fabrication and device performance.

1. The fabrication of molecular FETs could be with low cost, high controllability, and reproducibility. As we know, to fabricate a silicon MOSFET at the nanoscale, lithography and etching technology with extremely high resolution (< 10nm) is required, which may greatly increase the cost of IC fabrication. Moreover, the variations in lithography and etching can seriously affect the device performance. For CNTFETs (see previous paragraph), although the high-resolution lithography may not be needed for the device fabrication, the variations (i.e., the chirality and diameter of a CNT) from tube to tube could affect the controllability and reproducibility of the circuits. In contrast to silicon MOSFETs and CNTFETs, a molecular FET with numerous identical molecules might be realized at quite low cost by using the self-assembly technology [29]. The FET channel length is naturally equal to the length of the molecules so that the process variations would be effectively suppressed.
2. The molecular transistor has special physical properties that may be exploited to enhance the device performance. First, there could be no dopants in a molecular transistor. The type (n or p) of the FET can be determined by the gate work function [30]. As a result, the scattering inside the channel would be reduced. Considering the extremely short channel length of a molecular FET, transport inside the channel could be ballistic. Second, molecules are flexible and tunable,

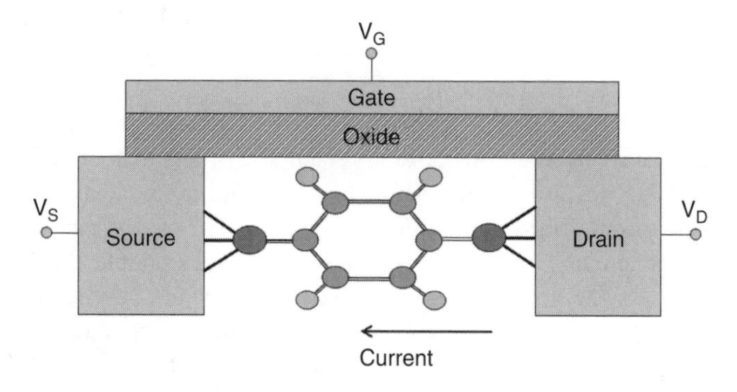

FIGURE 4.6 A schematic structure of the organic molecular transistor.

so it may be possible to control the molecules' shape (conformation) by a gate voltage [31]. This opens up the possibility of a molecular relay with a subthreshold swing better than the thermal-emission limit, $2.3k_BT/q$ (60 mV/dec at room temperature) [31]. Initial studies, however, show that thermal fluctuations of the flexible molecule are a serious issue [31].

As for any other nanotransistor, the organic molecular FET has its own challenges. Due to its extremely short channel, the molecular FET may seriously suffer the 2D electrostatics (so-called short channel effects, SCEs) (e.g., a 3-nm channel length may require a 0.2-nm equivalent oxide thickness to achieve good electrostatics, which is very difficult to realize in practice) [32]. The relatively low drive current may also limit its application as a logic circuit element. Therefore, the optimization of device performance becomes important for the future application of molecular transistors.

4.5.3 MOSFETs with New Channel Materials and Semiconductor Nanowire Transistors

A well-designed transistor should have an efficient gate control and good transport property (high channel mobility), so to improve the device performance of silicon MOSFETs, researchers are trying to exploit new channel materials and new gate geometry configurations.

Extensive experimental work [33–35] has been done on germanium and strained silicon, promising new channel materials that could provide higher mobility for both electrons and holes. On the other hand, silicon nanowire transistors are also being explored [6,36,37]. Such a 1D structure provides a possibility to make tri-gate or gate-all-around transistors that offer the best gate control. (Because the device physics of those transistors is similar to that of the silicon MOSFET, we do not discuss the details here.) With new channel materials or new gate geometry configurations, it may be possible to scale the MOSFET beyond the scaling limit of the planar silicon MOSFET.

4.6 Beyond the FET

Nanotechnology will not only provide the fabrication techniques to build nanoscale FETs, but also make it possible to realize some quantum-effects devices with special applications in the future electronics. Indeed, the most promising applications of molecular electronics may not be to replace Si MOSFETs but, instead, to complement CMOS with new capabilities. This section discusses two examples: the single electron transistor and the spin transistor.

4.6.1 Single-Electron Transistors

For future nanoscale transistors, the total number of electrons in the channel may be approximately 10, but a single-electron transistor (SET) is not just a smaller version of the same device. To produce a single electron transistor, the size of the "island" between the source and drain must be small enough so that the change in voltage due to a single electron is large compared to the thermal energy:

$$q^2/2C_G >> k_BT \tag{4.4}$$

Reliable room temperature operation requires an island size of less than about 1 nm, the size of a small molecule. In addition, we also require that the source and drain be weakly coupled to the gated island, which is usually accomplished by introducing tunnel barriers at the two contacts. When these conditions are met, some unique I-V characteristics result [38]. For example, the number of electrons on the island changes in discrete steps as the gate voltage increases, and a "Coulomb blockade" prevents current flow until V_{DS} exceeds a critical value. The critical voltage for conduction is periodic in gate voltage. Single electron transistors have been investigated for applications in digital systems, but they have several limitations [38]. The voltage gain is low and so is the drive current (because the tunnel junctions introduce a large series resistance). As might be expected, they are also extremely sensitive to stray background charges. Certain hybrid SET/MOSFET circuits, however, combine single (or few) electron devices and CMOS transistors and have interesting possibilities for memory [38].

4.6.2 Spin Transistors

The operation of a conventional transistor is based on the charge that electrons carry, but electrons also carry spin, a fundamental unit of magnetic moment. The electron's spin is the basis for magnetic memories, but it is also conceivable that spin could be modulated by a gate to realize new types of devices [39]. For example, if the source and drain were ferromagnetic, then spin-polarized electrons might be injected into a semiconductor. If they retain their spin as they propagate across the channel, they could easily exit the ferromagnetic drain, but it may be possible to rotate the electron spins by a gate voltage thereby preventing them from exiting through the drain and contributing to the drain current.

Devices of this type have not yet been demonstrated, but current research is examining how to combine ferromagnetic metals and semiconductors, how to inject spin-polarized electrons into the semiconductor, and how to maintain the spin polarization once the electrons are in the semiconductor [40]. If devices of this type could be realized, they promise faster switching and lower switching energy than conventional electrostatic MOSFETs. Eventually, it may be possible to manipulate the spins of individual electrons (single electron spin transistors), which could lead to the realization of quantum computers.

4.7 From Microelectronics to Nanoelectronics

Nanoelectronics is not simply a smaller version of microelectronics; things change at the nanoscale. At the device level, silicon transistors may give way to new materials such as organic molecules or inorganic nanowires [41]. At the interconnect level, microelectronics uses long, fat wires, but nanoelectronics seeks to use short nanowires [41]. Fundamentally new architectures will be needed to make use of simple, locally connected structures that are imperfect and are comprised of devices whose performance varies widely.

We believe that 21st-century silicon technology has evolved into a true nanotechnology. Critical dimensions are already below 100 nm. The materials used in these silicon devices have properties that differ from the bulk. Nanoscale silicon transistors have higher leakage, lower drive current, and exhibit more variability from device to device. New circuits and architectures will need to be developed to accommodate such devices. It matters little whether the material is silicon or something else; the same issues face any nanoelectronics technology. It is likely that many of the advances and breakthroughs at the circuits and systems levels that will be needed to make nanoelectronics successful will come from the silicon design community.

Developing an understanding of how devices operate at the nanoscale is a good reason to support nanoscience research. Another reason is that devices to complement silicon technology might be discovered. For example, carbon nanotube FETs could be exquisite singlemolecule detectors, and SETs could be integrated with MOSFETs for high-density memory applications. Another possibility is molecular structures that improve the performance of a CMOS platform. For example, ballistic CNTs could be high performance interconnects and efficient at heat removal. Nanowire thermoelectric cooling could lower chip temperature and increase performance [42]. Therefore, research on nanoelectronics will prove to be a good investment for several reasons.

The successful development of nanoelectronics will require a partnership between science and engineering. It was the same for semiconductor technology. The scientific community developed the understanding of semiconductor materials and physics and the engineering community used this base to learn how to design devices, circuits, and systems. Figure 4.7 summarizes this partnership. Science works in the nanoworld with individual atoms, molecules, nanoscale structures and devices, and assembly processes. Systems engineers work in the macroworld on complex systems with terascale device densities. In the middle are the device and circuit engineers. They must learn to think and work at the nanoscale to build devices and circuits that can connect to the macroworld. Their job is to hide the complexity of

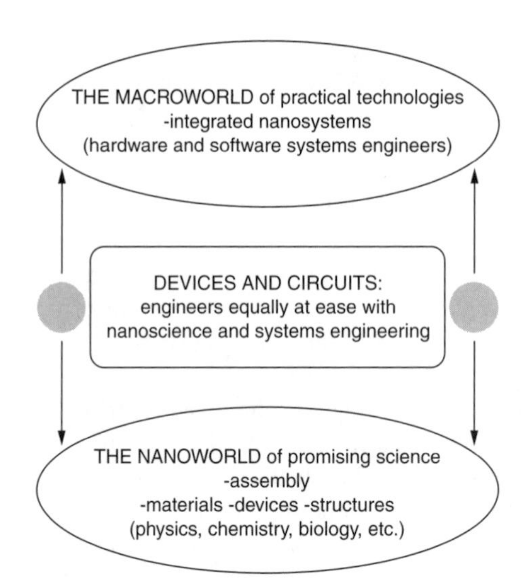

FIGURE 4.7 Science, engineering, and nanoelectronics.

the nanoscale device by packaging it in a form that systems engineers can use (e.g., a compact circuit model). To turn the promise of nanoscience into practical technologies, it is essential that the systems engineering community be engaged in the effort.

4.8 Conclusion

This chapter has introduced the emerging field of nanoelectronics — the new concepts and physical phenomena in the nanoscale MOSFET, the scaling limits of silicon MOSFETs, novel nanoscale FETs, quantum-effects devices with special applications, and the future of the electronics research. The scenario that we have outlined is an evolutionary one, but exponential evolution for another two to three decades would have a revolutionary impact on society. It is also true that it is hard to predict the future. Remember that the transistor was developed for a very specific purpose — to replace the vacuum tube; the integrated circuit was an unexpected bonus. The march of science and technology has carried us to the nanoscale; it is where the important questions are and where unforeseen breakthroughs may occur. Our march toward nanoelectronics is unstoppable. Who knows where it may lead.

4.9 Acknowledgments

It is our pleasure to acknowledge the contributions of Dr. Supriyo Datta, whose insights have deepened our understanding of conduction at the molecular scale. We also thank Sayed Hasan for providing the figures for Section 4.4. Our thanks also go to the sponsors of our work: the Semiconductor Research Corporation, the National Science Foundation, the Army Research Office (ARO) Defense University Research Initiative in Nanotechnology, and the Microelectronics Advanced Research Corporation (MARCO) Focused Research Center in Materials, Structures, and Devices (which is funded at the Massachusetts Institute of Technology, in part by MARCO under contract 2001-MT-887 and Defense Advanced Research Projects Administration (DARPA) under grant MDA972-01-1-0035).

References

[1] S. Datta, *Electronic Transport in Mesoscopic Systems,* Cambridge University Press, Cambridge, UK, 1999.

[2] M.S. Lundstrom, Elementary scattering theory of the Si MOSFET, IEEE *Electron. Dev. Lett.*, 18, 361–363, 1997.
[3] C. Collier et al. Electronically configurable molecular-based logic gates, *Science*, 285, 391–394, 1999.
[4] J. Chen, M.A. Reed, A.M. Rawlett, and J.M. Tour, Large on-off ratios and negative differential resistance in a molecular electronic device, *Science*, 286, 1550, 1999.
[5] P.L. McEuen, M.S. Fuhrer, and H. Park, Single-walled carbon nanotube electronics, IEEE *Trans. Nanotechnol.*, 1, 78–85, 2002.
[6] Y. Cui, and C.M. Lieber, Functional nanoscale electronic devices assembled using silicon nanowire building blocks, *Science*, 291, 851–853, 2001.
[7] National Nanotechnology Initiative, http://www.nano.gov, 2004.
[8] Semiconductor Industry Association (SIA), *The International Technology Roadmap for Semiconductors*, 2001.
[9] B. Yu, L. Chang, S. Ahmed, H. Wang, S. Bell, C.-Y. Yang, et al. FinFET Scaling 10-nm gate length, *IEEE Int. Electron. Devices Meeting (IEDM), Tech. Dig.*, pp. 251–254, San Francisco, CA, Dec. 2002.
[10] B. Doris, M.I.E. Ong, T. Kanarsky, Y. Zhang, R.A. Roy, O. Dokumaci, et al. Extreme scaling with ultra-thin silicon channel MOSFET's (XFET), *IEEE Int. Electron. Devices Meeting (IEDM), Tech. Dig.*, pp. 267–270, San Francisco, CA, Dec. 2002.
[11] Y. Yaur and T.H. Ning, *Fundamentals of Modern VLSI Devices*, Cambridge University Press, Cambridge, UK, 1998.
[12] D.J. Frank, R.H. Dennard, E. Nowak, P.M. Solomon, Y. Taur, and H.S.P. Wong, Device scaling limits of Si MOSFETs and their application dependencies, *Proc. IEEE*, 89, 259–288, 2001.
[13] B. Doyle, B. Boyanov, S. Datta, M. Doczy, S. Hareland, B. Jin, et al. Tri-gate fully depleted CMOS transistors: fabrication, design, and layout, *2003 Symp. on VLSI Technol.*, Kyoto, Japan, June 2003.
[14] S.H. Lo, D.A. Buchanan, Y. Taur, and W. Wang, Quantum-mechanical modeling of electron tunneling current from the inversion layer of ultra-thin-oxide nMOSFET's, *IEIEEE Electron. Dev. Lett.*, 18, 5, 209–211, 1997.
[15] W.K. Shih, E.X. Wang, S. Jallepalli, F. Leon, C.M. Maziar, and A.F. Taschjr, Modeling gate leakage current in nMOS structures due to tunneling through an ultra-thin oxide, *Solid-State Electron.*, 42, 6, 997–1006, 1998.
[16] N. Yang, W.K. Henson, J.R. Hauser, and J.J. Wortman, Modeling study of ultrathin gate oxides using direct tunneling current and capacitance-voltage measurements in MOS devices, *IEEE Trans. Electron Dev.*, 46, 7, 1464–1471, 1999.
[17] J. Wang, Y. Ma, L. Tian, and Z. Li, Modified airy function method for modeling of direct tunneling current in metal-oxide-semiconductor structures, *Applied Physics Letters*, 79, 12, 1831–1833, 2001.
[18] J. Wang and M. Lundstrom, Does source-to-drain tunneling limit the ultimate scaling of MOSFETs? *IEEE Int. Electron. Devices Meeting (IEDM), Tech. Dig.*, pp. 707–710, San Francisco, CA, Dec. 2002.
[19] K. Natori, Ballistic metal-oxide-semiconductor field effect transistor, *J. Appl. Phys.*, 76, 8, 4879–4890 1994.
[20] Z. Ren, R. Venugopal, S. Datta, M. Lundstrom, D. Jovanovic, and J.G. Fossum, The ballistic nanotransistor: a simulation study, *IEEE Int. Electron. Devices Meeting (IEDM), Tech. Dig.*, pp. 715–718, Dec. 2000.
[21] A. Rahman and M. Lundstrom, A compact scattering model for the nanoscale double-gate MOSFET, *IEEE Trans. Electron. Dev.*, 49, 3, 481–489, 2002.
[22] A. Rahman, J. Guo, S. Datta, and M. Lundstrom, Theory of ballistic nanotransistors, *IEEE Trans. Electron. Dev.*, 50, 9, 1853–1864, 2003.
[23] J. Wang and M. Lundstrom, Ballistic transport in high electron mobility transistors, *IEEE Trans. Electron. Dev.*, 50, 7, 1604–1609, 2003.
[24] S. Datta, Nanoscale device modeling: the Green's function method, *Superlattices and Microstructures*, 28, 253–278, 2000.

[25] K. Likharev, Electronics below 10nm, *Giga and Nano Challenges in Microelectron.*, J. Greer, A. Korkin, and J. Lanbanowski, Eds., North-Holland, 2003.
[26] S. Hasan, J. Wang, and M. Lundstrom, Device design and manufacturing issues for 10nm-scale MOSFETs: a computational study, *Solid State Electronics,* 48, 6, 867–875, 2004.
[27] S. Rosenblatt, Y. Yaish, J. Park, J. Gore, V. Sazonova, and P.L. McEuen, High-performance electrolyte-gated carbon nanotube transistors, *Nano Letters,* 2, 869–872, 2002.
[28] R. Saito, G. Dresselhaus, and M. Dresselhaus, *Physical Properties of Carbon Nanotubes,* Imperial College Press, London, 1998.
[29] J.M. Tour, A.M. Rawlett, M. Kozaki, Y.X. Yao, R.C. Jagessar, S.M. Dirk, et al. Synthesis and preliminary testing of molecular wires and devices, *Chemistry-A European J.*, 7, 23, 5118–5134, 2001.
[30] A. Ghosh, personal communication, May 2003.
[31] A.W. Ghosh, T. Rakshit, and S. Datta, Gating of a molecular transistor: electrostatic and conformational, *Nano Letters,* 4, 4, 565–568, 2004.
[32] P. Damle, T. Rakshit, M. Paulsson and S. Datta, Current-voltage characteristics of molecular conductors: two versus three terminal, *IEEE Trans. Nanotechnology,* 1, 3, 145–153, 2002.
[33] C-O. Chui, S. Ramanathan, B.B. Triplet, P.C. McIntyre, and K.C. Saraswat, Germanium MOS capacitors incorporating ultrathin high-κ gate dielectric, *IEEE Electron. Device Lett.*, 23, 8, 473–475, 2002.
[34] J. Hoyt, H. Nayfeh, S. Eguchi, I. Aberg, G. Xia, T. Drake, et al. Strained silicon MOSFET technology. Invited paper, *IEEE Int. Electron. Devices Meeting (IEDM), Tech. Dig.*, pp. 23–26, San Francisco, CA, Dec. 2002.
[35] K. Rim, S. Narasimha, M. Longstreet, A. Mocuta, and J. Cai, Low field mobility characteristics of sub-100nm unstrained and strained Si MOSFETs, *IEEE Int. Electron. Devices Meeting (IEDM), Tech. Dig.*, pp. 43–46, San Francisco, CA, Dec. 2002.
[36] T. Saito, T. Saraya, T. Inukai, H. Majima, T. Nagumo, and T. Hiramoto, Suppression of short-channel effect in triangular parallel wire channel MOSFETs, *IEICE Trans. Electron.*, E85-C (5), 2002.
[37] H. Majima, Y. Saito, and T. Hiramoto, Impact of Quantum mechanical effects on design of nano-scale narrow channel n- and p-type MOSFETs, *IEEE Int. Electron. Devices Meeting (IEDM), Tech. Dig.*, pp. 733–736, Washington, D.C., Dec. 2001.
[38] K. Likharev, Sub-20-nm electron devices, *Advanced Semiconductor and Organic Nano-Technologies, Part 1,* H. Morkoc, Ed., Academic Press, New York, 2003.
[39] S. Datta, and B.A. Das, Electronic analog of the electro-optic modulator, *Appl. Phys. Lett.*, 56, 665–667, 1990.
[40] D. Awschalom, M.E. Flatte, and N. Samarth, Spintronics, *Scientific American,* 67–73, June 2002.
[41] C.M. Lieber, The incredible shrinking circuit, *Scientific American,* 285, 3, 58–65, 2001.
[42] A. Seabaugh, T. Blake, B. Brar, T. Broekaert, R. Lake, F. Morris, and G. Frazier, Transistors and tunnel diodes for analog/mixed signal circuits and embedded memory, *IEEE Int. Electron. Devices Meeting (IEDM), Tech. Dig.*, pp. 429–432, San Francisco, Dec. 1998.

5

Advanced Research in On-Chip Optical Interconnects

Ian O'Connor
Frédéric Gaffiot
Ecole Centrale de Lyon

5.1 The Interconnect Problem

Due to continually shrinking feature sizes, higher clock frequencies, and the simultaneous growth in complexity, the role of interconnect as a dominant factor in determining circuit performance is growing in importance. The 2001 International Technology Roadmap for Semiconductors (ITRS) [1] shows that by 2010, high-performance integrated circuits (ICs) will count up to 2 billion transistors per chip and work with clock frequencies of the order of 10 GHz. Coping with electrical interconnects under these conditions will be a formidable task. Timing is already no longer the sole concern with physical layout: power consumption, cross talk, and voltage drop drastically increase the complexity of the trade-off problem. With decreasing device dimensions, it is increasingly difficult to keep wire propagation delays acceptable. Whereas dielectric constants below 2 (around 1.7–1.8) can be achieved using nanoporous silicon oxycarbide (SiOC)-like or organic (SiK-type) materials with an "air gap" integration approach, integration complexity is higher and mechanical properties are weaker. In addition, the use of ultra low-k materials is physically limited by the fact that no material permittivity can be less than 1 — that of

0-8493-1941-2/05/$0.00+$1.50

air. Thus, even with the most optimistic estimates for resistance-capacitance (RC) time constants using low-resistance metals, such as copper and low-k dielectrics, global interconnect performance required for future generations of integrated circuits (ICs) cannot be achieved with metal. Furthermore, because IC power dissipation is strongly linked to switching frequency, tomorrow's architectures will require power over the 100-W mark to be able to operate in the 10-GHz range and above. At this level, thermal problems will jeopardize system performance if not strictly controlled.

5.1.1 Analysis of Electrical Interconnect Performance

The overall device scaling factor *s* makes it possible to determine the performance of interconnects. Each process shrink has a large impact on electrical parameters of metallic interconnections. Before deep submicron (DSM) nodes (the threshold is widely accepted as being around the 0.35-μm technology node), the gate delay was higher than the interconnect delay, such that each shrink led to an improvement of the maximum working speed of a system by a factor of *1/s*. In the present DSM era, however, global interconnect delay has become larger than gate delay and, consequently, interconnect has become the dominant factor determining speed.

Sakurai and Tamaru's equation (Equation 5.2) gives the propagation delay of a signal transmitted from an emitter gate to a receiver gate.

$$t_d = R_{out}\left(C_{out} + C_L\right) + R_{out}cl_W + 0{,}4\left[\left(crl_W^2\right)^{1,6} + \tau_{tof}^{1,6}\right]^{1/1,6} + 0{,}7rl_wC_L \tag{5.1}$$

In this expression, R_{out} and C_{out} are, respectively, the output resistance and output capacitance of the emitter gate; C_L is the input capacitance of the receiver gate; *r*, *c*, and l_w are the lineic resistance, the lineic capacitance, and the length of the link between the emitter and the receiver; and τ_{tof} is the time of flight (i.e., the length of the line divided by the speed of the electromagnetic field).

In the case of local and intermediate interconnects, this equation reduces to:

$$t_d = R_{out}\left(C_{out} + C_L\right) + R_{out}cl_W \tag{5.2}$$

Equation 5.2 shows that the delay time is a combination of the gate output resistance, interconnect, and load capacitances. Gate sizing makes it possible to reduce delay by increasing gate strength, at the cost of increased area and power consumption.

In the case of global links, Sakurai's formula shows that the delay time in the line becomes predominant. To limit the delay time in the metallic line, global links are routed on the upper metal layers where it is possible to increase the width and the thickness of the line, and thus to reduce the lineic resistance. Reverse scaling (by reducing the thickness of the metal layer less than the scale factor) is commonplace, leading to high aspect ratios: Gate sizing makes it possible to minimize t_d, and it is possible to show that t_d varies with l_w^2. This increase of the delay time with the second power of the line length cannot be avoided. Repeater insertion makes it possible to make the delay vary with l_w, but this of course comes at the cost of a very large number of repeaters. In this scenario therefore, a relatively high percentage of silicon real estate and IC power consumption is devoted to interconnect instead of to data processing functions.

Subsequently, the problem facing us is that evolutionary solutions will not be sufficient to meet the performance roadmap. To tackle the issues developed previously, radically different interconnect approaches displaying a highly improved data-rate-to-power ratio must be developed. At present, the most prominent ideas are the use of integrated radio frequency or microwave interconnects [3], three-dimensional (3D) (nonplanar) integration [4], and optical interconnects [5]. This chapter focuses on the latter concept.

5.1.2 The Optical Alternative

A promising approach to the interconnect problem is the use of an optical interconnect layer. Such a layer could empower an enormous bandwidth increase, immunity to electromagnetic noise, a decrease in the power consumption, synchronous operation within the circuit and with other circuits, and reduced immunity to temperature variations. Important constraints when developing the optical interconnect layer are the fact that all fabrication steps have to be compatible with future IC technology and that the additional cost incurred remains affordable. Difficulties expected are obtaining a large enough optical-electrical conversion efficiency, reducing the optical transmission losses while allowing for a sufficient density of photonic waveguides on the circuit and reduction of the latency while operating above the 10-GHz mark. Sections 5.3, 5.4, and 5.5 describe, respectively, the issues involved in photonic waveguides, active devices, and optoelectronic conversion circuits.

5.1.3 Identified Applications

Optical links can be categorized into three broad domains, for which various analyses have been carried out and applications identified: single wavelength point-to-point (1-1 link); single and multiple wavelength broadcast (1-n link); multiple wavelength bus and switching (n-n link). The latter category is rather new and is discussed in Section 5.8.

5.1.3.1 Point-to-Point (1-1) Links

Today, complex chips typically need hundreds or thousands of global links [6]. The basic idea behind using point-to-point optical links consists of replacing electrical global links with optical ones. Research has been carried out on analyzing the benefits of introducing optical interconnect in critical data-intensive links, such as CPU-memory buses in processor architectures [7]. These analyses showed that point-to-point links do not present a sufficiently high performance gain to warrant their widespread use in future technologies. In essence, the bandwidth/power ratio for point-to-point optical links is higher than for electrical wires, but not high enough, when interface circuit power is taken into consideration. Instead, it is preferable to apply architectural modifications in order to enable bottlenecks to be overcome (in the given example application, the solution was to add more cache memory), even at the expense of greater silicon area and power. The benefits of optical interconnect in terms of physical cost in this situation are, in the long run, not viable for industrial manufacturers because the entire manufacturing process (from design to fabrication) would have to be changed: a very costly course of action. This proves that for optical interconnect to be accepted as a real alternative to metallic interconnect, performance gains of at least one order of magnitude must be demonstrated through circuit and device research advances, as well as through application targeting.

5.1.3.2 Broadcast (1-n) Links

Another and potentially more profitable application of optical interconnect technology is in clock distribution networks (CDN) [8]. To operate at high frequencies, CDNs require several hundreds of repeaters to drive the metallic tracks over the entire chip, resulting in using a high portion of overall IC power (up to 40–50%). This mode of operation also leads to stringent constraints on the design of the clock tree because an unbalanced tree will result in serious clock skew and, consequently, system failure. An electrical alternative is global clock distribution at a relatively low frequency and local clock multiplication to generate the required clock speed. Disadvantages of this approach include interzone synchronization and clock multiplication lock time. By replacing the electrical clock distribution tree by an optical one, the need for repeaters or clock multiplier circuits would be eliminated, thus reducing power consumption and clock skew; however, it would be illusory to believe that the optical clock signal could be routed down to the single-gate level: optoelectronic interface circuits are of course necessary and consume power. An example system realizing a clock distribution function, illustrated in Figure 5.1, requires a single photonic source coupled to a symmetrical waveguide structure routing to a number of optical receivers. At the receivers, the high-speed optical signal is converted to an electrical one and provided to local

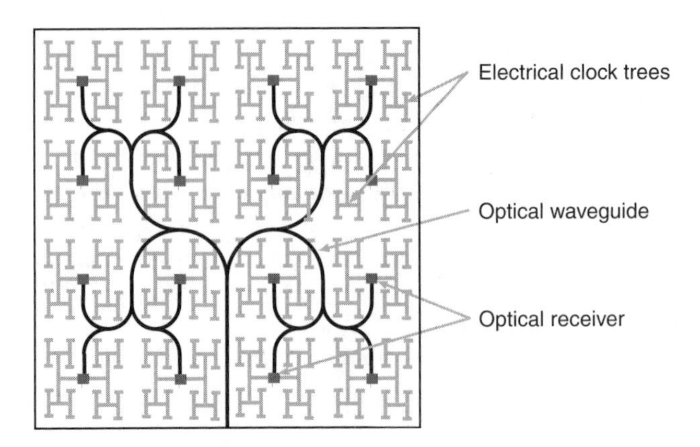

FIGURE 5.1 1-16 point optical clock distribution tree.

electrical networks. The number of clock distribution points is a particularly crucial parameter in the overall system.

5.2 Top-Down Link Design

5.2.1 Technology

Various technological solutions may be proposed for integrating an optical transport layer in a standard complementary metal oxide semiconductor (CMOS) system. The first choice is to specify where this optical layer has to be placed. Then one has to choose the different materials used for the active devices and the passive transport layer.

5.2.1.1 Materials

Materials have to be chosen with different constraints:

- Efficient light detection. Obviously, the active devices are of fundamental importance to the power budget of the link. Optics is suitable only if, for a given throughput, the global power consumption of the whole link is lower than the power consumption of classical metallic links. The quantum efficiency of the active devices is of prime importance in this context. In addition, particular attention has to be paid to the receiver: the signal to noise ratio determines the minimal optical power at the detector.
- Efficient signal transport. Attenuation and compactness are the main parameters for the choice of the passive waveguide. Technological compatibility with mature existing technologies (to ensure the required reproducibility and homogeneity of the device parameters).
- Technological compatibility with standard CMOS processes. An industrial solution is conceivable only if the optical process is completely separated from the CMOS process (the development cost of a new CMOS technology is so high that it seems very difficult to propose a solution which would require a fundamental rethink of IC fabrication processes).

Different materials are available for the realization of the optical passive guides but we focus here on silicon/silica waveguides. Silicon is an excellent material for wavelengths above 1.2 μm, and monomode waveguiding with attenuation as low as 0.8 dB/cm has been proven [9]. Moreover, the high refractive index difference between silicon and silica makes it possible to realize passive structures with dimensions compatible with DSM technologies (for example, it is possible to realize monomode waveguides less than 1 μm wide). The realization of silicon/silica waveguides is (at least in principle) compatible with a standard CMOS process. The choice of silicon waveguides leads to the use of wavelengths greater than 1.2 μm. To capitalize on the maturity of devices and concepts developed by the telecommunications industry, the choice of the wavelength is in practice limited to 1.3–1.55-μm windows.

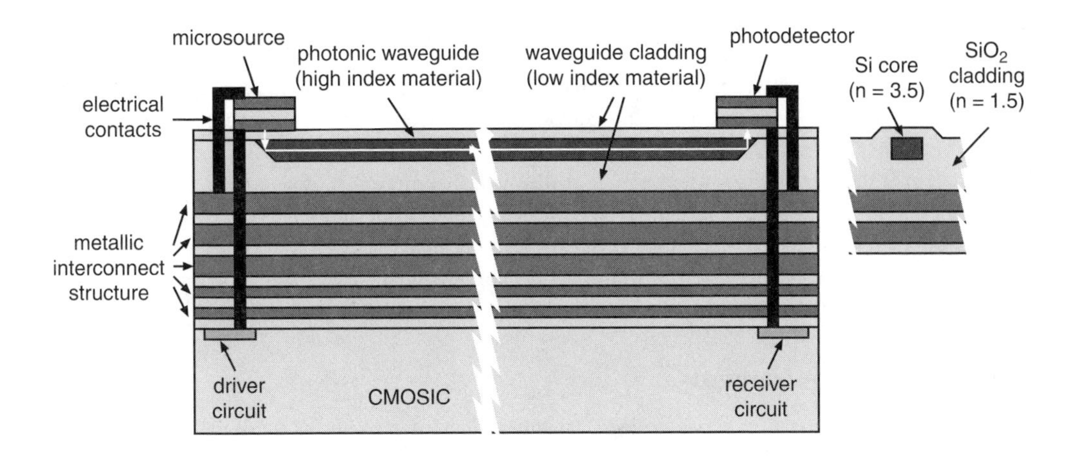

FIGURE 5.2 Cross section of hybridized interconnection structure.

5.2.1.2 Hybrid or Monolithic

The use of silicon waveguides makes it possible to imagine either monolithic (planar) or hybrid (3D) integration of the optical subsystem with CMOS systems. It is believed that the former solution is not realistic.

The integration of silicon waveguides at the front end of the CMOS process (i.e., before fabrication of the metallic interconnection layers) is certainly possible but other considerations have to be taken into account. At the transistor level, the routing of the waveguide is extremely difficult and requires routing space at the IC level. Further, the problem of the active devices remains: silicon-based sources cannot yet (and for the foreseeable future) be considered mature, while the growth of III-V devices on silicon faces strong technological barriers. The use of external sources and detectors bonded by flip-chip is unrealistic due to the high number of individual bonding operations required. In addition, silicon-based devices can only work at low wavelengths (850 nm), which translates to higher attenuation in the waveguides. This solution requires an extraordinary mutation in the CMOS process and, as such, is highly unattractive from an economic point of view.

Hybrid integration of the optical layer on top of a complete CMOS IC is much more practical and offers more scope for development. The source and detector devices are no longer bound to be realized in the host material. Figure 5.2 is a cross section of how a complete "above IC" photonic layer could be realized. The photonic source shown can be on- or off-chip: it seems likely that for some near-term applications, such as clock distribution, it is better to target off-chip signal generation for thermal reasons, even if it means higher assembly costs. It should be noted that this solution also applies to multichip module (MCM) technology. The optical process is completely independent from the CMOS process, which is appealing from an industrial point of view. Disadvantages of this approach include the more complex electrical link between the CMOS subcircuits (source drivers and detector amplifiers) and, inevitably, more advanced technological solutions for bonding.

In the system depicted in Figure 5.2, the microsource is coupled to the passive waveguide structure and provides a signal to an optical receiver (or possibly to several, as in the case of a broadcast function). At the receiver, the high-speed optical signal is converted to an electrical signal and, subsequently, distributed by a local electrical interconnect network.

To form a planar optical waveguide, silicon is used as the core and SiO_2 as the cladding material. Si/SiO_2 structures are compatible with conventional silicon technology and transparent for 1.3–1.55-μm wavelengths. Such waveguides with high relative refractive index difference $\Delta \approx (n_1^2\text{-}n_2^2)/2n_1^2$ between the core ($n_1 \approx 3.5$ for Si) and claddings ($n_2 \approx 1.5$ for SiO_2) allow the realization of a compact optical circuit, with bend radius of the order of a few μm [10]. To avoid modal dispersion, improve coupling efficiency, and reduce loss, single-mode conditions are applied to the waveguide dimensions. For a wavelength of 1.55 μm, this means a waveguide width of 0.3 μm.

The main criterion in evaluating the performance of digital transmission systems is the resulting bit error rate (BER), which may be defined as the rate of error occurrences. Typically, the BER figure required by Gigabit Ethernet and by Fiber Channel is 10^{-12} or better. For an on-chip interconnect network, a BER of 10^{-15} is acceptable. It should be noted here that BER is not commonly considered in IC design circles, and for good reason: metallic interconnects typically achieve BER figures better than 10^{-45}. Future operating frequencies are likely to change this, however, because the combination of necessarily faster rise and fall times, lower supply voltages, and higher cross talk increases the probability of wrongly interpreting the signal that was sent.

5.2.2 Design Requirements

To make a reasoned comparison between electrical and optical interconnect, a set of design requirements must be established. We have already mentioned BER; we must add to this the ubiquitous power/speed/area trinity found in any digital system. Power and speed can be compared directly, while area (in the 3D scenario) is more difficult to evaluate because we are essentially aiming at adding a photonic layer of the same size as the chip itself. What is important therefore is the average achievable area/bit ratio.

To evaluate and optimize link performance criteria correctly, predictive models and design methodologies are required. Concerning the power aspects, the aim is to establish the overall power dissipation for an optical link at a given data rate and BER. The receiver essentially conditions the calculation because the BER defines the lower limit for the received optical power. This lower limit can then be used to calculate the required power coupled into waveguides by optical sources, the required detector efficiency (including optical coupling), and acceptable transmission losses. Power can then be estimated from source bias current and photoreceiver front-end design methodologies.

For integration density aspects, source and detector sizes must be taken into account, while the width, pitch, and required bend radius of waveguides is fundamental to estimating the size of the photonic layer. On the circuit layer, the additional surface due to optical interconnect is in the driver and receiver circuits, as well as the contact and via stack to the photonic layer. The circuit layout problem is compounded by the necessity of using clean supply lines (i.e., separate from digital supplies) to reduce noise (for BER).

The data rate is essentially governed by the bandwidth of the photoreceiver: high modulation speed at the source is generally more easily attainable than similar detection speed at the receiver. This is essentially due to the photodiode parasitic capacitance at the input of the transimpedance amplifier.

Apart from these concerns, functional aspects also have to be considered. For example, using the same signal to drive two nodes is not trivial (as is the case in electrical interconnect) because the layout of a 1-2 splitter is crucial to the equal distribution of power to each node. More fundamentally, dividing the power has a direct influence on the power required at the source to achieve the lower power limit at the receiving nodes.

5.3 Passive Photonic Devices for Signal Routing

5.3.1 Waveguides

Optical system performance depends on the minimum optical power required by the receiver and on the efficiency of passive optical devices used in the system. The total loss in any optical link (represented in Figure 5.3) is the sum of losses (in decibels) of all optical components:

$$L_{total} = L_{CV} + L_B + L_Y + L_{CR} \tag{5.3}$$

where L_{CV} is the coupling coefficient between the photonic source and optical waveguide, L_W is the rectangular waveguide transmission loss, L_B is the bending loss, L_Y is the Y-coupler loss, and L_{CR} is the

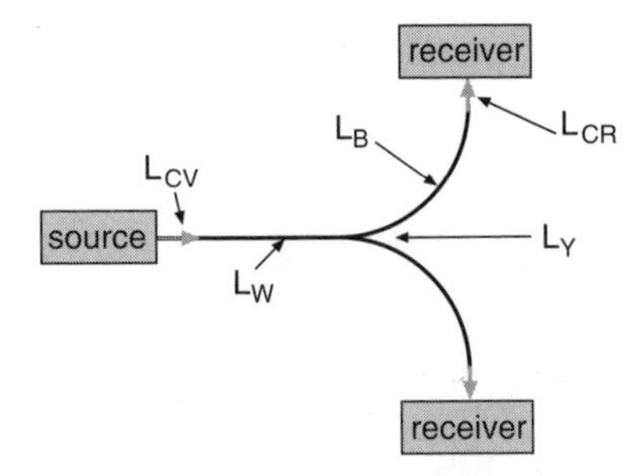

FIGURE 5.3 Losses in an optical link.

coupling loss from the waveguide to the optical receiver. To provide an unambiguous comparison in terms of dissipated power between optical and electrical on-chip interconnect networks, it is necessary to incorporate all of these quantities.

In the present technology, several methods are used to couple the beam emitted from the laser into the optical waveguide. In the proposed system, we assumed 50% coupling efficiency, L_{CV}, from the source to a single mode waveguide.

Transmission loss, L_W, describes the attenuation rate of the optical power, as light travels in the waveguide. Due to small waveguide dimensions and large index change at the core/cladding interface in the Si/SiO_2 waveguide the sidewall scattering is the dominant source of losses. To calculate the attenuation coefficient we used the Payne formula [11] associated with the Effective Index Method [12, 13]. For the waveguide fabricated by Lee et al. [9] with roughness of 2 nm, the calculated transmission loss is 1.3 dB/cm.

The bending loss L_B is highly dependent on the refractive index difference Δ between the core and cladding medium. For low Δ, the bending loss is very high, which prevents increasing the packing density. In Si/SiO_2 waveguides, Δ is relatively high and, therefore, due to this strong optical confinement, bend radii as small as a few μm may be realized. To assess the bending loss L_B we use the Marcuse method [14]. Figure 5.4 shows that the bending losses associated with a single mode strip waveguide are negligible if the radius of curvature is bigger than 2 μm.

The Y-junction loss L_Y depends on the reflection and scattering attenuation into the propagation path and surrounding medium. Different Y-branch structures have been analyzed by several methods [15,16]. For high index difference waveguides, the losses for the Y-branch are significantly smaller than for the low-Δ structures, and the simulated losses are less than 0.2 dB per split [17].

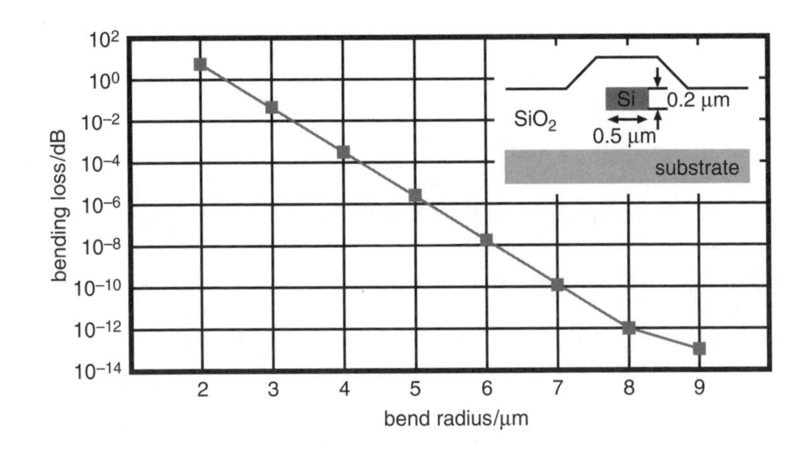

FIGURE 5.4 Simulated bending loss for Si/SiO_2 strip waveguide.

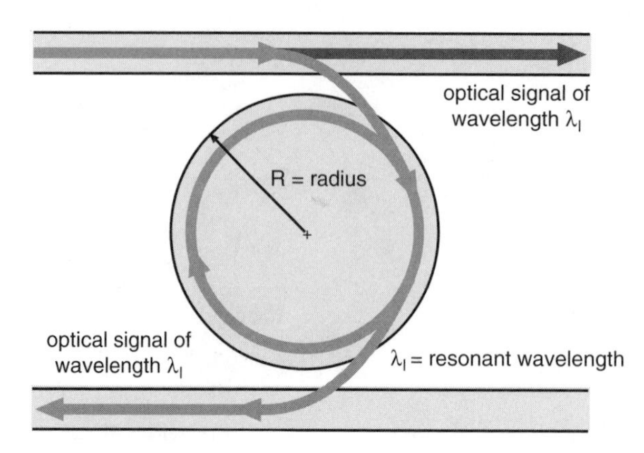

FIGURE 5.5 Micro-disk realization of an add-drop filter.

Using currently available materials and methods it is possible to achieve an almost 100% coupling efficiency from waveguide to optical receiver. In the proposed system, the coupling efficiency L_{CR} from the waveguide to the optical receiver is assumed to be 87% [18].

5.3.2 Resonators

Microdisks are resonating structures and are most commonly used in "add-drop" filters (so-called because of their capacity to add or subtract a signal from a waveguide based on its wavelength). The filter itself (Figure 5.5) is composed of one or more identical disks evanescently side-coupled to signal waveguides. The electromagnetic field is propagated within the structure only for modes corresponding to specific wavelengths, where these resonant wavelength values are determined by geometric and structural parameters (i.e., substrate and microdisk material index; thickness and radius of microdisk).

The basic function of a microresonator can be thought of as a wavelength-controlled switching function. If the wavelength of an optical signal passing through a waveguide in proximity to the resonator does not correspond to the resonant wavelength, then the electromagnetic field continues to propagate along the waveguide and not through the structure. If, however, the signal wavelength is close enough to the resonant wavelength (i.e., tolerance is of the order of a few nm, depending on the coupling strength between the disk and the waveguide), then the electromagnetic field propagates around the structure and then out along the second waveguide. Switching has occurred, based on the physical properties of the signal.

An obvious application of this device is in optical crossbar networks. More elaborate N × N switching networks have been devised [19], but experimental operation has yet to be proven. The advantages of such structures lie in the possibility of building highly complex, dense, and passive on-chip switching networks.

5.3.3 Photonic Crystals

Photonic crystals are nanostructures composed of, in the two-dimensional (2D) case, ultra-small cylinders periodically arranged in a background medium. 3D photonic crystals also exist but are much more difficult to fabricate from a technology point of view. Typically, for 2D photonic crystals, the cylinders are realized in a low-index material (such as SiO_2 or air), the background being a high-index material (such as Si). For light of certain wavelengths, such structures have a photonic band gap, leading to optical confinement. By introducing line defects (i.e., by removing one or more rows of cylinders), single-mode waveguides can be created. Other functions can be created using photonic crystals, such as couplers [20], multiplexers, demultiplexers, microresonators (using point defects instead of line defects), and even lasers. Photonic crystals are certainly good candidates for microscale optical integrated circuits due to their small size (a typical value for waveguide pitch is 0.5 μm) and massive fabrication potential; however, attenuation is an order of magnitude higher than that of planar waveguides (6 dB/mm [21]), although good progress has recently been made in this area.

5.4 Active Devices for Signal Conversion

5.4.1 III-V Sources

Fundamental requirements for integrated semiconductor lasers are small size, low threshold lasing operation, and single-mode operation (i.e., only one mode is allowed in the gain spectrum). From the viewpoint of mode field confinement and mirror reflection, two types of microcavity structures exist: multiple reflection (VCSELs and photonic crystals) and total internal reflection (microdisks). An overview of microcavity semiconductor lasers can be found in Baba [22].

5.4.1.1 Vertical Cavity Surface Emitting Lasers (VCSELs)

VCSELs are without doubt the most mature emitters for on-chip or chip-to-chip interconnections. As their name indicates, light is emitted vertically at the surface, by stimulated emission via a current above a few microamperes.

The active layer is formed by multiple quantum wells surrounded by III-V compound materials, and the whole forms the optical cavity of the desired wavelength. Above and below are Bragg reflectors, with deep proton implant to confine the current injected via the anode.

VCSELs are intrinsically single-mode due to their small cavity dimensions. They also have a very low threshold current and low divergence. Further, arrays of VCSELs are easy to fabricate; however, the internal cavity temperature can become quite high, and this is important because both wavelength and optical gain are dependent on the temperature.

Commercial VCSELs, when forward biased at a voltage well above 1.5 V, can emit optical power of the order of a few mW around 850 nm, with an efficiency of some 40%. Threshold currents are typically in the mA range. It is clear that effort is required from the research community if VCSELs are to compete in the on-chip optical interconnect arena, to increase wavelength, efficiency, and threshold current. Long wavelength and low-threshold VCSELs are only just beginning to emerge (e.g., a 1.5-μm, 2.5-Gb/s tuneable VCSEL [23] and an 850-nm, 70-μA threshold current, 2.6-μm diameter CMOS-compatible VCSEL [24] have been reported).

5.4.1.2 Microsources

Integrated microlasers differ from VCSELs in that light emission is in-plane to be able to inject light directly into a waveguide with minimum loss. Such devices, to be compatible with dense photonic integration, must satisfy the requirements of small volume, high optical confinement, low threshold current and emission in the 1.3–1.6-μm range. This wavelength implies the necessary use of indium phosphide (InP) and related materials, which leads to heterogeneous integration: bonding issues arise, which are covered in Section 5.6.

The structure of a microdisk laser is depicted in Figure 5.6 [25]. Upper and lower posts support the active region of the disk. Small cavity volume and strong optical confinement through semiconductor/air boundaries leads to low threshold currents. Current injection via the top contact causes carriers to diffuse to the disk edge and, consequently, produce optical gain. Lasing oscillation is generated by "whispering gallery" modes (so-called because of how the energy is distributed) rounding inside the disk edge. These modes, defined by the disk radius and representing the emission wavelengths, can be calculated using finite difference time domain (FDTD) simulations.

Photonic crystals can also be used as microsources. Although they are potentially smaller than microdisks and with better control of emission directivity and coupling, their mode behavior is complex and difficult to evaluate, and structures designed for lightwave frequencies are difficult to fabricate and to characterize. Research is ongoing in this area.

5.4.2 Detectors

Conventional positive-intrinsic-negative (PIN) photodiodes have relatively small area per unit capacitance values, meaning that the optical responsivity bandwidth product is low. This is a problem for high-speed operation in optical interconnect because transimpedance amplifier interface circuits cannot sup-

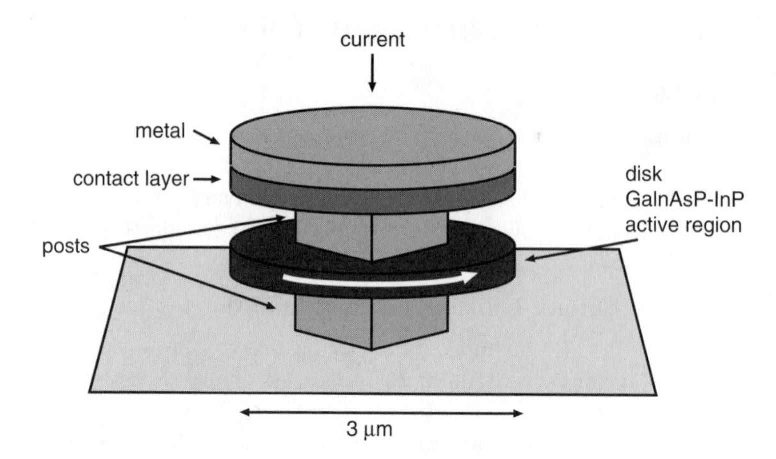

FIGURE 5.6 Structure of a microdisk laser.

port high photodetector capacitance values. Current research is focusing on thin-film, metal-semiconductor-metal (MSM) photodetectors, due to their improved area per unit capacitance [26].

5.5 Conversion Circuits

Between electronic data processing and photonic data transport lie crucial building blocks to the optical interconnect solution: high-speed optoelectronic interface circuits. On the emitter side, the power dissipated by the source driver is largely governed by the bias conditions required for the source itself. Advances in this area thus follow, to a large extent, improvements resulting from device research. On the receiver side however, things are rather different: most of the receiver power is due to the circuit. Only a small fraction is required for the photodetector device. The objective therefore is to attain the maximum speed/power ratio using dedicated circuit design methodologies.

5.5.1 Driver Circuits

The basic current modulation configuration of the source driver circuit is illustrated in Figure 5.7. The source is biased above its threshold current by M_2 to eliminate turn-on delays, and because the bias current value is the main contributing factor to emitter power, reducing the source threshold current is a primary device research objective. Figures of approximately 40 μA [27] have been reported. Device M_1 serves to modulate the current flowing through the source and, consequently, the output optical power injected into the waveguide. As with most current-mode circuits, high bandwidth can be achieved because the voltage over the source is held relatively constant and parasitic capacitances at this node have reduced influence on the speed.

5.5.2 Receiver Circuits

The classical structure for a receiver circuit is illustrated in Figure 5.8: a transimpedance amplifier (TIA) converts the photocurrent of a few μA into a voltage of a few mV; a comparator generates a rail-to-rail signal; and a data recovery circuit eliminates jitter from the restored signal.

Of these, the TIA is arguably the most critical component because it has to cope with a generally large photodiode capacitance situated at its input. Bandwidth/power ratio maximization can be achieved in several ways:

- Parametric optimization. For a given transimpedance structure, find the combination of component parameters necessary for maximum bandwidth.

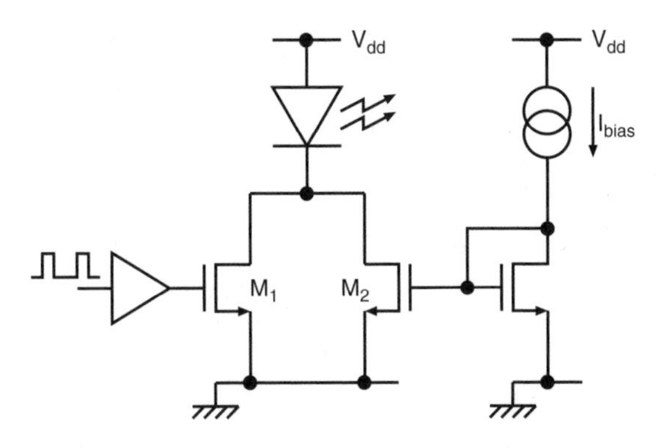

FIGURE 5.7 Basic current modulation source driver circuit.

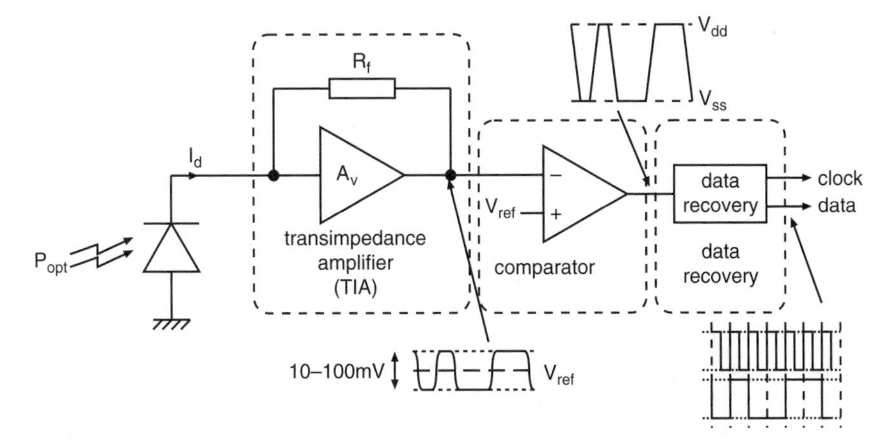

FIGURE 5.8 Typical photoreceiver circuit.

- Structural modification. For a given preamplifier architecture, make structural modifications, usually by adding elements, such as inductors for shunt peaking [28] or capacitors as artificial loads or feedback [29].
- Architectural exploration. Use complex architectures such as bootstrap or common-gate input stages [30].

The basic transimpedance amplifier structure in a typical configuration is depicted in Figure 5.9 [31]. The bandwidth/power ratio of this structure can be maximized by using small-signal analysis and mapping of the individual component values to a filter approximation of the Butterworth type, which gives analytical equations for the static transimpedance gain (Z_{g0}), the pole angular frequency (ω_0), and the pale quality factor (Q):

$$Z_{g0} = \frac{R_0 - R_f A_v}{1 + A_v} \tag{5.4}$$

$$\omega_0 = \frac{1}{R_0 C_y} \sqrt{\frac{1 + A_v}{M_f (M_x + M_m + M_x M_m)}} \tag{5.5}$$

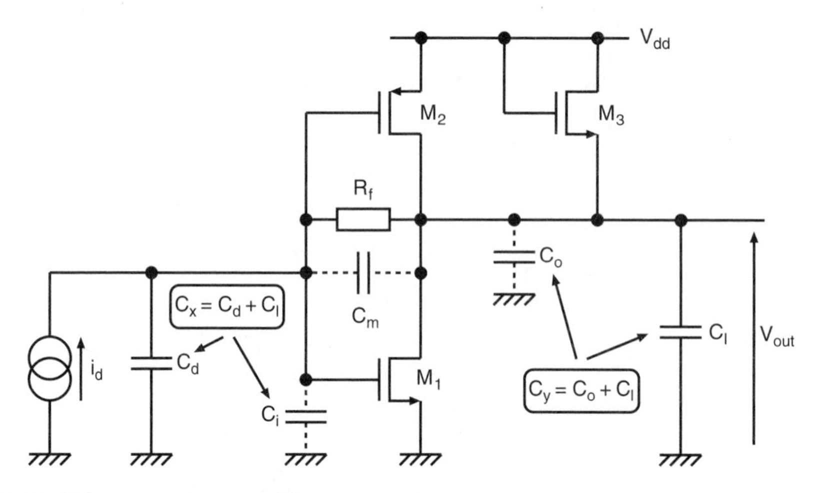

FIGURE 5.9 CMOS transimpedance amplifier structure.

$$Q = \frac{\sqrt{M_f\left(M_x + M_m(1+M_x)\right)(1+A_v)}}{1+M_x\left(1+M_f\right)+M_m M_f\left(1+A_v\right)} \tag{5.6}$$

where the multiplying factors $M_f = R_f/R_o$, $M_i = C_x/C_y$, and $M_m = C_m/C_y$ are introduced, normalizing all expressions to the time constant $\tau = R_o C_y$. By rearranging these equations, it is then possible to develop a synthesis procedure that, from desired transimpedance performance criteria (Z_{g0}, bandwidth and Q) and operating conditions (C_d, C_l), generates component values for the feedback resistance R_f and the voltage amplifier (A_v and R_o) (voltage gain A_v and output resistance R_o).

Taking into consideration the physical realization of the amplifier, those with requirements for low-gain and high-output resistance (high R_o/A_v ratio) are the easiest to build, and require the least quiescent current and area. Figure 5.10(a) shows a plot of this quantity against the TIA specifications (bandwidth and transimpedance gain) for $C_x = C_d = 500$ fF and $C_y = C_l = 100$ fF.

Approximate equations for the small-signal characteristics and bias conditions of the circuit allow a first-cut sizing of the amplifier. The solution can then be fine-tuned by numerical or manual optimization, using simulation for exact results [32].

Using this methodology and predictive BSIM3v3 models for technology nodes from 180 nm down to 70 nm [33], we generated design parameters for 1-THzΩ transimpedance amplifiers to evaluate the evolution in critical characteristics with technology node. Figure 5.10(b) shows the results of transistor level simulation of fully generated photoreceiver circuits at each technology node. According to traditional "shrink" predictions, which consider the effect of applying a unit-less scale factor of *1/s* to the geometry of metal-oxide semiconductor (MOS) transistors, the quiescent power and device area should decrease by a factor of $1/s^2$. Between the 180-nm and 70-nm technology nodes, $s^2 \approx 6.61$, which is verified through the sizing procedure. This methodology also allows us to find a particular specification to a given tolerance, as shown in Figure 5.10(c), which gives the active area and power of the generated TIA for bandwidths of 1GHz–5GHz (with $Z_{g0} = 1\text{k}\Omega$ and $Q = 1/\sqrt{2}$).

5.6 Bonding Issues

Connection of the optical interconnect network and the electronic IC is a nontrivial aspect to the whole optical interconnect concept. Probably the most effective and proven technique is flip-chip bonding [34]. This involves the depositing of gold solder bumps on either the electronic or photonic IC, then alignment and, finally, bonding, usually using thermocompression. At the wafer-scale bonding level, advanced machines are capable of precision alignment down to the order of 1 μm. In such cases, the solder bump

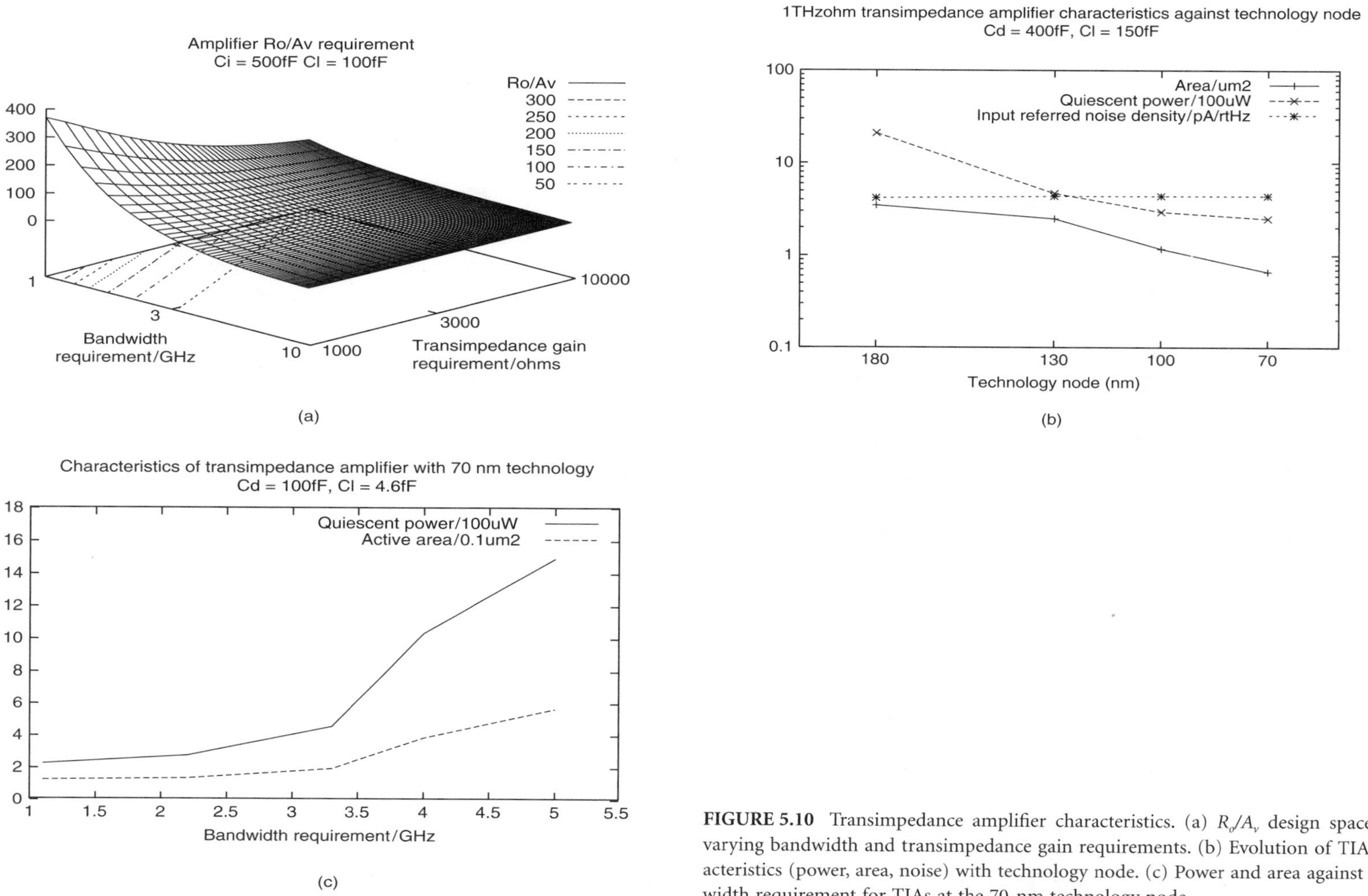

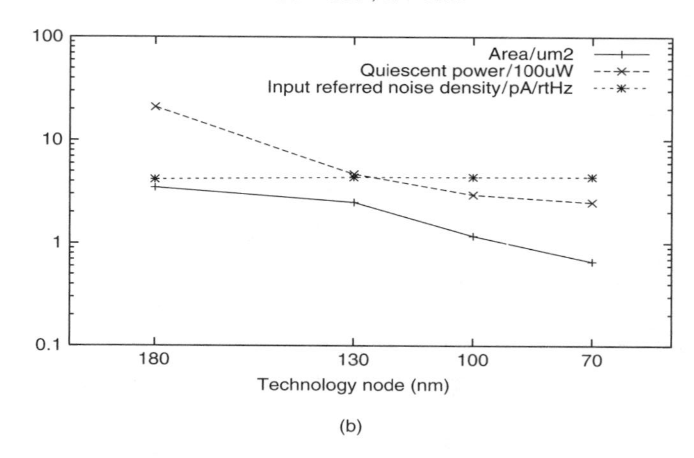

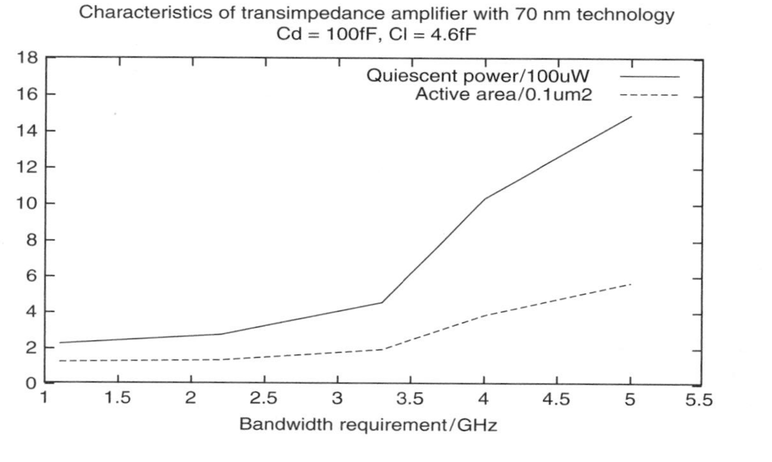

FIGURE 5.10 Transimpedance amplifier characteristics. (a) R_o/A_v design space with varying bandwidth and transimpedance gain requirements. (b) Evolution of TIA characteristics (power, area, noise) with technology node. (c) Power and area against bandwidth requirement for TIAs at the 70-nm technology node.

can be made small (under 10 μm in diameter) so that the total capacitance of the link, including pads, is of the order of a few tens of fF, compatible with high-speed interface circuit operation.

More futuristic ideas concern epitaxial integration, where the III-V material is grown directly onto the silicon substrate. This is possible, but the temperatures involved are usually rather high (800°C). However, recent research [35] has successfully demonstrated hydrophilic wafer bonding of an InP micro-disk laser onto a silicon wafer, at room temperature, by means of SiO_2 layers on both InP and silicon substrates. The highest temperature in this process was 200°C (for annealing, to increase bonding energy), which is compatible with CMOS IC fabrication steps.

5.7 Link Performance (Comparison of Optical and Electrical Systems)

To provide a clear comparison in terms of dissipated power between the optical and electrical interconnect networks it is necessary to estimate the electrical power dissipated in both systems. As an example, the power dissipated in clock distribution networks was analyzed in both systems at the 70-nm technology node. Power dissipation figures for electrical and optical clock distribution networks (CDNs) were calculated based on the system performance summarized in Table 5.1 and Table 5.2. The power dissipated in the electrical system can be attributed to the charging and discharging of the wiring and load capacitance and to the static power dissipated by the buffers. To calculate the power, we used an internally developed simulator, which allows us to model and calculate the electrical parameters of clock networks for future technology nodes.

The first input to this program is the set of technology parameters for the process of interest, particularly the feature size, dielectric constant, and metal resistivity according to the ITRS roadmap. In the next step, the resistance, capacitance, and inductance values for a given metal layer are calculated, as well as the electrical parameters of minimum size inverters. Based on this information, it is then possible to determine the optimal number and size of buffers needed to drive the clock network. For such a system, the program creates the SPICE netlist where the interconnect is replaced by resistance-capacitance (RC) or resistance-inductance-capacitance (RLC) distributed lines coupled by buffers

TABLE 5.1 Electrical System Performance

Technology [μm]	0.07
V_{dd} [V]	0.9
T_{ox} [nm]	1.6
Chip size [mm^2]	400
Global wire width [μm]	1
Metal resistivity [μΩ-cm]	2.2
Dielectric constant	3
Optimal segment length [mm]	1.7
Optimal buffer size [μm]	90

TABLE 5.2 Optical System Performance

Wavelength λ [μm]	1.55
Waveguide core index (Si)	3.47
Waveguide cladding index (SiO_2)	1.44
Waveguide thickness [μm]	0.2
Waveguide width [μm]	0.5
Transmission loss [dB/cm]	1.3
Loss per Y-junction [dB]	0.2
Input coupling coefficient [%]	50
Photodiode capacitance [fF]	100
Photodiode responsivity [A/W]	0.95

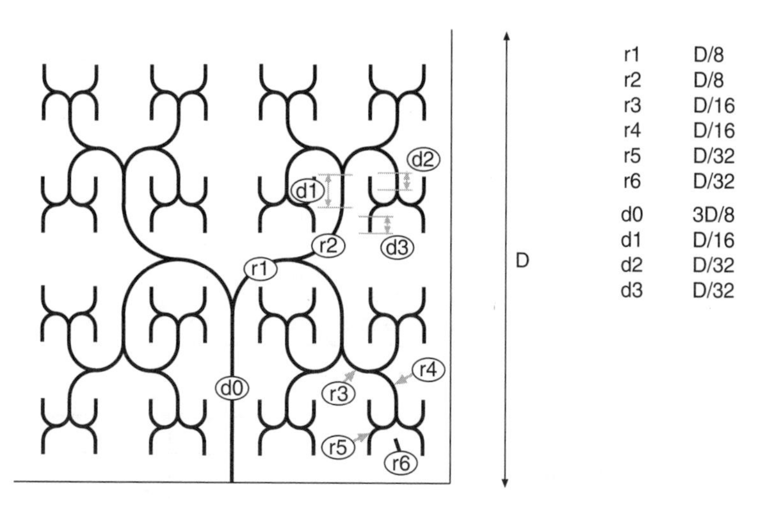

FIGURE 5.11 Optical H-tree network with 64 output nodes. $r_1..r_6$ are the bend radii and $d_0..d_3$ are the lengths of straight lines linked to the chip width D.

designed as CMOS inverters. Berkeley BSIM3v3 [33] parameters were used to model the transistors used in the inverters. The power dissipated in the system is extracted from transistor-level simulations.

Two main sources of electrical power dissipation exist in the optical clock distribution network (Figure 5.11):

1. Power dissipated by the optical receivers
2. Energy needed by the optical source to provide the required optical output power

To estimate the electrical power dissipated in the system, we used the methodology given in Figure 5.12. We assumed the use of an external VCSEL source, instead of an integrated microsource. The global optical H-tree was optimized to achieve minimal optical losses. The bend radii are designed to be as large as possible. For 20-mm die width and 64 output nodes in the H-tree at the 70-nm technology mode, the smallest radius of curvature (r_5, r_6 in Figure 5.11) is 625 μm, which leads to negligible pure bending loss.

First, based on the given photodiode parameters (C_d, R, I_{dark}), the method described in Section 5.5 is used for the design of the transimpedance amplifier. Next, for a given system performance (BER) and for the noise signal associated with the photodiode and transimpedance circuit, we calculate the minimum optical power required by the receiver to operate at the given error probability, using the Morikuni formula [36] in the preamplifier noise calculations.

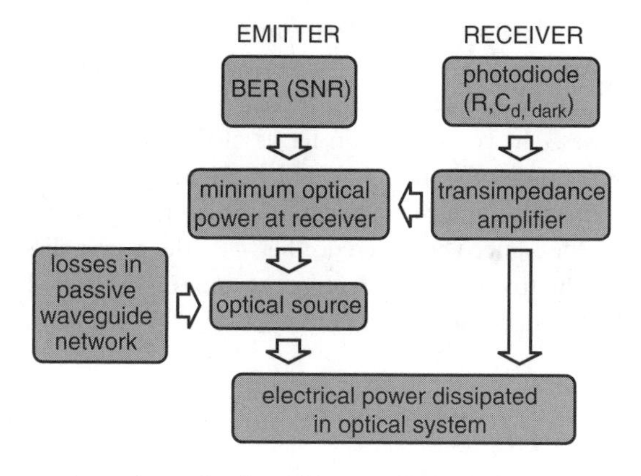

FIGURE 5.12 Methodology used to estimate the electrical power dissipation in an optical clock distribution network.

TABLE 5.3 Optical Power Budget for 20-mm Die Width at 3 GHz

Number of nodes in H-tree	4	8	16	32	64	128
Loss in straight lines [dB]	1.3	1.3	1.3	1.3	1.3	1.3
Loss in curved lines [dB]	1	1.31	1.53	1.66	1.78	1.85
Y-dividers [dB]	6	9	12	15	18	21
Loss in Y-couplers [dB]	0.4	0.6	0.8	1	1.2	1.4
Output coupling loss [dB]	0.6	0.6	0.6	0.6	0.6	0.6
Input coupling loss [dB]	3	3	3	3	3	3
Total optical loss [dB]	12.3	15.8	19.2	22.5	25.8	29.1
Min. receiver power [dBm]	−22.3	−22.3	−22.3	−22.3	−22.3	−22.3
Laser optical power [mW]	0.1	0.25	0.5	1.1	2.30	4.85

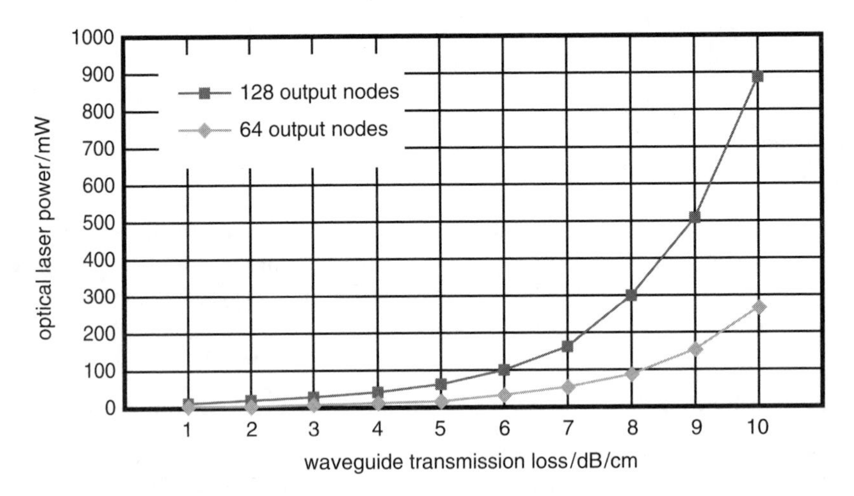

FIGURE 5.13 VCSEL optical output power required by the H-tree to provide a given BER for varying waveguide transmission loss.

To estimate the optical power emitted by the VCSEL, we took into account the previously calculated minimal required signal by the receiver and the losses incurred throughout passive optical waveguides. The electrical power dissipated in optical clock networks is the sum of the power dissipated by the number of optical receivers and the energy needed by the VCSEL to provide the required optical power. The electrical power dissipated by the receivers has been extracted from transistor-level simulations. To estimate the energy needed by the optical source, we use the laser light-current characteristics given by Amann et al. [37].

For a BER of $10^{-15,}$ the minimal power required by the receiver is −22.3 dBm (at 3 GHz). Losses incurred by passive components for various nodes in the H-tree are summarized in Table 5.3.

Figure 5.13 plots the optical power emitted by the VCSEL necessary to provide a given BER for various waveguide transmission losses. The comparison in terms of dissipated power between the optical and electrical global clock distribution networks is plotted in Figure 5.14. It can be seen that the power dissipated by the electrical system is highly dependent on the operating frequency. In the optical system, however, it remains almost the same. The difference between the power dissipated in both systems is clearly higher if we increase the frequency and number of nodes in H-trees. For a classical 64-node H-tree at 5GHz frequency, the power consumption in the optical CDN should be 5 times lower than in an electrical network.

5.8 Research Directions

Integrated optical interconnect is one potential technological solution to reduce the power required to move volumes of data between circuit blocks on integrated circuits, but it only makes sense to use this

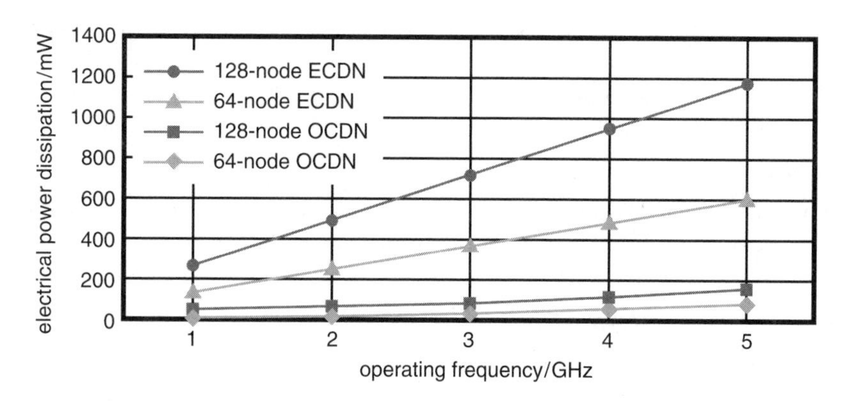

FIGURE 5.14 Electrical power dissipated by electrical (ECDN) and optical (OCDN) clock distribution networks for varying operating frequency.

technology for global high-speed data links. In addition, if the use of this technology implies a hard breach in terms of process technology and design methodologies, then architectural solutions may be an easier way to reduce power, by optimizing layout for the application such that the need for global high-speed data links is alleviated.

It is difficult to predict between these two scenarios. Future ICs will probably make use of advances in both areas. One parameter that is likely to make the difference in favor of optical interconnect is new research into the possibility of on-chip wavelength division multiplexing (WDM).

5.8.1 Network Links

Network, or n-n optical links, where wavelength routing may be used, would be targeted at (a) optical buses and possibly (b) reconfigurable networks. System architectures are moving rapidly toward platform-based designs, whereby every functional block on the chip (digital signal processors (DSP), analog/radio frequency [RF], video processors, memory, etc.) interfaces to an interconnect network architecture for data communication. Global system on chip (SOC) communication is around several tens of Gb/s. Commercial solutions for SoC development platforms are now based on bus architectures [38,39]: metallic interconnect architectures rely heavily on wide (64/128 bits) buses, as well as frequent use of switch boxes to dynamically define a communication route between two functional internet protocol (IP) blocks. Again, because the order of distance of communication is the chip die size, systematic use of repeaters (over the buses or within the actual switch boxes) is necessary and increases power consumption.

In the future, limitations (e.g., latency due to line delay, nonscalability, time sharing, and nonreconfigurability) of bus-based architectures will appear: future architectures of integrated systems will require new concepts for on-chip data exchange. The ever increasing number of transistors in a chip will lead to such complexity that IP reuse will be mandatory: a system is designed by integrating some hundreds of predesigned complex functional blocks, with the designer concentrating mainly on the organization of data transfer between these blocks.

A number of innovative interconnect architectures, often called networks on chip (NoC), have been recently proposed to overcome the limitations of bus-based platforms [40–42]. NoC architectures look much more like switching telecommunication networks than conventional bus-based architectures. Depending on the target application (multiprocessor SoCs or systems in which different functional blocks process heterogeneous signals), NoCs may have different structures, such as rings, meshes, hypercubes, or random networks. Latency, connectivity, global throughput, and reconfigurability constitute the main performance indicators of these networks.

Integrated optics may constitute an effective and attractive alternative for NoC. The superiority of optical interconnects in long distance links is established, and, possibly, some advantages of optical propagation may overcome the limitations of classical technologies for data exchanges at the integrated system scale. Some of the physical advantages of optical interconnects may be of a prime interest for NoC: flat frequency response (i.e., signal attenuation does not depend on frequency), limitation of cross talk, no repeaters, and power consumption. Above all, however, wavelength division multiplexing (WDM) may offer new and appealing solutions, such as optical buses and reconfigurable networks.

5.8.1.1 Optical Buses

As in telecommunications, WDM provides a route to very high data rates, even if the individual devices cannot be modulated much faster than electrical bus data rates. A single waveguide could be used to replace a 64-bit bus, for example, where each individual signal makes use of a distinct wavelength.

5.8.1.2 Reconfigurable Networks

By using a WDM approach, reconfigurable networks could be realized in the optical domain, leading to power reduction and higher integration density. Switch boxes, a key element for reconfigurable networks, could also be realized by using compact micro-resonators (about $10 \times 10\ \mu m^2$), capable of selecting and redirecting a signal based on its wavelength. Such networks would be entirely passive (i.e., no power would be required to transport the data, whatever the communication route necessary). Such a scheme, however, would imply a shift in the routing paradigm from a centralized arbiter acting on the switch boxes to one acting on the block interfaces to select the wavelength(s) to be used. In addition, tunable and thermally stable microlasers would be required.

5.9 Acknowledgments

The authors thank Dr. Xavier Letartre for his valuable discussions on active integrated devices for optical interconnect.

References

[1] Semiconductor Industry Association, *International Technology Roadmap for Semiconductors, 2001 ed.*, http://public.itrs.net, 2001.

[2] Sakurai, T. and Tamaru, K. Simple formulas for two- and three-dimensional capacitances, *IEEE Tran. Electron. Devices*, 30, 183, 1983.

[3] Chang, M.F. et al. RF/wireless interconnect for inter- and intra-chip communications, *Proc. IEEE*, 89, 456, 2001.

[4] Banerjee, K. et al. 3-D ICs: A novel chip design for improving deep-submicrometer interconnect performance and systems-on-chip integration, *Proc. IEEE*, 89, 602, 2001.

[5] Miller, D.A.B. Rationale and challenges for optical interconnects to electronic chips, *Proc. IEEE*, 88, 728, 2000.

[6] Davis, J.A., De, V.K., and Meindl, J.D. A stochastic wire-length distribution for gigascale integration (GSI) — Part I: derivation and validation, *IEEE Trans. Electron. Dev.*, 45, 580, 1998.

[7] Collet, J.H. et al. Architectural approach to the role of optics in mono- and multi-processor machines, *Applied Optics*, 39, 671, 2000.

[8] Friedman, E.G. Clock distribution networks in synchronous digital integrated circuits, *Proc. IEEE*, 89, 665, 2001.

[9] Lee, K.K. et al. Fabrication of ultralow-loss Si/SiO_2 waveguides by roughness reduction, *Optics Lett.*, 26, 1888, 2001.

[10] Sakai, A., Hara, G., and Baba, T. Propagation characteristics of ultrahigh-Δ optical waveguide on silicon-on-insulator substrate, *Japanese J. Appl. Phys. — Part 2*, 40, 383, 2001.

[11] Payne, F.P. and Lacey, J.P.R. A theoretical analysis of scattering loss from planar optical waveguides, *Optical Quantum Electron.*, 26, 977, 1994.

[12] Nishihara, H., Haruna, M., and Suhara, T. *Optical Integrated Circuits*, McGraw-Hill, New York, 1988.

[13] Kim, C.M., Jung, B.G., and Lee, C.W. Analysis of dielectric rectangular waveguide by modified effective-index method, *IEEE Electron. Lett.,* 22, 296, 1986.
[14] Marcuse, D. *Light Transmission Optics,* Van Nostrand Reinhold, New York, 1972.
[15] Chu, F.S. and Liu, P.L. Low-loss coherent-coupling Y branches, *Optics Lett.,* 16, 309, 1991.
[16] Rangaraj, M., Minakata, M. and Kawakami, S. Low-loss integrated optical Y-branch, *IEEE J. Lightwave Technol.,* 7, 753, 1989.
[17] Sakai, A., Fukazawa, T., and Baba, T. Low-loss ultra-small branches in a silicon photonic wire waveguide, *IEICE Trans. Electron.,* E85-C, 1033, 2002.
[18] Schultz, S.M., Glytsis, E.N., and Gaylord, T.K. Design, fabrication, and performance of preferential-order volume grating waveguide couplers, *Applied Optics-IP,* 39, 1223, 2000.
[19] Little B.E, Chu S.T., Pan W., and Kokubun Y. Microring resonator arrays for VLSI photonics, *IEEE Photonics Tech. Lett.,* 12, 323, 2000.
[20] Martinez, A., Cuesta, F., and Marti, J. Ultrashort 2-D photonic crystal directional couplers, *IEEE Photonics Tech. Lett.,* 15, 694, 2003.
[21] Notomi, M. et al. Structural tuning of guiding modes of line-defect waveguides of silicon-on-insulator photonic crystal slabs, *IEEE J. Quantum Electron.,* 38, 736, 2002.
[22] Baba, T. Photonic crystals and microdisk cavities based on GaInAsP-InP system, *IEEE J. Selected Topics in Quantum Electron.,* 3, 808, 1997.
[23] Filios, A. et al. Transmission performance of a 1.5-mm 2.5-Gb/s directly modulated tunable VCSEL, *IEEE Photonics Tech. Lett.,* 15, 599, 2003.
[24] Liu, J.J. et al. Ultralow-threshold sapphire substrate-bonded top-emitting 850-nm VCSEL array, *IEEE Photonics Lett.,* 14, 1234, 2002.
[25] Fujita, M., Sakai, A., and Baba, T. Ultrasmall and ultralow threshold GaInAsP-InP microdisk injection lasers: design, fabrication, lasing characteristics and spontaneous emission factor, *IEEE J. Selected Topics in Quantum Electron.,* 5, 673, 1999.
[26] Cho, S.Y. et al. Integrated detectors for embedded optical interconnections on electrical boards, modules and integrated circuits, *IEEE J. Selected Topics in Quantum Electron.,* 8, 1427, 2002.
[27] Fujita, M., Ushigome, R., and Baba, T. Continuous wave lasing in GaInAsP microdisk injection laser with threshold current of 40μA, *IEEE Electron. Lett.,* 36, 790, 2000.
[28] Mohan, S.S. et al. Bandwidth extension in CMOS with optimized chip inductors, *IEEE J. Solid-State Circuits,* 35, 3, March 2000.
[29] Kuo, C.W. et al. 2 Gbit/s transimpedance amplifier fabricated by 0.35μm CMOS technologies, *IEEE Electron. Lett.,* 37, 1158, 2001.
[30] Graeme, J. *Photodiode Amplifiers,* McGraw-Hill, New York, 1996, chap. 4.
[31] Ingels, M. and Steyaert, M.S.J. A 1-Gb/s, 0.7-μm CMOS optical receiver with full rail-to-rail output swing, *IEEE J. Solid-State Circuits,* 34, 971, 1999.
[32] O'Connor, I. et al. Exploration paramétrique d'amplificateurs de transimpédance CMOS à bande passante maximisée, *Proc. Traitement Analogique de l'Information, du Signal et ses Applications,* 73, Paris, September 12–13, 2002.
[33] Cao, Y. et al. New paradigm of predictive MOSFET and interconnect modeling for early circuit design, *Proc. Custom Integrated Circuit Conf.,* Orlando, FL, May 21–24, 2000.
[34] Krishnamoorthy, A.V. and Goossen, K.W. Optoelectronic-VLSI: Photonics integrated with VLSI circuits, *IEEE J. Selected Topics in Quantum Electron.,* 4, 899, 1998.
[35] Seassal, C. et al. InP microdisk lasers on silicon wafer: CW room temperature operation at 1.6 μm, *IEEE Electron. Lett.,* 37, 222, 2001.
[36] Morikuni, J.J. et al. Improvements to the standard theory for photoreceiver noise, *IEEE J. Lightwave Technol.,* 12, 1174, 1994.
[37] Amann, M.C., Ortsiefer, M. and Shau, R. Surface-emitting laser diodes for telecommunications, *Proc. Symp. Opto- and Microelectronic Devices and Circuits,* Stuttgart, March 10–16, 2002.
[38] IBM, The core connect bus architecture, http://www-306.ibm.com/products/coreconnect, 1999.

[39] VSI Alliance, http://www.vsi.org.
[40] Benini, L. and De Micheli, G. Networks on chip: a new SoC paradigm, *IEEE Computer*, 35, 70, 2002.
[41] Guerrier, P. and Greiner, A. A generic architecture for on-chip packet-switched interconnections, *Proc. Design, Automation and Test in Europe 2000*, 250, Paris, France, March 27–30, 2000.
[42] Dally, W.J. and Towles, B. Route packets, not wires: on-chip interconnection networks, *Proc. 38th Design Automation Conf.*, Las Vegas, June 18–22, 2001.

II

Low-Power Circuits

6

Modeling for Designing in Deep Submicron Technologies

Daniel Auvergne
Philippe Maurine
Nadine Azémard
LIRMM, University of Montpellier

6.1 Introduction

Much effort is being devoted to metal-oxide semiconductor (MOS) device modeling in deep submicron process [1–6]. Most of this effort addresses the modeling of short-channel and narrow-width effects to derive an accurate model of the threshold voltage and of the velocity saturation effect of the field dependent mobility. This has resulted in very complex modeling equations of the device performances. If these equations, validated on experimental silicon, can be used to accurately simulate analog designs with limited number of devices, they remain completely unpractical for digital design involving a huge number of components.

However, the performance of complementary metal-oxide semiconductor (CMOS) digital circuits depends on relatively few parameters, such as the threshold voltage and the drive current. In that case, a simple model of the average current available in any digital structure is sufficient to get an accurate representation of the delay and power performance of a design, if the model can easily be calibrated on the process. The objective of this chapter is to derive a design-oriented model of the delay and power consumption performances of digital structures. This model must give the explicit sensitivity of the performance parameters to the process, to the design and control conditions, as well as to the temperature and the value of the supply voltage. It is demonstrated here that using an engineering model [7] allows fulfillment of this objective, with a useful representation of the design space.

0-8493-1941-2/05/$0.00+$1.50

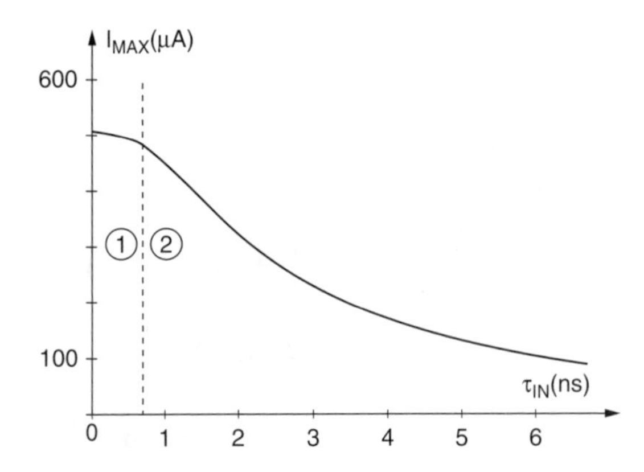

FIGURE 6.1 Sensitivity to the input ramp duration of the inverter maximum current value.

6.2 Current Modeling

Since the pioneering works of Mead and Rem [8] and Veendrick [9], it is well recognized that the speed and the total dynamic power consumption (including the short-circuit component) of any structure are directly related to the value of its switching current, which depends on the structure size and topology. As a result, the use of an accurate and design parameter explicit modeling of the switching current is the only way to define alternatives for low-power designing power-efficient circuits.

This design-oriented modeling can be obtained into two steps: by determining the maximum available current in the structure, and then considering the input slope effect, which defines the variation of this maximum value with the input control conditions.

6.2.1 Maximum Switching Current

An analytical direct current model to be used for hand calculations by designers has been presented in [7], where it is shown that considering transverse and high field effects on the carrier mobility, a drift velocity saturation current can be defined as:

$$I_{SAT} = K \cdot W \cdot (V_{GS} - V_T) \tag{6.1}$$

where $K = \kappa . v_{SAT} . C_{OX}$ is the product of the short-channel effect factor κ, by the carrier saturation speed and the oxide capacitance, W is the transistor width, and V_T the threshold voltage value. Note that Equation 6.1 corresponds to the Sakurai and Newton equation [10], for $\alpha = 1$.

Let us now evaluate the maximum value of this current on a simple CMOS structure, such as an inverter. The switching of an inverter is a dynamic process of which characteristics depend on the time duration of the voltage ramp applied to its input. Figure 6.1 represents the sensitivity to the input ramp duration of the maximum switching current of an inverter. In region 1, the current has a maximum value during all the discharge process, while in region 2, the maximum current value decreases when the input ramp duration increases.

The complete analysis of the input ramp effect, on both the speed and power performances, allows definition, as suggested in Maurine et al. [11], two switching ranges — the fast and slow input ranges in which a specific evaluation of the current must be performed.

6.2.2 Fast Input Range

The inverter maximum current value can easily be obtained from Equation 6.1 considering the maximum value of the controlling voltage, $V_{GSN} = V_{DD}$, which is instantaneously applied on the N transistor. In that case, the maximum current available in the N transistor is used to discharge the output node. We obtain:

TABLE 6.1 Relative Discrepancy between Simulated and Calculated Values of I_{MAX}

τ_{LN}/T_{HLS}	$F_o = 1$	$F_o = 3$	$F_o = 5$	$F_o = 10$	$F_o = 15$	$F_o = 20$
0.15	3%	2%	2%	1%	2%	1%
1	4%	3%	3%	3%	3%	2%
2	6%	5%	5%	5%	5%	4%
4	3%	2%	2%	0%	2%	1%
6	5%	3%	3%	2%	4%	1%
8	2%	1%	2%	3%	4%	3%
10	8%	2%	3%	5%	5%	5%
12	4%	3%	5%	6%	6%	6%
14	4%	5%	6%	7%	7%	7%
16	5%	6%	7%	8%	9%	8%
18	6%	7%	9%	9%	10%	9%
20	10%	9%	10%	10%	10%	10%

$$I_{MAX}^{Fast} = K_N \cdot W_N \cdot (V_{DD} - V_{TN}) \tag{6.2}$$

with an equivalent expression for the P transistor. In this expression K_N, is the conduction factor, previously defined, which can be directly calibrated on the process.

6.2.3 Slow Input Range

In this range, the transistor is still in saturation, but its driving voltage is lower than in the preceding part. Considering a nearly symmetric variation of the current around the maximum [11], its maximum value can be obtained as:

$$I_{MAX}^{Slow} = \sqrt{\frac{K_N \cdot W_N \cdot V_{DD}^2 \cdot C_L}{\tau_{IN}}} \tag{6.3}$$

which clearly exhibits the input ramp duration and load dependency of the current. This is illustrated by the nonlinear variation of I_{MAX}, observed in Figure 6.1. Table 6.1 gives an illustration of the accuracy obtained, with respect to Spice simulations, in determining the maximum current value (Equation 6.2 and Equation 6.3) in an inverter ($W_N = 1$ μm, $W_P = 2.2$ μm) implemented in a 0.18-μm process. As demonstrated in a large range of loading and controlling conditions, a very good agreement between simulated and calculated values is obtained.

6.2.4 Extension to Gates

For an n input gate the current available is input vector dependent, and the current can be evaluated as one or m times ($1 \le m \le n$) the current available in an inverter of identical size. It has been shown [12] that an array of series-connected transistors is equivalent to a current generator with a current capability reduced by the current reduction factor (DW), compared with an inverter of identical size:

$$DW = \frac{I_{Inverter}}{I_{Array}} \tag{6.4}$$

Detailed expressions of this coefficient have been given in Maurine et al. [11] for fast and slow input ranges and for controls applied on the top, bottom, or intermediate transistor of the array.

In Table 6.2, an example of validation obtained on a NAND2 (0.18 μm) illustrates the sensitivity of the reduction factor value to the input transition time.

TABLE 6.2 Sensitivity to the Input Transition Time of the Reduction Factor Value of a NAND2, for Top and Bottom Controls

	NAND2					
	Red_{Top}			Red_{Bot}		
τ_{IH}/t_{HLS}	Cal.	Sim.	Δ%	Cal.	Sim.	Δ%
1	1.54	1.54	0%	1.54	1.53	1%
2	1.54	1.50	3%	1.54	1.52	1%
3	1.54	1.50	3%	1.50	1.49	4%
8	1.24	1.28	3%	1.24	1.20	4%
13	1.24	1.24	0%	1.10	1.18	7%
18	1.24	1.23	1%	1.03	1.17	13%
20	1.24	1.22	2%	1.00	1.16	15%

TABLE 6.3 Comparison of the Reduction Factor Values in the Fast and Slow Input Range

CMOS 0.18 μm	Red^{Fast}	Red^{Slow}	Refs. [16, 22–25]
INV	1.00	1.00	1
NAND2	1.55	1.20	2
NAND3	2.10	1.48	3
NAND4	2.65	1.78	4
NOR2	1.93	1.39	2
NOR3	2.96	1.72	3
NOR4	3.76	1.94	4

Table 6.3 presents the variation of the DW reduction factor value with the gate complexity, for a top control of the series-connected transistors. For the process under study (0.18 μm), the values of DW^{Slow} obtained for NAND and NOR gates with 2, 3, and 4 inputs, respectively, are found quite different from the values obtained for fast input transition time conditions or from a direct reduction based on the number of serially connected transistors.

6.3 Definition of Metric for Performance

Using the preceding expressions for evaluating the maximum value of the switching current of CMOS structures, it is possible to obtain closed form expressions of the output transition time, the propagation delay, and the short-circuit power component.

6.3.1 Metric for the Transition Time

The output or input transition time (τ_{OUT}, τ_{IN}, respectively) is one of the fundamental performance parameters. It directly controls the value of the propagation delay and that of the short-circuit power component. It is defined as the time spent by the cell output (input) voltage to switch between the supply rail values. Its value is structure current capability (I_{MAX}) and output load dependent. Considering a linear variation of the output voltage, a simple first-order expression of the output transition time can be obtained from the charge conservation law initially introduced by Mead and Conway [13]:

$$\tau_{OUT} = \frac{C_L \cdot V_{DD}}{I_{MAX}} \tag{6.5}$$

where τ_{out} represents the time spent by the output voltage to swing over the full supply voltage value, V_{DD} is the variation of the output node voltage, C_L the output loading capacitance, and I_{MAX} the maximum value of the switching current. In this expression, currently used in roadmaps to predict the speed of extrapolated processes, the driving element is considered as a current generator supplying a constant current to the output loading capacitance. The output transition time value can directly be obtained by replacing I_{MAX}, from Equation 6.5, by one of the expressions (Equation 6.2 and Equation 6.3) previously introduced.

This gives for the fast input control range:

$$\tau_{OUT}(fall) = \frac{C_L \cdot V_{DD}}{I_{NMAX}} = C_L \cdot \frac{V_{DD}}{K_N \cdot W_N \cdot (V_{DD} - V_{TN})}$$

$$\tau_{OUT}(rise) = \frac{C_L \cdot V_{DD}}{I_{PMAX}} = C_L \cdot \frac{V_{DD}}{K_P \cdot W_P \cdot (V_{DD} - V_{TP})} \tag{6.6}$$

where the switching current has been obtained from Equation 6.2 considering a step input voltage.

To introduce a metric for the transition time, let us consider an ideal inverter (free from parasitic capacitance and I/O coupling), implemented with identically sized and minimum length, N and P transistors ($W_N = W_P$). Equation 6.6 becomes:

$$\tau_{OUTHL} = \frac{2Cox \cdot L_{MIN} \cdot V_{DD}}{K_N \cdot (V_{DD} - V_{TN})} = 2\tau_{ST}$$

$$\tau_{OUTLH} = 2\tau_{ST} \cdot R_\mu \tag{6.7a}$$

where R_μ is an indicator of the current imbalance between the N and P transistors.

$$R_\mu = \frac{K_N}{K_P} \cdot \frac{V_{DD} - V_{TN}}{V_{DD} - V_{TP}} \tag{6.7b}$$

For a general output load, we obtain:

$$\tau_{OUT}^{Fast}(fall) = \tau_{ST} \cdot \frac{C_L}{C_N} = \tau_{ST} \cdot (1+k) \cdot \frac{C_L}{C_{IN}}$$

$$\tau_{OUT}^{Fast}(rise) = R_\mu \cdot \tau_{ST} \cdot \frac{C_L}{C_P} = R_\mu \cdot \tau_{ST} \cdot \frac{(1+k)}{k} \cdot \frac{C_L}{C_{IN}} . \tag{6.8}$$

In these equations, k, C_{IN}, and C_L are, respectively, the configuration ratio, the input capacitance of the switching structure, and the total load evaluated at the output of the structure. As shown for a switching inverter controlled by a step-input voltage, the output transition time by unit load (C_{IN}) depends only on the configuration ratio of the inverter and on a technological factor τ_{ST}, which is a characteristic of the process speed [9]. This factor can be used, as well as a metric for the transition time.

For a slow input control, the reduction of the maximum current due to the input ramp effect has to be considered. From Equation 6.3 and Equation 6.5, we obtain:

$$\tau_{OUT}^{Slow}(Inv) = \sqrt{\frac{(V_{DD} - V_{TN})}{V_{DD}}} \cdot \sqrt{\tau_{OUT}^{Fast} \cdot \tau_{IN}} \tag{6.9}$$

Considering the full input transition range, the expression of the inverter output transition time is obtained from the maximum value of the Equation 6.8 and Equation 6.9:

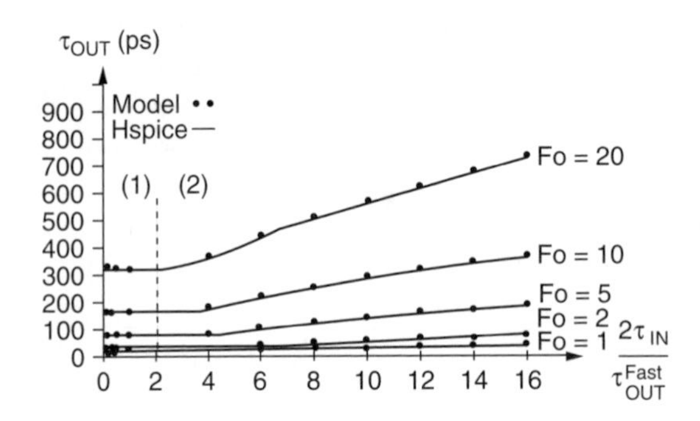

FIGURE 6.2 Inverter output transition time sensitivity to the load; Equation 6.1 and Equation 6.2 identify the fast and slow input ramp domains ($W_N = 1\ \mu m$, $W_P = 2.5\ \mu m$, $L_{MIN} = 0.18\ \mu m$).

$$\tau_{OUT}(Inv) = MAX\left\{ \begin{array}{l} \tau^{Fast}_{OUT}, \\ \sqrt{\frac{(V_{DD} - V_{TN})}{V_{DD}}} \cdot \sqrt{\tau^{Fast}_{OUT} \cdot \tau_{IN}} \end{array} \right\} \tag{6.10}$$

In Figure 6.2 we compare the calculated and simulated input transition time sensitivity of an inverter output transition time.

In the same way, the transition time of gates is directly deduced from the inverter expressions corrected by the DW given in Equation 6.4:

$$\tau^{Fast}_{OUT}(Gate) = DW^{Fast}_{Top} \cdot \tau^{Fast}_{OUT}(Inv)$$

$$\tau^{Slow}_{OUT}(GateTop) = DW^{Slow}_{Top} \cdot \tau^{Slow}_{OUT}(INV) \tag{6.11}$$

As shown in Equation 6.11, the output (input) transition time expression can easily be obtained as the product of three terms:

1. A technological factor common to all the structures
2. A symmetrical factor (logical effort of Sutherland et al. [20]), characteristic of the implemented logical function and of the internal configuration ratio ($k = W_P/W_N$)
3. The ratio of load to input cell capacitance (electrical effort of Sutherland et al. [20]) that characterizes the cell environment

If the first factor characterizes the process and the second one is library specific, designing for low power requires a lot of effort to optimize at physical level the third factor.

6.3.2 Metric for the Process

As shown in Equation 6.7, in the fast input control domain, the output transition time of an ideal symmetrical inverter, loaded by an identical one, is a direct characteristic of the process and of the current difference between N and P transistors:

$$\frac{\tau_{OUT}(fall)}{2} = \frac{Cox \cdot L_{MIN} \cdot V_{DD}}{K_N \cdot (V_{DD} - V_{TN})} = \tau_{ST}$$

$$\frac{\tau_{OUT}(rise)}{\tau_{OUT}(fall)} = R_\mu \tag{6.12}$$

TABLE 6.4 τ_{ST} Value Evolution with the Process

L (μm)	Cox (fF/μm²)	V_{DD} (V)	τ_{ST}calc. (ps)	τ_{ST}sim (ps)	V_{TN} (V)
0.13	13.3	1.2	4.05	4.05	0.60
0.18	7.85	1.8	4.51	4.57	0.54
0.25	6.91	2.5	7.00	6.70	0.64
0.35	4.30	2.85	10.8	11.5	6.62
0.50	2.80	2.85	15.9	16.4	0.73
0.80	2.30	5.0	24.8	23.8	1.22
1.00	1.73	5.0	28.0	26.3	0.82
1.20	1.38	5.0	33.0	30.7	0.70

In Table 6.4 we compare, for different processes, the τ_{ST} values simulated or measured on ring oscillators to the values calculated from Equation 6.12. The calculated values have been obtained from the value determined on the 0.25-μm process, used as a reference, by updating the L_{MIN}, V_{DD}, and V_T values of the corresponding process. The good agreement between the calculated and the measured values gives evidence of the interest in using τ_{ST} as a metric for defining or predicting the process speed performance.

6.3.3 Supply Voltage and Temperature Sensitivity

As shown in Equation 6.12, τ_{ST} gives the explicit supply voltage sensitivity of the transition time.

The temperature sensitivity can easily be included in the model considering both the mobility and the threshold voltage variations described in Sze [14] and Power et al. [15]:

$$K_\theta = K_{nom} \cdot \left(\frac{\theta_{nom}}{\theta}\right)^{XT}$$
$$V_T(\theta) = V_{Tnom} - \delta(\theta - \theta_{nom}) \tag{6.13}$$

where K and V_T are, respectively, the conductivity factor and the threshold voltage; θ_{nom} and θ represent, respectively, the reference and the targeted temperature; and XT and δ are the temperature coefficients of the mobility and of the threshold voltage.

Combining Equation 6.12 and Equation 6.13, a general expression for the τ_{ST} supply voltage and temperature sensitivity can be obtained:

$$\frac{\tau_{ST}(V_{DD},\theta)}{\tau_{STnom}} = \left(\frac{\theta}{\theta_{nom}}\right)^{XT} \left(\frac{V_{DD}}{V_{DDnom}}\right) \cdot \frac{1}{\dfrac{V_{DD} - V_{Tnom} + \delta(\theta - \theta nom)}{V_{DDnom} - V_{Tnom}}} \tag{6.14}$$

In this expression, the different parameters X_T and δ can directly be determined from specific simulation conditions to be defined in the next section.

6.3.4 Metric for the Delay

A realistic delay model must be input slope dependent and must distinguish between falling and rising signals. As developed in Jeppson [16], considering the input-to-output coupling effect, the input slope effect can be introduced in the propagation delay as:

$$t_{fall}(i) = \frac{v_{TN}}{2}\tau_{INrise}(i-1) + (1 + \frac{2C_M}{C_M + C_L})\frac{\tau_{OUTfall}}{2}(i)$$
$$t_{rise}(i) = \frac{v_{TP}}{2}\tau_{INfall}(i-1) + (1 + \frac{2C_M}{C_M + C_L})\frac{\tau_{OUTrise}}{2}(i) \tag{6.15}$$

where $\tau_{IN(fall,rise)}$ is the duration time of the input signal, generated by the controlling gate. C_M is the coupling capacitance between the input and output nodes [17], that can be evaluated as one half the input capacitance of the P(N) transistor for input rising (falling) edge, respectively or directly calibrated from electrical simulations. Indexes (i) and (i-1) refer to the location of the cell in a gate array.

As shown in Equation 6.15, the considered input transition time is short enough to assume that the output switching of the gate still occurs under a constant value of the current. This assumption justifies the use of the transition times for evaluating the delay. Otherwise, in the Slow input control range, the Slow ramp expression of the internal cross talk [18] must be used in Equation 6.15, and consideration of the short-circuit current [19] must be given. This results in a more complex expression out of the scope of this part.

Considering Equation 6.8, Equation 6.10, and Equation 6.15, it can be concluded that both the transition time and the propagation delay can be accurately evaluated using few parameters, which can be obtained from the design specifications and calibrated on the process, as shown later.

6.3.4 Metric for the Short-Circuit Power Dissipation

The dynamic power dissipation of static CMOS structures contains two terms:

1. An external term, the switching component, required to charge or discharge the different capacitances involved in the design
2. An internal term, the short-circuit component, which appears when both the N and P array of transistors are conducting

This last component is directly proportional to the common part of the current flowing between the supply rails. In the previous sections, it was shown that from the modeling of the switching current of CMOS structures, it was possible to define metrics for evaluating the transition time and propagation delay values. Now, from the modeling of the short-circuit current, metrics for evaluating the short-circuit power component can be defined.

The typical waveforms of the switching and-short-circuit currents flowing in an inverter controlled respectively by a fast and slow input ramps are given in Figure 6.3. The analysis of these typical waveforms shows that:

- In the fast input range, the short-circuit current, and thus the short-circuit power, is negligible with respect to the switching current.
- In the slow input range, the amplitude and the duration of the short-circuit current have values comparable to that of the switching current.

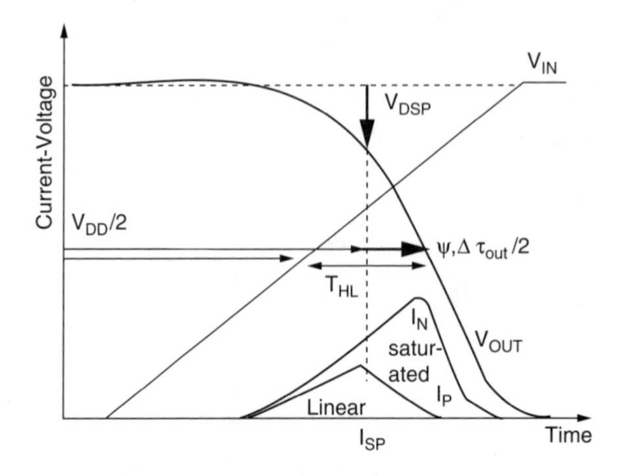

FIGURE 6.3 Illustration of the switching waveform of an inverter.

It is therefore obvious that, to reduce the short-circuit power dissipation, it is of prime importance to design CMOS circuit with a right control of the signal slopes along combinatorial paths. This is particularly true considering noncritical paths that are usually implemented with minimum drive gates.

Thus, to reduce, at gate level, the short-circuit power dissipated by noncritical paths, two steps have to be applied:

1. To develop a design-oriented model of the short-circuit power dissipation; such a model and the corresponding metric are introduced in the following
2. To deduce from this model a gate-sizing criterion to control effectively the signal slopes. This application of the model is given in Section 6.6

In his seminal work, Veendrick [9] first introduced a design-oriented model of the short-circuit power consumption. This model has been developed for micronic technology and unloaded structures, resulting in an upper bound of the short-circuit component evaluation. This model is extended here to submicronic process, considering realistically loaded structures; however, it is still assumed that:

- The short-circuit current waveform is symmetrical with respect to its maximum value, $I_{SC}{}^{MAX}$.
- The short-circuit current varies linearly between the times t_{OV} (end of the overshoot) and $t_{SP,N}$ (occurrence of $I_{SC}{}^{MAX}$, Figure 6.3).

Then evaluating, for an input rising edge, the maximum short-circuit current flowing between the supply rails [19] gives:

$$I_{SC}^{MAX} = \left\{ \psi_1 \cdot \frac{\tau_{IN}}{\tau_{OUT}} + \psi_2 \right\} . W_P \tag{6.16}$$

and the charge Q_{SC} supplied to the load by the P transistor as:

$$Q_{SC} = \frac{1}{2} \cdot (1 - v_{THP} - v_{TN}) \cdot \left\{ \psi_1 \cdot \frac{\tau_{IN}}{\tau_{OUT}} + \psi_2 \right\} . W_P \tag{6.17}$$

where v_{THP} (v_{TN}) is the normalized threshold voltage of the P (N) transistor working in linear (saturated) mode, and ψ_1 and ψ_2 are unique parameters to be calibrated on the process. As an example $\psi_1 = 10$ μa/μm and $\psi_2 = 2$ μa/μm, for a 0.18-μm process.

The energy short-circuit component can now be defined in the same way as the switching component:

$$E_{SC} = \frac{1}{2} \cdot C_{SC} \cdot V_{DD}^2 \tag{6.18}$$

Calculating the total charge transferred during the short-circuit period gives the following expression:

$$C_{SC} = \frac{Q_{SC}}{Vdd} = \frac{1}{2 \cdot Vdd} \cdot (1 - v_{THP} - v_{TN}) \cdot \left\{ \psi_1 \cdot \frac{\tau_{IN}}{\tau_{OUT}} + \psi_2 \right\} . W_P \tag{6.19}$$

with an identical expression for a falling input edge.

Equation 6.19 is of great interest for comparing the energy dissipated by the short-circuit process (E_{SC}) to the energy required to discharge (charge) the output load (E_{CL}). The following expression holds for an input rising edge:

$$R = \frac{E_{SC}}{E_{CL}} = \frac{\frac{1}{2} \cdot C_{SC} \cdot V_{DD}^2}{\frac{1}{2} \cdot C_L \cdot V_{DD}^2} = \frac{C_{SC}}{C_L} < 1 \tag{6.20}$$

In this expression, R represents the inefficiency of the gate in terms of energy consumption. Its value is necessarily lower than 1 because the short-circuit current waveform is included in the driving current waveform (Figure 6.3).

Note here (Equation 6.8, Equation 6.11, and Equation 6.19) that C_{SC} depends on the input and output transition time values, and C_L completely defines the output transition time. In this case, R appears as a good metric for evaluating the slope control on the design.

A value of R close to 1 indicates that the gate is badly controlled, while a value close to 0 indicates a good input slope control.

Considering Equation 6.19 and Equation 6.20, it clearly appears that R may reach significant values (empirically 0.3 or 0.4), if the designers do not control properly the signal slopes along combinatorial paths. Such values of R can be obtained in combinatorial noncritical paths where minimum size gates are used without proper control of the input slope. This may result in significant extra power dissipation on noncritical paths. Consequently, it could be of interest to develop a sizing criterion to properly control the signal slope along combinatorial paths in order to minimize this useless short-circuit power consumption.

6.4 Application to a Standard Cell Library

In a standard industrial approach, timing performance verification is obtained using a tabular method. The performance of each gate on a path, for each loading and control condition, is deduced from an interpolation between a set of predefined values. These values are determined from electrical simulations performed for a limited number of design conditions, such as load, input transition time, supply voltage value, and operating temperature. Characterizing each edge of the transition time and the propagation delay of each library cell, for typically five loading and input ramp conditions, involves 100 simulations. Then considering the process corners, defined for three supply voltage values (V_{max}, V_{nom}, V_{min}) and three temperature values (T_{max}, T_{nom}, T_{min}), the characterization of a logic function imposes 900 simulations by drive strength of this function. This huge number of simulations just allows representing the design space with five loading and controlling conditions. Intermediate conditions must then be interpolated using a linear characteristic equation (e.g., $f(\tau_{IN},C_L) = A\tau_{IN} + BC_L + C\tau_{IN}CL + D$). In submicron process, the transition time and the propagation delay exhibit a nonlinear variation with respect to the control and loading conditions that depends on each particular operating point imposed on the different combinatorial paths. This nonlinear range must clearly be located in the design space to adequately choose the simulation points to be inserted in the lookup table.

A continuous representation of the timing performance of a CMOS library will be introduced in the next paragraph to define the output transition time and propagation delay sensitivities of the cells to the design space parameters, such as the load, the input transition time, the supply voltage and the temperature values.

6.4.1 Continuous Representation of Standard Cell Performance

While considering Equation 6.10 to Equation 6.15, it clearly appears that in the fast input range, τ_{OUT}^{Fast} characterizes an inverter (gate) structure and its load. Considering the sensitivity of the different expressions to the input slope, τ_{OUT}^{Fast} can be used as an internal reference of the structure output transition time. In this condition, the following expression can be written:

$$\frac{\tau_{outfall}}{\tau_{outfall}^{Fast}}(Inv) = Max\left[1 \quad \sqrt{\frac{V_{DD}-V_{TN}}{V_{DD}}} \cdot \sqrt{\frac{\tau_{INrise}}{\tau_{outfall}^{Fast}}}\right] \tag{6.21}$$

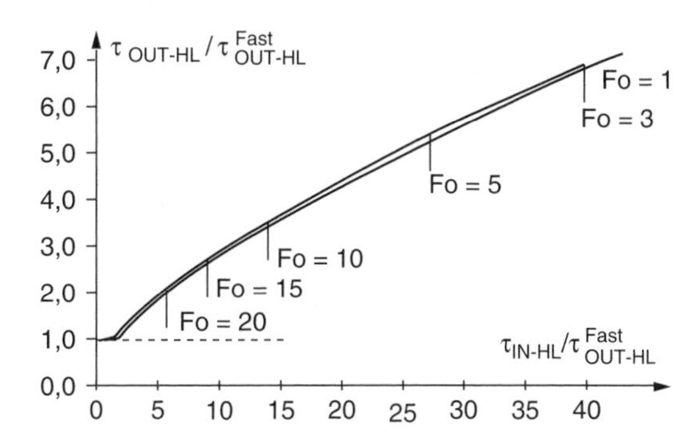

FIGURE 6.4 Full representation of the output transition time variation of the seven inverters of a 0.25-μm library.

Normalizing the output transition time with respect to $\tau_{outfall}^{Fast}$, used as a reference, the resulting expression only depends on the input transition time and is configuration ratio and load independent. Similar results can be obtained for gates and the representation of propagation delay.

This is illustrated in Figure 6.4 where the output transition time variation of the complete family of inverters of a 0.25-μm library is represented. As expected, all the curves pile up on the same one, representing the output transition time sensitivity to the input transition time. The final value for each specific cell is then directly obtained from the evaluation of τ_{OUT}^{Fast} given in Equation 6.11, which contains the structure and load dependency.

6.4.2 Calibration Procedure

From the preceding equations and considering the variation displayed in Figure 6.4, it appears that the output transition time and the propagation delay of all the gates of a library can be characterized with a reduced set of electrical simulations. The calibration of the parameters can be performed as follows.

1. The τ_{ST} value is obtained from the output transition τ_{HL} (falling edge) of a heavily loaded inverter (with a known configuration ratio k) controlled by a fast input ramp ($\tau_{IN} < \tau_{OUT}$).
2. R is obtained from the value of the ratio τ_{LH}/τ_{HL}.
3. For a small load, the variation of the apparent τ_{ST} value determines the value of C_{par} and C_M.
4. In the slow input range ($\tau_{IN} > \tau_{OUT}$) at constant load, varying τ_{IN} determines the input slope sensitivity (Equation 6.6).
5. Using the inverter as a reference, the gate parameters k and DW are directly determined from the ratio τ_{Gate}/τ_{Inv}.
6. Equation 6.9 completely determines the supply voltage sensitivity.
7. The temperature sensitivity parameters, XT and δ, are obtained from the preceding steps realized at different temperature values.

6.4.3 Validation

The validation of this representation has been done on a 0.13-μm library. The target is to get a continuous characterization of the timing performance with a robust identification of the design space (fast, slow input control range) including the temperature and supply voltage sensitivity. Only simple gates are considered, such as Inverter with seven different drives, NAND, and NOR gates with two and three inputs and five different drives. Initially the timing performance (transition time and propagation delay) of all these elements has been characterized from electrical simulations. They are available in tables (TLF, STF) that give, for each edge of the transition time and the propagation delay of each element and for three temperature and supply voltage values, the corresponding performance for five different values of the load and the input transition time.

Following the procedure described in Section 6.3, the values of the technology parameters are determined on the different tables, thus allowing to plot in Figure 6.5 to Figure 6.8 the variation of the transition time and the propagation delay of each logic family. As shown, the performance variation of all the elements of each family can be represented by one curve, as predicted by Equation 6.10.

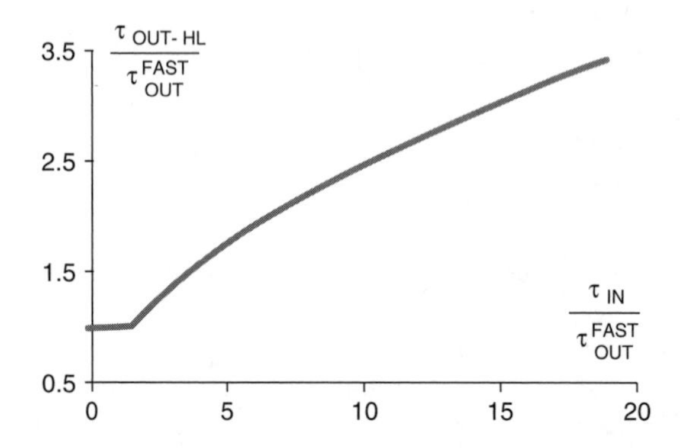

FIGURE 6.5 Output transition time representation of the seven inverters of a 0.13-μm process.

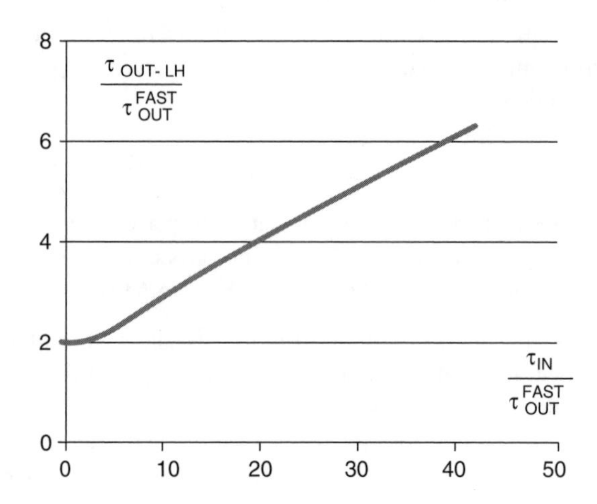

FIGURE 6.6 Output transition time representation of the five NAND2 of a 0.13-μm process.

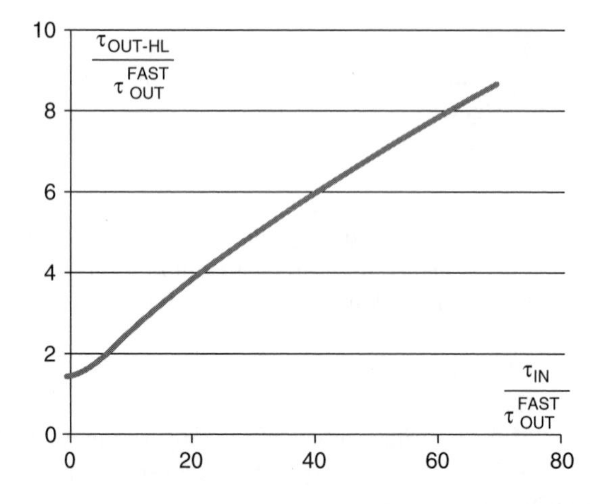

FIGURE 6.7 Output transition time representation of the five NOR2 of a 0.13-μm process.

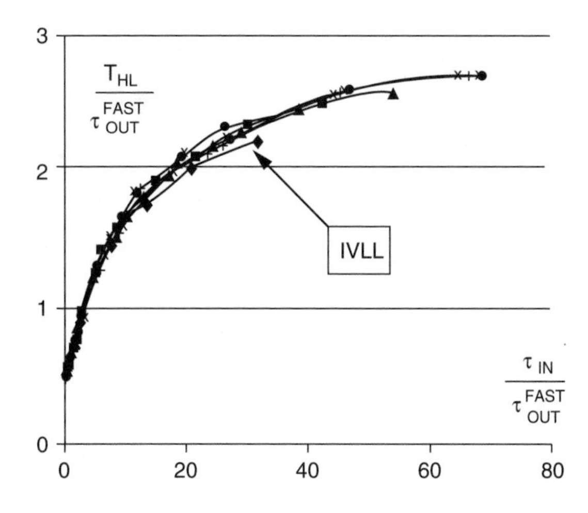

FIGURE 6.8 Propagation delay representation of the seven inverters of a 0.13-μm process.

TABLE 6.5 Voltage and Temperature Sensitivity of the τ_{ST} Parameter, XT=1.65, δ = 2.10⁻³

		V_{PP} (V)					
		1.08		1.2		1.32	
τ_{ST}		Model	Simul.	Model	Simul.	Model	Simul.
Temp. (°K)	233	4.26	3.92	3.86	3.56	3.02	3.3
	298	4.59	4.56	4.05	4.05	3.69	3.66
	398	5.16	5.45	4.85	4.93	4.62	4.58

Table 6.5 compares the τ_{ST} supply voltage and temperature sensitivity calculated from Equation 6.9 and Equation 6.11 with the value deduced from the simulated values on the lookup tables. As shown, an excellent agreement is obtained between calculated and simulated values for all the considered supply voltage and temperature range. These variations can then be completely represented by (11):

$$\frac{\tau_{ST}(V_{DD},\theta)}{\tau_{ST}} = \left(\frac{\theta}{298}\right)^{1.65} \cdot \left(\frac{V_{DD}}{1.2}\right) \cdot \frac{1}{\dfrac{V_{DD} - 0.62 + 2\cdot 10^{-3}(\theta - 298)}{0.58}} \tag{6.22}$$

where the different coefficients have been directly determined, following the calibration procedure given in the preceding part.

6.5 Application to Low-Power Design

6.5.1 Rule for Slope Control

To minimize the short-circuit power consumption, this section presents a gate-sizing criterion for properly controlling the input slope. Let us consider the structure depicted in Figure 6.9, where C_A models a parasitic (including routing) capacitance. The main challenge here is to control the value of the input transition time value τ_{IN}, allowing the reduction of the short-circuit power dissipated by stage (i).

This can be accomplished by increasing the size of stage (i-1); however, this results in an increase of the energy required for its control ($1/2 \cdot C(i\text{-}1).V_{DD}\Sigma$). This means that, in the same way as for delay optimization, a trade-off must be defined between the reduction of the short-circuit energy consumption of stage (i) and the increase of the energy required to control stage (i-1). The optimal sizing of stage (i-1), for minimizing the total energy consumption E_{TOT} of the structure (Figure 6.9) can be obtained from:

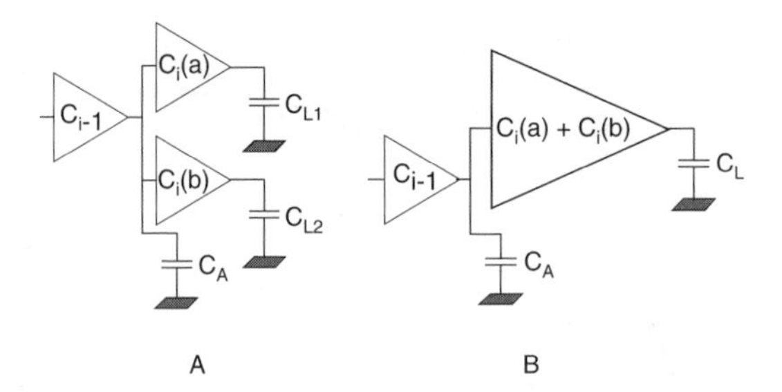

FIGURE 6.9 An example of divergence and its equivalent structure.

$$E_{TOT} = \frac{1}{2}C(i-1)\cdot V_{DD}^2 + \frac{1}{2}\left(C(i)+C_{SC}(i-1)\right)\cdot V_{DD}^2 + \frac{1}{2}\left(C_L + C_{SC}(i)\right)\cdot V_{DD}^2 \tag{6.23}$$

where C(k), C_{SC}(k) represent respectively the input capacitance and the short-circuit capacitance of stage k.

In this expression, the first term represents the energy required to control stage (i-1), the second term, the energy consumption of stage (i-1), and, finally, the last term is the energy consumption of stage (i).

Assuming that the stage (i-1) is controlled in such a way that its inefficiency R_{i-1} is minimized, Equation 6.23 becomes:

$$E_{TOT} = \frac{1}{2}C(i-1)\cdot V_{DD}^2 + \frac{1}{2}C(i)\cdot V_{DD}^2 + \frac{1}{2}\left(C_L + C_{SC}(i)\right)\cdot V_{DD}^2 \tag{6.24}$$

where only the first term can be reduced by a specific slope control. Searching analytically for the optimal value of C(i-1) gives:

$$C_{OPT}^{5/2}(i-1) \approx \frac{3}{2}\cdot A \cdot \frac{C^{3/2}(i).\left(C(i)+C_A\right)^{3/2}}{C_L^{1/2}} \tag{6.25}$$

where A is a process dependent parameter defined by:

$$A = \frac{\left(1-v_{THN}-v_{TP}\right).\left(\psi_1^{HL} + R_?.\psi_1^{LH}\right).\tau_{ST}}{2.V_{DD}.C_{OX}.L_{GEO}} \tag{6.26}$$

6.5.2 Application

The application of the sizing criterion (Equation 6.26) to an inverter tree is almost straightforward, processing backward from the output to the input of the tree; however, both the problems of divergence branches and of the output drivers have to be considered.

In minimizing the total power dissipated in an inverter tree, it appears that the optimal sizing of the output drivers depends strongly on the load content. For example, in optimizing the logic that drives a register or next gates, it can be considered that the output load is an active load or the sum of active and passive loads. Therefore, the sizing of the output driver has to be performed using Equation 6.25. If the output driver controls a passive load, however, no short-circuit power dissipation occurs in the load, and the driver must be sized at the minimum value satisfying the delay constraint.

The case of divergence branches presents a difficulty because the sizing criterion developed in the preceding section does not allow predicting the optimal sizing of the (i-1). The adopted solution is based on the fact that the power is an additive characteristic of the structure. To justify this approach, let us consider the structure represented in Figure 6.9.

The sizing criterion (Equation 6.26) supplies the optimal value of C_{i-1} only if $C_{L1} = C_{L2}$, in which case the two inverters can be lumped in a unique inverter with an input gate capacitance equal to $C_i(a) + C_i(b)$. In a general configuration, however, C_{L1} and C_{L2} have different values.

Nevertheless, as the short-circuit power dissipation is a decreasing function of C_L, the two inverters (a) and (b) are modeled by a unique inverter (Figure 6.9b) loaded by C_L=MAX(C_{L1},C_{L2}) to avoid any overestimation of the short-circuit power dissipated by (a) and (b).

6.5.3 Validation

This sizing heuristic, based on the sizing criterion defined by Equation 6.25, has been applied to an inverter tree represented in the Figure 6.10. The total power dissipated in the different implementations has been obtained from SPICE simulations.

Figure 6.11 illustrates the power gain and loss values obtained when comparing the proposed sizing solution to a minimal surface implementation. Different values of the parasitic routing capacitance P are considered to illustrate the sensitivity of the result to the parasitic content of the load.

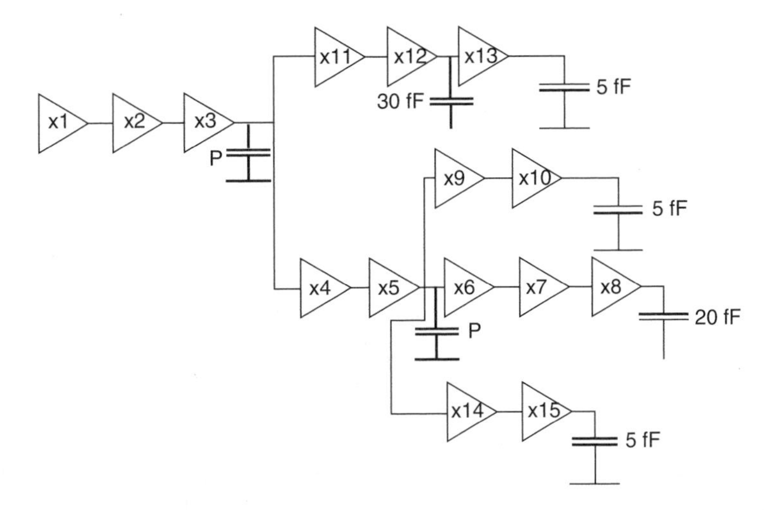

FIGURE 6.10 Representation of the inverter tree configuration used to validate the sizing criterion (Equation 6.25).

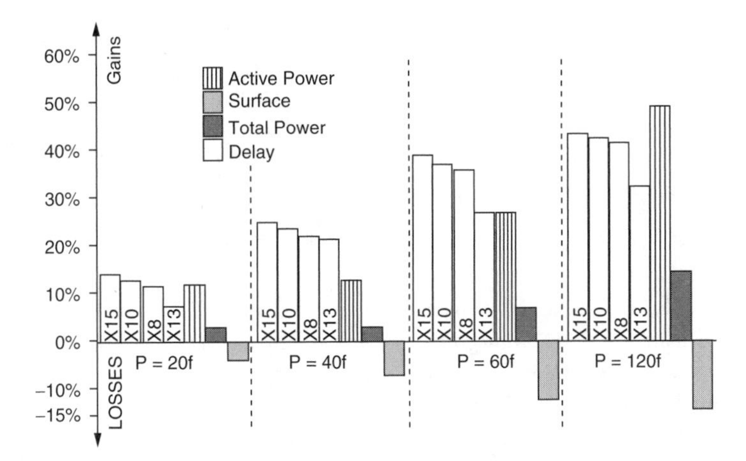

FIGURE 6.11 Gain and loss in delay, power, and area obtained on the inverter tree for different values of the parasitic capacitance $P = P_{3,4} = P_{5,6}$.

As shown, depending on the parasitic content of the inverter tree, the gain in power and speed is ranging from 3 to 15% and 13 to 45%, respectively.

The speed increase can be easily justified after a detailed analysis of the simulation results. For the considered example, the application of the sizing criterion increases the size of the stages X12, X5, and X3. This induces a reduction of the ramp duration applied at the input of the stage X14, X11, X9, X6, and X4 reducing their switching delays.

This ramp control allows to size noncritical paths at minimum dynamic power consumption. Moreover, this sizing method improves the speed, compared with a minimum size implementation. As a result, this solution can be recommended as an initial solution to be implemented before any critical path optimization.

6.6 Conclusion

Due to the fast evolution of the CMOS process, associated to the increasing complexity of the structures to be managed, it becomes necessary to define metric for performance allowing designers to use easy but robust indicators to evaluate alternatives at all the steps of the design flow. Using an analytical model to evaluate the maximum switching current value, a simple but accurate design oriented representation of the performance of CMOS logic was obtained. Simple and closed form formula for the output transition time, the propagation delay, and the short-circuit power component were derived. Metrics to characterize the speed of the CMOS process, as well as its sensitivity to the supply voltage and the temperature were defined.

Clear evidence was given that the transition time can be used as a simple and robust indicator for evaluating the cell performance and for defining the conditions of load and control satisfying the imposed constraints. The definition of the transition time through parameters, which are characteristics of the process, the structure and the load, gives to the designer opportunity to characterize a cell library in terms of load and critical transition time, and to improve lookup table centering in the useful design space.

A new way for a continuous representation of these performances was introduced, allowing modeling of the complete load and inputting ramp sensitivity by one curve. A method to calibrate the parameters of this representation was given, which that was completely validated on a 0.13-μm process for different temperature and supply voltage conditions.

Considering the power dissipation as a critical design parameter, a sizing criterion for minimizing the switching power dissipation component has been presented. The latter has been obtained by lowering the short-circuit component through a control of the gate input transition time. Using an analytical model of the short-circuit power dissipation and of the output transition time, it has been demonstrated that a sizing condition minimizing the short-circuit component, can be defined. Application has been given to general inverter configurations in various loading conditions. Gain in power and speed as large as 15 and 45% can be obtained, with respect to minimal size implementations.

These indicators also give facilities in controlling the load and input transition time distribution in combinatorial paths, which is, at the physical level, the most efficient way to manage the speed to power trade-off for circuit optimization.

References

[1] A. Chatterjee, C.F. Machala, and P. Yang, A submicron DC MOSFET model for simulation of analog circuits, *IEEE Trans. on CAD of Integrated Circuits and Syst.*, vol. 14, no. 10, pp. 1193–1207, 1995.

[2] A.F. Tasch, The challenges in achieving sub-100nm MOSFETS, *1997 Int. Conf. on Innovative Syst.*, pp. 53–60, Austin, TX.

[3] B.L. Austin, K.A. Bowman, X. Tang, and J.D. Meindl, A low-power transregional MOSFET model for complete power-delay analysis of CMOS gigascale integration, *11th Int. ASIC Conf*, pp. 125–129, Rochester, NY, 1998.

[4] S.H. Jen and B. Sheu, A compact unified MOS DC current model with highly continuous conductance for low-voltage ICs, *IEEE Trans. on CAD of Integrated Circuits and Syst.*, vol. 17, no. 2, pp. 169–172, 1998.

[5] Y. Cheng, K. Chen, K. Imai, and C. Hu, A unified MOSFET channel charge model for device modeling in circuit simulation, *IEEE Trans. on CAD of Integrated Circuits and Syst.*, vol. 17, no. 8, pp. 641–644, 1998.

[6] T. Skotnicki, Analysis of the silicon technology roadmap – how far can CMOS go? *C. R. Acad. Sci. Paris*, t.1, Série IV, pp. 885–909, 2000.

[7] K.-Y. Toh, P.-K. Ko, and R.G. Meyer, An engineering model for short-channel MOS devices, *IEEE J. Solid-State Circuits*, vol. 23, no. 4, pp. 959–958, 1988.

[8] C. Mead and M. Rem, Minimum propagation delays in VLSI, *IEEE J. Solid-State Circuits*, vol. SC17, no. 4, pp. 773–775, 1982.

[9] H.J.M. Veendrick, Short-circuit power dissipation estimation for CMOS logic gates, *IEEE J. Solid-State Circuits*, vol. 19, no. 4, pp. 468–473, 1984.

[10] T. Sakurai and A.R. Newton, A simple MOSFET model for circuit analysis, *IEEE Trans. on Electron. Devices*, vol. 38, no. 4, pp. 887–894, April 1991.

[11] P. Maurine, M. Rezzoug, N. Azemard, and D. Auvergne, Transition time modeling in deep submicron CMOS, *IEEE Trans. on CAD of Integrated Circuits and Syst.*, vol. 21, no. 11, pp. 1352–1361, Nov. 2002.

[12] P. Maurine, M. Rezzoug, and D. Auvergne, Output transition time modeling of CMOS structures, *ISCAS '01*, Sydney, Australia, May 2001, pp. V-363–V-366.

[13] C. Mead and L. Conway, Introduction to VLSI systems, in *Addison-Wesley Series in Computer Science, 2nd ed.*, Addison-Wesley, Reading, MA, 1980.

[14] S.M. Sze, *Physics of Semiconductor Devices*, John Wiley & Sons, New York, 1983.

[15] J.A. Power et al., An investigation of MOSFET statistical and temperature effects, *Proc. IEEE 1992 Int. Conf. on Microelectronic Test Structures*, vol. 5, March 1992.

[16] K.O. Jeppson, Modeling the influence of the transistor gain ratio and the input-to-output coupling capacitance on the CMOS inverter delay, *IEEE J. Solid-State Circuits*, vol. 29, pp. 646–654, 1994.

[17] J. Meyer, Semiconductor device modeling for CAD, chap. 5, G.K. Herskowitz and R.B. Schilling, Eds. Mc–Graw-Hill, New York, 1972.

[18] S. Turgis and D. Auvergne, A novel macro-model for power estimation in CMOS structures, *IEEE Trans. on CAD of Integrated Circuits and Syst.*, vol. 17, no. 11, pp. 1090–1098, Nov. 1998.

[19] P. Maurine, R. Poirier, N. Azemard, and D. Auvergne, Switching current modeling in CMOS inverter for speed and power estimation, *DCIS '01*, pp. 618–622, Porto, Portugal, November 2001.

[20] I. Sutherland, B. Sproull, and D. Harris, *Logical Effort: Designing Fast CMOS Circuits*, Morgan Kaufmann Publishers, Inc., San Francisco, CA, 1999.

7

Logic Circuits and Standard Cells

Christian Piguet
CSEM & LAP-EPFL

7.1 Introduction

Today, digital logic design is performed by using standard cell libraries and place and route computer-aided design (CAD) tools; however, many different logic styles have been and continue to be proposed for general-purpose and specialized standard cell libraries. Low power is even more important than speed and silicon area, but it is increasingly difficult to achieve in very deep submicron technologies as well as for specialized libraries for self-timed or cryptographic applications. This chapter summarizes logic design styles, stressing low power design issues. It also describes new and emerging logic styles for specialized libraries.

7.2 Logic Families

To achieve a very low dynamic power, a large number of logic families have been proposed and used in various designs. This section describes some of these logic families, which are assumably the most interesting regarding low-power designs. Various comparisons are proposed, but to be fair, some specific circuits are used for the comparison.

7.2.1 Static CMOS Logic

Static CMOS is the older and still most used logic family. It is still considered the most simple and robust logic style [1]. Each CMOS gate is constructed with two dual N-ch and P-ch networks, connected respectively between V_{ss} and V_{dd} and the gate output. Any logic Boolean function can be designed by

0-8493-1941-2/05/$0.00+$1.50

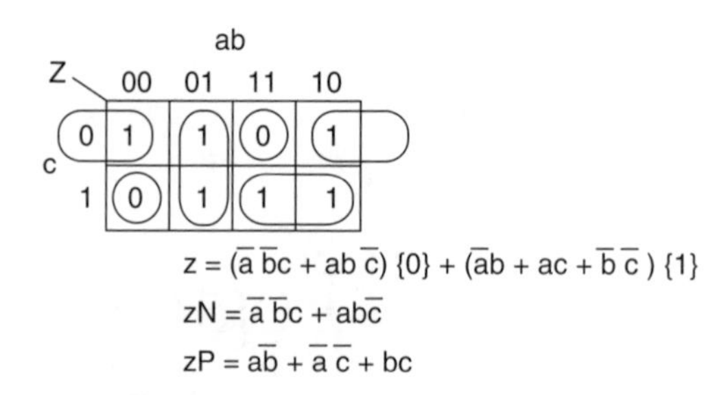

ab
Z
00 01 11 10
c
0
1
$z = (\bar{a}\,\bar{b}c + ab\,\bar{c})\,\{0\} + (\bar{a}b + ac + \bar{b}\,\bar{c})\,\{1\}$
$zN = \bar{a}\,\bar{b}c + ab\bar{c}$
$zP = a\bar{b} + \bar{a}\,\bar{c} + bc$

FIGURE 7.1 CMOS gate synthesized by the "separated simplification" method.

connecting, in series or parallel, CMOS transistors in the two N-ch and P-ch networks, if inputs are available in true and complemented forms.

The design of a CMOS gate is generally performed by synthesizing the N-ch network by taking the "0" cubes in the Karnaugh map of the Boolean function. The P-ch network is then derived as the dual network by connecting in series (parallel) the transistors that are in parallel (series) in the N-ch network. Figure 7.1 illustrates the synthesis of a Boolean function given by a Karnaugh map, in which the z symmetrical equation contains two terms, the first one with the "0" cubes indicated by {0} and the second term with the "1" cubes indicated with {1}. The metal-oxide semiconductor (MOS) structure of the N-ch and P-ch networks are then designed as zN and zP, by taking the first term {0} as such and by inverting each letter in the second term {1}, as P-ch transistors are conducting when they have a "0" on their gate. In the zN and zP expressions, AND operators mean a serial connection of transistors, while OR operators mean a parallel connection.

This method has been introduced in order to be capable of having the two N-ch and P-ch expressions as sums of products. As described in the next paragraph, it results in the so-called "branch-based" logic style that provides some advantages with respect to layout regularity, better performances in speed and power and a better testability [2,3].

7.2.2 Branch-Based Logic

In the "branch-based logic" [2], logic cells are designed exclusively with branches composed of transistors in series connected between a supply line and the gate output (Figure 7.2). The number of MOS in series is limited to three for speed performances. The main advantage of such an implementation is the layout density. For instance, the symbolic layout of the non-branch-based P-ch network (Figure 7.2) contains two supplementary contacts with two drain parasitic capacitances that can be removed in the more compact branch-based implementation. The symbolic layout of Figure 7.2 comes from the logical equation $S = (B + C)\ (A\overline{C} + \overline{A}\ D)$. If implemented as such, the P-ch network is presented at the top of Figure 7.2. If a Karnaugh map is designed, the minimum number of blocks of "1" that are necessary is three, resulting in the branch-based implementation with six transistors shown in the middle of Figure 7.2.

Figure 7.2 also describes a very regular geometrical branch-based layout consisting of three branches. It provides no diffusion interruption, common drain for two branches, a minimal number of contacts and few metal connections. This is not the case, for instance, for the implementation presented at the top of Figure 7.2, where a product of sum has to be implemented with a supplementary wire. This branch-based technique, first introduced to reduce parasitic capacitances for achieving low power [1], is also beneficial for high-speed logic such as fast adders in silicon-on-insulator (SOI) technology [3].

An obvious drawback of this technique is the possible increase of the number of transistors when realizing an MOS network of complex CMOS gates with a sum of products (a transistor controlled by the same input in two parallel branches is repeated); however, this problem is not so serious. A standard cell library does not contain a large number of complex gates. The most used cells are simple gates, flip-flops, latches, and multiplexers. In a 200-cell library, compared with non-branch-based logic, some cells contain one supplementary transistor (e.g., XOR, AOI, OAI, latch with reset, D-flip-flop [DFF],

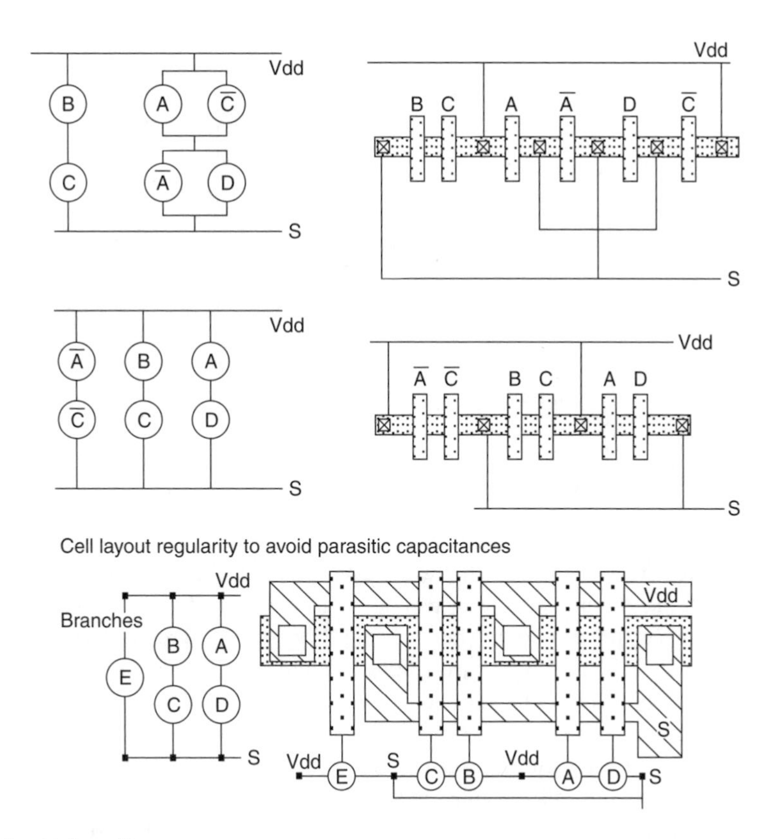

FIGURE 7.2 Branch-based layout.

and frequency dividers), and a few cells contain two supplementary transistors (e.g., latch and DFF with set/reset).

7.2.3 Transmission Gates

Transmission gate-based design has been largely used, ever since the CMOS technology introduction, because the MOS transistor is a very good switch. Transmission gates use two complementary transistors, as a single N-ch pass transistor, for instance, presenting a small gate-source V_{GS} when it has to conduct V_{dd} from input to output, reaches only V_{dd}-VT (threshold voltage). In the same situation, the complementary P-ch transistor has a full gate-source V_{GS} and provides V_{dd} at the output of the transmission gate. Rules of thumb are often used for the design of these transmission gate-based cells. This paragraph shows that the same basic methodology introduced for "branch-based" logic can be applied to transmission gates or pass-transistor circuits.

A transmission gate, controlled by an input variable, connects or not another input variable to the gate output. This means that some inputs are connected to transistor sources and not only to the gates of transistors. Compared to the branch-based style, for which sources of transistors or branches are always connected to V_{ss} {0} or to V_{dd} {1}, in transmission gate designs, sources of transistors or branches can also be connected to input variables. Thus, some cubes in the Karnaugh map are not only "0" cubes indicated by {0} or "1" cubes by {1}, but also some cubes identified by {input}. The content of the cubes is not "0" or "1," but is identical to a given input variable (or the complemented input). As a result, some cubes containing both "0" and "1" can be chosen, provided that the arrangement of "0" and "1" are identical to a given input.

Figure 7.3 provides an example for which one cube is a conventional cube with {0} and for which two other cubes are transmission cubes containing both "0" and "1." From Figure 7.3, it can be observed that

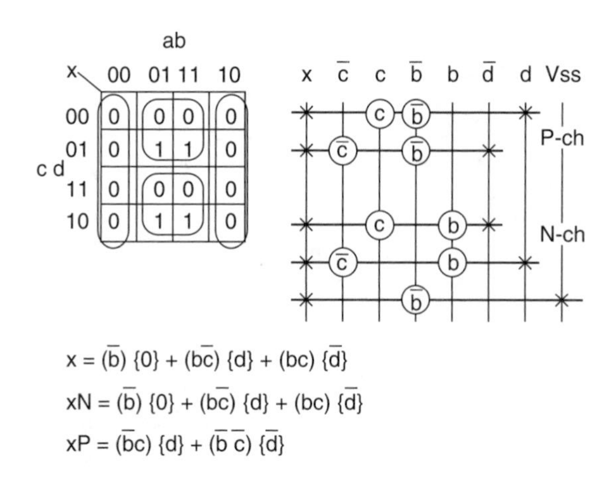

FIGURE 7.3 Synthesis of a complex CMOS gate with transmission gates.

the top cube content is identical to the input variable d, while the other cube content is identical to the complement of the input variable d. As such, the designer can write the symmetrical equation of x by a first term representing the "0" cube {0} and by two other terms for which the only difference is that they refer to input variables (i.e., {d} and $\{\bar{d}\}$). The N-ch and P-ch networks are then derived using the same rules (i.e., the terms with {0} and {input} are selected without any change for the N-ch network xN, while the terms with {1} and {input} are selected for the P-ch network xP by inverting each letter in the expression of the cubes). The transmission cubes give a contribution in both N-ch and P-ch networks, as the designer wants to get a transmission gate-based circuit. If only N-pass transistors are required, only the contribution of the transmission cubes in the N-ch network is necessary. The example of Figure 7.3 demonstrates that both cube types {0} or {1} and {input} can be simultaneously synthesized. The resulting circuit will therefore contain some branches connected to V_{ss} (and V_{dd}) and other branches connected to input variables.

The symbolic schematic of the resulting circuit (Figure 7.3), designed in such a way that branches are highlighted, shows, for instance, that two branches are connected to input d (i.e., one P-ch branch controlled by c and $\bar{b}$ and one N-ch branch controlled by $\bar{c}$ and b). These two branches do implement two transmission gates connected in series.

Transmission gate-based circuits do have a smaller number of transistors compared to static CMOS logic. In terms of layout density, however, depending on the layout style, the cell area is often very similar. Some circuits are advantageously designed as transmission gate-based design, such as XOR gates and adders (based on XOR gates), but other basic cells in a library can be designed in static CMOS logic without any penalty.

7.2.4 N-Pass Logic

Before CMOS became the mainstream technology, N-MOS logic was extensively used with a depleted transistor as the load device. N-MOS pass transistor logic was also used for many cells resulting in a very low transistor count; however, in N-MOS logic, the gate output produces a V_{dd}-VT voltage. Although it was not a problem many years ago with V_{dd} at 5.0 V, it is a major drawback today with supply voltages close to or below 1.0 V.

To keep the transistor count low, N-pass logic can be used with an output keeper or a restoring transistor as illustrated in Figure 7.4 for an XOR gate. This logic style is called single pass transistor logic (SPL). The keeper device is used to force full V_{dd} at the output when the inverted output is "0." This logic style appears to be interesting only for multiplexers and XOR gates, explaining why it is generally benchmarked for adders and multipliers. This SPL logic could compete with static CMOS at high V_{dd}, but not at low V_{dd}, where SPL is slower and consumes much more than static CMOS.

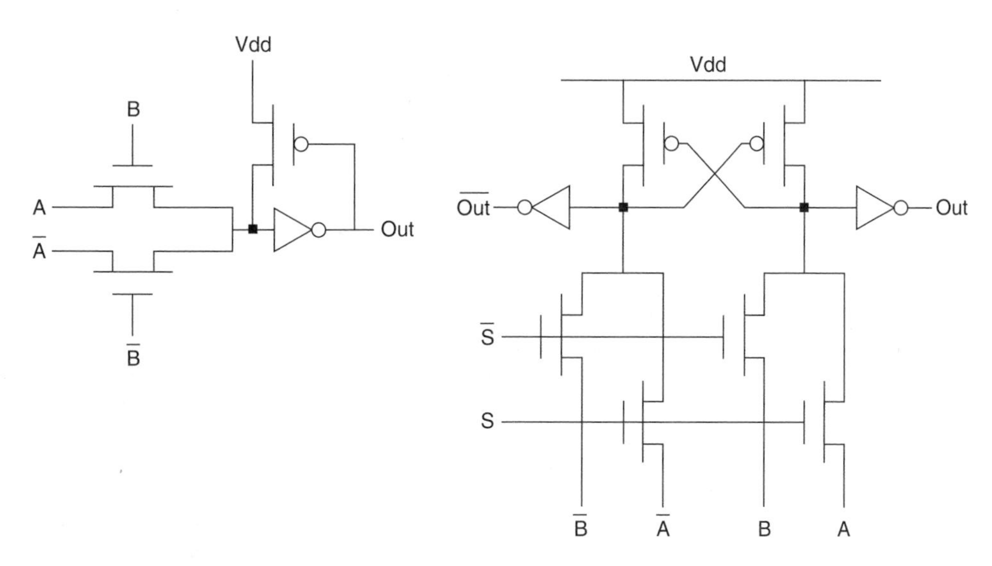

FIGURE 7.4 SPL XOR gate and CPL (dual-rail) 2:1 multiplexer.

TABLE 7.1 Comparison for Full Adder [6]

	Delay (ns)		Power (mW)		Power * Delay	
Logic Family	3.3 V	1.5 V	3.3 V	1.5 V	3.3 V	1.5 V
CMOS	1.89	7.88	32.9	6.4	1.00	1.00
CPL	1.39	8.33	34.1	6.0	0.76	0.99

Figure 7.4 presents another similar logic style called complementary pass transistor logic (CPL) and shows a 2:1 multiplexer. This CPL style is dual rail logic, as both true and complemented outputs are provided. The number of interconnect wires is therefore increased. A comparison (Table 7.1) with static CMOS shows some advantages in speed at high V_{dd}, but not better performances at low V_{dd} [4], the transistor count being similar to static CMOS. Many other logic styles are inspired from SPL and CPL, such as dual rail differential cascode voltage switch logic (DCVSL) [5], for which the two N-ch networks are not designed as N-pass logic but as conventional N-ch networks. Two cross-coupled P-ch MOS are used as load devices, similarly to two P-ch MOS of the CPL logic shown in Figure 7.4. It is a ratioed logic because the N-ch networks have to fight against the P-ch devices.

7.2.5 Dynamic Precharged Logic

Dynamic precharged logic is used to avoid the realization of P-ch networks to limit the transistor count in very complex logic gates. In single rail implementation, it is furthermore impossible to use resistive load devices if power consumption is an issue. This is why the P-ch network is replaced by a precharged P-ch transistor, which is used to precharge the gate output to V_{dd} in a first precharge phase. A second precharge N-ch transistor is used to cut off the N-ch network during the precharge phase. In the evaluation phase, the precharge N-ch MOS is conducting while the P-ch MOS is cut off. If the N-ch network is on, the gate output is switched to Vss, which is the right state for a logic gate with on N-ch network. If the N-ch network is off, the gate output keeps dynamically its precharged "1" state stored in the parasitic output capacitance.

Simple precharged logic gates cannot be connected in series, as the outputs of the first gates all precharged to "1" and connected to the inputs of the second gate result in a conducting N-ch network of the second gate. If the common precharge signal reaches the second gate a few nanoseconds before the first gate, the second gate can be discharged erroneously. Implementing a delayed precharge signal

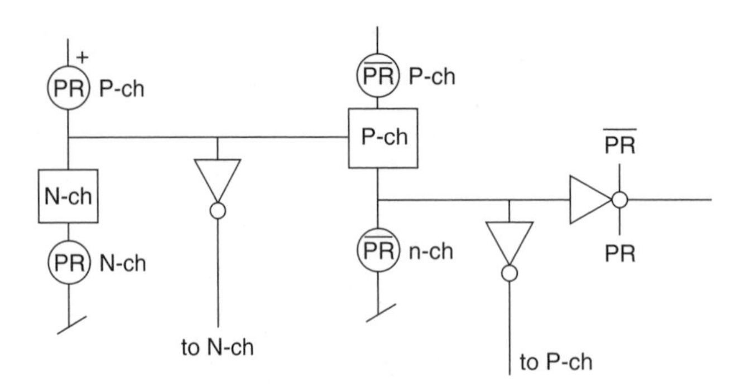

FIGURE 7.5 Precharged NORA gate.

for the second gate can solve this problem. Another solution is to implement a second gate with a P-ch network and precharged to "0" or to invert systematically the outputs of the precharged gates.

These solutions are used for dynamic DOMINO and NORA logic styles [7], as illustrated in Figure 7.5. A DOMINO gate is a conventional precharged gate with an inverted output, in such a way that DOMINO gates can be connected in series. The acronym "NORA" means "NO Race." This logic style combines DOMINO gates and precharged gates using alternate N-ch and P-ch networks (Figure 7.5). Furthermore, a NORA gate implements a dynamic latch (a simple tri-state gate) at its output. This latch is used for memorizing the output information and therefore to remove any hazard. Such logic has to work in pipeline due the output latch.

Dynamic logic is also used for precharged dual rail DCVSL logic [5]. By replacing the P-ch load devices with P-ch precharged transistors, the outputs in the evaluation phase switch necessarily to a complemented state (i.e., one output to "0" and the other to "1"). Consequently, for each computation, one output signal is always switched to "0," implying the same activity and the same power consumption for each computation. It is therefore more difficult to trace variation in power consumption during execution, which may be useful, for instance, for cryptographic applications attacked by differential power attack (DPA). Regarding power consumption, however, the activity of dual-rail gates is dramatically increased to 100%, as one of the outputs precharged to V_{dd} always switches to V_{ss} for each computation.

7.2.6 Memory Elements

The design of low-power flip-flops is crucial for the design of low-power circuits, as a digital block contains many memory elements. Besides power consumption issues, these elements have to be designed in such a way that they avoid any hazard for any gate delay. Master-slave structures, when true and complemented clocks respectively drive the master and the slave, present such a hazard [8]. Race-free flip-flops have therefore been designed to obtain speed-independent cells. The method [9], based on the fact that basic building blocks of CMOS are inverting or negative gates, result in structures that require inputs including the clock only in true forms, similar to the true single phase clock (TSPC).

For instance, using the method described in [10], a race-free DFF has been designed, containing only NAND gates and the single clock is connected to only four transistors to reduce the clock capacitance (Figure 7.6). Another interesting feature of this DFF is a weak sensitivity to the clock input slope. If a wire delay occurs on the clock fork connected to the two X and Y gates, this delay has to be shorter than three gate delays (e.g., gate delays Y, NM, and M) to guarantee a correct behavior.

7.2.7 Double-Edge Triggered Flip-Flops

Using the parallelization scheme proposed in [11], a flip-flop can be parallelized to obtain a double-edge triggered (DET) flip-flop clocked at half the master frequency [12]. A conventional DFF is implemented with two latches in series. Its parallelization results in two latches in parallel with an output multiplexer

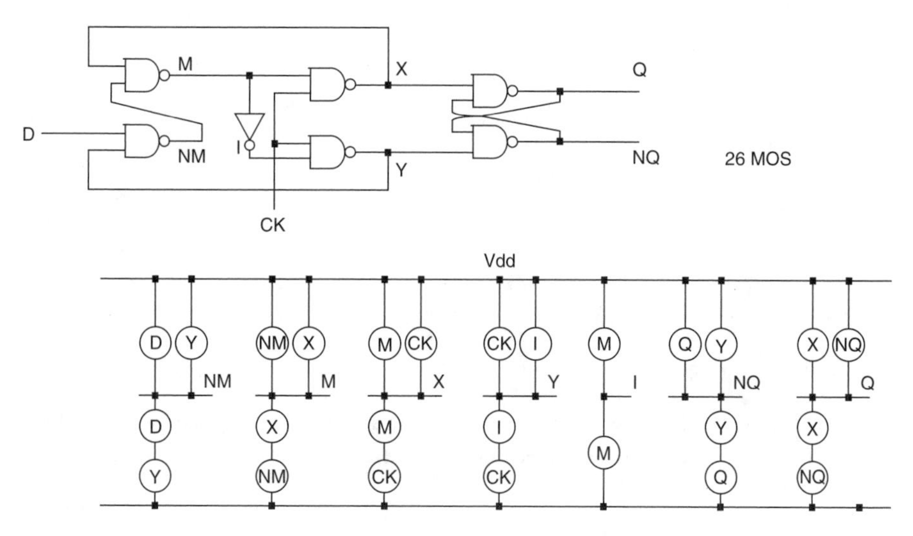

FIGURE 7.6 Race-free NAND-based DFF.

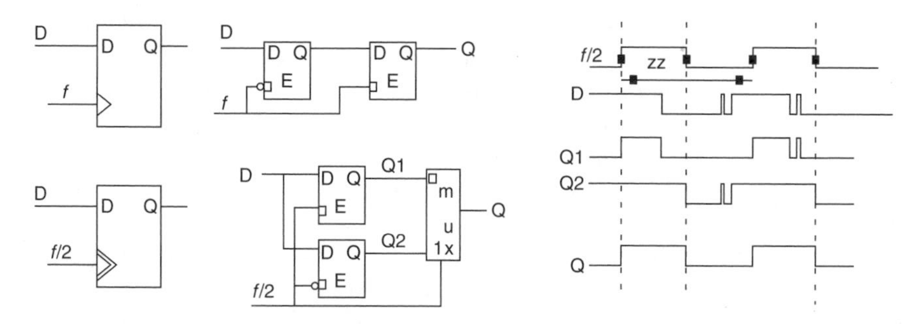

FIGURE 7.7 DFF parallelization.

(Figure 7.7). Such a flip-flop is sensitive to both edges of the input clock signal $f/2$ [13]. Its use in synchronous systems results in a master clock reduced of a factor 2. Any finite state machine may be implemented with double edge DFFs to reduce the input frequency by a factor 2 and the power consumption of the clock tree.

The choice of a logic family for a low-power design is generally not an issue. In systems on chip (SoC) design, a standard cell library is used. Its basic cells are generally designed in static CMOS. The controversy still holds about the benefits of transmission gates in some technologies, such as SOI [14]. For very fast microprocessors, some dynamic logic styles are used to achieve the huge speed required, but the layout is partially handcrafted. The design of fast and low-power flip-flops is still an issue for microprocessors [15]. For standard cell libraries, however, as the number of cells is increasingly reduced, conventional flip-flops instead of DET are used.

7.3 Low-Power and Standard Cell Libraries

The power consumption is today the major issue in the design of integrated circuits for portable devices. Design methodologies at different abstraction levels, such as systems, architectures, logic design, basic cells, as well as layout, must take into account the power consumption. The main goals of such design methods are V_{dd} reduction, activity reduction, as well as, reduction of parasitic capacitance [11,16]. It is well-known that most of the power can be saved at the highest levels; however, these choices are strongly application dependent. At the lowest levels, for instance, a low-power standard cell library, only a smaller factor in power reduction can be achieved, but the resulting library can be used for any design.

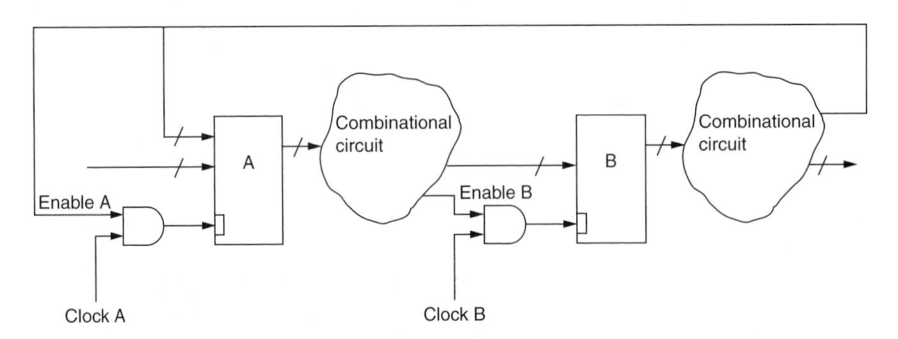

FIGURE 7.8 Latch-based design and gated clocks.

For standard cell libraries, besides the parasitic capacitance reduction for instance achieved by branch-based logic, V_{dd} reduction and gated clocks techniques are the most successful techniques used to reduce the power consumption.

7.3.1 Gated Clocks

The gated clock technique is extensively used in the design of low-power circuits [16,17]. It consists to gate the clock of sub-circuits that are in idle mode or that have just to keep their data as such. Arithmetic and logic units (ALU) of microcontrollers, for instance, have been designed with input and control registers that are loaded only when an ALU operation has to be executed. During the execution of another instruction (e.g., branch, load/store), these registers are not clocked avoiding any transition into the ALU. Gated clocks are also used to gate the clock of finite state machines [17] when the next state is identical to the present state.

Some logic synthesizers introduce gated clocks automatically; however, they could also gate clocks that have to be always active, which is useless. It is preferable to describe in very high speed hardware description language (VHDL) the necessary code to gate a clock and to introduce it only if it is useful. The most critical problem is to prevent the synthesizer from optimizing the clock gating "AND" gate with the rest of the combinational logic (Figure 7.8). This can be easily done manually by the designer by placing these AND gates in a separate level of hierarchy of his design or placing a "don't touch" attribute on them [18].

7.3.2 Latch-Based Designs

When designing a digital block using a standard cell library, the clocking scheme is extremely important for speed and power consumption. Generally, a single clock design is chosen, using master–slave flip-flops, well supported by the CAD tools; however, the clock tree synthesis is more and more difficult to achieve for avoiding a too large clock skew, resulting in large and power consumer buffers. Thirty percent of the total power could be in the clock circuits.

Latch-based designs with several nonoverlapping clocks have been proposed to solve this problem [18], and it has been demonstrated that they are more reliable at very low supply voltage. Conservative nonoverlapping clocks are used (i.e., an Ø1 clock pulse for the first period of the master clock CK (generated by an on-chip oscillator) and a second Ø2 pulse for the second period of the master clock). Therefore, the clock skew has to be shorter than half a period of the clock CK; however, such a scheme requires two clock cycles of the master clock CK to execute a single operation clocked by Ø1 and Ø2.

Latch-based designs provide several advantages over single clock master–slave flip-flop designs. In the design of a microcontroller, the power consumption can be reduced by about a factor of 2 [18]. The constraint with respect to the clock skew can be relaxed for both the Ø1 and Ø2 clock trees. This allows the synthesizer and router to use smaller clock buffers and to simplify the clock tree generation, which will reduce the power consumption of the clock tree.

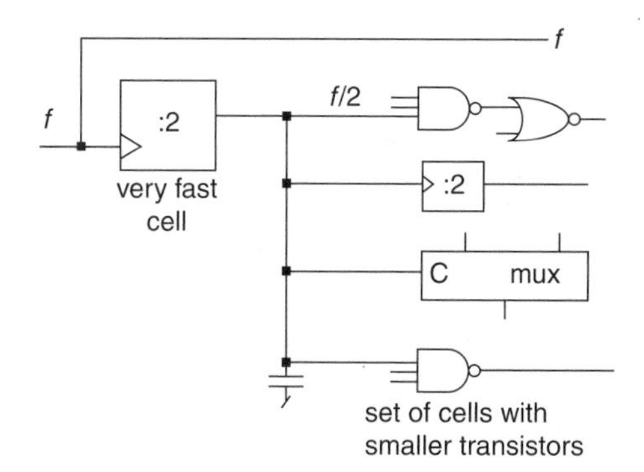

FIGURE 7.9 Several cell categories.

The latch design provides a smaller logic depth in combinational circuits and additional time barriers, which stop the transitions, avoid unneeded propagation of the signal, and thus reduce glitch power consumption. Using latches can also reduce the number of MOS of a design, for instance, in a register bank. With latches, the master part of the register bank can be common for all the registers, which gives a single master and many slaves, achieving a register bank area reduced by a factor of 2.

Using latches for pipeline structure is also very good for power consumption when using such a scheme in conjunction with clock gating. Figure 7.8 depicts a simple and safe way of generating enable signals for clock gating. This method gives glitch-free clock signals without the adding of memory elements, as it is needed with master–slave flip-flop clock gating [17]. Logic synthesizers very nicely handle the latch-based design methodology if the designer writes the description of the clock gating in his VHDL code.

7.3.3 Cell Drives

Standard cell libraries have to provide several drive versions of the same Boolean function (i.e., low-power, high-speed, and very high-speed cells as well as many buffer drives). This allows the designer or the logic synthesizer to place very high-speed cells on the critical path. However, if very high-speed cells are used outside the critical path, these cells will contain oversized transistors, which increase the power consumption and slow down the critical path. Figure 7.9 illustrates a simple example in which a first very fast cell is loaded with many other cells that are not on the critical path. If these cells contain oversized transistors, the load capacitance of the first very fast cell is increased, resulting in decreased speed. If the other cells are low-power cells with small transistors, however, the speed of the first cell will be higher and the power consumption reduced.

7.3.4 Complex Gate Decomposition

Complex gate decomposition is necessary if the number of transistors in series must be limited, as was the case in the proposed branch-based style [2]. As presented in Figure 7.10, the result is that simple gates with more than three inputs are decomposed into several simpler gates. This results in more transistors for the same Boolean function, but the total delay is reduced.

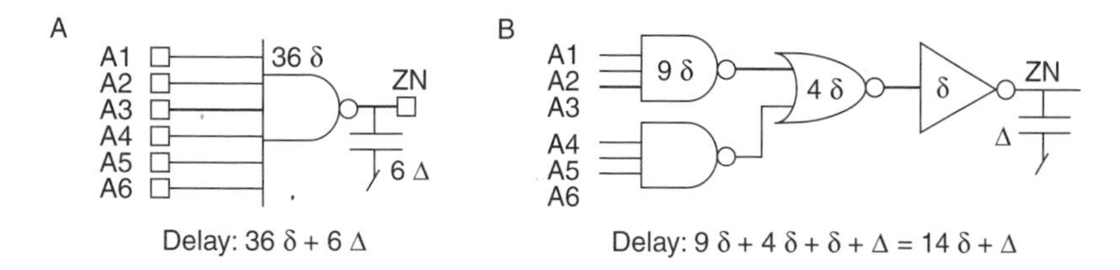

FIGURE 7.10 NAND6 decomposition.

TABLE 7.2 Gate Delay Comparison

Logic Gate	Delay
NAND6 (not decomposed)	0.70 ns
NAND6 (decomposed)	0.42 ns
NOR6 (not decomposed)	1.81 ns
NOR6 (decomposed: 2*NOR3 + NAND2)	0.65 ns
NOR6 (decomposed: 3*NOR2 + NAND3)	0.53 ns

According to a rough gate delay model [19], an N-input NAND gate contains a branch with N transistors in series, resulting in an increased internal resistance of $N^*\delta$. Furthermore, the internal parasitic capacitance is also increased roughly by a factor N (N drain capacitances). The internal delay of an N-input gate is therefore $N^{2*}\delta$. The load delay of an N-input gate is $N^*\Delta$ as the output capacitance must be charged or discharged by a branch with N transistors in series. Therefore, the total delay of an N-input gate is: delay = $N^{2*}\delta + N^*\Delta$. For a 6-input NAND gate, the total delay is $36\ \delta + 6\ \Delta$. If such a 6-input gate is decomposed as illustrated in Figure 7.10, the critical path of such a gate is made of three simple gates in series (e.g., a 3-input gate, a 2-input gate, and an inverter), resulting in a shorter total delay of $14\ \delta + \Delta$.

Table 7.2 gives some results in a 0.7-μm technology for both NAND6 and NOR6 gates with small-sized transistors (W = 2.2 μm). The load capacitance has been considered as six gate capacitances of similar logic gates. Obviously, the delay reduction for decomposed gates is better for NOR gates, which present P-ch transistors in series. At the same V_{dd} and for the same transistor sizes, a decomposed gate presents higher power consumption, as the simple gates could switch without an output transition.

7.3.5 Standard Cell Libraries

Standard cell libraries often provide a huge number of cells, up to 300 or even 500. A new approach is proposed, more or less similar to RISC vs. CISC processor architectures, which is based on a limited set of standard cells [20]. The number of functions for the new library has been reduced to 22 and the number of layouts to 92. It can be seen that the ratio between the number of layouts and the number of functions is larger (92/22 = 4.2 instead of 220/60 = 3.6 for the previous library). This means that the number of cell and buffer drives is larger. For speed and power optimization achieved by the logic synthesizer, the increased ratio of layouts to functions, as presented previously, is beneficial.

It appears obvious that the logic synthesizer could do a better job if the number of cells in the library is large. With a larger choice, it should be possible to provide a better solution, but this is not the case. Experiments in Table 7.3 and Table 7.4 demonstrate that the delay of some operators is significantly reduced with the new library resulting in a very small increase in silicon area (Table 7.3) and that the silicon area is reduced at the same speed with the new library (Table 7.4). These results show that the logic synthesizer is more efficient because it has a limited set of well-chosen cells and cell sizing adapted to the considered logic synthesizer. With significantly fewer cells than conventional libraries, the synthesizer is not lost in some optimization loops due to a too large choice of cells.

TABLE 7.3 Delay Comparison (synthesis for maximum speed, 0.5-μm process)

	Old Library		New Library	
	Delay [ns]	μm²	Delay [ns]	μm²
32-bit multiplier	16.4	907 K	12.1	999 K
Floating-point adder	27.7	510 K	21.1	548 K
CoolRISC ALU [18]	1.08	140 K	7.7	170 K

TABLE 7.4 Silicon Area Comparison (synthesis for a given delay, 0.5-μm process)

	Old Library		New Library	
	Delay [ns]	μm²	Delay [ns]	μm²
32-bit multiplier	17.1	868 K	17.0	830 K
Floating-point adder	28.1	484 K	28.0	472 K
CoolRISC ALU [18]	11.0	139 K	11.0	118 K

The design of the library has been based on keeping only very fast cells (i.e., to remove all the cells with 3 P-ch transistors in series and to have a very limited number of cells with 2 P-ch transistors in series). The number of cell layouts for the same function has been increased; however, it is not a simple increase from, for instance, sizing D1 (small transistors), D2, and D3 (medium-sized transistors) to D1, D2, D3, D4, and D5 (very large transistors). The cell sizing performed takes into account how the synthesizer uses the considered cells. The third consideration is based on buffer insertion (i.e., the combination of a given cell and of a buffer to replace complex gates).

Such a strategy must be checked through many experiments. The choice of the 22 functions was performed with a large number of experiments with and without a specific cell, and then the decision was made to either insert this cell in the library or not. Similar experiments were performed with various sizing and buffering of the cells. At the end, only 22 functions and 92 layouts were kept in the new library.

Furthermore, as the number of layouts is drastically reduced, it takes less time to design a new library for a more advanced process. Substantial time can also be saved for the library characterization, which is often the most time-consuming activity in library design. Reducing the number of layouts from 220 to 92 is a significant advantage. The reduction of the number of cells implies removal of complex gates from the library, forcing the logic synthesizer to decompose complex gates, which, as described, is beneficial in terms of speed.

It will also be a crucial point in future libraries for which more versions of the same function will be required while considering static power problems. The same function could be realized, for instance, with low or high VT for double VT technologies, or with several cells such as a generic cell with typical VT, a low-power cell with high VT and a fast cell with low VT.

7.3.6 Static Power

Static power is a dramatic issue for deep submicron technologies. Due to lower and lower supply voltages, the threshold voltages are also significantly reduced to keep some speed and the leakage is increased exponentially with the VT reduction [21].

Several circuit techniques have been proposed, which partially solve this issue. For standard cell libraries, three techniques directly affect the design of the library cells:

1. Multi-threshold CMOS (MTCMOS)
2. Stacked transistors
3. Dynamic threshold MOS (DTMOS)

The MTCMOS technique [22] is based on a technology offering two low and high VT for each MOS transistor. The low VT devices have to be used only on the critical path. Very fast cells of the library are therefore designed with low VT transistors on their critical path. Generally, only 10% of the transistors of a digital block are low VT devices, resulting in leakage reduction of about a factor of 10. Another method implements more stacked transistors [23], however, speed is impacted. This method implies to design logic cells of a library taking into account the trade-off between speed (the smaller number of MOS in series) and leakage (the larger number of MOS in series). The third technique, DTMOS [24], originally proposed for SOI technology, implements a connection between the transistor gate and the body of the transistor. It results in increased VT when the transistor is cut off.

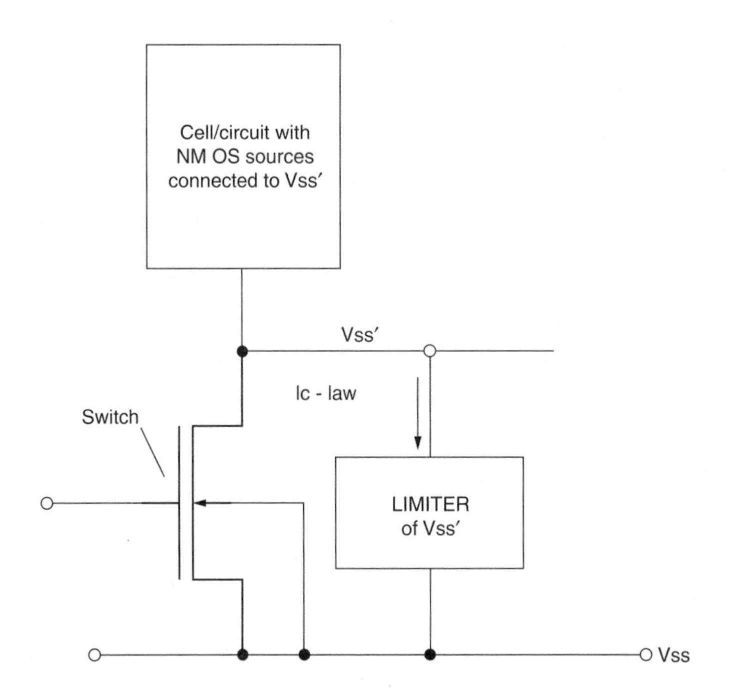

FIGURE 7.11 Switch in supply line with a limiter circuit.

Other well-known techniques, such as variables VT (variable threshold CMOS [VTCMOS] or self adjusted threshold scheme [SATS]) or switches in supply lines can be implemented at the fine grain (library cell) or large grain (block) levels. The former applies a well bias voltage to increase VT in idle or weak activity modes, while the second technique cut off the supply lines in idle modes. When cutting off transistors in idle modes, the resulting voltage of the considered cell or block may be so low that data in memory elements is lost. This is why some techniques, as illustrated in Figure 7.11, implement a limiter circuit in such a way that the resulting voltage is sufficient to keep the data in memory elements [25, 26]. Furthermore, when the switch is cut off, the N-MOS threshold voltages of the considered circuit are increased due to transistor source bias (V_{ss} is higher than the body of transistors).

Digital design is based on standard cell libraries, but it is also strongly impacted by the logic synthesis. As such, it is interesting to consider if synthesized digital architectures, such as pipeline, parallel, and asynchronous, may be better for reducing leakage. A very low activity factor does not provide a good ratio between dynamic and static power, as an idle circuit does present the same leakage than a very active circuit. To reduce the total power consumption, an attractive goal could be to have fewer, yet more active, transistors to perform the same logic function. If a given logic function is performed with 10,000 gates with an activity factor of 1%, this means that on average 100 gates are switching in a clock period. If the same logic function could be implemented with 1,000 gates, keeping the same number of switching gates (100), the activity will be 10%, with the same dynamic power but with a leakage reduced by a factor of 10 due to the reduction of the total number of gates.

Improved use of the gate switching is also dependent on the duty factor (Chapter 16) (i.e., the switching duration of a given logic gate over the clock period duration). The duty factor is defined as $\alpha = f^*Td$ (1/f: clock period; Td: switching duration). If there are many gates connected in series, as depicted in Figure 7.12, this duty factor is either equal to or less than $\alpha = 1/LD$ (LD: logical depth). Furthermore, as depicted in Figure 7.12 for five gates, a significant part of the clock period may be unused, without any switching, resulting in a smaller α than $\alpha = 1/LD$.

To compare digital architectures regarding leakage, the following design parameters can be used: activity factor *a*, duty factor α, the number N of logic gates, the capacitance C per gate, and the ratio of I_{OFF}/I_{ON}. The dynamic and static energies can be defined as:

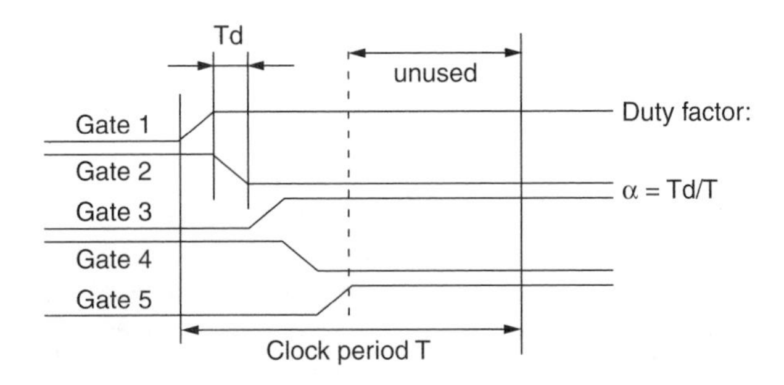

FIGURE 7.12 Duty factor and logic gates connected in series.

$$Edyn = a * N * C * V_{dd}^2$$

$$Estat = (1/f) * N * V_{dd} * I_{OFF}$$

Considering that the clock period is fully used, and introducing the fmax frequency (product of α by fmax of a logic gate) fmax = $\alpha * I_{ON}/(C*V_{dd})$ into the expression of the static energy:

$$Estat = (1/(\alpha * I_{ON})) * N * C * V_{dd}^2 * I_{OFF}$$

Therefore, the total energy becomes:

$$Etot = (a + 1/\alpha * I_{OFF}/I_{ON}) * N * C * V_{dd}^2$$

$$Etot = (a + LD * I_{OFF}/I_{ON}) * N * C * V_{dd}^2$$

This last equation shows that the static energy is proportional to the number LD of gates connected in series. A large LD implies a small α and a smaller use of the available logic gates. It has been shown [33] that the optimum of the total power consumption is roughly at 50% of dynamic and 50% of static power (i.e., Edyn = Estat). This allows defining:

$$a = LD * I_{OFF}/I_{ON}$$

or

$$I_{ON}/I_{OFF} = LD/a = 1/(\alpha * a)$$

This expression indicates that relatively small values of I_{ON}/I_{OFF} are possible if LD is small and activity a relatively large. For I_{ON}/I_{OFF} = 100, LD = 10, and a = 10%. These values result in very small VT and V_{dd} as well as very low total power consumption at a reasonable speed [33]. It also means that LD=100 and a=1% will result in I_{ON}/I_{OFF} =10'000, value for which VT and V_{dd} could not be reduced significantly. It is therefore necessary for digital synthesis to achieve very low values of I_{ON}/I_{OFF}, and, consequently, small LD and high activity. Clearly, pipeline architectures (small LD) are better than nonpipelined [34], asynchronous architectures could also be interesting to avoid unused part of the clock period (Figure 7.12), but high activity architectures are more difficult to design, as low activity has been a goal in recent years to reduce dynamic power.

7.4 Logic Styles for Specific Applications

The design of cell libraries is largely considered independent of the applications (i.e., any library can be used for any application). This assumption no longer holds, as certain specific applications require special cell libraries. Some examples are described here, such as self-timed design, cryptographic applications, and fault-tolerant logic.

7.4.1 Library Cells for Self-Timed Design

Self-timed logic (i.e., digital circuits without any master clock) has been introduced to solve the problem of the clock tree synthesis, which has proven more difficult and power-consuming [27,28]. In SoC design, the master clock cannot be propagated through the chip without having clock skew larger than the clock period. Globally asynchronous locally synchronous (GALS) SoC architectures have been proposed to solve this problem, but another solution could be the complete removal of the clock by designing pure asynchronous logic blocks.

Several asynchronous techniques have been proposed, at the block and/or cell levels. They are based on handshaking (i.e., a local control of the data shifted in a pipeline). This control logic is largely based on C-Muller elements, generally not proposed by conventional libraries. For asynchronous design, it would be beneficial to have this C gate as a library cell. Another logic style that is used in self-timed architectures is dynamic DCVSL dual rail logic. During the precharged phase, both outputs are "1," an invalid state. After evaluation, the valid state is reached ("01" or "10"), indicating that the operation is completed. This signal is consequently used to start the next operation (request) and to acknowledge the previous pipeline stage (acknowledge). In a manner similar to global clocks, in which rising edges are used for synchronization, rising and falling edges of these "request" and "acknowledge" signals are also used in self-timed logic. A static logic family, called "event logic," can also be designed while using the signal edges for which only these edges are the events of interest. Two different protocols are used:

1. Two phases protocol, for which a rising edge has the same significance as a falling edge
2. Four phases protocol, for which only the rising edge is taken into account while the falling edge means only reset

Figure 7.13 depicts the four phases protocol, in which the rising edge of the "request" signal occurs when data is ready (valid data). It is important to note that the "request" signal is *not* generated by the control logic itself (in this case, a critical race can occur between the data and the "request" wire), but by the data, using a specific code, such as the dual rail code or other codes such as one hot, parity or Berger codes. In this way, no critical race exists between data and the "request." Such a scheme is called quasi delay insensitive (QDI). When the data has been used, the "acknowledge" rising edge occurs and the data can be resettled (empty data). Then the two signals are resettled and a new cycle can be started.

The basic logic gates of "event logic" are quite different from conventional logic. For instance, the AND gate is a C-element or Muller gate. This gate is switched on if the two inputs switch to "1" and switched off if the two inputs switch off. This means that this gate has a memory (i.e., if only one input goes to "0" after that the two inputs switch on, the gate keeps its output = 1, so it is not a combinational AND). A library of "Event Logic" standard cells can be designed using the dual rail approach. The code for "1" is Z0 = 0 and Z1 = 1, while a "0" is represented by Z0 = 1 and Z1 = 0. From a Karnaugh map, for instance, the "OR" function in Table 7.5, the two outputs Z0 and Z1 can be designed as the sum of the minterms of each cube. Figure 7.14 illustrates the logic implementation of the OR function using C-

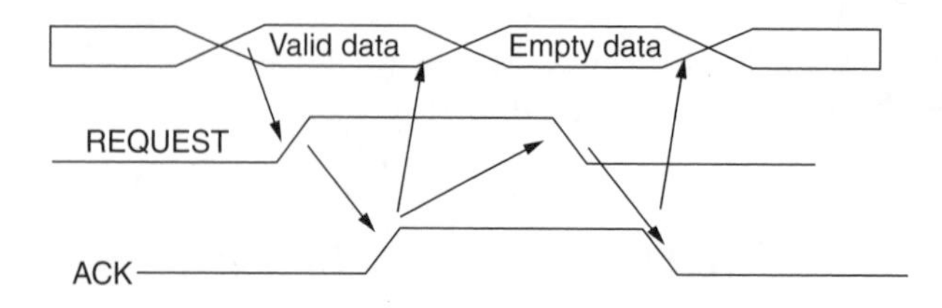

FIGURE 7.13 Four phases protocol.

TABLE 7.5 Karnaugh Map of the OR Function

Z	A = 0	A = 1
B = 0	0	1
B = 1	1	1

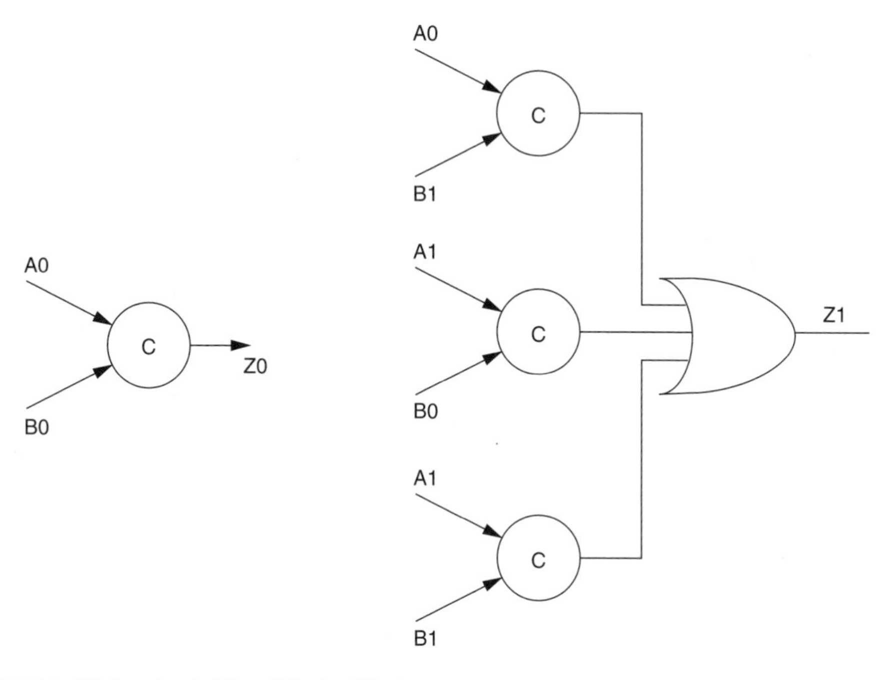

FIGURE 7.14 OR function in "Event" dual-rail logic.

elements. Any combinational circuit can be designed using this approach starting from its Karnaugh map. The cost in terms of number of transistors is significantly increased.

$$Z0 = A0^{*}B0$$

$$Z1 = A0^{*}B1 + A1^{*}B0 + A1^{*}B1$$

7.4.2 Library Cells for Cryptographic Applications

Application-specific integrated circuits (ASICs) for smart cards can be attacked by differential power attacks (DPA), which trace the power consumption and identify operations that are data-dependent after removal of the power consumption, which is data-independent. In this way, secret keys could be obtained. Consequently, DPA-resistant circuits will be of crucial importance in the future.

DPA was demonstrated in 1999 [29], thus conveying that it is possible to examine the power consumed by the circuit when processing data or executing instructions. By analyzing the variation in power consumption and the data processed, an attacker can discover the secure information being processed and the keys hidden in the circuit. The attacker can analyze a single power trace (SPA) or can perform a statistical analysis of many collected power traces (DPA). These power traces will provide an average power trace that represents the data-independent power trace. By having a given power trace with a given hidden key, the attacker can subtract the average power trace and determine the difference that represents only the data-dependent power. By comparing the difference to the simulated power traces, the attacker can quite easily deduce the secure key of the considered circuit.

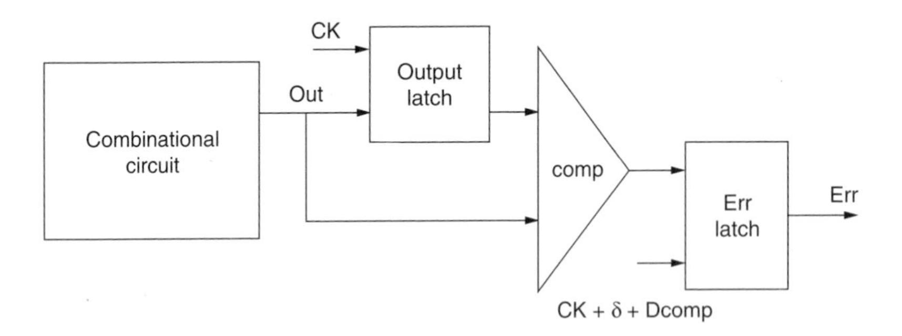

FIGURE 7.15 SEU fault-tolerant logic.

The DPA attack is because executed operations or instructions present power consumption that is data-dependent. A very general goal is to find circuit implementations that are not sensitive to operands regarding the power. Some circuit techniques, such as current steering logic or dual rail DCVSL logic, are known to be less data-dependent. Self-timed logic does not have a global clock, so there is not a global timing signal for use as a reference and the analysis and correlation of power traces of power consumption are more difficult [30]. It is difficult to determine when an operation or instruction starts and stops. Because power issues are crucial for smart cards, however, these logic families do not meet this requirement. Furthermore, recent research has revealed weaknesses in the basic DCVSL scheme and suggested improvements [31]. In dual-rail DCVSL logic, parasitic capacitances are different in the two N-ch networks implementing the dual Boolean functions and, therefore, the power consumption is. A new sense amplifier based logic (SABL) logic family is introduced in [31], for which the output will charge the same capacitance for each clock, even if the output transition is 1-1 or 0-0. This SABL logic is based on the differential strong arm flip-flop (SAFF); it consumes, however, two times the power of a static CMOS. Balanced "Event Logic" can also be used, by adding into Figure 7.14, for instance, a dummy OR gate for z0.

7.4.3 SEU-Tolerant Logic

Another dramatic problem in deep submicron technologies is due to soft errors, which may arise due to particles that can discharge a logic node. Several years ago, only dynamic nodes were an issue, for instance, in dynamic RAM memories (DRAMs), and mainly for space applications. With technology scaling, however, such as V_{dd} and parasitic capacitance scaling, the charge at each node becomes smaller and smaller. It is therefore becoming easier for a particle to discharge such a node even on earth, resulting in a glitch that could affect the circuit behavior.

Techniques have been proposed to create a soft error (single event upset [SEU]) tolerant logic. It is based on duplication of the logic, while observing if the two results are the same. If not, the executed operation is repeated until the SEU has disappeared. Such logic duplication is very expensive in terms of silicon area and power consumption. This is why some of the most interesting techniques are based on timing redundancy [32] (i.e., the operation is executed in a single combinational block), but the result is stored in a first latch at the rising edge of the CK, and in an extra latch at CK + δ. If the SEU occurs after the CK edge, the two stored data will be different and the SEU detected. Figure 7.15 depicts an implementation in which even the extra latch is removed by performing the comparison at CK + δ. The latch contains the data memorized at the CK edge, while the other input of the comparator is the data at CK + δ. To obtain an SEU fault tolerant logic, it is possible to transform the VHDL description of any circuit into a VHDL description containing cells as depicted in Figure 7.15 [32].

7.5 Conclusion

For deep submicron technologies, robustness is the main challenge after low power consumption. Therefore, robust logic styles, such as the old but very simple static logic style, are the best candidates for future libraries [35]. Leakage reduction will be the most difficult problem to solve regarding low power, as well as leakage reduction in active modes. As more constraints must be satisfied for specific applications, specialized libraries will emerge, and, ultimately, these are expected to lead to application specific libraries.

References

[1] T.G. Noll, E. De Man, Pushing the performances limits due to power dissipation of future ULSI chips, *IEEE Int. Symp. on Circuits and Syst., ISCAS '92,* San Diego, CA, May 10–13, 1992, pp. 1652–1655.

[2] J-M. Masgonty et al. Technology- and power-supply-independent cell library, *IEEE CICC '91,* San Diego, CA, May 12–15, 1991, Conf. 25.5.

[3] A. Nève et al. Design of a branch-based 64-bit carry-select adder in 0.18 μm partially depleted SOI CMOS, *Proc. ISLPED '02,* Monterey, CA, August 12–14, 2002, pp. 108–111.

[4] S. Nikolaidis and A. Chatzigeorgiou, Circuit-level low-power design, Chapter 4 in *Designing CMOS Circuits for Low-Power,* D. Soudris, C. Piguet, and C. Goutis, Eds., Kluwer Academic Press, Dordrecht, 2002.

[5] L.G. Heller et al. Cascode voltage switch logic: a differential CMOS logic family, *Proc. ISSCC 1984,* San Francisco, CA, February 12–14, pp. 16–17.

[6] R. Zimmermann and W. Fichtner, Low-power logic styles: CMOS versus pass-transistor logic, *IEEE JSSC,* Vol. 37, No. 7, pp. 1079–1090, July 1997.

[7] N.F. Goncalvez and H.J. De Man NORA: a racefree dynamic CMOS technique for pipelined logic structure, *IEEE J. Solid-State Circuits,* SC-18, No. 3, June 1983, p. 261.

[8] C. Piguet, Supplementary condition for STG-designed speed-independent circuits, *Electron. Lett.,* Vol. 34, No. 7, April 2, 1998, pp. 620–622.

[9] C. Piguet and J. Zahnd, Signal-transition graphs-based design of speed-independent CMOS circuits, *ESSCIRC '98,* Den Haag, The Netherlands, September 21–24, 1998, pp. 432–435

[10] C. Piguet, Robustness of asynchronous sequential standard cells in a synchronous environment, *AINT '2000,* Delft, The Netherlands, July 1920, 2000.

[11] A.P. Chandrakasan, S. Sheng, and R.W. Brodersen, Low-power CMOS digital design, *IEEE J. of Solid-State Circuits,* Vol. 27, No. 4, April 1992, pp. 473–484.

[12] C. Piguet et al. Logic design for low-voltage/low-power CMOS circuits, *1995 Int. Symp. on Low-Power Design,* Dana Point, CA, April 24–26, 1995, pp. 117–122.

[13] R. Hossain et al. Low-power design using double edge triggered flip-flops, *IEEE Trans. on Very Large-Scale Integr. Syst.,* Vol. 2, No. 2, June 1994, p. 261.

[14] M. Belleville and O. Faynot, Low-power SOI design, *Proc. PATMOS 2001,* Yverdon, Switzerland, September 26–28, 2001, pp. 8.1.2–8.1.10.

[15] V. Stojanovic and V.G. Oklobdzija, Comparative analysis of master-slave latches and flip-flops for high-performance and low-power systems, *JSSC,* Vol. 34, No. 4, April 1999.

[16] C. Piguet, Low-power and low-voltage CMOS digital digital design, *Elsevier Microelectron. Eng.,* 39, 1997, pp. 179–208.

[17] L. Benini et al. Saving power by synthesizing gated clocks for sequential circuits, *IEEE Design and Test of Computers,* Vol. 11, No. 4, pp. 32–41, 1994.

[18] C. Arm et al. Double-latch clocking scheme for low-power I.P. cores, *Proc. PATMOS 2000,* Goettingen, Germany, September 13–15, 2000, pp. 217–224.

[19] C. Piguet et al. Low-power low-voltage digital CMOS cell design, *Proc. PATMOS '94,* Barcelona, Spain, Oct. 17–19, 1994, pp. 132–139.

[20] J-M. Masgonty et al. Low-power low-voltage standard library cells with a limited number of cells, *PATMOS 2001,* Yverdon, Switzerland, September 26–28, 2001, pp. 9.4.1–9.4.8.

[21] T. Sakurai, Perspectives on power-aware electronics. Plenary talk 1.2, *Proc. ISSCC 2003,* San Francisco, CA, Feb. 9–13, 2003, pp. 26–29.

[22] S.V. Kosonocky et al. Enhanced multi-threshold (MTCMOS) circuits using variable well bias, *Proc. ISLPED '01,* August 6–7, 2001, Huntington Beach, CA, pp. 165–169.

[23] S. Narendra et al. Scaling of stack effect and its application for leakage reduction, *Proc. ISLPED '01,* Huntington Beach, CA, August 6–7, 2001, pp. 195–200.

[24] F. Assaderaghi et al. A dynamic threshold voltage (DTMOS) for ultra-low voltage operation, *IEEE IEDM Tech. Dig., Conf. 33.1.1,* pp. 809–812, 1994.

[25] T. Enomoto, Y. Oka, H. Shikano, and T. Harada, A self-controllable-voltage-level (SVL) circuit for low-power high-speed CMOS circuits, *Proc. ESSCIRC 2002,* Forence, Italy, September 24–26, 2002, pp. 411–414.

[26] S. Cserveny, J-M. Masgonty, and C. Piguet, Stand-by power reduction for storage circuits, *Proc. PATMOS 2003,* Torino, Italy, September 10–12, 2003, pp. 229–238.

[27] Asynchronous circuits and systems, *Special Issue of IEEE Proc.,* Vol. 87, No. 2, February 1999.

[28] J. Sparso and S. Furber, *Principles of Asynchronous Circuit Design, A Systems Perspective,* Kluwer Academic Publishers, Dordrecht, 2001.

[29] P. Kocher, Differential power analysis, Advanced in Cryptology–Crypto 99, Springer LNCS, Vol. 1666, pp. 388–397.

[30] M. Renaudin and C. Piguet, Asynchronous and locally synchronous low-power SoCs, *DATE 2001,* Münich, Germany, March 13–16, 2001, pp. 490–491.

[31] C. Tiri et al. A dynamic and differential CMOS logic with signal independent power consumption to withstand differential power analysis on smart cards, *Proc. ESSCIRC 2002,* Florence, Italy, September 2002, pp. 403–406.

[32] L. Anghel and M. Nicolaidis, Cost reduction and evaluation of a temporary faults detecting technique, *Proc. DATE 2000,* Paris, France, March 27–30, 2000, pp. 591–598.

[33] C. Heer et al. Designing low-power circuits: an industrial point of view, *PATMOS 2001,* Yverdon, September 26–28, 2001.

[34] C. Piguet et al. Techniques de circuits et méthodes de conception pour réduire la consommation statique dans les technologies profondément submicroniques, *Proc. FTFC '03,* Paris, France, May 15–16, 2003, pp. 21–29.

[35] M. Allam et al. Effect of Technology Scaling on Digital CMOS Logic Styles, *Proc. IEEE CICC 2000,* Conf. 19.1, Orlando, Florida, May 21–24, 2000, pp. 401–408.

8

Low-Power Very Fast Dynamic Logic Circuits

Jiren Yuan
Lund University

8.1 Introduction

To achieve a high throughput usually means more power consumption in a given CMOS technology because the dynamic power consumption is proportional to the activity ratio. It implies that at a low activity ratio static logic circuits may consume less power compared with clocked dynamic logic circuits. For a given logic function, however, high-speed or a short-propagation delay does not necessarily mean high-power consumption, if highly efficient dynamic logic circuits with low power-delay products are used. Such low-power and very fast dynamic circuits are introduced in this chapter. A simple way to distinguish dynamic logic from static logic is to see whether the logic states are still correctly maintained, as for the static circuits, or destroyed, as for the dynamic circuits when the clock is turned off. This is because dynamic logic circuits need to be regularly refreshed for the charge stored on the logic nodes while static logic circuits need not.

Although clocking is used in all synchronous circuits, it is used only as synchronization for static logic circuits while as both synchronization and refreshment for dynamic logic circuits. Note that both complementary logic circuits and precharged logic circuits can be either static or dynamic, depending on whether the logic states are locked, for example by a cross-coupled loop, or not.

Section 8.2 focuses on the basic synchronizing components, single-clock latches and flip-flops, with comparisons in power and delay. Section 8.3 presents high throughput logic styles based on these components. Section 8.4 demonstrates examples of very fast dynamic CMOS functional circuits. Section 8.5 discusses the future of dynamic logic when the leakage current becomes a serious problem in deep submicron technologies. The conclusion is given in Section 8.6.

0-8493-1941-2/05/$0.00+$1.50

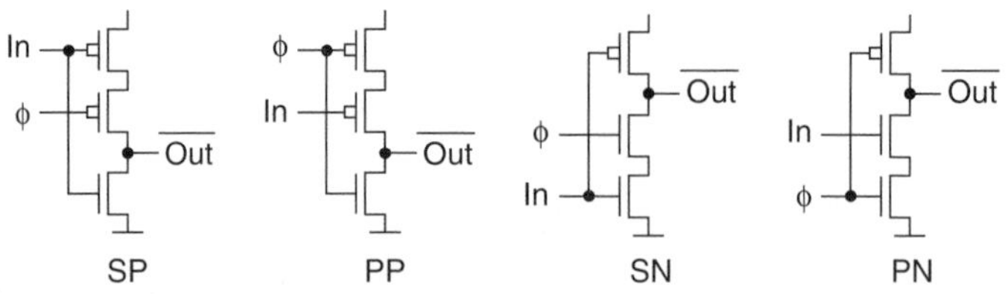

FIGURE 8.1 Basic stages in TSPC. (© 2004 IEEE.)

8.2 Single-Clock Latches and Flip-Flops

Latches and flip-flops controlled by clock(s) are the fundamental blocks of a synchronous system. It is well-known that $P_D = C_L V_{DD}^2 f_C A$, where P_D is the dynamic (usually the dominating) power consumption, C_L the load capacitance, V_{DD} the power supply voltage, f_C the clock frequency, and A the activity ratio. Clock is considered a fully active signal with a reference activity ratio of 1.0. Clocking strategy and the types of latches and flip-flops used for this system thus have a significant impact on its power consumption. Regarding the dynamic power consumption, a smaller number of clock wires and a smaller number of clocked devices will likely result in lower power dissipation. Based on this principle, we prefer to have as few clocked devices as possible and to use a single clock if it does not mean more clocked devices.

8.2.1 TSPC Latches and Flip-Flops

The true single phase clock (TSPC) circuit technique [1,2] uses only a single clock and two to three clocked transistors in each latch without local inversion of the clock as such an inversion requires more clocked devices. The basic stages SP, PP, SN, and PN in TSPC are depicted in Figure 8.1, where the first letter represents the logic style (S for nonprecharged and P for precharged), and the second represents the type of clocked devices (P for p-type and N for n-type). Stages SP and PP are identical except the exchange of data and clock inputs, the same for stages SN and PN. Two cascaded SP stages (nontransparent when clock is high) or SN stages (nontransparent when clock is low) become a p-type or n-type nonprecharged TSPC latch, respectively (see Figure 8.2(a)). A PP stage followed by an SP stage or a PN stage followed by an SN stage become a p-type or n-type precharged TSPC latch, respectively (see Figure

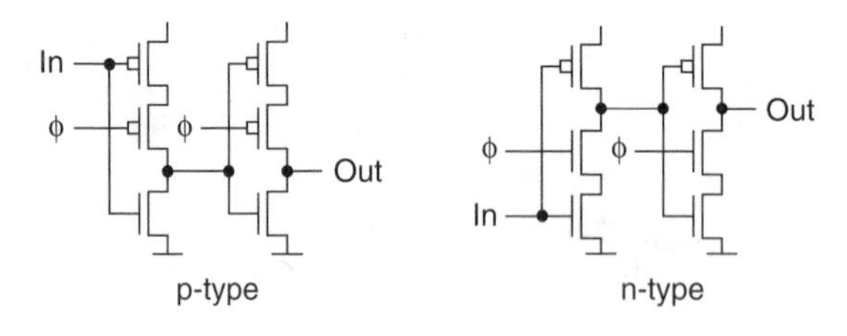

(A) Non-precharged TSPC latches.

(B) Precharged TSPC latches.

FIGURE 8.2 TSPC latches. (© 2004 IEEE.)

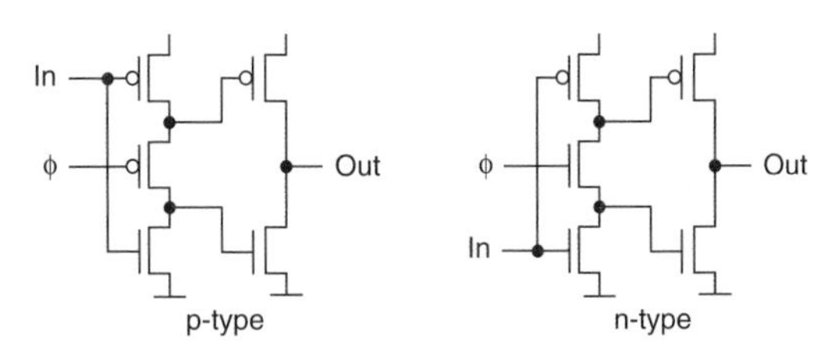

FIGURE 8.3 TSPC split-output latches. (© 2004 IEEE.)

8.2(b)). A TSPC nonprecharged flip-flop consists of two cascaded nonprecharged latches, a p-type and an n-type, and it becomes positive edge-triggered when the p-type is before the n-type, or negative edge-triggered otherwise. A TSPC precharged flip-flop is formed by a nonprecharged TSPC latch followed by a precharged TSPC latch in an opposite type, and it becomes positive edge-triggered when the non-precharged TSPC latch is a p-type, or negative edge-triggered otherwise. The nonprecharged TSPC latches and flip-flops are superior in low-power performance [3].

To reduce power consumption, it is possible to use only a single clocked transistor for each latch. Figure 8.3 depicts such latches in p-type and n-type. They are so-called TSPC split-output latches in which the output of the first stage is split [2]. Edge triggered flip-flops can be built by cascading the split-output latches. As the number of clocked devices is at its minimum, the power spent on the clocked node is minimized. Because the clocked transistor propagates both high state and low state, however, one of the two states will not have a full swing. The clocked transistor should be properly sized when the supply voltage is low. In submicron technologies, the TSPC split-output latches can still be used due to reduced threshold voltages.

A very efficient and fast TSPC flip-flop using only nine transistors, based on a nonclassic flip-flop concept, is depicted in Figure 8.4(a), which gives an inverted data output [2]. The nonclassic flip-flop concept is illustrated in Figure 8.5(b), in comparison with the classic flip-flop concept depicted in Figure 8.5(a). A flip-flop transfers the input to the output only when the right clock edge comes and must be nontransparent otherwise. In a classic flip-flop the master and slave are completely nontransparent in its latching phase regardless the input logic states. In a nonclassic flip-flop, the master may be transparent in its latching phase for either a high or a low input, but the slave (which is in its nonlatching phase) must be nontransparent for the output of the master. In the example given in Figure 8.5(b), the master is transparent for a high input in its latching phase but the slave is nontransparent for the low output of the master although the slave is in its nonlatching phase. In such a way, the flip-flop is still nontransparent, which is exactly the case for the nine-transistor TSPC flip-flop depicted in Figure 8.4(a). The master is

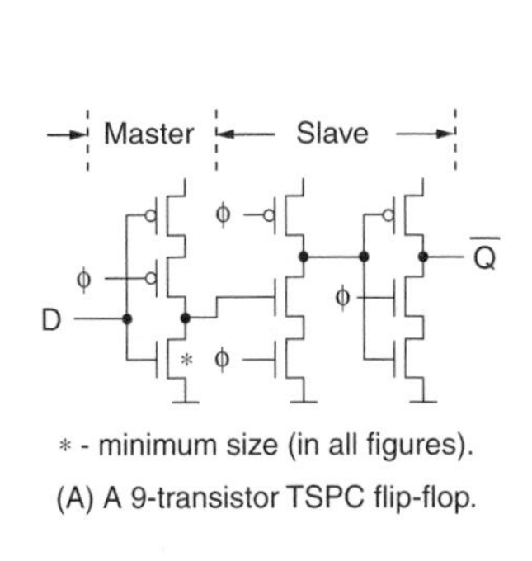

∗ - minimum size (in all figures).

(A) A 9-transistor TSPC flip-flop.

(B) A flip-flop with all clocked devices connected to power or ground.

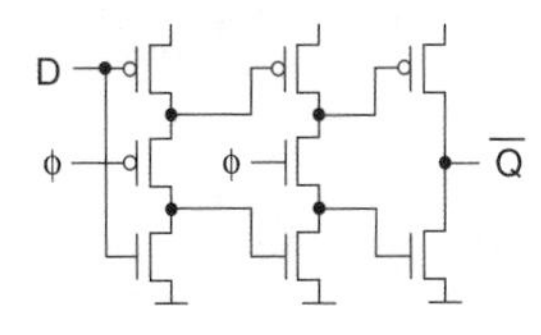

(C) A flip-flop with 2 clocked devices and totally 8 transistors.

FIGURE 8.4 Efficient nonclassic single-clock flip-flops. (© 2004 IEEE.)

Clock phase	Input	(a) Classic flip-flop Master	(a) Classic flip-flop Slave	(b) Non-classic flip-flop Master	(b) Non-classic flip-flop Slave
High (Low)	Low				
	High				
Low (High)	Low				
	High				

FIGURE 8.5 Illustration of classic and nonclassic flip-flop concepts.

a p-type half-latch with only an SP-stage and transparent for a high input. When the clock is high (latching phase), it gives a low output. The slave, an n-type precharged latch, however, is nontransparent for the low input after the PN stage finishes its evaluation. The condition of finishing evaluation increases the required hold time but just slightly because the evaluation of PN stage takes very short time. The speed increase, however, is significant because the p-type master latch with two SP stages is slower than the n-type precharged slave latch, and the removal of one of the two SP stages balances the delays of both latches. In the same time, the power consumption is reduced due to the removal of three transistors, especially the clocked p-transistor. A similar circuit but in a different transistor stacking order, illustrated in Figure 8.4(b), was published in 1973 [4] and its advantage in speed optimization was addressed in Huang and Rogenmoser [5].

Because all clocked devices are connected to power and ground rails, they can be sized without excessive loading to improve speed, though care must be taken for the charge sharing between nodes A and B and between nodes C and D. In another example, a nonclassic flip-flop can be built from a split-output SP stage and an n-type split-output latch (or an SN stage and a p-type split-output latch), reducing the number of total transistors to eight and the number of clocked devices to two, which is shown in Figure 8.4(c). For a high clock, the SP stage is half-transparent, but the split outputs respectively to the p- and n-transistors in the following latch stage can never make them transparent, although the latch is in its nonlatching phase. Another method using flow tables and signal transition graphs (STG) has been presented in Piguet [6] and Piguet and Zahnd [7] to design similar circuits including dynamic flip-flops, aiming at race-free (or, today, speed-independent [SI]) circuits.

To reduce the hold time of the flip-flop in Figure 8.4(a), a 10-transistor TSPC flip-flop illustrated in Figure 8.6(a) can be used [8]. Only the hold time for a low input needs to be reduced. The added nMOS transistor controlled by the precharged node signal will firstly increase the delay for a high input to a low output and secondly make the single stage master completely nontransparent (i.e., a full latch) without any additional clock or clocked device. A similar counterpart but with the single stage full latch

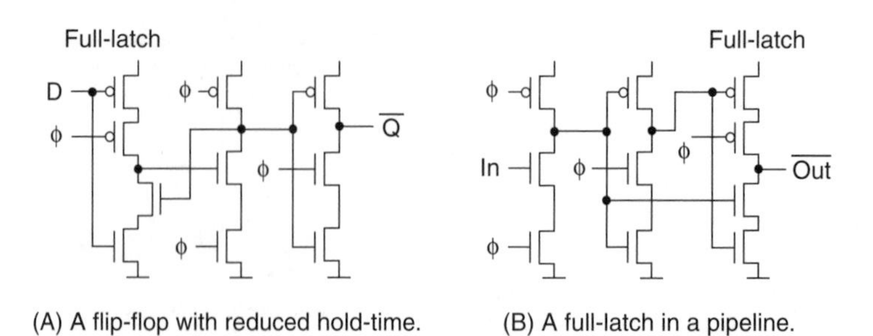

(A) A flip-flop with reduced hold-time. (B) A full-latch in a pipeline.

FIGURE 8.6 Single-stage full latches using the precharged node signal. (© 2004 IEEE.)

at the output is shown in Figure 8.6(b). The two circuits can be used for a TSPC double pipeline [9,10] to improve its robustness, which will be discussed later.

8.2.2 Differential Single-Clock Latches and Flip-Flops

It is necessary to avoid precharge for low-power applications, especially for a circuit with a low activity ratio. The aforementioned nonprecharged latches and flip-flops are therefore preferred in this case. To obtain differential outputs, however, an inverter has to be added for all single-ended latches and flip-flops, which consumes additional power. Differential latches and flip-flops can produce complementary outputs without an additional inverter. One example is the CVSL dynamic latches [11] illustrated in Figure 8.7. The problem with a cross-coupled differential latch is that it is sensitive to the ratio between p- and n-transistors, especially for the p-type latch (see the chart in Figure 8.7). This problem can be avoided by using dynamic ratio insensitive (DRIS) differential latches in which there is no fighting between p- and n-transistors [8]. The p-type latch of this kind is shown in Figure 8.8 along with the comparison between CVSL and DRIS. All latches and flip-flops introduced so far are dynamic (i.e., they have to be refreshed above a minimum clock rate). Static latches and flip-flops can accept a zero clock rate (clock off for low power) without losing data. The fully static counterpart to CVSL and DRIS latches are the random-access memory (RAM)-type [11] and SRIS latches [8], respectively, depicted in Figure 8.9. SRIS p-latch is faster than its RAM-type counterpart due to the absence of fighting.

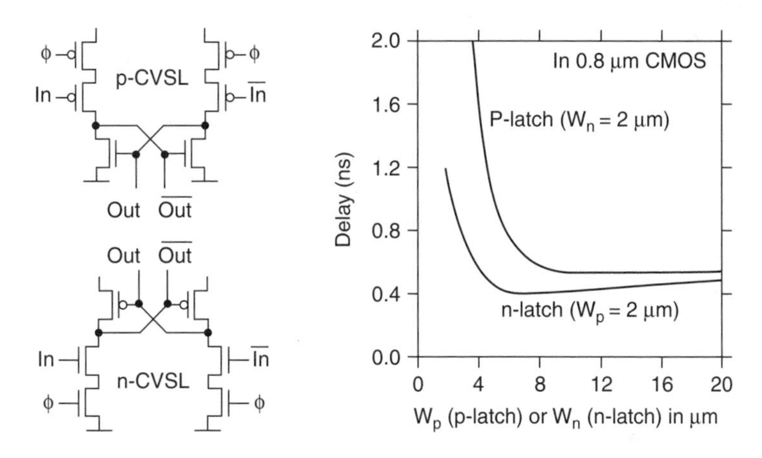

FIGURE 8.7 CVSL latches and their ratio sensitivities. (© 2004 IEEE.)

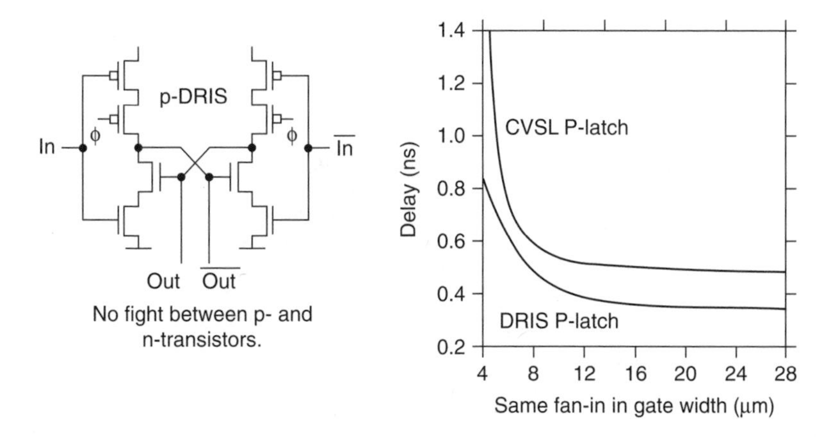

FIGURE 8.8 DRIS p-latch. (© 2004 IEEE.)

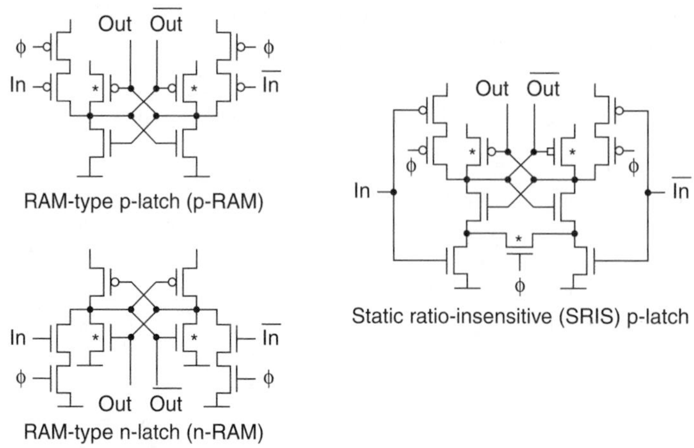

RAM-type p-latch (p-RAM)

RAM-type n-latch (n-RAM)

Static ratio-insensitive (SRIS) p-latch

FIGURE 8.9 Static RAM-type latches and static ratio-insensitive p-latch. (© 2004 IEEE.)

It is also possible to use only a single clocked transistor for a differential latch to reduce power consumption, as the p-type and n-type dynamic single-transistor clocked latches (version 1) depicted in Figure 8.10, named p-DSTC1 and n-DSTC1 latches [8]. They have the advantages of minimized clock load, minimized input load, and available differential outputs. Caution is needed for using DSTC lathes. There is a risk of charge sharing between the two output nodes when the two inputs have overlapped periods or glitches during which both input transistors are conductive. The same as the p-CVSL latch, p-DSTC1 latch is severely transistor ratio sensitive and slow, which will be handled later by its second version. To handle the charge sharing, static single-transistor clocked (SSTC) latches can be used [8]. Its n-type (n-SSTC) is depicted in Figure 8.10, which was developed from the RAM-type latches but with only a single clocked device. The two added n-transistors compared to n-DSTC1 need only the minimum size for reducing the load. The full output differential swing, if degraded by the charge sharing through two input transistors, could be effectively recovered by the cross-coupled inverter pair, which makes the SSTC latch highly robust.

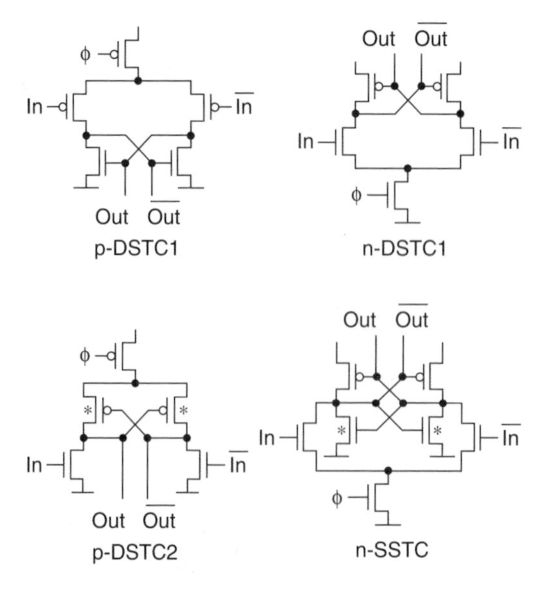

FIGURE 8.10 Single transistor clocked differential latches. (© 2004 IEEE.)

To break the speed bottleneck, the p-DSTC1 latch can be replaced by the p-DSTC2 latch [8], shown in Figure 8.10. The p-DSTC2 latch looks similar to the n-DSTC1 latch but with the clocked transistor being p-type and at the top of the latch. The p-DSTC2 latch is not a full latch but half-transparent when the clock is high, and a low-to-high input transition will result in a high-to-low output transition. If the p-DSTC2 latch (as the master) is followed by an n-type differential latch (as the slave) (e.g., an n-DCST1 or n-SSTC latch) it becomes a positive edge-triggered nonclassic flip-flop. The high-to-low output transition from the master will not propagate through the slave because the slave's input n-transistor will not respond to a high-to-low transition as long as the data propagation in the slave finishes. The advantages are obvious. First, the input transistors of both latches are n-type, resulting in low fan-in and high speed due to all logic in nMOS. Second, the p-DSTC2 latch is less sensitive to transistor ratio. Third, the two flip-flops are highly robust because the p-DSTC2 latch produces nonoverlap and glitch-free signals to the n-DSCT1 or n-SSTC latch. Finally, the flip-flop with the n-SSTC latch is semi-static, allowing the clock to standby at low state.

8.2.3 Power-Delay Comparison

The worst delay and power consumption of the aforementioned flip-flops are compared with classic and conventional solutions. For this purpose, the dynamic flip-flop with p-classic and n-classic latches, the dynamic flip-flop with p-C²MOS and n-C²MOS latches, and the static flip-flop with classic latches are presented in Figure 8.11(a), Figure 8.11(b), and Figure 8.11(c), respectively. The comparison is done by using the parameters of a 0.8 μm CMOS process through SPICE simulations to extract the worst delay and power consumption values under different activity ratios. It is not easy to fairly compare different circuits. To make the comparison as fair as possible, the following conditions are applied. The minimum length, 0.8 μm, is used for all transistors. For the widths, Wp = 6 μm, Wn = 3 μm, and Wmin = 2 μm (the transistor marked with *). For ratio sensitive n-type latches, such as n-CVSL, n-RAM and n-DSTC1 and n-SSTC latches, Wp = 4 μm, Wn = 6 μm, and Wmin = 2 μm. For ratio sensitive p-type latches, such as p-CVSL and p-RAM latches, Wp = 12 μm, Wn = 3 μm, and Wmin = 2 μm. The load to each flip-flop is the input capacitance of two inverters with Wp = 6 μm and Wn = 3 μm. In the simulation, both the power for driving the inputs of each flip-flop and the power for driving the output load are included in the power consumption. At an activity ratio of 0.25, the values of power consumption and worst delay are listed in Table 8.1. The power-delay products for all flip-flops with activity ratios from 0.0 to 0.5 are plotted in Figure 8.12. The minimum delay is given by the flip-flop with p-DSTC2 and n-DSTC1. At an

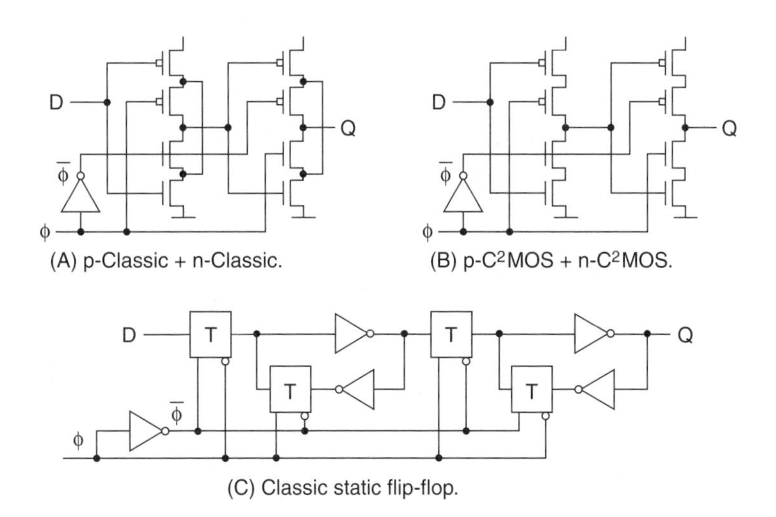

FIGURE 8.11 Three flip-flops for comparison.

TABLE 8.1 Comparison of Flip-Flops

Type	No.	Master + Slave	Power (μW)	Delay (ns)
Dynamic	1	p-CVSL + n-CVSL	699.4	0.691
	2	p-C²MOS + n-C²MOS	491.8	0.950
	3	p-Classic + n-Classic	512.4	0.776
	4	(SP + SP) + (PN + SN)	404.3	0.835
	5	SP + (PN + SN)	331.6	0.832
	6	(SP + SP) + (SN + SN)	317.6	0.802
	7	p-DSTC2 + n-DSTC1	313.1	0.717
Static	8	Classic + Classic	668.8	1.008
	9	p-RAM + n-RAM	685.4	0.673
	10	p-DSTC2 + n-SSTC	393.5	0.705

Note: Activity ratio = 0.25 and load = two inverters, in 0.8 μm CMOS.

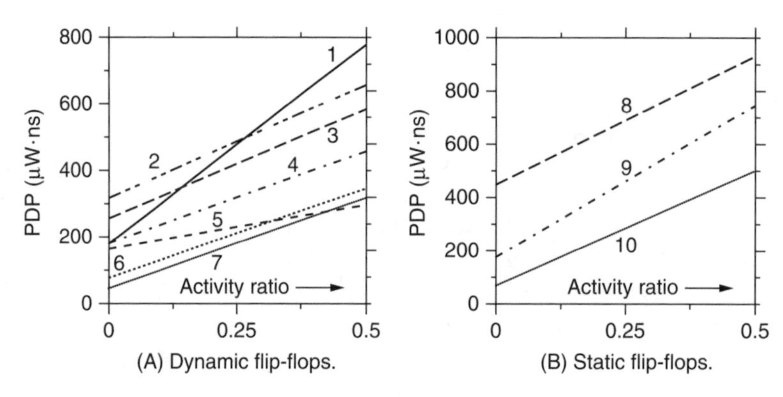

FIGURE 8.12 Comparison of power-delay products.

activity ratio of 0.25, the minimum power consumption is given by the flip-flop with p-RAM and n-RAM. As far as power-delay products are concerned, the best dynamic flip-flop is made of p-DSTC2 and n-DSTC1, and the best static (including semi-static) flip-flop is made of p-DSTC2 and n-SSTC.

8.3 High-Throughput CMOS Circuit Techniques

8.3.1 TSPC Pipeline

TSPC flip-flops can be used as edge-triggered elements in a synchronous pipeline. Its short setup-time, hold-time, and propagation delay contribute to high speed. Complementary logic stages can be placed between two TSPC latches in the pipeline. More efficiently, the logic gates can be embedded within TSPC latches [8], as depicted in Figure 8.13(a) and Figure 8.13(b). The previously mentioned pipeline can be divided into p-blocks and n-blocks, as depicted in Figure 8.13(c). A p-block consists of a p-type latch, which may embed logic, associated with the complementary logic stages before and after the p-latch, and it is the same for an n-block but with an n-type latch instead. The blocks must be connected with p-type and n-type latches alternately. Feedback is allowed but must also follow the rule from p-type to n-type or vice versa. In such a pipeline, p-blocks are the speed bottlenecks, especially when logic gates are included in the p-blocks or embedded in the p-latch with many stacked p-transistors. Therefore, in order to achieve a high throughput, logic gates are preferably placed in the n-blocks, leaving the p-block as a passing stage or with very simple logic. The nonclassic concept may be used to simplify the p-block to just a single SP stage if directly (or indirectly after an even number of complementary stages) followed by an n-type precharged latch. An all n-logic true-single-phase dynamic CMOS circuit technique was proposed in Gu and Elmasry [12] to speed up the p-block, in which the logic embedded in a p-type precharged latch uses n-transistors instead of p-transistors.

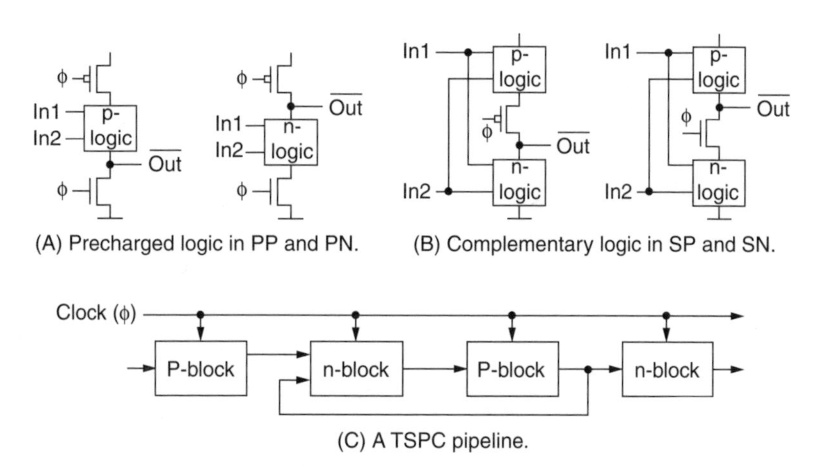

FIGURE 8.13 Logic embedded in latch stages in a TSPC pipeline.

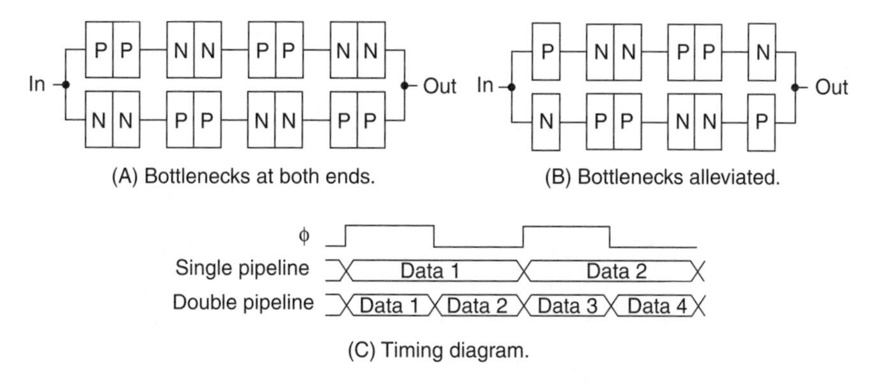

FIGURE 8.14 TSPC double pipelines.

8.3.2 TSPC Double Pipeline

Synchronous elements in a pipeline are usually triggered by a single clock edge. In a double pipeline, however, both edges of a clock are utilized for achieving high throughput and efficiency [9]. The data rate at the input and output of a double pipeline is at twice the clock rate. Internally, each pipeline works as normal, and data can be cross-connected between the two lines as long as following the n-to-p or p-to-n rule. As illustrated in Figure 8.14(a), two TSPC pipelines starting and finishing with opposite types of blocks can be such a double pipeline, and the two input-connected input blocks and the two output-connected output blocks become a multiplexer and a demultiplexer respectively. Because the input and output blocks have to work at a double data rate, they are the speed bottlenecks. It is therefore preferred to have the double pipeline configured as shown in Figure 8.14(b) (i.e., single-stage latches at both ends). This can be done by using the single-stage full latches [10], depicted in Figure 8.6(a) and Figure 8.6(b), which narrows the forbidden windows of low-to-high and high-to-low data transition by almost half, increasing speed and robustness. To reduce power consumption at a given data rate, a low-swing clock double-edge triggered flip-flop was proposed in Kim and Kang [13], in which both edges of a low swing clock are used to trigger a single flip-flop to reduce overall clock rate and associated power consumption.

8.3.3 Clock-and-Data Precharged Circuit Technique

All the preceding circuits are aiming at a high throughput regardless of the latency or the number of operating clock cycles for a final output. In many applications, however, the decision has to be made in one clock cycle. A technique named clock-and-data precharged dynamic (CDPD) circuit technique may

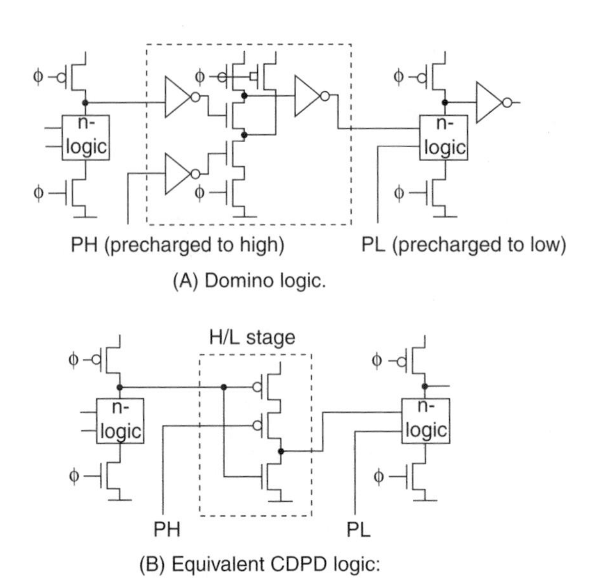

FIGURE 8.15 Domino logic and its equivalent CDPD logic.

offer an alternative for a fast one-clock-cycle decision and in the same time reduce the power consumption [14]. Domino logic is often used for logic calculations with a large depth as the logic parts can be distributed along the domino chain and are all in nMOS. As illustrated in Figure 8.15(a), however, an inverter has to be placed between two precharged stages to prevent an erroneous high-output to the next stage at the beginning of evaluation. Moreover, charge sharing may occur between the output node and the intermediate nodes so extra precharging transistors have to be used. As illustrated in Figure 8.15(b), all contents in the dashed line box can be replaced by only three transistors in CDPD technique, and no clocked transistor is contained in it. This CDPD block is named an H/L (high-to-low) stage in which the output is precharged to low by a high data input, and the NOR function is simply fulfilled by the two p-transistors. An H/L stage can be followed by an L/H (low-to-high) stage in which the output is precharged to high by a low data input. An n-type CDPD chain can be formed by the original domino precharged stages along with the H/L and L/H stages in between, as illustrated in Figure 8.16(a). It needs

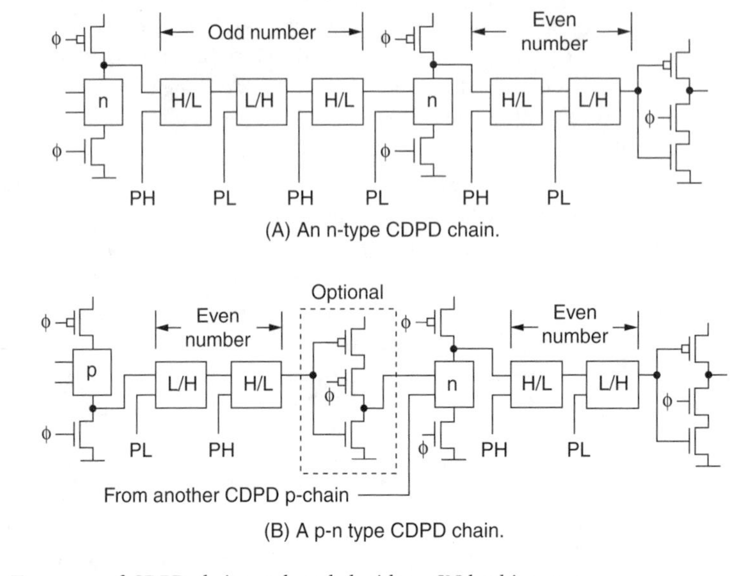

FIGURE 8.16 Two types of CDPD chains each ended with an SN latching stage.

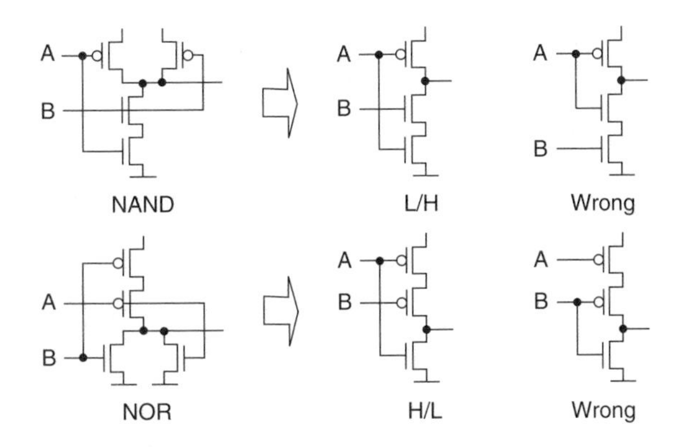

FIGURE 8.17 NAND and NOR gates transferred into L/H or H/L stages.

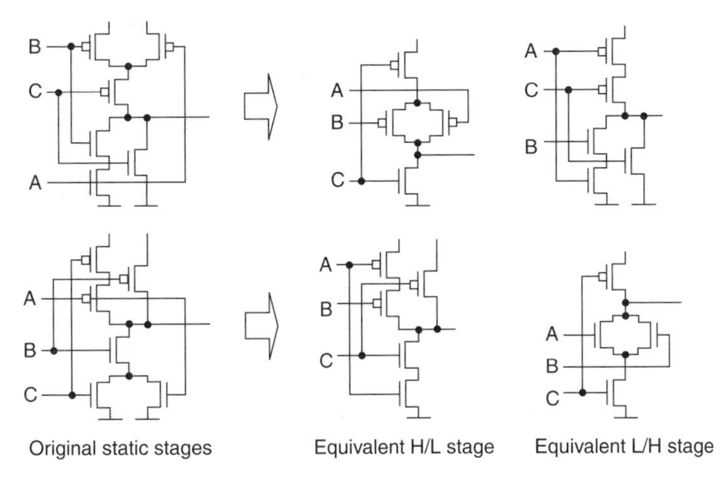

FIGURE 8.18 Logic gates transferred into either L/H or H/L stages.

an odd number of CDPD stages between two domino precharged stages, and an even number of CDPD stages between a domino stage and an output latch.

A number of advantages can be cited. First and second, all domino inverters are removed, and the number of clocked devices is minimized, reducing unnecessary power consumption. Third, the skewed precharging of CDPD stages effectively reduces the peak current. A p-n type CDPD chain is presented in Figure 8.16(b), and the rules can be found in the figure.

The p-n type CDPD chain has additional advantages. First, the logic operations are completed in both high and low clock periods so each duty cycle of the clock is fully utilized. Second, not only the number of clocked devices but also the number of latch stages is reduced. As indicated in Figure 8.16(b), the latch before the n-type precharged stage is optional, only depending on the need of inversion. Cares and skills are needed for designing CDPD stages to avoid erroneous results, as illustrated in Figure 8.17. A "NAND" function can be simplified in an L/H stage but is directly used for an H/L stage, while a "NOR" function can be simplified in an H/L stage but is directly used for an L/H stage. The wrong connections, which will result in charge sharing, should be avoided. Generally, complementary gates are simplified differently in an H/L or an L/H stage, (see Figure 8.18). In the worst-case scenario, complementary gates can be directly used for either an L/H or an H/L stage.

8.3.4 United Connection Rules of TSPC and CDPD Stages

It is important to follow the connection rules for constructing TSPC and CDPD circuits, and computer-aided design (CAD) tools should be able to check the correctness of the circuit connection according to

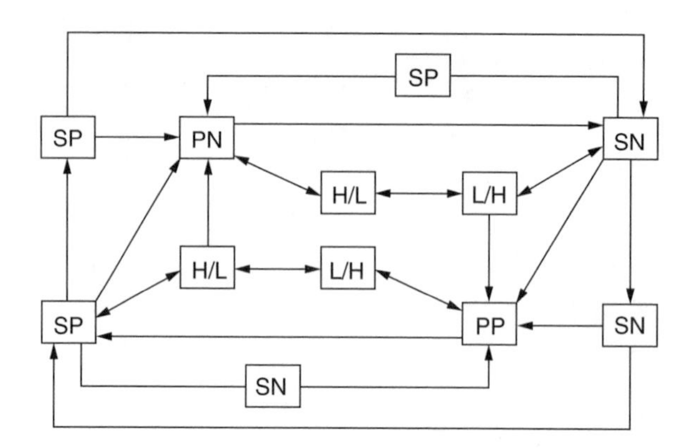

FIGURE 8.19 Unified connection rules of TSPC and CDPD stages.

the rules. If connections are correct, the circuit will undoubtedly work but the target function and speed have to be checked by simulation. The unified connection rules of TSPC and CDPD stages are illustrated in Figure 8.19. For example, SP→SP→SN→SN represents a TSPC nonprecharged flip-flop, and SP→SP→PN→SN becomes a TSPC precharged flip-flop. Nonclassic flip-flops are represented by the connections of SP→PN→SN (positive edge-triggered) and SN→PP→SP (negative edge-triggered). The connection rules between CDPD and TSPC stages are also clearly included.

8.4 Fast and Efficient CMOS Functional Circuits

The CMOS functional circuits introduced in this section are featured with high efficiency and high speed. High efficiency leads to a small number of both clocked and logic-operating devices, resulting in low-power consumption, and high speed offers a large delay margin that can be used for trading power at a lower supply voltage.

8.4.1 Dividers and Ripple Counters

A very fast divider can be constructed simply by connecting the output and the input of the nonclassic nine-transistor TSPC flip-flop depicted in Figure 8.4(a). Its dynamic version is shown in Figure 8.20 while its semi-static version is presented in Figure 8.21, respectively. The transistor widths in 2-μm CMOS, given in Figure 8.20, are optimized for speed. An 8-bit ripple counter built from the dynamic divide-by-two stage reached an input frequency of 750 MHz [15]. The semi-static divider in Figure 8.21 can be used in a very long ripple counter where the frequencies get very low in later stages. Note that the narrow pulse signal (out 1) should be fed to the next stage so that the condition for a dynamic circuit will be always satisfied for all stages in the ripple chain, while the 50% duty-cycle signal (out 2) used for bit-output.

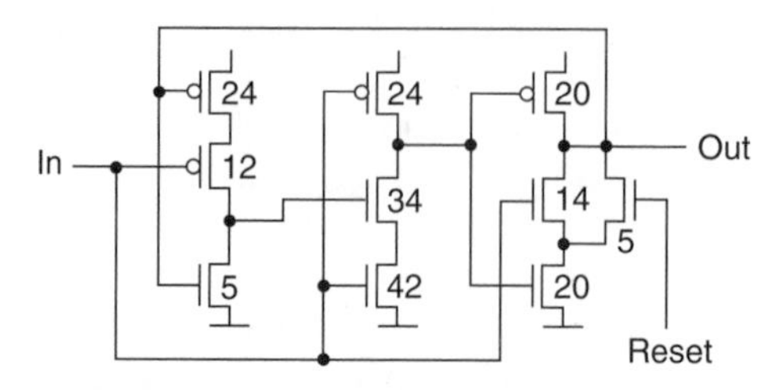

FIGURE 8.20 A dynamic divide-by-two circuit (D-1/2).

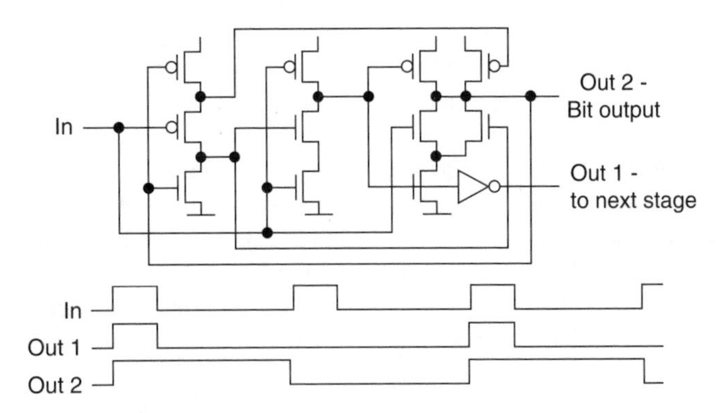

FIGURE 8.21 A semi-static divide-by-two circuit.

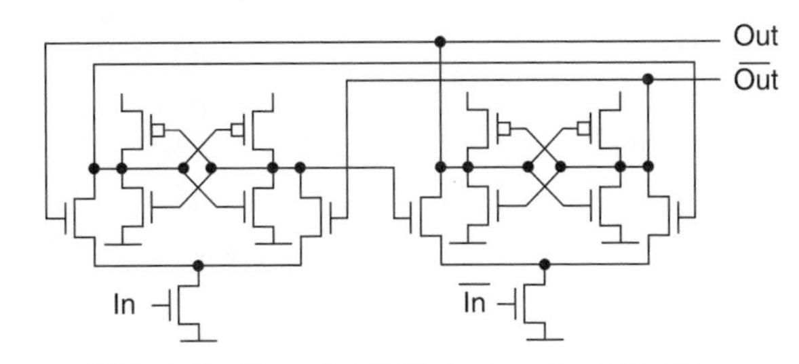

(A) A static differential divide-by-two circuit (S-1/2).

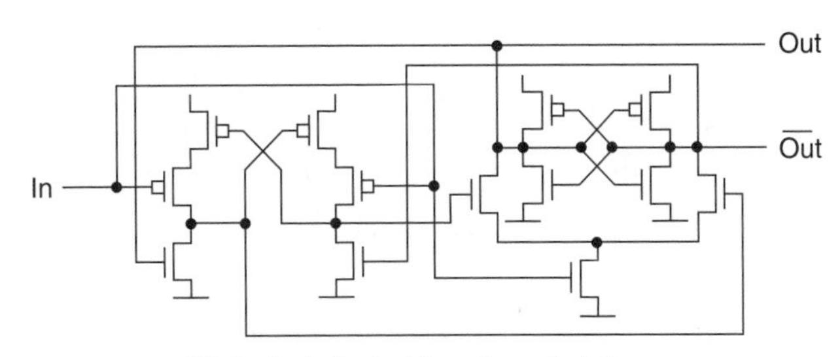

(B) A single-in double-out semi-static divide-by-two circuit (HS-1/2).

FIGURE 8.22 Differential divide-by-two circuits.

A differential divider offers differential outputs, which is sometimes a quite useful feature. In favor of speed, two n-SSTC latches can be cascaded to form a divider instead of using a p-SSTC and an n-SSTC, by using differential output signals available from the previous stage, named an S-1/2 stage and presented in Figure 8.22(a). For a single ended input, a semi-static differential divider may be used, named an HS-1/2 stage and presented in Figure 8.22(b). Two divider chains, one constructed by four D-1/2 stages with a buffer between stage 1 and stage 2 and another one constructed by an HS-1/2 stage followed by three S-1/2 stages, were constructed in IBM's partially scaled 0.1-μm CMOS process [16]. The measured input frequencies achieved 16.6 GHz for the dynamic divider chain and 12.5 GHz for the static divider chain [17].

8.4.2 Synchronous Counter

A TSPC synchronous counter is depicted in Figure 8.23 as an example showing how the carry-logic can be arranged in a p-block in favor of speed while using the dynamic divider as the toggle stage with the carry control function embedded [15]. The transistor widths optimized for speed are valid in a 3-μm technology. An 8-bit synchronous counter of this kind in the 3-μm technology was measured to reach a clock rate of 200 MHz. For a very long counter, however, the carry propagation becomes a serious speed bottleneck even with the parallel carry-logic depicted in Figure 8.23. A so-called backward carry propagation topology, in contrast to the conventional forward carry propagation, can be used to break the limit [18]. The principle block diagram of a backward carry propagation synchronous counter is presented in Figure 8.24. In a conventional counter, the worst-case scenario happens at output 0111 … 111, and the 0→1 flip of LSB has to be propagated through the whole chain of "AND" gates to MSB to enable the next output 1000 … 000. In Figure 8.24, however, when the out is 0111 … 110, the carry propagation is almost finished, and when the 0→1 flip of LSB comes all bits are ready for next output simultaneously. A more practical architecture is presented in Figure 8.25, mixing backward with forward carry propagation, at a lowest area and power penalty. The interface between the two propagation strategies depends on the counter length. Generally, a few bits using the backward carry propagation are enough.

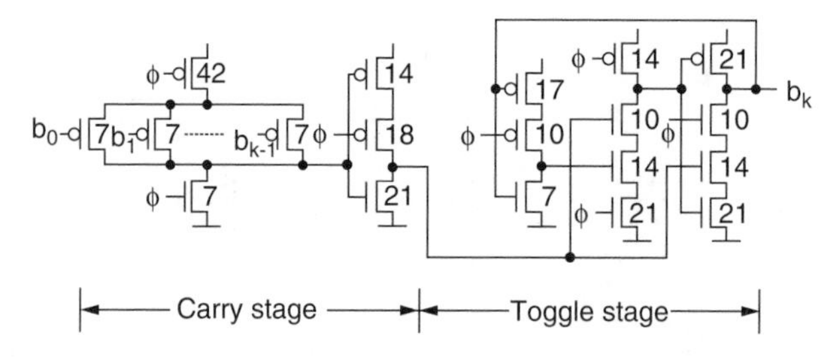

FIGURE 8.23 A bit-slice of a synchronous counter with parallel carry-logic in TSPC.

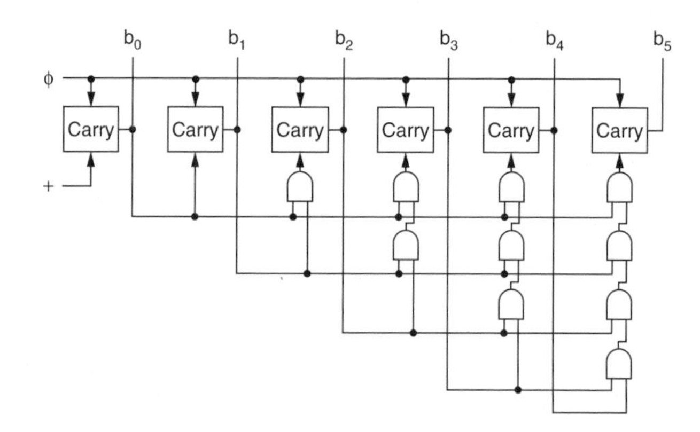

FIGURE 8.24 A synchronous counter with fully backward carry propagation.

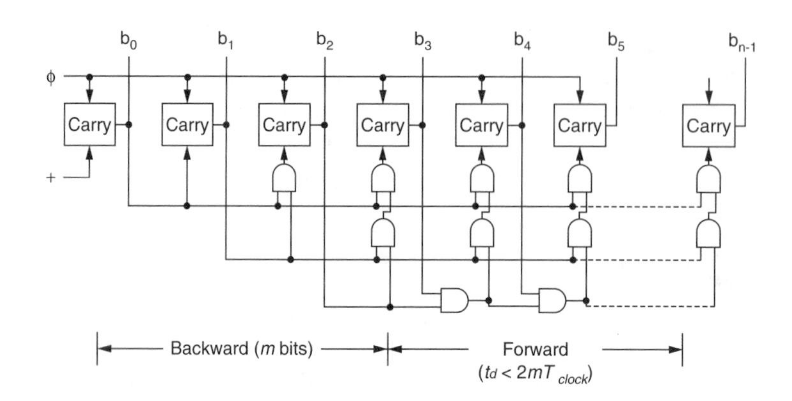

FIGURE 8.25 A synchronous counter mixing backward and forward carry propagations.

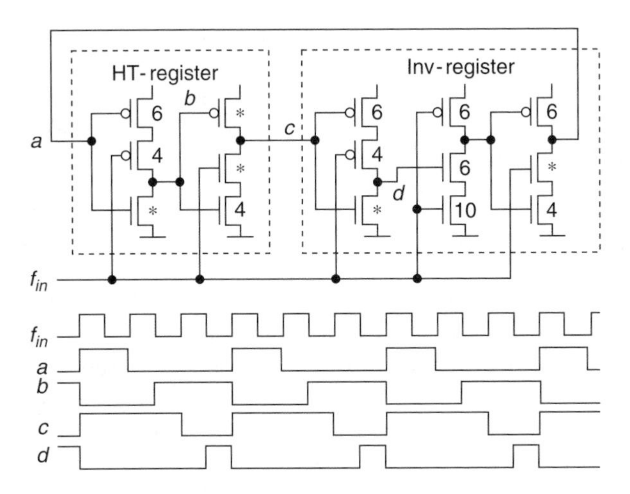

FIGURE 8.26 A dynamic nonbinary divider (1/3).

8.4.3 Nonbinary Divider and Prescaler

A nonbinary divider is usually constructed by a synchronous counter plus a decoding logic (i.e., when the output code reaches the target dividing ratio, the counter is reset). Such a topology can offer any dividing ratio; however, a synchronous counter is slow and the decoding logic adds additional delay. It was found that an SP stage followed by an SN stage in TSPC becomes a half-transparent register (HT-register) (i.e., registering a low-input [imposing a clock cycle delay] but no-register function for a high-input [transparent]). This feature can be utilized for constructing a nonbinary divider [19]. A divide-by-three circuit is depicted in Figure 8.26 along with the waveforms at different nodes. At node *b,* a symmetric waveform is obtained at a frequency of $f_{in}/3$, assuming the input clock is symmetric. If (n-2) HT-registers are used, it becomes a divide-by-n circuit. Because no decoding and no carry propagation exist, this circuit can work at the same speed as a 1/2 divider. The long propagation delay due to many cascaded transparent stages can be solved by a few speed-up transistors, see Yuan and Svensson [19]. Note that in a nonbinary divider the output is still edge-triggered (i.e., there is no skew between input and output as could be a problem for a ripple counter. Although a single nonbinary divider can offer any dividing ratio, it is more efficient to cascade two or more than two nonbinary dividers to achieve a high dividing ratio. For example, a 1/3 divider cascaded by a 1/7 divider becomes a 1/21 divider, and so on. The nonbinary divider is extremely useful for prescalers as the needed operating speed is often very high and the dynamic feature is usually not a problem. A dual-modulus prescaler, divided by either n or (n-1), is presented in Figure 8.27 for the purpose of frequency synthesis, where the "Inv-register"

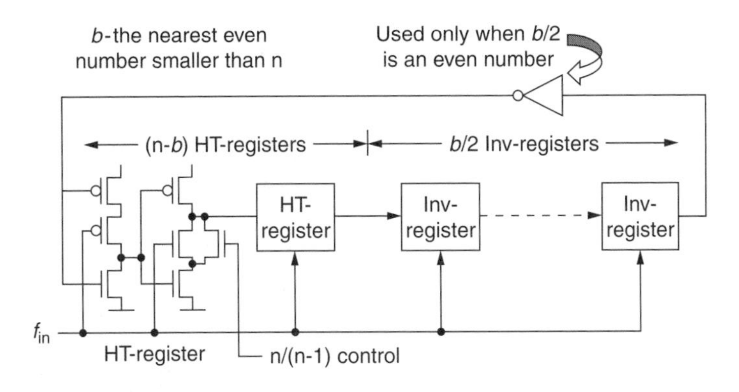

FIGURE 8.27 A divide-by-n/(n-1) prescaler.

represents the 9-transistor TSPC flip-flop. The control of divide-by-n and divide-by-(n-1) is extremely simple, only a single n-transistor in one of the (n-b) HT-registers, making this circuit highly attractive. When the input of the transistor is high, this HT-register becomes fully transparent. Other techniques in dual-modulus prescalers based on the modification of the nine-transistor TSPC flip-flop can be found in Chang et al. [20] and Yang et al. [21].

8.4.4 Adder and Accumulator

The core part of an adder is the "XOR" logic. A highly efficient pipelined XOR gate in TSPC is shown in Figure 8.28. The basic topology is to implement the XOR function in two steps respectively in a p-block and an n-block and to embed logic into latches [22]. The logic diagram is given at the upper part while the circuit diagram is given at the lower part of the figure. The NAND, OR, and AND functions are respectively embedded in an SN stage, an n-type precharged latch and a p-type precharged latch. The connection exactly follows the rule mentioned previously. The efficiency comes from two facts. First, the nonclassic principle is applied to the pair of SN and precharged p-type latch, so a single SN stage is used to embed the NAND function in favor of speed. Second, both the OR function in the n-type precharged latch and the AND function in the p-type precharged latch use parallel transistors which is also in favor of speed. The pipelined XOR gate can be directly cascaded to deliver the sum output for a full adder, and the sum can be fed back to its own input for accumulation. An accumulator can be therefore configured efficiently by using the pipelined XOR gate, and one of the bit-slices is given in Figure 8.29. A 24-bit pipelined accumulator in 1.2-μm CMOS for a numerically controlled oscillator based on the topology achieved a clock rate of 700 MHz [23].

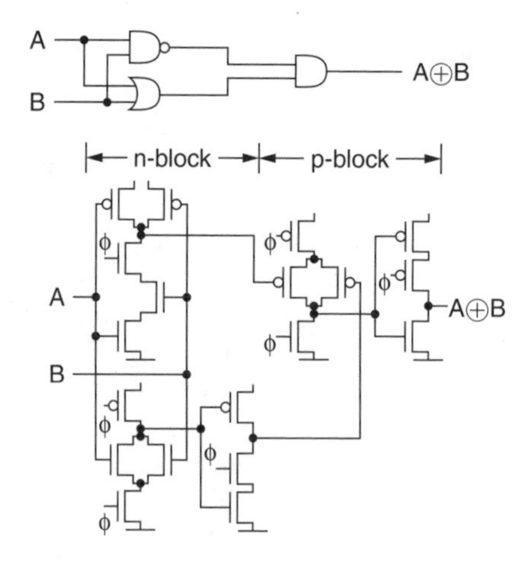

FIGURE 8.28 A pipelined XOR gate in TSPC.

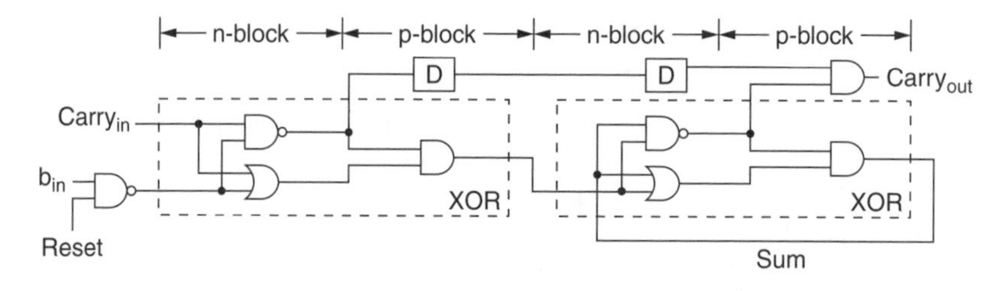

FIGURE 8.29 A bit-slice of an accumulator using the pipelined XOR in TSPC.

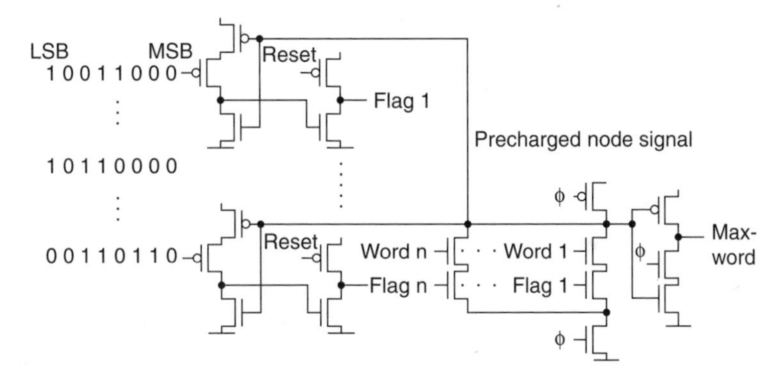

FIGURE 8.30 A bit-serial and word-parallel max/min selector.

8.4.5 Bit-Serial Comparator and Sorter

Dynamic logic may greatly simplify a complex circuit function and minimize the number of devices, resulting in low power without sacrificing speed or even with improved speed. One example is the bit-serial word-parallel maximum/minimum selector [24] in Figure 8.30. The main part of the selector is an n-type TSPC precharged latch embedding the selecting logic with AND functions (two n-transistors in stack) in parallel. The purpose to show this circuit is to emphasize that the precharged node signal can be used to effectively simplify the configuration. This signal is used as a second "clock" for a number of parallel PP stages. Each PP stage receives an input word. The precharged node signal of each PP stage is again used for the nMOS input of the flag logic stage, which consists of only an n- and a p-transistor. The p-transistor sets the flag high before the new input words start, and therefore all flags are high from the beginning. All word inputs start with MSB and are compared digits by digits. If all digits are the same, no matter zero or one, all flags are kept high. If partial digits become zero, the input with a zero digit will make the output of the PP stage high and the flag low to disable this input. In the end, only the maximum input is left. During comparison and selection, the output never stops. A minimum selector can be easily completed by inverting the input words and, of course, inverting the output again.

A bit-serial and word-parallel compare-and-swap cell [24] is presented in Figure 8.31. The maximum selector is used along with a minimum selector, which uses an inverted flag and the logic opposite to the maximum selector. When the two digits are equal, both go to the outputs, and when the two digits are different, the upper output will be "one" and the lower output will be "zero." In the same time, the smaller input will be disabled in the maximum selector, while the larger input will be disabled in the minimum

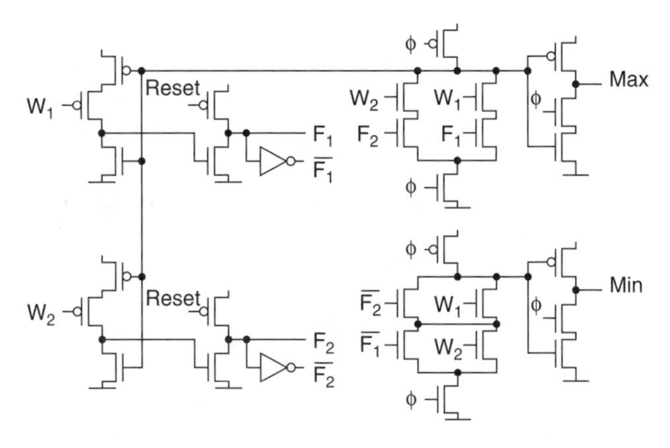

FIGURE 8.31 A bit-serial compare-and-swap cell.

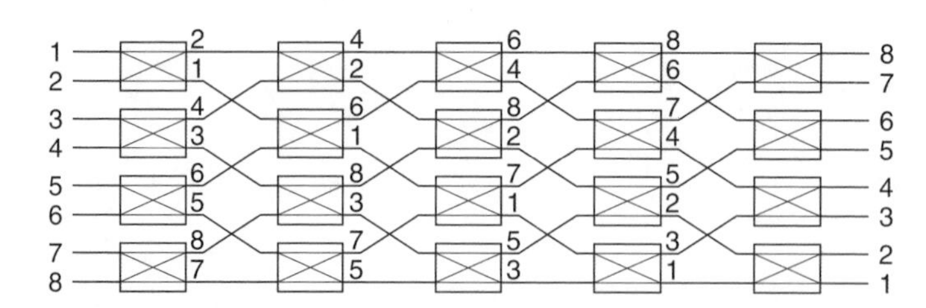

FIGURE 8.32 A bit-serial pipelined sorter using the compare-and-swap cell.

selector. It must be noted that the complete cell is an n-block in TSPC, and to cascade two such cells, a p-type latch must be used in between. An 8-input bit-serial and word-parallel sorter is depicted in Figure 8.32, where each box represents the compare-and-swap cell plus a p-type latch. This pipelined sorter can achieve a very high data throughput.

8.5 The Future of Dynamic Logic

It is well-known that dynamic CMOS logic circuits suffer from the leakage current, which puts a limit on the minimum possible clock frequency. This became a serious problem in deep submicron technologies. When V_{DD} is scaled down, V_{TH}, the threshold voltage, has to be scaled down to not lose the advantage of scaling, which is associated with the exponential increase in subthreshold leakage current [25]. It not only limits the use of dynamic logic but also dominates the overall power consumption. Advanced leakage control methods will become indispensable for future technologies [26].

Fortunately, some techniques have already emerged. One technique uses the self-reverse-biasing effect of stacked transistors, called stacking-effect, to reduce the standby leakage current [27]. This effect was successfully used in a low-leakage, gated-ground cache in which two off transistors connected in series reducing the leakage current by orders of magnitude [26]. It means that the topology and design methods have to change to meet the new challenge. Of course, to use static logic may avoid the reliability problem due to the leakage current; however, the reason to use dynamic logic has been its efficiency and high speed over its static counterpart under the same or a lower power budget. Therefore, dynamic logic may still find its position in deep submicron technologies if new techniques reducing the subthreshold leakage current are discovered by active researches in this field.

8.6 Conclusion

The comparisons in Table 8.1 and Figure 8.12 demonstrate that carefully designed dynamic CMOS flip-flops and latches — nonprecharged or precharged, single-ended or differential — present obvious short delays and small power-delay products over traditional dynamic and static flip-flops at different activity ratios. The strategies of single clock, low count of clocked devices, fewer transistors, nonclassic concept, and simple configurations lead to the results. The latches and flip-flops can be used in a pipeline or a double pipeline, resulting in a very high data throughput. When logic is embedded in an n-block, the number of devices and the overall capacitance is significantly less than what is necessary for a complementary logic, gaining high speed and low-power advantages. Advanced circuit topologies presented in the examples of functional circuits make the dynamic logic circuits highly attractive; however, precautions have to be taken. The robustness of dynamic (especially precharged) logic against power and ground noises is not as good as complementary logic. The node leakage current is another problem that has to be handled in deep submicron technology, and new leakage control techniques have to be found to make the dynamic logic circuits continuously attractive.

References

[1] Y. Ji-ren, I. Karlsson, and C. Svensson, A true single-phase-clock dynamic CMOS circuit technique, *IEEE J. Solid-State Circuits,* vol. SC-22, pp. 899–901, October 1987.

[2] J. Yuan and C. Svensson, High-speed CMOS circuit technique, *IEEE J. Solid-State Circuits,* vol. 24, pp. 62–70, February 1989.

[3] C. Svensson and D. Liu, Low-power circuit techniques, *Low-Power Design Methodol.,* J.M. Rabaey and M. Pedram, Eds., Kluwer Academic Publishers, Dordrecht, 1996, chap. 3.

[4] H. Oguey and E. Vittoz, CODYMOS frequency dividers achieve low-power consumption and high frequency, *Electron. Lett.,* pp. 386–387, August 23, 1973.

[5] Q. Huang and R. Rogenmoser, Speed optimization of edge-triggered CMOS circuits for Gigahertz single-phase clocks, *IEEE J. Solid-State Circuits,* vol. 31, pp. 456–465, March 1996.

[6] C. Piguet, Logic synthesis of race-free asynchronous CMOS circuits, *IEEE J. Solid-State Circuits,* vol. 26, pp. 271–380, March 1991.

[7] C. Piguet and J. Zahnd, Electrical design of dynamic and static speed-independent CMOS circuits from signal transition graphs, *ACiD Workshop,* Newcastle, U.K., January 18–19, 1999.

[8] J. Yuan and C. Svensson, New single clock CMOS latches and flip-flops with improved speed and power savings, *IEEE J. Solid-State Circuits,* vol. 32, pp. 62–69, January 1997.

[9] M. Afghahi and J. Yuan, Double edge-trigged flip-flop for high-speed CMOS, *IEEE J. Solid-State Circuits,* vol. 26, pp. 1168–1170, 1991.

[10] J. Yuan and C. Svensson, Fast and robust CMOS double pipeline using new TSPC multiplexer and demultiplexer, *Proc. 2nd Int. Conf. on ASIC,* pp. 271–274, October 1996.

[11] N. Weste and K. Eshraghian, *Principles of CMOS VLSI Design, 2nd ed.,* Addison-Wesley, Reading, MA, 1993, chap. 5.

[12] R.X. Gu and M.I. Elmasry, All-N-logic high-speed true-single-phase dynamic CMOS logic, *IEEE J. Solid-State Circuits,* vol. 31, pp. 221–229, February 1997.

[13] C. Kim and S-M. Kang, A low-swing clock double-edge triggered flip-flop, *IEEE J. Solid-State Circuits,* vol. 37, pp. 648–652, May 2002.

[14] J. Yuan, C. Svensson, and P. Larsson, New domino logic precharged by clock and data, *Electron. Lett.,* vol. 29, pp. 2188–2189, December 1993.

[15] J. Yuan, Efficient CMOS counter circuits, *Electron. Lett.,* vol. 24, pp. 1311–1313, October 1988.

[16] Y. Taur et al. CMOS scaling into the 21st century: 0.1 μm and beyond, *IBM J. Res. Dev.,* vol. 39, pp. 245–260, January/March 1995.

[17] J. Yuan and C. Svensson, Multigigahertz TSPC circuits in deep submicron CMOS, *Physica Scripta,* vol. T-79, pp. 283–286, 1999.

[18] P. Larsson and J. Yuan, Novel carry propagation in high-speed synchronous counters and dividers, *Electron. Lett.,* vol. 29, pp. 1457–1458, August 1993.

[19] J. Yuan and C. Svensson, Fast CMOS nonbinary divider and counter, *Electron. Lett.,* vol. 29, pp. 1222–1223, June 1993.

[20] B. Chang, J. Park, and W. Kim, A 1.2 GHz dual-modulus prescaler using new dynamic D-type flip-flops, *IEEE J. Solid-State Circuits,* vol. 31, pp. 749–752, May 1996.

[21] C-Y. Yang et al. New dynamic flipflops for high-speed dual-modulus prescaler, *IEEE J. Solid-State Circuits,* vol. 33, pp. 1568–1571, October 1998.

[22] J. Yuan, C. Svensson, F. Lu, and H. Samueli, A high speed pipelined CMOS accumulator for implementing numerically controlled oscillators, *Proc. ISCAS '90,* vol. 1, pp. 113–116, May 1990.

[23] F. Lu, H. Samueli, J. Yuan, and C. Svensson, A 700-MHz 24-bit pipelined accumulator in 1.2-um CMOS for application as a numerically controlled oscillator, *IEEE J. Solid-State Circuits,* vol. 28, pp. 878–886, August 1993.

[24] J. Yuan and K. Chen, Bit-serial realization of maximum and minimum filters, *Electron. Lett.,* vol. 24, pp. 485–486, April 1988.

[25] E. Vittoz and J. Fellrath, CMOS analog integrated circuits based on weak inversion operation, *IEEE J. Solid-State Circuits*, vol. SC-12, pp. 224–231, June 1977.

[26] A. Agarwal, H. Li, and K. Roy, A single V_t low-leakage gated-ground cache for deep submicron, *IEEE J. Solid-State Circuits*, vol. 38, pp. 319–328, February 2003.

[27] Y. Ye, S. Borkar, and V. De, A new technique for standby leakage reduction in high-performance circuits, *Symp. on VLSI Circuits, Dig. Tech. Papers*, pp. 40–41, 1998.

9

Low-Power Arithmetic Operators

Arnaud Tisserand

INRIA LIP Arénaire

9.1 Introduction

Arithmetic operators are among the most used basic blocks in digital integrated circuits. They are the core of functional units such as arithmetic and logic units (ALUs), integer multipliers, floating-point units (FPUs), or multimedia units. They also play a part in memory address generation units and in some controllers. In a complex system, such as a system on chip (SoC), the power consumption of those operators is often smaller than the power consumption of memories and buses. Nevertheless, some significant energy savings can be achieved at the arithmetic level without performance penalty. Furthermore, it is important to reduce power wastage wherever possible.

This chapter presents some low-power consumption aspects of the main arithmetic operators and representations of numbers used in high-performance circuits. A basic knowledge on computer arithmetic is assumed. More details on arithmetic algorithms and number systems can be found in reference computer arithmetic books such as Ercegovac and Lang [1] and Koren [2].

The optimization of a circuit can be done at different levels: system, algorithm, architecture, circuit, and technology. Arithmetic operators have no concern with the system and technology levels. Algorithm, architecture, and circuit levels widely impact the design of arithmetic operators and vice versa. More precisely, there is a complex trade-off among:

- The number system(s) used to represent the data (i.e., width and number coding)
- The algorithms used to compute the mathematical operations (i.e., evaluation methods, speed/area trade-offs, and fused operations)

0-8493-1941-2/05/$0.00+$1.50

- The characteristics of the data (i.e., accuracy, signal activity, and space/time correlations, etc.)
- Some circuit constraints (i.e., specific cells in the standard library and logic style, etc.)

This chapter focuses on standard-cell complementary metal oxide semiconductor (CMOS) circuits. As discussed in the previous chapters of this book, the power consumption of CMOS circuits can be decomposed into three main parts:

$$P = P_{switching} + P_{short-circuit} + P_{leakage}$$

The switching power $P_{switching}$ is due to the charge and discharge of the capacitors driven by the circuit. The short-circuit power $P_{short\text{-}circuit}$ is caused by the simultaneous conductance of P and N transistor networks. The leakage power $P_{leakage}$ is due to the leakage current that flows in the circuit such as sub-threshold or reverse-biased PN junction leakages for instance. The sum of the switching and the short-circuit powers is called the dynamic power, while the leakage power is called the static power.

The activity into an operator is caused by two kinds of signal transitions. The useful activity due to the input transitions that produce the internal transitions required to perform the computation. The redundant activity (also called glitching activity) is caused by different delays from the inputs to the same output and circuit defects. Specific circuit styles or careful placement and routing of the gates can significantly reduce the glitches, but important energy savings can also be achieved on the useful activity by using specific number systems or optimized evaluation algorithms. Some methods for reducing both kinds of activity are presented.

This chapter focuses on the dynamic power and mainly on the switching contribution, as it is the largest one for arithmetic operators. The static power has a smaller contribution. The methods presented in other chapters for reducing the dynamic power or the static power can be used on arithmetic operators; however, the goal of this chapter is to present some energy savings that can be simply achieved at the arithmetic level.

The two's complement representation of integers and fixed-point numbers is the most used number system in digital circuits. Unless it is explicitly mentioned, we will assume two's complement representation of numbers in this chapter. Suppose $X = x_{n-1}x_{n-2}\ldots x_1x_0$ is an n-bit two's complement number, then x_{n-1} denotes the sign of the number and the magnitude of X is given by $-x_{n-1}2^{n-1} + \sum_{i=0}^{n-2} x_i 2^i$.

The chapter is organized as follows. Section 9.2 and Section 9.3 focus on the two main integer operations: addition and multiplication. Finally, Section 9.4 briefly presents other operations, number systems, and possible evolutions of low-power arithmetic.

9.2 Addition

Addition is the most-used arithmetic operation in microprocessors, digital signal processors (DSPs), and many digital circuits. It is the core operation in some computation units such as ALUs, but it also plays a part in multipliers, dividers, FPUs, and address generation units.

Assuming two's complement representation of numbers, the addition and the subtraction operations are very close. In the following, we denote addition both operations (i.e., addition and subtraction).

9.2.1 1-Bit Addition Cells

Most of current digital circuits are designed using standard cell libraries and place and route tools. The type of the basic cells (complexity, drive strength, etc.) and their characteristics are of first importance. Besides the various logic gates, flip-flops, and latches, a few cells are dedicated to 1-bit addition: the half-adder (HA) and the full-adder (FA) cells. The purpose of these cells, also called counters, is to *count* the number of "1" at the inputs and to output their sum using a standard binary representation (i.e., $X = \Sigma_{i=0}^{n-1} x_i 2^i$).

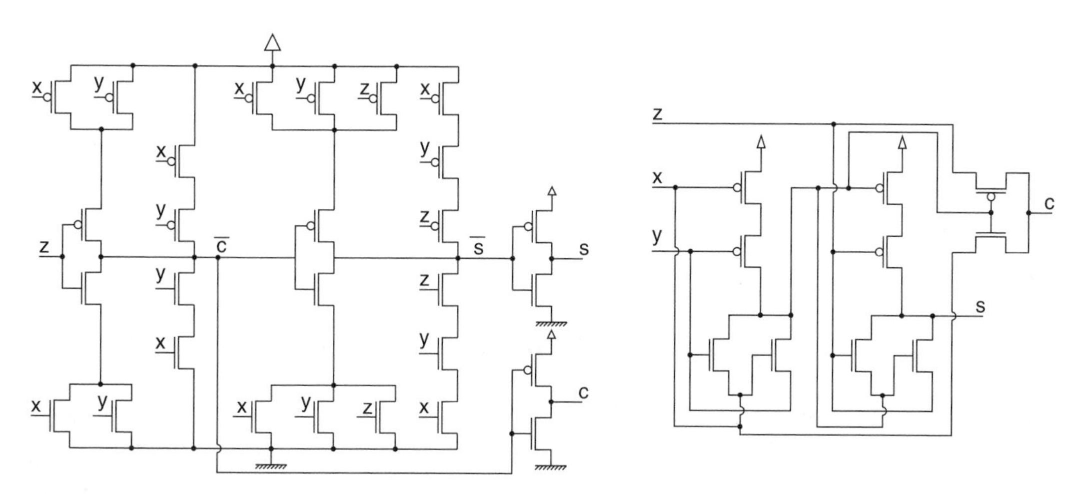

FIGURE 9.1 28 and 10-transistor implementations of the FA cell.

The half-adder is a (2,2) counter. The sum of the two input bits x and y is represented by the two output bits s (sum) and c (carry) such as $s + 2c = x + y$. A logic implementation of the HA cell gives $s = a \oplus b$ and $c = ab$. The full-adder is a (3,2) counter. The sum of the three input bits x, y, and z is represented by the two output bits s (sum) and c (carry) such as $s + 2c = x + y + z$. A logic implementation of the FA cell gives $s = a \oplus b \oplus c$ and $c = ab + ac + bc$. In an FA cell, the third input z is also called the carry-in c_{in} bit and the output c the carry-out c_{out}. Other counter cells are sometimes in libraries, such as the "4-to-2" cell used for reduction trees in multipliers (see Section 9.3).

Several possible circuit styles are used for the implementation of the HA and FA cells: pure CMOS, transmission gate, and complementary pass-transistor logic (CPL). The circuit implementation style of the gates is not an arithmetic parameter but it widely impacts the efficiency of larger operators. Depending on the circuit style, the transistor count of the FA cell may vary between 10 and 28. Figure 9.1 presents two implementations of the FA cell at the transistor level with different circuit characteristics.

In Shams and Bayoumi [3], a 16-transistor FA cell is presented. Energy savings up to 30% are achieved for a 16-bit carry select adder based on this cell compared with a version based on a previous 14-transistor FA cell without any speed penalty. Therefore, it shows that the power reduction is not just a problem of transistor number. A recent article presents a complete circuit style comparison for FA cell design in sub-micron technologies and many details about transistor sizing of those cells [4].

9.2.2 Sequential Adder

The simplest addition architecture is based on a linear array of FA cells as depicted in Figure 9.2. It is known as sequential adder or ripple carry adder (RCA). This adder is the slowest useful adder, but it is also the smallest.

Due to the very simple structure of the sequential adder, its power consumption has been the subject of many studies. For instance, Guyor and Abou-Samra [5] present a formal model of the sequential adder activity (and also for some other simple adders). In accordance with experimental results, this model shows that the average activity overhead (glitches) is about 50% for a sequential adder.

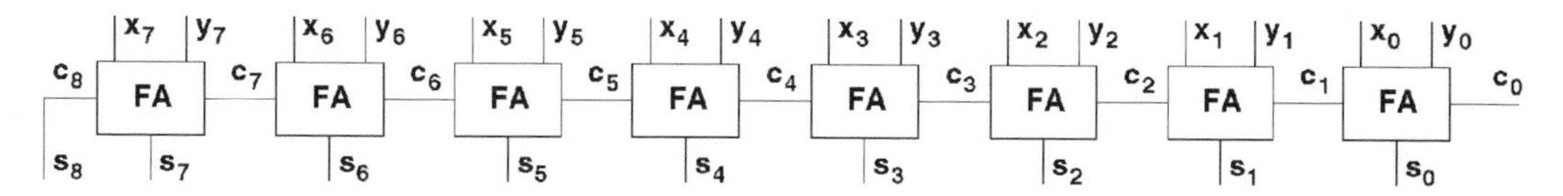

FIGURE 9.2 8-bit RCA or sequential adder.

TABLE 9.1 Bit-level Propagation and Generation of Carries

x	y	c_{out}	s	Carry Transfer
0	0	0	c_{in}	generate 0
0	1	c_{in}	$\overline{c_{in}}$	propagate
1	0	c_{in}	$\overline{c_{in}}$	propagate
1	1	1	c_{in}	generate 1
$x = y$		x	c_{in}	generate
$x \neq y$		c_{in}	$\overline{c_{in}}$	propagate

9.2.3 Propagate and Generate Mechanisms

One key point in the addition process is the computation of the carry-in bit for each rank (i.e., the $c_{in,i}$ values for all i in $\{0,1,\ldots n-1\}$). Once $c_{in,i}$ is known, the sum bit at rank i can be easily computed using $s_i = a_i \oplus b_i \oplus c_{in,i}$. As illustrated in Table 9.1, the carry-out bit c_{out} can be computed without need of the carry-in bit *cin* for some values of x and y. In this table, if x and y are different; the carry-in bit is propagated to the carry-out bit. If the input bits x and y are equal, the carry-out bit is generated without any need on the value of the carry-in. Sometimes, the generation of a carry-out bit equal to 0 is called a kill or absorption. Generation is then reserved for the case $c_{out} = 1$.

Three bits are defined to compute the carry-out bit at each rank. Those bits are: the propagate bit p, the generate bit g, and the kill bit k. Some of those bits are used in adders, such as carry lookahead or Manchester chain adders. A possible logic implementation of those bits is given by equation. Notice that the bits p and g can be computed using an HA gate ($p = c$ and $g = s$).

$$p = x \oplus y, \quad g = xy, \quad k = \overline{x}.\overline{y} = \overline{x + y} \tag{9.1}$$

9.2.4 Carry Select Adder

The main idea in a carry select adder (CSeA) is to split a sequential adder into two parts and performing the computation of the most significant bit (MSB) part with the two possible carry-in bits in parallel and selecting the right one using the carry-out bit of the least significant bit (LSB) part. Recursively applied, this method leads to a logarithmic time adder at the algorithmic level. As an example, Figure 9.3 depicts a 4-bit CseA, but CSeAs are not so fast in practice because of high fan out problems. This type of adder is used in combination with a faster scheme in some multiple size operators.

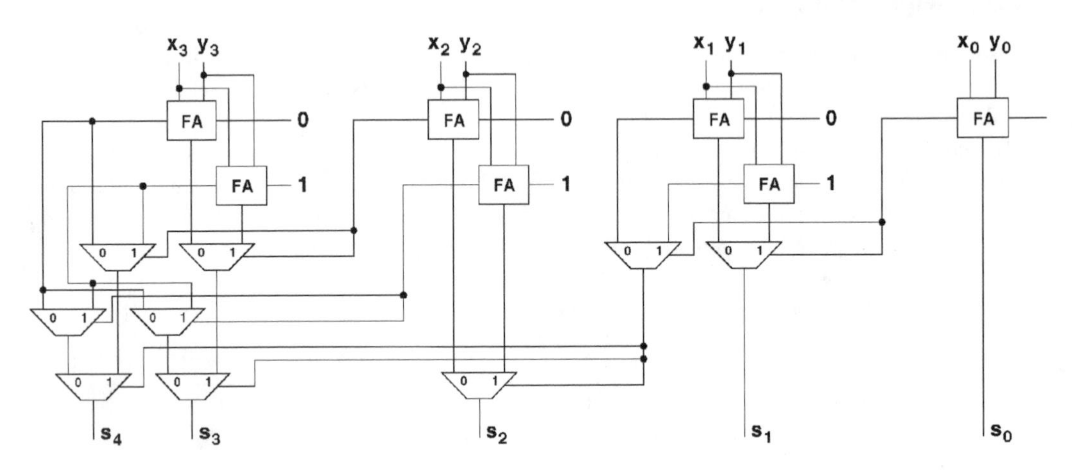

FIGURE 9.3 4-bit CSeA.

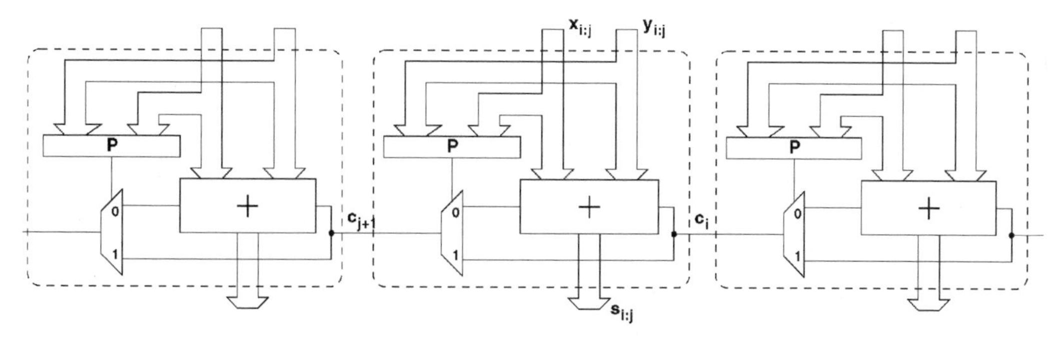

FIGURE 9.4 Principle of 1-level CSkAs.

9.2.5 Carry Skip Adder

The carry skip adder (CSkA) is based on a linear structure of blocks of sequential adders and additional logic used to skip blocks when all ranks propagate the carry inside the block. There is a skip over a block when $\prod_{i \in block} p_i = 1$. They are constant or variable block widths CSkAs. The principle of carry skip adders is depicted in Figure 9.4. The speed of those adders is $O(\sqrt{n})$ for n-bit operands. Although this adder has not the highest theoretical speed, it is widely used for fast but not critical additions because of its small area and regular layout. A hierarchical application of the carry-skip scheme can be used. This leads to multiple-level carry skip adders. In practice, a simple 1-level solution with variable block widths leads to simple and quite efficient adders.

9.2.6 Logarithmic Adders

Several kinds of logarithmic adders are available, such as carry-lookahead adders (CLAs) or parallel-prefix adders. The CLA is one special case of parallel prefix. This type of adder is widely used in fast circuits because of its high speed. The principle of CLAs is to compute the values $c_{in,i}$ using propagate/generate trees in parallel for all ranks instead of trying to propagate them as fast as possible. At rank i, a carry-in equal to 1 occurs in the following cases:

- Rank i-1 generates a carry-out equal to 1 (i.e., $g_{i-1} = 1$)
- Or rank i-1 propagates a carry generated at rank i-2 (i.e., $p_{i-1} = g_{i-2} = 1$)
- Or ranks i-1 and i-2 propagate a carry generated at rank i-3 (i.e., $p_{i-1} = p_{i-2} = g_{i-3} = 1$)

  ⋮
- Or ranks i-1 to 0 propagate the adder carry-in $c0$ equal to 1 (i.e., $p_{i-1} = p_{i-2} = \ldots = p_1 = p_0 = c_0 = 1$)

Therefore, all the carry-in bits can be computed using the relation:

$$c_i = g_{i-1} + p_{i-1}g_{i-2} + p_{i-1}p_{i-2}g_{i-3} + \ldots + p_{i-1}\cdots p_1 g_0 + p_{i-1}\cdots p_0 c_0 \tag{9.2}$$

The architecture of CLAs is based on the three following steps for all rank i:

1. Parallel computation of (g_i, p_i)
2. Parallel computation of c_i using Equation 9.2
3. Parallel computation of $s_i = a_i \oplus b_i \oplus c_i = p_i \oplus c_i$

The example of the computation of the carries in a 4-bit CLA is depicted Figure 9.5. Only small CLAs can be built because of their fan-in and fan-out limitations. Small CLAs, such as 4-bit block, are used to build larger adders. Various circuit styles have been evaluated on 32-bit CLAs in Ko et al. [6]. Very efficient implementations of such a small CLA can be achieved using AND-OR complex gates.

Parallel prefix adders are based on the same kind of structure: performing the computation of the carry-in bits for all ranks using partially shared generate/propagate trees. The sharing of the different

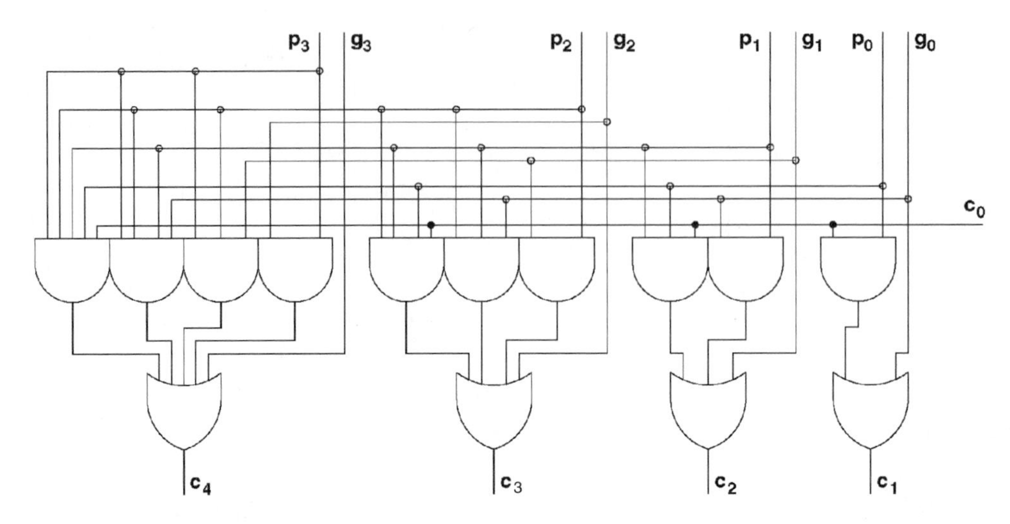

FIGURE 9.5 Computation of the carries in a 4-bit CLA.

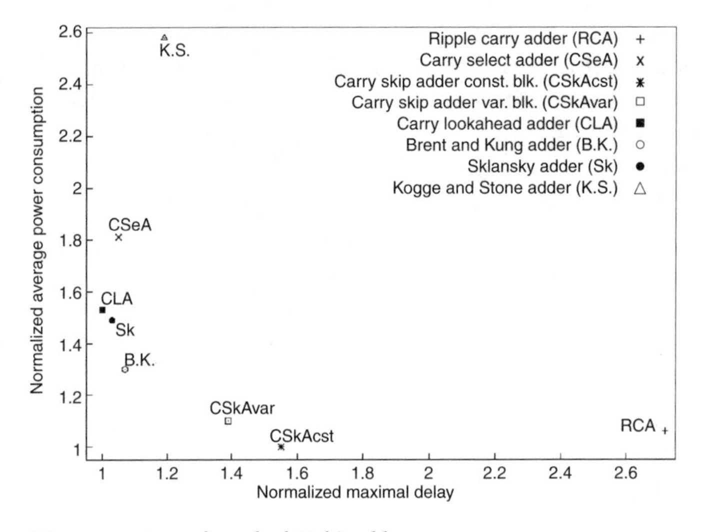

FIGURE 9.6 Power-delay comparison of standard 32-bit adders.

trees can be done using different schemes. This leads to the various types of parallel-prefix adders. The most well-known parallel prefix adders are: Brent and Kung, Sklansky, and Kogge and Stone. For a complete discussion on parallel prefix adders design, we refer the reader to Zimmerman [7].

9.2.7 Power/Delay Comparison

There are only a few papers on accurate SPICE-level power/delay comparison of the previous adders. For instance, Nagendra et al. [8] discusses numerous details about former technologies. Figure 9.6 presents a synthesis of accurate comparisons of various adders.

9.2.8 Redundant Adders

The use of redundant number systems allows constant time addition. The carry-save (CS) number system is widely used. This is a radix-2 representation in which the digits belong to {0,1,2}. The representation is considered redundant because some numbers have several representations. For instance, the integer 6 can be represented in CS using $(0110)_{cs}$ or $(0102)_{cs}$. In CS, each digit x of $\{0,1,2\}$ is represented by the

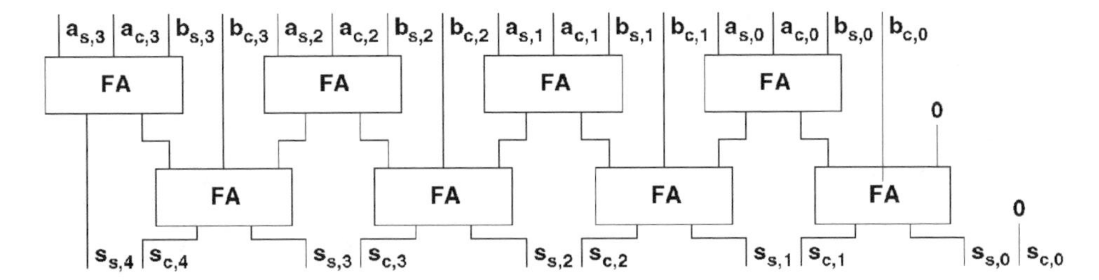

FIGURE 9.7 A 4-digit CSA.

sum of two bits: $x = x_c + x_s$ where x_c and x_s are in $\{0,1\}$. As an example, a four-digit carry-save adder (CSA) is depicted in Figure 9.7. We refer the reader to the books cited in the introduction for a comprehensive presentation of redundant number systems.

The conversion from a value represented using a redundant number system into a value represented using a nonredundant number system can be done using a nonredundant addition. In the case of CS numbers for instance, the conversion can be done using a standard fast adder. The sum vector and the carry vector of the CS representation are considered as the two nonredundant inputs in the adder.

Although a redundant number system allows to compute the addition of two arbitrary large numbers in a constant time, it has some drawbacks. First, more bits are required to represent the values than using a nonredundant number system. Second, due to the possible multiple different representations of a value, the comparison operation is quite complex. Therefore, redundant addition is mainly used as an internal representation. One frequent use of redundant adders is to perform the sum of three or more values. For instance, the sum of k values can be done using a $\log(k)$ -level tree of CSAs and a fast, nonredundant adder at the end. The adder tree produces the sum of k values represented in CS. The conversion into a nonredundant number system is done using a standard adder.

Nagendra et al. [8] gives the power/delay comparison of CSAs and CLAs. For 32-digit operands, the CSA is three times faster than the CLA and its average power consumption is 20% lower compared to the CLA just for the arithmetic part.

The comparison of redundant adders, such as CSAs and standard adders, sometimes called carry-propagate adder (CPA), is not straightforward. Indeed, redundant number systems require more bits than nonredundant ones for the same interval of representable values. This leads to storage or additional bus resources. For instance, in the case of a n-digit CSA, $2n$ bits are required. One good point about redundant adders is the fact that the glitching power of those operators is very limited due to their structure in comparison to CPAs, but this gain should be very small in front of the additional power consumption due to the storage or communication bus overhead.

9.3 Multiplication

Multiplication is the most area consuming arithmetic operation in high-performance circuits. This large area is required to implement high-speed multipliers. We refer the reader to Flynn and Oberman [9] for a comprehensive presentation of speed optimizations of multipliers. These large and complex operators also have the highest power consumption among all the arithmetic operators. Therefore, there are many research works on low-power design of high-speed multipliers.

Multiplication involves two basic operations: the generation of the partial products and their sum. Therefore, there are two possible ways to accelerate the multiplication: reduce the number of partial products or speed up their sum. Both solutions can be applied simultaneously. Reducing the number of partial products also reduces the time required to perform their sum.

High-speed multiplication have no concern with sequential or array multipliers, which are too slow. In this section, we only present tree based parallel multipliers. In practice, most of those high-performance multipliers are based on the three following steps:

- Parallel generation of the partial products
- Tree reduction of the partial products into a redundant sum (e.g., carry-save tree)
- Conversion of the redundant sum into a nonredundant representation using a fast adder

Several power optimization methods can be applied to multiplier design: circuit styles, improved computation algorithms, specific internal number systems. This section presents such optimizations for each basic step of high-speed parallel multipliers.

We denote X and Y the two n-bit operands of the multiplication where X is the multiplier and Y the multiplicand. We assume that the operands and the result are represented using the standard unsigned binary number system (i.e., $X = \Sigma_{i=0}^{n-1} x_i 2^i$). Unless it is explicitly mentioned, we look at the $2n$-bit full-width product $P = X \times Y$. In some applications, only the most significant bits of P are required, then we talk about truncated multiplication.

The modifications required to handle signed numbers in two's complement representation are small, and they do not affect significantly the power consumption. Therefore, we only focus on unsigned multiplication. More details on signed multiplication can be found in the books mentioned in the introduction.

9.3.1 Partial Products Generation

The generation of the partial products consists in the computation of all the $p_{i,j} = x_i \times y_j$ for all i and j in $\{0,1,\ldots,n-1\}$. The basic architecture is based on n^2 AND gates array. This generation appears to be straightforward, but the algorithmic description hides some load problems. Indeed, in an $n \times n$-bit multiplier, each input bit x_i or y_j is used as an input in n different gates. For instance x_i will be used to produce the n partial products $p_{i,0}, p_{i,1}, \ldots, p_{i,n-1}$. Therefore, the load on those signals can be very high in large multipliers. A careful buffering and gate sizing is then required.

Using a standard radix-2 coding of both n-bit operands, there are $n \times n$ partial product bits or n partial product n-bit words. For each bit of the multiplier X, two possible partial product words are available: 0 and Y. The reduction of the number of the partial products can be done using a high-radix recoding of one operand. Usually the multiplier is the recoded operand. Using a radix-4 recoding of the multiplier, the number of partial products reduces to $n/2$. For a radix-2^k recoding of the multiplier, only n/k partial products are required.

The recoding is performed by examining several bits of the multiplier X for each partial product. Using a naive radix-4 recoding, an examination of two bits of the multiplier X leads to four possible partial products: 0, Y, $2Y$, and $3Y$. Of course, this recoding is not efficient in practice because of the generation of the value $3Y$, which requires a preliminary addition. A better radix-4 recoding uses the partial products $-2Y$, $-Y$, 0, Y, and $2Y$. Multiplication by two is done by shifting one bit left. Negation is done by complementing and adding a correction +1 in the LSB.

The modified Booth's recoding is a radix-4 recoding that generates $n/2$ partial products for n-bit multipliers. It examines three bits of the multiplier X with an overlap of one bit. Booth's recoding is based on a careful application of the identity: $2^{i+k} + 2^{i+k-1} + 2^{i+k-2} + \cdots + 2^i = 2^{i+k+1} - 2^i$. Table 9.2 presents the truth table of the modified Booth's recoding used for all odd values of i, namely $= 1,3,5,\ldots$, to produce the recoded multiplier X'. Only one nonnull value of the recoded multiplier X' exists for each line. This ensures that only $n/2$ partial products are generated.

Based on recoding schemes, the complete generation of the partial products requires two types of cells: recoding cells and partial product generator cells. In some rich standard cell libraries, there are specific gates to implement those functions. Indeed, those functions can be easily implemented using AOI complex gates. In the case of the radix-4 modified Booth's recoding applied to an $n \times n$-bit multiplier, there are:

- $n/2$ recoding cells that generate the operation signals, such as $-2Y$, $-Y$, 0, $+Y$, and $+2Y$, from three bits of the multiplier X

TABLE 9.2 Modified Booth's Recoding Transformations

x_i	x_{i-1}	x_{i-2}	x'_i	x'_{i-1}	Comment	Operation
0	0	0	0	0	string of 0's	+0
0	0	1	0	1	end of 1's	$+Y$
0	1	0	0	1	a single 1	$+Y$
0	1	1	1	0	end of 1's	$+2Y$
1	0	0	$\bar{1}$	0	beginning of 1's	$-2Y$
1	0	1	1	$\bar{1}$	a single 0	$-Y$
1	1	0	0	$\bar{1}$	beginning of 1's	$-Y$
1	1	1	0	0	string of 1's	+0

- $n/2 \times n$ partial product generator cells that actually generate the partial products using the y_j's and the operation signals

Higher radices can be used to recode the multiplier, but this leads to very complex recoders. In practice, it seems that a radix-4 modified Booth recoding offers a good trade-off between circuit complexity and number of partial products reduction.

If the basic functions used for the generation of the partial products are quite simple, their implementation require to be careful at the layout level (see Abu-Khater et al. [10], for instance). Indeed, many signals in the generation of the partial products have to be generated at the "right" time in order to avoid spurious transitions. For instance, recoding cells drive large loads and glitches on the corresponding wires would lead to higher power consumption and slower circuits. This is a challenging problem for standard place and route tools.

9.3.2 Reduction Trees

Once all the partial products have been generated, the second basic operation to perform in the multiplication is the sum of those partial products. In high-speed multipliers, this is performed using a tree of redundant additions. As described in the previous section, a fast way to sum up several numbers is the use of a tree of redundant additions such as a CS tree.

An h-level CS tree can reduce $n(h)$ nonredundant inputs to a CS sum. The "CS function" $n(h)$ is defined by $n(h) = \lfloor 3n(h-1)/2 \rfloor$ and $n(0) = 2$. Some useful values of this function are presented in Table 9.3. This means that to reduce 24 partial products for instance, a 6-level CS tree should be used.

CS trees and Wallace's trees are very close. In the multiplier architecture proposed by Wallace, a recursive decomposition of the computation is performed but the reduction of the different terms is done using CS trees. Some other full-adder based reduction trees can be used such as Dadda's trees or those generated using fast reduction algorithms (see Stelling et al. [11], for instance).

To improve the reduction step performance, higher order counters can be used. In Goldovsky et al. [12], a reduction tree based on (3,2), (5,3), and (7,4) counters is presented. Several logic and circuit level optimizations can be used when considering such counters instead of simple FA cells. The (5,3) counter is also called 4-to-2 compressor. Possible implementations of the 4-to-2 compressor are presented in Figure 9.8. Using such a cell leads to faster reduction trees. The power dissipation is also reduced because of the more regular layout and the smaller number of signal transitions in the tree. Typical power reduction is around 30%. One of the main problems that remain is the efficient placement of complex and irregular structures such as trees using standard electronic design automation (EDA) tools.

TABLE 9.3 Some Useful Values of the CS Function for Multiplier Design

h	1	2	3	4	5	6	7	8	9	10	11
$n(h)$	3	4	6	9	13	19	28	42	63	94	141

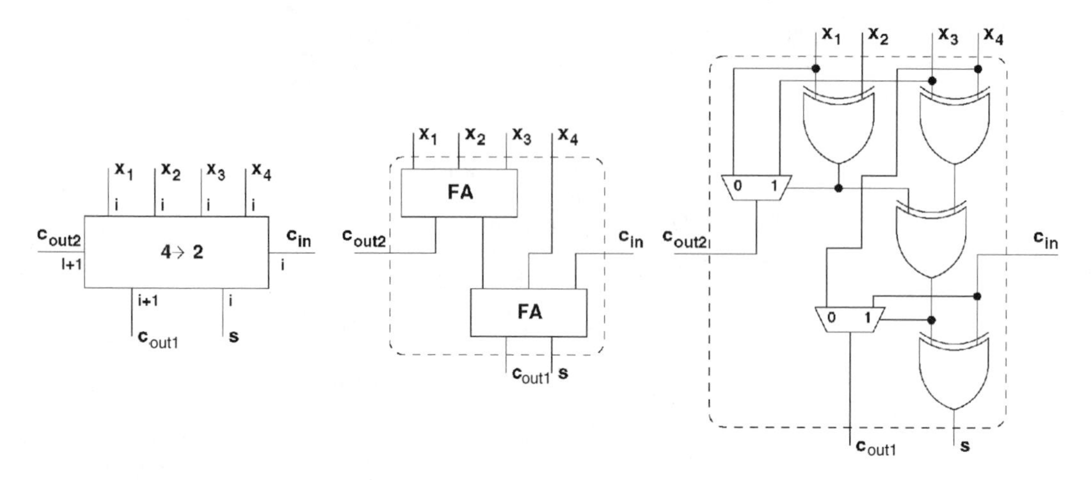

FIGURE 9.8 Some implementations of a 4-to-2 compressor.

9.3.3 Final Addition

The last step in the high-speed multiplication process is the conversion of the redundant sum produced by the reduction tree into a nonredundant representation. This step is also called assimilation of the carries. As described in Section 9.2, this conversion can be performed using a standard nonredundant adder.

Most adders are built assuming a uniform arrival time profile for their inputs (i.e., all the input signals are stable at the beginning of the computation). Nevertheless, the arrival times of the reduction tree outputs are nonuniform. A typical arrival time profile of the reduction tree outputs is presented Figure 9.9. This profile is expressed in XOR gate delays in the case of a 53×53-bit multiplier.

Based on these timing characteristics, the final adder structure can be optimized. The addition in region 1 can be done using a slow adder, such as an RCA. In region 2, a fast adder is required because the bits of this region arrive late. A CLA is often used in this region. The carry-out of region 1 can be used to select between the output of the adder of region 2 and this value plus one. High-speed adders such as CLA or parallel prefix adders can be modified to compute $a + b$ and $a + b + 1$ in the same time with a very small area overhead. Finally, the addition in region 3 can be done using a fast but simple adder such as a carry select adder. Using a CSeA for this last region allows to simply integrate the carry-out of region 2.

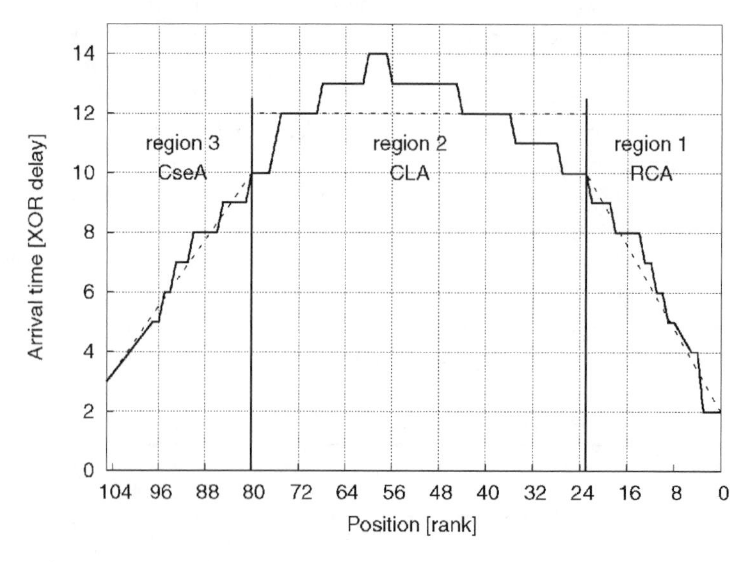

FIGURE 9.9 Typical arrival time of reduction tree output of a 53×53-bit multiplier.

The optimization of the final adder in a high-speed multiplier has two main advantages. First, high-speed adders require large area, and then by using slower and smaller adders for region 1 and 3, the area of the whole multiplier is improved. Second, the use of a structure that computes as late as possible avoids some spurious transitions. A power reduction around 20% is achieved using an optimized final adder instead of a standard high-speed one.

9.3.4 Fused Multiply and Add

The fused multiply and add operation $P = X \times Y + Z$ was introduced in the 1980s in DSPs to accelerate some signal processing algorithms. In many algorithms, such as filters, vector products, convolutions, or fast Fourier transforms, a multiplication is often followed by an addition in which one operand is the previously computed product. Therefore, a lot of modern general processors or DSPs have built-in hardware multiply-add and/or multiply-accumulate capabilities.

The basic idea in fast multiply-add units is to consider the third operand Z as an additional line of partial products. This leads to significant speed improvements compared to a simple concatenation of the multiplier and the adder. Indeed, for some values k, a reduction tree with h levels can reduce k or $k + 1$ partial products at the same speed. For instance, in the case of a 4-level CS tree, 9 up to 12 partial products can be reduced to a CS value in the same delay (see Table 9.3).

Although the fusion of the multiplication and the addition into a single operation leads to significant speed improvements, similar power consumption reductions can be achieved. The multiply-add operation has a smaller circuit and no intermediate storage of the product $X \times Y$ is required. Up to 40% power dissipation reduction can be achieved compared with two separate operations. The only drawback of the fused multiply-and-add unit is the need for fetching a third operand. In practice, the third operand is the output of a local accumulator to avoid heavy modifications of the processor register file.

9.3.5 Truncated Multiplication

In some applications, the $2n$-bit product produced by the multiplier is rounded to the n most significant bits to avoid growth in word size. Truncated multiplication is a method used to produce an accurate rounded evaluation of the n MSB of the product without computing most of the least significant bits. Several methods have been proposed to minimize the error produced by omitting some of the LSBs. In Schulte and Stine [13], post-layout simulations demonstrate that truncated 32-bit multipliers dissipate 40% less power than standard high-speed multipliers with a ±1 LSB error on 32-bit result.

9.3.6 Square

Some algorithms such as norm computation in image processing applications or Viterbi decoders involve a lot of square functions. Any standard multiplier can be used for computing those squares, but dedicated operators lead to significantly improved performances.

The optimization of a squaring unit is based on some simple identities applied to the generation of the partial products. As an example, we look for the square $P = X^2$ of an unsigned n-bit integer X represented using the standard binary format. The generation of the partial products leads to the values $x_i x_j$ for all i and j in $\{0,1,\ldots,n-1\}$. Several identities can be applied to these partial products:

- The term $x_i x_i$ reduces to x_i
- The sum of the two terms $x_i x_j + x_j x_i$ in a given column can be replaced by $2x_i x_j$ in the next column (of higher weight)
- Another simplification is possible using $x_i x_{i-1} + x_i = 2x_i x_{i-1} + x_i \bar{x}_{i-1}$ (i.e., two partial products in a column are replaced a by one partial product in the same column and another one in the next column [of higher weight])

The first simplification suppresses n partial products. The second one halves the number of partial products of the form $x_i x_j$, where $i \neq j$ (among the n^2 possible partial products, there are $n^2 - n$ such products). Finally, the last simplification decreases by one the height of highest column of partial products. It also reduces the length of the final adder.

Based on these simplifications, a dedicated square operator is significantly faster, smaller, and dissipates less power than a standard multiplier. Up to 60% power reduction can be achieved by using such an optimized squarer instead of a standard high-performance multiplier. Similar optimized operators can be designed for the cube function $P = X^3$.

9.4 Other Operations, Number Systems, and Constraints

Integer addition and multiplication are the two main blocks in arithmetic circuits. Most of the other operations or representations involve those two basic blocks. In this last section, we briefly present some low-power design aspects of other operations, number systems and some technology evolutions and constraints that may impact arithmetic operators.

9.4.1 Division and Square Root

Because of the lack of efficient division unit in former processors, some numerical and signal processing algorithms have been modified to avoid the use of division. Today, most of current processors have more or less complex hardware support for division. Some processors integrate a complete dedicated division unit while some other processors use functional iteration methods based on multiplication and table look-up. Even DSPs have dedicated instructions to speed-up basic division algorithms. There are a lot of division algorithms. A complete survey on division algorithms and implementations can be found in Oberman and Flynn [14]. Two main methods are used for high-speed division: digit recurrence and functional iteration.

The class of digit recurrence division algorithms is the simplest and most widely implemented. In those algorithms, a fixed number of quotient digits is produced in every iteration. The basic version of this algorithm is the "paper-and-pencil" algorithm taught at school, with just small modifications to use a radix-2 representation instead of a radix-10 one. A very efficient form of this algorithm is the SRT division (for Sweeney, Robertson, and Tocher). A comprehensive presentation of digit recurrence algorithms and implementations can be found in Ercegovac and Lang [15].

The SRT iteration is based on the following residual recurrence:

$$w[j+1] = rw[j] - q_{j+1}d$$

where j is the iteration number, r the radix, x the dividend, d the divisor, and q_{j+1} the new digit of the quotient, with an initial residual $w[0] = x$. To simplify the product $r\,w[j]$ the radix is chosen as a power of two. At each iteration, a new digit of the quotient is produced by a table lookup addressed by a few most significant digits of $w[j]$ and d. A redundant representation of the residual allows constant time subtraction and simplifies the selection of the new quotient digit. A complete presentation of the complex parameter space of the SRT division algorithms is out of the scope of this chapter. We refer the reader to the references given above.

One complete reference is available on low-power design of SRT dividers: Nannarelli and Lang [16]. Many optimizations are used in this chapter: retiming and path equalization to reduce the spurious transitions, modification of the internal redundant representation to reduce the number of flip-flops, gates with low drive capability, dual voltage supplies for the gates that are not in the critical path, clock gating, and "switch off" the power supply of inactive blocks. Based on all these optimizations, an impressive up to 60% power reduction can be achieved for a 53-bit radix-4 divider without speed penalty.

The other widely used class of division algorithms is based on functional iteration. In these algorithms, the multiplication is the fundamental operation. The most well-known algorithm is the Newton–Raphson method. It is based on the following iteration:

$$x_{i+1} = x_i(2 - x_i d)$$

where x_0 is an approximation of $1/d$ produced by a lookup in a small table. This iteration converges quadratically toward $1/d$ under some assumptions. The multiplication of the dividend by the reciprocal of the divisor finishes the division $q = x \times 1/d$. The cost of one iteration is two multiplications and one addition (performed in one's complement). The functional iteration has two main advantages over digit recurrence. First, due to the quadratic convergence of the functional iteration method, the number of digits of the quotient doubles at each iteration. This leads to very fast divisions. Second, the multiplier can be used as a shared unit. This avoids the need of dedicated division unit. The only hardware requirement is a small lookup table for the initial approximation of the divisor reciprocal (usually addressed by 6 to 10 MSB of the divisor).

Unfortunately, to our knowledge, there is no accurate comparison of digit recurrence and functional iteration algorithms for low power aspects. This task is very complex in practice. Indeed, as the functional iteration solution does not have a stand-alone unit, a processor model is required to perform the comparison. It appears, however, that the functional iteration method may have higher power consumption because of the use of the heavily loaded wires for the data transfers between the multiplier unit and the register file.

Both digit recurrence and functional iteration algorithms can be modified to efficiently compute square-roots.

9.4.2 Elementary Functions Evaluation

The evaluation of the elementary functions (e.g., sine, cosine, exponential, logarithm, and arctangent) requires quite complex algorithms. A complete presentation of the various algorithms used for evaluating these functions is given in Muller [17].

Three main classes of algorithms are available for the evaluation of the elementary functions: polynomial or rational approximations, shift-and-add methods, and table-based methods. The algorithms based on polynomial or rational approximations do not require specific hardware support. The shift and add algorithms such as the famous CORDIC are similar to SRT division algorithms, but their implementations are quite complex in practice. The last class is based on table look-up and addition, see Dinechin and Tisserand [18] for efficient implementations of this method.

It appears that the last class could lead to power efficient implementations because of their simple architecture for moderate precision (up to 32 bits). But for higher precision, the choice of the best method is still open. To our knowledge, there is no general and accurate comparison of these algorithms with respect to low-power considerations. This task is very complex, but it seems to be a motivating challenge for the research on computer arithmetic in the next years.

9.4.3 Floating-Point Arithmetic

Besides the fixed-point notation, the other widely used representation of real numbers is the floating-point number system and especially the IEEE 754 standard. In this system, a real number X is represented using the following coding:

$$X = (-1)^{s_x} \times m_x \times 2^{e_x}$$

where s_x is the sign bit, m_x the fixed-point mantissa, and e_x the integer exponent. The mantissa is a value in the interval [1,2). The exponent is a biased integer value (i.e., the stored exponent is $e_x + b$ where b is a given integer constant). All the characteristics of the IEEE 754 floating-point representations can be found in the books mentioned in the introduction.

The algorithms used to perform the arithmetic operations in the floating-point number system are more complex than in the fixed-point system. Indeed, they require some shifts, comparisons, and additive corrections to perform the alignment of the mantissa, the normalization, and the rounding. Usually, several prediction schemes are used to accelerate the computations. This leads to large and complex operators.

To reduce the power consumption of the floating-point operators, two ways have been investigated: optimization of the algorithms and reduction of the operands width. Most of the solutions presented for the integer operators can be used for floating-point operators, but additional improvements can be achieved on the shifts, the comparisons, and the corrections. The optimization of the length of the mantissa and the exponent leads to significant power reduction, but it requires a complex evaluation of the specifications of the algorithms.

9.4.4 Logarithmic Number System

In the logarithmic number system (LNS), the real numbers are represented using a sign bit and a fixed-point approximation of the logarithm of their absolute value. The value zero requires a specific representation, usually this is done by a dedicated bit. In the radix-2 LNS, the main operations can be performed using:

$$\log_2(a \times b) = \log_2 a + \log_2 b$$

$$\log_2(a \div b) = \log_2 a - \log_2 b$$

$$\log_2(a + b) = \log_2 a + \log_2(1 + 2^{\log_2 b - \log_2 a})$$

$$\log_2(a - b) = \log_2 a + \log_2(1 + 2^{\log_2 b - \log_2 a})$$

$$\log_2(a^2) = 2 \times \log_2 a$$

$$\log_2(\sqrt{a}) = \frac{\log_2 a}{2}$$

where the functions $\log_2(1+2^x)$ and $\log_2(1-2^x)$ are usually tabulated or evaluated by specific hardware operators. Based on these equations, it is clear that the LNS can lead to significant performance improvements in applications that involve a lot of multiplications, divisions, squares, or square roots. The main drawback in the implementation of the LNS is the cost of the huge tables used for the addition and subtraction functions.

In some signal processing applications, up to 30% power reduction is achieved using an LNS DSP instead of a fixed-point one. In Paliouras and Stouraitis [19], an interesting comparison between LNS and fixed-point circuit activity is investigated. It shows that to up to 50% power reduction can be achieved using the logarithmic number system.

9.4.5 Technology Evolution

The static power used to be very low and neglected in the past. In current technologies, the leakage power contribution increases significantly. The only thing that can be done at the arithmetic level is to use smaller operators, but this chapter has emphasized that large areas are often required to implement high-speed operators. For higher static to dynamic power ratios, we may have to change the algorithms used for the evaluation of arithmetic operations. For instance, sequential algorithms may be preferable to wide parallel ones.

References

[1] M.D. Ercegovac and T. Lang. *Digital Arithmetic.* Morgan Kaufman Publishers, San Francisco, CA, 2003.

[2] I. Koren. *Computer Arithmetic Algorithms, 2nd ed.* A.K. Peters Ltd., Natick, MA, 2001.

[3] A.M. Shams and M.A. Bayoumi. A novel high-performance CMOS 1-bit full-adder cell. *IEEE Trans. on Circuits and Syst. — II: Analog and Digital Signal Processing,* 47(5):478–481, May 2000.

[4] M. Alioto and G. Palumbo. Analysis and comparison on full adder block in submicron technology. *IEEE Trans. on Very Large Scale Integration (VLSI) Syst.,* 10(6):806–823, December 2002.

[5] A. Guyot and S. Abou-Samra. Modeling power consumption in arithmetic operators. *Microelectron. Eng.,* 39:245–253, 1997.

[6] U. Ko, P.T. Balsara, and W. Lee. Low-power design techniques for high-performance CMOS adders. *IEEE Trans. on Very Large Scale Integration (VLSI) Syst.,* 3(2):327–333, June 1995.

[7] R. Zimmerman. Binary adder architectures for cell-based VLSI and their synthesis. Ph.D. thesis, Swiss Federal Institute of Technology (ETH), Zurich, Hartung-Gorre Verlag, 1998.

[8] C. Nagendra, M.J. Irwin, and R.M. Owens. Area-time-power trade-offs in parallel adders. *IEEE Trans. on Circuits and Systems — II: Analog and Digital Signal Processing,* 43(10):689–702, October 1996.

[9] M.J. Flynn and S.F. Oberman. *Advanced Computer Arithmetic Design.* Wiley Interscience, New York, 2001.

[10] I.S. Abu-Khater, A. Bellaouar, and M.I. Elmasry. Circuit techniques for CMOS low-power high-performance multipliers. *IEEE J. Solid-State Circuits,* 31(10):1535–1546, October 1996.

[11] P.F. Stelling, C.U. Martel, V.G. Oklobdzija, and R. Ravi. Optimal circuits for parallel multipliers. IEEE Trans. on Computers, 47(3):273–285, March 1998.

[12] A. Goldovsky, B. Patel, M. Schulte, R. Kolagotla, H. Srinivas, and G. Burns. Design and implementation of a 16 by 16 low-power two's complement multiplier. *IEEE Int. Symp. on Circuits and Syst.,* pp. 345–348, 2000.

[13] M.J. Schulte and J.E. Stine. Reduced power dissipation through truncated multiplication. *IEEE Alessandro Volta Memorial Workshop on Low-Power Design,* pp. 61–69, 1999.

[14] S.F. Oberman and M.J. Flynn. Division algorithms and implementations. *IEEE Trans. on Computers,* 46(8):833–854, August 1997.

[15] M.D. Ercegovac and T. Lang. *Division and Square-Root Algorithms: Digit-Recurrence Algorithms and Implementations.* Kluwer Academic, Dordrecht, 1994.

[16] A. Nannarelli and T. Lang. Low-power divider. *IEEE Trans. on Computers,* 48(1):2–14, January 1999.

[17] J.-M. Muller. *Elementary Functions: Algorithms and Implementations.* Birkhauser, Boston, MA, 1997.

[18] F. de Dinechin and A. Tisserand. Some improvements on multipartite tables methods. In N. Burgess and L. Ciminiera, Eds., *IEEE 15th Int. Symp. on Computer Arithmetic ARITH15,* pp. 128–135, Vail, CO, June 2001.

[19] V. Paliouras and T. Stouraitis. Low-power properties of the logarithmic number system. In N. Burgess and L. Ciminiera, Eds., *IEEE 15th Int. Symp. on Computer Arithmetic ARITH15,* pp. 229–236, Vail, CO, June 2001.

10

Circuits Techniques for Dynamic Power Reduction

Dimitrios Soudris
Democritus University

10.1 Introduction

Power consumption has emerged as a very significant design parameter, which should be taken into consideration by the designer. Market-driven aggressive demands and technology-related limitations have steered researchers to try to invent new design techniques and methodologies to confront the power requirements. In particular, the field of personal computing devices (i.e., laptops, palmtops, as well as video- and audio-based multimedia products), wireless communication systems (i.e., personal digital assistants and mobile phones), home entertainment (i.e., consumer set-top boxes and video games) and wearable computers are some from the plethora of the market products, which are becoming increasingly popular. On the other hand, physical and technology issues related to chip packaging, cooling, signal integrity, threshold-voltage fluctuations, and variable supply voltages are some challenging problems, which should be studied and solved.

10.2 Dynamic Power Consumption Component

The dynamic power dissipation, P_{dyn}, is caused by the charging and discharging of parasitic capacitances in the circuit. We illustrate the computation of the dynamic dissipation through the example of a CMOS

0-8493-1941-2/05/$0.00+$1.50

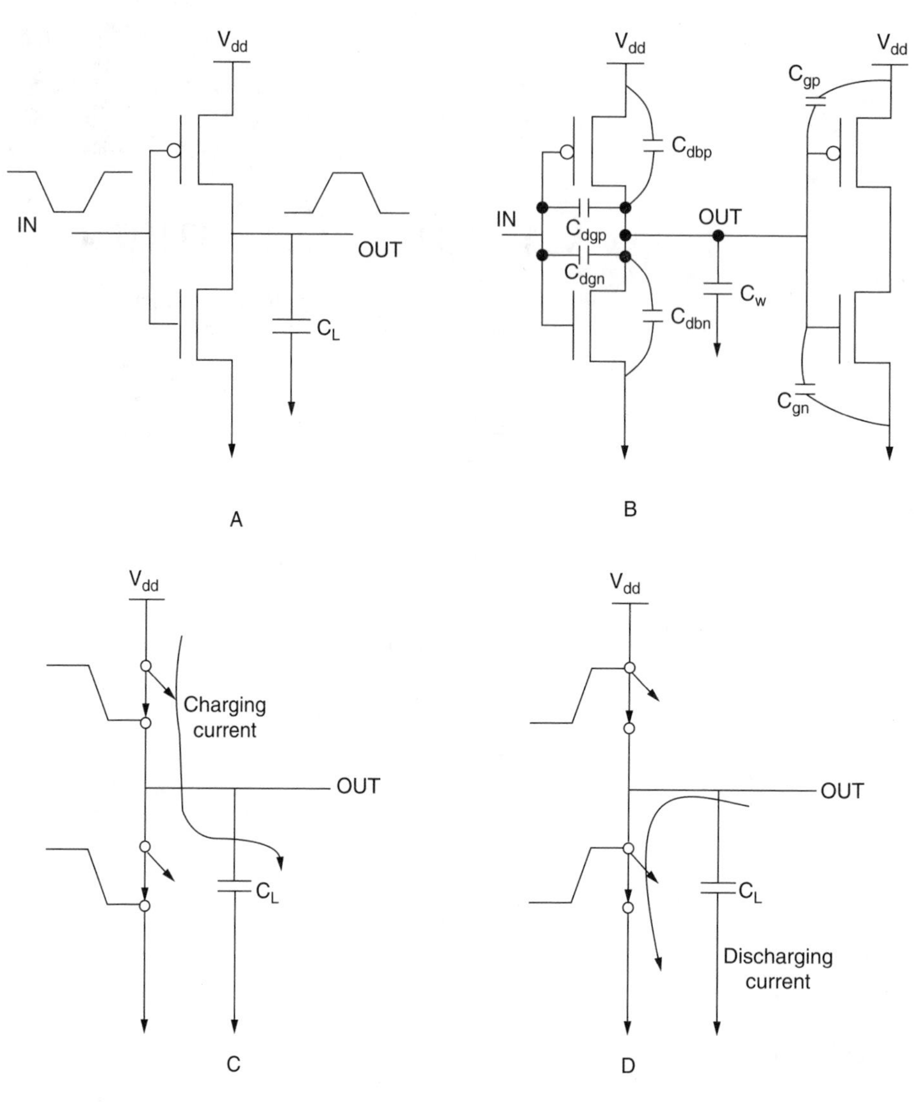

FIGURE 10.1 The function of CMOS inverter: (a) CMOS inverter, (b) capacitor C_L elements, (c) charging phase, and (d) discharging phase.

inverter driving a load capacitor C_L, as it is shown in Figure 10.1(a). The load capacitance C_L depicted in Figure 10.1(b) consists of the gate capacitance of subsequent inputs attached to the inverter output C_{gp} and C_{gn}, interconnect wire capacitance C_W, and the diffusion capacitance on drains of the inverter transistors C_{dgn}, C_{dgp}, C_{dbn}, and C_{dbp}. The total P_{dyn} comprises the sum of two power components: the first one occurs during low-to-high the output transition (i.e., charging phase), while the second one during high-to-low transition (i.e., discharging phase). More specifically, for every low-to-high output transition in a digital CMOS gate, the capacitance C_L on the output node incurs a voltage change ΔV, drawing energy of $C_L \langle \Delta V \langle V_{dd}$ joules from the supply voltage [8]. During this process, one-half of the energy is stored in the capacitor, whereas the second half is dissipated in the PMOS and interconnect wire. In the case of simple inverter, it holds $\Delta V = V_{dd}$, and thus, the power consumption is given by:

$$P_{dyn} = C_L \cdot V_{dd}^2 \cdot a_{0 \to 1} \cdot f \tag{10.1}$$

where $a_{0\to1}$ is an activity factor that represents the average fraction of clock cycles in which a low-to-high transition occurs, and f is the clock frequency. Similarly, a high-to-low transition dissipates the energy stored on the capacitor C_L in NMOS transistor, pulling the output low. Consequently, the total dynamic power consumption is given by the golden formula:

$$P_{dyn} = C_L \cdot V_{dd}^2 \cdot a \cdot f \tag{10.2}$$

Here, we focus on the circuit level techniques for reducing the P_{dyn} power component. The remaining two power components are analyzed and appropriate techniques are discussed in other chapters. It should be stressed that the circuit techniques described for reducing dynamic power consumption may have impacts on the performance and silicon areas as well as on the remaining power components.

10.2.1 Power Reduction Approaches

Equation (10.3) calculates the dynamic power consumption, P_{dyn}, of CMOS logic gates. It can be easily inferred that P_{dyn} is proportional to the load capacitance, C_L, the square of V_{dd}, the switching activity, a, and the clock frequency, f. Consequently, power consumption reduction can be achieved by:

- Reducing of output capacitance, C_L
- Reducing of supply voltage, V_{dd}
- Reducing of switching activity, a
- Reducing of clock frequency, f

Thus, a designer should devise new techniques aiming at the decrease of each abovementioned parameter or any combination of them. A very popular low strategy concerns the reduction of the *switched capacitance* or *effective capacitance*, C_{eff}, which is defined as the product of output capacitance times switching activity (i.e., $a \cdot C_L$).

Generally, the two main low-power reduction strategies concern the reduction of supply voltage and the switched capacitance. The reason is that we consider the throughput rate of a low powered-designed circuit remains the same with an existing circuit. In particular, the reduction of power supply voltage has the major impact on the power consumption due to the quadratic dependence of V_{dd}. Although such reduction is usually very effective, the circuit delay increases and system throughput degrades. In addition, the shift of industry from a supply voltage to a smaller one is quite expensive and slow due to, for instance, the compatibility issues of input/output signals with the peripheral circuits. In contrast, the reduction of the switching activity or the capacitance for a certain technology depends mainly on the designer's creativity. Thus, someone can reuse an existing silicon technology achieving satisfactorily level of power consumption without the need for purchasing new technology libraries, which may lead to design cost reduction. In other words, a designer may proceed to a more advance silicon technology only if he or she has explored all the possibilities for realizing a circuit with an existing technology considering the design cost and time-to-market constraints. The reduction of switching activity requires among others a detailed analysis of signal transition probabilities, careful redesign of circuit nodes with high activity, balanced paths, and selection of appropriate logic style. The capacitance load can be reduced by, for instance, technology scaling, transistor resizing, and logic family selection.

10.3 Circuit Parallelization

Circuit parallelization has been proposed to maintain, at a reduced V_{dd}, the throughput of logic modules that are placed on the critical path [8,17,23,24]. It can be achieved with M parallel units clocked at f/M. Results are provided at the nominal frequency f through an output multiplexer controlled at f (Figure 10.2(a)). Each unit can compute its result in a time slot M times longer (Figure 10.2(b)), and can therefore be supplied at a reduced supply voltage. If the units are datapaths or processors [23], the latter have to

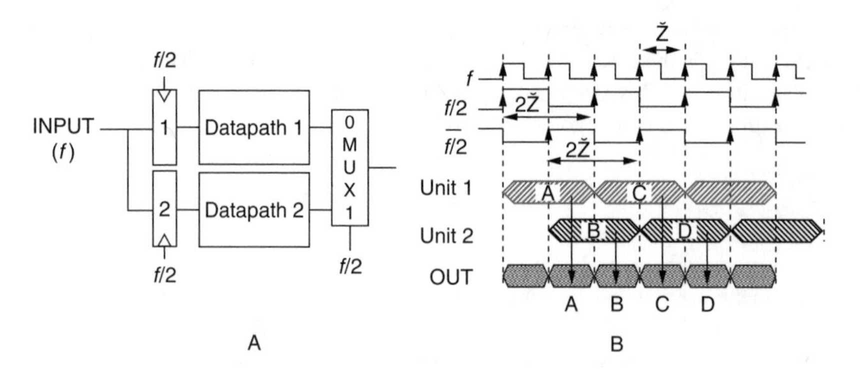

FIGURE 10.2 (a) Datapath parallelization concept, and (b) timing diagram.

TABLE 10.1 8-bit Adder Power Simulation With the CoolChip Library

2 μm Technology	F	V_{dd} (V)	Power	%
8-bit adder	f = 7 MHz	4.5	540 μW	100
2-// 8-bit adder	f/2 = 3.5 MHz	4.5	760 μW	140
2-// 8-bit adder	f/2 = 3.5 MHz	3.0	339 μW	63
2-// 8-bit adder	f/2 = 3.5 MHz	2.5	235 μW	44

be duplicated, resulting in an M times area and switched capacitance increase. Applying the well-known power formula, one can write:

$$P = M \cdot C_{eff} \cdot f / M \cdot V_{dd}^2 = C_{eff} \cdot f \cdot V_{dd}^2 \tag{10.3}$$

Table 10.1 presents the reduction of power of an 8-bit adder. One could deduce that power is saved only if V_{dd} is reduced. As operating frequency is reduced, however, the use of cells with smaller or unsized transistors results in a power reduction. Furthermore, some parallelized logic modules do not require M-unit duplication. It is the case, for instance, for memories [25], in which each unit contains $1/M$ data or instructions, resulting in the same total area to store the information and in the same C_{eff} or smaller C'_{eff} total switched capacitance, if cells with unsized transistors are used (Figure 10.3). In such a case, the power is the following:

$$P = C'_{eff} \cdot f / M \cdot V_{dd}^2 \tag{10.4}$$

At first order, power could be saved even if V_{dd} is not reduced; however, some overhead has to be considered, such as the address registers duplication and the output multiplexer (Figure 10.3). If this overhead is not too expensive, such a parallelization scheme has to be considered for logic modules that are not on the critical path. At a low V_{dd}, the latter are working without parallelization. At the same low V_{dd}, power could be saved if they are parallelized at the cost of a small overhead. Memories, shift registers, and serial-parallel converters provide interesting examples.

10.3.1 Memory Parallelization

In a parallelized module, operations of the execution units or data accesses in memories are performed in an overlapped or interleaved fashion (Figure 10.2(b)). Therefore, the result is provided with an M-1 latency delay compared to a nonparallel architecture. One can see on the timing diagram of Figure 10.2(b)

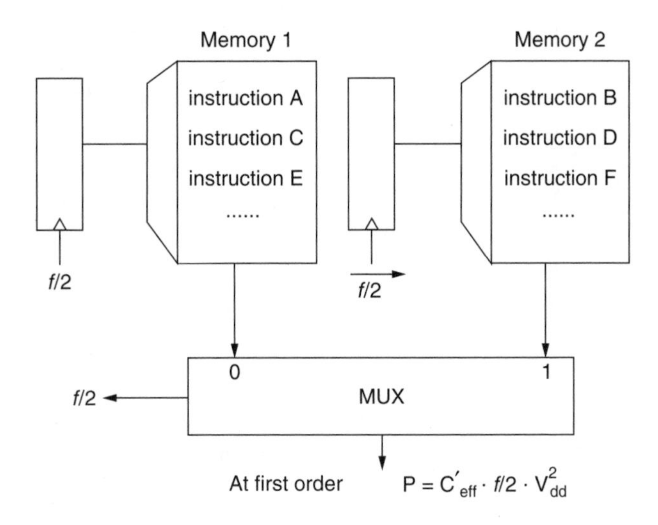

FIGURE 10.3 Memory parallelization.

that the output multiplexer can be controlled at $f/2$. The operation or access of a Unit 2 is started before the completion of the operation of Unit 1. Therefore, M successive computations do not have to be dependent on each other.

Controllers with a fixed sequence of commands without any branch instruction, or specialized processors for special linear computation, or random-access memories (RAMs) used to store coefficients for programmable finite impulse response (FIR) filters, can be parallelized according to the structure of Figure 10.3. It can be used, for instance, for transcoders in which several lookup tables (i.e., read-only memory [ROM]) are connected in parallel; however, parallel memories are difficult to use if branch instructions are used. Interleaved or parallelized memories (Figure 10.4) with branch instructions were used in the 1960s for computers [14]. With, for instance, 32 memory modules and an access time of 10 cycles ($f/10$), the probability to insert a branch delay is reduced as 10 successive instructions are, most of the time, stored in different modules.

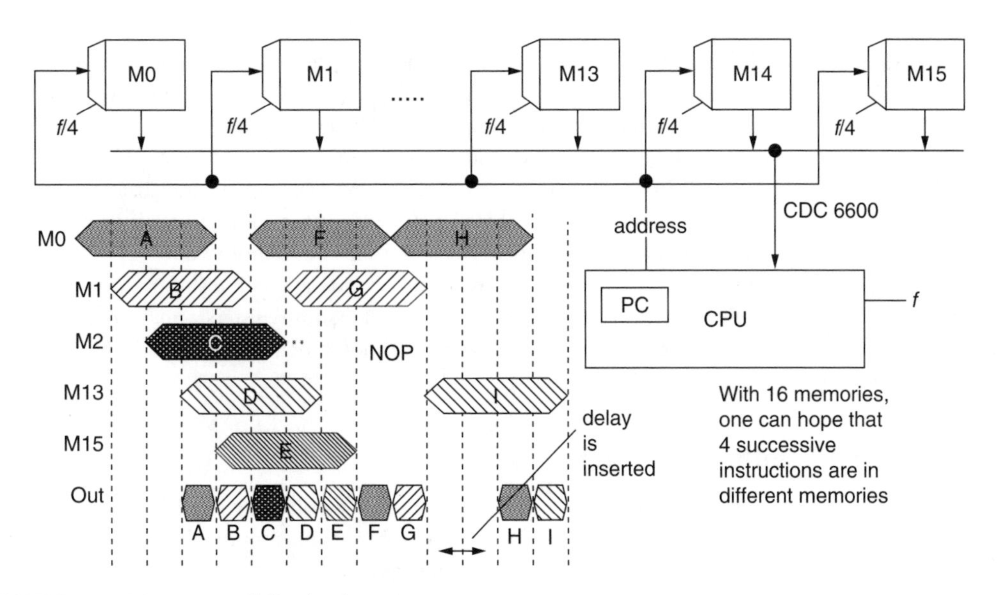

FIGURE 10.4 Memory parallelization in computers.

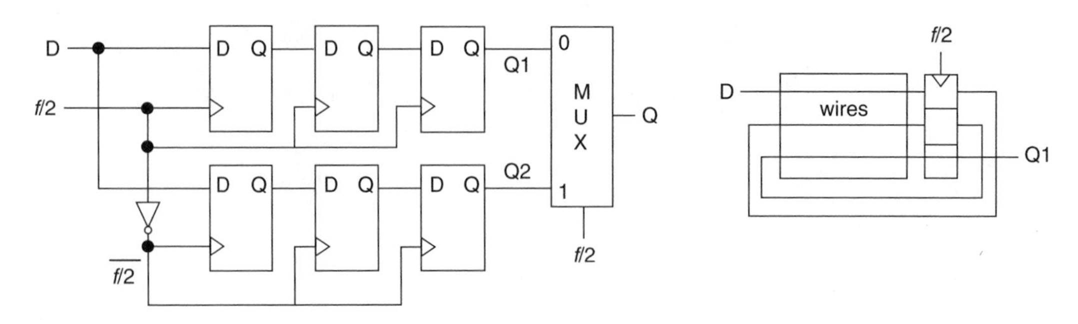

FIGURE 10.5 Parallelized shift register.

TABLE 10.2 Power Simulation with the CoolChip Library for: (i) Nonparallelized 16-bit Shift Register and (ii) 2- and 4-Parallelized 16-bit Shift Registers

2 μm Technology	f (MHz)	V_{dd} (V)	Power (μW)	%
16-bit SR	f = 33	4.5	1535	100
2-// 16 bit SR	f/2 = 16.5	4.5	887	58
4-// 16 bit SR	f/4 = 8.25	4.5	738	48
16-bit SR	f = 33	3.2	797	100
2-// 16 bit SR	f/2 = 16.5	3.2	448	56
4-// 16 bit SR	f/4 = 8.25	4.0	585	83

10.3.2 Parallelized Shift Register

Figure 10.5 depicts a parallelized shift register. Such a concept has been proposed for CCD serial memories [14,21]. The input is successively provided to the upper or to the lower half shift register at a reduced frequency, while the output multiplexer restores the output at the frequency f. No latency exists because the combinatorial circuit of the state machine "shift register" is implemented by simple wires, resulting in no associated delay. The total number of D-flip-flops (DFFs) is the same as in the nonparallelized shift register [24,25].

For the nonparallelized shift register, the maximum frequency is limited by the delays of the latches of the DFF. For the parallelized shift register, the maximum frequency is limited by one latch delay and the output multiplexer delay. Thus, the maximum frequency of the parallelized structure is the same as the classic structure (an f_{max} = 100 MHz classic shift register can be replaced by an $f/2$ = 50 MHz parallelized shift register, but it is impossible to increase $f/2 > 50$ MHz). Such a parallelization does not provide faster shift registers. It is therefore impossible to reduce V_{dd} if the shift register is on the critical path. For shift registers, which are not on the critical path, one can reduce both f and V_{dd}.

Table 10.2 presents the power consumption of nonparallelized and parallelized shift registers, depending on the degree of parallelism. Such a comparison is only valid for shift registers, which are not at their frequency limits, however, because an 8- or 4-parallelized cannot provide the same throughput as a nonparallelized shift register.

10.3.3 Serial-Parallel Converter

Figure 10.6 depicts a parallelized structure of a 16-bit serial-parallel converter in which the 1-bit input is successively loaded in four 4-bit shift registers clocked at $f/4$. Power consumption is reduced by a factor of four with the same throughput. Because no output multiplexer exists, the maximum frequency of such a structure can be much higher than the nonparallelized serial-parallel converter.

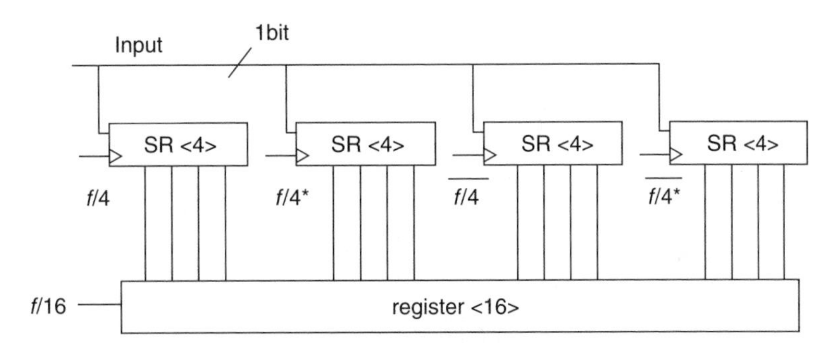

FIGURE 10.6 Parallel-serial converter.

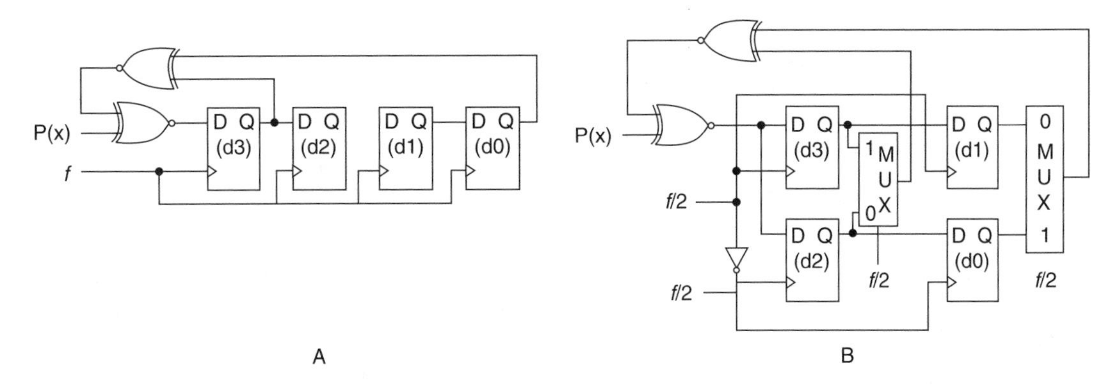

FIGURE 10.7 Linear feed-back shift register.

10.3.4 Linear Feed-Back Shift Registers

Shift register parallelization can be used for linear feed-back shift registers [19] with as many output multiplexers as the number of inputs of the XOR tree. Figure 10.7 gives an example with two output multiplexers. The example in [19] is an *M*-parallelized M-bit shift register with *M*-input simplified multiplexers. In addition, a parallelized LFSR architecture was used for the development of a division algorithm [16] and for the implementation of steam ciphers in cryptography [11].

10.3.5 Double-Edge Triggered Flip-Flop

Figure 10.8(a) and Figure 10.8(b) as well as Figure 10.8(c) to Figure 10.8(e) show the schematic and various circuit designs of single-edge triggered flip-flop (SET-FF) and double-edge triggered flip-flop (DET-FF), respectively. A classic SET-FF is implemented with two latches in series, while its parallelization results in two latches in parallel with an output multiplexer (i.e., derivation). A DET-FF is triggered on both rising and falling edge of a clock pulse. Using DET-FF the clock frequency, *f*, can be halved for the same throughput rate, thus reducing the power dissipation on the clock distribution network. Although many alternative DET-FF designs have been proposed, they have not been used extensively, due to the increased silicon area (i.e., increased input capacitance and number of transistors). This implies a larger number of internal nodes, which is strongly dependent on the input signal transition probability, *a*. It was proved [26] that if the switching activity *a* is low, significant power savings may be achieved, while high activity *a* may lead to increased power consumption. In addition, DET-FFs exhibit increased glitching activity compared with SET-FFs. SPICE simulation results show power savings around 10% using DET-FFs at the expense of a reduction of 10% in performance.

In Chung et al. [9], a detailed comparative study of five existing DET-FFs in terms of performance (i.e., latency^{-1}), total power consumption and power × delay product (PDP) is given in Table 10.3. It is

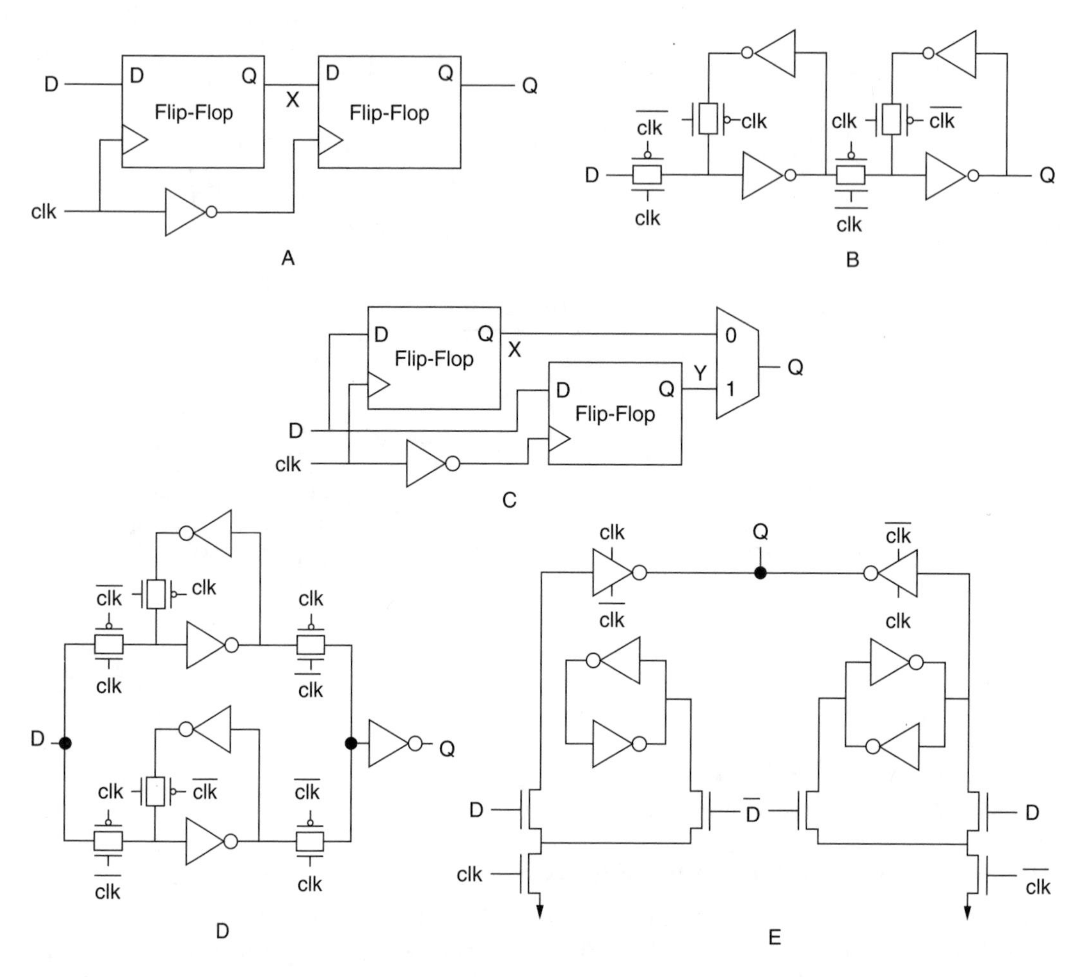

FIGURE 10.8 Flip-Flops: (a) block diagram of a single-edge triggered flip-flop (SET-FF), (b) circuit design of a SET-FF, (c) block diagram of a double-edge triggered flip-flop (DET-FF), (d) circuit design of a DET-FF [18], and (e) circuit design of a DET-FF [9].

TABLE 10.3 Comparison Results of DET-FF in Terms of Power Consumption, Latency, and Power-Delay Product

Type DET-FF	Clock Power (μW)	Data Power (μW)	Internal Power (μW)	Total Power (μW)	Latency (ps)	PDP (fJ)
[22]	17.6	65.6	241.7	324.9	245.4	79.6
[18]	17.0	4.6	153.4	175.0	312.3	54.7
[11]	23.2	11.6	131.4	166.2	262.2	43.6
[26]	30.0	13.4	194.5	237.8	235.3	56.0
[9]	18.1	10.9	189.4	218.4	230.5	50.3

assumed 0.18-μm technology and supply voltage of 1.8 volts. Notice that the total power consumption consists of three components:

1. Internal power dissipation
2. Data power
3. Local clock power, where the contribution of the internal power is over the 70% of total power consumption

Specifically, the first component concerns the power consumed inside a DET-FF including the power consumed for driving C_L. Thus, the power optimization techniques should concern the careful design of DET topology reducing capacitance or switching activity.

10.4 Voltage Scaling-Based Circuit Techniques

Because dynamic power is proportional to V_{dd}^2, even a small reduction in supply voltage causes a quadratic decrease in power consumption. However, a supply voltage reduction influences circuit's delay negatively. To preserve a constant system throughput using lower supply voltages, there exist three main approaches:

1. To redesign the circuits exploiting the principles of parallelism and pipelining
2. To reduce the threshold voltage, V_{th}, in order to compensate V_{dd} reduction
3. To assign lower V_{dd} to noncritical paths

10.4.1 Multiple Voltages Techniques

To preserve performance, while also reducing power consumption, a dual-V_{dd} approach can be used. The main concept is to assign the high V_{dd}, V_{ddH} to the gates that belong to the critical path, while the low V_{dd}, V_{ddL} is assigned to off-critical path remaining gates; however, the designer should be very careful to avoid the creation of static current. More specifically, the output of V_{ddL} gates cannot be fed directly to V_{ddH} gates because the output of a V_{ddL} gate can never be raised higher than V_{ddL}. Therefore, if connected to a V_{ddH} circuit, static current flows due to the pMOS in the V_{ddH} circuit are never being completely cut-off (Figure 10.9) [29].

To remove the static current, one possible solution is the use of level converters placed between the V_{ddL}- and V_{ddH}-supplied gates, which may increase area and power. To alleviate level converters' power and area overhead, one approach is to insert the level shifting function of a flip-flop circuit (FFLC) depicted in Figure 10.10. More specifically, the master latch is the same as a conventional flip-flop, while the slave latch also realizes the level-conversion function. This results in the power of the flip-flop being less than that of V_{ddH} flip-flop, while increasing delay slightly [30].

Layout is another important issue when dealing with multiple supply voltages. V_{ddL} and V_{ddH} cells should be separated because they have different n-well voltages. Generally, a row-by-row separation, depicted in Figure 10.11 is used due to high performance and applicability to both standard-cell and gate-arrays. Novel algorithms have been developed for optimal assignment of cells to the layout rows with V_{ddH} and V_{ddL} supply voltages [29].

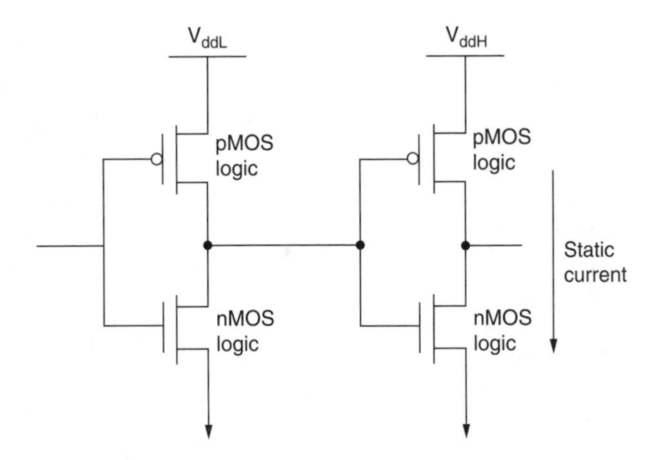

FIGURE 10.9 Dual supply voltages assignment concept.

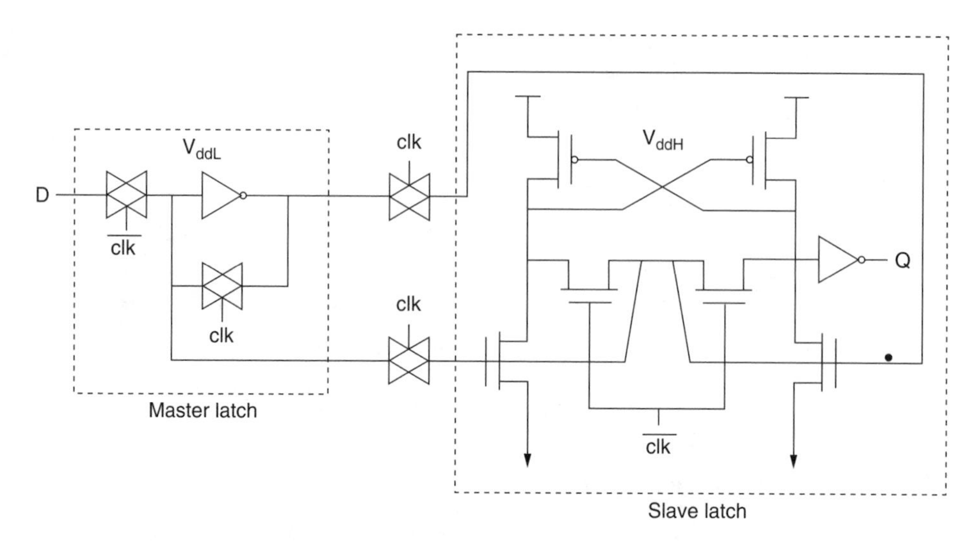

FIGURE 10.10 Circuit design of flip-flop with level-conversion function (FFLC) [30].

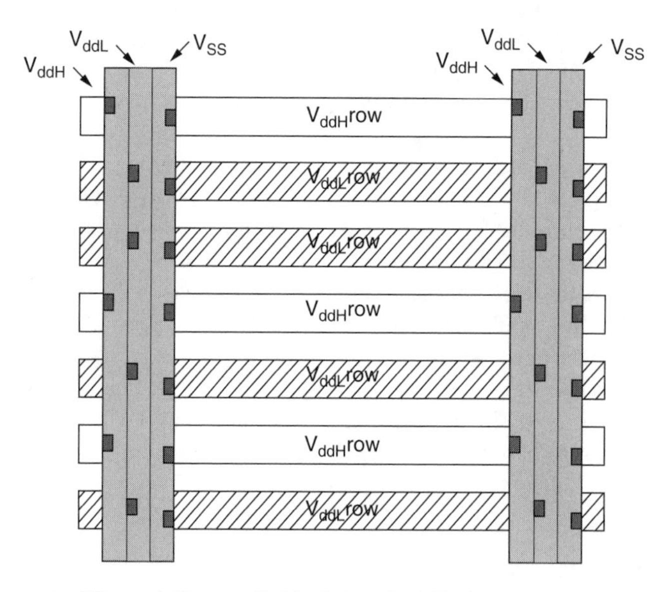

FIGURE 10.11 Placement of V_{ddH} and V_{ddL} supplied logic in a dual-V_{dd} layout.

Clustered voltage scaling (CVS) is one structure proposed to implement dual-V_{dd} design. With the CVS technique, all cells should be placed with a certain order starting from the primary inputs to primary outputs through V_{ddH}-supplied gates, V_{ddH}-supplied gates and level converters, as shown in Figure 10.12. This structure leads to clusters of V_{ddH} cells and V_{ddL} cells. By introducing level converters only at the end of a path, FFLC can be used and a minimal number of them is needed. To assign the V_{ddL} cells, a depth-first-search algorithm is used from the primary outputs to the primary inputs. As each cell is visited, an attempt is made to replace it with a V_{ddL} cell. If it can be replaced, the algorithm continues, otherwise the traversal is stopped. Dealing with multiple fan-outs can be tricky. To replace a cell, all of the cells in the fan-out of that cell should be replaced with V_{ddL} cells.

An extended CVS technique was developed [30], where the main difference compared with CVS is the fact it allows placement of a level converter even between logic cells. This can be useful in the case where a gate has multiple inputs and only one critical path. Before a level converter is inserted between logic cells, the insertion is checked to see if it does indeed reduce power consumption.

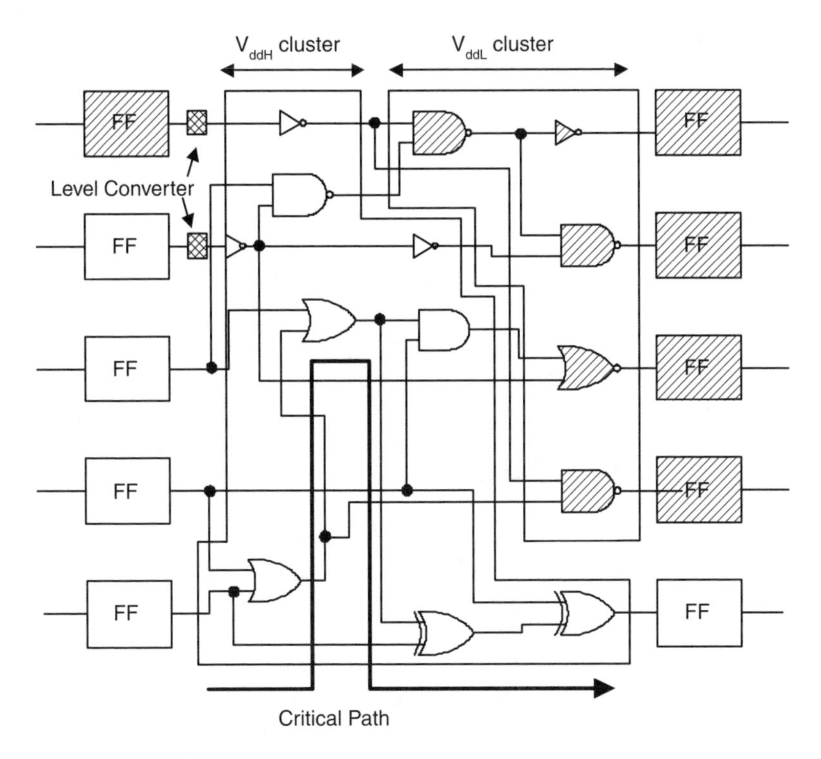

FIGURE 10.12 The concept of CVS technique.

In the case of multiple power supplies and multiple threshold voltages, theoretical models have been developed by Kuroda [15], which are used to determine "rules of thumb" for optimal multiple V_{dd} and V_{th}. These rules can be summarized as follows:

Multiple Voltages

For $\{V_{dd1}, V_{dd2}\}$:

$$\frac{V_{dd2}}{V_{dd1}} = 0.5 + 0.5\frac{V_{th}}{V_{dd1}}$$

For $\{V_{dd1}, V_{dd2}, V_{dd3}\}$:

$$\frac{V_{dd2}}{V_{dd1}} = \frac{V_{dd3}}{V_{dd2}} = 0.6 + 0.4\frac{V_{th}}{V_{1}}$$

For $\{V_{dd1}, V_{dd2}, V_{dd3,} V_{dd4}\}$:

$$\frac{V_{dd2}}{V_{dd1}} = \frac{V_{dd3}}{V_{dd2}} = \frac{V_{dd4}}{V_{dd3}} 0.7 + 0.3\frac{V_{th}}{V_{dd1}}$$

Multiple V_{th}

For $\{V_{th1}, V_{th2}\}$:

$$V_{th2} = 0.1 + V_{th1}$$

For $\{V_{th1}, V_{th2,} V_{th3}\}$:

TABLE 10.4 Comparison Results for Multiple V_{dd} and V_{th}

Approach	Technique	Power Reduction
[15][1,2]	Multiple V_{dd}	50%
[15][1]	Multiple V_{dd}, low- V_{th} device	30%
[15][1]	Multiple V_{th}, high- V_{th} device	15%
[29]	Dual- V_{dd}	10–20%

[1] Theoretical results. No real circuits were used
[2] Level converters power consumption not included

$$V_{th2} = 0.06V_{dd} + V_{th1}, \quad V_{th3} = 0.07V_{dd} + V_{th2}$$

For $\{V_{th1}, V_{th2}, V_{th3}, V_{th4}\}$:

$$V_{th2} = 0.04V_{dd} + V_{th1}, \quad V_{th3} = 0.05V_{dd} + V_{th2}, \quad V_{th4} = 0.05V_{dd} + V_{th2}$$

Table 10.4 gives some comparison results between different techniques based on multiple supply voltages. Despite the strong dependence of power consumption on supply voltage, Table 10.4 indicates that the power savings arising from the adoption of multiple supply voltages technique may be insignificant, due to the use of additional level converters.

Although the reduction of V_{dd} has a large impact on power consumption due to its quadratic dependence, leakage power may be the best candidate for reducing power especially for systems with nonuniform load and many standby periods. Consequently, a designer should be very careful when he or she attempts to reduce power consumption.

10.4.2 Low Voltage Swing

The low-swing voltage design technique aims at the power reduction on a long interconnect wire (i.e., large capacitance) through the use of reduced voltage swing on the wire. Given the fact that we are working in a specific design process (i.e., capacitance, frequency clock, and supply voltage remain unchanged) it is proved that on-chip lower supply voltages can be achieved [7,37], using specially designed circuits or DC-DC converters. We will discuss the concept of reduced voltage swing and the associated circuit implementations as well as its impact on P_{dyn}.

A typical form of signaling is the classical two inverters-based configuration scheme with rail-to-rail signal swing, as shown in Figure 10.13(a). The two CMOS inverters correspond to the driver and receiver circuit of the signal. In addition, intermediate repeaters/buffers are frequently used to improve signal characteristics [1]; however, the rail-to-rail swing (i.e., full swing, is an inefficient energy design approach). A possible solution is based on the reduction of the signal swing at the output driver and over the interconnect wire. Generally, the reduced voltage swing may increase not only the circuit performance, but also the major gain coming from the reduced dynamic power consumption, which can be significant in the case of large load capacitances (i.e., long wire).

Figure 10.13(b) depicts a typical circuit structure where a low swing design technique can be applied. In particular, having a long bus (or interconnection) (i.e., a large capacitive load, specially designed circuits for converting the normal swing signal to a low swing voltage and vice versa are placed at the interconnection ends. In this scheme, power savings can be achieved by two ways:

1. The charge needed for charge/discharge of C_L is smaller.
2. The driver size can be reduced because the current to be delivered by the driver to charge/discharge C_L in a certain time is smaller than the full-swing case.

The design of an efficient low-swing scheme has become a difficult problem with the deep-submicron process technology, due to very small supply voltage, V_{dd}, and threshold voltage, V_{th}.

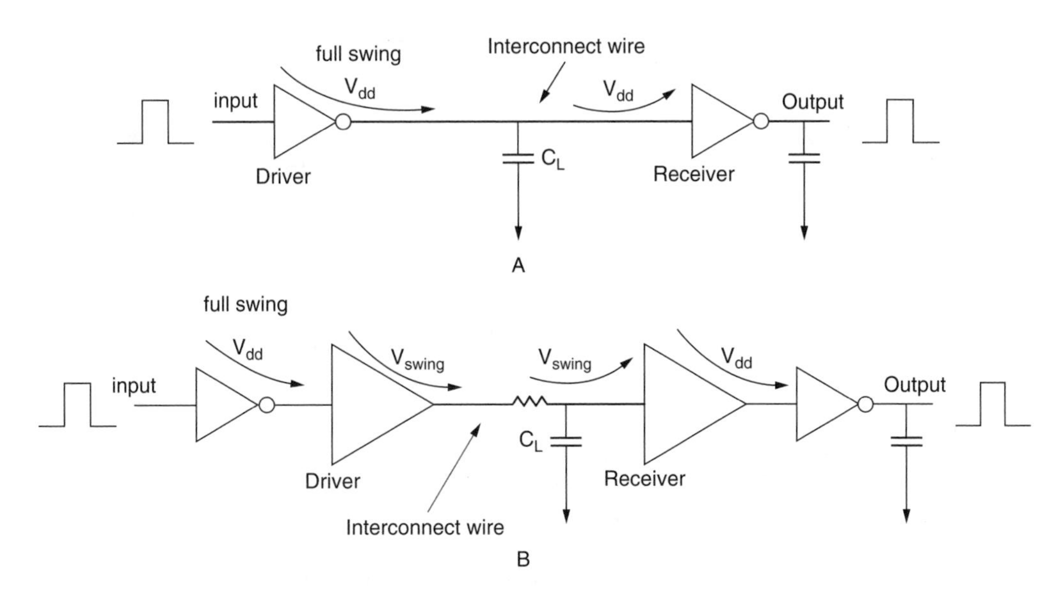

FIGURE 10.13 (a) Two-inverters-based signaling configuration scheme, and (b) typical reduced-swing circuit with a driver and receiver, and the long interconnect wire.

To design efficient low-swing-like circuits, the following parameters should be taken into account:

Energy. The dynamic energy consumption, E_{dyn}, of a interconnect wire in one cycle is given by:

$$E_{dyn} = a \cdot C_L \cdot V_{dd} \cdot V_{swing} \tag{10.5}$$

where a is the switching activity of the signal to be transmitted, and V_{swing} is the voltage across the wire.

It has been proven [32] that the selection of static drivers is preferable due to lower signal switching activity, and the supply voltage of the chosen driver should be as low as possible. The key challenge is how to detect a "one" signal at the receiver end. Novel low-swing designs using threshold voltage drops reduce the energy consumption by half order of magnitude. Additional power savings up to 4 to 6 times order of magnitude can be achieved by using very small supply voltages (from on-chip DC-DC converters).

Design Complexity. To meet certain design constraints, the designer should pay attention to how easy or complex the driver/receiver circuit design is and how much silicon area requires a new design. In addition, among other design issues related to the use of extra DC-DC converters for realizing different supply voltages and the use of only single-ended signaling schemes should be considered.

Delay. The use of voltage swing $V_{swing} < V_{dd}$ increases the propagation delay through a long interconnect wire. However, if a designer with careful architectural design can hide the increased bus delay and place a latch at the receiver, the low-swing transceiver can provide significant reduction in energy consumption.

A plethora of efficient low-swing-based techniques was reported [32]. The existing reduced voltage swing techniques can be classified as dynamic and static depending on the existence of a precharge phase (e.g., use of clock signal) during logic operation or not, respectively. Furthermore, depending on the chosen signaling approach, a low-swing circuit can implement either a single-ended or a differential signaling technique. Specifically, using the former signaling technique, a receiver detects an absolute change in a voltage in a single wire, while with the latter technique, a receiver should detect a relative difference in voltage between two wires.

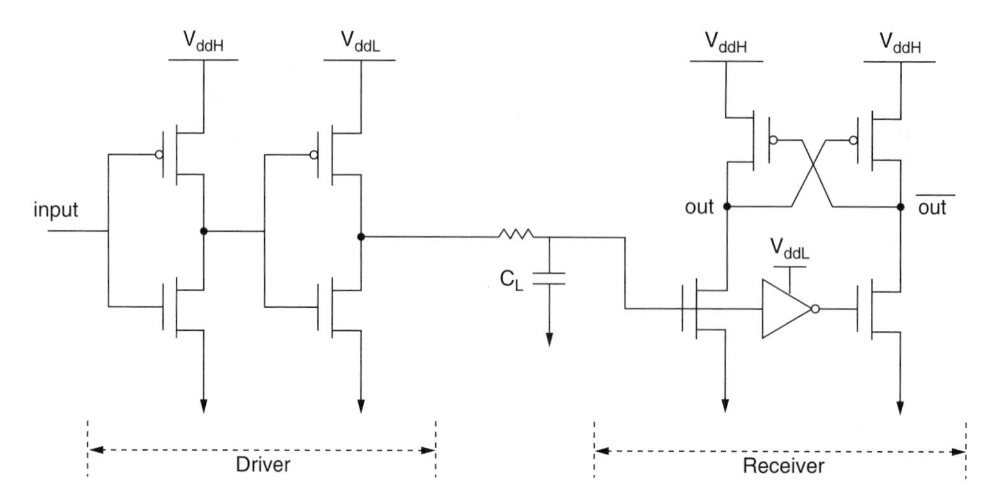

FIGURE 10.14 The architecture of the conventional level converter.

Several low-swing interfaces have been proposed in the literature; conventional level converter (CLC), differential interconnect (DIFF) [7], pulsed-controlled driver (PCD) [10], charge intershared bus (CISB) [12], charge-recycling bus (CRB) [31], capacitive-coupled level converter (CCLC) [32], level-converting register (LCR) [32], and pseudo-differential interconnect (PDIFF) [32]. It is beyond the scope of this chapter to describe in detail the architecture, the operation, and the advantages and disadvantages of each low-swing scheme; however, some comments about a typical static and differential swing scheme are provided for CLC and DIFF [7]. In particular, the CLC scheme shown in Figure 10.14 represents the conventional way for converting a full-swing signal to a low-swing one and vice versa. Moreover, the driver's circuit uses an additional supply voltage V_{ddL} to drive the load capacitance (i.e., interconnect). This voltage value is the voltage swing on the interconnect wire. From Table 10.5 it can be deduced that the noise margin remains in an acceptable level, because the receiver exhibits a differential behavior, while the circuit delay increases. Figure 10.15 illustrates the circuit of DIFF [7]. Differential signaling is an attractive choice due to its high common-mode noise rejection, leading to a very small signal swing.

Detailed comparison results of existing low-swing circuit techniques in terms of energy consumption, delay, energy × delay product, voltage swing, and SNR value are given in Table 10.5. The CMOS scheme of the first row represents the conventional full swing scheme, and it is assumed V_{dd} = 2 volts, C_L = 1pF. All swing schemes can achieve energy savings starting from 50% to a factor of seven. The PDIFF scheme provides the optimal solution with a very small energy consumption, acceptable performance, and perfect noise immunity level, employing a very low swing voltage of 0.5 V. Furthermore, a qualitative comparison of the plethora of low-swing techniques is depicted in Table 10.6.

TABLE 10.5 Low-Swing Techniques [32]

Approach	Energy (pJ)	Delay (ns)	Energy–Delay Product	Voltage Swing (V)	SNR
CMOS	11.6	2.1	24.5	2	1.52
CLC	4.4	3.1	13.6	1.1	1.24
DIFF [7]	3.0	2.7	8.1	0.25	1.64
PCD [10]	3.5	2.0	7.0	0.5	0.70
CISB [12]	3.5	4.4	15.4	0.25	0.66
CRB [13]	3.1	3.1	10.9	0.25	0.74
CCLC [32]	2.67	2.6	6.94	0.8	0.81
LCR [32]	2.44	2.59	6.32	0.8	1.23
PDIFF [32]	1.92	2.4	4.6	0.5	1.92

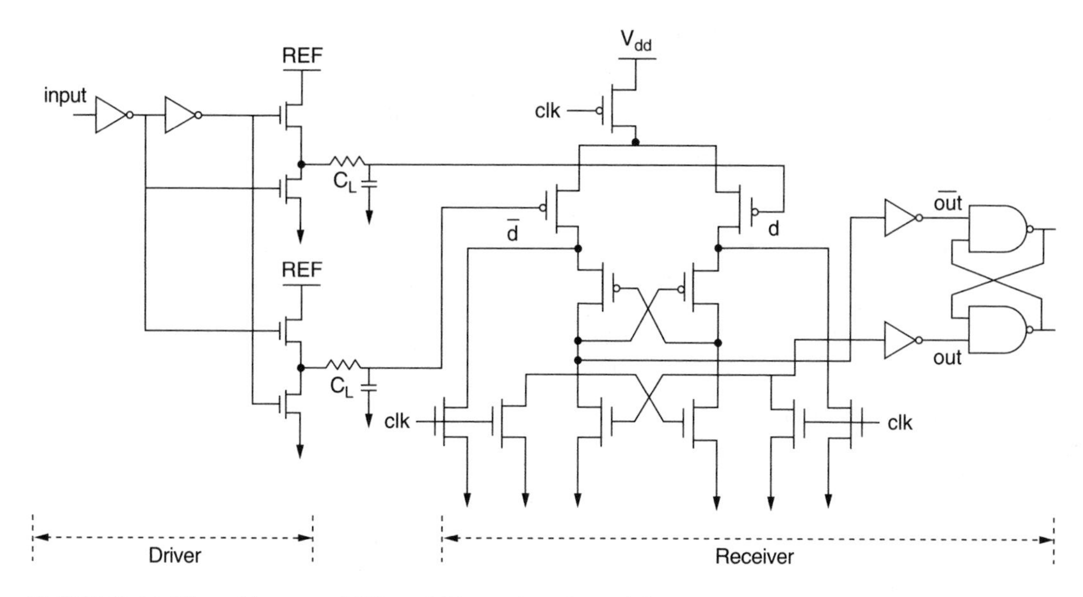

FIGURE 10.15 The architecture of differential low-swing scheme [7].

TABLE 10.6 Qualitative Comparison of Low-Swing Techniques

	CMOS	CLC	DIFF	PDC	CISB	CRB	CCLC	LCR	PDIFF
Extra power supplies	√				√	√			√
Reference voltages	√					√			
Multiple V_{th}	√		√	√	√	√	√	√	√
Voltage scaling	√	√	√	√	√	√	√	√	√
Low power		√	√	√	√	√	√	√	√
Low delay		√	√	√			√	√	√
Good SNR	√	√	√				√	√	√
Area penalty	√	√		√			√	√	√
Interconnect	Single	Single	Double	Single	Single	Double	Single	Single	Single

10.5 Circuit Technology-Independent Power Reduction

During a design process, it is possible to have only the behavioral of circuit (e.g., Boolean expression, state equations) with a few gates or flip-flops of a circuit. Although sometimes the actual implementation technology is not selected, it is possible to employ design techniques, which target circuit capacitance and switching activity reduction. Such low-power techniques may also reduce the total design time cost because a designer may have a good power consumption estimate during the early design phases.

10.5.1 Precomputation

This optimization technique is based on selectively precomputing the output logic values of a circuit one clock cycle before they are required, and then use the precomputed values to reduce the internal switching activity of the combinational logic in the successive clock cycle [2].

A simple example of this idea-based structure is shown in Figure 10.16, where the inputs of sequential block A have been partitioned into two sets, corresponding to registers R_1 and R_2, and the output of block A is the input of register R_3. The Boolean functions g_1 and g_2 serve as the predictor functions according to the following equations:

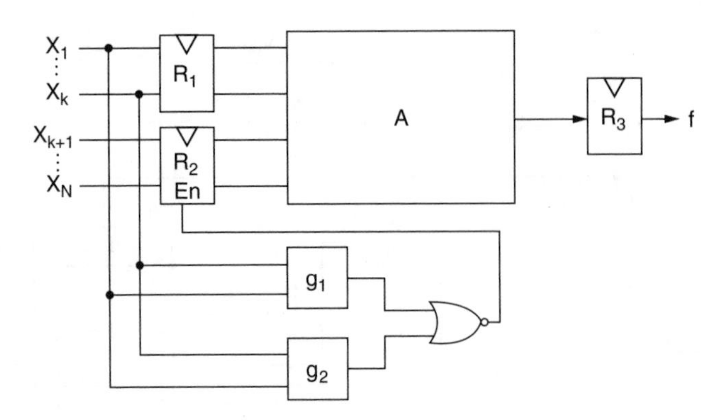

FIGURE 10.16 Precomputation structure for sequential circuits.

$$\begin{aligned} g_1 = 1 &\Rightarrow f = 1 \\ g_2 = 1 &\Rightarrow f = 0 \end{aligned} \tag{10.6}$$

The physical meaning of Equation (10.6) is that if function g_1 or g_2 equals 1, the value of the output function, f, is fully determined; however, the Boolean variables of functions g_1 and g_2 are a subset of the input signals of block A. Thus, the remaining signals, $(X_{k+1},\ldots, X_N)$, can be frozen. It must be stressed that it is not allowed for both g_1 and g_2 to be evaluated to 1.

Consequently, if the logic level of g_1 or g_2 is high, during clock cycle T, then the enable signal of register R_2 is low. Thus, the outputs of R_2, during clock cycle $T + 1$, are not changed. Because the output of R_1 is updated, the function f is evaluated correctly. As a subset of the inputs of block A changes, the switching activity of this block is reduced, implying a remarkable power saving. Because functions g_1 and g_2 occupy extra area and consume additional power, attention should be paid to the construction of g_1 and g_2 and an appropriate trade-off analysis should be performed.

10.5.2 Retiming

A novel method for reducing the power consumed in pipelined sequential circuits has been proposed in [20]. The method is based on retiming, which is a technique that repositions the flip-flops of the circuit resulting in the minimization of either the area or the delay of the circuit.

Because a flip-flop output makes at most one transition when the clock is asserted, the idea is to place a flip-flop in a circuit node with high glitching activity and high load capacitance. Thus, glitches are not propagated to the transitive fan-out of the node resulting in a reduction of the total switching activity, as shown in Figure 10.17.

However, attention should be paid because the switching activity of some nodes of the circuit may be changed due to retiming, which may result in an increase of the power consumption. In addition, the number of the used registers should be minimized because the power dissipation of the registers and clock line are not negligible. Finally, attention should also be paid to preserve the timing behavior of the circuit when retiming for low power is performed.

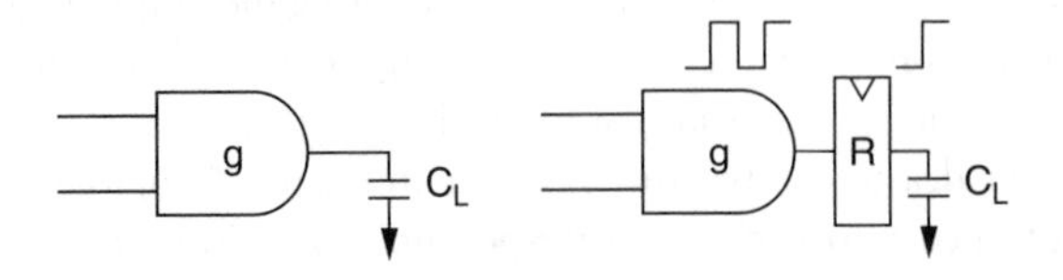

FIGURE 10.17 Retiming for low-power.

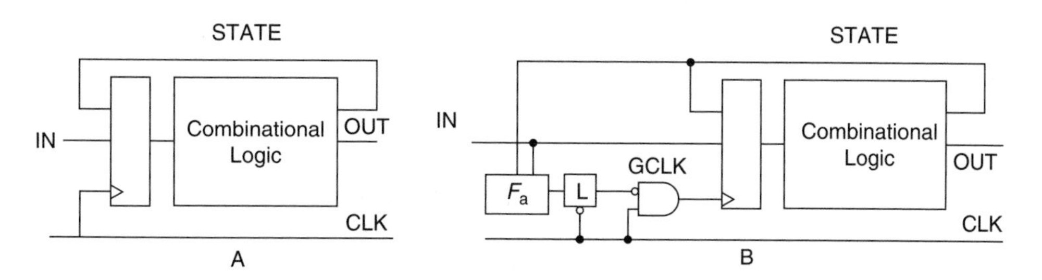

FIGURE 10.18 (a) Single clock, flip-flop based FSM, and (b) gated-clock version.

10.5.3 Synthesis of FSMs with Gated Clocks

This technique, presented in Benini et al. [3–6], refers to the power reduction that can be achieved in finite-state machine (FSM) circuits by using gated clocks. The key idea is that during the operation of an FSM there are conditions where the state and the output of the FSM do not change. Thus, clocking the circuit in the corresponding time intervals wastes power both in the combinational logic and in registers. Thus if we can detect the idle conditions of the FSM, we can also stop the clock during the corresponding time intervals. The benefit of using a gated-clock is twofold: first, when the clock is stopped, no power is consumed by the combinational logic because its inputs remain unchanged. Second, no power is consumed by the flip-flops and gated-clock line.

The flip-flop-based architecture is modified by setting a new activation signal, *GCLK*, as illustrated in Figure 10.18. The purpose of this signal is to selectively stop the local clock for the FSM when the machine is idle. The combinational circuit, F_a, which provides the activation signal, uses as its inputs the primary inputs and the state lines of the machine. It has been found that the application of this technique to such circuits provides power savings ranging between 10 and 30%. Because the function F_a consumes power, it is recommended to select a subset of all idle conditions such that this subset takes place with high probability during the circuit operation.

10.6 Circuit Technology-Dependent Power Reduction

The physical implementation of circuit behavior (e.g., Boolean function) may be differentiated by the chosen technology. In other words, the realized circuit topology and the chosen circuit components (e.g., from a library) may result in circuit designs with different hardware features (e.g., chain topology vs. tree topology), which affect the circuit capacitance and switching activity. The next paragraphs describe a series of low-power techniques, which achieve power savings through the reduction of a or C_L (Equation 10.2).

10.6.1 Path Balancing

The way the gates of a logic circuit are interconnected can strongly affect the overall switching activity, and hence the power dissipation. For example, timing skew between signals in a circuit can cause spurious transitions (glitches) resulting in extra power. To reduce the possible spurious activity in a circuit, delay of all true paths that converge at each gate must be balanced, as depicted in Figure 10.19, where the logic

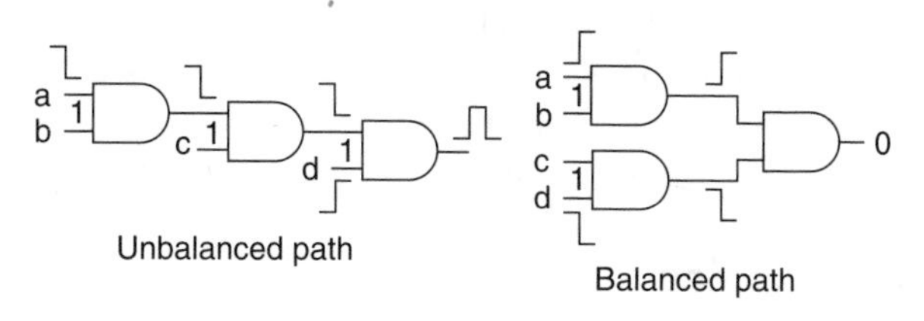

FIGURE 10.19 Path balancing for glitching reduction.

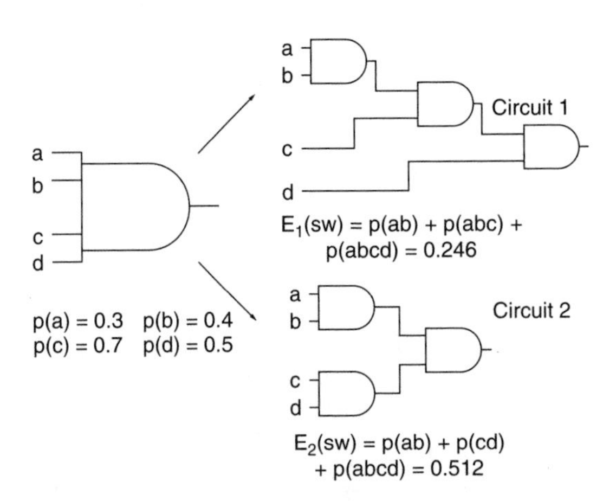

FIGURE 10.20 Technology decomposition for minimizing switching activity.

function $f = abcd$ is implemented in two alternative ways (i.e., chain structure and tree structure). In addition, notice that the tree implementation of function f provides glitches elimination, thus reducing effectively the total power dissipation.

Path balancing can be achieved before technology mapping by selective collapsing and logic decomposition or after technology mapping by delay insertion and pin reordering. The advantage of this technique is that by selectively collapsing the fan-ins of a node, the arrival time at the output of the node can be changed. Logic decomposition and extraction can be performed to minimize the level difference between the inputs of the nodes that are driving high capacitive nodes. Additionally, by inserting variable-delay buffers in a circuit, the delays of all paths in the circuit can be made equal. The issue in delay insertion is to use the minimum number of delay elements to achieve the maximum reduction in glitching activity. Path delays may sometimes be balanced by an appropriate signal to the pin assignment. This is possible, because the delay characteristics of CMOS gates vary as a function of the input pin that is causing a transition at the output.

10.6.2 Technology Decomposition

The next step during logic synthesis of a network is to convert the network to another, which only contains two-input AND/NAND and inverter gates. This step, named technology decomposition, is very useful for network synthesis and is carried out before the mapping of the network, according to the current cell library, takes place. Therefore, a decomposition scheme that minimizes the total switching activities of the network is a good starting point for power-efficient technology mapping.

Given the switching activity at each input of a node, Tsui et al. [28] suggested a technique for AND decomposition of this node, which reduces the total switching activity in the resulting two-input AND structure under a zero-delay model. The idea is to inject the high switching activity inputs into the decomposition model as late as possible, as shown in Figure 10.20, where two different decomposition structures for the four-input AND gate are depicted.

Note that signal d, which has the highest switching activity, is injected last in configuration A, thus implying better power performance for this configuration. This technique has been found as being optimal for dynamic CMOS circuits, but also produces very good results for static CMOS circuits. In general, the low-power technology decomposition procedure reduces the total switching activity in the circuits by 5% over the conventional balanced tree decomposition method.

10.6.3 Technology Mapping

Technology mapping refers to the process of binding a given Boolean network to the gates included in a target cell library. In Lin and Man [17], Tiwari et al. [27], and Tsui et al. [28], some design techniques

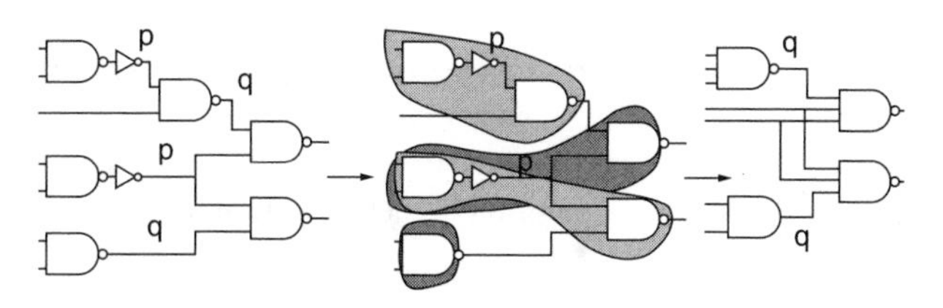

FIGURE 10.21 Technology mapping for minimizing switching activity.

for low-power consumption during technology mapping have been proposed. The main concept is to hide nodes with high switching activity inside the gates, thus they can drive smaller load capacitance, as presented in Figure 10.21.

According to Tsui et al. [28], the whole process consists of two steps. The first step requires the computation of power-delay curves (i.e., power consumption vs. arrival time) of all nodes in the network. The second step produces the mapping solution according to the previous curves and the required times at the primary inputs. This method has been proven to imply an 18% power savings at the expense of a 16% increase in area, without any penalty in network performance. In other words, we can say that the power-delay mapper reduces the number of high switching activity sub-networks at the expense of increasing the number of them having low switching activity. In addition, it reduces the network average load.

Although the approach mentioned previously refers to mapping for zero-delay circuits, an extension to a real-delay model is considered in Tsui et al. [28], resulting in optimum power solutions. According to [28], every point on the power-delay curve of a specific node uniquely defines a mapped subnet from the circuit inputs up to the node. The principle is to compute each such point with the probability waveform for the node in the corresponding mapped subnet. Thus, the total power cost, owing to steady-state transitions and hazards, of a candidate match can be calculated from the computed power-delay curves at the inputs of the gate and the power-delay characteristics of the gate itself.

10.7 Conclusions

The dynamic power consumption is a dominant power component for the current and future design technologies. Dynamic power substantially increases in nanometer technologies because of increased number of on-chip functions as well as a prolonging trend on getting higher clock frequencies. A multi-objective approach for reducing dynamic power consumption should combine multiple supply and threshold voltages with flexible gates from suitable cell libraries and efficient signaling schemes. Two design strategies can be adopted to reduce dynamic power. The first strategy concerns the supply voltage reduction, where substantial power savings can be achieved due to its quadratic dependence (i.e., $P \propto V_{dd}^2$). The second strategy concerns the capacitance or switching activity reduction, which is very useful when the design process is fixed. Four different sets of low-power design techniques were presented. More specifically, circuit techniques based on the principle of parallelism, techniques that use multiple supply voltages and low on-chip voltage swing, and techniques that are circuit technology-dependent and technology-independent. The key challenges to using multiple voltage supplies on a chip are minimizing area cost, placing logic cells under appropriate clustering constraints, as well as using dual power rails and efficient cell libraries that are capable of assigning the appropriate threshold voltage to each cell.

References

[1] V. Adler and E. Friedman, Repeater design to reduce delay and power in resistive interconnect, *Trans. Circuits and Systems — II,* Vol. 45, May 1998, pp. 607–616.

[2] M. Alidima, J. Monteiro, S. Devadas, A. Ghosh, and M. Papaefthimiou, Precomputation-based sequential logic optimization for low power, *IEEE Trans. on VLSI,* Vol. 2, No. 4, pp. 426–435, Dec. 1994.

[3] L. Benini, G. De Micheli, E. Macii, M. Poncino, and R. Scrasi, Symbolic synthesis of clock-gating logic for power optimization of synchronous controllers, *ACM Trans. on Design Automation of Electron. Syst.,* Vol. 4, No. 4, pp. 351–375, Oct. 1999.

[4] L. Benini, G. De Micheli, A. Lioy, E. Macii, G. Odasso, and M. Poncino, Synthesis of power-managed sequential components based on computation kernel extraction, *IEEE Trans. on CAD,* Vol. 20, No. 9, pp. 1118–1131, Sept. 2001.

[5] L. Benini, P. Siegel, and G. De Micheli, Saving power by synthesizing gated clocks for sequential circuits, *IEEE Design Test of Comput.,* pp. 32–41, Winter 1994.

[6] L. Benini and G. De Micheli, Automatic Synthesis of low-power gated-clock finite-state machines, *IEEE Trans. on CAD,* Vol. 15, No. 6, pp. 630–643, June 1996.

[7] T. Burd and R.W. Brodersen, *Energy Efficient Microprocessor Design,* Kluwer Academic Publishers, Boston, 2002.

[8] A.P. Chandrakasan and R.W. Brodersen, *Low-Power Digital CMOS Design,* Kluwer Academic Publishers, Boston, 1995.

[9] W. Chung, T. Lo, and M. Sachdev, A comparative analysis of low-power low-voltage dual-edge triggered flip-flops, *Trans. on VLSI Syst.,* Vol. 10, No. 6, Dec. 2002, pp. 913–918.

[10] R. Colshan and B. Jaroun, A novel reduced swing CMOS BUS interface circuit for high-speed low-power VLSI systems, *Proc. of Int. Symp. on Circuits and Syst. (ISCAS),* 30 May 1994, London, UK, Vol. IV, pp. 351–354.

[11] J. Goodman and A.P. Chandrakasan, Low-power scalable encryption for wireless systems, *Wireless Networks,* 4, 1998, pp. 55–70.

[12] M. Hiraki et al., Data-dependent logic swing internal bus architecture for ultra low-power LSI's, *IEEE J. Solid-State Circuits,* Vol. 30, Apr. 1995, pp. 397–402.

[13] R. Hossain, L. Wronski, and A. Albicki, Low-power design using double edge triggered flip-flops, *Trans. on VLSI Syst.,* Vol. 2, No. 2, June 1994, pp. 261–265.

[14] J.P. Hayes, *Computer Architecture and Organization,* McGraw-Hill, New York, 1978, p. 382.

[15] T. Kuroda, Low-power CMOS design challenges, *IEICE Trans. on Electron.,* Vol. E84-C, Aug. 2001, pp. 1021–1028.

[16] H.-J. Kwon and K. Lee, A new division algorithm based on lookahead of partial-remainder (LAPR) for high-speed/low-power coding applications, *IEEE Trans. of CAS-II,* Vol. 46, No. 2, Feb. 1999, pp. 202–209.

[17] B. Lin and H. De Man, Low-power driven technology mapping under timing constraints, *Proc. ICCAD,* 1993, pp. 421–427.

[18] R.P. Llopis and M. Sachdev, Low-power, testable dual-edge triggered flip-flops, *Proc. Int. Symp. Low-Power Electronics and Design,* 1996, pp. 341–345.

[19] M. Lowy, Low-power spread spectrum code generator based on parallel shift registers, *1994 IEEE Symp. on Low-Power Electron.,* San Diego, CA, Oct. 10–12, 1994, pp. 22–23.

[20] J. Monteiro, S. Devadas, and A. Ghosh, Retiming sequential circuits for low power, *Proc. ICCAD,* Nov. 7–11, Santa Clara, CA, pp. 398–402, 1993.

[21] G. Panigrahi, The implications of electronic serial memories, *Computer,* July 1977, pp.18–25.

[22] M. Pedram, Q. Wu, and X. Wu, A new design of double-edge triggered flip-flops, *Proc. ASP-DAC '98 Asian and South Pacific Design Automation Conf.,* Feb. 10–13, 1998, Yokohama, Japan, pp. 417–421.

[23] C. Piguet, J.-M. Masgonty, V. von Kaenel, and T. Schneider, Logic design for low-voltage/low-power CMOS circuits, *1995 Int. Symp. on Low-Power Design,* Dana Point, CA, Apr. 23–26, 1995, pp. 117–122.

[24] C. Piguet, Logic design for low-power CMOS circuits. Invited talk at TENCON '95, Hong-Kong, Nov. 7–10, 1995, pp. 299–302.

[25] T. Schneider, V. von Kaenel, and C. Piguet, Low-voltage/low-power parallelized logic modules, *Proc. PATMOS '95,* Paper S4.2, Oldenburg, Germany, Oct. 4–6, 1995, pp. 147–160.

[26] A. Antonio, G.M. Strollo, E. Napoli, and C. Cimino, Analysis of power dissipation in double edge-triggered flip-flops, *Trans. on VLSI Syst.*, Vol. 8., No. 5, Oct. 2000, pp. 624–629.

[27] V. Tiwari, P. Ashar, and S. Malik, Technology mapping for low-power in logic synthesis, *Integration, the VLSI J.*, July 1996.

[28] C.-Y. Tsui, M. Pedram, and A. Despain, Power-efficient technology decomposition and mapping under extended power consumption model, *IEEE Trans. on CAD,* Vol. 13, No. 9, Sept. 1994.

[29] K. Usami and M. Horowitz, Clustered voltage scaling technique for low-power design, *Proc. Int. Symp. on Low-Power Design,* Apr. 1995, pp. 3–8.

[30] K. Usami and M. Igarashi, Low-power design methodology and applications utilizing dual supply voltages, *Proc. Asia and South Pacific Design Automation Conf.*, Jan. 25–28, 2000, Yokohama, Japan, pp. 123–128.

[31] H. Yamauchi et al., An asymptotically zero power charge–recycling bus architecture for battery-operated ultra-high data rate ULSIs *IEEE J. Solid-State Circuits,* Vol. 30, Apr. 1995, pp. 423–431.

[32] H. Zhang, G. Varghese, and J. Rabaey, Low-swing on-chip signaling techniques: effectiveness and robustness, *Trans. on VLSI Syst.,* Vol. 8, No. 3, June 2000, pp. 264–272.

11

VHDL for Low Power

Amara Amara
ISEP

Philippe Royannez
Texas Instruments

11.1 Introduction

The purpose of this chapter is to provide front-end designers with guidelines and good design practices for writing efficient register transfer logic (RTL) code from a low-power standpoint. It is suitable for engineers that are already familiar with RTL coding for synthesis, but are not necessarily aware of low-power techniques.

RTL-level techniques are very efficient because hardware description language (HDL) programmers are knowledgeable about the circuit architecture and functionality. They can ask and answer questions such as: Why should we clock a register if the input data has not changed? Why should we update a data in a register or on a heavily loaded bus if this data is not used by anybody? A lot of power can be saved based on this information, but obviously, this can not be observed at the standard cell library level or at the technology level. It must be done at the RTL level.

After a brief reminder of basic coding rules and techniques, we address different types of blocks like operative parts or control logic. For each type of block, we introduce the appropriate techniques including clock gating, finite-state machine (FSM) state assignment, bus encoding, and conditional computing to optimize RTL code and to obtain a significant power consumption reduction after synthesis. In this chapter, we only consider the case of RTL synthesis for silicon complementary metal oxide semiconductor (CMOS) digital circuits. All examples are coded in VHDL, and synthesis script examples use synopsys DC.

0-8493-1941-2/05/$0.00+$1.50

11.2 Basics

11.2.1 Power Consumption

RTL synthesis has been a major improvement in the field of digital integrated circuit (IC) design. Besides the higher reusability and the better verification methodology, RTL-based design has definitely changed the way designers consider a digital circuit. Indeed, most of them do not see a circuit as a netlist of electronic components anymore but instead as a high-level software functional description. In that sense, RTL coding has improved the productivity, but, on the other hand, it has introduced a major disconnect between the front-end design and the real electronic devices where the power is burnt. For many years, however, this has not been a major issue. RTL synthesis was mostly timing driven with several iterations to optimize area and fix timing violations. Power was not a real concern, but with CMOS process scaling and ever increasing switching speeds, the power density can reach tens of W/cm^2 in today's digital ICs. Moreover, the exploding market of the portable electronic devices is also driving very strongly the need for low-power solutions. Therefore, the problem must also be tackled at the RTL stage, and RTL programmers cannot ignore power anymore. In particular, they need to bear in mind the underlying circuitry where it is consumed. Thus, let us briefly summarize the sources of power consumption.

Power is either static or dynamic. The static power consumption is due to MOS sub-threshold leakage and to a lesser degree extends to gate-induced drain leakage (GIDL), gate leakage, and diode leakage. The static current consumption used to be in the range of μA. Without leakage-reduction techniques, it is now in the range of mA and, in some cases, can account for more than 50% of the total power consumption. Temperature makes the picture even worse.

The dynamic power is due to the switching of CMOS gates. Ideally, this power is the well-known $P_D = \alpha CV^2 f$, where α is the switching activity. This includes the clock distribution network consumption and the parasitic power due to glitches. Often neglected, however, the latter can account for up to 15%.

Metal-oxide semiconductor (MOS) transistors do not switch instantaneously, and signals do not have zero transition time. There is always a short amount of time where both the pull-up and pull-down paths of a CMOS gate are on simultaneously, thus creating a parasitic current that is wasted. This additional power consumption sometimes called "short-circuit" power depends on the input and output transition times, on the output capacitance load and on the transistor characteristics. P_{Short} can account for 10% of the total dynamic power. Unlike the P_D term, this consumption decreases with the output capacitance load.

11.2.2 RTL Coding Applicability to Power Reduction

Static power-reduction techniques include multiple V_T, multiple V_{DD}, back biasing, and power supply scaling or switching, among others. These techniques affect the MOS process technology or the global system architecture. They are not really implemented at the RTL level. Thus, static power reduction is out of the scope of this chapter.

The dynamic power $P_D = \alpha CV^2 f$ can be optimized by reducing each factor. The power supply, V, as well as the operating frequency, f, is not handled at the RTL level, and voltage and frequency scaling are covered by other chapters of this book. The capacitance factor, C, is mostly dependent on the process technology and the standard cell library. The fan-out can be controlled by the dc_shell command set_max_fanout, but RTL code has a limited impact on the C factor. Thus, the RTL techniques will be used mainly to reduce the switching activity α and, to a lesser extent, the C factor.

As far as P_{Short} is concerned, this wasted power can be minimized by controlling the rise and fall times. For instance, we can use the set_max_transition directive for synthesis and check for post-synthesis reports as well as post-backend report. This constraint is not specific to power reduction. Slow nodes affect signal integrity, reliability, and timing. They should be avoided anyway. RTL coding has very little impact on this part of the power consumption.

To summarize, RTL level techniques are not applicable to reduce every type of power consumption. Therefore, the techniques presented in this chapter focus on dynamic power consumption reduction. This reduction will be mainly due to a better management of the switching activity.

The various strategies to reduce this switching activity often use the same types of basic techniques that are reviewed briefly in the next subsections.

11.2.3 Latch Inference

Latch insertion is a common practice to suppress or reduce unnecessary switching; however, latches, as memory elements, can cause race condition, logic hazard, or metastability. It is mandatory that the latch-enable command signals are spike-free, and have the appropriate setup and hold margins with regard to the data inputs. It is also recommended that these latches be initialized with the common hardware reset used by the other sequential elements of the block.

To infer a latch, it is recommended that either an incomplete IF clause or an explicit instantiation is used (see the VHDL example next).

```
EN_T <= EN or scan_mode;
LT: process (RST, EN_T)
begin
if (RST='1') then
       Q<='0';
elsif (EN_T='1') then
       Q<=D;
end if;
end process LT;
```

RTL designers should keep in mind that latch insertion can affect the testability, the static timing analysis (STA), and the equivalence checking. To avoid unwanted latches, check that your sensitivity lists are complete, that IF and CASE statements have default clauses, and that the variable contents are always initialized before being used. In addition, check the inference reports and the postsynthesis results against the report_reference and the all_registers options.

11.2.4 Direct Component Instantiation

Design for low power might require a very predictable and reproducible synthesis mapping. It is therefore very common to directly instantiate in the RTL code-specific cells, such as metal programmable delay buffers and clock-gating cells of clock tree buffers; however, this is at variance with the RTL reusability principle. To avoid this problem, it is a good practice to keep these direct inferences in some separate wrapper around the reusable RTL code. A generic component name should be used in the architecture part of the VHDL code. Then, for each target library, the explicit component can be defined in the specific configuration part of the VHDL description.

```
configuration my_block_cfg of my_block is
  for my_block_arch
  for INST0: generic_special_cell_name
  use configuration WORK.target_lib_special_cell_name_cfg
  end for;
  end for;
end my_block_cfg;
```

11.2.5 Explicit-State Encoding

In the RTL approach, the designer focuses on the high-level description and relies on the synthesis tool to implement the functionality. Most often, the FSM states or any symbols declared as enumerated types are encoded automatically. For low-power optimization, however, the designer might need to precisely control this encoding. This can be done either at the synthesis scrip or at the RTL level. With synopsys DC, the encoding is controlled by the set_fsm_state_vector and set_fsm_encoding commands.

To define the encoding at the RTL level, various synthesis tools support several attributes, but because there is no standardization, we recommend avoiding the enumeration type, such as that in the following portable code:

```
type state_type is std_ulogic_vector(1 downto 0);
constant S0: state_type :="01";
constant S1: state_type:="10";
signal curr_st, next_state : state_type;
```

11.3 Glitch Reduction

Glitches are due to converging combinatorial paths with different propagation delays. Let us consider the simple example depicted in Figure 11.1. A 32-bit adder is followed by an XOR-tree that counts the number of "1" in the sum and gives the parity. In the case of the -1 +1 addition the parity bit will oscillate many times before stabilizing to the valid state "0." Because the final and initial results are the same, all activity on the output node brings no information yet consumes both dynamic and short circuit power. This oscillation can also propagate to other combinatorial blocks and generate activity that is even more spurious. This propagation will stop either with a sequential element or by pulse swallowing once the transition times will become shorter than the intrinsic gate propagation delay. Glitches are not an issue for power consumption only. Because of the parasitic capacitive coupling, glitches also affect the signal integrity and the timing closure with effects like dynamic cross talk of driver weakening.

11.3.1 Gate-Level Control

To interrupt the propagation of glitches, a first idea is to pipeline the design, but this very efficient method comes at the expense of additional registers, latency, control logic, and clock tree distribution. The clock tree and registers will consume both static and dynamic power, and the designer must find the best compromise. Moreover, pipelining is not always possible because it delays the output data delivery by one or more clock cycles. In some cases, the architecture change requirements can go up to the compiler and the real time operating system (RTOS), which is often not possible. An intermediate solution is to subdivide a clock cycle into two or more phases. The multiple phase clocks can be used to mask the datapath signals with simple AND gates or with latches. A common implementation is the well-known two-phase master–slave latch logic; however, the overhead in terms of clock generation and distribution, static timing analysis, and design complexity must be carefully evaluated before using such complex clocking schemes.

Another approach consists in balancing the delay between different combinatorial paths. Delay cells are directly instantiated in the RTL and fine-tuned at the place and route step. Because delays vary with process variation and temperature, this method is difficult to implement. With CMOS device scaling and advanced technologies, those techniques are not recommended except for some full-custom, high-performance blocks.

Glitches activity can also be reduced by using sum of products style of Boolean equations. It is even more convenient to handle it at the synthesis step using the flattening options of the synthesis tool with, for instance, set_flatten true. These options prioritize the speed (i.e., logic depth) and thus reduce glitches; however, it is, in most cases, at the expense of area and dynamic power consumption.

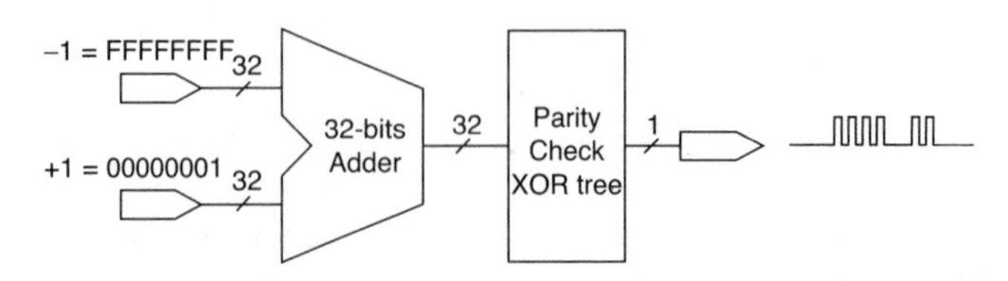

FIGURE 11.1 Example of glitches generation.

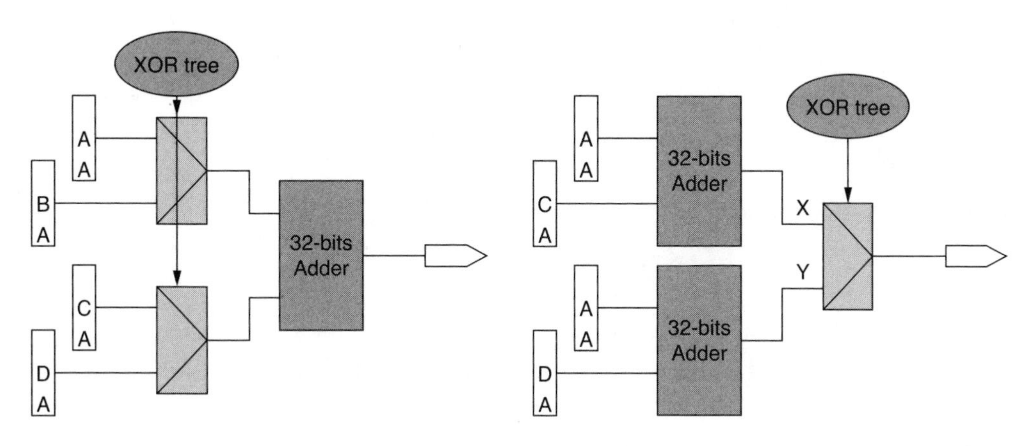

FIGURE 11.2 Glitch reduction by block reordering.

In case of a high-speed logic block, we can also use a monotonic design like domino logic. This type of logic is glitch-free, but requires a dedicated library of cells and an additional clock signal. To map the RTL code, we can again use direct instances or synthesis scripts to control the inferences (e.g., set_dont_use and set_use_only).

11.3.2 Block-Level Control

The last technique presented consists of rearranging the logic structure. To illustrate this approach, let us consider the block diagram in Figure 11.2. We use the previous XOR tree to select either A or B as the first operand of a 32-bit adder and C or D as the second operand. Because A, B, C, and D come from registers, they are stable data; but if the control signal of the multiplexers is oscillating, then the operands of the adder are unstable and propagate glitches which consume power. If we use two adders to compute X and Y sums first and then multiplex them, then adders see stable inputs and have much less power due to glitches. This reduction comes at the expense of one additional 32-bit adder block. In addition, note that synthesis tools are able to detect the two adders and, after a resource allocation step, could move back to the single-adder structure. To prevent this, the set_dont_touch attribute on net X and Y might be useful.

11.4 Clock Gating

Clock gating, which is probably one of the most well-known low-power techniques, is very effective in reducing the power consumption in digital circuits. The goal of this technique is to disable or suppress transitions from propagating to parts of the clock path (i.e., flip-flops, clock network, and logic) under a certain condition computed by clock-gating circuits. The savings are mainly due to the switching capacitance reduction in the clock network and the switching activity in the logic fed by the storage elements because unnecessary transitions are not loaded when the clock is not active.

Clock gating (CG) is illustrated in Figure 11.3. A block CG, which inhibits the clock signal when the idle condition is true, is associated with each sequential functional unit.. The clock signal is computed by function F_{cg}. CLK is the system clock and CLKG the gated clock of the functional unit. Clock-gating techniques have been successfully implemented in many microprocessors [1,2].

11.4.1 Flip-Flop-Based Design

Many implementations have been proposed for function CG. The simplest one uses an AND or an OR gate (Figure 11.4(a) and Figure 11.4(b)), but is not efficient because of the possible spikes at the output of the gate. An alternative and better solution is given in Figure 11.5 and is based on a latch L transparent

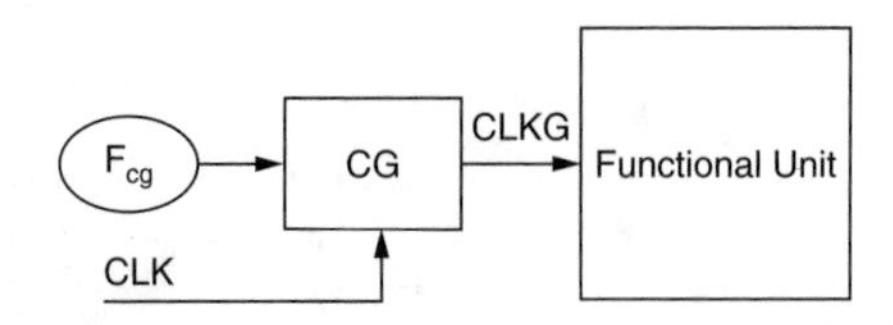

FIGURE 11.3 Clock-gating principle.

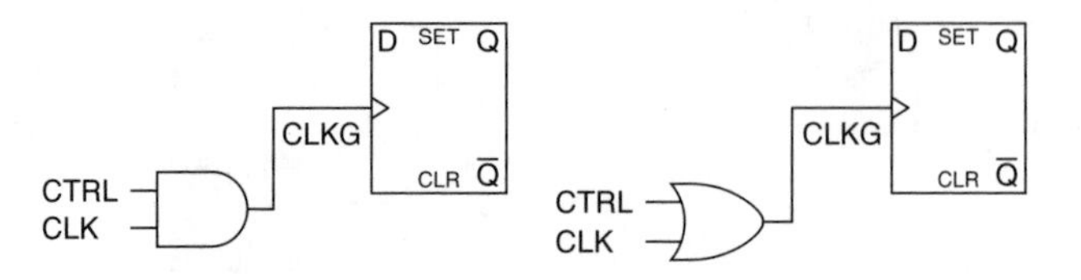

FIGURE 11.4 AND/OR CG block implementation.

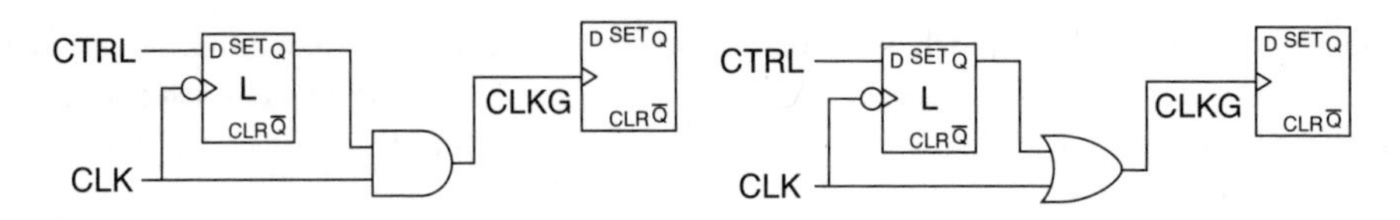

FIGURE 11.5 LATCH/AND/OR CG block implementation.

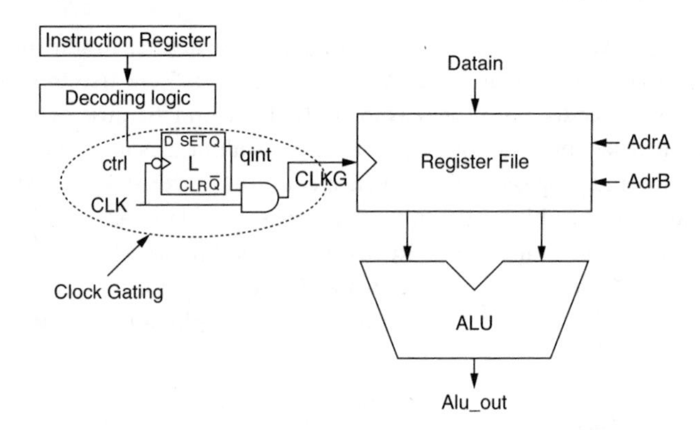

FIGURE 11.6 Clock gating example.

(when the clock is low) and an AND gate. With this configuration, the spurious transitions generated by function F_{cg} are filtered.

As an example, the circuit given in Figure 11.6 illustrates the clock gating of a datapath register file.

The clock-gating file and the register file must be physically close to reduce the impact on the skew and to prevent unwanted optimizations during the synthesis phase. They can be modeled by two separate processes in the same hierarchical block, synthesized, and then inserted into the parent hierarchy with a "don't touch" attribute. The following VHDL code describes the file register and its clock-gating circuit:

```
library ieee;
use ieee.std_logic_1164.ALL;
entity CG_RF_e is
port(
clk   : in std_logic;
adrA  : in std_logic_vector(4 downto 0);
adrB  : in std_logic_vector(4 downto 0);
datain: in std_logic_vector(7 downto 0);
```

```
wr    : in std_logic;
ctrl  : in std_logic;
A,B   : out std_logic_vector(7 downto 0));
end CG_RF_e;
architecture CG_RF_a of CG_RF_e is
signal clkg : std_logic;
type ram is array (0 to 31) of std_logic_vector(7 downto 0);
signal RF : ram;
begin
CG : Process (CLK, CTRL)
variable qint : std_logic;
begin
if clk = '0' then qint := ctrl; end if;
     CLKG <= (not Qint) and CLK;
end process;
process (CLKG)
begin
if CLKG = '1' then
          if WR ='1' then
               RF(conv_integer(adrA)) <= datain;
          else
               A <= RF(conv_integer(adrA));
               B <= RF(conv_integer(adrB));
          end if;
end if;
end process;
end;
```

In some applications, conditionally executed parts of the VHDL code can be identified and separated. Clock gating can be applied for each part. This technique has been proposed by Raghavan et al. [3]. The following VHDL code gives an example illustrating this technique. The initial process (P0) in the architecture listing2_1_a has been transformed into three processes (i.e., P1, P2, P3) as depicted in architecture listing2_2_a. A glitch-free load signal c_load is generated and combined to the clock, clk, to generate the gated clock, clkg, to be used by process P2.

```
Library ieee;
use ieee.std_logic_1164.ALL;
entity listing2_e is
port(
clk   : in std_logic;
load  : in std_logic;
A, B, C, E: in std_logic_vector(7 downto 0);
X, D, Z : out std_logic_vector(7 downto 0));
end listing2_e;
architecture listing2_1_a of listing2_e is
begin
P0 : process (clk)
begin
if (clk'event and clk='1')then
          X <= A + B;
          D <= E;
          if (load='1') then
```

```
                Z <= C;
            end if;
end if;
end process;
end;
architecture listing2_2_a of listing2_e is
signal Gclk : std_logic;
begin
P1 : process (clk)
begin
    if clk'event and clk='1' then
        X <= A + B;
        D <= E;
    end if;
end process;
P2 : process (Gclk)
begin
    if Gclk'event and Gclk='1' then
        if (load='1') then
            Z <= C;
        end if;
    end if;
end process;
P3: process (clk, load)
Variable c_load: std_logic;
begin
if clk = '0' then
        c_load <= load;
end if;
Gclk <= clk and c_load;
end process;
end;
```

In some designs, enabled flip-flops are used as shown in Figure 11.7(a). It is well-known that this kind of flip-flops are area and power-consuming, but their advantage compared with gated-clock-based design is that testability can be easily implemented and clock skew is more manageable. This kind of structure can be easily transformed into a gated clock structure (see Figure 11.7(b)). It is noteworthy that this transformation leads to important savings in area and power consumption. The following VHDL code gives the description of enabled flip-flop and its corresponding gated clock version:

```
library ieee;
use ieee.std_logic_1164.ALL;
entity EDFF_e is
```

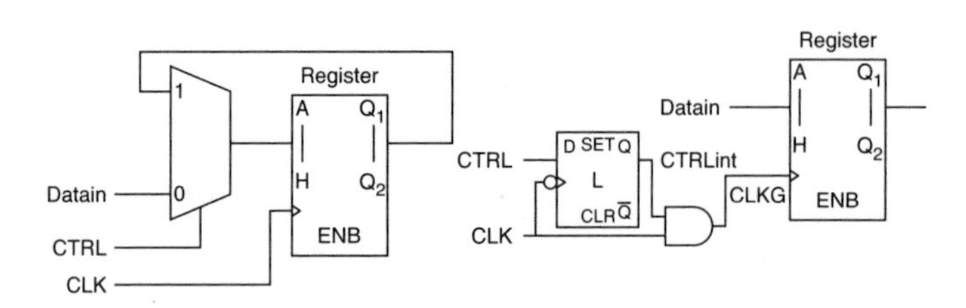

FIGURE 11.7 Enabled (a) to gated clock transformation (b).

```
port(
clk   : in std_logic;
ctrl  : in std_logic;
datain: in std_logic_vector(7 downto 0);
Q     : out std_logic_vector(7 downto 0));
end EDFF_e;
architecture EDFF_a of EDFF_e is
begin
process (clk)
begin
if clk'event and clk = '1' then
          if (ctrl='1') then
                Q <= datain;
          end if;
end if;
end process;
end;
library ieee;
use ieee.std_logic_1164.ALL;
entity CG_DFF_e is
port(
clk   : in std_logic;
ctrl  : in std_logic;
datain: in std_logic;
Q     : out std_logic);
end CG_DFF_e;
architecture CG_DFF_a of CG_DFF_e is
signal clkg : std_logic;
begin
process (clkg)
begin
if clkg'event and clkg = '1' then
          Q <= datain;
end if;
end process;
process (clk, ctrl)
variable ctrl_int: std_logic;
begin
if clk = '0' then
          ctrl_int := ctrl;
end if;
clkg <= clk and ctrl_int;
end process;
end;
```

11.4.2 Issues in Clock Gating of DFF-Based Design

11.4.2.1 Timing Issues

The clock gate (i.e., AND or OR) must not alter the waveform of the clock other than turning the clock on or off. Unfortunately, introducing clock gating may result in setup time or hold time violations. Moreover, in most power design flows [4,5], the clock gating is inserted before the clock tree synthesis.

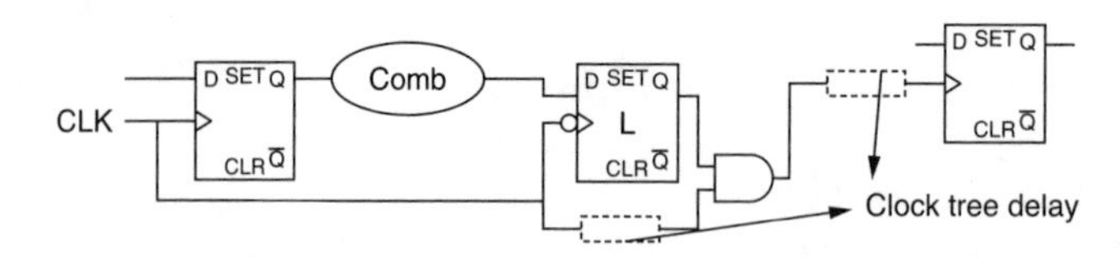

FIGURE 11.8 Timing issues in clock gating.

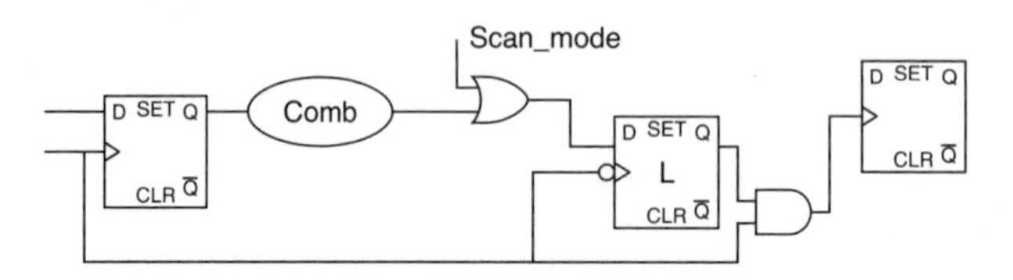

FIGURE 11.9 Testability issues in clock gating.

The existing synthesis tools by setting some variables allow the designer to specify these critical times before synthesis. When choosing these times, the designer has to estimate the delay impact of the clock tree from the clock gate to the gated register as depicted in Figure 11.8.

11.4.2.2 Testability Issues

Clock gating introduces multiple clock domains in the design, and this will affect the testability of the circuit. One way to improve the testability of the design is to insert a control point, which is an OR gate as indicated in Figure 11.9, controlled by an additional signal scan_mode. Its task is to eliminate the function of the clock gate during the test phase and thus restores the controllability of the clock signal.

11.4.2.3 Computer-Aided Design (CAD) Issues

Determining which flip-flops should be grouped for clock gating is an issue. Two techniques have been proposed:

1. Hold condition detection [6]. Flip-flops that share the same hold condition are detected and grouped to share the clock-gating circuitry. This method is not applicable to enabled flip-flops.
2. Redundant-clocking detection [7]. The method is simulation-based. Flip-flops are grouped with regard to the simulation traces to share the clock-gating circuitry. It is obvious that this method cannot be automated.

11.4.3 Latch-Based Design

In some applications, latch-based designs are preferred to D Flip Flop (DFF)–based designs. The basic concept is that a DFF can be split into two latches, and each one is clocked with an independent clock signal. The two clocks are nonoverlapping clocks as presented in Figure 11.10. Combinational network is usually inserted between the two latches to build a pipelined datapath (Figure 11.11). The main advantage is that this kind of design supports greater clock skew before failing than a similar DFF-based design. The second advantage is that time borrowing is achieved naturally in the pipelined datapath.

The clock gating is easy to implement. Figure 11.11 depicts a simple and robust way to do it [8]. A simple AND gate is used to generate the gated clock. This configuration is glitch-free because the control

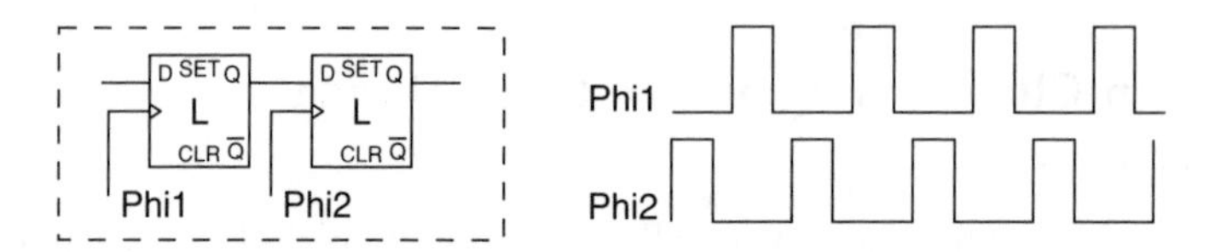

FIGURE 11.10 Master-slave latch and nonoverlapping clock concepts.

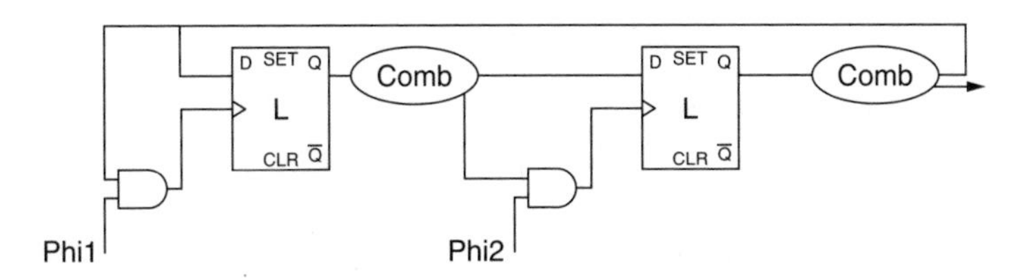

FIGURE 11.11 Clock gating of latch-based design.

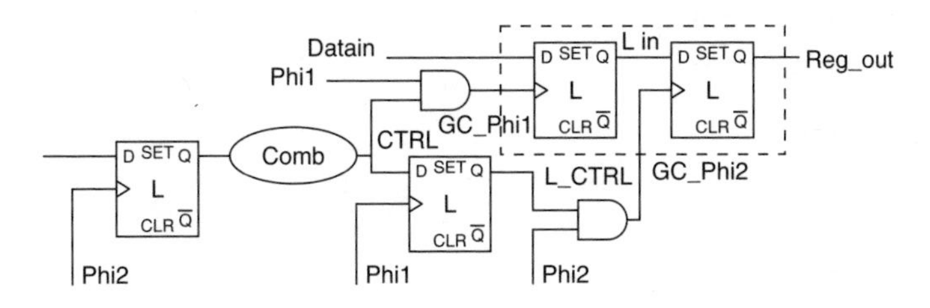

FIGURE 11.12 Clock gating in latch-based datapath.

signal, generated when Phi1 is high, is stable and remains stable when Phi2 goes high. In the case of registers, given the fact that the control signal is coming either from a latch clocked by Phi2 or a combinational function in which inputs are clocked by Phi2, it is necessary to add a latch clocked by Phi1 to delay the control signal as indicated in Figure 11.12 [9]. Notice that the AND gate must be handled carefully.

To prevent any optimization by the synthesis tool, the gate must be placed in a separate hierarchy level and assigned a "don't touch" attribute, for example. The code that follows this paragraph gives a VHDL description of a latch-based 32-bit register bank with the clock gating of Figure 11.12. Block CG contains the clock gating circuits (i.e., latch, 2 AND) in a separate hierarchy. This block can be assigned the "don't touch" attribute. Block RB contains the register bank, and, finally, GRB includes the structural description of the gated clock register bank.

```
Library ieee;
use ieee.std_logic_1164.ALL;
Entity RB is
Port (Phi1, Phi2 : in std_logic;
datain: in std_logic_vector(31 downto 0);
Reg_out: out std_logic_vector(31 downto 0));
End RB;
architecture Register_Bank of RB is
signal L_In: std_logic_vector(31 downto 0);
begin
process (GC_Phi1, datain)
begin
if (GC_Phi1 = '1') then
L_In <= datain;
end if;
end process;
process (GC_Phi2, L_In)
begin
if (GC_Phi2 = '1') then
Reg_out <= L_In;
end if;
end process;
```

```
end Register_Bank;
library ieee;
use ieee.std_logic_1164.ALL;
entity GC is
port(
CTRL, Phi1, Phi2: in std_logic;
GC_Phi1, GC_Phi2: out std_logic);
end GC;
architecture Gated_Clock of GC is
signal L_CTRL: std_logic;
begin
process (Phi1, CTRL)
begin
      If (Phi1 = '1') then
            L_CTRL <= CTRL;
end if;
end process;
GC_Phi1 <= CTRL AND Phi1;
GC_Phi2 <= L_CTRL AND Phi2;
end Gated_Clock;
library ieee;
use ieee.std_logic_1164.ALL;
entity GRB is
Port (
Phi1, Phi2 : in std_logic;
ctrl  : in std_logic;
datain: in std_logic_vector(31 downto 0);
Reg_out: out std_logic_vector(31 downto 0));
end GRB;
architecture Gated_Clock_Register_Bank of GRB is
component RB
port (Phi1, Phi2: in std_logic;
      datain: in std_logic_vector(31 downto 0);
      Reg_out: out std_logic_vector(31 downto 0));
end Component;
component GC
port (CTRL, Phi1, Phi2: in std_logic;
      CG_Phi1, CG_Phi2: out std_logic);
end Component;
signal GC_Phi1, GC_Phi2: std_logic;
begin
RB_instance: RB port map
      (GC_Phi1, GC_Phi2, datain, Reg_out);
GC_instance: GC port map
      (CTRL, Phi1, Phi2, GC_Phi1, GC_Phi2);
end Gated_Clock_Register_Bank;
```

11.4.4 Issues in Latch-Based Design

One of the design issues related to latch-based clock gating has been reported in Arm et al. [8]. In fact, the synthesis tool finds timing loops going through the control paths. As we can see from Figure 11.11,

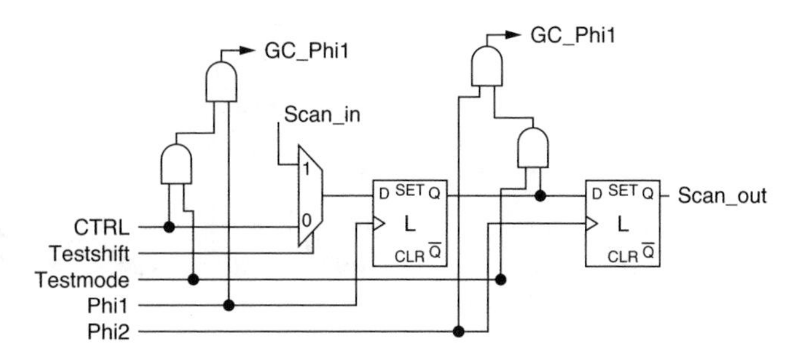

FIGURE 11.13 Clock-gating block with testability improvement.

the clock enable signal of each latch depends on the value computed by the logic fed by the other latches. As mentioned in Arm et al. [8], at least one of these paths can be cut using tool-dependent attributes (i.e., set-disable-timing for Synopsys design compiler).

The second design issue is related to testability. Figure 11.13 [9] illustrates the modification of clock-gating block to improve the testability of latch-based design.

11.5 Finite-State Machines

Finite-state machines (FSMs) are very common parts of digital systems. They are intensively used to generate signal sequences, to check an input signal sequence or to control datapath parts. The basic structure of an FSM is a state register and two logic blocks. The input (or "next-state") logic block computes the next state as a function of the current state and of the new set of inputs. The output logic block generates the outputs as a function of the current state (for a Moore FSM). The power can be burnt either in the logic blocks or in the clock distribution to the flip-flops of the state register. We present here various techniques to minimize this power consumption, using explicit state encoding and clock gating. The RTL coding of these techniques have been presented in the previous sections.

11.5.1 Gated-Clock FSM

The basic idea of gated-clock FSM is that it is not useful to have switching activity in the next-state logic or to distribute the clock if the state register will sample the same vector [20]. Let us take a simple, yet very common, example.

Figure 11.14 depicts a state machine that interacts with a timer-counter to implement a very long delay of thousands of clock cycles before executing a complex but very short operation (in the DO_IT state). We can use the clock-gating techniques to freeze the clock and the input signals as long as the ZERO flag from the time-out counter is not raised. This idea is efficient because this FSM spends most of the time in the WAIT state. It can be even more efficient if we assume that the FSM is used to control a very large datapath which outputs will not be used in the WAIT state. We can gate the clock or mask the inputs of this datapath and, therefore, avoid dynamic power consumption during all the countdown phases.

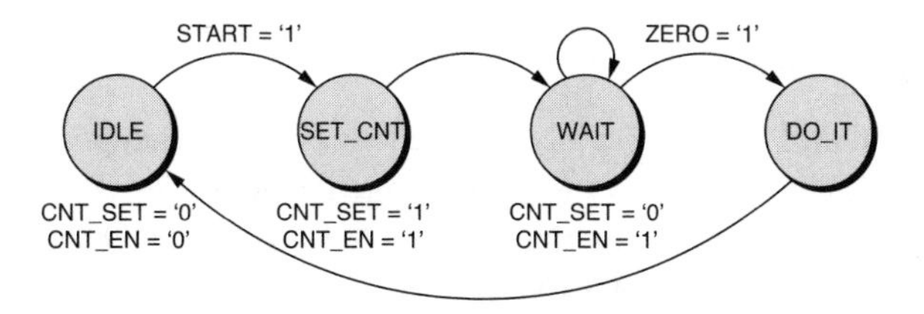

FIGURE 11.14 Example of FSM where the clock gating is easy and efficient.

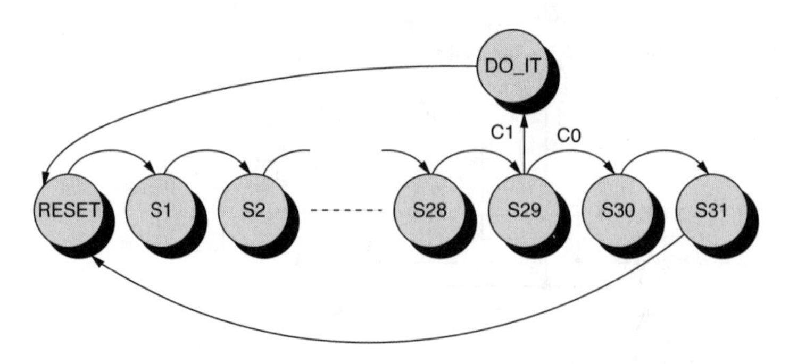

FIGURE 11.15 Example of FSM with gray encoding.

Software programmers are used to the 90/10 rule that states that 90% of the time a program loops over only 10% of the code. Actually, this rule is often valid for the FSM as well. It is the RTL designer's task to try to extract these small subparts of the FSM, isolate them, and then freeze the rest of the logic that is large and that most of the time does not achieve any useful computation.

11.5.2 State Encoding

This kind of technique uses the state encoding to reduce the logic activity in the input and output combinatorial blocks. One simple idea is to use an encoding that minimizes changes from one state to another if this transition is very likely to happen. In other words, we should minimize the hamming distance of the transition with high probability. This requires tools that propagate transition probabilities on the FSM inputs and calculate the probability of each transition. Again, we can make the parallel with software because this is similar to the branch prediction techniques. Although this probability estimation can be difficult, very common cases exist where this can be applied easily.

In the example depicted in Figure 11.15, states from RESET to S29 are chained sequentially with 100% probability of transition. Therefore, a gray encoding is the best choice. If we assume that condition C0 has a much lower probability than C1, the gray encoding should be not be incremented from S29 to S30 and S31.

However, what we gain in the next-state logic might be lost in the output logic activity. The designer has to find the best trade-off. If we consider now the power reduction on the output logic, we can also choose a judicious state encoding. A very common choice is the "one hot" encoding to optimize speed, area, and power for the output logic [19]. This approach is only valid for a small FSM (i.e., less than 8 to 10 states) because of the large state register and the increasing complexity of the next-state logic. For a larger FSM, a case-by-case analysis is needed. A good practice is to group states that generate the same outputs and assign them codes with minimum hamming distance.

A simple example is given in Figure 11.16, where an FSM is used to recognize the sequence "BEEFBEEF" on the input I and generates a flag Y = 1 if the sequence is complete. The encoding proposed achieves both a minimum "next-state logic" activity due to the "gray-like" encoding as well as no power consumption at all in the output logic because the orthogonal encoding defines the most significant bit of the state register as the flag Y itself [22].

11.5.3 FSM Partitioning

Often, FSM can be partitioned into smaller pieces. The idea here is to decompose a large FSM into several simpler FSMs with smaller state registers and combinatorial logic blocks. Only the active FSM receives clock and switching inputs. The others are static and do not consume any dynamic power [21]. Let us illustrate this technique with a simple example.

We consider a large FSM that includes a small subroutine, which is used very often in a real application scenario. We can easily partition the big FSM into two parts and isolate the subroutine loop. We add a

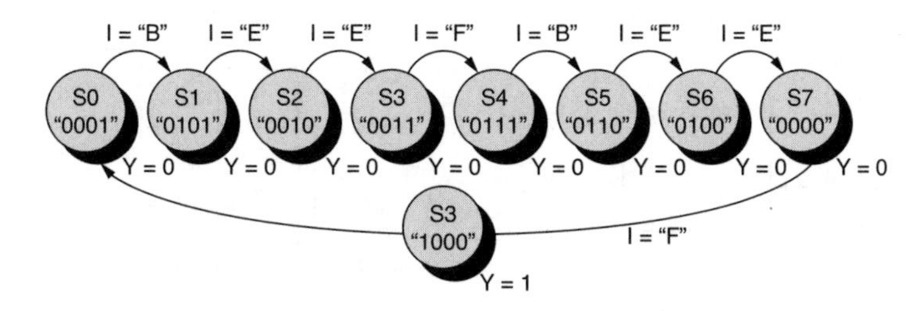

FIGURE 11.16 Example of FSM with zero-output logic.

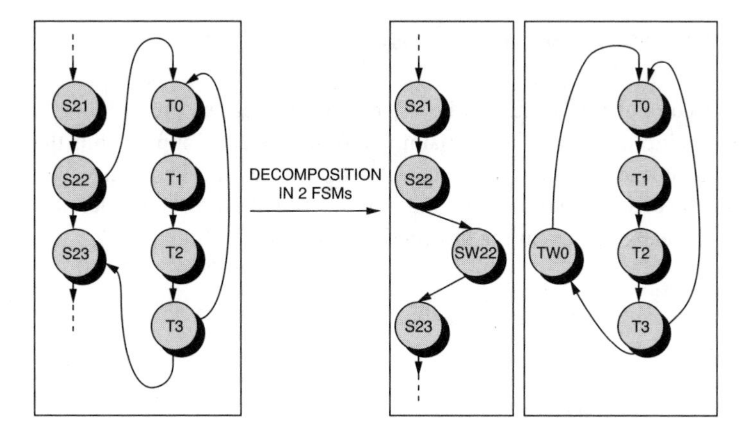

FIGURE 11.17 Example of FSM decomposition.

wait state, SW22 and TW0, between the entry and exit points of the subroutine in both FSMs. The new FSMs are mutually exclusive; when one is operating, the other one remains in its wait state. In such a state, clock and inputs can be gated to prevent any dynamic power (power supply could even be switched off to save leakage). The power savings is even higher if we can isolate very small subsets of states where the initial FSM remains most of the time.

11.6 Datapaths

An important amount of energy may be wasted in the datapath due to switching activity that does not contribute to the functionality of the circuit. Different techniques have been proposed to suppress or reduce dynamically this energy. Among these techniques, precomputation logic [10], guarded evaluation [11], and control-signal gating [12] techniques are widely used by low-power circuit designers. These techniques can be used early in the design flow (i.e., at the RTL level).

11.6.1 Precomputation Design Techniques

The principle of precomputation is to identify a logic condition on some inputs of a combinational circuit for which the output does not vary. Figure 11.18(a) gives a generic example of such a circuit. The inputs of the combinational logic f(X) are partitioned into precomputed inputs and gated inputs. If the output Y is independent of the gated inputs, then the function g generates a control signal for the register R2 that freezes its outputs. Yeap [13] describes a systematic method to derive the function g, but unfortunately, for given inputs, partitioning the solution is not unique, and the designer has to find the one that gives the best power-performance-area trade-off. Many implementation alternatives of precomputation logic are given in Yeap [13]. Figure 11.18(b) is a simple and realistic example of precomputation logic. It is a binary comparator that computes A > B (see the VHDL code that follows this paragraph).

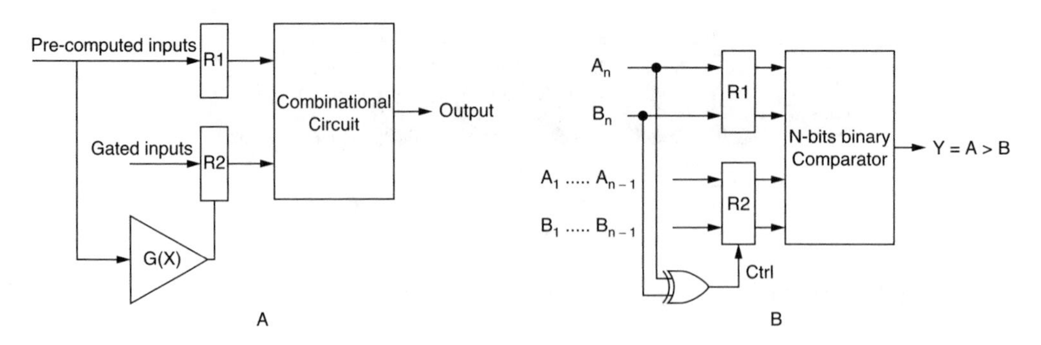

FIGURE 11.18 (a) Precomputation logic, and (b) application to a comparator.

In this case, the precomputation condition is very easy to derive. The precomputed inputs are A_n and B_n, and the most significant bits and the remaining bits are the gated inputs, thus the precomputation is a simple XOR gate. It is obvious that the designer needs to have some knowledge on the input statistics to apply efficiently the precomputation techniques. In practice, the selection of R1, R2, and the precomputation function depends heavily on the designer's experience.

```
Library ieee;
use ieee.std_logic_1164.ALL;
entity PC_Comp is
port (
A, B: in Std_Logic_Vector(31 downto 0);
Clk : in Std_Logic;
Y : out Std_Logic);
end PC_Comp;
architecture b32Comp of PC_comp is
signal Ctrl : std_logic;
signal A_R1, B_R1 : Std_Logic;
signal PC_A_R2, PC_B_R2 : Std_Logic_Vector(30 downto 0);
begin
Ctrl <= A(31) Xor B(31);
Y <= '1' when ((A_R1 & PC_A_R2) > (B_R1 & PC_B_R2)) else '0';
R1 : process (Clk)
begin
      if (Clk'event and Clk ='1') then
            A_R1 <= A(31);
            B_R1 <= B(31);
      end if;
end process;
R2 : process (Clk)
begin
      if (Clk'Event and Clk ='1') then
            if (Ctrl = '0') then
                  PC_A_R2 <= A(30 downto 0);
                  PC_B_R2 <= B(30 downto 0);
            end if;
      end if;
end process;
end b32Comp;
```

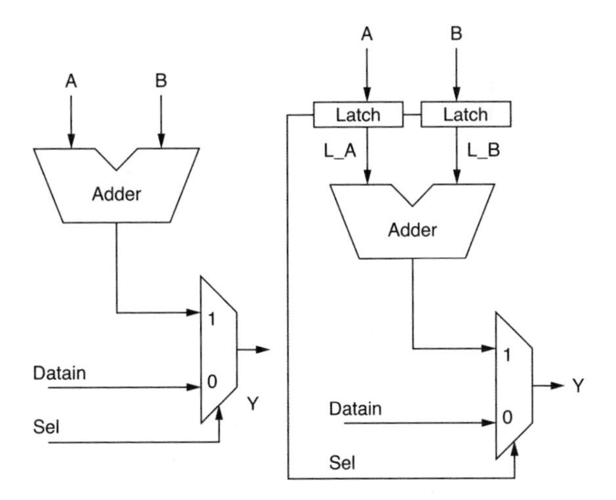

FIGURE 11.19 (a) Original circuit, and (b) its guarded evaluation version.

11.6.2 Guarded Evaluation Design Techniques

The technique is applicable to embedded combinational blocks from which outputs are in idle condition. Transparent latches are inserted at all the inputs of the embedded block. Control circuitry is added to determine the idle condition, which is then used to disable the latches. Figure 11.19 is a simple example that illustrates this technique. The arithemetic and logic unit (ALU) output may or may not be used depending on the condition selection of the multiplexer. If it is not used, the latches preserve the previous output values of the ALU. It is obvious that for wide buses, the area and power dissipation overhead should be nonnegligible. Following is the VHDL description of the Figure 11.19(b) circuit:

```
library ieee;
use ieee.std_logic_1164.ALL;
entity GE_Alu is
port(
A, B, Datain : in Std_Logic_Vector(31 downto 0);
Y : out Std_Logic_Vector(31 downto 0);
Sel : in Std_Logic);
end GE_Alu;
architecture Garded_Evaluation_Alu of GE_Alu is
signal L_A, L_B : Std_Logic_Vector(31 downto 0);
begin
process (Sel, A, B)
begin
      if Sel ='1' then
            L_A <= A;
            L_B <= B;
      end if;
end process;
Y <= (L_A + L_B) when (Sel ='1') else Datain;
end Garded_Evaluation_Alu;
```

11.6.3 Control-Signal Gating Design Techniques

The techniques we presented previously all aim to reduce the switching activity in a datapath module. The control-signal technique takes advantage of a fine granularity analysis to reduce the switching activity

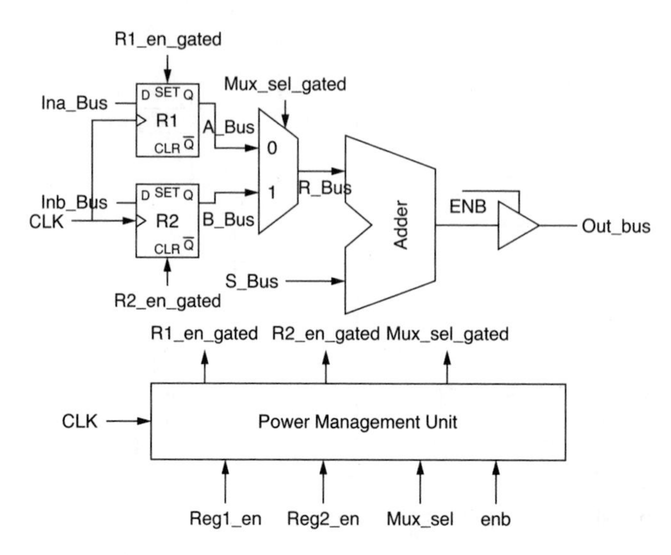

FIGURE 11.20 Control-signal gating technique.

in the datapath buses. The method is based on the observability don't care concept (ODC) [14] to detect when a bus is not used and to stop the propagation of the switching activity through the module(s) driving the bus. We can find in Kapadia et al. [12] a good formulation of the concept and its application to gating the datapath control signals. Figure 11.20 illustrates a datapath example. When enb is not active, mux_sel, reg1_en, and reg2_en can be gated, leading to a 100% switching activity reduction in R_Bus, A_bus, and B_Bus. When mux-sel is active, either reg1_en and reg2_en can be gated depending on the value of mux-sel. The gating conditions for the datapath of Figure 11.20 have been derived in Kapadia et al. [12] and are given next.

```
R1_en_gated = reg1_en AND (not(mux_sel OR (not enb)))@(T+1)
R2_en_gated = reg2_en AND (not(not mux_sel OR not enb))@(T+1)
(mux_sel_gated) @T = (mux_sel_gated) @(T-1) if ((not enb) @(T+1) = = True)
```

The suffix @T means the value of a variable or a function at the current clock cycle, @T-1 is the value one clock cycle before, and, finally, @T+1 is the value at the next clock cycle.

These equations can be implemented in a power management unit as depicted in Figure 11.20. The power management unit (PMU) generates all the gated control-signals for the datapath. The VHDL description of the PMU is given next.

```
library ieee;
use ieee.std_logic_1164.ALL;
entity PMU is
port (
Reg1_en, Reg2_en, Mux_sel, Enb, Clk : in Std_Logic;
R1_en_gated, R2_en_gated, Mux_sel_gated : out Std_Logic);
end PMU;
architecture Power_Management_Unit of PMU is
signal Enb_int, R1_en_tmp, R2_en_tmp : Std_Logic;
begin
R1EG : process (Clk)
begin
if (Clk'Event and Clk='1') then
R1_en_tmp <= NOT(mux_sel OR (NOT Enb));
end if;
```

```
R1_en_gated <= R1_en_tmp AND reg1_en;
end process;
R2EG : process (Clk)
begin
if (Clk'Event and Clk='1') then
R2_en_tmp<= NOT(NOT(mux_sel) OR (NOT Enb));
end if;
R2_en_gated <= R2_en_tmp AND reg2_en;
end process;
MSG : process (Clk)
begin
if (Clk'Event and Clk='1') then
Enb_int <= NOT Enb;
            if (Enb_int = '0') then
mux_sel_gated <= mux_sel;
            end if;
end if;
end process;
end Power_Management_Unit;
```

11.7 Bus Encoding

Advanced systems are typically characterized by wide and long buses, which consume a large amount of power mainly due to large capacitance and a significant switching activity. Many techniques have been proposed to deal with this issue at different design levels: low-swing bus [15], charge recycling bus [16], bus pipelining [17], bus multiplexing, and bus encoding techniques. The latter are more suitable for VHDL coding for low power. We will focus on one of them and briefly present other bus encoding techniques.

11.7.1 Bus Invert Encoding

Bus invert encoding is suitable for a set of parallel and synchronous signals such as internal buses in modern system on chip (SoC) architectures [18]. The idea behind is very simple: Before sending the data, the emitter compares its current value with the previous one and decides whether to send it or to send its inverted value along with a polarity signal. A bank of XOR gates at the sending and receiving ends inverts the bus data if necessary. Figure 11.21 depicts the bus encoding architecture, and the corresponding VHDL code is given next.

```
library ieee;
use ieee.std_logic_1164.ALL;
```

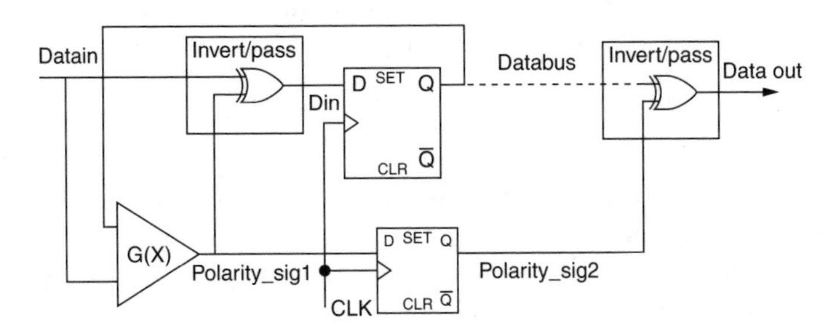

FIGURE 11.21 Bus invert encoding scheme.

```
entity BIE is
port(
Datain : in Std_Logic_Vector(31 downto 0);
Dataout : out Std_Logic_Vector(31 downto 0);
CLK : in Std_Logic);
end BIE;
architecture Bus_Invert_Encoding of BIE is
signal Din, Databus: Std_Logic_Vector(31 downto 0);
signal polarty_sig1, polarity_sig2: Boolean;
begin
Polarity_sig1 <= Polarity_Gen(datain, databus);
– Polarity_Gen is a procedure that returns True
– if the bus value has to be inverted
process (CLK)
begin
     if CLK'Event and CLK = '1' then
          if Polarity_sig1(datain,databus) == True; then
Databus <= not Datain;
else
Databus <= Datain;
          end if;
          polarity_sig2 <= polarity_sig1;
     end if;
end process;
Dataout <= polarity_sig2 xor Databus;
end Bus_Invert_Encoding;
```

11.7.2 Other Bus Encoding Techniques

In the bus invert approach, we have to calculate the hamming distance between two consecutive codes and transmit an additional polarity signal, which changes the interfaces between the emitter and the receiver. An improvement consists of making the guess that if the most significant bit (MSB) is "1," the inverted code should be transmitted. The MSB = "1" can be used as the polarity information. This technique is efficient when the vector to transmit is a 2's complement arithmetic data bus, with MSB being the sign bit.

Another common technique takes advantage of the fact that, very often, the value transmitted on a bus (an address bus, for instance) is simply the previous value with an increment. Therefore, the lines can remain the same (i.e., no power consumption) as long as the codes are consecutive, which is mentioned to the receiver by an additional control signal.

Finally, we can also use the knowledge of the set of symbols or probability of sequences to encode them in order to reduce the switching activity on the bus. For instance, if the sequence "0101" ⇒ "1010" occurs 90% of the time, we can save power by recoding "0101" into "0000" and "1010" into "0001."

11.8 Conclusion

The RTL coding step is not too early in the design flow to address power consumption optimization. For each source of consumption and each type of digital block, appropriate solutions can be implemented. Although the theory behind some of these techniques can be complex, they are often easy to implement. RTL designers should be aware of these techniques and use their knowledge of the system not only to optimize the speed performance, but also to reduce the unnecessary switching activity.

11.9 Acknowledgments

The authors thank Dr. Zinai Karima and Dr. Thomas Ea for their helpful comments and suggestions.

References

[1] G. Gerosa et al., A 2.2-W 80-Mhz superscalar RISC microprocessor, *IEEE J. Solid-State Circuits,* vol. 29, no. 12, pp. 1440–1454, Dec. 1994.
[2] C. Piguet et al., Low-power design of 8-bit embedded CoolRISC microcontroller cores, *IEEE J. Solid-State Circuits,* vol. 32, no. 7, July 1997.
[3] N. Raghavan, V. Akella, and S. Bakshi, Automatic insertion of gated clocks at register transfer level, *Proc. 12th Int.l Conf. on VLSI Design,* January 1999.
[4] *Power Compiler Design Manual,* Synopsys Ltd.
[5] PowerTheater, SequenceDesign Ltd.
[6] F. Theeuwen and E. Seelen, Power reduction through clock gating by symbolic manipulation, *Proc. Symp. Logic and Architecture Design,* Dec. 1996, pp. 131–136.
[7] M. Ohnishi, A. Yamada, H. Noda, and T. Kambe, A method of redundant clocking detection and power reduction at RT-level design, *Proc. 1997 Int. Symp. Low-Power Electronics and Design,* Monterey, CA, Aug. 1997, pp. 184–191.
[8] C. Arm, J.-M. Masgonty, and C. Piguet, Double-latch clocking scheme for low-power IP cores, *PATMOS 2000,* Goettingen, Germany, September 13–15, 2000.
[9] T. Schneider, *VHDL: Méthodologie de Design et Techniques Avancées,* Dunod, Paris, France, 2001.
[10] M. Alidina, J. Monteiro, S. Devadas, A. Gosh, and M. Papaefthymiou, Precomputation-based sequential logic optimization for low power, *Proc. 1994 Int. Comput.-Aided Design,* San Jose, CA, Nov. 1994, pp. 74–81.
[11] V. Tiwari, S. Malik, and P. Ashar, Guarded evaluation: pushing power management to logic synthesis/design, *Proc. Low-Power Design Symp.,* Dana Point, CA, Apr. 1995, pp. 221–226.
[12] H. Kapadia, L. Benini, and G. De Micheli, Reducing switching activity on datapath buses with control-signal gating, *IEEE J. Solid-State Circuits,* vol. 34, pp. 405–414, Mar. 1999.
[13] G.K. Yeap, *Practical Low-Power Digital VLSI Design,* Kluwer Academic Publishers, Dordrecht, 1998.
[14] G. De Micheli, *Synthesis and Optimization of Digital Circuits,* McGraw-Hill, New York, 1994.
[15] M. Hikari, H. Kojima, et al., Data-dependent logic swing internal bus architecture for ultralow-power LSIs, *IEEE J. Solid-State Circuits,* vol. 30, no. 4, pp. 397–402, Apr. 1995.
[16] H. Yamauchi, H. Akamatsu, and T. Fujita, An asymtomatically zero-power charge-recycling bus architecture for battery-operated ultra-high data rate ULSIs, *IEEE J. of Solid-State Circuits,* vol. 30, no. 4, pp. 423–431, Apr. 1995.
[17] L. Benini, Designing advanced NoCs architectures, *Int. Seminar on Application-Specific Multi-Processor SoC,* Chamonix, France, July 7–11, 2003.
[18] M. Stan and W. Burleson, Bus-invert coding for low-power IO, *IEEE Trans. on VLSI Syst.,* vol. 3, no. 1, pp. 49–58, Mar. 1995.
[19] C. Tsui and M. Pedram, Low-power state assignment targeting two and multi-level logic implementation, *ACM/IEEE Int. Conf. on CAD,* pp. 82–87, Nov. 1994.
[20] L. Benini and G. DeMicheli, Transformation and synthesis of FSMs for low-power and gated-clock implementation, *ACM/SIGDA ISLP '95,* Apr. 1995.
[21] L. Benini, G. DeMicheli, and F. Vermulen, Finite-state machine partitioning for low power, *IEEE ISCAS '98,* pp. 5–8, May 1998.
[22] R. Shelar and M.P. Desai, Orthogonal partitioning and gated-clock architecture for low-power realization of FSMs, *IEEE ASIC/SOC '2000,* pp. 266–270, Sept. 2000.

12

Clocking Multi-GHz Systems

Vojin G. Oklobdzija
University of California—Davis

12.1 Introduction

The clock speed has been rising rapidly, doubling every 3 years as plotted in Figure 12.1. Currently, the highest microprocessor clock frequency is slightly above 3 GHz, while that number is changing rapidly upward. It is expected that we will reach 10 GHz in the next 5 years, and by the year 2010, the processors will be running at frequencies beyond 10 GHz. At that clock rate, several challenges may force us to reexamine standard approaches to clocking.

As the clock speed increases, the number of logic levels in the critical path diminishes. In today's high-speed processors, instructions are executed in one cycle, which is driven by a single-phase clock. In addition, the pipeline depth is increasing to 15 or 20 to accommodate the speed increase. Today, 10 levels of logic in the critical path are common; however, the amount of logic between the two stages is decreasing further. Thus, any overhead associated with the clock system and clocking mechanism that is directly and adversely affecting the machine performance is critically important.

At today's frequencies, the ability to absorb clock skew and to use a faster clocked storage element (CSE) results in direct and significant performance improvements. These improvements are very difficult to obtain through architectural techniques or micro-architecture levels. As the clock frequency reaches 5 to 10 GHz, traditional clocking techniques will be stretched to their limits. New ideas and new ways of designing digital systems are required.

12.1.1 Clock Distribution

The two most important timing parameters affecting the clock signal are: clock skew and clock jitter.

Clock skew is a spatial variation of the clock signal as distributed through the system. It is caused by the various RC characteristics of the clock paths to the various points in the system, as well as different loading of the clock signal at different points on the chip. In addition, we can distinguish between global clock skew and local clock skew. These are both equally important in high-performance system design.

0-8493-1941-2/05/$0.00+$1.50

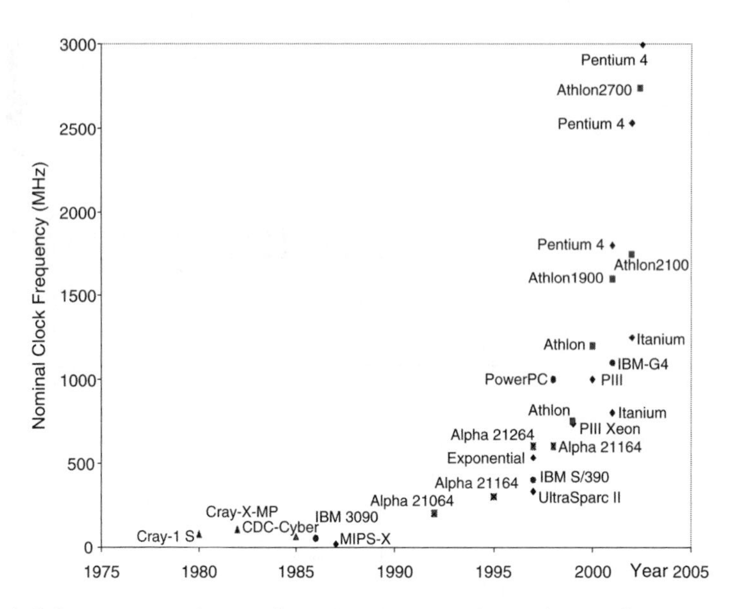

FIGURE 12.1 Clock frequency over the years for various representative machines and processors.

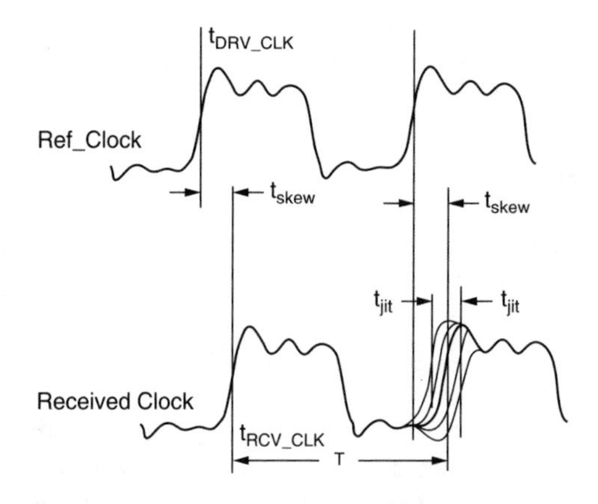

FIGURE 12.2 Clock parameters: period, width, clock skew, and clock jitter.

Clock jitter is a temporal variation of the clock signal with regard to the reference transition (i.e., reference edge) of the clock signal as illustrated in Figure 12.2.

Clock jitter represents edge-to-edge variation of the clock signal in time. As such, clock jitter can also be classified as long-term jitter and edge-to-edge clock jitter, which defines the clock signal variation between two consecutive clock edges. In the course of high-speed logic design, we are more concerned about edge-to-edge clock jitter because this phenomenon affects the time available for the logic operation.

Typically, the clock signal has to be distributed to several hundreds of thousands of the CSEs. Therefore, the clock signal has the largest fan-out of any node in the design, which requires several levels of amplification. Consequently, the clock system alone can use up to 40 to 50% of the power of the entire very large-scale integration (VLSI) chip [1,9]. We also must assure that every CSE receives the clock signal precisely at the same moment in time.

12.2 Clocking Considerations in Sequential Systems

A traditional view of the finite state machine (FSM) is represented by the Huffman model, which consists of combinational logic (CL) and CSEs. In this model, the next state, which is determined by the present

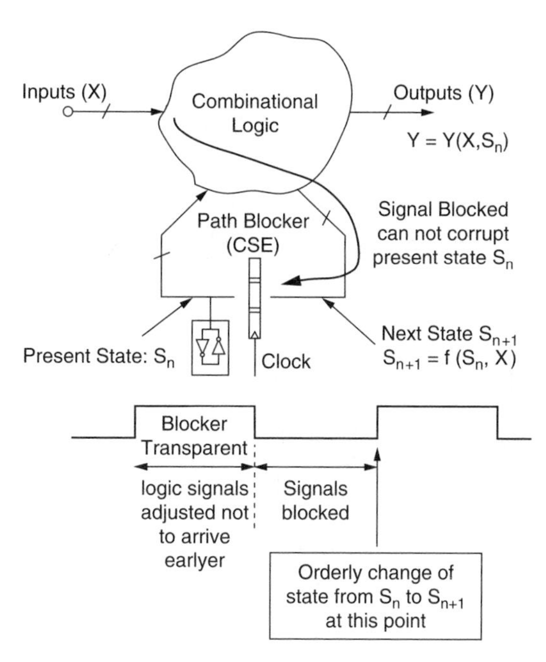

FIGURE 12.3 Different view of an FSM (Huffman model).

state and the input (as in the case of Mealy machine), is stored into the CSE by the triggering mechanism of the clock (i.e., edge or level). Following this model, we are used to thinking that the purpose of the CSE is to "hold" or "memorize" the state. This view is further supported by the level sensitive scan design (LSSD) methodology, which uses the storage elements to "scan-out" the state of the machine during the test and debug mode.

We want to present a slightly different view. The purpose of the CSE is to prevent the corruption of the next state as illustrated in Figure 12.3.

This model is broader and can represent wave pipelining [2], for example. In the case of wave pipelining, the signal is blocked from corrupting the present state S_n by a sheer delay of the wire. It simply cannot arrive in time, therefore, no blocking is necessary; however, this model also reveals problems of wave pipelining technique. Ideally, all the signals should arrive at the same point in time, which is not possible. Therefore, the fast-path problem becomes more difficult to control and stringent requirements are necessary. Thus, the system will run the risk of corrupting the state after several cycles.

The case of skew-tolerant domino logic [5,6], illustrated in Figure 12.4, conforms to the model presented in Figure 12.3.

Blocking of the signal is accomplished by the precharge phase of the clock. For example, while clock Φ_2 is "low" (precharge), data from stage 1 cannot be passed onto the stage 2. Only after the precharge phase has elapsed and clock Φ_2 has returned to "high" value can data from the stage 1 be passed to Stage 2. This transfer has to be completed while the clock Φ_1 is "high." Obviously, the speed of this logic is determined by precise matching of the clocks. This is accomplished by having the clock signal travel along the data-path, while delaying the clock for the amount of time needed in the logic stage generates the local clocks. In some way, this is similar to the clocking used in the early mainframe computers [3].

12.2.1 Clocked Storage Elements

The function of a CSE (flip-flop or latch) is to block the signal path, thus preventing it from corrupting the present state. In addition, it may be used to capture the state information and preserve it as long as it is needed by the digital system. It is not possible to define a storage element without defining its relationship to the *clock*.

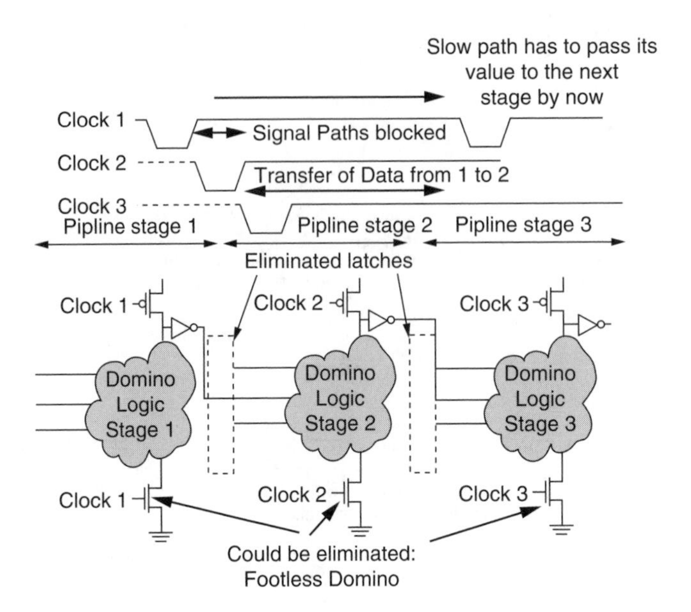

FIGURE 12.4 Skew-tolerant domino logic: no explicit latches.

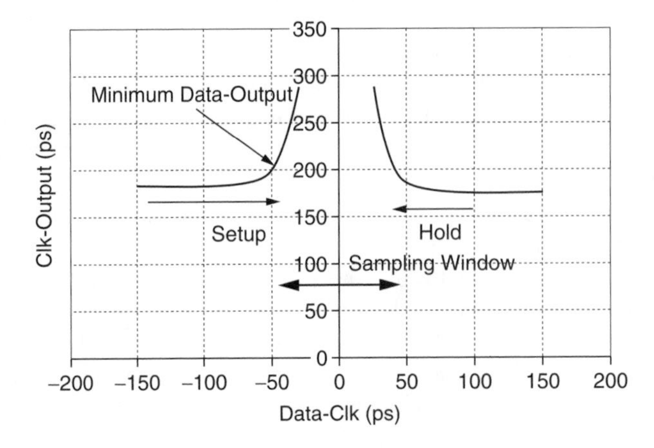

FIGURE 12.5 Setup and hold time behavior as a function of clock-to-output delay [8].

12.2.1.1 Timing Parameters

Data and clock inputs of a CSE must satisfy basic timing restrictions to ensure correct operation [4]. Fundamental timing constraints between data and clock inputs are quantified with setup and hold times, as illustrated in Figure 12.5 [8]. Setup and hold times define time intervals during which input has to be stable to ensure correct flip-flop operation. The sum of setup and hold times define the sampling window of the CSE. The sampling window is the period in which the CSE is sampling, and data is not allowed to change.

12.2.1.2 Setup and Hold Time Properties

Failure of the CSE due to the setup and hold time violations is not an abrupt process. This failing behavior is depicted in Figure 12.6.

Two opposing requirements exist with respect to the change of data signal as the locking event is approaching [8]:

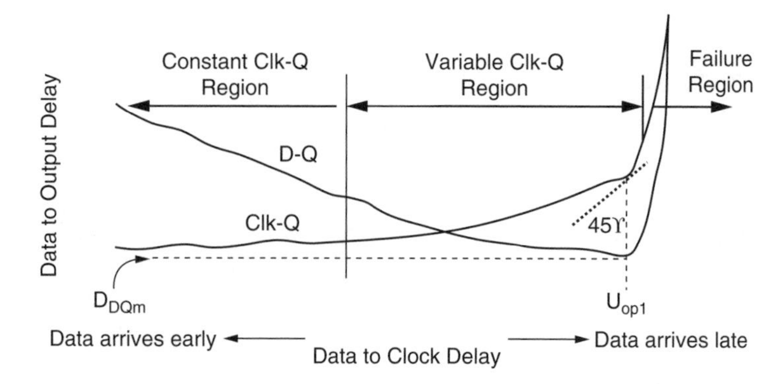

FIGURE 12.6 Setup and hold time behavior as a function of Data-to-Output delay.

1. The change should be kept farther from the failing region (Figure 12.6) for the purpose of design reliability.
2. The change should be as close as possible to increase the time available for the logic operation.

In industry, setup and hold times are specified as points in time when the Clk-Q (t_{CQ}) delay raises for an arbitrary number (commonly 5 to 20%). This reason is not valid. If we pay attention to D-Q (t_{DQ}) delay (instead of Clk-Q), we see a different picture. Despite the increase in Clk-Q delay, there are still benefits of getting closer to the locking event because D-Q delay (representing the time taken from the cycle) is reduced [16].

12.2.2 Time Borrowing and Absorption of Clock Uncertainties

Even if data arrives past the clock edge, the delay contribution of the storage element is still smaller than the amount of delay passed onto the next cycle. This allows for more time for useful logic operation. This is known as time borrowing or cycle stealing [7]. To understand the full effects of delayed data arrival, we have to consider a pipelined design where the data captured in the first clock cycle is used as input in the next clock cycle (see Figure 12.7).

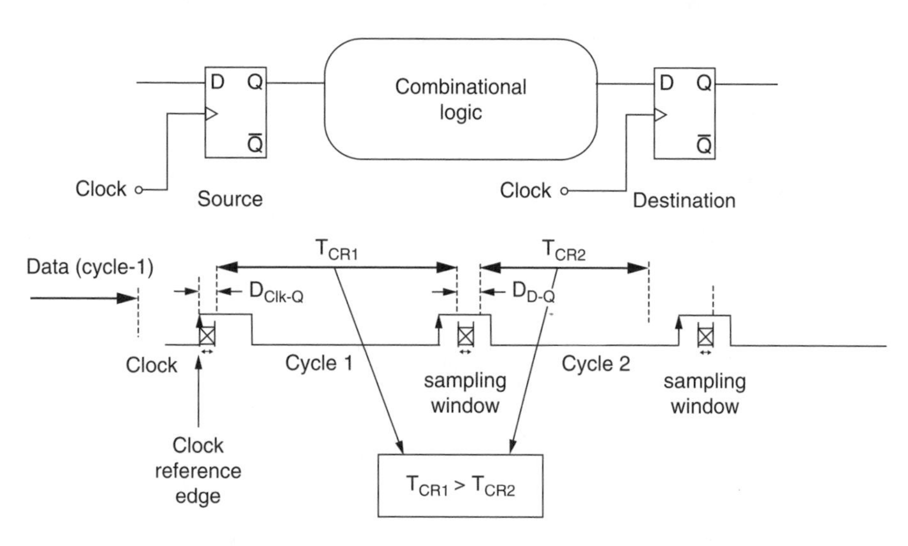

FIGURE 12.7 "Time borrowing" in a pipelined design.

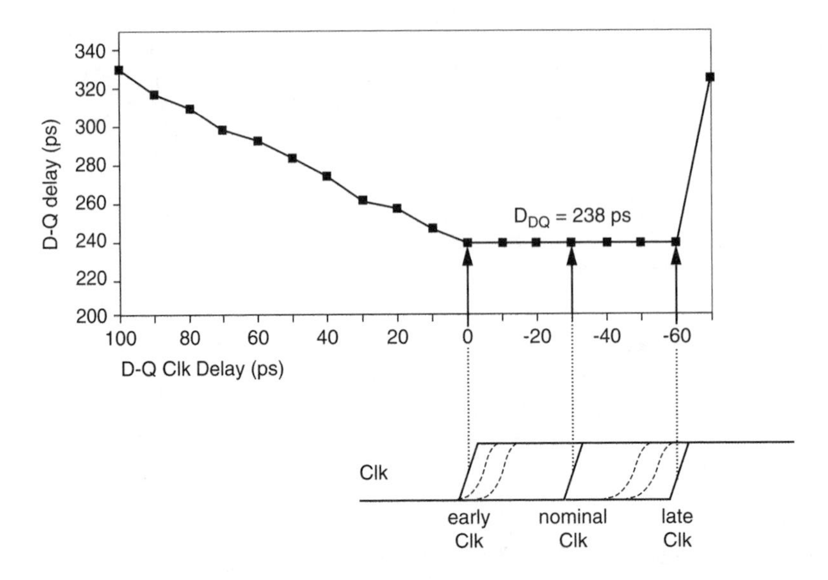

FIGURE 12.8 Clock skew absorption property: data-to-output delay vs. clock arrival time.

The sampling window is defined as the period in which CSE is sampling, and data is not allowed to change. The amount of time for which the T_{CR1} was stretched did not come for free. It was simply taken away (i.e., stolen or borrowed) leaving less time in the next cycle (Cycle 2) for T_{CR2}. As a result, of late data arrival in Cycle 1, less time is available in Cycle 2. Thus, a boundary between pipeline stages is somewhat flexible. This feature not only helps accommodate a certain amount of imbalance between the critical paths in various pipeline stages, but it helps in absorbing the clock uncertainties: skew and jitter.

Thus, time borrowing is one of the most important characteristics of today's high-speed digital systems. Absorption of the clock jitter is depicted in Figure 12.8 [7], and the effect on data arrival in the following cycle is illustrated in Figure 12.9. We observe how moderate amounts of clock uncertainties can be effectively absorbed, while the absorption property diminishes as the clock uncertainties become excessive.

The benefits of the "flat" data-to-output characteristic are presented in Figure 12.8 and Figure 12.9. We create the flat characteristic by expanding the transparency window of the CSE. Widening of the transparency window is equivalent to increasing the separation between the two reference events in time:

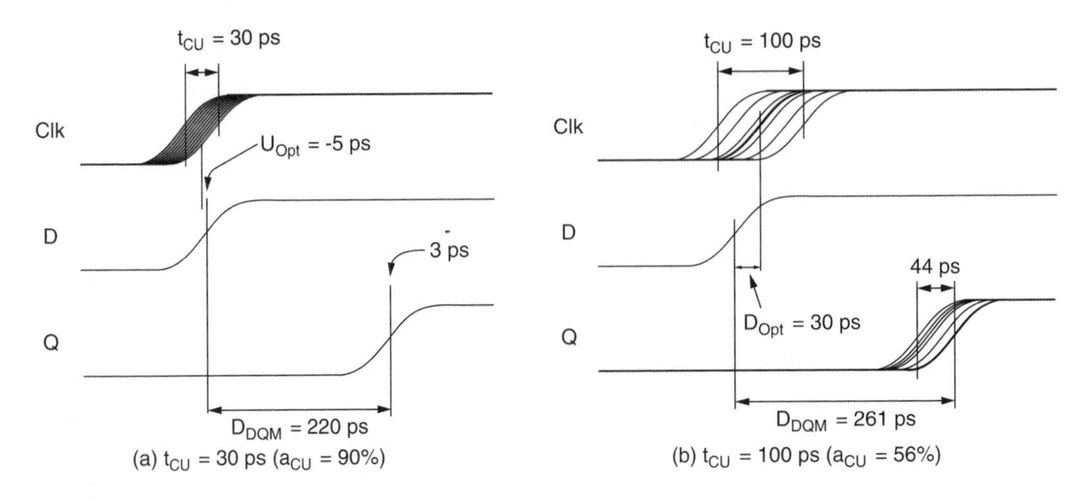

FIGURE 12.9 Effects of clock uncertainties to data arrival in the next cycle [20].

one that opens and other one that closes the CSE. In effect, the storage element behaves as a transparent latch for the short amount of time after the active clock edge [10]. Widening the transparency window can be achieved by intentionally creating wider capturing pulses of flip-flop and pulsed latch [11], or overlapping master and slave clocks in the master–slave latch. A consequence of increasing the transparency window is that the failure region of the data-to-output characteristic is moved away from the nominal clock edge. This results in the negative setup time but at the expense of increasing the hold time of the storage element. Large hold time makes fast path requirement harder to meet. Thus, the design for clock uncertainty absorption is often traded for a longer hold time. In many cases, however, these two requirements are not contradictory because different types of storage elements are used in fast and slow paths. The maximal clock skew that a system can tolerate is determined by CSEs. If the clock-to-output delay of a CSE is shorter than the hold time required and no logic exists in between two storage elements, a race condition can occur. A minimum delay restriction on the clock-to-output delay is given by:

$$t_{CLK-Q} \geq t_{hold} + t_{skew} \tag{12.1}$$

If this relation is satisfied, the system is immune to hold time violations.

The clock uncertainty absorption property shows how the propagation delay of a CSE is changing if the arrival of the reference clock is uncertain. Applying the clock uncertainty to a CSE is equivalent to holding reference clock arrival fixed and allowing data arrival to change.

More generally, uncertainty absorption should be treated as degradation of data-to-output delay for uncertain data-to-clock delay. As such, it can be used to describe the timing of the CSE if used in time borrowing, in exactly the same way if used for clock uncertainty absorption. Therefore, a "soft clock edge" designates a storage element where the output follows both early and late arrivals of the input, allowing slower stages to borrow time from the faster subsequent stages.

The time-borrowing capability and the clock uncertainty absorption are not mutually exclusive. They can be traded off for each other. Figure 12.10 illustrates a case where a wide transparency window, denoted

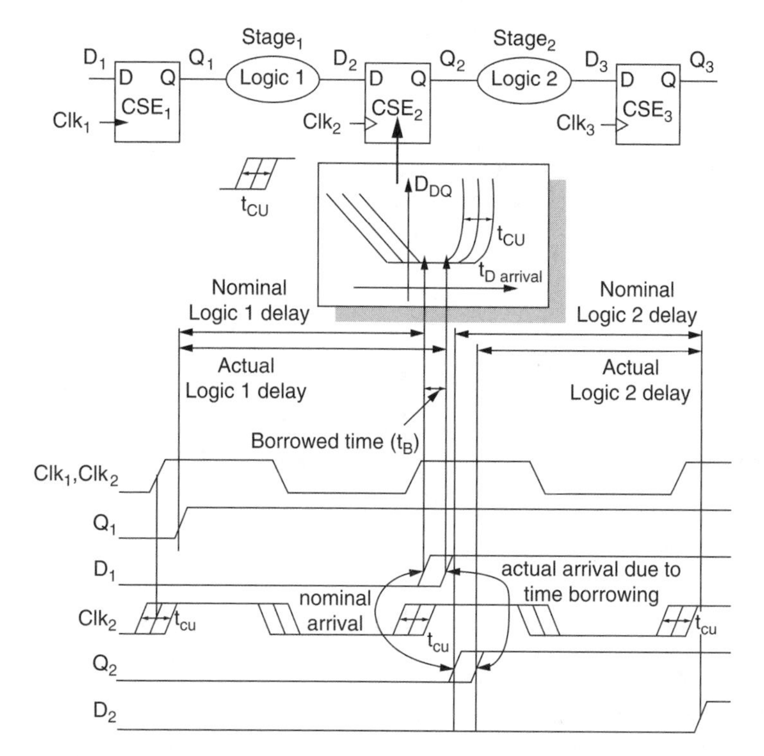

FIGURE 12.10 Time borrowing with uncertainty-absorbing CSEs [16].

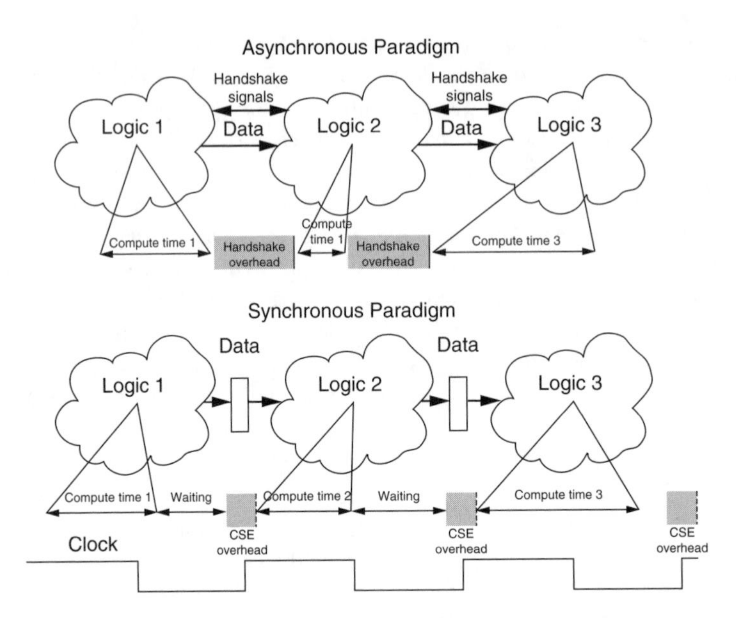

FIGURE 12.11 Data transfer in an asynchronous system vs. a synchronous system.

as a flat data-to-output characteristic, is used to absorb both the clock uncertainties, t_{CU}, and to borrow time, t_B, from the surrounding stages. Combinational logic of stage 1 takes more time than nominally assigned, and it borrows a portion of the cycle time from stage 2. In general, the storage element may not be completely transparent (i.e., data-to-output characteristics are not completely flat). The combination of clock uncertainty t_{CU} and time borrowing t_B causes an increase in the data-to-output delay of the flip-flop ΔD_{DQ}.

It should be noted that the practical values of the total borrowed time are approximately the width of the transparency window of the storage element and are, in any event, shorter than the hold time.

12.3 Asynchronous Systems

As the clock frequency increases, synchronous systems are facing serious problems such as the lack of ability to precisely control the clock, nonscaling clock uncertainties, wire delays, and the simple fact that the signal may need one or more clock cycles to reach its destination. Thus, asynchronous system design has been revisited.

The overhead imposed on the synchronous system by the clock uncertainties and CSE properties is simply traded for the overhead imposed by the handshake signaling in the asynchronous system (see Figure 12.11). Thus, the question really is: which one of the two can be designed so that it imposes lesser penalties on the data transfer as the speed of the logic keep rapidly increasing? As of today, it makes logical sense to use synchronous design in local domains, which can be clocked synchronously without considerable difficulties. Data transfer lasting several clock cycles could be accomplished using asynchronous communication. This opinion is supported by the fact that at 10 GHz or more, it would take several clock cycles to cross from one chip edge to another, as well as the fact that an entire processor in a 1-billion-transistor chip would occupy only a small portion of the chip.

12.4 Globally Asynchronous Locally Synchronous Systems

Following industry projections, VLSI chips will contain 1 billion transistors before the year 2010; however, the number of transistors used to build the logic of a single processor has not been increasing at the same rate. On the contrary, that number has remained relatively constant. Table 12.1 lists some

TABLE 12.1 Logic Transistors in Representative RISC Processors

Feature	Digital 21164	MIPS 10000	PowerPC 620	HP 8000	Sun UltraSparc
Frequency	500 MHz	200 MHz	200 MHz	180 MHz	250 MHz
Pipeline stages	7	5–7	5	7–9	6–9
Issue rate	4	4	4	4	4
Out-of-order execution	6 loads	32	16	56	none
Register renaming (int/FP)	none/8	32/32	8/8	56	none
Total transistors	9.3 M	5.9 M	6.9 M/	3.9 M*	3.8 M
Logic transistors	1.8 M	2.3 M	2.2 M	3.9 M	2.0 M

Source: Microprocessor Report 1998 issues.

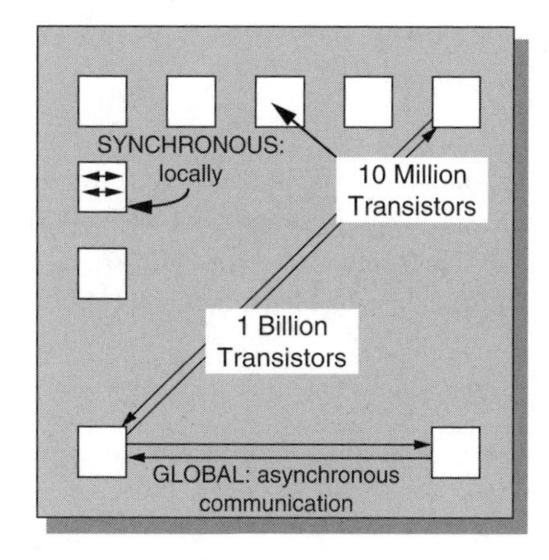

FIGURE 12.12 Projection of a 1-billion-transistor VLSI chip.

of the transistor numbers for a sample of typical super-scalar RISC architecture processors around the year 2000.

Figure 12.12 provides an illustration of a 1-billion-transistor chip. If we project the speed of the chip at that time and compare it with the projection for the speed of the interconnect, it becomes obvious that synchronous design on the entire chip may not be possible. Several clock cycles would be necessary for the signal to cross from one side of the chip to the other. From the design standpoint, it is also obvious that the future 1-billion-transistor chip will contain multiple cores in either multiprocessor or system on chip (SoC) arrangement. Therefore, globally asynchronous and locally synchronous clocking is considered as a promising technique for the SoC design.

In such a system, a number of independently synchronized modules communicate between each other through an asynchronous handshake mechanism. It is projected that interconnect effects will be manageable within a local synchronous module; therefore, a synchronous design would continue to be a viable option for the processor core.

Globally asynchronous and locally synchronous systems contain several independent synchronous blocks that operate with their own local clocks and communicate asynchronously with each other. The main feature of these systems is the absence of a global timing reference and the use of several distinct local clocks, or clock domains, possibly running at different frequencies.

This methodology is also viable when various intellectual property (IP) blocks are integrated in a single chip in an SoC environment because proven IP blocks can be reused without any modifications while relying on asynchronous interface between blocks. Such design preserves the benefit of synchronous design while avoiding problems caused by global wiring, especially a global clock signal.

Most conventional microprocessor designs are synchronous in their construction; that is, they have a global clock signal that provides a common timing reference for the operation of all the circuitry on the chip. On the other hand, fully asynchronous designs built using self-timed circuits do not have any global timing reference.

The globally asynchronous, locally synchronous design is not a novel approach, given that such a concept has long been used in mainframe computer systems, and this represents its logical migration into the VLSI chip.

12.5 Conclusion

Clocking for high-performance and low-power systems represents a challenge given the rapid increase in clock frequency, which has already reached multiple GHz rates. We expect that current clocking techniques will hold up to 10 GHz. Afterward, the pipeline boundaries will start to blur and synchronous design will be possible only in limited domains on the chip. A mix of synchronous and asynchronous design may emerge even in digital logic. This may represent the next design challenge in complex chips.

To Probe Further

For complete analysis of clocking, please see Oklobdzija [16]. Overviews of low-power circuit design techniques are available in Kuroda and Sakurai [9] and Oklobdzija [15]. Good references for asynchronous clocking including articles by Hauck and Sutherland [12,13]. Globally asynchronous, locally synchronous systems are described in the article by Hemani et al. [14].

References

[1] P.E. Gronowski, et al., High-performance microprocessor design, *IEEE J. Solid-State Circuits*, Vol. 33, No. 5, May 1998.

[2] W.P. Burleson, M.Ciesielski, F. Klass, and W. Liu, Wave-pipelining: a tutorial and research survey, *IEEE Trans. of Very Large-Scale Integration (VLSI) Systems*, Vol. 6, No. 3, September 1998.

[3] L.W. Cotten, Circuit implementation of high-speed pipeline systems, *AFIPS Proc., Fall Joint Comput. Conf.*, pp. 489–504, 1965.

[4] S.H. Unger and C.J. Tan, Clocking schemes for high-speed digital systems, *IEEE Trans. on Computers*, Vol. C-35, No. 10, October 1986.

[5] D. Harris and M.A. Horowitz, Skew-tolerant domino circuits, *IEEE J. Solid-State Circuits*, Vol. 32, No. 11, November 1997.

[6] D. Harris, et al., Opportunistic Time-Borrowing Domino Logic. U.S. Patent No. 5,517,136. Issued May 14, 1996.

[7] H. Partovi et al., Flow-through latch and edge-triggered flip-flop hybrid elements, *IEEE Int. Solid-State Circuits Conf. (ISSCC), Dig. of Tech. Papers*, San Francisco, CA, February 8–10, 1996.

[8] V. Stojanovic and V.G. Oklobdzija, Comaparative analysis of master-slave latches and flip-flops for high-performance and low-power VLSI systems, *IEEE J. Solid-State Circuits*, Vol. 34, No. 4, April 1999.

[9] T. Kuroda and T. Sakurai, Overview of low-power VLSI circuit techniques, *IEICE Trans. on Electron.*, E78-C, No. 4, April 1995, pp. 334–344. Invited paper, special issue on low-voltage low-power integrated circuits.

[10] F. Klass et al., A new family of semidynamic and dynamic flip-flops with embedded logic for high-performance processors, *IEEE J. Solid-State Circuits*, Vol. 34, No. 5, pp. 712–716, May 1999.

[11] T. James, S. Narendra, Z. Chen, S. Borkar, M. Sachdev, and V. De, Comparative delay and energy of single edge-triggered and dual edge-triggered pulsed flip-flops for high-performance microprocessors, *Proc. 2001 Int. Symp. on Low-Power Electron. and Design*, Huntington Beach, CA, August 6–7, 2001.

[12] S. Hauck, Asynchronous design methodologies: an overview, *Proc. IEEE*, Vol. 83, No. 1, January 1995.

[13] I.E. Sutherland, Micropipelines, *Commun. ACM,* Vol. 32, No. 6, June 1989.
[14] A. Hemani, T. Meincke, S. Kumar, A. Postula, T. Olsson, P. Nilsson, J. Oberg, P. Ellervee, and D. Lundqvist, Lowering power consumption in clock by using globally asynchronous locally synchronous design style, *Proc. 36th Design Automation Conf.,* June 21–25, 1999.
[15] V.G. Oklobdzija, Ed., *High-Performance System Design: Circuits and Logic,* John Wiley & Sons/IEEE Press series on microelectronics systems, 1999.
[16] V.G. Oklobdzija, V. Stojanovic, D. Markovic, and N. Nedovic, *Digital System Clocking: High-Performance and Low-Power Aspects,* John Wiley & Sons, New York, January 2003.

13

Circuit Techniques for Leakage Reduction

Kaushik Roy
Amit Agarwal
Chris H. Kim
Purdue University

13.1 Introduction

Semiconductor devices are aggressively scaled each technology generation to achieve high integration density while the supply voltage is scaled to achieve lower switching energy per device. To achieve high performance, however, commensurate scaling of the transistor threshold voltage (V_{th}) is needed. Scaling of transistor threshold voltage is associated with exponential increase in subthreshold leakage current [1]. Aggressive scaling of the devices in the nanometer regime not only increases the subthreshold leakage, but also has other negative impacts such as increased drain-induced barrier lowering (DIBL), V_{th} roll-off, reduced on-current to off-current ratio, and increased source-drain resistance [2]. DIBL increases the dependency of V_{th} on channel length. A small variation in channel length might result in large V_{th} variation, which makes device characteristics unpredictable. To avoid these short-channel effects (SCE), oxide thickness scaling and higher nonuniform doping needs to be incorporated [3] as the devices are scaled in nanometer regime. The International Technology Roadmap for Semiconductors (ITRS) predicts gate oxide thickness of 1.2 to 1.6 nm for sub-100nm CMOS [4]. The low oxide thickness gives rise to high electric field, resulting in considerable direct tunneling current [5]. This current destroys the classical infinite input impedance assumption of metal-oxide semiconductor (MOS) transistors and thus affects circuit performance severely. Higher doping results in high electric field across the p-n junction (source-substrate or drain-substrate), which causes significant band-to-band tunneling (BTBT) of electrons from the valence band of the p-region to the conduction band of the n-region. Peak halo doping (P+) is restricted such that the BTBT component is maintained reasonably small compared with the other leakage components.

0-8493-1941-2/05/$0.00+$1.50

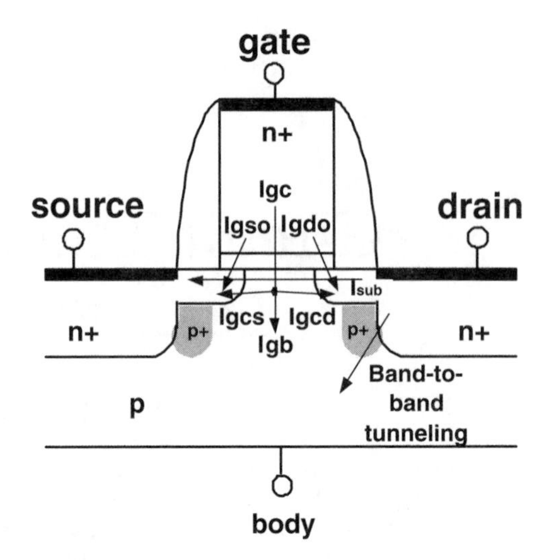

FIGURE 13.1 Leakage components in MOSFET.

This chapter proposes different integrated circuit techniques to reduce overall leakage in both logic and cache memories. A spectrum of circuit techniques including dual V_{th}, variable V_{th}, dynamically varying the V_{th} during runtime, sleep transistor, natural stacking, and multiple/dynamic supply circuits are reviewed. Based on these techniques, different leakage tolerant schemes for logic and memories are summarized.

13.2 Leakage Components

A metal-oxide semiconductor fluid-effect transistor (MOSFET) of the nanometer regime has three dominant components of leakage:

1. Subthreshold leakage, which is the leakage current from drain to source (I_{sub} in Figure 13.1).
2. Direct tunneling gate leakage, which is due to the tunneling of electron (or hole) from the bulk silicon through the gate oxide potential barrier into the gate.
3. The source/substrate and drain/substrate reverse biased p-n junction BTBT leakage; this leakage component is expected to be large for sub-50-nm devices [6].

Other components of leakage current described in Roy et al. [7], such as gate-induced drain leakage (GIDL) and impact ionization leakage, are not expected to be large for regular nanoscale CMOS devices.

13.2.1 Subthreshold Leakage

Subthreshold or weak inversion conduction current between source and drain in a MOS transistor occurs when gate voltage is below V_{th} [8]. Weak inversion typically dominates modern device off-state leakage due to the low V_{th} that is used. The weak inversion current can be expressed based on the Equation (13.1) [8].

$$I_{subth} = Ae^{\frac{q}{nkT}(V_{GS}-V_{TH0}-\gamma V_{SB}+\eta V_{DS})}(1-e^{-\frac{qV_{DS}}{kT}}) \tag{13.1}$$

where

$$A = \mu_0 C'_{ox} \frac{W}{L_{eff}} \left(\frac{kT}{q}\right)^2 e^{1.8}$$

V_G, V_D, V_S, and V_B are the gate voltage, drain voltage, source voltage, and body voltage of the transistor, respectively. Body effect is represented by the term γV_{SB}, where γ is the linearized body effect coefficient; η is the DIBL coefficient, representing the effect of V_{DS} on threshold voltage; C_{ox} is the gate oxide capacitance; μ_0 is the zero bias mobility; and n is the subthreshold swing coefficient of the transistor. This equation shows the exponential dependency of subthreshold leakage on V_{TH0}, V_{GS}, V_{DS} (due to DIBL), and V_{SB}. Each of the leakage reduction techniques described in the latter sections utilizes these parameters in a MOSFET to achieve a low leakage state.

13.2.2 Gate Leakage

Gate direct tunneling current is due to the tunneling of electron (or hole) from the bulk silicon through the gate oxide potential barrier into the gate. The direct tunneling is modeled as

$$J_{DT} = A\left(V_{ox}/T_{ox}\right)^2 exp\left(\frac{-B\left(1-\left(1-V_{ox}/\phi_{ox}\right)^{3/2}\right)}{V_{ox}/T_{ox}}\right) \tag{13.2}$$

where J_{DT} is the direct tunneling current density, V_{ox} is the potential drop across the thin oxide, ϕ_{ox} is the barrier height of tunneling electron, and t_{ox} is the oxide thickness [9]. The tunneling current increases exponentially with decrease in oxide thickness. It also depends on the device structure and the bias condition [10]. Figure 13.1 describes the various components of gate tunneling in a scaled NMOS device [11]:

- Edge-direct tunneling (EDT) components between gate and source-drain extension (SDE) region (I_{gso} and I_{gdo})
- Gate-to-channel current (I_{gc}), part of which goes to source (I_{gcs}) and rest goes to drain (I_{gcd})
- Gate-to-substrate leakage current (I_{gb})

Tunneling current increases with the increase in voltage drop across oxide (V_{ox}). The voltage across oxide in different regions (i.e., channel, gate-source overlap, and gate-drain overlap region) depends on biasing of the nodes representing the region.

13.2.3 Source/Substrate and Drain/Substrate P-N Junction Leakage

Drain and source-to-well junctions are typically reverse biased, causing p-n junction leakage current. A reverse biased p-n junction leakage has two main components: One is minority carrier diffusion/drift near the edge of the depletion region, and the other is due to electron-hole pair generation in the depletion region of the reverse biased junction.

In the presence of a high electric field (> 10^6 V/cm), electrons will tunnel across a reverse biased p-n junction. A significant current can arise as electrons tunnel from the valence band of the p-region to the conduction band of the n-region. Tunneling occurs when the total voltage drop across the junction is greater than the semiconductor band-gap. Because silicon is an indirect band gap semiconductor the BTBT current in silicon involves the emission or absorption of phonons.

In an NMOS device, when the drain or source is biased at a potential higher than that of the substrate, BTBT current flows through the drain-substrate or source-substrate junction. If both *n*- and *p*-regions are heavily doped, which is the case for scaled MOSFETs using heavily doped shallow junctions and halo doping for better SCE, BTBT significantly increases and becomes a major contributor to the total off-state current. Substantial increase in BTBT current is observed at high reverse biases. Reducing substrate doping near the substrate-drain/source junction is an effective way to reduce the BTBT current; however, this increases the SCE leading to considerable increase in the subthreshold current. Although circuit techniques specifically targeted at reducing BTBT have not been reported, forward substrate biasing can

TABLE 13.1 Circuit Techniques to Control Leakage in Logic

	Run Time Techniques	
Design Time Techniques	Standby Leakage Reduction	Active Leakage Reduction
Dual-V_{th} [12,13,14,15,16,17]	Natural Stacking [20,21,22,23]	DVS [37]
	Sleep Transistor [24,25,26,27,28,29,30,31]	
Multiple Supply Voltage [19]	VTCMOS [32,33,34,36]	DVTS [38,39]

be used to reduce BTBT in a MOSFET (because electric field reduces with reduction in the reverse bias across the junction).

13.3 Circuit Techniques to Reduce Leakage in Logic

Because circuits are mostly designed for the highest performance — for instance, to satisfy overall system cycle time requirements — they are composed of large gates and highly parallel architectures with logic duplication. As such, the leakage power consumption is substantial for such circuits; however, not every application requires a fast circuit to operate at the highest performance level all the time. Modules in which computation is bursty in nature (e.g., functional units in a microprocessor or sections of a cache) are often idle. It is of interest to conceive of methods that can reduce the leakage power consumed by these circuits. Different circuit techniques have been proposed to reduce leakage energy utilizing this slack without impacting performance. These techniques can be categorized based on when and how they utilize the available timing slack (Table 13.1) (e.g., dual V_{th} statically assigns high V_{th} to some transistors in the noncritical paths at the *design time* to reduce leakage current). The techniques, which utilize the slack in *runtime*, can be divided into two groups depending on whether they reduce standby leakage or active leakage. Standby leakage reduction techniques put the entire system in a low leakage mode when computation is not required. Active leakage reduction techniques slow down the system by dynamically changing the V_{DD} or V_{th} to reduce leakage when maximum performance is not needed. In the active mode, the operating temperature increases due to the switching activities of transistors. This has an exponential effect on subthreshold leakage (Equation (13.1)), making it the dominant leakage component during active mode and aggravating the leakage problem.

13.4 Design Time Techniques

Design time techniques exploit the delay slack in noncritical paths to reduce leakage. These techniques are static; once they are fixed, they cannot be changed dynamically while the circuit is operating.

13.4.1 Dual Threshold CMOS

In logic, a high V_{th} can be assigned to some transistors in the noncritical paths to reduce subthreshold leakage current, while the performance is not sacrificed by using low V_{th} transistors in the critical path(s) [12]. No additional circuitry is required, and both high performance and low leakage can be achieved simultaneously. Figure 13.2(a) illustrates the basic idea of a dual V_{th} circuit. The path distribution of dual V_{th} and single V_{th} standard CMOS for a 32-bit adder is illustrated in Figure 13.2(b). Dual V_{th} CMOS has the same critical delay as the single low V_{th} CMOS circuit, but the transistors in the noncritical paths can be assigned high V_{th} to reduce leakage power. Dual threshold CMOS is effective in reducing leakage power during both standby and active modes. Many design techniques have been proposed, which consider upsizing of high V_{th} transistor [13–15] in dual V_{th} design to improve performance. Upsizing of high V_{th} transistor affects switching power and die area that can be traded off against using a low V_{th} transistor that increases leakage power.

Domino logic can be susceptible to leakage — especially wide OR domino gates. Low threshold evaluation logic reduces noise immunity. Thus, for scaled technologies, domino may require larger keeper

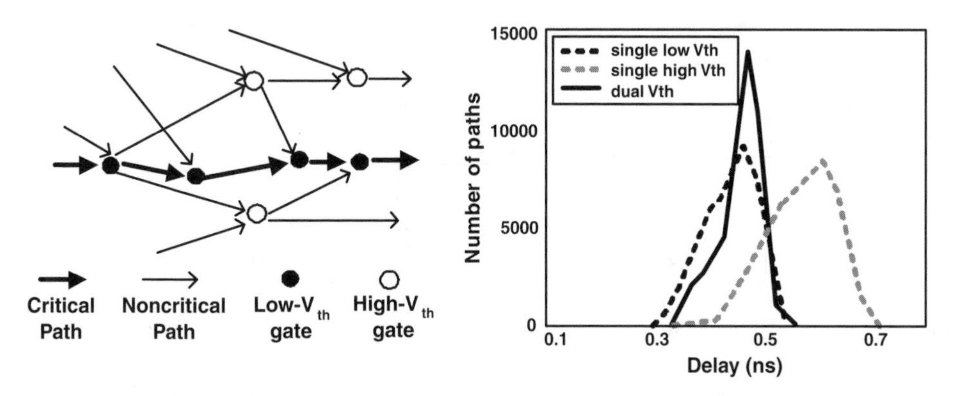

FIGURE 13.2 (a) A dual V_{th} CMOS circuit and (b) path distribution of dual V_{th} and single V_{th} CMOS.

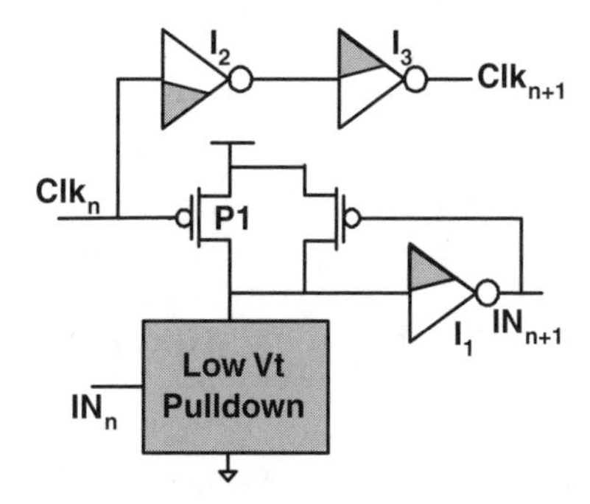

FIGURE 13.3 Dual V_{th} Domino gate [16] with low V_{th} devices shaded.

transistors that, in turn, can affect speed. Figure 13.3 depicts a typical dual V_{th} domino logic [16] for low leakage noise immune operations. Because of the fixed transition directions in domino logic, one can easily assign low V_{th} to all transistors that switch during the evaluate mode and high V_{th} to all transistors that switch during precharge modes. When a dual V_{th} domino logic stage is placed in standby mode, the domino clock needs to be high (evaluate) to shut off the high V_{th} devices (e.g., P1, I2 PMOS, and I3 NMOS). Furthermore, to ensure that the internal node remains at solid logic ZERO, which turns off the high V_{th} keeper and I1 NMOS, the initial inputs into the domino gate must be set high.

Instead of changing the channel doping profiles, a higher t_{ox} can be used to obtain a high V_{th} device for dual threshold CMOS circuits. In order to suppress the SCE, the high t_{ox} device needs to have a longer channel length as compared with the low t_{ox} device. Multiple t_{ox} CMOS (MoxCMOS) can optimize the power consumption due to subthreshold leakage, gate oxide tunneling leakage as well as switching power. An algorithm for selecting and assigning optimal transistor oxide thickness is derived in Sirisantana et al. [17]. The simulation results on IEEE International Symposium on Circuits and Systems (ISCAS) benchmark circuits for 70-nm technology show that the total power consumption of MoxCMOS circuits can be reduced by an average of 34% with over 70% reduction in gate oxide tunneling leakage compared with standard CMOS circuits.

13.4.2 Multiple Supply Voltage

Supply voltage scaling was originally developed for switching power reduction. It is an effective method for reducing switching power because of the quadratic dependency of switching power on supply voltage. Supply voltage scaling also helps reduce leakage power because the subthreshold leakage due to GIDL and DIBL decreases as well as the gate leakage component when the supply voltage is scaled down. In a 1.2-V, 0.13-μm technology, it is demonstrated that the supply voltage scaling has impacts in the orders of V^3 and V^4 on subthreshold leakage and gate leakage, respectively.

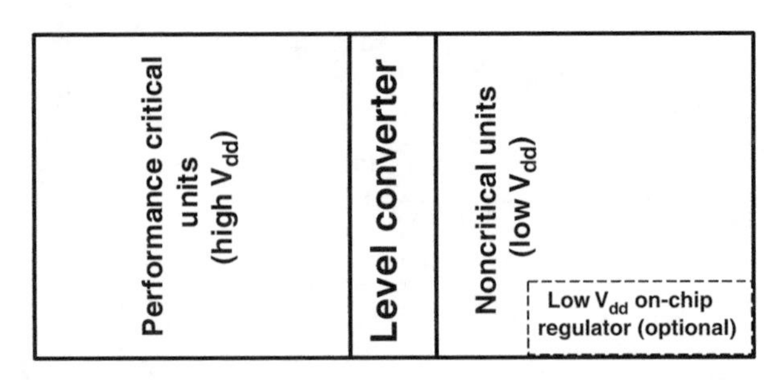

FIGURE 13.4 Two-level multiple supply voltage scheme [18].

To achieve low-power benefits without compromising performance, two methods of lowering supply voltage can be employed: static and dynamic (see Section 13.6) supply scaling. Static supply scaling is a multiple supply voltage approach [18] as depicted in Figure 13.4. Critical and noncritical paths or units of the design are clustered and powered by higher and lower supply voltages, respectively [19]. Because the speed requirements of the noncritical units are lower than the critical ones, supply voltage of noncritical unit clusters can be lowered without degrading system performance. Whenever an output from a low V_{DD} cluster has to drive an input to a high V_{DD} cluster, a level conversion is needed at the interface. The secondary voltages may be generated off-chip or regulated on-die from the core supply.

13.5 Runtime Standby Leakage Reduction Techniques

A common architectural technique to keep the power of fast, hot circuits within bounds has been to freeze the circuits — place them in a standby state — any time when they are not needed. Standby leakage reduction techniques exploit this idea to place certain sections of the circuitry in standby mode (low leakage mode) when they are not required.

13.5.1 Leakage Control Using Transistor Stacks (Self-Reverse Bias)

Leakage currents in NMOS or PMOS transistors depend exponentially on the voltage at the four terminals of transistor (Equation (13.1)). Figure 13.5 illustrates the variation of I_{DS} with respect to V_{GS} (V_G is tied to "0"). Increasing V_S of NMOS transistor reduces subthreshold leakage current exponentially due to the following three effects:

- Gate-o-source voltage becomes negative, thus the subthreshold current reduces exponentially.
- Negative body to source potential causes more body effect resulting in increased threshold voltage and thus reducing the subthreshold leakage.
- Drain-to-source potential decreases, resulting in less DIBL and thus lower subthreshold leakage.

This effect is also called self-reverse biasing of transistor. The self-reverse bias effect can be achieved by turning off a stack of transistors [20]. Turning off more than one transistor in a stack raises the internal

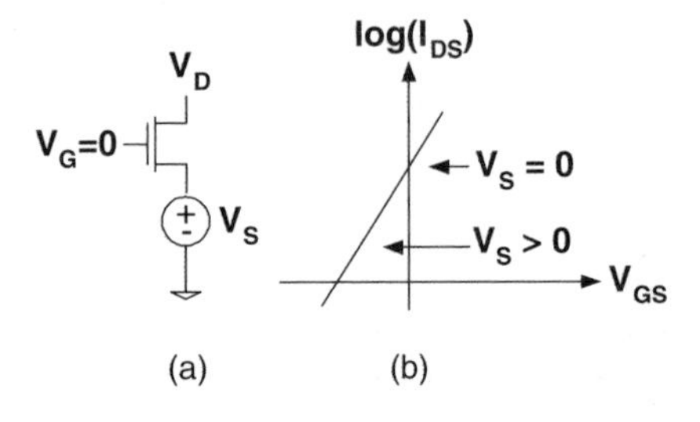

FIGURE 13.5 Leakage control using self-reverse bias.

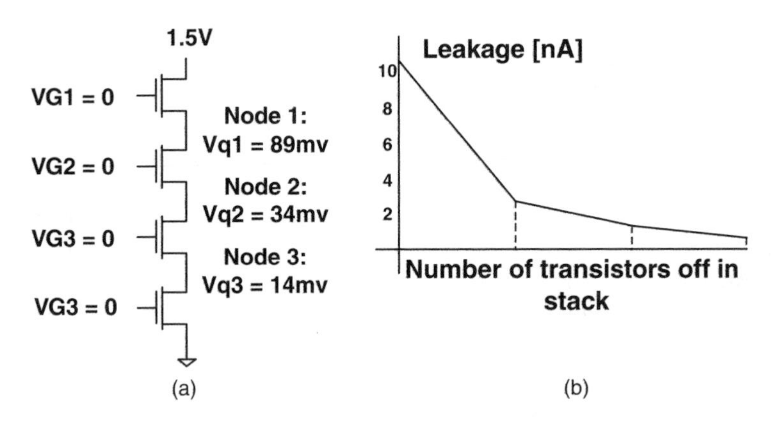

FIGURE 13.6 (a) Effect of transistor stacking on source voltage and (b) leakage current vs. number of transistors off in stack.

voltage (source voltage) of the stack, which acts as reverse biasing the source. Figure 13.6(a) depicts a simple pull-down network of a four input NAND gate. This pull-down network forms a stack of four transistors. If some of the transistors are turned off for a long time, the circuit reaches a steady state where leakage through each transistor is equal and the voltage across each transistor settles to a steady state value. In a case where only one NMOS device is off, the voltage at the source node of off transistor would be virtually zero because all other on transistors will act as short circuit. Thus, there is no self-reverse biasing effect, and the leakage across the off transistor is large. If more than one transistor is off, the source voltages of the off transistor, except the one connected to ground by on transistors, will be greater than zero, and the leakage will be determined mainly by the most negatively self-reverse biased transistor (because subthreshold leakage is an exponential function of gate-source voltage). The voltages at the internal nodes depend on the input applied to the stack. Figure 13.6(a) shows the internal voltages when all four transistors are turned off. These internal voltages make the off transistors self-reverse biased. The reverse bias makes the leakage across the off transistor very small. Figure 13.6(b) depicts the subthreshold leakage current vs. number of off transistors in a stack. A large difference in leakage current exists between one off transistor and two off transistors. Turning off three transistors does improve subthreshold leakage, however, there is a diminishing return.

It is evident from the preceding discussion that the leakage through logic gates depends on the applied input vector. Functional blocks such as NAND, NOR, or other complex gates readily have a stack of transistors. Maximizing the number of off transistors in a natural stack by applying proper input vectors can reduce the standby leakage of a functional block. A model and heuristic is proposed in [21] to estimate leakage and to select the proper input vectors to minimize the leakage in logic blocks. Table 13.2 presents the quiescent current flowing into different functional blocks for the best and worst case input vectors. All the results are based on HSPICE simulation using 0.18-μm technology with V_{DD} = 1.5 V. Results show

TABLE 13.2 Input Vector Control

Circuit	Input Vector	Iddq (nA)	Comments
4 input NAND	ABCD=000	0.60	Best
	ABCD=111	24.1	Worst
3 input NOR	ABC=111	0.13	Best
	ABC=000	29.5	Worst
Full adder	A,B,Ci=111	7.8	Best
	A,B,Ci=001	62.3	Worst
4 bit ripple adder	A=B=0000,Ci=0	91.3	Best
	A=B=1111,Ci=1	94.0	Best
	A=B=0101,Ci=1	282.9	Worst

that application of proper input vector can be efficient in reducing the total subthreshold leakage in the standby mode of operation [22].

Recent studies [23] demonstrate that as gate leakage is becoming a significant component of leakage, the input vector control technique using a stack of transistors needs to be reinvestigated to effectively reduce the total leakage. It has been reported that with gate leakage, the traditional way of using stacking fails to reduce leakage and in the worst case might increase the overall leakage. The gate leakage depends on the voltage drop across different regions of transistor (Section 13.2). Having "00" as the input in a two transistor stack has a high voltage drop across the gate-drain overlap region of the first transistor increasing the gate leakage, which may dominate the total leakage at room temperature. (Gate leakage is insensitive to temperature whereas subthreshold leakage is a strong function of temperature and increases with temperature [7].) Forcing inputs to "10" reduces this gate leakage component at the cost of subthreshold leakage. In scaled technology where gate leakage dominates the total leakage, using "10" might produce more savings in leakage as compare to "00." A three-transistor stack (NMOS) with input "100" can improve total leakage compare to "000" inputs for similar stack where subthreshold is the major component of leakage. The source/substrate and drain/substrate junction BTBT leakage is a weak function of input voltage, and thus it can be neglected from the analysis [6].

13.5.2 Sleep Transistor

This technique inserts an extra series-connected transistor (sleep transistor) in the pull-down/pull-up path of a gate and turns it "off" in the standby mode of operation [24]. During regular mode of operation, the extra transistor is turned on. This provides substantial savings in leakage current during standby mode of operation. As depicted in Figure 13.6(b), stacking of two off devices can significantly reduce leakage as compared with a single off device. Due to the extra stack transistor (sleep transistor), however, the drive current of forced-stack gate will be lower resulting in increased delay. Thus, this technique can only be used for paths that are noncritical. If the V_{th} of the sleep transistor is high, extra leakage saving is possible. This circuit topology is known as MTCMOS (multi-threshold CMOS) (Figure 13.7) [25].

In fact, only one type (i.e., either PMOS or NMOS) of high V_{th} transistor is sufficient for leakage reduction. The NMOS insertion scheme is preferable, because the NMOS on-resistance is smaller at the same width, and thus it can be sized smaller than a corresponding PMOS [26]. MTCMOS can be easily implemented on already existing circuits. A 1-V DSP (digital signal processing) chip for mobile phone applications has been recently developed using the MTCMOS scheme [27]; however, MTCMOS can only reduce leakage power in standby mode and the large inserted sleep transistors can increase the area and

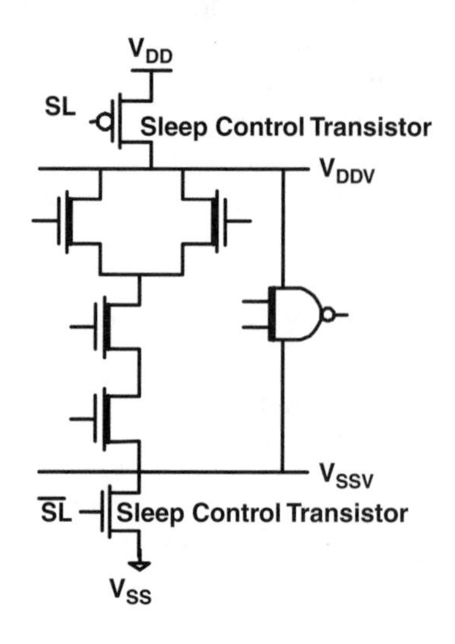

FIGURE 13.7 Schematic of MTCMOS circuit [25] with low V_{th} devices shaded.

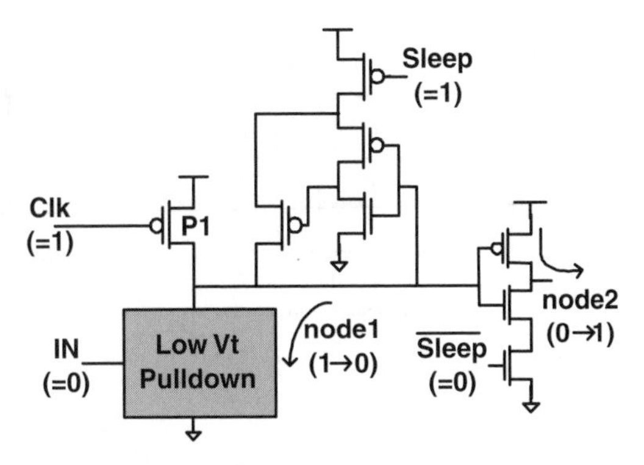

FIGURE 13.8 Domino gate with sleep transistor [31].

delay. Moreover, if data retention is required in standby mode, an extra high V_{th} memory circuit is needed to maintain the data [28]. Instead of using high V_{th} sleep transistors, a super cut-off CMOS (SCCMOS) circuit uses low V_{th} transistors with an inserted gate bias generator [29]. For the PMOS (NMOS) insertion, the gate is applied to 0 V (V_{DD}) in the active mode, and the virtual V_{DD} (V_{SS}) line is connected to the supply V_{DD} (V_{SS}). In standby mode, the gate is applied to V_{DD} + 0.4 V (V_{SS} − 0.4 V) by using the internal gate bias generator to fully cut off the leakage current. Compared with MTCMOS where it becomes difficult to turn on the high V_{th} sleep transistor at very low supply voltages, SCCMOS circuits can operate at very low supply voltages. Recent designs have been proposed using low-V_{th} devices for the sleep transistor to minimize the performance and area impacts [30].

In Figure 13.8, two small sleep transistors are added to conventional CMOS domino gate to save leakage [31]. In standby mode, clock is left high and sleep signal is asserted. If the data input were high, node 1 would have been discharged. If the data input was low, node 1 would be high, but leakage through NMOS dynamic pull-down stack would slowly discharge the node to ground. The NMOS sleep transistor is added to prevent any short circuit current in the static output logic while the dynamic node discharges to ground. Node 2 would rise as static pull-up turns on which would cause the NMOS transistors in the pull-down stacks of the following domino gates to turn on, accelerating the discharge of their internal dynamic nodes. Because sleep transistors are not in the critical path (evaluation path), minimal performance loss is incurred.

13.5.3 Variable Threshold CMOS (VTCMOS)

Variable threshold CMOS (VTCMOS) is a body-biasing design technique [32]. Figure 13.9(a) depicts the VTCMOS scheme.To achieve different threshold voltages, a self-substrate bias circuit is used to control the body bias. In the active mode, a zero body bias (ZBB) is applied. While in standby mode, a deep reverse body bias (RBB) is applied to increase the threshold voltage and cut off the leakage current. This scheme has been implemented in a two-dimensional discrete cosine transform core processor [32]. Furthermore, in active mode, a slightly forward substrate bias can be used to increase the circuit speed while reducing the SCE [33]. Providing the body bias voltage requires routing a body bias grid and this adds to the overall chip area. Keshavarzi et al. reported that RBB lowers integrated circuit (IC) leakage by three orders of magnitude in a 0.35-µm technology [34]. More recent data, however, demonstrates that the effectiveness of RBB to lower I_{OFF} decreases as technology scales due to the exponential increase in band-to-band tunneling leakage at the source/substrate and drain/substrate p-n junctions due to halo doping in scaled devices [34]. For scaled technologies, forward body biasing (FBB) can be used together with RBB to achieve better current drive with less SCE [35].

Raising the NMOS source voltage while tying the NMOS body to ground can produce the same effect as RBB. Forward body biasing can also be realized by applying a negative source voltage with respect to the body, which is tied to ground. Figure 13.9(b) illustrates the circuit diagram of this technique [36]. The main advantage is that it eliminates the need for a deep N-well or triple-well process because substrate

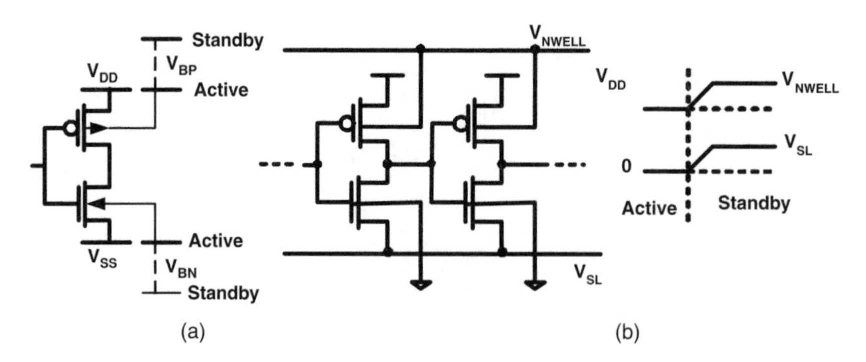

FIGURE 13.9 (a) Variable threshold CMOS [32], and (b) realizing body biasing by changing the source voltage with respect to body voltage, which is grounded [36].

of the target system and the control circuitry can be shared. The source voltage of the PMOS should also be raised if the charge in the storage node is to be kept constant while the NMOS source voltage is raised. A charge pump circuit is required if the body of the PMOS is to be raised higher than the source of the PMOS for RBB. In cases where V_{DS} can be further reduced, additional leakage improvement is possible. The smaller V_{DS} raises the transistor V_{th} (DIBL mechanism) and substantially reduces the subthreshold leakage component. GIDL component and gate leakage are also reduced due to the smaller gate-to-drain/source voltages.

13.6 Runtime Active Leakage Reduction Techniques

Not every application requires a fast circuit to operate at the highest performance level all the time. Active leakage techniques exploit this idea to intermittently slow down the fast circuitry and reduce the leakage power consumption as well as the dynamic power consumption when maximum performance is not required.

13.6.1 Dynamic V_{dd} Scaling (DVS)

Dynamic supply scaling overrides the cost of using two supply voltages (static supply scaling), by adapting the single supply voltage to the performance demand. The highest supply voltage delivers the highest performance at the fastest designed frequency of operation. When performance demand is low, supply voltage and clock frequency are lowered, just delivering the required performance with substantial power reduction. Implementing DVS in a general-purpose microprocessor system includes three key components:

1. An operating system that can intelligently vary the processor speed
2. A regulation loop that can generate the minimum voltage required for the desired speed
3. A microprocessor that can operate over a wide voltage range

Figure 13.10 depicts a DVS system architecture [37]. Control of the processor speed must be under software control, as the hardware alone may not distinguish whether the currently executing instruction is part of a computation-intensive task or a nonspeed-critical task. Supply voltage is controlled by hard-wired frequency-voltage feedback loop, using a ring oscillator as a critical path replica. All chips operate at the same clock frequency and same supply voltage, which are generated from the ring oscillator and the regulator.

13.6.2 Dynamic V_{th} Scaling (DVTS)

Similar to the dynamic VDD scaling (DVS) scheme, a dynamic V_{th} scaling (DVTS) scheme can be used to reduce the active leakage power in sub-100-nm generations where leakage power accounts for a large fraction of the total power consumption even during runtime. When the current workload is less than

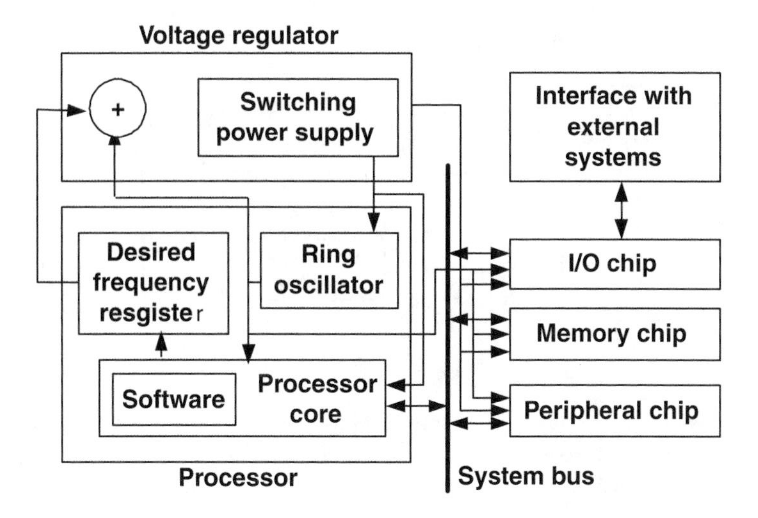

FIGURE 13.10 Dynamic voltage scaling architecture [37].

the maximum, the operating system commands a lower clock frequency to the hardware. Based on the given reference clock frequency, the DVTS hardware raises the transistor V_{th} via RBB to reduce the runtime leakage power dissipation. In cases when there is no workload at all, the V_{th} can be increased to its upper limit using body biasing, to significantly reduce the standby leakage power. "Just enough" throughput is delivered for the current workload by tracking the optimal V_{th} while leakage power is considerably reduced by intermittently slowing down the circuit.

Figure 13.11 plots the power consumption of DVTS and DVS systems for a speculative 70-nm process technology (only subthreshold leakage is considered) where leakage power accounts for 52% of total power dissipation (T = 125°C) [38]. By reducing the clock frequency without changing the V_{DD} or V_{th}, total power decreases in proportion to the operating frequency. This is because dynamic power is a linear function of clock frequency. The leakage power does not change with clock frequency, which makes 52% of total power wasted even when the clock frequency is zero. By dynamically scaling the V_{DD} together with the clock frequency, the speed requirement can be met while consuming significantly less power. Because the leakage power is dominant, DVTS appears to be comparable to DVS in saving total power for this technology. Figure 13.11 demonstrates that when the desired clock rate is 30% of the maximum operation frequency, 92% total energy savings can be achieved using DVTS. The following discussions address the merits and issues related to DVTS system designs.

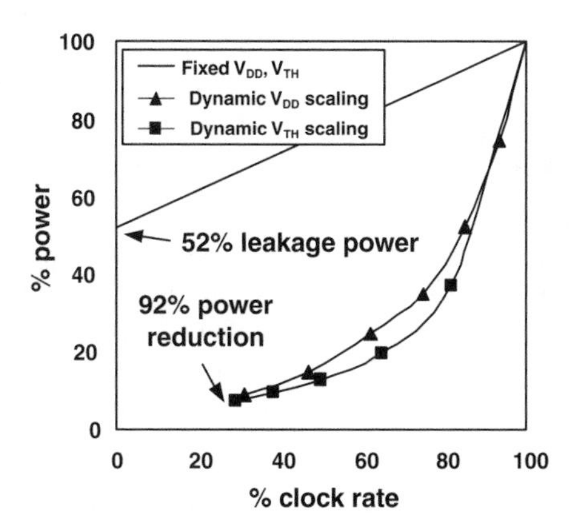

FIGURE 13.11 Total power vs. clock frequency for DVTS scheme and DVS scheme (BPTM, 70 nm, V_{DD} = 0.9 V, V_{th} = 0.15 V [38]).

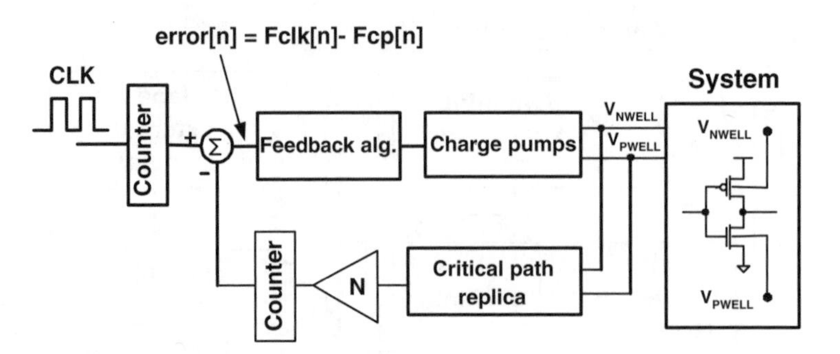

FIGURE 13.12 Dynamic V_{th} scaling system proposed in Kim and Roy [39].

- Simple hardware. Charge pumps are a simple solution for boosting voltages where current demand is low. No external inductors are needed and power consumption is very low compared with buck converters, which are used for DVS systems. Charge pumps are used in DVTS systems to generate the body bias voltages, which are outside the supply rail.
- Transition energy overhead. Results VTCMOS demonstrate that the energy overhead for a 120-K-transistor test chip in 0.3μm triple well technology consumes 10 nJ per V_{th} transition [32]. In case the V_{th} transition occurs frequently, transition energy overhead for DVTS systems becomes nonnegligible.
- Substrate noise. Charge pumps generate an unregulated body bias voltage due to the absence of external inductors. Any fluctuation in body bias will induce noise in logic.
- Process complexity. PMOS and NMOS body biases of the DVTS control circuit must be isolated from the target system in order to function as a reference. Thus, deep N-well or triple well technology is needed for the DVTS systems. The overall cost penalty by using these advanced processes is less than 5% [32].

Several different DVTS system implementations have been proposed in literature [39,40]. Figure 13.12 shows a DVTS hardware that uses continuous body bias control to track the optimal V_{th} for a given workload. A clock speed scheduler, which is embedded in the operating system, determines the (reference) clock frequency at runtime. The DVTS controller adjusts the PMOS and NMOS body bias so that the oscillator frequency of the critical path replica tracks the given reference clock frequency. The error signal, which is the difference between the reference clock frequency and the oscillator frequency, is fed into the feedback controller. The continuous feedback loop can also compensate for process, supply voltage, and temperature variations. A simpler method called "V_{th} hopping scheme," which dynamically switches between low V_{th} and high V_{th} depending on the performance demand, is proposed in [40]. The schematic diagram of the V_{th} hopping scheme is depicted in Figure 13.13. Compared with the continuous body bias control in Figure 13.12, the discrete control has two levels of V_{th}. If control signal VTHlow_Enable is asserted, the transistors in the target system are forward body biased and the V_{th} is low. When performance can be traded off for lower power consumption, VTHhigh_Enable is asserted and a high V_{th} is applied. The operating frequency of the target system is set to f_{CLK} when V_{th} is low and to $f_{CLK/2}$ when the V_{th} is high. An algorithm that adaptively changes the V_{th} depending on the workload is also verified and applied to an MPEG4 video encoding system. In future technology generations, the effectiveness of RBB is expected to be low due to the worsening SCE and increasing band-to-band tunneling leakage at the source/substrate and drain/substrate junctions. FBB can be applied together with RBB to achieve a better performance-leakage trade-off for DVTS systems.

13.7 Circuit Techniques to Reduce Leakage in Cache Memories

Figure 13.14(a) illustrates the seven available terminals in a conventional 6T SRAM cell; V_{SL}, V_{PWELL}, V_{NWELL}, V_{DL}, V_{WL}, V_{BL}, and V_{BLB}. Various SRAM cell architectures have been proposed in the past where

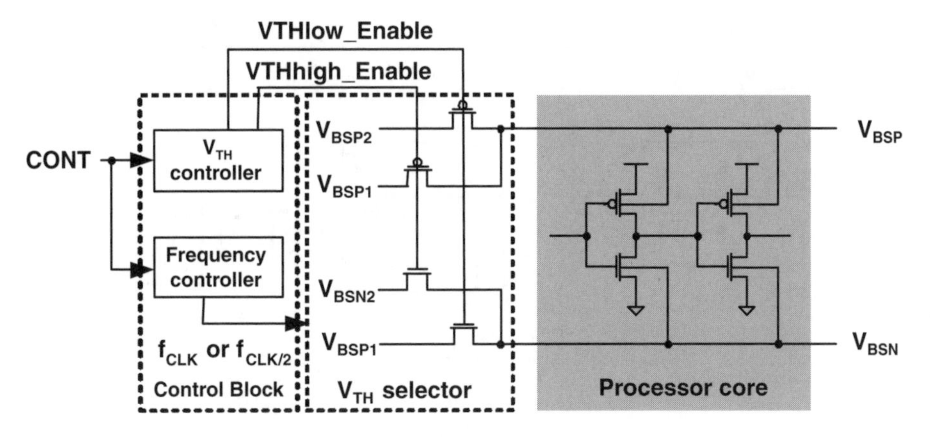

FIGURE 13.13 V_{th} hopping scheme proposed in Nose et al. [40].

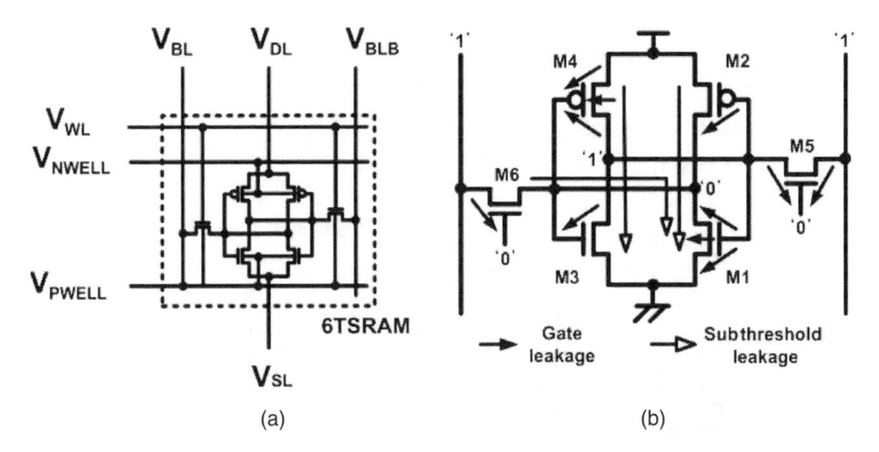

FIGURE 13.14 (a) Seven terminals of the 6-T SRAM cell, and (b) dominant leakage components in a 6-T SRAM.

one or more of the seven terminal voltages are controlled during standby mode for reducing the leakage components shown in Figure 13.14(b). Each technique exploits the fact that the active portion of a cache is very small, which gives the opportunity to put the large idle portion in a low-leakage sleep mode.

Effectiveness and overhead of each technique are evaluated based on the following discussions. First, the impact of the technique on various leakage components should be considered. Although subthreshold leakage still continues to dominate the I_{OFF} at high temperatures, ultra-thin oxides and high doping concentrations have led to a rapid increase in direct tunneling gate leakage and BTBT leakage at the source and drain junctions in the nanometer regime. Each leakage reduction technique needs reevaluation in scaled technologies where subthreshold conduction is not the only leakage mechanism. Second, the impact of the leakage reduction technique on SRAM read/write delay should be considered. Third, the transition latency/energy overhead should be taken into account, because of the limited time and energy budget for the mode transition. Last, the leakage reduction technique should not have a noticeable impact on SRAM cell stability or soft error rate (SER). Based on these discussions, the different low-leakage SRAM cells are summarized in Table 13.3.

A source biasing scheme raises the source line voltage (V_{SL}) in sleep mode [41–45], which reduces subthreshold leakage due to the three effects described in Section 13.5. The gate leakage in the cell is also reduced due to the relaxed signal rail, V_{DD}–V_{SL} (Section 13.2) [45]. An extra NMOS has to be series connected in the pull-down path to cut off the source line from ground during sleep mode, and this, in turn, imposes an extra access delay. The reduced signal charge in sleep mode also causes the soft error rate (SER) to rise, which requires additional error correction coding circuits [44].

TABLE 13.3 Low-Leakage SRAM Cell Techniques

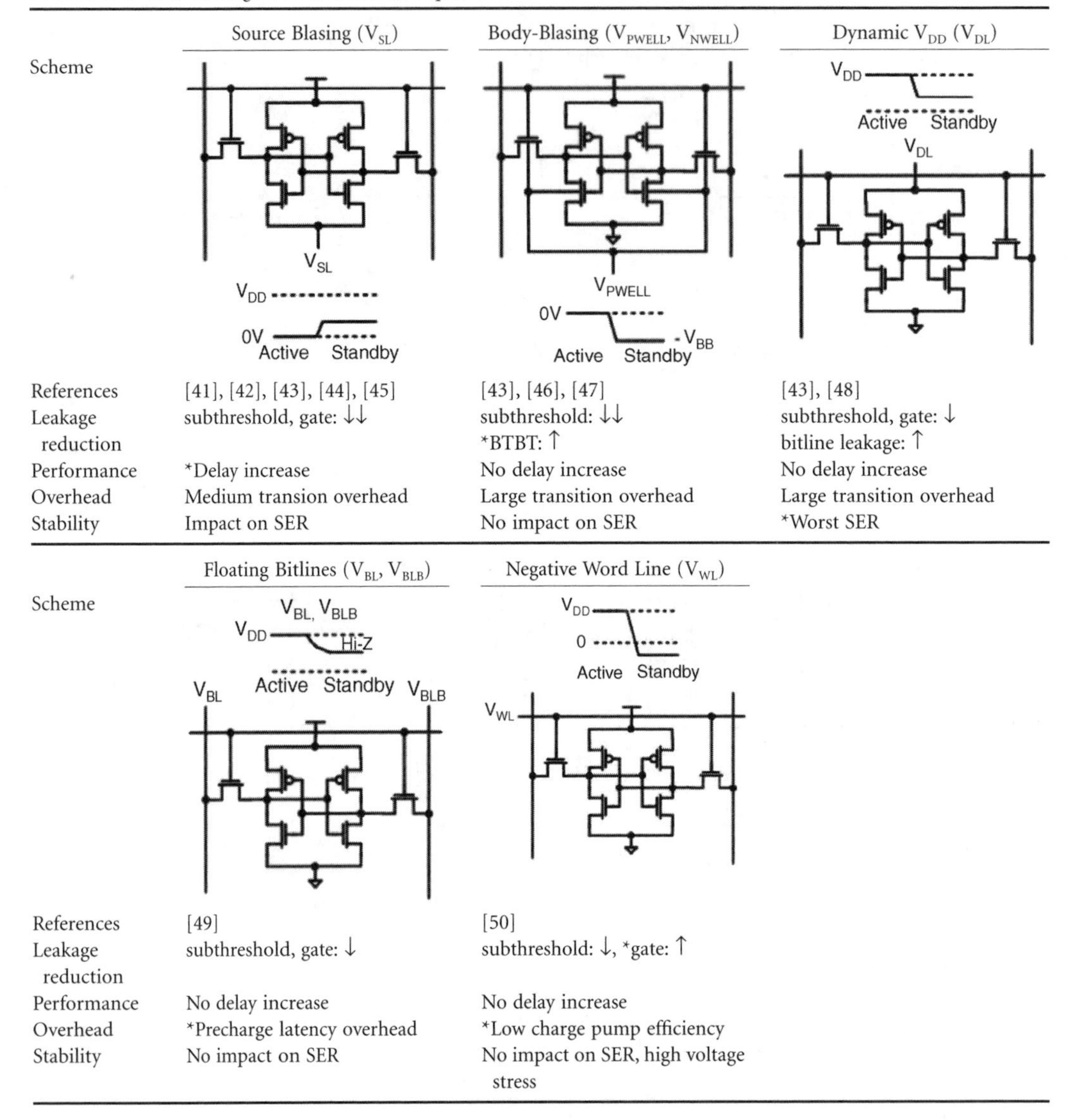

	Source Blasing (V_{SL})	Body-Blasing (V_{PWELL}, V_{NWELL})	Dynamic V_{DD} (V_{DL})
Scheme			
References	[41], [42], [43], [44], [45]	[43], [46], [47]	[43], [48]
Leakage reduction	subthreshold, gate: ↓↓	subthreshold: ↓↓ *BTBT: ↑	subthreshold, gate: ↓ bitline leakage: ↑
Performance	*Delay increase	No delay increase	No delay increase
Overhead	Medium transion overhead	Large transition overhead	Large transition overhead
Stability	Impact on SER	No impact on SER	*Worst SER

	Floating Bitlines (V_{BL}, V_{BLB})	Negative Word Line (V_{WL})
Scheme		
References	[49]	[50]
Leakage reduction	subthreshold, gate: ↓	subthreshold: ↓, *gate: ↑
Performance	No delay increase	No delay increase
Overhead	*Precharge latency overhead	*Low charge pump efficiency
Stability	No impact on SER	No impact on SER, high voltage stress

Reverse body-biasing (RBB) the NMOS (or PMOS) can reduce subthreshold leakage via body effect, while not affecting the access time by switching to zero body-biasing (ZBB) in the active mode [43,46,47]. A large latency/energy overhead is imposed for the body-bias transition due to the large V_{BB} swing and substrate capacitance. This scheme becomes less attractive in scaled technologies because the body coefficient decreases with smaller dimensions, and the source/drain junction BTBT leakage becomes enhanced by RBB.

Supply voltage is lowered in a dynamic V_{DD} SRAM (DVSRAM) [43,48], which, in turn, reduces the subthreshold, gate, and BTBT leakage. This scheme requires a smaller signal rail (V_{DL}–V_{GND}) compared with the SBSRAM for equivalent leakage savings. Although there is no impact on delay in the active mode, the large V_{DD} swing between sleep and active mode imposes a larger latency/energy transition overhead than the SBSRAM. Moreover, the greatest drawback of the DVSRAM is that it increases the bitline leakage in the sleep mode because the voltage level in the stored node also drops as the V_{DD} is lowered. Therefore, this scheme is not suitable for dual V_{th} designs where the speed-critical access transistors may already be using low V_{th} devices with high leakage levels.

A technique that biases the bitlines to an intermediate level has been proposed to reduce the access transistor leakage via the DIBL effect [49]. Because only the access transistors benefit from the leakage reduction, the overall leakage savings is moderate. Unlike the three previously mentioned techniques, this scheme has to be applied to the entire subarray because the bitline is shared across different cache lines. The main limitation comes from the fact that there is a precharge latency whenever a new subarray is accessed. This would mean that an architectural modification is required in order to resolve the multiple hit times in case the precharge instant is not known ahead of time.

The negative word line scheme [50] pulls down the V_{WL} to a negative voltage during standby in order to avoid the subthreshold leakage through the access transistors. However, it has issues such as increase in gate leakage and higher voltage stress in the access transistors. Although this technique has no impact on performance or SER, a power loss occurs due to the generation of a negative bias using charge pumps. This becomes more serious as the supply voltage is scaled.

References

[1] Borkar, S., Design challenges of technology scaling, *IEEE Microelectron.,* 19(4), 23, 1999.

[2] Brews, J., *High-Speed Semiconductor Devices,* Sze, S.M., Ed., John Wiley & Sons, New York, 1990, chap. 3.

[3] Roy, K. and Prasad, S.C., *Low-Power CMOS VLSI Circuit Design*, Wiley Interscience Publications, New York, 2000, chap. 5, 224.

[4] International Technology Roadmap for Semiconductors (ITRS), 2001 ed., Semiconductor Industry Association, http://public.itrs.net/Files/2001ITRS/Home.htm.

[5] Taur, Y. and Ning, T.H., *Fundamentals of Modern VLSI Devices,* Cambridge University Press, New York, 1998, chap. 2, 94.

[6] Mukhopadhyay, S. and Roy, K., Accurate modeling of transistor stacks to effectively reduce total standby leakage in nano-scale CMOS circuits, *Symp. of VLSI Circuits,* June 12–14, 2003.

[7] Roy, K., Mukhopadhyay, S., and Mahmoodi-Meimand, H., Leakage current mechanisms and leakage reduction techniques in deep-submicron CMOS circuits, *Proc. IEEE,* 2003.

[8] Taur, Y. and Ning, T.H., *Fundamentals of Modern VLSI Devices,* Cambridge University Press, New York, 1998, chap. 3, 120.

[9] Schuegraf, K. and Hu, C., Hole injection SiO_2 breakdown model for very low voltage lifetime extrapolation, *IEEE Trans. on Electron. Devices,* 41, 761, 1994.

[10] Choi, C., Nam, K., Yu, Z., and Dutton, R.W., Impact of gate direct tunneling current on circuit performance: a simulation study, *IEEE Trans. on Electron. Devices,* 48, 2823, 2001.

[11] Cao, K. et al., BSIM4 gate leakage model including source drain partition, *IEDM Tech. Dig.,* 815, 2000.

[12] Wei, L., Chen, Z., Johnson, M., Roy, K., Ye, Y., and De, V., Design and optimization of dual threshold circuits for low voltage low power applications, *IEEE Trans. on VLSI Syst.,* 16, 1999.

[13] Karnik, T. et al., Total power optimization by simultaneous dual-V_t allocation and device sizing in high-performance microprocessors, In *ACM/IEEE Design Automation Conf.,* 486, June 10–14, 2002.

[14] Pant, P., Roy, K., and Chatterjee, A., Dual-threshold voltage assignment with transistor sizing for low-power CMOS circuits, *IEEE Trans. on Very Large-Scale Integration (TVLSI) Syst.,* 9, 390, 2001.

[15] Sirichotiyakul, S. et al., Stand-by power minimization through simultaneous threshold voltage selection and circuit sizing, *Proc. 36th ACM/IEEE Conf. on Design Automation,* 436, June 21–25, 1999.

[16] Kao, J.T. and Chandrakasan, A.P., Dual-threshold voltage techniques for low-power digital circuits, *IEEE J. of Solid-State Circuits,* 35, 1009, 2000.

[17] Sirisantana, N., Wei, L., and Roy, K., High-performance low-power CMOS circuits using multiple channel length and multiple oxide thickness, *Proc. 2000 Int. Conf. on Computer Design,* 227, Sept. 17–20, 2000.

[18] Krishnamurthy, R.K., Alvandpour, A., De, V., and Borkar, S., High-performance and low-power challenges for sub-70-nm microprocessor circuits, *Proc. IEEE Custom Integrated Circuits Conf.*, 125, May 12–15, 2002.

[19] Takahashi, M. et al., A 60-mW MPEG4 video CODEC using clustered voltage scaling with variable supply-voltage scheme, *IEEE J. of Solid-State Circuits,* 33, 1772, 1998.

[20] Ye, Y., Borkar, S., and De, V., A new technique for standby leakage reduction in high performance circuits, *IEEE Symp. on VLSI Circuits,* 40, June 11–13, 1998.

[21] Chen, Z., Wei, L., Johnson, M., and Roy, K., Estimation of standby leakage power in CMOS circuits considering accurate modeling of transistor stacks, *IEEE Int. Conf. on Comput.-Aided Design,* Aug. 10–12, 1998.

[22] Chen, Z., Wei, L., Keshavarzi, A., and Roy, K., IDDQ testing for deep submicron ICs: challenges and solutions, *IEEE Design and Test of Comput.*, 24, 2002.

[23] Mukhopadhyay, S., Neau, C., Cakici, T., Agarwal, A., Kim, C.H., and Roy, K., Gate leakage reduction for scaled devices using transistor stacking, *IEEE Trans. on Very Large-Scale Integration Syst.*, 2003.

[24] Johnson, M.C., Somasekhar, D., and Roy, K., Leakage control with efficient use of transistor stacks in single threshold CMOS, *Proc. ACM/IEEE Design Automation Conf.*, 442, June 21–25, 1999.

[25] Mutoh, S. et al., 1-V Power supply high-speed digital circuit technology with multi-threshold voltage CMOS, *IEEE J. Solid-State Circuits,* 30, 847, 1995.

[26] Kao, J., Chandrakasan, A., and Antoniadis, D., Transistor sizing issues and tool for multi-threshold CMOS technology, *Proc. of ACM/IEEE Design Automation Conf.*, 495, June 9–13, 1997.

[27] Mutoh, S. et al., A 1-V multi-threshold voltage CMOS DSP with an efficient power management for mobile phone application, *Dig. Tech. Papers EEE Int. Solid-State Circuits Conf.,* 168, Feb. 8–10, 1996.

[28] Shigematsu, S. et al., A 1-V high-speed MTCMOS circuit scheme for power-down applications, *IEEE J. Solid-State Circuits,* 32, 861, 1997.

[29] Kawaguchi, H., Nose, K., and Sakurai, T., A CMOS scheme for 0.5-V supply voltage with pico-ampere standby current, *Dig. Tech. Papers IEEE Int. Solid-State Circuits Conf.,* 192, Feb. 5–7, 1998.

[30] Tschanz, J. et al., Dynamic-sleep transistor and body bias for active leakage power control of microprocessors, *IEEE Int. Solid-State Circuit Conf.*, 2003.

[31] Heo, S. and Asanovic, K., Leakage-biased domino circuits for dynamic fine-grain leakage reduction, *Symp. on VLSI Circuits,* 316, June 13–15, 2002.

[32] Kuroda, T. et al., A 0.9-V, 150-MHz, 10-mW, 4-mm^2, 2-D discrete cosine transform core processor with variable-threshold-voltage scheme, *Dig. of Tech. Papers IEEE Int. Solid-State Circuits Conf.,* 166, Feb. 8–10, 1996.

[33] Oowaki, Y. et al., A sub-0.1um circuit design with substrate-over-biasing, *Dig. of Tech. Papers of IEEE Int. Solid-State Circuits Conf.,* 88, Feb. 5–7, 1998.

[34] Keshavarzi, A., Hawkins, C.F., Roy, K., and De, V., Effectiveness of reverse body bias for low power CMOS circuits, *Proc. 8th NASA Symp. on VLSI Design,* 231, Oct. 1999.

[35] Keshavarzi, A. et al., Forward body bias for microprocessors in 130nm technology generation and beyond, *Symp. on VLSI Circuits,* 312, June 13–15, 2002.

[36] Mizuno, H. et al. An 18-μA standby current 1.8-V, 200-MHz microprocessor with self-substrate-biased data-retention mode, *IEEE J. Solid-State Circuits,* 34, 1999.

[37] Lee, S. and Sakurai, T., Run-time voltage hopping for low-power real-time systems, *Proc. IEEE/ACM Design Automation Conf.*, 806, June 5–9, 2000.

[38] UC Berkeley device group, Predictive technology model, http://www-device.eecs.berkeley.edu/~ptm /.

[39] Kim, C.H. and Roy, K., Dynamic V_{th} scaling scheme for active leakage power reduction, *Design, Automation and Test in Europe,* 163, March 5–8, 2002.

[40] Nose, K. et al., V_{th}-hopping scheme for 82% power saving in low-voltage processors, *Proc. IEEE Custom Integrated Circuits Conf.*, 93, May 6–9, 2001.

[41] Agarwal, A., Li, H., and Roy, K., A single-V_t low-leakage gated-ground cache for deep submicron, *IEEE J. of Solid-State Circuits,* 2003.
[42] Yamauchi, H. et al., A 0.8V/100MHz/sub-5mW-operated mega-bit SRAM cell architecture with charge-recycle offset-source driving (OSD) scheme, *Symp. on VLSI Circuits,* 126, June 13–15, 1996.
[43] Bhavnagarwala, A.J., Kapoor, A., and Meindl, J.D., Dynamic threshold CMOS SRAMs for fast, portable applications, *ASIC/SOC Conf.,* 359, 2000.
[44] Osada, K. et al., 16.7fA/cell tunnel-leakage-suppressed 16Mb SRAM for handling cosmic-ray-induced multi-errors, *Int. Solid-State Circuits Conf.,* 302, 2003.
[45] Agarwal, A. and Roy, K., Noise-tolerant cache design to reduce gate and subthreshold leakage in nanometer regime, In *Int. Symp. Low-Power Electronics and Design (ISLPED2003),* Aug. 25–27, 2003.
[46] Kawaguchi, H., Itaka, Y., and Sakurai, T., Dynamic leakage cut-off scheme for low-voltage SRAMs, *Symp. on VLSI Circuits,* 140, June 11–13, 1998.
[47] Kim, C.H. and Roy, K., Dynamic V_t SRAM: a leakage-tolerant cache memory for low-voltage microprocessors, *Int. Symp. on Low-Power Elecron. and Design,* 251, Aug. 12–14, 2002.
[48] Flautner, K. et al., Drowsy caches: simple techniques for reducing leakage power, *Int. Symp. on Comput. Architecture,* 148, May 25–29, 2002.
[49] Heo, S. et al., Dynamic fine-grain leakage reduction using leakage-biased bitlines, *Int. Symp. on Comput. Architecture,* 137, May 25–29, 2002.
[50] Itoh, K., Fridi, A.R., Bellaouar, A., and Elmasry, M.I., A deep sub-V, single power-supply SRAM cell with multi-V_t, boosted storage node and dynamic load, *Symp. on VLSI Circuits, Dig. of Tech. Papers,* 132, June 13–15, 1996.

14

Low-Power and Low-Voltage Communication for SoCs

Christer Svensson
Linköping University

14.1 Introduction

Integrated circuits (ICs) mainly consist of transistors and interconnects. Normally, we are more interested in the transistors and how they are combined to form logic gates, flip-flops, memories, and other functional units. Interconnects are easily overlooked because they are just nodes in a circuit diagram; but interconnects are responsible for all communication between logical gates, functional units, and subsystems and are, therefore, of crucial importance. In reality, interconnects are one or several wires with various lengths, which connect transistors and blocks over various distances. Their behavior strongly depends on their lengths. When discussing power consumption, interconnect tends to dominate the power consumption, due to their large total capacitance [1–3]. In Liu and Svensson [2], 30 to 40% of the power consumed by a chip (input/output, I/O, excluded) is estimated to be related to interconnect and an additional 40% to the clock distribution (of which some half is related to the wires). In Chandra et al. [3], about 70% of the power of a high-performance chip (microprocessor) is estimated to be related to interconnect and clock in the 180-nm technology node. It is, therefore, well motivated to consider interconnect power separately.

Very short wires, connecting transistors inside small blocks, as for example simple logic gates, are most easily treated as parasitic capacitance added to other parasitic capacitances of similar size, as gate or drain capacitances [4]. For interconnect between blocks, we normally see a broad distribution of wire lengths

0-8493-1941-2/05/$0.00+$1.50

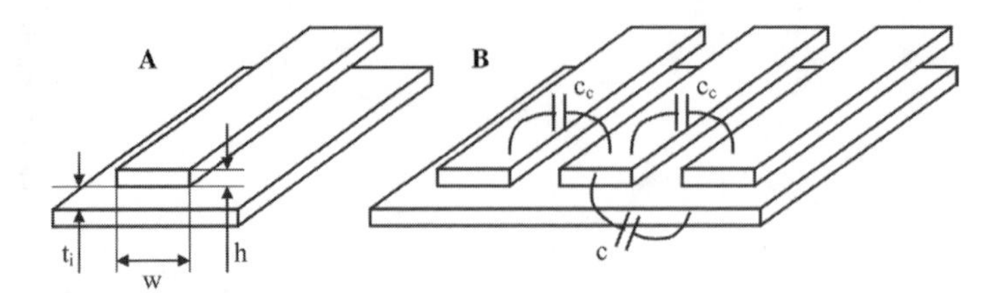

FIGURE 14.1 Single and multiple microstrip wire.

[4]. Here, interconnect capacitance easily dominates over other capacitances, making interconnect power consumption substantial.

When looking outside the chip, we normally have quite large capacitive or resistive loads, making the I/O power substantial. As much as 60% of the total chip power consumption can be related to I/O [2]. Because the outside load is largely controlled by the properties of interconnect between chips (on the circuit board), we will include a treatment of I/O power consumption. Outside the chip, interconnect may become electrically long, with the result that they behave more as transmission lines than as capacitive lines [4], making their behavior quite different, also in a power perspective. For large enough clock frequencies, on-chip wires may also approach transmission line behavior in the future.

The continuous scaling of IC processes also introduces changes in the metal stack [5]. Most important changes are an increased aspect ratio (metal height/width) of the wires and an increased number of metal layers, with larger dimensions of the upper layers. The increased aspect ratio leads to an increased wire-to-wire capacitance, compared to wire-to-ground capacitance, which will increase power consumption. The increased number of metal layers offers new opportunities, as the upper layers have much lower loss than the lower layers.

We start with a brief discussion of the properties of wires and interconnects, then describe power modelling of interconnects, and, finally, we discuss various methods to reduce interconnect power consumption.

14.2 Basics of Wires

14.2.1 General

The simplest model of a wire is the microstrip, that is a metal strip on top of a ground plane (Figure 14.1(a)). Such a wire can be modelled through its capacitance to ground per unit length, c, its resistance per unit length, r and its inductance per unit length, l. In the simplest case, very short wires, capacitance is sufficient as a model. For longer wires, we need to add resistance; for very long wires, inductance will also become important. For the microstrip, we may approximate the capacitance per unit length as [6]:

$$c = c_{pp} + c_{fringe} = \frac{\varepsilon_i (w - h/2)}{t_i} + \frac{2\pi\varepsilon_i}{\log(2 + 4t_i/h)} \tag{14.1}$$

where c_{pp} is the “plate capacitor” capacitance per unit length, c_{fringe} is the fringing capacitance per unit length, w is the wire width, t_i is the insulator thickness, h is the wire thickness, and ε_i is the insulator dielectric constant.

More generally, the wire is surrounded by many other wires (Figure 14.1(b)). We then have capacitances not only to ground, but also to the neighboring wires, the coupling capacitances per unit length, c_c. A reasonable model here is to use the principle of superposition of signals. We thus assume that all wires except the actual one are grounded, and describe it through its total capacitance to its surroundings. The effect of any signal on a neighboring wire is then treated by calculating the crosstalk from each neighboring wire to the actual one, and add this voltage to the voltage on the actual wire. Most important is thus the total capacitance per unit length, c_{tot}, which may replace c in Equation 14.1. In some cases, crosstalk effects may increase power consumption above the superposition model described above (see Section 14.3.1). The situation becomes even more complex if we have no ground plane or other metal in parallel with the actual wire. In such cases, the return current may take a complex path leading to a large inductance, which is quite hard to predict. Such situations are not discussed here.

For longer wires, we also need to consider the series resistance per unit length, r, given by

$$r = \frac{\rho}{A} \tag{14.2}$$

where ρ is the metal resistivity and A = hw the wire cross section. For longer wires at high speed, we may need to include also the inductance per unit length, l. In such a case, the wire behaves as a transmission line (assuming a ground plane or regular return path) [4,7]. A transmission line has the following properties: it is considered to carry two waves, one in each direction. Each wave moves with velocity v_d, given by:

$$v_d = \frac{c_0}{\sqrt{\varepsilon_r}} \tag{14.3}$$

where c_0 is the velocity of light in vacuum, and ε_r is the relative dielectric constant of the dielectric. Furthermore, for each wave its characteristic impedance, Z_0, relates voltage and current. Z_0 is real at high frequencies. An ideal transmission line has no resistance and transports any wave without attenuation or distortion. A transmission line with resistance is said to be lossy and will attenuate each wave as $e^{-\beta L}$, where β is the attenuation factor:

$$\beta = \frac{r}{2Z_0} \tag{14.4}$$

For $\beta L > 1$, we may consider the wire an RC-wire, as described previously. For $\beta L < 1$, we may consider it a lossy transmission line. A lossy transmission line also exhibits skin effect, which is the resistance becomes frequency dependent, making the wire distort high-speed signals [7].

Transmission lines with low loss should normally be terminated, that is connected to the impedance Z_0, at least at one of its ends. The reason for this is to avoid reflections. If a wave travels in x-direction along the wire and meet Z_0, it will be completely absorbed; however, if it instead meets, for instance, $Z = \infty$, then a new wave in $-x$ direction is generated (to fulfill the voltage to current relation), which travels in the opposite direction on the wire. This is also termed "full reflection." When the reflected wave reaches the terminated transmitter, it will however be absorbed with no further action. From this, we can understand that if the wire is incorrectly terminated in both ends, we will have waves bouncing back and fourth several times, thus leading to uncontrolled "ringing." If the wire is terminated in one end, we control the signal much better. To ensure full control, we often prefer to terminate the wire in both ends.

On-chip wires can normally be treated as RC-lines. Very long global lines in thick upper level metal may sometimes be considered lossy transmission lines [7]. Off-chip lines (on printed circuits boards for example) normally behave as lossy transmission lines.

14.2.2 Interconnect Delays

Any wire-carrying signal is driven by some circuit. The simplest driver is the inverter. We will therefore use a simple inverter model when discussing driven wires in this section. The simple inverter is described by its Thevenin equivalent (Figure 14.2). For a short wire with length L, we may describe the wire with its total capacitance, $C_w = c_{tot}L$, and we may then estimate the delay to:

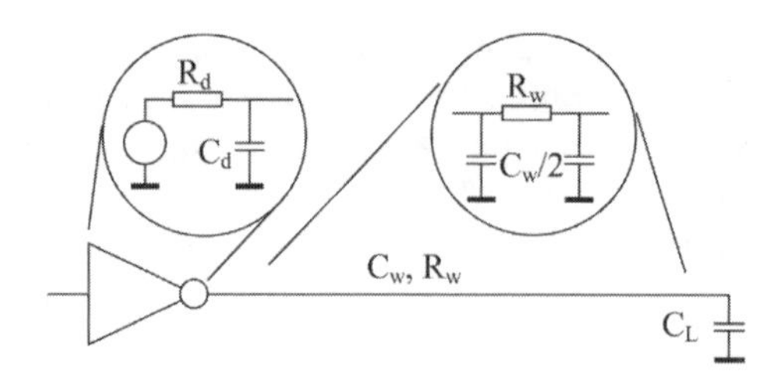

FIGURE 14.2 Inverter driver and wire with lumped element models in the inserts.

$$t_d = R_d(C_d + C_w + C_L)\log(2) \tag{14.5}$$

where R_d and C_d are the driver output resistance and capacitance, respectively, and C_L is the load capacitance.

For a longer wire, including wire resistance $R_w = rL$, we may approximate the delay to [4]:

$$t_d = \left(R_d(C_d + C_w + C_L) + R_w\left(\frac{C_W}{2} + C_L\right)\right)\log(2) \tag{14.6}$$

This expression includes a term, R_wC_w, which is proportional to wire length squared, L^2 (as each of R_w and C_w is proportional to wire length). This leads to very long delays for long wires.

In the case of crosstalk from neighboring wires an increased delay may occur [8]. Consider the case of a positive transition of amplitude ΔV on the actual wire interacting with a negative transition of the same amplitude on its neighbor. The coupling capacitance between the two wires, $C_c = c_cL$, thus change its voltage from ΔV to $-\Delta V$, thus $2\Delta V$, making the transient current through the capacitor double comparing to the noncrosstalk case. The capacitance thus appears as a capacitance to ground of value $2C_c$, increasing wire delay. For two neighbors (in a bus) we may experience a worst-case capacitance of $4C_c$, to be considered in worst-case delay calculations. This phenomenon is similar to the Miller effect in inverters and affects power consumption (see Section 14.3.1).

For a transmission line, finally, the interconnect delay is given by the velocity of light in the actual dielectric:

$$t_d = \frac{L}{v_d} \tag{14.7}$$

14.2.3 Wires with Repeaters

As noted earlier, wire delay grows fast with wire length. One way to mitigate this is to introduce repeaters along the wire [8]. Let us divide the wire into m sections of length $L_s = L/m$ each and attach a repeater (an inverter) at each section (Figure 14.3). Minimum delay occurs for a certain section length, given by:

$$L_{sopt} = \sqrt{\frac{t_{p1}}{0.38rc}} \tag{14.8}$$

where t_{p1} is the delay of an inverter driving another similar inverter.

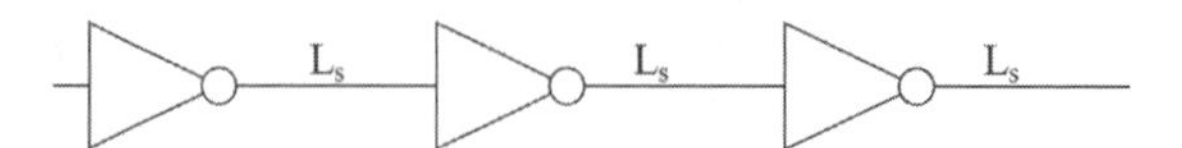

FIGURE 14.3 Wire with repeaters.

The minimum delay occurs at an inverter size (compared to the minimum size) given by:

$$s_{opt} = \sqrt{\frac{R_d c}{r C_d}} \tag{14.9}$$

and will have the value:

$$t_{d\min} = \left(1.02 + \frac{1.38}{\sqrt{1+\gamma}}\right)\sqrt{rct_{p1}}\, L \tag{14.10}$$

Here, R_d and C_d are the minimum inverter output resistance and input capacitance respectively, and γ is the ratio between the inverter input capacitance and its output capacitance. The total delay of the wire with repeaters is thus proportional to L instead of L^2, as described previously. The total capacitance of a section of the wire, C_s is the sum of the wire capacitance, $c_{tot}L_{sopt}$ and the inverter capacitances, sC_d and γsC_d. Calculating C_s/L_{sopt} yields:

$$\frac{C_s}{L_{sopt}} = c + 0.74\sqrt{(1+1/\gamma)}c \tag{14.11}$$

The total switching capacitance of the wire including the optimized repeaters is thus larger than the wire itself (c), leading to corresponding higher power consumption. With a typical value of $\gamma = 1$, the capacitance overhead is about 100%.

14.3 Power Consumption Related to Interconnect

14.3.1 Basics

As described earlier, short interconnects are just described by their total capacitance, C_w. For longer interconnects that are still electrically short, we describe them as RC-lines or by an RC-lumped model. In both cases, power consumption occurs as in logic, which is the interconnect capacitance is charged by the supply voltage and then discharged, making the power consumption related to the interconnect given by:

$$P_w = \frac{1}{2}\alpha f_c C_w \Delta V^2 \tag{14.12}$$

where α is the signal activity (the probability that the signal will change per clock cycle), f_c is the clock frequency, and ΔV the signal voltage swing. Here, we assumed that the driver is an ideal inverter, that is a switch connected either to ΔV or to ground. Equation (14.12) is also valid for a switch with series resistance (e.g., an inverter where the transistors have series resistance) driving an open wire. We included only C_w in the expression. For the full interconnect including driver and load, we should include the capacitances of these as well.

The fact that power consumption depends on signal activity (described by α in Equation (14.12)) leads to severe difficulties in power prediction and, therefore, in power optimization. Any prediction and optimization must consider signal activity, which depends on actual data statistics and therefore on actual architectures and the applications run on these. One consequence is that total interconnect length is not sufficient for power estimation. Instead, we need individual wire lengths and individual signal activities [9].

For wires with crosstalk, the worst-case power consumption may be larger than predicted by Equation (14.12), due to the Miller effect discussed previously. If there is a transition with opposite polarity on a neighboring wire, the effective value of the coupling capacitance, C_c, may double, thus increasing the power consumption through a larger C_w in Equation (14.12). This effect was analyzed in [10], where also methods to reduce the effect was discussed. The effect depends on the correlation between neighboring signals and is quite sensitive to the exact timing relation between the two transitions (see Sasaki et al. [11]).

For transmission lines, we have two main cases: a wire that is terminated to Z_0 at the far end and a wire that is open at the far end. The terminated case is very simple, as the input impedance to a terminated wire simply is Z_0. The wire thus behaves as a resistor with resistance Z_0. Assuming it is driven by a driver with output impedance Z_0, we get a power consumption of

$$P_w = \frac{V_{dd}^2}{4Z_0} \tag{14.13}$$

where V_{dd} is the supply voltage, and we assumed equal probability for ones and zeros. Note that the voltage swing is smaller than the supply voltage in this case, $\Delta V = V_{dd}/2$. The open case was treated in Svensson [7]. We may understand the behavior in terms of forward and reflected waves. If the driver output changes from low to high at $t = 0$, it creates a forward wave of amplitude $V_{dd}/2$ in the wire (assuming a driver output impedance of Z_0). This wave is reflected at the far end of the wire, creating a backward wave of amplitude $V_{dd}/2$. After time $2T_d$, where T_d is the wire delay, the voltage at the wire input becomes V_{dd}. If the input driver still is driving high, the current becomes zero and the total charge driven into the wire is $2T_dV_{dd}/Z_0$. This charge can be shown to be equal to the wire capacitance charged to V_{dd}. If the driver instead has changed to driving zero after time $2T_d$ (which may occur if the electrical wire length T_d is larger than half the data symbol length, T_s), the backward wave is terminated to ground and will partly discharge the wire. We can imagine that for a long open wire, the whole data sequence sent will return after time $2T_d$, and depending on which state the driver is in (output connected to V_{dd} or ground), the return current will either cancel the V_{dd} current or go to ground. The average current consumption therefore depends on the correlation between sent data at times t and $t+2T_d$. In Svensson [7], this average was calculated for random data. Thus, we have:

$$P_w = \frac{1}{4} f_c C_w V_{dd}^2 \qquad 2T_d < T_s \tag{14.14}$$

$$P_w = \frac{V_{dd}^2}{8Z_0} \qquad 2T_d > T_s \tag{14.15}$$

We may then conclude the transmission line case as follows (assuming a random sequence of binary data and a driver with output impedance Z_0). For transmission lines terminated by Z_0 in the far end, the power consumption is simply the same as that of a resistor of value $2Z_0$ (driver resistance in series with load resistance). For an open transmission line, we have two cases: for the electrically short line, the power consumption is the same as of a capacitor of value C_w. For an electrically long line, it is half of the terminated line case with the same supply voltage (note, however, that voltage swing in the terminated case is $V_{dd}/2$, and, in the open case, is V_{dd}) (see Figure 14.4).

14.3.2 Power Consumption Related to Drivers and Repeaters

Because wire capacitance often is quite large, the driver must be upsized to facilitate short delay and fast rise-time. For driving large loads, we normally use a tapered inverter chain in order to minimize delay. This means that we have an upsized inverter to drive the wire and then a multistage predriver to drive the upsized inverter. The power consumption related to the driver itself (i.e., on top of the power

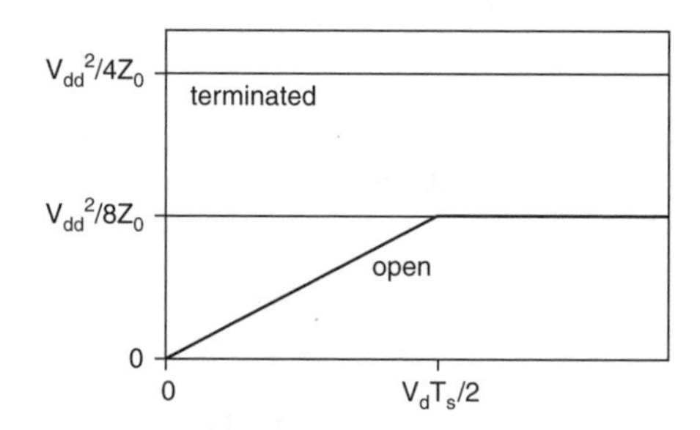

FIGURE 14.4 Power consumption of transmission lines vs. length for binary symbols with length T_s. Two cases are depicted: a line terminated by Z_0 and an open line.

related to its load) also becomes significant and proportional to the load capacitance. The dynamic portion of the driver power consumption can be expressed through the total switched capacitance of the driver, C_d [12]:

$$C_d = \frac{(1+1/\gamma)}{(f-1)}(C_w + C_L) \tag{14.16}$$

where f is the tapering factor (size ratio between following inverters in the chain; the value for minimum delay is about 3.5). This leads to a power overhead of about 80% of the power consumed by wire and load ($C_w + C_L$). Drivers are also vulnerable to additional power consumption due to short-circuit power, because of their large loads. In the case of drivers aimed for driving external loads (I/O), there may exist additional constraints, asking for more complex circuits than a simple inverter [13]. There may be needs for accurate output impedance, for tri-state outputs, for wired-or capability, for differential signals, for mitigating short-circuit current and for rise-time control. Such additional constraints normally leads to a larger power consumption compared with the simple inverter model discussed previously, either directly, through for example voltage loss in the circuit, or indirectly, through increased transistor sizes leading to larger capacitance and larger predrivers. These issues are further discussed in Section 14.4.5.

As mentioned earlier, repeaters are often used on chip to optimize wire delays. Repeaters are also utilized to mitigate crosstalk in long wires. The dominating power consumption in well-designed repeaters is dynamic power, thus following Equation (14.12) (assuming electrically short wires) with C_w replaced by the sum of C_w and the total switching capacitance of the repeaters:

$$C_{tot} = \frac{C_s}{L_s}L \tag{14.17}$$

As mentioned earlier, C_{tot} is about 100% larger than C_w for optimized delay. However, the delay minimum is very shallow, so in order to save power it is preferable to have smaller repeaters at longer distances than optimum, which saves considerable power at a small delay penalty [14].

14.3.3 Power Related to Precharged Buses

Sometimes wires are precharged instead of driven statically [8]. This may speed up a wire with large capacitance. It also mitigates the detrimental effect of coupling capacitance (i.e., the Miller effect) on both delay and power consumption. The reason for this is that signals on neighboring wires are monotonous, that is they cannot change in opposite direction. However, precharged nodes always have a large activity, about 0.5, independent of the signal activity, α, thus increasing the power consumption compared to a static bus [12].

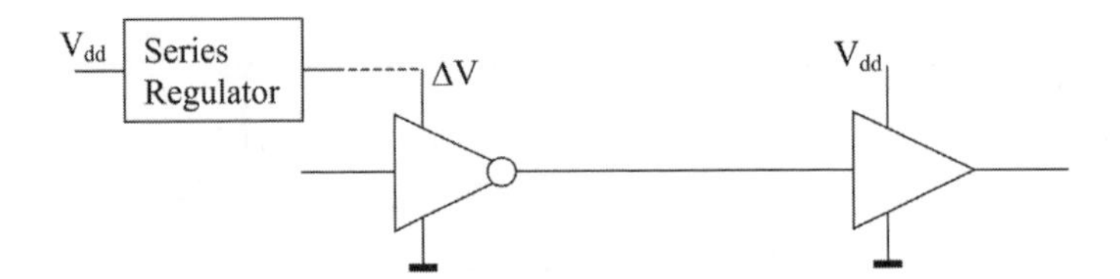

FIGURE 14.5 Transmitter, wire, and receiver.

14.4 Strategies for Power Savings in Interconnect

14.4.1 Introduction

As mentioned in the introduction to this chapter, power consumption related to interconnect is substantial. It may therefore be very profitable to find methods to reduce interconnect-related power consumption. An obvious method is of course to limit long distance communications. This is a very important route to reduced power, but an architectural issue and therefore outside the scope of this chapter. It may be important to minimize wire capacitance through changes in the fabrication process (Equation (14.12) and Equation (14.14)). Such work is ongoing, by searching for dielectrics with lower dielectric constants. At the same time, however, scaling tends to increase capacitance through changes in wire aspect ratios leading to larger coupling capacitances [5].

A very efficient power savings method is to reduce the signal voltage swing (Equation (14.12)). This method normally leads to some delay penalty, which may be hard to accept in high performance systems. Particularly in wires with repeaters, reduced swing is tricky and leads to delay penalties. A possible route is to avoid repeaters by utilizing the lower loss in an upper level, thicker, metal layer [15,16]. If repeaters must be used, their delay may be traded for lower power consumption.

Another powerful method for power saving is to reduce the product of data activity and capacitance. One way to accomplish this is to consider data activity during floorplanning and routing, thus facilitating power optimization based on wire activity/length product. Another way is to reduce data activity on buses, and minimize the amount of crosstalk-related power consumption. This can be accomplished through coding. Several coding methods have been evaluated for power saving in data buses. Utilizing precharged buses could be seen as a special case of data coding.

All methods mentioned can be utilized for on-chip as well as off-chip wires. Off-chip interconnect is, however, more vulnerable to noise, often motivating special solutions. Finally, some more exotic methods are charge redistribution, charge recovery, and adiabatic methods. Let us discuss each of these possibilities below.

14.4.2 Reduced Voltage Swing

Reduced voltage swing is often an effective way to save power in logic. This is true also for wires as can be seen in Equation (14.12) through Equation (14.15). If we consider interconnect between two logic blocks, we could reduce the voltage at the transmitter, thus saving power of the wire. To drive the logic at the receiving side, however, we normally need to restore the reduced voltage to the normal logic levels again [8]. We thus need an amplifying receiver that causes a delay penalty and needs power. For high data rates, we may find an optimum voltage swing, for which the total power (wire and receiver) is minimum [15,17]. For lower data rates, this optimum voltage becomes very small and will instead be limited by noise.

Let us study a complete link with driver, interconnect, and receiver (Figure 14.5). The simplest possible driver is an inverter, driven by a reduced supply voltage ΔV. Let us for simplicity assume that one voltage level is equal to ground, which means that the interconnect voltage is 0 or ΔV. The current consumption of the driver when driving an electrically short wire is then:

$$I_D = \frac{1}{2}\alpha f_c C_{tot} \Delta V \tag{14.18}$$

where C_{tot} is the total load of the driver (i.e., driver output capacitance, wire capacitance, and receiver input capacitance). The power consumption now depends on from where we take the current. If the current is taken from a separate power supply of voltage ΔV, then the power consumption becomes:

$$P_w = I_D \Delta V = \frac{1}{2} \alpha f_c C_w \Delta V^2 \tag{14.19}$$

If, on the other hand, the current is taken from the ordinary supply at voltage V_{dd}, assuming that ΔV is generated from V_{dd} through a lossless series regulator, we get:

$$P_w = I_D V_{dd} = \frac{1}{2} \alpha f_c C_w V_{dd} \Delta V \tag{14.20}$$

The latter case also describes circuits where the voltage swing reduction is obtained through a series diode or MOS diode (for example reducing the swing from V_{dd} to $V_{dd}-V_T$ by an MOS diode in series with V_{dd}).

We thus have a transmitter (for simplicity assumed to be an inverter) and a receiver amplifier (Figure 14.5), which should amplify the reduced swing signal to full logic swing (assumed to be V_{dd}). The amplifier speed and gain will be limited by the gain-bandwidth product (or f_T) of the actual process. This leads to a substantial delay in the receiver if the gain is large (swing is low). Furthermore, the amplifier must be designed to fit the actual gain needed. It turns out that the number of amplifier stages and the transistor sizes of the amplifier depend on the gain needed, resulting in a power consumption that increases with gain and speed requirements [17]. A lower swing therefore saves power in interconnect but leads to a larger power consumption of the receiver amplifier. Therefore, we may see an optimum voltage swing corresponding to minimum total power consumption [17]. As an example, a 4-mm long open wire, run at 5 Gb/s, using an amplifier in a 0.18-μm process and using V_{dd} = 1.8 V as supply voltage show a minimum power consumption of 0.7 mW at a voltage swing of 200 mV [15]. Several implementation schemes have been proposed for low-swing interconnect [8,12,15].

If the interconnect use repeaters, a reduction in voltage swing is less obvious. We then need a repeater that accept the reduced swing input and generate the same reduced swing at the output. As repeaters normally are used to mitigate delay, the new repeater should not increase delay. The simplest solution to this problem is to use an inverter with a reduced supply voltage of ΔV as repeater. It will then automatically adjust to as well the reduced input voltage as to the reduced output voltage. We will then still follow Equation (14.8) through Equation (14.11), but we must note that the reduced supply voltage leads to increased values of t_{p1} and R_d because of a smaller effective gate voltage and, therefore, a reduced current driving ability of the transistors. Reducing voltage swing thus leads to an increase in L_{sopt} (which is good as we get less repeaters) and an increased minimum delay (Equation (14.10)). Some simulations in a 0.13-μm process indicate that a swing of 75% of the supply voltage leads to a power saving of 23% (assuming that V_{dd} is used as transmitter supply) at a delay penalty of 15% [18]. Larger power savings leads to considerably larger delay penalty, which is often not accepted. An alternative solution to the simple inverter could consist of a standard receiver amplifier followed by a standard driver. Such a solution will have considerable delay, however, because the amplifying receiver always has a large delay [15].

14.4.3 Reduced Interconnect Activity

For all electrically short interconnects, dynamic power consumption dominates, making the power consumption proportional to the logical activity on the interconnect. It is therefore important to reduce the length of interconnects with large activity. This can be accomplished by having the wire activity control the place and route process. The effect of such an optimization procedure was investigated in Prabhakaran et al. [19], using a few benchmark circuit examples. It was then shown that a reoptimization toward a minimum power goal (replacing a minimum latency goal) resulted in power savings of 25 to 70% with a latency penalty of 10 to 50% and no area penalty.

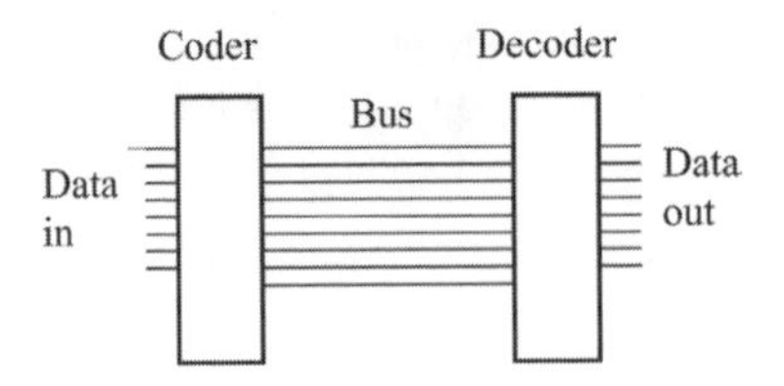

FIGURE 14.6 Using coding on a bus.

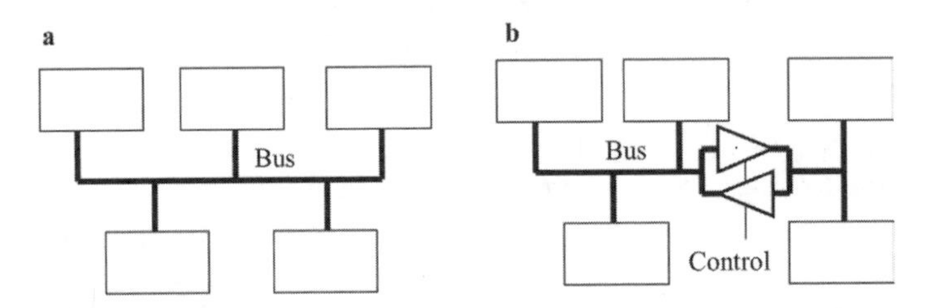

FIGURE 14.7 Going from a common bus (a) to a split bus (b).

Another method for reducing the power consumption is to reduce the logical activity on long interconnects (e.g., long on-chip or off-chip buses) [20]. Such a reduction of activity can be accomplished through coding (Figure 14.6). A simple example is the bus-invert-coding applied to an n-bit bus. The principle is to add one wire to the bus (the invert line, I) to give room for coding. On the rest of the bus we transmit the data, D, by the codeword C. For each new piece of data, D_{next}, we measure the Hamming distance to the previous codeword $C_{previous}$. If the Hamming distance is less than n/2, C_{next} is equal to D_{next} and I = 0, else C_{next} is equal to the inverted value of D_{next} and I = 1. The decoder on the receiver side then recovers D again. In this way, we minimize the number of transitions between two consecutive codes on the bus, thus minimizing the power consumption. Stan and Burleson [20] showed that this is an optimal solution for random data if we add just one extra wire, and finds that the power saving is 25% at best. The best savings occur for small bus widths and is reduced to about 15% at n = 16. In all cases, the bus-invert-coding reduces the maximum supply current with 50%, which is also a very important accomplishment.

In addition, introducing the effect of mutual capacitances (i.e., the Miller effect) makes coding even more efficient for power savings. Sotiriadis and Chandrakasan [10] demonstrated a power savings of 15 to 40% for buses with c_c/c varying from zero to infinity. Another way to mitigate the extra power consumption caused by the Miller effect is to code the bus in such a way that signals on neighboring wires are monotonous, that is we use a precharged bus. The drawback with such an approach is that precharging always increases the total activity (as also some of the lines which do not change data, will charge and discharge). Still, a combination of precharging and coding can give performance benefits without power penalty [21].

Considerably more power savings can be accomplished by context-dependent coding. If we have *a priori* knowledge of the bus traffic, we may use this knowledge to adapt the coding to the traffic. One example is memory bus coding for processor systems, where we know that addresses often are sequential. In such cases, coding at an architectural level may save as much as 64 to 85% [22]. In general, various coding and other architectural optimizations are quite effective for power savings [23].

Buses often use much power. One reason for this is that the whole bus is excited for each transaction, even if the transaction concerns blocks that are close together (Figure 14.7). One way to save power is to divide the bus into several segments, so all short-range transactions only excites part of the bus. Simple simulations indicate power savings of 16 to 50%, depending on the characteristics of the data transfer among the modules and the configuration of the split bus [24].

14.4.4 Power Savings in Drivers and Repeaters

The power overhead related to the wire driver may be quite large (80% or more) in high performance systems. This power overhead can be reduced with a delay penalty by increasing the tapering factor, f.

Increasing f from 3.5 to 9, for example, reduces the power overhead from 80 to 25% at a delay penalty of 20% [12].

As mentioned earlier, interconnect that is delay-optimized by repeaters has a power overhead of about 100%. As the number of repeaters is expected to increase with scaling, this problem will increase in the coming process generations. The delay-minimum is quite shallow, however, so there is a good opportunity to reduce power consumption with a relatively small delay penalty [14]. By optimizing power consumption for a given delay penalty, it has been demonstrated that repeater power dissipation can be decreased by 50%, with a very limited delay penalty of 5% [14]. Allowing a larger delay penalty facilitates larger power savings. A more drastic solution is, of course, to avoid repeaters by using an upper metal layer for long interconnects [15,16].

14.4.5 Off-Chip Interconnect

Off-chip interconnect is traditionally run at moderate speeds (i.e., a few hundred Mb/s) at the same voltage swing used by the ICs. They are, therefore, electrically short. The capacitance of these interconnects consists of the wire capacitance itself (C_w) and the capacitance of the loads. Board wire capacitance is of the order of 100 pF/m [13] and a chip input capacitance of the order of 1 pF. Normally, several loads are allowed, so the output driver is specified for quite a large capacitive load (e.g., 50 to 100 pF). The large specified load combined with a large voltage swing makes the off-chip interconnect have very large maximum power consumption, as mentioned in the introduction, 14.1. In addition, the peak current needed is particularly problematic in the case of I/O because the I/O current surge cannot be absorbed by decoupling capacitors either on-chip or off-chip, as the current pass the chip edge. Because of the large current spikes on the driver supply voltage, this supply is often separated from the supply for the rest of the circuit to protect the circuit from the I/O noise.

The wire power consumption discussed previously refers to a simple driver in the form of an inverter. This case also represents the minimum power case (for a given voltage swing). Quite often, I/O circuits are more complicated than the simple inverter, thus giving rise to larger power consumption. Let us discuss some of these circuits and the reasons for their use.

Often, we want several I/O drivers to drive the same wire (like a common external bus). This is accomplished either with a tri-state driver or with an open drain driver. In the tri-state case, we could use gating transistors in series with the driving transistors (Figure 14.8(a)) or we could use one p-MOS and one n-MOS transistor driver with some additional logic (Figure 14.8(b)). The second solution is preferable from the power consumption point of view, as we can keep the output transistors of minimum size. (In the first case, both output transistors must have double width to keep the output resistance.) Similarly, avoiding short-circuit power in the driver by "break-before-make-action" can be solved by logic in front of the driver transistors [13]. An open-drain output (Figure 14.8(c)) is simple but consumes static power (that is current is continuously consumed when data on the wire is low). Thus, assuming equal probabilities for ones and zeros on the wire:

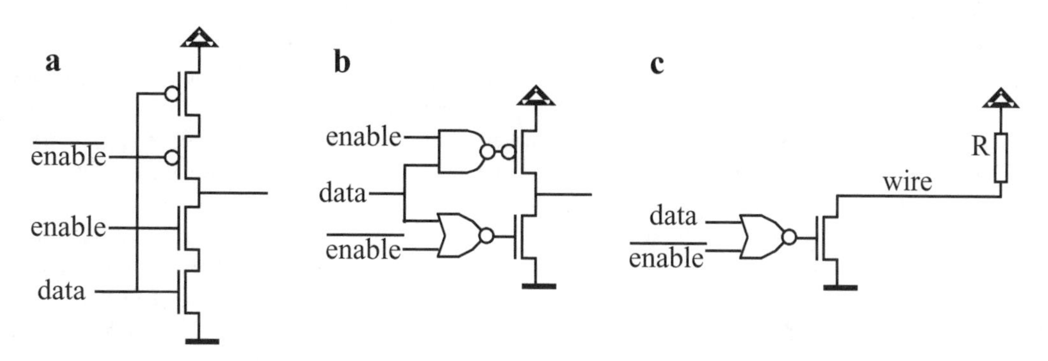

FIGURE 14.8 Tristate drivers (a and b) and open drain driver (c).

$$P = \frac{V_{dd}^2}{2R} \quad (14.21)$$

On the other hand, it saves transistor size, particularly because p-transistors are normally two to three times wider than n-transistors.

Another class of drivers is the current-mode drivers. Such a driver looks exactly as the open-drain driver, but the transistor operates as current generator and the voltage swing is reduced, $\Delta V = I_0 R < V_{dd}$, where I_0 is the current driven by the transistor. The size of the current from the transistor can be controlled by a current-mirror [13]. The power consumption thus becomes:

$$P = \frac{V_{dd} I_0}{2} = \frac{V_{dd} \Delta V}{2R} \quad (14.22)$$

For high data rates, reflection may become a problem. This problem is far more pronounced for off-chip communication than for on-chip wires of two reasons. First, off-chip wires are much longer than on-chip wires. Second, off-chip wires have much less loss and, therefore, behave more like ideal transmission lines. To mitigate reflections, we need to terminate the wire in at least one end. Using an inverter-driver and an open wire allows such a termination without power penalty (except if the transistors need to be wider to get the right on-resistance) by just adjusting the driver output impedance to Z_0. The nonlinear behavior of the transistor on-resistances may be unacceptable, however, calling for a linear resistor in series with each transistor [25]. Then the transistors must be much wider to have smaller resistance, thus increasing their capacitance, which leads to larger power consumption. Still, using a self-terminating driver like this is the most economical solution in terms of power savings. Using a current-mode driver also allows termination, as we can make the resistor in the far end equal to Z_0. The power-consumption is still given by Equation (14.22). In this case, the current-generator driver is considered to have an output impedance that is much larger than Z_0 (wire open on transmitter side). Sometimes, we want to have the wire terminated in both ends. For example, this can be obtained by a current driver with a load resistor of Z_0 in both ends of the wire. The total resistance is then $Z_0/2$, doubling the power consumption compared with Equation (14.22):

$$P = \frac{V_{dd} \Delta V}{Z_0} \quad (14.23)$$

Because of the large power consumption of off-chip interconnect, a reduced voltage swing on these is very profitable. The risk when using low voltage swing is that the interconnect becomes vulnerable to noise. Therefore, most low swing off-chip interconnect use differential signaling [13], which is much less sensitive to noise. A modern standard using differential signaling is LVDS [26]. Because LVDS is aimed for high speeds, it utilizes terminated wires.

A simple way to achieve low swing is to use current-mode drivers as described previously. The drawback is a higher power consumption than necessary. The most power-efficient way is to use a voltage-mode driver with a separate supply voltage as discussed earlier in this paper (Equation (14.13) with $V_{dd} = \Delta V$). Let us exemplify the saving opportunity by using low voltage swing from an example from Svensson [17]. Assuming a terminated wire (terminated by 50 Ω), a 0.13-μm process with a 1.3-V ordinary supply and with separate transmitter supply we arrive to a total power consumption of 0.25 mW at 10Gb/s and at an optimum swing of 120mV. This could be compared with an experimental example of a differential LVDS link in a 0.25-μm process with 2.5-V supply. Here, a total power consumption of 1 mW at 1 Gb/s and 100 mV swing was observed, corresponding to 0.5 mW per wire [27]. A similar experiment, with the voltage swing related to ground, demonstrates 0.8 mW per wire at 200 mV swing [28]. A full swing

(2.5 V) single interconnect, while still terminated to 50 Ω, would have a power consumption of 16 mW (Equation (14.13)).

14.4.6 Charge Recovery Techniques

When considering electrically short wires, which can be modeled as capacitors, the power dissipation is not really occurring in the capacitor, but in the driver. Therefore, could we avoid this dissipation? The answer is yes. If we charge the capacitor slowly with a voltage ramp, no energy is lost during charging. It is then possible to recover this charge back into the power source by discharging the capacitor to a down-ramping supply. This is the adiabatic or energy-recovery principle [29]. Such principles, together with various principles for charge reuse by redistribution, can be utilized for power savings in interconnects [30–32] (see also Chapter 15).

14.5 A Comment about Optical Interconnect

It has been suggested that optical interconnect consumes less power than electrical interconnect [33]. This was based on the observation that also a small photocurrent from a photodiode can give rise to a large voltage, comparable to the logical swing, if the impedance level is large enough. On the other hand, a full logical swing on a wire causes large dynamic power consumption; however, this conclusion did not consider the opportunity for reduced voltage swing on the electrical interconnect and is therefore wrong. A more accurate comparison between optical and electrical interconnect should be done at the same signal-to-noise ratio and consider the receiver power consumption in both cases. This was done in Berglind et al. [34] with the conclusion that electrical interconnect use less or the same power compared with an optical interconnect. A similar analysis performed in Yoneyama et al. [35] considers very long interconnect and estimates the breakeven length, below which electrical interconnect is superior. They find that electrical interconnect is superior for distances less than about 5 m at a data rate of 20 Gb/s. An extrapolation to 100 Gb/s indicates that electrical interconnect is superior up to about 3 m in length. This latter observation demonstrates the most important difference between electrical and optical interconnect, that is optical interconnect is superior for long distances because of a very low attenuation. Then, there may also be other benefits with optical interconnect, such as less electromagnetic emission and less sensitivity to electrical noise, as well as drawbacks as technical complexity and alignment problems.

14.6 Conclusion

The power consumption related to on-chip and off-chip communication is substantial. It is, therefore, very profitable to seek power reductions of interconnect. We have described the most important properties of wires and their design, including the impact of drivers, receivers, and repeaters. We have further described how to model power consumption of wires, including long transmission lines. Power saving can be accomplished by various methods, from the architectural level to the physical level. Reducing the need for communication between distant blocks based on the choice of system architecture is probably the most efficient method. Other architectural methods include utilizing coding on buses to save transitions. Such methods may save about 50% power. On the physical level, floor planning and layout based on communication activity may save up to 70% of power. Reduction of the voltage swing on the interconnects is very effective, particularly for the off-chip case. Savings can be anything from a few percent to maybe 20×. Power savings using voltage swing reduction always leads to some delay penalty, which often can be kept quite small. In the case of off-chip interconnect it may also be very profitable to choose a low-power circuit technique for the wire driver. Some further savings can be achieved through pure technological changes, such as low dielectric constant insulators; however, this savings is expected to barely compensate for capacitance increase caused by scaling. Finally, several more advanced techniques have been proposed, such as charge recovery (adiabatic) techniques or optical interconnects. Among these techniques, the charge recovery technique has potential, whereas optical interconnect is not expected to lead to power savings.

References

[1] H.B. Bakoglu, *Circuits, Interconnections, and Packaging for VLSI,* Addison-Wesley, Reading, MA, 1990.

[2] D. Liu and C. Svensson, Power consumption estimation in CMOS VLSI chips, *IEEE J. Solid-State Circuits,* vol. 29, pp. 663–670, June 1994.

[3] G. Chandra, P. Kapur, and K.C. Saraswat, Scaling trends for the on-chip power dissipation, *Proc. IEEE 2002 Int. Interconnect Technol. Conf.,* pp. 154–156, 2002.

[4] J.M. Rabaey, A. Chandrakasan, and B. Nicolic, *Digital Integrated Circuits, 2nd ed.,* Prentice Hall, Upper Saddle River, New Jersey, 2003, chapter 4.

[5] K. Soumyanath, S. Borkar, C. Zhou, and B.A. Bloechel, Accurate on-chip interconnect evaluation: a time-domain technique, *IEEE J. Solid-State Circuits,* vol. 34, pp. 623–631, May 1999.

[6] Corrected version of Equation (4.2) in Rabaey et al. [4].

[7] C. Svensson, Electrical interconnects revitalized, *IEEE Trans. VLSI Syst.,* vol. 10, p. 777, Dec. 2003.

[8] J.M. Rabaey, A. Chandrakasan, and B. Nicolic, *Digital Integrated Circuits, 2nd ed.,* Prentice Hall, Upper Saddle River, New Jersey, 2003, chapter 9.

[9] A. Alvandpour, P. Larsson-Edefors, and C. Svensson, GLMC: interconnect length estimation by growth-limited multifold clustering, *Proc. Int. Symp. on Circuits and Syst.,* vol. 5, pp. 465–468, 2000.

[10] P.P. Sotiriadis and A. Chandrakasan, Low-power bus coding techniques considering inter-wire capacitances, *IEEE Custom Integrated Circuit Conf.,* pp. 507–510, 2000.

[11] Y. Sasaki, M. Sato, M. Kuramoto, F. Kikuchi, T. Kawashima, H. Masuda, and K. Yano, Crosstalk delay analysis of a 0.13-μm node test chip and precise gate-level simulation technology, *IEEE J. Solid-State Circuits,* vol. 38, pp. 702–708, 2003.

[12] C. Svensson and D. Liu, Low-power circuit techniques, in J. M. Rabaey and M. Pedram, Eds., *Low-Power Design Methodologies,* Kluwer, Dordrecht, 1996.

[13] W.J. Dally and J.W. Poulton, *Digital Systems Engineering,* Cambridge University Press, Cambridge, 1998.

[14] P. Kapur, G. Chandra, and K.C. Saraswat, Power estimation in global interconnects and its reduction using a novel repeater optimization methodology, *Proc. 39th Design Automation Conf.,* pp. 461–466, 2002.

[15] P. Caputa and C. Svensson, Low-power, low latency global interconnect, *15th Annu. IEEE Int. ASIC/SOC Conf.,* pp. 394–398, 2002.

[16] C. Svensson and P. Caputa, High-bandwidth, low-latency global interconnect, *Proc. SPIE, VLSI Circuits and Syst.,* vol. 5117, pp. 126–134, May 2003.

[17] C. Svensson, Optimum voltage swing on on-chip and off-chip interconnects, *IEEE J. Solid-State Circuits,* vol. 36, p. 1108, July 2001.

[18] C. Svensson, unpublished data.

[19] P. Prabhakaran, P. Banerjee, J. Crenshaw, and M. Sarrafzadeh, Simultaneous scheduling, binding and floorplanning for interconnect power optimization, *Proc. 12th Int. Conf. on VLSI Design,* pp. 423–427, Jan. 1999.

[20] M.R. Stan and W.P. Burleson, Bus-invert coding for low-power I/O, *IEEE Trans. VLSI Syst.,* vol. 3, pp. 49–58, March 1995.

[21] M. Anders, N. Rai, R.K. Krishnamurthy, and S. Borkar, A transition-encoded dynamic bus technique for high-performance interconnects, *IEEE J. Solid-State Circuits,* vol. 38, p. 709, May 2003.

[22] Y. Aghaghiri, F. Fallah, and M. Pedram, BEAM: bus encoding based on instruction-set-aware memories, *Proc. Asian and South Pacific Design Automation Conf.,* pp. 3–8, 2002.

[23] E. Macii, M. Pedram, and F. Somenzi, High-level power modeling, estimation, and optimization, *IEEE Trans. Comput.-Aided Design,* vol. 17, p. 1061, Nov. 1998.

[24] C.-T. Hsieh and M. Pedram, Architectural power optimization by bus splitting, *IEEE Trans. Comput.-Aided Design,* vol. 21, pp. 408–414, 2002.
[25] M. Haycock and R. Mooney, 3.2 GHz 6.4 Gb/s per wire signaling in 0.18-μm CMOS, *48th Int. Solid-State Circuits Conf., Dig. Tech. Papers,* pp. 62–63, Feb. 2001.
[26] ANSI/TIA/EIA-644 LVDS Standard. See also J. Goldie, The many flavors of LVDS, http://www.national.com/nationaledge/feb02/flavors.html
[27] S. Hirsch and H.-J. Pfleiderer, CMOS receiver circuits for high-speed data transmission according to LVDS standard, *Proc. SPIE, VLSI Circuits and Syst.,* vol. 5117, pp. 238–244, 2003.
[28] M. Hedberg and T. Haulin, I/O family with 200 mV to 500 mV supply voltage, *44th Int. Solid-State Circuits Conf., Dig. Tech. Papers,* pp. 340–341, Feb. 1997.
[29] W.C. Athas, L.J. Svensson, J.G. Koller, N. Tzartzanis, and E.Y.-C Chou, Low-power digital systems based on adiabatic-switching principles, *IEEE Trans. VLSI Syst.,* pp. 398–407, Dec. 1994.
[30] M. Hiraki, H. Kojima, H. Misawa, T. Akasawa, and Y. Hatano, Data-dependent logic swing internal bus architecture for ultralow-power LSIs, *IEEE J. Solid-State Circuits,* vol. 30, pp. 397–401, April 1995.
[31] H. Yamauchi, H. Akamatsu, and T. Fujita, An asymptotically zero power charge recycling bus architecture for battery-operated ultrahigh data rate ULSIs, *IEEE J. Solid-State Circuits,* vol. 30, pp. 423–431, April 1995.
[32] L.J. Svensson, W.C. Athas, and R.S.-C. Wen, A sub-CV2 pad driver with 10-ns transition time, *IEEE Int. Symp. on Low-Power Electron. and Design,* pp. 105–108, 1996.
[33] D.A.B. Miller, Optics for low-energy communications inside digital processors: quantum detectors, sources and modulators as efficient impedance converters, *Opt. Lett.,* vol. 14, p. 146, Oct. 1996.
[34] E. Berglind, L. Thylén, B. Jaskorzynska, and C. Svensson, A comparison of dissipated power and signal-to-noise ratios in electrical and optical interconnects, *J. Lightwave Technol.,* vol. 17, p. 68, Jan. 1999.
[35] M. Yoneyama, K. Takahata, T. Otsuji, and Y. Akazawa, Analysis and application of a novel model for estimating power dissipation of optical interconnections as a function of transmission bit error rate, *J. Lightwave Technol.,* vol. 14, pp. 13–22, Jan. 1996.

15

Adiabatic and Clock-Powered Circuits

Lars Svensson
Chalmers University

15.1 Introduction

The integrated digital circuitry ubiquitous in the electronic equipment that surrounds us is mainly implemented with complementary metal-oxide semi-conductor (CMOS) technologies. The commonly used logic styles operate in binary voltage mode, where each logic gate drives its output to one of the end points of the available voltage range. Simple circuit styles exist where the supply voltage rails define the voltage range. These full-swing logic styles, including the well-known static-CMOS and domino styles, dominate CMOS logic; special-purpose circuits occupy important niches such as memories. In summary, full-swing voltage-mode CMOS logic styles have been extremely successful, both technically and in terms of market share. In this chapter, they will sometimes be referred to as "conventional" logic.

A lower limit on the dynamic power dissipation of a capacitively loaded conventional logic gate is easy to calculate. Each transition will at least cause a dissipation that depends only on the load capacitance and the voltage swing. Any attempt to seriously reduce this dissipation must reduce the number of transitions, the load capacitance, the voltage swing, or some combination of these.

This chapter describes *adiabatic charging,* a family of techniques to design logic and other switching circuits which circumvent this lower limit of dynamic power dissipation. The principle of adiabatic charging is wide-reaching — it is grounded in very generic models of the switching elements — but implementations have so far been completely dominated by CMOS. Future implementations may benefit from the availability of other manufacturing technologies.

15.2 The Adiabatic-Charging Principle

Consider the conventional, capacitively loaded CMOS inverter depicted in Figure 15.1(a). For the purpose of this example, the idealized resistive-switch network depicted in Figure 15.1(b) can represent the inverter. (Because this discussion concerns lower limits for the dissipation, short-circuit currents and other second-order effects are ignored.) When the input, V_{in}, is pulled low, the pMOS device turns on, and the linear load capacitance (C) is charged from 0 to V. The charging process causes energy dissipation in the pMOS device, because the charge experiences a potential drop on its way from the supply node

0-8493-1941-2/05/$0.00+$1.50

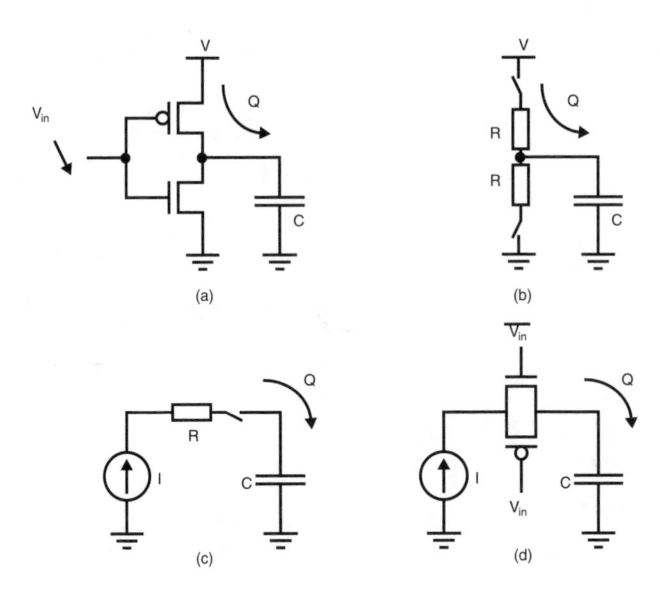

FIGURE 15.1 (a) A static CMOS inverter charging a capacitance C to the supply voltage V. (b) A simplified resistive-switch representation of the same circuit. (c) Resistive-switch representation of constant-current charging of C. (d) CMOS implementation of the same circuit.

to the load. At the outset, no charge is yet stored in C, so the potential drop is V; at the end of the transition, the potential drop is zero. The average potential drop traversed by the charge, V_{avg}, is therefore V/2. The amount of charge stored, Q, is equal to CV. The energy dissipation is therefore $E_{conv} = V_{avg}\ Q = (V/2)\ (C\ V) = CV^2/2$: a familiar result. The node energy, E_{cap}, which is stored on C after charging, is also $CV^2/2$. This energy will dissipate during the subsequent discharging of C.

It is clear from the expression for E_{conv} that dissipation is reduced if V or C are made smaller. At first sight, it seems that nothing can be done if both C and V are fixed. Closer inspection reveals, however, that the dissipation would be reduced if V_{avg} could somehow be reduced below V/2. Actually, the energy dissipation could be reduced to an arbitrary degree, if V_{avg} were reduced commensurately.

Consider now the idealized resistive-switch network in Figure 15.1(c). The constant supply voltage level, V, has been replaced by a constant current source, I; the voltage drop across the switch is IR throughout the charging, so the energy dissipation is $E_{curr} = IR\ CV$. The charging current, I, is equal to CV/T, where T is the charging time used to charge C from 0 to V. When this expression is substituted for I, the result is the expression $E_{curr} = (RC/T)\ CV^2$.

We may now make several observations:

- E_{curr} can be lower than E_{conv}, if T is long enough. Actually, E_{curr} may be made arbitrarily small by further extending the charging time. In the limit, charge is moved onto C with no dissipation (i.e., with no heat exchange with the environment). This observation is the motivation for the term "adiabatic charging" [1].
- It can easily be demonstrated that the constant-current charging is the most energy-efficient way to charge a linear capacitance though a resistance in a given time, be it short or long. (Note, however, that the constant voltage drop across R must be maintained throughout the charging. For short-enough T, it would be much larger than V.)
- A lower path resistance R brings a lower dissipation. This result is in contrast to the conventional case: the expression for E_{conv} does not contain R.
- If the current direction is reversed, C is discharged through the same path through which it was charged. The node energy, E_{cap}, which was moved onto C during charging, is then removed, again with a dissipation that depends inversely on the charging time. The node energy minus the

dissipated part is recovered by the current source, and may be reused in the next charging. This mechanism is the motivation for the term "energy-recovery CMOS" [2].

A simple implementation of the resistive-switch network of Figure 15.1(c), in terms of metal-oxide semiconductor (MOS) devices, is depicted in Figure 15.1(d). The switch and the resistance R are implemented as a transmission gate. This choice allows the control signals for the switch to swing from 0 to V, just like the output. A single device of either kind would need a higher-swing control signal to allow the output to be charged all the way from 0 to V.

In conventional static-CMOS circuits, such as the inverter in Figure 15.1(a), a supply-voltage reduction brings lower dynamic dissipation. The dissipation of the circuit of Figure 15.1(d), however, has a minimum value at a certain voltage swing [1]. Further reduction of the swing will bring a dissipation increase caused by the reduced gate-to-channel voltages of the transmission-gate devices; E_{curr} will rise with the on-resistance of the transmission gate.

The constant-current generator presents implementation problems. It is not clear how to build individual controllable constant-current generators for each capacitive load in a large circuit without wasting more power than was gained by introducing them. Thus, all adiabatic-charging circuits presented to date use some approximation of the constant-current source. A time-dependent voltage source that generates periodic positive- and negative-going linear voltage ramps will create current waveforms similar to those generated by the current source, if the (RC/T) factor in the expression for E_{curr} is small enough. In addition to the operating power, the ramp signals naturally provide timing information to the circuits, and, therefore, are often referred to as power-clocks or simply as clock signals. This is the motivation for the term "clock-powered circuits" [22].

15.3 Implementation Issues

The adiabatic-charging principle outlined in the previous section describes how dynamic energy dissipation may be reduced below the $CV^2/2$ per switching event required for conventional switching circuits. Many engineering problems must be solved to utilize the principle in the design of logic circuits. As already indicated, approximately constant currents must be generated and distributed to those circuit nodes that are to be charged. It is not trivial to accomplish this current generation; most published approaches have centralized the current generation in an adiabatic power supply (APS) and solved the logic-design problem separately. The timing of charge and discharge phases of the logic circuits ties together the APS design with the logic and pipeline design. All these subproblems must be solved in any realization of the adiabatic-switching principle; an efficient adiabatic logic style that lacks an efficient APS will not provide a low-power solution.

Two main approaches dominate the adiabatic-logic styles presented to date. The more ambitious approach aims to recover the node energies of *all* circuit nodes, including nodes inside logic gates. The other, more pragmatic approach applies the adiabatic principle to nodes with large capacitance and, therefore, large node energies. The APS designs for these approaches will be somewhat different, as described next.

15.3.1 Adiabatic Logic

Section 15.2 described how adiabatic techniques could be used to reduce the energy dissipation caused by charging a capacitive load. This section presents the incremental construction of a logic style, which allows fully adiabatic operation [1]. The construction starts from the buffer circuit in Figure 15.1(d), but adds a clamp device to ensure that the output is securely grounded when it is not to be charged. The resulting buffer circuit is depicted in Figure 15.2(a).

The transmission gate used to connect the load to the APS requires a dual-rail control signal, as do all logic gates based on transmission gates. It is not admissible to introduce a conventional inverter to derive one of these control signals from the other. At each transition, the inverter would dissipate energy, which would not scale with the transition time, as required for fully adiabatic operation. In a logic style

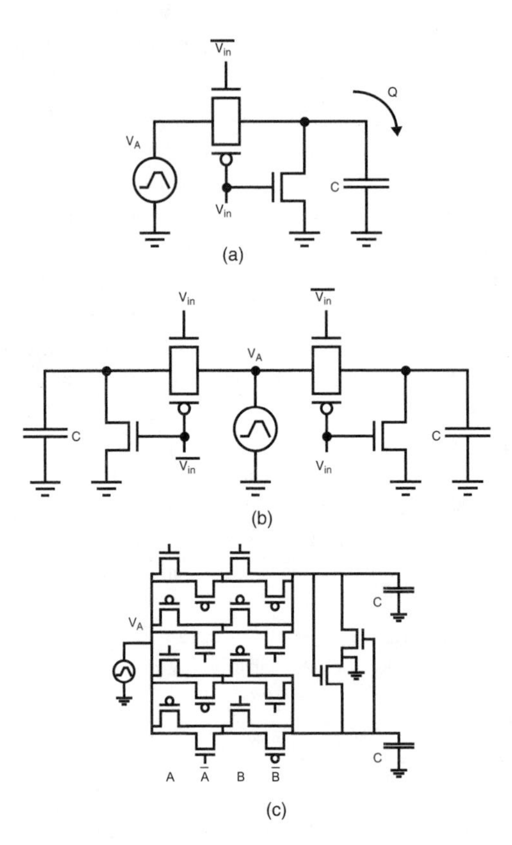

FIGURE 15.2 (a) The adiabatic buffer of Figure 15.1(d), with an additional clamp device. The current generator has been replaced with a time-dependent voltage source, V_A. (b) A dual-rail version of the same circuit. (c) A dual-rail EXOR circuit.

based on transmission gates, therefore, all logic gates must generate dual-rail signals. The dual-rail version of the buffer is depicted in Figure 15.2(b).

The circuit of Figure 15.2(b) is trivially extendable to implement a logic function other than duplication of the input signal to the output. Figure 15.2(c) presents an EXOR gate with two inputs. The function controlling the clamping of an output to ground must be the inverse of the function implemented by the transmission-gate network driving that output. Cross-coupled clamp devices, as depicted in the figure, can replace a full pull-down network controlled by the inputs to the gate. The combinational gate of Figure 15.2(c) can drive its output fully adiabatically, but because of the transmission gates, its device count is twice that of a corresponding static-CMOS counterpart.

Signal timing presents a difficult problem. For fully adiabatic operation to be possible, the inputs must be held static throughout the charging and discharging of the load. It would be possible to charge the load capacitances, latch the result, and discharge the load; but conventional latches invariably cause nonadiabatic dissipation [3] and are therefore not permissible. Instead, a discharge path for each load capacitance, separate from the charge path, must be used to relax the static-input requirement (Figure 15.3). The inputs must now be held static throughout the charging of the load, whereas the subsequent discharge is controlled by another set of signals. It is tempting to try to derive these control signals directly from the gate outputs; but such signals will, by necessity, not be stable throughout the discharge. Any schemes based on the same idea are bound to fail for thermodynamic reasons [4]. The solution is more complex: each stage in a fully adiabatic pipeline must generate signals that control the charging path of the following stage, as well as signals that control the discharging path of the previous stage in the pipeline. Such a pipeline is illustrated in Figure 15.4; the figure is still simplified, in that all signals, including the voltage ramp signals, must be dual-rail-encoded [1]. For the approach to be workable, all combinational

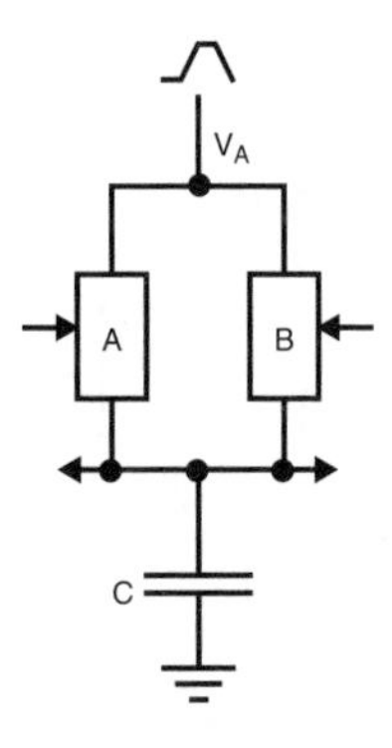

FIGURE 15.3 The capacitance C is charged though one path, A, and later discharged through another path, B. The inputs to the path A must be stable throughout the charging, whereas the inputs to path B must be stable throughout the discharging.

functions used in the pipeline must be invertible: there must exist an inverse function, which can compute the inputs to the original function from its results. The resulting pipeline is reversible and runs backward if the clocks are reversed in time. This reversibility property is characteristic of all fully adiabatic pipelines.

The complexity of two mutually inverse function blocks are usually similar, so the separate discharge path doubles the amount of hardware. Additionally, the restriction to use only invertible functions adds overhead which may be quite large. In a benchmark experiment, a fully reversible, bit-level-pipelined three-bit adder required 20 times as many devices as a conventional one, and 32 times the silicon area [5]. Other styles of reversible logic have less overhead [6,7], but none are as compact as the conventional, nonadiabatic logic styles. Nevertheless, it is possible to design fully reversible processors [8], which would dissipate very little power when operated slowly enough.

The hardware overhead of reversible logic has encouraged many researchers to seek ways to apply adiabatic techniques also for nonreversible logic. Such solutions will not be fully adiabatic, so their dissipation will not scale to the very lowest levels; but at higher performance levels, they may well offer lower dissipation than a corresponding reversible implementation. For a fair evaluation, any such partially adiabatic logic style should be benchmarked against logic styles according to best conventional practice, and the conventional control case should be optimized for power (typically by supply-voltage selection) at the same performance level as the adiabatic style.

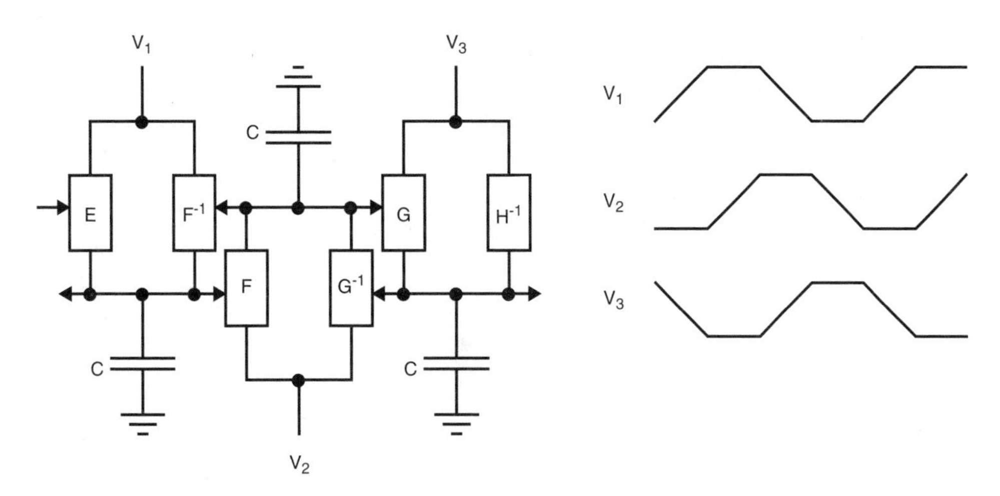

FIGURE 15.4 Node capacitances in this reversible pipeline are charged and discharged through two different paths. Each discharge path implements the inverse logic function to the charge path of the next pipeline stage. Thus, the discharge path has full information of the state of the variables to erase.

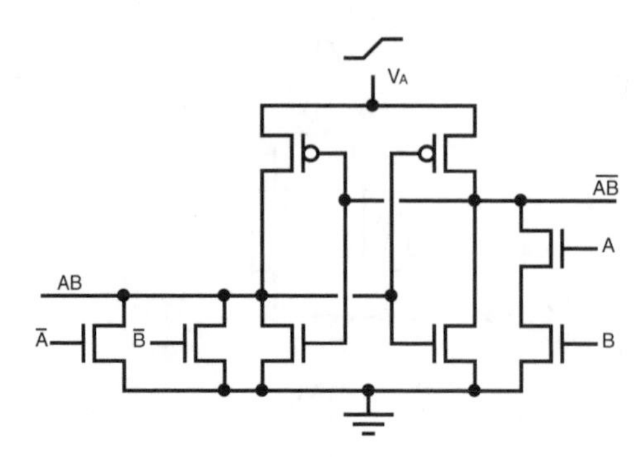

FIGURE 15.5 A partially adiabatic logic gate, based on two cross-coupled inverters. One of the two output nodes is held low for each combination of input values. The other output node is charged from the APS through one of the pMOS devices when the voltage ramp is applied.

Most of the partially adiabatic logic styles presented to this date use cross-coupled devices connecting two nodes that form the true and inverted outputs of a gate. The gate settles in one of two stable states when a voltage ramp is applied. The behavior of the outputs depends on a small imbalance between the two nodes, which in turn has been caused by the input values. Figure 15.5 illustrates an early example of the type [3,9]. The figure presents an AND/NAND gate. When the clock signal rises, the right output will be held low iff both inputs are high; the output not held low will follow the clock signal and be charged adiabatically once the corresponding pMOS device has turned on. The change from one stable state to the other (when the output changes from the value in the previous cycle) is associated with a nonadiabatic dissipation of approximately $C\ V_{th}^2$, where V_{th} is the device threshold voltage and C is the driven capacitance. The clock signal must be held high to provide static inputs for the next pipeline stage while its clock is ramped up. A pipeline of such gates can be operated with four identical clock signals ninety degrees out of phase [9].

Newer partially adiabatic logic styles seek to minimize the number of separate clock signals needed. Several examples have been demonstrated to work with a single sinusoidal clock. Further reduction in the number of clocks appears unlikely.

15.3.2 Adiabatic Buffering

The previous section outlined some of the issues to be addressed when designing adiabatic-charging CMOS logic circuits. Although many of the problems have been solved in principle, no consensus has been reached on how best to design circuits that recover most of the charge and energy of the internal nodes of logic gates. This section will describe a more limited approach, where mainly nodes that contribute a large amount of dynamic power (that is, nodes with a large capacitance and a high switching frequency) are driven with adiabatic-switching techniques. In contrast, most nodes inside blocks of combinational logic have a rather small capacitance; many of the complications of adiabatic switching can be circumvented if such combinational blocks are implemented with conventional circuitry.

The largest node capacitances in a conventional chip (aside from the supply and ground nodes) typically belong to the clock distribution network. A continuously running, global clock signal is particularly suitable to energy recovery: the clock node capacitance may be readily resonated with on- or off-chip inductances. Because the charge path is identical from cycle to cycle, there is no need for a resistive switch to be inserted to direct the charge flow. The result would be a sinusoidal clock signal.

Present-day low-power digital chips use clock gating as an essential tool for power minimization. By disabling the clock signal for an unused subblock, the designer ensures that any flip-flops served by the clock stay idle and therefore consume no dynamic power. Combinational logic blocks whose inputs emanate from these flip-flops are also kept in an idle state. The power, which would have been used to distribute the disabled clock, is also saved with clock gating.

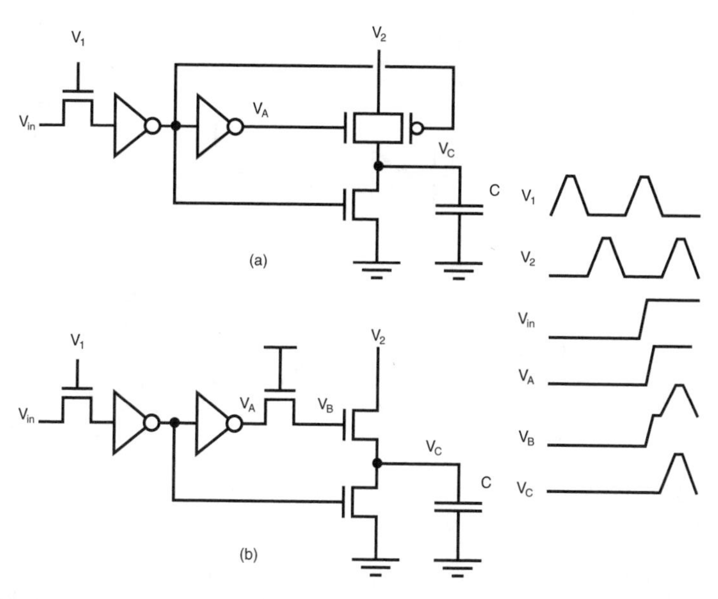

FIGURE 15.6 (a) A latch and energy-recovery driver circuit for two-phase, nonoverlapping clocks. (b) The Athas-Tzartzanis latch and energy-recovery driver, where a bootstrapped nMOS device replaces the transmission gate of (a).

Clock gating is still possible with a resonant clock driver. The subblock clock network must be connected to the global clock network through a low-resistance switch, such as the one given in Figure 15.1(d). The same principle may be further extended. A high-capacitance signal line will be charged and discharged if it is connected to the clock network for an appropriate time. If the RC time constant of the load capacitance and the connecting-switch resistance is small compared with the rise time of the clock, the driven signal will faithfully follow the clock signal. Figure 15.6(a) depicts a combined latch and energy-recovery driver circuit, suitable for this purpose. It is intended for two-phase, nonoverlapping clocks. The input value is latched on the falling edge of phase 1 and buffered through the inverter pair. When the latched value is low, the clamp device holds the output at ground potential, and the transmission gate devices are turned off. When the latched value is high, the transmission gate connects the load to V_2 in time for the phase-2 clock pulse.

The transmission-gate pMOS device must be quite wide to maintain a low switch resistance at the peak voltage. Consequently, its gate capacitance (which is driven without the benefit of energy recovery) will be uncomfortably large. A variation of the circuit, introduced by Tzartzanis and Athas [10] and further developed by Athas et al. [2], is depicted in Figure 15.6(b). When the latched value is low, it works as the circuit in Figure 15.6(a). When the latched value is high, the second inverter will charge the gate of the bootstrap device through the isolation device. The positive edge of phase 2 will be capacitively coupled from the channel of the bootstrap device to the boot node, which will rise, immediately turning off the isolation device. The boot node will now follow the phase-2 waveform through its positive and negative transitions, if the isolation device has been properly sized with respect to the bootstrap device (the boot-node voltage is subject to capacitive voltage splitting). In the process, the capacitive load will first be charged from the phase-2 clock line and will then deliver its charge back to the same clock line, all through the bootstrap device.

It is instructive to study the design of the Athas–Tzartzanis energy-recovery latch, or "ER latch." Its simplicity and efficiency stems from several considerations. The bootstrapping technique allows a single nMOS device to replace the transmission gate, which saves both power and layout area. The drawback is that the boot node rises to a voltage larger than that of the clock. With a full-swing clock, careful analysis is needed to ensure long-term survival of the substrate-diffusion junction at the boot node. Furthermore, the latch produces pulse rather than level: it will transfer a clock pulse to its output when the input signal is high. If a high value is latched repeatedly, several clock pulses will be transferred to

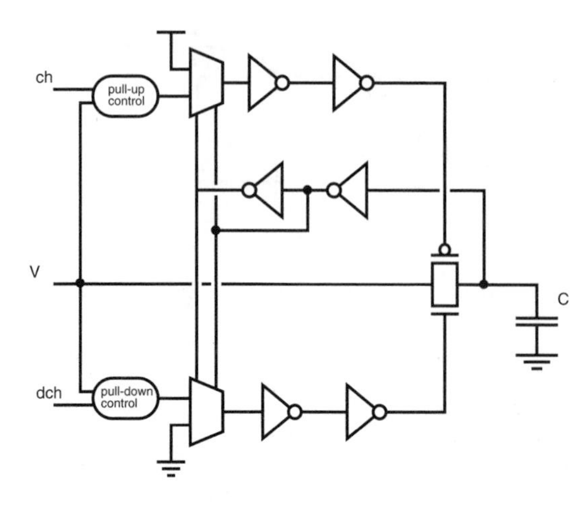

FIGURE 15.7 Energy-recovery driver according to Kim [11]. Signals marked ch and dch control whether the load, C, is to be charged or discharged (see Figure 15.6).

the output, with each pulse causing energy dissipation. Such a signaling scheme is not a good choice when the receiving circuits are implemented in a conventional static-CMOS style; dynamic, domino-style circuits are more suitable, and need no footer devices because their input signals are already gated with the clock signal.

Figure 15.7 is another circuit proposed for selectively connecting data lines to the clock line. This circuit, introduced by Kim et al. [11], retains the transmission gate of Figure 15.1(d) but adds a feedback path, which allows the driver to release the driven node after a one-way transition. The benefit is that a long run of a single value will cause dissipating transitions only at the beginning and the end of the run. (The output node seems to be left floating between these transitions, which could however easily be corrected with two clamp devices connecting the output to the supply and ground nodes, respectively.) The circuit is clearly more complex than the Athas–Tzartzanis ER latch. The transistor-count overhead and the extra parasitic capacitance at the output indicate that the Athas–Tzartzanis ER latch may be used with smaller loads, whereas the Kim ER driver would likely be more efficient at long runs of high values.

Both these drivers have been used in implementations of static memories [11,12], an application that offers rich opportunities for signal buffering due to the large-capacitance bit lines and word lines. The implementations differ in the details: Tzartzanis uses a current-mode sense amplifier, whereas Kim uses a voltage-mode design; Kim uses energy recovery for bit- and word-line drivers only, whereas Tzartzanis uses the principle also for row decoders and internal data buses. Despite all these differences, both designs reach similar speeds in simulation (Kim: 300 MHz for a 256 × 256 memory in 0.35 μm; Tzartzanis: 200 MHz for a 256 × 256 memory in 0.8 μm) and offer similar energy savings (Tzartzanis: 2.4×–4.2×; Kim: 2.66×) when compared with a conventional control case.

Neither design uses adiabatic charging to read values out of the memory cells. It would seem that a memory cell could connect a resonantly driven word line with a bit line, depending on the value stored in the memory cell, maybe with circuits similar to those in Figure 15.8. This approach would, however, require the connecting switch to be enabled already when the word line is driven, or nonadiabatic dissipation would result. The switches of the unselected words must however *not* be enabled because the pulse on the bit line could travel backwards through these switches and onto other unrelated word lines, as presented in the figure. The requirement to preenable the switches in only the selected word leads to the introduction of an auxiliary word line, which would go high before the power-delivery word line. The memory cell itself would be quite large, as the readout circuitry in Figure 15.8 would need to be augmented with enable logic.

This problem points to a general design difficulty with the energy-recovery circuit styles described here: they are ill-suited to circuit nodes with large fan-ins (such as memory bit lines). The reason is that the switch that connects a gate output to a power-clock is bidirectional by design. Charge is supposed

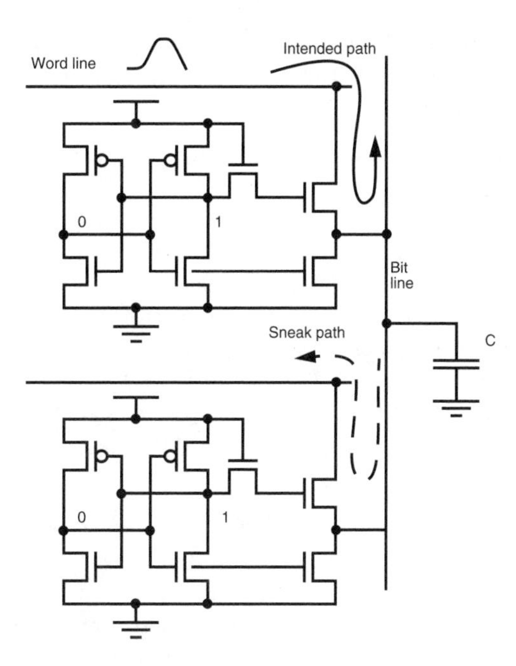

FIGURE 15.8 A first attempt at a memory-cell readout circuit that is able to charge the bit lines adiabatically. The upper cell is selected when its word line is adiabatically driven. A bootstrapped driver as in Figure 15.6 drives the bit line when the cell content is as indicated. The bootstrap device in the lower cell, however, is also enabled, allowing the pulse to enter the second word line.

to be able to flow both ways through the switch. Many such switches are connected to a node with a large fan-in. It will often be necessary to explicitly turn off all but one of these switches before driving the fan-in node. This requirement significantly complicates both circuits and timing. In the case of memory readout, another low-dissipation alternative is readily available because memory sense amplifiers can work with small differential swings. Thus, there appears to be little to gain in terms of dissipation by recovering bit-line node energies during read operations.

Clock lines and memory bit-lines (during write operations) are good candidates for adiabatic driving, because of their large capacitance. Outside the confines of a single chip, more opportunities appear. A trace capacitance on a printed circuit board (PCB) is likely to be larger than any on-chip capacitance (with the exception of a clock net in a large microprocessor). Some energy-recovery drivers targeted to this application are described at the end of Section 15.3; they do not work by tying the output to a clock line transition, but rather charge the load gradually in several steps.

Large capacitances and a nonnegotiable voltage swing can also be found in liquid-crystal display (LCD) drivers. The column drivers of an active-matrix LCD dominate the dissipation of the display (with the possible exception of the backlight). Because the speed requirements are modest, the column drivers would seem to be prime candidates for adiabatic charging, but the situation is complicated by the requirement for driving the column line to a controllable voltage level. Two solutions have been published, both targeting chip-sized rather than laptop-sized displays. Ammer et al. [13] distributes digital pixel values to each column driver and a common, adiabatically generated voltage ramp to all columns. Initially, all columns are connected to the voltage ramp; each column is then disconnected from the ramp when a counter determines that the proper voltage has been reached. Lal et al. [14] distributes an analog video signal, which is sampled at each column; the sampled value is used to control the final voltage on the column.

Throughout this section, it has been implicitly assumed that the driven load can be considered purely capacitive. Adiabatic charging was invented for such cases, but node energies may be recovered also when the load has a significant resistive component, as is likely when driving a long signal line off chip or on chip. In these cases, the formulae for the dissipation are more complex than those given in Section 15.2 because of the wire-resistance influence. The qualitative behavior of the energy dissipation can be summarized in the following points:

- The switch resistance, R_S, and the wire resistance, R_W, must both be much smaller than T/C_W for the dissipation to approach 0.
- If T is approximately $R_W C_W$, the wire properties limit the amount of energy recovery possible. It will be useful to reduce R_S to a certain degree, but further reductions are to no avail. In contrast, all switch-resistance reduction is useful when the load is purely capacitive.
- For a wire with a uniform resistance and capacitance per unit of length, a maximum length exists at which some given percentage of its energy can be recovered. This maximum length depends on the charging time T. Thus, global signals in large, fast chips must be split in parts and adiabatically rebuffered if most of their node energy is to be recovered.

15.3.3 Adiabatic Power Supplies

Adiabatic switching can reduce overall power dissipation only if some part of the switching-circuit node energies can be recovered and reused. As exemplified in the previous sections, the switching circuits (which may be logic circuits or simply drivers for large capacitances) must be designed with this requirement in mind. Additionally, the recovered node energies must somehow be stored in the APS while waiting to be reused. The two principal alternatives for energy-storing circuit elements are inductors and capacitors. Both have been used for APS energy storage; some examples are described next. Other suggestions include transmission lines [15] and piezo-electric resonators [16].

A conceptually simple method to repeatedly deliver energy to a capacitive load is to connect the capacitance to an inductance, thus forming an LC resonance circuit. Once the circuit has been excited, the energy will oscillate between the inductance and the capacitance, with a frequency proportional to the inverse of the square root of the LC product. The sine-shaped voltage waveform on the capacitance will be damped because the inevitable resistive losses will convert part of the circuit energy to heat in each cycle of oscillation. A sustained near-sinusoidal waveform may be produced if the energy is replenished regularly, as indicated in Figure 15.9. This simple arrangement, which requires a logic style compatible with a single-phase sinusoidal clock, was used by Maksimovic et al. [17]; a more evolved version was presented by Ziegler et al. [18].

The LC circuit of Figure 15.9 displays some properties that are inherent to all simple LC-resonance APSs. First, the frequency of operation is set by the total driven capacitance C and the resonance inductor L. It may be possible to operate the APS at a frequency slightly different from its self-resonance frequency, but at a reduced efficiency.

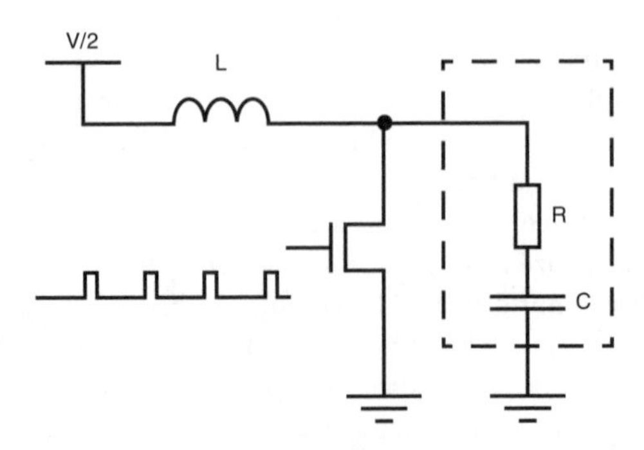

FIGURE 15.9 A simple adiabatic power supply (APS). The logic circuit, in the dashed box, is represented as an RC link, where the C corresponds to the node capacitances and the R corresponds to the on-resistances of the logic gates. The RC link is periodically shunted by the nMOS device. When the control signal frequency agrees with the LC resonance, the voltage across the RC link approximates a sine wave with a peak-to-peak value of V.

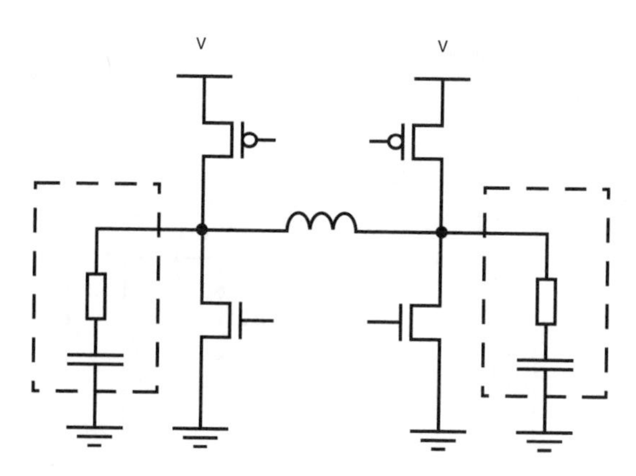

FIGURE 15.10 A dual-rail version of the circuit in Figure 15.9. Each pMOS device is used to tie one adiabatic power rail to V when the other is tied to ground by its nMOS device.

Second, the overall efficiency is determined by the Q value of the LC resonance circuit, which in turn is set by the average resistance in series with the driven capacitances. If L can be increased without introducing more parasitic resistance, efficiency will be increased, but at the cost of a lower frequency of operation: with the same voltage swing, the average charging currents follow the frequency, and because $P \sim I^2R$, power will fall as the square of the frequency.

Third, if the driven capacitance would change instantaneously, as would be typical for a design using single-rail logic or clock gating, both the self-resonance frequency and the amplitude of the waveform could be immediately affected. (To avoid the amplitude change, the auxiliary capacitance must be switched into or out of the circuit when the inductor current is zero; also, the capacitance must be charged to the same voltage as the in-circuit capacitances when it is switched in.) The changes in frequency and amplitude are related to the square root of the relative capacitance change, which suggests that a "ballast" capacitance could be added in parallel with the payload to keep the variations small. This approach works, but adds to the overall capacitance driven and therefore to the dissipation.

The almost-sinusoidal waveforms of the circuit in Figure 15.9 are easy to produce, but they restrict the choice of logic styles available to the designer. A two-phase version of the same circuit generates a sinusoidal waveform and its inverse, using only one inductor, as presented in Figure 15.10 [19]. (Large inductors with high Q-values cannot presently be integrated on a silicon chip, so with current packaging techniques, the number of inductors must be kept small to reduce cost.) With slightly modified control-signal timing, the same circuit can produce a good approximation of the "ideal" clock signal for many adiabatic logic styles: linear voltage ramps interspersed with periods of constant voltage. When the circuit is operated in this mode, the switches first connect the inductor across the supply rails, allowing a current to build up. When the switches are released, the current continues to flow and moves charge from one of the load capacitances to the other. When both load capacitances have reached the opposite supply voltage, the inductor is connected to the rails again, but now in the opposite direction. The inductor current will shrink, change direction, and build up again, and another cycle can start. Clearly, this circuit requires equal capacitances to deliver equal rise times at both ends of the inductor.

As described previously, logic circuits of very high efficiency require that the input voltages of a logic gate are held constant while charge is transported to and from its output. Logic styles suitable for reversible computing therefore require several interleaved clock signals. Several circuits such as that in Figure 15.10 can be used to provide four, six, eight, or more equally spaced clock signals. It is also possible to generate any number of such clock signals with only one inductor, if only two signals transition at the same time [7].

An important property of the circuit in Figure 15.10 may not be immediately obvious. The inductor is connected with switches to a constant voltage to allow current to build up. This current will cause ohmic losses in the switches. Wider switch devices have lower resistances and thus cause proportionally lower ohmic losses. It would seem that to minimize the losses, the devices should be chosen as wide as

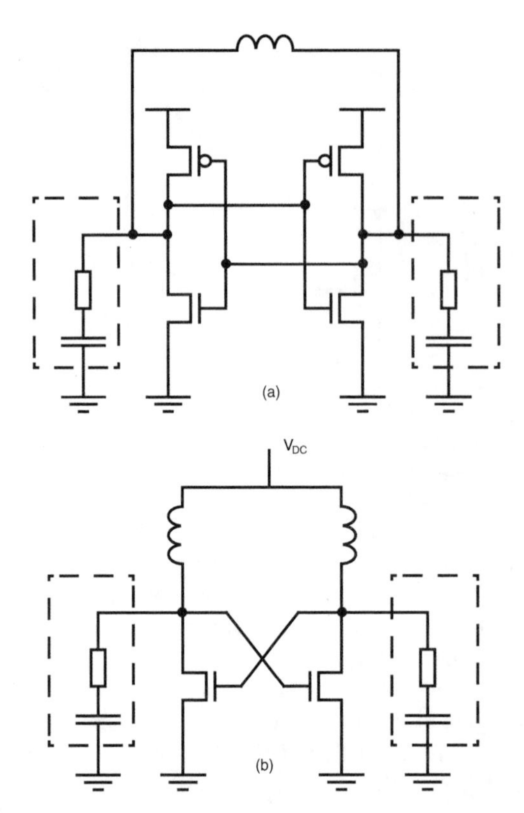

FIGURE 15.11 (a) The APS of Figure 15.10, but with resonant driving of the gates of the inductor switches. (b) Simplified version of (a), which avoids crowbar current through the inductor switches.

practically possible; but the gate capacitance is also proportional to device width, and the circuits driving these gate capacitances will dissipate proportionally larger amounts of energy per cycle. This relationship changes the system-level power dependence on transition time from $P \sim T^{-1}$ to $P \sim T^{-1/2}$ [1]. To avoid this degradation, the driving signals for the ohmic switches must actually be resonantly driven, as in Figure 15.11(a) [19]. A simplified version of this circuit, depicted in Figure 15.11(b), does not suffer from crowbar current; it generates waveforms where sinusoidal "blips," suitable for the Athas–Tzartzanis ER latch of Figure 15.6(b), alternate on the two output lines [20]. The peak voltage is approximately 3 times the DC voltage.

The ohmic switches can be made much wider when they are resonantly driven. A first-order analysis indicates that the overall dissipation is minimized when their gate capacitance is equal to the total payload capacitance. Thus, the switches act as "ballast" capacitances, minimizing the influence of fluctuations in the payload capacitance. The price paid is that the system is now free-running, without control signals, and its frequency cannot be easily controlled.

All APS designs described previously use inductors for energy storage. Inductors have several practical drawbacks: high-Q inductors cannot be integrated on silicon chips; timing errors can cause ringing or damaging voltage spikes; the inductor-based APS solutions suffer from jitter when subjected to variable capacitive loads. It is possible to avoid these drawbacks by using capacitors for energy storage [21]. Consider the circuit in Figure 15.12, where the tank capacitors C_{Ti} are charged to the evenly distributed voltages (i/N) V. When the load C_L is charged from 0 to V, it is connected to each of these tank capacitors in sequence and receives from each of them a charge C_LV/N before finally being connected to the supply voltage V. To discharge C_L, it is again connected to all the tanks, but in opposite sequence, and then finally connected to ground. Each tank capacitor receives during the discharge procedure the same amount of charge as it provided during the charge procedure, so the tank capacitor voltages are self-

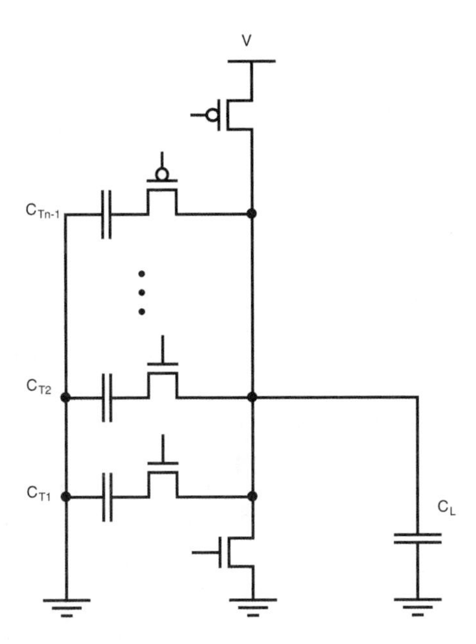

FIGURE 15.12 A stepwise driver for the load capacitance CL. The load is connected to each of the large "tank" capacitors, C_{Ti}, in sequence during charging, and in reverse sequence during discharging. The tank capacitor voltages converge to i (V/N).

sustaining. Additionally, the tank capacitances converge automatically towards the evenly distributed voltages when the circuit is exercised, so no auxiliary circuits are needed to maintain the tank voltages. The reduced dissipation is due to the smaller voltage drop traversed by the charge. In addition, it is clear that only a charge C_LV/N is drawn from the supply line for each cycle: a reduction from the conventional case by a factor of N. The energy drawn from the supply line, C_LV^2/N, is also reduced by the same factor.

The circuit in Figure 15.12, known as a stepwise driver, may be used as an APS to drive adiabatic logic circuits. It is, however, not as energy-efficient as an inductor-based circuit with the same transition time, because of all the switches that must be turned on and off for each step (which itself causes additional dissipation). This overhead limits the useful number of steps to below 10 in almost all cases.

The most natural place for a stepwise driver is not as an APS for a computing circuit, but rather as an off-chip pad driver: such circuits rarely use the maximum speed possible in the technology, because the resulting high current derivatives would cause large voltage spikes across the package inductances. A stepwise driver of three steps can offer 50% lower dissipation with little performance impact [21]. It may also be introduced without system-level redesign, as would often be required for the more ambitious inductive solutions.

15.4 Conclusion

Adiabatic and energy-recovery techniques offer new possibilities to trade dynamic power dissipation for delay in switching circuits. In some cases, the voltage swing is fixed for external reasons, such as in display drivers and micro-mechanical actuators, and in pad drivers that must produce industry-standard voltage levels. In these cases, adiabatic switching is the only known technique that allows this fundamental trade-off to be made at the circuit level.

Most published research in adiabatic-circuit techniques has focused on novel logic styles. The application examples for each style have been few, however, and many of them have been arithmetic blocks of various types (adders, multipliers, etc.). A wider range of application examples would make it easier to understand where each of these the adiabatic techniques outperform the conventional ones.

Some system-level experiments have been carried out where a moderately complex digital system has been implemented with energy-recovery techniques. The AC-1 and MD-1 microprocessors [2,22] used the less-ambitious approach to energy recovery: no attempt was made to recover signal energies inside logic gates. The results were encouraging, in that the processor-APS combinations displayed power levels under resonant drive, which were significantly lower than for conventional control cases. The improvements were, however, not large enough to cause an immediate paradigm shift in digital circuit design. Further system-level experiments, building on these experiences, should be able to improve significantly on these initial results.

The very lowest dynamic dissipation figures for a digital logic block can be reached only if all logic functions are invertible and connected to form a reversible apparatus. The circuit overhead for building fully reversible logic pipelines in present-day CMOS appears large enough to be prohibitive in most cases. Logic-style breakthroughs may still occur which would reduce the overhead; but it is equally important not to overlook the power-supply influence on the overall dissipation. It is fruitless to build a reversible-logic chip from which 99.999% of the node energies can be recovered if the power supply is only 99% efficient. In addition, device leakage through nominally off devices in the logic circuits themselves must be better analyzed for ultimate-low-power claims to be credible.

References

[1] W.C. Athas, L.J. Svensson, J.G. Koller, N. Tzartzanis, and E.Y.-C. Chou. Low-power digital systems based on adiabatic-switching principles. *IEEE Trans. on VLSI Systems,* Vol. 2, No. 4, pp. 398–407, Dec. 1994.

[2] W.C. Athas, N. Tzartzanis, L.J. Svensson, and L. Peterson. A low-power microprocessor based on resonant energy. *IEEE Journal of Solid-State Circuits,* Nov. 1997, pp. 1693–1701.

[3] J.G. Koller and W.C. Athas. Adiabatic switching, low-energy computing, and the physics of storing and erasing information. *Proc. of the Workshop on Physics and Computation, PhysCmp '92,* Oct. 1992, IEEE Press, 1993.

[4] R. Landauer. Irreversibility and heat generation in the computing process. *IBM J. Res. Dev.,* Vol. 5, pp. 183–191, 1961.

[5] W.C. Athas and L. Svensson. Reversible logic issues in adiabatic CMOS. *Proc. 1994 Workshop on Physics and Computation,* Nov. 1994.

[6] S. Younis and T.F. Knight. Asymptotically zero-energy split-level charge recovery logic. *Proc. Int. Workshop on Low-Power Design,* Napa, CA, 1994, pp. 177–182.

[7] J. Lim, D.-G. Kim, and S.-I. Chae. nMOS reversible energy recovery logic for ultra-low-energy applications. *IEEE J. Solid-State Circuits,* June 2000, pp. 865–875.

[8] C. Vieri. Reversible computer engineering and architecture. Ph.D. thesis, Massachusetts Institute of Technology, Cambridge, MA, June 1999.

[9] J.S. Denker. A review of adiabatic computing. *Symp. on Low-Power Electronics,* San Diego, CA, Oct. 10–11, 1994, pp. 94–97.

[10] N. Tzartzanis and W.C. Athas. Design and analysis of a low-power energy-recovery adder. *Proc. 5th Great Lakes Symp. on VLSI Design,* Buffalo, NY, Mar. 16–18, 1995, pp. 66–69.

[11] J. Kim, C.H. Ziesler, and M.C. Papaefthymiou. Energy recovering static memory. *Proc. 2002 IEEE ISLPED,* Monterey, CA, Aug. 12–14, 2002, pp. 92–97.

[12] N. Tzartzanis, W. Athas, and L. Svensson. A low-power SRAM with resonantly powered data, address, word, and bit lines. *Proc. ESSCIRC 2000,* Stockholm, Sweden, Sep. 19–21, 2000.

[13] J. Ammer, M. Bolotski, P. Alvelda, and T.F. Knight, Jr. A 160 × 120 pixel liquid-crystal-on-silicon microdisplay with an adiabatic DACM. *ISSCC 1999 Dig. of Tech. Papers,* pp. 212–213.

[14] R. Lal, W. Athas, and L. Svensson. A low-power adiabatic driver system for AMLCDs. *Symp. on VLSI Circuits Dig. of Tech. Papers,* Honolulu, June 15–17, 2000, pp. 198–201.

[15] S. Younis and T. Knight. Non-dissipative rail drivers for adiabatic circuits. *Proc. 16th Conf. on Advanced Research in VLSI, 1995,* Los Alamitos, CA, March 27–29, 1995, pp. 404–414.

[16] P. Solomon and D. Frank. The case for reversible computation. *Proc. Int. Workshop on Low-Power Design,* Napa, CA, April 24–27, 1994, pp. 93–98.

[17] D. Maksimovic et al. Clocked CMOS adiabatic logic with integrated single-phase power-clock supply. *IEEE Trans. VLSI Syst.,* Vol. 8, No. 4, Aug. 2000, pp. 460–463.

[18] C.H. Ziesler, S. Kim, and M.C. Papaefthymiou. A resonant clock generator for single-phase adiabatic systems. *Proc. Int. Symp. on Low-Power Electronics and Design,* Huntington Beach, CA, Aug. 6–7, 2001, pp. 159–164.

[19] A. Dickinson and J.S. Denker. Adiabatic dynamic logic. *IEEE J. Solid-State Circuits,* Vol. 30, No. 3, Mar. 1995, pp. 311–315.

[20] W.C. Athas, L.J. Svensson, and N. Tzartzanis. A resonant clock driver for two-phase, almost-non-overlapping clocks. *Proc. IEEE ISCAS '96,* Atlanta, GA, May 12–15, 1996.

[21] L.J. Svensson, W.C. Athas, and R.S.-C. Wen. A sub-CV^2 pad driver with 10-ns transition time. *Proc. Int. Symp. on Low-Power Electronics and Design,* Monterey, CA, Aug. 12–14, 1996.

[22] W. Athas, N. Tzartzanis, W. Mao, L. Peterson, R. Lal, K. Chong, J.-.S Moon, L. Svensson, and M. Bolotski. The design and implementation of a low-power clock-powered microprocessor. IE*EE J. Solid-State Circuits,* Nov. 2000, pp. 1561–1570.

16

Weak Inversion for Ultimate Low-Power Logic

Eric A. Vittoz
CSEM

16.1 Introduction

In digital circuits, power is needed to charge the load capacitance C of each logic node at the switching frequency f. This dynamic power consumption can be expressed as

$$P_{dyn} = fC\Delta VV_B \tag{16.1}$$

where V_B is the supply voltage and ΔV the logic voltage swing, smaller or equal to V_B. Thus, the dynamic power can be reduced by reducing ΔV, but this gate voltage swing is needed to ensure a sufficient current ratio I_{on}/I_{off} in the transistors producing the transitions. Indeed, the on-current I_{on} must be large enough to ensure transitions at the required speed, and I_{off} should be as small as possible to limit the static power consumption $P_{stat} = I_{off}V_B$ between transitions.

The swing ΔV needed to achieve a given value of I_{on}/I_{off} can be reduced by reducing the gate voltage overhead, until it becomes minimum when weak inversion is reached. Logic circuits based on transistors

0-8493-1941-2/05/$0.00+$1.50

operated in weak inversion (also called subthreshold) therefore offer minimum possible operating voltage, and thereby minimum P_{dyn} for a given P_{stat}. This is only possible, however, if the threshold voltage of the transistors can be precisely adapted to this very low value of supply voltage V_B. The feasibility of CMOS inverters with supply voltages as low as 200 mV was already demonstrated more than 30 years ago [1], with the possibility of reducing it to 100 mV if fast surface state would be negligible (which has become true for more than 20 years). However, the minimum channel length was still on the order of 5 μm, limiting the maximum frequency to just a few hundred kHz. Therefore, the idea was buried for several decades dominated by the struggle for maximum speed. It has been revived recently [2] and applied to complete subsystems operated below 200 mV [3,4]. In the meantime, weak inversion was used extensively for very low-power analog circuits [5–7], and a special model was developed to better describe the behavior of a MOS transistor from weak to strong inversion [8,9]. This chapter relies on this experience of weak inversion and on this model to derive the analytical results needed to optimize such low-voltage digital circuits and to identify their ultimate limits.

16.2 MOS Model in Weak Inversion and Basic Assumptions

The drain-to-source current I_{DS} of n-channel MOS transistors operated in weak inversion can be expressed as [6,8,9]:

$$I_{DS} = I_S \exp\frac{V_{GS} - V_T}{nU_T}\left[1 - \exp\frac{-V_{DS}}{U_T}\right] \tag{16.2}$$

where $n > 1$ is the slope factor (practically always below 1.6), V_{GS} and V_{DS} the gate to source and drain to source voltages, V_T the gate to source threshold voltage and I_S the specific current given by

$$I_S = 2n\mu C_{ox} U_T^2 W / L \tag{16.3}$$

where μ is the carrier mobility, C_{ox} the gate oxide capacitance per unit area, $U_T = kT/q$ and W/L the width-to-length ratio of the channel. The threshold voltage V_T depends on the source-to-substrate voltage V_{SB} according to

$$V_T = V_{T0} + (n-1)V_{SB} \tag{16.4}$$

where V_{T0} is the threshold voltage for $V_{SB} = 0$. Because V_{SB} must be larger than about $-4U_T$ (source junction reverse biased or only slightly forward biased, to avoid parasitic bipolar effects), V_T can only be increased or just slightly decreased with respect to V_{T0}.

By introducing the saturation current for $V_{GS} = 0$:

$$I_0 = I_S \exp\frac{-V_T}{nU_T} = I_S \exp\frac{-(V_{T0} + (n-1)V_{SB})}{nU_T} \tag{16.5}$$

which is also controllable by V_{SB}, Equation (16.2) is reduced to

$$I_{DS} = I_0 \exp\frac{V_{GS}}{nU_T}\left[1 - \exp\frac{-V_{DS}}{U_T}\right]. \tag{16.6}$$

The same equations are valid for p-channel transistors if the sign of current and voltages is inverted. Thus, I_{DS}, V_{GS}, V_{DS} become I_{SD}, V_{SG}, V_{SD}, and V_T is the threshold value of V_{SG}.

For the following analysis, two basic assumptions will be made about the process:

1. The native threshold V_{T0} for p- and n-channel transistors does not exceed $4U_T$, even in the worst case of process and temperature variation. Practical ways of adjusting the value of V_T according to Equation (16.4) will be discussed in Section 16.8.
2. True twin wells (often called triple-well) are available, to allow separate adjustment of p- and n-channel V_T.

16.3 Static CMOS Inverter

Consider the simple CMOS inverter of Figure 16.1 with input voltage V_i and output voltage V_o. We will assume that the local p- and n- substrates are properly biased with respect to the $V+$ and $V-$ rails of the power supply to control V_T and I_0, according to Equation (16.4) and Equation (16.5). To simplify the analysis, we will further assume that the two types of transistors are symmetrical, with same values of n and I_0.

By normalizing currents and voltages according to

$$x_k = V_k / U_T \text{ and } y_k = I_k / I_0 \tag{16.7}$$

the normalized values of current I_p and I_n flowing through the two transistors can be expressed from Equation (16.6) as

$$y_n = \mathrm{e}^{x_i/n}(1 - \mathrm{e}^{-x_o}) \tag{16.8}$$

$$y_p = \mathrm{e}^{(x_B - x_i)/n}(1 - \mathrm{e}^{x_o - x_B}) \tag{16.9}$$

The inverter is only loaded by a capacitance C, thus the static transfer function is obtained by equating these two currents, which yields:

$$x_i = \frac{x_B}{2} + \frac{n}{2}\ln\left[(1 - \mathrm{e}^{x_o - x_B}) / (1 - \mathrm{e}^{x_o})\right] \tag{16.10}$$

which can be inverted to give

$$x_o = x_B + \ln\frac{1 - G + \sqrt{(G-1)^2 + 4G\mathrm{e}^{-x_B}}}{2} \quad \text{where } G = \mathrm{e}^{(2x_i - x_B)/n} \tag{16.11}$$

This transfer function is plotted in Figure 16.2(a) for n = 1.6 and several values of x_B.

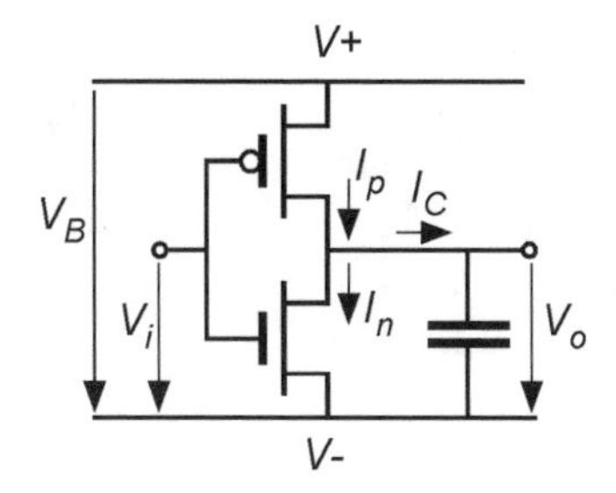

FIGURE 16.1 CMOS inverter.

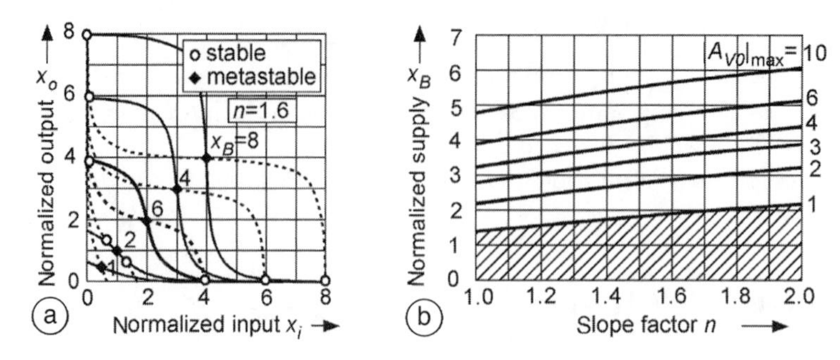

FIGURE 16.2 Stables states: (a) transfer function; (b) voltage required for gain.

The plots are repeated with horizontal and vertical axes permuted, to represent two inverters connected as a flip-flop. If x_B is not too small, there are two stable solutions corresponding to the logic states. Now the latter only exist if the maximum value of the inverter gain $|A_{V0}|$ exceeds 1. Differentiating Equation (16.10) or Equation (16.11) yields

$$-A_{V0} = \frac{2(1-e^{x_o-x_B}-e^{-x_o}-e^{-x_B})}{n(2e^{-x_B}-e^{x_o-x_B}-e^{-x_o})} \tag{16.12}$$

the maximum of which occurs for $x_o = x_i$:

$$|A_{V0}|_{max} = (e^{x_B/2}-1)/n \text{ or } x_B = 2\ln(n|A_{V0}|_{max}+1) \tag{16.13}$$

The latter is plotted in Figure 16.2(b) as a function of n. The lowest curve with $|A_{V0}|_{max} = 1$ depicts the absolute minimum voltage required for bistability. For a realistic value of $n = 1.6$, this absolute minimum is 1.91 ($V_B \cong 50$ mV at ambient temperature).

The normalized high value x_H and low value x_L of the stable points can be calculated by iteration in a series of inverters described by Equation (16.11), driving them by any value outside the metastable state. The result is represented in Figure 16.3(a) for $n = 1.6$.

The static current I_{stat} flowing through both transistors at the stable states can easily be calculated by introducing the values of x_H and x_L as input and output voltages in Equation (16.8) or Equation (16.9). The result is plotted in Figure 16.3(b). It shows that this static current is only slightly larger than I_0.

For $n = 1.6$ or less, $x_B = 4$ ($V_B = 100$mV at ambient temperature) is sufficient to ensure a swing almost equal to V_B and a static current virtually equal to I_0.

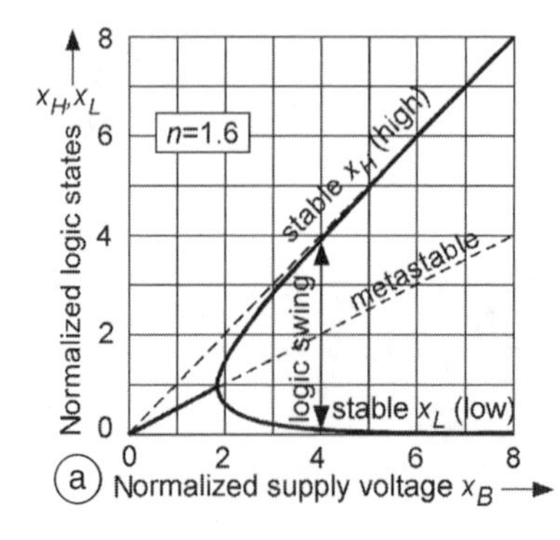

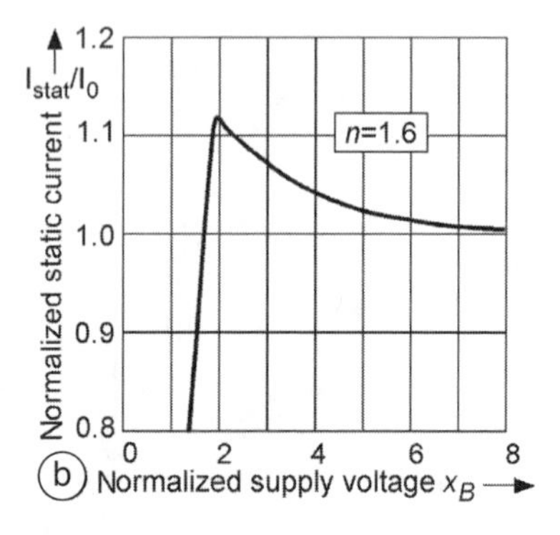

FIGURE 16.3 Logic states: (a) evolution with supply voltage; (b) static current.

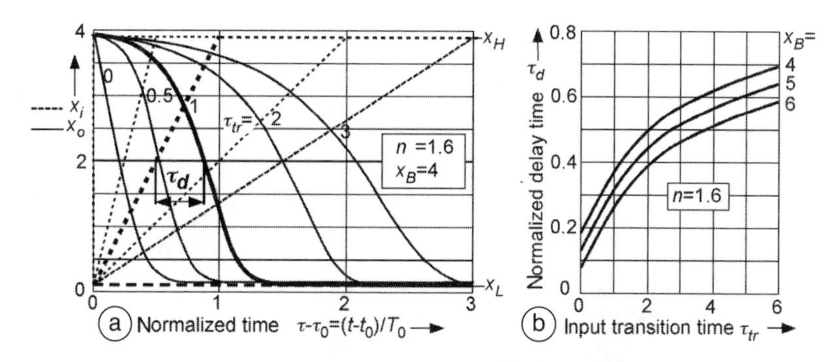

FIGURE 16.4 Gate transition: (a) input and output transitions; (b) delay time.

16.4 Dynamic Behavior of the CMOS Inverter

16.4.1 State Transition

Referring again to Figure 16.1, the normalized current flowing into capacitor C during a transition is obtained from Equation 16.8 and Equation 16.9:

$$y_C = I_C / I_0 = y_p - y_n = \mathrm{e}^{(x_B - x_i)/n}(1 - \mathrm{e}^{x_o - x_B}) - \mathrm{e}^{x_i/n}(1 - \mathrm{e}^{-x_o}) \tag{16.14}$$

The evolution of the output voltage of the inverter is thus obtained by integrating I_C into C:

$$x_o = x_{o0} + \frac{I_0}{CU_T}\int_{t_0}^{t} y_C(x_i, x_o)dt = x_{o0} + \int_{\tau_0}^{\tau} y_C(x_i, x_o)d\tau \tag{16.15}$$

$$\text{where } \tau = t / T_0 \text{ and } T_0 = CU_T / I_0 \tag{16.16}$$

are the normalized time and the characteristic time, respectively, and x_{o0} is the initial value of x_o at $t = t_0$. This evolution depends on that of the input voltage $x_i(t)$.

It is plotted in Figure 16.4(a) for various constant slopes of $x_i(t)$ characterized by their normalized transition time $\tau_{tr} = T_{tr} / T_0$ between the two logic states $x_i = x_L$ and $x_i = x_H$. The delay time $T_d = \tau_d T_0$ defined on this figure is minimum for a step input and increases with input transition time T_{tr}. This variation is plotted in Figure 16.4(b) for various values of supply voltage x_B.

16.4.2 Currents and Charges

During transitions, neither of the two transistors is as blocked as in the static states, therefore, some additional current y_{sc} flows directly through them. For a rising input, this short-circuit current corresponds to an increase of y_p, the normalized current flowing through the p-channel device, as can be seen in Figure 16.5(a).

Integrating this surplus current during the whole transition gives the short-circuit charge per transition Q_{sctr}, or its normalized value:

$$q_{sctr} = \frac{Q_{sctr}}{CU_T} = \int_{before}^{after} (y_p - y_{stat})d\tau \tag{16.17}$$

which is also presented in the figure, amplified by a factor of 10. This charge is delivered by the power supply, in addition to the main charge Q_{Ctr} flowing from the capacitor, given by:

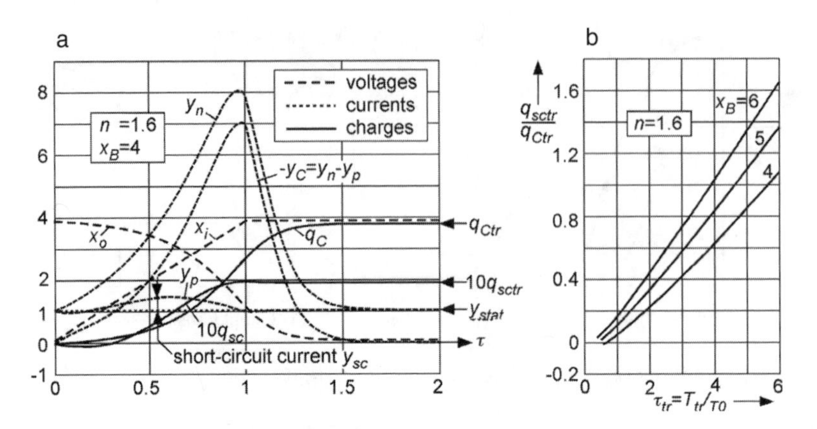

FIGURE 16.5 Transition of the inverter: (a) currents and charges; (b) proportion of short-circuit charge.

$$q_{Ctr} = \frac{Q_{Ctr}}{CU_T} = \int_{before}^{after} (y_n - y_p)d\tau \tag{16.18}$$

The total charge delivered by the power supply for each cycle of up and down transitions (and added to that due to static current I_{stat} is

$$Q_{tr} = Q_{Ctr} + 2Q_{sctr} \tag{16.19}$$

The calculated charge ratio Q_{sctr}/Q_{Ctr} as a function of the normalized input transition time $T_{tr} = \tau_{tr} T_0$ is plotted in Figure 16.5(b) for various values of normalized supply voltage x_B. It increases approximately proportionally to τ_{tr}, but remains below 20% for $\tau_{tr} < 1$ $(T_{tr} < T_0)$.

16.5 Behavior of the Inverter for Standard Transitions

16.5.1 Definition and Delay Time

In large digital circuits, the inputs of most gates are outputs of other gates. If all gates have same fan-in and approximately same capacitive load, then they all tend to the same transition behavior. Such standardized transitions can be emulated in a long chain of inverters identical to the single inverter analyzed in Section 16.4. The result is illustrated in Figure 16.6(a) for a cascade of eight stages.

With the first stage driven by a rising step, the transitions become perfectly standardized after just a few stages. The standard delay time for a pair of gates $2\tau_d = 2T_d / T_0$ can then be defined as presented in the figure: such a definition would still be valid if the inverters are not be symmetrical. The variation of

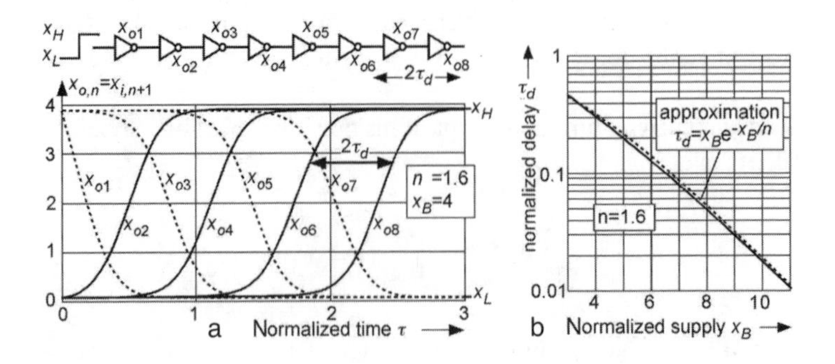

FIGURE 16.6 Cascade of inverters: (a) transitions of successive stages; (b) standard delay time.

this delay with supply voltage, calculated between x_{o6} and x_{o8} is represented in Figure 16.6(b). An approximative analytical expression of T_d can be obtained as follows: assuming that, at each transition, the whole on-current I_{on} of a fully saturated transistor driven by the full supply voltage V_B is flowing through capacitor C and changes the voltage across it by V_B, then:

$$T_d = \frac{CV_B}{I_{on}} = \frac{CV_B}{I_0 \exp\frac{V_B}{nU_T}} \text{ or } \tau_d = x_B e^{-x_B/n} \tag{16.20}$$

where the saturated on-current I_{on} is obtained from Equation (16.6) for $V_{DS} >> U_T$.

This expression is also plotted on Figure 16.6(b), demonstrating that is is a relatively good approximation. Because the exponential increases much faster than its argument, the delay time decreases approximately exponentially with increasing supply voltage, this for a constant characteristic time $T_0 = CU_T/I_0$. As can be seen, T_d is always smaller or much smaller than T_0 for such standard transitions.

16.5.2 Currents and Charges

The various voltages, currents, and charges of a standard transition calculated at the eighth stage of the cascade of inverters are plotted in Figure 16.7(a) for a particular small value of normalized supply voltage x_B.

As a consequence of the short standard transition time, the short-circuit current y_{sc} (now shown amplified by a factor of 100) is always a very small fraction of current y_C in the capacitor. It results in a short-circuit charge q_{sctr} much smaller than the charge q_{Ctr} flowing in or out of the capacitor, as can be expected from Figure 16.5(b) for values of $\tau_d << 1$. This proportion, represented in Figure 16.7(b) for various values of normalized supply voltage x_B, is always smaller than 1% for $x_B > 4$. The dynamic power consumption in homogeneous circuits with standard transitions is therefore very close to that due to the charge supplied to the capacitor:

$$P_{dyn} = f(Q_{Ctr} + Q_{sctr})V_B \cong fQ_{Ctr}V_B = fC(V_H - V_L)V_B \cong fCV_B^{\,2} \tag{16.21}$$

where f is the frequency of transitions.

The short-circuit charge is always a very small fraction of the charge flowing in the capacitor. As a result, the short-circuit current can be practically neglected, the only current flowing (permanently) through the two transistors being I_{stat}.

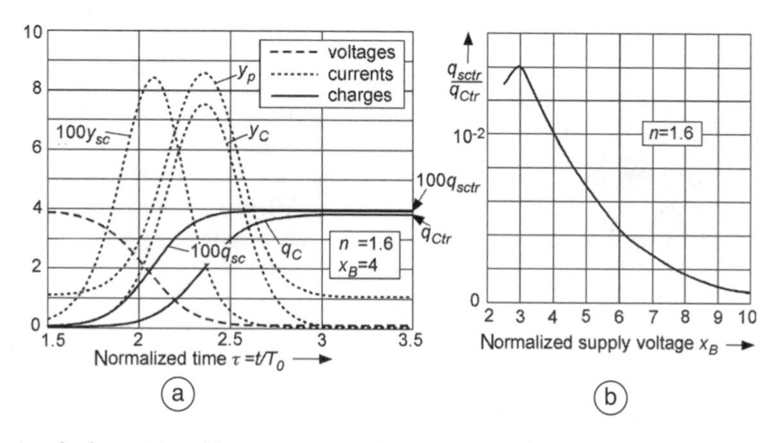

FIGURE 16.7 Standard transition: (a) currents and charges; (b) proportion of short-circuit charge.

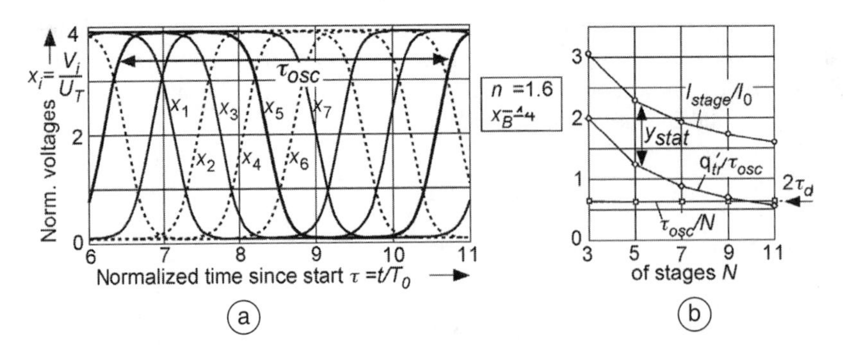

FIGURE 16.8 Ring oscillator: (a) voltages of 7-stage ring; (b) current per stage and period of oscillation.

16.5.3 Ring Oscillator

The dynamic behavior of inverters can also be examined by connecting an odd number N of them in closed loop to form a ring oscillator. The normalized output voltages of the different stages are plotted in Figure 16.8(a) for seven stages.

Because each stage must switch up and down once per period of oscillation T_{osc}, the period of oscillation for N stages should be given by

$$T_{osc} = 2NT_d \text{ or } \tau_{osc} = 2N\tau_d \tag{16.22}$$

As plotted in Figure 16.8(b), T_{osc} is slightly larger for the shortest ring ($N = 3$) because the swing does not reach the full static swing $x_H - x_L$, but Equation (16.22) becomes valid for longer rings. Figure 16.8(b) also shows that the average current I_{stage} consumed by each stage is given by

$$I_{stage} = Q_{tr} / T_{osc} + I_{stat} = (q_{tr} / \tau_{osc} + y_{stat})I_0 \tag{16.23}$$

Again, this equation becomes exact when the ring is long enough to achieve full static swing. When N is increased, corresponding to a decrease of the frequency of oscillation, the activity of each stage decreases, resulting in an increased proportion of static current I_{stat}.

16.5.4 Power-Delay Product

As was shown in Figure 16.3(b), the static current I_{stat} consumed in each stable state is negligibly higher than I_0, the saturation current for $V_{GS} = 0$. Therefore, the static power consumption per inverter can be simply approximated by

$$P_{stat} = V_B I_{stat} = V_B I_0 \tag{16.24}$$

Furthermore, it has been shown in Figure 16.7(b) that, for standard transitions obtained in an homogeneous system, the short-circuit charge Q_{sctr} is only a negligible fraction of the charge Q_{Ctr} flowing in the capacitor. The dynamic power consumption can thus reasonably be approximated as in Equation (16.21), and the total power P by:

$$P = P_{stat} + P_{dyn} = V_B I_0 + fCV_B^2 \tag{16.25}$$

Now, the frequency of operation f cannot be larger than $1/(2T_d)$. For $f = 1/(2T_d)$, the inverter (or in general the gates) are always in transition. A duty factor α can thus be defined as:

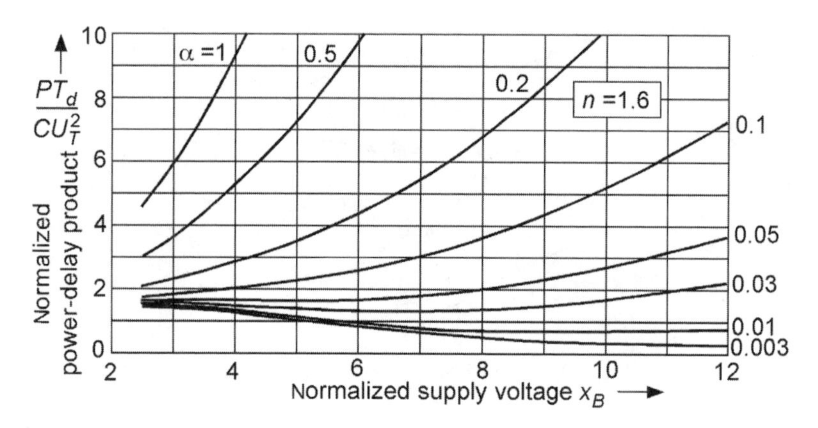

FIGURE 16.9 Power-delay product.

$$\alpha = 2fT_d \leqq q1 \tag{16.26}$$

Introducing Equation (16.26) in Equation (16.25) with the definitions from Equation (16.7) and Equation (16.8) yields the power-delay product:

$$PT_d = CU_T^{\,2}\left(x_B\tau_d(x_B) + \frac{\alpha x_B^2}{2}\right) \tag{16.27}$$

where the first term results from the static power and the second from the dynamic power. This product can be calculated as a function of the normalized supply voltage x_B by using the previous calculations of $\tau_d(x_B)$ represented in Figure 16.6(b). Results are plotted in Figure 16.9 for various values of duty factor α.

Using the approximation Equation (16.20) for $\tau_d(x_B)$, the power-delay product can be expressed from Equation (16.27) as

$$PT_d = CU_T^2 x_B^2\left(e^{-x_B/n} + \frac{\alpha}{2}\right) \tag{16.28}$$

giving results very close to those of Figure 16.9. Using this expression with the definition of α by Equation (16.26) of α to replace the gate delay T_d by the (average) frequency of operation f, the power to frequency ratio can be expressed as:

$$\frac{P/f}{C(nU_T)^2} = (x_B/n)^2\left(\frac{2}{\alpha}e^{-x_B/n} + 1\right) \tag{16.29}$$

where the first term is again the static power and the second term the dynamic power. This result is represented in Figure 16.10 as a function of the normalized supply voltage x_B divided by n, for various values of duty factor α.

Except for duty factors approaching unity, these curves have a minimum for an optimum value of x_B. This optimum value is obtained by differentiating Equation (16.29) with respect to x_B and equating the result to zero. This yields:

$$\alpha = (x_{Bopt}/n - 2)e^{-x_{Bopt}/n} \tag{16.30}$$

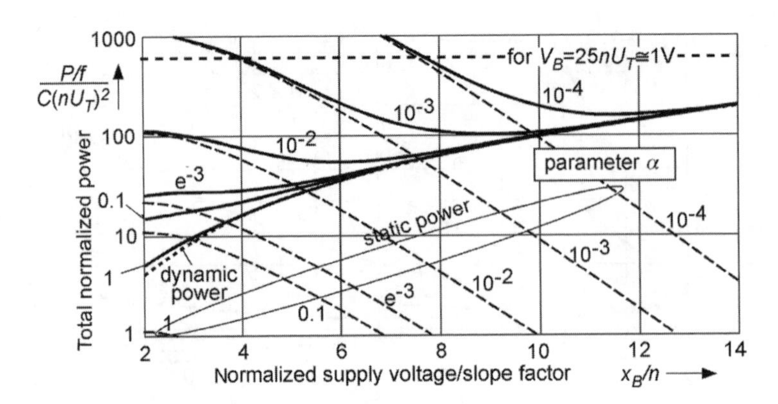

FIGURE 16.10 Power/frequency ratio.

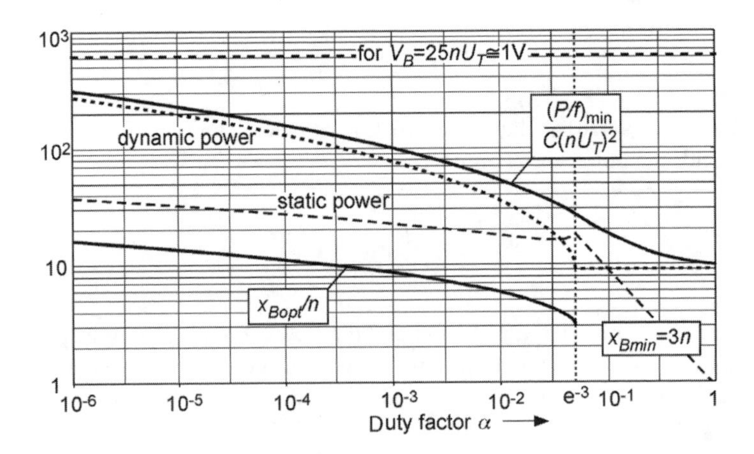

FIGURE 16.11 Optimum value of supply voltage and corresponding minimum power/frequency ratio.

which only has a solution for $\alpha \geqq e^{-3}$. Figure 16.10 shows that this optimum is not very critical. It can be obtained by numerically inverting Equation (16.30). The result can then be introduced into Equation (16.29) to provide the minimum possible value of P/f for a given duty factor α plotted in Figure 16.11.

Because no optimum value of x_B is available, for $\alpha > e^{-3}$, its minimum value must be fixed to ensure a logic swing close to the supply voltage, according to Figure 16.3(a). A value of $x_B = 3n$ has been chosen for this figure. The dynamic power for a supply voltage of 25 nU_T (approximately 1 V) is also indicated on the figure. As can be seen, a 50-fold reduction would be possible for $\alpha > 0.3$; the possible gain compared with 1-V operation is smaller for more realistic lower values of the duty factor, but it remains larger than a factor 10 for $\alpha > 3/1000$. For $\alpha << 1$, the minimum power occurs for $P_{stat} << P_{dyn}$. Indeed, according to Equation (16.20), increasing I_0 does not allow to reduce V_B significantly if delay time T_d must be maintained constant.

16.5.5 Minimum Delay Time in Weak Inversion

The degree of inversion of a transistor can be characterized by the inversion coefficient IC [6] defined as

$$IC = I_{DSsat} / I_S \tag{16.31}$$

where I_S is the specific current of the transistor given by Equation (16.3). Introducing this definition in the approximation of the delay time given by Equation (16.20) results in

$$T_d = \frac{CV_B}{I_S IC_{on}} = BV_B / IC_{on} \tag{16.32}$$

where $B = C/I_S$ has a minimum value given by the process. The only possibility to reduce T_d is thus to increase the inversion coefficient in the on-state of the transistor; but weak inversion and its features according to the model introduced in Section 16.2 is limited to $IC < 1$. Therefore, the minimum delay time and the maximum average frequency achievable in weak inversion are simply given by

$$T_{dmin} = BV_B \text{ and } f_{max} = \frac{\alpha}{2BV_B} \text{ (in weak inversion)} \tag{16.33}$$

16.6 Effect of Entering Moderate and Strong Inversion

16.6.1 Transistor Model

With the definitions of currents and voltages introduced in Section 16.2, the following simple expression can be used to describe the drain current in and above weak inversion [6,8,9]:

$$I_{DS} = I_S\left(\ln^2\left[1+\exp\frac{V_{GS}-V_T}{2nU_T}\right] - \ln^2\left[1+\exp\frac{V_{GS}-V_T}{2nU_T}\exp\frac{-V_{DS}}{2U_T}\right]\right) \tag{16.34}$$

which reduces to Equation (16.2) for weak inversion when the exponential terms are much smaller than unity. In saturation, the second term is negligible, and this equation becomes

$$IC = \frac{I_{DSsat}}{I_S} = \ln^2\left[1+\exp\frac{V_{GS}-V_T}{2nU_T}\right] \tag{16.35}$$

This equation can be inverted to provide

$$V_{GS} = V_T + 2nU_T \ln(e^{\sqrt{IC}} - 1) \tag{16.36}$$

16.6.2 Required Voltage Swing

The gate voltage swing required to obtain a ratio K between the on-current I_{on} and the off-current I_{off} of a transistor can be obtained by application of Equation (16.36):

$$\Delta V_{GS} = 2nU_T\left[\ln(e^{\sqrt{IC_{on}}} - 1) - \ln(e^{\sqrt{IC_{on}/K}} - 1)\right] \tag{16.37}$$

This equation is plotted in Figure 16.12(a) for various values of K. It can be seen that the required swing is minimum and constant in weak inversion (as can be expected from the exponential behavior), but increases drastically when the maximum inversion coefficient IC_{on} is increased beyond unity.

Solving Equation (16.37) with respect to current ratio K yields

$$K = \frac{IC_{on}}{\ln^2\left[1+\exp\frac{-\Delta V_{GS}}{2nU_T}(\exp\sqrt{IC_{on}} - 1)\right]} \tag{16.38}$$

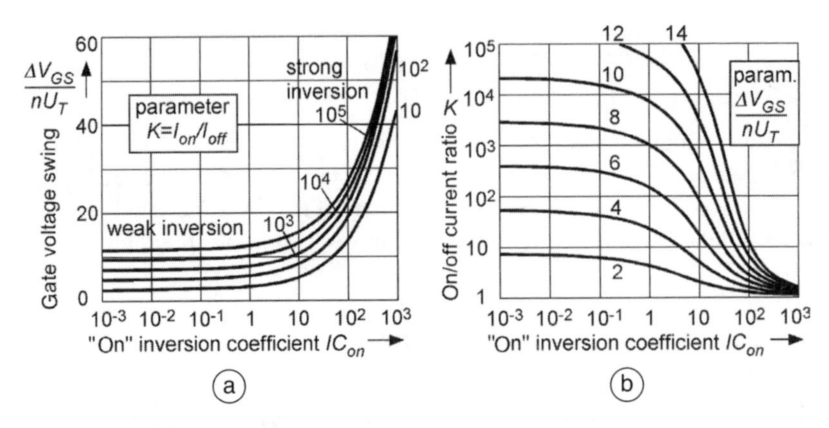

FIGURE 16.12 Gate voltage swing required for an on/off current ratio *K*.

which is plotted in Figure 16.12(b). This representation clearly demonstrates that the on/off current ratio produced by a given value of gate voltage swing is drastically reduced when the inversion coefficient is increased beyond unity. Thus, if the transistors are pushed into strong inversion to further reduce the delay time T_d according to Equation (16.32), the swing and thus supply voltage V_B must be increased to maintain the static power. The dynamic power is thus increased and a part of the expected reduction of T_d is lost.

16.6.3 Degeneration of Logic States

Assuming that the two types of transistors have the same specific current I_S and the same slope factor n, the current flowing through each of them in a CMOS inverter operated above weak inversion can be expressed by means of Equation (16.34):

$$I_n / I_S = \ln^2\left[1+\exp\frac{x_i - x_T}{2n}\right] - \ln^2\left[1+\exp\frac{x_i - x_T - nx_o}{2n}\right] \quad (16.39)$$

$$I_p / I_S = \ln^2\left[1+\exp\frac{x_B - x_i - x_T}{2n}\right] - \ln^2\left[1+\exp\frac{(1-n)x_B - x_i - x_T + nx_o}{2n}\right] \quad (16.40)$$

where $x_T = V_T/U_T$ is the normalized threshold voltage. The maximum value IC_{on} of the inversion coefficient is reached when the transistors are saturated with the gate voltage equal to the supply voltage. Thus, from Equation (16.35):

$$IC_{on} = \ln^2\left[1+\exp\frac{x_B - x_T}{2n}\right] \text{ or } x_T = x_B - 2\ln(e^{\sqrt{IC_{on}}} - 1) \quad (16.41)$$

which can be introduced in Equation (16.39) and Equation (16.40) to eliminate x_T. These two equations can then be used to calculate a long chain of inverters to obtain the values of the low- and high-logic states x_L and x_H and that of the static current I_{stat}. The results are presented in Figure 16.13 for two values of normalized supply voltage.

As can be seen, for $x_B = 4$ (part a of the figure), the logic states degenerate rapidly when the inversion factor IC_{on} is increased beyond unity; they vanish for $IC_{on} > 5$. Moreover, the static current I_{stat} increases drastically, which will increase the static power dissipation. If x_B is doubled to 8 (part b of the figure), more speed can be obtained by increasing IC_{on} beyond unity, but the dynamic power will be increased by a factor 4. This demonstrates again that the minimum power-delay product increases significantly as soon as the devices are operated beyond weak inversion to obtain more speed than the approximate limit given by Equation (16.33).

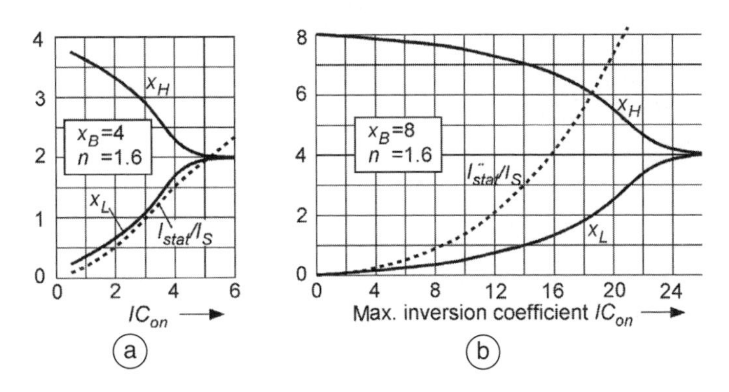

FIGURE 16.13 Degeneration of stable states and increase of static current above week inversion.

16.7 Extension to Logic Gates and Numerical Examples

To build useful logic circuits, the simple inverter of Figure 16.1 must be replaced by a logic gate, with at least two inputs. Let us assume that all gates of the system have the same number N of inputs, which means that the fan-in and fan-out of each gate are both equal to N. Let us further assume that N has the reasonable value 2 or 3, and that positive logic is used, in which each NAND gate is built by connecting N n-channel transistors in series and N p-channel transistors in parallel, as depicted in Figure 16.14.

If all transistors have the same (minimum) dimensions, then such a gate is approximately equivalent to the symmetrical inverter assumed in the previous analysis because the mobility of electrons is two to three times larger than that of holes, and the positive current I_C charging load capacitor C is normally that of only one p-channel transistor (except if several input transit simultaneously). According to Equation (16.3), the equivalent specific current I_S is then given by

$$I_S = \frac{2nW\mu_n C_{ox} U_T^2}{NL} \tag{16.42}$$

The equivalent capacitance C of the loaded gate can be expressed as

$$C = 2N(C_G + C_D) + C_{int} \tag{16.43}$$

where C_G and C_D are the gate capacitance and drain capacitance of a single transistor, and C_{int} is the interconnection capacitance.

Values of I_S and C given by Equation (16.42) and Equation (16.43) are evaluated in Table 16.1 for two typical technologies with $N = 3$. All transistors are of minimum size, and it is assumed that $C_{int} = C/2$.

Numerical values of the most important results derived previously are also given in this table. As can be seen, the minimum energy per transition (P/f for $\alpha = 1$ and $V_B = 4U_T$) is only about 0.23 fJ for

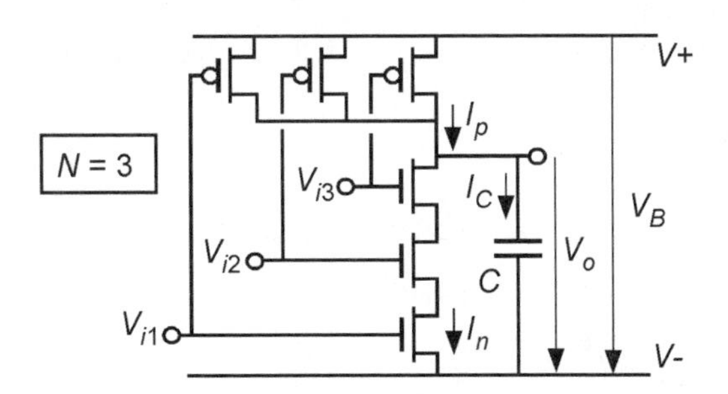

FIGURE 16.14 3-input CMOS NAND gate.

TABLE 16.1 Numerical Values for Two Different Processes ($N = 3$)

Parameter	Process A	Process B	Unit	Reference
L_{min}	500	180	nm	
U_T	25	25	mV	
n	1.5	1.3		
I_S	200	400	nA	equ. (16.42)
C	20	4	fF	equ. (16.43)
$C(nU_T)^2$	$2.8.10^{-17}$	$4.2.10^{-18}$	J	
B	1.10^{-7}	1.10^{-8}	V/s	equ. (16.32)
P/f for $\alpha = 1$ and $V_B = 4U_T$	$2.28.10^{-16}$	$4.37.10^{-17}$	J	equ. (16.29)
$(P/f)min$ for $\alpha = 10^{-2}$ and $V_B = V_{Bopt} = 6nU_T$	$1.46.10^{-15}$	$2.20.10^{-16}$	J	Fig. 16.11
P_{dyn}/f at $V_B = 1V$	$2.00.10^{-14}$	$4.00.10^{-15}$	J	
T_{dmin} in weak inversion for $V_B = 4U_T$	10	1	ns	equ. (16.33)
f_{max} for $\alpha = 1$ and $V_B = 4U_T$	50	500	MHz	equ. (16.33)
f_{max} for $\alpha = 10^{-2}$ and $V_B = V_{Bopt}$	0.22	2.56	MHz	equ. (16.33)
P_{min} at f_{max} above	32.5	56.3	nW	Fig. 16.11

conservative process A, and is further reduced by a factor 5 with the more advanced process B. If the duty factor α is reduced to 1%, the minimum equivalent energy per transition is increased by a factor 5 to 6, due to the increased importance of static power and to the necessary increase of supply voltage. Still, these values are 14 to 18 times lower than the dynamic power of the same gate operated at 1 V.

The minimum delay time in weak inversion is respectively 10 and 1 ns for the two processes, corresponding to a maximum possible clock frequency of 50 MHz and 500 MHz. If the duty cycle α is only 1%, the maximum average frequency is reduced to 220 kHz, respectively 2.56 MHz; it is decreased more than proportionally to α because V_B must be increased to limit the static power.

As discussed in Section 16.6, increasing the average frequency beyond this limit for weak inversion rapidly increases the necessary equivalent energy per transition.

16.8 Practical Considerations and Limitations

The ideal situation discussed in the previous sections is based on some basic assumptions. It is now necessary to examine if and how these assumptions can be made valid in practice.

16.8.1 Low-Voltage Power Source

Very low power is possible in weak inversion because the supply voltage V_B can be reduced to a very small value. If the duty factor is large ($\alpha > e^{-3}$), this value should be at least about $4U_T$ to ensure a full logic swing, according to Figure 16.3. For smaller α, the power consumption for a given average frequency f of transition is minimum for an optimum value of V_B that depends on α, as plotted in Figure 16.11. This optimum ranges from 4 to $15U_T$ for realistic values of α.

To take advantage of the scheme, the supply voltage must thus be adapted to this low value by means of a high-efficiency voltage converter. Although V_B should ideally be proportional to U_T and therefore proportional to the absolute temperature (PTAT), a fixed value could be used because the optimum is not very critical as demonstrated by Figure 16.10. Different supply voltages V_B should be used for blocks with different levels of duty factor α.

16.8.2 Low-Threshold and Threshold Adjustment

Once $P/f = 2PT_d/\alpha$ has been minimized by using the optimum value of V_B, the total power consumption P can be minimized by adjusting T_d to the maximum value compatible with f. Examination of the approximation (Equation (16.20)) of T_d shows that because C is imposed by the process, the only remaining possibility is to adjust I_0, which itself depends on threshold voltage V_T according to Equation

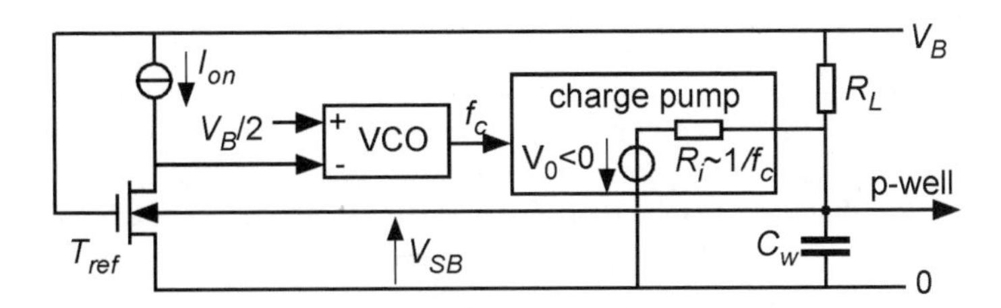

FIGURE 16.15 Principle of threshold adjustment for n-channel transistors [10].

(16.5). The latter can be adjusted by the source-to-bulk voltage V_{SB} according to Equation (16.4). More precisely, what must be adjusted by V_{SB} is the on-current I_{on}, in order to eliminate the exponential dependency on V_B exhibited by Equation (16.20). A possible implementation of such an adjustment is depicted in Figure 16.15 [10].

The on-current of reference transistor T_{ref} is compared with the required value of I_{on}. This produces a voltage that controls the frequency f_c generated by a VCO. The output of the VCO drives a capacitive charge pump, which produces a negative voltage V_0. The internal resistance R_i of this capacitive charge pump is inversely proportional to control frequency f_c. The output of the charge pump is connected to the p-well of T_{ref} and of all the transistors to be controlled, and is loaded by a resistor of fixed value R_L. As a result, the value of V_{SB} depends on frequency f_c and the loop is closed. It stabilizes when the on-current of T_{ref} (and that of all transistors to be controlled) is equal to the required value I_{on}. The large well-to-substrate capacitance C_w stabilizes the loop.

The effective value of threshold V_T can only be slightly reduced by $V_{SB} < 0$ (until a parasitic bipolar is activated by the forward-biased source-to-bulk junction). Thus, most of the possible adjustment is an increase of V_T by $V_{SB} > 0$. Because the required value of V_T is never more that about $10U_T$ (a few hundred mV), the nominal native threshold $V_{T0} = V_T(V_{SB} = 0)$ should be close to zero.

As depicted by Figure 16.10 and Figure 16.11, the minimum value of P/f is proportional to n^2; however, a value of n sufficiently above unity is needed to control V_T according to Equation (16.4).

The scheme must be symmetrically duplicated to control all p-channel devices in their separate n-well (true twin well is needed). The entire complementary scheme must be repeated for each different value of supply voltage V_B, if any.

16.8.3 Symmetry and Matching

The entire preceding analysis has assumed a perfect p/n symmetry of the gates. This situation can be approached by using NAND gates in a positive logic (see Figure 16.14) to approximately compensate for the difference of hole and electron mobility. The residual asymmetry causes a difference between rise and fall time, which may result in a reduced value of maximum frequency. The short-circuit charge Q_{sc} might also be increased, while remaining essentially negligible for homogeneous gates (see Figure 16.7(b)). Further adjustment of symmetry can be obtained by adapting the width of p- or n-channel devices, at the cost of an increased minimum power.

The mismatch of currents in weak inversion is dominated by the effect of threshold mismatch [6], with

$$\sigma(\Delta I / I) = \sigma(\Delta V_T) / (nU_T) \qquad (16.44)$$

This spread of currents may reach several tens of percent from gate to gate, resulting in a proportional reduction of maximum frequency. Because mismatch is approximately proportional to $1/(WL)^{1/2}$, it may be reduced by increasing gate width W, but this would drastically increase capacitance C and thereby the power consumption. Increasing length L might have a lesser effect on power (if the drain capacitance dominates), but would drastically reduce the speed.

16.8.4 Process Scaling and Short-Channel Effects

Scaling of a process amounts to reducing all its geometrical dimensions by a factor S. In order to maintain constant fields, this geometrical scaling should be associated with a down-scaling of voltages and an up-scaling of doping concentrations by the same factor S. Now, because in most cases the absolute temperature T cannot be scaled down, $U_T = kT/q$ remains constant. As a result, approaching and even entering weak inversion should come naturally with scaled-down processes.

Such a constant-field scaling should bring no qualitatively new effect, until the apparition of quantum effects. Because all capacitors would be scaled-down by S, the power-delay product PT_d should be reduced by the same factor according to Equation (16.28). In weak inversion, the channel is equipotential and the current is carried by diffusion. As long as this is true, the mobility remains essentially constant and specific current I_S given by Equation (16.3) scales up with S. Therefore, the minimum delay time T_{dmin} given by Equation (16.33) should be reduced by S^2.

In practice, submicron scaling has been more aggressive in striving to reach the maximum possible speed by reducing voltages slower than dimensions. The fields have therefore increased dramatically, resulting in the need for special means to avoid high-field effects. As a result, some basic characteristics such as the slope n in weak inversion have been degraded in some processes. Whether this trend will be pursued with further scaling is not clear nowadays.

The gate current due to tunneling becomes a very important problem with the very thin oxide of deep submicron processes; however, this problem should be strongly alleviated at supply voltages of just a few hundred mV.

The evolution of threshold mismatch with scaled-down processes depends on the dominant cause of mismatch. If it is due to random fluctuations of the fixed interface charge, and if these fluctuations can be maintained constant in absolute value, then the increase of mismatch due to the reduction of gate area should be compensated by the increase of C_{ox}.

16.8.5 System Architecture and Applications

As explained in Section 16.5, all gates of the same digital block should be homogeneous in speed to keep the short-circuit charge per transition Q_{sctr} negligible. This is not different for strong inversion logic operated at a supply voltage larger than the sum of p and n thresholds [11,12], but new architectural approaches are needed to fully exploit the potential of weak inversion logic.

Traditional approaches in CMOS digital design exploit the fact that a CMOS gate consumes power only during transitions. Thus, the total power consumption can be minimized by minimizing the overall switching activity [13], independently of the total number of gates because idling gates consume negligible power. On the contrary, idling gates should be avoided in weak inversion logic because the minimum power increases when duty factor α is decreased as shown by Figure 16.11.

Ideally, the delay time T_d of each gate given by Equation (16.20) should be adjusted so that the gate is always in transition, corresponding to $\alpha = 1$. This is, of course, not possible in practice, but Figure 16.10 and Figure 16.11 demonstrate that a considerable improvement with respect to 1-V operation is still possible for α as low as 0.01. Architectures allowing the largest possible value of α should be developed. For example, small logic depth, pipelined and possibly asynchronous architectures should be favored. Large groups of gates should have similar values of T_d and α, so that the number of separate threshold adjustment loops of Figure 16.15 can be limited.

Dynamic adjustment of speed (adjustment of threshold voltage in time) might be considered, including sleep modes with a drastic reduction of I_0, but this is expected to complicate the implementation of the adjustment loop, especially if fast wake-up is required. Such a dynamic adjustment might be the only way to deal with the very low duty factor of read/write memory blocks.

Weak inversion logic represents the ultimate limit of the present trend to lower the supply voltage in order to reduce the power consumption and/or to limit the electric fields in scaled-down processes. Indeed, with the low threshold voltages characteristic of deep submicron processes, the residual channel current I_0 of blocked transistors increases according to Equation (16.5). Therefore, a reduction of the the average idling time of gates (increase of α) is becoming a priority to avoid a situation where power consumption is dominated by its static component.

Because the maximum frequency is considerably lower than for strong inversion, weak inversion logic is best applicable to systems that do not require high local speed. It seems ideally suited to implement massively parallel digital architectures, for example, in applications related to image processing [7].

16.9 Conclusion

Digital circuits may in principle be operated at a supply voltage as low as $4U_T$ (100 mV at ambient temperature), while maintaining well-defined stable states (Figure 16.3) and a sufficient on/off current ratio (Figure 16.12), thanks to the exponential behavior in weak inversion. However, this requires a special CMOS process with values of native threshold voltages close to zero and separate wells (truly twin wells) for both types of transistors, in order to electrically adjust these threshold voltages against process spreading and temperature variations.

The dynamic power consumption is reduced with the square of the supply voltage, thus by as much as 100 compared with a 1-V operation. Static power consumption cannot be neglected anymore, as is the case for any CMOS digital circuits using low-threshold devices. The relative importance of this static power depends on the duty factor α , defined as the average percentage of time the gates are in transition. Still, the overall power reduction (with respect to 1 V) remains a factor larger than 30 for $\alpha = 0.1$ and larger than 10 for $\alpha = 0.01$ (see Figure 16.10 and Figure 16.11).

Despite the limited drain current density available in weak inversion, the maximum frequency of operation may reach several hundred MHz for a deep submicron process ($0.18\mu m$). Moreover, the very low voltage of operation is expected to prevent or to alleviate most high-field effects that plague short-channel devices operated at higher voltages.

The approach is substantially different from traditional CMOS digital circuits. Therefore, new architecture should be developed to best exploit its potential for low power. It appears best suited for very low-power digital systems running at moderate clock frequency, such as parallel image processing.

Even if it is not used directly, weak inversion logic represents the lower limit of supply voltage and power-delay product that are achievable with a given process at a given temperature. It can therefore be used as a reference to assess the merit of any digital circuit.

It is interesting to notice that nature has also found this voltage limit in its processing capability because the action potentials in neurons is also in the range 50 to 100 mV (2 to $4U_T$). The local speed of neurons is much lower, however, because of the lower mobility of ions. Still, it is compensated by a huge degree of parallelism that results in fast and low-power capabilities for processing very complex global tasks.

References

[1] R.M. Swanson and J.D. Meindl, Ion-implanted complementary MOS transistors in low-voltage circuits, *IEEE J. Solid-State Circuits,* vol. SC-7, April 1972, pp. 146–153.

[2] H. Soeleman and K. Roy, Ultra-low power digital subthreshold logic circuits, *Proc. Int. Symp. on Low-Power Electronics and Design,* pp. 94–96, 1999.

[3] J. Burr and J. Shott, A 200-mV self-testing encoder/decoder using Stanford ultra-low-power CMOS, *ISSCC Dig. of Tech. Papers,* 1994, pp. 84, 85, 316.

[4] M. Miyazaki et al., A 175-mV multiply-accumulator unit using an adaptive supply and body bias (ASB) architecture, *ISSCC Dig. of Tech. Papers,* 2002, pp. 58, 59, 444.

[5] E. Vittoz and J. Fellrath, CMOS analog integrated circuits based on weak inversion operation, *IEEE J. Solid-State Circuits,* vol. SC-12, June 1977, pp. 224–231.

[6] E. Vittoz, Micropower techniques, *Design of VLSI Circuits for Telecommunications and Signal Process.,* chapter 3, J. Franca and Y. Tsividis, Eds., Prentice Hall, Englewood Cliffs, NJ, 1994.

[7] E. Vittoz, Low-power design: ways to approach the limits, *ISSCC Dig. of Tech. Papers,* 1994, pp. 14–18.

[8] E. Vittoz, Very low power circuit design: fundamentals and limits, *Proc. Int. Symp. on Circuits and Syst.,* 1993, vol. 2, pp. 1439–1442.

[9] C. Enz, F. Krummenacher, and E. Vittoz, An analytical MOS transistor model valid in all regions of operation and dedicated to low-voltage and low-current applications, *Analog Integrated Circuits and Signal Process.,* vol. 8, July 1995, pp. 83–114.

[10] V. von Kaenel et al., Automatic adjustment of threshold and supply voltage for minimum power consumption in CMOS digital circuits, *Proc. IEEE Symp. on Low-Power Electron.,* San Diego, CA, 1994, pp. 78–79.

[11] J.R. Burns, Switching response of complementary-symmetry MOS transistor logic circuits, *RCA Review,* Dec. 1964, pp. 627–661.

[12] H.J.M. Veendrick, Short-circuit dissipation of static CMOS circuitry and its impact on the design of buffer circuits, *IEEE J. Solid-State Circuits,* vol. SC-19, August 1984, pp. 468–473.

[13] A.P. Chandrakasan and R.W. Brodersen, Minimizing power consumption in digital CMOS circuits, *IEEE Proc.,* vol. 83, April 1995, pp. 498–523.

17

Robustness of Digital Circuits at Lower Voltages

Harry Veendrick
Philips Research Labs

17.1 Introduction

Over the last two decades, there has been a change in the drive for the continuous scaling of devices and circuits. Figure 17.1 plots the scaling dependence of different parameters.

The presented diagram is divided into two regions: constant-voltage scaling and constant-field scaling. The constant-voltage scaling timeframe reflects the period in which (C)MOS devices were still supplied by a constant voltage of 5 V. After a certain point in time (in the diagram: about 1997 for volume production) the supply voltage was reduced at the same pace as the transistor's channel length, thereby keeping a constant field across the transistor channel: constant-field scaling. During the first period, in each new technology node, the average fabrication costs increased with a factor of $s^{-0.5}$ (about 1.2 times; $s \approx 0.7$), while the intrinsic speed improvement as obtained from the technology scaling was about a factor of s^{-2} (≈ 2). In the same period, the power efficiency (= $1/\tau D$-product: how much speed do I get per watt?) improved only by a factor of s^{-1} (≈ 1.4). Therefore, the rise in fabrication costs could easily be compensated by the speed and density increase in that period.

After that period, the voltage scales at the same pace as the technology feature sizes and causes a reduction in the speed increase to only a factor of s^{-1}. The fabrication costs now also increase with the same factor, however, meaning that the speed increase has become less of a drive for further scaling. Nevertheless, voltage scaling has one big advantage: it improves the power-efficiency increase to a factor of s^{-3} (≈ 2.8) per technology node. This means that if a function consumes a certain amount of power in a given technology node, it will consume a factor of s^{-3} less power in the next technology node. Or,

0-8493-1941-2/05/$0.00+$1.50

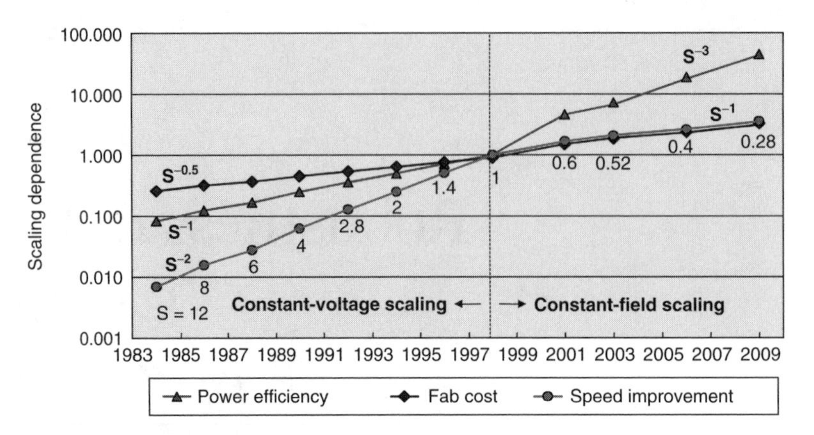

FIGURE 17.1 From speed drive to power-efficiency drive.

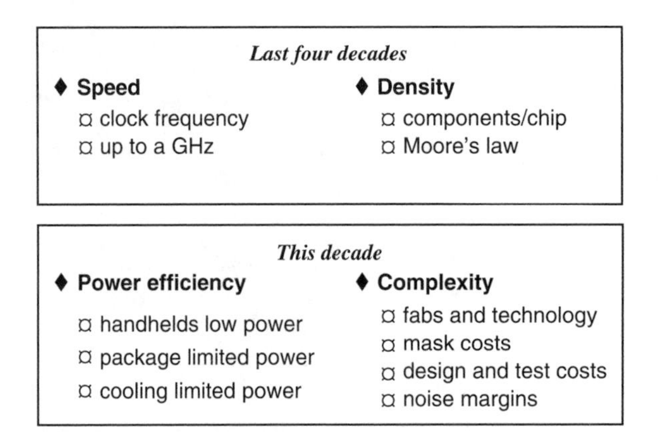

FIGURE 17.2 Summary of changing scaling trends.

to put it another way, in the next technology node, we can have about 2.8 times more functionality (2 times the number of transistors running at 1.4 times the speed) at the same power consumption. Therefore, power efficiency has become the major performance drive in this decade. Figure 17.2 is a rough summary of these scaling trends over the past and near future.

Whereas the speed and density improvements drove the semiconductor scaling process according to Moore's law during the last couple of decades, today, the performance focus has changed toward power efficiency. Yet, the complexity has become another focus to manage (read: limit) the excessive increase of the costs in all semiconductor disciplines. This holds for the production costs (i.e., fabs and the lithography), the mask costs, and, finally yet importantly, the design and test costs. The increased design costs are not only due to the higher complexity of the circuits, it is also a result of increased noise and reduced noise margins. This noise margin reduction is an issue that we, as integrated circuit (IC) designers, should be worried about mostly. Today, many so-called deep-submicron (DSM) effects manifest themselves more as the minimum feature sizes and the voltages are reduced. These DSM effects cause an increase of physical design aspects that have to be taken into account, mostly during the back-end of an IC design. Figure 17.3 depicts a heterogeneous system on chip (SoC) including both system-design and physical-design aspects. As an example, we take the global bus/switch-matrix. Whereas the protocol and bandwidth are typical system design aspects of a bus, signal propagation and cross talk are more physics related. In designing or choosing the I/O pads, the bandwidth and the interface standard are system-design aspects, while di/dt noise and EMC behavior have a closer relation to the physical layer. An overall conclusion from this SoC representation is that, as we scale further, more of these physical aspects pop up and start threatening the robustness of operation of the IC.

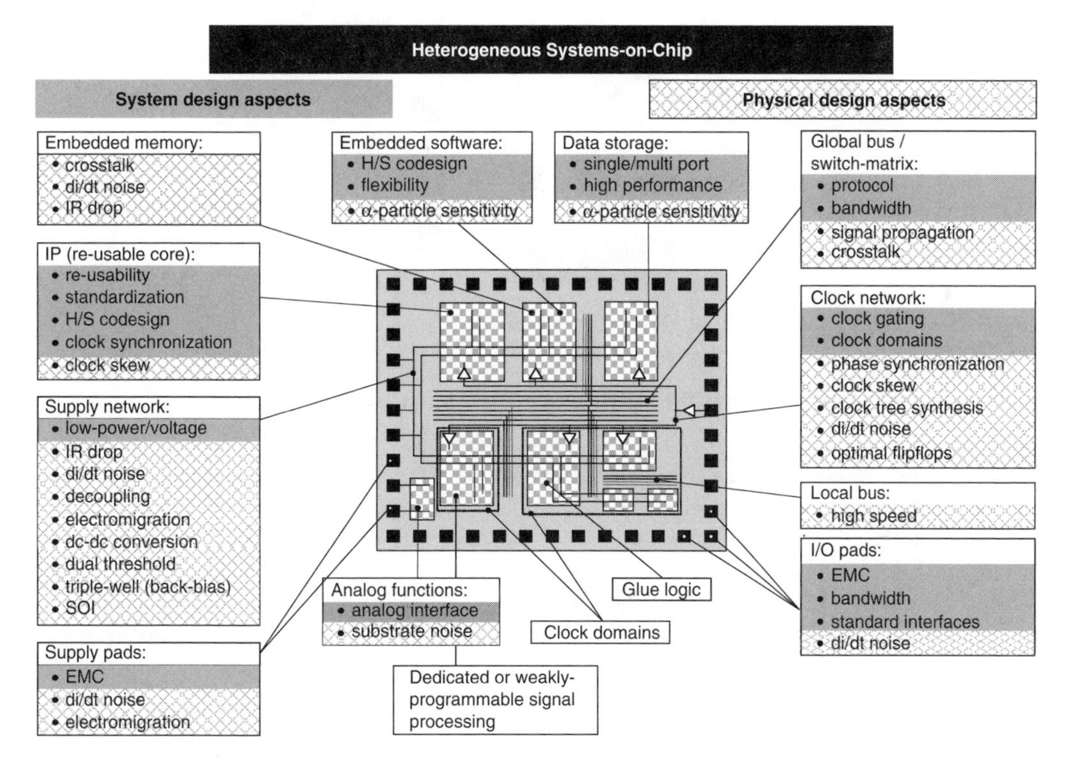

FIGURE 17.3 System and physical design aspects in an SoC [1].

This chapter deals with the robustness of digital circuits in relation to the continuous scaling process. It covers most topics related to signal integrity (e.g., cross talk, signal propagation, voltage drop, supply and substrate noise, soft-errors, and electromagnetic compatible [EMC]) as well as such reliability issues as electromigration, electro-static discharge (ESD), latch-up, hot-carrier effect, and negative temperature bias instability (NBTI). The reducing signal integrity is a result of two conflicting effects: the increase of noise and the reduction of the noise margins (V_{dd} and V_t). The next section, therefore, begins with a discussion on noise issues and ways to maintain signal integrity at a sufficiently high level. A continuous reduction of the noise margins also has a severe impact on the quality of the IC test. The increasing discrepancy between chip operation during test and in the application will result in more customer returns and design spins. The next paragraph will therefore also include some remarks on the effect of scaling on test coverage and complexity. The third paragraph in this section is devoted to trends in reliability and ways to maintain it. Paragraph four presents some conclusive remarks and deals with different scaling scenarios. It presents scaling tables for constant-voltage scaling, constant-field scaling, and constant-size scaling and focuses on the challenges that DSM effects imply on the robustness of operation of future ICs.

17.2 Signal Integrity

Signal integrity indicates how well a signal maintains its original shape when propagating through a combination of circuits and interconnections. On-chip effects from different origin may influence this shape. Signals can be influenced by switching of nearby neighbors (cross talk), by voltage changes on the supply lines (voltage drop and supply noise), by local potential changes in the substrate (substrate noise), or when the signal node is hit by radioactive or cosmic particles (soft-error). In addition, the speed at which a signal propagates through bus lines is heavily affected by the switching behavior of neighboring bus lines.

The next subsections will focus on each of these signal-integrity topics individually and present ways to limit the noise level or the influence of the potential noise sources that threaten the signal integrity.

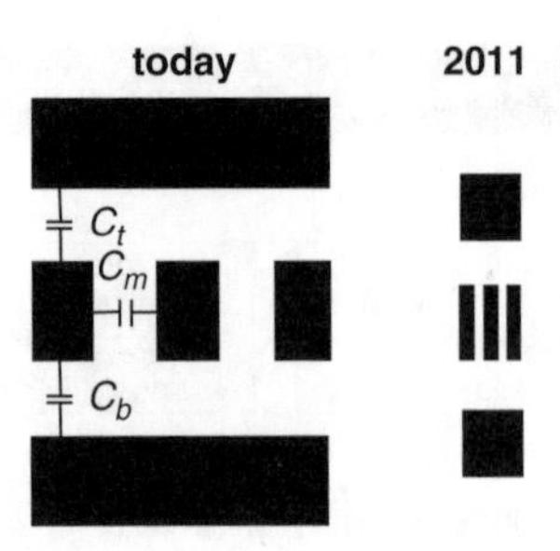

FIGURE 17.4 Expected scaling of metal track width and spacing.

17.2.1 Cross Talk and Signal Propagation

Due to the scaling of the transistors, their density has almost doubled every new technology node for more than four decades already. This requires the metal lines (width and spacing) to be scaled in the same order to be able to connect this increasing number of devices per unit of area. Per unit of area, however, the total length of the interconnections in one metal layer only increases with a factor of 1.4. This means that additional metal layers are needed to allow a high-density connection of all logic gates. The metal layers are also used to supply the current from the top metal layer all the way down to the individual devices. As will be discussed in the subsection on electro-migration, the current density also increases with a factor of 1.4 every new technology node, meaning that the thickness of the metal layers cannot be scaled at the same pace as the width and spacing. Consequently, the mutual capacitance between neighboring signal lines is dramatically increasing.

Figure 17.4 presents two cross sections of three parallel metal lines: one in a 120-nm CMOS technology and the other one in a 35-nm process (year 2011). It clearly shows that the bottom (C_b) and top capacitances (C_t) reduce while the mutual capacitances (C_m) increase. This increase in mutual capacitance has dramatic effects on the performance and robustness of integrated circuits. The first one is the growing interference between two neighboring interconnect lines, which is usually referred to as cross talk. The second one is the growing signal propagation delay because of the increasing *RC* times across the interconnect. Third, the increased interconnect capacitances also affect the overall IC's power consumption. We will discuss each one of these effects in more detail now. Figure 17.5 depicts the trend in the cross talk over several technology nodes.

The used model refers to two minimum-spaced interconnect wires in the same metal layer. A signal swing ΔV_{M1} on metal track M1, causes a noise pulse ΔV_{M2} on a floating metal track M2, as defined by:

$$\Delta V_{M2} = \Delta V_{M1} * Cm/(Cm + Cground) \tag{17.1}$$

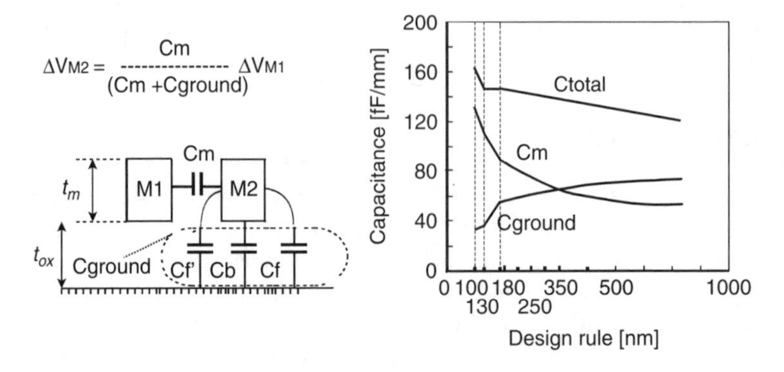

FIGURE 17.5 Interconnect capacitances across various technology nodes [1].

TABLE 17.1 Capacitance Values for Second Metal Layer in Different Technology Nodes

	180 nm CMOS	130 nm CMOS	90 nm CMOS
C_m	89	110	132
C_{ground}	58	36	32
C_{total}	147	146	164
$\Delta V_{M2} =$	$0.6 * \Delta V_{M1}$	$0.75 * \Delta V_{M1}$	$0.8 * \Delta V_{M1}$

Table 17.1 lists the capacitance values for different technology nodes. The bottom row in this table presents the amount that one signal propagates into the other one through cross talk. For the 90-nm node this means that 80% of the switching signal propagates into its floating neighbors. Because of this, all floating lines (e.g., precharged bit lines in a memory and tri-state buses are very susceptible to cross talk noise).

Even nonfloating (driven) lines in digital cores are becoming increasingly susceptible to cross talk causing spurious voltage spikes in the interconnect wires. Traditional design flows only deal with cross talk analysis in the back-end part, to repair the violations with manual effort, after the chip layout is completed. Because timing and cross talk are closely related, they should be executed concurrently with the place-and-route tools. The introduction of multi-V_{dd} and multi-V_t pose a challenge for the physical synthesis and verification tools because both design parameters affect timing and signal integrity.

In memory design, scaling poses other challenges to maintain design robustness. The layout of a static random-access memory (SRAM), for example, includes many parallel bit lines and word lines at minimum spacing in different metal layers. It is clear that these will represent many parasitic capacitances; however, there might be even more mutual capacitance between the various contacts (pillars) than between the metal tracks. Memories in DSM technologies therefore require very accurate three-dimensional (3-D) extraction tools in order to prevent that the silicon will, unexpectedly, run much slower than derived from circuit simulations.

Next to the cross talk between metal wires, the signal propagation across metal wires is also heavily affected by scaling. In a 32-bit bus, for example, most internal bus lines (victims) are embedded between two neighbors (aggressors). The switching behavior of both aggressors with respect to the victim causes a large dynamic range in signal propagation across the victim line. In case both aggressors switch opposite from the victim, the signal propagation across the victim lasts about six times longer than in case the aggressors and victim all switch in the same direction, for 20-mm long bus lines in a 180-nm CMOS technology. Figure 17.6 plots the increasing propagation delay (in nano-seconds) with the technology node for a 20-mm long bus line, embedded between two quiet (nonswitching) aggressors.

Although the introduction of copper with the 120-nm node provides some relief in the increase of the propagation delay, it will only help for about one technology node. This means that in the 120-nm node, with an aluminum backend, the interconnect propagation delay would reach the same order of magnitude as the 90-nm node with a copper backend. The diagram also indicates that the propagation delay will

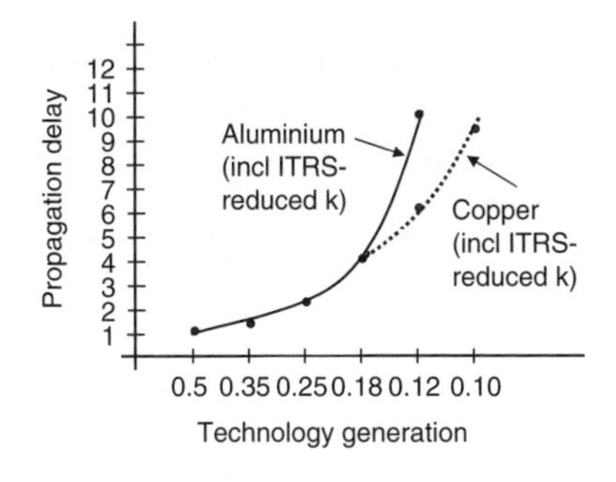

FIGURE 17.6 Propagation delay versus technology node in case aggressors are quiet.

further increase. This requires different design architectures, in which the high-speed signals are kept local. Such architectures must allow latency in the global communication or communicate these global signals asynchronously (i.e., islands of synchronicity; globally asynchronous, locally sychronous (GALS)).

In the preceding discussions, self- and mutual inductances where not taken into account; however, with the advances in speed and clock frequencies, the influence of these inductances becomes increasingly pronounced. The resistances of most of today's signal lines still exceed the values of inductance by more than one order of magnitude. For one reason this is because the resistance increases every technology node. The second reason is that the inductance contribution is linearly proportional to the frequency. As a rule of thumb, we can state that at 10 GHz, the inductance contribution to the total impedance of a metal wire reaches about the same value as the resistance contribution. This means that we need to change from an *RC* interconnect model to an *RLC* model for designs that exceed 1 GHz (at this frequency the inductance value is about 10% of the resistance value and can thus no longer be neglected). Generally, two effects determine the difference in accuracy between an *RC* and an *RLC* model: the damping factor and the ratio between the input signal rise time and the signal propagation speed across the line. Therefore, even in designs that do not yet reach 1GHZ, the wider metal lines (with lower resistance) (e.g., in clock distribution networks) and upper metal layers can exhibit significant inductive effects [2]. Because also the rise times of signals on interconnect lines are reducing with the advance of the technologies, RLC models need to be included in our computer-aided design (CAD) tools soon, in order to avoid inaccurate performance predictions or underestimate signal integrity effects, which may also lead to a reliability problem.

Finally, a number of methods exist to reduce cross talk and/or improve signal propagation. We will summarize them here, without discussing them in more detail:

- Use fat wires to reduce track resistance.
- Increase spacing to reduce mutual capacitance.
- Use shielding between the individual bus lines.
- Use staggered repeaters to compensate for noise.
- Use tools that can detect, replace, and reroute critical nodes.
- Use current sensing or differential signaling for improved speed and noise compensation.

17.2.2 Supply and Ground Bounce

Every new technology node allows us to almost double the number of transistors. Next to this, the bus widths have also gradually grown over the last couple of decades: from 4-bit in the late 1970s to 64-bit, or even 128-bit, today. The interface to a 256-Mbit DDR SDRAM, for instance, requires communicating 32 data bits — about 23 address bits plus a few additional control bits — totally adding up to some 60 parallel bits. In addition, due to the increased speed requirements, more flip-flops/pipelines are used within the logic blocks. All these individual trends contribute to a dramatic increase of simultaneously switching activity in an IC causing huge currents (i) and current peaks (δi). These currents cause voltage drop across the resistance (R) of on-chip supply network, while the current peaks cause relatively large voltage drops across the self-inductances (L) in the supply path. As is discussed in the previous subsection, most of the self-inductance is still in the bond wires and the package leads, instead of in the on-chip metal supply lines.

Another trend that keeps pace with technology advances is the reduction in switching times (δt) of the logic gates and driver circuits. The combination of these two trends leads to a dramatic increase of $\delta i/\delta t$, which term is mainly responsible for the supply and ground bounce generated on chip.

In total, we can summarize the voltage drop by:

$$\Delta V = i.R + L.\delta i/\delta t \tag{17.2}$$

The impact of this voltage drop on the behavior of the chip is twofold. First, the average supply voltage throughout the complete clock period determines the speed of a circuit. Let V_{dd} be the nominal supply voltage of a chip. Most commonly, this means that the chip is specified to operate within a 5 to 10% margin in this supply voltage. In case of a 0.18-μm CMOS design, this means that it should operate between 1.65 V and 1.95 V. So, in the application, the IC should operate correctly, even at 1.65 V. Because the logic synthesis is done using the gate delays specified at this lower voltage, an additional voltage drop ΔV within the chip could be disastrous for proper functionality. In other words, the designer should limit the total average voltage drop within stringent limits to assure the circuit operates according to the required frequency spec. It is commonly accepted that this voltage drop is limited to just a small percentage of the supply voltage (less than 5%). Second, ΔV introduces noise into the supply lines of the IC. The current is supplied through the V_{dd} supply lines and leaves the circuit through the V_{ss} ground lines. When the impedances of the supply and ground lines are identical, which is most commonly the case, the introduced bounce on the respective lines show complementary behavior and are identical in level. The total inductance (L) consists of on-chip contributions of the supply and ground networks and off-chip contributions of the bond wires, package leads, and board wires. Usually, the damping effect of high resistive narrow signal wires reduces the effect of on-chip inductive coupling. To reduce the contribution of the first term in the Equation (17.2), however, the supply and ground networks require wide metal tracks in the upper metal layers with very low sheet resistance. Particularly for designs operating at GHz frequencies, inductance in IC interconnects is therefore becoming increasingly significant.

The supply noise can be reduced in several ways. When using n supply pads for the supply connection, which are more or less homogeneously distributed across the IC periphery, the self-inductance will reduce to L/n. Both the use of a low-resistive supply network and multiple supply pads, however, contribute to a reduction of the overall impedance of the supply network. Because the bond wires, package leads, and board wiring, all act as antennae, the resulting increase of the current peaks ($\delta i/\delta t$) lead to a dramatic rise of interference with neighboring ICs on the board and may cause EMC problems in the system. Therefore, it is also required to keep the peak currents local within the different cores on the IC. In other words, it is necessary to lower the global $\delta i/\delta t$ contribution in the Equation (17.2) as well. The use of staggered driver turn-on, to limit the amount of simultaneous switching activity, as well as encouraging the use of "slow" clock transients will directly contribute to a lower $\delta i/\delta t$. Another measure to limit the global $\delta i/\delta t$ is the use of decoupling capacitors within each of the different cores. Figure 17.7 depicts two implementations of decoupling capacitor cells [1]. Figure 17.7(a) is a complementary set of transistors connected as an nMOS and pMOS capacitor, directly between V_{dd} and V_{ss}. Figure 17.7(b) is a "tie-off" cell used as decoupling capacitor. In several applications, a tie-off cell supplies dummy V'_{dd} and V'_{ss} potentials to circuits, which, for reasons of ESD, are not allowed to have an input directly connected to the V_{dd} and V_{ss} rails. The channel resistances R_n and R_p (Figure 17.7(c)) of the nMOSt and pMOSt, respectively, serve as additional ESD protection for the transistor gates connected to the V'_{ss} and V'_{dd}. This advantage can also be exploited when we use this cell only as a capacitor cell between V_{dd} and V_{ss}, however, without using the dummy V'_{dd} and V'_{ss} terminals. When a supply dip occurs, the charge stored on the gate capacitance C_n (C_p) of the nMOSt (pMOSt) must be supplied to the V_{dd} (V_{ss}) in a relatively

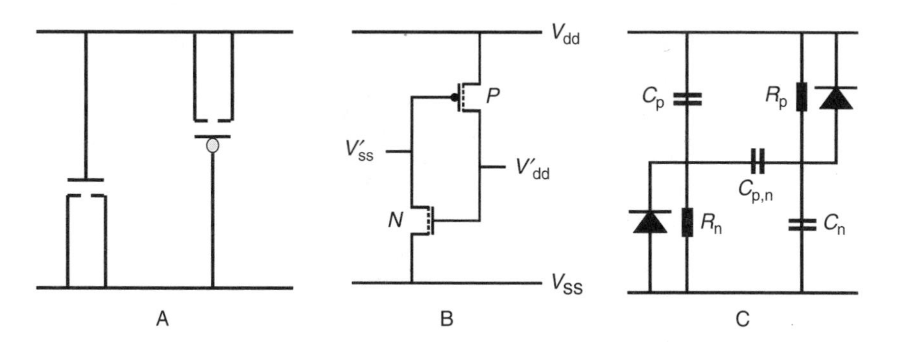

FIGURE 17.7 (a) normal decoupling capacitor, (b) tie-off cell decoupling capacitor, and (c) equivalent circuit.

short time, which puts some constraints to the value of R_n (R_p). Therefore, decoupling capacitor cell *b* presents a better ESD behavior compared with cell *a*.

These decoupling capacitors are charged during steady state (e.g., at the end of the clock period when the total switching activity has almost or completely come to an end). The additional charge, stored in these capacitors is then redistributed to the supply network during moments of intense switching, particularly at the clock transient that initiates the next signal propagation through the logic paths. These decoupling capacitor cells are designed as standard cells and are usually available in different sizes. The amount of decoupling capacitance that needs to be added in each core depends on the number of flip-flops in it and on the switching activity of its logic. The switching activity α is defined as the average number of gates that switch during a clock cycle. When a logic core has an activity factor of $\alpha = 1/3$, it means that the average gate switches one out of every three clock periods. Different algorithms require different logic implementations, which show different switching activities. It is known that average telecom and audio algorithms show less switching activity ($0.05 < \alpha < 0.3$) than an average video algorithm ($0.15 < \alpha < 0.4$), for example. The operating frequency of a logic block is a major component in determining the total amount of decoupling capacitance that needs to be added in a logic core. The higher the frequency, the less voltage drop can be allowed and the more decoupling capacitance needs to be added. As an example, the total additional decoupling capacitance in an average logic block, performing a video algorithm, running at 350 MHz in a 0.12-μm CMOS core in a digital chip, may occupy about 10 to 15% of its total area. When the standard-cell block utilization is less than 85%, this amount of decoupling capacitance fits within the empty locations inside a standard-cell core. In mixed analog/digital ICs, however, this amount could grow dramatically because the noise in these ICs is more restricted by the sensitivity of the analog circuits. Due to further scaling, δi will increase, while the δt will just do the opposite, requiring an increasing amount of decoupling capacitance every new technology node.

17.2.3 Substrate Bounce

Substrate noise is closely related to the ground bounce. On a mixed analog/digital IC, usually the digital circuits are responsible for most of this bounce, while the analog circuits are most sensitive to it (Figure 17.8). The substrate bounce has several contributors. The transistor substrate current injection is responsible for only a few mV. Junction and interconnect capacitances account for several tens of mV. The highest noise levels (several hundred mV), however, are introduced through the current peaks in the supply network, also causing the previously discussed supply noise.

In most CMOS circuits, it is common practice to connect the substrate to the V_{ss} rail, meaning that the ground bounce that is generated in the V_{ss} rail is directly coupled into the substrate. This is even a bigger problem, when the chip is realized on epitaxial wafers (see Section 17.3.4) with a low-ohmic substrate because it propagates the noise through the substrate to the analog part almost instantaneously and with hardly any loss of amplitude. Because the noise margins reduce with reducing supply voltages, the use of high-ohmic substrates is becoming increasingly important. Triple-well technology allows improved isolation of analog circuits from digital cores. The use of a silicon-on-insulator (SOI) technology

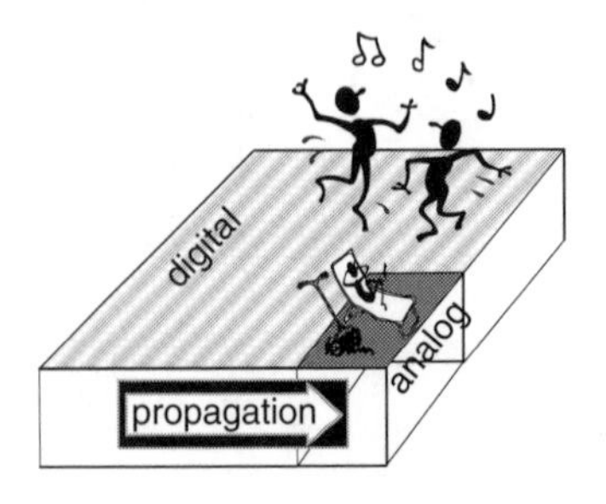

FIGURE 17.8 Symbolic representation of a mixed analog/digital IC.

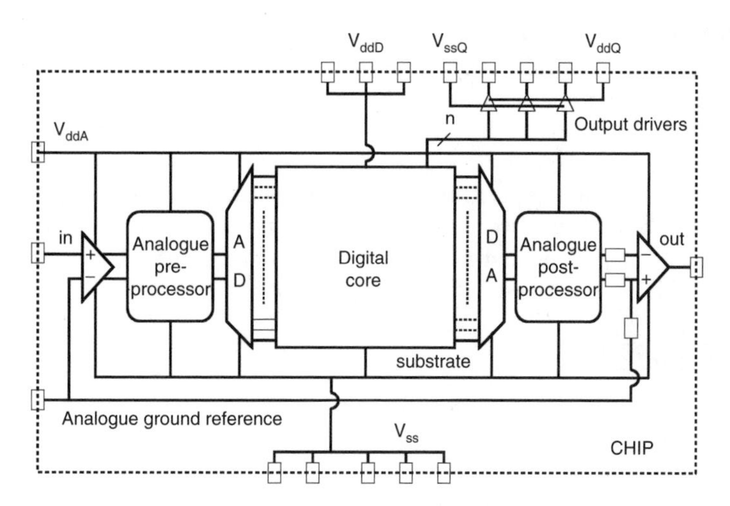

FIGURE 17.9 Proposed supply connections in a mixed analog/digital IC.

allows even a complete separation of the analog and digital circuits. Several other measures exist to reduce the level of substrate bounce. First, the measures that help reduce the supply and ground bounce, as discussed in the previous sub-section, are also beneficial for substrate bounce reduction. Second, a physical separation of the core and I/O supply nets from the analog supply net, according to Figure 17.9, prevents the relatively large noise introduced in these nets to propagate directly into the analog net [3].

The figure also illustrates that most digital and analog circuits share the same ground (V_{ss}) because it also serves as a reference for the communicated signals. Usually, the impedance of the internal and external V_{dd} and V_{ss} networks are almost symmetrical, meaning that they have equal widths and the same number of bonding pads. An increase in the impedance of the V_{dd} network with respect to the impedance of the V_{ss} network would increase the bounce in the V_{dd} supply network, while reducing it in the V_{ss} ground network. Because the analog and digital V_{dd} where separated anyway, this additional digital supply bounce is not coupled into the analog V_{dd}. Because the analog and digital circuits share the same ground, the lower V_{ss} ground bounce also reduces the substrate bounce. Therefore, to increase the margins and robustness of mixed analog/digital ICs, it may be advantageous to dedicate more supply pads to V_{ss} and less to the V_{dd}. Finally, particularly in the case of high-ohmic substrates, circuits with the highest switching activities, and driving strengths (e.g., I/O pads, clock drivers, and drivers with a high fan-out) must be located as far away from the analog circuits as possible.

17.2.4 EMC

The problem of supply and ground bounce caused by large current changes is not restricted to on-chip circuits. High current peaks may also introduce large electromagnetic disturbances on a printed-circuit board (PCB) because of the electromotive force and threatens the off-chip signal integrity. Because bonding pads, package, and board wiring act as antennae, they can "send" or "receive" an electromagnetic pulse (EMP), which can dramatically affect the operation of neighboring electronic circuits and systems [4].

When realizing EMC circuits and systems, the potential occurrence of EMPs must be prevented. The use of only one or a few pins for supply and ground connections of complex high-performance ICs is one source of EMC problems. Even the location of these pins is very important with respect to the total value of the self-inductance. The use of three neighboring pins for V_{dd}, for instance, results in an electromagnetic noise pulse that is twice as large as when these supply pins were equally divided over the package. The best solution is to distribute the power and ground pins equally over the package in a sequence such as V_{dd}, V_{ss}, V_{dd}, and V_{ss}. Bidirectional currents compensate each other's electromagnetic fields in the same way as twisted pairs do in cables. Another source of EMC problems is formed by the outputs. They can be many (about 60 for the address and data bits in a 256-Mbit DDR SDRAM interface), contain relatively large drivers with high current capabilities, and often operate at higher voltages than the cores. Actually, each output requires a low-inductance current return path, such that the best position for an output is right between one pair of V_{dd} and V_{ss} pads. This results in the smallest electromagnetic disturbances at PCB level and reduces the supply noise at chip level. Because this is not very realistic in many designs, however, more outputs will be placed between one pair of (V_{dd} and V_{SS}) supply pads. The limitation of this number is the designer's responsibility (simulation). In addition, the $\delta i/\delta t$, generated by these outputs, must be limited to what is really needed to fulfill the timing requirements. Finally, all measures that reduce on-chip supply and ground bounce, also improve the electromagnetic compatibility of the chip and result in a more robust and reliable operation.

17.2.5 Soft Errors

Because of the continuous shrinking of devices on an IC, the involved charges on the circuit nodes have been scaled down dramatically. Particles, independent of their origin, do have an increasing impact on the behavior of these shrinking devices. Several categories of particles can be distinguished, which all generate free electron-hole pairs in the semiconductor bulk material [5]:

- Alpha particles, originating from radioactive impurities (mainly uranium and thorium) in materials; these materials can be anything near the chip: solder, package, or even some of the materials used in the production process of an IC (metals or dielectrics). These so-called *α-particles* can create many electron-hole pairs along their track.
- High-energy cosmic particles, particularly neutrons, can even fracture a silicon nucleus. The resulting fragments cause the liberation of large numbers of electron-hole pairs.
- Low-energy cosmic neutrons, interacting with boron-10 (^{10}B) nuclei; when a ^{10}B nucleus breaks apart, an α-particle and a lithium nucleus are emitted, which are both capable of generating soft errors.

In all cases, the generated electrons and holes can be captured by capacitors (in dynamic logic and DRAMs) and may flip states of both dynamic and static storage circuits (e.g., memories, latches, and flip-flops). The resulting incorrect state is called a soft error because the next clock period its data may have been restored, meaning that the flipped state has not caused permanent damage to any of the circuit nodes. In addition, in static CMOS logic (or SRAM cells) the total charge of a node is an important criterion for the possibility of flipping its state after being hit by an ionizing particle. In a first-order approximation, the total critical charge (Q_{crit}) needed to flip a circuit node to its complementary state is defined by:

$$Q_{crit} = C.V \tag{17.3}$$

where V equals the supply voltage and C the total capacitance of the node. Due to the continuous scaling of sizes and voltages, both C and V reduce with about a factor of 0.7 (which is the average scale factor between two successive technology nodes). Furthermore, design complexity and memory size increase, whereas, conversely, the charge collection efficiency reduces. As a net result, the soft-error sensitivity of integrated circuits dramatically increases with scaling. Particularly the large capacity, densely packed (embedded) memories show an increasing failure in time (FIT). The previous considerations particularly hold for dynamic circuits and DRAMs. In static storage cells (e.g., SRAM cell, latch, or flip-flop), the critical charge is not only dependent on the capacitance of the nodes in these cells, but also on the drive strengths of the transistors that try to maintain the logic state. In this case, the critical charge varies with the width of the transient current pulse induced by a particle hit.

Several measures can prevent or limit the occurrence of soft errors:

- Careful selection of purified materials (i.e., package, solder, and chip manufacture) with low α-emission rates
- Usage of a shielding layer, most commonly polyimide; this layer must be sufficiently thick (~ 20 μm) in order to achieve about three orders of magnitude reduction of the soft-error rate (SER) caused by α-particles. This does not help against cosmic particles.

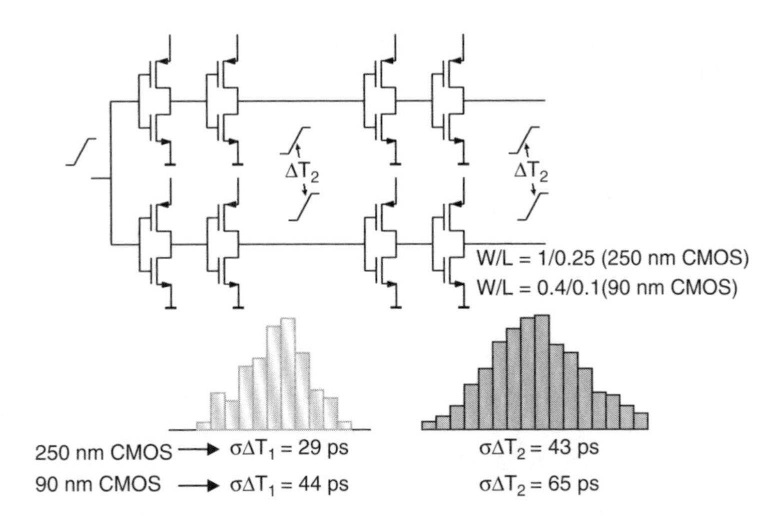

FIGURE 17.10 Spread in signal arrival times due to transistor mismatch.

- SER hardening of the circuits by changing memory cells, latches, and flip-flops
- Usage of process options or alternative technologies, such as SOI, to reduce the volume in which charges are generated along a particle track
- Inclusion of error-detection/correction circuits or making the designs fault tolerant

Currently, much effort is being put into the evaluation and prevention of soft-errors, particularly in systems containing large amounts of densely packed memories.

17.2.6 Transistor Matching

Matching of transistors means the extent to which two identical transistors (i.e., identical in type, size, and layout topology) show equal device parameters, such as β and V_t. Particularly in analog circuits (a memory is also an analog circuit) where transistor pairs are required to have a very high level of matching [6], the spread in V_t due to the doping statistics in the channel of the MOS transistors results in inaccurate or even anomalous circuit behavior. For minimum transistor sizes (area), this effect increases every new IC process generation, such that both the scaling of the physical size and the operating voltage of analog CMOS circuits lag one or two generations behind the digital CMOS circuits. In addition, for digital CMOS circuits (logic), matching of transistors is becoming an important issue, resulting in different propagation delays of identical logic circuits. Figure 17.10 depicts two identical inverter chains (e.g., in a clock tree), but due to the V_t spread, they show different arrival times of the signals at their output nodes. For circuits in a 90-nm CMOS technology, this time difference is in the order of several gate delays. Particularly for high-speed circuits, for which timing is a critical issue, transistor matching and its modeling is of extreme importance to maintain design robustness at a sufficiently high level.

17.2.7 Statistical Timing Analysis

In the preceding subsection, the influence of device parameter spread with respect to transistor matching is discussed; however, process-induced parameter spread in both the device and interconnect structures are also increasingly challenging chip-level timing behavior and analysis. Transistors vary in relation to oxides, doping, V_t, width and length. Interconnects vary in relation to track width, spacing, and thickness and dielectric thickness. So far, this spread was included in simulators in the so-called worst-case, nominal, and best-case parameter sets to provide sufficient design margins. For example, in worst-case timing analysis it is assumed that the worst-case path delay equals the sum of the worst-case delays of all individual logic gates from which it is built. This produces pessimistic results, incorrect critical paths and over-design. Static timing analysis is a means to optimize and estimate timing across the chip. Current static timing analysis tools use the previously mentioned deterministic values for gate and wire delays, which are appropriate for inter-die parameter variations, but does not account for in-die variations. Particularly these in-die variations show significant impact on the overall timing behavior. Delay faults caused by noise sources (e.g., cross talk and supply noise) are also unpredictable with respect to the induced delay. Statistical timing analysis is therefore needed in order to cope with these local variations, which cause random gate and wire delays.

An objective of statistical timing analysis is to find the probability density function of the signal arrival times at internal nodes and primary outputs. Traditionally statistical timing analysis has suffered from extreme run times. Related research is therefore focused to reduce run times [7,8]. Statistical timing analysis is just taking off. For the 90-nm technology node and below, statistical timing analysis is considered necessary, particularly for the complex and higher performance categories of ICs.

17.2.8 Signal Integrity Summary and Trends

From the previous subsections, it can be seen that all noise components increase because of scaling and integrating more devices onto the same die area. At the same time that noise levels in digital CMOS ICs increase with scaling, the noise margins reduce due to reducing supply voltages (Figure 17.11). Because they deal with large current peaks, high-performance ICs, such as the PowerPC (IBM, Motorola), the Pentium (Intel), and the α-chip (DEC/Compaq/HP), have faced signal-integrity problems already in the early 1990s. The average application-specific integrated circuit (ASIC), however, consumes more than a factor of ten less power (and current) and therefore faces these problems a couple of technology generations later in time.

When a certain noise level has reached a limit, a design or technology measure is required to reduce the noise level.

Examples of technology measures are:

- The use of copper instead of aluminum allows reduction of the metal height, thereby reducing the cross talk (see Section 17.2).
- The use of low-k dielectrics in the back-end of the technology has the same effect.

Examples of design measures are:

- The increase of space between long signal lines (buses) also reduces the cross talk.
- The use of on-chip decoupling capacitors reduces supply, ground, and substrate bounce.

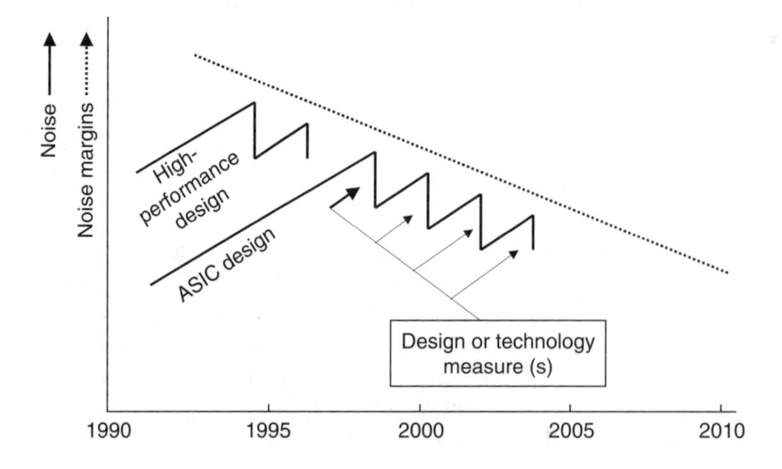

FIGURE 17.11 Noise and noise margin trends over the past and current decade.

Whatever technology or design measure is taken, it only fulfills the requirements in that technology node. The next technology node offers twice the number of transistors, which can intrinsically switch about a factor of 1.4 times faster. This results in a huge increase in the noise levels. In addition, the noise margin has reduced. Therefore, in every new technology node, it becomes more difficult to limit the noise within shrinking boundaries. In other words, the line (in Figure 17.11) that represents the increasing noise must be bended in the direction of the line that represents the reducing noise margins. This can only be obtained by applying more and more design and/or technology measures. In example: in today's ASIC designs, the decoupling capacitors occupy between 5 to 15% of the total area within a standard-cell block. It is expected that this number will have increased to almost 50% by the end of this decade, which means that, by that time, 50% of all transistor equivalents on a chip is needed to support the other 50% in their functional and storage operations. This is yet another factor that adds up to the already rocketing semiconductor development costs.

Another increasingly important topic is the relation between signal integrity and test. Because noise has the tendency to increase, while noise margins reduce (again Figure 17.11), there is not much room left for a reliable operation of an IC. Different operating vectors introduce different local and global switching activities. In many complex ICs, the operation and switching activity during testing are different from the operation and switching activity in the application. As a result, the noise, generated during a test, is different from the noise generated in the application. Because of the reducing noise margins, this increasing discrepancy between "test noise" and "application noise" cause devices that were found correct during testing to operate incorrectly in the application. This is because, in many cases, scan tests are performed to verify the IC's functional operation. These tests are mostly performed locally and in many cases at different frequencies causing a lower overall switching activity and less noise than in the application. On the other hand, depending on the design, different scan chain tests may run in parallel, synchronous and at the same frequency, causing much more simultaneous switching and noise than in the application. These ICs may be found to operate incorrectly during testing while showing correct functional behavior in the application.

Because the voltages continue to decrease, this trend is expected to continue, at least until the end of this decade. Provisions should therefore be taken in the designs, such that, during test, inactive IP cores should run dummy operations in order to emulate application activity. This poses additional challenges to the design, increases its complexity, and adds up to the total development costs.

17.3 Reliability

The continuous scaling of both the devices and interconnect has severe consequences for a reliable operation of an IC. Reliability topics, such as electro-migration, hot-carrier effects, NBTI, latch-up, and ESD are all influenced by a combination of physical and electrical parameters: materials, sizes, dope, temperature, electrical field, and current density. Improving reliability therefore means choosing the right materials, the right sizes and doping levels, as well as preventing excessive electrical fields, temperatures, and currents. This section discusses the effects of scaling on each of the aforementioned reliability issues.

17.3.1 Electromigration

The increase in current density associated with scaling may have detrimental impact not only on circuit performance, but also on the IC's reliability. High currents, flowing through the metal lines, may cause metal atoms to be transported through the interconnection layers due to the exchange of sufficient momentum between electrons and the metal atoms. For this effect, which causes a material to physically migrate, many electrons are required to collide with its atoms. Because of this physical migration of material from a certain location to another location, we get open circuits or voids (Figure 17.12(a)) on locations where the material is removed, and hillocks (Figure 17.12(b)) on locations where material is added. This "electromigration" effect damages the layer and results in the eventual failure of the circuit. Electromigration may therefore dramatically shorten the lifetime of an IC. Preventing excessive current

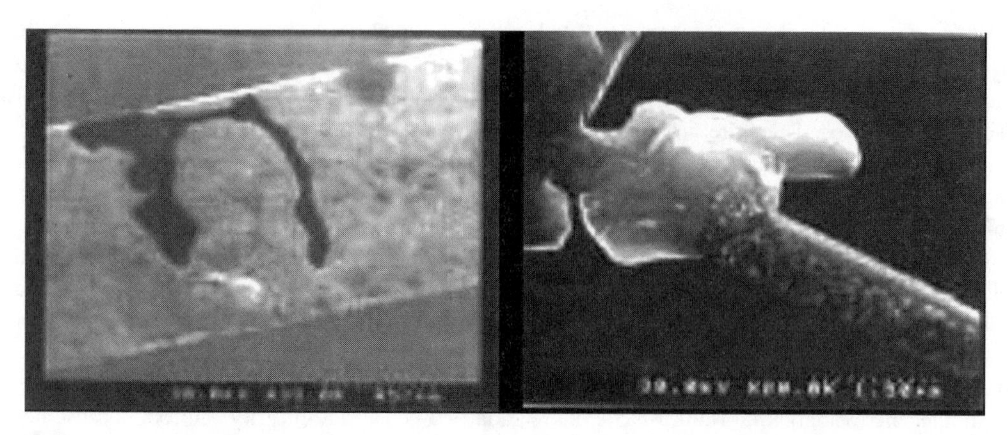

FIGURE 17.12 Electromigration damage in metal interconnect lines: voids (left); hillocks (right). (Courtesy of R. Frankovic, X. Pang, and G.H. Bernstein, University of Notre Dame, Indiana).

densities eliminates the impact of electromigration. Electromigration design rules are therefore part of every design kit. These rules specify the minimum required metal track width for the respective metal (e.g., aluminum or copper) for a certain desired current flow at given temperatures. Electromigration effects increase with temperature because of the temperature dependence of the diffusion coefficient. This dependency causes a reduction of the maximum allowed current density (J_{max}) at higher temperatures in on-chip interconnect. The required metal width for electromigration roughly doubles for every 20°C to 25°C increase in temperature. Because most IC data sheets indicate a maximum ambient temperature of around 70°C or higher, the real worst-case junction temperature of the silicon itself may exceed 100°C in many applications. Therefore, it is common design practice to use the value for J_{max} at 125°C.

The minimum allowed width W_{em} of a metal wire with height H, to carry a current I, according to this electromigration requirement, is then equal to:

$$W_{em} = I/(J_{max} \cdot H) \tag{17.4}$$

Table 17.2 lists some parameter values, which are characteristic for metal layers in 0.18-μm and 0.12-μm CMOS technologies.

Because most of the currents on an IC flow through the supply lines, it is obvious that these are often implemented in the upper metal layer(s), which usually have a larger height (Table 17.2). Similarly, currents through contact holes and vias must be limited to eliminate electromigration-induced damage of the contact conductor. A typical maximum current density value for a 0.2 × 0.2-μm contact or via in a 0.12-um CMOS technology is around 5 mA/μm². The increase in the aspect ratios of the contacts and vias, in combination with a reduction of maximum currents through them, makes them an incremental part of the overall IC reliability.

TABLE 17.2 Metal Characteristics for 0.12-μm and 0.18-μm Bulk-CMOS Technologies

Technology and metal layer	R_{sheet}	H	J_{max} @ 125°C
0.18 μm CMOS second metal (aluminium)	72 mW/□	550 nm	2.3 mA/μm²
0.18 μm CMOS upper metal (aluminium)	35 mW/□	900 nm	2.3 mA/μm²
0.12 μm CMOS second metal (copper)	85 mW/□	350 nm	3.5 mA/μm²
0.12 μm CMOS upper metal (copper)	26 mW/□	900 nm	3.5 mA/μm²

The continuous scaling of feature sizes and voltages (constant-field scaling) by about a factor of 0.7, every new technology node, did not change the intrinsic power density of most standard-cell designs. Due to the reduction in supply voltage, however, the supply current per unit area of logic increases with about a factor of 1.4 every generation. This puts severe constraints to maintaining electromigration reliability across complex designs.

Due to the expected increase in currents through the metal layers, more Joule heating is expected in these layers. This, in combination with low-k dielectrics, which show a higher thermal resistance, made designers to start worrying about this so-called "wire self-heating" mechanism; however, the width of a metal wire is not only specified by the appropriate electromigration requirements, but also by the maximum allowed voltage drop across the wire in order to limit speed loss of the connected circuit(s). Suppose an active logic block draws a supply current of 100 mA. When this block is located nearby the supply pads of the chip, the width of the supply lines is determined only by the electromigration requirement for this 100-mA current. When this block is near the center of the chip, for instance, at a 5-mm distance from the supply pads, the supply lines must be much wider in order to limit the voltage drop across it. Therefore, above a certain distance from the supply pads, the width of the metal (and thus its cooling area) grows with its length, keeping the voltage drop across the line constant. As a result, the resistance of the line (and thus its total I^2R Joule heating) will also be constant. In other words, the maximum wire self-heating occurs in wires with length equal to a cross-over length L_{co}, which is defined to be the length at which the metal-width required by electromigration is identical to the width required by the maximum allowed voltage drop. In Veendrick [9], it is shown that for 0.18-μm and 012-μm bulk-CMOS technologies, wire self-heating in supply lines causes only a limited temperature rise of the wires of just a few degrees. This temperature rise is by far negligible compared to the temperature rise due to the power consumption of the silicon part of the chip. From this result, it can be concluded that wire self-heating in supply lines should not be a real issue in current (and near future) properly designed CMOS VLSI chips.

17.3.2 Hot-Carrier Degradation

When carriers in the MOS transistor channel are given enough energy, these carriers collide with the substrate atoms and generate electron-hole pairs. These, in turn, are also accelerated and may collide with substrate atoms. This so-called impact ionization may cause large substrate currents, device breakdown and/or degradation of the silicon-to-gate oxide interface. Electrons actually collide with the gate oxide. When electrons achieve sufficient energy, they may cross this silicon-to-silicon-dioxide (Si/SiO_2) interface barrier (with a barrier energy of about 3.1 eV for electrons and 3.8 eV for holes) and will then be injected into the gate oxide. Injected carriers lead to the degradation of the Si/SiO_2 interface (i.e., electrically active interface defects are generated), to the generation of defects in the gate oxide film and to charge trapping in the oxide interface (both preexisting and newly generated). Oxide charge trapping and interface state generation induce a shift of the transistor threshold voltage and cause a degradation of the device drive current. This effect is called the hot-carrier effect (HCE) and leads to degraded device performance and reliability problems. Due to the lower mobility of holes with respect to electrons in the transistor channel, impact ionization in p-channel metal-oxide semiconductor field-effect transistors (MOSFETs) is much less. Therefore, the hot-carrier effect is much more severe in n-type MOSFETs.

Theoretically, in a 0.18-μm CMOS technology with a supply voltage of 1.8 V, an electron can only get an energy level of 1.8 eV during its flow through the channel from source to drain. This is less than the previously mentioned barrier energy to create hot electrons. Due to multiple collisions, however, some electrons may collect more energy than the required barrier energy and become "hot." From these considerations, it was generally accepted that when supply voltages are reduced, the chance to generate hot carriers in the transistor channel would reduce as well and the hot-carrier effect was expected to eventually disappear totally.

With the continuous scaling process, critical-dimension (CD) control becomes more difficult leading to transistors with different channel lengths showing different hot-carrier behavior. Shorter channel lengths easier introduce punch-through. Punch-through prevention requires different doping profiles around sources and drains, with increased doping levels. This has some negative effects on the hot-carrier behavior.

When voltages across the transistor are scaled at the same pace as the transistor feature sizes, the electrical fields remain almost constant, and the chance for impact ionization would hardly change. Particularly now, however, with 90-nm and smaller CMOS technologies, the effective channel length is scaling faster than the supply voltage, so that the increase in electrical field may lead to increased impact ionization. Because of this, hot-carrier effects may manifest themselves again more in sub 100-nm technologies than in the last couple of technology nodes, especially in the early development phase due to bad transistor drain engineering. Assuming the transistor is stressed under a worst-case condition (i.e., V_g such that the substrate current is maximal), the hot-carrier lifetime is described by a well-accepted empirical expression (Takeda) as:

$$\tau_{\mathrm{drift}} = A.\, L_{eff}^{\,C}.\, \mathrm{e}^{B/Vds} \tag{17.5}$$

where τ_{drift} represents the lifetime (usually at 10% degradation), L_{eff} the effective channel length, and A, B, and C are process-related coefficients. Practical values for B and C, in current technologies, are 60 and 10, respectively. It is clear that the hot-carrier lifetime reduces with decreasing channel length and increasing voltage. Therefore, when we scale the supply voltage with the same factor as the feature sizes, still this lifetime may increase, dependant on the constants A, B, and C.

An additional effect is that for future technologies the silicon dioxide will be replaced by high-k dielectrics. Most of them, however, have a significantly lower barrier [10] and the hot-carrier effects are not just slowly fading away due to reducing supply voltages below the barrier. Results from literature [11,12] stress the importance of continuous monitoring deep submicron technologies for hot carrier degradation, in order to maintain functional reliability at a sufficiently high level.

17.3.3 Negative Bias Temperature Instability (NBTI)

NBTI is a result of a negative bias applied to the gate of a p-channel MOS transistor with respect to the bulk. The mechanism is temperature activated. NBTI results in the degradation of many transistor parameters (drive current, trans-conductance, and threshold voltage), but the threshold voltage appears to be the most degrading one. NBTI was first reported in 1967, but the attention devoted to this mechanism has been escalating over the last couple of years, due to the introduction of gate-oxide nitridation [13] that enhances NBTI and the fact that other oxide wear-out mechanisms, such as HCE and oxide breakdown, were expected to become less severe as the gate oxide scales down. NBTI is strongly process dependent. It has been reported that a higher nitrogen concentration in the oxide [13], boron penetration [14], and plasma processing can enhance NBTI, while fluorine incorporation in the gate dielectric is beneficial against NBTI [15]. The physical nature of the wear-out mechanism induced by NBTI is not fully understood yet. The most accepted models imply positive charge build-up in the oxide-to-bulk and at the Si/SiO_2 interface (donor-like interface states) [16,17].

Whereas hot-carrier injection mostly affects n-channel MOSFETs and depends on the transistor channel length, NBTI mostly affects the pMOS transistor and is only slightly dependent on the transistor geometry, although it has also been reported that in shorter channel devices NBTI can be more severe [18]. Furthermore, the NBTI does not imply a current flow in the transistor channel and can occur at zero drain to source bias. This would mean that NBTI stress could even occur in the standby mode. Design configurations in which matched p-channel MOSFET pairs are subjected to unbalanced stress are reported as most sensitive to NBTI degradation because the threshold voltages of the transistor pair change differently with the stress [19]. Also matched p-channel MOSFET pairs operated symmetrically can lead to reliability fails due to NBTI when the transistors are subjected to different biases in power-down mode. Burn-in can also be a source of NBTI-induced circuit fails, due to the involved high temperature.

Even when an IC is produced in different fabs that run the same process, it may perform differently with respect to NBTI, because not all individual processing steps are identical. NBTI is therefore a technology issue, but critical design configurations, such as matched p-channel MOSFET pairs subjected to unbalanced stress, in either operation or power-down mode, should be avoided. For NBTI there is not a well-accepted model. Assuming a power-law dependence on the stress voltage (field), then the change in *Vt* is proportional to:

$$\Delta V_t = D.\, F_{ox}{}^m \tag{17.5}$$

where D is a process dependant parameter, F_{ox} represents the electrical field across the oxide and m a coefficient dependant on the dielectric material and the dielectric thickness [an approximate value is $m \approx 4$].

V_t shifts of 50 mV and more have been reported, so designers need to be convinced to build enough tolerance in their designs. The occurrence of NBTI can be lowered when a device is not subjected to voltage overshoot and/or high temperatures, either from its own heat dissipation or from its application environment. Therefore, reduced power consumption is also beneficial to reduce the chance for NBTI stress.

17.3.4 Latch-Up

The presence of nMOS and pMOS transistors in a CMOS process leads to the creation of parasitic thyristors, as shown in Figure 17.13. In this figure, R_1 and R_2 represent the substrate and n-well resistances, respectively.

Relatively high currents through the bipolar transistors will create relatively high voltages in the substrate and/or n-well. When a sufficiently high positive voltage is present somewhere in the substrate (e.g., at position A), it will switch on the parasitic NPN transistor, or a local voltage (e.g., at position B) within the n-well that is sufficiently lower than the V_{dd} will switch on the parasitic PNP transistor. When both transistors conduct, they are connected into a feed-forward loop, which means that they enhance each other's conduction state, which will finally be latched (maintained) in the thyristor. This state can only be recovered when the supply is completely switched off. This undesirable effect is called "latch-up" and leads to incorrect circuit behavior or even damage. Inductive effects or coupling capacitances may also cause the node connected to the drain to have overshoots and/or undershoots, thus forward biasing the drain substrate junction, which may initiate latch-up. This requires a controlled start up of ICs.

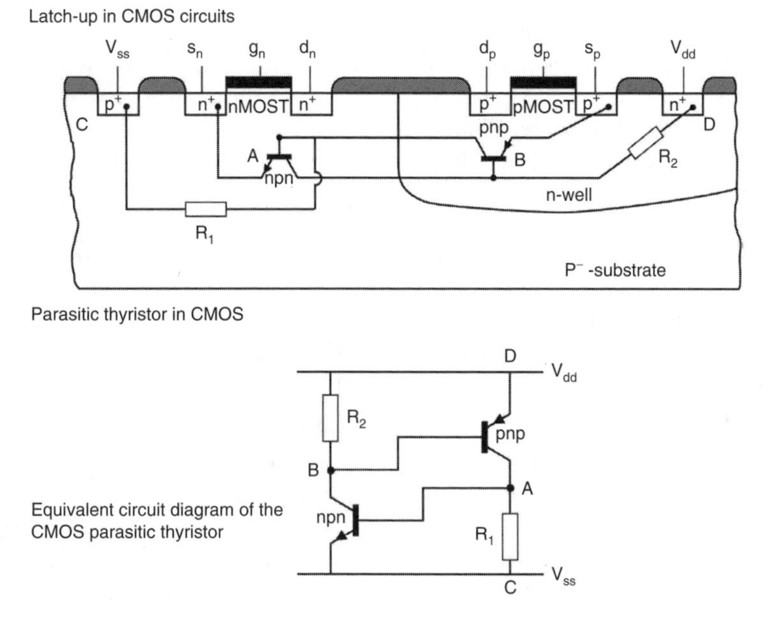

FIGURE 17.13 Parasitic thyristor in CMOS and its equivalent circuit diagram.

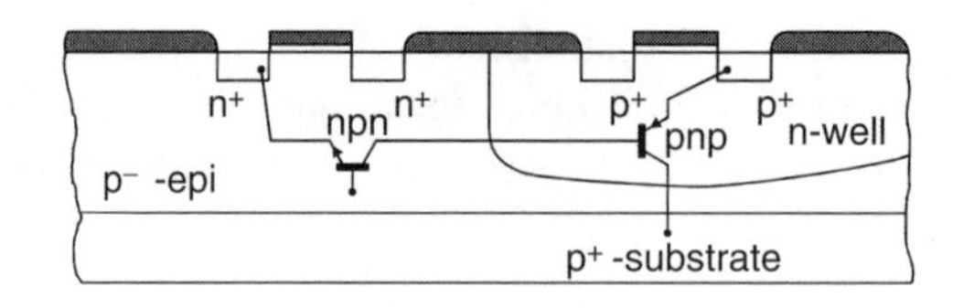

FIGURE 17.14 Cross section of a wafer with a thin p⁻-epi-layer on a thick p⁺-substrate.

Latch-up in CMOS circuits can be avoided by applying the following technological and/or design remedies:

- Minimize the substrate and/or n-well resistances. This can be done in two ways. One is the use of many substrate and n-well contacts in the design, which will reduce the values for *R1* and *R2*, respectively. The parasitic thyristor is then unlikely to turn on. Reducing both resistances by increasing the substrate and n-well doping is not an option because it also changes the threshold voltages and overall transistor behavior. A good alternative is the use of so-called epitaxial wafers (Figure 17.14).
- Epitaxy is a layer of single-crystalline silicon deposited/grown onto a single-crystalline silicon wafer. The crystalline structure of the substrate is reproduced in the growing material. This epitaxial layer, in which the devices are formed and whose thickness is usually between 1 to 5 μm, can be doped, as it is deposited, to the required doping type and concentration (usually with a resistivity of ≈ 10–20ΩCM) while continuing the substrate's crystalline structure. Therefore, we can create a thin p⁻-epitaxial layer on top of a p⁺-substrate. Because the current wafer thickness is between 200 and 700 μm, the p⁺-substrate is relatively thick and has a low resistivity (≈ 0.01–0.05 ΩCM). Such low-ohmic substrates show very low values for R_1. A large part of the PNP collector current will therefore flow through this substrate and only a small part will flow into the base of the NPN transistor. This transistor can no longer be turned on easily and is then largely excluded from the latch circuit. Epitaxial wafers with low-ohmic substrates have been massively used for CMOS products in 0.25-μm and older technologies. Due to decreasing supply voltages and increasing noise levels, the combination of analog and digital circuits onto one single chip has made its design a difficult and cumbersome task. Particularly the substrate noise sensitivity of analog circuits requires a good isolation from the digital noise "generators," which is why a high-ohmic substrate is preferred for mixed analog digital circuits.
- The use of guard rings is another way to make strong (low-ohmic) connections of local substrate and/or n-well areas to V_{ss} and V_{dd}, respectively. Moreover, the distance between n-type and p-type areas is also a matter of concern during the design phase and is particularly of interest in I/O circuits, which are usually supplied by higher voltages. Guard rings are more effective on high-ohmic substrates.
- Apply a back-bias voltage to the substrate. When the p⁻-substrate in Figure 17.13 is connected to a negative voltage instead of to V_{ss}, the base voltage V_A of the NPN transistor will be lowered. Therefore, this transistor can no longer be turned on easily. This technique is more a theoretical option and is not frequently used for latch-up prevention.
- Use SOI technology to completely isolate the nMOS from the pMOS transistors. In this technology, the NPN and PNP transistors are completely isolated from one another and so the connections to create latching thyristor circuits are missing.

The application of one or more of the preceding remedies has increased latch-up immunity to a very high level. The highest chance of occurrence for latch-up is during testing. Standard testing requirements include immunity to 100 mA or more, depending on what the IC can and should withstand from an application point of view. This means that with epi-wafer material, 100 mA can be supplied to the output of an output buffer (driver) even though no output transistor is conducting. This current, then, directly flows into the substrate, thereby raising the substrate voltage and possibly turning the thyristor on (Figure

17.13). In practice, some latch-up tests are done with 150–200 mA at a maximum ambient rated temperature for the device.

In future technologies, the latch-up phenomenon is likely to disappear inside electronic circuits, as the supply voltages will be reduced in every new technology node. At the chip I/Os, however, the requirements on latch-up remain relatively high because many applications still require a higher interface voltage (e.g., 1.8 V, 2.5 V, or 3.3 V). More on latch-up basics can be found in Troutman [20].

17.3.5 Electro-Static Discharge (ESD)

ICs are exposed to many possible sources of damage, both during and after the manufacturing process. The principle cause of damage is ESD, due to the transfer of charge between bodies at different electrical potentials. ESD pulse durations are very short and normally range from 1 to 200 ns, but they may introduce very large power spikes. The high impedance of MOS input circuits makes them particularly vulnerable to physical damage when they are exposed to these spikes. This may result from operations during the fabrication process or from handling (un)packaged dies and bonding. It may also occur during testing and maintenance or in the application. Although only a few devices or connections may be severely damaged, many more may suffer damage that is not immediately apparent. These latent failures will result in customer returns, which is one of the biggest worries of semiconductor vendors. Thus, an ESD is one of the most important factors that determine the reliability of an IC.

The damage caused by an ESD is irreversible. The human body is one of the main sources responsible for ESD. Just by walking on a carpet on a low-humidity day, for instance, a person, wearing shoes with highly insulating soles can build up a voltage in excess of 30,000 V. The resulting charge can then be transferred via an ESD to an electronic circuit during touching. It is also very important that precautions need to be taken to prevent ESD damage during IC fabrication. In addition, protective measures must be included in an IC's design to ensure that it can withstand acceptably large ESD pulses. On-chip MOS protection circuits are used to increase the immunity of an IC to ESD pulses. These circuits are designed to provide input and output circuits with low-impedance shunt paths, which prevent the occurrence of excessive voltages on the chip.

17.3.5.1 ESD Test Models and Procedures

ESD sources are emulated in several different ways. The human-body model is currently the most popular industry model and simulates the direct transfer of electrostatic charge from the human body to a test device. It is internationally accepted as a standard (JEDEC Standard No. 22-A114-B). Figure 17.15 is a human-body test setup. The basic requirement for this model, in combination with the parasitics (L) of the tester interface cables, is to generate ESD pulses with rise times between 10 to 15 ns.

The test is normally done on an ESD tester. This human-body model has not changed much over the last decade. A 100-pF capacitor is charged to the test voltage, and then discharged through a 1.5-KΩ resister across any combination of pins A and B (Table 17.3) of the device under test (DUT). The chip may consist of several supply (V_{dd}) and ground (V_{ss}) domains. Each domain may be supplied by more than one pin. The V_{ss} and V_{dd} in Table 17.3 refer to just one of the respective pins of a supply domain. In other words: each pin is then tested with respect to all grounded V_{ss} and V_{dd} domains and not to all grounded V_{ss} and V_{dd} pins, to save test time. Each signal pin is also tested with respect to all other grounded signal pins. The maximum test voltage ranges from 2 kV to 8 kV and depends on the application area of the chip. Because production environments are well controlled, a maximum voltage of 2 kV is usually

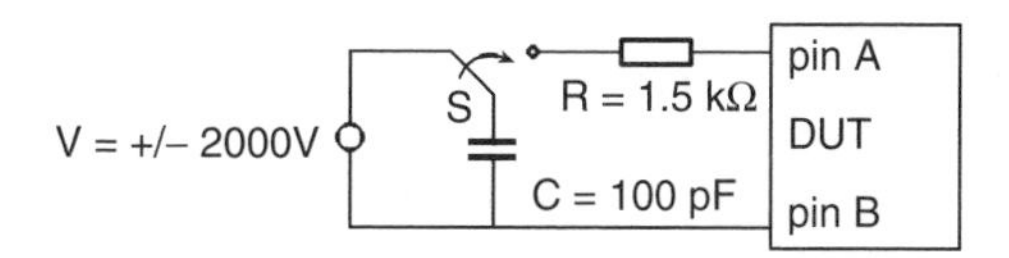

FIGURE 17.15 A typical equivalent circuit based on the human-body model.

TABLE 17.3 Different ESD Test States

State	DUT	
	Pin A	Pin B
1	input	V_{ss}
2	V_{ss}	input
3	input	V_{dd}
4	V_{dd}	input
5	output	V_{ss}
6	V_{ss}	output
7	output	V_{dd}
8	V_{dd}	output
9	input	output
10	output	input
11	V_{ss}	V_{dd}
12	V_{dd}	V_{ss}

required. However, because more and more IC pins can be touched in daily life (i.e., plug-ins such as USB ports, chip cards, SIM cards, memory sticks, and flash cards), the ESD-test requirements tend to increase. The 8-kV requirement is no longer the exception. The devices are classified when meeting a particular sensitivity criterion. A class-2 device, for instance, has passed the 2 kV, but fails after exposure to an ESD pulse of 4 kV (see the previously mentioned standard at http://www.jedec.org).

Generally, three to five positive and negative pulses are applied at 300ms intervals in all test states. Stressed pins are tested after application of each ESD pulse series. If no failure is observed for a sequence through the pins, then the ESD voltage level is increased by 100V and the sequence is repeated. The process continues until a failure occurs or the required maximum voltage is reached. The ESD is complete when a failure is observed or when all pins on the DUT have been stressed as described. Generally, the following (example) criteria may be used to determine failure:

- Incorrect functional operation or a violation of the device specifications
- A change of more than 5% in the forward voltage drop and breakdown voltage in the diode characteristic
- An increase of more than 10% in the I_{ddq} leakage current

Another standardized and popular ESD test model is the machine model, which emulates the rapid direct transfer of electrostatic charge, from a charged conductive object (tool or equipment) to a test device. Compared with the human-body model of Figure 17.15, the machine model specifies a discharge of a 200-pF capacitor through a 0.75-μH inductor. Due to the absence of the current limiting resistor, this model is considered more severe, and tests are run at lower voltages. The charged-device model is an alternative ESD test set up, which is most commonly used to emulate rapid electrostatic charge transfer during packaging and assembly. More details on the latter two models can be found in http://www.esd-lab.com/others.htm [21] or directly from the JEDEC Web site: http//:www.jedec.org.

17.3.5.2 On-Chip ESD Protection Circuits

Although much ESD and ESD-protection knowledge has been built over the last couple of decades, the design of on-chip ESD protection circuits is both scientific and experimental. This is because in every new semiconductor node, device architectures and feature sizes (e.g., width, spacing, and oxide thickness) have changed with respect to the previous node, which requires new protection solutions. Usually, several alternative protection circuits are explored in each new technology node and often semiconductor process development goes hand in hand with ESD protection development.

The purpose of a protection circuit is that it provides a low-ohmic shunt path in parallel with the MOS input and output transistors during the occurrence of an ESD pulse. MOS input protection circuits usually comprise a voltage spike filter and diode clamps. Because MOS inputs are connected to high-ohmic transistor gates, the protection of input circuits is more critical than that of output circuits. Output pads are connected to drain areas. Usually these drain areas are relatively large, because outputs usually have to drive large capacitance (10–50 pF) and the complementary drain junctions act as intrinsically available diode clamps. Of course, also the outputs must fulfill certain ESD design rules.

The behavior of MOS protection circuits depends very much on their size and layout and on various process parameters. Each manufacturing process has its own specific design rules for ESD protection circuits. Therefore, the design of such circuits should be done in cooperation with specialists in the field of protection devices.

Future technologies, particularly those for high-performance designs, require different substrates such as SOI or silicon germanium (SiGe). SOI technologies need a different approach for the development of ESD protection devices because their devices are built on an isolating substrate. The implementation of ESD protection diodes on SOI needs to change from the high-perimeter bulk CMOS diodes to an SOI lateral-gated diode structure. SiGe technology has become another important alternative for high-speed communications and wireless applications. Because the change in material and mobility will also influence ESD, developing an ESD strategy for SiGe circuits will be very challenging. More about ESD can be found in Ameraskera and Duvvury [22].

17.3.6 Charge Injection during the Fabrication Process

Many IC processing steps use plasma or sputter-etching techniques, in which charge particles are collected on conducting surface materials (e.g., polysilicon, metals). This, so-called antenna effect can create significant electrical fields across the thin gate oxides which can be stressed to such an extent that the transistor's reliability can no longer be guaranteed. It can also cause a threshold-voltage shift, which affects the matching behavior of transistor pairs in analog functions. It is industry practice to introduce additional "antenna design rules" to limit the ratio of antenna area to gate-oxide area. The back-end design tools can handle these design rules by limiting the maximum wire (antenna) length in the different metal layers. In addition, protection diodes are used in the library cells to shunt the transistor gates. Due to the trend in gate-oxide thickness scaling, the appearance of the antenna effect is expected to increase. The use of high-k gate dielectrics in building the transistor stack would therefore also be beneficial to reduce this antenna effect.

17.3.7 Reliability Summary and Trends

Most of the previously discussed reliability topics depend on size, doping profiles and levels, voltages, temperatures, and device materials. Scaling requires a change in many of these parameters and will therefore have dramatic effects on the reliability of CMOS devices and circuits. Moreover, in technologies with channel lengths below 45 nm, the transistors are expected to be built from a completely different stack of materials as compared with today's high-volume products. SOI and/or SiGe will probably replace the bulk-silicon substrate; due to the high leakage current, the SiO_2 gate oxide is expected to be replaced by a high-k dielectric and, because of gate depletion, a metal gate may replace the polysilicon gate. This has an additional impact on the reliability of the devices and vice versa. Maintaining reliability at a sufficiently high level will put severe demands on this new transistor stack and makes the choice for the right materials a very difficult and cumbersome one.

17.4 Conclusion

For many previous technology generations the supply voltage has been constant and equal to 5V. The scaling process over that period was called constant-voltage scaling. Over the last decade, the advances in CMOS technology were not just related to scaling of the devices and the minimum features sizes, but also of the supply voltages. This is called constant-field scaling. If a chip is fabricated in a certain technology, then it should be operated at the nominal supply voltage for which that technology and libraries are developed. Certain applications, particularly those driven by low-power requirements, need further reduced supply voltages. This can be seen as maintaining the size and reducing only the supply

TABLE 17.4 Different Scaling Scenarios

	Topic	Relation	Scaling Factor ($s \approx 0.7$) $p \neq s \neq q$	$p=1$	$p=s$	$s=q=1$
Basic parameters	Voltages	V	p	1	s	p
		V_t	q	q	q	1
	Feature sizes	W, L, T_{is}, dist, t_{ox}(EOT)	s	s	s	1
	Devices per unit area	÷1/A	$1/s^2$	$1/s^2$	$1/s^2$	1
	Transistor bias current	i	p	1	s	p
	Average current/unit area	I	p/s^2	$1/s^2$	1/s	p
	Capacitance	$C = \varepsilon_0\varepsilon_r A/t_{is}$	s	s	s	1
	Metal resist. (top metals)	$R = \rho\ell/(t_m W)$ ($t_m \approx$const)	1	1	1	1
Performance	Gate delay τ, (÷1/f)	CV/i	s	s	s	1
	Power dissipation/gate, D	CV^2f	p^2	1	s^2	p^2
	Power-delay product, τD	÷CF^2	p^2s	s	s^3	p^2
	Power density, P	CV^2f/A	p^2/s^2	$1/s^2$	1	p^2
	Subthr. leakage current	Expon. with V_t and V	$s^{-1}12^{10(1-q)Vt+0.1(p-1)V}$	$s^{-1}12^{10(1-q)Vt}$	$s^{-1}12^{10(1-q)Vt+0.1(s-1)V}$	$12^{0.1(p-1)V}$
	Gate leakage current	Expon. with V and t_{ox}	$s^2 \cdot 10^{5(1-s)t_{ox}+2\log p}$	$s^2 \cdot 10^{5(1-s)t_{ox}}$	$s^2 \cdot 10^{5(1-s)t_{ox}+2\log s}$	$10^{2\log p}$
Reliability	Electromigr. (curr. dens.)	I = P/V	p/s^2	$1/s^2$	1/s	p
	Latch-up (for Vdd >> 1V)	÷V/dist	p/s	1/s	~1	p
	ESD susceptibility	÷$1/t_{ox}$	1/s	1/s	1/s	1
	Hot-carrier lifetime	f(V,V/distance)	$s^C \cdot e^{B/V(1/p-1)}$	—	—	$e^{B/V(1/p-1)}$
	NBTI V_t-shift	δVt ÷ f(V,L,t_{ox})	$(p/s)^m$	—	—	p^m
Signal integrity	Cross-talk/unit length	÷1/dist.	1/s	1/s	1/s	1
	Induct. noise/unit area	(di/dt)/A	p/s^3	$1/s^3$	$1/s^2$	p
	Voltage drop/unit length	IR/ℓ	p/s^3	$1/s^3$	$1/s^2$	p
	Soft-error rate (÷1/Q)	Q = CV	~1/ps	1/s	$1/s^2$	1/p
	Noise margin	V_{dd} and V_t	p and q	1 and q	s and q	p and 1
			With velocity saturation	↑ constant-voltage scaling	↑ constant-field scaling	↑ constant-size 'scaling'

voltage. This scaling scenario will be referred to as constant-size scaling. It is obvious that these different scaling scenarios have a different impact on the basic transistor parameters and on the performance and robustness of CMOS ICs. Table 17.4 shows how the transistor performance, reliability and signal integrity parameters depend on the scaling factor *s* for the sizes (an average value for $s \approx 0.7$ between successive technology generations) and *p* for the voltages and the impact of the different scaling scenarios, when we continue the scaling process as we did for more than four decades now. It means that no dramatic design and technology measures/changes have been taken into account.

The first scaling column ($p \neq s$) demonstrates how a parameter scales, when the voltages scale with a different factor than the sizes. In the constant-voltage scaling column ($p=1$), only the sizes scale, while the voltages are kept constant. In the constant-field-scaling column ($p=s$), both the sizes and the voltages scale with the same factor. Finally, in the constant-size scaling column ($s=1$), only the supply voltage scales, while keeping the sizes (= technology) constant.

In the table, the carrier mobility degradation due to velocity saturation is taken into account. It means that it is assumed that the transistor bias current (i) has a linear instead of the quadratic relation with the voltage. In understanding the table, a few more assumptions need to be explained. First, the thickness of the upper metal layers, which are commonly used for supply lines is assumed to stay almost constant and does not scale with s. This is because this thickness has hardly been scaled over many technology generations and is not expected to scale much further, because of electromigration requirements. However, metal layers 2 to 5 (or 8, depending on the technology node and metal layer options), which are used for routing, are assumed to almost scale with *s*. This is done, in combination with a slowly decreasing dielectric constant, to reduce the mutual wire capacitance. This not only reduces crosstalk, but it also helps to reduce the active power consumption per gate. It is also assumed, that the size of the cores (standard-cell blocks) remain almost constant as well. Therefore, the total capacitance per logic gate, which is defined by the fan-in capacitance of the connected logic gates, and the capacitance of the metal interconnections, is also assumed to scale with *s*.

The expression for the scaling of the subthreshold leakage current (per transistor) is based on the subthreshold slope, which is assumed to be 80mV/decade for bulk CMOS. This means that the subthreshold leakage current increases with about a factor of 12 for every 100mV reduction in the threshold voltage. The relation with the voltage scaling factor *p* originates from the drain-induced barrier lowering (DIBL) effect on the Vt, which is assumed to have a linear relation with the change in Vt ($\Delta Vt = -\gamma Vds = -0.1V$, where γ represents an emperically determined constant, which is assumed to stay close to 0.1).

For the gate leakage current scaling expression (also per transistor), it is assumed that it increases by a factor of 10 for every 0.2nm reduction of the gate-oxide thickness and it also increases by a factor of 10 for every doubling of the supply voltage.

For several parameters, the relation with the scaling factors is not completely clear.

Particularly the expressions for hot-carrier lifetime and NBTI depend heavily on the technology node, by the values used for B, C, and m.

Since these values only hold for one technology node, the expressions cannot be used to reflect the scaling trends and are therefore not included in the constant-voltage and constant-field scaling columns. In 45nm CMOS technology the device architecture is expected to completely change [e.g. (double) metal gate, high-k dielectric, strained silicon, etc.], which may dramatically affect NBTI as well as most of the other reliability parameters.

Latch-up depends on both voltage and size scaling. When only the supply voltage is reduced, the chance for latch-up is also reduced. If only the sizes shrink, however, the latch-up is expected to increase due to the smaller n^+ to p^+ spacings. Therefore, in the constant-field scaling column, a scaling factor of ≈ 1 is assumed for latch-up.

When going from the 90nm CMOS technology node to the 65nm and 45nm nodes, the voltages, particularly for low-leakage applications, no longer, or only hardly scale with the feature sizes. This means that, for these applications, we are slowly moving back from the constant-field scaling column towards the constant-voltage scaling column in Table 17.4. This has severe consequences, particularly for the active power consumption, the reliability and signal integrity topics. Parameters, such as the power and current density, as well as the inductive noise and the voltage drop all increase dramatically and will have severe design, package and application consequences. The drive for low leakage in standby operation then becomes a real burden to limit the power consumption during active operation.

Creative solutions, both in technology and design, are needed to keep the IC's robustness at a sufficiently high level in order to extend Moore's law for yet another decade; however, this will lead to a major increase of the complexity and total development and production costs of an IC. It is the author's opinion that, for many applications, the 32nm CMOS technology node, plus or minus one generation, is expected to be the last economically viable one.

17.5 Acknowledgment

The author thanks Dr. Andrea Scarpa for reviewing the hot-carrier and NBTI sections, Dr. Theo Smedes for the ESD and latch-up sections, Dr. Yuang Li for the section on electromigration, and, finally, Dr. Dick Klaassen and Dr. Ronald van Langevelde for discussions on the scaling table.

References

[1] H.J.M.Veendrick, *Deep-Submicron CMOS ICS: FROM Basics to ASICs, 2nd ed.*, Dec. 2000, Kluwer Academic Publishers, Dordrecht, 2000.

[2] Y.I. Ismail, On-Chip Inductance Cons and Pros, *IEEE Trans. on VLSI Syst.*, vol. 10, no. 6, Dec. 2002.

[3] B. Nauta and G. Hoogzaad, How to deal with substrate noise in analog CMOS circuits, *European Conf. on Circuit Theory and Design*, Budapest, September 1997.

[4] H.B. Bakoglu, *Circuits, Interconnections, and Packaging for VLSI*, Addison-Wesley, Reading, MA, 1990.

[5] E. Dupont et al., Embedded Robustness IPs for transient-error-free ICs, *IEEE Design Test of Comput.*, vol. 19, no. 3, pp. 56–70, May/June 2002.

[6] M. Vertregt, Embedded analog technology, *IEDM Short Course on System-on-a-Chip Technology*, Dec. 5, 1999.

[7] A. Agarwal et al., Statistical timing analysis using bounds, *DATE*, March 2003.

[8] J.-J. Liou et al., Fast statistical timing analysis by probabilistic event propagation *DAC 2001*, June 2001, Las Vegas, NV.

[9] H.J.M. Veendrick, Wire self-heating in supply lines on bulk-CMOS ICs, *ESSCIRC 2002 Dig. of Tech. Papers*, pp. 199–202, Sept. 2002.

[10] G.D. Wilk et al., High-k dielectrics: current status and materials properties considerations, *J. Applied Physics*, vol. 89, no. 10, pp. 5243–5275, May 15, 2001.

[11] A. Kottantharayil, Low-voltage hot-carrier issues in deep-sub-micron MOSFETs, Thesis, Universitat der Bundeswehr (München), 2001, http://137.193.200.177/ediss/kottantharayil-anil/inhalt.pdf, June 21, 2004.

[12] S. Mahaptra et al., Device scaling effects on hot-carrier induced interface and oxide-trapped charge distributions in MOSFETs, *IEEE Trans. on Electron. Devices*, vol. 47, no. 4, April 2000.

[13] K. Kushida-Abdelghafar et al., *Appl. Physics Lett.*, vol. 81, no. 23, 2002.

[14] Y. Hiruta et al., Interface state generation under long-term positive-bias temperature stress for a p^+ poly gate MOS structure, *IEEE TED 36*, p. 1732, 1989.

[15] T. B. Hook et al., The effect of fluorine on parametric and reliability in a 0.18-μm 3.5/6.8-nm dual gate oxide CMOS technology, *IEEE TED*, vol. 48, no. 7, p. 1346, 2001.

[16] C.E. Blat et al., Mechanism of negative-bias-temperature instability, *Journal of Applied Physics (JAP)*, vol. 69, no. 3, 1991.

[17] Ogawa et al., Interface-trap generartion at ultrathin (4–6 nm) interfaces during negative-bias temperature aging, *JAP 77*, 3, 1995.

[18] A. Scarpa et al., Effect of the process flow on negative-bias-temperature-instability, *Proc. 8th Int. Symp. on Process- and Plasma-Induced Damage*, p. 142, 2003.

[19] P. Chaparala et al. NBTI in dual gate oxide PMOSFETs, *Proc. 8th Int. Symp. on Process- and Plasma-Induced Damage*, p. 138, 2003.

[20] R.R. Troutman, *Latchup in CMOS Technology*, Kluwer Academic Publishers, Dordrecht, 1986.

[21] http://www.esdlab.com/others.htm, June 21, 2004.

[22] A. Ameraskera and C. Duvvury, *ESD in Silicon Integrated Circuits*, John Wiley & Sons, New York, 2002.

III

Low-Power Processors and Memories

18

Techniques for Power and Process Variation Minimization

Lawrence T. Clark
Arizona State University

Vivek De
Intel Labs

18.1 Introduction

For more than a decade, integrated circuit (IC) power has been steadily increasing due to higher integration and performance enabled by process scaling. As shrinking transistor dimensions are fabricated, and as the absolute value of the dimensions diminish, greater device variations must be addressed. Until recently, increased power was driven primarily by active switching power. Threshold voltages must be decreased to maintain performance at the lower supply voltages required by thinner oxides, however, raising drain to source leakage exponentially. Steeper doping gradients and higher electric fields increase other leakage components, giving rise in sub-0.25-μm generations to DC leakage currents that may limit overall power and performance in future chips. This comes on top of still increasing active power dissipation, driven by architectural changes such as greater parallelism and deeper pipelining. The latter

0-8493-1941-2/05/$0.00+$1.50

implies fewer gates per stage and in turn, requires more aggressive circuit techniques such as domino, which can also increase active power. Having fewer logic stages increases the susceptibility to process variations. Finally, as scaling requires lower voltages, in-die, and system-level voltage variations are also increasingly problematic.

The focus of this chapter includes the design implications of increasing device variation and leakage. The mechanisms are a direct result of basic physics and will continue to grow in importance over time, requiring design effort to mitigate them. Variation in microprocessor frequency has been dealt with by "speed binning" whereby faster dies are separated and sold at a premium. Dies with inadequate speed or excessive standby current are discarded. These yield considerations are important for robust design. We also discuss design techniques, notably the application of body bias and supply voltage adjustment, which can help deal with both variation and average leakage, as well as active power. Examples from fabricated designs demonstrating the efficacy of the techniques are discussed.

18.2 Integrated Circuit Power

Increasing leakage currents are a natural byproduct of transistor scaling and comprise a significant portion of the total power since the 0.25-μm-process generation. By the 90-nm technology node, it can contribute over a fifth of the total IC power on high-performance products [1]. The profusion of battery powered "hand-held" devices introduced in recent years (e.g., cell phones and personal digital assistants) has made power management a first-order design consideration. These sections focus on circuit design approaches to alleviate leakage power using reverse body bias (RBB) "Drowsy" mode when an IC is in a standby mode and later, in Section 18.9, optimizing the active power by dynamic voltage management (DVM). Although other implementations are briefly discussed, the bulk of the discussion describes the specific implementation on the 0.18-μm XScale microprocessor cores intended for system on chip (SoC) applications [2].

18.2.1 Active Power and Delay

The total power of a static CMOS integrated circuit is given by

$$P_{tot} = P_{dyn} + P_{static} + P_{short\text{-}circuit} \tag{18.1}$$

representing the dynamic power (i.e., that due to charging and discharging capacitances during switching) the static leakage power, and the "short-circuit" or crowbar power due to both P and N transistors being on simultaneously during a switching event, respectively. The latter term tracks with the active power and is generally on the order of 5% or less for well-designed circuits. It is typically ignored, as it will be here. The dynamic power of a digital circuit follows the well-known

$$P_{dyn} = a/2\ C\ V_{dd}^2\ F \tag{18.2}$$

where C is the switched capacitance, V_{dd} is the power supply voltage, F is the operating frequency, and a is the switching activity factor measured in transitions per clock cycle. Leveraging the V_{dd}^2 dependency is consequently the most effective method for lowering digital system power; however, the switching speed of a digital circuit with a fixed input slope and fixed load is given by Chen and Hu [3]:

$$T_{delay} = K\ V_{dd}/(V_{dd} - V_t)^{\alpha} \tag{18.3}$$

where α^* is typically 1.1 to 1.5 for modern velocity saturated devices, tending toward the former for NMOS and the latter for PMOS [4], and K is a constant depending on the process. To first order, this

* α is typically used as in the literature.

delay dependency on voltage can be treated as linear. The concept of DVM is to limit the V_{dd} and frequency such that the application latency constraints are met, but the energy to perform the application function is minimized by following the square law dependency of Equation (18.2) instead of linearly tracking F. The chosen frequency F, representing the reciprocal of the worst-case path delay, is constrained by Equation (18.3) for a given supply voltage.

18.2.2 Leakage Power

Leakage power sources are numerous [5], with the primary contributor historically being transistor off state drain to source leakage (I_{off}). For modern processes having gate dielectric thicknesses under 3 nm, gate leakage I_{gate} is becoming a larger contributor but is generally smaller than I_{off}, particularly at high temperatures, given the stronger temperature dependency of 8–12×/100°C for I_{off} vs. approximately 2×/100°C for I_{gate}. I_{off} increases on scaled transistors because, to maintain performance, V_t must be lowered to compensate for decrease in V_{dd}. This increases the leakage according to

$$I_{off} \propto e^{-V_T/(S/\ln 10)} \tag{18.4}$$

where S is the subthreshold swing given by

$$S = \frac{kT(\ln 10)}{q}\left(1 + \frac{C_D}{C_{OX}}\right) \tag{18.5}$$

where k is the Boltzmann constant, T is the temperature in Kelvin, q is the elementary charge, C_D is the depletion layer capacitance, and C_{OX} is the gate oxide capacitance. Noting that C_D is nonvanishing, the subthreshold swing parameter S is essentially a fixed parameter for Si MOSFETs, typically 80–100 mV/decade depending upon the process at room temperature. Referring to Equation (18.4), it is obvious that lowering V_t affects the I_{off} exponentially.

For gate oxide thicknesses below 3 nm, quantum mechanical (direct band-to-band) tunneling current becomes significant. This leakage is extremely voltage dependent, increasing approximately with V^3 [6]. It also increases dramatically with decreasing thickness (e.g., increasing 10× for a change from 2.2 nm to 2.0 nm [7]). Gate-induced drain leakage (GIDL) at the gate-drain edge is important at low current levels and high applied voltages. It is most prevalent in the NMOS transistors where it is about two orders of magnitude greater than for PMOS devices. For a gate having a 0-V bias with the drain at V_{dd}, significant band bending occurs in the drain region, allowing electron-hole pair creation. Essentially, the gate voltage attempts to invert the drain region, but because the holes are rapidly swept out, a deep depletion condition occurs [8]. The onset of this mechanism can be lessened by limiting the drain to gate voltage. It can be exacerbated by high source or drain to body voltages. Diode area leakage components from both the source-drain diodes and the well diodes are generally negligible with respect to I_{off} and GIDL components. This is also improved by compensation implants intended to limit the junction capacitance. However, transistor scaling requires increasingly steep (often halo) doping profiles increasing band-to-band tunneling (BTBT) currents at the drain to channel edge, particularly as the drain to bulk bias is increased. This component may also limit use of RBB on sub-0.18-μm processes. Controlling these leakages will be key to effective use of body biasing and will require careful circuit design as well as appropriate transistor architecture.

18.3 Process Selection and Rationale

Thinner oxides are required to allow transistor length scaling while maintaining channel control. These scaled oxides require lower supply voltages to limit electric fields to a reliable value. Additionally, to maintain performance at lower voltage by retaining gate overdrive V_{dd}–V_t it is necessary to lower V_t. For handheld battery powered devices, V_t must be chosen to balance standby power with active power dissipation for maximum battery life. Absent clever design to mitigate leakage, the duty cycle between standby and active operation for the given application determines the optimal threshold voltage [9]. This leads to considerable divergence in future processes and considerable power constraints to scaling processes used for portable devices [10]. One of the purposes of circuit techniques to limit active and standby power is to help widen the allowable V_t and process performance range. Handheld battery lifetime requires IC standby currents below 500 μA requiring total leakage under 100 pA/μm of transistor width. This implies a V_t over 500mV, independent of supply voltage, increasing active power at the same performance level.

Figure 18.1 plots the simulated power vs. performance for a microprocessor operating at different frequencies on processes with different V_t, assuming complete flexibility in the supply voltage or DVM (i.e., the voltage is chosen such that it is just sufficient to meet the processor frequency). The curves are based on the transistor performance metric described in Thompson [11] and normalized to the microprocessor performance with V_t of 390 mV (solid line in both plots) and 500 mV (dashed line in both plots). Figure 18.1(a) emphasizes the active power, which depicts the greater overall performance available from the lower V_t process. Note the improved power vs. the linear characteristic that would be obtained by scaling frequency alone. Figure 18.1(b) plots the log scale power for the low frequency ranges. At low frequencies, it is assumed that the power supply voltage cannot be scaled below a minimum value due to circuit functionality constraints. This value is 0.6 V for the 390-mV and 0.7 V for the 500-mV processes. Below the minimum operating voltage, the clock frequency is lowered resulting in a linear, instead of quadratic power savings. The break between square law and linear behavior is evident in the log scale plot of Figure 18.1(b). It is apparent that the lower V_t process has a higher leakage, as indicated by the zero frequency point, while it has a lower active power at the same frequency. It is also capable of higher overall performance. The lower active power is the result of reaching a given performance at a lower voltage, and its benefit was presented in Equation (18.2). The dotted line in Figure 18.1(b) demonstrates that with the addition of RBB Drowsy mode, the higher-performance process is power competitive at low effective frequencies with the slower process. The methods for achieving this comprise Section 18.4 and Section 18.5.

Nonstate-retentive sleep modes also incur power penalties. The present logical state must be saved before sleep and restored upon resuming active operation, requiring a low standby power storage medium. The data movement requires time and power that must be amortized by the leakage power savings achieved in the time in sleep. This can preclude frequent use. If the storage is off-chip, the higher IO voltages and off-chip capacitances increase the power penalty. A number of schemes, ranging from "greedy" to timeout based, have been proposed for determining when to enter a low-power state. The key considerations are achieving low energy cost to entry and exit, as well as low latency to awaken and respond to input.

18.3.1 Effective Frequency

For compute intensive applications, the active power dominates as illustrated in Figure 18.1. The leakage power is of interest when the compute demands are modest, for instance when a processor is waiting for user input or in a cell phone, in the intervals between contacts with the cell. The former can be expected to be multiple times per second, and the latter less than once per second [12]. The total computed cycles per second is very low, although the frequency of the part might be higher as described next. Here, the term "effective frequency" is used to mean the number of cycles of computation accomplished over a given period. The actual frequency may vary during that time, according to whether the processor is running or is in a low-power Drowsy mode. Effective frequency is a measure of the average actual work performed by the processor. For example, assume that the processor receives interrupts at an average frequency determined by the application, for instance, from keystrokes on a keypad. Each interrupt awakens the processor where it computes for a number of cycles required to process the input (e.g., add it to the display buffer). The computational requirements might be quite different depending on the type of interrupt that is being serviced — it may be a command to sort mail messages. The effective frequency

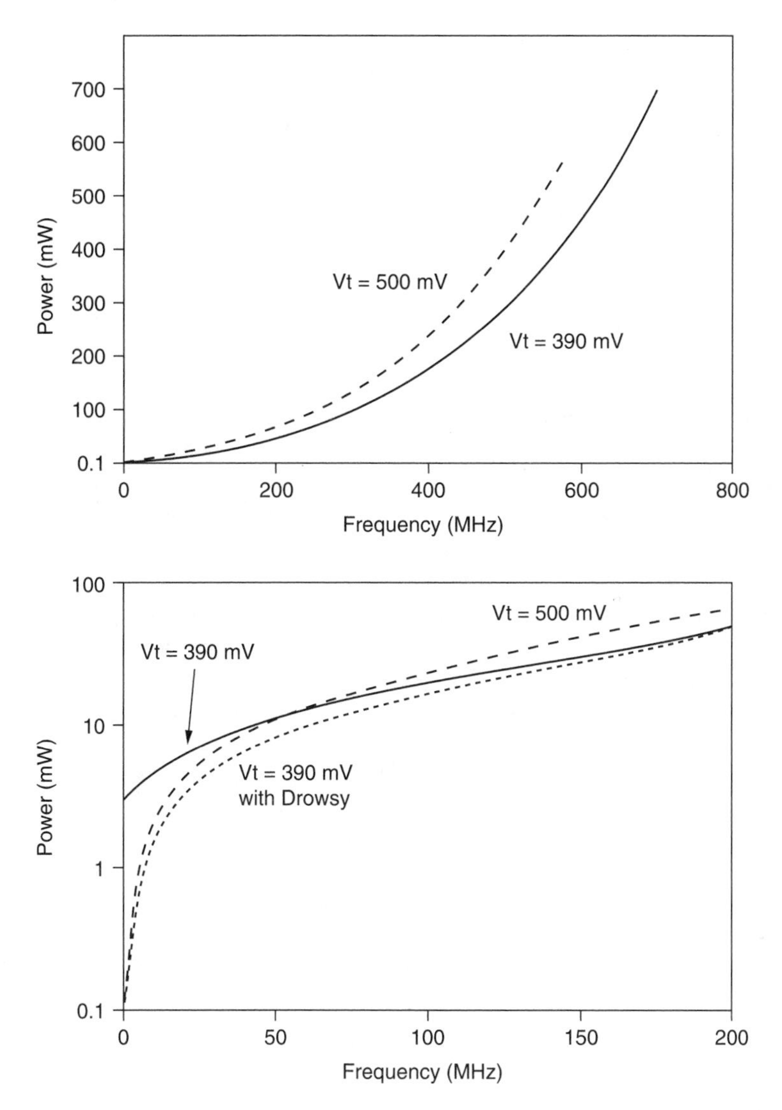

FIGURE 18.1 (a) The effect of V_t on power vs. frequency, and (b) the low frequency, leakage dominated power levels. In the upper plot, the low V_t with Drowsy is coincident with the non-Drowsy.

is then the total long-term average number of useful clocks per unit time (i.e., the number of instructions per interrupt times the number of interrupts). For example, with a 100-Hz interrupt rate and 100,000 instructions per interrupt, the processor will have an effective frequency of (100 Hz * $100(10)^3$) = 10 MHz, although the clock rate may be much higher (e.g., 300 MHz).

18.4 Leakage Control via Reverse Body Bias

RBB has been suggested for leakage control for some time [13,14]. Essentially, this leverages the well known body effect, that raises the V_t of a transistor having a source voltage above the bulk, as commonly occurs in the upper transistors of an NMOS stack during switching. Although normally a designer's bane that reduces circuit speed, it can be used to advantage because

$$V_t = V_{FB} + \gamma\sqrt{\phi_s - V_{bs}} - K_2(\phi_s - V_{bs}) - \eta V_{ds} \tag{18.6}$$

where γ is the body effect coefficient, which, along with K_2 models nonuniform doping [15]. These coefficients represent the efficacy of a change in the source to body voltage in modulating I_{off}. η is the drain induced barrier lowering (DIBL) coefficient, which represents the ability to Control V_t by applying drain bias. Drain and body bias also affect the subthreshold slope.

RBB to modulate leakage has a number of advantages:

1. It is a circuit design approach.
2. It does not adversely affect the active performance.
3. It is state retentive.

The first point allows this approach to be utilized on any process. Longer channel lengths generally have a stronger body effect [16], under designer control at the resolution of the drawing grid. The second assumes that the implementation does not incur a significant IR drop or alternatively, it allows improved active power at the same standby current level vs. a device not so equipped. The final point is the advantage over "sleep" modes where the power supply is completely disconnected. With RBB, data is not lost when entering and exiting the low-power state — important in that it allows the power control to be transparent to the operating system and application software and saves significant energy. It is frequently difficult to predict *a priori* how long a device will be in a standby state, particularly when this depends upon user interaction. Retaining precisely the state of the IC before the entrance, as well as minimizing any power penalty to enter or exit the low-power mode makes the mode usable more frequently.

Body bias was used to limit leakage on a 1.8 V microprocessor implemented in a dual-well 0.25-μm process described in Mizuno et al. [17]. This device used separate supplies for both the NMOS and PMOS bulk connections. A strong negative bias greater than 1 V was applied to the NMOS bulk via a charge pump and the PMOS bulks (N wells) were connected to the 3.3-V power supply rail during standby. Hundreds of local switches, distributed across the device, apply the body bias and provide a low impedance bulk connection, at the expense of routing the controls and supplies throughout the layout. This strong biasing is inappropriate for smaller geometry processes, where more abrupt doping and thinner oxides increase second order effects. This implementation of RBB increases GIDL, which can thus be the limiting leakage mechanism. Direct BTBT leakage in the source diodes of sub-130-nm halo doped transistors can be increased to also limit total standby current by reverse biasing the junctions. Consequently, to use RBB effectively on processes beyond 0.25 μm, it will need to be comprehended in the transistor design and the RBB operation should use the lowest effective voltages.

18.4.1 RBB on a 0.18-μM IC

The Intel 80200 microprocessor is an implementation of the XScale microarchitecture implemented in a 0.18-μm process. Although sold commercially as a high-performance embedded device, it was also used as a development vehicle to develop Drowsy mode [18] circuitry and techniques. This mode utilizes RBB as well as V_{dd}–V_{ss} collapse to limit leakage power, achieved via large supply gating transistors that allow the source to be raised. They also allow full collapse of the core voltage, which produces the non-state-retentive "sleep" mode, essentially the classical multi-threshold CMOS (MTCMOS) approach to leakage control [19]. The manner in which RBB is applied, utilizing lower source to bulk voltages while collapsing the V_{dd}–V_{ss}, alleviates second order components. Drowsy mode retains state in all storage elements on the die and is exited on any interrupt. Sleep mode is not state-retentive, requiring a "cold-start." Consequently, asserting reset instead of an interrupt terminates it. The Drowsy implementation and results are described in detail in the following sections.

18.4.2 Circuit Configuration

The circuit configuration is depicted in Figure 18.2. Power pads are on the V_{dd}, $V_{dd(IO)}$, and $V_{ss(GND)}$ pins. Large N channel devices M1 provide V_{ss} to the active circuitry during active operation. Simultaneously, large P channel devices M2 provide clamping of the N well ($V_{dd(SUP)}$) providing the PMOS bulk connection

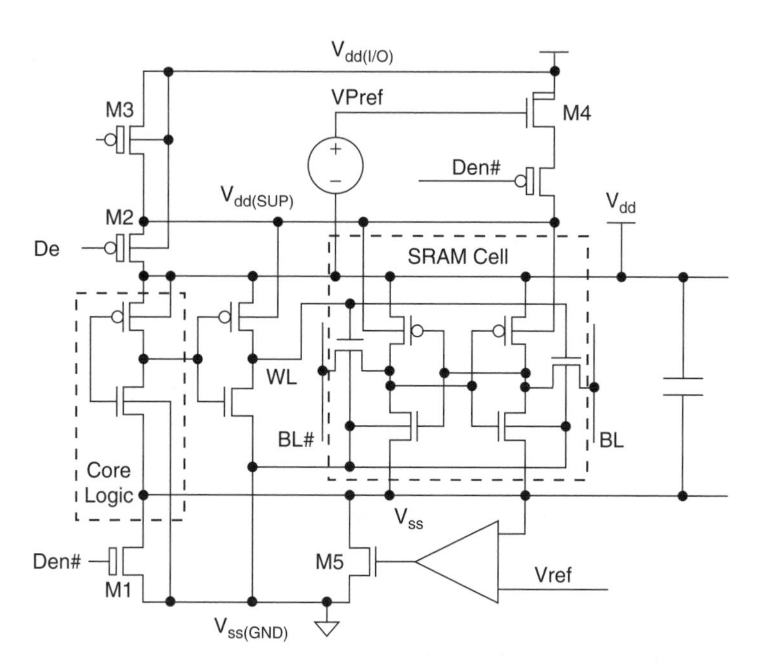

FIGURE 18.2 Circuit configuration for RBB Drowsy mode.

to V_{dd}. These transistors must be thick oxide because they are exposed to high voltages as indicated — here, the thick gate IO transistors are used. The PMOS clamping transistors carry no DC current and are 15 mm in total width. The NMOS clamp transistors carry the entire power supply current during operation and must do so with minimal IR drop. They are 85 mm in total width, which is less than 2% of the total transistor width of the microprocessor. This high ratio between the rest of the core is indicative of the low activity factor achieved by the design and relies on adequate on-die decoupling capacitance to provide instantaneous current demand. To this end, total of 55 nF of decoupling capacitance was interspersed among the active circuitry.

For sleep as well as Drowsy modes, the transistors comprising M1 are in cutoff. In the former cases, the core V_{ss} is allowed to float to V_{dd}, and power consumption is dominated by the leakage current through the NMOS clamp devices. The clamp devices should be high V_t to minimize this current because they do not have body bias applied. In Drowsy mode, to apply body bias to the NMOS devices, V_{ss} is allowed to rise toward V_{dd} but regulated to avoid losing state. Raising the NMOS source voltage instead of decreasing the NMOS body voltage is advantageous because it does not require a twin-tub or triple-well process, nor charge pump circuitry. It also lowers the I_{off} by the η coefficient of Equation (18.6) as well as limiting GIDL components because drain to bulk voltage is not increased. Because gate current is strongly affected by the drain to gate voltage, it is substantially reduced on processes with thin oxides. Another regulator provides a high voltage to the $V_{dd(SUP)}$ node, to reverse body bias the PMOS transistors. In the static random access memory (SRAM), the word-lines are driven to $V_{ss(GND)}$ as presented in Figure 18.2. This places a negative gate-to-source bias on the SRAM pass devices as presented, lowering the SRAM current a further 40%. This may not be desirable for thin oxides as it can increase the gate leakage component beyond the I_{off} savings.

Simulated waveforms of the V_{ss} and $V_{dd(SUP)}$ nodes are plotted in Figure 18.3, at 110°C. Minimal overshoot can be discerned in the figure. In Drowsy mode, the V_{ss} node rises to approximately 650 mV, with some PVT variation. $V_{dd(SUP)}$ is driven to 750 mV above V_{dd}. At room temperature, the V_{ss} node takes approximately half a msec to rise because it is pulled toward V_{dd} solely by leakage. The advantage of this passive V_{ss} rise is that movement of this highly capacitive node is limited if Drowsy is exited soon after entrance, limiting the power cost of using this mode. No energy is explicitly expended to enter the mode because it achieved by transferring charge from the core nodes to the V_{ss} node, instead of supplying

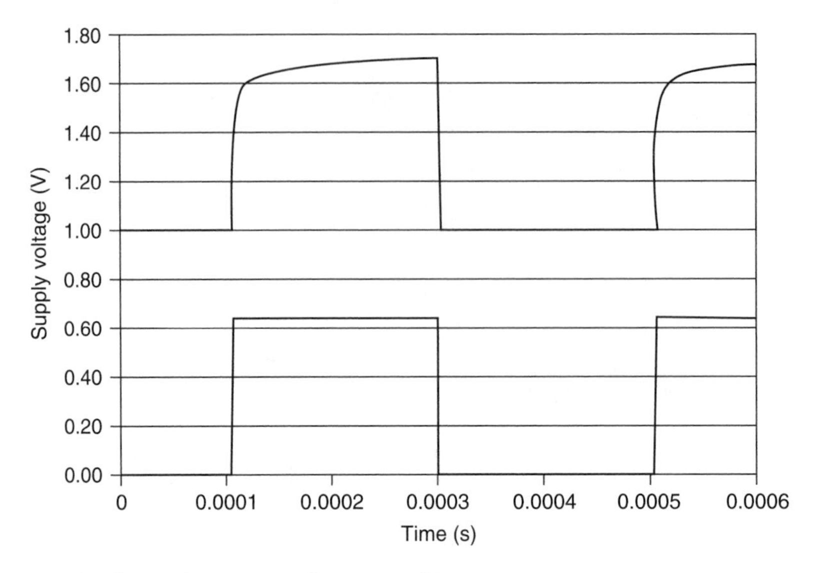

FIGURE 18.3 Simulated V_{ss} and $V_{dd(SUP)}$ waveforms at 110°C.

it from the IC power pins. This is not possible on the PMOS bulk node. This regulator circuit is designed with limited drive, as that node is less capacitive at 5 nF, and with low current demand, generally just the diode contributions of the N-well and PMOS source-and-drain diodes.

18.4.3 Layout

Application of any body bias requires separate bulk and source supplies for both P and N transistors. This design opts for minimal intrusion due to the separate body connections. The power supply clamping transistors are provided in the pad ring only, occupying otherwise empty (or IO decoupling capacitor) space within the supply pins. Because the core was over 4000 μm per side, circuits could be over 2000 μm from the nearest clamp. Additionally, the bulk connections are routed sparsely through the logic circuitry, limiting the density impact. This is feasible because these provide no DC power, making resistance less important. A two-layer routing grid with 50 μm between bulk supplies was utilized. The substrate is highly doped, providing an effective short circuit between V_{ss} (ground) rails and limiting noise due to switching. N-wells are intentionally contiguous, forming a grid at the substrate level for $V_{dd(SUP)}$.

18.4.4 Regulator Design

The V_{ss} regulator comprises Figure 18.4(a) and strictly limits the regulator overhead power. The output voltage must be essentially constant over 3 decades of current demand at all process, temperature, and voltage (P, V, T) corners (see Section 18.6). At the high end, when entering the low-power state directly from high frequency operation the die may be hot, where MOS drain to source leakage may be over 100× the RBB low temperature leakage and must be provided to avoid collapsing logic state. As expected, the amplifier compares the voltage on V_{ss} with a reference voltage. A PMOS stack simulating a resistor string, which allows it to vary with power supply variations, generates this reference. In this manner, higher supplies allow larger body bias — this flexibility was desirable for a test device. The resistor stack current is under 100 nA and is continuously biased in all modes. The regulator is a three-stage amplifier with an NMOS output transistor M5. Three stages were required due to the bias conditions and low current requirements to keep the regulator power consumption less than 5% of the total standby power at the typical process corner. The output transistor is sized to provide the full IC leakage current at high temperature and the worst-case process corner. The first stage is a differential operational transductance amplifier (OTA), while the second buffer stage provides increased voltage output range and current drive to the gate of M5. The first and second stages combined use less than 4 μA at typical operating conditions.

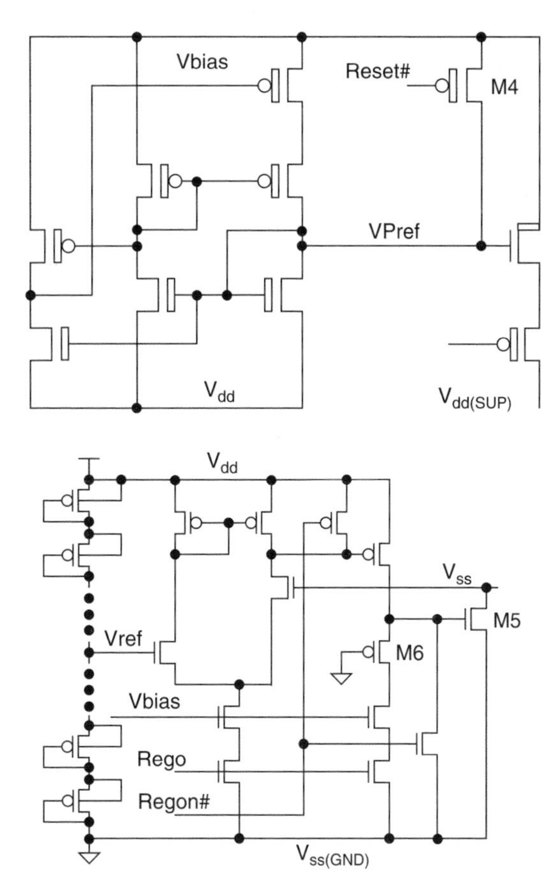

FIGURE 18.4 V_{ss} (a) and $V_{dd(SUP)}$ (b) supply regulation circuits. All NMOS share substrate $V_{ss(SUP)}$, and all PMOS in (a) have V_{dd} and in (b) $V_{dd(SUP)}$ body connections.

At such low current levels, gain is limited, which improves stability, as discussed next. Slew rate also suffers, which makes the step response poor. To address this, the buffer stage includes the diode connected transistor M6, which, combined with proper sizing keeps transistor M5 from completely cutting off, except in sleep mode. The enables are evident in the figure.

Stability must be ensured at all P, V, T conditions and overshoot on V_{ss} must be limited. Entering the body bias state, which is essentially a voltage step on V_{ss} represents the worst-case stability condition. Adequate phase margin ensures stability of the system comprised of the regulator and V_{ss} node on the IC. Overshoot on V_{ss}, even momentarily, can cause state loss. The circuit poles may be approximated by the dominant terms to simplify the analysis. The V_{ss} node is controlled to first order by the output conductance of transistor M5, while the amplifier pole is dominant. The former pole is at approximately 670 kHz calculated from the small signal parameters, while the latter is at 9 kHz. The low gain of the amplifier produces a low unity gain bandwidth and greater than 60 degrees of phase margin at the typical process. Essentially, the highly capacitive V_{ss} node low-pass characteristic does not require high amplifier speed for stability.

To back-bias the PMOS devices, two schemes may be used. At low IO voltages (e.g., 1.8V), the PMOS transistor bulk node may be directly connected to this voltage via M3. For higher IO voltages diminishing leakage reduction does not offset the greater charge switched in raising the well voltage. Therefore, in this case, this voltage is regulated. The open loop regulator is depicted in Figure 18.4(b), which derives a constant voltage from the IO supply $V_{dd(IO)}$. It is worth noting that as long as circuit configurations that accumulate the gates of the PMOS transistors are avoided, high voltages may be applied to the bulk without oxide stress or damage. The regulator is a bootstrapped voltage reference driving a wide NMOS vertical drain transistor in a source follower configuration as presented in Figure 18.4(b). This device

(M4 in Figure 18.2) has a naturally low V_t and operates in subthreshold, providing a negligible voltage drop from the reference voltage to $V_{dd(SUP)}$ in operation. The vertical drain configuration allows the thin gate oxide device to tolerate high drain to gate voltages as in Clark [21].

The relatively high active current of the phase-locked loop (PLL) necessitates disabling it in Drowsy mode. Leaving standby mode requires the PLL to restart and lock, triggered by an external interrupt. Because this takes approximately 20 μs, the mode is usable often (e.g., between keystrokes). On the prototype, the lock time is set by a counter to enable deterministic testing. In actuality, the PLL lock time can be as low as 2 μs depending on voltage. Faster interrupt latencies can be supported by providing the PLL reference clock directly to the IC, while the PLL locks. Consequently, PLL lock-time need not affect interrupt latency or limit the applicability of Drowsy usage.

18.4.5 Limits of Operation

All memory, such as latches, need to be able to hold a "0" or a "1" with RBB applied. Although it is more difficult from a circuit aspect, holding state in all elements greatly simplifies logic design verification. As V_{dd} and V_{ss} collapse toward one another the transistors move from saturation into subthreshold, as the reverse body bias increases V_t and the increase in V_{ss} decreases V_{gs}. In subthreshold, these "on" transistors rapidly weaken with their current following the subthreshold slope. In a memory element, the voltage level of a node is maintained by an "on" transistor being able to supply enough current to overcome the leakage of all the attached "off" transistors. In normal, high V_{ds} operation, this is not a problem due to the large I_{on} to I_{off} ratio. As transistors reach subthreshold, the on current drops rapidly with V_{ds} (= V_{gs}) due to

$$I_{ds,sat} = \frac{\mu C_{ox} Z}{2L}\left(V_{gs} - V_t\right)^{\alpha} \tag{18.7}$$

becoming Equation (18.4) as the gate overdrive (V_{gs}–V_t) is reduced below 0. Ideally, V_{dd}–V_{ss} can be lowered to drive all of the transistors into subthreshold operation because the I_{on}/I_{off} ratio will scale for all transistors. Assuming an 80-mV/decade transistor subthreshold characteristic, over three decades of current difference between on and off transistors will be maintained with 250 mV of V_{ds}. Lowering the voltage too far on future ultra-small devices will reach thermodynamic constraints [22]. The relative size and strength of the N and P transistors, including local channel length and V_t variation must be considered. In practice, state loss depends upon many factors such as the type of latch, the transistor ratios, the logic state being held, the local transistor V_t and the temperature. Domino circuits, with the largest N to P (keeper) width ratios, are the first to fail.

The fail point as a function of the PMOS body voltage and NMOS body voltage as measured on silicon is presented in Figure 18.5. Because in this design the V_{ss} is referenced to the $V_{ss(GND)}$ supply node, V_{ss} is the applied NMOS body bias. Points lower on the vertical axis have higher NMOS V_t and further right have higher PMOS V_t. Measured parts retained state below the curve (Pass) and lost state above it after application of that level of reverse body bias (Fail). As V_{ss} is increased, the NMOS transistors have increasing reverse body bias applied to them, so "on" devices are in subthreshold. The right side of the curve represents a memory element failing as logic "0" is flipped to a "1." As $V_{dd(sup)}$ increases the PMOS transistors leakage is reduced, so that the amount of reverse body bias that can be applied to the NMOS transistors can be increased, continuing until a maximum value of $V_{dd(SUP)}$ and V_{ss} is reached. The left part of the curve represents the converse case where the PMOS transistors are weakened with respect to the NMOS. With a large $V_{dd(SUP)}$ applied "on" devices are in subthreshold and are eventually unable to supply enough current to overcome leakage from NMOS transistors. This left part of the curve represents a memory element holding a "1" flipping to a "0." The flat zone depicts the saturation of any body effect as voltage increases.

18.4.6 Measured Results

When the leakage current from the microprocessor is low the voltage on V_{ss} will not rise to the reference voltage because the regulator does not actively drive its output. At V_{dd} of 1.05 V, the regulator clamps at

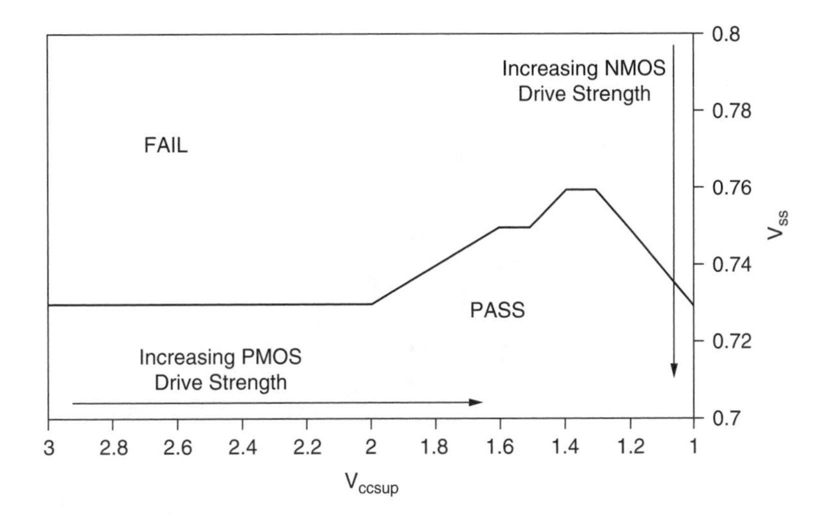

FIGURE 18.5 Shmoo plot of state retention with PMOS and NMOS body bias as parameters.

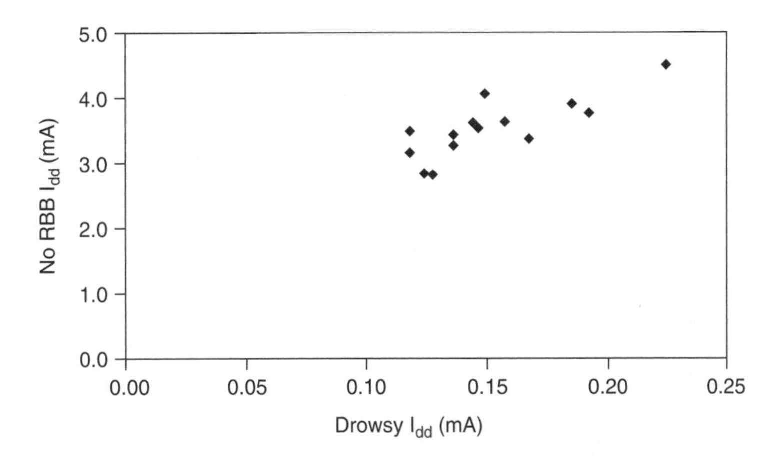

FIGURE 18.6 Standby current of the microprocessor with and without body bias.

the reference voltage, about 0.73 V for high leakage. The leakage current is reduced by a factor of over 25 across most devices when the body bias is applied. Figure 18.6 plots the no body-bias (NBB) standby vs. the RBB Drowsy mode current. Figure 18.7 gives the distribution of the current with reverse body bias for all die on one wafer. A wide variation, due to variations in the process (e.g., threshold voltage and channel length) as well as the regulator output, is evident.

18.5 System Level Performance

This section describes experiments using Drowsy mode to simulate low leakage by running an IC in short bursts of operation interspersed with time in the leakage control mode, time domain multiplexing (TDM) Drowsy mode [23]. The IC power is dominated by different components in different operating modes described in Section 18.1. First, the active power component dominates during intervals of operation. Second, there are the two primary leakage components, the active component, potentially multiplied by die heating, but still small compared with the active power, and leakage during Drowsy mode. Third, the PLL and clock-generation power, as well as that of the interrupt circuitry required to wake up the device from Drowsy mode. The former provide an active power component that runs during active operation and for 20 μs before each active interval. There is a small non-RBB leakage component

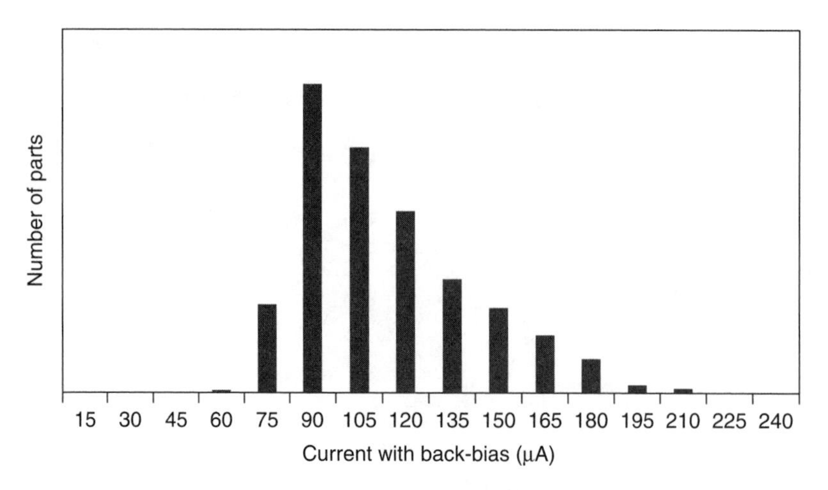

FIGURE 18.7 Standby current of the microprocessor with body bias.

during PLL startup, but the time is small enough to make this component negligible. Finally, the power cost of each power supply movement represents the "penalty" power of entering and exiting Drowsy mode. This low frequency high capacitance switching power is mitigated by the low-voltage swing utilized. The energy to transition the clamp transistors and their driver circuitry is small enough to be considered negligible because the total gate capacitance of the clamp transistors is 119 pF. Entry into the standby mode consumes no power on the V_{ss} node due to its being driven high by passive leakage of the core (i.e., redistribution of charge from the nodes within the core logic to V_{ss}). Power dissipation is incurred only when leaving the mode. The total energy cost of a single entry into the low-power mode is calculated to be 30.6 nJ from measurements.

The experiments were performed on a microprocessor board [24] at 1 V V_{dd} running IO at 100 MHz and with a core frequency of 300 MHz. Separate core and analog PLL supplies connected to external power supply and ammeter connections allowed these currents to be distinguished. The Drowsy circuitry allows high performance — the device under test was run on the board to 800 MHz at 1.55 V. Instantaneous current demand was measured, while the interrupt signal was asserted at the chosen interrupt frequency. Each interrupt runs code comprised of a simple loop, intended to be representative of the power that would be consumed by the typical instructions, which are generally quite similar [25]. The number of instructions to run at each interrupt is set by a loop count parameter. At the end of the loop, the IC returns to Drowsy mode. Subsequent interrupts wake the microprocessor and begin the loop anew. Due to branch prediction, the processor executes one instruction per clock in the loop (i.e., there are no stall cycles). State retention while Drowsy maintains the cached instructions, so there are no misses after the first compulsory ones. The operating voltage was adjusted based on the reading from a locally connected voltmeter in order to account for IR loss in the power supply leads. Figure 18.8 is a representative power measurement.

18.5.1 System Measurement Results

Drowsy power was measured to be 0.1mA at 1V on the IC used in these measurements. In a DC condition, the I_{sb} at room temperature (i.e., the standby core supply current with no clock running) was 2.8 mA at 1 V. The PLL consumes 6.6 mW at the same voltage. The processor was run at a number of interrupt frequencies and instruction per interrupt rates with the results plotted in Figure 18.9. As expected, the power shows a linear dependency on the effective frequency at high rates, where the active power dominates, while at low rates a floor due to leakage components is evident. The energy per instruction is calculated to be 0.5 nJ. All interrupt and instruction rates fall on the same curve as presented in the figure. Measurements made in "idle" mode, in which the PLL is kept active, no RBB is applied, but clocks that are gated at the PLL generate a relatively high power floor due to PLL power.

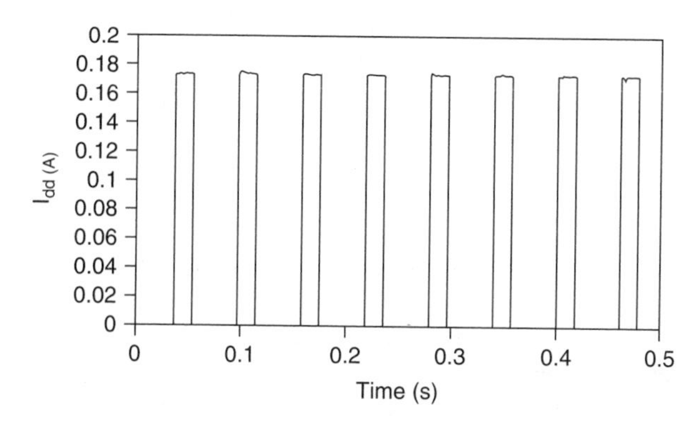

FIGURE 18.8 Current measurement with the time in standby and active modes evident.

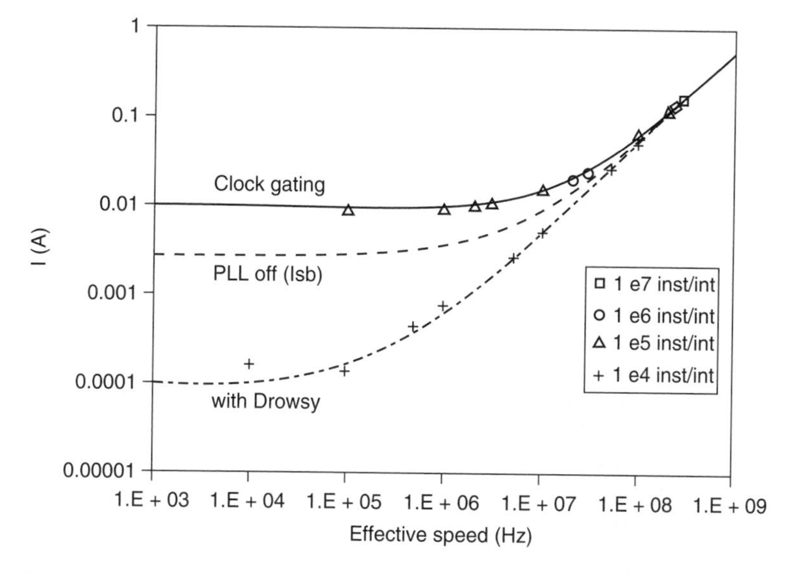

FIGURE 18.9 Current measurement depicting active and leakage power dominated frequencies.

The power savings of using Drowsy mode over clock gating alone is approximately 100×. This is over 25× less than the I_{sb} leakage power floor that would be obtained without Drowsy (e.g., by simply lowering the clock to a very low rate). Power is saved and the response time to external stimulus is improved by running in short bursts at high frequencies. Effective frequency allows direct comparison with devices running at lower frequency demonstrating that the efficacy of TDM Drowsy mode, matching the theoretical curve of Figure 18.1. By raising the V_t with RBB to achieve low standby power, it is combined with improved low voltage and higher maximum performance. The active power improvement can be estimated by considering the V_t increase required to match the I_{off} reduction and the required V_{dd} increase to achieve the same performance. By simulating the circuit metric mentioned previously, calibrated to the measured frequency vs. voltage performance of the microprocessor, a V_t increase of 110 mV (to a typical value of 500 mV) results in the same reduction. At this V_t, the same frequency at $V_{dd} = 0.75$ V is obtained by an increase to 0.86 V demonstrating an active power savings of 24% by using Drowsy mode instead of a higher V_t.

18.6 Process, Voltage, and Temperature Variations

Systematic and random variations in P, V, and T are posing a major challenge to future high-performance microprocessor design [26,27]. Technology scaling beyond 90 nm is causing higher levels of device parameter variations, which are changing the design problem from deterministic to probabilistic [28,29]. The demand for low power and thinner gate oxides causes supply voltage scaling. Then, voltage variations become a significant part of the overall challenge. Finally, the quest for growth in operating frequency is manifested in significantly high junction temperature and within die temperature variation.

18.6.1 Process Variation

Distributions of frequency and standby leakage current (I_{sb}) of microprocessors on a wafer are presented in Figure 18.10. The spread in frequency and leakage distributions is due to variation in transistor parameters, causing about 20× maximum variation in chip leakage and 30% maximum spread in chip frequency. This variation in frequency has led to the concept of "frequency binning" to maximize revenue from each wafer. Note that the highest frequency chips have a wide distribution of leakage, and for a given leakage, there is a wide distribution in the frequency of the chips. The highest-frequency chips with large I_{sb}, and low-frequency chips with too high I_{sb} may have to be discarded, thus affecting yield. Limits to maximum acceptable I_{sb} are dictated by total active power as affordable by cost-effective cooling and current delivery capabilities, as well as idle power required to achieve a target battery life in portable applications. The spreads in standby current and frequency are due to variations in channel length and threshold voltage, both within die and from die to die. That leakages are affected exponentially, while delay is affected approximately linearly is evident in their relative magnitudes. Figure 18.11 illustrates the die-to-die V_t distribution and its resulting chip I_{sb} variation. V_t variation is normally distributed and its 3-σ variation is about 30 mV in an 180-nm CMOS logic technology. This variation causes significant spreads in circuit performance and leakage. The most critical paths in a chip may be different from chip to chip. Figure 18.11 also presents the 20× I_{sb} variation distribution in detail.

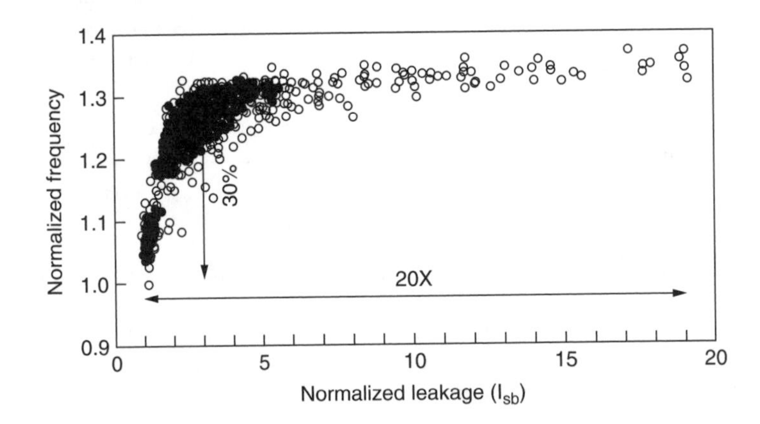

FIGURE 18.10 Leakage and frequency variations.

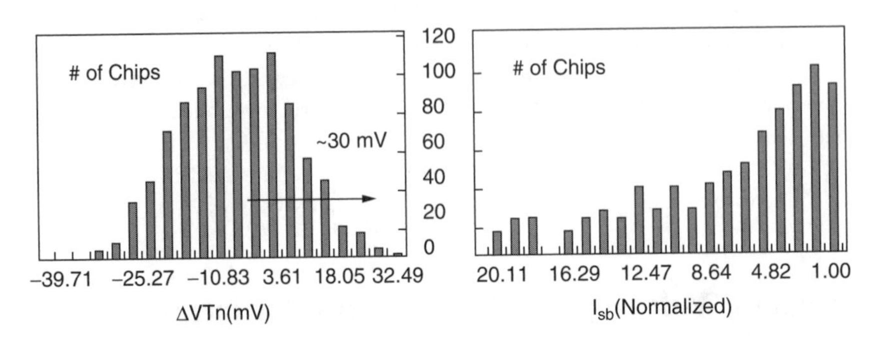

FIGURE 18.11 Die-to-die V_t, I_{sb} variation.

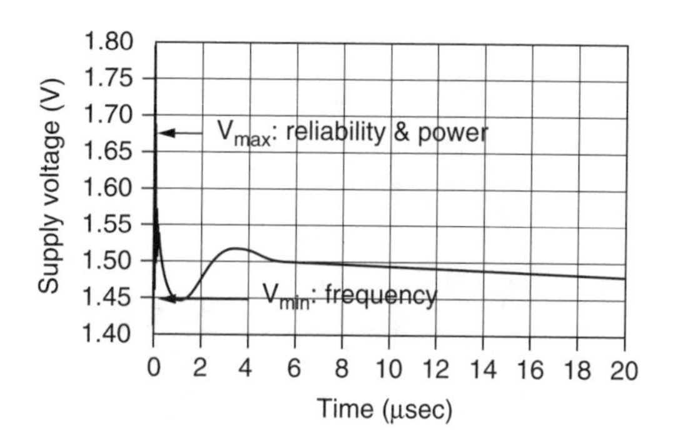

FIGURE 18.12 Supply voltage variation.

18.6.2 Supply Voltage Variation

Uneven and variable switching activity across the die and diversity of the type of logic, result in uneven power dissipation across the die. This variation results in uneven supply voltage distribution and temperature hot spots, across a die, causing transistor subthreshold leakage variation across the die. Supply voltage (V_{dd}) will continue to scale modestly by 15%, not by the historic 30% per generation, first, due to difficulties in scaling threshold voltage and second, to meet the transistor performance goals. Maximum V_{dd} is specified as a reliability limit for a process, and minimum V_{dd} is required for the target performance. V_{dd} variation inside the max–min window is plotted in Figure 18.12. This figure depicts a droop in V_{dd}, when IC current demand changes rapidly, which degrades the performance. This is the result of platform, package, and IC inductances and resistances that do not follow the scaling trends of CMOS process. Specifically, the time "0" point is relatively inactive, while a rapid change in power demand, by the processor leads to the large supply droop pictured. This problem is increased by good low-power design (e.g., clock gating). Power delivery impedance does not scale with V_{dd} and ΔV_{dd} has become a significant percentage of V_{dd}.

18.6.3 Temperature Variation

Figure 18.13 illustrates the thermal image of a leading microprocessor die with hot spots as high as 120°C. Within die temperature fluctuations have existed as a major performance and packaging challenge for

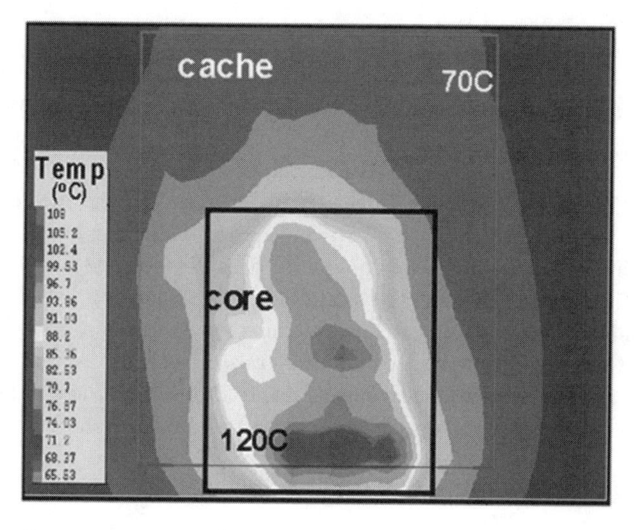

FIGURE 18.13 Within die temperature variation.

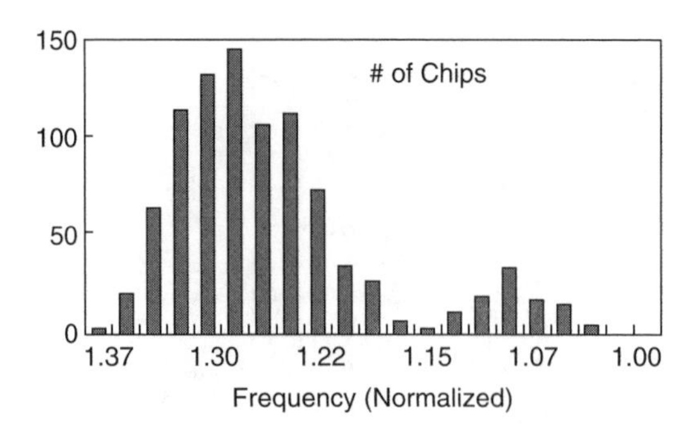

FIGURE 18.14 Die-to-die frequency variation.

many years. Both the device and interconnect performance have temperature dependence, with higher temperature causing performance degradation. Additionally, temperature variation across communicating blocks on the same chip may cause performance mismatches, which may lead to logic or functional failures. Because these thermal variations are the result of uneven local heating, they can be ignored in standby, where lower power dissipation creates minimal heating. Additionally, it can be assumed that the die temperature equals that of the ambient, typically room temperature.

18.7 Variation Impact on Circuits and Microarchitecture

A primary consequence of the P, V, T variation manifests itself as maximum operating frequency (F_{max}) variation. Figure 18.14 presents the distribution of microprocessor dies in 180-nm technology across a frequency range. The data is taken at a fixed voltage and temperature, and thus this F_{max} variation is caused by the process variations discussed previously. This frequency distribution has serious cost implications associated with it — low performing parts need to be discarded which in turn affects the yield and hence the cost. The P, V, T variations consequently impact all levels of design. For instance, products that have only one operating frequency of interest (e.g., networking devices that either do or do not meet a specific standard) must be designed conservatively. Frequently this means designing all circuits to the worst-case P, V, T corner. This section highlights some of the impact that process has on circuit and microarchitecture design choices.

18.7.1 Design Choice Impact

Dual-V_t circuit designs [30,31] can reduce leakage power during active operation, burn-in and standby. Two V_t are provided by the process technology for each transistor. High-V_t transistors in performance critical paths are either upsized or are made low-V_t to provide the target chip performance. Because upsizing has limited benefit in gate-dominated paths, as capacitive load is added at the same rate as current drive, lower V_t can be beneficial. Larger transistor sizes increase the relative probability of achieving the target frequency at the expense of switching power. Increasing low-V_t usage also boosts the probability of achieving the desired frequency, but with a penalty in leakage power. It was demonstrated in Karnik et al. and Tschanz et al. [30,31], that by carefully employing low-V_t devices, 24% delay improvement is possible to trade off leakage and switching power components, while maintaining the same total power. However, a design optimized for lowest power by careful assignment of transistor sizes and V_t values is more susceptible to frequency impact due to within-die variations because they sharpen the path delay distributions making a larger number of paths and transistors critical.

18.7.2 Microarchitecture Choice Impact

The number of critical paths that determine the target frequency vary depending on both microarchitecture and circuit design choices. Microarchitecture designs that demand increased parallelism and/or

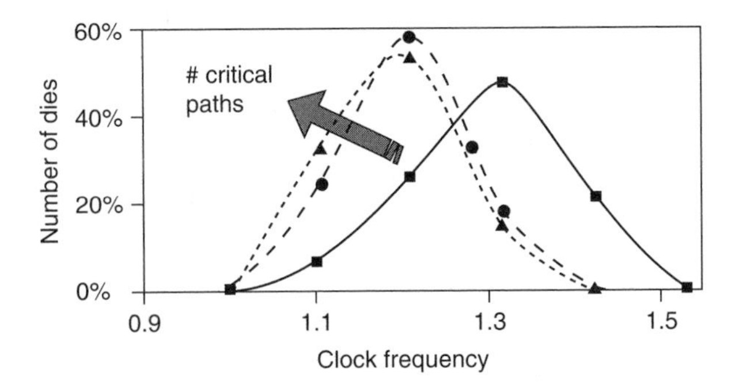

FIGURE 18.15 Die-to-die critical path distribution.

functionality require increase in the number of critical paths. Designs that require deeper pipelining, to support higher frequency of operation, require increase in the number of critical paths and decrease in the logic depth. The impact process variation has on these choices are described next. Test chip measurements in Figure 18.15 demonstrate that as the number of critical paths on a die increases, within-die delay variations among critical paths cause both mean (μ) and standard deviation (σ) of the die frequency distribution to become smaller. This is consistent with statistical simulation results [26] indicating that the impact of within-die parameter variations on die frequency distribution is significant. As the number of critical paths exceeds 14, there is no noticeable change in the frequency distribution. So, microarchitecture designs that increase the number of critical paths will result in reduced mean frequency because the probability that at least one of the paths is slower will increase.

Historically, the logic depth of microarchitecture critical paths has been decreasing to accommodate a 2× growth in the operating frequency every generation, faster than the 42% supported by technology scaling. As the number of logic gates per pipeline stage that determine the frequency of operation reduces, the impact of variation in device parameter increases. Measurement on 49-stage ring oscillators demonstrated that σ of the within-die frequency distribution was 4× smaller than σ of the device saturation current distribution [26]; however, measurements on a test chip containing 16-stage critical paths demonstrate that σ of within die (WID) critical path delay distributions and NMOS/PMOS drive current distributions are comparable. Specifically, NMOS I_{dsat} $\sigma/\mu = 5.6\%$, PMOS I_{dsat} $\sigma/\mu = 3.0\%$, while the 16-stage delay $\sigma/\mu = 4.2\%$. The impact of process variation on the microarchitecture design choices can be summarized as follows: with either smaller logic depth or with increasing number of microarchitecture critical paths, performance improvement is possible. The probability of achieving the target frequency that translates to performance, however, drops due to the impact of within-die process variation.

18.8 Adaptive Techniques and Variation Tolerance

This section describes some of the research and design work to enhance the variation tolerance of circuits and microarchitecture and to reduce the variations by clever circuit and microarchitectural techniques. These techniques expand on those discussed previously by expanding the use of body bias from only RBB to include forward body bias (FBB) to reduce V_t and thereby improve circuit speed. Adjusting the voltage to the optimal required as determined at test time is introduced as another method to increase yield in the presence of variation.

18.8.1 Body Bias Control Techniques

Lowering V_t can improve device performance, with the commensurate increase in leakage and standby current (I_{sb}) as described earlier. One possible method to trade off performance with leakage power is to apply a separate bias to critical devices. In addition to application of RBB to reduce leakage, V_t can be modulated for higher performance by forward body bias (FBB). This method also reduces the impact

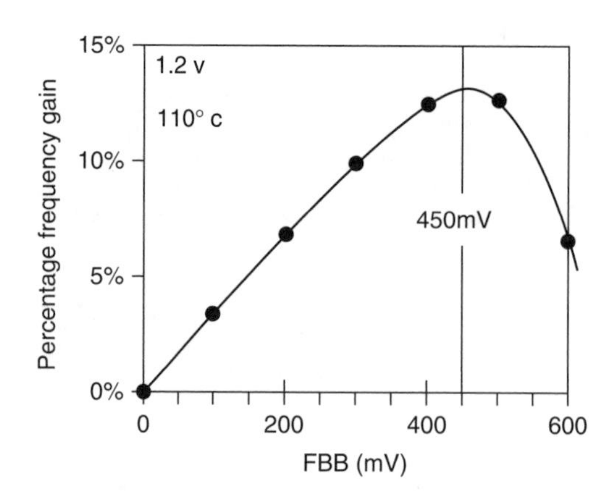

FIGURE 18.16 Optimal FBB for sub-90-nm generations.

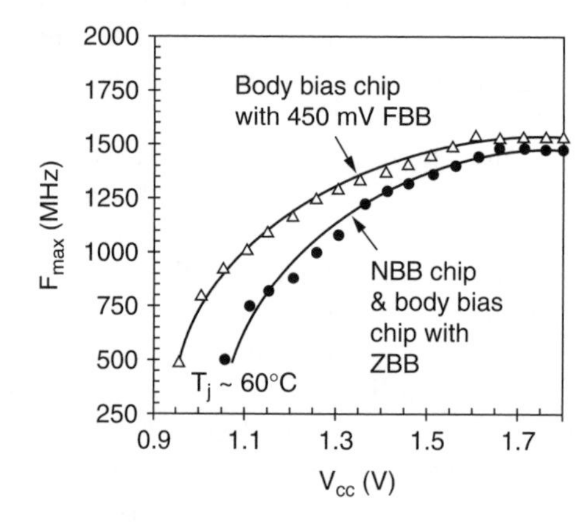

FIGURE 18.17 Forward body bias results.

of short channel effects, hence reducing V_t variations. Figure 18.16 plots the percentage frequency gain as a function of FBB. It was demonstrated empirically that 450 mV is the optimal FBB for sub-90-nm generations at high temperature [32]. A 6.6-M transistor communications router chip [33], with on-chip circuitry to provide FBB to PMOS transistors during active operation and zero body bias (ZBB) during standby mode, was implemented in a 150-nm CMOS technology. Performance of the chip is compared with the original design that has no body bias (NBB) in Figure 18.16. The maximum operating frequency (F_{max}) of the NBB and FBB router chips are compared from 0.9 V to 1.8 V V_{dd} at 60°C (see Figure 18.17). The FBB chip with forward body bias achieves 1GHz operation at 1.1 V, compared with 1.25 V required for the NBB chip, or 23% less switching power at 1 GHz. The frequency of FBB is 33% higher than NBB at 1.1 V. Area overhead supporting ABB was approximately 2%, while the power overhead was 1%.

RBB was also applied to the same device to reduce leakage. Figure 18.18 plots the leakage current for the worst-case channel length (L_{wc} dashed) and the nominal channel length (L_{nom} dotted) as a function of RBB. The measured full-chip leakage current is within these upper and lower leakage current bounds over a range of RBB values. The optimum RBB value derived from the measured chip for minimum leakage is 500 mV [34]. Higher RBB values cause the junction leakage current to increase and overall leakage power to go up because, as in Mizuno et al. [17], the V_{dd} was not collapsed; however, effectiveness of RBB reduces as channel lengths become smaller or V_t is lowered. Essentially, the V_t-modulation capability by RBB weakens as short-channel effects become worse or body effect diminishes due to lower channel doping.

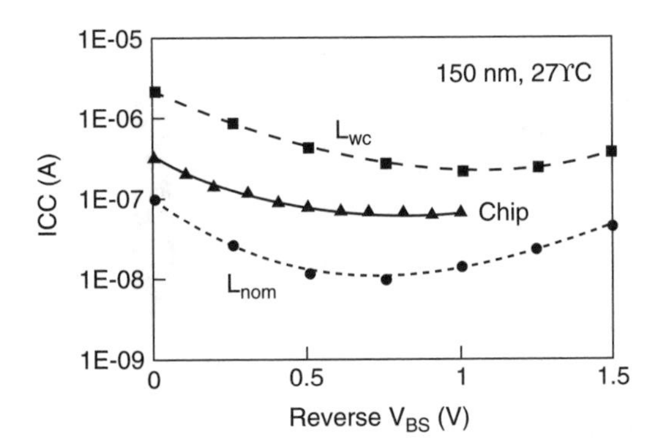

FIGURE 18.18 Leakage reduction by reverse body bias.

18.8.2 Adaptive Body Bias and Supply Bias

The previous two subsections presented the advantages of both FBB and RBB. It is possible to utilize both of these approaches as depicted in Figure 18.19. Due to the frequency spread in fabricated parts caused by process variations, the low frequency parts may be discarded for lower performance and the high frequency parts may be discarded for higher leakage power. As presented on the right side, devices can be adaptively biased to increase the performance of the slow parts by FBB and to decrease leakage power of the fast parts by RBB.

A test chip was implemented in a 150-nm CMOS technology to evaluate effectiveness of the adaptive body bias (ABB) technique for minimizing impacts of both die-to-die and within-die parameter variations on processor frequency and active leakage power [35]. The bias is based on a 5-bit digital code, which provides one of 32 different body bias values with 32mV resolution to PMOS transistors. NMOS body is biased externally across the chip. Bidirectional ABB is used for both NMOS and PMOS devices to increase the percentage of dies that meet both frequency requirement and leakage constraint. As a result, die-to-die frequency variations (σ/μ) reduce by an order of magnitude, and 100% of the dies become acceptable (see Figure 18.20). Bin 2 is the highest frequency bin, while Bin 1 is the lowest acceptable frequency bin — any dies that are slower than Bin 1 are discarded. Almost 50% of dies with NBB fell below Bin 1 but are recovered using ABB. In addition, 30% of the dies are now in the highest frequency bin allowed by the power density limit. WID-ABB (applying multiple bias values per die to compensate for within-die as well as die-to-die variation) reduces σ of the die frequency distribution by 50%, compared with ABB. In addition, almost all the dies are accepted in the highest possible frequency bin, compared with 30% for ABB. Another technique to increase yield in the high frequency bins, is to apply adaptive V_{dd}. Figure 18.21 presents the advantage of adaptive V_{dd} over fixed V_{dd}. Bin 3 is the highest

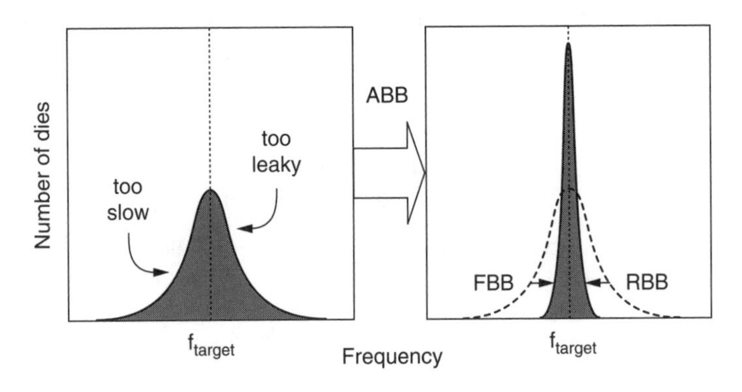

FIGURE 18.19 Target frequency binning by adaptive body bias.

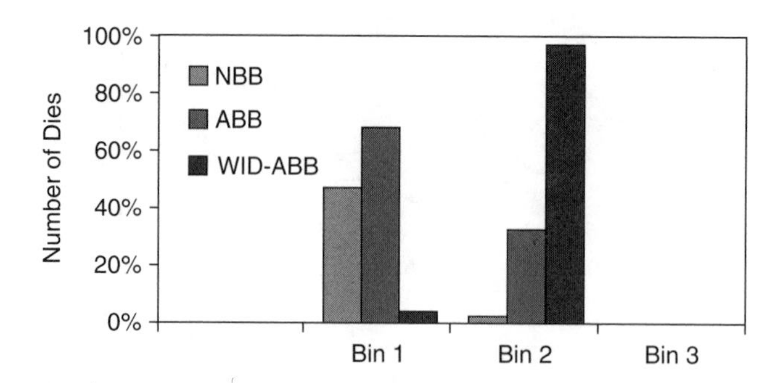

FIGURE 18.20 Adaptive body bias results.

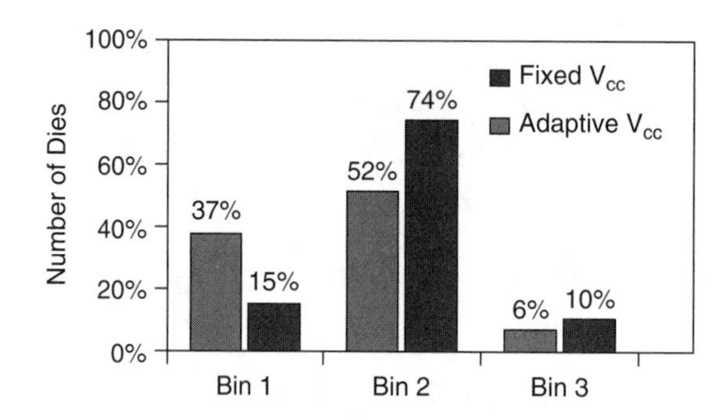

FIGURE 18.21 Bin improvement by adaptive V_{cc}.

frequency bin, while Bin 1 is the lowest acceptable frequency bin. The dark bars indicate that adaptive V_{dd} (V_{cc} in the figure) has pushed more than 20% dies from Bin 1 to Bin 2 and even Bin 3, as well as recovered those dies that fell below Bin 1.

18.9 Dynamic Voltage Scaling

Although adapting the power supply voltage to manufacturing variation was introduced previously, it may also be used to adjust power usage dynamically to the workload at hand. This section describes the dynamic variation of the power supply voltage V_{dd} appropriate to the instantaneous workload of the integrated circuit, commonly described as dynamic voltage scaling (DVS) or dynamic voltage management (DVM) [36]. The results are from system level measurements performed on the 80200 microprocessor. The basic premise is to adjust the frequency and voltage of the device to the lowest values that will simultaneously meet the required application throughput and the operating envelope of the processor. If the performance voltage curve of the device is violated (i.e., a circuit critical path is provided insufficient voltage to meet its timing constraints) then a circuit failure will occur. This implies that changing voltage and clocks must conform to two rules:

1. V_{dd} must be adjusted upward before initiating a frequency increase.
2. Frequency must be lowered before adjusting V_{dd} downward.

The power, voltage, and frequency measured on the processor are plotted in Figure 18.22. The large power range obtainable using DVM is evident, ranging from 6 mA with the clocks gated off to 1.4 W at 1 GHz.

18.9.1 Clock Generation

In conventional designs, the PLL must be allowed to relock to the new frequency when a new frequency is chosen because a divided version of the core clock itself is compared with the reference, as illustrated

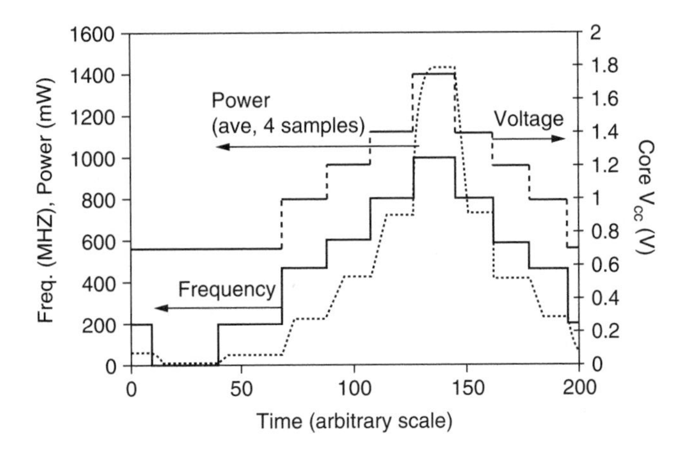

FIGURE 18.22 Frequency, voltage, and power vs. time using DVM.

in Figure 18.23(a). Because the PLL may generate clocks that are shorter than the chosen frequency during this time, clocks to the logic core must be gated off while relocking the PLL. The PLL relock time is predictable and often fixed (by comparing with a counter representing the worst-case lock time — 20 μs previously) to simplify specification and testing. V_{dd} adjustment and initiation of a frequency change may be coincident if the time to change V_{dd} is predictable and consistent (e.g., slew rate limited). In this case, the time to reach the specified voltage is dependent on the starting voltage. This clock change time introduces latency to achieving the lower power. An energy cost also occurs from moving the highly capacitive supply voltage. The latter is unavoidable, but the former can be mitigated in two ways. First, the processor can be supplied with the reference clock during clock changes. This can maintain some, generally lower performance, but allows computation to continue and avoids a "dead zone" where interrupts cannot be taken. Second, a more sophisticated PLL divider scheme allows "on the fly" changes in clock rate.

This approach, illustrated in Figure 18.23(b) keeps the PLL running at a consistent maximum frequency for all voltage and frequency configurations. This requires a separate power supply connection for the PLL, which is virtually required to keep the analog PLL supply isolated from noise. Typically, this supply is separated and is additionally filtered or regulated to improve the clock jitter component due to supply

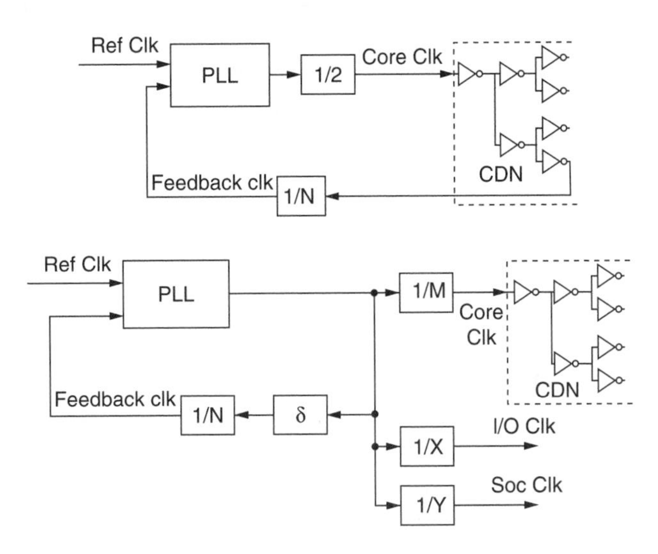

FIGURE 18.23 (a) Conventional PLL and clock generation, and (b) scheme to allow speed changes without performance penalty. The 1/M divider in (b) is dynamically adjustable.

noise. Here, the PLL supply is not scaled with the logic core. The PLL power is not strongly dependent on the VCO frequency and is a small fraction of the overall active power, so the penalty of not scaling the PLL V_{dd} is acceptable. Given the clock performance benefit accrued by regulating the PLL supply, a fixed PLL supply is the preferred approach.

In a conventional design, the core or IO clock is fed back to the phase-frequency detector (PFD) of the PLL as depicted in Figure 18.23(a). Because the point is to lock the internal clock edges to the external reference clock, this feedback is from the end of the clock distribution network (CDN) to include the insertion delay. The only insertion delay to match is that of the feedback divider. To allow on-the-fly clock changes, the clock dividers are configured as depicted in Figure 18.23(b). The feedback divider reduces the VCO frequency to that of the reference clock independent of the core clock divisor chosen. The core clock divisor can then be changed dynamically within certain constraints. First, no clock glitches can be allowed. Second, the clock changes must be predictable to allow consistent behavior when transferring data across domains, as well as for testing and validation. More dividers to other clock domains are likely in a large SoC design, as discussed. It is important to have the same insertion delay for all of the clocks, so that the version returning to the PLL tracks the others. In practice, some latitude in insertion delays can be allowed, which will show up as systematic additions to the off-chip or inter-domain clock skew. Finally, the mechanism for crossing from one domain to another must be independent of the actual frequencies. In practice, this is provided by generating separate signals from the same clock divider circuits, which anticipate coincident edges constituting allowed domain crossings. Ideally, the maximum core speed is half that of the PLL voltage controlled oscillator (VCO) because a 2× divide easily provides a 50:50 duty cycle clock.

18.9.2 Experimental Results

The important factor in implementing dynamic voltage management is the amount of work performed (e.g., the number of instructions required) instead of the number of clocks and some indicator of whether or not the machine is busy. Modern systems use an interrupt-driven model, whereby the processor will enter a low-power state, pending being awakened via interrupt, when there is no useful work to perform. Many operations are available, however, particularly memory accesses and IO, which have significant latency. Increasing core to memory frequency ratios exacerbates this. Consequently, to effectively utilize DVM, it is necessary to detect when the processor is constrained by such operations, which effectively limit the number of instructions per clock (IPC) below the peak value.

As an example, two experiments were performed using an 80200 running a modified Linux operating system (OS) kernel to monitor the work performed two different ways. The first merely determined if the scheduled tasks had been completed early, while the second determined the actual IPC using the on-core performance monitors. The interval between adjustments was 10 ms.

The experiments performed the following at the end of each interval:

1. OS only using time-slice utilization
```
If (task finished early)
Lower the voltage and frequency
Else
Raise the voltage and frequency.
```
2. OS using time slice utilization and core performance monitors (Sampled number of instructions executed and number of data dependencies every 2 ms):
```
If (work performed increases)
Raise voltage and frequency
Else
Lower the voltage and frequency
```

In the first case, whether or not a task completed early provided a coarse assessment of the needed computational power, and it was assumed that the future demands would be similar. In the second case, the actual work was determined. It can be inferred that the latter approach may also provide a more quantitative estimate of how much the frequency should be raised and lowered. It also allows the power to be lowered, essentially matching the processor clock to memory or system clock ratios more appropriately to a given workload, automatically detecting and minimizing the energy consumption in memory bound cases.

18.10 Conclusions

Higher digital IC power and parameter variations are an inevitable consequence of scaling and promise to increase further in the future. This chapter has described some of the presently important as well as emerging limiting mechanisms. Various design techniques that mitigate these issues were discussed. These techniques rely on leveraging the often-neglected bulk terminal as well as careful selection of the supply voltage to both the specific device as manufactured as well as the computing task at hand. It has been demonstrated that the overhead of using these techniques, although nonnegligible, is modest. As transistor scaling forces future products into increasingly difficult cost, power, and performance trade-offs, we can expect to see greater reliance on these, as well as other design schemes to enable still further scaling.

References

[1] S. Borkar, Obeying Moore's law beyond 0.18 micron, *Proc. 13th Annu. ASIC/SOC Conf.*, pp. 13–16, 2000.

[2] L. Clark et al., An embedded 32b microprocessor core for low-power and high-performance applications, *IEEE J. Solid-State Circuits,* 36, p. 1599, 2001.

[3] K. Chen and C. Hu, Performance and V_{dd} scaling in deep submicrometer CMOS, *JSSC,* 33, pp. 1586–1589, Oct. 1998.

[4] Y. Taur et al., *Fundamentals of Modern VLSI Devices,* Cambridge University Press, U.K., 1998.

[5] A. Keshavarzi, S. Narenda, S. Borkar, C. Hawkins, K. Roy, and V. Dey, Technology scaling behavior of optimum reverse body bias for leakage power reduction in CMOS ICs, *Proc. ISLPED,* pp. 252–254, 1999.

[6] R. Krishnamurthy et al., High-performance and low-power challenges for sub-70-nm microprocessor circuits, *CICC Proc.,* pp. 125–128, 2002.

[7] H. Wong, D. Frank, P. Solomon, H. Wann, and J. Welser, Nanaoscale CMOS, *Proc. IEEE,* 87, pp. 537–570, 1999.

[8] S. Wolf, *Silicon Processing for the VLSI Era: Volume 3 — The Submicron MOSFET,* Lattice Press, Sunset Beach, CA, 1995.

[9] R. Gonzalez, B. Gordon, and M. Horowitz, Supply and threshold voltage scaling for low-power CMOS, *IEEE J. Solid-State Circuits,* 32, pp. 1210–1216, Aug. 1997.

[10] D. Frank, Power constrained CMOS scaling limits, *IBM J. Res. Dev.,* 46, 2/3, p. 235, 2002.

[11] S. Thompson, Technology performance: trends and challenges, IEDM short course, *IEDM '99 Tutorial,* Washington, D.C., 1999.

[12] H. Holma and A. Toskala, Eds., *WDCMA for UMTS: Radio Access for Third-Generation Mobile Communications,* John Wiley & Sons, New York, 2001.

[13] S. Thompson, I. Young, J. Greason, and M. Bohr, Dual threshold voltages and substrate bias: keys to high performance, low-power 0.1-μm logic designs, *VLSI Tech. Symp. Dig.,* pp. 69–70, 1997.

[14] M. Horiguchi, T. Sakata, and K. Itoh, Switched-source-impedance CMOS circuit for low standby subthreshold current giga-scale LSIs, *IEEE JSSC,* 28, pp. 1131–1135, Nov. 1993.

[15] B. Sheu, D. Scharfetter, P. Ko, and M. Jeng, BSIM: berkeley short-channel IGFET model for MOS transistors, *IEEE JSSC,* 22, pp. 558–566, Aug. 1987.

[16] L. Clark, N. Deutscher, S. Demmons, and F. Ricci, Standby power management for a 0.18-μm microprocessor, *Proc. ISLPED,* pp. 7–12, 2002.

[17] H. Mizuno et al., An 18-μA standby current 1.8-V, 200-MHz microprocessor with self-substrate-biased data-retention mode, *IEEE J. Solid-State Circuits,* 34, 1999, p. 1492.
[18] S. Yang et al., A high-performance 180-nm generation logic technology, *Proc. IEDM,* pp. 197–200, 1998.
[19] M. Morrow, Microarchitecture uses a low-power core, *IEEE Computer,* p. 55, April, 2001.
[20] S. Mutoh, T. Douseki, Y. Matsuya, T. Aoki, S. Shigematsu, and J. Yamada, 1-V power supply high-speed digital circuit technology with multithreshold-voltage CMOS, *IEEE J. Solid-State Circuits,* 30, 1995, pp. 847–854, Aug. 1995.
[21] L. Clark, A high-voltage output buffer fabricated on a 2-V CMOS technology, *VLSI Circuit Symp. Dig.,* pp. 61–62, 1999.
[22] R. Swanson and J. Meindl, Ion-implanted complementary MOS transistors in low-voltage circuits, *IEEE JSSC,* SC-7, pp. 146–153, April, 1972
[23] L. Clark, M. Morrow, and W. Brown, Reverse body bias for low effective standby power, *IEEE Trans. VLSI,* Sept. 2004.
[24] BRH Reference Platform specifications are available at http://www.adiengineering.com/products-BRH.html.
[25] M. Osqui, Evaluation of software energy consumption on microprocessors, Master's thesis, Massachusetts Institute of Technology, Cambridge, MA, Oct. 2001.
[26] K. Bowman et al., Impact of die-to-die and within-die parameter fluctuations on the maximum clock frequency distribution for gigascale integration, *IEEE J. Solid-State Circuits,* 37, pp. 183–190, Feb. 2002.
[27] S. Borkar, Parameter variations and impact on circuits and microarchitecture, *C2S2 MARCO Review,* March 2003.
[28] G. Sery et al., Life is CMOS: why chase the life after? *DAC 2002,* pp. 78–83.
[29] T. Karnik et al., Sub-90nm technologies — challenges and opportunities for CAD, *ICCAD 2002,* pp. 203–206.
[30] T. Karnik et al., Total power optimization by simultaneous dual-V_t allocation and device sizing in high performance microprocessors, *DAC 2002,* pp. 486–491.
[31] J. Tschanz et al., Design optimizations of a high-performance microprocessor using combinations of dual-V_t allocation and transistor sizing, *VLSI Circuits Symp. 2001,* pp. 218–219.
[32] J. Tschanz et al., Dynamic-sleep transistor and body bias for active leakage power control of microprocessors, *ISSCC 2003,* pp. 102–103.
[33] S. Narendra et al., 1.1-V 1-GHz communications router with on-chip body bias in 150-nm CMOS, *ISSCC 2002,* pp. 270–271.
[34] A. Keshavarzi et al., Effectiveness of reverse body bias for leakage control in scaled dual V_t CMOS ICs, *ISLPED 2001,* pp. 207–210.
[35] J. Tschanz et al., Adaptive body bias for reducing impacts of die-to-die and within-die parameter variations on microprocessor frequency and leakage, *ISSCC 2002,* pp. 422–423.
[36] T. Burd, T. Pering, A. Stratakos, and R. Broderson., A dynamic voltage scaled microprocessor system, *IEEE J. Solid-State Circuits,* 35, pp. 1571–1580, Nov. 2000.

19

Low-Power DSPs

Ingrid Verbauwhede
University of California—Los Angeles

19.1 Introduction

Mobile wireless communications show an incredible growth, as illustrated in Figure 9.1. It is estimated that by the year 2010, wireless phones will surpass wire line phones, each having a worldwide penetration of more than 20%. The market for digital signal processors (DSPs) has a growth rate of 40%. In 1966, it was a $2 billion market, and by 1999 it had grown to a $4.4 billion market. After a dip in 2001–2002, the forecast for 2004 is $7.7 billion and a predicted $17 billion by 2008 [28]. More than 60% of all DSP shipments are used in cellular phones [28]. In the industrialized world, the numbers are even more impressive: in a small country like Belgium with a population of 10 million, more than 2 million cell phones are sold every year compared with approximately 600,000 PCs [6].

Power optimization can be done at several levels of abstraction: technology level, circuit level, gate level, architectural level, algorithm level, and system level. Multiple chapters in this book are devoted at each of these abstraction levels: at the technology level is the usage of multiple threshold voltages, a low V_t for the logic circuits and a high V_t for the memory circuits. At the circuit level, a designer has the choice of using complementary static CMOS instead of high-speed dynamic logic. At the logic level, gated clocks and power down of unused modules will reduce the power consumption. At the architectural level, an optimization of the processor components such as the datapath and the memory architecture will reduce power. At the system level, the selection of variable voltages, idle and sleep modes, etc. will contribute to the reduction of power.

The focus of this chapter is on the power and energy reduction because of optimizations at the architectural and micro-architectural level. Indeed, by tuning the processor components to the application field, a huge amount of power can be saved. This means *all* processor components, and includes the datapaths, the memory architecture, the bus network, and the control architectures, which includes instruction set design.

The first successful DSP processors were introduced in the early 1980s. Many good overview papers are available that describe the evolution of these processors and the special features to support signal processing applications [10,16,17]. Examples in this category are the Texas Instruments TMS320C1x, C2x, C5x, series or the Lucent DSP16A and DSP1600 series. This chapter focuses on the evolution of

0-8493-1941-2/05/$0.00+$1.50

DSP processors during the last couple of years, especially the special features in the processors to support the demands from wireless communications.

Until recently, the same DSP processors are used both in the mobile terminal (i.e., the actual cell phone) and in the base stations; however, a trend starts to emerge to place different processors in the mobile terminal and the base stations. The main drivers for the processors in the mobile are cost and very low energy consumption. This leads to processors that have a very compact but complex instruction set (CISC) and work with domain-specific or application-specific coprocessors. High-performance processors need to be included in the base stations, and they tend to become more compiler-friendly because the software complexity requires it. Thus, the success of very large instruction word (VLIW) processors and modified VLIW processors for the base station applications.

It is insightful to first define the meaning of million instructions per second (MIPS) and million operations per second (MOPS). Most traditional DSP processors belong to the class of CISC processors. This means that in 1 instruction, typically 16 bits wide, several operations, sources, and destinations for the operations are coded. For instance, in one dual-multiply-accumulate instruction of the Lode processor, six different operations are performed: two memory-read operations, two address calculations, and two multiply-accumulate (MAC) operations [30]. Assuming the processor runs at 100 MHz, this corresponds to 100 MIPS and 600 MOPS. If the multiply and adds are considered two operations, this becomes 800 MOPS. Similarly, in one dual MAC instruction on the Lucent 16210, seven different operations are executed: one three-input addition, two multiplications, two memory reads, and two address pointer updates. This corresponds to 700 MOPS.

CISC type processors are usually compared on the amount of MIPS. Sometimes, to make things confusing, the two multiply-accumulate operations are counted separately (usually by marketing or sales people). Therefore, it might be a 100-MHz processor, advertised for "200 MIPS."

One instruction of a VLIW processor consists of a set of small (e.g., six or eight), primitive instructions, issued in parallel. It is customary to multiply the clock frequency of these processors by the number of parallel units and define these as MIPS or MOPS. The processor described in Weiss et al. [33] uses a VLIW variation combined with SIMD properties, to reach 3000 MOPS with a 100 MHz clock. The processor in Igura et al. [14] runs at 50 MHz and is described as an 800-MOPS solution.

To make a fair comparison between processors, we will use the MIPS terminology when referring to the clock frequency and count the primitive operations for both the CISC and VLIW machines as MOPS.

A second insight is the means of measuring performance of DSP processors. Instead of comparing processors based on GHz or MOPS, DSP processors are usually compared on the number of instructions to get the job done. Therefore, the goal is to minimize the number of instructions, also expressed in MIPS, (to make it even more confusing). For instance, the MIPS for several speech coding standards on a SH-DSP are reported in Baji et al. [5]: the simplest full-rate GSM speech codec requires 3.1 MIPS. A half-rate coder already requires 23 MIPS.

Section 19.2 introduces the driving application — in this case, wireless mobile communications. Section 19.3 identifies the most important computation-intensive functions, and gives the DSP approach for a low-power solution. Section 19.4 discusses the integration of DSP processors and coprocessors in systems on chip (SoCs). Conclusions are formulated in Section 19.5.

19.2 The Application Driver

DSP processors are made to support hard real-time signal processing applications. This translates in the rule that 10% of the code is executed 90% of the time, and 90% of the code is executed 10% of the time. The code that is executed all the time tends to sit in tight loops, of which every instruction or clock cycle counts. DSP processors are compared based on the number of instructions and the number of clock cycles it takes to execute basic DSP kernels.

The main building blocks of a wireless terminal are depicted in Figure 19.1. The computation-intensive functions can be subdivided in two main categories. The first is associated with the communication

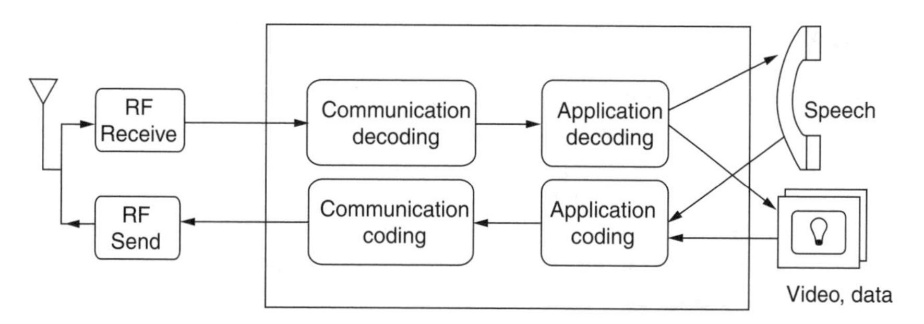

FIGURE 19.1 Application overview.

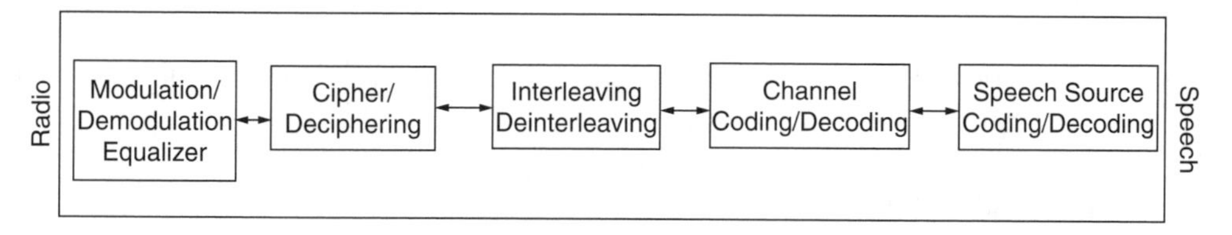

FIGURE 19.2 DSP functions of a second-generation communication system.

processing, also called baseband processing. The second is associated with the application processing, also called source coding. The main baseband building blocks of second-generation cellular phones, such as for GSM, GSM+, and IS-95, are illustrated in Figure 19.2 [11,12,23]. About half of the processing functions are at the physical layer, implementing the modulation/demodulation, the equalizer, and the channel coder and decoder. The other half of the processing occurs at the application level. For second-generation phones, this means the speech coder.

All functions of the system in Figure 19.2 can be implemented in one state-of-the-art DSP processor running at a clock frequency between 80 MHz to 150 MHz. The differentiation between the processors and implementations sits in either the power consumption or the extra features that are included in the processor, such as noise cancellation or equalizers that are more advanced.

Third-generation (3G) cellular wireless standards put higher demands on the modem functions as well as the application functions. 3G systems support not only speech, but also data, image, and video communication. These advanced applications require more processing power from the DSP. At the same time, the advanced applications put higher demands on the coding algorithms, requesting improved bit-error-rates. Thus, equalizers that are more advanced as well as coding algorithms that are more advanced, such as turbo coding algorithms, are used. This is combined with a higher bandwidth requirement.

The blocks with the largest computational requirements are the following:

- Filters (FIR, IIR), autocorrelations, and other "traditional" signal processing functions.
- Convolutional decoders based on the Viterbi algorithm.
- To support data processing requirements in 3G systems, turbo coders are introduced.
- On the application side, efficient codebook search and max–min search, etc. for speech coders and vector search algorithms are required.
- Image and video decoding is the next highly computation-intensive function requiring efficient implementation. Major examples are JPEG and MPEG.

The next subsections discuss how different DSP processors have special architecture features to support the most commonly required computational building blocks. Some of these features are tightly coupled to the DSP processor architecture and integrated in the instruction set. We call these tightly coupled coprocessor units. Some features run on separate building blocks through a bus or memory mapped interface. In this case, jobs are delegated to the coprocessors. We call these loosely coupled coprocessors.

19.3 Computation-Intensive Functions and DSP Solutions

Power consumption in CMOS circuits is mainly dynamic switching power (assuming that the leakage power is well under control, which is a separate topic). Thus, the goal is to avoid unnecessary switching in the processor and limiting the switching to actions necessary to create the outcome of the algorithm. As an example, a multiplication and addition operation is fundamental to calculate an FIR filter, assuming that the multiplication can be done without glitching power; however, the instruction read, decode, and memory accesses can be considered as "overhead." This overhead is not present in a full custom application specific integrated circuit (ASIC) that only performs an FIR filter.

A processor has four fundamental components: datapaths to calculate the algorithm and three supporting building blocks, including control (e.g., all the instruction read/decode logic), storage, and interconnect. To reduce power in a DSP processor, one should look at the supporting processor blocks and reduce or tune them toward the application domain. This will reduce the unnecessary overhead power.

This concept is illustrated in the next section, with several computationally intensive functions running on DSP processors.

19.3.1 FIR Implementation

The basic equation for an N tap FIR equation is the following:

$$y(n) = \sum_{i=0}^{i=N-1} c(i) \cdot x(n-i)$$

When this equation is executed in software, output samples *y(n)* are computed in sequence. This means that to compute one output sample, there are N multiply–accumulate operations and 2N memory read operations to fetch the data and the coefficients. N is the number of taps in the filter. It is well-known that DSP processors include datapaths to execute multiply accumulate operations in an efficient way [17]. Therefore, we focus on the memory architecture, which is a much more fundamental design issue for DSP processors.

19.3.1.1 Memory Architectures

On a traditional von Neumann architecture, 3N access cycles are needed to compute one output: for every tap one needs to fetch one instruction, read one coefficient, and read one data sample sequentially from the unified memory space. Already early on, DSP processors were differentiated from von Neumann architectures because they implemented a Harvard or modified-Harvard architecture [16,17]. The main characteristic is the use of two memory banks instead of one common memory space in the von Neumann architecture. The Harvard architecture has a separate data memory from program memory. This reduces the number of sequential access cycles from three to two because the instruction fetch from the program memory can be done in parallel with one of the data fetches. The modified Harvard architecture improves this even further. It is combined with a "repeat" instruction. In this case, one multiply–accumulate instruction is fetched from program memory and kept in the one instruction deep instruction cache. Then, the data access cycles are performed in parallel: the coefficient is fetched from the program memory in parallel with the data sample being fetched from data memory. This architecture is found in all early DSP processors and is the foundation for all following DSP architectures. It is an illustration of the "tuning" of the processor components to the application, in this case the memory architecture and the control logic.

The newer generation of DSP processors has even more memory banks, accompanying address generation units and control hardware, such as the repeat instruction, to support multiple parallel accesses. The execution of a 32-tap FIR filter on the dual MAC architecture of the Lucent DSP 16210 is depicted in Figure 19.3. The corresponding pseudo code is the following:

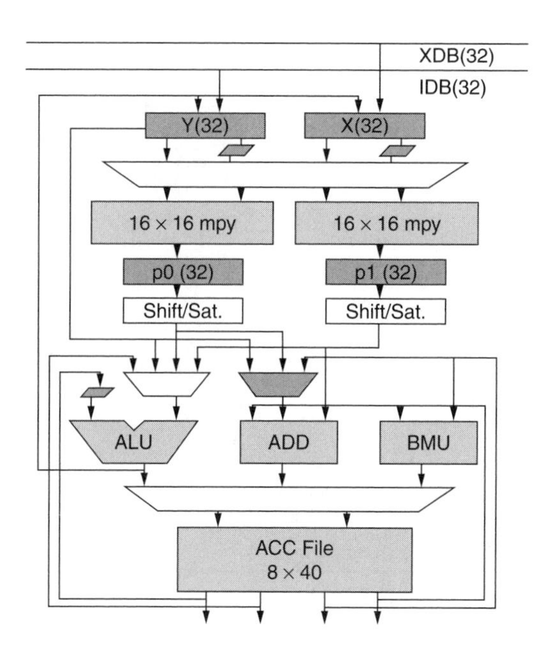

FIGURE 19.3 Lucent/Agere DSP16210 architecture.

```
do 14 {//one instruction !
a0=a0+p0+p1
p0=xh*yh p1=xl*yl
y=*r0++ x=*pt0++
}
```

This code can be executed in 19 clock cycles with only 38 bytes of instruction code. The inner loop takes one cycle to execute and as can be seen from the assembly code, seven operations are executed in parallel: one addition with three inputs, two multiplications, two memory reads, and two address pointer updates.

The difficult part in the implementation of this tight loop is the arrangement of the data samples in memory. To supply the parallel datapaths, two 32-bit data items are read from memory and stored in the X and Y register, as illustrated in Figure 19.3. Then, the data items are split in an upper half and a lower half and supplied to the two 16 × 16 multipliers in parallel. It requires a correct alignment of the data samples in memory, which is usually tedious work done by the programmer because compilers are not able to handle this. A similar problem exists in single instruction multiple data (SIMD) instructions on general-purpose microprocessors. If the complete word length of the memory locations is used, it requires a large effort from the programmer to align the smaller subwords (e.g., at the byte level) into larger words (e.g., 32-bit integers). A similar data alignment approach is used in Kabuo et al. [15]. Instead of two multipliers, only one multiplier working at double the frequency is used, but the problem of alignment of data items in memory remains. This approach will not reduce the total amount of bits read from memory; only the number of instructions (control overhead) is reduced.

To reduce the amount of data read from memory, more local reuse of the data items is needed. This is illustrated with the Lode architecture [30]. In this example, a delay register is introduced between the two MAC units as illustrated in Figure 19.4. This halves the amount of memory accesses. Two output samples are calculated in parallel as indicated in the pseudo code of Table 19.1. One data bus will read the coefficient from memory; the other data bus will read the data sample from memory. The first MAC will compute a multiply-accumulate for output sample $y(n)$. The second multiply–accumulate will compute in parallel on $y(n + 1)$. It will use a delayed value of the input sample. In this way, two output samples are computed at the same time.

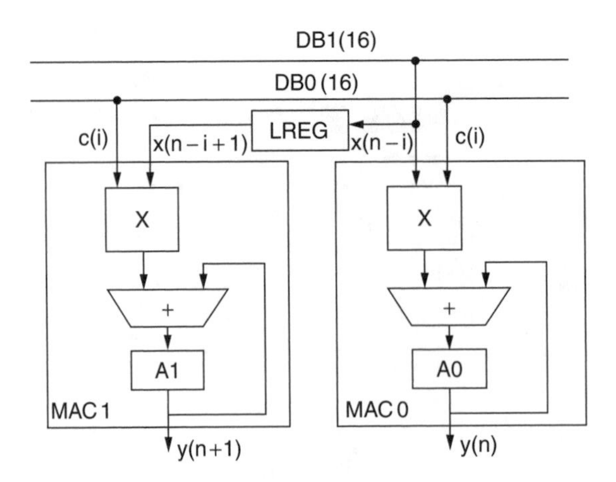

FIGURE 19.4 Lode's dual MAC architecture with delay register.

TABLE 19.1 Pseudo Code for FIR Implementation

```
y(0) = c(0)x(0) + c(1) x(-1) + c(2) x(2) + ... + c(N-1) x(1-N):
y(1) = c(0) x(1) + c(1) x(0) + c(2) x(-1) + ... + c(N-1) x2-N):
y(2) = c(0) x(2) + c(1) x(1) + c(2) x(0) + ... + c(N-1) x(3-N):
...
y(n) = c(0) x(n) + c(1) x(n-1) + c(2) x(n-2) + ... + c(N-1) x(n-(N-1)):
```

This concept of inserting a delay register can be generalized. When the datapath has P MAC units, P-1 delay registers can be inserted and only $2N/(P + 1)$ memory accesses are needed. These delay registers are pipeline registers and thus if more delay registers are used, more initialization and termination cycles need to be introduced.

The TI TMS320C55x [24] is a processor with a dual MAC architecture and three 16-bit data busses. To supply both MACs with coefficients and data samples, the same principle of computing two output samples at the same time is used. One data bus will carry the coefficient and supply this to both MACs, the other two data busses will carry two different data samples and supply this to the two different MACs [3]. Figure 19.5 illustrates this.

Table 19.2 summarizes the different implementations. Note that most energy savings are first obtained from reducing the amount of memory accesses and, second, from reducing the number of instruction cycles. Both are considered overhead. Indeed, the total energy associated with the MAC operations is fixed because an N tap FIR filter requires N multiply-accumulate operations. A dual MAC computes two

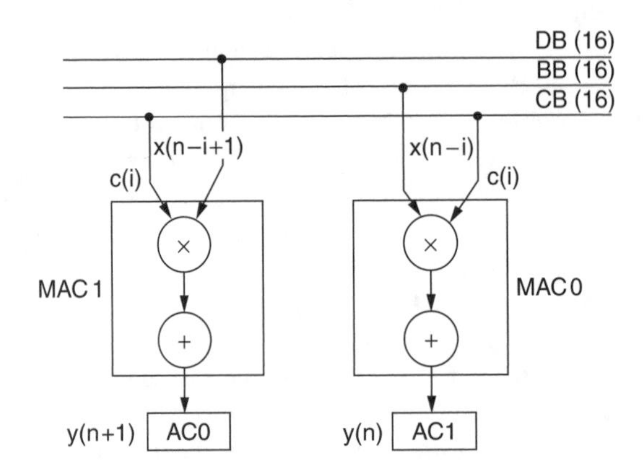

FIGURE 19.5 Dual MAC with three data buses.

TABLE 19.2 Energy Evaluation for an N Tap FIR Filter

DSP	Data Memory Access	MAC Operations	Instruction Cycles	Instructions
Von Neumann	2N	N	3N	2N
Harvard	2N	N	2N	2N
Modified Harvard	2N	N	N	2 (repeat instruction)
Dual Mac	2N	N	N/2	2 (same)
Dual Mac with 3 data busses	1.5N	N	N/2	2
Dual Mac with 1 delay reg	N	N	N/2	2 (same)
Dual Mac with P delay reg	2N/(P + 1)	N	N/(P+1)	2

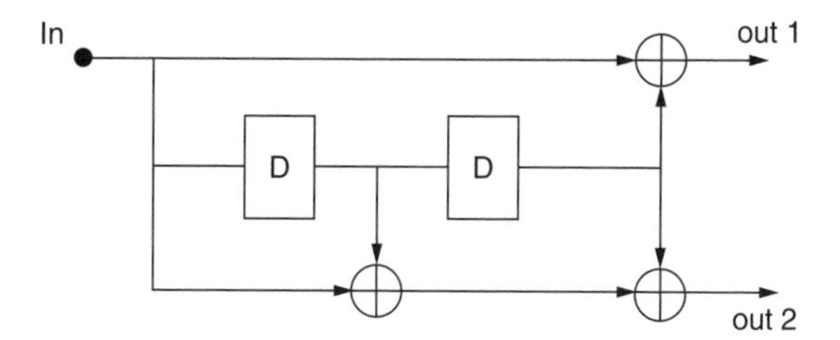

FIGURE 19.6 Example convolutional coder.

MAC operations in parallel and thus the instantaneous power could even be higher in this case. Of importance for battery-operated handsets is, however, the total energy drawn from the supply.

19.3.2 Viterbi Acceleration

The Viterbi decoders are used as forward error correction (FEC) devices in many digital communication devices, not only in cellular phones but also in digital modems and many consumer appliances that require a wireless link. The Viterbi algorithm is a dynamic programming technique to find the most likely sequence of transitions that a convolutional encoder has generated.

Most practical convolutional encoders are rate 1/n (which means that one input bit generates n coded output bits). A convolutional encoder of "constraint length K" can be represented as a finite state machine (FSM) with K-1 memory bits. This means that the FSM has 2^{K-1} possible states, also called trellis states. If the input is binary, there are two possible next states starting from a current state because the next state is computed from the current state and the input bit. This is illustrated in Figure 19.6 with a simple example of a coder with constraint length K = 3, number of states 4. The generator function is $G(D) = [1 + D^2\ 1 + D + D^2]$.

The task of the Viterbi decoding algorithm is to reconstruct the most likely sequence of state transitions based on the received bit sequence. This approach is called the "most likelihood sequence estimation." To compute this most likely path, a trellis diagram is constructed, as illustrated in Figure 19.7. It will compute from every current state, the likelihood of transitioning to one out of two next states.

This leads to the kernel of the Viterbi algorithm, called the Viterbi butterfly. From two current states, two next states are reached. The basic equations executed in this butterfly are:

$$d(2i) = \min\{d(i) + a, d(i + s/2) - a\}$$

$$d(2i + 1) = \min\{d(i) - a, d(i + s/2) + a\}$$

19.3.2.1 Memory Architecture

For power and performance efficiency, DSP processors will include special logic for an efficient implementation of these two equations, mostly called an "add-compare-select" (ACS) operation. One needs

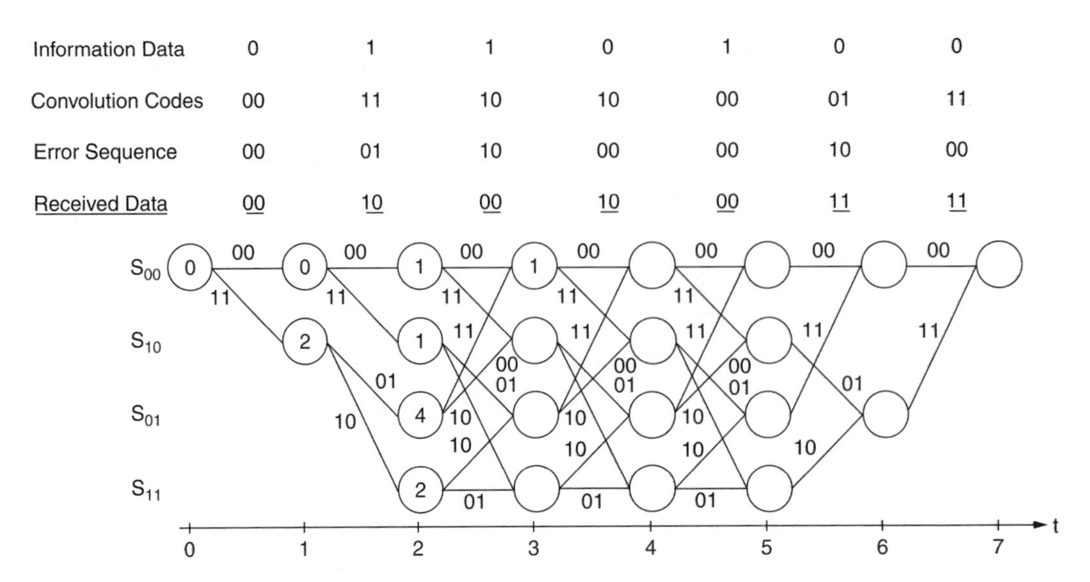

FIGURE 19.7 Example Viterbi trellis diagram.

to add or subtract the branch metric from states i and $i + s/2$, compare them and select the minimum. Similarly, state $2i + 1$ is updated. The first main power reduction comes from the butterfly arrangement because it reduces the amount of memory accesses by half; however, it does slightly complicate the address arithmetic.

19.3.2.2 Datapath Architecture

DSP processors have special hardware and instructions to implement the ACS operation in the most efficient way. The Lode architecture uses the two MAC units and the ALU to implement the ACS operation, as dipicted in Figure 19.6. The dual MAC operates as a dual add/subtract unit. The ALU finds the minimum. The shortest distance is saved to memory and the path indicator (i.e., the decision bit is saved in a special shift register A2). This results in 4 cycles per butterfly [30].

The Texas Instruments TMS320C54x and the Matsushita processor described in Okamoto et al. [20] use a different approach, which also results in four cycles per butterfly. Figure 19.8(b) illustrates this. The ALU and the accumulator are split into two halves (much like SIMD instructions), and the two halves operate independently. A special compare, select, and store unit (CSSU) compares the two halves, selects the chosen one, and writes the decision bit into a special register TRN. The processor described in Okamoto et al. [20] describes two ACS units in parallel. To illustrate the importance of an efficient implementation of the ACS butterflies, consider the IS-95 cellular standard. The IS-95 standard uses a rate 1/2 convolutional encoder with a constraint length of 9 [23], which corresponds to 256 states or 128 butterflies. It has a window size of 192 samples. This corresponds to $128 \times 192 \times (ACS)$ operations. The most efficient implementation requires four cycles per butterfly. This still corresponds to close to 100 MIPS. One should note that without these specialized instructions and hardware, one butterfly requires 15 to 25 or more instructions, which results in a factor 5 to 10 increase in number of instructions to calculate a complete Trellis diagram.

19.3.2.3 Datapath Support

The hardware support for the Viterbi algorithm on the 16210 also allows for the automatic storage of decision bits from the ACS computations. This functionality can be switched on or off. When the built-in comparison function `cmp1()` is called, the associated decision bit is shifted into the auxiliary register `ar0`. This auxiliary register is a special shift register to move decision bits in at the LSB side. During the trace back phase, its bits are used to reconstruct the most likely path. Each ACS takes two cycles (one

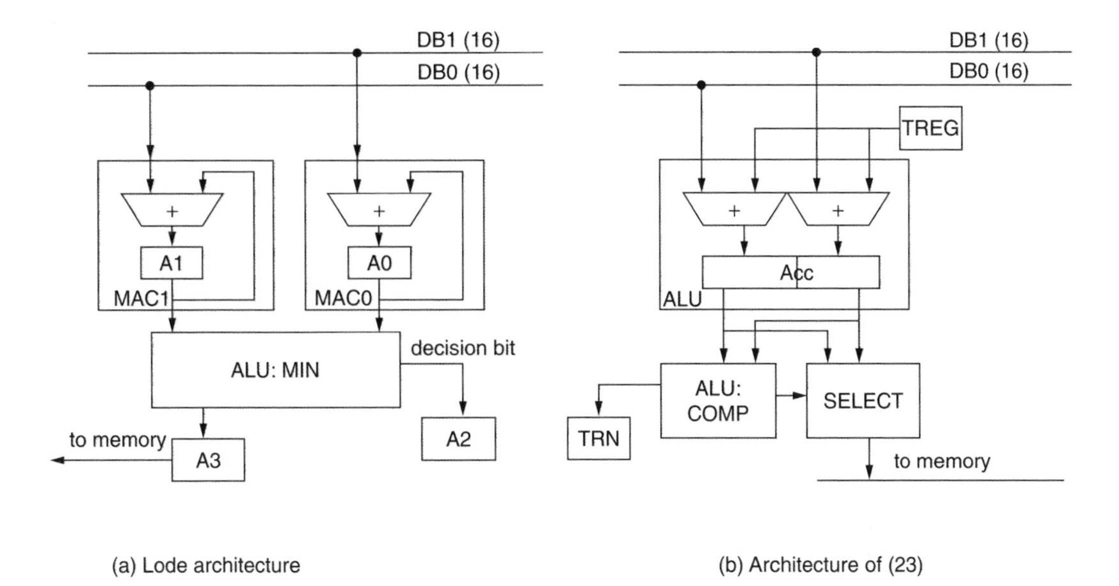

(a) Lode architecture (b) Architecture of (23)

FIGURE 19.8 Add-compare-select (a) on the Lode architecture and (b) on C54x architecture and on the architecture of Okamoto and co-workers [20].

TABLE 19.3 Pseudo Code for the Viterbi Butterfly on the DSP 16210

```
do 8 {
      a0=a4+y  a1=a5=y  *r3++=a0h
      a2=a4y   a3=a5+y  *r5++=a2h
      a0=cmpl  (a1, a0)  yh=*r0  r0=r1+j  j=k  k=*  pt1++
      a2=cmpl  (a3, a2)  a4  5h=*pt0++
}
*r2++=ar0
```

for the additions, one for the compare/select), and thus a single butterfly takes a total of four cycles. The code segment in Table 19.3 performs the butterfly computations.

19.3.2.4 Control Architecture

The 16210 has hardware looping support, and there is only a single cycle required to initialize this looping support before the loop executes with zero overhead. When decoding a standard GSM voice channel, which has a constraint length of 5 or 16 states in the trellis, the `ar0` register is filled with 16 decision bits after the 8 butterflies are processed. Thus, with a single memory access, the decision bits can be stored in memory and the next symbol pair can be processed. This is an efficient use of memory bandwidth. For codes with higher constraint lengths and thus more states, the code segment can be executed multiple times with each decision bit word written to memory as required.

19.3.3 Turbo Decoding

Although convolutional decoding remains a top priority (the decoding requirement for EDGE has been identified as greater than 500 MIPS), the performance needed for turbo decoding is an order of magnitude greater. We therefore describe the turbo decoders needed in 3G systems. Turbo decoding (see Figure 19.9) is a collaborative structure of soft-input/soft-output (SISO) decoders with the inclusion of interleaver memories between decoders to scatter burst errors [7]. Either soft-output Viterbi algorithm (SOVA) [13] or maximum a posteriori (MAP) [4] can be used as SISO decoders. Within a turbo decoder, the

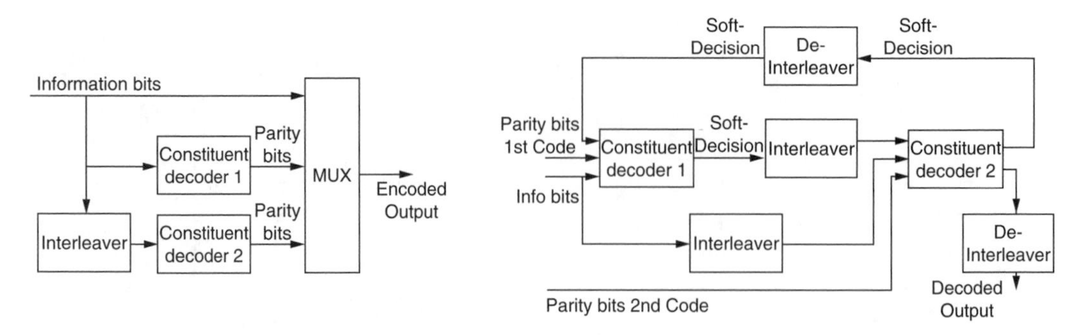

FIGURE 19.9 Turbo encoder and Turbo decoder.

two decoders can operate on the same or different codes. Turbo codes are included to provide coding performance to within 0.7 dB of the Shannon limit (after a number of iterations).

The Log-MAP algorithm can be implemented in a manner very similar to the standard Viterbi algorithm. Perhaps the most important difference between the algorithms when they are implemented is the use of a correction factor on the new "path metric" value (the alpha, beta, and log-likelihood ratio values in Log-MAP) from each ACS, which is dependent on the difference between the values being compared. This is typically implemented using a lookup table, with the absolute value of the difference used as an index into this table and the resulting value added to the selected maximum before it is stored.

19.3.3.1 Datapath Architecture

The C55x DSP processor includes explicit instructions to support the turbo decoding process. This is illustrated in Figure 19.10. A new instruction, the *max_diff* (ACx, ACy, ACz, ACw) is introduced [24]. It makes use of the same ALU and CSSU unit as the Viterbi instructions. Again, the ALU is split into two 16-bit halves. This processor has four accumulator registers compared with two in the previous generation. All four accumulator registers are split in half. The two differences, between ACx(H) and ACy(H) and between ACx(L) and ACy(L), are stored in the ACw halves. The maximum of the ACx(H) and ACy(H) is stored in ACz(H); the maximum of the ACx(L) and ACy(L) is stored in ACz(L). Two special registers are used to store the path indicators, TRN0 and TRN1.

The preceding modifications support the requirements for wireless baseband processing. To also improve the performance for multimedia, a tightly coupled mechanism of instruction extension and

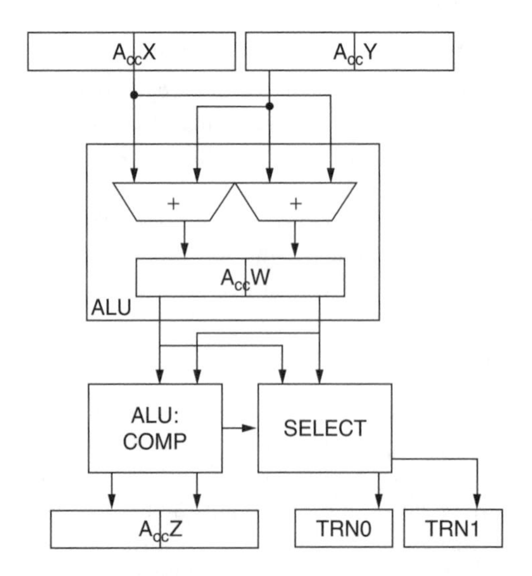

FIGURE 19.10 Turbo decoding acceleration on the C55x.

TABLE 19.4 Examples of Low-Power Programmable DSP Processors

Reference	MOPS	Technology	Threshold Voltages	Power	Standby-Power
Mutoh [19]	26	0.5 μm	Two	2.2 mW/MHz (at 1 V, 13.2 MHz)	350 μW
Lee [18]	300	0.25 μm	Two	0.21 mW/MHz (at 1 V, 63 MHz)	4.0 mW
Shiota [26]	NA	0.25 μm	Two	0.26 mW/MHz (at 1 V max, 50 MHz)	100 μW
Igura [14]	800	0.25 μm	One	2.2 mW/MHz (at 1.5 V, 50 MHz)	NA
Zhang [35]	240	0.25 μm	One	0.05 mW/MHz (at 1 V, 40 MHz)	NA

Note: NA = not available.

hardware acceleration is added to the C55x processor [9]. A special set of instructions is provided, with sources and destinations that are shared with the regular instructions. These special instructions have one set of opcodes: `copr()`. This avoids explosion of the instruction code size. Then, the application specific integrated processor (ASIP) designer has the choice to define the functionality of the hardware accelerator units addressed by these `copr()` instructions. Typical examples are video processing extensions for mobile applications [22].

Table 19.4 summarizes several low-power DSP processors. Notice that all operate with dual threshold voltages. In addition, note that although the clock frequency is not spectacularly high, the MOPS efficiency is very high for each of these processors.

19.4 DSPs as Part of SoCs

The previous section presented several modifications to the processor architecture to optimize the architecture toward the application domain of mobile wireless communications. It included examples of modifications to the memory architecture, datapath architecture, control architecture, instruction set, and bus architecture. So far, these modifications are tightly coupled (i.e., it reflects directly in the instruction set). The optimized instruction sets result in very compact program sizes and very efficient code. Yet, it is also very hard to produce efficient embedded software. The specialized CISC type instructions are extremely hard to be recognized by the compiler. Thus, the approach usually results in hand-optimized assembly code libraries for the computation-intensive functions.

In addition, the demands of next generation mobile applications are not satisfied by these instruction set modifications alone. More applications and multiple applications in parallel (and on demand) are running on the battery-operated devices.

Because of this, we see two distinct trends: one is in the direction of more powerful, but also more energy-hungry, processors used in the infrastructure. The other trend is in the direction of ultra-low power DSP solutions used in the handheld, battery operated terminals. Processors used in the basestation infrastructure are more compiler-friendly [1]. One popular type is the class of VLIW processors that are developed for wireless communications. Some examples are the TIC6x processor [2], the Lucent/Motorola Starcore [27], the ADI TigherSharc [21]. The main advantage of these processors is that they are compiler-friendly: efficient compiler techniques are available. The main disadvantage is that the program size is large and thus creates a large memory overhead [31]. This makes them mostly attractive to base station infrastructure.

It is interesting to note that some CISC features have reappeared: specialized instructions or loosely coupled coprocessors been added to the base VLIW architecture to improve the performance and to reduce the power consumption. A first example is the TIC6x processor, to which loosely coupled Viterbi and turbo coding coprocessor units are added [2]. A second example is the Starcore processor to which some specialized instructions are added [21]. Because the main driver is base station infrastructure (i.e., the baseband part of the application), there is no explicit support for the application side of the system, such as speech or video processing.

Because one processor is not sufficient to process the multiple and widely varying applications, a second, more energy efficient trend, is the addition of specialized coprocessors or accelerator units to the main processor on the SoC. This can take several forms. Initially, an SoC had multiple but almost identical processor units, as in Igura et al. [14]. This processor contains four identical DSP processors. Global tasks (coarse-grain) are assigned to the DSP processors in a static manner. Alignment of the tasks is provided by synchronization routines and interrupts. It is demonstrated that a video codec (H.263) and a speech codec (G.723) can run at the same time within the 110-mW power budget. Memory accesses (internal, shared, and external) consume half of the power budget, which indicates again that the memory architecture and the match of the application to the memory architecture is crucial.

A second form is an SoC with heterogeneous processor units. An example of this is the OMAP architecture [22]. It consists of a specialized DSP processor, the TMS320C55x DSP and a microcontroller, an ARM9xx CPU. The microcontroller is used for the control flow, including running an operating system, user interfaces, and so on. The DSP is used for the number crunching signal processing tasks. As discussed before, it is highly optimized for the communication signal processing, and through its extension possibilities toward multimedia applications. Thus, the OMAP is a result of several strategies: domain specific instruction sets, tightly coupled instruction set acceleration through coprocessor instructions, loosely coupled coprocessors, and multiple processors on one SoC. The global flow of data as well as the corresponding interconnect architecture and memory architecture are still fixed.

This leads to a third form. To combine flexibility with energy efficiency, it is our opinion that the SoC architecture should consist of multiple heterogeneous building blocks connected together by a reconfigurable interconnect architecture. We call this a RINGS (reconfigurable interconnect for next generation systems) architecture [32], illustrated in Figure 19.11. Each of the building blocks is optimized for its specific application domain, represented by an application domain pyramid. Within an application pyramid, the reconfiguration or reprogrammability level can be determined individually. For instance,

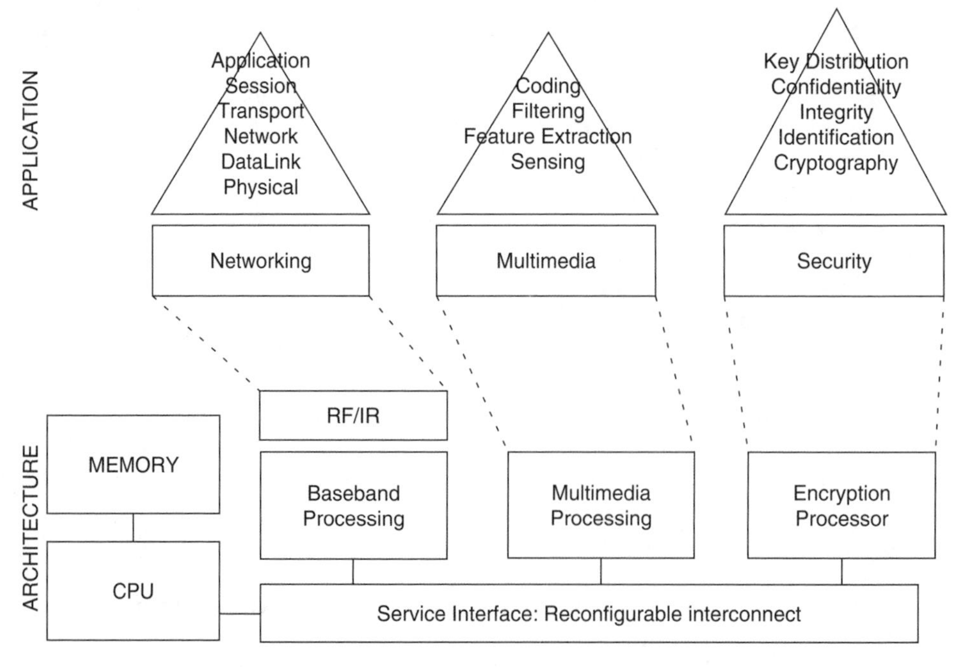

FIGURE 19.11 Generic RINGS architecture.

a baseband pyramid can be realized with a programmable DSP processor, augmented with a few coprocessors. A security pyramid can be realized with a small FSM and lowly programmable crypto acceleration engines. A network protocol stack will need a highly programmable central processing unit (CPU) approach. Multimedia applications are probably best served by a dataflow approach and a chain of hardware acceleration units, and so on. Thus, the transit line between hardware and software can be positioned at different levels in different pyramids. The top level is a general system application (in software) that connects the different pyramids together.

At the bottom, the communication is provided by means of a flexible interconnect. Reconfiguration of the interconnect is also crucial for the MAIA processor [35]. This processor contains an ARM microcontroller and hardware acceleration units (e.g., MAC, ALU, and AGU). The ARM core controls and decides the reconfiguration of the interconnect. To optimize the energy flexibility trade-off a two-level hierarchical mesh network is chosen. A local mesh connects local tightly coupled units. A global mesh with a larger granularity is provided at the top level. This global mesh has switchboxes that connect both globally and downwards to the local level. Another example is the DM310 digital media processor [29]. To obtain low power, it has dedicated coprocessors for image processing, a programmable DSP processor for audio processing, and an ARM processor to process system level tasks.

At the physical level, this is a typical example of a time and space division-based interconnect. To improve density, combined with a larger degree of programability and energy efficiency, we propose to use frequency and code division access to the interconnect medium [32]. This can be combined with the space and time division. From the programmer's viewpoint, the actual physical implementation should be hidden and a programming model should be available, that allows to model different interconnect paradigms and gives the uses the possibility to perform the energy flexibility trade-offs.

A RINGS architecture allows the platform to be changed as the target changes. This approach has been proven successful for an embedded fingerprint authentication system [25]. We are currently working on applying the same design methodology to accelerate multimedia applications for wireless embedded systems.

19.5 Conclusion and Future Trends

Low power can only be obtained by tuning the architecture platform to the application domain. This chapter presents multiple examples to illustrate this for the domain of signal processing and, more specifically, to support the signal processing algorithms for wireless communicating devices. At the same time, demand for flexibility is increasing. Thus, the designer must try to balance these conflicting requirements by providing flexibility at the right level of granularity and to the right components. It is extremely important to realize that this tuning involves *all* components of a processor: the datapaths, the instruction set, the interconnect, and the memory strategy. Traditional DSPs are CISC machines with an adapted modified Harvard interconnect and memory architecture (coming in many flavors). With increasing demands, coprocessors are added to these architectures. As SoCs grow in complexity, however, the architecture becomes one where one integrated device will contain multiple heterogeneous processors. Each processor supports an application domain and its programmability is tuned to the domain. The different components are connected together by a reconfigurable interconnect paradigm. This reconfigurable interconnect poses several research challenges are different abstraction levels: physical realization, the modeling at a higher abstraction level and the reconfigurable programming at compile time and run time.

19.6 Acknowledgments

The author acknowledges the following DSP processor experts: Chris Nicol, Dave Garrett, Wanda Gass, Mihran Touriguian, and Katsuhiko Ueda. The author also acknowledges the contributions of Frank M.C. Chang and Patrick Schaumont.

References

[1] B. Ackland and P. D'Arcy, A new generation of DSP architectures, *Proc. IEEE CICC '99*, Paper 25.1, pp. 531–536, May 1999.

[2] S. Agarwala et al. A 600-MHz VLIW DSP, *IEEE J. Solid-State Circuits*, Vol. 37, No. 11, pp. 1532–1544, Nov. 2002.

[3] D. Alter, Efficient implementation of real-valued FIR filters on the TMS320C55x DSP, Application Report SPRA655, April 2000, available from www.ti.com.

[4] L. Bahl, J. Cocke, F. Jelinek, and J. Raviv, "Optimal decoding of linear codes for minimizing symbol error rate," *IEEE Trans. Information Theory*, Vol. IT-20, pp. 284–287, Mar. 1974.

[5] T. Baji, H. Takeyama, and T. Nakagawa, Embedded-DSP superH family and its applications, *Hitachi Review*, Vol. 47, No. 4, pp. 121–127, 1998.

[6] Belgen kopen opnieuw meer gsm's/Belgians buy again more cells phones, *De Tijd*, Sept. 17, 2003.

[7] C. Berrou, A. Glavieux, and P. Thitimajshima, Near Shannon limit error-correcting coding and decoding: turbo-codes (1), *Proc. ICC '93*, Vol. 2, pp. 1064–1070, May 1993.

[8] M. Bickerstaff, D. Garrett, T. Prokop, C. Thomas, B. Widdup, G. Zhou, L. Davis, G. Woodward, C. Nicol, and R.-H. Yang, A unified turbo/Viterbi channel decoder for 3GPP mobile wireless in 0.18-mm CMOS, *IEEE J. Solid-State Circuits*, Vol. 37, No. 11, pp. 1555–1564, Nov. 2002.

[9] J. Chaoui, K. Cyr, S. de Gregorio, J.-P. Giacalone, J. Webb, and Y. Masse, Open multimedia application platform: enabling multimedia applications in third-generation wireless terminals through a combined RISC/DSP architecture, 2001. *Proc. (ICASSP '01). 2001 IEEE Int. Conf. on Acoustics, Speech, and Signal Processing, Volume: 2,* May 7–11, 2001.

[10] W. Gass and D. Bartley, Programmable DSPs, in *Digital Signal Processing for Multimedia Systems*, Marcel Dekker Inc., 1999, chap. 9.

[11] A. Gatherer, T. Stelzler, M. McMahan, and E. Auslander, DSP-based architectures for mobile communications: past, present, and future, *IEEE Commun. Mag.*, pp. 84–90, January 2000.

[12] A. Gatherer and E. Auslander, *The Application of Programmable DSPs in Mobile Communications*, John Wiley & Sons, New York, 2002.

[13] J. Hagenauer and P. Hoeher, A Viterbi algorithm with soft-decision outputs and its applications, *Proc. Globecom '89*, Nov. 1989, pp. 47.1.1–47.1.7.

[14] H. Igura, Y. Naito, K. Kazama, I. Kuroda, M. Motomura, and M. Yamashina, An 800-MOPS, 110-mW, 1.5-V, parallel DSP for mobile multimedia processing, *IEEE J. Solid-State Circuits*, Vol. 33, pp. 1820–1828, Nov. 1998.

[15] H. Kabuo, M. Okamoto, et al., An 80-MOPS peak high-speed and low-power consumption 16-bit digital signal processor, *IEEE J. Solid-State Circuits*, Vol. 31, No. 4, pp. 494–503, 1996.

[16] P. Lapsley, J. Bier, A. Shoham, and E. Lee, *DSP Processor Fundamentals*, IEEE Press, 1997.

[17] E.A. Lee, Programmable DSP processors: part I and II, *IEEE ASSP Mag.*, Oct. 1988 and Jan. 1989.

[18] W. Lee et al. A 1-V programmable DSP for wireless communications, *IEEE J. Solid-State Circuits*, Vol. 32, No. 11, Nov. 1997.

[19] S. Mutoh, S. Shigematsu, Y. Matsuya, H. Fukuda, and J. Yamada, A 1-V multi-threshold voltage CMOS DSP with an efficient power management technique for mobile phone application, *IEEE Int. Conf. on Solid-State Circuits*, Paper FA 10.4, pp. 168–169, Feb. 1996.

[20] M. Okamoto, K. Stone, T. Sawai, H. Kabuo, S. Marui, M. Yamasaki, Y. Uto, Y. Sugisawa, Y. Sasagawa, T. Ishikawa, H. Suzuki, N, Minamida, R. Yamanaka, and K. Ueda, A high-performance DSP architecture for next generation mobile phone systems, *1998 IEEE DSP Workshop*.

[21] A. Olofsson and F. Lange, A 4.32-GOPS 1- general-purpose DSP with an enhanced instruction set for wireless communications, *Proc. ISSCC*, pp. 54–55, Feb. 2002.

[22] M. Peresse, K. Djafarian, J. Chaoui, D. Mazzocco, and Y. Masse, Enabling JPEG2000 on 3-G wireless mobiles through OMAP architecture, *Proc. Acoustics, Speech, and Signal Processing, 2002 (ICASSP '02)*, Vol. 4, May 13–17, 2002, pp. IV-3796–IV-3799.

[23] T. Rappaport, *Wireless Communications, Principles & Practices,* IEEE Press, New York and Prentice Hall, New Jersey, 1996.

[24] TMS320C55x DSP Mnemonic Instruction Set Reference Guide, document SPRU374C, June 2000, available from www.ti.com.

[25] P. Schaumont and I. Verbauwhede, Domain-specific codesign for embedded security, *IEEE Comput. Mag.,* pp. 68–74, April 2003.

[26] T. Shiota, I. Fukushi, R. Ohe, W. Shibamoto, M. Hamaminato, R. Sasagawa, A. Tsuchiya, T. Ishihara, and S. Kawashima, A 1-V, 10.4-mW low-power DSP core for mobile wireless use, *1999 Symp. on VLSI,* Paper 2-2, 1999.

[27] Starcore launched first architecture, *Microprocessor Report,* Vol. 12, No. 14. p. 22, Oct. 1998.

[28] W. Strauss, DSP Market Bulletin, Forward Concepts, June 2, 2004, available from www.forward-concepts.com, June 2004.

[29] D. Talla, C. Hung, R. Talluri, F. Brill, D. Smith, D. Brier, B. Xiong, and D. Huynh, Anatomy of a portable digital mediaprocessor, *IEEE Micro.,* Vol. 24, Issue 2, pp. 32–39, March–April 2004.

[30] I. Verbauwhede and M. Touriguian, A low-power DSP engine for wireless communications, *J. VLSI Signal Process., Vol.* 18, pp. 177–186, 1998.

[31] I. Verbauwhede and C. Nicol, Low-power DSPs for wireless communications, *Proc. Int. Symp. on Low-Power Electron. Design (ISLPED 2000),* pp. 303–310, July 2000.

[32] I. Verbauwhede and M.-C.F. Chang, Reconfigurable interconnect for next-generation systems, *Proc. ACM/Sigda 2002 Int. Workshop on System Level Interconnect Prediction (SLIP 2002),* Del Mar, CA, April 2002, pp. 71–74.

[33] M. Weiss, F. Engel, and G. Fettweis, A new scalable DSP architecture for system on chip (SoC) domains, *Proc. IEEE ICASSP Conf.,* May 1999.

[34] J. Williams, K.J. Singh, C.J. Nicol, and B. Ackland, A 3.2-GOPs multiprocessor DSP for communication applications, *Proc. IEEE ISSCC 2000,* San Francisco, February 2002, Paper 4.2.

[35] H. Zhang, V. Prabhu, V. George, M. Wan, M. Benes, A. Abnous, and J. Rabaey; A 1-V heterogeneous reconfigurable DSP IC for wireless baseband digital signal processing, *IEEE J. Solid-State Circuits,* Vol. 35, pp. 1697–1704, Nov. 2000.

20

Energy-Efficient Reconfigurable Processors

Raphaël David
Sébastien Pillement
Olivier Sentieys
ENSSAT/University of Rennes
IRISA Laboratory

20.1 Introduction

Rapid advances in silicon technology and embedded computing bring two conflicting trends to the electronics industry. On the one hand, high-performance embedded applications dictate the use of complex battery-powered devices. The evolution of the battery capacity being significantly lower than that of the application complexity, energy efficiency becomes a critical issue in the design process of these systems. On the other hand, these systems have to be flexible enough to support rapidly evolving applications, restricting the use of domain-specific architectures. These trends have led to the reconfigurable computing paradigm [1,2].

Formally, configuring permits the adjustment of something or a change in the behavior of a device so that it can be used in a particular way. This definition leads to a very large design space bounded by bit-level reconfigurable architectures and by von Neumann-style processors. Common execution (i.e., reconfiguration) schemes can be extracted for different paradigms in this design-space [3,4], on the basis of the processing primitive granularity. On one side of the design space, bit-level reconfiguration is used in field programmable gate-array (FPGA). They provide bit-level reconfigurability and typically use mesh topology for their interconnection network. They allow designers to fully optimize the architecture at the bit level. The flexibility of these devices is associated with a very important configuration data volume and with performance and energy overheads. On the opposite side, system-level reconfiguration corresponds to instruction-based processors, including digital signal processors (DSP). They achieve flexibility through a set of instructions that dynamically modify the behavior of statically connected components. Their performance is limited by the amount of operator parallelism. Futhermore, their power-hungry data and instruction access mechanisms lower their energy efficiency.

0-8493-1941-2/05/$0.00+$1.50

In between, to increase the optimization potential of programmable processors without the bit-level reconfigurable architecture drawbacks, the functional-level reconfiguration has been introduced for reconfigurable processors. In such architectures, functional units as well as their interconnection network are reconfigurable and handle worldwide data. Most of these architectures use two-dimensional network topologies, usually hierarchical [5], for communications between functional units. In this context, numerous approaches have been proposed, such as DReAM [6], Morphosys [7], Piperench [8], FPFA [9], RaPiD [10], or Pleiades [11]. The main concern of these architectures is to introduce flexibility while maintaining high performance and reducing reconfiguration cost.

Reconfigurable architectures such as the Chameleon [12] have demonstrated their efficiency on implementing 3G base stations. More generally, reconfigurable architectures have demonstrated their efficiency on computation-hungry signal processing applications. Unfortunately, energy efficiency has been rarely a topic of interest in the reconfigurable framework. In this chapter we focus on the energy/flexibility trade-off for high-performance reconfigurable architectures.

Section 20.2 presents the energy efficiency criterion and highlights energy wastes in the reconfigurable design space as well as the opportunities to reduce energy consumption. Section 20.3 presents the DART architecture implementing energy aware design techniques and innovative reconfiguration schemes. Finally, Section 20.4 discusses the implementation results of a key application of next-generation mobile communication systems.

20.2 Energy Efficiency of Reconfigurable Architectures

20.2.1 Problem Definition

The energy efficiency (E.E.) of an architecture can be quantified by considering the number of operations it processes per second when consuming one mW. This parameter can be defined by Equation (20.1) [13]:

$$E.E. = \frac{N_{OP} . F_{clk}}{A_{Chip} . \alpha . C_N . F_{clk} . V_{DD}^2} \quad \left[\frac{MOPS}{mW}\right] \tag{20.1}$$

where *NOP* is the number of operations computed at each cycle and *Fclk* the operating frequency [*MHz*]. *AChip* is the total area of the chip [*mm2*], *CN* the normalized capacitance by area unit [*mF/mm2*], α the average activity, and *VDD* the supply voltage [*V*]. The product *NOP.Fclk* thus represents the computation power of the architecture and is given in millions of operations per second (MOPS). The product $A_{Chip} . \alpha . C_N . F_{clk} . V_{DD}^2$ gives the power consumed during the execution of the *NOP* operations. *AChip* parameter is obtained by Equation (20.2):

$$A_{Chip} = N_{opr} . A_{opr} + A_{mem} + A_{ctrl} \tag{20.2}$$

where *Aopr* is the average area per operator, *Nopr* the number of operator in the design. *Nopr.Aopr* thus represents the operator area in the design. *Amem* is the memory area and *Actrl* the area of the control and configuration management resources.

These two equations can be used to find out which parameters could best be optimized to design an energy efficient architecture. The $N_{OP} . F_{clk}$ product has to cover the needs of the implemented application (i.e., the architecture has to be powerful enough to compute the application). Consequently, N_{OP} and F_{clk} need to be jointly optimized. The normalized capacitance mainly depends on the technology. So, its optimization was not studied for this work.

The definition of an energy aware architecture dictates the optimization of the remaining parameters: average operator area, storage, and control resource area as well as activity through the circuit and of course clock frequency and supply voltage. To define an optimal *delay* × *power* product, parallelism inherent to the implemented system must finally be fully exploited.

According to the energy efficiency criterion, application-specific integrated circuits (ASIC) can be considered as the ultimate solution. Because they are dedicated only for one specific processing, no computational unit is larger or more complicated than it has to be. In such devices, the operator area (A_{opr}) is thus minimized. Moreover, no architectural mechanisms have to be introduced to support flexibility (i.e., there is no need to fetch and decode instructions). The execution is fully deterministic and all known optimizations, such as using wires instead of shifters, can be used. The circuit is controlled thanks to a finite state machine and the control area of the chip A_{ctrl} is also minimized.

In such designs, processing parallelism can be fully exploited. By increasing the parallelism level, the operating frequency along with the supply voltage can be reduced, and therefore an optimal energy delay product can be achieved. Moreover, because there is no resource waste, design area is reduced. Finally, clock distribution energy waste can also be minimized by defining several clock domains.

With these devices, data accesses are fully determined at the synthesis time. Thus, data can be placed as near as possible to the functional units that will handle them. A memory hierarchy can also be defined in order to minimize the energy consumed by data transactions within the architecture. Furthermore, optimizations such as first-in first-out (FIFO) memory instead of static random-access memory (SRAM) can be used. Consequently, memory area (A_{mem}) is reduced along with energy.

Beside classical high-performance and low-energy consumption constraints, flexibility becomes a major concern to the development of multimedia and mobile communication systems. This dictates the use of programmable or even reconfigurable devices [14]. The next section discusses energy efficiency optimization techniques that can be applied in the case of reconfigurable processors.

20.2.2 Energy Efficiency Optimization

20.2.2.1 Energy in Computations

An architecture is considered as energy efficient only if its operators are the main source of energy consumption. Nevertheless, the optimization effort for these components has also to be important. Programmable processors integrate in their datapath, general-purpose functional units designed to perform a large variety of computations. They are thus significantly more complicated than they need to be, and are a source of energy waste. Moreover, if their bit-width is larger than the data length used in the algorithm, additional energy is wasted.

On the contrary, in bit-level reconfigurable architectures, each operator is built to execute only one operation. The very fine granularity of the computation primitive (e.g., look-up tables) dictates the association of numerous cells. Consequently, the power dissipated in such an operator is mainly issued from the interconnection network (60 to 70%) [15,16]. Even if the operators are tailored-made to execute only one operation on fixed-size data, they are inefficient from an energy point of view.

To reduce energy waste, the amount of operations supported by functional units has to be limited. A functional decomposition of these units leads to the isolation of its different parts by using latches. In this case, only transistors useful to the execution consume dynamic power.

Many application domains handle several data sizes during time (e.g., 8, 11, 13, and 16 bits). To support all these data sizes, very flexible functional units have to be designed. Consequently, latency and energy penalties occur. Another alternative is to optimize functional units only for a subset of these data sizes by designing sub-word parallelism (SWP) operators [17]. This technique consists of dividing an operator working on N-bit data to allow the execution of k operations in parallel on part of the input data of N/k length. Integrating such operators allows to increase computation power while keeping the consumed energy per operator nearly constant during processings with data-level parallelism. Therefore, SWP can increase energy efficiency of the architecture.

20.2.2.2 Exploiting the Parallelism

To minimize energy consumption, supply voltage has to be reduced aggressively. To compensate the associated performance loss, concurrent execution must be supported. Digital signal processing algorithms provide several levels of parallelism that can be exploited to achieve this objective.

Operation- or instruction-level parallelism (ILP) is inherent to every computation algorithm. Although it is constrained by data dependencies, its exploitation is generally quite easy. It requires the introduction of several functional units working independently with each other. To exploit this parallelism, the architecture controller has to be able to jointly specify, to several operators, the operation to be executed.

Thread-level parallelism (TLP) represents the number of processings which may be executed concurrently to implement an algorithm. The TLP is far more complicated to exploit than ILP. TLP strongly varies from one application to another, and even more between two descriptions of the same application. To exploit this parallelism while guaranteeing a good computation density, the architecture must be able to adapt its organization of processing resources [18]. The trade-off between ILP and TLP must thus be adapted to the application to be executed. This can be realized by organizing the architecture into a hierarchy, the lowest level of which being a set of functional units (e.g., a datapath). Each datapath should be able to be controlled independently to implement a particular thread of the application. On the contrary, the datapaths should be interconnected as a single resource exhibiting a large amount of ILP.

Application or algorithm parallelism can be considered as an extension of thread parallelism. The goal is here to identify the applications that may be implemented concurrently on the architecture. On the opposite of threads, applications executed in parallel are working on distinct data sets. To exploit this kind of parallelism, a second level of hierarchy needs to be added to the architecture. The architecture may be divided into clusters working independently on several applications. These clusters have their own control, storage, and processing resources.

20.2.2.3 Reducing the Control Overhead

In a reconfigurable processor, two types of information are needed to manage the architecture: the configuration data which specify the hardware structure of the architecture (operators, logic, interconnections), and the control which manages the data transactions within the architecture. Distributing the configuration and control information has a significant impact on performance and energy efficiency of the system. This is mainly issued from the configuration and control data volume needed to execute an application and to the reconfiguration frequency.

The architectural paradigms included in the reconfigurable design space have very different strategies to distribute these information. Bit- and system-level reconfigurable architectures have the two most extreme reconfiguration schemes. On the one hand, a very large amount of configuration data (several thousands or millions of bits) is distributed in FPGA architecture. The reconfiguration cost is very high but once specified, there is no control overhead. On the other hand, programmable processors eliminate the overhead linked to the specification of the datapath because it is fixed. The control cost of the architecture is very important, however, and corresponds to fetch and decode instructions at each cycle.

The 80/20 rule asserts that 80% of the execution time is consumed by 20% of the program code [19]. Few portions of source code are thus executed during long periods of time. These blocks of code are described as regular and are typically loop kernels during which a same computation pattern is used for long time. Between these blocks of regular code, instructions follow one another without particular order and in a nonrepetitive way. These portions of code are described as irregular. Because of their lack of parallelism, they present few optimization opportunities.

To minimize the architecture control cost, the distribution strategy can be adapted to the implemented processing. For this purpose, regular and irregular processings have to be distinguished to define two reconfiguration modes. The first one is used to specify the architecture configurations which will allow optimal implementations of regular processings. The second reconfiguration mode is used to specify the control information that will allow to execute irregular processings. By reducing the amount of reconfiguration targets, functional-level reconfigurable architectures limit the configuration data volume associated to the datapath structure specification.

To reduce even more the configuration data volume, redundancy in datapath can also be exploited. It allows to distribute simultaneously the same configuration information to several targets, whenever these targets execute the same processing.

20.2.2.4 Reducing the Data Access Cost

Data access cost also has a significant impact on the energy efficiency of the architecture. It depends on the amount of memory accesses and on the cost of one memory access. In programmable processors, each computation step dictates register file accesses. These architectures cannot completely exploit spatial and temporal locality of data because all the data have to be stored in registers. Thanks to bit- or functional-level reconfiguration, operators may be interconnected to efficiently exploit the locality of data. Spatial locality is used by directly connecting operators. Producer and consumer of data can be directly connected, no memory transfers are necessary to store intermediate results. Temporal locality can be exploited thanks to one-to-all connections. That kind of connection allows the transfer of one data element toward several targets in a single transaction and to skip redundant data accesses. This temporal locality may moreover be exploited thanks to delay chains, which reduce the amount of memory accesses when several samples of a same signal are concurrently handled in an application.

Defining a memory hierarchy reduces the data access cost while providing a high memory bandwidth [20]. This hierarchy has to combine high capacity, high bandwidth, and energy efficiency.

Because multi-port memories are characterized by high-energy consumption, it is more efficient to integrate several single-port memories. High-bandwidth and low-energy constraints thus require the integration of a large amount of small memories. Moreover, to provide a quite important storage space, a second level of hierarchy can be added. Finally, to reduce memory management costs, the address generation task is distributed along with the memories.

Associating flexibility with high-performance and energy efficiency, is a critical issue for embedded applications. Beside the dynamically reconfigurable device Xc6200 from Xilinx [21], numerous research projects have contributed to the simplification of the reconfiguration process to increase performance and flexibility (e.g., Singh et al. [7], Goldstein et al. [8], Cronquist et al. [10], and Callahan et al. [22]). Despite the energy optimization potential of reconfigurable architectures, few projects have integrated this constraint. In Abnous and Rabaey [11], the authors propose a low-power reconfigurable processor. Because it is a domain-specific platform, its flexibility is limited. Furthermore, the validation of this platform has only been proposed for medium-complexity application domain, such as speech coding [23].

The next section discusses a reconfigurable processor associating energy efficiency, high performance and flexibility. This architecture is based on the optimization mechanisms presented in this section.

20.3 The DART Architecture

DART is a hierarchical architecture supporting the different levels of parallelism. To exploit task parallelism, DART has been broken up into clusters. Distinct tasks can be processed concurrently by clusters because each of them has its own control and storage resources. At the system level, tasks are distributed to clusters by a controller. This controller supports the real-time operating system which assigns tasks to clusters according to urgency and resources availability constraints. The system level of DART also includes shared memories (data, configuration) and an I/O block which allows its interfacing with external components through a standard bus (e.g., AMBA and VCI).

This section first describes the architecture of DART clusters. Next, the processing primitives are presented. Finally, dynamic reconfiguration and development tools are discussed.

20.3.1 Cluster Architecture

Each cluster of DART (Figure 20.1) integrates two types of processing primitives: some reconfigurable datapath (RDP) used for arithmetic processing and an FPGA core processing data at the bit level. The RDPs, detailed in the next section, are reconfigurable at the functional level to optimize the interconnections between arithmetic operators according to the calculation pattern. The FPGA core is reconfigurable at the gate level to efficiently support bit-level parallelism of processings (e.g., generation of Gold or Kasami codes in wideband code division multiple access (WCDMA) or channel coders [24]). Using these two kinds of operators allows an architecture to be defined in adequacy with the algorithm for a

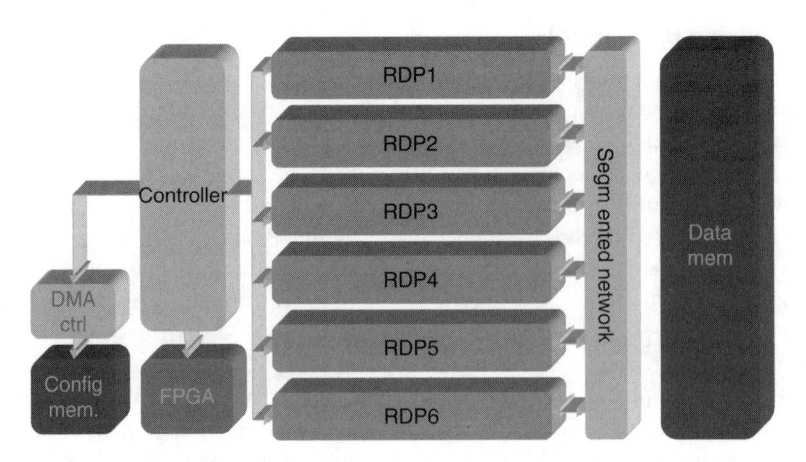

FIGURE 20.1 Architecture of a DART cluster.

large set of applications. Experiments have demonstrated that integrating one FPGA core and six RDPs in each cluster of DART delivers enough calculation power.

The RDPs are interconnected thanks to a segmented mesh network. Depending on the parallelism level of the application, the RDPs can be interconnected to compute in a highly parallel fashion to support high ILP, or can be disconnected to work independently on different threads. The segmented network allows dynamic adaptation of the instruction- and thread-level parallelism of the architecture, depending on the processing needs.

This hierarchical organization of DART allows not only the distribution of control but also that of the processing resources. Thus, it is possible to efficiently connect a very large number of resources without being too penalized by the interconnection cost. The processing resources distribution allows the definition of a hierarchical interconnect network which is significantly more energy efficient for complex design than a typical global interconnection networks [5]. With this kind of network, the lowest level of the resource hierarchy is completely connected while the higher levels communicate via the segmented network.

Moreover, thanks to the flexibility of this topology, the resulting architecture becomes a better target for the development tools.

All the processing primitives (i.e., the FPGA and RDPs) access the same data memory space and their reconfigurations are managed by a controller. To minimize the associated control overhead, reconfigurations of the FPGA are realized via a DMA controller. The cluster controller has only to specify an address bound to the DMA controller which will transfer the data from a configuration memory towards the FPGA. Besides that, the cluster controller also manages the reconfiguration of the RDPs via instructions. Its architecture is similar to that of a typical programmable processor, but it distributes configurations instead of instructions. Consequently, it does not have to access an instruction memory at each cycle. Fetch and decode operations are only realized when a reconfiguration occurs and are hence very infrequent. This drastic reduction of the instruction memory readings and decodings leads to very significant energy savings (cf. Section 20.4).

20.3.2 RDP Architecture

The arithmetic processing primitives in DART are the RDPs (Figure 20.2). They are organized around functional units and memories that are interconnected according to a very powerful communication network. Every RDP has four functional units (two multipliers/adders and two ALUs) handling double-precision 16-bit data, followed by a register. They support SWP (sub-word parallelism) processings and have been designed with low-power concerns [25].

The functional units are dynamically reconfigurable (see next section) and are working on data stored in four small local memories. On the top of each memory, four local controllers (the *AGi* on the top of

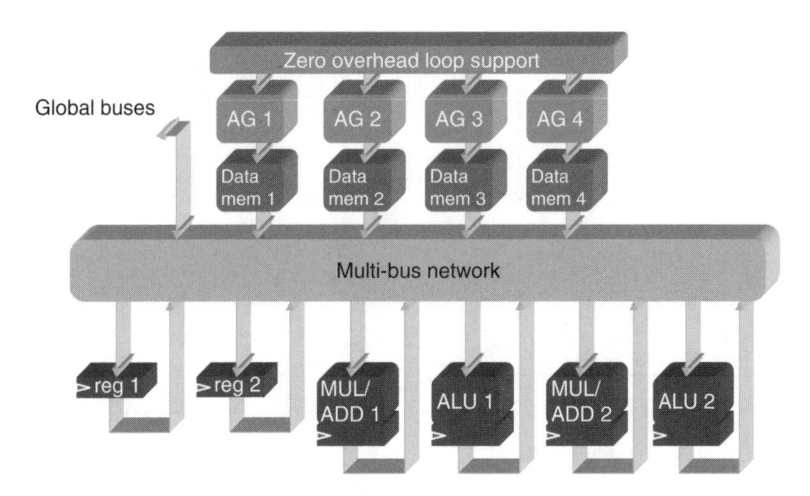

FIGURE 20.2 Architecture of an RDP.

Figure 20.2) are in charge of providing the addresses of the data handled inside the RDPs. These local controllers are like tiny reduced instruction-set computer (RISC) processors and support a large set of addressing patterns. The four local controllers of each RDP share a zero-overhead loop support. In addition to the memories, two registers are also available in every RDP. These registers are used to build delay chains, and hence to realize time data sharing.

All these resources are connected through a completely connected network. The hierarchical organization of DART permits these connections to be kept relatively small and hence to limit their energy consumption. Thanks to this network, each resource of the RDP can communicate with every other resource and hence, the datapath can be optimized for every calculation pattern. Moreover, this flexibility eases data sharing. Indeed, because a memory can simultaneously be accessed by several functional units, some energy savings can also be achieved. The upper left part of Figure 20.2 depicts the connections with global buses that allow the connection of several RDPs to implement massively parallel processing.

20.3.3 Dynamic Reconfiguration

One of the main features of DART is to support two RDP reconfiguration modes which ensues from the 80/20 rule (see Section 20.2). During regular processing, the RDPs are dynamically reconfigured to be adapted to the calculation pattern. This reconfiguration — hardware reconfiguration — may take a few cycles, but is used for long period of time. On the contrary, during irregular processing, the calculation pattern is changing very often. In that case, the reconfiguration time has to be minimized, and the RDPs structure is modified thanks to software reconfiguration. Another important feature of a DART cluster is to exploit the redundancy in the RDPs to minimize the configuration data volume.

20.3.3.1 SCMD Concept

A portion of code is usually qualified as regular when it is used for a long period of time, and applied to a large set of data, without being suspended by another processing. Loop kernels support this qualification because their computation patterns are maintained during all the loop iterations. Instruction-level parallelism of such regular processing is often exhibited by compilation techniques such as loop unrolling or software pipelining [26]. With such techniques, the computation pattern of the loop kernel is repeated several times, which leads to a highly regular architecture. If this loop kernel is implemented on several RDPs, their configuration might be redundant. Specifying several times the same configuration is an energy waste, we then introduce a concept called single configuration multiple data (SCMD). It may be considered as an extension of SIMD (single instruction multiple data), in which several operators execute the same operation on different data sets. Within the framework of SCMD, the configuration data sharing is no longer limited to the operators but is extended to the RDPs.

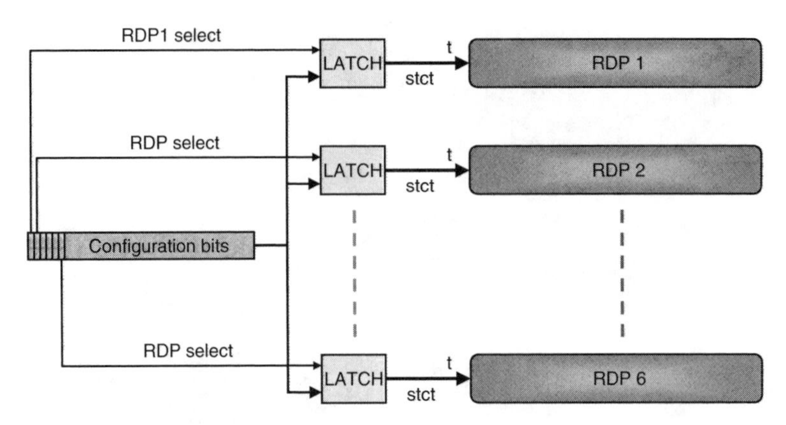

FIGURE 20.3 SCMD implementation for DART.

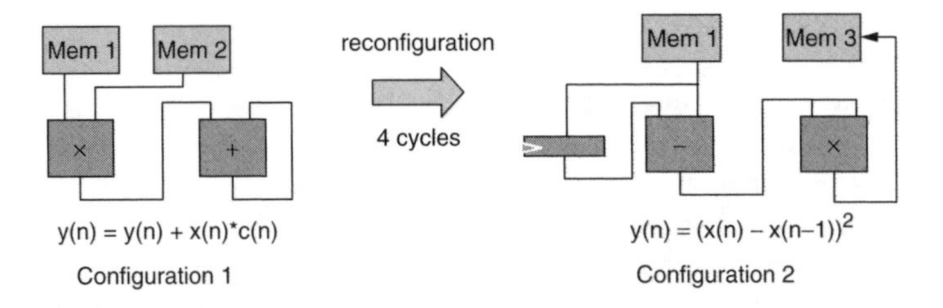

FIGURE 20.4 Hardware reconfiguration example.

The SCMD concept allows the simultaneous configuration of several RDPs. Practically, a field is concatenated to the configuration instructions to specify the targets of the configuration bits. With six RDPs, six bits have to be added to the instruction. Each RDP validates the configuration instructions according to the value of its select bit (Figure 20.3). This allows the reduction of the configuration data volume for regular computations, where there is a lot of redundancy between the RDP configurations.

20.3.3.2 Hardware Reconfiguration

During regular processing, a complete flexibility of the RDPs will be allowed by the full use of the functional-level reconfiguration paradigm. By allowing the modification of the way in which functional resources and memories are interconnected, the architecture can be optimized for the computation pattern that has to be implemented. With six RDPs, the configuration data volume for a cluster is 826 bits. According to the regularity of the computation pattern and the redundancy of the RDP configurations (which influences the SCMD performances), between three and nineteen 52-bit instructions will be required to reconfigure all the RDPs and their interconnections. Once these configuration instructions have been specified, no other instruction readings and decodings have to be done until the end of the loop execution.

This kind of configuration can for example be illustrated by Figure 20.4. The datapath is optimized at first in order to compute a digital filter based on multiply-accumulate operations. Once this configuration has been specified, the data-flow computation model is maintained as long as this computation pattern is used. At the end of the computation, after a reconfiguration step which needs four cycles, a new datapath is specified in order to be in adequacy with the calculation of the square of the difference between $x(n)$ and $x(n\text{-}1)$. Once again, no control is necessary to end this computation.

20.3.3.3 Software Reconfiguration

For irregular processing, which implies frequent modifications of the RDP configuration, a software reconfiguration has also been defined. To be able to reconfigure the RDP in one cycle with an instruction of reasonable size, their flexibility has been limited. In that case, DART uses a read-modify-write behavior,

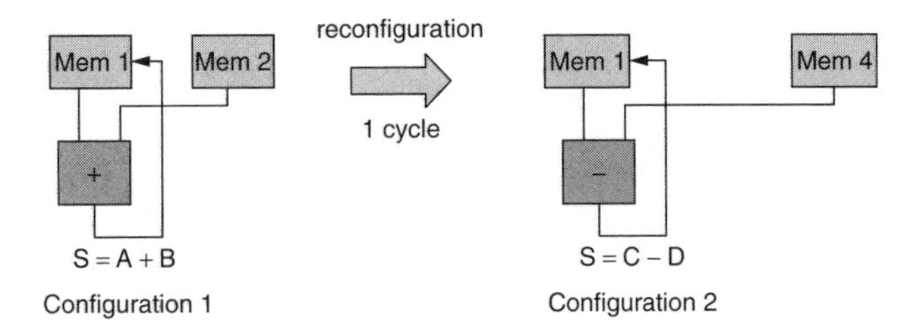

FIGURE 20.5 Software reconfiguration example.

such as that of very long instruction word (VLIW) processors. For each operator used, the data are read in memory, computed, and then the result is stored back to the memory at each cycle.

This software reconfiguration thus concerns only the functionality of the operators, the size of the data and their origin. Thanks to these flexibility limitations, the RDP may be reconfigured at each cycle with only one 52-bit instruction. This is illustrated on Figure 20.5, which represents the reconfiguration needed to replace an addition of data stored in the memories 1 and 2 by a subtraction on data stored in the memories 1 and 4.

Thanks to these two reconfiguration modes and to the SCMD concept, DART supports every kind of processing while being able to be optimized for the critical (i.e., regular) ones. These two types of reconfiguration can moreover be mixed without any constraint, and have a great influence on the development methodology.

20.3.4 Development Flow

To exploit the computation power of DART, an efficient development flow is the key to enhance the status of the architecture. Hence, a compilation framework, which allows the exploitation of the previously mentioned programming models, has been defined. It is based on the joint use of a front-end allowing the transformation and the optimization of a C code [27], a retargetable compiler [28], and an architectural synthesis tool [29]. As in most development methodologies for reconfigurable hardware, the key of the problem has been to distinguish the different kinds of processing. This approach has already been used with success in the program-in chip-out (PICO) project developed at HP labs in order to distinguish regular codes, implemented in systolic array, and irregular codes, executed in a VLIW processor [30]. Other related works such as the Pleiades project [31] or GARP [32] are also distinguishing regular processings and irregular ones. Massively parallel processings are implemented on circuits respectively reconfigurable at the functional and at the bit level, and irregular codes are executed on a RISC processor.

The development flow allows the user to describe its applications in the C language. This high-level description is translated at first into control and data flow graph (CDFG), from which some automatic transformations (e.g., loop unrolling and loop kernel extractions) [33] are done to optimize the execution time. After these transformations, the distinction between regular and irregular codes and data manipulations permits to translate, thanks to the compilation and the architectural synthesis, a high-level description of the application into binary executable codes for DART [34]. A cycle-accurate bit-accurate simulator developed in SystemC finally allows to validate the implementation and to evaluate its performance and energy consumption.

20.4 Validation Results

This section presents significant results stemming from a WCDMA receiver implementation on DART. Energy distribution between the different components of the architecture is also discussed. Performance and energy efficiency of DART is finally compared with typical reconfigurable architectures and programmable processors.

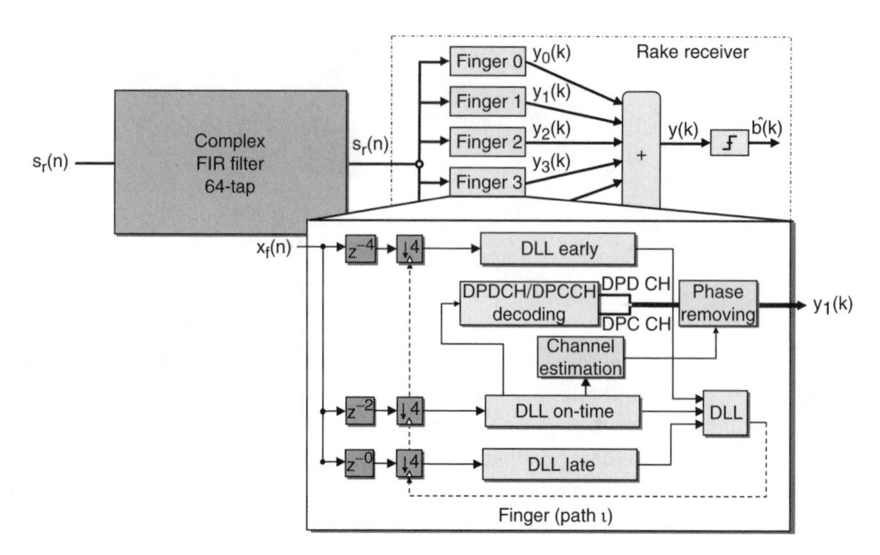

FIGURE 20.6 WCDMA receiver synoptic.

20.4.1 Implementation of a WCDMA Receiver

WCDMA is typically considered as one of the most critical applications of next-generation telecommunication systems [35]. A synoptic of this receiver is given in Figure 20.6. Within a WCDMA receiver, real and imaginary parts of data received on the antenna, after demodulation and digital-to-analog conversion, are filtered at first thanks to two real FIR (finite impulse response) filters. These two 64-tap filters operate at a high frequency (15.36 MHz), which lead to a tremendous complexity of 1966 millions of MAC per second (MMACS). Next, a rake receiver has to extract the usable information in the filtered samples and to retrieve the transmitted symbol. Because the transmitted signal reflects in obstacles like buildings or trees, the receiver get several replica of the same signal with different delays and phases. By combining the different paths, the decision quality is drastically improved and consequently, a rake receiver is constituted of several fingers which have to despread one part of the signal, corresponding to one path of the transmitted information. The decision is finally done on the combination of all these despreaded paths. The complexity of this complex despreading is 184.3 MOPS for six paths. To improve the decision quality, amplitude and delay of each path have to be estimated and removed from the signal. The synchronization between the received signal and the codes internally generated, i.e., the delay estimation and removing, is done in two times. The first part of this processing operates at a high-frequency (chip rate: F_c = 3.84 MHz) and has a complexity of 331 MOPS. The second one operates at the symbol frequency (F_s), which depends on the required bit-rate, and has a low complexity (e.g., 1.3 MOPS for a spreading factor of 256). Finally, the channel estimation is a low-complexity process and also operates at F_s.

Therefore, five configurations of the architecture may be distinguished: filtering, chip-rate synchronization, symbol-rate synchronization, channel estimation, and complex despreading. They follow one another on the architecture as depicted in Figure 20.7.

DART clusters have been designed under a 1.9-V, 0.18- μm technology. The synthesis has led to an operating frequency of 130 MHz. Running at this frequency, DART is able to provide up to 3120 MMACS per cluster, on 8-bit data. Thanks to the cycle-accurate bit-accurate simulator, the overall energy consumption of the architecture is evaluated from the activity of the different modules (e.g., functional units, memories, interconnection networks, control and registers) and their average energy per access. The latter has been estimated at the gate level.

Thanks to the configuration data volume minimization, reconfiguration stages are very short, and thus represent only 0.05% of the overall execution time. The effective computation power proposed by DART on these applications is 6.2 giga operations per second (GOPS). In such conditions, the processing of a

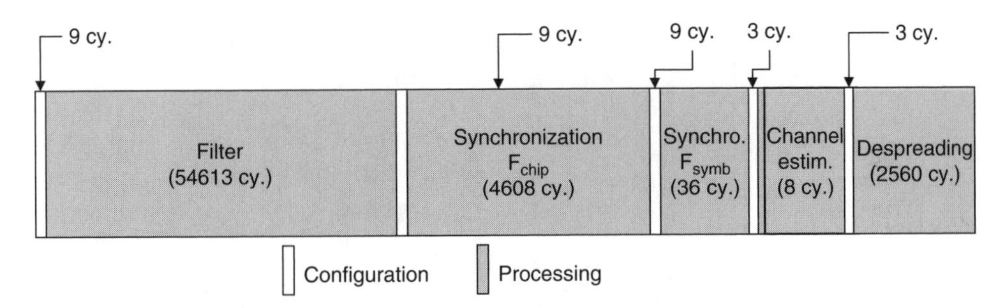

FIGURE 20.7 DART reconfigurations during the processing of one slot.

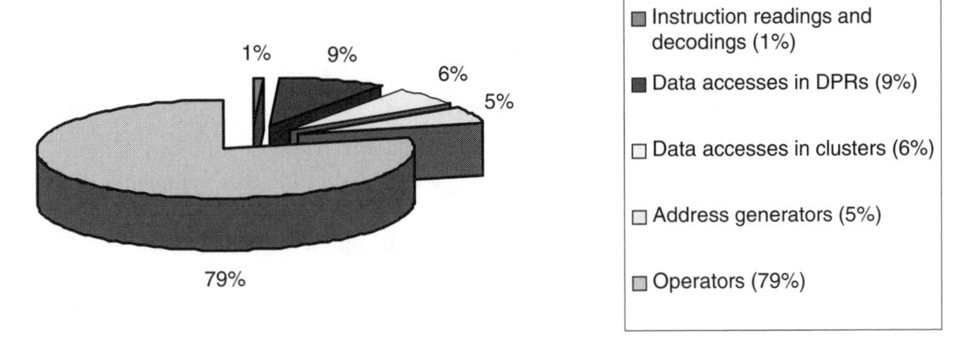

FIGURE 20.8 Power consumption distribution in a DART cluster during the processing of the WCDMA receiver.

WCDMA receiver on DART leads to a cluster usage rate of 72.6%. This performance level is done possible by the flexibility of the DART interconnection network, which allows a nearly optimal use of the RDP internal processing resources.

20.4.2 Energy Distribution in DART

The average energy efficiency of DART during the implementation of this WCDMA receiver is 38.8 MOPS/mW. Figure 20.8 represents the power consumption distribution between the different components of the architecture. We can notice that the main part of the cluster consumption comes from the operators (79%). Thanks to the configuration data volume minimization and to the reconfiguration frequency reduction, the energy overhead associated to the control of the architecture is negligible. During this processing, only 0.9 mW are consumed to fetch and decode control information, that is to say less than 0.8% of the 114.8 mW needed for the processing of a WCDMA receiver.

The power consumption issuing from data accesses is also reduced (20% for memory accesses and address generation). This is notably due to the minimization of the energy cost of local memory access, obtained by the definition of an appropriate memory hierarchy. In the same time, one-to-all connections allow to significantly reduce the amount of data memory accesses. In particular, on filtering and complex despreading applications, which exploit a thread-level parallelism, the simultaneous use of several functional units with a same data-flow allows to drastically reduce the number of accesses to the data memory. The use of delay chains also allows to benefit from data temporal locality and to skip a lot of data memory accesses. For this WCDMA receiver, the joint use of delay chains and of one-to-all connections permit to save 46 mW, representing a 32% reduction of the overall consumption of a cluster.

20.4.3 Performance Comparisons

This section compares the performances of DART with bit- and system-level reconfigurable architectures. The first considered architecture is the Virtex Xcv200E FPGA from Xilinx [36]. This choice is justified

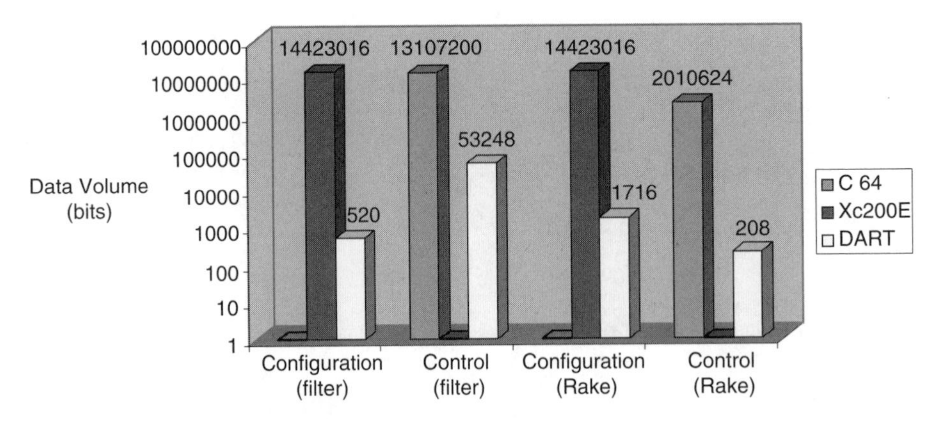

FIGURE 20.9 Configuration and control data volume of the Xc200E, the C64x, and DART for filter and rake receiver.

by the need to study the reconfiguration cost. This component has been dimensioned to have a computation power strictly corresponding to the FIR filters implementation. Two configurations will thus follow one another on the FPGA: the filter and the rake receiver. The configuration data volume for this circuit is about 1.4 Mbits. As DART, it is distributed on a 0.18- μm technology. The second considered architecture is the TMS320C64x digital signal processor (DSP) from Texas Instruments. This DSP is a VLIW architecture able to exploit an ILP of eight as well as data-level parallelism thanks to SWP capabilities [37]. This processor is distributed on a 0.12-μm technology. With a 720-MHz clock frequency, it may deliver up to 5760 MOPS.

Configuration and control cost has a critical impact on the performance and the energy efficiency of the system. Figure 20.9 represents the information volume needed for these two operations during the filtering and the rake receiver applications, for the C64, the Xc200E, and DART. Figure 20.9 clearly illustrates the conceptual divergences between the bit-level reconfiguration and the system-level reconfiguration. In the case of FPGA, a very large amount of information is distributed to the component before executing the application. The reconfiguration cost is here very important but once specified, the reconfiguration has no influence on the execution time. On the other hand, system-level reconfigurable architectures do not need to configure the datapath structure. On the contrary, the architecture control cost is critical. It corresponds, for the C64, to a 256-bit instruction reading and decoding at each cycle. DART allows the trade-off of these two operations, and thus minimizes energy waste. The hardware configuration between the processing is limited to the distribution of few hundreds of bits, while the architecture control is limited to the specification of reset instructions.

These considerations, relative to the architecture management, partly explain the results appearing in Figure 20.10, which represents the computation time associated to the three architectures according

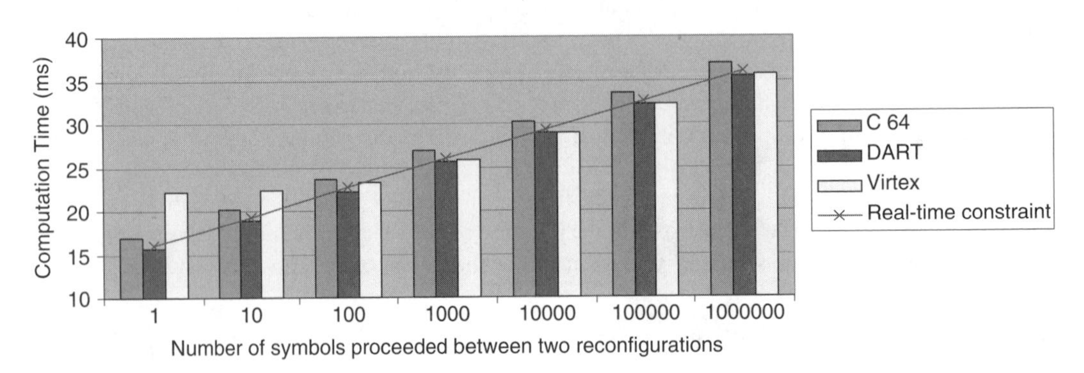

FIGURE 20.10 Xc200E, TMS320C64x, and DART performance on a WCDMA receiver.

to the number of symbols processed between two reconfigurations. The normalized real-time deadline represented on Figure 20.10 demonstrates that the DSP performance does not allow the implementation of the WCDMA receiver in real time, even when SWP is fully exploited. The FPGA is able to support the real-time constraint when the number of symbols processed between two reconfigurations exceeds 150. In such conditions, the configurations have to be stable during at least 10 ms because the reconfiguration time of this component is 2.7 ms. This remark highlights the impact of the reconfiguration overhead. In such conditions, it is essential to filter all the samples of a frame (153,600 samples), to store the filtered data into the memory, then to reconfigure the component to decode these data.

The power consumption associated with the implementations on the FPGA has been estimated at the gate level with the Xpower tool from Xilinx. The FPGA consumes 670 mW during the filtering and 180 mW during the rake receiver. In other words, the energy efficiency of the Xcv200E is 5.8 MOPS/mW during the filters and 2.9 MOPS/mW during the rake receiver.

An important drawback of the implementation with the WCDMA receiver on the Xc200E FPGA thus comes from the large delay separating data receiving and data decoding. This temporal shift exceeds 10 ms and might be unacceptable in mobile applications. Another problem associated with this solution comes from the volume of temporary data. The need to store the filtered samples before to decode them implies the use of a large amount of memory. To store a frame, 1.2 Mbits of memory are needed, which exceeds the storage capacity of the Xc200E. It is thus necessary to use external memory which implies an important energy overhead.

It has to be noticed that these drawbacks can be overcome by using larger chips. For example, the Xcv1000E, from the same FPGA family, allows the implementation of the WCDMA receiver in one configuration. In that case, no reconfiguration occurs and the real-time constraints can always be achieved. Obviously, this solution leads to a drastic increase of the device cost and to an energy efficiency reduction (about 20%).

The DSP power consumption has been estimated thanks to the results presented in Texas Instruments [38]. We have estimated this consumption to be 1.48 watt during the filters and 1.06 watt during the rake receiver. The energy efficiency of this architecture is therefore 2.6 MOPS/mW during the filters and 1.8 MOPS/mW during the rake receiver.

By minimizing the energy waste related to the architecture control and to the data accesses, DART is able to execute nearly 39 MOPS for each mW consumed. Unlike high-performance DSP and FPGA, flexibility does not come with a significant energy efficiency reduction.

20.5 Conclusions

This chapter discussed how to improve energy efficiency in flexible architectures. In such context, reconfigurable processors offer opportunities. They allow energy waste in control, storage as well as in computation resources to be reduced by adapting their datapath structure and by minimizing reconfiguration data volume. The association of key concepts, and of an energy aware design, lead to the definition of the DART architecture. Innovative reconfiguration schemes allow to deal concurrently with high-performance, flexibility, and low-energy constraints. We have validated this architecture by presenting implementation results of a WCDMA receiver. A computation power of 6.2 GOPS combined with an energy efficiency of 40 MOPS/mW demonstrate its potential in the context of multimedia mobile computing applications.

20.6 Acknowledgments

This project was conducted in collaboration with STMicroelectronics and UBO, and received funding from the French government. The authors would like to thank Dr. Tarek Ben Ismail and Dr. Osvaldo Colavin from STMicroelectronics, as well as Professors Bernard Pottier and Loïc Lagadec from UBO for their contributions.

References

[1] S. Hauck. The roles of FPGAs in reprogrammable systems. *Proc. IEEE*, 86:615–638, April 1998.

[2] E. Sanchez, M. Sipper, J.O. Haenni, J.L. Beuchat, A. Stauffer, and A. Perez-Uribe. Static and dynamic configurable systems. *IEEE Trans. on Computers*, 48(6):556–564, 1999.

[3] J. M. Rabaey. Reconfigurable processing: the solution to low-power programmable DSP. In *Int. Conf. on Acoustics, Speech, and Signal Processing (ICASSP)*, April 1997.

[4] A. Dehon. Reconfigurable architectures for general-purpose computing. Ph.D. thesis, Massachusetts Institute of Technology, Artificial Intelligence Laboratory, Cambridge, MA, October 1996.

[5] H. Zhang, M. Wan, V. George, and J. Rabaey. Interconnect architecture exploration for low-energy reconfigurable single-chip DSPs. *Int. Workshop on VLSI*, April 1999.

[6] J. Becker, T. Pionteck, and M. Glesner. DReAM: a dynamically reconfigurable architecture for future mobile communication applications. *Int. Workshop on Field Programmable Logic and Applications (FPL '00)*, pp. 312–321, Villach, Austria, August 2000. Lecture Notes in Computer Science 1896.

[7] H. Singh, G. Lu, M. Lee, E. Filho, and R. Maestre. MorphoSys: case study of a reconfigurable computing system targeting multimedia applications. *Int. Design Automation Conf.*, pp. 573–578, Los Angeles, CA, June 2000.

[8] S. Goldstein, H. Schmit, M. Moe, M. Budiu, and S. Cadambi. PipeRench: a coprocessor for streaming media acceleration. *Int. Symp. on Comput. Architecture (ISCA '99)*, Atlanta, GA, May 1999.

[9] G. Smit, P. Havinga, P. Heysters, and M. Rosien. Dynamic reconfiguration in mobile systems. *Int. Conf. on Field Programmable Logic and Applications (FPL '02)*, pp. 171–181, Montpellier, France, September 2002. Lecture Notes in Computer Sciences 2438.

[10] D. C. Cronquist, P. Franklin, C. Fisher, M. Figueroa, and C. Ebeling. Architecture design of reconfigurable pipelined datapath. *Advance Research in VLSI (ARVLSI '99)*, pp. 23–40, Atlanta, GA, March 1999.

[11] A. Abnous and J. Rabaey. Ultra low-power specific multimedia processors. *In VLSI Signal Processing IX*. IEEE Press, November 1996.

[12] Chameleon Systems. Wireless base station design using reconfigurable communications processors. Technical report, 2000.

[13] B. Brodersen. Wireless systems-on-a-chip design. *Int. Symp. on Quality Electronic Design (ISQED '02)*. Invited paper, San Jose, CA, March 2002.

[14] R. Hartenstein, M. Hertz, Th. Hoffman, and U. Nageldinger. Generation of design suggestions for coarse-grain reconfigurable architectures. *Int. Workshop on Field Programmable Logic and Applications*, Villach, Austria, August 2000. Lecture notes in Computer Science 1896.

[15] K.K.W. Poon. Power estimation for field programmable gate arrays. Master's thesis, University of British Columbia, Vancouver, Canada, 2002.

[16] L. Shang, A.S. Kaviani, and K. Bathala. Dynamic power consumption in Virtex-II FPGA family. *Int. Symp. on Field Programmable Gate Arrays (FPGA '02)*, pp. 157–164, Monterey, CA, February 2002.

[17] J. Fridman. Sub-word parallelism in digital signal processing. *IEEE Signal Process. Mag.*, 17(2):27–35, March 2000.

[18] J.P. Wittenburg, P. Pirsh, and G. Meyer. A multithreaded architecture approach to parallel dsps for high-performance image processing applications. *Workshop on Signal Process. Syst. (SIPS '99)*, Taipei, Taiwan, October 1999.

[19] G. Stitt, B. Grattan, J. Villarreal, and F. Vahid. Using on-chip configurable logic to reduce system software energy. *Symp. on Field-Programmable Custom Computing Machines (FCCM '02)*, Napa, CA, September 2002.

[20] S. Wuytack, J.Ph. Diguet, F. Catthoor, and H. De Man. Formalized methodology for data reuse exploration for low-power hierarchical memory mappings. *IEEE Trans. on VLSI Syst.*, 6(4):529–537, December 1998.

[21] Xilinx. *Xilinx 6200 Preliminary Data Sheet*. San Jose, CA, 1996.

[22] T.J. Callahan, J.R. Hauser, and J. Wawrzynek. The Garp architecture and C compiler. *IEEE Comput.*, 33(4):62–69, April 2000.

[23] X. Zhang and K.W. Ng. A review of high-level synthesis for dynamically reconfigurable FPGAs. *Microprocessors and Microsystems*, 24:199–211, 2000.

[24] E. Dinan and B. Jabbari. Spreading codes for direct sequence CDMA and wideband CDMA cellular network. *IEEE Commun. Mag.*, 36(9):48–54, September 1998.

[25] R. David, D. Chillet, S. Pillement, and O. Sentieys. DART: a dynamically reconfigurable architecture dealing with next-generation telecommunications constraints. *Int. Reconfigurable Architecture Workshop (RAW '02)*, Fort Lauderdale, FL April 2002.

[26] P. Faraboshi, J.A. Fisher, and C. Young. Instruction scheduling for instruction level parallel processors. *Proc. IEEE*, 89(11):1638–1659, November 2001.

[27] R. Wilson et al. SUIF: an infrastructure for research on parallelizing and optimizing compilers. Technical report, Computer Systems Laboratory, Stanford University, Stanford, CA, May 1994.

[28] F. Charot and V. Messe. A flexible code generation framework for the design of application specific programmable processors. *Int. Symp. on Hardware/Software Codesign*, Rome, Italy, May 1999.

[29] O. Sentieys, J.P. Diguet, and J.L. Philippe. A high-level synthesis tool dedicated to real time signal processing applications. *European Design Automation Conf. (EURODAC '95)*, Brighton, U.K., September 1995.

[30] R. Schreiber, S. Aditya, S. Mahlke, V. Kathail, B. Ramakrishna Rau, D. Cronquist, and M. Sivaraman. PICO-NPA: high-level synthesis of non-programmable hardware accelerators. Technical report HPL-2001-249, Hewlett-Packard Laboratories, Palo Alto, CA, 2001.

[31] M. Wan. Design methodology for low-power heterogeneous digital signal processors. Ph.D. thesis, University of California at Berkeley, Berkeley Wireless Design Center, 2001.

[32] J. Hauser. Augmenting a microprocessor with reconfigurable hardware. Ph.D. thesis, University of California at Berkeley, 2000.

[33] A. Fraboulet, K. Godary, and A. Mignotte. Loop fusion for memory space optimization. *Int. Symp. on Syst. Synthesis (ISSS '01)*, Montreal, Canada, October 2001.

[34] R. David, D. Chillet, S. Pillement, and O. Sentieys. A compilation framework for a dynamically reconfigurable architecture. *Int. Conf. on Field Programmable Logic and Applications*, pp. 1058–1067, Montpellier, France, September 2002. Lecture Notes in Computer Science 2438.

[35] T. Ojanpera and R. Prasad. *Wideband CDMA for Third-Generation Mobile Communication.* Artech House Publishers, London, 1998.

[36] Xilinx. *VirtexE Series Field Programmable Gate Arrays.* Xilinx, San Jose, CA, July 2001.

[37] Texas Instruments. *TMS320C64x Technical Overview.* Texas Instruments, Dallas, TX, February 2000.

[38] Texas Instruments. *TMS320C6414/15/16 Power Consumption Summary.* Application report, spra811a, Dallas, TX, March 2002.

21

Macgic, a Low-Power Reconfigurable DSP

Flavio Rampogna
Pierre-David Pfister
Claude Arm
Patrick Volet
Jean-Marc Masgonty
Christian Piguet
CSEM SA

21.1 Introduction

Low-power programmable digital signal processors (DSP) can nowadays be found in a broad range of battery-operated consumer devices, such as MP3/CD/DVD players, or in the ubiquitous cellular phone. The trend being in the software implementation of more complex signal processing algorithms, very high-performance programmable DSP microprocessors having both a very high computational power capability and a very low-power consumption will be required in the near future to seamlessly implement such algorithms.

21.1.1 DSP Architectures Evolution

The first programmable DSPs were relatively simple microprocessors, specialized in the handling of very specific data formats: either fixed-point or floating-point, depending on their architecture [9,10,13,17,19]. These processors were very efficient in transferring data between the memory and their data processing unit, as well as in the processing of the data itself. The data processing unit was typically optimized for the handling of multiply-and-accumulate (MAC) operations between two data words read from two different memories. Memory accesses were indirect, and most DSPs supported modulo indirect addressing modes especially useful in convolutions or for implementing circular buffers. Sometimes, a special reverse-carry addressing mode was also available, and was useful for the reordering of the data in fast-Fourier-transform (FFT) computations. The address computation hardware was typically located into address generation units (AGUs). An AGU usually contains a set of specialized index, offset and modulo registers.

0-8493-1941-2/05/$0.00+$1.50

Historically, programmable DSPs implemented a very limited and specialized set of registers: temporary registers, accumulators, and AGU registers. This nonorthogonality of the architecture and the limited resources made these processors very difficult for a high-level language compiler to generate program code. To fully exploit the processing power and available instruction-level-parallelism (ILP) of these DSPs, they had to be programmed in assembly language, which, very often, was quite a tedious task.

In the last few years, the tendency has been to limit the number of specialized registers by implementing large sets of general-purpose registers that can be used as operands for most, if not all, instructions, and to have the hardware providing support for multiple data types [6,11,12,14,15,16,18]. The latest high-performance DSP architectures generally provide a very high data transfer bandwidth that can be exploited by a large number of parallel processing units. There are still different kinds of processing units, specialized for a given kind of processing. Several general-purpose ALUs are typically available, as well as a branch/loop unit, and specialized address generation ALUs. These recent architectures provide a relatively good hardware support for high-level language compilers. Code generation is indeed eased by the availability of multiple relatively basic parallel processing elements (ALUs), by the sufficiently large number of general-purpose registers, and by the support of standard data types (i.e., chars, integers, long integers, and floating-point).

In some modern DSPs, and to reduce both the power consumption and the hardware complexity of the circuit, the instruction-level parallelism (ILP) made available by these architectures is made explicit [12,16] (i.e., operations to be executed are grouped together into clusters, which are typically between 128 and 256 bits wide and may contain between 4 to 8 operations). Within a cluster it is sometimes possible to specify what are the operations that can be executed in parallel, and which ones need to be executed sequentially. The simplest approach, however, being to have all operations in a cluster executed in parallel and a direct and simple mapping between available hardware resources and operations coding in a cluster. In this approach, the scheduling of operations execution has to be explicitly specified by the programmer (or the compiler), and not chosen by the hardware, such as in superscalar architectures.

An alternative to programmable DSPs comes from configurable but nonprogrammable DSP coprocessors [7], optimized and specialized for the computation of a very specific signal processing task, such as FFT, FIR, IIR, Viterbi decoding, or image motion estimation computation. Such coprocessors may use the system's direct memory access (DMA) mechanisms to fetch the data needed to compute their algorithm, or implement their own memory address generation mechanisms. Future high-performance DSP systems may well include one or more of these coprocessors, together with one or more programmable DSPs or general-purpose microprocessors. Indeed, use of coprocessors may easily improve a system's performance by an order of magnitude or even more, by allowing very efficient parallel implementation of specific algorithms (e.g., Viterbi decoder or FFT computation).

21.1.2 Parallelism, Instruction Coding, Scheduling, and Execution

Today's high-performance DSP microprocessors can often execute up to eight different operations in parallel coded in a single instruction word (e.g., four MAC, two data address computations together with two memory accesses, one branch, and one bit manipulation operation). The packing of a large number of parallel operations into instruction word(s) can be performed using different approaches, leading to different DSP architectures.

A first possible approach consists in keeping all parallel operations as separate and independent instructions, and defining an instruction-set in which instructions are relatively small in terms of the number of bits required for the coding of an operation. The processor would read multiple instruction words from the memory at once and schedule their execution. Scheduling could either be automatically performed by the processor's hardware, as in a superscalar microprocessor, or predefined by the programmer or the compiler (Figure 21.1(a)). If the scheduling is predefined, the architecture is called static superscalar. Explicit scheduling information has to be encoded in an instruction's opcode (Figure 21.1(b)). The kind of explicitly provided scheduling information could be: Execute this instruction in parallel or in sequence with the previous instruction.

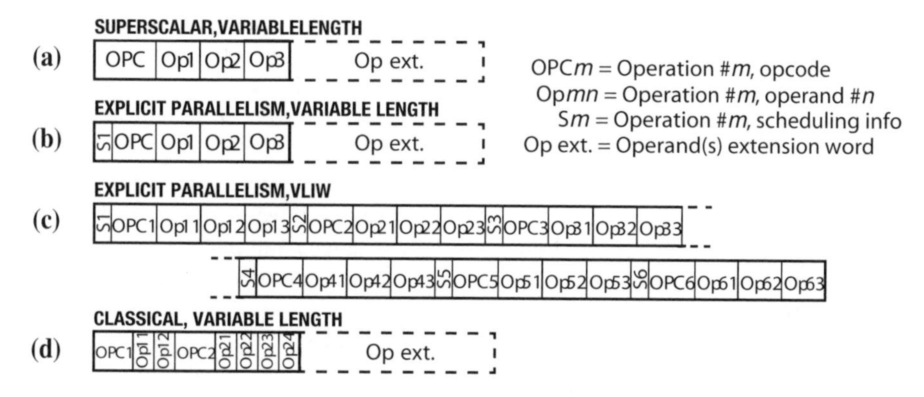

FIGURE 21.1 Parallelism and instruction coding.

A second, slightly different, approach may consist in the explicit coding of different parallel operations into a single very large instruction word (VLIW), typically of 128 bits or more. The processor fetches such a large instruction and executes all parallel operations contained in it. The operation execution scheduling is explicit and the simplest possible scheduling mechanism would be that all operations coded into the VLIW instruction are executed in parallel. Such an approach could prove to be quite memory wasting, however, especially in situations where low parallelism is actually available in a program. To solve this problem, a more advanced scheduling mechanism may be implemented by encoding some additional information on the need for parallel or sequential execution to each operation coded in a VLIW instruction (Figure 21.1(c)). By using this additional scheduling information, and by an appropriate ordering of operations within an instruction word, it is possible to fulfill any instruction execution scheduling needs, while still keeping an optimum code memory density because no-operations (NOPs) are not required to fill up VLIW instructions.

A third possible approach consists in using instruction words of relatively small size, typically between 24 to 64-bit, and by packing very few parallel operations in such words (Figure 21.1(d)). This is the approach that has originally been followed in the first programmable DSP microprocessors, but which is still widely used today [11,14,17,19]. In this approach, an instruction word consists of up to four different operations that are executed in parallel. It is indeed common to find DSP architectures allowing for the encoding of an ALU operation together with the encoding of two indirect memory accesses and address indexes updates operations in a single instruction word. The limitation of this approach, when applied to high-performance DSPs, is that these relatively small instruction words do not allow for the encoding of a very large number of parallel operations, therefore limiting the maximum available parallelism that can be programmatically exploited at the instruction level. To overcome part of this limitation it is possible to define operations that actually perform a unique computation on multiple data. These operations are of the single-instruction-multiple-data (SIMD) category [8]. For example, an MUL4 operation allows performing four multiplications on four pairs of data. As for VLIW instruction packing, here too, if little parallelism can be extracted from a program, the available parallel operations of an instruction word cannot be fully exploited and are to be replaced with NOPs, thus needlessly increasing program code size. By using variable-length instruction words, it is possible to reduce such a program memory occupation overhead, while unfortunately somewhat complicating the instruction fetch and decoding hardware's task.

21.1.3 High-Performance for Low-Power Systems

Power consumption reduction in a system-on-chip (SoC) can be achieved at different levels, ranging from the careful analysis of an application's needs, to appropriate algorithms selection, as well as a good knowledge of precision requirements for the data to be processed. Selection of appropriate hardware architecture for implementing these algorithms, taking into account the DSP processor core(s), the memory subsystem (i.e., internal/external RAMs, ROMs, and DMA availability) and the data acquisition chain (i.e., ADC/DAC, and I/Os), may allow a significant power reduction [20]. In addition, the use of an appropriate semiconductor technology together with a good trade-off between the hardware's computational power, and operating voltage and frequency may allow for a large power consumption reduction.

Modern high-performance DSPs and general-purpose microprocessor circuits and systems usually implement multilevel memory access hierarchies: typically, two cache-memory levels followed by a high-speed internal RAM or external (S)DRAM memory containing data and instructions. The memory is usually seen as unified for the programmer; data and instructions can be intermixed. Internally, however, the DSP uses multiple independent memory busses that actually implement distinct memory spaces. Generally, a DSP implements a program memory bus and one or two data memory busses. Each memory bus is typically connected to a specific level-1 (L1) cache memory. Unification of memory spaces may occur after the L1 cache or after the level 2 (L2). In very low-power systems, where memory needs and maximum operating speed is not very high, caches can be avoided and the memory spaces may remain distinct. This, together with a simpler hierarchical memory subsystem may help reduce power consumption. The power consumption of the memory subsystem can quite easily be reduced simply by placing the most often accessed data into smaller memories, and the less often accessed data in larger ones because smaller memories are faster and consume less energy per access than larger ones.

21.1.4 DSP Performance and Reconfigurability

With the increasingly high costs for accessing advanced semiconductor technologies, and with implementation of more computationally demanding signal processing algorithms, a modern SoC implementing programmable DSP(s) should be as efficient and generic as possible to allow for the implementation of the larger possible number of applications. Therefore, a DSP core has to be as power efficient as specialized hardware, and be retargetable to different algorithms and applications without any significant loss of performance. Fortunately, in some applications, performance vs. power-consumption vs. reconfigurability goals may be met by the implementation of an appropriate programmable signal processing architecture. For example, a system's computational performance may be increased by allowing multiple parallel processing units to compute an algorithm on different data, different parts of an algorithm on pipelined data, or different algorithms on identical or different data. The maximum achievable parallelism depends on the algorithms to be executed and available hardware resources: processors, coprocessors, and memories.

Power consumption of a system can be reduced by appropriate selection of power supply voltage, operating circuit frequency, and available parallelism. Indeed, by increasing the execution parallelism, the timing constraints are relaxed and the circuit's operating frequency can be decreased, which allows to lower its operating voltage and therefore to reduce its dynamic power consumption.

If an SoC system has to be reconfigured to support a new application, or multiple applications, reconfiguration can be achieved at various levels, the main being at the program code level by the programming of a new application's algorithms. The program has then to be stored into a reprogrammable memory, such as an EEPROM. Sometimes, an external serial EEPROM chip can be used to initialize the content of an internal RAM at reset-time, and thereby allowing the configuration of the system. Additional reconfiguration levels are obtained when the program actually reconfigures the SoC's hardware: coprocessors, direct memory access (DMA) hardware, peripherals, or even the actual DSP processor [1,3,4,5].

The runtime reconfiguration of a programmable DSP core may be achieved in different ways. The Macgic DSP architecture allows the programmer to reconfigure and use a small set of extended instructions, which allow a fine-grain control over specific datapaths: the address generation unit (AGU) and the data processing unit (DPU) datapaths. This fine-grain control increases the programmatically exploitable hardware parallelism. Such extended instructions are typically used in algorithm's kernels to significantly speed up their execution, while keeping the program code density performance of the DSP at a good level.

21.2 Macgic DSP Architecture

21.2.1 General Architecture

Macgic is a low-power programmable DSP core for SoC designs. It can be used as a stand-alone DSP, or as a coprocessor for any general-purpose microprocessor or DSP. It has been designed to be efficient for a broad range of DSP applications. This is done by providing the designers with the possibility to tailor some Macgic features to best fit an application class (e.g., audio, video, or baseband radio). In particular, the various specifiable word sizes (e.g., data and address), the DSP modularity, and the instruction-set specialization are key features allowing Macgic to be very efficient both in terms of processing speed and energy consumption. This customization of the DSP must be performed before hardware synthesis.

Figure 21.2 presents the Macgic DSP architecture. The DSP core is made of four distinct operating units, each playing a specific role in the architecture: The program sequencing unit (PSU) handles branches, exceptions, and instruction fetch. The host and debug unit (HDU) handles data transfers with a host microprocessor and the debugging of Macgic programs through a specific debugging bus. The data move unit (DMU), containing the X and Y address generation units (AGU), handles data transfers between registers and between registers and external data memories. The data processing unit (DPU) handles the processing of the data.

Macgic uses relatively small (32-bit) instruction words. The data word size can be freely specified (e.g., dw = 12 to 32 bits) before synthesis. The DSP implements two distinct data memory spaces (X, Y). Concurrent accesses to these two memory spaces are supported. Up to four data words per memory space can be transferred between the DSP and the external memory per clock cycle. Macgic is a load/store architecture [8] and implements two banks of eight wide general-purpose (GP) registers, one bank per memory space. A wide register can store four data words. A GP register can be accessed either as a single data word, as half-wide, or as wide data words. Data processing operations can access up to 16 data words per clock cycle, from up to four wide GP registers, two per data space. The program and data address space sizes can be independently specified (paw, daw = 16 to 32 bits) before synthesis.

Complex addressing modes are made available by the two customizable and software reconfigurable address generation units (AGUs). The data processing unit (DPU) can also be customized and specialized before synthesis. Extended operations of the DPU can be software reconfigured. An HDU allows the control of Macgic from a host microprocessor or from a remote software debugger. The HDU also allows the exchange of data with the host microprocessor through specific data transfer FIFOs and registers.

21.2.2 Program Sequencing Unit

The program sequencing unit (PSU) is responsible for handling the instruction fetch, the global instruction decoding and the execution of branches, subroutine calls, hardware loops, and exceptions. This unit handles external interrupt requests as well as internal software exceptions. Eight prioritized and vectorized external interrupts requests lines are available to an external interrupt controller.

An external hardware stack stores the return address (of subroutines, exceptions) and the loop status when needed. The number of hardware loops, subroutines and interrupts that can be nested is given by the size of the hardware stack, which is a customization parameter.

The PSU contains eight 16-bit flag registers (IN, EX, EC, PF, HD, DM, PA, and PB). Two of these registers are actually controlled by the DPU (PA and PB), one by the DMU (DM), one by the HDU (HD) and the remaining ones by the PSU. DPU flags are typically the Z, N, C, and V flags of each ALU.

The PSU handles the conditional execution of operations in a manner similar to the conditional branches (i.e., operations are executed or not depending on the value of a flag taken from one of the eight flag registers). It implements hardware loops and instruction repetition mechanisms. Hardware loops automatically handle the iteration counting and branching to the beginning of the loop at the end of an iteration. Only one clock cycle is necessary to initialize a loop and there are no additional clock cycles penalties during its execution. Hardware loops can help reduce the clock cycles count of an

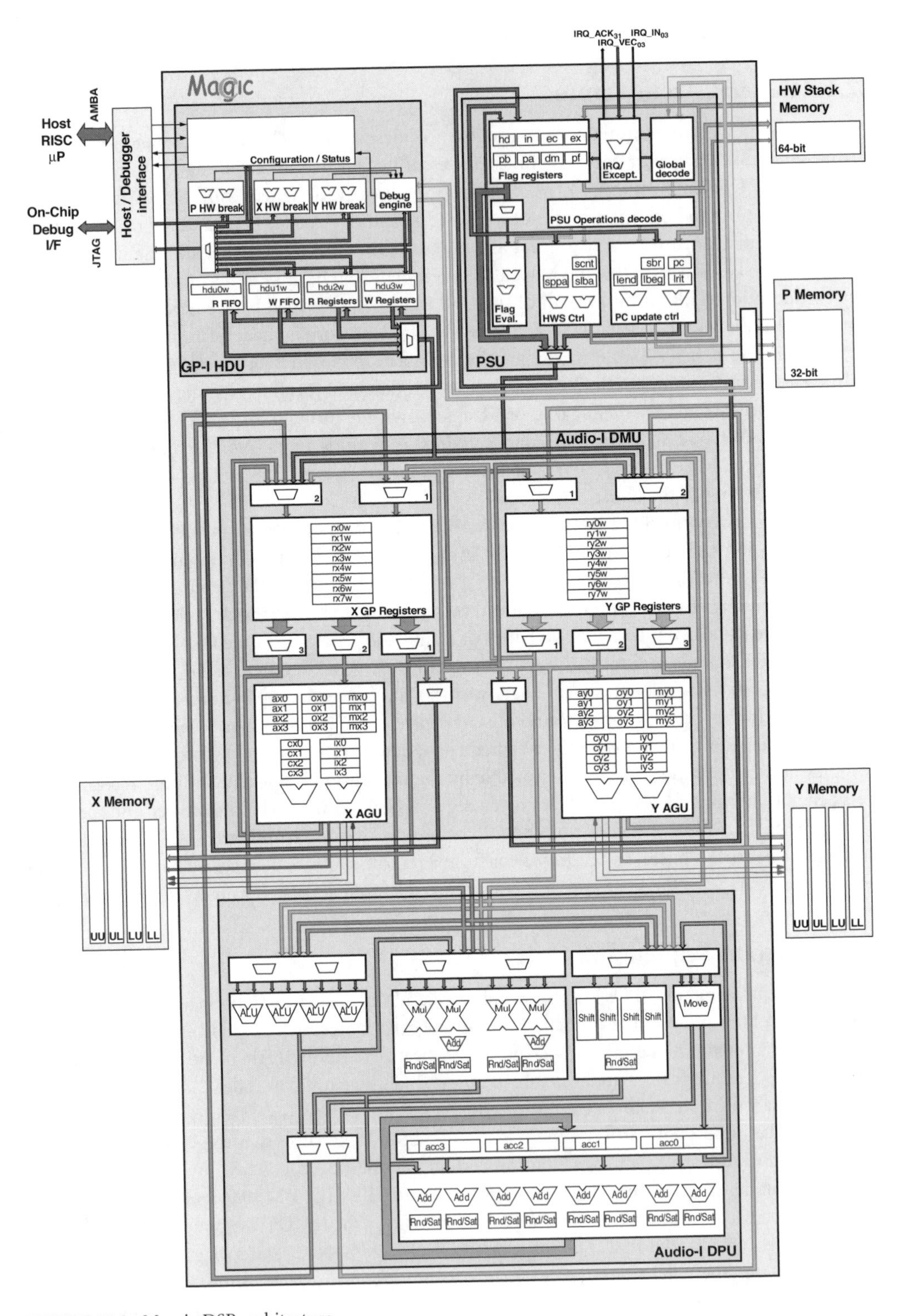

FIGURE 21.2 Macgic DSP architecture.

algorithm's execution in a significant manner, particularly for small loops that are iterated a large number of times. In instruction-repeat operations, the instruction to be repeated is fetched only once from the program memory, thus saving unnecessary, power-consuming, program memory accesses.

21.2.3 Data Move Unit

The DMU implements the data transfer mechanisms of the Macgic DSP. Data can be transferred between the DMU and the external memory, as well as between the DMU and the other units: DPU, HDU, and PSU. All data transfers use at least one GP register, either as a source or as a destination register.

The large number of data busses between the DMU and the DPU allows a very high data transfer bandwidth between these units: up to 16 data words can be transferred from the DMU to the DPU per clock cycle, and up to 8 data words from the DPU to the DMU.

Two address generation units (AGUs) are available in the DSP: one per data memory space. The AGUs are used to generate addresses for data memory accesses. Each AGU has four index register sets. In addition to the traditional base address, offset, and modulo registers, the configuration and extended-instruction registers allow the configuration and customization of the AGUs to best fit the targeted algorithms memory addressing needs. The two independent AGUs allow concurrent accesses to the two memory spaces.

21.2.4 Data Processing Unit

The data processing unit (DPU) implements the data processing capabilities of the Macgic DSP. Because the DSP architecture is modular, new DPUs can be developed and specialized to obtain the best possible performance for the class of algorithms to be executed. The first implementation of this unit has been a general-purpose one but slightly specialized towards audio processing. This first Audio-I DPU implements four ALUs, four multipliers, four shifters, together with four accumulator registers and their associated adders. The accumulator registers (width: 2 dw + 8 bits) allow storing the results of multiply-and-accumulate operations.

The DPU can handle data either as fixed-point, signed, or unsigned integer, depending on the operation selected. It implements round-to-nearest rounding, and a saturation mechanism can be enabled to ensure accurate computations.

21.2.5 Host and Debug Unit

The HDU is the link between Macgic and an external host microprocessor or a software debugger. This unit implements both data transfers and software debugging mechanisms.

The host microprocessor or debug interface access the HDU through the host/debug bus. A set of registers is then available and allows configuring and controlling the HDU. For this bus, the HDU acts as the slave and the host microprocessor or debug interface as the master.

The HDU allows the transfer of data between a host/debug bus master and the HDU registers. For this, two FIFOs are available, one per data transfer direction, as well as two groups of four registers, one per data transfer direction. The depths of the FIFOs are customizable. Flow-control mechanisms have been implemented. Writing or reading data into or from the FIFOs or registers can, for example, trigger the generation of an event to the bus master, or of an exception to Macgic.

In addition to data transfer mechanisms, the HDU implements a set of hardware breakpoints engines, one per memory space. Each engine allows monitoring accessed memory addresses and is able to generate a breakpoint if either an address range or a single address matches a given memory access kind: read, write, read or write.

The hardware breakpoint engines use the HDU debug engine to actually implement the breakpoint. The debug engine allows controlling the Macgic program execution. It allows to stop the DSP, execute instructions step-by-step, insert instructions in the DSP pipeline, access the program memory (e.g., to

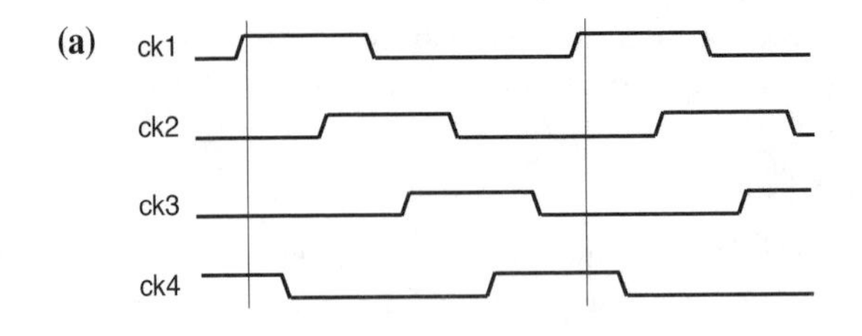

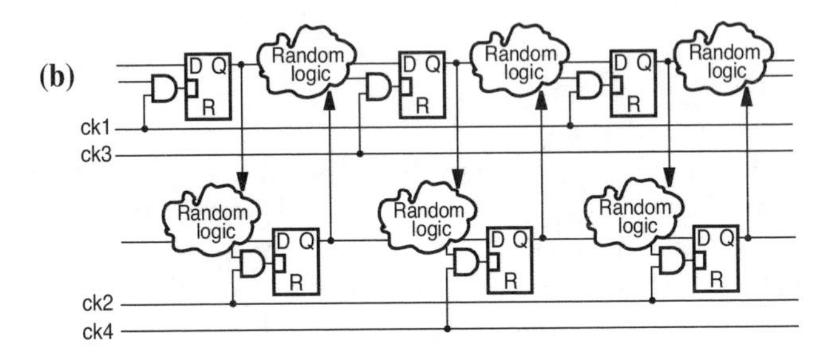

FIGURE 21.3 (a) Macgic DSP clock signals; (b) clock gating and pipeline.

place software breakpoints instructions or to download a program), or to get information on the DSP processor state.

21.2.6 Clocking Scheme

The Macgic DSP core uses four clocks signals ck1 to ck4. These signals must be nonoverlapping by pairs (i.e., ck1 must not overlap ck3, and ck2 must not overlap ck4). The DSP uses latches as data storage elements instead of flip-flops. The Macgic DSP uses these two pairs of nonoverlapping signals (ck1/ck3 and ck2/ck4) to implement the various pipeline stages and clock gating. Figure 21.3(a) depicts the partial overlapping of the four clock signals. Figure 21.3(b) illustrates how the level-sensitive storage elements implement the pipeline stages as well as clock gating. Clock-gating signals are generated either from signals latched during the previous clock phase or from two clock phases before the clock phase they are supposed to enable. They must be stable before the activation of the clock signal they enable.

The appropriate use of level-sensitive storage elements [2], makes the hardware less sensitive to clocks jitter than by using edge-sensitive storage elements, therefore allowing to implement more robust circuits, capable of working under very extreme operating conditions (e.g., voltage, temperature, and technology corner). With this approach, any trade-off between maximum clock frequency and power consumption related to jitter minimization in clock distribution trees can be made, without actually compromising the good working of the circuit — only the achievable maximum operating frequency.

The large number of clock phases enables a finer control over the pipeline, the clock gating mechanism, and simplify the generation of clean I/O data and control signals on the various DSP's external busses.

21.2.7 Pipeline

To simplify the description of the pipeline, the following clock-phase notation *c.n* is used. Where *c* represents the clock cycle number (c = 1..x) and n the clock phase number (n = 1..4). When $n = 1$, it means that the clock signal ck1 is asserted, when $n = 2$ that clock signal ck2 is asserted, and so on.

The Macgic DSP has been targeted to very lower-power applications. To keep both the design complexity and the power consumption to acceptable levels, the DSP pipeline depth has been made relatively short. Figure 21.4 depicts the various pipeline stages. Most instructions are typically executed in only three clock cycles.

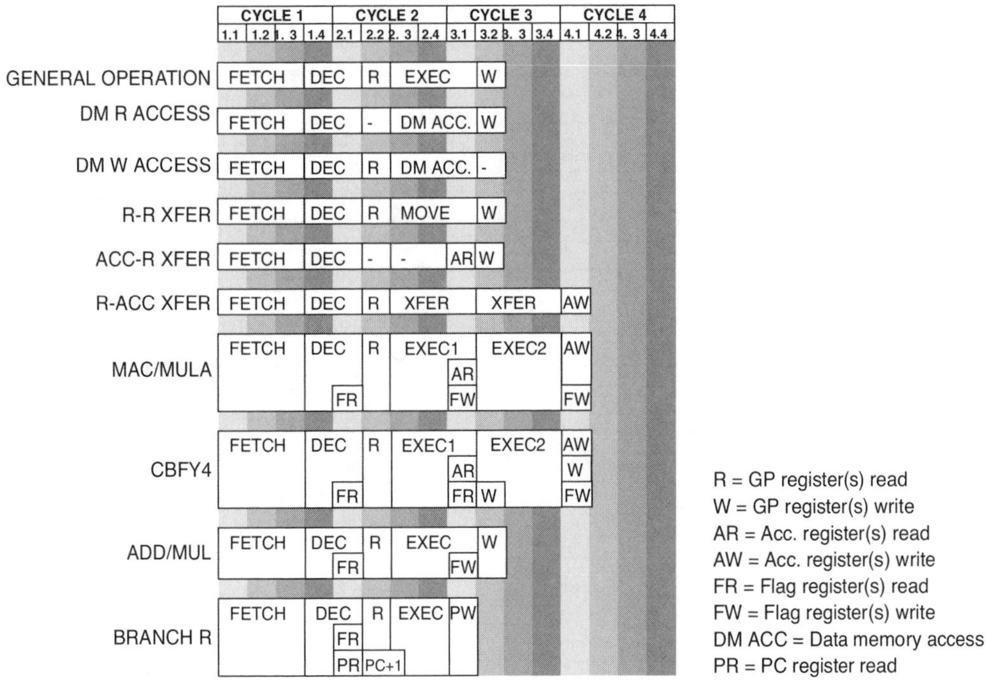

FIGURE 21.4 Macgic DSP pipeline (Audio-I DMU, Audio-I DPU, GP-I HDU).

In the Macgic DSP, the PC, the flag registers, and the accumulators are updated during phase 1 of each clock cycle, and the GP registers during clock phase 2. The program memory is accessed during phases 2 and 3 of each clock cycle, and the data memory during phases 4 and 1, while the hardware stack memory is accessed during the clock phase 3.

The delay between the reading and the writing-back of a register is typically of one clock cycle. This makes the pipeline transparent to the programmer, which greatly eases assembly-language programming. Only branches and a few instructions that are executed in four clock cycles need a special attention by the programmer. Branches necessitate a one-cycle delay slot (i.e., the instruction immediately following a branch is always executed). The instructions that write a result in a GP register with an additional latency cannot be immediately followed by instructions exploiting such a result. Unrelated operations or NOPs need to be inserted for the duration of the latency, before the result can actually be exploited.

One instruction is fetched per clock cycle, except if the pipeline has to be stalled by a program or data memory access wait-state, which delays the fetching and execution of subsequent instructions. Customized DPU/DMU or HDU instructions may, if needed, request a stall of the pipeline or a delaying of exception handling. The PSU handles the fetching and partial decoding of instructions. Fetched instructions are first partially decoded in the PSU, and the category of the operation(s) is determined. Then, the operation is dispatched to the appropriate unit for further decoding and execution. The PSU is not actually fully aware of the whole DSP instruction-set and pipeline. Completely independent and arbitrarily long execution pipelines can therefore be implemented in the DMU, DPU, and HDU. The pipeline can therefore vary from one implementation of a given unit to another implementation of the same unit. (e.g., short pipeline fixed-point hardware is implemented in one unit vs. long pipeline floating-point DPU hardware in the other one).

21.2.8 Instruction-Set

Macgic DSP instructions are 32 bits wide. This relatively small instruction size helps keeping the program memory power consumption to acceptable levels. A 32-bit instruction word fits one or two operations.

First operation		Second operation (executed in parallel with first operation)	
PSU-L	PSU Long operations (jumps, subroutine call, loop, etc.)	None	
PSU-M	PSU flag move operations	None	
DMU-L	DMU long operations (move immediate, direct memory)	None	
DPU-L	DPU long operations	None	
HDU-L	HDU long operations	None	
PSU-C	PSU conditional execution operation	**PSU-S**	PSU short operations (register-register transfers)
		PSU-M	PSU flag move operations
		DMU-S	DMU short operations (register-register transfers)
		DPU-P	DPU parallel operations
		DMU-P	DMU parallel operations (two indirect memory accesses)
PSU-P	PSU parallel operations (flag clear/set/invert)	**DMU-S**	DMU short operations
		DPU-P	DPU parallel operations
		DMU-P	DMU parallel operations
PSU-S	PSU short operations (HW stack PUSH/POP, reg. move)	**DMU-P**	DMU parallel operations (registers data move, indirect memory accesses)
DMU-S	DMU short operations (registers data move)		
DPU-P	DPU parallel operations		
HDU-P	HDU parallel operations		

FIGURE 21.5 Macgic instruction word operations categories coding.

Macgic DSP operations are split into several categories. Operations contained in an instruction word are executed in parallel. Figure 21.5 illustrates the available instruction-level parallelism. Additional parallelism can further be encoded within an operation (e.g., extended, SIMD, vectorial, or specialized operations).

PSU operations are "built-in" and cannot be customized nor removed from the instruction-set of the Macgic DSP. Hardware support is provided for nested hardware loops and instruction repeat.

Branches can be either direct or indirect. In case of indirect addressing, either GP registers or the software branch register (SBR) can be used as index. Program memory addressing can be either absolute or PC-relative. Branches can be conditional. The condition is the value of a flag. There are operations for handling and processing flags. Flags can be set, cleared, inverted. A Boolean expression evaluation can take expressions of up to three flags as operands, perform AND/OR/XOR operations on them and save the Boolean result into a flag in the PF flags register.

The Audio-I DMU makes a comprehensive set of data move operations available. These data transfers can move either single or wide data words. Up to two, parallel wide registers data can be moved into two wide GP registers in a single DMU-S or P operation. Up to two, wide data words can be transferred between the two memory spaces and two GP registers in a single DMU-P operation. In addition to single-word or wide data moves, half-wide or word-specific data moves are available. Immediate data moves are also available.

The Audio-I version of the AGU implements three types of indirect data memory access operations basic, predefined, and extended. Basic operations implement a simple set of very common DSP addressing operations. Up to three predefined operations can be configured for each index register through the appropriate programming of a configuration register. Predefined operations allow access to more powerful addressing modes than the ones made available by the basic operations. Extended operations further extend the complexity of addressing modes and operations that the AGU can perform. Up to four extended operations are available per AGU.

The actual operation performed by an extended operation is configured through an extended operation register. Extended operations may help reducing the number of clock cycles necessary to compute a specific address computation, potentially saving precious clock cycles in key parts of time-consuming algorithms.

The DPU is responsible for the processing of the data in Macgic. This unit can be customized to best fit the targeted class of algorithms and application. Two categories of DPU operations are available: DPU-P and DPU-L operations. DPU-P operations can be executed in parallel with PSU-P or DMU-P operations. In the Audio-I DPU, four kinds of DPU-P operations have been implemented. Standard DPU operations, such as the classical MAC, ADD, SUB, MUL, CMP, AND, OR, and XOR, are available. Computations on complex numbers are also supported. Use of SIMD operations, such as MAC4, ADD4, SUB4, and MUL4, may speed-up the computation performance by a factor up to four. The same is true for vector-oriented operations such as MACV, MSUBV, ACCV, MINV, and MAXV, which usually take multiple input values and compute a single result in a single clock cycle. Specialized or customized operations allow for the speed-up of some targeted algorithms. In the Audio-I DPU, special instructions for FFT computations, IIR and FIR filtering, function interpolation, bit-stream creation and decoding, min and max searches, and data clipping have been implemented. As an example, in a baseband-oriented DPU, specialized operations for the implementation of Viterbi or turbo decoders can easily speed-up the algorithm's performance from a factor 2 to > 30 over classical software implementation of such algorithms, depending on the additional hardware used to implement such specialized operations. Audio-I DPU-L operations allow performing two independent DPU operations in parallel (e.g., four MUL and four ADD). More than 170 data processing operations have been implemented in the Audio-I DPU. This extensive number of operations can be further completed/customized, if needed, to better match application-specific algorithms needs. As for the AGU, if a high level of parallelism is needed, and heterogeneous data processing operations should be executed in parallel, a limited set of reconfigurable extended DPU operations can be made available.

A few examples of Macgic Audio-I instructions are given next.

```
irepeat 16
   mac4      acc,rx0w,ry3w  ||  movpx2p rx0w,(ax2,  pr0)  ry3w,(ay1,iy2)
cmacc acc0,acc1,rx0w.l,ry0w.l  ||  movb2p rx0w,(ax0)+ ry0w,(ay0)+
loop ry7, end_radix4_fft_loop
   cbfy4a0 acc,ry0w.l                              ||  movpxp rx2w,(ax0,pr0)
   cbfy4a1 rx0w,acc,rx1w.l,ry1w.l,rx5w.l,ry4w.l  ||  movpxp rx3w,(ax0,pr1)
   cbfy4a2 rx0w,acc,rx2w.l,ry2w.l,rx5w.u,ry4w.u  ||  movpxp (ax1,pr0),rx0w
   cbfy4a3 rx0w,ry5w.l,rx4w,ry4w,acc,rx3w.l,ry3w.l,rx4w,ry5w.l  ||  movpxp (ax1,pr0),rx0w
end_radix4_fft_loop:
```

The irepeat operation allows repeating the execution of the next instruction the given number of times. The cmacc operation takes the complex conjugate of the second operand, performs a complex multiplication, and accumulates the complex result into the specified accumulators. The loop operation allows repeating a specified sequence of operations for a given number of times. The cbfy4 operations are specialized instructions for FFT/IFFT computation. The various movpxp and movbp operations are data move operations that usually perform data memory accesses, or just index registers updates when no source or destination registers are specified.

21.3 Macgic DSP Reconfiguration Mechanisms

Two reconfiguration mechanisms have been developed for the audio versions of the AGU and DPU. Both mechanisms use a similar principle. Given the relatively small instruction word of the Macgic DSP (32-bit), the degree of ILP available to the DSP programmer may be relatively limited, unless the powerfulness of SIMD, and of specialized operations can be exploited. To provide the programmer with additional programming capabilities, a set of extended, software reconfigurable, DMU-P and DPU-P operations are made available by the AGUs and the DPU.

21.3.1 Address Generation Unit Reconfiguration

Each AGU permits the reconfiguration of four extended operations. An extended operation allows to both perform an address computation, to access the data memory using an indirect addressing, and to save address computation results into up to three AGU registers. In a DMU-P operation, two 3-bit fields, one per AGU, specify the kind of operation to be performed by the AGU: either a predefined AGU operation or an extended AGU operation. Examples of use are:

```
macv acc,rx0w,ry3w || movpx2p (ax2, pr0),rx0w (ay1,iy2)
clra acc
irepeat 16
    macv acc,rx1w,ry2w || movpx2p rx1w,(ax2, ix0) ry2w,(ay1,iy3)
```

The programmer shall specify the use of an extended operation by typing the extended operation register Is*n* (*s* = X,Y, *n* = 0..3) after the index register. Predefined operations are specified in a similar manner, by typing PR*m* (*m* = 0..2) after the index register to be used. The mnemonic for predefined and extended operations is movpxp or movpx2p.

As an example, a single extended AGU operation may perform the following computations in parallel:

```
rx2w <= XDM[ax1+ox3], ax2<=(ax0+ox3)%mx1, ox1<= ox2+mx3
```

or

```
ax1 <= ClearLSBs(ax2,ox3), ox2<=ox2>>2, mx2<= mx2>>2
```

Figure 21.6 illustrates the various parts of an AGU datapath and its reconfiguration capability. Direct addressing operations, as well as basic and predefined indirect addressing operations are internally remapped into extended operations.

Figure 21.6(a) illustrates the selection of two address registers, two offset registers, and two modulo registers that will be read. The first address, offset and modulo register can either be selected by the value specified in the instruction's opcode, or by a value specified in the extended operation' configuration register. The second address offset and modulo registers can independently be selected from the configuration register. The six values read from the AGU registers are made available to two ALUs. The first one (Figure 21.6(b)) is responsible for computing the actual address to be accessed in the data memory, while the second one (Figure 21.6(c)), which allows more complex operations to be performed is rather used to compute post-modified addresses that will be used the next time a memory access will be performed. The results coming from these two ALUs can be input to a third ALU (Figure 21.6(d)) for further post-processing. Table 21.1 lists the various operations available for each of the three ALUs present in an AGU. Up to three results can be saved into the AGU registers: one in an address register, one in an offset register, and one in a modulo register (Figure 21.6(e)). Figure 21.6(f) gives the format of an extended AGU operation configuration register. Four of such registers are available in each AGU, one per extended operation.

21.3.2 Data Processing Unit Reconfiguration

Eight extended DPU operations are made available by the DPU. An extended DPU operation configuration word is 128-bit wide. An extended operation is invoked through a specific DPU mnemonic: dpxop.

```
dpxop n,Value1,Value2
```

The first operand of the extended operation is the operation number (0..7). Two 3-bit operand values can then be specified. The meaning of these additional operands depends on the configuration of the extended operation. Figure 21.7 presents the configuration possibilities.

Up to four wide GP registers can be read from the DPU. Figure 21.7a illustrates how the GP registers are selected. The values read from the registers are made available as four values XR1, XR2, YR1, and YR2. Figure 21.7b illustrates the four virtual processing units. The operations that these virtual units can perform are given in Table 21.2. The source operands for each operation can be freely chosen from any of the four wide values XR1, XR2, YR1, and YR2. Each virtual operation can be conditionally executed, using a flag value as condition. The result of each virtual operation can be saved into any of the two wide

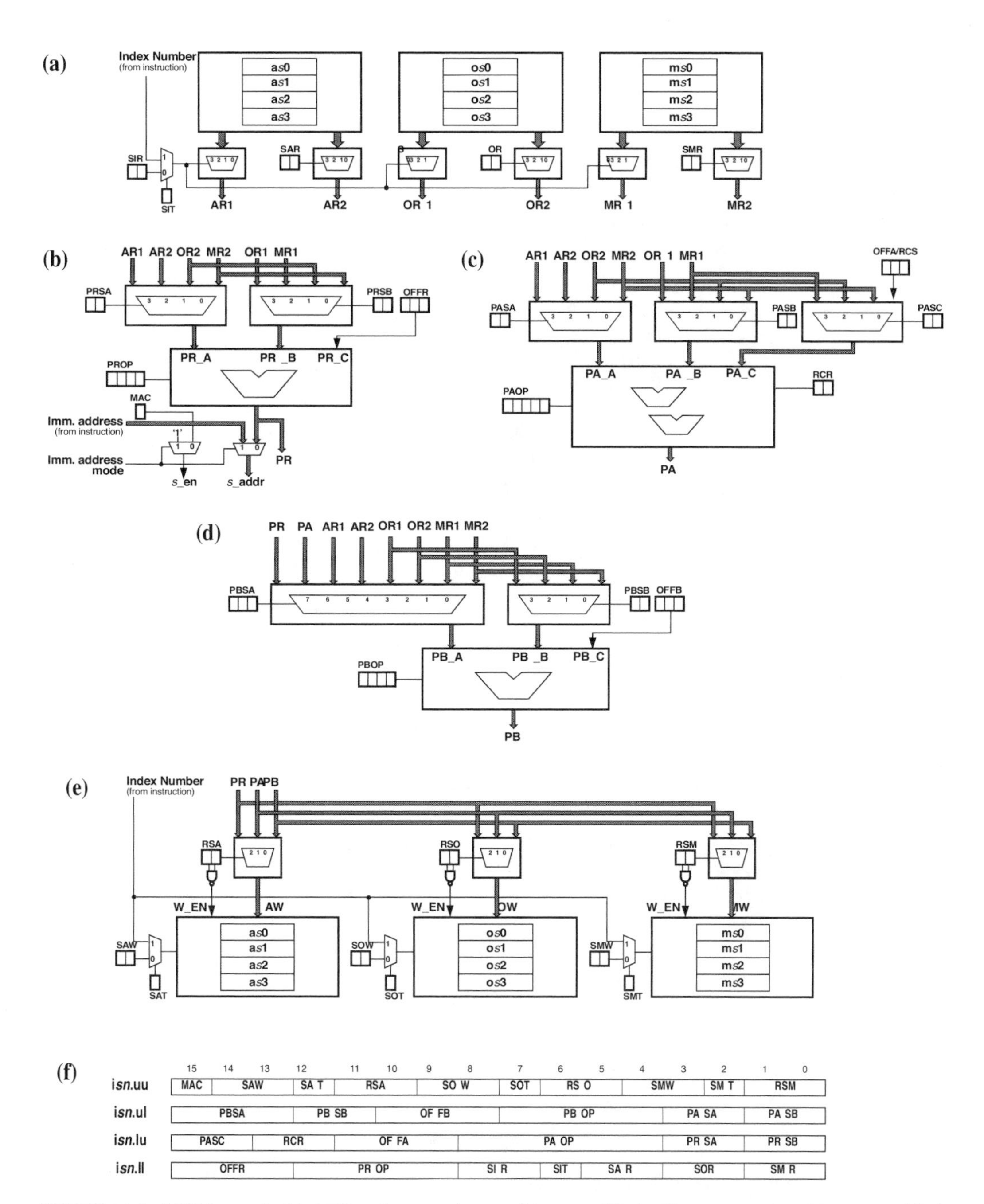

FIGURE 21.6 AGU datapath: (a) AGU registers read ports; (b) premodified address generation; (c) post-modified A address computation; (d) post-modified B address computation; (e) AGU registers write ports; (f) extended AGU operation configuration.

output values XW, YW. The two wide output results can be saved back into GP registers. Figure 21.7c illustrates the GP registers' write-selection mechanism. Sixteen 64-bit extended operation registers are made available to the programmer, two per extended operation. Figure 21.7d is an extended DPU operation configuration register. Eight of such registers are available, one per extended operation.

As an example, a single extended DPU operation can perform the following computations in parallel, in a single clock cycle:

TABLE 21.1 AGU Datapath, Example of Address Generation Operations

Value	PROP	PAOP	PBOP
0	A	C	C
1	B	-C	-C
2	A+B	A+C	A+C
3	A-B	A-C	A-C
4	B-A	B+C	B+C
5	A+OFFR	B-C	B-C
6	B+OFFR	A+B	A+B
7	A+(B>>1)	A-B	A-B
8	A-(B>>1)	B-A	B-A
9	A+(B>>2)	B<<(C AND 7)	A<<(C AND 7)
A	A-(B>>2)	B>>(C AND 7)	A>>(C AND 7)
B	A+(B<<1)	ClearLSBs(A,B)	ClearLSBs(A,B)
C	A-(B<<1)	RevCarryInc(A,RCR,C AND 7)	Reserved
D	A+(B<<2)	RevCarryInc(B,RCR,C AND 7)	Reserved
E	A-(B<<2)	(A+B)%C	Reserved
F	OFFR	(A-B)%C	Reserved
10	-	(A+C)%B	-
11	-	(A-C)%B	-
12	-	A+(B>>(C AND 7))	-
13	-	A-(B>>(C AND 7))	-
14	-	B+(A>>(C AND 7))	-
15	-	B-(A>>(C AND 7))	-
16	-	A+(B<<(C AND 7))	-
17	-	A-(B<<(C AND 7))	-
18	-	B+(A<<(C AND 7))	-
19	-	B-(A<<(C AND 7))	-

```
if (pf7) {//execute if flag 7 of PSU flag register PF is set
   Rx2w.uu <= rx7w.uu * rx6w.uu; Rx2w.µl <= rx7w.µl * rx6w.µl;
   Rx2w.lu <= rx7w.lu * rx6w.lu; Rx2w.ll <= rx7w.ll * rx6w.ll;
}
if (npf3) {//execute if flag 3 of PSU flag register PF is cleared
   Ry5w.uu <= ry3w.uu + ry2w.uu; Ry5w.µl <= ry3w.µl + ry2w.µl;
}
if (pa1) {//execute if flag 1 of DPU flag register PA is set
   Ry5w.lu <= rx7w.lu>>2; Ry5w.ll <= rx7w.ll>>2;
}
if (npa1) {//execute if flag 1 of DPU flag register PA is cleared
   Ry5w.lu <= rx7w.lu; Ry5w.ll <= rx7w.ll;
}
```

21.4 Performance Results

Some computation performance results of the Audio-I version of the DMU and DPU are given in Table 21.3. The results are expressed in clock cycles and are given for the algorithms' kernels. Additional clock cycles overhead may exist if these algorithms are to be placed into subroutines.

The complexity of the Audio-I version of the DSP in number of transistors count is about 750k for a dw = 24-bit implementation, and about 550 k for a dw = 16-bit implementation.

The simulated power consumption of the Macgic Audio-I DSP (dw = 24-bit) is about 1.2 mW/MHz at 1.8 V in the TSMC 0.18-µ technology, and only about 0.25 mW/MHz at 0.9 V (e.g., CSEM's CSELIB 6 standard cells library and Synopsys' power compiler).

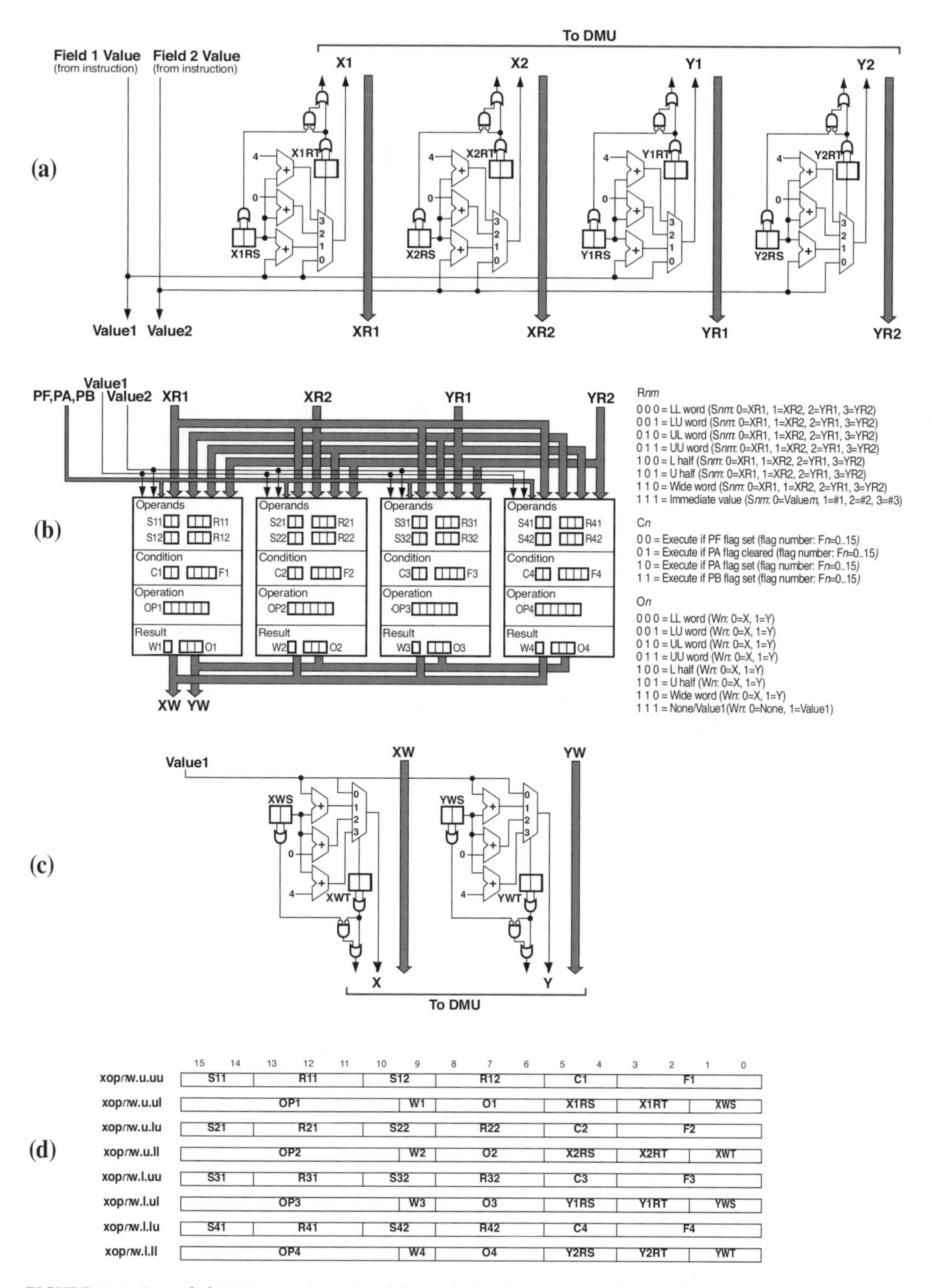

FIGURE 21.7 Extended DPU operations virtual datapath: (a) GP registers read ports; (b) extended DPU operations; (c) GP registers write ports; (d) extended DPU operation configuration.

TABLE 21.2 Example of Extended DPU Operations

Value	OP1	OP2	OP3	OP4	Value	OP1	OP2	OP3	OP4
00	NOP	NOP	NOP	NOP	20	SHR	SHR	SHR	SHR
01	ABS	ABS	ABS	ABS	21	SHRU	SHRU	SHRU	SHRU
02	ACCA	ACCA	ACCA	ACCA	22	SUB	SUB	SUB	SUB
03	ADD	ADD	ADD	ADD	23	SUBA	SUBA	SUBA	SUBA
04	ADDA	ADDA	ADDA	ADDA	24	SUBC	SUBC	SUBC	SUBC
05	ADDC	ADDC	ADDC	ADDC	25	SUBN	SUBN	SUBN	SUBN
06	ADDN	ADDN	ADDN	ADDN	26	ADD2	ADD2	SHL2	SHL2
07	AND	AND	AND	AND	27	MUL2	MUL2	MOV2	MOV2
08	CHKB	CHKB	CHKB	CHKB	28	SUB2	SUB2	SHR2	SHR2
09	CLRA	CLRA	CLRA	CLRA	29	MULS2	MULS2	SWAP2	SWAP2
0A	CLRB	CLRB	CLRB	CLRB	2A	ADDC2	ADDC2	SHRU2	SHRU2
0B	CMP	CMP	CMP	CMP	2B	MULU2	MULU2	SEL2	SEL2
0C	CMPA	CMPA	CMPA	CMPA	2C	SUBC2	SUBC2	SHLW	WCONC1
0D	CMPM	CMPM	CMPM	CMPM	2D	MULF2	MULF2	SHRW	WCONC2
0E	CMPU	CMPU	CMPU	CMPU	2E	MAX	MAX	BSEX	WCONC3
0F	DSQA	DSQA	DSQA	DSQA	2F	MAXU	MAXU	BSEXF	WCONCU
10	EXOR	EXOR	EXOR	EXOR	30	MAXM	MAXM	BSEXU	WCONCL
11	INVB	INVB	INVB	INVB	31	MIN	MIN	BSINSF	WCONCUS
12	MAC	MAC	MAC	MAC	32	MINU	MINU	BSINSF	WCONCLS
13	MOV	MOV	MOV	MOV	33	MINM	MINM	BSRD	WINS
14	MSUB	MSUB	MSUB	MSUB	34	CMUL	CMUL	BSRDF	WEXTR
15	MUL	MUL	MUL	MUL	35	CCONJ	CCONJ	BSRDU	TRSPL
16	MULA	MULA	MULA	MULA	36	CNORM	CNORM	EXTRAR	TRSPU
17	MULF	MULF	MULF	MULF	37	ACCV	MACV	EXTRA	CNTRMSB
18	MULS	MULS	MULS	MULS	38	CMP2	CMP2	CNTRMSB	CNT0MSB
19	MULU	MULU	MULU	MULU	39	CMPM2	CMPM2	CNT0MSB	CNT0LSB
1A	NEG	NEG	NEG	NEG	3A	CMPU2	CMPU2	CNT0LSB	MOV4
1B	NOT	NOT	NOT	NOT	3B	ADD4	MUL4	SHL4	SWAPHL
1C	OR	OR	OR	OR	3C	SUB4	MULS4	SHR4	SWAPIHL
1D	SEL	SEL	SEL	SEL	3D	ADDC4	MULU4	SHRU4	SWAPE
1E	SETB	SETB	SETB	SETB	3E	SUBC4	MULF4	CLR2	CLR4
1F	SHL	SHL	SHL	SHL	3F	Reserved	Reserved	Reserved	Reserved

TABLE 21.3 Macgic DSP Performance Results (Audio-I DMU, DPU, GP-I HDU)

Algorithm	Number of Clock Cycles (kernel)
Vector SUM (vector size = N)	~N/4
Vector addition	~N/2
Vector multiplication	~N/2
Dot product	~N/2
Vector normalization	~N/4
Minimum/maximum of a vector	~N/4
Matrix multiplication (matrix size = N × M)	~(N*M)*(N/4+2)
Matrix transportation	~N*M*(5/16)
Convolution (or FIR filter)	~1/4 per tap
Complex convolution	~1 per tap
Complex convolution, with conjugate of second operand	~1 per tap
IIR (biquad)	1 per stage
Complex FFT/1FFT radix 4 (size 64, optimized)	~220
Complex mixed-radix FFT/1FFT (size 2048, not optimized)	~14.6 k
Complex mixed-radix FFT/1FFT (size 4096, not optimized)	~29.1 k
Complex mixed-radix FFT/1FFT (size 8192, not optimized)	~66.2 k
MP3 decoder (not optimized)	~340 per stereo sample

21.5 Conclusions

Macgic has been developed in VHDL language and can therefore be synthesized to virtually any CMOS technology. It can provide a huge computational power and can be tailored to best suit an application's needs. Power consumption has been minimized by extensive use of clock gating. Design robustness and power consumption reduction has been obtained through use of level-sensitive storage elements (latches), which helps reducing constraints on clock distribution trees.

A set of software development tools has been specifically developed for the Macgic DSP to fully support its parameterization capabilities. These tools include an integrated development environment, a powerful macro-assembler as well as a source-level software debugger. A C++ phase-accurate pipelined model of the DSP is available and can be integrated in various cosimulations environments (e.g., Matlab/Simulink, System C, COSSAP/CoCentrics, and ModelSim). An FPGA implementation of the DSP also permits a quick prototyping of systems using the Macgic DSP.

A new DPU specialized for baseband data processing, as well as a simplified DMU and DPU for less computation-intensive applications, are in development.

References

[1] Õ. Paker, J. Sparso, N. Haandbaek, M. Isager, and L.S. Nielsen, A heterogeneous multi-core platform for low-power signal processing in systems-on-chip, *ESSCIRC 2002,* Firenze, Italy, pp. 73–76.

[2] C. Arm, J.-M. Masgonty, and C. Piguet, Double-latch clocking scheme for low-power IP cores, *PATMOS 2000,* Goettingen, Germany, September 13–15, 2000, pp. 217–224.

[3] J.M. Rabaey Reconfigurable computing: the solution to low-power programmable DSP, *ICASSP '97,* Munich, Germany, April 21–24, 1997, pp. 275–278.

[4] A. Abnous and J. Rabaey, Ultra-low-power domain-specific multimedia processors, in *VLSI Signal Processing IX,* IEEE Signal Processing Society 1996, Piscataway, NJ, pp. 461–470. http://www.ieee.org/organizations/society/sp.

[5] A. Abnous, K. Seno, Y. Ichikawa, M. Wan, and J. Rabaey, Evaluation of a low-power reconfigurable DSP architecture, in *IPPS/SPDP '98 Workshops, Vol. 1388 of Lecture Notes in Computer Science,* pp. 55–60, Springer-Verlag, Heidelberg, 1998.

[6] P. Kievits, E. Lambers, C. Moerman, and R. Woudsma, REAL DSP technology for telecom baseband processing, *ICSPAT '98,* 1998.

[7] P. Mosch, G.V. Oerle, S. Menzl, N. Rougnon-Glasson, K.V. Nieuwenhove, and M. Wezelenburg, A 720-μW, 50-MOPs, 1V DSP for a hearing aid chip set, *ISSCC 2000,* pp. 238–239, February 2000.

[8] M.J. Flynn, *Computer Architecture: Pipelined and Parallel Processor Design,* Jones & Bartlett, Boston, 1995.

[9] *TMS320C1x User's Guide (Rev. C),* Texas Instruments, SPRU013C, 1991. http://focus.ti.com/lit/ug/spru013c/spru013c.pdf

[10] *TMS320C3x User's Guide (Rev. E),* Texas Instruments, SPRU031E, 1997. http://focus.ti.com/lit/ug/spru031e/spru031e.pdf

[11] *TMS320C5x User's Guide (Rev. D),* Texas Instruments, SPRU056D, 1998. http://focus.ti.com/lit/ug/spru056d/spru056d.pdf

[12] *TMS320C64x CPU and Instruction Set Reference Guide,* Texas Instruments, SPRU189F, 2000. http://focus.ti.com/lit/ug/spru189f/spru189f.pdf

[13] *ADSP-2100 Family User's Manual, Third Edition,* Analog Devices, 1995. http://www.analog.com/Processors/Processors/ADSP/technicalLibrary/manuals/16BitIndex.html

[14] *Blackfin Processor Instruction Set Reference,* Analog Devices, P/N 82-001991-01, 2003. http://www.analog.com/Processors/Processors/blackfin/technicalLibrary/manuals/blackfinIndex.html#Processor%20Manuals

[15] *ADSP-BF535 Blackfin Processor Hardware Reference,* Analog Devices, P/N 82-000410-13, 2003. http://www.analog.com/Processors/Processors/blackfin/technicalLibrary/manuals/blackfinIndex.html#Processor%20Manuals

[16] *ADSP-TS101 TigerSHARC Processor Programming Reference,* Analog Devices, Part No. 82-001991-01, 2003. http://www.analog.com/Processors/Processors/tigersharc/technicalLibrary/manuals/tigersharcIndex.html#Processor%20Manuals

[17] *DSP56000/DSP56001 Digital Signal Processor. User's Manual,* Motorola, DSP56000UM/AD Rev. 1. http://e-www.motorola.com/files/dsp/doc/inactive/DSP56000UM.pdf

[18] *Starcore SC140 DSP Core Reference Manual,* Motorola/Agere, MNSC140CORE/D Rev. 3, 2001. http://e-www.motorola.com/files/dsp/doc/ref_manual/MNSC140CORE.pdf

[19] *DSP1611 DIGITAL SIGNAL PROCESSOR Information Manual,* Agere, MN02-016AUTO, 2001. http://www.agere.com/client/docs/MN02016.pdf

[20] I. Verbauwhede and M. Touriguian. A low-power DSP engine for wireless communications, *J. VLSI Process.,* 18, pp. 177–186, 1998.

22

Low-Power Asynchronous Processors

Kamel Slimani
Joao Fragoso
Mohammed Es Sahliene
Laurent Fesquet
Marc Renaudin
TIMA Laboratory

22.1 Introduction

This chapter gives an overview of the techniques based on the asynchronous technology to design low-power circuits, particularly low-power processors. The asynchronous microprocessors designed and fabricated during the last two decades are reviewed. As it is well-known, reducing the power consumption of integrated systems requires applying many different dedicated techniques, at different levels. This chapter is focused on the micro-architecture or structural level of asynchronous circuits. It also discusses how asynchronous processors can run dedicated power-aware operating systems, leading to a significant reduction of the power consumed by embedded systems — reduction that cannot be achieved using synchronous processors.

The first section covers fundamental power-reduction techniques that can be applied at the structural/micro-architecture level of a complex asynchronous circuit. The presentation is considering the design of low-power datapaths, pipelines, and control structures. In the second section, adequate design methodologies are presented, which take advantage of asynchronous circuits' properties for low power and target the power-reduction techniques presented in the previous section. The third section reviews fabricated asynchronous processors and highlights their performances with respect to low power. In this research domain, impressive prototypes were designed and fabricated, from 8-bit complex instruction set computer (CISC) microcontrollers to 32-bit reduced instruction set computer (RISC) machines. Finally, the last section demonstrates that asynchronous processors surpass synchronous processors when

0-8493-1941-2/05/$0.00+$1.50

designing low-power embedded systems using an operating system based on dynamic voltage scheduling algorithms. This original work illustrates how the software can exploit the key properties of asynchronous hardware to achieve a significant reduction of the power consumption.

22.2 Power Reduction Techniques in Asynchronous Circuits

22.2.1 Datapaths

Asynchronous datapaths have potentially low-power operation due to two main reasons [1]:

1. No global clock exists. In fact, clock is a major source of power consumption.
2. Asynchronous systems do nothing when their inputs have no data, so inactive asynchronous circuits shut themselves off.

The asynchronous circuit style (i.e., protocol and data coding) has a strong impact on the power consumption, however. A two cycle protocol (i.e., nonreturn-to-zero [NRZ]) allows to transmit and to process data in each handshaking cycle. In another way, a four cycle protocol (or return-to-zero [RZ]), which implies a more simple circuit, inserts a bubble (invalid data) into the protocol to finish the handshaking. This RZ phase consumes energy and no data is processed or transmitted.

The asynchronous bundled-data datapaths are similar to synchronous datapaths consequently all synchronous techniques to reduce power-consumption can be used in the asynchronous design.

On the other hand, delay insensitive (DI) data encoding circuits do not require a request line. Requests are generated using a transition in all digits to be sent. Therefore, delay insensitive circuits must be hazard-free because transitions are interpreted as requests. In this way, delay insensitive circuits do not waste energy in spurious transition. Additionally, the number of transitions on the circuit input lines is no longer data-dependent. This feature simplifies the energy estimation [2] because the switching activity can be evaluated with no care about data being processed, provided that implemented algorithms are not data-dependent.

Thus, reducing DI datapath energy consumption mainly means reducing the number of gate switching to process data. The widely used dual-rail data encoding codes a binary value on two wires. An efficient way to reduce energy consumption is to increase the radix. 1-of-M data encodings hold more information in a single transition and more information can be processed/transmitted with less gate transitions [3]. For example, an n-bit adder power consumption can be reduced by a factor two replacing the dual-rail encoding by the 1-of-4 data encoding [4].

In addition, busses energy is saved when using large radix together with one-hot encoding. It also enables to protect the busses from cross-talk effects. Indeed, when a transition occurs on a bus wire, the neighbor wires are by definition quiet [5,6]. Finally, asynchronous circuits are delay insensitive, thus repeaters can be efficiently replaced by asynchronous buffers. Obviously, n-of-m data encoding could be used and the designer has to exploit the solution space to choose the appropriate power/area/speed trade-off [7].

At the architectural level, there are other techniques to save energy-complexity in datapaths structures as presented in Manohar [8] and Park et al. [9]. Although synchronous datapaths power consumption is generally data-dependent, DI asynchronous circuits may loose this ability. For example, in a simple synchronous 32-bit array multiplier, the energy consumption depends on the operand wideness, but a given asynchronous multiplier consumes almost the same power for all operations.

Table 22.1 lists the results of benchmark programs for the 32-bit multiplier of microprocessors [10,11]. It is easy to see that an asynchronous implementation would waste a lot of energy using a constant 32-bit data width. In Manohar [8], a new class of number representation is presented. The number representation uses a self-delimited data encoding that allows dynamically adapting the operation width while keeping delay insensitivity.

Additionally, in specific applications such as finite impulse response (FIR) Filters, the knowledge about the data dependency can be used to improve the architecture power efficiency [12]. Finally, when decomposing gates to limit maximal fan-in, the input transition probability knowledge could be taken

TABLE 22.1 Benchmark Programs Data Pattern

Architecture	Benchmark Program	Number of Multiplication	Average Bit Length	
			Multiplicand	Multiplier
SPARC v8.0	lisp	2 K	5.32	8.00
	gcc	240 K	5.81	3.67
	ijpeg	82 M	7.69	11.24
Simplescalar	lisp	15 K	1.54	0.43
	gcc	1 M	6.69	9.35
	perl	42 M	0.01	4.01
	go	13 M	3.70	4.41
	ksim	5 M	11.43	12.88

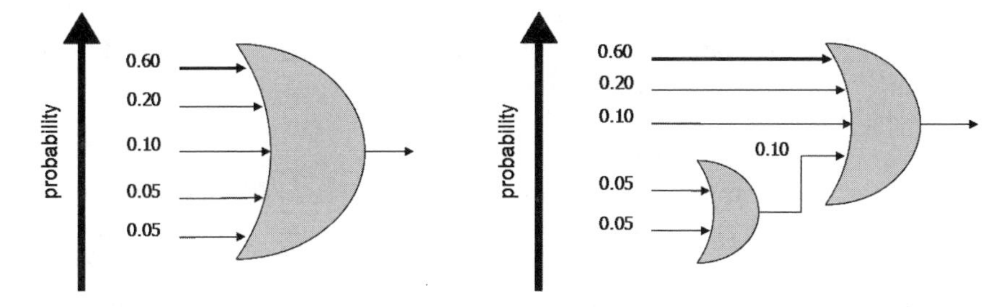

FIGURE 22.1 An OR-5 gate decomposition example.

into account to reduce the average number of transitions. Figure 22.1 is an example of OR-5 decomposition with maximal fan-in equal to four and all input mutually exclusive. The three most probable inputs (i.e., 60%, 20%, 10%) are connected to the output gate, resulting in a unique gate switching in 90% of the cases.

The use of all these techniques improves energy-aware asynchronous systems potential.

22.2.2 Pipelines

Several techniques have been proposed and introduced to improve the performance of circuits such as processors. Pipelining is a common and popular way to increase the throughput of a circuit.

In synchronous designs, pipelining is achieved by inserting registers within combinational stages and by adjusting the global clock accordingly. Whereas in asynchronous circuits, pipelining is performed only by inserting latches/buffers between stages, no global synthesis has to be done. It looks more complicated in synchronous than in asynchronous because all the circuit has to be resynthesized in synchronous to fulfill the new global synchronization for a functional correctness.

The ease of performing a deeper pipeline in asynchronous must be carried out with care. Indeed, inserting latches to improve the throughput of a circuit induces additional hardware affecting consequently the power consumption of a circuit. It is the eternal trade-off between speed and power consumption. Nevertheless, an upper bound of buffers can be added into a design to reach the maximum throughput. Beyond this number of buffers, adding more buffers increases the power consumption overhead without improving the performance of the circuit. More dramatically, adding an excessive amount of buffers leads to a decrease of performance because the delays of buffers affect the throughput [13].

Some models have been proposed to determine the optimum number of buffers to be added in an asynchronous pipeline to get the maximum throughput. An interesting experiment was proposed by [13]. It has demonstrated that performances of an asynchronous ring can vary between a blocked-state when the number of buffers is lower than the minimum to a maximum throughput at the optimum number of buffers. Beyond this optimum number of buffers, the performance decreases implying an increase of the energy dissipation.

Other models such as $E\tau^2$ [14] energy-time metric is used to determine the upper bound of buffers to be added in an asynchronous pipeline to get the highest throughput [15]. Fixing the optimal number of buffers in a pipeline is also called slack matching [16]. Slack matching can be applied without affecting the correctness of the circuit wherever no arbitration structure exists. These parts of the circuit, which do not include arbitration, are called slack elastic.

Another source of energy waste is observed when branch instructions are executed and taken. As a rule, the deeper the pipeline the higher the energy waste when a branch is taken. This is due to the presence of as many instructions in the pipeline as pipeline stages, so the energy dissipation is directly proportional to the number of instructions that must be discarded in the pipeline. To reduce this waste of energy, solutions based on branch prediction can be used. When the prediction is close to the reality, the energy overhead is significantly reduced. The branch prediction hardware consumes energy for every instruction, however, so it is questionable whether the total energy consumption of the whole execution is reduced. An elegant solution to this problem, which exploits the elasticity of asynchronous pipelines, is to control the pipeline depth by collapsing pipeline stages together. This is performed dynamically, so the processor can switch from a fully pipelined configuration to a low depth pipeline configuration very quickly during an instruction execution [17]. Dynamic configuration of pipeline stages has the potential to significantly save energy. Decreasing the pipeline depth slows down the processor but also saves energy because fewer "speculative" instructions are fetched and decoded. It is based on selectively making some latches transparent, which join pipeline stages together. In micropipeline asynchronous circuits, dynamic configuration of pipeline stages is achieved by the mean of reconfigurable latch controllers [18] that can be either "permanently transparent" (collapsed) or "normal." These new latch controllers were used in AMULET3 [19].

22.2.3 Control Structures

When the datapath is nonlinear (i.e., in the presence of control structures such as multiplexers), the complexity of the circuit is higher because several inputs can be propagated to one output. These architectures are expensive in terms of power consumption if no care is taken in their implementations. Strategies are proposed to reduce the power consumption of these components by unbalancing the structure of choice in favor of the cases with the highest probabilities [20]. The probabilities are established by simulation and are application-dependent. This optimization is a step further towards the power consumption reduction of a circuit. Indeed, the power dissipation overhead lies on the enormous forks that connect several inputs to one output. The idea for reducing the power consumption of these structures is to decompose these huge forks into a dissymmetric tree of small forks. This decomposition allows the signals with the highest probabilities to cross a reduced amount of smaller gates, thus reducing the loads switched. This decomposition is similar to the one proposed in Figure 22.1.

The same reasoning can be applied to any control structure, such as demultiplexers, for which one input is sent to one among several possible outputs.

22.3 Design Methodologies for Low Power

Several universities develop their own design flows and use them in order to demonstrate the efficiency of their design methodologies. Research in asynchronous logic has lead to interesting circuits that demonstrate the effectiveness of asynchronous logic. The results obtained stimulate the research of asynchronous logic. Among some design flows developed in some universities, we can point out Tangram [21], which was developed at Eindhoven University in collaboration with Philips Research Lab in Eindhoven, the Netherlands. This design flow was used to implement a low-power asynchronous 80C51 [22]. Tangram was a good inspiration source for the AMULET group at the University of Manchester, U.K., to develop Balsa [23]. Balsa is a tool that can generate bundled data circuits. A DMA controller [24] integrated in AMULET3i was a good example to demonstrate the potential of the Balsa synthesis tool.

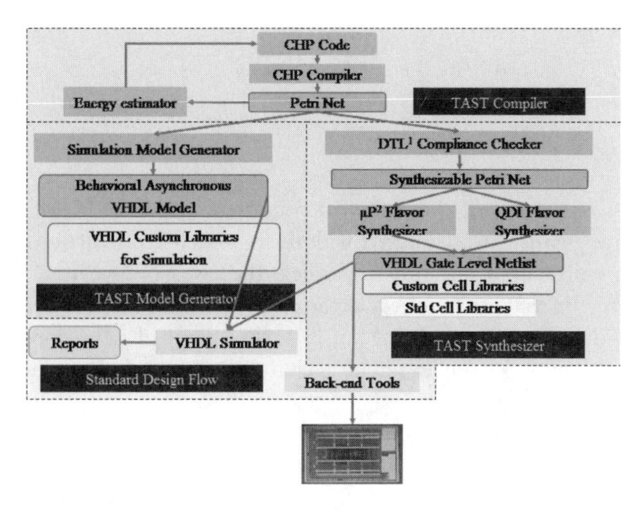

FIGURE 22.2 Design flow in TAST.

Recent works have been accomplished in Balsa to generate quasi-delay insensitive (QDI) architecture for secure applications. An asynchronous implementation of the commercial advanced RISC machine (ARM) microprocessor named SPA [25] was achieved to illustrate the efficiency of this synthesis tool. Finally, we introduce TIMA asynchronous synthesis tools (TAST) [26] (Figure 22.2). It was developed at TIMA laboratory in Grenoble, France, within the concurrent integrated systems (CIS) group. TAST is a complex and complete tool for designing asynchronous circuits. It takes a high-level language, which is based on VHDL and CHP developed by Alain Martin at the California Institute of Technology (Caltech, Pasadena, CA). TAST enables the designer to perform functional simulation in VHDL or C, and to synthesize the communicating hardware processes (CHP) specification into an asynchronous micropipeline or QDI circuit.

One of the interesting features of the TAST design flow is the trace estimator. It gets information on both the circuit activity and the energy consumption during a specific simulation [27,27a]. The trace estimator, which is an interesting option that can guide the designer to achieve a low-power asynchronous circuit, performs it. By following a rigorous strategy of design (it is possible to add labels in the CHP specification to monitor one or several particular instructions), the designer can get information on power consumption and can nicely exploit them to design low-power circuits.

The activity estimation tool allows the designer to achieve a profiling of the code to get code coverage information of a CHP description. The aim of this code profiling is to identify parts of the description that are never used or seldom used.

The activity estimation of a circuit allows the designer to identify parts of a circuit that are the most frequently used. For an asynchronous microprocessor for example, it is interesting to know which units are often used: the register bank, the ALU, the load/store unit, and so on. It is possible to get information at a fine grain with a higher accuracy. For instance, it is interesting to know which registers in a register bank are the most frequently accessed and, therefore, optimize the hardware accordingly. For an ALU, this information on activity can guide the designer to choose the best operator architecture (the operator that is solicited most often will have a low-power architecture).

Information on channels is useful to choose the best encoding. The activity of digits in a channel is useful to see which digits of a channel are rarely, if ever, used. According to this analysis, the channels can be sliced into a set of subchannels that are conditionally active (channel width adaptation).

Finally, statistics on choice structures are essential to privilege cases with the highest probabilities. The statistics obtained can guide the designer to unbalance the structure of choice in favor of the cases with the highest probabilities [20]. For example, if the guard G0 has the highest probability, the structure of choice could be unbalanced as follows:

```
[ G0  =>  I0                    [ G0  =>  I0
@ G1  =>  I1 =>                 not G0  =>[ G1  =>  I1
@G2  =>  I2                               @ G2  =>  I2
]                                             ]
                                        ]
```

The mutually exclusive structure of choices (written in pseudo CHP code) on the left hand side is a regular structure of three guards G0, G1, and G2 with their respective instructions I0, I1, and I2. On the right-hand side, it is an unbalanced structure to privilege the case G0. The effect on the final circuit by decomposing the structure in that way is to reduce the power consumption of the execution of the guard G0. Indeed, every time the guard G0 is executed, it costs the energy of a two-input multiplexer instead of a structure with three inputs. A complex circuit including complex choice structures having unequal probabilities results in large energy savings.

The activity estimation tool is an efficient tool to guide the designer for power reduction. Moreover, TAST goes beyond by offering to the designer the possibility to get an estimation of the energy in terms of the number of transitions. Commercial tools exist, but the estimation of the energy is done at the gate level [28] or at the transistor level [29], thereby occurring too late in the design flow and taking too much time to be simulated. This is the reason why the trace estimator tool in TAST allows the designer to get power estimation at the synthesizable CHP specification level. This estimation is done without performing the synthesis of the circuit, but it is based on how TAST does the synthesis of asynchronous circuits. Furthermore, this is a C-simulation, thus the simulation and the estimation are very fast, contrary to a gate simulation, which is generally a VHDL gate simulation involving gate delays (using VITAL libraries, for example).

This power consumption estimator is a good way to check whether an optimization is judicious, and indicates the relative gain between two different implementations of the same circuit.

22.4 Asynchronous Processors: A Review

Important studies have been conducted on asynchronous logic to prove the efficiency in terms of power reduction. Thanks to several design methodologies that have emerged from these studies, some designs around the world have been achieved and are very encouraging for the future. Among some circuits, a short list of asynchronous microprocessors, which have marked the last 15 years of intensive research, is presented.

22.4.1 CAP

Alain Martin's group at Caltech designed the first microprocessor entirely implemented in asynchronous logic at the end of the 1980s. This microprocessor is known as the Caltech asynchronous processor (CAP) [30]. This simple processor was designed to prove the feasibility of asynchronous circuits by using the methods of program transformations developed at Caltech [31].

CAP is a 3-stage pipeline, 16-bit RISC processor with 16 registers of 16 bits; it is implemented in dual-rail QDI architecture using a 4-phase communication protocol. The circuit is synthesized and mapped on static complex gates. It contains 20,000 transistors. The circuit is physically manufactured in MOSIS SCMOS 1.6-μm technology.

Performances of the processor at different supply voltages have been recorded: 26 MIPS for a power consumption of 1.5 W at 10V; 18 MIPS for a power consumption of 225 mW at 5 V; and 5 MIPS for a power consumption of 10.4 mW at 2 V.

22.4.2 MiniMIPS

More recently, the work done at the Caltech on asynchronous design methodologies has lead to the design and fabrication of a powerful asynchronous MIPS R3000 prototype: MiniMIPS [16]. One of the goals of this project was to apply the theory based on the $E\tau^2$ metric [14,32], devoted to the estimation of time and energy efficiency of computation.

The synchronous MIPS architecture is based on three pipeline stages: PC address computation, instruction fetch, and instruction execution; however, the asynchronous MiniMIPS is very finely pipelined inside these stages in order to increase the performances without sacrificing the power consumption. A slack matching was achieved in many parts of the circuit to obtain the maximum throughput (this slack matching was done without affecting the correctness of the circuit because it was achieved on slack elastic parts).

In addition, improvement in completion mechanism was performed on the datapath. Indeed the datapath was decomposed into small units (4 bits wide) that generate their own completion signals. As a result, the completion tree is smaller, thereby reducing the delay and the power consumption of the circuit.

MiniMIPS is very innovative for the year of its invention (1997) because it includes a bypass mechanism and exceptions management. The processor was manufactured in 0.6-μm SCMOS technology, and the performances are 150 MIPS for a power consumption of 1 W at 2 V and 280 MIPS for a power consumption of 7 W at 3.3 V.

22.4.3 AMULET1, 2, 3

AMULET (asynchronous microprocessor using low energy and technology) is a project born in late 1990 at the University of Manchester. The purpose of this project was to demonstrate the potential of the asynchronous technology to reduce the energy consumption.

The advanced RISC machine microprocessor (ARM) was chosen to prove the feasibility of designing a fully asynchronous commercial processor.

The first version was AMULET1 [33] released in early 1993. The realization of this processor required 5 men-year. The processor has been implemented using Sutherland micropipeline [34], and using a 2-phase communication protocol (nonreturn-to-zero). The result obtained for this circuit was not very successful — the performances measured were lower than those of the original ARM6. These unexpected results are partly due to an unsuitable deep pipeline. The power consumption of AMULET1 is 83 mW for a performance of 9 K dhrystones. For the original ARM6, the power consumption is 75 mW for a performance of 14 K dhrystones at 10 MHz.

The experience acquired with the realization of AMULET1 lead to a second version, AMULET2 [35], in 1996, which was higher performance than the former. The architecture is still based on Sutherland micropipeline. The 2-phase communication protocol was replaced by the 4-phase communication protocol, however, which has the advantage of being faster and less consuming despite the multiplication by two of the number of transitions. The 2-phase communication protocol consumes more than the 4-phase because of the use of a complex double edge sensitive logic. Improvement was also achieved by reducing the unsuited deep pipeline of AMULET1, which was partly responsible for the extra power dissipation.

AMULET2 integrates a "branch target cache" for branch prediction, a mechanism of register blocking in the form of a FIFO and a mechanism of forwarding. Logic blocks, such as the shifter and the multiplier, are bypassed when they are not necessary to reduce energy waste. Furthermore, AMULET2 includes a "halt" mechanism to stop all activity in the circuit. The circuit was implemented in a CMOS 0.5-μm technology. The performances of AMULET2 are 40 MIPS for a power consumption of 150 mW, which is better than the ARM710 but less than the ARM810. AMULET2 was incorporated in AMULET2e with flexible RAM memory.

A third version, AMULET3 [19], was designed in 2000. It is an asynchronous 32-bit RISC processor that is competitive with the synchronous ARM9TDMI core.

The datapath is a full custom design, and the standard library was provided by ARM Limited. The control part was synthesized using Petrify [36]; Petrify takes a circuit in form of a STG and transforms it into a speed independent (SI) circuit. AMULET3 uses latch controllers [18] to reduce the pipeline when branch instructions are executed. The technology used in AMULET3 is 0.35 μm and the power supply voltage is 3.3 V. The performances recorded for AMULET3 are 120 MIPS with a power con-

FIGURE 22.3 DRACO layout.

sumption of approximately 155 mW. These performances are very similar to the synchronous ARM9TDMI core.

AMULET3 demonstrated that asynchronous technology is competitive in terms of power consumption. It was integrated in AMULET3i [37] with an asynchronous bus named MARBLE [38], a DMA controller (synthesized using Balsa [23]) and RAM and ROM memories. AMULET3i is used for commercial applications; the first application is for a DRACO control system (Dect [digital enhanced cordless telecommunications] RAdio COmmunication controller) depicted in Figure 22.3.

22.4.4 Asynchronous 80C51

The CISC asynchronous 80C51 microcontroller [22] was designed at Eindhoven University of Technology in collaboration with Philips Research Laboratories in 1995. The circuit was implemented in a relatively short time of 6 months.

The asynchronous version is completely compatible with the original processor. The goal of this duplication was to demonstrate the benefits of an asynchronous implementation by comparing directly the original synchronous processor with the asynchronous implementation.

The circuit was specified in high-level language using Tangram [21], and was compiled in a netlist using the Tangram compiler. The compilation of a Tangram program is done in two stages in which the handshake circuits represent the intermediate form. The handshake circuits are implemented using the 4-phase protocol and single-rail architecture.

The transparency of the Tangram compiler allowed the designers to bring optimizations at the Tangram program specification. In addition, a suitable combination of bus-structure and point-to-point communication contributed to save extra energy dissipation.

The processor was mapped onto a standard cell library and fabricated in CMOS 0.5-μm technology. It integrates a DC-DC converter for voltage scaling (see Section 22.5).

The performances obtained at a supply voltage of 3.3 V are 4 MIPS for a power consumption of 9 mW. It is noted that the energy consumption was reduced by a factor of 4 compared with the original synchronous version. These results prove the benefits brought by the asynchronous logic in terms of power consumption and low electromagnetic emission. This processor has been used in a pager; the layout of the circuit is depicted Figure 22.4.

22.4.5 Lutonium

Lutonium [38a] is a recent asynchronous 8051 microcontroller developed by the California Institute of Technology. Although the 8051 is an irregular CISC processor, which is not really well suited for low-power, it is very popular and is often found in applications where minimizing the energy is important.

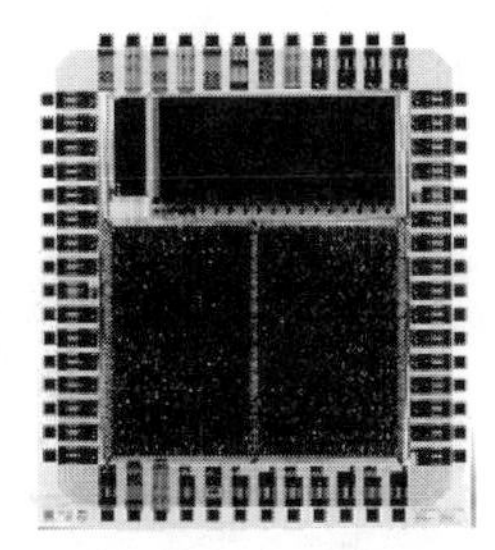

FIGURE 22.4 Asynchronous Philips 80C51.

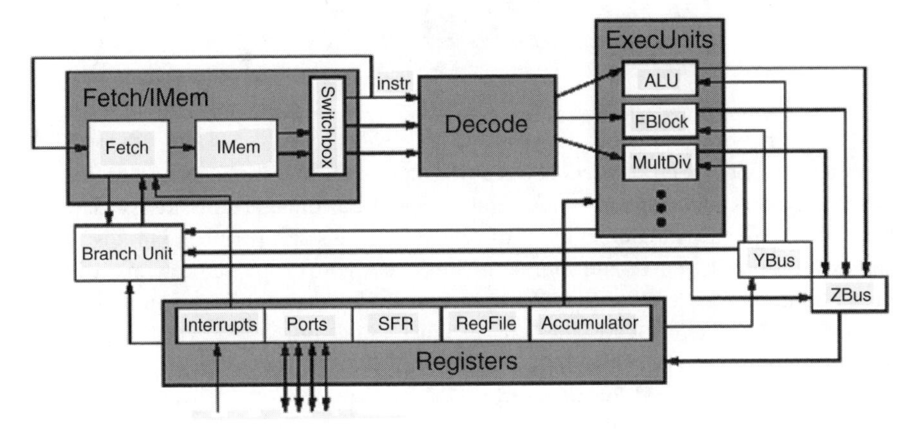

FIGURE 22.5 Lutonium architecture.

TABLE 22.2 Lutonium Performances

1.8 V	200 MIPS	100.0 mW	500 pJ/inst	1800 MIPS/W
1.1 V	100 MIPS	20.7 mW	207 pJ/inst	4830 MIPS/W
0.9 V	66 MIPS	9.2 mW	139 pJ/inst	7200 MIPS/W
0.8 V	48 MIPS	4.4 mW	92 pJ/inst	10900 MIPS/W
0.5 V	4 MIPS	170.0 mW	43 pJ/inst	23000 MIPS/W

Lutonium was implemented for an optimal $E\tau^2$ [14] parameter. It is highly pipelined to increase the speed. The instruction decoder is more complex than the MiniMIPS [16] and is centralized in one component to decrease the size and the number of communication channels in the circuit. Moreover, the lutonium performs a deep-sleep mode in which almost all switching activities are stopped. The advantage of asynchronous implementation is that Lutonium can wake up instantly from the deep-sleep mode. Moreover, Lutonium introduces a segmented bus-control protocol to save energy: useful digits only are sent (especially for control structures as presented in section 22.2.2).

The QDI logic is used for its robustness property at low voltages. Lutonium has been inplemented in SCN018 (a 0.18-μm CMOS library) by Taiwan Semiconductor Manufacturing Company (TSMC) supplied by MOSIS; the nominal supply voltage is 1.8 V. Figure 22.5 depicts the architecture of Lutonium.

The performances of Lutonium are shown in Table 22.2.

22.4.6 MICA

Microcontrôleur asynchrone (MICA) is a quasi-delay insensitive asynchronous 8-bit microcontroller CISC machine, designed by the CIS group of TIMA in collaboration with France Telecom research and development in Grenoble, France, and STMicroelectronics in Grenoble in 2000 [39,40]. It integrates two

FIGURE 22.6 MICA microcontroller and measured performances.

different register-files: eight 8-bit registers devoted to data, and eight 16-bit registers devoted to pointers (including the program counter and the stack pointer). Specific arithmetic units are associated with each register file enabling concurrent computations of data and addresses. A peripheral unit is also included, supporting six 8-bit parallel ports, and four serial links (using a two-phase delay insensitive protocol compatible with the RISC asynchronous ASPRO processor). Moreover, the microcontroller integrates 16 Kbytes of RAM and 2 Kbytes of ROM. The latter includes a built-in-self-test, which is executed at reset according to the boot mode selected (eight modes are available).

The MICA processor was tested functional from 3 V down to 0.6 V (2.5 V is the nominal voltage of the 0.25 μm CMOS technology used). It is noticeable that the chip only consumes 800 μW at 1 V, still delivering a computational power of 4.3 MIPS. At 0.8 V, the chip consumes less than 400 μW (Figure 22.6).

The microcontroller core was designed using the so-called quasi-delay insensitive (QDI) logic. A four-phase protocol was used in conjunction with an n-rail encoding. This chip was a vector for developing new skills in the design of standard-cell based QDI asynchronous circuits. The design of MICA was focused on two correlated concerns: designing distributed asynchronous finite state machine and designing for low power.

To reduce the power consumption of the microcontroller, the number and the energy-cost of communication actions occurring during the execution of each instruction, as well as the number of sequential steps to perform each instruction were minimized. In other words, instead of designing the architecture around a big central sequencer, the sequencer implementation was distributed all over the architecture and as much as possible. The asynchronous logic is particularly well suited to satisfy such a design approach because, by nature, the sequencing of an asynchronous circuit is performed by multiple local sequencers implementing handshaking communications and local treatments.

Thus, the architecture of MICA was designed as a distributed system, each part providing specific services. For example, the two register files, the status register, and the memory integrate local units, which manage the memory resources. These modules implement functions such as "read," "write," "read then write back" or even more complex functions such as: read a byte, increment/decrement the pointer/address, and read another byte (Copy and Push & Load instructions, for example, use these features). Adopting such an approach significantly simplifies the design of the main sequencer of a CISC microprocessor like MICA. It then minimizes the power consumed by the main sequencer, where the consumption associated with each instruction is the direct image of its complexity. In fact, complex instruction implementation does not penalize simple instruction implementation at the main sequencer level. Moreover, such a distributed approach minimizes the power consumed by communications because the minimum number of transactions occurs through busses (memory accesses for example).

Because of the low-power constraint and because computational power was not a priority for the targeted applications, a minimum number of pipeline stage was introduced. This does not avoid parallel execution of instruction subparts, but simply means that parallel execution of instructions is not supported. In some cases, however, successive instructions may partially overlap.

Finally, at the signal level, communication channels are using a low-power data encoding. Instead of using dual-rail coding, N-rail coding (also called "one hot"), is used (i.e., one out of the N wires is active during a transaction). Therefore, the different parts of the architecture are controlled by the means of channels using 5-rail to 12-rail data encoding, which minimizes the number of transitions per communication action, and thus minimizes the dynamic power consumption. The datapaths (8-bit and 16-bit) are entirely designed with 4-rail encoded data, requiring radix-4 logic/arithmetic processing units. The register files are also designed with 4-rail encoded data. Instead of bit-registers, they are built of digit-registers, each digit representing four values.

MICA was successfully used for the design of a contact-less smart-card chip, called MICABI [39], which integrates an on-chip coil connected to a power reception system and an emitter/receiver module compatible with the ISO 14443 standard. The benefit brought by an asynchronous processor in such a system on chip (SoC) is that design constraints are relaxed [39,41,42], especially concerning the software, the analog, and the radio-frequency parts.

22.4.7 ASPRO

ASynchronous PROcessor (ASPRO) [43] is a 3-stage pipeline 16-bit RISC asynchronous microprocessor. Its design started at the Ecole Nationale Supérieure de Télecommunications de Bretagne antenne de Grenoble, in Grenoble, France, in collaboration with France Telecom R&D and STMicroelectronics in 1998. The team then joined TIMA [40] in 1999. ASPRO is a regular RISC microprocessor, which decodes the instructions in order and completes them out of order; it contains 16 registers of 16 bits.

The design flow and the circuit style are an original application of the method of Alain Martin. The motivations were to get experience in the design of complex asynchronous QDI circuits using standard cells and to demonstrate that asynchronous design techniques could improve very large-scale integration (VLSI) systems in term of speed and power consumption.

ASPRO was implemented using multi-rail encoding and a 4-phase communication protocol. The synthesis was partially performed by hand and it targeted the standard cell library provided by the founder STMicroelectronics.

The processor integrates two distinct on-chip memories for instructions and data. It was manufactured in 0.25-μm CMOS technology by STM. It can deliver 24 MIPS for a power consumption of 20 mW at 1 V and 140 MIPS for a power consumption of 350 mW at 2.5 V (including consumption of the memories). The CPU integrates three independent sources of supply voltages (from 0.9 to 3.0 V) for dynamic voltage scaling purposes (see Section 22.5). The final physical circuit is pictured in Figure 22.7.

FIGURE 22.7 ASPRO.

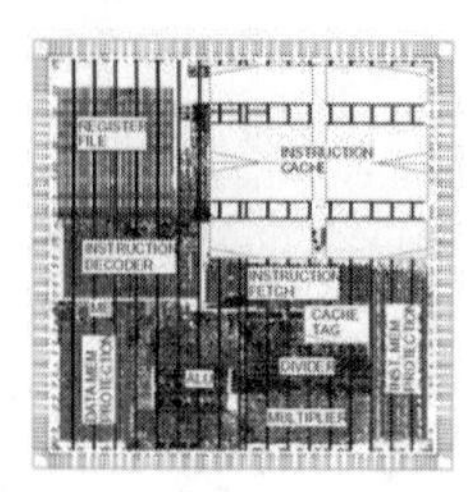

FIGURE 22.8 TITAC-2 layout.

22.4.8 TITAC-2

The realization of an asynchronous processor TITAC-2 [44] was invented by the Tokyo Institute of Technology in Tokyo, Japan. It is an asynchronous version of the MIPS R2000 processor (some instructions are missing or are modified). It is a 5-stage pipeline 32-bit RISC processor.

TITAC-2 is implemented using a scalable-delay insensitive (SDI) architecture [44]. SDI is based on QDI architecture with an unbounded delay model (i.e., no upper bound is assumed on the gate and wire delays). The benefits of this model are visible in terms of speed. The SDI circuits can effectively run faster than the equivalent QDI circuits.

To get the maximum performance without affecting the consumption of the processor, pipelining in TITAC-2 was fixed by simulation.

The data are coded into dual-rail and the communication is done using a 4-phase communication protocol. TITAC-2 was manufactured using a 0.5-µm CMOS technology and synthesized using a standard cell library.

The performances obtained are 52.3 VAX MIPS for a V2.1 test dhrystone and consumes 2.11 W at 3.3 V and 20°C. Figure 22.8 illustrates the layout of the circuit.

22.4.9 Conclusion

The results obtained for these microprocessors are in reward of all the work accomplished in asynchronous logic over the last 15 years of research. The asynchronous 80C51 and the AMULET3i processors are used in commercial applications. These two asynchronous processors, which are a reliable duplication of synchronous commercial processors, have demonstrated the advantages of asynchronous logic in comparison with the synchronous logic in terms of power reduction. The MiniMIPS prototype demonstrated that high performance is achievable using the asynchronous technology, while keeping the relative power consumed rather low. Finally, the MICA microcontroller and the MICABI contact-less system-on-chip demonstrated the benefits brought by the asynchronous technology to design mixed-signal circuits requiring low voltage, low power, and low noise.

All these prototypes have demonstrated that asynchronous logic has attained a maturity level that is commercially viable today. It is, however, clear that an ever greater reduction of asynchronous processors power consumption is possible, by combining the different techniques introduced in Section 22.2 and by applying design methodologies devoted to low power as suggested in Section 22.3.

22.5 Power Reduction Techniques at the System Level

22.5.1 Introduction

To go a step further in reducing the power consumption let us now jointly consider the hardware and software parts of an integrated system. Several hardware and software techniques have been developed over the last years to manage the electrical consumption of a system. Nevertheless, as devices become much more powerful and sophisticated, power requirements increase continuously [45]. Therefore, new power management techniques have been investigated at hardware and software levels [46,47]. In this section, a new method is considered that combines asynchronous processors and an operating system for low-power management. Although the literature on power management expounds extensive research on what to do at the hardware or software level, this approach investigates both levels simultaneously. In cooperation with a power management policy adapted to an asynchronous processor, the operating system (OS) adjusts the speed of the processor to the task requirements at runtime by controlling the processor operating voltage. This scheme exploits the ability of asynchronous processors to self-regulate their processing speed with respect to the supply voltage only [48].

22.5.2 Principles of Power Reduction with Operating Systems

For current CMOS integrated circuits, power dissipation is dominated for the moment by the switching power, even if the static leakage current induces more and more power losses in deep submicron technologies. This arises from the charging and discharging of the loading capacitance and is expressed for a synchronous processor as:

$$P = C\,V^2\,f \tag{22.1}$$

where C is the load capacitance, V is the supply voltage, and f is the clock frequency [49–51]. In the case of an asynchronous processor, the relation reads:

$$P = C\,V^2\,s \tag{22.2}$$

where s is the instantaneous speed of the processor. These equations suggest that minimizing the load capacitance, reducing supply voltage (or speed), or slowing down the clock can reduce power consumption.

Although the load capacitance can be affected during chip design by minimizing on chip routing capacitance [52], voltage scalable processor and power controllable peripheral devices make it possible to reduce power at the operating system level. In cooperation with scheduling policies, operating systems can vary the processor's speed and voltage (dynamic voltage scaling) and put devices in low-power sleep states (dynamic power management) [53]. Indeed, a typical embedded system consists of processors, memories, communication links, and other peripheral devices that are not always used. Thus, an OS can control the power states of peripheral devices in the system according to the workload. When a device is idle, it can be put in a low-power sleeping state after a long enough idle period to compensate time and energy overhead of shutting down and waking up [54,55]. To determine whether a device can sleep, time-out, predictive, and stochastic policies have been developed [56,57]. In the time-out based policy, after a device is idle for a time-out value, it remains idle for at least a certain time. Predictive policy uses the past idle periods or both the current and the past idle period to predict the length of future idle periods eliminating the time-out period wasted energy. Choosing a policy for a given application depends on prediction accuracy, power savings, and resources requirements such as memory and computation [54,56,57]. A shutdown mechanism can be applied to put the processor into idle mode when not in use [54]; however, a more fruitful way to save power is to run slower at reduced voltage according to computational load demands [58].

A technique called dynamic voltage scaling (DVS) allows processors to dynamically alter their voltage and speed at runtime under the control of voltage scheduling algorithms [59,60]. These algorithms predict future workload and set the processor voltage and speed based on this prediction. The interval based voltage scheduler called PAST (bounded-delay limited PAST scheduling algorithm) [61] assumes that the processor utilization of the next interval will be like the previous one and updates the processor speed accordingly. If the last interval was busier than idle, speed is increased. Similarly, if it was mostly idle, speed is decreased. A comparative simulation of PAST and other proposed algorithms points out that a simple smoothing algorithm can be more effective than sophisticated prediction techniques [62]. Recent studies take the real-time constraints [63,64] into account. The processor speed is estimated considering workload prediction and task deadlines. The goal is to complete tasks before or on deadlines. Because the processor is often running at a reduced speed, however, all these studies assume missed deadlines as a trade-off between power saved and deadlines missed.

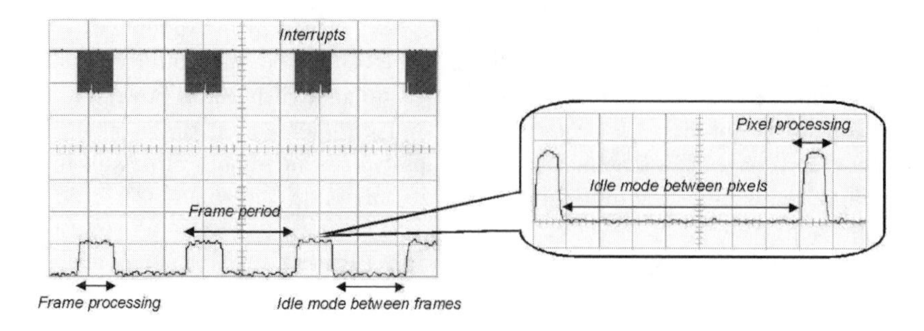

FIGURE 22.9 Fine grain idle mode with the ASPRO processor.

22.5.3 Low-Power Sleeping States

22.5.3.1 Synchronous Processors Idle Mode

Synchronous processors need a huge amount of time and energy to enter the idle mode or to wake up from a sleeping state. This constitutes a severe drawback to swap rapidly with low energy costs between the normal and the idle mode. Indeed, the transition time to a sleeping state currently takes several tenths of microseconds (for the lpARM [60] or the Crusoe [65,66]) to several tenths of milliseconds for the wake-up (with a StrongARM processor [64]). The time and energy overheads do not allow continuous starting and stopping of the processor. For minimizing these drawbacks, the synchronous processors own different levels of sleeping states.

22.5.3.2 Asynchronous Processors Idle Mode

Contrary to the synchronous processors, the asynchronous processors are well suited to exploit fine grain idle mode. Indeed, they only consume energy if the processor has data to process. If no data is processed, the processor is immediately set in idle mode until a request or interrupt wakes it up. The wake-up time overhead is equivalent to the time overhead of an interrupt. These specific properties of the asynchronous circuits allow the use of a fine-grain idle mode. This point is illustrated in Figure 22.9, which presents real measurements performed with a video application running on the ASPRO processor (see Section 22.4) [43,67]. A digital video camera sends images to ASPRO via a high throughput serial link. Between two consecutive frames, the processor is idle because it has no data to process. More astonishing, if the frame is zoomed, we can see that between two consecutive pixels the processor is idle too (see the balloon in Figure 22.9). The activity duty cycle is 0.333 for the frames and 0.125 for the pixels. This leads to a 95% energy reduction compared with a system without idle capabilities. This is not possible for a synchronous processor.

22.5.4 DVS for Synchronous Processors vs. DVS for Asynchronous Processors

22.5.4.1 Timing Model for Asynchronous Processor Speed Variation

The power supply voltage only drives the speed variation of asynchronous processors. The variation time t_v only depends on the DC-DC converter and the load capacitance of the processor. Time t_v could be modeled as a function of V_{dd1} and V_{dd2}, the supply voltages before and after the variation. A simple way to express t_v is to consider a linear expression:

$$t_v = k \cdot \left| V_{dd1} - V_{dd2} \right| \tag{22.3}$$

where k is a fitting parameter depending on the DC-DC converter and the load capacitance of the processor.

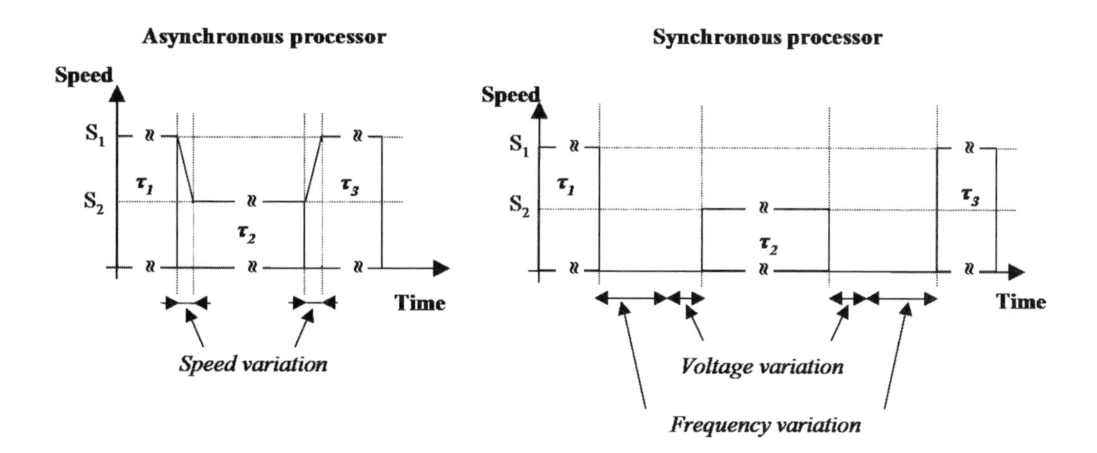

FIGURE 22.10 Task execution and speed variation for asynchronous and synchronous processors.

In the asynchronous case, we can notice that task processing is not stopped during the speed variation. Moreover, the speed can vary continuously and can be finely adjusted during the task execution. Figure 22.9 illustrates that the execution of tasks τ_2, and τ_3 is not stopped during the speed variation.

22.5.4.1.1 Timing Model for Synchronous Processor Speed Variation

In the synchronous case, the speed variation is controlled by the supply voltage and by the clock frequency. This operation needs two steps. In the case of speed reduction, the first step is used to slow down the frequency and the second step reduces the power supply voltage. In the case of speed increase, the first step increases the voltage and the second speeds up the frequency. The transition time, t_v, depends on the DC-DC converter and the phase-locked loop (PLL) that controls the clock frequency; t_v can be expressed as:

$$t_v = t_{DC\text{-}DC} + t_{PLL} \tag{22.4}$$

where $t_{DC\text{-}DC}$ is the transition time from the initial voltage to the final voltage, and t_{PLL} is the transition time to change the frequency. As modeled in Equation 22.3 t_v can be expressed as:

$$t_v = k_1\,|V_{dd2} - V_{dd1}| + k_2\,|f_2 - f_1| \tag{22.5}$$

where k_1 and k_2 are, respectively, the fitting parameter for the DC-DC converter and for the PLL. Frequencies f_1 and f_2 are the initial and the final frequency of the processor.

The frequency transition time t_{PLL} is constant in the case of the StrongARM processor [64] or variable if a Transmeta Crusoe is considered [65,66]. The frequency of synchronous processors does not vary continuously but by steps. For instance, the Crusoe processor frequency varies by 33 MHz steps. Moreover, a critical point (except for the lpARM), which drastically increases the inefficiency of synchronous processors, is that task processing is stopped during the speed variation. Figure 22.10 presents the timing model for the synchronous processor speed variation. Because the processor stops the processing during the transition time, the speed is set to zero for this duration.

22.5.4.1.2 DVS Additional Energy Costs for Synchronous Processors

Contrary to the asynchronous processors, the speed variation of synchronous processors has a cost in terms of energy and time. Indeed, synchronous processors are stopped, and they continue to spend energy during the speed variation, while asynchronous processors process their tasks and vary their speeds at the same time. During the speed variation, the energy spent could be expressed as the sum of the energetic cost of the power supply variation plus the energetic cost of the clock frequency variation.

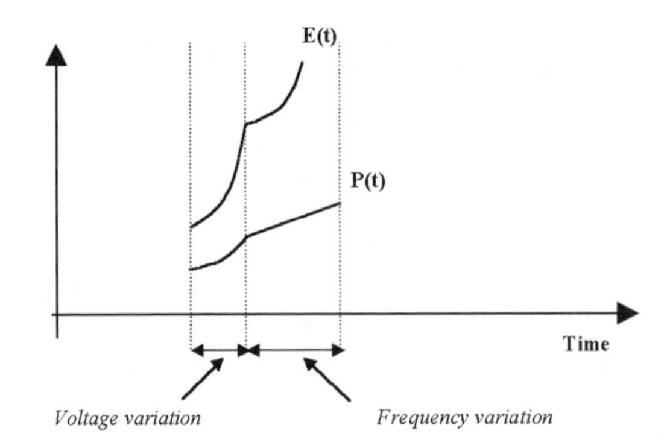

FIGURE 22.11 Energy and power evolution during the speed variation of a synchronous processor.

$$E = E_v + E_f \tag{22.6}$$

Referring to Equation 22.1 and Equation 22.6, the DVS energy cost (see Figure 22.11) is expressed as follows:

$$E = \alpha Cf \int_{t0}^{t1} V^2(t)dt + \alpha CV^2 \int_{t1}^{t2} f(t)dt \tag{22.7}$$

where α is a scaling factor and the intervals [*t0, t1*] and [*t1, t2*] are, respectively, the voltage variation time at fixed frequency and the frequency variation time at fixed voltage.

22.5.6 DVS Algorithms for Asynchronous Processors

A real-time system has often to manage periodic and sporadic tasks. Although periodic tasks are commonly used to process data from sensors and update the current state of the system, sporadic tasks are required to process asynchronous events; however, most of the voltage scheduling schemes presented in the literature consider systems with periodic tasks only. Few attentions have been dedicated to a system with sporadic tasks [63]. The following paragraphs consider both periodic and sporadic tasks.

22.5.6.1 Task Model Definition

Each task can be characterized by a triplet <NIi, Di, Ti>, where NIi is the number of instructions of the task, Di its deadline and Ti its period or its minimum inter-arrival time. We assume that:

- Tasks are independent and their parameters become known when arriving.
- Periodic tasks have deadlines equal to their periods and tasks periods are different.
- At the maximum supply voltage (at highest processor speed), all considered periodic and sporadic tasks can be processed.
- The overhead due to the context switching is negligible.

Because the computation can continue during the voltage switching, we assume that the overhead associated with the voltage scheduling is also negligible.

22.5.6.2 Sporadic Task Voltage Scheduling Algorithm

This subsection considers a case when only sporadic tasks arrive to the system. We assume that the ready time of each task is the instant of its arrival. When a new task τ_i arrives, an acceptance test is performed to determine whether the task can meet its deadline without causing any prior guaranteed tasks to miss their deadlines:

$$\left(\sum_{j\le i}\frac{\overline{NI_j}}{td_j - t}\right) \le S_{MAX} \tag{22.8}$$

where $\overline{NI_j}$ denotes the number of instructions still to be executed for the task τ_j, td_j denotes its deadline, t denotes the current time, and S_{MAX} denotes the highest processor speed in MIPS at the maximum supply voltage. If τ_i is accepted, it is inserted into a priority task queue according to the earliest-deadline-first (EDF) order, *and the voltage-scheduling algorithm updates the processor speed to complete all tasks in the task queue before or at their deadlines. This speed is given by:

$$S = \underset{l\in Q}{MAX}\left[\frac{\overline{NI_j}\sum\limits_{j\ne l/Pl\le Pj}\overline{NI_j}}{td_l - t}\right] \tag{22.9}$$

where Q denotes the stream of the sporadic tasks existing in the task queue, and P_k denotes the priority of task τ_k according to EDF policy. For each task τ_i in the task queue, including the new task, the voltage scheduler computes the required speed S_i to finish the task τ_i considering all priority tasks. Then, it sets the processor speed to the maximum value of S_i. The processor speed is updated whenever a task is added or removed from the task queue. Compared with the voltage scheduling proposed in Pering et al. [63] our algorithm avoids an overestimation of the processing requirements and leads to a higher power saving. This is because our approach takes only runnable and ready tasks into account to compute the operating voltage.

To illustrate the effectiveness of the proposed voltage scheduler, we consider three tasks as presented in Figure 22.12(a), with arrival time and deadlines assigned to each of them. Therefore, when no power reduction technique is applied, the processor runs at S_{MAX} and consumes P_{MAX}. All the speed and power figures reported are based on real measurements performed with an ASPRO motherboard supplied with different voltages.

Using shutdown techniques, the processor can be stopped on completion of task τ_2, woken up at task τ_3 arrival time, and stopped when this task completes. The consumed power is then 42% P_{MAX}. Because an asynchronous processor can instantly be stopped and woken up without any time overhead, all tasks meet their deadlines. In a synchronous system, this technique is ineffective. Because sporadic tasks have random arrival times, it is difficult to predict the future idle times. Thus, tasks can miss their deadlines. Furthermore, shutting down and waking up synchronous processors cause a time and energy overhead.

Voltage scheduling policy is more effective to reduce power consumption in comparison with the shutdown technique. When it is applied, the processor runs at variable voltage and speed as presented in Figure 22.12(b). In this example, the processor runs at 50% S_{MAX} until task τ_1 completes. Then the processor

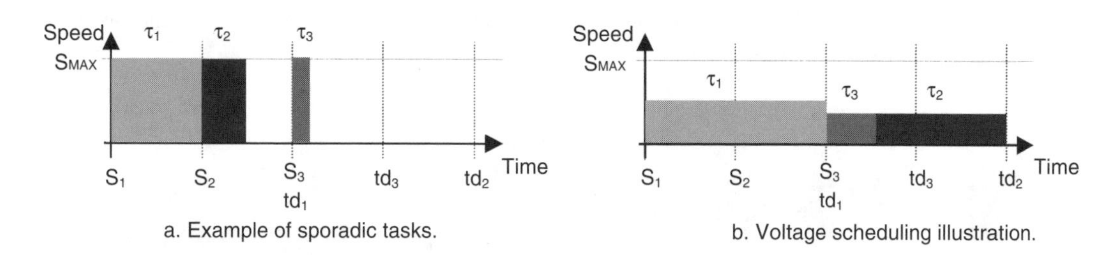

FIGURE 22.12 How sporadic task voltage scheduling reduces power consumption.

*Any scheduling policy is applicable.

TABLE 22.3 Example Periodic Tasks Set

Tasks	NI_j	D_j	T_j	Ready Time
τ_1	0.25×10^6	2	2	0
τ_2	1.0×10^6	5	5	0
τ_3	0.5×10^6	3	3	4

speed is reevaluated and set at 35% S_{MAX} for tasks τ_2 and τ_3. The consumed power is then 14% P_{MAX}. When no task is running, the processor enters a sleeping state in which it consumes no power.

22.5.5.3 Periodic Task Voltage Scheduling Algorithm

This subsection assumes that all tasks are periodic (i.e., no sporadic tasks arrive to the system). We also assume that at t = 0, a set of n periodic tasks is ready and sorted into *a priori*ty task queue according to the EDF order. The processor speed is set to:

$$S = \sum_{j=1}^{n} \frac{NI_j}{D_j} \tag{22.10}$$

When a new periodic task τ_i is added to the system, it is inserted into the priority task queue according to the EDF order, and the voltage-scheduling algorithm updates the processor speed to complete all tasks in the tasks queue on their deadlines. The new speed is given by:

$$S = \frac{NI_i}{D_i} + \sum_{j=1}^{n} \frac{\overline{NI_j}}{td_j - t} \tag{22.11}$$

where $\overline{NI_j}$ denotes the number of instructions still to be executed for the task τ_j, td_j denotes its deadline, and t denotes the current time. Similarly, if a periodic task is removed from the system, the processor speed is updated:

$$S = \sum_{j=1}^{n-1} \frac{\overline{NI_j}}{td_j - t} \tag{22.12}$$

Consider the tasks set in Table 22.3, with ready time and deadlines assigned to each task. Speed and power are normalized to their values at the maximum supply voltage: S_{MAX} and P_{MAX}.

In Figure 22.13(a), when the tasks are scheduled running at the highest processor speed, some waiting states exist between the end of a task and the arrival time of the next task. The system then wastes power waiting for the next task. Therefore reducing the supply voltage, such that the tasks make full use of the CPU time, can lower the processor speed.

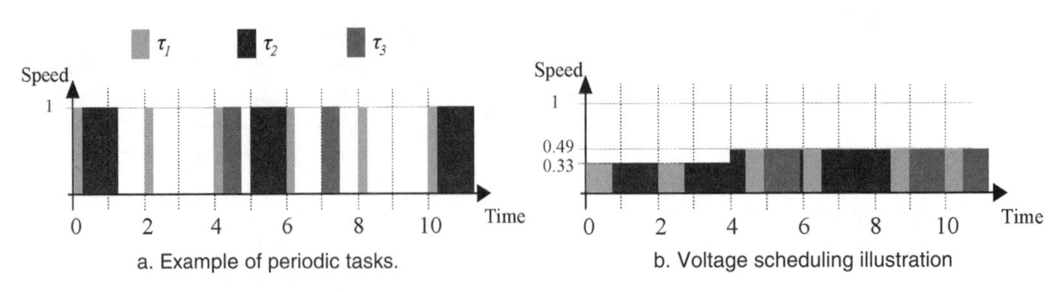

FIGURE 22.13 How periodic tasks voltage scheduling reduces power consumption.

In Figure 22.13(b), the processor speed is reduced to achieve power reduction. When starting, it is set to 33% S_{MAX} according to Equation 22.4. At t = 4, task $\tau 3$ becomes ready and the processor speed is updated, according to Equation 22.5, to 49% SMAX. The consumed mean power is then 14% P_{MAX}.

22.5.7 Conclusion

This section demonstrated the ability of asynchronous systems to jointly manage power at the hardware and software levels. Compared with their synchronous competitors, the asynchronous processors are easy to idle and to stop without time and energy overheads. Moreover, a scheduling policy adapted to asynchronous processors has been explained to take full advantage of their properties.

22.6 Conclusion

The goal of this chapter was to give to the reader a better and more global view of the properties and potentials of the asynchronous technology to reduce the power consumption of integrated systems. Starting from the power reduction techniques that can be applied at the structural level of asynchronous circuits, we moved to the software (OS) layers of an embedded system. At the hardware level, reducing the energy consists in reducing the number of transitions required by the computation. This assertion relies on asynchronous circuit properties: they are hazard-free (no energy wasted in spurious transitions) and event-driven (only operating parts consume energy). To minimize the number of transitions, hence the energy, we reviewed a set of techniques based on data encoding, logical and micro-architecture structures. At the system level, it was demonstrated that a relevant exploitation of the asynchronous hardware potentials by the software enables significant power savings that cannot be achieved by a synchronous/clocked implementation. As reported in this chapter, many asynchronous processors have been designed and manufactured. All these prototypes, even though quite complex, have involved low manpower. Regarding their performances, we want to point out that asynchronous processors operate at wide voltage range. Therefore their consumption can be adjusted within a two orders of magnitude range. Moreover, another key feature of asynchronous processors is their ability to promptly shutdown and wake-up.

This suggests adopting a very different point of view for the design of the hardware, the software as well as the running applications of an embedded system. In fact, instead of designing synchronous clock-driven systems, the asynchronous technology gives us the opportunity to design information-driven or event-driven systems. As for the power-aware OS presented in Section 22.5, it was demonstrated that the design of asynchronous algorithms implemented with asynchronous circuits has led to the design of ultra low-power systems, unattainable by clocked systems [68,69]. This extra power reduction originates from the exploitation of the knowledge available on the information to be processed, jointly by the hardware and by the software. This approach can be applied to the design of all subparts of SoCs, processors, peripherals, analog and RF parts, communication network, etc. Removing the clock and moving to asynchronous circuits enables to focus on information processing and design ultra low-power information-specific hardware and software.

References

[1] G. Birtwistle and A. Davis, Eds., *Asynchronous Digital Circuit Design*, Springer-Verlag, New York, 1995.

[2] P. Pénzes and A. Martin, An energy estimation method for asynchronous circuits with application to an asynchronous microprocessor, *Proc. of Design, Automation, and Test in Europe Conf.*, Mar. 2002, Paris, France.

[3] K. Stevens, Energy and performance models for clocked and asynchronous communication, *Proc. of 9th IEEE Int. Symp. on Advanced Res. in Asynchronous Circuits and Syst.*, May 2003. Vancouver, B.C., Canada. pp. 56–66.

[4] J. Fragoso, G. Sicard, and M. Renaudin, Power/area tradeoffs in 1-of-M parallel-prefix asynchronous adders, *Proc. of 13th Int. Workshop on Power and Timing Modelling, Optimization, and Simulation,* Sep. 2003, Turin, Italy.

[5] W.J. Bainbridge and S.B. Furber, Delay-insensitive system-on-chip interconnect using 1-of-4 data encoding, *Proc. of 7th IEEE Int. Symp. on Advanced Res. in Asynchronous Circuits and Syst.,* Mar. 2001, Salt Lake City, Utah, pp. 118–126.

[6] C. Piguet, M. Renaudin, and T.J.-F. Omnès, Special session on low-power systems on chip (SoC), *Proc. of Design, Automation, and Test in Europe Conf.,* Mar. 2001, Munich, Germany.

[7] W.J. Bainbridge, W.B. Toms, D.A. Edwards, and S.B. Furber, Delay-insensitive, point-to-point interconnect using M-of-N codes, *Proc. 9th Int. Symp. on Advanced Res. in Asynchronous Circuits and Syst.,* May 2003, Vancouver, B.C., Canada, pp. 132–140.

[8] R. Manohar, Width-adaptive data word architectures, *Proc. of 19th Conf. on Advanced Res. in VLSI,* Salt Lake City, Utah, March 2001, pp. 112–129.

[9] C.-H. Park, B.-S. Choi, D.-I. Lee, and H.-Y. Choi, Asynchronous array multiplier with an asymmetrical parallel array structure, *Proc. of 19th Conf. on Advanced Res. in VLSI,* 2001, pp. 202–212.

[10] D. Burger, T.M. Austin, and S. Benett, Evaluating future microprocessors: the simplescalar tool set, Tech. Rep. CS-TR-96-1308, University of Wisconsin—Madison, Jul. 1995.

[11] SUN Microsystem Laboratories, Introduction to shade, TR 415-960-1300, Revision A of 1, SUN Microsystem Laboratories, Mountain View, CA, Apr. 1992.

[12] V. A. Bartlett and E. Grass, A low-power asynchronous VLSI FIR filter, *Proc. 19th Conf. on Advanced Res. in VLSI,* 2001, pp. 29–39.

[13] T.E. Williams, Performance of iterative computation in self-timed rings, *J. VLSI Signal Proc.,* 7:17–31, Feb. 1994.

[14] A. Martin, M. Nyström, and P. Penzes, ET2: a metric for time and energy efficiency of computation, in *Power-Aware Computing,* R. Melhem and R. Graybill, Eds., Kluwer Academic Publishers, New York, 2002.

[15] J. Teifel, D. Fang, D. Biermann, C. Kelly, and R. Manohar, Energy-efficient pipelines, *Proc. 8th IEEE Int. Symp. on Advanced Res. in Asynchronous Circuits and Syst.,* Apr. 2002, pp. 23–33.

[16] A. Martin, A. Lines, R. Manohar, M. Nystroem, P. Penzes, R. Southworth, and U. Cummings, The design of an asynchronous MIPS R3000 microprocessor, *Proc. on Advanced Res. in VLSI,* Sep. 1997, pp. 164–181

[17] A. Efthymiou and J.D. Garside, Adaptive pipeline structures for speculation control. *Proc. 9th IEEE Int. Symp. on Advanced Res. in Asynchronous Circuits and Syst.,* May 2003. Vancouver, B.C., Canada, pp. 46–55.

[18] M. Lewis, J.D. Garside, and L.E.M. Brackenbury, Reconfigurable latch controllers for low-power asynchronous circuits, *Proc. ASYNC '99,* Apr. 1999. pp. 27–35.

[19] S.B. Furber, D.A. Edwards, and J.D. Garside, AMULET3: a 100-MIPS asynchronous embedded processor, *Proc. of Int. Conf. Computer Design (ICCD),* Sep. 2000.

[20] J. Tierno, R. Manohar, and A. Martin, The energy and entropy of VLSI computations, *Proc. 2nd IEEE Int. Symp. on Advanced Res. in Asynchronous Circuits and Syst.,* Mar. 1996.

[21] K. Van Berkel, J. Kessels, M. Roncken, R. Saeijs, and F. Schalij, The VLSI-programming language tangram and its translation into handshake circuits, *Proc. European Conf. on Design Automation (EDAC),* 1991, pp. 384–389.

[22] H. Van Gageldonk, D. Baumann, K. Van Berkel, D. Gloor, A. Peeters, and G. Stegmann, An asynchronous low-power 80c51 microcontroller, *4th Proc. Int. Symp. on Advanced Res. in Asynchronous Circuits and Syst.,* San Diego, CA, 1998, pp. 96–107.

[23] D. Edwards and A. Bardsley, Balsa: an asynchronous hardware synthesis language, *The Computer J.,* 45(1):12–18, 2002.

[24] A. Bardsley and D.A. Edwards, Synthesising an asynchronous DMA controller with Balsa, *J. Syst. Architecture,* 46:1309–1319, 2000.

[25] L.A. Plana, P.A. Riocreux, W.J. Bainbridge, A. Bardsley, J.D. Garside, and S. Temple, SPA — a synthesizable AMULET core for SMARTCARD applications, *Proc. of Int. Symp. on Advanced Res. in Asynchronous Circuits and Syst.*, Apr. 2002, pp 201–210.
[26] A.V. Dinh Duc, J.B. Rigaud, A. Rezzag, A. Sirianni, J. Fragoso, L. Fesquet, and M. Renaudin, TAST CAD tools: tutorial. Tutorial given at the *8th IEEE Int. Symp. on Advanced Res. in Asynchronous Circuits and Syst.*, Apr. 2002, Manchester, UK, TIMA internal report ISRN:TIMA-RR-02/04/01-FR, http://tima.imag.fr/cis.
[27] K. Slimani, Y. Remond, A. Sirianni, G. Sicard, and M. Renaudin, Estimation et optimisation de la consommation d'énergie des circuits asynchrones, *4ème journées francophones d'étude Faible Tension Faible Consommation (FTFC '2003)*, May 2003, Paris, France.
[27a] K. Slimani, Y. Remond, G. Sicard, and M. Renaudin, Test profiles and low energy asynchronous design methodology, PATHOS '04, Santorini, Greece, September 2004.
[28] http://www.synopsys.com/products/etg/powermill_ds.html.
[29] http://www.cadence.com/datasheets/spectre_cir_sim.html.
[30] A.J. Martin, S.M. Burns, T.K. Lee, D. Borkovic, and P.J. Hazewindus, The design of an asynchronous microprocessor, in *Decennial Caltech Conf. on VLSI*, C.L. Seitz, Ed., MIT Press, Cambridge, MA, 1989, pp. 351–373.
[31] S.M. Burns and A.J. Martin, Synthesis of self-timed circuits by program transformation, in *The Fusion of Hardware Design and Verification*, G.J. Milne, Ed., North Holland, Amsterdam, 1988. pp. 99–116.
[32] A.J. Martin, An asynchronous approach to energy-efficient computing and communication, *Proc. SSGRR 2000, Int. Conf. on Advances in Infrastructure for Electronic Business, Science, and Education on the Internet*, Aug. 2000.
[33] S.B. Furber, P. Day, J.D. Garside, N.C. Paver, and J.V. Woods, AMULET1: a micropipelined ARM, *Proc. IEEE Computer Conf. (COMPCON)*, Mar. 1994. pp. 476–485.
[34] I.E. Sutherland, Micropipelines, *Commn. ACM*, 32(6):720–738, Jun. 1989.
[35] S.B. Furber, J.D. Garside, and S. Temple, Power-saving features in AMULET2e, *Power-Driven Microarchitecture Workshop*, Jun. 1998.
[36] J. Cortadella, M. Kishinevsky, A. Kondratyev, L. Lavagno, and A. Yakovlev, Petrify: a tool for manipulating concurrent specifications and synthesis of asynchronous controllers, Technical report, Universitat Politècnica de Catalunya, Barcelona, Spain, 1996.
[37] J. Garside, W. Bainbridge, A. Bardsley, D. Edwards, S. Furber, J. Liu, D. Lloyd, S. Mohammadi, J. Pepper, O. Petlin, S. Temple, and J. Woods, AMULET3i: an asynchronous system-on-chip. *Proc. Int. Symp. on Advanced Res. in Asynchronous Circuits and Syst.*, Apr. 2000, pp. 162–175.
[38] W. Bainbridge and S. Furber, An asynchronous macrocell interconnect using MARBLE, *Proc. Int. Symp. on Advanced Res. in Asynchronous Circuits and Syst.*, 1998. pp. 122–132.
[38a] A.J. Martin, M. Nyström, K. Papadantonakis, P.I. Penzes, P. Prakash, C.G. Wong, J. Chang, K.S. Ko, B. Lee, E. Ou, J. Pugh, E.-V. Talvala, J.T. Tong, and A. Tura, The lutonium: A sub-nanojoule asynchronous 8051 microcontroller, *9th IEEE International Symposium on Asynchronous Systems & Circuits*, May 12–16, 2003, Vancouver, Canada.
[39] A. Abrial, J. Bouvier, M. Renaudin, P. Senn, and P. Vivet, A new contact-less smart card IC using on-chip antenna and asynchronous microcontroller, *J. Solid-State Circuits*, Vol. 36, 2001, pp. 1101–1107.
[40] TIMA laboratory, CIS group Web site: http://tima.imag.fr/cis/.
[41] M. Renaudin, Asynchronous circuits and systems: a promising design alternative, in microelectronics for telecommunications: managing high complexity and mobility (MIGAS 2000), special issue of the *Microelectron.-Eng. J.*, P. Senn, M. Renaudin, and J. Boussey, Guest Eds., Vol. 54, No. 1-2, December 2000, pp. 133–149.
[42] J. Kessels, G. den Besten, T. Kramer, and V. Timm, Applying asynchronous circuits in contactless Smart Cards, *Proc. Int. Symp. on Advanced Res. in Asynchronous Circuits and Syst.*, Eilat, Israel, April, 2000.

[43] M. Renaudin, P. Vivet, and F. Robin, ASPRO-216: A standard-cell QDI 16-bit RISC asynchronous microprocessor, *Proc. Int. Symp. on Advanced Res. in Asynchronous Circuits and Syst.*, 1998. pp. 22–31.

[44] A. Takamura, M. Kuwako, M. Imai, T. Fujii, M. Ozawa, I. Fukasaku, Y. Ueno, and T. Nanya, TITAC-2: an asynchronous 32-bit microprocessor based on scalable-delay-insensitive model, *Proc. Int. Conf. on Comput. Design (ICCD),* Oct. 1997, pp. 288–294.

[45] M. Gowan, L. Biro, and D. Jackson, Power considerations in the design of the alpha 21264 microprocessor, *Proc. 35th IEEE Design Automation Conf.,* San Francisco, CA, Jun. 1998, pp. 726–731.

[46] M. Es Salhiene, L. Fesquet, and M. Renaudin, Dynamic voltage scheduling for real-time asynchronous Systems, *12th Int. Workshop on Power and Timing Modeling, Optimization, and Simulation (PATMOS),* Sept. 2002, Sevilla, Spain.

[47] Y. Li, G. Patounakis, A. Jose, K. Shepard, and S. Nowick, Asynchronous datapath with software controlled on-chip adaptative voltage scaling for multirate signal processing application, *Proc. IEEE Int. Symp. on Advanced Res. in Asynchronous Circuits and Syst.,* May 2003, pp. 216–225.

[48] L. Nielsen, C. Niessen, J. Sparso, and J. Van Berkel, Low-power operation using self-timed circuits and adaptative scaling of the supply voltage, *IEEE Trans. on Very Large-Scale Integration (VLSI) Syst.,* 2(4):391–397, Dec. 1994.

[49] M. Pedram, Design technologies for low-power VLSI, in *Encyclopedia of Comput. Science and Technol., Vol. 36.* Marcel Dekker, New York, 1997, pp. 73–96.

[50] A. Chandrakasan, S. Sheng, and R. Brodersen, Low-power CMOS digital design, *IEEE J. Solid-State Circuits,* 27(4):473–484, Apr. 1992.

[51] T. Burd and R. Brodersen, Energy-efficient CMOS microprocessor design, *Proc. 28th IEEE Hawaii Int. Conf. on System Sciences,* Vol. 1, Jan. 1995, pp. 288–297.

[52] G. Smit and P. Havinga, A survey of energy saving techniques for mobile computers, Moby Dick technical report, University of Twente, The Netherlands, 1997.

[53] T. Simunic, L. Benini, A. Acquaviva, G. Glynn, and G. De Micheli, Dynamic voltage scaling and power managements for portable systems, *Proc. 38th IEEE Design Automation Conf.,* Las Vegas, NV, June 2001, pp. 524–529.

[54] L. Benini, A. Bogliolo, and G. De Micheli, A survey of design techniques for system-level dynamic power management, *IEEE Trans. on Very Large-Scale Integration (VLSI) Syst.,* 8(3):299–316, Jun. 2000.

[55] Y. Lu, L. Benini, and G. De Micheli, Operating-system-directed power reduction, *Proc. Int. Symp. on Low-power Electronic Design,* July 2000, Rapallo, Italy, pp. 37–42.

[56] Y. Lu and G. De Micheli, Comparing system-level power management policies, *IEEE Design Test of Comput.,* 2001, pp. 10–19.

[57] M. Srivasta, A. Chandrakasan, and R. Brodersen, Predictive system shutdown and other architectural techniques for energy efficient programmable computation, *IEEE Trans. on Very Large-Scale Integration (VLSI) Syst.,* 4(1):42–55, March 1996.

[58] K. Flautner, Automatic monitoring for interactive performance and power reduction, Ph.D. Dissertation, University of Michigan, Ann Arbor, 2001.

[59] T. Pering, T. Burd, and R. Broderesen, Dynamic voltage scaling and the design of a low-power microprocessor system, Power-Driven Microarchitecture Workshop, in conjunction with *Int. Symp. on Comput. Architecture,* June, 1998, Barcelona, Spain.

[60] T. Burd, T. Pering, A. Stratakos, and R. Brodersen, A dynamic voltage scaled microprocessor system, *IEEE J. Solid-State Circuits,* 35:1571–1580, Nov. 2000.

[61] M. Weiser, B. Welch, A. Demers, and S. Shenker, Scheduling for reduced CPU energy, *USENIX Symp. on Operating Syst. Design and Implementation,* Nov. 1994, Monterey, CA, pp. 13–25.

[62] K. Govil, E. Chan, and H. Wassermann, Comparing algorithms for dynamic speed-setting of a low-power CPU, *ACM Int. Conf. on Mobile Computing and Networking,* Nov. 1995, Berkeley, CA, pp. 13–25.

[63] T. Pering, T. Burd, and R. Brodersen, Voltage scheduling in the lpARM microprocessor system, *Proc. Int. Symp. on Low-Power Electronic Design,* July 2000, Rapallo, Italy, pp. 96–101.

[64] P. Kumar and M. Srivastava, Predictive strategies for low-power RTOS scheduling, *Proc. IEEE Int. Conf. on Computer Design: VLSI in Computers and Processors,* Sep. 2000, Austin, TX, pp. 343–348.
[65] M. Fleischmann, Crusoe LongRun power management, Transmeta Corporation, Santa Clara, CA, Jan. 2001.
[66] Transmeta Corporation, *Crusoe Processor System Design Guide,* http://www.transmeta.com /.
[67] M. Renaudin, P. Vivet, and F. Robin, ASPRO: an asynchronous 16-bit RISC microprocessor with DSP capabilities, *ESSCIRC '99,* Sep. 1999, Duisburg, Germany, pp. 28–31.
[68] E. Allier, G. Sicard, L. Fesquet, and M. Renaudin, A new class of asynchronous A/D converters based on time quantization, *Proc. 9th Int. Symp. on Advanced Res. in Asynchronous Circuits and Syst.,* May 2003, Vancouver, B.C., Canada, pp. 196–205.
[69] B. Galilée, F. Mamalet, M. Renaudin, and P.Y. Coulon, Watershed parallel algorithm for asynchronous processor array, *IEEE Int. Conf. on Multimedia and Expo (ICME),* August 26–29, 2002, Lausanne, Switzerland.

23

Low-Power Baseband Processors for Communications

Dake Liu
Eric Tell
Linköping University

23.1 Introduction

Three processors usually exist in a communication terminal system: the digital signal processing (DSP) baseband processor (BBP), the DSP application processor (APP), and the microcontroller (MCU).

Baseband signals are all signals in a radio system, which are not modulated onto the carrier wave. In a cellular phone, this means all signals except those in the radio frequency (RF) part of the phone. This chapter discusses processors for digital baseband signal processing, known as digital baseband DSP processors (DBBP). Figure 23.1 defines the basic partitioning of a radio communication system from both functional and hardware points of view. A DBBP plays an important role in both the transmitter and the receiver. In a transmitter, the DBBP converts the data from application sources to a format adapted to the radio channel. In a receiver, a DBBP recovers symbols from the distorted analog baseband signal and translates them to a bit stream with acceptable bit error rate (BER) for applications. Figure

0-8493-1941-2/05/$0.00+$1.50

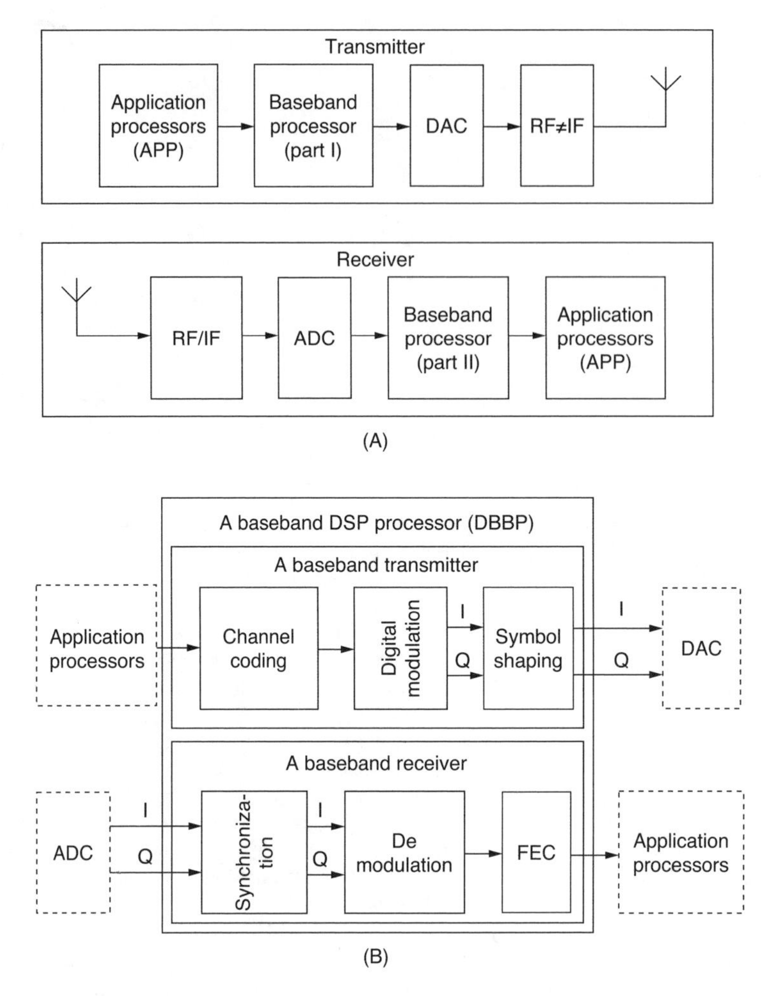

FIGURE 23.1 A radio communication transceiver.

23.1(b) describes functions in a DBBP. The most power-consuming parts are in the receiver, with functions such as synchronization, demodulation, and forward error correction. This chapter introduces the implementation of low-power DBBP. Detailed theory and knowledge of DSP for digital communications may be found elsewhere [2].

23.2 Digital Baseband DSP Processors (DBBP)

23.2.1 Function Coverage

This section gives an overview of the functions needed in a DBBP, using wireless local area network (LAN) as an application example. In principle, what we want to handle in the DBBP is the digital part of the physical layer, which includes all functions between analog to digital converter (ADC)/digital to analog converter (DAC) and the MAC (medium access control) layer interface. Figure 23.1(a) depicts the physical partitioning of a radio transceiver, and Figure 23.1(b) depicts the main functions handled by the DBBP.

23.2.2 The Transmitter

A transmitter performs three major functions: channel coding, digital modulation, and symbol shaping. Channel coding covers different methods for error correction (e.g., convolutional coding) and error detection (e.g., cyclic redundancy check (CRC)). Interleaving is used to minimize the effect of burst errors.

Digital modulation is the process of mapping a bit stream to a stream of complex samples. The first (and sometimes the only) step in the digital modulation is to map groups of bits to a specific signal constellation, such as binary phase shift keying (BPSK), quadrature phase shift keying (QPSK), or quadrature amplitude modulation (QAM). These are different ways of mapping groups of bits to the amplitude and phase of a radio signal). In most cases, a second step, domain translation, is applied. In an orthogonal frequency division multiplexing (OFDM) system (i.e., a modulation method where information is sent over a large number of adjacent frequencies simultaneously), an inverse fast fourier transform (IFFT) is used for this step. In a direct sequence spread spectrum (DSSS) system (e.g., direct sequence-code division multiple access [DS-CDMA], a "spread spectrum" method of allowing multiple users to share the RF spectrum by assigning each active user an individual "code"), each symbol is multiplied with a spreading sequence of ones and minus ones. The final step is symbol shaping which transforms the square wave to a band-limited signal using a finite impulse response (FIR) band-pass filter. This is necessary to make sure no parts of the transmitted signal are outside the permitted frequency band.

Channel coding and mapping functions operate on bit level (not on word level) and are therefore not suitable for implementation in a programmable processor. In a low-bandwidth transmitter, the symbol-shaping filter can be implemented in firmware. In a high-bandwidth baseband processor, however, a dedicated low-power, low-cost FIR filter circuit is needed.

23.2.3 Synchronization and Channel Equalization

Synchronization in the receiver can be divided into several steps. The first step includes detecting an incoming signal or frame, so called energy detection. In connection with this, operations, such as antenna selection and gain control, are also carried out. The next step is symbol synchronization, which aims to find the exact timing of the incoming symbols. All the preceding operations are typically based on complex auto- or cross-correlations.

In most cases, it is necessary that a receiver performs some kind of compensation for imperfections in the radio channel. This is known as channel equalization. In OFDM systems, this involves a simple scaling and rotation of each sub-carrier after the FFT. In a CDMA system, a so-called rake receiver is often used to combine incoming signals from multiple signal paths with different path delays. In some systems, least mean square (LMS) adaptive filters are used. Similar to synchronization, most operations involved in channel estimation and equalization employ convolution-based algorithms. These algorithms are not similar enough to share the same fixed hardware, but they can be implemented efficiently on a programmable DSP processor. If bandwidth and mobility is relatively low, the whole synchronization and equalization flow can be implemented in a programmable DSP processor. For higher bandwidth, high-speed processors with complex multiply and accumulate (MAC) units may be needed.

23.2.4 Demodulation and Forward Error Correction

Demodulation is the opposite operation of modulation. It involves an FFT in OFDM systems and a correlation with spreading sequence (so called despread) in DSSS systems. The last step of demodulation is to convert the complex symbol to bits according to the signal constellation.

Similar to channel coding, deinterleaving and channel decoding are not suitable for firmware implementation. In particular, Viterbi or turbo decoding, which are used for convolutional codes, are very demanding functions.

23.2.5 Comparison with a General DSP Processor

Different kinds of DSP processors are designed for different applications. A DBBP is designed especially for radio baseband signal processing, focusing on synchronization, modulation–demodulation, coding, and forward error correction. A DBBP can be implemented as an application-specific integrated circuit (ASIC) or as an application-specific instruction-set processor (ASIP). An architecture based on a programmable processor with surrounding accelerators is sometimes called a "centralized architecture" as

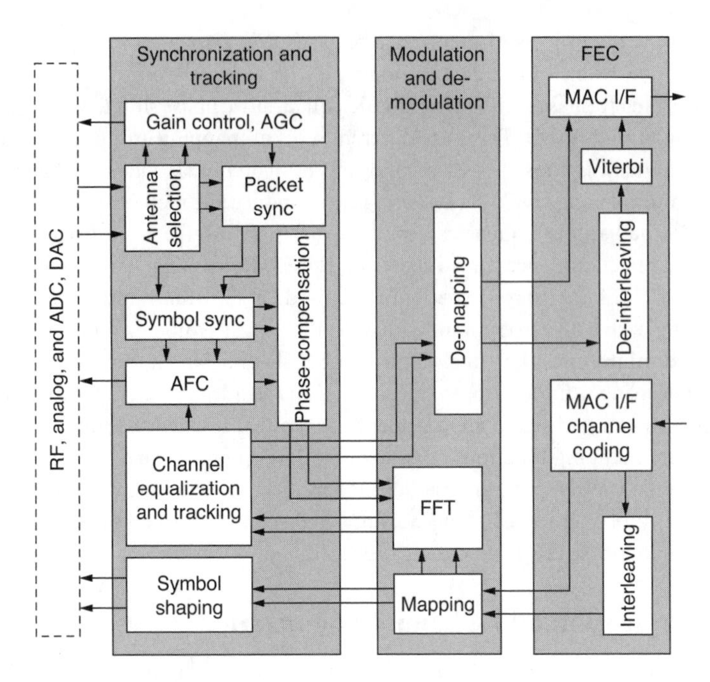

FIGURE 23.2 A baseband processor for IEEE802.11a.

opposed to a "distributed architecture," based on integration of custom circuits. Most current high-bandwidth baseband processors (for third-generation [3G] or wireless local area networks [WLAN]) are implemented as ASICs. One example of a DBBP for IEEE802.11a [4,13] is given in Figure 23.2.

To fulfill a certain level of flexibility and to cover multiple standards, modern low-power DBBPs are approaching a mix of ASIC and centralized ASIP architectures. Recently, the interest in software defined radio (SDR) solutions, which can cover as many standards as possible, has grown. Increased programmability in baseband processors has been the trend in both academia and industry; however, programmability in a DBBP is significantly different from the flexibility required in a general DSP processor. A general DSP processor [3] has to be flexible enough to cater to a large variety of DSP applications on the arithmetic level. Because of this, it has no support for acceleration of specific algorithms (except convolutions). A DBBP, on the other hand, gives dedicated algorithm acceleration for radio baseband DSP processing. For example, to increase performance and decrease control overhead, a baseband processor could have one dedicated instruction to carry out a complex vector operation that would require a subroutine of 20 instructions in a general DSP processor. Table 23.1 and Table 23.2 give the five most often used instructions for general DSP processors and DBBPs, respectively.

23.2.6 Classification of Baseband Processors

Different radio systems require different baseband DSP processors. Some systems, for example global system for mobile communications (GSM, which is the pan-European digital cellular radio standard) and wideband CDMA (WCDMA, which is a 3G mobile phone system) are full duplex (i.e., sending and receiving at the same time), and some, for example, WLAN, is half duplex (i.e., sending and receiving at different times). In some systems (e.g., digital audio broadcasting [DAB] and digital video broadcasting [DVB]), most devices are only receivers.

Requirements on DBBPs can be characterized by two variables, as presented in Figure 23.3. The first variable is the bandwidth. Computing power for synchronization, demodulation, and forward error correction (FEC) increases linearly or faster with the bandwidth. Therefore, the bandwidth has the largest impact on the power consumption of a baseband processor. Another important variable is the mobility. Higher mobility results in faster channel fading, which requires more processing to recover from channel

TABLE 23.1 Five Most Often Used Instructions of a General DSP Processor

Instructions	Functional Specification
Multiplication and accumulation	Accumulator <= accumulator + register 1 * register 2
Multiplication and round arithmetic in accumulator	Register 3 <= round (register 1 * register 2) Accumulator <= arithmetic operation (accumulator)
Memory to/from register	Memory [address pointer] <= (or =>) register
Register to/from accumulator	Accumulator <= (register 1, register 2) or Register <= round (accumulator)

TABLE 23.2 Most Often Used Instructions of a Baseband DSP Processor

Instructions	Functional Specifications
Conjugate complex convolution (auto correlation)	For I = 1 to N do (Complex Reg <=) Complex REG + V1[i] * Conjugate of V1(or 2)[i]
Complex convolutions	For I = 1 to N do {Complex REG <= Complex REG + V[I] * V2[i]}
Conjugate complex vector product	For I = 1 to N do (V3[i] <= V1[i] * Conjugate of V2[i])
Complex vector product	For I = 1 to N do (V3[i] <= V1[i] * V2[i])
Lookup table	REG2 <= Memory [Segment + REG1]

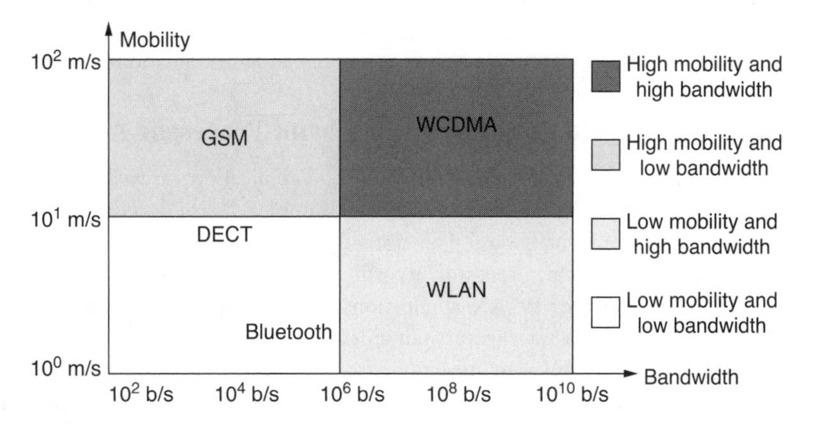

FIGURE 23.3 Classification of baseband DSP processors.

distortion. For a fast fading channel, the receiver may have to be updated for every received symbol; whereas with a slow fading channel, it may be enough to update the channel equalizer for every new packet. Because channel estimation and equalization often are among the most demanding operations in the receiver, the mobility requirements have a large impact on power consumption. Another parameter that should be considered is the dynamic range. This depends on the ratio of bit rate over symbol rate and on the distance over which transmissions may take place. With larger dynamic range, higher data precision is required in the receiver, which leads to higher power consumption.

23.3 Design of Low-Power Radio Baseband DSP Processors

23.3.1 Basic Principles for Low-Power Design

Dynamic power consumption of digital complementary metal oxide semiconductor (CMOS) circuits is $P = aCfV^2$, where a is the activity, C is the total load capacitance, f is the toggling frequency, and V is the supply voltage and the swing of the digital signal. Based on this formula, the principle of low-power design is to eliminate extra toggling, to decrease power supply voltage, and to minimize the overall capacitance. Elimination of extra toggling includes selecting a low-power datapath (data flow) architecture, selecting suitable precision for data processing units, eliminating control overhead, minimizing memory access, using operand stopping techniques, and designing for clock gating. Parallelization and pipelining are the major ways to reach low supply voltage. If the extra latency introduced by pipelining is not acceptable, extra hardware acceleration may be needed. If we disregard physical scaling, capacitances on silicon can be eliminated by shrinking/optimizing data and computing precision, by hardware acceleration, and by smart hardware multiplexing (sharing the same hardware by multiple functions in different time slots).

23.3.2 Trade-Off between Programmability and Fixed Function Hardware

At least three major advantages come from programmability: flexibility throughout the product lifetime, shorter debugging time, and a certain level of design error tolerance through the possibility of modifying firmware. It is also easier to share or multiplex a programmable device between multiple functions. On the other hand, extra power consumption is introduced by program memory and extra control circuits compared with an architecture where functions are direct mapped to fixed hardware. Due to the lack of hardware multiplexing, much more silicon is required by the direct mapped architecture. A trade-off between a programmable device and fixed function hardware is configurable devices that can be reused for similar functions. We know from previous sections that a programmable device is suitable for jobs related to synchronization and equalization. We also know that some other jobs, including modulation/demodulation and coding/decoding, are not suitable for firmware implementation. Instead, dedicated configurable circuits are a better solution for these jobs. We therefore conclude in this section that it is suitable to use a programmable device for synchronization and equalization and for miscellaneous jobs.

23.3.3 Nonprogrammable Low-Power Baseband Processor Architecture

Different kinds of architectures are used for different baseband applications, as well as for different trade-offs between power consumption and flexibility. FEC, for example, Viterbi decoders or turbo decoders, are always executed by dedicated hardware. The rest of the functions could be located in either a programmable or a nonprogrammable processor, according to product requirements. In this section, we discuss nonprogrammable DBBP for WLAN applications. One example is the baseband processor for HiperLAN2 and IEEE802.11a from interuniversitair micro-elektronica centrum (IMEC) Belgium [7]. Another example is the baseband and MAC processor from Atheros in Sunnyvale, California [6]. Non-programmable DBBPs are implemented by mapping algorithms to dedicated hardware. Because WLAN systems are a half-duplex, sending and receiving jobs are not processed simultaneously, so sharing hardware between the transmitter and the receiver is possible. Furthermore, because of their different execution times, sharing hardware between preamble and payload functions is also feasible. As another example of a nonprogrammable DBBP, Figure 23.4 gives an example of a converged DBBP for IEEE802.11a and IEEE802.11b [5].

The upper part of Figure 23.4 describes the IEEE802.11a functions divided into four data flows. The upper flow, which is the first to be activated, takes care of the so-called short pilot signals (10 equal symbols of 16 samples each). These are used mainly for energy detection, antenna selection, gain control, and synchronization tasks. The next flow operates on the so-called long pilots (2 equal symbols of 64 samples each), which are used for channel estimation and for fine-tuning the synchronization. Next, the main receiver flow takes care of the payload data symbols. The important steps here are I-Q-phase compensation (carried out by a rotor), FFT, channel equalization, demapping, deinterleaving, and forward error correction. After theses steps the resulting bit stream is handed over to the MAC layer. The final flow is the transmitter flow, which consists of forward error coding, interleaving, constellation mapping, IFFT, and the symbol-shaping filter. By multiplexing hardware of the four flows, the receiver filter is used by all receiver flows, and the FFT engine is used by the long pilot, payload reception, and transmitter flows. The same hardware is also used for both interleaving and deinterleaving.

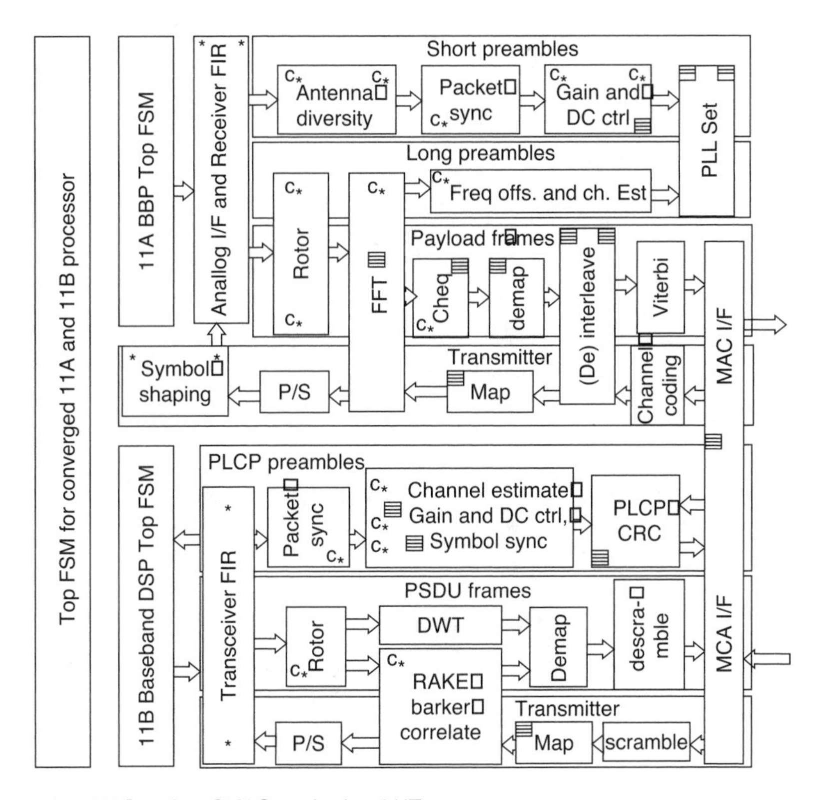

FIGURE 23.4 Mapping functions to nonprogrammable BBP architecture.

The lower part of Figure 23.4 depicts the IEEE802.11b functions, which are divided into three flows. The first flow operates on the preamble and takes care of energy detection, gain control, and synchronization. The second flow handles the payload data and has functions such as I-Q-compensation, channel equalization, despreading (which is either a correlation or a Walsh transform depending on the transmission mode), demapping, and descrambling. The third is the transmitter flow, which does scrambling, mapping, spreading, and symbol shaping. Possible candidates for hardware multiplexing are the receive filter and the scrambler/descrambler. The MAC layer protocol is identical for the two standards, so all MAC functions are also multiplexed.

The high degree of parallelization in this architecture enables low clock frequency (~ 50 MHz) and thereby makes it possible to lower the supply voltage. This architecture will give minimum power consumption at the price of no flexibility and high silicon cost. Architectures, such as the one in Figure 23.4, need approximately 17 complex multipliers, 15 lookup tables, and many other complex hardware blocks. Obviously, the total silicon cost will be high.

23.3.4 Programmable Baseband Processor (PBP) Architectures

In a converged baseband processor (e.g., a converged solution for GSM/GPRS/3G [1] or IEEE802.11a/b/g [8]), a custom circuit implementation will require more silicon area than a centralized ASIP solution because the hardware (HW) multiplexing does not cross between standards. To implement different standards with multiple modes of operation (i.e., preamble reception, payload reception, and transmission) and different data rates, a high degree of dynamic reconfigurability is required. The control will be

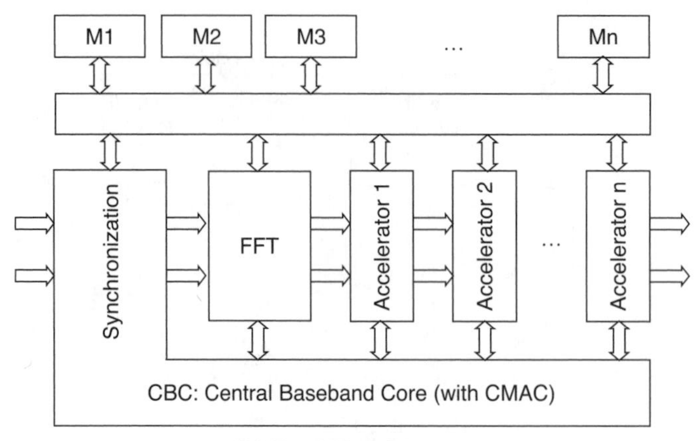

(A) The PBP architecture

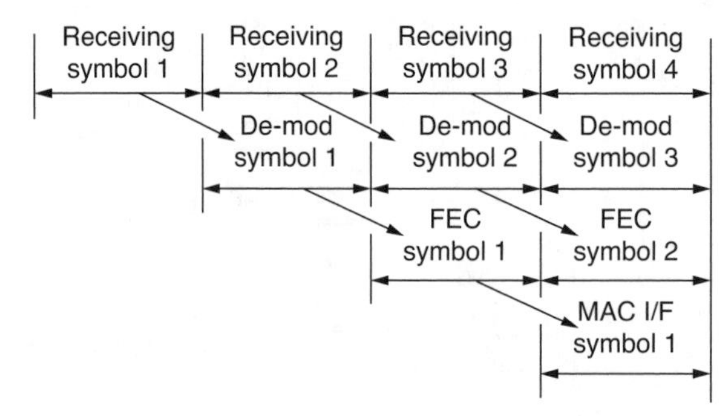

(B) The pipeline scheduling example of 11A payload process

FIGURE 23.5 A centralized PBP.

very complex. To reduce the complexity of the control path and reach the required flexibility, a programmable solution will be necessary.

Programmable solutions were investigated in Sengupta et al. [1] and Tell and Liu [8]. To support high millions of instructions per second (MIPS) capacity and low computing induced latency, centralized and scalable architectures are proposed by both academia and industry. By "centralized," we mean that a central programmable processor manages the DSP flow and some of the DSP functions. By "scalable," we mean that multiple accelerators and memories can be added. Figure 23.5(a) gives an example of such a solution. CBC in Figure 23.5(a) stands for "central baseband core." The CBC functions as the master of the system controlling the slave accelerator blocks. Accelerator and memory blocks are connected by a connection network, which is also configured by the CBC.

By analyzing the essential baseband algorithms listed in Figure 23.2 and the most often used instructions in Table 23.2, we have found that tasks suitable for CBC are the top DSP job flow management, complex vector computing, and lookup table functions. Therefore, a suitable CBC architecture will be a combination of a compact complex MAC and an arithmetic and logic unit (ALU) datapath. The complex MAC, which computes a complex multiplication, $(A_R + jA_I)^*(B_R + jB_I)$ in one clock cycle and complex accumulation in one clock cycle, supports complex vector computing (i.e., complex convolution, conjugate complex convolution, and complex vector dot product) [9]. The ALU will be used to support DSP job flow control. In the programmable baseband processor (PBP) of Figure 23.5, the CBC manages the following functions:

1. Running the top- and sublevel program flows, configuring the connection bus and the memory partitioning, initializing the accelerators, and implementing other miscellaneous controls.
2. Data quality control including scaling, antenna control, frequency offset estimation, and AGC/AFC to analog baseband.

TABLE 23.3 Kernel Instructions of the CBC

Operation Functions	Operand A	Operand B	Results
	Complex Arithmetic		
Normal MAC	Vector MEM A	Vector MEM B	Accumulator
Conjugate MAC	Vector MEM A	Vector MEM B	Accumulator
Vector energy	Vector MEM A	Vector MEM A	Accumulator
Vector product	Vector MEM A	Vector B	V-MEM C (A)
	Real Arithmetic		
LUT			Register file
±, compare	Register file	Register file	Register file
++, --	Register file	Register file	Register file
Shift/logic	Register file	Register file	Register file
	Flow and Control		
Configuration	Register/constant		Control register
Move			
Call/return			
Conditional jump			

3. Packet and symbol synchronization, channel estimation, and equalization — This requires complex vector computations including complex (conjugate) convolution, dot products, vector square-sum, and lookup tables (LUTs) for sine, arctangent, and 1/x.

The CBC runs two threads in parallel:

1. Program flow and miscellaneous jobs
2. Complex vector computations

The programmability gives several advanced features. The first feature is the excellent opportunity for convergence. For example, by adding an FFT accelerator and a rake receiver accelerator, both OFDM and CDMA standards are supported. This freedom of convergence decreases the time required for adapting to new market requirements. The second important feature is the hardware multiplexing. Most baseband algorithms can be implemented by the instructions in Table 23.2 on a complex multiply and accumulate (CMAC) in the CBC. The CBC can support most baseband functions with low silicon area cost if the firmware in the program memory can be synthesized into logic gates during the silicon backend design. For example, in a converged IEEE 802.11a/b/g receiver, approximately 2 complex MAC units will be sufficient, while the nonprogrammable solution may need approximately 17. The third advanced feature, and possibly the most important feature when it comes to implementation, is the relatively short time needed for system verification. Firmware verification is much easier than HW verification. Changing firmware can solve most small problems. The firmware design iteration time can be measured in minutes, compared with the long hardware iteration time, which could be about a day or more.

The instruction set of the CBC is given in Table 23.3. Because it is an ASIP, the instruction set of the CBC should be as simple as possible. Most complex mode arithmetic instructions are vector instructions. Real arithmetic instructions are used for table-based arithmetic acceleration and miscellaneous functions.

Figure 23.6 demonstrates how the functionality of an IEEE802.11a/b transceiver is allocated to a PBP. The fill patterns illustrate the functions allocated to different accelerators. For example, functions marked with the white fill pattern are allocated to the CBC and functions marked with horizontal lines are allocated to the FFT accelerator. To relax the load in the CBC, an extra CMAC is added as a configurable accelerator for vector processing. Also for example, part of the channel equalization could be allocated to this extra CMAC. The system clock frequency is relaxed by using a radix-4 FFT accelerator [12] and by off-loading CBC jobs to the extra CMAC. In the OFDM flow, accelerators could be used for FIR and analog/MAC interface, FFT, interleaver/deinterleaver, and FEC including Viterbi decoder, CRC, and

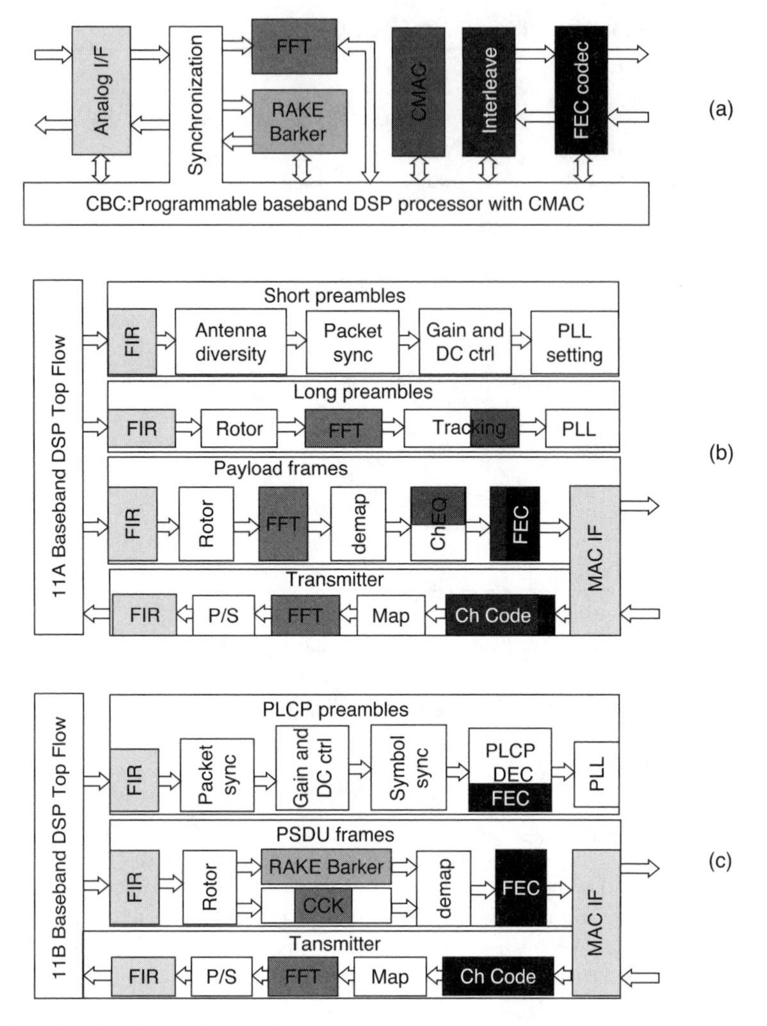

FIGURE 23.6 Mapping functions to PBP architecture.

scrambling. The DSSS transmission of IEEE std. 802.11b shares the FIR/interface and CRC/scrambling accelerators with the OFDM flow and uses an extra accelerator for spread/despread.

23.3.5 PBP Design Challenges

The main problem during the HW partitioning and integration is the memory access cost while delivering computing buffers from one processor/accelerator to another processor/accelerator. These extra memory accesses consume both extra power and execution time. Latency is especially a problem when jobs are divided and run in parallel at a lower clock frequency. Memory size and memory usage optimizations are important [8,17]. The most important memory size optimization in a baseband processor is the minimization of variable lifetime. General variable reuse techniques are very important when using cache as the computing buffer. When using only scratch pad memory for data stream processing in a baseband processor, most variable lifetimes are explicit and fixed according to the baseband data flow. What we need here is to minimize the memory access operations to minimize both the memory cost and run time. This optimization should be done during task level scheduling. In Abnous [15] and Tell and Liu [8], memory operations for delivering a computing buffer can be eliminated by reconfiguration of the memory connection network. When an algorithm is assigned to a computing unit (a processor or an accelerator) and a group of memories, one memory must be assigned as the result buffer. In the next stage, the memory connection network is reconfigured so that the result buffer of the previous unit becomes the input buffer for the new unit. In this way, the memory cost and the memory access time are minimized. This procedure, which we call "computing buffer delivery technique" [8], decreases both power consumption and latency. To implement this technique, a comprehensive methodology for system reconfiguration and integration is necessary. The research project Pleiades at the Department of Electrical Engineering and Computer Science (EECS) in the University of California-Berkeley, gives an architecture and methodology for reconfiguring and reconnecting computing devices and memories according to the dataflow at hand. In the Pleiades architecture, system reconfiguration and reconnection is performed via a system on chip connection network that includes a data network and a control network. The control network is used to configure the data network. In our architecture [8], the CBC controls the connections by writing a configuration vector to bus connection registers, which are addressable control and configuration registers.

Another major challenge is the design of the CBC-accelerator interface. The CBC is made for relatively general baseband jobs and is designed before the accelerators are specified. The interface must allow easy plug-in of accelerators of varying type without modification of the hardware. We need to send control and data from CBC to accelerators and results from accelerators to the CBC.

A bus architecture gives the required scalability and device address space. When we plug in an accelerator to the bus extension, the added hardware is seen as an ordinary execution unit of the processor. The design idea is to define a simple and robust protocol for accelerator integration. Giacalone [16] gives a good acceleration methodology. In the TIC55, scalability is given by instruction set architecture extension. This is also known as tightly coupled HW acceleration. In TIC55 and in our CBC, the extension is defined by a protocol accepted by both the processor and the accelerators. A set of special accelerator instructions is needed. Each accelerator instruction is divided into a common part and a custom part. The common part gives address and bus control, and must be decoded by both the processor and the accelerator. The custom part carries control codes that are only decoded by the accelerator. Because the CBC does not decode the custom control code, the processor can easily be upgraded by adding a new accelerator. In addition, no restrictions are made regarding the kind of accelerators that can be added.

23.3.6 Decreasing Supply Voltage

Lowering the supply voltage is the most effective way to decrease power because the reduction of power consumption follows the square rule; however, three constraints (i.e., the throughput required by the baseband specification, the limitations on the computing latency, and the limited choices of supply voltage on system level) have to be met while reducing the supply voltage. In this section, we aim for lowering the supply voltage thus satisfying throughput and computing latency requirements.

To decrease supply voltage, we should start by analyzing the system scheduling and finding the timing critical path. By accelerating the timing critical path, we get a chance of lowering the supply voltage. The following paragraph gives an example based on the system in Figure 23.6. We first allocate jobs to execution time slots following the constraints from the MAC layer specification. After job allocation according to the task pipeline definition in Figure 23.5(b), we find that the timing critical path is the demodulation step, which includes rotor, FFT, channel equalization, and demapping. We modify the schedule by moving the rotor to the receiving time slot (the receiving time slot now includes sampling, buffering, receiver FIR filter, and rotor). Next, we speed up the remaining operations in the demodulation step. By replacing the radix-2 FFT by a radix-4 FFT, the computing time is decreased from about 200 cycles to 52 cycles [12]. The channel equalization takes about 200 cycles to run in the CBC. Adding one more CMAC and allocating half of the channel equalization job to it decreases the cost for channel equalization to about 100 cycles. The demapping for 64-QAM takes about 1000 cycles if there is no instruction level acceleration. By specifying a special demapping instruction for 16-QAM and 64-QAM, the cycle cost for 64-QAM becomes about 144 cycles. Including about 200 cycles for miscellaneous jobs,

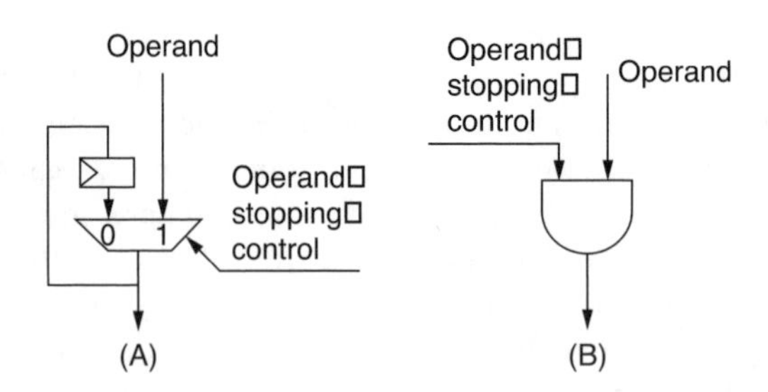

FIGURE 23.7 Operand-stopping technique.

the cycle cost of the payload processing has decreased from about 1644 cycles to about 498 cycles. This means that the system clock can be decreased from 416 MHz to 125 MHz. If the critical path at circuit level is given by the 10-bit CMAC, about 250 MHz can be reached using a 0.15-μm digital CMOS technology running on the nominal supply voltage. The fact that we only need 125 MHz means that we have a chance to lower the supply voltage by about 40%. The corresponding power reduction factor is 2.7. The actual power reduction factor is approximately two because of the extra power cost from the added hardware.

23.3.7 Eliminating Unnecessary Switching

Three possibilities are available for reducing logic switching: minimizing the memory access, minimizing the computing power, and minimizing the control power. Minimizing memory accesses has already been discussed in this chapter. About 10% power and execution time can be saved by eliminating the passing of computing buffers [8]. Operand stopping is a technique where operands are stopped from propagating through a bus to nonactive logic blocks. Two basic operand-stopping techniques are given in Figure 23.7. The method, given in Figure 23.7(a), stops the operand by keeping it in a register. It is used for stopping the logic toggling involved in single step instructions. The method given by Figure 23.7(b) stops the operand by masking it with AND or OR gates. It is used for stopping the logic toggling involved in vector mode instructions. By using operand-stopping technique, a datapath can be divided into several active regions. The operand stopping control signal could come from the decoded instruction.

Clock gating is another technique for achieving low-power consumption. During architecture and function level design, the hardware should be partitioned into blocks suitable for clock gating. The instruction decoding can then generate clock gating control signals to different blocks. Although a part of the circuit is not active, the clock to flip-flops in this part will be gated or shut down. It should be noted that the reset signal should be valid to every flip-flop even when the clock is gated. Therefore, asynchronous reset is preferred.

23.3.8 System-Level Power Management

If it is possible to use a controllable power switch or DC-DC converter for the WLAN subsystem, three levels of power management will be possible: power supply on-off control, system clock on-off control, and circuit clock on-off control. The supply on-off control and the system clock on-off control are managed on system level beyond the the programmable core of the baseband processor. Clock on-off control for circuits in a DBBP can be managed inside the DBBP or inside the accelerators. The control could be handled implicitly or explicitly. Explicit control is managed by running instructions such as no-operation (NOP) and sleep and wait for external interrupt (SLEEP). The instruction decoder manages implicit control. Because every instruction activates only certain parts of the processor, the instruction decoder can control clock gating and operand stopping inside the CBC. Power management is a general issue in low-power design, and detailed implementation techniques can be found in other chapters in the book.

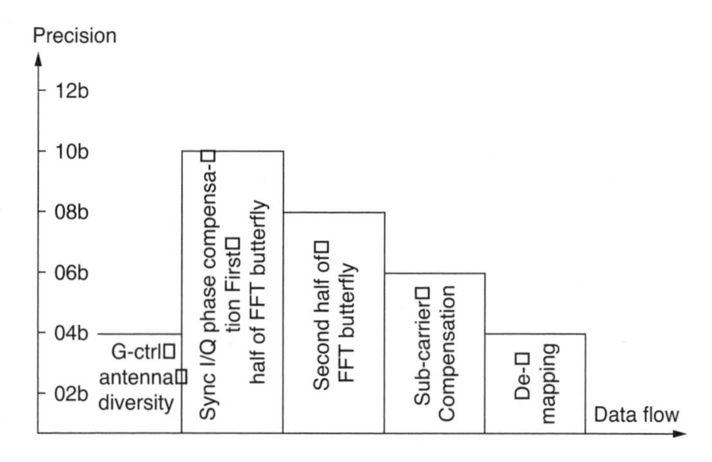

FIGURE 23.8 Variable precision in an IEEE 802.11a baseband processor.

23.4 Case Study One: Variable Data Length and Computing Precision

The power consumption increases with the data precision. In a nonprogrammable architecture, each step in an algorithm is assigned to a specific computing or storage unit. To save power, each unit is designed with the minimum acceptable data width. In a programmable BBP, however, the computing unit must be designed for the largest required precision.

To decrease the power consumption in the datapath of a programmable processor, dynamic precision firmware is introduced in the CBC [10]. In the algorithm design phase for the WLAN baseband firmware, we need to optimize precision for every step of all algorithms. When reduced precision can be used for a step of an algorithm, we will mask operands to a limited precision. Figure 23.8 demonstrates how different precision can be assigned for different algorithms in a receiver. (Data and computing precision is not specified after demapping because the remaining operations are bit level operations.) This chapter focuses on symbol processing because the precision of symbol processing algorithms has significant impact on the receiver quality.

We have estimated the switching power in a MAC unit with different masks applied and with different datapath precision. The results are given in Table 23.4. The right part of the table gives power consumption from 6 MAC units with different data width. The left part of the table gives the difference of the power consumption by masking the same MAC units, which is a 16-bit MAC with 32-bit accumulator. Note, for example, that replacing a 16-bit MAC operation with a 12-bit operation saves more than 50% power, although the same hardware is used. Furthermore, no major additional power saving can be made by replacing the 16-bit MAC unit by 12-bit hardware. We conclude that reducing the precision of computations can save a significant amount of power, even if a high-precision processor is used.

TABLE 23.4 Relative Power Consumption Measured from Masked MAC

Masked Operands on a 16-bit MAC		MAC Units with Different Precisions	
Mask Precision	Relative Power	Datapath Precision	Relative Power
16-bit (no mask)	1.00	16-bit (32-bit accumulator)	1.00
12-bit (mask 4-LSB)	0.47	12-bit (24-bit accumulator)	0.41
10-bit (mask 6-LSB)	0.31	10-bit (20-bit accumulator)	0.26
8-bit (mask 8-LSB)	0.18	8-bit (16-bit accumulator)	0.12
6-bit (mask 10-LSB)	0.09	6-bit (12-bit accumulator)	0.06
4-bit (mask 12-LSB)	0.04	4-bit (8-bit accumulator)	0.02

23.5 Case Study Two: Hardware Architecture for a Block Interleaver

23.5.1 Introduction

This case study introduces a low-power interleaver/deinterleaver architecture and compares it to the conventional hardware implementation. Excellent power saving was reached through significantly reduced toggling and power supply voltage reduction.

Interleaving is an operation often carried out as part of the channel coding in a radio system. The purpose of interleaving is to distribute transmitted bits in time, frequency, or both. Consecutive bits on the input stream should not be transmitted consecutively in time (or on the same frequency in an OFDM system). Interleaving reduces the effect of burst errors caused by fast fading. The requirements for interleaving depend both on the modulation and error correction schemes used, and on the channel characteristics.

One common method for interleaving is to use a block interleaver. A block interleaver operates on one block of data at a time, and no interleaving occurs between blocks. A block interleaver is typically implemented by writing the bits into a matrix row by row and then reading them column by column. Deinterleaving is simply the reversed operation — writing column by column and reading row by row. However, a real interleaver implementation could be more complicated. The following steps can define a more general block-interleaving algorithm:

1. Write bits to matrix row by row
2. Perform intra-row permutations
3. Perform intra-column permutations
4. Read bits columns by column

23.5.2 Traditional Interleaver Implementation

Figure 23.9(a) is a traditional interleaver implementation based on LUTs. This implementation can support any interleaving scheme. A sequential memory is used and a ROM LUT stores the specified interleaving sequence. For interleaving operation, the input bits are first written sequentially to the memory and then read in the order defined by the LUT. For deinterleaving, the bits are written according to the LUT and then read sequentially. The number of bits in the data memory is equal to the largest block size, and the number of words in the LUT is at least equal to the sum of all block sizes the interleaver should handle.

The main advantages of the traditional implementation are that it is simple, straightforward, and general. The main disadvantage of this implementation is that the bits have to be written and read one at a time, resulting in a high cycle cost. Another disadvantage is that the complete interleaving sequence of every interleaving scheme has to be stored explicitly in the LUT, making it relatively large if many standards have to be supported.

23.5.3 A New Block Interleaver Implementation

This section introduces a new implementation of a multi-standard block interleaver [14]. The purpose of the new implementation is to avoid the need for addressing individual bits in the memory and enable read and write of several bits in parallel. Generally, this scheme executes an interleaving of R rows and C columns in R + C clock cycles, while the old scheme needs two RC clock cycles. Figure 23.9(b) presents the main idea of the new architecture. It is based on a special matrix memory block where words are written to rows but read from columns.

A complete row can be written and a complete column can be read in one clock cycle. Intra-row permutations are carried out before the bits are stored to memory by simply reordering the bits on the input data bus. In the same way, inter-column permutations are carried out by reordering the bits on the output data bus after the data has been read.

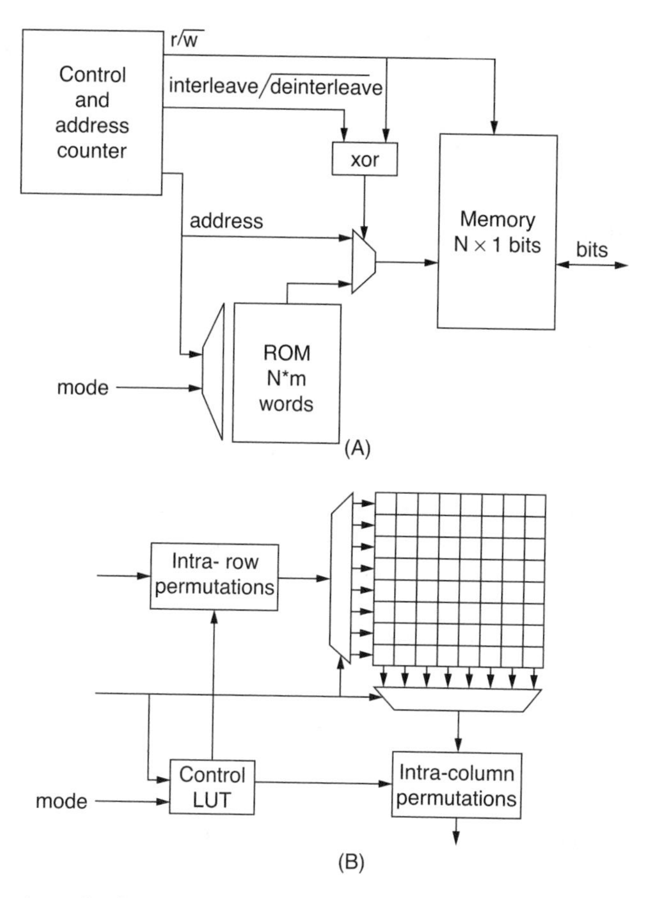

FIGURE 23.9 Two interleaver implementations.

If a small number of different permutation schemes are needed, the permutation blocks are a simple set of multiplexers. A more general interleaver may require a more intricate permutator.

A control block may be needed to set up the permutation block according to the current mode (and address). This could be implemented as a small LUT.

23.5.4 Hardware Implementation

An interleaver for the WLAN standard IEEE 802.11a has been implemented using the two previously discussed schemes. The IEEE 802.11a standard uses four different interleaving schemes with block sizes up to 16 × 18 bits and uses three different intra-column permutation schemes. The two implementations of the interleaver were both implemented in VHDL and synthesized to a 130-nm process. The result in terms of area and speed can be found in Table 23.5.

23.5.5 Power Issues

The new implementation is superior to the traditional one not only in performance, but also in power consumption for several reasons:

1. Most of the energy will be consumed in the data memory. Because we need to handle both 16 × 18 and 18 × 16 matrices (for interleaving and deinterleaving respectively), the number of physical memory cells is larger in the new scheme (324 vs. 288 in the old scheme). However, the number of active cells is the same in both schemes. Furthermore, because the new scheme uses much

TABLE 23.5 Interleaving implementation results

Feature	Old Architecture	New Architecture
Interleaving for 6–9 Mb/s transmission (BPSK, block size 3 × 16)	96 cycles	19 cycles
Interleaving for 36-54 Mb/s transmission (64-QAM, block size 18 × 16)	576 cycles	34 cycles
Approximated max frequency	250 MHz	500 MHz
Data memory area	0.0197 mm^2	0.0189 mm^2
Other area	0.0053 mm^2	0.0018 mm^2
Total area	0.0250 mm^2	0.0207 mm^2

smaller address decoders (18 addresses instead of 288) the power needed for addressing each bit is smaller in the new scheme.

2. The sequence ROM in the traditional scheme consumes significant power. For the 802.11a interleaver, the ROM size is 624 × 9 bits. (If the LUT in the new scheme was implemented in a ROM in the most naive way, it would need 128 × 4 bits, but it is possible to make it much smaller because the permutation does not depend on the address in three of the four modes.)
3. Because the new scheme needs fewer clock cycles, it can run at a much lower frequency, thereby the supply voltage can also be lower.
4. More power is also consumed in, for example, control logic in the traditional scheme, simply because it has to run for many more clock cycles.

23.6 Conclusion

A baseband DSP processor is an application specific processor dedicated for baseband signal processing. Requirements on a baseband processor vary a lot according to differences in dynamic range, channel fading, bandwidth, and applications (voice or data). Low-power design is essential in baseband processors for radio terminals. In addition to conventional low-power design rules, low-power design for baseband processors especially focuses on optimizing task level partitioning, scheduling, and parallelization. Minimizing memory size and memory accesses is essential. Functional acceleration is the key factor to achieve the balance between low-power consumption and flexibility. Variable precision, operand-stopping, and clock-gating techniques are also important ways of achieving low power.

References

[1] Sengupta, C. et al., The role of programmable DSPs in dual mode (2G and 3G) handset, in *The Application of Programmable DSPs in Mobile Communications,* Gatherer, A. and Luslander, E., Eds., John Wiley & Sons, New York, 2002, chap. 3, pp. 23-40.

[2] Gibson, J.D., *The Mobile Communication Handbook, 2nd ed.,* CRC Press, Boca Raton, FL, and IEEE Press, 1999.

[3] Lapsley, P., Bier, J., Shoham, A., and Lee, E.A., *DSP Processor Fundamentals, Architectures, and Features,* IEEE Press, 1997.

[4] IEEE Std. 802.11a-1999, High-speed physical layer in the 5-GHz band.

[5] IEEE Std. 802.11b-1999, High-speed physical layer extension in the 2.4-GHz band.

[6] Thomson, J. et al., An integrated 802.11a baseband and MAC processor, *Proc. ISSCC 2002,* San Francisco, CA, pp. 451-453.

[7] Eberle, W. et al., 80-Mb/s QPSK and 72-Mb/s 64-QAM flexible and scalable digital OFDM transceiver ASICs for wireless local area network in the 5-Ghz band, *IEEE J. Solid-State Circuits,* Vol. 36, 2001, pp. 1829-1838.

[8] Tell, E., Nilsson, A. (J), and Liu, D., A vector processor for converged baseband DSP, to be submitted to *WASP 2004,* Stockholm, Sweden.

[9] Huang, Y. and Chiueh, T., A sub-word parallel digital signal processor for wireless communication systems, *Proc. Asian-Pacific Conf. on ASIC 2002,*Taipei, pp. 287-290.
[10] Tell, E., Seger, O., and Liu, D., Operand masking for low-power baseband DSP firmware, submitted to IEEE NORCHIP 2004.
[11] Gunn, J.E., Barron, K.S., and Ruczczyk, W., A low-power DSP core-based software radio architecture, *IEEE, J. Selected Areas in Communications,* Vol. 17, April, 1999, pp. 574-590.
[12] Tell, E. and Liu, D., A converged hardware solution for FFT, DCT, and Walsh transform, *Proc. ISSPA 2003,* July, 2003, Paris, France.
[13] Heiskala, H. and Terry, J.T., *OFDM Wireless LANs: A Theoretical and Practical Guide,* Sams Publishing, Indianapolis, Indiana, 2002.
[14] Tell, E. and Liu, D., A hardware architecture for a multi-standard block interleaver, *Proc. ICCSC 2004,* Moscow, Russia, July 2004.
[15] Abnous, A., Low-power domain specific processor for digital signal processing, Ph.D. dissertation, University of California-Berkeley, 2001.
[16] Giacalone, J., Application-specific instruction-set architecture extensions for DSPs, in *The Application of Programmable DSPs in Mobile Communications,* Gatherer, A. and Luslander, E., Eds., John Wiley & Sons, New York, chap. 18, pp. 361-377.
[17] Catthoor, F. et al., *Custom Memory Management Methodology,* Kluwer Academic Publishers, Dordrecht, 1998.

24

Stand-By Power Reduction for SRAM Memories

Stefan Cserveny
Jean-Marc Masgonty
Christian Piguet
CSEM

24.1 Introduction

In processor-based systems on chip (SoCs), the memories limit most of the time the speed and are the main part of the power consumption. Much work has been done to improve their performances [1], however, new approaches are required to take into account the trend in scaled down deep sub-micron technologies toward an increased contribution of the static consumption in the total power consumption [2]. The main reason for this increase is the reduction of the transistor threshold voltages [3–7].

In a previous article [8], proposals were made for low-power SRAM and ROM memories working in a large range of supply voltages. With all bit-lines kept precharged at the supply value, however, the proposed techniques are not favorable for the static leakage. Simple solutions are available for a ROM: it can be switched off in the standby mode, or only the selected bit-lines are precharged before a read (with the corresponding speed penalty). For the six-transistor SRAM cell, however, enough supply voltage has to be present all the time to keep the stored information.

Negative body biasing increases the NMOS transistor threshold voltage and, therefore, reduces the main leakage component — the cutoff transistor subthreshold current. A positive source-body bias has the same effect and can be applied to the devices that are processed without a separate well, however, it reduces the available voltage swing and degrades the noise margin of the SRAM cell. Another important feature to be considered is the speed reduction resulting from the increased threshold voltage, which can be very severe when a lower than nominal supply voltage is considered.

This chapter presents an approach based on the source-body biasing method for the reduction of the subthreshold leakage, with the aim of limiting the normally associated speed and noise margin degradation by switching it locally. At the same time, this bias is limited at a value guaranteeing enough noise margins for the stored data.

Only the case of a SRAM is considered here, however, the same approach can be applied to any blocks containing storage circuits (e.g., flip-flops and registers).

0-8493-1941-2/05/$0.00+$1.50

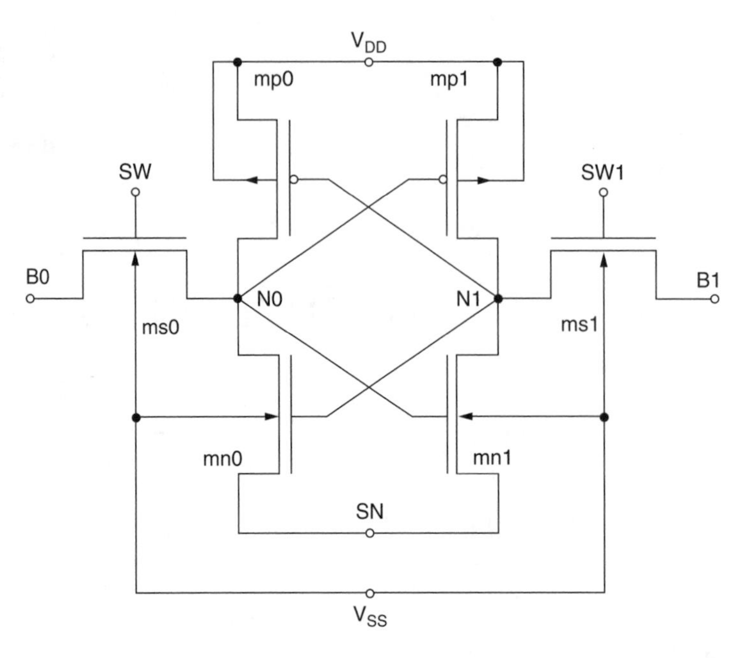

FIGURE 24.1 SRAM cell with separate NMOS source connection SN.

24.2 Leakage Reduction

In the deep sub-micron processes, as they scale down, there is a tendency toward a static leakage strongly dominated by the subthreshold current of the turned-off metal-oxide semiconductor (MOS) transistors. In even deeper nanometric technologies, an important tunneling gate leakage current exists [9]; however, it can be neglected in the preceding 100-nm technologies. Because this subthreshold leakage is much higher for the NMOS than that of the PMOS, only the NMOS transistors will be considered here for the leakage reduction (if necessary, however, the proposed methods can be applied as well for both). Notice also that the leaky NMOSs usually have no separate well in a standard digital process.

To allow source-body biasing in the six-transistor SRAM cell, the common source of the cross-coupled inverter NMOS (SN in Figure 24.1) is not connected to the body. Body pick-ups can be provided in each cell or for a group of cells, and they are connected to the V_{SS} ground.

In Figure 24.1, the possibility for separate select gate signals SW and SW1 has been illustrated, as for the asymmetrical cell described in Masgonty et al. [8] and Cserveny et al. [10], in which read is performed only on the bit-line B0 when only SW goes high while both are activated for write. Even if a symmetrical cell, which is selected for read and write with the same select word signal SW ≡ SW1, can also be considered for the leakage reduction techniques to be proposed, the asymmetrical six-transistor SRAM cell [10], presents several advantages.

With the asymmetrical cell, the dynamic power consumption at read is reduced because only one of the two bit-lines has a voltage swing. This consumption can be reduced even more by physically splitting these single bit-lines into sub-bit-lines as illustrated in the Figure 24.2. Because the sub-bit-lines are connected to a fraction of the cells in the column, the capacitive load to be discharged is significantly reduced.

With the asymmetrical cell, the functional voltage supply range is large, particularly for voltages well below the nominal process values because read on only one bit-line is performed without sense amplifiers, while both bit-lines are used for write. Not only are the problems related to sense amplifiers eliminated, even the power consumption compares favorably despite the full swing discharge of the sub-bit-lines where a "1" is read. Because only one select transistor is activated at read, the width (W) and length (L)

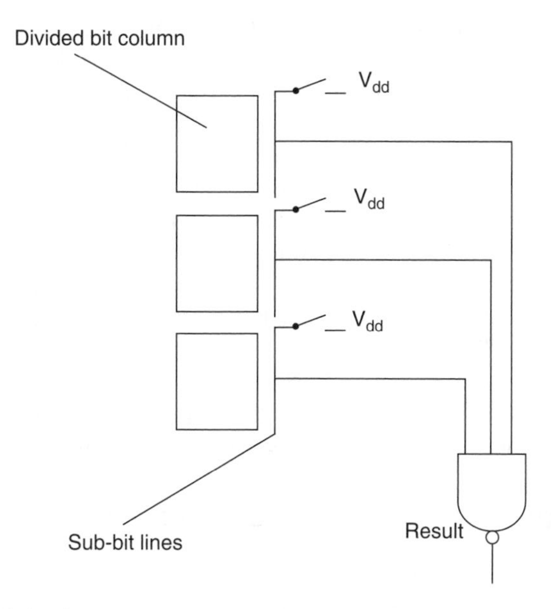

FIGURE 24.2 Physically split bit-lines principle: The sub-bit-lines are connected directly to a NAND gate, contrary to the speed reducing series transistor used when sub-bit-lines are connected to a main bit-line by a switch. In this example, three sub-bit-lines are considered.

of the transistors can be optimized differently on the two sides of the asymmetrical cell by taking advantage of the relaxed read noise margin constraint and the asymmetrical need for high driving capability.

A positive bias between the SN node and the V_{SS} ground (V_{SN}) will reduce the subthreshold leakage of a nonselected cell (SW and SW1 at V_{SS}) as plotted in Figure 24.3.

The result in Figure 24.3 is for a minimum size symmetrical cell in a 0.18-μm process (mn0, mn1, mp0, and mp1 have W = 0.22 μm and L = 0.18 μm only ms0 and ms1 are longer) for the state in which

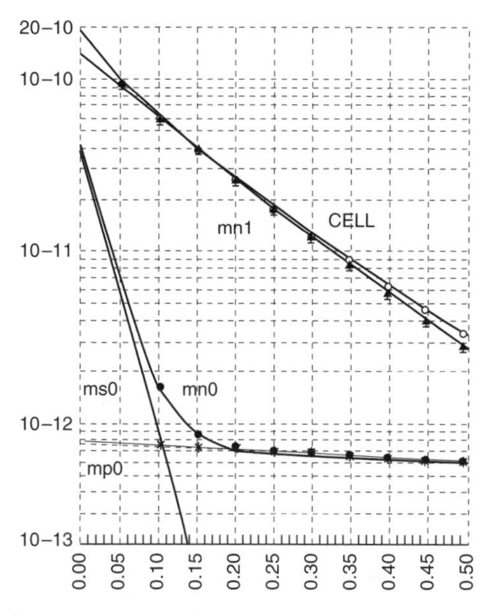

FIGURE 24.3 The simulated cell current (logarithmic scale in A) and its decomposition into the contributions of its transistors as a function of the source-body bias (linear scale in V).

the node N0 is low (at V_{SN}) and the node N1 is high (at V_{DD}). In this case the cell leakage is dominated by the transistor mn1, the NMOS which is cut off with $V_{GS} = 0$; its leakage is controlled by the body effect, the source-body voltage V_{BS} effect on its threshold voltage. The other branch contribution due to mp0, a less leaky PMOS, and to ms0, a longer NMOS with a gate-source voltage V_{GS} that becomes negative, is much less.

Furthermore, the ratio between the on (at VD = VG) and off (at $V_{GS} = 0$) currents in these deep submicron processes is worst at the minimum channel length, normally used in the digital designs; therefore, a somewhat longer transistor will be considered whenever possible and effective, without too much area penalty. According to the preceding analysis, such longer transistors are interesting mainly for the two inverter NMOS (mn0 and mn1), which, in the asymmetrical cell can be optimized differently also for low leakage.

The proposed method becomes less effective in the much deeper nanometric technologies; however, because there is less body bias effect to control the threshold voltage and the leakage due to the tunnelling gate, current and the gate induced drain leakage become increasingly important.

24.3 Noise Margin and Speed Requirements

A fixed bias V_{SN} on the source SN can reduce the subthreshold leakage of a nonselected cell as far as the remaining supply is enough to keep the flip-flop state, including the necessary noise margin.

For the selected cell in the active read/write mode with such a fixed V_{SN} bias, a speed loss is associated with the reduced available driving current and the noise margin at read is modified. This speed loss can be partially compensated adapting the V_{SN} value to the process corner because the corner that is worst for leakage is best for speed and vice versa. Nevertheless, this approach is hard to control over a large range of supply voltages because the speed/leakage relationship is a strong function of the supply voltage and temperature. Moreover, as depicted in the simulations of Figure 24.4, the noise margin at read is reduced at low supply voltages, which is a common situation for applications that need a supply value that is low or in a large range, such as low-power portable systems.

The noise margin, represented by the maximum size square that can be nested into the cross-coupled voltage transfer characteristics [11,12], is visibly reduced by the source-body bias $V_{DD} = 0.9$ V, a supply value well below the nominal 1.8 V for this process. Notice also that the crossing of the transfer characteristics changes from 3 to 5 points.

For the nonselected cell, important for the standby leakage, only the inverter transfer characteristics (without select MOS loading) come into account, and even if the V_{SN} increase reduces the noise margin,

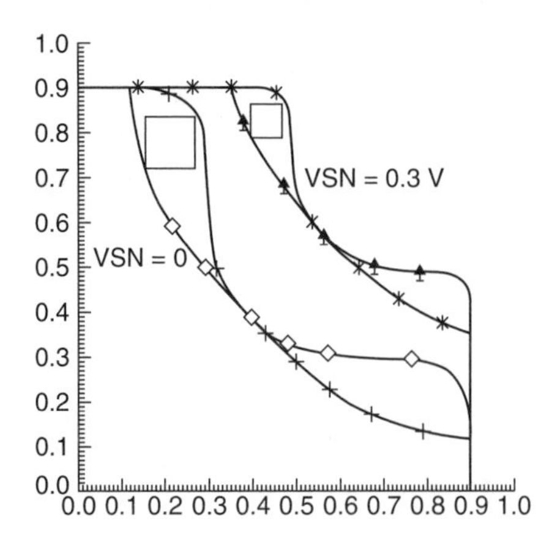

FIGURE 24.4 Worst-case (Nfast Pslow 125°C) read access static noise margin analysis for the minimum size symmetrical cell at $V_{SN} = 0$ and $V_{SN} = 0.3$ V for $V_{DD} = 0.9$ V.

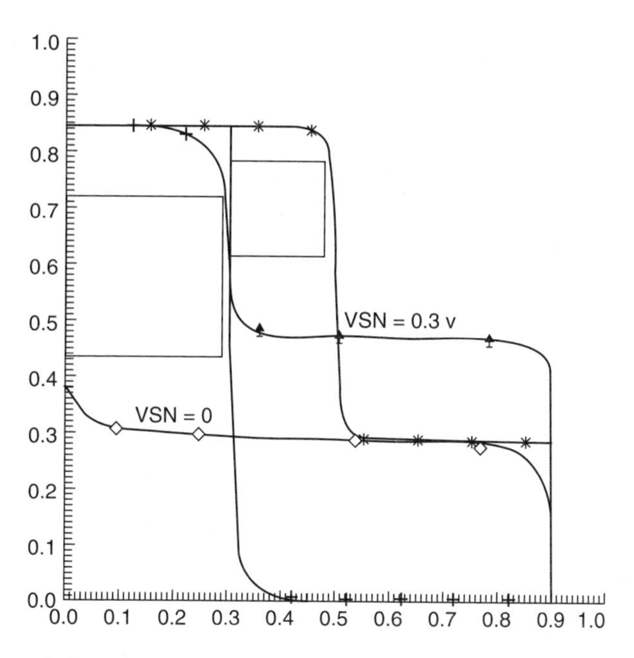

FIGURE 24.5 Worst-case (Nfast Pslow 125°C) standby static noise margin analysis for the minimum size symmetrical cell at $V_{SN} = 0$ and $V_{SN} = 0.3$ V for $V_{DD} = 0.9$ V.

it remains reasonably high if the V_{SN} does not increase too much, as plotted in Figure 24.5 for the same voltages discussed previously.

24.4 Locally Switched Source-Body Bias

The V_{SN} bias, useful for static leakage reduction, is acceptable in standby if its value does not exceed the limit at which the noise margin of the stored information becomes too small, but it degrades the speed and the noise margin at read. Therefore, it will be interesting to switch it off in the active read mode; however, the relatively high capacitance associated to this SN node, about six to eight times larger than the bit-line charge of the same cell, is a challenge for such a switching.

It is proposed here to partition the cell array into a number of groups, with the inverter NMOS sources of all cells inside a group being connected to a common terminal SN belonging to that group, and to locally assign a switch between these SN terminals and the ground, as illustrated in the Figure 24.6. When an active read or write operation takes place, the switch assigned to the group containing the selected word, connects the SN terminal of this group to ground. Therefore, in the active mode the performance of the cell is that of a cell without source bias. In standby, however, or if the group does not contain the selected word, the switch is open. With the switch open, the SN node potential increases reducing the leakage of the cells in that group, as described before, until the leakage of all cells in the group equals that of the open switch, which is slowly increasing with the SN potential (V_{DS} effect). Nevertheless, to guarantee enough noise margins for the stored state, the SN node potential should not become too high; this is avoided with a limiter associated to this node.

The group size and the switch design are optimized, compromising the equilibrium between the leakages of the cells in the group and the switch with the voltage drop in the activated switch and the selected group SN node switching power loss. The switch is an NMOS that has to be strong enough compared with the read current of one word, the selected word (i.e., strong compared with the driving capability of the cells defined by the select and inverter NMOS in series). At the same time, the NMOS switch has to be weak enough to leak without source-body bias as little as the desired leakage for all words in the group with source-body bias at the acceptable V_{SN} potential. On the other hand, to limit

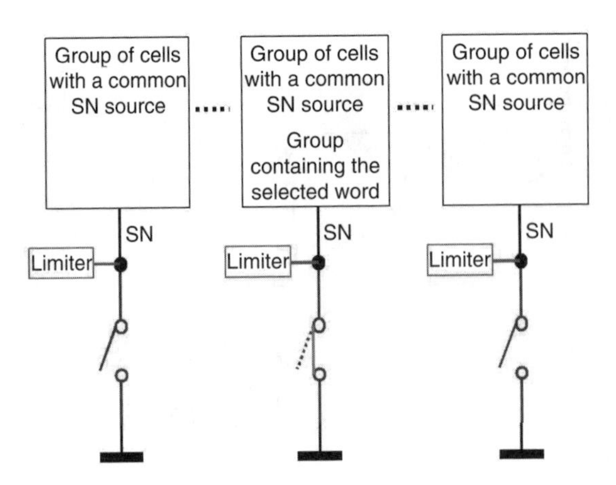

FIGURE 24.6 The limited and locally switched source-body biasing principle applied to a SRAM.

the total capacitive load on the SN node at a value keeping the power loss for switching this node much less than the functional dynamic power consumption, the number of words in the group cannot be increased too much. In particular, this last requirement demonstrates why the local switching is needed contrary to a global SN switching for the whole memory.

Figure 24.7 illustrates the most flexible approach to implement the described principle. Groups are built with a specific number of rows in a one-word column, the selected group of rows SRG containing the selected row SR. The switch size, built with one transistor for each bit, is adapted to the number of bits in a word. As proposed in Masgonty et al. [8], the divided word line blocks have the size of one word and the select column signal SC is activated at read and write for the column containing the selected word SW in the selected row SR.

In this implementation, with the number of rows in a group an integer power of 2 (e.g., 16 or 32), the selection of the group of rows (i.e., the SRG signal) is made by a simple partial row decoding. By combining this SRG signal with the one-word column selection signal SC, the group activation signal SG is generated. This SG signal closes the switch for that group (i.e., the NMOS transistors with the gate at SG). The limitation is obtained by connecting the SN node of the group to a VSN source with a NMOS

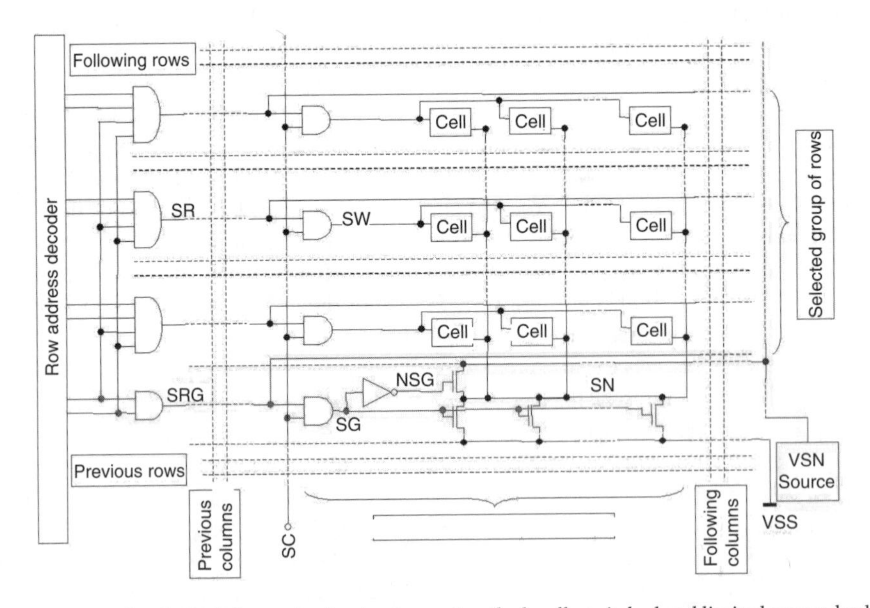

FIGURE 24.7 Example of a SRAM organization implementing the locally switched and limited source-body biasing.

pass transistor activated by the complement of the SG signal, thus connecting all groups except the activated one. Contrary to the fixed bias approach in which the whole active current goes through the biasing source, here the VSN source has to deliver or absorb only a very small current (i.e., the difference between the small leakages of the cells and the switches that are essentially equilibrated by design).

Other implementations were considered depending on the size and specific design constraints of the memory. It is quite interesting to activate the group only by address decoding, reducing the number of groups switching at the expense of a larger leakage in the group that remains selected; in particular, a group containing only all the words in a row can be activated directly by the select row signal SR. It is important to notice that the SN switching power loss depends on the group size and its relevance to the total power dissipation depends on the number of groups in the memory. Even more important is to notice that the limiter should only limit the SN potential increase. Figure 24.7 presents only one of the different implementations that have been considered to fulfill this function, however the VSN source can be replaced by an active or passive limiter circuit. Diode limiters, as considered in Enomoto et al. [13] for the typical case, are very interesting; however, they should be carefully designed and checked over all parameter corners.

24.5 Results

A cell implementing the described leakage reduction techniques together with all the characteristics of the asymmetrical cell described in Section 24.2 has been made in a 0.18-μm process. The SN node can be connected vertically or horizontally, and body pickups are provided in each cell for best source bias noise control. The inverter NMOS mn0 and mn1 use a 0.28-μm transistor length. Despite larger W/L of ms0 and mn0 on the read side used to take advantage of the asymmetrical cell for higher speed and relaxed noise margin constraint on their ratio, a further two to three times leakage reduction has been obtained besides the source-body bias effect plotted in Figure 24.3. The largest reduction is obtained where it is most desired — in the fast-fast process corner, the worst case for the leakage. The cell is a square with 2.8-μm sides — five times the upper level metal pitch allowing easier access blocks layout. It is 68% larger than the minimum size cell designed at the foundry with the special layout rules not accessible for other designs; however, the new cell is only 25% larger than a cell using the minimum size cell transistors if both layouts use the same standard rules. Put in another way, this further two to three times leakage reduction is equivalent with a reduction of 0.1 V to 0.15 V of the source-body bias needed for same leakage values. Overall, with V_{SN} near 0.3 V, the leakage of the cell has been reduced at least 25 times, however, more than 40 times for the important fast-fast worst case.

Figure 24.8 compares the read bit-line discharge delay for the group switched source-body bias with the fixed bias approach.

As expected, the equivalent V_{SN} bias for the same delay, depicted by the crossings in Figure 24.8, corresponds to the voltage drop in the switch. For the worst case considered here (all bits in the read word at one, that is, maximum current in the switch), the voltages at these crossings vary between 19 mV for SS (worst case for delay) and 28 mV for FF (best delay).

When the local group V_{SN} switching is compared with a SRAM using the same cell without source-body bias ($V_{SN} = 0$), the approximate 10% maximum speed loss, as well as the meaningless noise margin reduction due to these few mV, are not important.

On the other hand, the Figure 24.8 results illustrate the important speed improvement obtained with the local group switching as compared to the fixed V_{SN} bias; for example, at 0.3 V, the delays are 68 times shorter for the SS corner, the worst case for speed, and still 7.4 times shorter for the FF corner, the worst case for leakage, illustrating a very important speed increase. At the same time, the important reduction of the noise margin at read at low V_{DD} due to such a fixed V_{SN} bias (see Figure 24.3) is avoided.

The only area increase, compared with the previous design [8], is due to the group switches. If the switches are implemented as in Figure 24.7, the switches for two groups need an area about that of one row. Therefore, the area penalty for 16-word groups is about 3%. If the groups are organized in the rows, the area increase is even less.

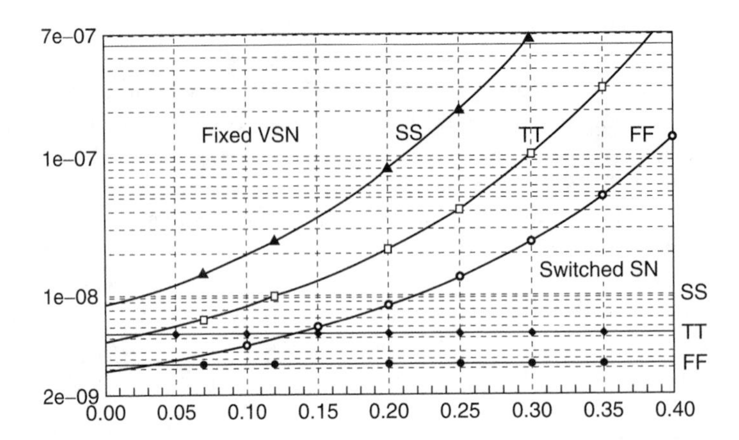

FIGURE 24.8 Simulated bit-line dicharge delay (in s) as a function of the V_{SN} source-body bias (in V) for the new cell and group switch design compared with the fixed V_{SN} bias result for the same cell at the slow-slow (SS), typical (TT), and fast-fast (FF) process corners, considering 256 rows, 25°C and all bits in the word at 1.

Following the strong leakage reduction in the cell array, it was important to design the remaining blocks of the memory to keep the total standby current of the memory still dominated by the cells. The standby leakage of the other blocks has been limited at well under 10% of the total standby total current of the memory by adapting some control signals for better low leakage of the blocks. For the most leaky elements, identified according to their functions, similar approaches have been used, such as source biasing, but without state retention requirements or longer than minimum length transistors. This did not require any extra area. The switches added for the bit-line pull-down transistors could be placed in the very large input signal buffers area already defined by the length of the select row and the height of the read/write blocks.

With the proposed locally switched source-body biasing applied to a SRAM, when compared with a SRAM without source-body bias, an important reduction of the standby leakage is obtained without any significant degradation of its speed or noise margin. Compared with the recently available SRAM with fixed source-body bias, for the same leakage reduction, the proposed SRAM improves significantly the speed and the noise margin at read.

24.6 Conclusion

Standby power reduction for storage circuits, which have to retain data, is obtained through limited locally switched source-body biasing. The standby leakage current is reduced by using a source-body bias not exceeding the value that guarantees safe data retention and less leaking nonminimum length transistors. This bias is short-circuited in active mode to improve the speed and the noise margin, especially for low supply voltages; however, this is made for a fraction of the circuit containing the activated part, allowing a trade-off between switching power and leakage. For a SRAM in a 0.18-μm process, the leakage is reduced more than 25 times without speed or noise margin loss.

References

[1] K. Itoh, Low-voltage memories for power-aware systems, Keynote Speech at *ISLPED '02,* Monterey, CA, August 12–14, 2002.

[2] H. Morimura et al., A shared-bitline SRAM cell architecture for 1-V ultra low-power word-bit configurable macrocells, *Proc. ISLPED 1999,* San Diego, CA.

[3] S. Kosonocky et al., Enhanced multithreshold (MTCMOS) circuits using variable well bias, *Proc. ISLPED '01,* pp. 165–169.

[4] S. Narendra et al., Scaling of stack effect and its applications for leakage reduction, *Proc. ISLPED '01,* pp. 195–200.
[5] C. Kim and K. Roy, Dynamic VT SRAM: a leakage tolerant cache memory for low-voltage microprocessors, *Proc. ISLPED '02,* pp. 251–254.
[6] N. Azizi et al., Low-leakage asymmetric-cell SRAM, *Proc. ISLPED '02,* pp. 48–51.
[7] C. Piguet et al., Techniques de circuits et méthodes de conception pour réduire la consommation statique dans les technologies profondément submicroniques. Invited paper, *Proc. FTFC '03,* Paris, May 15–16, 2003, pp. 21–29.
[8] J.-M. Masgonty, S. Cserveny, and C. Piguet, Low-power SRAM and ROM memories, *Proc. PATMOS 2001,* paper 7.4.
[9] F. Hamzaoglu and M. Stan, Circuit-level techniques to control gate leakage for sub-100-nm CMOS, *Proc. ISLPED '02,* pp. 60–63.
[10] S. Cserveny, J.-M. Masgonty, C. Piguet, and F. Robin, Random access memory, U.S. Patent 6,366,504 B1, April 2, 2002.
[11] E. Seevinck, F.J. List, and J. Lohstroh, Static-noise margin analysis of MOS SRAM cells, *IEEE J. Solid-State Circuits,* vol. 22, pp. 748–754, Oct. 1987.
[12] A.J. Bhavnagarwala, X. Tang, and J.D. Meindl, The impact of intrinsic device fluctuations on CMOS SRAM cell stability, vol. 36, pp. 658–665, April 2001.
[13] T. Enomoto, Y. Oka, H. Shikano, and T. Harada, A self-controllable-voltage-level (SVL) circuit for low-power high-speed CMOS circuits, *Proc. ESSCIRC 2002,* pp. 411–414.

25

Low-Power Cache Design

Vasily G. Moshnyaga
Koji Inoue
Fukuoka University

25.1 Introduction

Cache memories are the most area- and energy-consuming units in today's microprocessors. As the speed disparity between processor and external memory increases, designers try to put large multilevel caches on a chip to reduce the number of external memory accesses and thus boost the system performance. (See Table 25.1 for a survey of the on-die caches for several recent high-end microprocessors.) On-chip data and instruction caches are implemented using arrays of densely packed static RAM cells. The device count for the caches often exceeds the number of transistors devoted to the processor's datapath and controller. For example, the Alpha21364 [3] and PA-RISC Maco [5] microprocessors have over 90% of their transistors in RAM, with most of them dedicated for caches; the Itanium2 [1] has 80% in caches, the IBM G5 [7] has 72%, the PowerPC [8] has 71%, and Strong-ARM110 [9] has 70%. Due to the large load capacitance and high access rate, these caches account for significant portion of the overall power dissipation (e.g., 35% in Itanium2 [1]; 43% in Strong-ARM [9]). Therefore optimizing caches for power is increasingly important. Although much work on energy reduction has taken place in the circuit and technology domains [10,11], interest in cache design for power efficiency at the architectural level continues to increase. Architecture is the entry point in cache design hierarchy, and decisions taken at this level can drastically affect the efficiency of design.

This chapter describes architectural techniques appropriate for reducing power and energy in caches. Because power does not incorporate any notion of completing a given task, we could minimize the power by simply turning cache off or running it slowly. Our focus here is on techniques appropriate for reducing energy. Cache energy reduction is more difficult because it must be achieved under very strict timing requirements (i.e., without affecting performance). The chapter is organized as follows. First, we describe conventional cache design, and define sources of energy dissipation and degrees of freedom in the low-power design space. Then, we present an in-depth survey (and in many cases analyses) of cache energy-reduction techniques. We conclude by summarizing the major low-power design challenges that lie ahead.

0-8493-1941-2/05/$0.00+$1.50

TABLE 25.1 Survey of Recent High-Performance Microprocessors

Micro-Processor [reference]	Freq. (GHz)	Tech. (μm)	Die size (mm^1)	Trans. count (×10^6)	Power (W)	Cache Organization: Cache type	Size (KB)	Assoc. (ways)	Line Size (byte)	Latency (cycles)	Write update	Bandwidth (GBs)	Transistor count (% of total)
Itanium-2 McKinley [1]	1.0	0.18	421	221	130	L1-D	8	4	64	1	WT	16	
						L1-I	12op	4	n/a	1	n/a	32	
						L2	256	8	128	5i/6f	WB	32	80
						L3	3000	8	1024	12	WB	32	
Pentium 4	1.5	0.18	55	217	54	L1-D	12*	4	64	2i/6f	WT	44.8	
						L1-I	256	8	n/a	n/a	n/a	n/a	n/a
						L2	3000	8	128	7i/7f	WB	44.8	
Alpha 21364 [3]	1.0	0.18	397	100	150	L1-D	64	2	64	1	WT	19.2	
						L1-I	64	2	64	5i/6f	WT	19.2	93 (in fRAM)
						L2	1750	6	n/a	12	WB	16	
Ultra Sparc III [4]	1.0	0.18	244	23	80	L1-D	32	4	64	2i/6f	WT	16	
						L1-I	64	4	n/a	n/a	WT16	52	
						L1-pw	4	4	n/a	n/a	n/a	n/a	n/a
						L2-tags	88	n/a	n/a	n/a	n/a	n/a	
PA-RIS C Maco [5]	1.0	0.18	366	325	n/a	L1-D	1500	4	64	1	WT	n/a	
						L1-I	1500	4	64	1	WT	n/a	92 (in fRAM)
						L2-tags	1000	n/a	n/a	n/a	n/a	n/a	
Power4 (2 cores) [6]	13.0	0.18	400	n/a	125	L1-D	2*32	1	128	1	WT	416	
						L1-I	2*64	2	128	1	WT	416	n/a
						L2	1500	8	512	n/a	WB	n/a	
						L3-tags	176est.	8	n/a	n/a	WB	n/a	

Note: L1-D and L1-I denote level-1 data cache and level-1 instruction cache, respectively; L2 and L3 are level-2 and level-3 caches; pw is the prefetch cache; victim is the victim cache; WT is write-through; WB is write-back; 5i/6f defines latency of integer access and floating point access, respectively; n/a means not available.

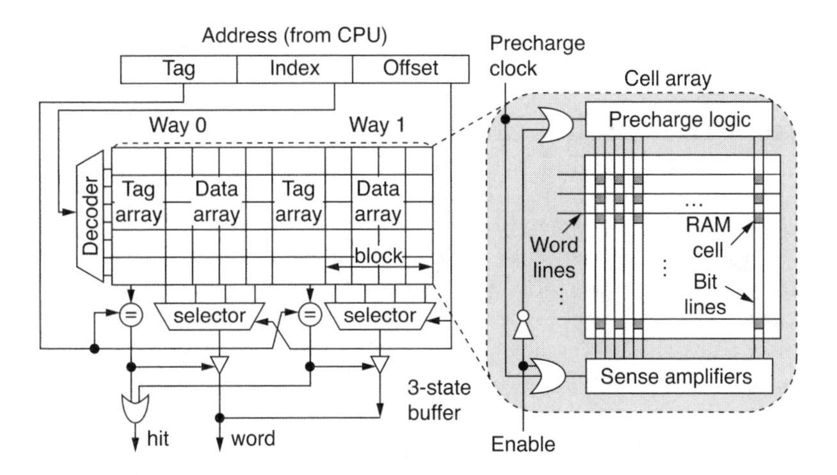

FIGURE 25.1 Organization of a two-way, set-associative cache.

25.2 Cache Organization

The dominant cache organization employed in modern microprocessors is m-way set-associative cache. Figure 25.1 exemplifies the organization for $m = 2$. A cache line stores a set of m memory blocks (several words each). Unlike traditional static RAM (SRAM), each data block in the cache is tagged with extra bits that indicate which address of main memory is actually being stored in the line, as well as other bits that indicate the validity and status of cached copy. A pair of tag array and data array is called way each consisting of S rows, or sets. If cache has only one way (i.e., $m = 1$), it is called direct-mapped; if the number of ways equals the number of blocks in the cache, the cache is called fully associative.

An m-way set-associative cache works as follows. Whenever processor initiates a memory access, it sends the physical address of the target word to the cache. The index bits of the address indicate cache line, while the least significant bits (or block offset) indicate position within the block. To read a datum, the cache activates in parallel all bits in the selected line, simultaneously comparing the upper portion of the address with m tags in the set. If a match occurs and data is valid, the data word from the way that hits is supplied to the processor. Otherwise, a cache-miss occurs, and the cache passes the address to a low-level memory to read the datum and replace it with a block within the cache. When the datum arrives, both the processor and the cache receive a copy. The cache then stores its copy with the appropriate address tag. On a write access, if the address being written to the memory has its copy in the cache, the cache updates value of the copy. Otherwise, the copy may be brought into the cache and then updated (write-allocate), or it may be updated in memory and not brought into the cache (write-no allocate).

The cache may employ two policies in handling a write to a block that is present in the cache: write-through, which updates both the cache and memory upon each write, and write-back, which writes the cache only. With write-through, read misses never produce copy-back writes to memory. With write-back, data are written to memory only when the data is removed from the cache (on cache miss), and an update has occurred. The cache checks the status bit and writes-back on miss only dirty (i.e., modified) blocks.

An m-way set-associative cache has as many as m alternative locations to write a new block on miss and, therefore, requires a specific policy (e.g., least recently used) for block replacement. A direct-mapped cache does not need such a policy because it has only one location from which to choose. If a program happens to repeatedly access words from two different blocks that map to the same line, however, the direct-mapped cache will continually swap the blocks, spending power and time. In comparison to the other alternatives (of the same cache size), direct-mapped cache is the fastest and the smallest, but has the highest miss rate. Set-associativity elevates the hit rate on expense of both the access time and hardware cost. For a fixed cache size, an m-way set-associative cache has as m-times as long word-lines and m times as more circuits for tag comparison and data selection. Therefore, caches with associativity $m > 32$ usually employ content-addressable memory (CAM) to store the tags and search them in parallel.

25.3 Factors Influencing Energy Consumption in Caches

There are two major components of energy dissipation in cache: dynamic energy, which is attributed to signal transitions in activated bit- and word-lines, sense-amplifiers, comparators, and selectors during reads/writes, and static energy, which is due to the total amount of leakage (or subthreshold) current, I_L, through inactive or OFF-transistors. The dynamic energy consumed per access is a sum of the energy spent on searching within the cache, an extra energy required for handling the writes and energy consumed by block replacement on cache miss. In CMOS circuits, energy consumption is proportional to the switching capacitance and the square of supply voltage, V. Thus, if we make N accesses to a CMOS cache that has miss rate of α, we will dissipate the following amount of energy:

$$\text{Energy} = N^*(C^*V^2 + \beta^*k^*E_m + \alpha^*2^*E_m) + V^*I_L \tag{25.1}$$

Here, C is the cache capacitance switched per access; β is the ratio of cache-writes to the total number of cache accesses, k is the write-policy dependent coefficient, and E_m is the energy required by one low-level memory access. Because β and E_m are independent to cache organization, optimizing cache for energy entails an attempt to minimize N, C, V, I_L, k, and α. This section briefly discusses these factors describing their relative importance, as well as the interactions that complicate the energy optimization process.

25.3.1 Miss Rate

Each external access requires many clock cycles and at least by two orders of magnitude more power than an on-chip access. Miss rate multiplies the number of external accesses by a factor of two and, therefore, has the highest optimization priority. Three parameters have an impact upon the cache miss rate: cache size, block size, and associativity:

1. Cache size. As cache size increases, the cache miss rate drops. A 64-KB cache, for example, has 6 times less misses than 1-KB cache for the 32-B block size and 10 times less misses for the 64-B block size [12]. Large on-chip caches reduce the overall energy consumption but cause a linear extension of bit-lines and more energy loss in the lines. This energy loss eventually may dominate other sources and result in energy increase, not savings. The cache energy, which is first lowered as cache size increases, starts to grow up as cache exceeds 32-KB in size [12]. In addition, caches larger than a few tens of KB are difficult to access in one cycle. If latency is not so important, the multi-cycle access provides a good opportunity to save energy by selectively enabling only relevant ways and sets in array.
2. Block size. With increase of block size, miss rate decreases due to enhanced spatial locality. As block grows in size, however, the width of data read to and from the cache also grows, leading to more energy dissipation in bit-lines, sense-amplifiers, and most important, in inter-memory traffic. Eventually, this energy cost may outweigh the energy savings of a smaller miss ratio because some instructions or data brought in on a miss will not be used. Usually, processors use smaller block sizes for low-latency and low-bandwidth L1 caches, and large blocks for high-latency and high-bandwidth L2 and L3 caches.
3. Set associativity. Miss rate decreases with more degrees of associativity though this effect diminishes with increasing cache size [13]. A four-way 64-KB set-associative cache (32-B block size) has a 14 and 44% miss rate advantage over a two-way set-associative cache and direct-mapped cache, respectively [14]; however, high associativity increases the cache access time. Although the energy consumption of each bit-line in an m-set associative cache is usually decreasing as associativity increases, the number of bit-transitions in cache usually grows with associativity. In addition, an *m*-way set-associative cache increases sense-amps power by *m* times. Meanwhile, sense-amps

consume less energy than the bit-lines in conventional caches. Therefore, associativity affects energy less than the cache size.

25.3.2 Write Policy

The write-through policy is simpler to implement than the write-back but requires more external accesses. With write-through, we access external memory the same number of times as we write to cache. With write-back, external memory is accessed only when block is replaced. Thus, we can simplify Equation 25.1 by using $k = 1$ for write-through and $k = \alpha$ for write-back. Because $\alpha << 1$, the write-back policy generally leads to better performance and energy-efficiency; however, it may cause temporal data inconsistency (i.e., when external memory and the cache associate different data with the same physical address). Problems may also arise when several processors with independent caches are sharing the same low-level memory. Regarding to write policy, there is also a choice whether to use write-allocate or write-no allocate on a cache-miss. Write-allocate appears to be a better choice for power because subsequent writes (or reads) to the allocated block may be done directly in the cache.

25.3.3 Cache Accesses Rate

High access rate (N) directly increases the number of tag checks and data reads, magnifying the cache energy consumption proportionally. If cache is not large enough to capture the majority of code and data used in relatively long execution period, the average amount of memory transfers will also grow up elevating both the fetching time and total energy dissipation. A common policy to reduce cache accesses is to use separate L1 data and instruction caches and increase memory hierarchy. An extra level in cache hierarchy introduces an extra delay and, therefore, is used only when latency increase is acceptable.

25.3.4 Switching Capacitance Per Access

In conventional caches, bit-lines, word-lines, and sense-amplifiers are the major contributors to the switching capacitance per access. In an *m*-set associative SRAM cache of size S, block size of L bytes, and tag size of T bits, the number of bit-lines is proportional to $2^*(m^*8^*L + T)$ and the number of word-lines to $S/(m^*L)$. Thus, we have: $C \propto 2^*(m^*8^*L + T)^*S/(m^*L)$. This provides us with four options to lower switching capacitance: decreasing the set-associativity, m, reducing the block size, L, reducing the cache size, S, and shrinking the tag bit-width, T. Note that all these options, except T, inversely affect the miss rate and, therefore, are viable only when cache switching activity dominates energy consumption.

25.3.5 Voltage

Because supply voltage (V) has a square impact on energy, voltage scaling offers the most effective means to minimize energy dissipation. Unfortunately, we pay a performance penalty for voltage reduction with delays drastically increasing as V approaches the threshold voltage Vt of the devices. Recent advances in technology have scaled both V and Vt aggressively, while maintaining a single clock access latency to L1 caches; however, reducing Vt increases leakage current, I_L, and makes the leakage term in Equation (25.1) appreciable. Another important concern for low V-low Vt regime is the fluctuation in Vt. As V approaches 1V, a Vt variation of ±0.15V causes delay changes by a factor of three. Such a large variation in nominal delay values cannot be tolerated. This sets a major limitation on how V can be reduced unless the Vt fluctuation is diminished to the level of ±0.05V [15]. Thus, lowering the threshold has a limited option for countering the effect of reducing V.

25.3.6 Leakage

Reducing energy consumption in CMOS circuits by lowering the supply voltage increases leakage energy dissipation. Leakage current I_L is independent of the circuit activity but dependent on the device area and temperature. Therefore, it occurs as long as power is supplied to the CMOS device. The on-chip L1

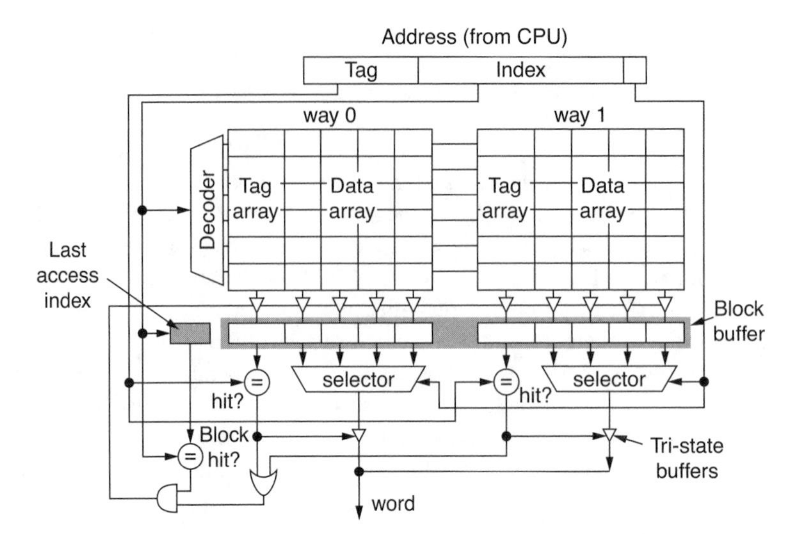

FIGURE 25.2 Block-buffered cache.

and L2 caches are one of the main candidates for leakage reduction because they utilize a significant fraction of chip transistors, most of which are inactive during long periods. At 0.13-μm technology, leakage energy already accounts for 30% of L1 cache energy and as much as 80% of L2 energy [16]. As technology moves below 0.1 μm, leakage energy increases exponentially, dominating the total energy used by the CPU [17]. Therefore, issues to reduce leakage without affecting performance become very important for the future low-energy cache design.

25.4 Energy Reduction Techniques

25.4.1 Reducing Cache Access Rate

When accessing a cache it is very likely that a requested word is confined to the block or one of the blocks in the set, that was last accessed. Thus, we can avoid some references to the cache by placing the most recently fetched block into a buffer and then reading-out the block data directly from the buffer without activating cache [18,19]. Figure 25.2 illustrates a structure of block-buffered cache. The cache checks first whether the line address of the current access is currently resident in the buffer. If there is a block hit, then the cache data is read from the block buffer without accessing the cache arrays. The normal access, including bit-line precharging and row-access decoding is performed only when a block-buffer miss occurs. Although effectiveness of this method strongly depends on the spatial locality of the access pattern and the block size, it can save 40 to 50% of energy over nonbuffered cache [19].

The main disadvantage of the block buffering is cache-access latency increase. To overcome this problem, Kamble and Ghose [20] proposed performing cache precharging and block-buffer access in parallel. Compared with the original block buffering [19], it spends some energy in precharging bit-lines on a block-buffer hit but does not affect cache latency.

A further extension to the block buffering is to use multiple (e.g., four) line buffers [21] or even a small direct-mapped level-0 (L0) cache, placed between processor and L1 cache [22,23]. This "filter" cache helps to omit many of the data accesses, so the L1 cache is not referenced as frequently. If data is found in the filter cache, energy is saved. If data is not found, however, an extra cycle is needed to access larger L1 cache. A 256-byte filter cache backed by a 32-KB L1 cache saves energy by 65% on average but causes a 29% performance degradation in comparison to a conventional cache due to access time increase on a filter cache miss.

Efficiency of using the L0 cache can be improved in two ways: by increasing data locality or by dynamic access management. The first approach puts in L0 cache only the most frequently executed instruction

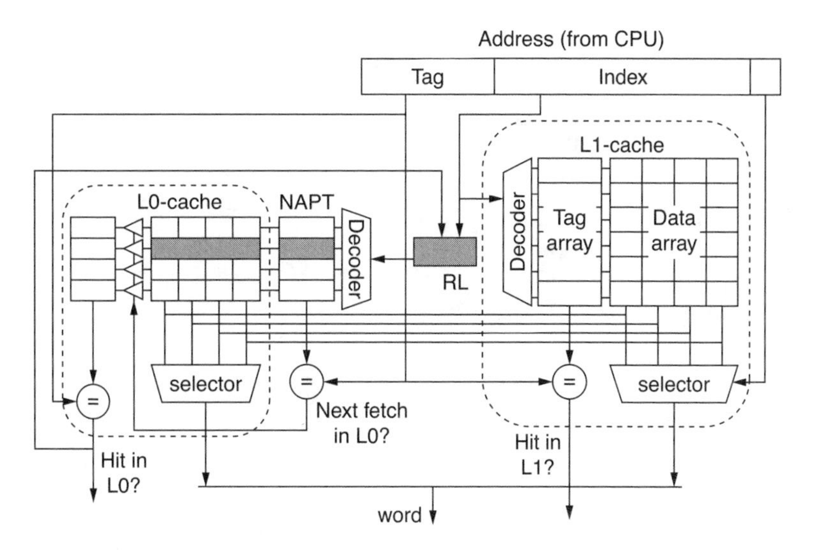

FIGURE 25.3 L0-cache bypassing hardware.

blocks or loops, identifying them based on branch prediction unit [24], loop instructions [26], or specific instructions inserted in the application code [25]. The second approach dynamically predicts the L0-cache miss and, if the prediction is correct, it accesses the L1-cache directly to reduce the miss penalty [27]. In small loops, consecutive addresses differ by a few least significant bits. Consequently, if current access hits (misses) in the L0 cache, the remaining accesses to the same block are likely to hit (miss) in the L0 cache. As illustrated in Figure 25.3, the approach adds to the caches a next address prediction table (NAPT), a register (RL) to store the index of the most recently accessed line in L0, and a comparator. The NAPT has the same number of entries as L0-cache, but stores only four lowest bits of the next address tag. Both L0-cache and the NAPT are read in parallel. If current access hits L0-cache and the NAPT entry matches the current address tag, then the next access is directed to L0 cache, and RL is updated. Otherwise, L1 cache is accessed, and the L0-cache is refilled.

25.4.2 Reducing Switching Capacitance Per Access

As discussed in Section 25.3, several degrees of freedom are inherent in cache organization: cache size, associativity, block size, and tag size. Reducing switching capacitance in cache can be achieved along different directions that exploit one or more of these freedoms. Most techniques utilize the same basic idea: divide the cache arrays into pieces, and then selectively activate on cache access only one that can hold the data. The partitioning can be determined structurally or behaviorally. The structural partitioning affects cache structure, while keeping the cache-access operation unchanged. In opposite, the behavioral partitioning modifies the cache-access sequence but leaves the cache structure unchanged.

25.4.2.1 Structural Partitioning

25.4.2.1.1 Word-Line Segmentation

Word-line segmentation is similar to column multiplexing used in SRAM memories [28]. Cells in each row are grouped into blocks, and a local word line, as depicted in Figure 25.4, accesses memory cells in each block. The local word-lines *LWLj* are connected to a global word *WLs* line through a transfer gate. Therefore, only cells in the activated block have their bit-line pair driven. With k local lines, the word line capacitance is reduced by $1/k$.

In caches, this method is usually applied by splitting the data array into sub-banks and using the low-order bits (block offset) of the address to disable sub-banks that are not accessed. Figure 25.4 illustrates a two-way set-associative cache divided in sub-banks. Because only one sub-bank is active per access, the

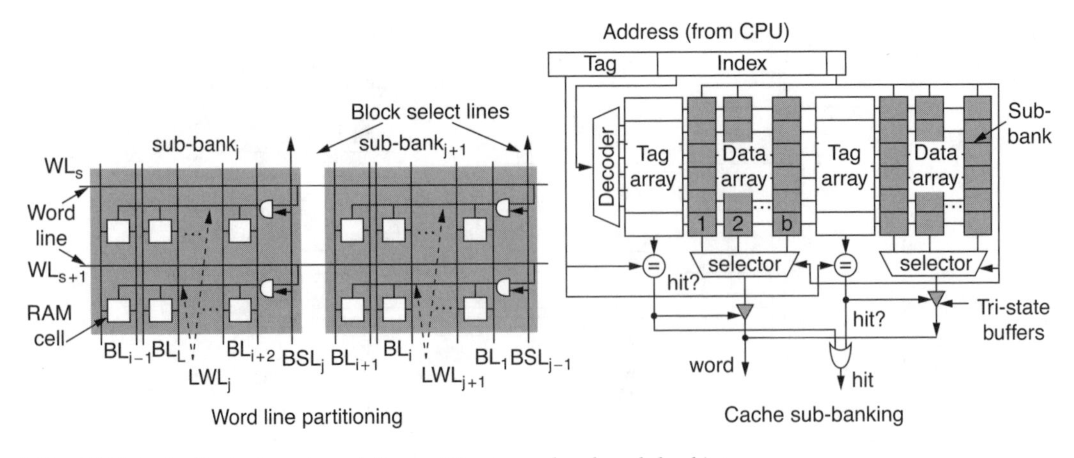

FIGURE 25.4 An illustration of word-line partitioning and cache sub-banking.

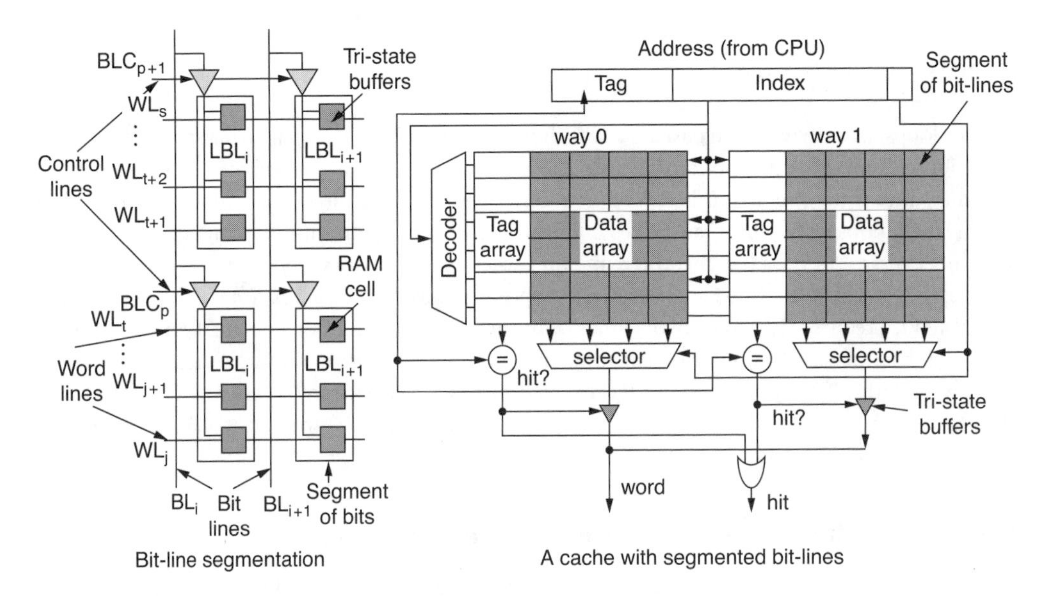

FIGURE 25.5 An illustration of bit-line segmentation.

more sub-banks in the cache, the better. For example, a 32-KB cache with a block size of 2, 4, 8, and 16 words per block saves energy by 46%, 63%, 80%, and 89%, respectively [19].

25.4.2.1.2 Bit-Line Segmentation

The idea of this method [28] is to split every column of bit-cells into independent segments, as depicted in Figure 25.5. An additional pair of local bit-lines (LBL) runs across the segments. The lines are connected to the global bit-lines (BL) through three-state buffers, which are activated/deactivated by the bit-line control lines (BLC). Before a read access, all segments are connected to the common lines, which are precharged as usual. The cache address decoder identifies the segment targeted by the row address issued to the array and isolates all but the targeted segment from the common bit-line. This reduces the effective capacitive loading on the common line and eventually lowers the energy.

25.4.2.1.3 Bit-Line Isolation

Along with the cache division into sub-arrays, holding each bit-line and output driver of each sub-array in a constant state of precharge when the array is not being accessed can reduce the switching activity. The lines remain in a precharge state unless a sub-array access is required. In the architecture, column switches are placed on bit-lines between memory cells and the sense amplifiers to isolate parasitic capacitor from of bit-lines from the active sense amplifiers [29]. The latch-type sense amplifiers engage after the column switches turn off. Thus, the sense amplifier load is reduced, and the switching capacitance decreases.

25.4.2.1.4 Multiple Cache Decomposition

The key feature of this technique [30] is to split the entire cache into independently addressable, small modules, each of which is an actual cache with its own cell array and peripheral circuits. A cache generally can be divided into *M* identical independently selectable modules, with each module further divided into K sub-banks. The sub-arrays may be created on any appropriate access boundary: byte, word, double-word, and so forth. The latency of this multi-divided cache architecture is equivalent to that of the single module and energy consumption is only 1/(M*K) of the regular, nondivided cache. The smaller the cache size, the bigger savings. A typical example is the SA-110 [9], which divides its 32-way associative 16-KB L1-I and L1-D-caches into 16 fully associative sub-arrays, so that only one eighth of the cache is enabled per access. With a 160-MHz target clock frequency, the SA-110 designers were able to maintain single cycle cache latency with this degree of associativity.

25.4.2.1.5 Cache Decomposition by Data Types

Locality of cache accesses increases when sub-caches are organized by data types. For example, a cache might have three modules: one for stack data only, another for global data, and the third for other data types [31]. The modules may be different in size (e.g., 4 KB, 4 KB, and 32 KB, respectively) and activated separately. When only the stack-module is referenced, the overall energy dissipated in cache can be reduced by 70% in comparison with a conventional cache.

25.4.2.2 Behavioral Partitioning

Current L1 set-associative caches are designed to operate in a single clock cycle. To ensure single-cycle operation, the caches probe all the data ways in parallel with the tag lookup, although the output only of a single matching way is used. The energy spent accessing the other ways is wasted. To save energy, designers make the way select in tag array to precede search in the data array [29]. Although this ordering can be efficiently used in multi-cycled L2 caches, it slows the access time of L1 caches by almost 60% [32]. Next, we present techniques that try to balance the speed-energy characteristics of L1 cache by dynamically optimizing the cache-access.

25.4.2.2.1 Phased Array Activation

The key idea of this method [33] is to drive the tag array and the data array by two different clocks: *C1* and *C2* (see Figure 25.6). If clock frequency is high, the difference between C1 and C2 is small. In this case, both ways in the data array start the access before the tag array dispatches the way select signal. Once the way select signals are produced in the tag array, unselected ways of the data array stop their access. If the clock frequency goes lower, the difference between C1 and C2 becomes long enough to allow generation of the way select signals to complete before the rising edge of C2. Thus, the only one way of data array is activated, resulting in low-energy operation. Although this so called "*automatic-power save architecture*" reduces the fastest access speed by a quarter, a low area overhead and large energy savings (almost 60%) make it very promising to adjust energy consumption to workload variation. The approach has been adopted in Hitachi's SH3 and Super-H microprocessors.

25.4.2.2.2 Way Prediction

Way prediction was originally proposed to reduce cache access time [34]. The idea is to predict the way in which data can be found prior to the cache access and probe the way instead of waiting on the tag array to provide the way number. Because only the predicted way is accessed, the method also offers immediate benefits for energy savings. A way-predicting scheme [35] speculatively chooses the most-recently used way to access, while disabling the nonaccessed ways. When prediction is correct, the cache completes operation in one cycle (see Figure 25.7), consuming single-way energy only. Otherwise, it searches the other ways in parallel and spends two clock cycles without any energy reduction. The way-

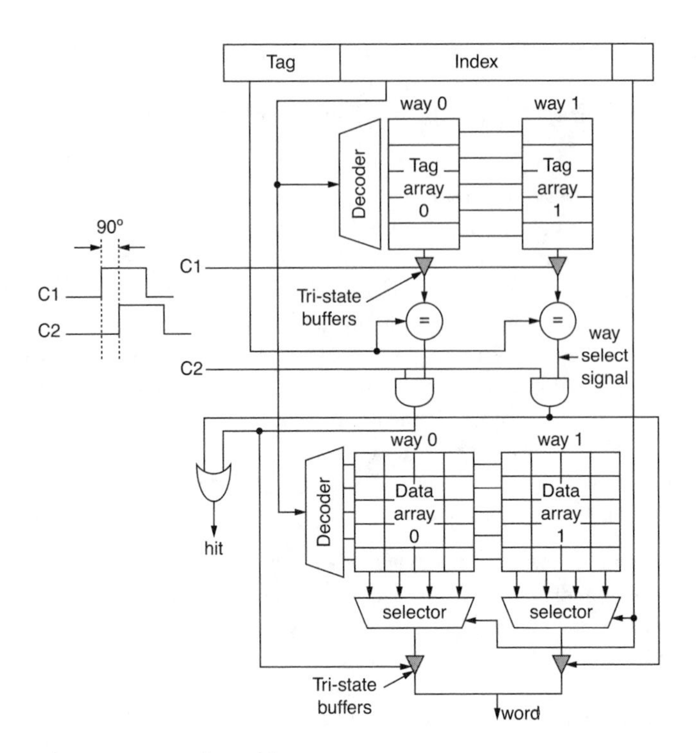

FIGURE 25.6 Automatic power-save cache architecture.

prediction improves the energy-delay product of cache by 60 to 70%; however, it incurs large performance loss on each misprediction.

Enforcing way prediction with other techniques can save cache energy without compromising performance. One approach is to unite way prediction with selective direct mapping [36]. Because 70 to 80% of blocks accessed in L1 D-cache are nonconflicting, allocating these blocks to direct-mapped positions can avoid both way prediction and tag checks. To identify conflicting blocks, the reactive cache [37] uses a list of recently replaced blocks, or victims. On a replacement, if the evicted block is present

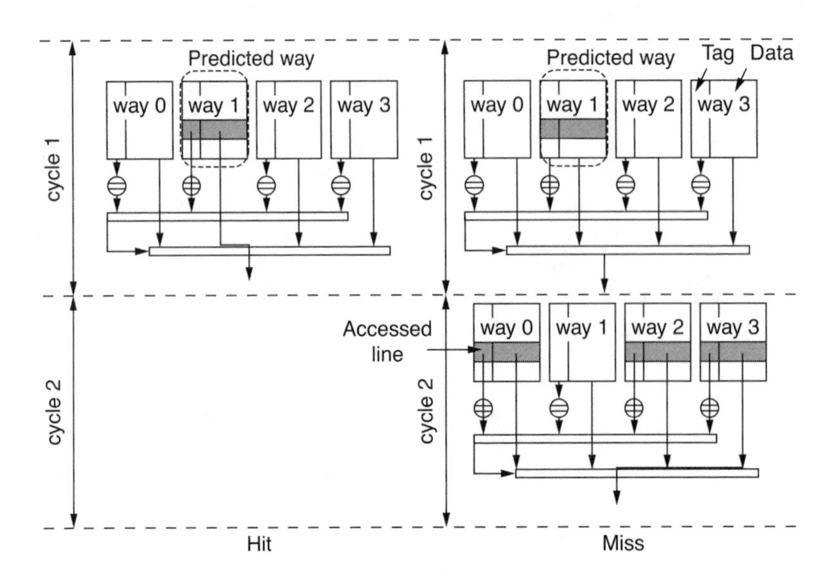

FIGURE 25.7 An illustration of a way-predicting scheme.

in the victim list, its entry's counter in the list is incremented; otherwise, a new victim list entry is created. Those blocks whose counter exceeds two are conflicting and placed in the set-associated position to avoid future conflicts. On access, the cache predicts whether the access is direct or set-associative. If prediction is correct, the cache probes only the matching data way, while a misprediction initiates a second probe of the correct data way. If a hit occurs in the direct-mapped way, the counter associated with the block is decremented. If a set-associative way hits, the counter is incremented. In comparison with original way prediction, the approach achieves the same energy-delay reduction but at less than 3% performance overhead.

Another approach [38] combines way prediction with phased-cache access based on a pseudo-associative cache. The cache utilizes a steering table and two schemes of phased-cache access to predict which way to probe first. The first scheme probes only a predicted way and, if the probe fails, it activates all the ways at the second attempt. The second scheme activates all the tag arrays and data array of the predicted way. If the prediction fails, the data array of correct way is subsequently accessed. A similar idea is advocated in [39]. In this proposal, prediction logic operates in parallel with the translation look-aside buffer (TLB) lookup. Based on the prediction, the cache controller activates appropriate sub-cache by sending the physical address. If the selected sub-cache hits, the access ends. Otherwise, the controller looks for the next sub-cache to probe. The reprobing continues until the data is found or it is determined not to be on-chip sub-caches. The architecture reduces the number of external memory accesses, improving both the system energy and delay by 42 and 23%, respectively.

25.4.2.2.3 Selective Cache Ways

This method [40] is one of the first attempts to adapt cache associativity to workload variation as between applications as well as during execution of an individual application. The idea is to disable some cache ways for applications that do not require full cache associativity, while enabling them for applications in which high performance is necessary. Figure 25.8 is a two-way set-associative cache with selective ways. In addition to the standard cache modules, the structure includes a cache controller, gating hardware for disabling operation of particular ways, and a software visible cache way select register (CWSR). The CWSR is modified by the operating system (OS) and contains k bits, each of which signals the cache controller to enable/disable a particular way. A zero value of the CWSR bit nulls the corresponding set-

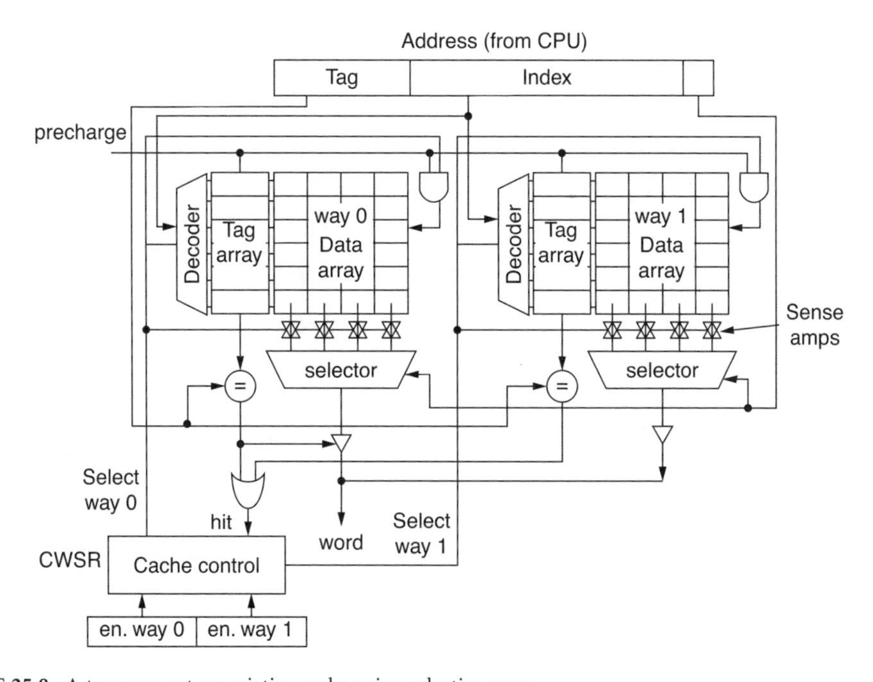

FIGURE 25.8 A two-way, set-associative cache using selective ways.

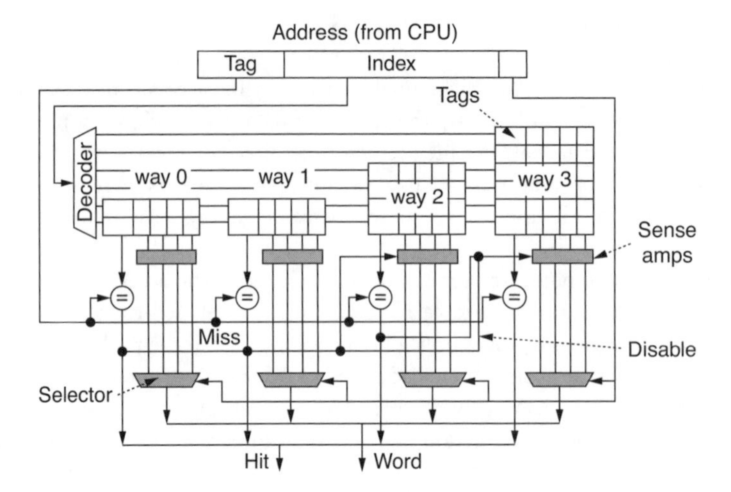

FIGURE 25.9 An asymmetric cache.

way signal thereby preventing a data way from precharging, bit selection, and firing the sense-amps. The OS monitors the performance degradation threshold and changes the number of ways enabled in the cache. Due to adaptive associativity, the cache energy can be reduced by 25 to 40% on average for a 64-KB, four-way set-associative cache with less than 2% loss in performance, and 10 to 25% for a 32-KB four-way cache. What it needs is specific software support for analyzing application requirements and modifying the CWSR. In addition, it requires an extra support for sharing data in disabled ways and maintaining its proper coherency state.

In conventional caches, some sets experience a higher access load and many more misses than others. Although most of the accesses tolerate quite a low associativity, frequently referenced addresses in programs require high associativity. Intuitively, we can reduce miss penalty by assigning more ways to the frequently used sets than others do (i.e., by making the cache asymmetric) [41]. Figure 25.9 gives an example. Extra associativity is shared by having two cache blocks from the large ways align with individual cache blocks of the smaller ways. In the asymmetric cache, the smaller ways are faster and consume less energy. A hit in a smaller way disables the sense amplifiers in all larger (or slower) ways. Because about half of the energy per access in dissipated on the data sense-amps, early hits on faster ways increase energy efficiency. Compared with the conventional cache, an asymmetric cache can save up to 17% with random replacement policy and 13% with least recently used (LRU) replacement policy. The hit latency is determined by the slowest way that produces a hit, plus overhead to route the data after the hit is detected.

25.4.2.2.4 Selective Cache Sets

Alternatively to the preceding approach, which alters set-associativity, this method [16,42] explores the second dimension in cache design, namely the number of cache sets, changing it in response to varying application demand for the cache size. Figure 25.10 is a two-way set-associative cache with dynamically resizable sets. The controller monitors cache operation in fixed-length intervals counting the number of misses. At the end of each interval, it increments/decrements the number of sets, depending on whether the miss count is lower/higher than a predefined threshold. The minimum number of sets achievable is a single sub-array per cache way. Because changing the number of sets alters both the required index and the tag bits, the cache includes a set-mask to indicate the number of index bits used in cache. Every time the cache is downsized, the mask shifts to the right, to enable fewer index bits (but more tag bits) and vice versa. The downsizing disables the high-order sets in the cache in groups of power of two. Because the cache maintains as many tag bits as required by the smallest number of sets, the tag array is larger than in the normal cache.

The method is orthogonal to selective cache ways and, therefore, can be combined into hybrid cache architecture to reduce energy dissipation. According to Yang [16], the method has better energy-delay

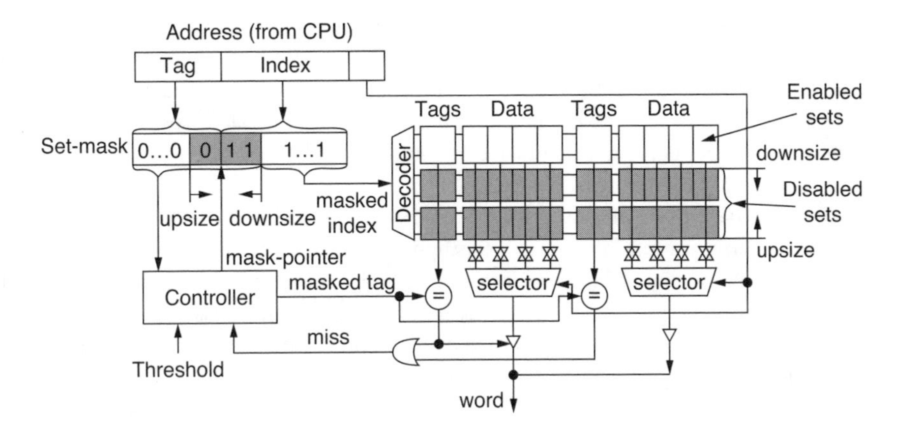

FIGURE 25.10 An organization of a two-way, set-associative cache with selective sets.

than the selective-ways when set-associativity is no more than four. At the same time, selective cache-ways is more beneficial for applications with small variation in cache size requirement. The hybrid cache has the best energy-efficiency but larger area.

25.4.2.2.5 Selective Line Sizing

Line (or block) length is the third dimension of cache optimization. The optimal line size, which results in minimal miss rate, varies with applications as well as within the application [43,44]. Most processors use only one cache line size and, therefore, often consume redundant energy. MIPS R3000 [45] supports multiple cache line sizes, but line size is configurable at boot time. It will be ideal if a cache line size can vary with application.

The line size can be controlled by software. If the compiler knows that the application only needs one word, only one word is fetched, and only that word and tag is activated in the cache. A cache can employ two fixed line sizes, changing them at runtime based on locality of data, as is done in Tang et al. [46]. The long "fetch size" is used for applications with good spatial locality and short fetch size for applications with good temporal locality. The line size is predicted for an interval of time by profiling the miss rate over a number of fetch sizes and selecting one with minimal miss rate. The dynamic profiling requires additional hardware and may affect performance on misprediction. If the prediction interval is much longer than the profiling time, however, the approach can benefit caches, especially those with large miss latencies.

Another example is the span cache [47]. In the RAM-based span cache line, each word can store both data and tag. To indicate whether the word is a tag or a datum an extra bit (t-bit) is attached to each word. The tags divide the set into spans of potentially different length. The length of line is determined by finding the next (t-bit) or the end of the set. A large overhead incurred by range-check and word-selection makes this cache quite challenging to build.

25.4.2.2.6 Reducing Switching Activity of Tag Checks

Conventional caches perform tag comparison on every access to detect whether the requested datum is within the cache. The tag operations associated with tag check are typically designed for a worst-case scenario and, therefore, utilize the entire effective address. In high associativity caches, the tag length approaches the length of the entire effective address, which results in energy expensive tag reads and comparisons. A k-way set-associative cache does k tag-checks in parallel even if only one way contains the requested data. If the same block is sequentially accessed, all the tag checks, except the first one, become unnecessary because all words within the same block have a common tag.

To avoid these redundant tag-checks, Panwar and Rennels [25] proposed testing whether the target instruction resides in the same cache block as the previous instruction. Although program counter can control this condition, it requires tagging each branch instruction to indicate whether the branch control is transferred outside the current line. A variant of this technique is to store the address index of the last line accessed within each bank and enable cache tags only if a different line is being accessed. In addition, a possible alternative is to memorize the tag of the last cache line that was accessed and compare it against the tag of the next memory access before enabling the tag search [48]. The main disadvantage of these solutions is that they introduce a wide compare operation into the critical path of every cache access, enlarging the delay of this latency-sensitive path.

Several techniques have been proposed to reduce tag-checks in I-cache without sacrificing performance. One is way-memorization [49]. Similar to way-prediction, the technique stores way information (links) within the cache but in addition maintains one "valid bit" per link to indicate that the link is correct. If the valid bit is set, the cache follows the link to locate the next instruction, thereby avoiding a tag check. Otherwise, it falls back to a regular tag comparison to find the target instruction and update the link. The following instruction fetches reuse the valid link. The approach is orthogonal to other methods and can be applied regardless of cache associativity. Compared with way-prediction, the way memorization can save an extra 13% of energy but requires a large area due to keeping the links inside the cache.

Another technique is a history-based tag-comparison [50]. The idea is to omit tag check if the target instruction has been already referenced in the past and no cache misses have occurred since that reference. To validate the condition, the cache records the history of accesses through execution footprints placed in an extended branch target buffer (BTB). An execution footprint indicates whether the target instruction block associated with the branch currently resides in the I-cache. A footprint is recorded when the corresponding block is referenced and erased whenever a cache miss occurs. On access, the cache checks a particular footprint and, if it is valid, avoids the tag check. Otherwise, it invalidates all the previously stored footprints, while setting a corresponding footprint associated with the newly referenced block. The technique reduces the cache energy by 17% with almost no impact on performance.

A compiler-driven tag check reduction for I-caches is reported in Witchel et al. [51]. Because the compiler often knows when the program is accessing the same piece of memory, the number of tag comparisons can be significantly reduced if the software guides hardware. The proposed directly addressed (DA) cache is augmented with a number of DA-registers to indicate exact locations of cache lines in the data array as well as status bits. The DA-registers are amenable by software, using extended versions of load, store, and jump operations. A load instruction, for example, causes a proper DA-register to record the location of the referenced line, as depicted in Figure 25.11. A tag-unchecked load or store instruction analyzes the DA-register and if it is valid, the cache line pointed to by the DA-register is accessed without any tag-check, and the word specified by the line offset is transferred to processor. Otherwise, hardware

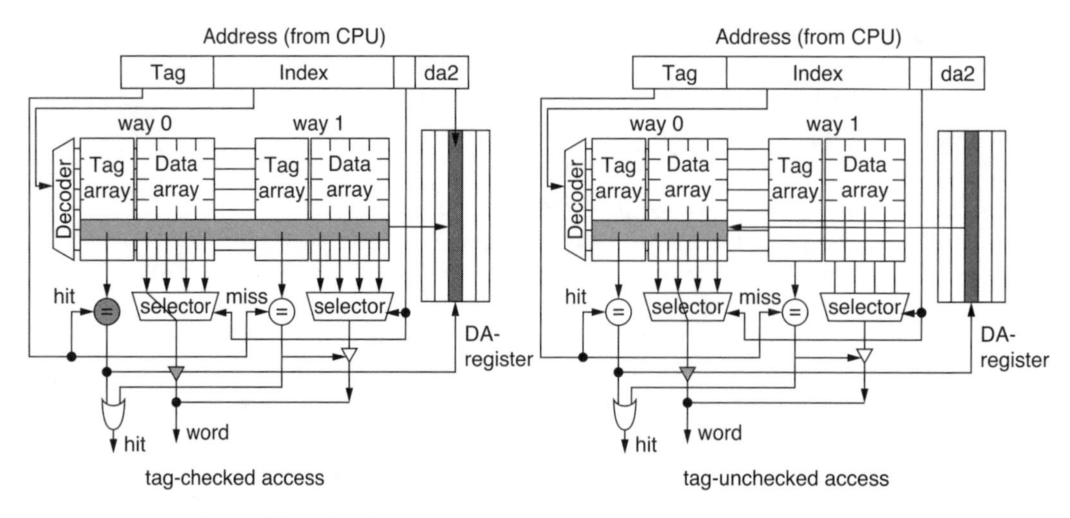

FIGURE 25.11 A direct-addressed cache.

performs the tag search and sets the DA-register. The technique saves energy by up to 40%; however, it might affect code compatibility due to the specific compilation scheme.

Another option in lowering the tag activity is to reduce the number of tag bits, which are read from the cache and used in the comparison. If referred locations are close in the address space, a small number of least significant bits can be enough to distinguish all the possible conflicting addresses in the line. The number of required tag bits can be determined for each loop by compiler, which inserts instructions to disable appropriate tag bits at the beginning of the loop and to enable the bits at the end of the loop, respectively [52]. During operation, this number is loaded to a control register which directly enables/disables particular bit-lines of the cache tag and gates the tag comparator cells. The method also can be applied for data caches if the compiler places the data sets within a region covered by one tag, or spanning more than one tag region by overlapping some of the data arrays [53]. The only problem is code compatibility.

25.4.2.2.7 Data Compression

Bit-line switching is the main source of energy consumption in cache memories. Sub-banking, segmented bit-lines, and hierarchical bit-lines decrease bit-line capacitance switched on each access by exploiting the address locality. Interestingly, data values in caches also exhibit high locality, being asymmetrically distributed across the bit-lines. Leveraging this uneven data distribution can increase energy savings.

Dynamic zero compression [54] is based on observation that over 70% of the cache values are zeros; so compressing these zeros can avoid redundant bit-line charging/discharging. The scheme adds an extra bit to each cache byte to indicate whether the byte contains all zeros. On a read access, the scheme prevents bit-line discharge by disabling the local word-line for each byte when the zero-indicator bit (ZIB) is set. If the ZIB is not set, the eight bits are read normally. On a write access, only the ZIB is written if byte is zero. Otherwise, both the data bits and the ZIB are written. The approach reduces the data cache energy by 26% and instruction cache energy by 10% under 9% area overhead and 7% clock penalty. It, however, is applicable to zero patterns only.

Besides zeros, just a few distinct values occupy the majority of cache locations [55]. These values are scattered somewhat uniformly across cache and remain almost the same during program run. We can exploit this phenomenon by dividing the data array into two parts, as depicted in Figure 25.12 [55]. The frequent values are stored in encoded form (2 to 7 bits) in the low-bit data array, while nonfrequent values are stored in unencoded form (32 bits) in both data arrays. A flag bit is attached to each word in the low-bit data array to indicate whether the word is encoded or not. On read access, the cache reads

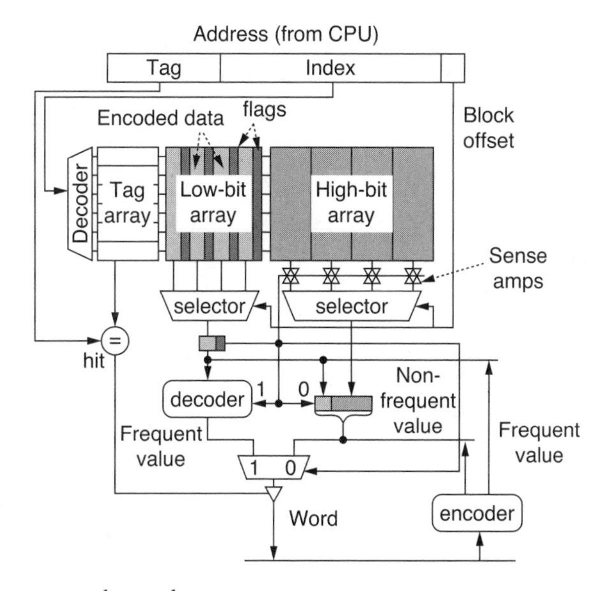

FIGURE 25.12 The frequent-value cache.

out the low-bit data array; if the flag points to an encoded word, then it decodes the word and completes the access without activating the high-bit array. Otherwise, it does not perform decoding but takes another clock cycle to read the remainder of the word from the second data array. The full value in this case is obtained by concatenating the two parts of the word. On write access, the incoming value is checked within the CAM-based encoder; if it is frequent, it is stored in cache in the encoded form jointly with the flag. Otherwise, the flag is stored with unencoded, original word. Although up to 32% energy saving has been reported for the eight-way 64-KB cache, the main challenge exists in implementation of the CAM-based data coding-decoding (codec) capable of detecting frequent values with low impact on performance and area.

Obviously, the compression in caches is neither limited to these two schemes nor to data arrays. According to Mahapatra et al. [56], existing compression techniques can reduce access times for L1 cache by 1/3, area by 2/3, and power consumption by 1/2.

25.4.3 Voltage Reduction

A widely used technique to limit the voltage swing in RAM is the word-line pulsing [57]. Pulses, short enough to read or write the cell arrays, activate the word-lines. The pulse level does not swing fully between V_{dd} and ground level: it is restricted to narrow range, and it is wide enough for correct sense-amplifier operation. This technique is effective in caches regardless of operating frequency.

Due to critical dependence of processor performance on cache operation, reducing voltage in caches without increasing the delay is very difficult. If cache is block-buffered, however, the supply voltage to buffer can be reduced. In block-buffered cache (Figure 25.2), the buffer and the arrays are operated by a single voltage that must be high enough to charge (discharge) the cache circuits in the clock-time interval. Because the circuit capacitance operated on the block-hit is less than that accessed on the block-miss, the block-buffered cache has an idle time on each block-hit. This idle time can be traded with voltage if the block buffer is operated by a dual voltage supply [58]. At the first phase of clock cycle 1, the cache precharges the memory arrays for read, decodes the address, and compares the address tag with the block-buffer tag. At the next phase, if there is a block-hit, the cache disables the arrays from the read port and enforces the low voltage. This voltage slows down the word selection from the block, filling the idle time with action. In opposition, when the block-miss is detected, the high voltage is selected to accelerate the operation. In this case, the cache reads out the selected block into the array output latches. In the next clock cycle, it performs tag comparison I (phase 1) and, if there is a hit, drives the matched block from the data array, copying it concurrently to the buffer and outputting the requested data word from the block (phase 2). The method maintains the performance of a single-voltage block-buffered scheme, while saving its energy consumption by up to 36% for programs with high data correlation.

25.4.4 Leakage Energy Reduction

Many techniques are used to reduce leakage energy dissipation at the circuit/technology level (see Chapter 3). Most of these techniques, however, affect circuit performance and, therefore, are not suitable to performance-critical circuits, such as caches [59]. Efforts have been put to devise micro-architectural techniques capable of reducing leakage energy dynamically (i.e., during program run) with less cost and area overhead. Usually, powering off unused SRAM devices reduces leakage energy. To shrink it dynamically we need a policy: when and what to switch off. Current microprocessors use a simple OS-driven policy: deactivate the entire processor when it enters a sleep mode.

A more intelligent solution is to disable cache sets that eventually become inactive during program execution [60]. The proposal combines selective cache sets [16] with a circuit technique, called *gated-V_{dd}*, which inserts an extra transistor in the supply voltage (V_{dd}) or the ground path (V_{ss}) of the cache's SRAM cells; the extra transistor is "turned on" in the active sets and "turn off" (or gated) in inactive sets. In comparison to conventional cache, the proposed dynamically resizable I-cache (or DRI cache) [42] reduces energy-delay by 62% with 4% latency increase.

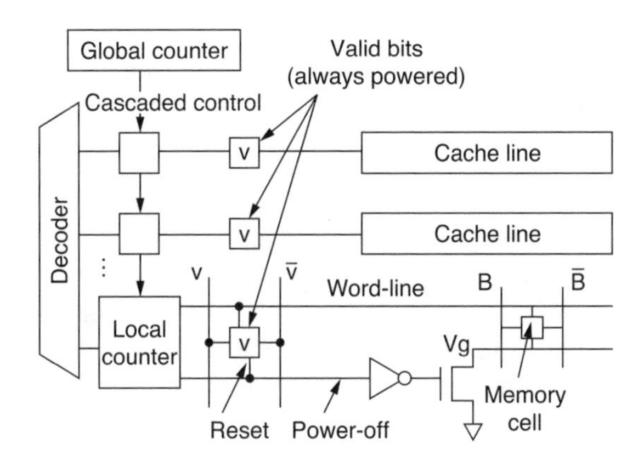

FIGURE 25.13 An illustration of cache-decay scheme.

Cache-line decay [61] exploits a similar idea but at a granularity of cache-line and without resizing. Motivated by the fact that almost half of cache lines remain unused for a long period of time, the technique turns-off the least recently used cache lines assuming that these lines will be not referenced in the future. In design, a counter is attached to each cache line, as depicted in Figure 25.13. The counter is incremented periodically at fixed time intervals (set by the global counter) and cleared on each access to the line. Once a line counter reaches its maximum count, the corresponding cache-line is invalidated, and its voltage supply is turned off via a gated-V_{dd} transistor. The scheme results in roughly 70% reduction in L1 data cache leakage energy.

The main drawback of these approaches is that the state of the cache line is lost when it is turned off. Reloading the line from L2 cache not only hurts performance, but could also diminish energy savings. To avoid these pitfalls, it is necessary to use adaptive algorithms and be conservative about which line is turned off. Another drawback of these schemes is dependence of arbitrary parameters that must be tuned per application to minimize the performance penalty. In the case of DRI cache, the number of active cache sets is increased or decreased based on the miss bound which must not exceed the application's typical miss rate. In case of cache line decay, the control parameter is line decay interval. Tuning such parameters by application profiling may be difficult because the results differ as within as well as across application. To overcome this limitation, Zhou et al. [62] suggests keeping the tag arrays active to dynamically monitor the cache miss rate and adjust the miss rate based on relative factor instead of an arbitrary, absolute value. The proposed adaptive mode control (AMC) examines the ratio of sleep misses to ideal misses, and if it is "too small," reduces the line decay interval to deactivate more lines. If the ratio is "too large," the line decay interval is increased. Otherwise, the interval remains constant. The hardware implementation of the AMC is relatively compact and requires two counters and a small logic to modify the new line-decay interval based on the miss ratio. On average, the method can turn off up to 73% of I-cache lines and 54% of D-cache lines with less than 2% performance loss.

Leakage reduction in caches can also be achieved by putting a cache line into a low-energy "drowsy" mode, as proposed in [63]. To implement drowsy caches, authors add a "drowsy bit" to each cache line and supply two voltages (V_{dd}-low and V_{dd}-high) to the cell array. The lines are put into drowsy mode either periodically (simple policy) or based on access statistics (i.e., only those lines which have not been accessed in a fixed time interval). Whenever cache is accessed, it reads the drowsy bit of corresponding line; if it indicates normal mode, the line is read without losing performance. If it indicates that the line is in drowsy mode, however, the line is woken up in the next cycle and then read. Compared to caches that use gated-V_{dd} technique, drowsy caches preserve line information and switch faster between the power modes. The penalty they pay for being wrong is a single clock cycle to wake up drowsy lines; however, drowsy caches depend on process variation and do not reduce leakage energy of L1-cache as much as the others. Even in drowsy mode, a cache consumes a quarter of the normal mode energy.

Several works exploit the fact that leakage of a block depends on input pattern and internal state [64,65]. To minimize leakage they apply, a combination of input patterns and internal state (so called sleep-vector) is applied to set internal latches into the correct state and turn inputs to the correct polarity. As Heo et al. [66] demonstrates, the leakage current from a bit-line into the memory cell depends on the stored value on that side of the cell. That is, there is no leakage if the bit line is at the same value as that stored in the cell. Therefore, if we assume that there are more zeros than ones in cache, we could reduce the bit-line leakage of inactive sub-bank by setting the true line to zero, while keeping the complement line precharged. If this assumption is incorrect, however, the "sleep vector" approach leads to additional energy dissipation. Another solution is to use leakage-biased bit-lines [66]. Instead of setting bit-lines of inactive sub-banks to a sleep-value, this scheme turns-off both the precharging transistors and sense-amplifiers, letting the bit-lines float to mid-rail. Because the precharge transistor switches exactly the same number of times as in conventional SRAM, the energy-performance overhead of the scheme is small; the precharge is only delayed until the sub-bank is accessed, and the wake-up latency is just that of the precharge phase. Using the scheme to deactivate SRAM read paths within I-caches saves over 24% of leakage energy and 22% of total I-cache energy at 0.07-μm processes. The cache, however, must be designed to deactivate blocks for multiple cycles and preferably to give them an early notice when to be reawakened.

25.5 Conclusion

We surveyed techniques that have been proposed to reduce energy consumption in caches. The presented survey is not comprehensive; instead, we have focused on the architectural optimizations for reducing the cache accesses, switching activity, voltage, and leakage. Even though a broad range of techniques have been proposed, many possible research directions remain.

First, most of the techniques presented here, except structural partitioning, have not been implemented. The challenge remains to integrate them into the design in which power plays as large a role as performance. Second, the cache design space is large and involves complicated tradeoffs. Exploring alternative hardware/software schemes to determine the less energy-consuming cache configuration for given performance constraint and voltage level is another big challenge. Third, static cache design cannot respond to variable applications and thus is inefficient from the energy reduction perspective. If we want to keep energy dissipation within the bounds of the future microprocessor generations, we need to move forward self-adjusting and adaptive cache architecture that can quickly and efficiently respond to the application change as well as varying data statistics. Another challenge is to develop architecture-driven techniques capable of performing cache reconfiguration based on application requirements on cache size or IPC. This is promising not only for applications where throughput can be traded for low energy, but also for reducing leakage. Fourth, leakage energy reduction by powering off unneeded portions of the cache induces additional off-chip accesses and increased latency. A definite challenge exists for exploring cache resizing trade-offs to carefully balance leakage energy with other energy components.

25.6 Acknowledgments

The research was supported in part by The Ministry of Education, Technology, Science, Sports, and Culture of Japan, Grant-in-Aid for Creative Basic Research (A) No.14GS0218, Grant-in-Aid for Scientific Research C (2) No.14580399, and Grant-in-Aid for Encouragement of Young Scientists (A) No.14702064.

References

[1] Naffziger, S.D., et al. The implementation of the Itanium2 microprocessor, *IEEE J. Solid-State Circuits,* Vol. 37, 11, 1448, 2002.

[2] Intel, A detailed look inside the Intel® NetBurst™ micro-architecture of the Intel Pentium4 processor, Nov. 2000, http://developer.intel.com/design/pentium4/manuals/248966.htm.

[3] Jain, A., et al. A 1.2-GHz alpha microprocessor with 44,8GB/s chip pin bandwidth, *IEEE Int. Solid-State Circuits Conf., Dig. of Tech. Papers,* 2001, 240.

[4] Heald, R., et al. A third-generation SPARC V9 64-b microprocessor, *IEEE J. Solid-State Circuits,* 37, 11, 1526, 2000.

[5] Johnson, D.J.C., HP's Maco processor, *Microprocessor Forum,* Oct. 2001.

[6] Behling, et al. The Power4 processor introduction and tuning guide, *IBM Redbooks,* Nov. 2001, available from http://www.redbooks.ibm.com.

[7] Northrop, G., et al. 600-MHz G5 S3/390 microprocessor, *1999 IEEE Int. Solid-State Circuits Conf., Dig. of Technical Papers,* 88, 1999.

[8] Alvarez, J., et al. 450-MHz Power PC microprocessor with enhanced instruction set and copper interconnect, *IEEE Int. Solid-State Circuits Conf., Dig. of Tech. Papers,* 96, 1999.

[9] Montanario, J., et al. A 160-MHz 32-b 0.5-W CMOS RISC microprocessor, *IEEE Int. Solid-State Circuits Conf., Dig. of Tech. Papers,* 1996.

[10] Chandrakasan, A. and Brodersen, R., *Low-Power Digital CMOS Design,* Kluwer Academic Publishers, Dordrecht, 1996.

[11] *Power-Aware Design Methodologies,* Rabaey, J. and Pedram, M., Eds., Kluwer Academic Publishers, Dordrecht, 2002.

[12] Shiue, W.-T. and Chakrabarti, C., Memory design and exploration for low-power, embedded systems, *J. VLSI Signal Processing,* 29, 167, 2001.

[13] Gary, S., Low-power microprocessor design, in *Low-Power Design Methodologies,* Rabaey I. and Pedram M., Eds., Kluwer Academic Publishers, Dordrecht, 1996, 255.

[14] Hu, Z., Martonosi, M., and Kaxiras, S., Improving cache power efficiency with asymmetric set-associative cache, *Proc. Workshop on Memory Performance Issues,* held jointly with *ISCA-2001,* Goteborg, Sweden, June 30–July 1, 2001.

[15] Kobayashi, T. and Sakurai, T., Self-adjusting threshold voltage scheme for low-voltage high-speed operation, *Proc. IEEE Custom Int. Circuits Conf.,* 1994, 271.

[16] Yang, S.-H., An integrated circuit/architecture approach to reducing leakage in deep-submicron high-performance I-caches, *Proc. 8th Int. Symp. on High-Performance Comput. Architecture,* 2001.

[17] Borkar, S., Design challenges of technology scaling, *IEEE Micro.,* 19(4): 23, 1999.

[18] Bunda, J., Athas, W., and Fussel, D., Evaluating power implications of CMOS microprocessor design decisions, *Proc. Int. Workshop on Low-Power Design,* 1994, 147.

[19] Su, C.L. and Despain, A.M., Cache design trade-offs for power and performance optimization: a case study, *Proc. Int. Symp. on Low-Power Design,* 1995, 69.

[20] Kamble, M.B. and Ghose, K., Analytical energy dissipation models for low-power caches, *Proc. Int. Symp. on Low-Power Electron. and Design,* 1997, 143.

[21] Ghose, K. and Kamble, M.B., Reducing power in superscalar processor caches using sub-banking, multiple line buffers and bit-segmentation, *Proc. Int. Symp. on Low-Power Design,* 1999, 70.

[22] Kin, J., Gupta, M., and Mangione-Smith, W., The filter cache: an energy-efficient memory structure, *Proc. 30th Annu. IEEE/ACM Int. Symp. on Microarchitecture,* 1997, 184.

[23] Bajwa, R.S., et al. Instruction buffering to reduce power in processors for signal processing, *IEEE Trans. on Very Large-Scale Integration Syst.,* 5(4): 417, 1997.

[24] Bellas, N., Hajj, I., and Polychronopoulos, C., Using dynamic cache management techniques to reduce energy in a high-performance processor, *Proc. 1999 Int. Symp. on Low-Power Electron. and Design,* 1999, 64.

[25] Panwar R. and Rennels, D., Reducing the frequency of tag compares for low-power I-cache design, *Proc. 1995 Int. Symp. on Low-Power Electron. and Design,* 1995, 57–62.

[26] Lee, L.-H., Moyer, B., and Arends, J., Instruction fetch energy reduction using loop caches for embedded applications with small tight loops, *Proc. Int. Symp. on Low-Power Electron. Design,* San Diego, CA, 1999, 267.

[27] Tang, W., Gupta, R., and Nicolau, A., Design of a predictive filter cache for energy savings in high performance processor architectures, *Proc. Int. Conf. on Comput. Design,* 2001, 68–73.

[28] Itoh, K., Low-Power Memory Design, in *Low-Power Design Methodologies*, Rabaey, J. and Pedram, M., Eds., Kluwer Academic Publishers, Dordrecht, 1996, 201.

[29] Hasegawa, A., et al. SH3: high code density, low power, *IEEE Micro.*, 1995, 11.

[30] Ko, U., Balsara, P.T., and Nanda, A.K., Energy optimization of multilevel cache architectures for RISC and CISC processors, *Trans. IEEE Very Large-Scale Integration (VLSI) Syst.*, 6(2): 1998, 299.

[31] Lee, H.S., and Tyson, G.S., Region-based caching: an energy-delay efficient memory architecture for embedded processors, *Proc. Int. Conf. on Compilers, Architecture, and Synthesis for Embedded Syst.*, 2000, 120.

[32] Wilson, S.J.E. and Jouppi, N.P., An enhanced access and cycle time model for on-chip caches. Technical report 93/5, Digital Equipment Corporation, Western Research Laboratory, Palo Alto, CA, 1994.

[33] Shimazaki, Y., An 8-mW, 8-KB cache memory using an automatic power-save architecture for low-power RISC microprocessors, *IEICE Trans. Electron.*, E79-C(12): 1693, 1996.

[34] Calder, B. and Grunwald, D., Next cache line and set prediction, *Proc. Int. Symp. on Computer Architecture,* 1995, 287.

[35] Inoue, K., et al. Way predicting set-associative cache for high performance and low energy consumption, *Proc. 1999 Int. Workshop on Low-Power Design,* 1999, 273.

[36] Powell, M., et al. Reducing set-associative cache energy via way prediction and selective direct-mapping, *Proc. 34th Annu. IEEE/ACM Int. Symp. on Microarchitecture,* 2001, 54–65.

[37] Batson, B. and Vijaykumar, T.N., Reactive associative caches, *Proc. 2001 Int. Conf. on Parallel Architectures and Compilation Techniques,* 2001, 49–60.

[38] Huang, M., et al. L1 data cache decomposition for energy efficiency, *Proc. Int. Symp. on Low-Power Electron. Design,* Huntington Beach, CA, 2001, 10.

[39] Kim, S., et al. Power-aware partitioned cache architecture, *Proc. Int. Symp. on Low-Power Electron. Design,* Huntington Beach, CA, 2001, 64.

[40] Albonesi, D.H., Selective cache ways: on-demand cache resource allocation, *Proc. 32th Int. Symp. on Microarchitecture,* 1999, 248.

[41] Hu, Z., Kaxiras, S., and Martonosi, M., Improving cache power efficiency with an asymmetric set-associative cache, *Proc. Workshop on Memory Performance Issues,* Goteborg, Sweden, June 30–July 1, 2001.

[42] Yang, S.-H., et al. Exploiting choice in resizable cache design to optimize deep-submicron processor energy-delay, *Proc. 8th Int. Symp. on High-Performance Comput. Architecture,* 2003, 151–161.

[43] Gonzales, A., Aliagas, C., and Valero, M., A data cache with multiple caching strategies tuned to different types of locality, *Proc. Int. Conf. on Supercomputing,* 1995, 338.

[44] Inoue, K., Kai, K., and Murakami, K., High-bandwidth, variable line-size cache architecture for merged DRAM/logic LSIs, *IEICE Trans. on Electron.*, E81C(9): 1438, 1999.

[45] MIPS Corporation. *MIPS R3000 Hardware Manual,* MIPS Corporation, Mountain View, CA.

[46] Tang, W., Veidenbaum, A.V., and Gupta, R., Architectural adaptation for power and performance, *Proc. 14th Annu. IEEE Int. ASIC/SOC Conf.,* Oct. 2001, 127–130.

[47] Witchel, E. and Asanovic, K., The span cache: software controlled tag checks and cache line size, in *Proc. Workshop on Complexity-Effective Design, ISCA-28,* Goteborg, 2001.

[48] Burd, T., Energy-efficient processor system design, Ph.D. thesis, University of California—Berkeley, May 2001.

[49] Ma, A., Zhang, M., and Arsanovic, K., Way memorization to reduce fetch energy in instruction caches, *Proc. Workshop on Complexity-Effective Design, ISCA-28,* Goteborg, Sweden, June 30–July 1, 2001.

[50] Inoue, K., Moshnyaga, V. G., and Murakami, K., A history-based I-cache for low-energy multimedia applications, *Proc. Int. Symp. on Low-Power Electron. and Design,* 2002, 148.

[51] Witchel, E., et al. Direct addressed caches for reduced power consumption, *Proc. 34th Annu. IEEE/ACM Int. Symp. on Microarchitecture,* Austin, TX, Dec. 1–5, 2001.

[52] Petrov, P. and Orailoglu, A., Energy-frugal tags in reprogrammable I-caches for application-specific embedded processors, *Proc. ACM CODES,* May 6–8, 2002, 181–186.

[53] Petrov, P. and Orailoglu, A., Data cache energy minimization through programmable tag size matching to applications, *Proc. IEEE Int. Symp. on System Synthesis,* Sept. 30–Oct. 3, 2001.

[54] Villa, L., Zhang, M., and Asanovic, K., Dynamic zero compression for cache energy reduction, *Proc. 33rd Annu. IEEE/ACM Int. Symp. on Microarchitecture,* Monterey, CA, Dec. 10–13, 2000, 214–222.

[55] Yang, J. and Gupta, R., Frequent value locality and its applications, *ACM Trans. on Embedded Comp. Syst.,* 2(3): 1, 2002.

[56] Mahapatra, N., et al. The potential of compression to improve memory system performance, power consumption and cost, *Proc. 22nd IEEE Int. Performance, Computing, and Communication Conf.,* Phoenix, AZ, Apr. 9–11, 2003.

[57] Mai, K., et al. Low-power SRAM design using half-swing pulse-mode techniques. *IEEE J. Solid-State Circuits,* 33, 1659, 1998.

[58] Moshnyaga, V.G. and Tsuji, H., Reducing cache energy dissipation by using dual voltage supply, *IEICE Trans. on Fundamentals,* E84-A(11): 2762, 2001.

[59] Hamzaoglu, F., et al. Dual-Vt SRAM cells with full-swing single ended bit-line sensing for high-performance on-chip cache in 0.13-um technology generation, *Proc. 2000 Int. Symp. on Low-Power Electron. and Design,* July 2000, 15–20.

[60] Powell, M.D., et al. Gated-V_{dd}: a circuit technique to reduce leakage in cache memories, *Proc. 2000 ACM/IEEE Int. Symp. on Low-Power Electron. and Design,* 2000, 90.

[61] Kaxiras, S., Hu, Z., and Martonosi, M., Cache decay: exploiting generational behavior to reduce cache leakage power, in *Proc. 28th Int. Symp. on Comput. Architecture,* June 30–July 4, 2001, 240–251.

[62] Zhou, H., et al. Adaptive mode control: a static-power-efficient cache design, *Proc. Int. Conf. on Parallel Architectures and Compilation Techniques,* Sept. 8–12, 2001, 61–70.

[63] Flautner, K., et al. Drowsy caches: simple techniques for reducing leakage power, *Proc. of the 29th Annu. Int. Symp. on Comput. Architecture,* Anchorage, AK, May 25–29, 2002, 148–150.

[64] Halter J.P. and Najm, F., A gate-level leakage power reduction method for ultra-low-power CMOS circuits, *Proc. IEEE Custom Integrated Circuits,* 1997, 457.

[65] Ye, Y., Borkar, S., and De, V., A technique for standby leakage reduction in high-performance circuits, *Proc. IEEE Symp. VLSI Circuits,* 1998, 40.

[66] Heo S., et al. Dynamic fine-grain leakage reduction using leakage-biased bitlines, *Proc. 29th Int. Symp. on Computer Architecture,* Anchorage, AK, May 25–29, 2002.

26

Memory Organization for Low-Energy Embedded Systems

Alberto Macii
Politecnico di Torino

26.1 Introduction

Memory design for multi-processor and embedded systems has always been a crucial problem, because system-level performance depends strongly on memory organization.

The proliferation of embedded systems, and the corresponding new chip and chip-set designs, have brought additional attention to storage units. Indeed, the heterogeneity of components and structures within embedded systems and the possibility of using application-specific storage systems have added a new dimension to memory design. Moreover, new degrees of freedom have been opened since the introduction of embedded memory arrays in different technologies, such as SRAMs, DRAMs, EEPROMs, and FLASH, and their realization on the same silicon substrate hosting the processing units.

Embedded systems are often designed under stringent energy consumption budgets, to limit heat generation and battery size. Because memory systems consume a significant amount of energy to store and to forward data, it is then imperative to balance energy consumption and performance in memory design. Contemporary system design focuses on the trade-off between performance and energy consumption in processing and storage units, as well as in their interconnections. Although memory design is as important as processor design in achieving the desired design objectives, the former topic has received less attention than the latter in the literature.

Two are the key issues in low energy memory design for embedded systems:

1. Reduce the energy consumed in accessing memories [12,14,28,29], which takes a dominant fraction of the energy budget of an embedded system for data-dominated applications [12].
2. Minimize the amount of energy consumed when information is exchanged between the processor and the memory by reducing the amount of required processor-to-memory communication bandwidth.

Regarding memory access optimization, the possibility of integrating one or more memories on the same die as the processor offers new opportunities for energy-efficient design. In fact, from one side,

0-8493-1941-2/05/$0.00+$1.50

accessing on-chip memory is much faster and energy-efficient than relying exclusively on off-chip memories [20,34]. On the other side, memory size and organization can be tailored to application requirements, and application-specific memory architectures can be developed to minimize memory energy for a given embedded application. Among the available solutions for memory hierarchy customization, memory partitioning has demonstrated very good potential for energy savings, as well as excellent suitability for usage in automated design environments. The principle of such a method is to subdivide the address space in several clusters and to map them to different memory banks that can be independently enabled/disabled.

Concerning communication bandwidth optimization, Burger [11] discussed that memory bandwidth is becoming more important as a metric for modern systems because of the increased instruction-level parallelism generated by superscalar or very long instruction word (VLIW) processors, and because of the density of integration, which allows shorter latencies. Unlike latency, but similar to energy, bandwidth is an average-case quantity, and is related to memory traffic. Therefore, not all solutions that reduce latency necessarily translate into bandwidth improvements, as for energy. A good way for decreasing the memory traffic, and memory energy as well, is to compress the information that is transmitted between two levels of the memory hierarchy.

This chapter focuses both on methods that aim at reducing the energy consumption by optimizing the memory hierarchy and on techniques that target the reduction of the energy consumed in memory transfers.

In particular, Section 26.2 surveys several effective memory partitioning approaches, especially focusing on methods that are suitable to be used in an automatic fashion, whereas Section 26.3 discusses solutions for information compression including both code and data.

26.2 Memory Partitioning

Within a given memory hierarchy level, energy consumption can be reduced by memory partitioning. The principle of memory partitioning is to subdivide the address space into several smaller and less energy consuming blocks, and to map such blocks to different physical memory banks that can be independently enabled and disabled. Memory partitioning by itself is a typical performance-oriented solution (because of the reduced latency due to accessing smaller blocks), and energy may be reduced only for some specific access patterns. What actually makes this class of techniques low-energy is the opportunity of selectively shutting down memory blocks that are not accessed, which has little or no effect on performance.

Arbitrary fine partitioning is prevented from the fact that an excessively large number of small banks is highly area inefficient, and imposes a severe wiring overhead, which tends to increase communication energy and performance. Partitioning-based techniques proposed in the literature differ in several aspects. First, the hierarchy level targeted for partitioning (from caches to off-chip memories). Second, the scope of partitioning: physical partitioning strictly maps the address space onto different, nonoverlapping memory blocks; logical partitioning allows some redundancy in the various blocks of the partition, with the possibility of storing addresses several times in the same level of hierarchy. Farrahi et al. [15] first studied memory partitioning to exploit sleep mode operation. This work is in the context of board-level memory optimization where memory blocks are large off-chip DRAMs that can be powered down when they are not storing live program variables, thereby eliminating memory refresh energy. Furthermore, it is assumed that activating an inactive memory incurs a significant energy cost. The technique presented by Farrahi et al. tries to cluster data into memories so that memory chips are transitioned in and out of the shutdown mode as little as possible. Several authors have analyzed partitioning of on-chip memories. In most cases, the various partitioning options at a given hierarchy level have been used as an additional dimension of the memory design space. For example, Su and Despain [33], Ko and Balsara [20], and Shiue and Chakrabarti [32] studied energy-efficient partitioned cache organizations, identifying cache sub-banking as an effective technique for reducing cache energy consumption.

Other solutions rely on hardware-based selective activation of individual ways for set-associative caches. Region-based caching proposes separate cache memories for stack data and global data, and a main cache for all other accesses. Clearly, non-accessed cache modules can be disabled for energy saving.

Coumeri and Thomas [13] studied embedded SRAMs, and described a partitioned SRAM model (called segmented configuration), providing energy models for partitioned memories as well. These techniques rely on a physical partitioning of the on-chip memory. Gonzàlez et al. [16] proposed logical partitioning, where the on-chip cache is split into a spatial and into a temporal cache to store data with high spatial and temporal correlation, respectively. This approach resorts to a dynamic prediction mechanism that can be realized without modification of the application code by means of a prediction buffer. Grun et al. [17] have extended the ideas behind this approach in the context of embedded systems for energy optimization. In their approach, data are statically mapped onto either cache, thanks to the high predictability of the access profiles in embedded programs. Depending on the application, data might be duplicated and thus be mapped to both caches.

Another class of logical partitioning techniques exploits buffer insertion along the I-cache or the D-cache or both to realize some form of cache parallelization. Such schemes can be regarded as a partitioning solution because the buffers and the caches are actually part of the same level of hierarchy. In this kind of architecture, data and instructions are explicitly replicated, and redundancy is an intrinsic feature of these approaches. Reducing the average cost of a memory access by increasing the cache-hit ratio saves energy. Another solution proposes the use of the buffer as a victim cache that is accessed on a main cache miss. In case of a buffer hit, the line is moved to the cache and returned to the CPU, while the replaced line in the cache is moved to the victim cache. In case of a buffer miss, the lower level of hierarchy is accessed and the fetched datum is copied into the main cache as well, while the replaced line in the cache is moved to the victim cache. In practice, the victim cache serves as an over-full buffer for the main cache. A similar approach has been introduced by Bahar et al. [2], where buffers are used for speculation: every cache access is marked with a confidence level, obtained by the processor state; the main cache contains misses with a high confidence level, while the buffers contains those with a low confidence level. Other techniques adopt a small associative buffer (e.g., 32 entries) in parallel to the L1-cache (called the noncritical buffer), used to "protect" the cache from being filled with noncritical (i.e., potentially polluting) data. Noncritical data are identified at run-time by monitoring the issue rate of the core. An alternative solution consists of filtering the data to be stored in the main cache through a small, highly associative cache close to the L1-cache. Unlike the victim cache (where data are kept before disposing them), the annex cache stores the data read from memory, which are copied into the main cache only on subsequent references to those data.

This idea of filtering cache accesses to reduce writes that are very likely to cause misses can be refined by selectively disabling the side buffers to save additional energy. Way-predicting caches attempt to minimize unnecessary way activation in set-associative caches, by predicting which way contains the data. Prediction is carried out based on memory access history.

26.2.1 Memory Partitioning for Low Energy

Moving from the observation drawn in Section 26.2 and finding inspirations from the existing memory partitioning techniques proposed in the literature, an automatic optimization methodology for on-chip memories to be used in embedded systems on chip (SoCs) is presented in Benini et al. [7]. Starting from the dynamic execution profile of an embedded application running on a given processor core, a multi-banked memory architecture optimally fitted to such a profile is synthesized. The rationale behind the approach we will describe in the following is to partition memory into multiple banks that can be independently accessed. Power-per-access is reduced as the size of a memory bank is decreased. On the other hand, as the number of banks increases, there is an unavoidable hardware overhead caused by:

1. Duplication of addressing and control logic
2. Increased communication resources required to transfer information

Such an overhead manifests itself in increased energy, access time, and area that prevent arbitrarily fine partitioning. Thus, finding an optimal partition with a tight constraint on the maximum number of memory banks is extremely important.

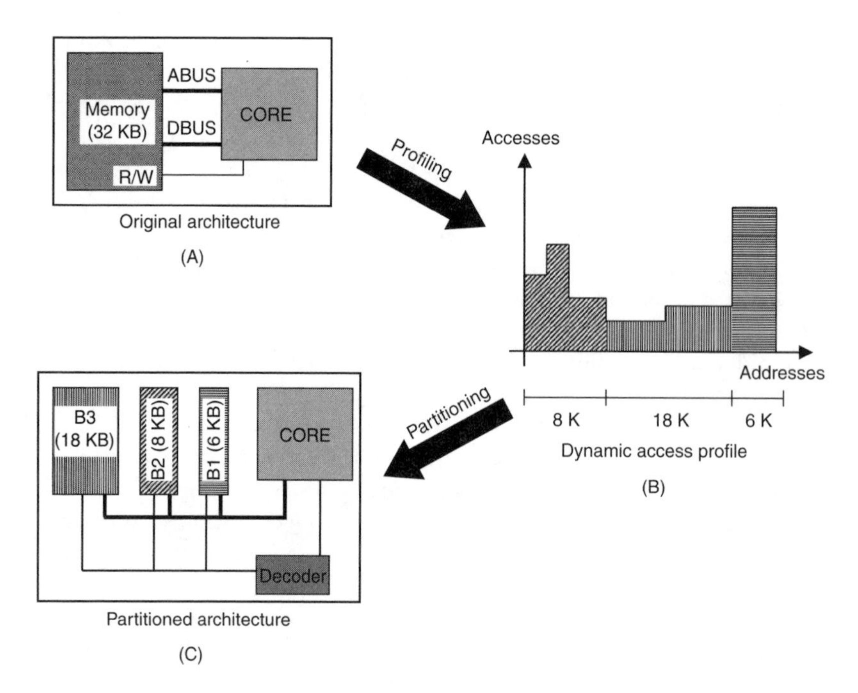

FIGURE 26.1 Example of memory partitioning.

Moving from the fact that in many nontrivial embedded applications the most frequently accessed addresses can fit into a relatively small memory space, the information that comes from the dynamic access profile, obtained through the simulation of the target application, is used to specify a range (a_{lo}, a_{hi}) of memory addresses that should be mapped onto the on-chip memory. For each address $a_{lo} \leq i \leq a_{hi}$, the profile gives the number of reads $r(i)$ and writes $w(i)$ to the address during the execution of a sample run of the target embedded application. Standard instruction-level simulators available for most processor cores can capture the profile.

In a traditional approach, all addresses in the range are mapped to a single memory array, the smallest array in the library, which is large enough to contain the specified range, as illustrated in Figure 26.1(a).

This solution is not optimal from the energy dissipation viewpoint. Assume, for the sake of illustration, that the dynamic access profile is that presented in Figure 26.1(b). A small subset of the addresses in the range has large $r(i)$ values for all its addresses (i.e., it is very "hot"; see the right horizontal-shaded slice of the profile). An energy-optimal partitioned memory organization is presented in Figure 26.1(c). It consists of three memories and a memory selection block (the decoder). The hot addresses are stored into a small memory (block B1), while two larger cuts (blocks B2 and B3) contain the other parts of the range. The average energy in accessing the memory hierarchy is decreased, because a large fraction of accesses is concentrated to a small, energy-efficient memory. Obviously, it is mandatory to account for the energy consumed in the entire partitioned memory system (i.e., the address and data buses), the decoder, and the control signals. In fact, these components introduce a nonnegligible overhead on energy consumption that may offset the advantages given by bank partitioning. Nevertheless, the obtained savings are significant especially when the access profile is highly nonuniform and high-access addresses are clustered into small banks (on average, for several embedded applications, energy savings are around 60% with respect to the traditional monolithic memory architecture).

The memory partitioning approach belongs to the class of memory optimization techniques that, moving from a given memory access trace and obtained by profiling an application, produce a customized memory hierarchy. The effectiveness of memory partitioning can be improved by adopting the concurrent optimization of memory access patterns and memory architecture. Techniques relying on such a concurrent optimization are the most powerful, yet also the most difficult to actuate. This fact is witnessed by the few solutions proposed in the literature, the most popular being the DTSE hardware/software (HW/SW) exploration framework [12]. One of the biggest difficulties in concurrent optimization lies in the fact that the two dimensions of the problem are regarded as orthogonal: architectural optimization is viewed as a purely hardware task, while the optimization of the access patterns is viewed as a purely software task. Recently, in Macii et al. [27], the access pattern optimization problem was revisited from an architectural perspective: the design of an application-specific memory architecture is concurrently carried out with the optimization of the access patterns, yet done through the introduction of proper hardware. In practice, the access patterns are modified on the fly, without any intervention on the software application running on the processor.

Thus, the idea is to combine the memory partitioning methodology with a technique, called address clustering, which consists of reorganizing (through extra hardware) the address trace fed to a memory block, in such a way that the potential for the memory partitioning engine is maximized. This is equivalent to modifying the memory access profile, yet in a very transparent fashion to the programmer.

The address clustering problem consists of finding a relocation of a proper subset of the address space that maximizes the locality of the dynamic trace, with the ultimate objective of minimizing the energy consumption of the memory architecture for the given trace, possibly under area and cycle time constraints.

The actual energy consumption of a partitioned memory architecture is determined by the outcome of the memory-partitioning algorithm of Benini et al. [7]; however, running the partitioning engine for each candidate clustering solution may become quite computationally expensive. It is thus necessary to devise a high-level cost function that can be used into an exploration framework to evaluate the suitability of a clustering solution.

The potential of memory partitioning is related to the locality of the trace. In Macii et al. [27], it was thus defined the density of a profile, C, as the maximum value of the cumulative number of accesses for a sliding window of size W over the trace. This metric is suitable for use in an exploration engine because it provides, through a low-effort analysis of an address trace, a quantification of which percentage of the total number of addresses can be covered by clustering W words. There is then a strict correlation between the number M of addresses to be clustered and the size W of the sliding window.

Figure 26.2 reports an example of address clustering. Figure 26.2(a) depicts the original profile (i.e., as obtained by profiling the application); here, two addresses (i and j) are illustrated with bold lines, together with the sliding window W. Figure 26.2(b) depicts one instance of the swap between i and j. This transformation increases the spatial locality by aggregating highly accessed memory locations. The information about the swap is recorded in a table, as the one of Figure 26.2(c).

Results obtained by applying address clustering before partitioning demonstrate advantages ranging from 5 to 55% with respect to energy consumption of the memory architecture generated by plain application of memory partitioning.

26.3 Memory Transfer Optimization

Although memory energy consumption is relevant, additional improvements can be achieved by combining the minimization of memory and bus energy, that is, of the whole memory-processor interface. Optimizing the processor-to-memory bandwidth is one possibility to simultaneously reduce memory and bus energy.

Bandwidth can be increased either by directly reducing the processor-to-memory traffic or by indirectly compressing the information transmitted through the interface.

In most cases, the main objective of the information compression is to provide a high compression ratio; instead, parameters like compression (and decompression) time and complexity are not seen as critical ones because software compression routines or dedicated hardware units perform their functions in environments where timing, resource, and energy constraints are relaxed. Think, for example, of data compression for hard-drives or transmission over communication links, where transfer rates are significantly slower than the time required by the HW or SW compressor/decompressor to transform data in

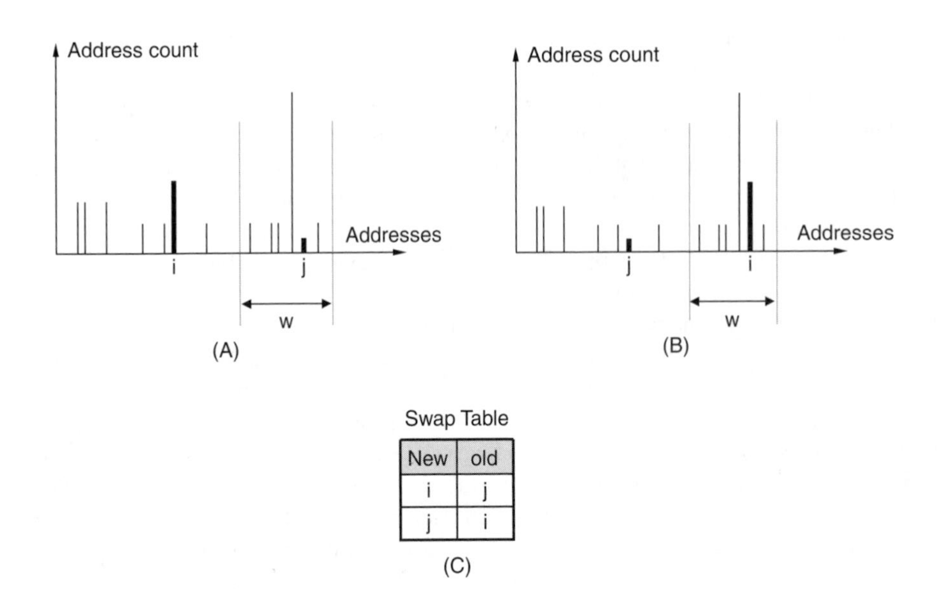

Swap Table

New	old
i	j
j	i

FIGURE 26.2 Example of address clustering.

the appropriate format. A different scenario must be faced when compression is applied to information being stored in memories or caches. Here, compression and decompression are subject to tight performance constraints because they should comply with data read and write speeds of modern, high-performance processors. As such, compression speed becomes the primary cost measure to be used for assessing the quality of a compression scheme. HW-assisted solutions are the only viable options in this context, where high compression ratios are traded for faster compression units. Besides speed, also size and complexity of HW compressors need to be controlled because modern SoCs call for implementations of such units near-by the core processors (for cache-compressed architectures) or between cache and memory (for memory-compressed systems). The problem of designing fast and compact HW compression units for usage in the memory path of a processor-based system has been studied extensively in the past (see, for example, Bunton and Borriello [10] and Lee et al. [22] for a survey of existing literature). Recently, HW memory compression has found its way in a number of commercial designs (see, for example, the MXT Pentium-based server by IBM [1]). Although reduction of memory and bus bandwidth has been, historically, the main motivation for resorting to HW-assisted memory compression, recent studies have demonstrated that this approach can also be exploited when the ultimate target is energy (or power) minimization of a processor-based system. In particular, successful attempts were made to limit energy consumption in systems containing embedded processors by reducing energy requirements of I-caches [5,26], program memory [6,36], and data memory [8].

In the next two subsections, we present a number of approaches that target the reduction of the memory-processor traffic and, as a consequence, the number of accesses to memory locations, exploiting memory compression. In particular, Subsection 26.3.1 ("Code Compression") discusses code compression techniques, whereas Subsection 26.3.2 ("Data Compression") focuses on data compression methodologies.

26.3.1 Code Compression

Currently, many embedded processors are based on high-performance RISC architectures, with on-chip cache [20] and full support for complex memory systems and peripheral controllers [30]. System integrators usually purchase these processors, as well as their software development environments, from third-party companies that specialize in embedded core design.

One of the key challenges in designing a complex system around a high-performance embedded RISC processor is to ensure sufficient instruction fetch bandwidth to keep the execution pipeline busy. The regularity of RISC instruction sets eases application and compiler development, but hinders code compaction. For this reason, designers and researchers have put significant effort in devising techniques for improving code density and reducing instruction-related costs, in terms of speed, area, and energy [3].

Numerous code compression techniques have been proposed for reducing instruction memory size in low-cost embedded applications (refer to Lefurgy [24] for an extensive set of references). The basic idea is to store programs in compressed form and decompress them on the fly at execution time. Later, researchers have realized that code compression can be beneficial for energy as well, because it reduces the energy consumed in reading instructions from memory and communicating them to the processor core [6,26,36].

Code compression leverages well-known lossless data compression techniques [4], but it is characterized by two distinctive constraints. First, it must be possible to decompress a program in relatively small blocks, as instructions are fetched, and starting from several points inside the program (i.e., branch destinations). Thus, traditional lossless techniques that decompress a stream starting from a single initial point are not applicable without changes. Second, the decompressor should be small, fast, and energy-efficient because the corresponding savings in memory size and energy must amortize its hardware, without compromising the performance.

For simple processors with no instruction cache, the hardware decompression block is either merged with the processor core itself or placed between program memory and processor. The first solution has been implemented in several commercial core processors, in the form of a "dense" instruction set, with short instructions (e.g., ARM Thumb [31] and MIPS16 [19] instruction sets). The second solution has been investigated in several articles [6,25,36]. Supporting restricted instruction sets requires changes to the core architecture, while an external decompressor does not. Furthermore, with an external decompressor it is possible to aggressively tailor code compression to a specific embedded application. Thus, external decompression is well suited for embedded designs employing third-party cores.

The basic assumption of the method presented in Yoshida et al. [36] is that the firmware running on a given embedded processor normally uses only a small subset of the instructions supported by the processor. By replacing such instructions with binary patterns of limited width (i.e., $\lceil \log_2 N \rceil$, where N is the number of distinct instructions appearing in the code), memory bandwidth usage can be reduced, thus decreasing the total energy. Notice that two k-bit instructions are said to be distinct if they differ by at least one bit.

The solution proposed in Yoshida et al. [36] does not require ad-hoc compilers; in fact, the original machine instructions can be automatically replaced by $\lceil \log_2 N \rceil$-bit instructions by means of a script after the subset of instructions actually used by the program is identified through execution profiling or instruction-level simulation, and the number $\lceil \log_2 N \rceil$ is determined. The original machine code can thus be compressed to reduce the memory bandwidth that is needed to execute the program. The so-called instruction decompression table and the related control circuitry can be designed and placed between the processor and the memory. Thus, the architecture of the core processor is left unchanged. This is a big plus for system designers employing third-party, off-the-shelf cores and microcontrollers that are either not disclosed (IP hard or soft macros) or not easy to modify.

The work in Benini et al. [6] describes a new technique that builds upon the method of Yoshida et al. [36] by overcoming its major limitation: if the number of instructions used by the embedded code gets large, so does the number of bits of the compressed instructions. Besides increasing the size of the instruction decompression table, this may excessively complicate the implementation of the controller that handles instruction fetching and decoding, especially when the bit-width of the compressed instructions is not compatible with the available memory-addressing scheme (e.g., bit-width different from a multiple of 8 on a byte-addressable memory).

Moving from the observation that the number of instructions used by most programs, although limited with respect to the total number of instructions supported by the processor, has a highly nonuniform statistical distribution. In other words, some instructions are usually much more used than others, it is convenient considering for compression only the instructions used by the embedded code with the highest execution probability. This solution allows fixing *a priori* the bit-width of the compressed instructions;

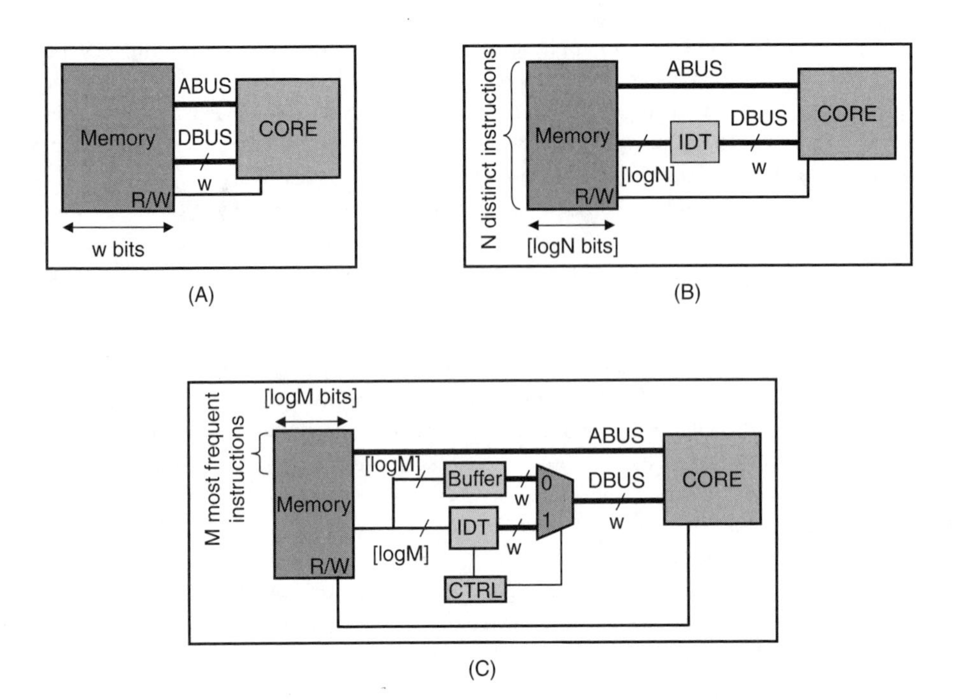

FIGURE 26.3 (a) Original architecture, (b) modifications proposed in Yoshida et al. [36], and (c) in Benini et al. [6].

in this way the size of the instruction decompression table is fixed and limited, the instruction fetching/decompression logic has reduced complexity, and the energy required to fetch the program from memory is minimized. Instruction decompression is performed on-the-fly by a hardware block located between processor and memory. No changes to the processor architecture are required because it always executes full-size instructions. Thus, this technique is well suited for systems employing IP cores whose internal architecture cannot be modified.

Figure 26.3(a) depicts a typical conceptual architecture of a processor-memory system, while Figure 26.3(b) and Figure 26.3(c) report the modified architectures proposed in Yoshida et al. [36] and Benini et al. [6], respectively.

In more advanced architectures that contain instruction caches, the decompressor can be placed either between the I-cache and the main memory (decompress on cache refill, or DCR architecture) or between the processor and the I-cache (decompress on fetch, or DF architecture). Both alternatives have been investigated in the recent literature [21,26].

The approach described in Larin and Conte [21] targets VLIW processors and compresses instructions using the Huffman algorithm. Basic blocks of compressed instructions are transferred and stored into the I-cache atomically. Compressed instructions are not aligned to cache line boundaries. On a cache access, two consecutive cache lines are decompressed and stored in a level-zero buffer. The following instructions are fetched in sequence from the buffer, until it is emptied, or a branch is executed. The hardware decompressor may have a very high cost; in fact, fetching and decoding two cache lines at a time imposes parallel Huffman decoding of multiple instructions in one clock cycle (even single-instruction Huffman decoding requires a quite large hardware block). Furthermore, branch targets addresses in compressed code are stored in a dedicated address remapping memory that must be accessed on every taken branch.

In Lekatsas et al. [26], it was demonstrated that from an energy and performance viewpoint, the DF architecture is superior to the DCR architecture, when decompressor overhead is small. The main reason for this effect is that instructions are stored in cache in a compressed fashion, effectively increasing cache capacity. Current silicon implementations of code compression are based on the DCR architecture

[18,23]; however, indicating that reducing decoding overhead is a non-trivial task that still entails significant challenges.

The main issue with the DF approach is that decompression is performed on every instruction fetch. In other words, the decompressor is on the critical path for the execution of every instruction, not only for cache refills. If its delay is not small, it may significantly slow down execution. Furthermore, it consumes energy on every instruction fetch, while in the DCR architecture it can be activated only on cache refills. Careful implementation of the decompression unit is thus key for making DF applicable in practice.

In Benini et al. [5], a novel DF architecture that focuses on reducing decoding overhead on energy and performance is proposed. This technique guarantees that storage requirements for the compressed program always decrease. Furthermore, the compression algorithm has been designed specifically for fast and low-energy decompression during cache lookup. Compressed instructions are always aligned to cache line boundaries, branch destinations are word-aligned, and instruction decompression is based on a single lookup into a small (and fast) memory buffer. The analysis is not limited to the architecture level, but present a complete implementation of the cache-decompressor block, including detailed analysis of its energy and delay. The achieved code size reductions are on average around 28%, while from the energy point of view the improvements vary a lot depending on cache size, original and compressed code size, dynamic memory access profile, and kind of adopted program memory (i.e., on-chip vs. off-chip). For example, for a 4 KB cache and an on-chip program memory, average energy savings are around 30%. This value grows to 50% for a system with a cache of the same size but an off-chip program memory.

26.3.2 Data Compression

The principle exploited in code compression (namely, reduction of memory traffic), can be extended to the case of data, with some additional difficulties.

Data compression techniques, together with hardware architectures, have been introduced recently in Benini et al. [8,9].

The approach of Benini et al. [8] relies on a fixed-dictionary scheme. It resorts to data profiling information to selectively compress cache lines before they are written back to the main memory, and to quickly decompress them when a cache refill operation is started. This solution is particularly suited to embedded systems, where the collection of data statistics to be used by compressor and decompressor is much more predictable than in general-purpose systems. The obtained reductions of memory traffic were around 42% for a significant number of benchmark programs. In this work, the authors consider systems with compressed main memory (i.e., information is stored in caches in uncompressed format), in which the compression HW is placed between caches and main memory.

A possible generalization of the approach of Benini et al. [8] to the case of general-purpose systems was sketched in Benini et al. [9]. The idea was that of avoiding the profiling step by doing online prediction of data statistics. Here, the architecture is able to adaptively update the compression dictionary according to the current data statistics. This improvement allows removal of the main limitation constituted by the need of off-line data profiling. In other words, the adaptive algorithm is applicable to systems where several programs need to be executed, and thus ad hoc data profiling information cannot be collected before the system is started.

In these approaches, the fundamental difference between code and data compression is that, for the latter, both compression and decompression are needed during the execution of a program, while for the former only decompression is required. This fact has far-reaching implications on the applicable compression algorithms and architectures; for instance, it rules out highly asymmetric schemes, where compression is much more involved than decompression.

Clearly, the energy cost of the compressor also needs to be accounted when evaluating the feasibility of an HW-based memory compression scheme. It was observed in Benini et al. [5] that the overhead introduced by the extra hardware is roughly proportional to the amount of storage space the compressor requires. As such, if care is given to the choice of the compression algorithm and to its implementation, the achieved bus and memory energy savings will offset the energy cost of the compressor.

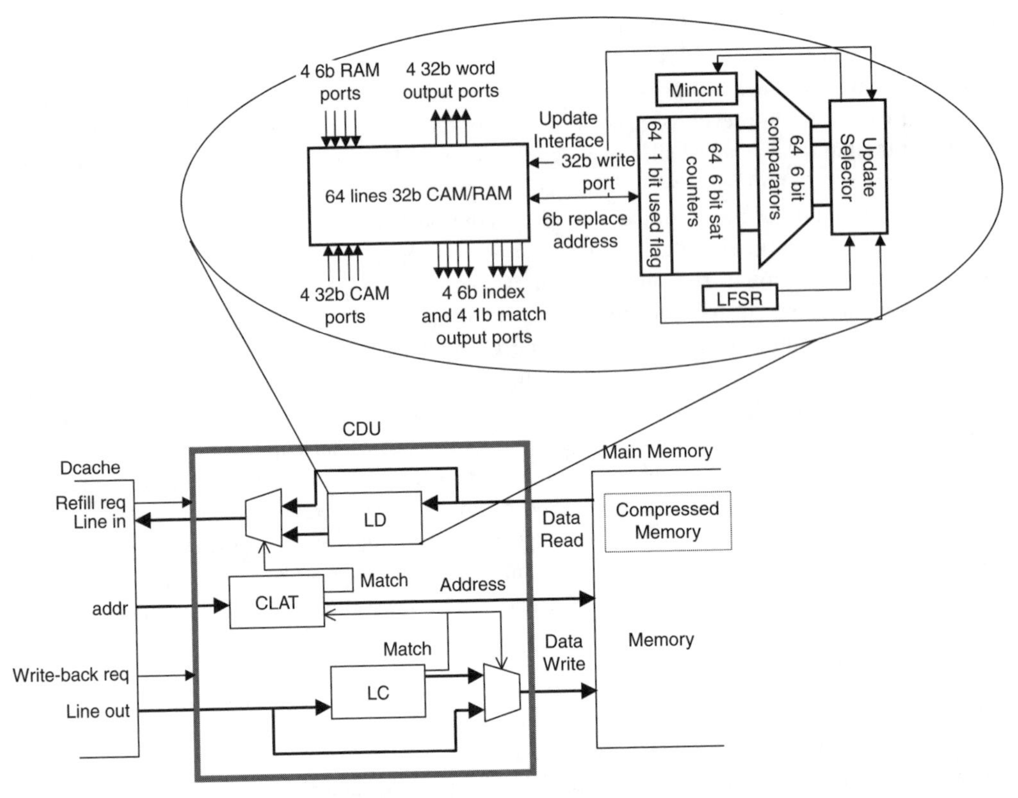

FIGURE 26.4 CDU architecture and implementation of a comp–decomp table.

Just to give to the reader the flavor of how the compressor can be implemented, Figure 26.4 reports a conceptual block diagram of the compression-decompression unit (CDU) adopted in Benini et al. [8,9] with the basic interface signals, together with a real hardware implementation of a compression–decompression table.

The CDU contains three major functional blocks, namely, the line compressor (LC), the line decompressor (LD), and the compressed line address table (CLAT). The compression–decompression table highlighted in Figure 26.4 is the basis of the LC and LD blocks.

The hardware implementation of the CDU is based on content-addressable memories and random-access memories (CAMs and RAMs, respectively), and it is mainly targeted toward energy minimization in the cache-bus-memory subsystem with a strict constraint on performance. As a result, average memory reductions evaluated on several benchmarks are around 23%, at no performance penalty (actually, on average, performance improved by 4%). Comparison to the memory traffic reductions achieved with the profile-driven method of Benini et al. [8] demonstrates clearly that the adaptive approach performs reasonably well in this respect (31% reductions, against 42% for the profile-driven solution).

A similar compression technique is the compressed cache presented in Yang et al. [35]. It is based on the frequent value locality, that is, the fact that very few values (typically small integers) account for usually around half of the total memory accesses, for most benchmarks. This allows storing the selected values in a compressed form. Compression/decompression is performed on the fly on accesses from/to the next hierarchy level.

26.4 Conclusions

Embedded systems are now becoming ubiquitous; in particular, they have large applicability in mobile, battery-operated personal communication systems, for which energy consumption is a major constraint.

Among the various contributors to the system's energy budget, memory plays a preeminent role; in fact, reading and writing data to the memory hierarchy is an operation that takes place very often, especially in data-dominated applications (e.g., video and audio playing). As such, memory energy optimization is one of the most promising and successful approaches to system power minimization.

References

[1] S. Arramreddy, et al. IBM X-Press memory compression technology debuts in a ServerWoks NorthBridge, *HOT Chips 12 Symp.*, Stanford University, Stanford, CA, August 2000.

[2] R.I. Bahar, G. Albera, and S. Manne, Power and performance trade-offs using various caching strategies, *ISLPED-98: ACM/IEEE Int. Symp. on Low-Power Electron. and Design,* pp. 64–69, Monterey, CA, August 1998.

[3] R. Bajwa, et al. Instruction buffering to reduce power in processors for signal processing, *IEEE Trans. on Very Large Scale Integration (VLSI) Syst.*, Vol. 5, No. 4, pp. 417–424, December 1997.

[4] T. Bell, J. Cleary, and I. Witten, *Text Compression,* Prentice Hall, New York, 1990.

[5] L. Benini, A. Macii, and A. Nannarelli, A code compression architecture for cache energy minimization in embedded systems, *IEEE Proc. — Comput. and Digital Techniques,* Vol. 149, No. 4, pp. 157–163, July 2002.

[6] L. Benini, A. Macii, E. Macii, and M. Poncino, Minimizing memory access energy in embedded systems by selective instruction compression, *IEEE Trans. on Very Large-Scale Integration (VLSI) Syst.*, Vol. 10, No. 5, pp. 521–531, October 2002.

[7] L. Benini, L. Macchiarulo, A. Macii, and M. Poncino, Layout-driven memory synthesis for embedded systems-on-chip, *IEEE Trans. on Very Large-Scale Integration (VLSI) Syst.*, Vol. 10, No. 2, pp. 96–105, April 2002.

[8] L. Benini, D. Bruni, A. Macii, and E. Macii, Hardware-assisted data compression for energy minimization in systems with embedded processors, *DATE-02: IEEE Design Automation and Test in Europe,* Paris, France, pp. 449–453, March 2002.

[9] L. Benini, D. Bruni, A. Macii, E. Macii, and B. Riccò, An adaptive data compression scheme for memory traffic minimization in processor-based systems, *ISCAS-02: IEEE Int. Symp. on Circuits and Syst.*, pp. IV-866–IV-869, Scottsdale, AZ, May 2002.

[10] S. Bunton and G. Borriello, Practical dictionary management for hardware data compression, *Commn. ACM,* Vol. 35, No. 1, pp. 95–104, January 1992.

[11] D.C. Burger, Hardware techniques to improve the performance of the processor/memory interface. Ph.D. dissertation, University of Wisconsin—Madison, 1998.

[12] F. Catthoor, S. Wuytack, E. De Greef, F. Balasa, L. Nachtergaele, and A. Vandecappelle, *Custom Memory Management Methodology Exploration for Memory Optimization for Embedded Multimedia System Design,* Kluwer, Dordrecht, 1998.

[13] S.L. Coumeri and D.E. Thomas, Memory modeling for system synthesis, *ISLPED-98: ACM/IEEE Int. Symp. on Low-Power Electron. and Design,* pp. 179–184, Monterey, CA, August 1998.

[14] S. Coumeri and D.E. Thomas, Memory modeling for system synthesis, *IEEE Trans. on Very Large-Scale Integration (VLSI) Syst.*, Vol. 8, No. 3, pp. 327–334, June 2000.

[15] A. Farrahi, G. Tellez, and M. Sarrafzadeh, Memory segmentation to exploit sleep mode operation, *DAC-32: ACM/IEEE Design Automation Conf.*, pp. 36–41, San Francisco, CA, June 1995.

[16] A. Gonzàlez, C. Aliagas, and M. Valero, A data-cache with multiple caching strategies tuned to different types of locality, *ICS-95: ACM Int. Conf. on Supercomputing,* pp. 338–347, Barcelona, Spain, July 1995.

[17] P. Grun, N. Dutt, and A. Nicolau, Access pattern based local memory customization for low-power embedded systems, *DATE-01: IEEE Design Automation and Test in Europe,* pp. 778–784, Munich, Germany, March 2001.

[18] IBM Corporation, *CodePack PowerPC Code Compression Utility, User's Manual Version 3.0,* IBM Corporation, Austin, Texas, 1998.

[19] K. Kissel, MIPS16: high-density MIPS for the embedded market. Technical report, Silicon Graphics MIPS Group, Mountain View, CA, 1997.

[20] U. Ko and P. Balsara, Energy optimization of multilevel cache architectures for RISC and CISC processors, *IEEE Trans. on Very Large Scale Integration (VLSI) Syst.,* Vol. 6, No. 2, pp. 299–308, June 1998.

[21] S. Larin and T. Conte, Compiler-driven cached code compression schemes for embedded ILP processors, *MICRO-32: 32nd Annual Int. Symp. on Microarchitecture,* pp. 82–92, Haifa, Israel, November 1999.

[22] J.-S. Lee, W.-K. Hong, and S.-D. Kim, Design and evaluation of a selective compressed memory system, *ICCD-99: IEEE Int. Conf. on Comput. Design,* pp. 184–191, Austin, TX, March 1999.

[23] C. Lefurgy, E. Piccinini, and T. Mudge, Evaluation of a high performance code compression method, *MICRO-32: 32nd Annu. Int. Symp. on Microarchitecture,* pp. 93–102, Haifa, Israel, November 1999.

[24] C. Lefurgy, Efficient execution of compressed programs. Doctoral dissertation, University of Michigan, Ann Arbor, MI, 2000.

[25] H. Lekatsas and W. Wolf, Code compression for embedded systems, *DAC-35: ACM/IEEE Design Automation Conf.,* pp. 516–521, San Francisco, CA, June 1998.

[26] H. Lekatsas, J. Henkel, and W. Wolf, Code compression for low-power embedded systems, *DAC-37: ACM/IEEE Design Automation Conf.,* pp. 294–299, Anaheim, CA, June 2000.

[27] A. Macii, E. Macii, and M. Poncino, Improving the efficiency of memory partitioning by address clustering, *DATE-03: IEEE Design Automation and Test in Europe,* pp. 18–23, Munich, Germany, March 2003.

[28] P. Panda and N. Dutt, *Memory Issues in Embedded Systems-on-Chip Optimization and Exploration,* Kluwer, Dordrecht, 1999.

[29] J. Rabaey and M. Pedram, *Low-Power Design Methodologies,* Kluwer, Dordrecht, 1996.

[30] S. Santhanam et al. A low-cost, 300-MHz, RISC CPU with attached media processor, *IEEE J. Solid-State Circuits,* Vol. 33, No. 11, pp. 1829–1839, November 1998.

[31] S. Segars, K. Clarke, and L. Goudge, Embedded control problems, thumb and the ARM7TDMI, *IEEE Micro,* Vol. 15, No. 5, pp. 22–30, October 1995.

[32] W. Shiue and C. Chakrabarti, Memory exploration for low-power, embedded systems, *DAC-36: ACM/IEEE Design Automation Conf.,* pp. 140–145, New Orleans, LA, June 1999.

[33] C.L. Su and A.M. Despain, Cache design trade-offs for power and performance optimization: a case study, *ISLPD-95: ACM/IEEE Int. Symp. on Low-Power Design,* pp. 63–68, Dana Point, CA, April 1995.

[34] T. Watanabe, R. Fujita, and K. Yanagisawa, Low-power and high-speed advantages of DRAM-logic integration for multimedia systems, *IEICE Trans. on Electron.,* Vol. E80-C, No. 12, pp. 1523–1531, December 1997.

[35] J. Yang, Y. Zhang, and R. Gupta, Frequent value compression in data caches, *MICRO-33: IEEE/ACM 33rd Int. Symp. on Microarchitecture,* pp. 258–265, Monterey, CA, December 2000.

[36] Y. Yoshida, B.-Y. Song, H. Okuhata, T. Onoye, and I. Shirakawa, An object code compression approach to embedded processors, *ISLPED-97: ACM/IEEE Int. Symp. on Low-Power Electron. and Design,* pp. 265–268, Monterey, CA, August 1997.

IV

Low-Power Systems on Chips

27

Power Performance Trade-Offs in Design of SoCs

Victor Zyuban
Philip Strenski
IBM Watson Research Center

27.1 Introduction

The design and implementation of processor cores is characterized by conflicting requirements of the ever increasing demand for higher performance and, usually, stringent power budget. Thus, compromises between performance and power need to be made early in the design cycle. In the design of a processor core, a very specific power budget typically exists, but power can be traded for performance in several ways.

At the system level, varying power supply is the most straightforward and well-understood method for controlling power. One advantage of this method is that power supply can typically be adjusted within a certain range even after the chip has been manufactured. In addition, scaling V_{dd} around the nominal value in application specific integrated circuits (ASIC) foundry technologies has a known cost which is "typically" 2% in energy per 1% in performance, although it can be anywhere from 0.5% to more than 3% in energy per 1% in performance. A notion of voltage intensity has recently been introduced [21] to quantify power–performance trade-offs through varying the power supply. Although a very powerful technique, scaling V_{dd} may have a relatively high performance cost for saving power in a processor core that does not meet its power budget, as discussed in Section 27.3 of this chapter.

Another method for making power–performance trade-offs is technology scaling, such as shrinking the oxide thickness and effective channel length. Although, such trade-offs are not generally available to average ASIC customers, some foundry technologies provide libraries and transistors with multiple threshold voltages, and some high-end microprocessor designs work with foundries (typically their own) to engineer these parameters effectively.

At the circuit level, power and performance can be traded by changing transistor sizes and power levels of ASIC cells, controlled by changing transistor tuning targets, or by restructuring logic to increase or

0-8493-1941-2/05/$0.00+$1.50

decrease the parallelism in circuits, for example, perform more computations in parallel in order to reduce the critical path. These trade-offs are controlled by either custom or logic designers or by running synthesis tools with different directives. Although more difficult to quantify, this method for power–performance trade-offs is at least as powerful as scaling the power supply. By scaling circuits, in "typical" designs that we analyzed, one percent in performance could be traded for from 0.5 to 5% in energy, or even higher, if the frequency target is too aggressive. The concept of *hardware intensity* was introduced in Zyuban and Strenski [21] to quantify power–performance trade-offs through scaling circuits.

Finally, scaling the processor core microarchitecture and, possibly instruction set architecture (ISA), is one more way for trading power and performance. This method involves changing machine organization, such as pipeline depth, issue width, the set of functional units, bypasses, number of ports and entries in queues and register files, the sizes of branch predictors, and other structures of the microarchitecture. Scaling microarchitecture is even a more powerful method for power–performance trade-offs than voltage and circuit scaling, but it is more difficult to quantify and optimize, and can only be used at early stages of the design. A concept of architectural complexity was introduced in Zyuban [19] to analyze power–performance trade-offs at the ISA and microarchitectural levels. It has been demonstrated that architectural complexity cannot only be measured but also set as a design target [10].

It was demonstrated in Zyuban and Strenski [21] that to develop an energy-efficient processor core, design decisions at all levels must be balanced in such a way that all forms of spending power, described above, have a similar marginal cost. The following sections summarize some of the most important formulas for balancing hardware intensity and power supply voltage derived in Zyuban and Strenski [21], and give a graphical interpretation of the major result. Then, the formula for balancing architectural complexity with voltage and hardware intensity, and the iterative process of refining the processor core architecture in the power–performance space are discussed in detail. Then, because making power-balanced decisions at the architectural level plays such an important role in the development of the energy-efficient processor cores, we give a derivation in Section 27.4 of a new form of the architectural energy-efficiency criterion that does not require evaluation of the relative changes in processor frequency. Section 27.5 discusses other power–performance metrics that are commonly used in the architectural community and some common mistakes made by architects when applying these metrics. Section 27.6 gives some examples of using the architectural energy-efficiency criterion, and Section 27.7 concludes the chapter.

27.2 Hardware Intensity

The concept of hardware intensity η was introduced in Zyuban and Strenski [21] as a quantitative measure of how aggressively the circuits in a processor are tuned to meet a target clock frequency (see related work in Brodersen et al. [1], Hofstee [9], and Oklobdzija et al. [14]). Hardware intensity shows the energy cost (in %) required to improve the delay D of a hardware macro by 1% through restructuring the logic and retuning the circuits, at a fixed power supply v,

$$\eta = -\frac{D\partial E}{E\partial D}\bigg|_{\substack{\text{fixed}\\ v}} \quad \text{or} \quad \eta = -\frac{\%E}{\%D}\bigg|_{\substack{\text{through}\\ \text{retuning}}} \tag{27.1}$$

Alternatively, we can define the hardware intensity η as a parameter in the cost function for optimizing hardware:

$$F_{cost}(E,D) = (E/E_0)(D/D_0)^{\eta} \quad \eta \geq 0, \tag{27.2}$$

where D is the critical path delay through the circuit, E is the average energy dissipated per cycle, D_0 and E_0 are the corresponding lower bounds that can be achieved through tuning and logic restructuring for

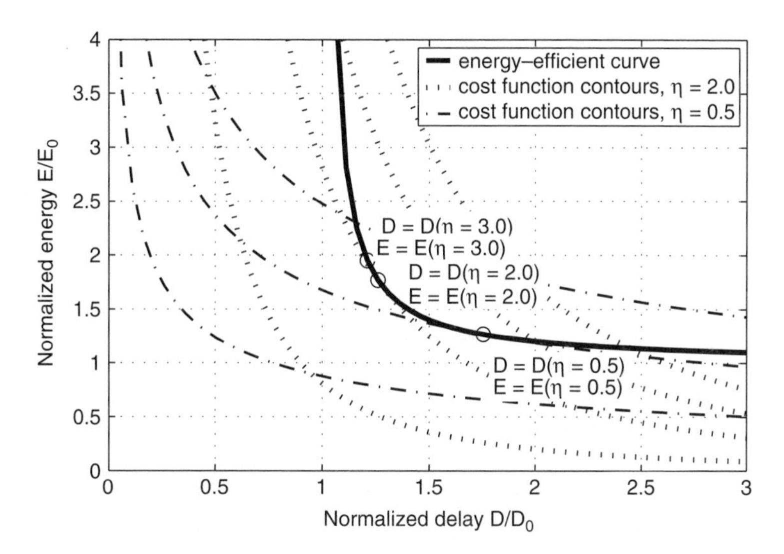

FIGURE 27.1 Typical energy-efficient curve and constant cost function contours for $\eta = 0.5$ and $\eta = 2.0$.

a fixed supply voltage. Under a very general assumption that the curvature of the energy-delay curve is larger than the curvature of the contour of the cost function (Equation (27.2)) at any point in the two tangent, $\frac{D^2}{E}\frac{\partial^2 E}{\partial D^2} > \eta(\eta+1)$, we can show that for any power supply voltage *v*, every point on the energy-delay curve corresponds to a certain value of hardware intensity η, $0 \le \eta < +\infty$. Then, the energy-delay curve in the energy-vs.-delay coordinates can viewed as a parameterized curve: $D = D(\eta, v)$, $E = E(\eta, v)$.

Figure 27.1 gives a graphical interpretation of the hardware intensity. The solid line plots a typical energy-delay curve for some hardware function. Dotted lines show several contours of the cost function (Equation (27.2)), for two values of hardware intensity η. Point (D,E) at which the energy-delay curve tangents the lowest of the contours $F_{cost}(E,D) = A$ (with the smallest value of A) corresponds to the implementation for this value of hardware intensity η. Taking advantage of the equivalence of the tangents to the energy-delay curve and the counter of the cost function, we get:

$$\left.\frac{\partial E}{\partial D}\right|_{\substack{\text{fixed} \\ \text{n}}} = \frac{\partial E(\eta, v)}{\partial \eta} / \frac{\partial D(\eta, v)}{\partial \eta} = -\frac{\partial F_{cost}}{\partial D} / \frac{\partial F_{cost}}{\partial E} = -\eta \frac{E}{D} \tag{27.3}$$

This establishes the equivalence of the two definitions of the hardware intensity in Equation (27.1) and Equation (27.2).

Then, by formally solving the problem of minimizing the delay as a function of η and *v*, subject to a constant energy constraint, the following relations were derived in Zyuban and Strenski [21] for the optimal balance between hardware intensity and power supply voltage:

$$\text{isolated macro} \quad \eta = \theta(v) \tag{27.4}$$

$$\text{composite macro} \quad \frac{w_j}{u_j}\eta_j = \theta(v) \quad 1 \le j \le M \tag{27.5}$$

$$\text{multi-stage pipeline} \quad \sum_i w_i \eta_i = \theta(v) \tag{27.6}$$

where w_i are energy weights of pipeline stages i in Equation (27.6), w_j and u_j are energy and delay weights of sub-blocks j in Equation (27.5), η_j are the hardware intensities in the corresponding sub-blocks, and θ is the voltage intensity defined as

$$\theta = \frac{E_v}{D_v} \quad E_v = \frac{v}{E}\frac{\partial E}{\partial v} \quad D_v = -\frac{v}{D}\frac{\partial D}{\partial v}.$$

Equation (27.4) has a simple interpretation, shown in Figure 27.2. The solid curve shows an energy-delay trade-off curve for variable hardware intensity at a fixed power supply, v = 1.5V, $\theta = 2$ in this example (hardware intensity energy-delay curve). The curve was fitted to simulation data [21] for an integer adder, obtained using the EinsTuner [6]. The dotted curves show the energy-delay curves for a fixed tuning point (fixed hardware intensity η) of the circuit, but varying power supply, plotted for an ideal $E \sim v^2$ and $f \sim v$ dependence, as commonly assumed in many studies. The dashed curves show simulated data (for a set of functional units (FUs), running PathMill and PowerMill), with 50 mV steps in the power supply marked with circles. The point at which the power supply energy-delay curve (dashed curve) tangents the hardware intensity energy-delay curve (solid curve) corresponds to the optimal balance between η and v.

To show this, suppose that the circuit is over-tuned, for example, $\eta = 4$. Then, retuning the circuit for a lower value of η will move the design point down the hardware intensity energy-delay curve. Increasing the power supply to recover the performance will move the design up the power supply energy-delay curve (dashed curve). Because the hardware intensity energy-delay curve (solid curve) is steeper than the power supply energy-delay curve, the same performance will be achieved at a lower energy. Similarly, if the circuit is under-tuned for, say, $\eta = 0.5$, then tuning the circuit for a higher η and then reducing V_{dd} to achieve the same critical path delay (if the faster operation is not needed) will result in a circuit operating at the same speed, but lower energy. Notice that this reasoning does not require that the curvature of the hardware intensity energy-delay curve be higher than that of the power supply energy-delay curve. Although this property was experimentally verified for a 0.13μ and older CMOS technologies it may or may not hold true in future technologies, depending on the dependence of the gate and subthreshold leakage currents on the power supply. If the curvature of the V_{dd} energy-delay curve becomes higher, the range in which energy and performance can be traded through adjusting V_{dd} will be more limited, but the optimality relation (Equation (27.4)) will still hold.

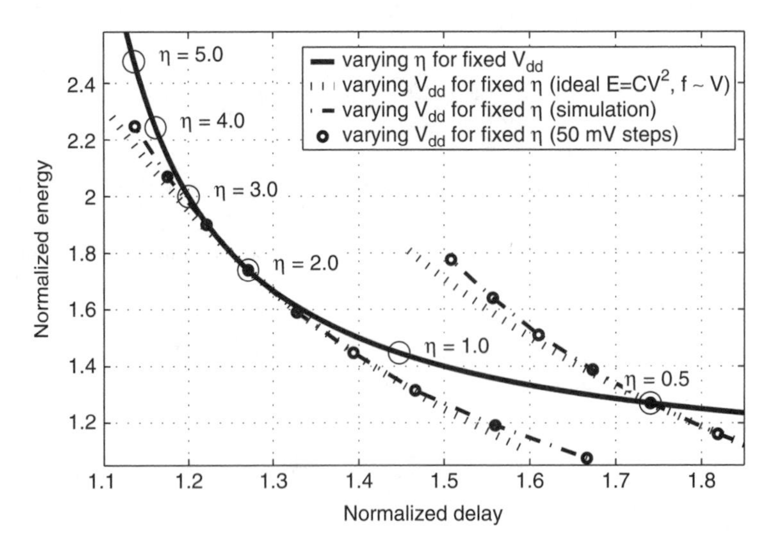

FIGURE 27.2 Graphical interpretation of the optimum hardware intensity balance for an isolated macro.

Although the optimal values of hardware intensities in different pipeline stages, η_i are different, it is useful to abstract for the higher-level microarchitectural analysis of energy-performance trade-offs a single aggregate quantity for hardware intensity η_{ag} that represents the whole processor, such that $\eta_{ag} = -\left.\frac{D\partial E}{E\partial D}\right|_v$, where D is the clock period, and E is the total average energy dissipated per cycle in the processor, $E = \sum E_i$. To derive an expression for η_{ag}, notice that increasing the clock cycle time by dD through retuning the circuits in all stages of the pipeline, increases the total energy of the pipeline by $dE = \sum dE_i = -\sum \frac{E_i}{D_i}\eta_i dD$, where the summation is performed over all stages of the pipeline. Since $D_i = D$ (all stages are tuned for the same delay), $\frac{dE}{E} = -\frac{dD}{D}\sum w_i\eta_i$, which means that the aggregate hardware intensity for a multi-stage pipeline is expressed through the hardware intensities of individual stages η_i as

$$\eta_{ag} = \sum_i w_i \eta_i \tag{27.8}$$

27.3 Architectural Complexity

Changing the processor architecture is another way to make trade-offs between performance and energy. Architectures that are more complex deliver higher architectural performance or instructions per cycle (IPC), but inevitably dissipate more energy per every executed instruction. Similar to building the optimal energy-delay curve in the circuit domain, an optimal energy-delay curve can be constructed in the architectural domain, as an envelope in the power-performance space of all feasible architectural alternatives [20]. Similar to Equation (27.1), the architectural complexity ξ can be defined as*

$$\xi = -\left.\frac{D\partial E}{E\partial D}\right|_{\text{fixed}\,\eta,v} \quad \text{or} \quad \xi = -\left.\frac{\%E}{\%D}\right|_{\text{through architecture}} \tag{27.9}$$

Similar to the optimal balance between η and v in the circuit domain, there exists an optimal balance between architectural complexity ξ and η and v in the unified architectural-circuit domain. By formally solving the problem of minimizing energy as a function of three variables $E = E(\xi, \eta, v)$, subject to a constant delay constraint $D(\xi, \eta, v) = D_0$, we arrive at**

$$\frac{\partial D}{\partial \eta}\frac{\partial E}{\partial v} = \frac{\partial D}{\partial v}\frac{\partial E}{\partial \eta} \quad \frac{\partial D}{\partial \xi}\frac{\partial E}{\partial v} = \frac{\partial D}{\partial v}\frac{\partial E}{\partial \xi} \tag{27.10}$$

Using Equation (27.1) and Equation (27.9), Equation (27.10) can be rewritten as

$$\xi = \eta = \theta \tag{27.11}$$

*It is assumed that the curvature of the architectural energy-delay curve is such that $\frac{D^2}{E}\frac{\partial^2 E}{\partial D^2} > \xi(\xi + 1)$ is satisfied at every point.

**The converse problem of minimizing D subject to constant E leads to the same equations.

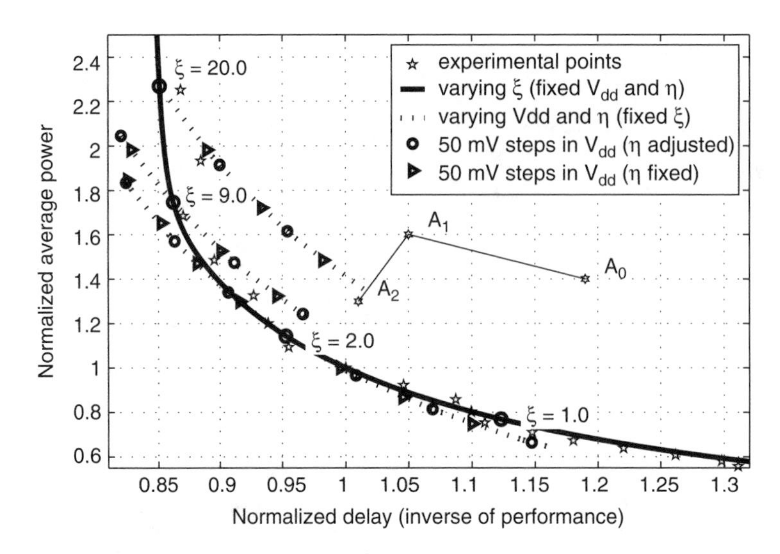

FIGURE 27.3 Graphical interpretation of the optimum architectural complexity balance.

Figure 27.3 gives a graphical interpretation of this relation. The solid curve shows the architectural energy-delay curve, plotted by curve-fitting the power-performance data reported in the optimal pipeline depth study [17]. Every data point, plotted as stars in Figure 27.3, represents a different pipeline depth, and we assume here that the processor microarchitecture was optimally tuned at every pipeline depth. The dotted curves in the figure show the circuit energy-delay curves for fixed ξ, with the power supply and hardware intensity varied simultaneously, so that the optimum balance (Equation (27.4)) is observed at each point. Circles mark 50-mV steps in V_{dd}, with the corresponding adjustment in η (Equation (27.6)). This data was obtained by simulating a set of representative circuits in a microprocessor [21]. For a reference, triangles show 50-mV steps in V_{dd} without adjusting η. Although the quality of the energy-delay trade-off of the fixed-η scaling is almost the same as that of the optimal v-η scaling, the span of the former is much smaller (larger change in V_{dd} is needed to achieve the same speed up or slow down).

The point at which the architectural energy-delay curve (solid curve) tangents the circuit energy-delay curve (dotted curve) is the point of the optimal balance between ξ, η, and v in Equation (27.11). To see this, suppose the architecture is over-designed (for instance, $\xi = 9$). Then, by reducing the architectural complexity, we can move the design point down the architectural energy-delay curve. Then the performance can be recovered by increasing V_{dd} (by 100 mV in this example) and tuning up circuits for higher η, according to Equation (27.11), which will move the design point up the circuit energy-delay curve. Because the circuit energy-delay curve is less steep than the architectural energy-delay curve, the same performance will be achieved at a lower power. Similarly, if $\xi < \eta$, for instance, $\xi = 1$, then increasing the architectural complexity to improve the architectural performance (moving up the architectural energy-delay curve) and reducing V_{dd} and η to save energy (moving down the circuit energy-delay curve) will result in the same performance at a lower power. As with hardware intensity, this relation is not dependent on assumptions about the relative curvatures of the various energy-delay trade-offs.

Although the nature of the energy-delay trade-offs at the architectural level is similar to that at the circuit level, one significant difference between them is that with recent advances in the circuit tuning techniques [6] all circuit-level implementations (provided an appropriate circuit topology is chosen) in a properly tuned processor can be assumed to be on the optimal energy-delay curve (Figure 27.2) (designs above the optimal energy-delay curve should be simply discarded), whereas getting the processor architecture that is on the architectural energy-delay curve (Figure 27.3) presents a significant challenge. The initial architectural proposal for a new processor is likely to be way off the optimal energy-delay curve. Multiple iterations of optimizing the architecture are required to transfer the design point to the optimal architectural energy-delay curve, and then, to the point of the optimum balance (Equation (27.11)), an iterative process illustrated by sequence $A_0 \rightarrow A_1 \rightarrow \cdots A_n$ in Figure 27.3. To make the methodology useful for comparing architectural configurations that are not necessarily on the optimal energy-delay curve, the definition of ξ was extended to designs above the optimal energy-delay curve and a more general form of the energy-efficiency criterion was derived in Zyuban [19], and generalized to include hardware intensity in Zyuban and Strenski [21]:

$$\frac{\eta_{ag}}{I}\frac{\Delta I}{\Delta\xi} - \frac{\eta_{ag}+1}{N}\frac{\Delta N}{\Delta\xi} > -\frac{\eta_{ag}}{f}\frac{\Delta f}{\Delta\xi}\bigg|_{\text{fixed}\eta,v} + \frac{1}{E}\frac{\Delta E}{\Delta\xi}\bigg|_{\text{fixed}\eta,v} \tag{27.12}$$

If the inequality holds, then the architectural feature under evaluation is energy-efficient, that is, after adopting it, the processor will deliver higher net performance at the same power budget, after appropriate retuning and, possibly, adjustment in the power supply voltage are done to meet the power budget. In this formula $\frac{\Delta f}{f\Delta\xi}$, $\frac{\Delta I}{I\Delta\xi}$, $\frac{\Delta E}{E\Delta\xi}$, and $\frac{\Delta N}{N\Delta\xi}$ are relative increments in the processor frequency, architectural performance IPC, average energy per instruction, and the dynamic instruction count arising from a modification at the architectural or microarchitectural level, evaluated for a fixed hardware intensity η_{ag} and power supply v. Thus, all deltas in Equation (27.12) have the meaning of partial derivatives with respect to the architectural complexity.

The terms $\frac{\Delta I}{I\Delta\xi}$ and $\frac{\Delta N}{N\Delta\xi}$ in Equation (27.12) can be measured by running the benchmark suite on an architectural simulator. Next we present a methodology for estimating the two remaining terms, $\frac{\Delta f}{f\Delta\xi}$ and $\frac{\Delta E}{E\Delta\xi}$, and derive a new form of the energy-efficiency criterion that does not require estimating the term Δf.

27.4 Energy-Efficiency Criterion

27.4.1 Frequency-Invariant Formulation

The key assumption in deriving the energy-efficiency criterion (Equation (27.12)) was that of the optimal tuning of circuits in every pipeline stage (Equation (27.4), Equation (27.5), and Equation (27.6)) for every architectural alternative, so that the aggregate hardware intensity of the processor η_{ag} (Equation (27.8)) is unchanged between designs implementing the architectural alternatives. This assumption imposes special rules on calculating $\frac{\Delta f}{f}$, and $\frac{\Delta E}{E}$, in particular, these relative increments must be calculated assuming that the processor pipeline is reoptimized after every modification to the microarchitecture to satisfy Equation (27.4), Equation (27.5), and Equation (27.6).

Suppose, an architectural feature under evaluation introduces an additional complexity in several (or all) stages of the pipeline, which leads to increments $\Delta D_i\big|_{noretune}$ in critical path delays in the corresponding pipeline stages, assuming that no retuning is done to recover the clock frequency. Suppose that the corresponding increments in average energies are $\Delta E_i\big|_{noretune}$. The increments $\Delta D_i\big|_{noretune}$ and $\Delta E_i\big|_{noretune}$ should be evaluated consistently with the initial hardware intensities of the corresponding stages. For example, logic added to stage i should be tuned (or assumed to be tuned) according to Equation (27.5). Then after adding the logic, the aggregate hardware intensity in pipeline stage i will not change. The delay and energy increments may be either positive or negative, and in those pipeline stages that are

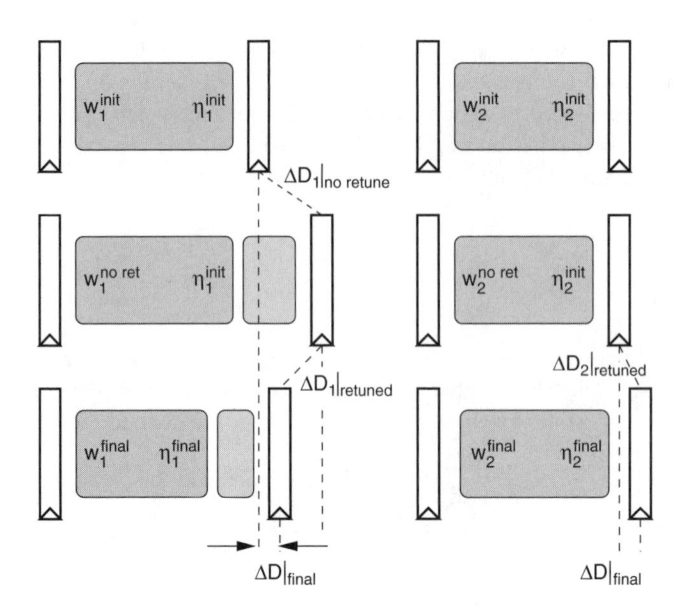

FIGURE 27.4 Retuning pipeline after architectural modification.

unaffected by the architectural modification, the delay and energy increments are zero, $\Delta D_i|_{noretune} = 0$, $\Delta E_i|_{noretune} = 0$, as shown in Figure 27.4.

Circuit designers usually have no difficulties estimating the "nonretuned" increments in delay and energy. For example, adding an execution bypass in 10FO4 pipeline results in increments in the critical path delay and average energy of the execution stage of the pipeline which are approximately $\left.\frac{\Delta D_{EX}}{D}\right|_{noretune} = 0.2$, and $\left.\frac{\Delta E_{EX}}{E_{EX}}\right|_{noretune} = 0.02$, whereas adding an extra read port to a multiported register file may result in $\left.\frac{\Delta D_{RF}}{D}\right|_{noretune} = 0.1$, and $\left.\frac{\Delta E_{RF}}{E_{RF}}\right|_{noretune} = 0.2$, with no impact in other stages of the pipeline.

To recover the clock frequency, circuits in those stages of the pipeline that are negatively affected by the architectural modification need to be tuned up for a higher hardware intensity. To restore the energy-optimal balance in the pipeline $\eta_{ag} = \theta$, circuits in all remaining stages need to be tuned down for a lower η, so that

$$\Delta\eta_{ag} = \sum_i \eta_i \Delta w_i + \sum_i w_i \Delta\eta_i = 0 \tag{27.13}$$

where $\Delta\eta_i$ is the increment in the aggregate hardware intensity in stage i as a result of retuning, $\Delta\eta_i = \eta_i^{final} - \eta_i^{initial}$, as illustrated in Figure 27.4, whereas Δw_i is the net increment in the corresponding energy weight, as a result of both adding hardware and subsequent retuning, $\Delta w_i = \frac{\Delta E_i}{E} - w_i \frac{\Delta E}{E}$.

We designate by $\Delta D_i|_{retune}$ and $\Delta E_i|_{retune}$ the increments in delay and energy in the pipeline stage i as a result of retuning the processor, whereas by $\Delta D_i = \Delta D$ and ΔE_i we designate the net increment in delay and energy in pipeline stage i as a result of both modifying the function and subsequent retuning:

$$\Delta D = \Delta D_i|_{noretune} + \Delta D_i|_{retune} \tag{27.14}$$

$$\Delta E_i = \Delta E_i\Big|_{no\,retune} + \Delta E_i\Big|_{retune} \tag{27.15}$$

Thus, the net delay and energy increments in every pipeline stage consist of increments due to a change in the functionality resulting from a microarchitectural modification, and additional increments as a result of retuning the circuits. The net delay increment ΔD does not need any index because all pipeline stages are assumed to have the same delay before and after the retuning, $D_i = D$. The relative increment in the maximum clock frequency is related to ΔD as

$$\left.\frac{\Delta f}{f}\right|_{\text{fixed}\,\eta,v} = -\frac{\Delta D}{D} \tag{27.16}$$

Assuming small changes in hardware intensities in all pipeline stages, and neglecting second order terms, the increments in energies $\Delta E_i\big|_{retune}$ because of the retuning can be expressed through the corresponding increments in delays $\Delta D_i\big|_{retune}$ as follows:

$$\Delta E_i\Big|_{retune} = -\eta_i \frac{E_i}{D} \Delta D_i\Big|_{retune} \tag{27.17}$$

Using Equation (27.14) and Equation (27.15), the final increments in energies can be expressed as

$$\Delta E_i = \Delta E_i\Big|_{no\,retune} - \eta_i \frac{E_i}{D}\left(\Delta D - \Delta D_i\Big|_{no\,retune}\right) \tag{27.18}$$

The total increment in energy of the whole pipeline, $\Delta E = \sum \Delta E_i$, is calculated by summing Equation (27.18) over all pipeline stages and taking advantage of Equation (27.8) and Equation (27.16):

$$\left.\frac{\Delta E}{E}\right|_{\text{fixed}\,\eta,v} = \left.\frac{\Delta E}{E}\right|_{no\,retune} + \sum_i \eta_i w_i \left.\frac{\Delta D_i}{D}\right|_{no\,retune} + \eta_{ag} \left.\frac{\Delta f}{f}\right|_{\text{fixed}\,\eta,v} \tag{27.19}$$

Substituting this expression into the earlier derived energy-efficiency criterion (Equation (27.12)), we notice that the term $\Delta f / f$ cancels out, since in both expressions it has the same meaning of a partial derivative with respect to architectural complexity ξ. Then, dropping $\Delta\xi$ in the denominators of all terms we arrive at the form of the energy-efficiency criterion that does not require estimating the increment in frequency:

$$\eta_{ag} \frac{\Delta I}{I} - (\eta_{ag} + 1)\frac{\Delta N}{N} > \left.\frac{\Delta E}{E}\right|_{no\,retune} + \sum_i \eta_i w_i \left.\frac{\Delta D_i}{D}\right|_{no\,retune} \tag{27.20}$$

where $\left.\frac{\Delta E}{E}\right|_{no\,retune}$ is the total increase in average energy dissipated per instruction, assuming no retuning, $\left.\frac{\Delta E}{E}\right|_{no\,retune} = \sum \left.\frac{\Delta E_i}{E}\right|_{no\,retune}$, summation being done over all stages in the pipeline, affected by the architectural modification.

Equation (27.20) is a more convenient form of the energy-efficiency criterion than Equation (27.12). According to Equation (27.20), to evaluate the energy-efficiency of an architectural feature, the architects must supply the relative gain (or loss) in the architectural performance $\frac{\Delta I}{I}$ and relative change in the dynamic instruction count $\frac{\Delta N}{N}$ that result from this feature. These estimates can be obtained by running an architectural simulator, or timer, like Turandot [11,12]. The second term, ΔN is nonzero if changes to the instruction set architecture (ISA) are considered, or compiler optimizations are analyzed for energy efficiency. It may also be nonzero if microarchitectural changes are considered in a speculative issue processor that impact the average number of instructions executed from mispredicted paths.

The architect needs to consult circuit designers to estimate the impact of the architectural feature under consideration on the average energy dissipated per instruction and the critical path delay through every stage of the pipeline affected by this architectural feature. A significant advantage of the derived formula is that in estimating the relative changes in energy and critical path delays the circuit designer does not need to worry about retuning the circuits to recover the frequency, or reducing the positive timing slack to save power in logic on paths that are no longer critical. Then, the relative increments in critical path delays are summed, multiplied by the appropriate energy weights and hardware intensities. The higher the energy weight w_i and the hardware intensity η_i of a part of the pipeline i affected by the architectural feature the higher the weight of the increase in the critical path delay through this part of the pipeline.

The energy weights w_i in Equation (27.20) are typically available as part of power budgeting at the early stages of the definition of the processor pipeline. The only additional data that is needed to use the energy-efficiency criterion are hardware intensities η_i in all blocks of the processor. Those quantities can be measured by static tuning tools, like the EinsTuner [6], based on the simulations of previous designs, or set as targets at early planning of the microarchitecture, similar to the way the power targets are budgeted.

Then Equation (27.20) is evaluated. If the inequality holds, then the architectural feature under evaluation is energy-efficient, that is, after adopting it, the processor will deliver higher net performance at the same power budget, after appropriate retuning and, possibly, adjustment in the power supply voltage are done to meet the power budget.

Figure 27.5 gives a graphical interpretation of the architectural energy-efficiency criterion (Equation (27.20)). Modifying the architecture from an alternative A to B is evaluated for energy efficiency

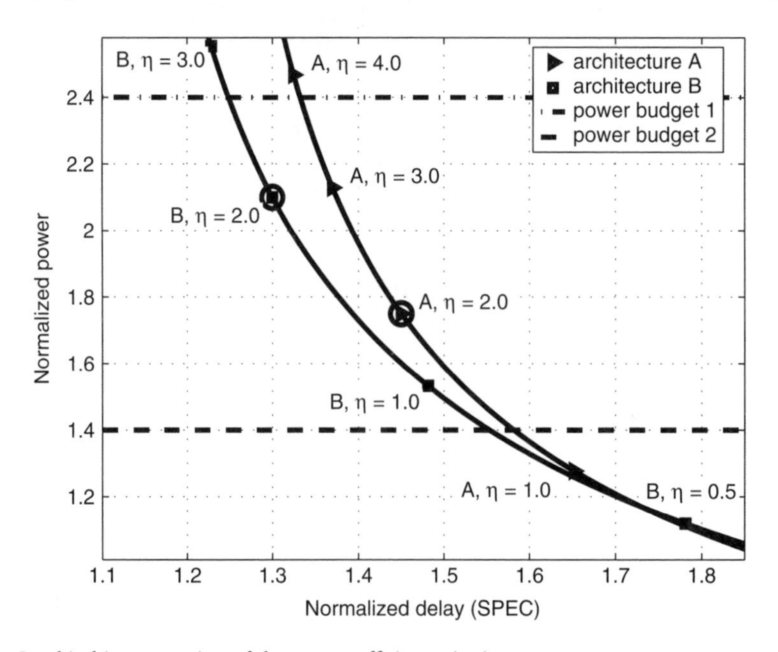

FIGURE 27.5 Graphical interpretation of the energy-efficient criterion.

using Equation (27.20). The points corresponding to the implementations of both architectural alternatives with the same hardware intensity $\eta = 2$ are marked with circles, and the curves passing through these circles show implementations of architectural alternatives *A* and *B* with different hardware intensities, assuming that the power supply is adjusted accordingly, to keep the optimum balance between the hardware intensity and power supply (Equation (27.6)). In other words, these curves show where the corresponding design points will move in the energy-delay space if the same architectures are implemented with more or less aggressive circuits.

If the inequality in Equation (27.20) evaluates as true for the architectural change from alternative *A* to *B*, then in the neighborhood of these points, the circuit energy-delay curve passing through point *B* is below (or left of) the circuit energy-delay curve passing through point *A*, as shown in Figure 27.5. This means that for any power budget, within a certain range of the initial points, implementations of architecture *B*, deliver higher performance than implementations of architecture *A*, for the same power budget.

Notice that the curves may intersect, as shown in Figure 27.5, which means that architectural alternative *B* is more energy efficient than *A* only within some range of the initial design points. This demonstrates the fact that an architecture optimized for a certain range in the power-performance space may not perform well outside of this range. For example, a high-performance core, scaled down to operate in the lower performance space may not be competitive with a core specifically optimized for the low-power, low-performance applications. Another conclusion from this analysis is that an accurate estimate of the available power budget for a core is essential for developing an energy-efficient architecture because only by knowing the power budget can we estimate the maximum value for architectural complexity, hardware intensity, and power supply that can be used in the core.

Figure 27.3 plots a possible outcome of overestimating the power budget available for a processor core at the microarchitecture definition stage, and choosing an overly high value for the architectural complexity. Suppose, at early design stages, the power budget is estimated to be at 2.2 and the microarchitecture of the processor core is optimized for the architectural complexity of $\xi = 20$. Suppose that at a circuit phase of the design it is discovered that the actual power budget is only 1.2. Since changing the architecture at this point would result in missing the product release schedule, the only way to bring the processor power under the budget is to redesign all circuits for a lower hardware intensity and reduce the power supply, or just reduce the power supply, if it is too late for redesigning the circuits. This will send the design point down the dashed curve passing through the point $\xi = 20$ in Figure 27.3, and leading to an almost 15% loss in performance compared to the design originally optimized for the power budget of 1.2 with the architectural complexity of $\xi = 2.0$. Thus, such late changes in the design may lead to a significant performance degradation, and scaling down the power supply to bring an overpowered processor core to the power budget may have a very high performance cost. This demonstrates the importance of accurately estimating the available power budget at early design stages, and disproves the notion, common in architectural community, that relative power estimates are always sufficient when proposing new architectural features.

27.5 Other Power Performance Metrics

Until energy efficiency criterion (Equation (27.20)) was introduced, the most popular power-performance metric used in the architectural community has been [2–5,7,8,13,15,18]

$$\frac{MIPS^{\gamma}}{Watt} \tag{27.21}$$

with the value of parameter γ ranging from $\gamma = 0$ to $\gamma = 3$, depending on the class of the microprocessor. As discussed in Zyuban [19], the prior art metric (Equation (27.21)) is a special case of the integral form of the derived metric (Equation (27.12)), with γ set to $\gamma = \eta_{ag} + 1$.

Another recent work proposed the following metric for evaluating architectural features [16]:

$$3\frac{\Delta IPC}{IPC} > \frac{\Delta Power}{Power} \tag{27.22}$$

Notice that Equation (27.22) is also a special case of Equation (27.12), with $\eta_{ag} = 2$, $\Delta f = 0$ and $\Delta N = 0$, because in clock-gated designs $Power \sim E \times IPC$, and $\frac{\Delta Power}{Power} = \frac{\Delta E}{E} + \frac{\Delta IPC}{IPC}$.

The main advantage of the derived criterion (Equation (27.20)) is that, in addition to being formally derived and being more general than the preceding metrics (Equation (27.21) and Equation (27.22)), all its terms have a clear meaning and an unambiguous method for estimating them as "naive" increments in energies and delays. On the other hand, the metrics in Equation (27.21) and Equation (27.22), though correct, may be confusing to use by an architect, because they hide important assumptions about the method for estimating increments in million instructions per second (MIPS) and Watts. In particular, estimating the term $\Delta Power$ may be ambiguous because it requires knowing both Δf and ΔE, which are interrelated and depend on the assumptions about the allowed change in the clock frequency and retuning the pipeline after modifying the architecture. When using these metrics, some architects assume that circuit designers will do whatever is needed to recover the extra delay due to an introduced architectural feature, and set $\Delta f = 0$, neglecting the increase in energy due to redesigning and retuning the circuits. Others calculate the extra delay introduced due to an added architectural feature and set $\frac{\Delta f}{f} = -\frac{\Delta D}{D}$, assuming that nothing can be done at the circuit level to recover the frequency, and neglecting that circuits in stages not affected by the change will have a timing slack and could be tuned down to save power. In both cases, the conclusion of applying the metrics in Equation (27.21) and Equation (27.22) may be incorrect. The next section presents a typical example of incorrectly using the metrics in Equation (27.21) and Equation (27.22).

27.6 Example: Adding an Execution Bypass

As an example, we evaluate the energy-efficiency of implementing an execution bypass in the integer unit (IU) of a microprocessor with a target cycle time of 10FO4, and an aggregate hardware intensity target of $\eta_{ag} = 2$ (shown in Figure 27.6). This microarchitectural feature affects only the register file (RF)

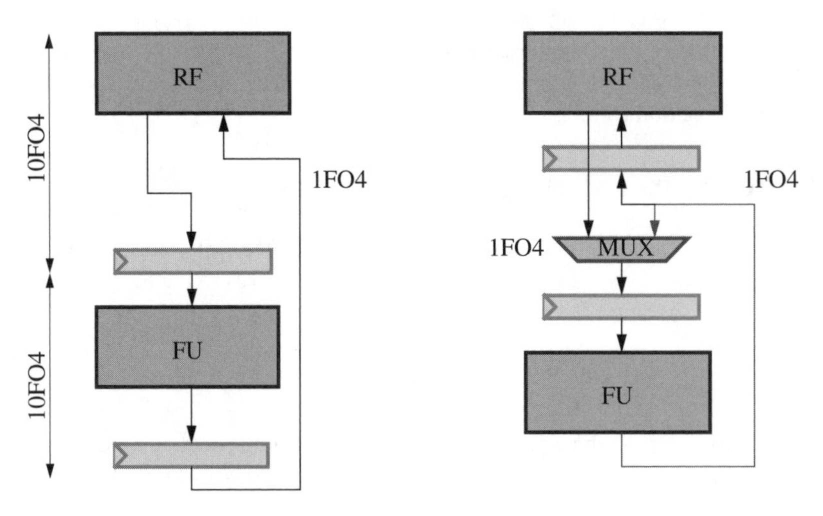

FIGURE 27.6

and execution (EX) stages of the processor. The delay insertion penalty of the bypass multiplexer in front of the latch is approximately 1FO4, and the delay of the bypass wire, including the rebuffering is approximately 1FO4. Then the relative "nonretuned" increments in the critical path delays through the register file and execution stages are $\left.\frac{\Delta D_{RF}}{D}\right|_{noretune} = 0.1$ and $\left.\frac{\Delta D_{EX}}{D}\right|_{noretune} = 0.2$.

Adding the bypass multiplexer and the bypass wires also introduces an energy overhead, dissipated whenever the IU is accessed. Based on simulation results, we estimate the relative energy overhead of the bypass wires and multiplexers as 5% of the average energy dissipated in the IU. Suppose the energy budget of the IU is 10% of the total energy dissipated by the microprocessor. Then, $\left.\frac{\Delta E}{E}\right|_{noretune} = 0.05 \cdot 0.1 = 0.005$.

Suppose, the aggregate hardware intensity of the microprocessor is $\eta_{ag} = 2.0$, but since pipelining the register file access and integer functional units has a high cost in IPC degradation, a higher value of hardware intensity is budgeted to them, $\eta_{RF} = \eta_{EX} = 3$. Also, suppose, $w_{RF} = 0.04$ and $w_{EX} = 0.06$. Then, using the criterion in Equation (27.20), we determine the relative increment in IPC that needs to be demonstrated to justify adding the execution bypass in the IU as $\frac{\Delta I}{I} > 2.7\%$.

Notice that if we used the $\frac{MIPS^3}{Watt}$ metric then, depending on the assumption about the change in frequency, the architect could arrive at different conclusions. If the view is taken that the circuit designers can do nothing to recover the frequency, then the metric in Equation (27.21) leads to the answer $\frac{\Delta I}{I} > 20\%$. On the other hand, if the view is taken that circuit designer will do whatever is needed to recover the frequency, and the architect does not need to worry about it, then the metric in Equation (27.21) leads to the answer $\frac{\Delta I}{I} > 0.25\%$. Similarly, the metric in Equation (27.22) leads to $\frac{\Delta I}{I} > 0.25\%$, assuming the frequency is unchanged. In both cases, the conclusions about the energy-efficiency of the IU execution bypass produced by straightforwardly applying the metrics in Equation (27.21) and Equation (27.22) are incorrect.

27.7 Conclusions

This chapter analyzed common approaches to trading power and performance in the design of processor cores for systems on chip, such as varying the power supply, hardware intensity, and architectural complexity. It was demonstrated that in order to develop an energy-efficient processor core, that is a core that delivers maximum performance at a strictly limited power budget, design decisions at all levels must be balanced in such a way that all forms of spending power have a similar marginal cost. A criterion for optimizing the core architecture was described which is useful for guiding the iterative architectural optimization process that leads to the optimal balance between the architectural complexity, hardware intensity and power supply. It was demonstrated that a single core may not be competitive in both high and low performance domains, and accurate estimates of the available power budget for a core are essential for developing an energy-efficient architecture, as opposed to making decisions based on relative power estimates only. It was also demonstrated that scaling down the power supply to bring an overpowered processor core to under the power budget might have a very high performance cost.

27.8 Acknowledgment

The authors thank Dr. Jaime Moreno and Dr. Kevin Warren for their management support.

References

[1] R. Brodersen, M. Horowitz, D. Markovic, B. Nikolic, and V. Stojanovic. Methods for true power minimization. In *Proc. ICCAD,* 35–42, November 2002.

[2] D. Brooks and P. Bose et al. Power-aware microarchecture: design and modeling challenges for next-generation microprocessors. *IEEE MICRO,* 20(6):26–44, November 2000.

[3] T. Burn and R. Broderson. Energy-efficient CMOS microprocessor design. *Proc. 28th Annu. Hawaii Int. Conf. on System Sciences,* 288–297, 1995.

[4] J. Burr and A. Peterson. Energy considerations in multchip module-based multiprocessors. *Proc. ICCD,* 593–600, 1991.

[5] A. Chandrakasan, S. Sheng, and R. Brodersen. Low-power CMOS digital design. *IEEE J. Solid-State Circuits,* 27(4):473–484.

[6] A. Conn et al. Gradient-based optimization of custom circuits using a static-timing formulation. *Proc. Design Automation Conf.,* 452–459, June 1999.

[7] R. Gonzales, B. Gordon, and M. Horowitz. Supply and threshold voltage scaling for low-power CMOS. *IEEE J. Solid-State Circuits,* 32(8):1210–1216, August 1997.

[8] R. Gonzales and M. Horowitz. Energy dissipation in general-purpose microprocessors. *IEEE J. Solid-State Circuits,* 31(9):1277–1283, September 1996.

[9] H.P. Hofstee. Power-constrained microprocessor design. In *Proc. IEEE on Computer Design,* 14–16, September 2002.

[10] J. Moreno et al. An innovative low-power, high-performance programmable signal processor for digital communications. *IBM J. Res. Dev.,* 47(2/3):299–327, 2003.

[11] M. Moudgill, P. Bose, and J.H. Moreno. Validation of Turnadot, a fast processor model for microarchitecture exploration. *Proc. IEEE Int. Performance, Computing, and Communications Conf. (IPCCC),* 451–457, February 1999.

[12] M. Moudgill, J.D. Wellman, and J.H. Moreno. Environment of PowerPC microarchitecture exploration. *IEEE Micro,* 19(3):9–14, May/June 1999.

[13] K. Nowka, P. Hofstee, and G. Carpenter. Accurate power efficiency metrics and their application to voltage scalable CMOS VLSI design, *IEEE Trans. on VLSI Syst.,* 2003.

[14] V.G. Oklobdija, B.R. Zeydel, D. Hoang, S. Mathew, and R. Krishnamurthy. Energy-delay estimation technique for high-performance microprocessor VLS1 adders. In *Proc. 16th IEEE Symp. on Computer Arithmetic,* 2003.

[15] P. Penzes and A. Martin. Energy-delay efficiency of VLSI computations. *Proc. Great Lakes Symp. on VLSI,* 104–107, April 2002.

[16] J. Rattner. Making the right-hand turn to power-efficient computing. Keynote speech, *35th Annu. Int. Symp. on Microarchitecture,* November 2002.

[17] V. Srinivasan et al. Optimizing pipelines for power and performance. *Proc. 35th Annu. Int. Symp. on Microarchitecture,* November 2002.

[18] M. Stan. Low-power CMOS with subvolt supply voltages. *IEEE Trans. on VLSI Syst.,* 9(2):394–400, April 2001.

[19] V. Zyuban. Unified architecture level energy-efficiency metric. *Proc. Great Lakes Symp. on VLSI,* 24–29, April 2002.

[20] V. Zyuban and P. Kogge. Optimization of high-performance superscaler architectures for energy efficiency. *IEEE Symp. on Low-Power Electron. and Design,* 84–89, August 2000.

[21] V. Zyuban and P. Strenski. Unified methodology for resolving power-performance trade-offs at the microarchitectural and circuit levels. *Proc. Int. Symp. on Low-Power Electron. and Design,* 166–171, August 2002.

28

Low-Power SoC with Power-Aware Operating Systems Generation

Sungjoo Yoo
Aimen Bouchhima
Wander Cesario
Ahmed A. Jerraya
TIMA Laboratory

Lovic Gauthier
FLEETS

28.1 Introduction

Recently, embedded software (SW) is becoming increasingly important in system-on-chip (SoC) design [1]. Two important characteristics of embedded SW in SoC are:

1. Interaction with physical components in hardware (via I/O and interrupt)
2. Concurrency required to handle the interaction with physical components as well as to better exploit the processor computing power

To satisfy the characteristics, embedded SW needs the operating system (OS) both for the hardware (HW) interaction and for multi-tasking.

Embedded SoCs have strict constraints in energy consumption. The main drains of energy in SoC are processor, memory, on-chip bus, to name a few. From the viewpoint of embedded SW, the energy consumption is decomposed into two parts: energy consumption by the application SW and by the operating system.

For the reduction in energy consumption by the application SW, there have been presented many techniques categorized into dynamic voltage scaling [2] and dynamic power management [3,4]. Recent analyses and experiments have demonstrated that the actual OS can consume a significant portion of energy [5,6].

To reduce the energy consumption of SoC, we need methods to reduce the energy consumption of the OS as well as methods to reduce that of the application SW. To reduce the energy consumption of the OS, we can consider two approaches. One is to change the usage style of OS (e.g., replacing polling

0-8493-1941-2/05/$0.00+$1.50

operations into interrupts). The other is to reduce the energy overhead of OS itself (e.g., minimizing the OS size).

In this chapter, we handle the problem of minimizing the energy overhead of OS itself. The sources of energy consumption of the OS are twofold. One is the energy consumption of memory portion that contains the OS code. The other is the energy consumption of processor where the OS executes. To reduce the energy consumption of both sources, a possible method will be to design a small OS specific to the given application SW. The small OS implements only the functionality necessary to the application SW. We call such an OS application-specific OS.

The application-specific OS reduces the energy consumption of the memory by reducing the OS code size. To understand the effects of small OS code, for instance, assume a network-on-chip processor that contains 100 processors [7] and that each processor needs an OS with a code size of 100 KB. In that case, only for the OS code, we need 10 MB of memory and the energy consumption in the memory occupied by the OS code can be significant. Because a small OS usually enables fast OS execution, it can also reduce the energy consumption of processor by reducing the number of processor clock cycles necessary to perform the same OS functions.

The main difficulty in designing application-specific OSs is the design time. For the design, we need to tailor the functionality of OS to that of application SW. For instance, if the application SW does not use semaphore, the OS needs to remove the semaphore functionality from its implementation. Because many different OS functionalities (e.g., scheduling, inter-process communication, I/O, and interrupt management) and different implementations of the same OS functionality exist (e.g., different implementations of interrupt management), if the tailoring process is performed manually, application-specific OS design can be prohibitively time-consuming. Such a long design time of application-specific OS is not acceptable in ever tightening time-to-market pressure. Thus, we need new methods to accelerate the design of application-specific OS. In this chapter, we present a novel method to automatically generate application-specific OSs. The presented method generates very small OSs comparable to the smallest hand-written commercial OSs.

This chapter is organized as follows. Section 28.2 presents a short review of energy consumption issues related to the OS. Section 28.3 introduces SoC architecture. Section 28.4 explains the presented method. Section 28.5 discusses the effectiveness of the presented method with experiments. Section 28.6 concludes this chapter.

28.2 Related Work

Although most of SoC applications require low-power OSs, applications requiring ultra low-power OS are sensor network [8] and network on chip [7,9]. In Hill [8], an OS called TinyOS is presented for the sensor network where each sensor node has significant OS usage for data acquisition and transmission. In network on chip where each of (up to) hundreds of processors may need its OS, the energy overhead of OS can become significant.

In Dick et al. [5], an analysis of energy consumption by the OS (uC/OS-II in this case) is presented. The analysis demonstrates that we can obtain a 27.5 to 42.8% reduction in total energy by changing RTOS usage (e.g., replacing application code using polling by interrupt code). In Tan et al. [6], a technique of transforming SW architecture is presented to reduce the energy consumption of OS (ARM Linux in this case). The transformation gives a 10.7 to 66.1% reduction in total energy consumption. In this method, for the reduction in energy consumption, the OS code does not change, but the application code changes; however, our method changes (i.e., minimize) the actual OS code to reduce the OS overhead. Thus, the method in Tan et al. [6] and ours complement each other.

The size of the smallest versions of full-featured commercial OS (e.g., micro-kernel + file system and memory management) ranges between 80 KB (VxWorks [10]) and 100 KB (QNX Realtime [11]). Several very small OSs (i.e., micro-kernel designs including TinyOS [8] [~ 3.7 KB], pOSEK [12] [~ 2 KB], Chorus OS [13] [~ 10 KB], Ariel [14] [~ 19 KB], etc.) have been presented. In terms of OS size, the presented method yields OS sizes, 1.6 to 7.7 KB, comparable to the handwritten micro-kernels for our examples.

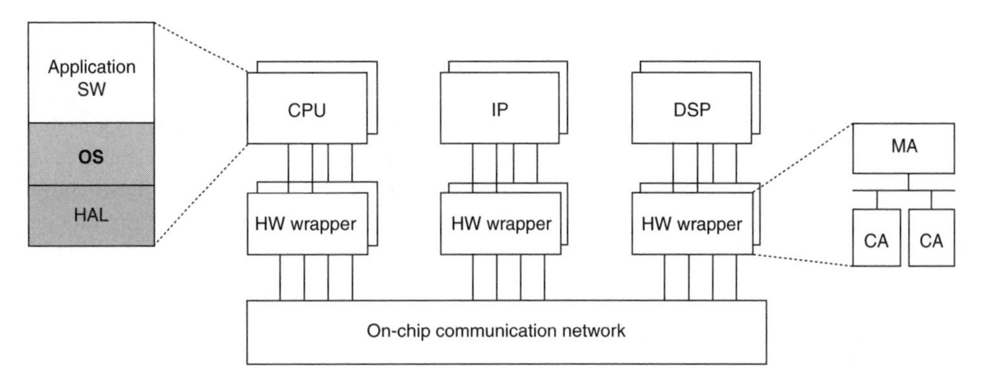

FIGURE 28.1 SoC architecture.

Compared with the existing micro-kernels, our contribution is to automatically generate small OSs without manual design.

28.3 Preliminary: SoC Architecture Generation

28.3.1 SoC Architecture

Figure 28.1 depicts a typical SoC architecture with heterogeneous processors (i.e., CPUs, DSPs, IPs, etc.) and the on-chip communication network (e.g., AMBA and packet/circuit switch network).

Embedded SW is generally organized as a stack of layers on top of the processor as depicted in Figure 28.1. The lowest SW layer called HW abstraction layer (HAL) provides the upper SW layer with an abstraction of underlying HW architecture. The OS provides the application SW with services such as task scheduling and synchronization, interrupt management, I/O, memory management, etc.

The processor is connected with on-chip communication network via an interface called HW wrapper. This chapter focuses on the OS generation in the SoC architecture. For further details of HW wrapper generation, refer to Lyonnard et al. [15] and Cesario et al. [16].

28.3.2 SoC Architecture Generation

The application-specific OS is generated as a part of SoC architecture generation. Figure 28.2 illustrates the SoC architecture generation flow called ROSES. From a high-level SoC specification called VADEL, the flow generates application-specific OS and HW wrappers.

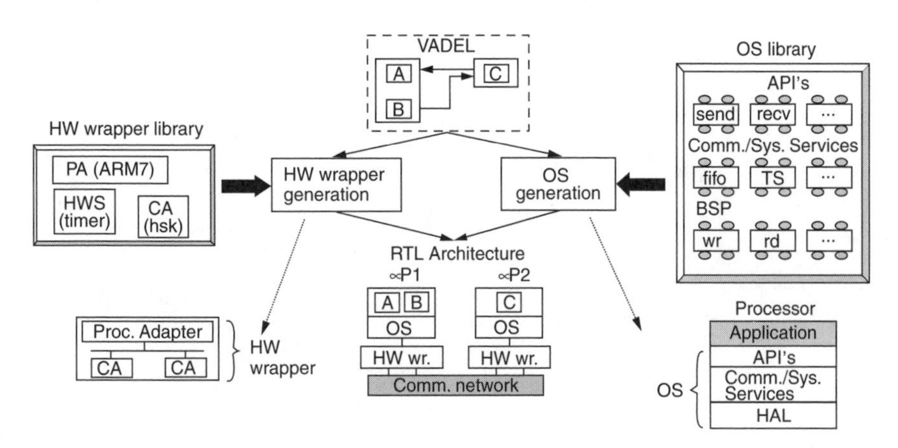

FIGURE 28.2 The flow of SoC architecture generation: ROSES.

In VADEL, the system is described with a network of hierarchical modules. Modules are connected with each other through its ports via communication channels. A module consists of behavioral parts and ports. Each module can be a leaf module or a hierarchical one composed of a network of module instances. We call a leaf module a *task*. For communication, ports provide module behavior with communication application programming interface (API) encapsulating communication details (i.e., communication protocol). In the real SoC implementation, the OS implements the API (illustrated in the OS library in Figure 28.2) in the form of service. Note that the module behavior (i.e., application SW code) communicates with the OS via the APIs. In fact, the API is a programming model used by the designer to write the application SW code. In our case, the API is extendable.

Architecture generation from the VADEL representation to SoC architecture consists of:

1. OS generation
2. HW wrapper generation

As illustrated in Figure 28.3, an OS consists of two types of services: communication/system services and HAL services. To generate application-specific OSs, only the services required by the application SW are selected from the OS library. Assembling HW components from the HW library, as illustrated in Figure 28.2, also generates the HW wrapper. In this paper, we focus on the OS generation flow. For the details of HW wrapper generation, refer to Lyonnard et al. [15] and Cesario et al. [16].

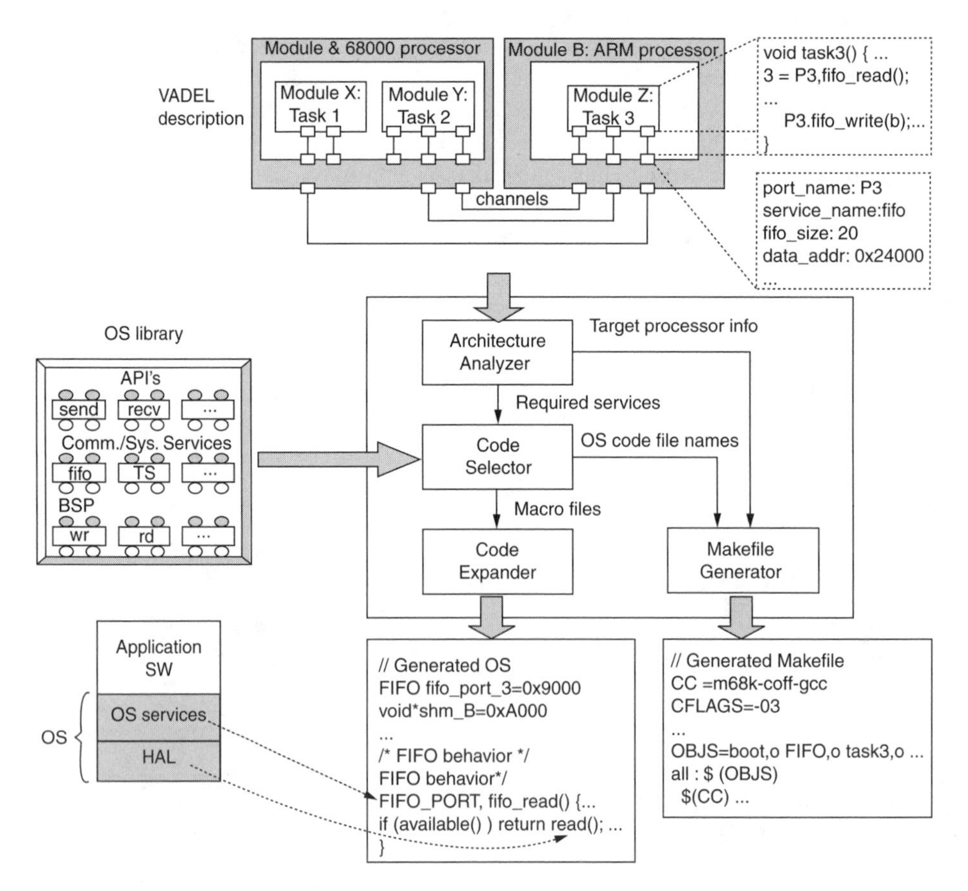

FIGURE 28.3 OS generation flow.

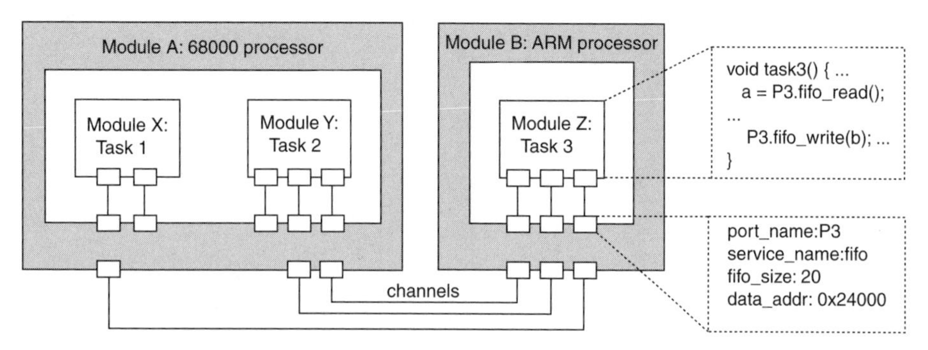

FIGURE 28.4 An example of VADEL description.

28.4 Automatic Generation of Application-Specific Operating Systems

Figure 28.3 depicts the flow of automatic OS generation. The input to the OS generation is the VADEL description that contains a network of hierarchical modules, communication between modules and parameters for OS generation.

As presented in the figure, the architecture analyzer takes the structural information and the parameters in the VADEL description. The code selector receives a list of services specific to the application and the architecture from the architecture analyzer and finds the full list of (original and deduced) services. The code expander generates the source code of OS. Then, the makefile generator gives makefiles. Thus, the outputs of the design flow are the source code of the generated OS and a makefile for each processor. To obtain the binary code to be downloaded onto the target processor memory, the designer runs a compilation of both generated OS and the application SW code using the generated makefile.

28.4.1 System Description Input

As the system description input, the flow takes a structural representation of communication in a hierarchical network of modules. Figure 28.4 is an example of hierarchical network of modules.

In the figure, two modules (tasks) X and Y are mapped on a 68,000 processor and the other module Z on an ARM7 processor. Each module (task or processor) has a set of ports (depicted as small rectangles in the figure). The port has parameters. Figure 28.4 is an example of port parameters (e.g., *port_name (P3), service_name (fifo),* and so forth). Each task has also parameters such as task priority.

Note that the OS is generated on a processor basis. Between modules, inter/intraprocessor communications are represented. For instance, there is an intraprocessor communication between modules X and Y (a line in the shaded region of the figure). In the figure, the three channels between module A and B exemplify interprocessor communication.

Figure 28.4 is an example of communication via the port in task 3. In the figure, the behavioral part of task 3 just calls a high-level communication function provided by port P3 (i.e., a port function), for example, *P3.fifo_read()* for communication via the port. In the VADEL description, we do not describe the behavior part of task, but list task source code files and the parameters of OS API that the task calls for.

28.4.2 OS Library

The OS library is the core of the OS generation tool. It provides small and flexible pieces of code that can be adapted and used to build the OS. The library contains two parts:

1. A code part containing the code
2. A description part

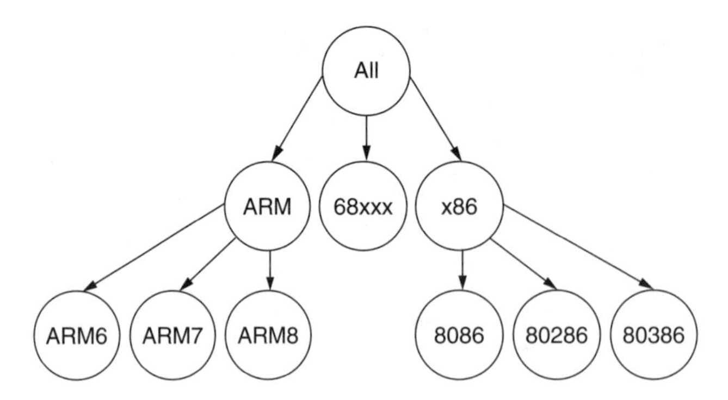

FIGURE 28.5 Processor dependency of OS code.

The code part is written in a macro language. The description part enables assembly of OS code pieces. Its main components are the OS element, the service, and implementations.

28.4.2.1 OS Element

An OS element represents a SW code piece of the generated OS. It provides services and may require other services. Depending on different target processors, an OS element can have several code pieces each of which is specific to a processor. We call such a processor-specific code piece an implementation. Note that each implementation code of an element can be described in a different language (for instance, C or assembly).

28.4.2.2 Service

A service can represent any kind of OS functionality. In our OS library, three types of services are used: communication, system, and HAL services. Communication services implement the API functions of communication between modules. For instance, the communication protocol FIFO is implemented as a communication service. Examples of system service are task scheduling service (e.g., function to schedule tasks in preemptive/nonpreemptive priority-based/round-robin schedulers) and timing service. Communication/system services use device driver services to implement their functionality. *HAL* services are specific to the processors where the application SW is running.

28.4.2.3 Implementations

Implementations of an OS element are organized in a tree relationship. An implementation can also be compliant with specific peripherals (mainly for driver descriptions).

Figure 28.5 is an example of such a tree relationship. In the figure, each oval represents a part of implementation of an OS element. For instance, an OS element, boot for ARM7 processor, can consist of:

1. A processor-independent code piece in C (the top oval in the figure)
2. ARM processor family-specific code piece in assembly (the oval denoted with ARM)
3. ARM7 processor-specific code piece in assembly

28.4.3 OS Code Generation

28.4.3.1 Architecture Analyzer

The architecture analyzer finds the following information from the system description input:

1. Application-specific services and their detailed parameters
2. Module-specific parameters
3. Modules interconnection topology

Application-specific OS services are extracted from the parameters of modules, channels, and ports in the system description input. For instance, if a port has a parameter for FIFO implementation, FIFO service is selected to be included into the generated OS. Further detailed parameters of required services are also found in the VADEL description. For instance, the address range of FIFO communication and the interrupt level of interrupt-driven port can be found. A list of required service names is sent to the code selector.

Module-specific parameters (e.g., task priority, processor speed, and processor type) are also found from module parameters. The type of processor is sent to the makefile generator to choose the right compiler and to the code expander to produce the OS code to the processor. The interconnection topology (e.g., point to point and multipoint) is used to deduce internal data structure used for communication and synchronization.

28.4.3.2 Code Selector

The code selector takes as input the list of required service names from the architecture analyzer. It uses the OS library to check service dependencies and finds all the OS elements that have dependency relation with the required services and that are compliant with the target architecture. The dependency relationship can be transitive. For instance, if a FIFO communication service used in the application code requires an interrupt handling service, an OS element providing the interrupt handling service should be also chosen for inclusion in the OS to be generated.

An OS element is compliant with an architecture if one of its implementations is compliant with the architecture and if all the services it requires can be provided by other compliant OS element. Because an OS element may require some services, the previous algorithm is repeated recursively to perform a kind of transitive closure. Sometimes, several OS elements compliant with the architecture may provide the same required service. In such a case, it is up to the user to choose the good one. After the OS element selection is done, the code selector sends the list of the code file names to the makefile generator and the macro file names to the code expander.

28.4.3.3 Code Expander

The code expander takes as input a list of macro file names from the code selector and parameters from the architecture analyzer. It generates the final OS code by:

1. Associating the right parameters to each OS element
2. Expanding the macro codes of OS elements to source codes by calling an external macro processing program

Figure 28.6 is an example of code expansion. In Figure 28.6(a), a macro-code section is listed for a part of the semaphore synchronization mechanism. First, the code expander fixes the input parameters: *SYNC*, which is an array of identifiers for the synchronization units and *Sem_P*, which indicates that the service function is required. Then, it performs the macro expansion. For instance, in Figure 28.6(b), the data structure for the semaphores is generated with its size (*SIZE* = 2) (i.e., the number of semaphores, its queues, and its states [zero]). Note that the synchronization queue uses another service (in fact, also to be expanded) *Soft_Wait* that provides wait and signal mechanism.

28.4.3.4 Makefile Generator

The makefile generator takes as input:

1. Processor type information from the architecture analyzer
2. A list of source codes of the OS (in C and assembly) from the code selector
3. A list of the application SW codes described in the VADEL description

It determines the right compiler and linker and generates a makefile (for each processor) that includes the two code lists of OS and application SW.

28.4.4 Application to Existing OSs

An existing OS can be integrated into the proposed flow of automatic generation (to be specific, in this case, automatic configuration) of application-specific OS. To explain the integration, we assume that the

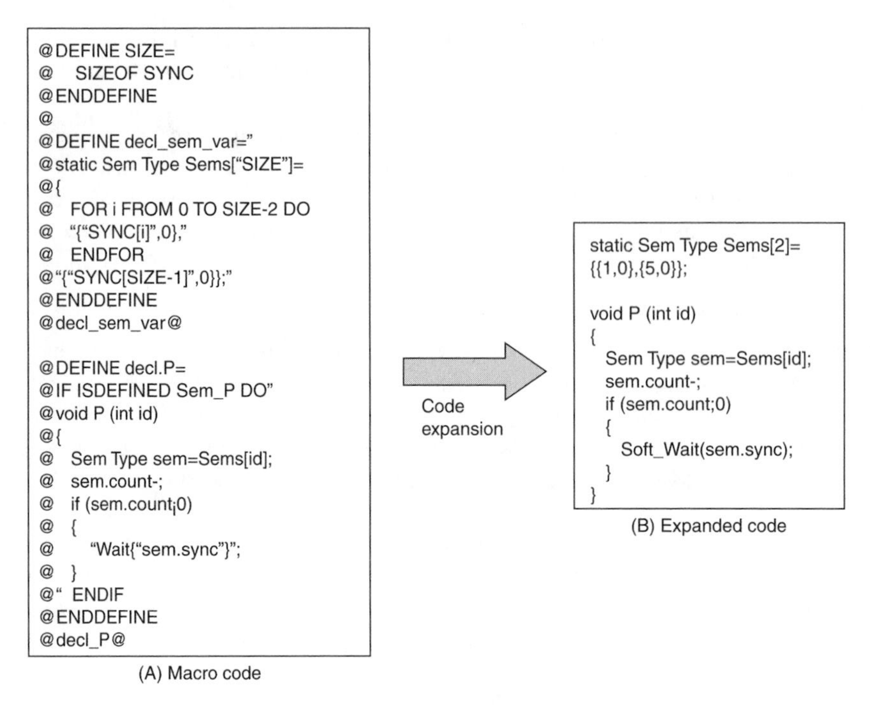

FIGURE 28.6 Code generation example.

existing OS supports OS configuration by "#" statements (i.e., configuration by defining required macros) without modifying the OS source code because most commercial OSs allow such a configuration. The integration can be done as follows:

1. Information of available OS services (e.g., API, macros defined for services) and dependency relationship between services in the existing OS are taken into the OS library.
2. To the OS generation flow in Figure 28.3, the designer gives the same system description input in VADEL.
3. The architecture analyzer performs the same operation (i.e., extracting required information such as services and target processor information) as described in this section.
4. The code selector finds all the required (derived) services from the OS library as explained in this section. Then, it selects, from the OS library, macro definitions corresponding to the required services instead of selecting macro code files as explained in this section. Note that in the case of automatic configuration of existing OS, the code expander does not generate the OS source code because the existing OS source code is not modified.
5. The makefile generator outputs a makefile with the selected macro definitions received from the code selector.

Note that compared with the original flow of automatic OS generation proposed in this chapter, in the case of integrating the existing OS into the flow, no change occurs in automatic execution of service extraction (by the architecture analyzer) and makefile generation. The code quality (i.e., size and execution time) of the automatically configured OS depends on the configuration granularity of OS services in the existing OS.

28.5 Experiments

We applied the proposed method to two system examples: a token-ring system and a very high data-rate digital subscriber line (VDSL) framer system.

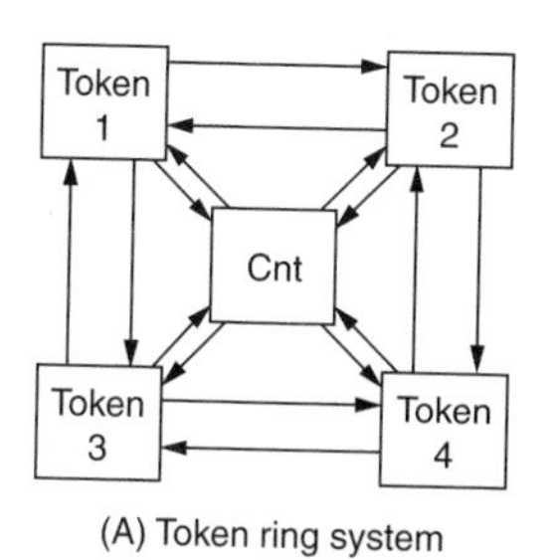

(A) Token ring system

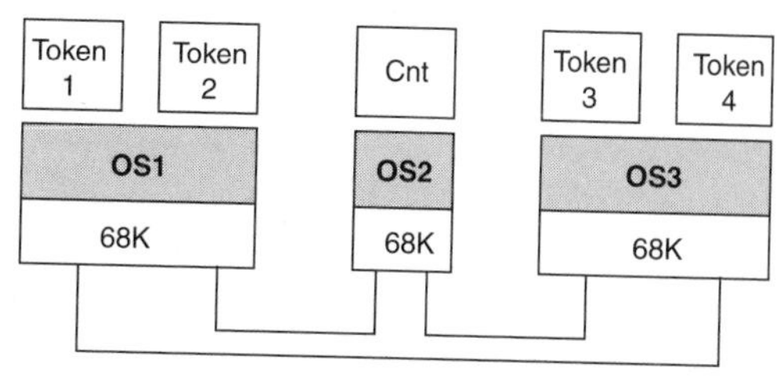

(B) Three generated OSs

FIGURE 28.7 A token ring example.

28.5.1 Token-Ring Example

Token-ring system (1245 lines in SystemC) consists of four tasks (called *token*) that exchange tokens with each other and one counter task (called *Cnt*) that counts the number of tokens exchanged.

Figure 28.7(a) depicts the interconnection of tasks in the example. As presented in the figure, four *token* tasks make a bidirectional ring connection with each other. *Cnt* task is connected to all *token* tasks.

In our experiment, we implement the system example on a multi-processor target architecture with three 68 K processors depicted in Figure 28.7(b). In the figure, four *token* tasks are mapped to two processors (two tasks on each processor) and *Cnt* task is mapped to the other processor. In the system description input, we assigned equal priority to all the tasks.

For the communication channel between modules, we specified one word communication with non-blocking write/blocking read. In the example, the size of transferred data (i.e., counter value and token) is one word.

First, the system description input in VADEL is read into the architecture analyzer. Then, the code selector selects the following OS services for the two OSs (OS1 and OS3 in the figure) of the two processors each of which runs two *token* tasks:

1. Round-robin scheduler service with preemption service because tasks have the same priority
2. A timer service because the round-robin scheduler with preemption is used
3. Nonblocking write (called *exoutd*) and blocking read services (called *exinb*)

The code expander generates the OS source code that handles two tasks of equal priority and two communication service functions (*exoutd* and *exinb*). For the processor where only *Cnt* task is mapped, the same communication services are selected for OS2; however, no scheduler and timer services are selected because only one task exists on the processor.

We obtained three binary executables for three processors after running compilation with the generated OS codes and makefiles. We validated the system implementation in cosimulation with three instruction set simulators of 68,000 processor and a VHDL simulator. As the result, in this experiment, the generated OSs give very small code sizes: 797 lines (90% in C and 10% in assembly), compiled and linked to 1.86 KB for each of two processors with two Token tasks and 1.62 KB for the processor with one *Cnt* task. In terms of performance, it gives 83 instruction cycle latency in the channel read operation from interrupt trigger to the end of single-word data access.

28.5.2 VDSL Example

The design presented in this section was taken from the implementation of a VDSL modem using discrete components [17]. The block diagram for this prototype implementation is illustrated in Figure 28.8. The subset we will use in the rest of this paper is shaded in the figure. It is composed of two ARM7 processors and a part of the datapath, *TX_Framer*, described at the RT-level.

On the two processors, parallel tasks are running. The control over the three modules (two processors and *Tx_Framer*) is fully distributed. All three modules act as masters when interacting with their envi-

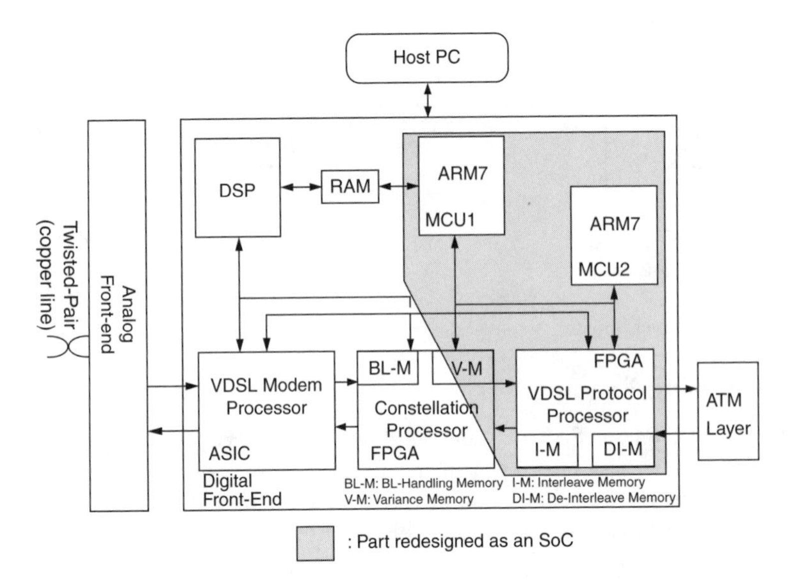

FIGURE 28.8 Entire VDSL system.

ronment. Additionally, the application includes some multipoint communication channels (shared memory services) requiring sophisticated OS services.

Figure 28.9 illustrates the VADEL description of VDSL example. Modules **VM1** and **VM2** correspond to the ARM7s on Figure 28.8 and module **VM3** represents *TX_Framer* block.

Given the VADEL description, our flow in Figure 28.2 generates two OSs for two ARM7 processors and three HW wrappers (i.e., two for two ARM7 processors and one for the Tx Framer). The architecture generation takes only a few minutes on a Linux PC 500 MHz. Figure 28.10 depicts the RTL architecture obtained after OS and HW wrapper generation. For further details of HW wrapper generation, refer to Cesario et al. [16].

Each generated OS is customized to the set of tasks executed on each of the processors. For example, SW tasks running on **VM1** access the OS using an API composed of two functions: pipe for communication with **VM2**, and signal to modify the task scheduling on runtime. The OS contains a round-robin

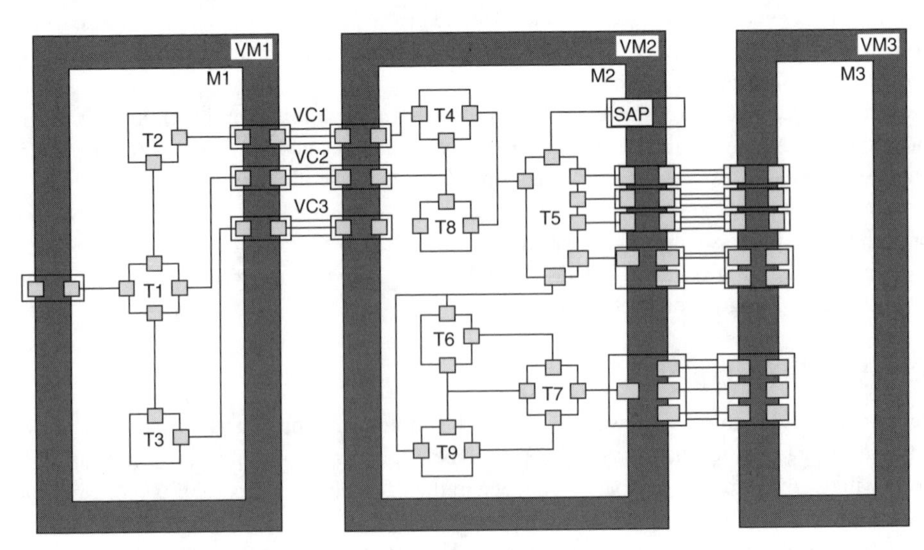

FIGURE 28.9 VADEL description of VDSL example.

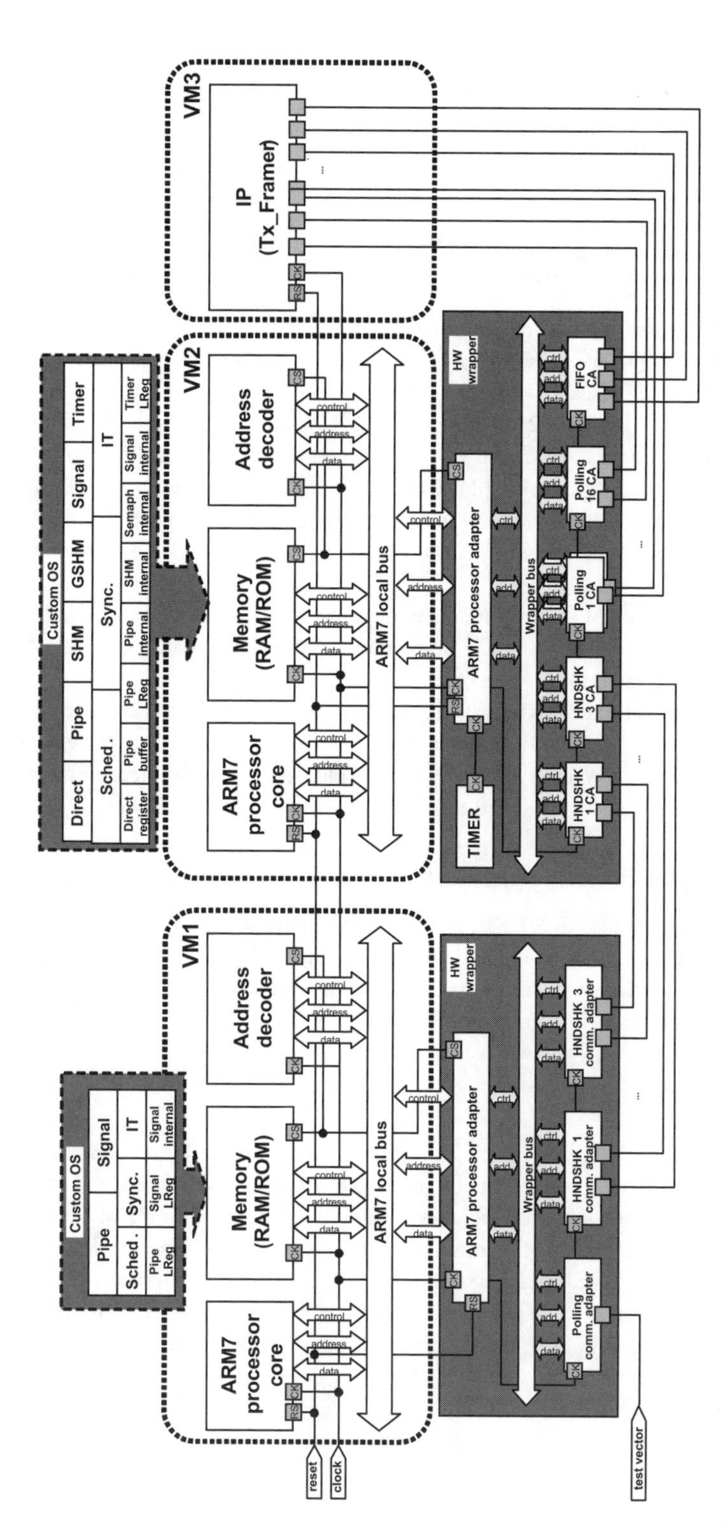

FIGURE 28.10 Generated SoC architecture.

TABLE 28.1 Results for OS Generation

OS Results	No. of Lines C	No. of Lines Assembly	Code Size (bytes)	Data Size (bytes)
VM1	968	281	3,829	500
VM2	1,872	281	6,684	1,020
Context switch (cycles)				36
Latency for interrupt treatment (cycles)				59 (OS) + 28 (ARM7)
System call latency (cycles)				50
Resume of task execution (cycles)				26

scheduler (*Sched*) and resource management services (*Sync, IT*). The HAL contains low-level code to access the HW (e.g., *Pipe LReg* to access *HNDSHK* CA in the HW wrapper as presented in Figure 28.10) and some low-level kernel routines.

The HW wrapper for processor **VM2** includes a TIMER block because task *T5* (see Figure 28.9) must wait 10 ms before starting its execution. When the timer expires, the TIMER block generates an HW interrupt. The task can configure this block using Timer API provided by the OS. The OS for **VM2** provides a more complex API. *Direct* API is used to write/read to/from the configuration/status registers inside *TX_Framer* block; *SHM* and *GSHM* are used to manage shared-memory communication between tasks.

Application code and generated OS (in C and assembly) are compiled and linked together to execute on each ARM7 processor. The HW wrapper can be synthesized using RTL synthesis. Table 28.1 presents the results regarding the generated OSs.

The OS sizes (4.3KB for VM1 and 7.7KB for VM2) presented in Table 28.1 are comparable to the sizes of commercial micro-kernels (2 to 10 KB). In terms of OS runtime, as listed in the table, context-switch takes 36 cycles, latency for the HW interrupt is 59 cycles (plus 4 to 28 cycles needed by the ARM7 to react), latency for system calls is 50 cycles, and task reactivation takes 26 cycles.

28.5.3 Gain Compared with Conventional OSs

In terms of OS size, our results (1.6 to 7.7 KB) presented in this section are comparable to the smallest micro-kernels (2 to 10 KB as explained in Section 28.2). Compared with the conventional small OSs, our contribution is to automatically generate small OSs thereby enabling to explore the OS design space, especially to obtain low-power OS implementations.

28.6 Conclusion

This chapter presented the problem of reducing the energy consumption by the OS. To minimize the energy overhead causes by the OS, we presented a method to design application-specific OSs with sizes that are minimal enough to provide the OS services required by the application SW. Because the application-specific OS reduces the overhead of memory usage and processor runtime, it can reduce the energy consumption caused by the OS itself. To resolve the important problem of application-specific OS design (i.e., slow design cycle) our method generates automatically application-specific OSs. The experiments demonstrate that the automatically generated OSs yield OS sizes (1.6 to 7.7 KB) comparable to those of commercial handwritten micro-kernels (2 to 10 KB). Fast design of small-size OSs will enable the exploration of the low-power implementations of OSs.

References

[1] Semiconductor Industry Association, *ITRS 2001 Edition,* available at http://public.itrs.net/Files/2001ITRS/Home.htm.

[2] P. Pillai and K. Shin, Real-time dynamic voltage scaling for low-power embedded operating systems, *Symp. on Operating Syst. Principles,* Oct. 2001.

[3] C. Pereira, V. Raghunathan, S. Gupta, R. Gupta, and M. Srivastava. An SW architecture for building power-aware real-time operating systems, Technical Report No. 02-07, University of California–Irvine, March 2002.

[4] Y.H. Lu, et al. Operating system directed power reduction, *Int. Symp. on Low-Power Electron. and Design,* 2000.

[5] R.P. Dick, G. Lakshiminarayana, A. Raghunathan, and N.K. Jha, Power analysis of embedded operating systems, *Design Automation Conf.,* June 2000.

[6] T.K. Tan, A. Raghunathan, and N.K. Jha, SW architectural transformations: a new approach to low-power embedded SW, *Design, Automation, and Test in Europe (DATE),* March 2003.

[7] L. Benini and G. De Micheli, Networks on chips: a new SoC paradigm, *IEEE Comput.,* 35(1), pp. 70–80, 2002.

[8] J. Hill. An SW architecture supporting networked sensors. Master's thesis, EECS, University of California–Berkeley, Dec. 2000.

[9] E. Rijpkema, K.G.W. Goossens, A. Radulescu, J. Dielissen, J. van Meerbergen, P. Wielage, and E. Waterlander, Trade-offs in the design of a router with both guaranteed and best-effort services for networks on chip, *Design Automation and Test Conf. in Europe,* March 2003.

[10] VxWorks, available at http://www.windriver.com/products/vxworks5/index.html.

[11] D. Hildebrand, An architectural overview of QNX, available at http://www.qnx.com/literature/whitepapers/archoverview.html.

[12] pOSEK, A super-small, scalable real-time operating system for high-volume, deeply embedded applications, available at http://www.isi.com/products/posek/index.htm.

[13] Chorus operating system open source, available at http://www.experimentalstuff.com/Technologies/ChorusOS/index.html.

[14] Microware ariel technical overview, http://www.microware.com/ProductsServices/Technologies/ariel_technology_brief.html.

[15] D. Lyonnard, S. Yoo, A. Baghdadi, and A.A. Jerraya, Automatic generation of application-specific architectures for heterogeneous multiprocessor system-on-chip, *Design Automation Conf.,* 2001.

[16] W.O. Cesario, D. Lyonnard, G. Nicolescu, Y. Paviot, S. Yoo, L. Gauthier, M. Diaz-Nava, and A.A. Jerraya, Multiprocessor SoC platforms: a component-based design approach, *IEEE Design and Test of Comput.,* Vol. 19, No. 6, Nov.-Dec., 2002.

[17] M. Diaz-Nava and G.S. Okvist, The zipper prototype: a complete and flexible VDSL multi-carrier solution, *J. ST Microelectronics,* Sept. 2001.

29

Low-Power Data Storage and Communication for SoC

Miguel Miranda
Erik Brockmeyer
Tycho van Meeuwen
Cedric Ghez
IMEC

Francky Catthoor
IMEC and Katholiek University

29.1 Introduction

Despite the recent architectural advances for multimedia aiming at improving computational efficiency (e.g., sub-word parallel data level processing, reconfigurable computing, and networks on chip), the dominance in data storage and transfer of these systems still remains as one of the main bottlenecks for power and speed efficient implementations. The reason is the ever-increasing gap in speed and energy between the memory and the data processing subsystems.

To cope with such a gap, the addition of more layers to the memory hierarchy, from where data can be efficiently accessed from smaller, faster, and more energy efficient memories becomes mandatory. This is true both for systems on chip (SoC) platforms based on random access memories as well as for cache memories; however, the potential efficiency offered by these multi-layer memory organizations becomes attainable on condition that the storage and transfer of data between the different layers is done in an optimal manner.

Memory hierarchy layers can contain software controlled scratch-pad memories or caches. To guarantee an efficient transfer of data along the memory hierarchy, often requires that smaller copies of the data be made from the larger data arrays, which can be stored in the smaller layers [1,2]. Those copies must be selected such that they minimize the overall transfer cost. In this context, any transfer of data from a higher layer to the current one is considered an overhead for the current layer.

0-8493-1941-2/05/$0.00+$1.50

This happens most efficiently under full software control (e.g., a number of static random access memories (SRAM) used as a scratch-pad memories) because a global view on the transfer can be obtained at design time. In this case, copy operations should be explicitly present in the application code. This is mostly possible for design time analyzable applications that are characterized via Pareto curves collecting all optimal energy trade-offs for the different execution times (see Section 29.3). The decision of the selected Pareto point can also be made at design time for purely static applications.

Many real-life applications are dynamic in nature, however, and they cannot be completely characterized at design time. Traditionally, to cope with this dynamism, HW-controlled caches are used instead of SW-controlled scratch-pad memories. In this case, the hardware cache controller will make the copies of signals at the moment they are accessed (and the copy is not present yet in the cache); however, this is inefficient because data present in the cache and required in the near future can be (wrongly) evicted to accommodate new fetched data. This evicted data, when needed again by the processor, will have to be brought to the cache for a second time, thus leading to transfer and power overhead. To minimize such overhead, architects tend to use bigger caches with hardware controllers implementing complex mapping policies; however, this is not efficient for power given the extra overhead in every single access even when these are cache hits. According to Wilton and Jouppi [3], a 1-KByte (KB), four-way associative cache is 4 to 5 times more inefficient in terms of energy per access than a scratch-pad memory of the same size and 2 to 3 times more than a (one-way associative) direct mapped cache (DM-cache); the latter is only 60% less efficient than a scratch-pad memory of the same size. This overhead is due to accessing the tag array of the cache which size (thus energy overhead) considerably increases with the associativity factor.

For dynamic applications, a much better approach is to characterize the different run-time conditions of the application at design time and to generate for each a Pareto curve as demonstrated in Marchal et al. [4]. This is quite design time-consuming, but it can be sped up by appropriate tools (see, e.g., Section 29.3 and Marchal et al. [4]). In this case, the selection of the Pareto operation point will be made at run-time, which can be done very efficiently in the middleware layer and not at design-time, as is the fully static case, although this still allows employing SRAMs. When the application becomes even more data-dependent, however, the number of Pareto curves to be stored would become too high. In that case, the HW controller of a cache can still help to obtain better run-time results, starting from the code with copy candidates which are "locked" in the cache whenever appropriate (Pareto) design time analysis is available. The cache will then implement the right copy of data according to the selected Pareto point, but the HW controller takes over whenever no appropriate locking choice is available.

In this approach, it is crucial that we use energy-efficient cache memories with simple controller mechanisms and low tag overhead that are still SW-controlled (such as the use of address-lock mechanisms, by-passable transfers and SW-controlled write-backs). An example of such cache memories is a lockable DM-cache. DM-caches are very attractive for low energy operation but if not well steered from the SW, they can lead to an excess in misses, thus in power. Fortunately, an effective way to reduce this overhead in misses can be achieved by a careful placement of the data in the main memory. Nevertheless, this is only useful on condition that the additional overhead involved in this placement is kept limited. Therefore, if good design time analysis and decisions are available, the code can be written such that the controller is forced to make the right decision [5]. For this purpose, we have developed an array interleaving data-layout approach that involves a relatively small amount of overhead in required storage size. This approach is very efficient in reducing data-layout related conflict misses, which are known to be the major contribution in DM-caches. This scheme interleaves signals in the main memory to assign different signals to own locations in the cache. This method can be complemented with an array padding technique for fine-tuning the placement of signals in the cache.

29.2 Related Work

Optimizing the memory hierarchy for performance using software controlled memory layers has been a well explored topic [6–9]; however, several recent articles also address the energy-related issues [10–13], and Benini and Micheli [10] and Panda et al. [12] have published good surveys.

In Grun et al. [14], a clustering of the data sets into different memory types is proposed. These different clusters have certain memory type preferences and are assigned accordingly; however, the mapping is suboptimal, especially for the regular accesses, because it is based on average characteristics and does not allow accurate predictions. Only then, performance is measured by simulation. In addition, Panda et al. [15] make a distinction between caches and scratch-pad memories; however, no real layer assignment is made. For that purpose, the technique presented in Steinke et al. [16] assigns data (and instructions) to the scratch-pad memory; however, no consideration is made to benefit from data reuse [1,2] and memory inplace optimizations [19].

The closest work to our approach is presented in Kandemir and Choudhary [13]. It analyzes and exploits the temporal locality by inserting local copies. Their layer assignment builds a separate hierarchy per loop nests and then combines them into a single hierarchy; however, a global view of the assignment and lifetime of the arrays and copies is required for real life applications having imperfect nested loops. Moreover, no overhead estimation is made, which makes it impossible to trade-off copy sizes (see Section 29.3) vs. array sizes in a certain layer. Similarly, the work published Masselos et al. [17] lacks a global application view.

On the other hand, access trace based analysis techniques [18] have limited optimization capabilities. The quality of the analysis depends on the preceding compilation step. For instance, from an access profile point of view, all elements of an array are accessed equally while a small data reuse copy could be present. As a result, the search space cannot be explored properly.

To the best of our knowledge, no previous work has combined data reuse and inplace opportunities in a systematic way, leading to a technique for real life applications that is not based on simulation. Our approach allows finding the optimal assignment in a predictable way for both memories and hardware caches. Moreover, it decides the trade-offs in a controlled manner instead of evaluating them afterward by adding the relevant estimations.

For HW-controlled cache memories, source-level program transformations have been proposed to modify the execution order. This can greatly improve the cache performance of these applications [19–21], but still a significant number of cache misses are present in the experimental results. To remedy this, loop blocking has been first proposed primarily for improving the cache performance [22], but we observe that for multimedia applications, a significant amount of cache misses remain due to conflict misses. Similarly, storage order optimizations [19,23] are very helpful in reducing the capacity misses. Thus, mostly conflict cache misses related to the suboptimal data-layout remain. Array padding had been proposed earlier to reduce conflict misses [24–26]. These approaches are useful for reducing the inter-array conflict misses to some extent; however, existing approaches do not eliminate the majority of the conflict misses yet. Besides the studies conducted by Kandemir et al. [20], Panda and Dutt [25], and Burger et al. [27], very little has been done to measure the impact of data-layout transformations on the cache performance. Thus, there is a need to investigate additional data-layout organization techniques to reduce these cache misses [5].

29.3 SW-Controlled Memory Hierarchy Optimization

By exploiting data reuse information [1], a part of an array can be copied from a higher to a lower layer from where it is read multiple times. As a result, energy can be saved by ensuring that most accesses take place on the smaller copy signal and not on the larger (potentially also bigger, thus more energy consuming) original array.

Many different opportunities exist for deciding a data reuse copy. These are called copy candidates (CCs). Only when it has been decided to instantiate a CC do we call it a copy (although the copy candidates will be instantiated as arrays in the application, we reserve the name array for the "original array"). A relation exists between the size of a CC and the number of transfers from the higher layer to the local (potentially smaller) layer needed to update that one with new data. These transfers, hereafter called copy writes (CWs) are for cache memories the equivalent to minimal capacity misses according to our compiler centric cache miss classification (see Section 29.4). Read operations from the CCs are called copy read operations (CRs). Write operations needed to initialize the original arrays (e.g., with data coming from the application's test-bench) will be called compulsory writes.

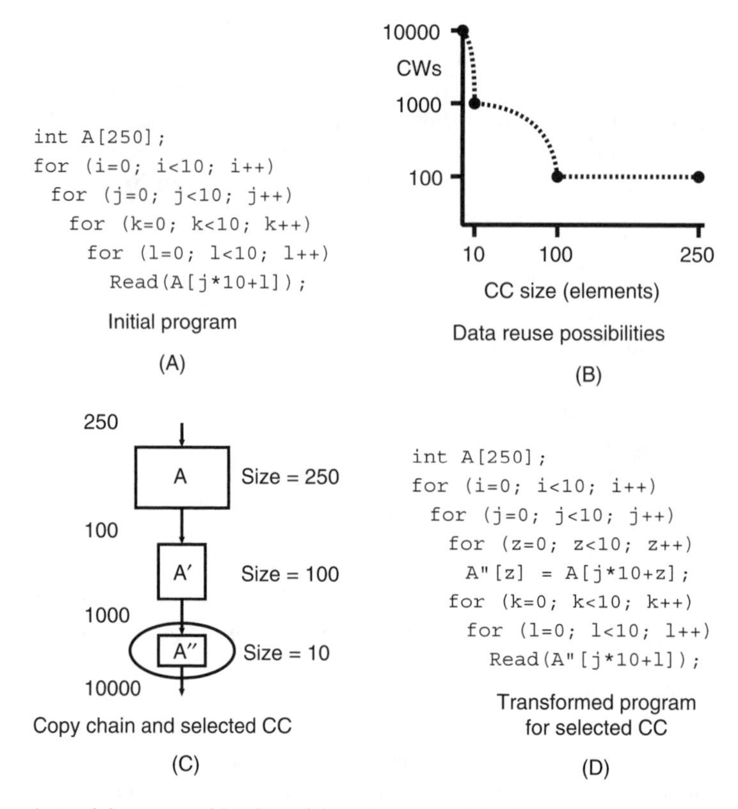

FIGURE 29.1 Analysis of the temporal locality of data: data reuse behavior.

Figure 29.1 presents an illustrative example of a loop nest with one reference to an array A with size 250. Originally, that array is accessed 10,000 times and several CCs are possible as depicted in Figure 29.1(B). For instance, we could select a CC A″ of size 10 by transforming the source code as illustrated in Figure 29.1(D) code. This is done by adding an explicit CC and corresponding copy statement in front of the k-loop. This copy statement will be executed 100 times, resulting in 1000 CWs to the array A. This CC selection corresponds to the second leftmost point plotted in Figure 29.1(B). Note that the good temporal locality of array A does not influence the amount of CWs from the next level. In theory, any CC size ranging from one element to the full array size is a potential candidate. In practice, however, only a limited set of CCs lead to efficient solutions. All possible CCs are plotted in Figure 29.1(B). The most promising CC sizes and CWs are kept and added to a data reuse chain as depicted in Figure 29.1(C). In this case, these are exactly those that have a relation to the loop bounds. That data reuse chain is completed with all 250 compulsory writes to array A.

The preceding example considers only a one array with a single read operation. In practice, however, several arrays exist in the code, each with one or more read operations, thus a data reuse chain will be associated to each read operation. To globally incorporate these issues, all data reuse chains of a given array must be combined in a data reuse tree. This concept is illustrated in the upper left part of Figure 29.2, where both data reuse trees for array A and array B are given. Array B is assumed to have only one read and it has no tree but a single chain associated. On the other hand, array A is assumed to have two references where one of them has no promising CC (indicated by the leftmost branch of the tree). More details on identification of data reuse chains and trees can be found in Achteren et al. [1] and Kandemir and Choudhary [13].

Obviously, larger CCs could retain the data longer and could therefore avoid more CWs. On the other hand, the larger the CC the bigger would be the memory required to hold the data and thus the bigger the overhead in energy per access will also be for each CR operation. Thus, a careful balance is required here between the number of CR and CW operations, and the layer implementation and associated energy

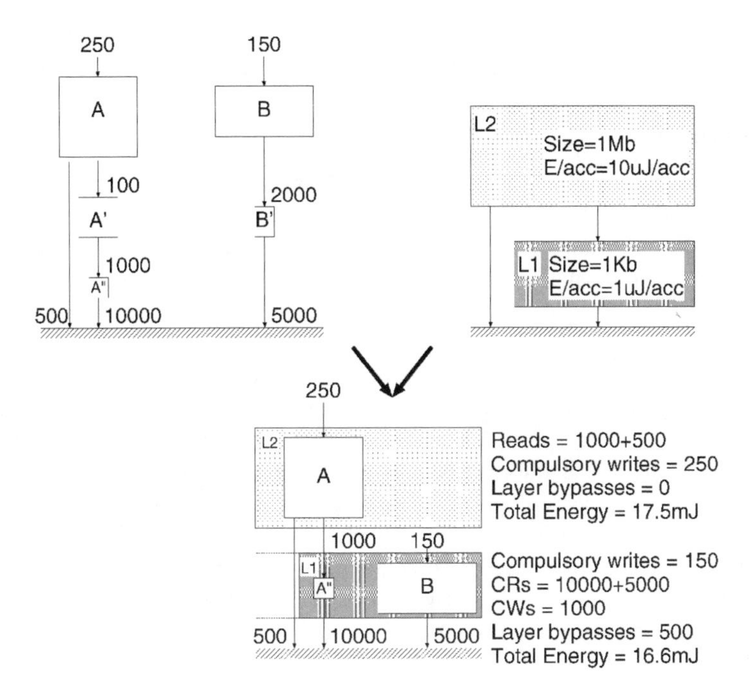

FIGURE 29.2 MHLA problem definition.

per access. This exploration is actually considered in the next step, when CCs and arrays must be mapped to the data memory hierarchy.

29.3.1 Memory Hierarchy Layer Assignment Techniques

We consider a generic SoC memory hierarchy template. It may contain multiple memory layers *Li*, each layer with multiple memory partitions (e.g., software controlled SRAMs, cache memories or a mixture of both) and off-chip (synchronous) dynamic random access memories ((S)DRAMs). Still, each partition can be organized on multiple memory modules, all of the same type but potentially having different sizes and number of ports.

In general, the energy consumed by the data memory hierarchy can be optimized for a target template by a careful selection of the set of CCs and arrays of the application, and by assigning these to the different layers of the memory hierarchy. This memory hierarchy layer assignment (MHLA) step results on an energy-efficient mapping of the different signals of the application to the memory hierarchy, and it is dependent on the different platform memory parameters. The goal is to select the mapping selection with the lowest energy possible where the total energy is the sum of the energy of all memory partitions.

The energy associated to each memory partition can be estimated by the number of accesses to this, multiplied by the energy per access of the partition (which can be potentially different depending on whether this is a read or write operation). On the one hand, the energy per access is a function of the memory size, partition type, and other memory parameters; thus it can be easily modeled in a memory library [23]. On the other hand, the number of accesses is the sum of compulsory writes, CRs, and CWs operations. This cost function is accurate in case of scratch-pad memories. For cache memories, the overhead in miss count should in principle be incorporated as well. As we will discuss in Section 29.4, however, that cost function is still valid for DM-caches on condition that additional optimizing memory data-layout transformations are applied to the application, a posteriori in a subsequent step. This is because the most important miss contributions are already being incorporated in the actual access count, namely the compulsory and minimal capacity misses that are by definition equivalent to compulsory writes and copy writes operations, respectively. For DM-caches, the major miss type contribution would be coming from data-layout conflict misses where the overhead of these in the overall access count can be kept relatively low by the use of optimizing data-layout transformations [5] (see Section 29.4).

29.3.2 Illustration of the MHLA Techniques

The MHLA process and its mapping results are depicted in Figure 29.2. We assume CC A″ and array B are selected to be stored in L1 and array A in L2. This is a valid selection because we assume it fulfills the memory layer size constraints. Assuming this, on one hand we have for layer L2, 250 compulsory writes needed for the first initialization of array A. On the other hand, 500 reads are needed for reading A via the read operation having no promising CC associated (see upper left of Figure 29.2) and 1000 CRs needed for the update of A″ in layer L1.

Similarly in layer L1, 150 compulsory writes are needed for the initialization of array B, and 1000 CWs for updating CC A″. On the other hand, 10,000 CRs from CC A″ and 5000 CRs from array B will take place from the same layer. Note that the 500 reads from array A not having promising CC will not affect the activity of layer L1 for this architecture because a bypass from layer L2 is foreseen. This may not be the case for hardware controlled caches because all accesses have to pass through the cache when no bypass is foreseen. Also note that for caches no explicit copies are introduced in the code; however, the cache controller can be enforced to make the desired copy by a proper memory layout (see Section 29.4).

The assignment of arrays and CCs to the layers of the memory hierarchy must be performed globally to effectively minimize the energy consumption. The size of one copy must be traded for the size of another copy because the assignment must fulfill the maximum layer size constraints or because activity in that layer must be kept low for energy reduction.

A simple illustrative example of the MHLA balance having multiple references is given in Figure 29.3 for an example assuming only two read operations. The two curves at the left of the figure illustrate the amount of CWs for different CC of increasing sizes, thus it depicts the Pareto trade-off between CWs and CC-size. Only the evolution of CWs with CC-size is given because, in all cases, the number of CRs from the layer will not change. Array B has a maximum of 6000 CWs when no CC is selected and this can be reduced to 500 CWs when the largest CC is selected for layer assignment. Similarly, array A has more than 2000 CWs when the smallest CC is selected and less than 1000 when the largest CC is selected. The number of write operations to the layer is minimal when selecting the largest CC for both arrays; however, this assignment is not valid because the total required size for accommodating both CCs does not meet the maximum layer size constraint pictured at the right-hand side of Figure 29.3. There, all

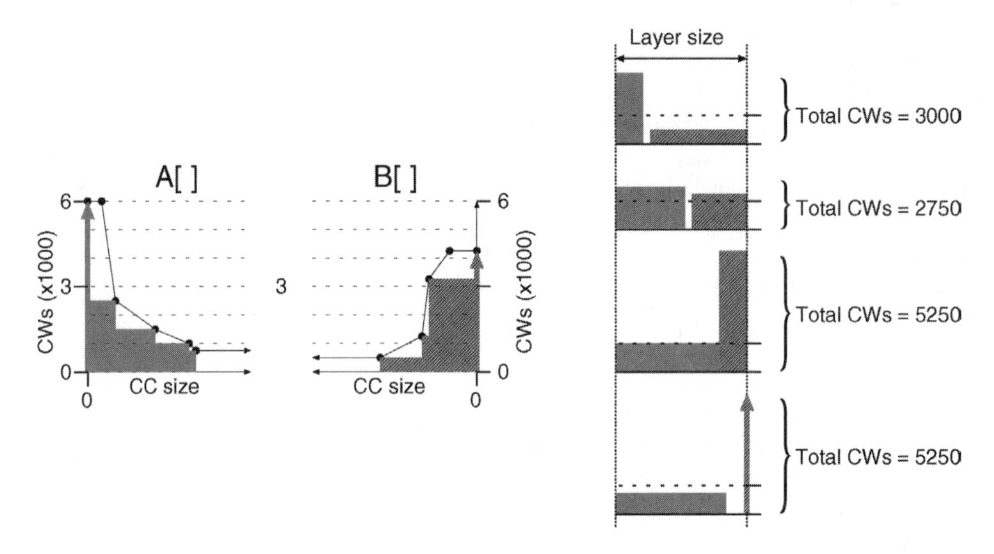

FIGURE 29.3 MHLA trade-offs for the selection of copy candidates.

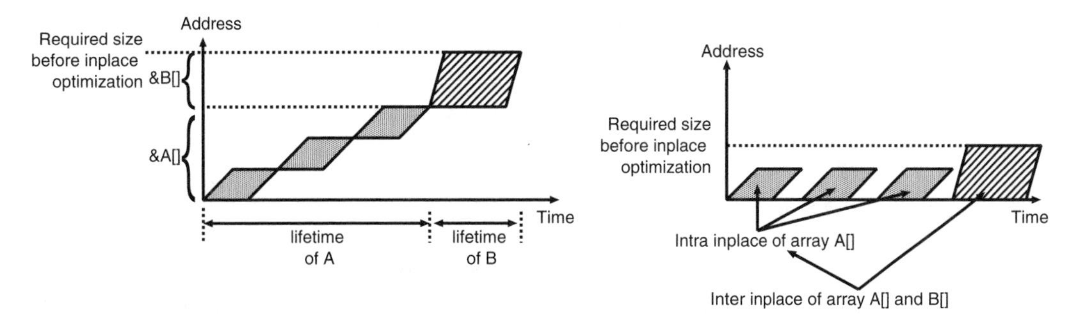

FIGURE 29.4 Illustration of inter/intra-inplace optimization for memory size estimation.

feasible combinations of CC selection to this layer are presented, together with the total number of CWs operations associated to the choice. The upper solution combines the largest CC of array B and the largest feasible CC of array A. The total number of 3000 CWs is the sum of the individual CCs. While constructing the other solutions, some space of the CC of array B is traded for the CC of array A. As a result, the number of CWs for A reduces and the number of CWs for B increases. In this simple example, it is easy to see that the optimal choice is the second in the row with 2550 CWs.

In general, however, realistic applications will have too many possibilities to still find the best solutions manually. Moreover, that explodes the search space when considering more the one layer for the mapping because unnecessary copies from one layer to the next one must be avoided. Additionally, the problem must be considered over the memory hierarchy globally because CWs of one layer can be traded for CWs on another layer. Thus at IMEC, formalized techniques have been developed [28], and these have been implemented on a research prototype tool to support this step.

29.3.3 Lifetime Analysis for Memory Partition Size Estimation

The exploitation of limited lifetime allows having smaller layers or storing more data in an equal sized layer. Both can have a huge impact on energy and performance. Especially, the short lifetime of the CCs should be considered carefully. In addition, it can be expected that a technique without inplace estimation could lead to suboptimal decisions for larger applications because most arrays will have a relatively smaller lifetime as the application complexity increases.

Figure 29.4 illustrates how to exploit the limited lifetimes of the signals to reuse memory locations. At the left of the figure, the plot illustrates how elements of arrays A and B are used in time. The shaded areas indicate which elements are used and for how long. Clearly, the declared size of array A could be reduced by a factor of three by reusing the same locations due to the fact the shaded areas are not overlapping on time. This type of memory location reuse is called intra-inplace [19]. Similarly, the elements of array B do not overlap in lifetime with any of the elements of array A. Thus, the complete array B could also reuse the locations of array A. This type of memory location reuse is called inter-inplace [19]. The results of both inplace opportunities are depicted at the right-hand side of Figure 29.4.

We have implemented a low complexity inter-inplace estimation technique in our prototype tool for a realistic estimation of the required storage in the partitions of the memory layers. That is based on tracking what signals are simultaneously alive in the innermost loops of the application. As a result, we only have to update the storage size of those inner loops that span the lifetime of the CC. This is accurate enough because, typically, most CCs have a very short lifetime and its size is already known from the data reuse analysis phase.

29.3.4 Relation to Other Steps of the DTSE Design Methodology

MHLA is usually applied as part of IMEC's data transfer and storage exploration (DTSE) script [29] developed to optimize the data transfer and storage issues in data-dominated applications. Two phases can be distinguished in this script.

First, a platform independent phase performs program transformations to improve locality of the reference and the required storage size [30]. If this phase is skipped, the approach still works, but the results will typically be far less optimal if the original temporal locality of the code is poor [29]. This will have a clear impact in quality of the search space due to less opportunities for CCs.

The second phase maps the application to the target platform. This is performed in four decoupled steps. MHLA is the first of the platform dependent steps, and it starts by determining for each data set a layer and a memory partition type, thus allowing detailed timing information about the array references be obtained.

This information is required for the next step that optimizes the required memory access parallelism for meeting the timing constrains. Techniques, such as Wuytack et al. [31] and Grun et al. [32], could be used here because certain accesses must still happen in parallel to meet the target timing budget. The conflicting memory accesses must be either stored in different memories or in a dual port memory. In the third memory allocation and assignment step, arrays and CCs assigned to the memory partitions are assigned to the various memory modules within each partition obeying the required parallelism [18,33].

A final data-layout step decides on the storage layout of the data inside the memory. This would further minimize the required size of the memory partition and also avoid the overhead in conflict misses in case of HW-controlled caches (see Section 29.4).

29.4 Case Studies for MHLA Exploration

Two real-life application drivers having different characteristics are selected to illustrate the impact of MHLA decisions on the energy consumed in the memory hierarchy.

The first driver, a quad-tree structured difference pulse code modulation (QSDPCM) algorithm, is an application from the video compression application domain consisting of a few large arrays exposing high temporal locality in the memory references, thus it is largely dominated by data reuse opportunities. It involves an inter-frame compression technique for video images based on a hierarchical motion estimation step, and a quad-tree based encoding of the motion compensated frame-to-frame difference signal [34].

The second driver is a wireless receiver for digital audio broadcast (DAB) containing many smaller arrays, with almost no data reuse opportunities present. The transmission system is based on an orthogonal frequency division multiplex (OFDM) transportation scheme using up to 1536 carriers [35].

For the target platform, we assume two layers of memory hierarchy, the local being implemented using a scratch-pad memory. Although as we will discuss in Section 29.4, our approach is not limited to these memory types, and it is also valid for HW- and SW-controlled caches using a varying number of layers. The used energy model is based on a real memory library. Because relative energy figures are sufficient for the exploration, these are presented relative to an off-chip memory of 1MByte (MB) of fix size. The largest on-chip memory considered is 16 KByte, which is a factor 3 less energy consuming than the off-chip memory. This is because the energy model used is slightly super logarithmic, so a memory that is 256× larger consumes 8.6× more energy per access. This energy model is also used for L0 and L1 in both drivers. This relation demonstrates a very conservative factor in energy per access between off- and on-chip memories.

29.4.1 The QSDPCM Driver

A global view of the QSDPCM's main signals and their data reuse chains is given in Figure 29.5. Many data reuse opportunities exist for the QSDPCM application as can be seen from the many data reuse chains. Figure 29.6(A) shows the energy contribution of the L1 memory (bottom bar) and the main memory (top bar). When increasing the layer size, the energy decreases because fewer accesses occur on main memory (due to the introduction of more CR operations). On the other hand, the reduction in CWs in L1 does not decrease much for a L1 size larger than 640 Bytes (B). Therefore, the L1 energy per

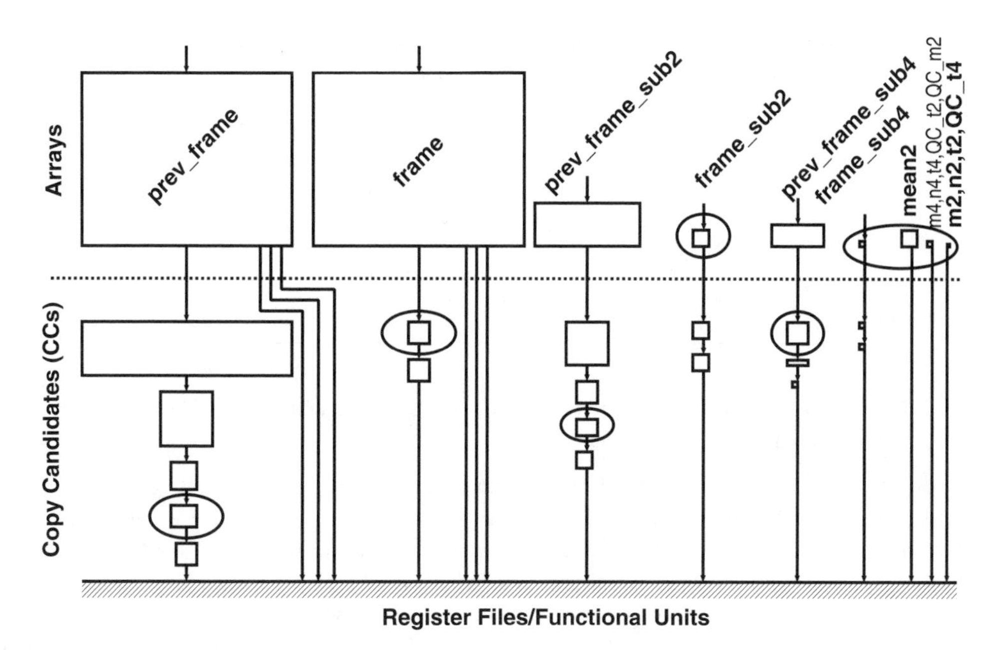

FIGURE 29.5 Assignment of arrays and CCs to the memory hierarchy for QSDPCM. Arrays and CCs marked inside circles have been selected for storage in the local memory.

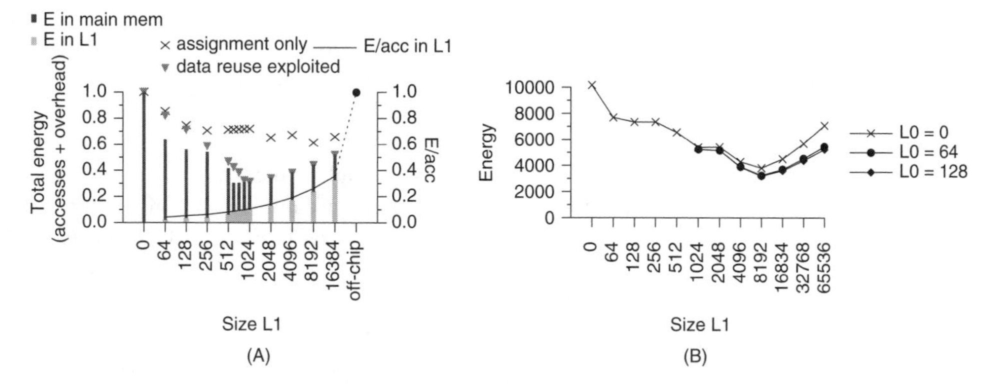

FIGURE 29.6 Memory layer size exploration for energy for both QSDPCM and DAB drivers: (A) varying L1 layer size for QSDPCM (solid line plots); (B) varying L0 (RF) and L1 layer sizes for DAB.

access penalty due to the larger L1 size is not compensated by the potentially less amount of CWs. Thus, the overall energy consumption increases from that point. Thus, an L1 of 640B is the optimal layer size for this application. The optimal assignment of arrays and CCs corresponding to this point is given in Figure 29.5. Arrays and CCs selected as the optimal ones to be stored in L1 are indicated by circles. The rest of arrays are stored in main memory. Nonselected CCs are not stored because copy operations are not needed for these.

To consider the effect of the number of hierarchy layers in energy, we have also reexplored the options by inserting an extra L0 layer. However, this has not resulted in a significant reduction in overall energy because the optimal L1 size for a two-layer platform is already small enough to exploit the data-reuse opportunities present in this application. This is also exposed by comparing our approach with an array-level assignment technique, thus without the introduction of CCs (indicated by crosses in Figure 29.5). In this case, a factor 2 in energy is gained.

TABLE 29.1 Energy Values, Optimal Layer Size, and Numbers of Accesses for the Three-Loop Transformed Versions of QSDPCM

	Nonmerged	Fully Merged	Partially Merged
Total Energy	71.97	47.17	46.47
L1 Energy	13.79	14.32	14.29
L2 Energy	58.18	32.85	32.17
Layer size for L1 (Bytes)	742	748	802
Layer size for L2 (Bytes)	114048	66352	63360
Number of L1 accesses ($\times10^6$)	1.14	1.19	1.19
Number of L2 accesses ($\times10^3$)	542	306	300

29.4.2 Global Loop Transformations for Improved Memory Hierarchy Utilization in QSDPCM

For the QSDPCM, we have also evaluated the gains in overall energy by disabling the inplace mapping heuristics mentioned in the earlier section for an accurate estimation of the storage size in L1. This case is indicated in Figure 29.5 by triangles. The use of the heuristic allows a L1 size of 640B instead of the 1KB required otherwise; however, the overall energy gain in this case is rather limited (only another 3.2%).

The limited effect of the inplace mapping optimization techniques is due to the overall loop structure of the code (with several loop nests each traversing an image frame), thus requiring intermediate buffers to store data between loop nests. However, it is possible to largely change the lifetime of the data, and thus the size of these buffers, by globally transforming the loop structure of the application source code as illustrated by Catthoor et al. [29].

To overcome this limitation and to study the effect that an optimized storage size would have in the memory hierarchy, we have implemented loop-merging transformations in the initial code because they eliminate the need for intermediate buffers and, potentially, reduce the lifetime of the data. All these optimizing transformations are part of the global storage cycle budget distribution (SCBD) step in the DTSE script [29,36].

Three loop-transformed versions of the QSDPCM have been implemented: the original version with all loops (hereafter called nonmerged), a version with most loops merged into a global one (fully merged); and an intermediate version of the last one where only selected loops have been merged together (partially merged). Table 29.1 gives an overview of the results obtained on each code version.

Clearly, the best solution for energy is for the partially-merged version. In all cases, the energy contributions from the L1 are very similar. This is somehow coincidental because the arrays and copies assigned to L1 are very different in all three cases and the amount of L1 accesses could have been more different; however, the L2 energy contribution is much smaller in the fully and partially merged code versions. This is due to the big reduction of the number of accesses to this layer. In this case, the total energy gain is rather small with the partially merged version, but exploring other solutions may provide better results. An important conclusion appears contrary to what intuitively one would expect, the most local code (fully merged) is not always the best one for energy, and an exploration of the possible transformation options is therefore needed.

For the fully-merged and the partially-merged codes, an exploration on the L1 size has been also performed during the MHLA step. It appears that for both versions the best hierarchy for energy is with an L1 memory of 1KB. This appears to be a good trade-off between the amount of CW operations (important for small L1 memories) and the energy per accesses (that increases with the layer size). The same size of 1KB for the nonmerged code has been kept to fairly compare all three versions.

29.4.3 The DAB Driver

Similar to the previous driver, an L1 size exploration is performed and presented in the top curve of Figure 29.6(B). For this driver, the minimum energy consumed is found for an L1 size of 8KB; however,

TABLE 29.2 MHLA Results for DAB for the TriMedia TM1300 Processor

	Estimated Load/Stores	Actual Load/Stores	Required Registers	Data Cache Misses	CPU Cycles
Mild register usage	17,152	18,408	22	2895	91,552
Aggressive register usage	11,232	11,837	70	763	47,341

the introduction of an extra L0 layer has brought additional gains in energy in this case. By implementing this L0 layer as a register file (RF) with varying size from 64 to 128B and reexploring the assignment, an extra 15% in energy is gained, while the optimal size for layer L1 remains unchanged. The reason is an additional reduction in 25% of the memory accesses in the OFDM block now becoming RF operations.

The most energy efficient assignment decision has been used to map the DAB application to the TriMedia processor. This processor is selected because it has a data memory hierarchy that matches the optimal architecture. The processor has an L0 layer of 128 registers, 16KB cache and 8MB of SDRAM; however, the exploitation of the L0 register file has to be carefully evaluated. On one hand, the number of data load stores will decrease as the size of L0 increases because more data remains in the L1. On the other hand, the higher register pressure might counteract this gain as register spilling is required to schedule all instructions given the actual RF size available in the processor. In addition, the unrolling transformations, required to keep the data in the RF, needs more instruction cache space.

The trade-off between the load stores, spilling, and instruction cache is given in Table 29.2. The native TriMedia simulator is used for the evaluation of three differently transformed implementations having a relatively more aggressive L0 register usage (second row). The large reduction of 34% in memory accesses has of course a large impact on data memory energy and performance (as presented in Table 29.2). In addition, the prediction of the MHLA technique has been very close to the actual number of accesses. After an examination of the analysis report provided by the compiler, the small difference between MHLA's estimated activity and the simulation results can be explained by the few register spilling operations.

Of course, the same result could be obtained manually by an experienced designer using a combined trial and error and simulation process; however, MHLA can significantly save design time by predicting beforehand the good solution. Moreover, this prediction gives the designer a target minimum for which to optimize. No time is wasted in finding more opportunities that actually will not pay off.

29.5 SW-Controlled Cache Miss Optimizations

Before introducing SW optimizations for HW and SW controlled caches, in this section we revise the traditional classification [37] of cache misses from a compiler viewpoint. This is the first step needed before developing design-time optimization techniques for cache memories. The definition of a cache miss does not change though. In the most general case a cache miss is any transfer of data from main memory to the main memory taking place whenever a load/store operation issues for the data that is not present in the cache. Transfers from the cache to main memory are write-backs and these are not focus of this work.

29.5.1 Compiler-Centric Cache Miss Classification

In computer architecture literature [37], cache misses have been traditionally divided in three categories: compulsory, capacity, and conflict miss types; however, that classification is purely architecture oriented and is clearly insufficient for compiler optimizations because they do not properly expose the nature of the miss type. For that purpose, we have defined a new taxonomy for data cache misses which spans seven categories. These are compulsory, minimal capacity, block prefetch, block allocate, associativity conflict, replacement, and data-layout conflict misses. This new taxonomy, while isolating the different sources of overhead for cache misses, is much better suited for modern SW-controlled caches than previously proposed ones [37].

To study the nature of the miss contribution, we start by assuming two caches for comparison purposes: a real cache (our target cache) and an ideal cache. The ideal cache is a memory of infinite size with a single data element per line, implementing a fully associative mapping organization, with an optimal replacement policy and following an optimal block allocate policy. Unlike the write-around or fetch-on-write block allocate policies, we consider an optimal block allocate policy in which a write miss ensures that no block of data is unnecessarily allocated in a cache line when all data elements of that line are to be updated before the line is to be replaced. As for the optimal block replacement policy [37], the optimal block allocation policy cannot be implemented via a hardware mechanism because that would imply predicting the behavior of the algorithm in the future with respect to the actual state of the cache.

29.5.1.1 Compulsory Misses

Assuming our definition of ideal cache, a compulsory miss happens whenever a new data element of the application's test bench is needed by the application's algorithm and this data needs to via the cache or local memory for the first time. Any data written from the algorithm back to memory (back to the test bench) can be considered a compulsory write-back. All miss types not yet dealing with prefetching issues must be accounted at the data level. This applies both to compulsory and minimal capacity misses.

Compulsory misses are independent of the physical memory organization parameters (such as the cache memory size) and controller policy issues (such as the replacement policy). Thus, these are truly platform independent and their contribution depends solely on the application characteristics. This is the only miss type requiring a change in the execution order of the application to be avoided. This can be done by means of global data-flow [23] or global loop transformations [29]. All the rest of the cache misses are data-layout related for a given execution order, thus the way to avoid these is by transforming the placement of data in main memory. This miss type is valid for HW-controlled caches, SW-controlled scratch-pad memories and mixes, and it is one of the major (but largely unavoidable) contributors to overall miss count for DM-caches.

24.5.1.2 Minimal Capacity Misses

After assuming an ideal cache, platform constraints can start to be gradually introduced in our target cache, while still assuming optimal controlling policies. In this context, the first platform dependent, controller policy independent miss type is the minimal capacity miss and expressed as the additional contribution in misses in an ideal cache which is constrained in size (by same capacity in bytes) as the real cache. Due to this size limitation, some data values are discarded and have to be retrieved later. If the cache is full, the data value to discard is the furthest to be used in the future (optimal replacement policy). Minimal capacity misses are significant contributors to the overall miss count, especially for small cache sizes; however, the MHLA techniques discussed in Section 29.3 are effective in minimizing the overhead of these, thus enabling the efficient utilization of small cache spaces.

24.5.1.3 Block Prefetch Misses

This miss type is related to the line size of the cache. By definition, a block is the group of data fetched from main memory and that is allocated altogether in the same cache line. A block prefetch miss is the next miss contribution for an ideal cache with the same number of lines and same line size as the real one. It characterizes the miss occurring from a data that is required in the very near future (such as data being subsequently fetched in the innermost loop of an application) and that is not present in the block that has been most recently loaded in cache. This miss contribution gives an indication on how inefficiently the memory layout adapts to the ideal situation where all data prefetched in a cache line is actually read before this line is to be flushed due to cache capacity limitations or other reasons.

Together with block prefetch misses, minimal capacity misses depend on the actual physical memory organization parameters (such as number of lines and line size), thus these are platform dependent miss types although still independent on the controller-policy of the cache (such as replacement policy and mapping organization). These miss types are valid for purely HW and SW controlled caches and mixes when prefetching is exploited in the purely SW controlled case.

24.5.1.4 Block Allocate Misses

At this point, the geometry of the cache is now fully fixed and the effect in miss rate due to a nonoptimal controlling policy is considered. These are platform and controller-policy dependent misses, and they are valid miss types for purely HW controlled caches but less relevant for mixes and irrelevant for purely SW controlled caches (pure compiler controlled policies).

In this context, the first platform and controller policy dependent miss type is the block allocate miss. This is the next miss contribution resulting from the (nonoptimal) allocation policy of the real cache. Some caches (especially the HW controlled types) fetch the block from main memory on a block write miss (e.g., following a fetch-on-write block allocate policy). This is done to ensure the consistency between the copy of the data in the cache and the value stored in memory. This extra read operation (block allocate miss) should only be needed in case the block is replaced before all its data elements have been updated; in any other case this transfer is considered redundant, thus a miss.

Block prefetch and block allocate miss types can be clustered in a common category (block miss) to collect all miss effects related to the organization of data in the cache in lines and the transfer of data in blocks. The impact of block misses can also be high when the effect of SDRAM page misses (in main memory) is incorporated; however, this is not the focus of this chapter but a topic for future research work.

24.5.1.5 Associativity Conflict Misses

This is the next miss contribution (measured at the block level) due to nonideal policies, and it results from the limited associativity of the real cache. A cache that is not fully associative will limit the replacement freedom, even when the mapping organization is exploited by using an optimal memory data-layout. For instance, if the same block has to be loaded in the cache several times (due to any of the previous miss effects), the block may be allocated in different places against what the optimal replacement policy will actually select. Indeed, a particular data-layout could eventually avoid the first allocations of the target block that will conflict with an existing one in the cache, but not for the subsequent allocations of the same block. The contribution of this miss type in the overall miss count for DM-caches is limited due to the fixed mapping organization.

24.5.1.6 Replacement Misses

This is the next miss contribution resulting from the (nonoptimal) replacement policy (typically least recently used or LRU [37]) of the real cache. An optimal replacement strategy uses the capacity of the cache and misses can be avoided because this policy exploits knowledge of the future accesses. For instance, data required in the near future will not be replaced. With a nonoptimal replacement policy, however, it can happen that data are flushed just before being required, which is the source of extra misses. Therefore, a suboptimal replacement strategy increases the replacement misses. In advanced multimedia caches, the replacement policy is software controlled and in DM-caches there is no replacement policy due to its fixed mapping organization. Therefore, these misses can be clearly ignored in our context.

24.5.1.7 Data-Layout Conflict Misses

Misses in this last category are purely devoted to isolate the effects due to a nonoptimal utilization of the cache mapping organization. At this point, both the geometry and controller policy of the cache are fully fixed and the only misses left are related to a nonoptimal utilization of the cache mapping organization. Therefore, data-layout conflict misses are the last miss contribution, thus by definition these are any misses that could be avoided by choosing an optimal (conflict oriented) memory data-layout (e.g., using array interleaving, array padding, selective gap placement, array merging, or any mix of these). Two types of data-layout conflict misses exist: inter-array when elements of different arrays conflict in the same cache line and, intra-array when the situation happens for elements of the same array.

These misses are typically the largest contributors to miss rate in DM-caches, once a good memory layer assignment has been selected (as we have achieved with the techniques of Section 29.3). Data-layout techniques for the elimination of these misses are the topics addressed in rest of this chapter.

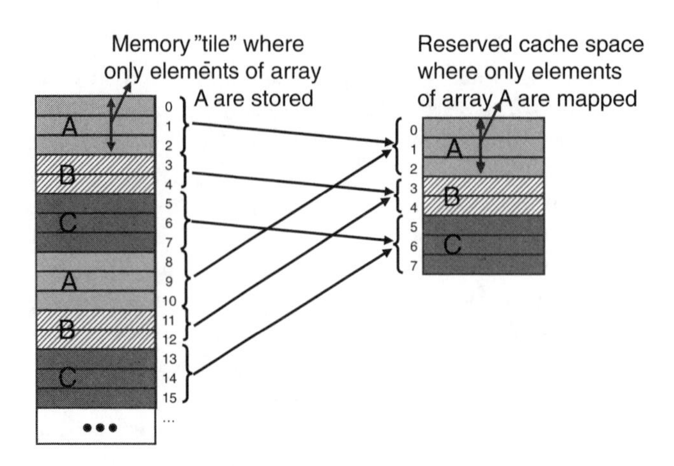

FIGURE 29.7 Illustration of how memory tiles reserve spaces in cache.

24.5.2 Data-Layout Transformations for Conflict Miss Reduction

To maintain the decisions taken during the MHLA step, we must ensure that the overall miss contribution follows the estimated number of CWs at that level. Therefore, it is needed to ensure that the data-layout conflict miss contribution can be minimized to a large extent. This means that for each array a certain amount of space should be reserved in the cache to hold all data of the the selected CC without this being flushed because of conflicts in the cache lines.

The necessary spaces can be reserved in the (lockable) cache by using the array-interleaved data-layout approach for data-layout conflict misses presented in Kulkarni [38]. In this chapter the original approach is improved further by considering not only the inter-array data-layout conflict misses but also the intra-array component and the influence of the MHLA decisions in minimal capacity misses.

Our data-layout approach aims at an interleaved placement of arrays in main memory such that every array is distributed in memory as data "tiles." It assumes a cache with fixed mapping organization as the one found in DM-caches. For explanation purposes, we will also consider (without loss of generality) cache lines of the same size as each memory location containing at most one array element. The concept is illustrated in Figure 29.7. In this case, due to the fixed mapping organization of the DM-cache, every element of arrays A[], B[], or C[] gets a fixed cache location, which is defined by the fixed mapping organization of the DM-cache (e.g., $C_{addr} = M_{addr} \% C_{size}$, where C_{addr} and M_{addr} are the cache and memory locations, and C_{size} is the size of the cache). The interleaved nature of our data-layout approach ensures that the tiles of the same array are evenly distributed in memory at equal address distances. This is achieved by an index transformation (e.g., $index\%TS + index \div TS \times C_{size} + O_{ffset}$, with *index* being the index expression of the array reference, *TS* the tile size, and O_{ffset} the offset of the original array placement). For this data layout, the cache controller will always assign the same cache space to elements of the same array (see Figure 29.7). As a consequence, each array will always get disjoint address spaces in the cache, thus elements of different arrays will never conflict in the same cache line, thus eliminating any possible inter-array data-layout conflict miss effect.

In addition, this data organization allows for an address range of the cache that is always reserved to the elements of a particular array. We can use this property to ensure that a fixed amount of the cache size is always devoted to hold data of that array. Therefore we can ensure that the optimal MHLA decisions, relating the amount of memory space reserved for a particular CC, are also guaranteed even when the control is done in HW and not only in SW. This is done by choosing the tile size of each array to be equal to the size of the selected CC (see Section 29.3).

We must also consider the intra-array effect, however, because elements from the same array may still conflict in the same cache line. This situation happens because different elements of a working set (e.g., the elements of an array belonging to the copy selected by the cache controller) could be located within

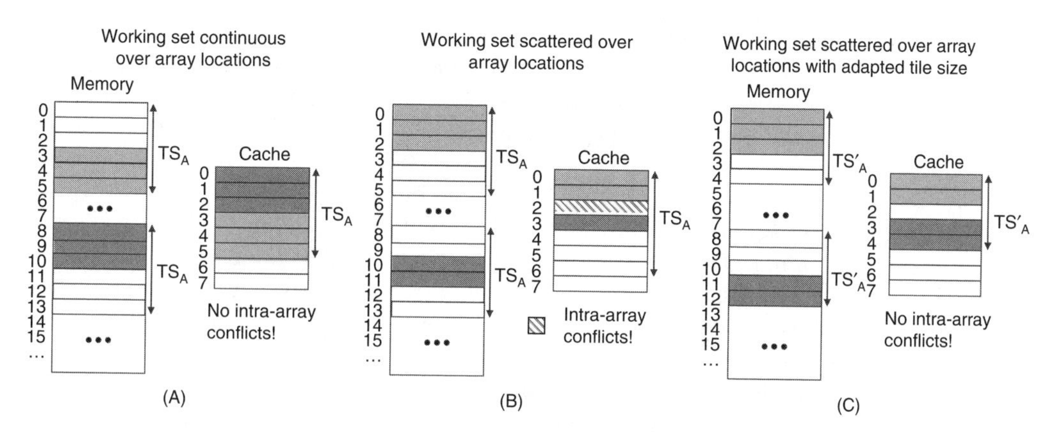

FIGURE 29.8 Illustration of how intra-array, data-layout conflict misses occur in an array-interleaved data-layout: (A) situation without intra-array conflict misses for a Working Set (WS) over continuous array locations; (B) situation with (potential) intra-array conflict misses for a WS with locations scattered all over the array; (C) adaptation of the tile size to avoid intra-array conflict misses for situation (B).

the array at distances larger than the actual tile size, thus leading to cache mapping situations with potentially conflicting cache lines. On the other hand, there are also situations when the mapping organization of the cache causes no conflict misses. This is typically the case when the different elements of the working set are accessed at consecutive locations within the array, thus they will never overlap in cache. Figures 29.8(A) and 29.8(B) illustrate both concepts.

These data-layout conflict situations due to intra-array effects can be avoided by ensuring that the elements of the working sets within a memory tile are evenly distributed over the available cache locations. In fact, when studying the miss contributions for both minimal capacity and data-layout conflict misses for different tile-sizes, a seemingly random behavior is observed: very high and very low miss counts are possible for small variations of the tile size. Indeed, changes in the tile size can modify the cache mapping decisions for the affected array, thus some sizes could lead to more conflict situations that others and vice versa. In general, the larger the tile size, the less the probability to have conflicts in the cache lines. Still, it is possible to find small enough tiles for which the amount of conflicts is still minimized. Moreover, these variations can be made small enough such as it is possible to trade few minimal capacity misses for a large amount of data-layout conflict misses. This is done by adjusting the tile size, thus avoiding a "peak" and selecting one of the neighborhood "valleys." Figure 29.9(A) illustrates the effect that the tile

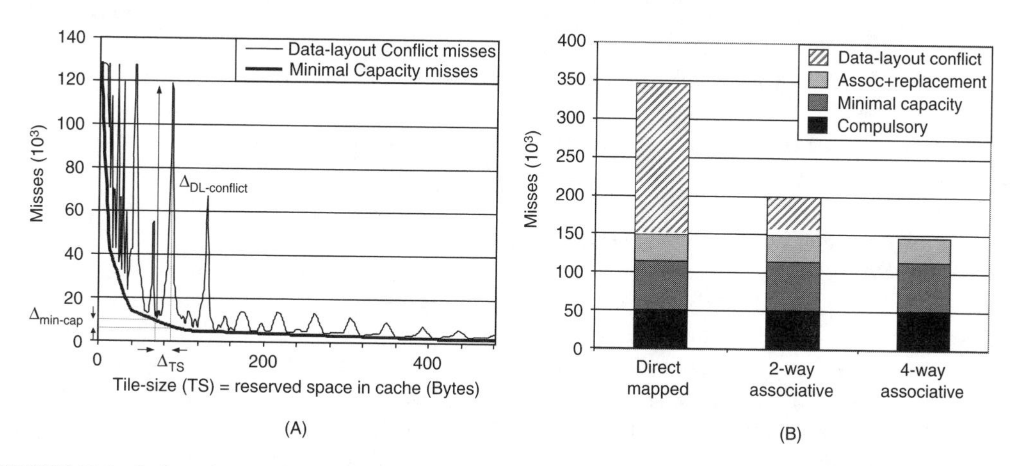

FIGURE 29.9 Cache miss simulations for QSDPCM: (A) miss count contributions due to minimal capacity and data-layout conflict misses in a DM-cache associated to one of the arrays in the QSDPCM when changing the tile size; (B) simulation results for the complete application with breakdown on different miss type contributions.

size has in miss rate for one of the arrays of the QSDPCM driver. The locations of these "valleys" can be avoided by ensuring that cache-lines are evenly used by the elements of the working set. This situation happens when the combination between tile size and the distances between elements of the working sets are co-prime numbers (e.g., numbers not having a greatest common divisor other than 1) as illustrated in Figure 29.8(C).

Finally, the mapping between elements of the working set and the cache lines can be influenced not only by adjusting the tile sizes but also by changing the size of the dimensions of the arrays (e.g., by using array padding techniques [39]). Globally considered, changing the tile size affects the mapping organization by modifying the space reserved in cache, while changing the padding decision also affects the mapping organization by modifying the memory address. However, a bigger tile size for one array decreases the amount of cache lines available for other arrays in the cache. Thus, these decisions must be globally considered during the MHLA step. A drawback to the use of array padding techniques is that they would definitely introduce holes in the array and those holes will become unused locations in main memory, thus the memory occupation would be far less efficient than adjusting the tile size. In fact, a trade-off exists in sacrificing cache or memory space for reducing the conflict misses. It is desirable that both memory and cache costs are globally minimized though. A global exploration is needed because the choice for a tile size for one array influences the choice for other arrays. Such exploration can be performed for the complete application in which the additional space in either cache or main memory is expressed; however, the details for that extension fall outside the scope of this chapter.

24.5.3 Case Study for Data-Layout Transformations

To evaluate the impact of data-layout transformations in cache miss rate, we have used the QSDPCM driver introduced in Section 29.3.

During the memory-hierarchy layer assignment step (see Section 29.3) an optimal L1 layer of 1KB has been selected based on the estimated amount of accesses to the different layers and the energy-consumption per memory access. During that exploration, the sizes of the arrays and CCs are used as initial tile sizes within the cache, and these are passed to the data-layout phase as first estimation. In the fine-tune phase, the tile sizes are slightly changed and some array dimensions padded marginally (e.g., by extending their declaration size). This is done to optimize the data-layout for intra-array conflict-misses.

Note there is no need to optimize those signals whose complete top-level array is selected to be stored in the cache. In this case, each element has its own location in the cache, elements of this signal will never compete for the same location. In this case, there are four signals (prev_frame, frame, prev_sub2_frame, and prev_sub4_frame) with a selected copy-candidate smaller than the top-level array (see Figure 29.5). This potentially results in conflict misses that should be avoided. The refinement of the initial tile size as indicated by MHLA allows avoidance of the data-layout conflict miss contributions and retention of the miss contribution due to minimal capacity misses decided by the precedent MHLA step.

Simulations have been performed to measure the impact of the data-layout optimizations for the complete QSDPCM application. The main results are given in Figure 29.9(B). The two bottom bars of the graph represent the (unavoidable) miss contribution due to compulsory and minimal capacity misses. To have a reference of how good the performance is when an optimal data-layout is applied, the miss rates of the DM-cache are compared to a two-way and a four-way associative cache with the least recently used replacement policy. Those caches have a more "intelligent" controller, and they should be able to resolve (part of) the conflict misses; however, the DM-cache with an optimized data-layout can still outperform the two-way associative cache (without data-layout optimizations) and become as efficient as a four-way associative cache. Note that a four way associative cache is about two to three times less energy efficient than the DM-cache of the same size (see Section 29.1). Thus, the combination of MHLA techniques (for an optimal sizing of the local memory layer) together with the use of data-layout optimization techniques (for the efficient use of DM-caches) provides a low-power alternative for platform architects and application designers of application domain specific solutions.

29.6 Conclusions

In this chapter, we have presented design time optimization techniques for an optimal dimensioning of the memory hierarchy for low-power operation. This is based on data-to-memory hierarchy layer assignment decisions that consider information resulting from the temporal locality and lifetime characteristics of the application data. This is true both for HW and SW controlled memories (e.g., scratch-pad and cache memories). For HW-controlled caches, however, because of the mapping organization of the hardware controller, additional data-layout organization techniques are needed to avoid the overhead in additional data transfer resulting from conflicts in the cache lines.

References

[1] T. Achteren, R. Lauwereins, and F. Catthoor. Systematic data reuse exploration techniques for non-homogeneous access patterns. *Proc. 5th ACM/IEEE Design and Test in Europe Conf. (DATE),* pp. 428–435, Paris, France, Apr. 2002.

[2] I. Issenin, E. Brockmeyer, M. Miranda, N. Dutt, Data reuse analysis technique for software-controlled memory hierarchies, *Proc. 7th ACM/IEEE Design and Test in Europe Conf. (DATE),* pp. 202–207, Paris, France, Feb. 2004.

[3] S.J.E. Wilton and N.P. Jouppi, CACTI: an enhanced cache access and cycle time model, *IEEE J. of Solid-State Circuits,* Vol. 31, No. 5, pp. 677–688, May 1996.

[4] P. Marchal, J.I. Gomez, D. Bruni, F. Catthoor, M. Prieto, L. Benini, and H. Corporaal, SDRAM-energy-aware memory allocation for dynamic multi-media applications on multi-processor platforms. *Proc. 6th ACM/IEEE Design and Test in Europe Conf. (DATE),* Munich, Germany, pp. 516–521, March 2003.

[5] C. Kulkarni, M. Miranda, C. Ghez, F. Catthoor, and H. De Man, Cache-conscious data layout organization for embedded multimedia applications, *Proc. 4th ACM/IEEE Design and Test in Europe Conf. (DATE),* Munich, Germany, pp. 686–691, March 2001.

[6] C. Ancourt et al. Automatic data mapping of signal processing applications. *Proc. Int. Conf. on Application-Specific Array Processors,* pp. 350–362, Zurich, Switzerland, July 1997.

[7] J. Anderson, S. Amarasinghe, and M. Lam. Data and computation transformations for multiprocessors. *5th ACM SIGPLAN Symp. on Principles and Practice of Parallel Programming,* pp. 39–50, Aug. 1995.

[8] M. Kampe and F. Dahlgren. Exploration of spatial locality on emerging applications and the consequences for cache performance. *Proc. Int. Parallel and Distributed Processing Symp. (IPDPS),* pp. 163–170, Cancun, Mexico, May 2000.

[9] H.-B. Lim and P.-C. Yew. Efficient integration of compiler-directed cache coherence and data prefetching. *Proc. Int. Parallel and Distributed Processing Symp. (IPDPS),* pp. 331–339, Cancun, Mexico, May 2000.

[10] L. Benini and G. De Micheli, System-level power optimization techniques and tools, *ACM Trans. on Design Automation for Embedded Syst. (TODAES),* Vol. 5, No. 2, pp. 115–192, Apr. 2000.

[11] L. Benini, A. Bogliolo, and G. Micheli. A survey of design techniques for system-level dynamic power management, *IEEE Trans. on VLSI Syst.,* pp. 299–316, 2000.

[12] P. Panda, F. Catthoor, N. Dutt, K. Danckaert, E. Brockmeyer, C. Kulkarni, A. Vandecappelle, and P.G. Kjeldsberg, Data and memory optimizations for embedded systems, *ACM Trans. on Design Automation for Embedded Syst. (TODAES),* Vol. 6, No. 2, pp. 142–206, Apr. 2001.

[13] M. Kandemir and A. Choudhary. Compiler-directed scratch-pad memory hierarchy design and management. *39th ACM/IEEE Design Automation Conf.,* pp. 690–695, Las Vegas, NV, June 2002.

[14] P. Grun, N. Dutt, and A. Nicolau. Apex: access pattern based memory architecture exploration. *14th Int. Symp. on Syst. Synthesis,* pp. 25–32, Montreal, Canada, Oct. 2001.

[15] P.R. Panda, N.D. Dutt, and A. Nicolau. Data cache sizing for embedded processor applications. *Proc. 1st ACM/IEEE Design and Test in Europe Conf. (DATE),* pp. 925–926, Paris, France, Feb. 1998.

[16] S. Steinke, L. Wehmeyer, B.-S. Lee, and P. Marwedel. Assigning program and data objects to scratch-pad for energy reduction. *Proc. 5th ACM/IEEE Design and Test in Europe Conf. (DATE)*, pp. 409–415, Paris, France, Apr. 2002.

[17] K. Masselos et al. Memory hierarchy layer assignment for data re-use exploitation in multimedia algorithms realized on predefined processor architectures. *8th IEEE Int. Conf. on Electron., Circuits and Syst. (ICECS)*, pp. 285–288, Oct. 2001.

[18] L. Benini, L. Macchiarulo, A. Macii, and M. Poncino. Layout-driven memory synthesis for embedded system-on-chip. *IEEE Trans. on VLSI*, pp. 96–105, September 2002.

[19] E. de Greef, Storage size reduction for multimedia applications. Doctoral dissertation, ESAT/KUL, Belgium, Jan. 1998.

[20] M. Kandemir, J. Ramanujam, and A. Choudhary, Improving cache locality by a combination of loop and data transformations, *IEEE Trans. on Comput.*, Vol. 48, No. 2, pp. 159–167, Feb. 1999.

[21] C. Kulkarni, F. Catthoor, and H. De Man, Cache transformations for low power caching in embedded multimedia processors, *Proc. Int. Parallel Processing Symp. (IPPS)*, Orlando FL, pp. 292–297, Apr. 1998.

[22] M. Lam, E. Rothberg, and M. Wolf, The cache performance and optimizations of blocked algorithms, *Proc. 4th Int. Conf. on Architectural Support for Prog. Lang. and Operating Syst. (ASPLOS)*, Santa Clara, CA, pp. 63–74, Apr. 1991.

[23] F. Catthoor, S. Wuytack, E. de Greef, F. Balasa, L. Nachtergaele, and A. Vandecappelle, *Custom Memory Management Methodology — Exploration of Memory Organisation for Embedded Multimedia System Design*, Kluwer Academic Publishers, Boston, 1998.

[24] N. Manjiakian and T. Abdelrahman, Fusion of loops for parallelism and locality. Technical report CSRI-315, Computer Systems Research Institute, University of Toronto, Ontario, Canada, Feb. 1995.

[25] P. Panda and N. Dutt, Low-power mapping of behavioral arrays to multiple memories, *Proc. IEEE Int. Symp. on Low Power Design*, Monterey, CA, pp. 289–292, Aug. 1996.

[26] P.R. Panda, N.D. Dutt, and A. Nicolau, Efficient utilization of scratch-pad memory in embedded processor applications, *Proc. 5th ACM/IEEE Design and Test in Europe Conf. (DATE)*, Paris, France, March 1997.

[27] D.C. Burger, J.R. Goodman, and A. Kagi, The declining effectiveness of dynamic caching for general purpose multiprocessor. University of Wisconsin Computer Sciences Tech. Report CS-TR-95-1261—Madison, WI, 1995.

[28] E. Brockmeyer, M. Miranda, F. Catthoor, and H. Corporaal, Layer assignment techniques for low-power in multi-layered memory organisations, *Proc. 6th ACM/IEEE Design and Test in Europe Conf. (DATE)*, Munich, Germany, pp. 1070–1075, March 2003.

[29] F. Catthoor, K. Danckaert, C. Kulkarni, E. Brockmeyer, P.G. Kjeldsberg, T. Van Achteren, and T. Omnes, *Data Access and Storage Management for Embedded Programmable Processors*, Kluwer Academic Publishers, Boston, 2002.

[30] F. Catthoor, K. Danckaert, S. Wuytack, and N. Dutt, Code transformations for data transfer and storage exploration preprocessing in multimedia processors, *IEEE Design and Test of Comput.*, Vol. 18, No. 2, pp. 70–82, June 2001.

[31] S. Wuytack, F. Catthoor, G. Jong, B. Lin, and H. Man. Flow graph balancing for minimizing the required memory bandwidth. *Proc. 9th ACM/IEEE Int. Symp. on System-Level Synthesis (ISSS)*, pp. 127–132, La Jolla, CA, Nov. 1996.

[32] P. Grun, N. Dutt, and A. Nicolau. Mist: an algorithm for memory miss traffic management. *Proc. IEEE Int. Conf. on CAD*, pp. 431–437, Santa Clara, CA, Nov. 2000.

[33] P. Slock, S. Wuytack, F. Catthoor, and G. Jong. Fast and extensive system-level memory exploration for ATM applications. *Proc. 10th ACM/IEEE Int. Symp. on System-Level Synthesis (ISSS)*, pp. 74–81, Antwerp, Belgium, Sep. 1997.

[34] P. Strobach. QSDPCM — a new technique in scene adaptive coding. *Proc. 4th Eur. Signal Processing Conf. (EUSIPCO)*, pp. 1141–1144, Grenoble, France, Sep. 1988.

[35] Radio broadcasting systems; digital audio broadcasting to mobile, portable and fixed receivers. Standard RE/JTC-00DAB-4, ETSI, ETS 300401, May 1997.

[36] K.C. Shashidar, A. Vandecappelle, and F. Catthoor, Low-power design of turbo decoder module with exploration of energy-performance trade-offs, *Workshop on Compilers and Operating Systems for Low Power (COLP '01)* in conjunction with *Int. Conf. on Parallel Architecture and Compilation Techniques (PACT),* Barcelona, Spain, pp. 10.1–10.6, Sept. 2001.

[37] D. Patterson and J. Hennessey, *Computer Architecture: A Quantitative Approach,* Morgan Kaufmann, San Francisco, CA, 1996.

[38] C. Kulkarni. Cache optimization for multimedia applications. Doctoral dissertation, ESAT/KUL, Belgium, 2001.

[39] P.R. Panda, H. Nakamura, N.D. Dutt, and A. Nicolau, A data alignment technique for improving cache performance, *Proc. IEEE Int. Conf. on Comput. Design,* Santa Clara, CA, pp. 587–592, Oct. 1997.

30

Networks on Chips: Energy-Efficient Design of SoC Interconnect

Luca Benini
University of Bologna

Terry Tao Ye
Giovanni De Micheli
Stanford University

30.1 Introduction

The challenge of designing systems on chip (SoCs) is related to both the large scale of integration and the small transistor features in the upcoming silicon technologies. Moreover, SoCs will be applied in many embedded systems, where reliability of operation and low-energy consumptions are key figures of merit.

It is a common belief that SoCs are designed using preexisting components, such as processors, controllers, and memory arrays. Design methodologies have to support component reuse in a plug-and-play fashion to be effective.

We think that the most critical factor in system integration will be related to the communication scheme among components. The implementation of on-chip communication largely affects the system correctness, reliability, and energy consumption. Indeed, technology trends foresee an increase in device density and frequency of operation, which both correlate to higher power consumption. Voltage down-scaling will mitigate the energy cost at the expenses of reduced signal integrity. Thus, future system designs will be based on a balancing act between performance, reliability, and energy consumption. This chapter will analyze this trade-off in the domain of on-chip component interconnection.

The challenges for on-chip interconnect stem from the physical properties of the interconnection wires. Propagation delays on global wires — spanning a significant fraction of the chip size — will carry signals whose propagation delay will exceed the clock period. Thus, signals on global wires will be pipelined. At the same time, the switched capacitance on global wires will constitute a significant fraction of the dynamic power dissipation. Moreover, estimating delays accurately will become increasingly harder, as wire geometries may be determined late in the design flow. Thus, the need for latency insensitive design is critical. The most likely synchronization paradigm for future chips is globally asynchronous locally synchronous (GALS), with many different clocks.

0-8493-1941-2/05/$0.00+$1.50

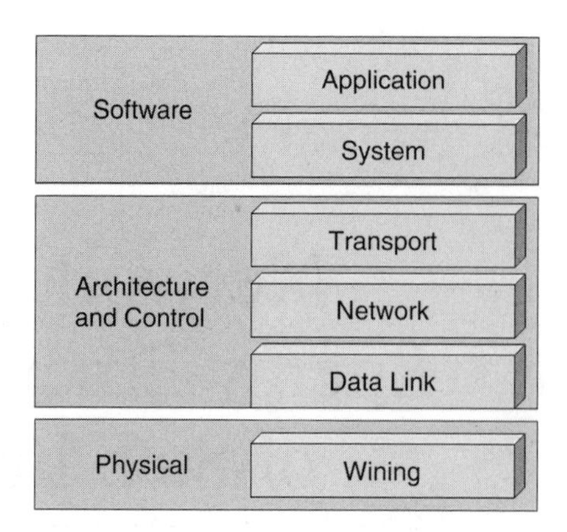

FIGURE 30.1 Micro-network stack.

SoC design will be guided by the principle of consuming the least possible power. This requirement matches the need of using SoCs in portable battery-powered electronic devices and of curtailing thermal dissipation, which can make chip operation infeasible or impractical. Energy considerations will impose small logic swings and power supplies, most likely below 1 V. Electrical noise due to crosstalk, electromagnetic interference (EMI), and radiation-induced charge injection (soft errors) will be likely to produce data upsets. Thus, the mere transmission of digital values on wires will be inherently unreliable.

To cope with these problems, we use network design technology to analyze and design SoCs modeled as micro-networks of components. The SoC interconnect design analysis and synthesis is based upon the micro-network stack paradigm, which is an adaptation of the protocol stack [30] (Figure 30.1) used in networking. This abstraction is useful for layering micro-network protocols and separating design issues belonging to different domains.

SoCs differ from wide-area networks because of local proximity and because they exhibit much less nondeterminism. In particular, micronetworks have a few distinctive characteristics, namely, energy constraints, design-time specialization, and low communication latency. This chapter addresses specifically the first problem.

Whereas computation and storage energy greatly benefits from device scaling (smaller gates, smaller memory cells), the energy for global communication does not scale down. On the contrary, projections based on current delay optimization techniques for global wires [15,28,29] demonstrate that global communication on chip will require increasingly higher energy consumption. Thus, communication-energy minimization will be a growing concern in future technologies. Furthermore, network traffic control and monitoring can help in better managing the power consumed by networked computational resources. For instance, clock speed and voltage of end nodes can be varied according to available network bandwidth. The emphasis on energy minimization creates a sleuth of novel challenges that have not been addressed by traditional high-performance network designers.

Design-time specialization is another facet of the SoC network design. Whereas macroscopic networks emphasize general-purpose communication and modularity, in SoCs networks, these constraints are less restrictive, because most on-chip solutions are proprietary. This degree of freedom can be used effectively to design low-energy communication schemes.

30.2 Micro-Networks: Architectures and Protocols

Much literature is available about architectures for macroscopic networks and, more specifically, for single-chip multi-processors [6,12,17]. These architectures can be classified by their topology, structure, and parameters. The most common on-chip communication architecture is the shared medium architecture, as exemplified by the shared bus. Unfortunately, bus performance and energy consumption are deeply penalized by the scaling up of the number of end nodes. Point-to-point architectures, such as mesh, torus, and hypercube, have been demonstrated to scale up despite a higher complexity in their design. Examples of recent micro-network architectures include Octagon [17], which is a direct network (i.e., with routers attached to the end node) consisting of eight nodes arranged as the vertices of an octagon and connected via the octagon sides and diameters. Octagon has the properties that any two nodes can be reached in two hops, and that more octagons can be connected by sharing an end node. The Nostrum network is a two-dimensional indirect mesh network (i.e., with routers separate from end nodes) [19]. The network performs the routing function and acts as the network interface for each node processor as well. Another example of a recent micro-network is SPIN [12], which is also an indirect network, and is built with a 4-ary fat-tree topology.

Communication in any given architecture is regulated by protocols, which are designed in layers. We analyze next specific issues related to the different layers of abstraction outlined in the micro-network stack in a bottom-up way.

30.2.1 Physical Layer

Global wires are the physical implementation of the communication channels. Physical layer signaling techniques for lossy transmission lines have been studied for a long time by high-speed board designers and microwave engineers [2,10].

Traditional rail-to-rail voltage signaling with capacitive termination, as used today for on-chip communication, is definitely not well suited for high-speed, low-energy communication on future global interconnects [10]. Reduced swing, current-mode transmission, as used in some processor-memory systems, can significantly reduce communication power dissipation while preserving speed of data communication.

Nevertheless, as the technology trends lead us to use smaller voltage swings and capacitances, the upset probabilities will rise. Thus, the trend toward faster and lower-power communication may decrease reliability as an unfortunate side effect. Reliability bounds as voltages scale can be derived from theoretical (entropic) considerations [14] and can be measured by experiments on real circuits.

We conjecture that a paradigm shift is needed to address the aforementioned challenges. Current design styles consider wiring-related effects as undesirable parasitics, and try to reduce or cancel them by specific and detailed physical design techniques. It is important to realize that a well-balanced design should not over-design wires so that their behavior approaches an ideal one, because the corresponding cost in performance, energy-efficiency, and modularity may be too high. Physical layer design should find a compromise between competing quality metrics and provide a clean and complete abstraction of channel characteristics to micro-network layers above.

30.2.2 Data Link, Network, and Transport Layers

The data-link layer abstracts the physical layer as an unreliable digital link, where the probability of bit upsets is nonnull (and increasing as technology scales down). Furthermore, reliability can be traded off for energy [14]. The main purpose of data-link protocols is to increase the reliability of the link up to a minimum required level, under the assumption that the physical layer by itself is not sufficiently reliable.

An additional source of errors is contention in shared-medium networks. Contention resolution is fundamentally a nondeterministic process, because it requires synchronization of a distributed system, and for this reason it can be considered as an additional noise source. Generally, nondeterminism can be virtually eliminated at the price of some performance penalty. For instance, centralized bus arbitration in a synchronous bus eliminates contention-induced errors, at the price of a substantial performance penalty caused by the slow bus clock and by bus request/release cycles.

Future high-performance, shared-medium on-chip micro-networks may evolve in the same direction as high-speed local area networks, where contention for a shared communication channel can cause errors, because two or more transmitters are allowed to concurrently send data on a shared medium. In this case, provisions must be made for dealing with contention-induced errors.

An effective way to deal with errors in communication is to packetize data. If data is sent on an unreliable channel in packets, error containment and recovery is easier, because the effect of errors is contained by packet boundaries, and error recovery can be carried out on a packet-by-packet basis. At the data link layer, error correction can be achieved by using standard error correcting codes (ECC) that add redundancy to the transferred information. Error correction can be complemented by several packet-based error detection and recovery protocols. Several parameters in these protocols (e.g., packet size and number of outstanding packets) can be adjusted depending on the goal to achieve maximum performance at a specified residual error probability and/or within given energy consumption bounds.

At the network layer, packetized data transmission can be customized by the choice of switching and routing algorithms. The former establishes the type of connection while the latter determines the path followed by a message through the network to its final destination. Popular packet switching techniques include store-and-forward, virtual cut-through, and wormhole. When these switching techniques are implemented in on-chip networks, they will have different performance metrics along with different requirements on hardware resources.

- Store-and-forward (SAF) routing inspects each packet's content before forwarding it to the next stage. While SAF enables more elaborated routing algorithms, (e.g., content-aware packet routing), it introduces extra packet delay at every router stage. Furthermore, SAF also requires a substantial amount of buffer spaces because the switches need to store multiple complete packets at the same time. Because on-chip storage resources (i.e., static random access memory (SRAMs) and dynamic random access memory (DRAMs)) are very expensive in terms of area and energy consumption, SAF approaches are not appropriate for on-chip communications.
- Virtual cut-through (VCT) routing can forward a packet to the next stage before its entirety is received by the current switch. Therefore, VCT switching reduces the store-and-forward delays. When the next stage switch is not available, however, the entire packet still needs to be stored in the buffers of the current switch.
- Wormhole routing was originally designed for parallel computer clusters [11] because it achieves the minimal network delay and requires fewer buffers. In wormhole routing, each packet is further segmented into flits (flow control unit). The header flit reserves the routing channel of each switch, the body flits will then follow the reserved channel, and the tail flit will later release the channel reservation.

One major advantage of wormhole routing is that it does not require the complete packet to be stored in the switch while waiting for the header flit to route to the next stages. Wormhole routing not only reduces the store-and-forward delay at each switch, but it also requires much less buffer space. Because of these advantages, wormhole routing is an ideal candidate switching technique for on-chip interconnect networks [7].

In wormhole routing, one packet may occupy several intermediate switches at the same time. Thus, it may block the transmission of other packets. Deadlock and livelock are the potential problems in wormhole routing schemes [9,11].

At the transport layer, algorithms deal with the decomposition of messages into packets at the source and their assembly at destination. Packetization granularity is a critical design decision, because the behavior of most network control algorithms is very sensitive to packet size. Packet size can be application-specific in SoCs, as opposed to general networks. In general, flow control and negotiation can be based on either deterministic or statistical procedures. Deterministic approaches ensure that traffic meets specifications, and provide hard bounds on delays or message losses. The main disadvantage of deterministic techniques is that they are based on worst cases, and they generally lead to significant under-utilization of network resources. Statistical techniques are more efficient in terms of utilization, but they cannot provide worst-case guarantees. Similarly, from an energy viewpoint, we expect deterministic schemes to be more inefficient than statistical schemes, because of their implicit worst-case assumptions.

30.2.3 Software Layers

Current and future systems on chip will be highly programmable, and therefore their power consumption will critically depend on software aspects. Software layers comprise system and application software. The system software provides us with an abstraction of the underlying hardware platform, which can be leveraged by the application developer to exploit the hardware's capabilities safely and effectively.

Current SoC software development platforms are mostly geared toward single microcontroller with multiple coprocessor architectures. Most of the system software runs on the control processor, which orchestrates the system activity and farms off computationally intensive tasks to domain-specific coprocessors. Micro-controller–coprocessor communication is usually not data-intensive (e.g., synchronization and reconfiguration information), and most high-bandwidth data communication (e.g., coprocessor–coprocessor and coprocessor–IO) is performed via shared memories and direct memory access (DMA) transfers. The orchestration activities in the micro-controller are performed via runtime services provided by single-processor real time operating systems (RTOSs) (e.g., VxWorks, Micro-OS, and Embedded Linuxes), which differentiate from standard operating systems in their enhanced modularity, reduced memory footprint, and support for real-time-scheduling and bounded time-interrupt service times.

Application programming is mostly based on manual partitioning and distribution of the most computationally intensive kernels to data coprocessors (e.g., very long instruction word (VLIW) multimedia engines, digital signal processors, etc.). After partitioning, different code generation and optimization toolchains are used for each target coprocessor and the control processor. Hand-optimization at the assembly level is still quite common for highly irregular signal processors, while advanced optimizing compilers are often used for VLIW engines and fine-grained reconfigurable fabrics. Explicit communication via shared memory is usually supported via storage classes declarations (e.g., noncacheable memory pages) and DMA transfers from and to shared memories are usually set up via specialized system calls which access the memory-mapped control registers of the DMA engines.

Even from this cursory analysis, the poor scalability in a network on chip (NoC) setting of current software abstractions and runtime environments is evident. In our view, the most critical issues are the following:

- Confining the OS onto a single centralized micro-controller is a sensible choice for small-to-medium scale and asymmetric multi-processing architectures, but this choice is bound to create a performance bottleneck and significant power overhead as architectures become more symmetric and scale up in complexity and parallelism, resulting in a significant energy inefficiency. This is because all centralized control functions and policies will require communication (often under tight real-time constrains) to all peripheral processors. Even worse, a centralized OS would need to continuously collect information on all system components to maintain an updated system state snapshot. The cost in performance and power of system control and monitoring is significant and could either lead to an over budgeting of NoC resources (e.g., dedicated control channels) if quality-of-service guarantees (e.g., bounded control message delivery delay) must be provided, or to uncertain and unreliable operation in case of a best-effort network service.
- The manual and ad-hoc partitioning and workload distribution procedure is too slow and error-prone in parallel, large-scale applications and target architectures. Furthermore, the lack of communication analysis tools may lead to highly inefficient task mappings. From a performance viewpoint, a communication sub-optimal task mapping leads to reduced throughput and/or high latency. The energy implications can be even more serious, because in many cases reduced performance is caused by local congestion, which is a high-occupancy condition for network resources and implies high power consumption. Thus, energy efficiency decreases quadratically (high power and low performance).
- Current programming styles are based on a shared memory paradigm, which is quite natural and well suited for tightly coupled, small-scale clusters. Unfortunately, shared memory abstraction tends to hide the cost and unpredictability of communication, which are destined to grow in an

NoC setting. Furthermore, DMA burst transfers, often advocated as a mean to increase throughput in memory transfers, do increase the risk of starvation and the variance in delivery time of short, sporadic messages. From an energy viewpoint, we need to raise the level of awareness of programmers on the energy cost in accessing shared memories, especially when a single memory is shared among many multiple processors. Such a cost is only in part due to pure memory array access. Significant overheads are associated with communication and contention resolution.

In our view, software issues are among the most critical and less understood in NoC. We believe that the full potential of on-chip networks can be effectively exploited only if adequate software abstractions and programming aids are developed to support them.

30.3 Energy-Efficient Micro-Network Design

This section delves into a few specific instances of energy-efficient, micro-network design problems. In most cases, we also outline specific solutions that have been proposed in the literature, even though it should be clear that many design issues are open and significant progress in this area is expected in the near future.

30.3.1 Physical Layer

At the physical layer, low-swing signaling is actively investigated to reduce communication energy on global interconnects [36]. In the case of a simple CMOS driver, low-swing signaling is achieved by lowering the driver's supply voltage V_{dd}. This implies a quadratic dynamic power reduction (because $P_{dyn} = K\,V_{dd}^2$). Unfortunately, swing reduction at the transmitter complicates the receiver's design. Increased sensitivity and noise immunity are required to guarantee reliable data reception. Differential receivers have superior sensitivity and robustness, but they require doubling the bus width. To reduce the overhead, pseudo-differential schemes have been proposed, where a reference signal is shared among several bus lines and receivers, and incoming data is compared against the reference in each receiver. Pseudo-differential signaling reduces the number of signal transitions, but it has reduced noise margins with respect to fully differential signaling. Thus, reduced switching activity is counterbalanced by higher swings and determining the minimum-energy solution requires careful circuit-level analysis.

Dynamic voltage scaling has been recently applied to busses [26,33]. In Worm et al. [33], the voltage swing on communication busses is reduced, even though signal integrity is partially compromised. Encoding techniques are used to detect corrupted data that is retransmitted. The retransimission rate is an input to a closed-loop DVS control scheme, which sets the voltage swing at a trade-off point between energy saving and latency penalty (due to data retransmission).

Another key physical-layer issue is synchronization. Traditional on-chip communication has been based on the synchronous assumption, which implies the presence of global synchronization signals (i.e., clocks) that define data sampling instants throughout the chip. Unfortunately, clocks are extremely energy-inefficient, and it is a well-known fact that they are responsible for a significant fraction of the power budget in digital integrated systems. Thus, postulating global synchronization when designing on-chip micronetworks is not an optimal choice from the energy viewpoint. Alternative on-chip synchronization protocols that do not require the presence of a global clock have been proposed in the past [3,37], but their effectiveness has not been studied in detail from the energy viewpoint.

30.3.2 Data-Link Layer

At the data-link layer, a key challenge is to achieve the specified communication reliability level with minimum energy expense. Several error recovery mechanisms developed for macroscopic networks can be deployed in on-chip micronetworks, but their energy efficiency should be carefully assessed in this context. As a practical example, consider two alternative reliability-enhancement techniques: error-correcting codes and error-detecting codes with retransmission. A set of experiments involved applying error correcting and detecting codes to an AMBA bus and comparing the energy consumption in four cases [4]:

1. Original unencoded data
2. Single-error correction
3. Single-error correction and double-error detection
4. Multiple-error detection

Hamming codes were used. Note that in case 3, a detected double error requires retransmission. In case 4, using (n,k) linear codes, there are $(2^n - 2^k)$ error patterns of length n that can be detected. In all cases, some errors may go undetected and be catastrophic. Using the property of the codes, it is possible to map the mean time to failure (MTTF) requirement into bit upset probabilities, and thus comparing the effectiveness of the encoding scheme in a given noisy channel (characterized by the upset probability) in meeting the MTTF target.

The energy efficiency of various encoding schemes varies: we summarize here one interesting case, where three assumptions apply. First, wires are long enough so that the corresponding energy dissipation dominates encoding/decoding energy. Second, voltage swing can be lowered until the MTTF target is met. Third, upset probabilities are computed using a white Gaussian noise model [13]. Figure 30.2 gives the average energy per useful bit as a function of the MTTF (which is the inverse of the residual word error probability). In particular, for reliable SoCs (i.e., for MTTF = 1 year), multiple-error detection with retransmission is demonstrated as more efficient than error-correcting schemes. We refer the reader to Bertozzi et al. [4] for results under different assumptions.

Another important aspect affecting the energy consumption is the media access control (MAC) function. Currently, centralized time-division multiplexing schemes (also called centralized arbitration) are widely adopted [1,8,32]. In these schemes, a single arbiter circuit decides which transmitter accesses to the bus for every time slot. Unfortunately, the poor scalability of centralized arbitration indicates that this approach is likely to be energy-inefficient as micronetwork complexity scales up. In fact, the energy cost of communicating with the arbiter, and the hardware complexity of the arbiter itself scales up more than linearly with the number of bus masters.

Distributed arbitration schemes as well as alternative multiplexing approaches, such as code division multiplexing, have been extensively adopted in shared-medium macroscopic network, and are actively investigated for on-chip communication [34]. Research in this area is just burgeoning, however, and significant work is needed to develop energy-aware media-access-control for future micronetworks.

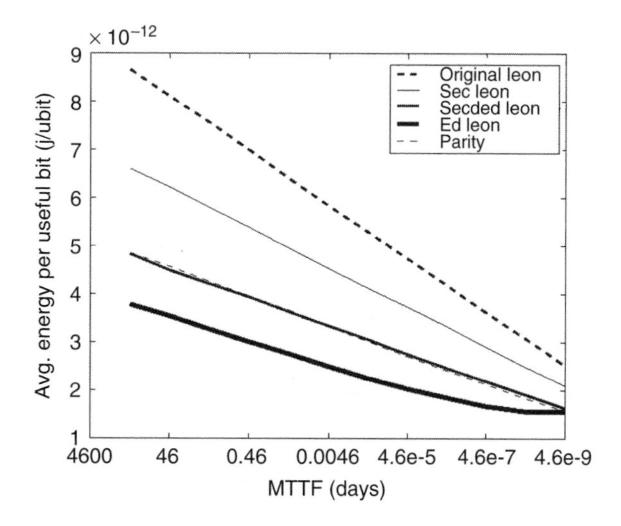

FIGURE 30.2 Energy efficiency for various encoding schemes.

30.3.3 Network Layer

Switching and routing for on-chip micro-networks affect heavily performance and energy consumption, whereas contention plays an important role. On one hand, contention delays packet transmission. On the other hand, resolving contention requires packets to be stored temporarily on the storage elements (on-chip SRAMs or DRAMs), which will increase power consumption significantly.

30.3.3.1 Contention-Look-Ahead Routing

A contention-look-ahead routing scheme is the one where the current routing decision is helped by monitoring the adjacent switches, thus possibly avoiding or reducing blockages and contention in the coming stages.

A contention-aware routing scheme is described in Nilsson [23]. The routing decision at every node is based on the "stress values" (the traffic loads of the neighbors) that are propagated between neighboring nodes. This scheme is effective in avoiding "hot spots" in the network. The routing decision steers the packets to less congested nodes.

To solve the contention problems in the wormhole routing schemes, we propose a contention-look-ahead routing algorithm that can "foresee" the contention and delays in the coming stages using a direct connection from the neighboring nodes. We use a two-dimensional mesh on-chip multiprocessor network to further explain and implement this routing algorithm. The processors are connected directly to each other in a tile-array formation, similar to that proposed by Dally and Toles [7]. Each processor tile performs packet routing and arbitration independently. The major difference from Nilsson [23] is that information is handled in flits, and thus packets with large or variable sizes can be handled with limited input buffers. Furthermore, because it avoids contention between packets and requires much less buffer usage, the proposed contention-look-ahead routing scheme can greatly reduce the network power consumption.

30.3.3.2 Wormhole Contention-Look-Ahead Algorithm

At every intermediate stage, there may be many alternate routes to go to the next stage. We call the route that always leads the packet closer to the destination a profitable route. Conversely, a route that leads the packet away from the destination is called misroute [11] (Figure 30.3). In mesh networks, profitable routes and misroutes can be distinguished by comparing the current node ID with the destination node ID.

Profitable routes will guarantee a shortest path from source to destination. Nevertheless, misroutes do not necessarily need to be avoided. Occasionally, the buffer queues in all available profitable routes are full, or the queues are too long. Thus, detouring to a misroute may lead to a shorter delay time. Under these circumstances, a misroute may be more desirable.

Any packet entering an intermediate switch along a path finds a set C of output channels to exit. As an example, for a two-dimensional mesh, $C = \{\text{North, South, East, West}\}$. We further partition C into profitable routes P and misroutes M. We define the buffer queue length of every profitable route $p \in P$ as Q_p. Similarly, we define the buffer queue length of every misroute $m \in M$ as Q_m.

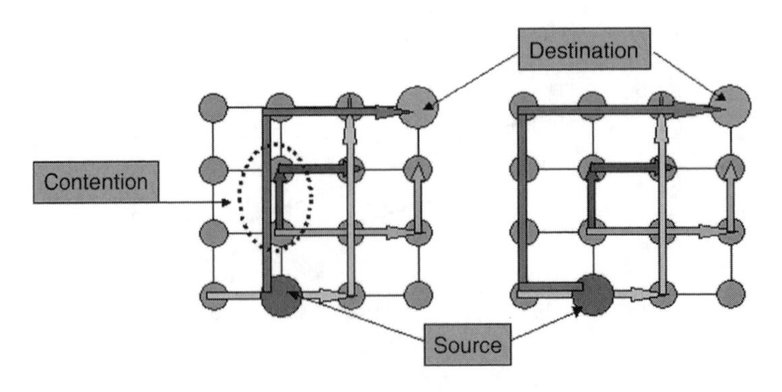

FIGURE 30.3 Profitable route and misroute.

Assume the flit delay of one buffer stage is D_B, and the flit delay of one switch stage is *DS*. The delay penalty to take a profitable and a misroute is defined as D_{profit} and $D_{misroute}$, respectively, in the following equation (Equation 30.1).

$$D_{profit} = min(D_B \times Q_p) \forall p \in P \tag{30.1}$$

$$D_{misroute} = min(D_B \times Q_m + 2D_S) \forall m \in M \tag{30.2}$$

In a mesh network, when a switch routes a packet to a misroute, the packet moves away from its destination by one switch stage. In the subsequent routing steps, this packet needs to get back on track and route one more stage back toward its destination. Therefore, the delay penalty for a misroute is $2 \times D_S$. The delay D_S can be estimated beforehand, and, without loss of generality, we assume the same D_S value for all switches in the network.

If all profitable routes are available and waiting queues are free, the packet will use profitable routing decision. If the buffer queues on all of the profitable routes are full or the minimum delay penalty of all the profitable routes is larger than the minimum penalty of the misroutes, it is more desirable to take the misroute (Equation 30.3):

$$(D_{profit} \leq D_{misroute})(Q_p \leq Q_{p_{max}} \quad \forall p \in P) ? ProRoute\colon Misroute \tag{30.3}$$

where Q_{pmax} is the maximum buffer queue length (buffer limit). This routing algorithm is heuristic, because it can only "foresee" one step ahead of the network. It provides a local best solution but does not guarantee the global optimum.

30.3.3.3 Network Power Consumption

This routing scheme was simulated with RSIM, a multiprocessor instruction level simulator, using 16 reduced instruction set computer (RISC) processors connected in a 4×4 (4-ary 2-cube) mesh network. Control wires that deliver the input queue information to the adjacent switches also connect adjacent processors.

The contention-look-ahead routing algorithm is compared with dimension-ordered routing — a routing scheme that always routes the packets on one dimension first, upon reaching the destination row or column, then switch to the other dimension until reaching the destination. Dimension-ordered routing is deterministic and guarantees shortest path, but it cannot avoid contention. The comparison is performed on four benchmarks: quicksort, fft, lu, and sor. These benchmarks are ported from Stanford SPLASH suite [27] and running on the RSIM simulation platform.

On-chip network power consumption comes from three contributors:

1. The interconnect wires
2. The buffers
3. The switch logic circuits

A network power consumption estimation technique is proposed by Ye et al. [35], and we will use it in the experiments.

The contention-look-ahead routing will reduce the power consumption on the buffers because it can "foresee" the contention in the forthcoming stages and shorten the buffer queue length. Figure 30.4(a) presents the averaged buffer power reduction of different benchmarks. The reduction is more significant under larger buffer sizes. This is because larger buffers will consume more power, and the power consumption is more sensitive with contention occurrence.

The power consumption on the interconnect can be estimated by counting the average number of hops a packet travels from source to destination. Dimension-ordered routing always steers the packets along the shortest path. In comparison, our proposed routing scheme may choose the misroute when contention occurs. Therefore, the contention-look-ahead routing has larger average hop count per

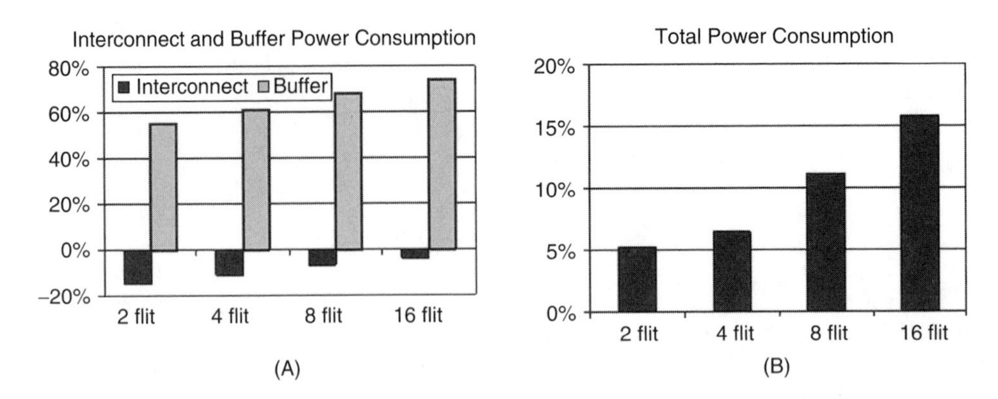

FIGURE 30.4 Power consumption comparison on interconnect wires and buffers.

packet than the dimension-ordered routing, and consequently consumes more power on the interconnects. Figure 30.4(A) depicts the proposed routing scheme consumes more power (presented as negative values) with smaller buffer size, this is because smaller buffer sizes will cause more contention and induce more misroutes.

The contention-look-ahead routing switch needs more logic gates than dimension-ordered routing. From synopsys power compiler simulation, the proposed switch circuit consumes about 4.8% more power than dimension-ordered switch. Combining the power consumption on the interconnects and buffers, the total network power consumption is depicted in Figure 30.4(B). It presents the total network power reduction compared with dimension-ordered routing. The reduction is more significant with larger buffer sizes (15.2% with 16-flit buffers).

30.3.3.3.1 Transport Layer

Above the network layer, the communication abstraction is an end-to-end connection. The transport layer is concerned with optimizing the usage of network resources and providing a requested quality of service. Clearly, energy can be considered as a network resource or a component in a quality-of-service metric. An example of transport-layer design issue is the choice of information decomposition into packets or flits, as well as the choice of packet size. Energy efficiency can be heavily impacted by this decision. Next, we will use the shared-memory multi-processor system on chip (MPSoC) as a case study to analysis the packet size trade-offs both qualitatively and quantitatively.

A typical shared-memory MPSoC architecture is illustrated in Figure 30.5. The MPSoC power consumption originates from three sources:

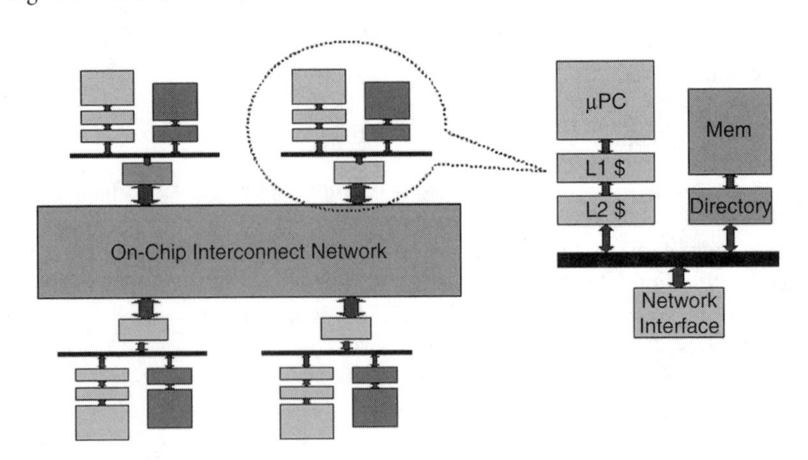

FIGURE 30.5 MPSoC architecture.

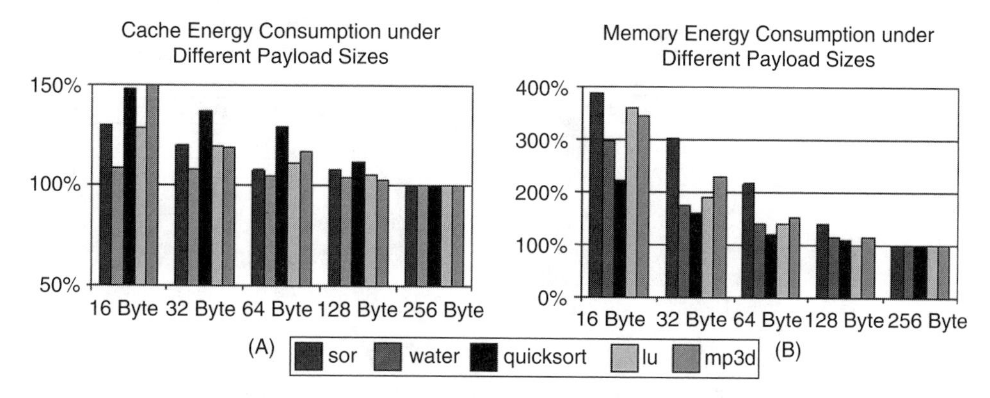

FIGURE 30.6 Cache and memory energy decrease as packet payload size increases.

1. The node processor power consumption
2. The cache and shared memory power consumption
3. The interconnect network power consumption

We will start first from the cache and memory analysis.

30.3.3.4 Cache and Memory Power Consumption

Whenever there is a cache miss, the cache block content needs to be encapsulated inside the packet payload and sent across the network. In shared-memory MPSoC, the cache block size correlates with the packet payload size. Larger packet sizes will decrease the cache miss rate, because more cache content can be updated in one memory access. Consequently, both cache energy consumption and memory energy consumption will be reduced. This relationship can be observed in Figure 30.6. It depicts the energy consumption by cache and memory under different packet sizes. The energy in the figure is normalized to the value of 256 Bytes, which achieves the minimum energy consumption.

30.3.3.5 Interconnect Network Power Consumption

The power consumption of packetized dataflow on MPSoC network is determined by three factors. The effects of these factors are summarized and listed next:

1. The number of packets on network. Packets with larger payload size will decrease the cache miss rate and consequently decrease the number of packets on the network. This effect can be observed in Figure 30.7(A). It gives the average number of packets on the network (traffic density) at one clock cycle. As the packet size increases, the number of packets decreases accordingly.

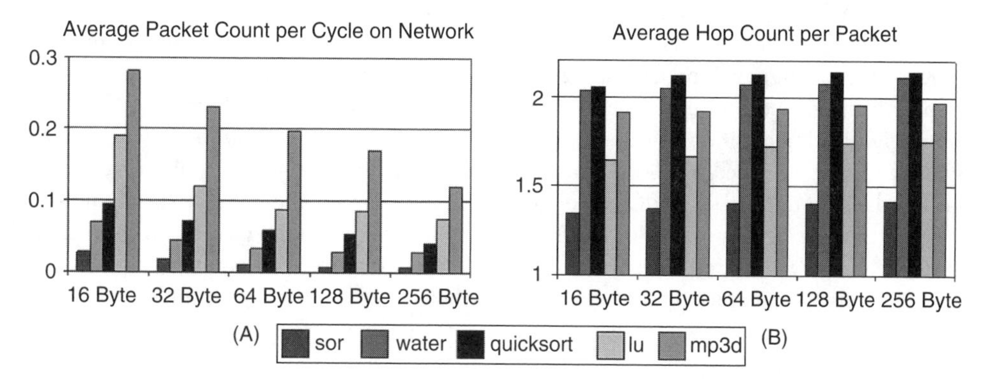

FIGURE 30.7 Packet count and hop count per packet under different payload sizes.

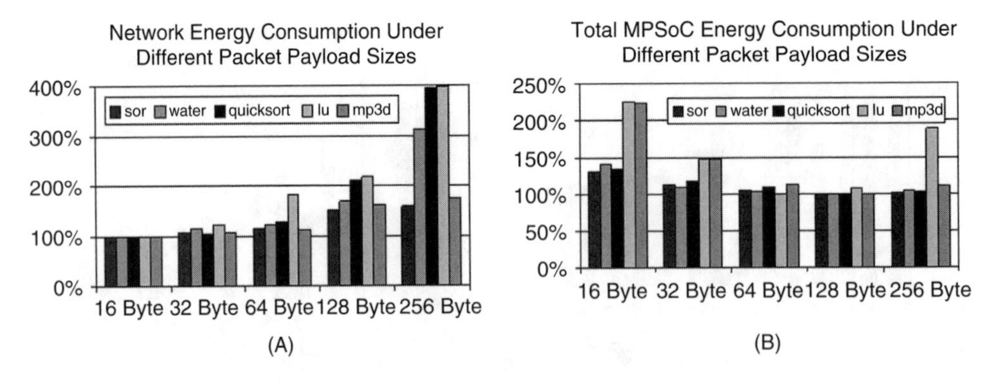

FIGURE 30.8 Network and total MPSoC energy consumption under different packet payload sizes.

2. The energy consumed by each packet on one hop. Larger packet size will increase the energy consumed per packet, because there are more bits in the payload.
3. The number of hops each packet travels. Larger packets will occupy the intermediate node switches for a longer time, and cause other packets to be rerouted to longer paths. This leads to more contention that will increase the total number of hops needed for packets traveling from source to destination. Figure 30.7(B) illustrates the effect. As packet size (payload size) increases, average hop count per packet increases as well.

Actually, increasing the cache block size will not decrease the cache miss rate proportionally. Therefore, the decrease of packet count cannot compensate for the increase of energy consumed per packet caused by the increase of packet length. Larger packet size also increases the hop counts on the datapath. Figure 30.8(A) presents the combined effects of these factors. The values are normalized to the measurement of 16 Bytes. As packet size increases, energy consumption on the interconnect network will increase.

The total energy dissipated on MPSoC comes from noncache instructions (instructions that do not involve cache access) of each node processors, the caches and the shared memories as well as the interconnect network. The overall results are given in Figure 30.8(B). From this figure, we can see that the total MPSoC energy will decrease as packet size increases. When the packets are too large, however, as in the case of 256 Bytes in the figure, the total MPSoC energy will increase. This is because when the packet is too large, the increase of interconnect network energy will outgrow the decrease of energy on cache and memories. In our simulation, the noncache instruction energy consumption does not change significantly under different packet sizes.

30.3.3.5.1 Application and System Layer

As hinted in Section 30.2, software layers are critical for the NoC paradigm shift, especially when energy efficiency is a requirement. As outlined in the previous sections, NoCs have the potential for overcoming many of the energy bottlenecks of current integrated architectures (i.e., globally shared communication and storage blocks), but only if programming abstractions, development tools, and system software help programmers understand communication-related costs and how to cope with them.

From a high-level application viewpoint, multi-processor SoC platforms can be viewed as networks of computing nodes equipped with local storage. Computation and storage are highly energy efficient if confined to the local resources within a node. Communication cost should be made explicit throughout all steps of the code development flow. Software analysis tools should help designers in identifying communication bottlenecks and code optimizers should heavily emphasize communication cost reduction. Many effective techniques have been devised in the area of parallel programming for large-scale supercomputers, and there is good potential for leveraging these experiences. It is important, however, to point out three key differences:

1. Target MPSoC architectures are much more heterogeneous than general-purpose parallel computers.

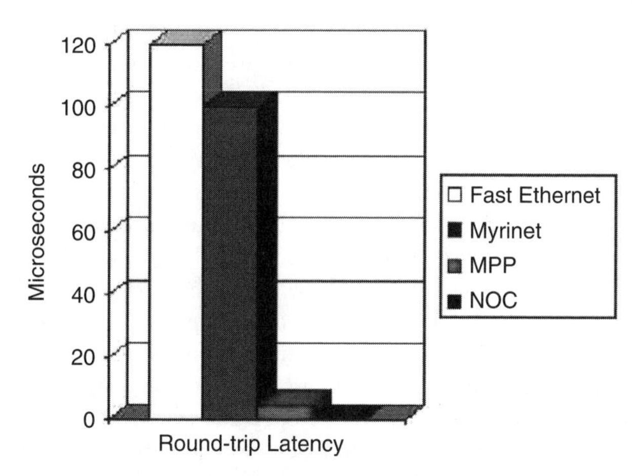

FIGURE 30.9 Interconnect latency for different parallel machines.

2. Physical link latency, albeit significant, is not nearly as dominant in on-chip networks as it is in macroscopic parallel computers (see Figure 30.9).
3. Energy constraints are extremely tight, while they have never been significant in traditional parallel machines.

The following paragraphs outline four critical areas for the evolution of energy efficient software layers in current and future NoCs. We survey seminal contributions and identify critical needs.

1. Programming abstractions. Developing adequate abstractions for NoC programming is a critical objective. Dataflow programming abstractions, such as streams [18] and Kahn networks [22], are based on a model of computation that matches very well NoC architectures. With these abstractions, communication is made explicit starting from the early steps of application development, because data flow is explicitly represented. At these levels, energy efficiency can be pursued by minimizing redundant communication, and by carefully balancing local computation and communication costs. A critical need in this area is the definition of hardware platform dependent high-level metrics, such as energy per local operation and energy per transmitted bit, which can help in first-cut exploration of the communication vs. computation trade-off during algorithm development. Unfortunately, even though many digital signal processing and multimedia applications are developed starting from dataflow models, numerous legacy applications use more traditional programming styles, where tasks are not clearly decoupled and communication is implicitly performed through memory. Leveraging the existing code basis, without compromising performance and energy efficiency, is today an open challenge.
2. Task-level analysis and optimization. A number of interesting opportunities are open for high-level optimization tools, which can help designers mapping data-flow specifications onto target hardware platforms. Consider, for instance, task splitting and merging (i.e., distributing the computation performed by a task among two or more computational nodes and collapsing two or more tasks onto the same node), task allocation, as well as communication splitting and merging over available physical NoC links. Even though a few of these problems have been explored in preliminary works [16,25], we critically need high-level energy models and analysis tools to explore techniques for increasing energy efficiency. It is important to acknowledge that energy-optimal solutions can differ significantly from performance-optimal ones in this context. Consider, for instance, a situation where available computational resources are used to achieve marginal performance benefits (e.g., a task is speculatively executed), at a price of significantly increased power consumption.

3. Code optimization. Classical code optimization, at the single task level, will still hold an important place in future NoC software development. In this area we view as critical further developments of two types of code optimizers: tools for parallelism extraction from a single task or a legacy applications [31]; tools that reduce the memory footprint and improve access locality for both code and data [24]. The first class of tools represents enabling technology that enables the reuse of legacy software as well as task splitting. The second class has a critical role in reducing "implicit" communication (as opposed to "explicit" data flow communication) from/to large background memories, which ensues because of large working sets that do not fit into local node memory. Another critical area in code optimization is the development of highly efficient communication primitives, possibly with significant dedicated hardware support. The latency and energy consumption associated with software handling of communication primitives (e.g., message send or receive) in traditional parallel programming libraries are simply unacceptable in an NoC setting [5].
4. Distributed operating systems. Intuitively, the operating system support NoC operation cannot be centralized. Truly distributed embedded OSes are required [5,6] to create a scalable runtime system. In addition to traditional functions (i.e., scheduling, interrupt handling), the NoC OS should natively support power management. End-nodes (processing elements) in SoC micronetworks will most likely be power-manageable "voltage islands" [20], with individually controllable clock speeds and supply voltages. One of the key tasks of the system software will be to control the voltage islands power states. We can envision a network-centric approach, where components send messages to neighbors to request state changes [21]. Such requests are originated and serviced at the system software levels. For example, an image processor can be required to raise its service levels before receiving a stream of data. In this case, the system software supports policies that accept requests from other components and perform transitions according to such requests.

30.4 Conclusions

The challenges of designing SoCs in 50- to 100-nm technologies available in the second part of this decade include coping with design complexity and providing reliable, high-performance operation and minimizing energy consumption. Starting from the observation that interconnect technology will be the limiting factor for achieving the operational goals, we envisioned a communication-centric view of design. We focused on energy efficiency issues in designing the communication infrastructure for future SoCs. We described several open problems at various layers of the communication stack, and we outlined basic strategies to effectively tackle the energy efficiency challenge for on-chip communication networks.

References

[1] P. Aldworth, System-on-a-chip bus architecture for embedded applications, *IEEE Int. Conference on Comput. Design*, pp. 297–298, 1999.
[2] H. Bakoglu, *Circuits, Interconnections, and Packaging for VLSI*, Addison-Wesley, Reading, MA, 1990.
[3] W.J. Bainbridge and S.B. Furber, Delay insensitive system-on-chip interconnect using 1-of-4 data encoding, *IEEE Int. Symp. on Asynchronous Circuits and Syst.*, March 2001, pp. 118–126.
[4] D. Bertozzi, L. Benini, and G. De Micheli, Low power error resilient encoding for on-chip data busses, *Proc. Int. Conf. on Design and Test Europe*, Paris, France, March 2000, pp. 102–109.
[5] D. Bertozzi, F. Poletti, L. Benini, and A. Bogliolo, Performance analysis of arbitration policies for SoC communication architectures, *J. Design Automation for Embedded Sys.*, Vol. 8, June–Sept. 2003, pp. 189–210.
[6] W.O. Cesario, D. Lyonnard, G. Nicolescu, Y. Paviot, S. Yoo, L. Gauthier, M. Diaz-Nava, and A.A. Jerraya, Multiprocessor SoC platforms: a component-based design approach *IEEE Design andTest of Comput.*, Vol. 19 No. 6, Nov.–Dec., 2002
[7] W. Dally and B. Towles, Route packets, not wires: on-chip interconnection networks, *Proc. 38th Design Automation Conf.*, June 2001, pp. 684–689.

[8] B. Cordan, An efficient bus architecture for system on chip design, *IEEE Custom Integrated Circuits Conf.*, May 1999, pp. 623–626.
[9] W.J. Dally and H. Aoki, Deadlock-free adaptive routing in multicomputer networks using virtual channels *IEEE Trans. on Parallel and Distributed Syst.*, April 1993, pp. 466–475.
[10] W. Dally and J. Poulton, *Digital System Engineering*, Cambridge University Press, New York, 1998.
[11] J. Duato, S. Yalamanchili, and L. Ni, *Interconnection Networks: An Engineering Approach,* IEEE Computer Society Press, Washington, D.C., 1997.
[12] P. Guerrier and A. Greiner, A generic architecture for on-chip packet-switched interconnections, *Proc. Int. Conf. on Design Automation and Test in Europe,* March 2000, pp. 250–256.
[13] R. Hegde and N. Shanbhag, Toward Achieving Energy Efficiency in Presence of Deep Submicron Noise, *IEEE Trans. on VLSI Syst.*, pp. 379–391, Vol. 8, No. 4, August 2000.
[14] R. Hegde and N. Shanbhag, Toward achieving energy efficiency in presence of deep submicron noise, *IEEE Trans. on VLSI Syst.*, pp. 379–391, Vol. 8, No. 4, August 2000.
[15] R. Ho, K. Mai, and M. Horowitz, The future of wires, *Proc. IEEE,* April 2001, pp. 490–504.
[16] J. Hu and R. Marculescu, Energy-aware mapping for tile-based NOC architectures under performance constraints, *Proc. ASP Design Automation Conf.*, Jan. 2003, pp. 233–239.
[17] F. Karim, A. Nguyen, and S. Dey, On-chip communication architecture for OC-768 network processors, *Proc. 38th Design Automation Conf.*, June 2001, pp. 678–683.
[18] B. Khailany et al. Imagine: Media Processing with Streams, IEEE *Micro* vol. 21, no. 2, pp. 35–46, 2001.
[19] S. Kumar, A. Jantsch, J. Soininen, M. Forsell, M. Millberg, J. Oberg, K. Tiensyrij, and A. Hemani, A network on chip architecture and design methodology, *Proc. IEEE Computer Society Annual Symp. on VLSI,* April 2002, pp. 105–112.
[20] D. Lackey, P. Zuchowski, T. Bednar, D. Stout, S. Gould and J. Cohn, Managing power and performance for systems on chip design using voltage islands, *ICCAD – Int. Conf. on Computer-Aided Design,* Nov. 2002, pp. 195–202.
[21] A. Laffely, J. Liang, P. Jain, N. Weng, W. Burleson, and R. Tessier, Adaptive systems on a chip (aSoC) for low-power signal processing, *35th Asilomar Conf. on Signals, Syst., and Comput.*, Nov. 2001, pp. 1217–1221.
[22] P. Lieverse, P. van der Wolf, K. Vissers, and E. Deprettere, A methodology for architecture exploration of heterogeneous signal processing systems *J. VLSI Signal Process. for Signal, Image and Video Technol.*, Vol. 29, No. 3, pp. 197–207, 2001.
[23] E. Nilsson Design and implementation of a hot-potato switch in a network on chip, M.S. thesis, Department of Microelectronics and Information Technology, Royal Institute of Technology, Stockholm, Sweden, June 2002.
[24] P. R. Panda, N. D. Dutt, A. Nicolau, F. Catthoor, A. Vandecappelle, E. Brockmeyer, C. Kulkarni, and E. de Greef, Data memory organization and optimizations in application-specific systems, *IEEE Design and Test of Comput.*, Vol. 18, No. 3, May–June 2001.
[25] A. Pinto, L Carloni, and A. Sangiovanni-Vincentelli, Constraint-driven communication synthesis,' *Design Automation Conf.*, June 2002, pp. 783–788.
[26] L. Shang, L.-S. Peh, and N.K. Jha, Dynamic voltage scaling with links for power optimization of interconnection networks, *HPCA — Proc. Int. Symp. on High-Performance Computer Architecture,* Anaheim, CA, February 2003, pp. 91–102.
[27] J.P. Singh, W. Weber, and A. Gupta, SPLASH: Stanford parallel applications for shared-memory *Computer Architecture News,* Vol. 20, No. 1, March 1992, pp. 5–44.
[28] D. Sylvester and K. Keutzer, A global wiring paradigm for deep submicron design, *IEEE Trans. on CAD/ICAS,* Vol. 19, No. 2, February 2000, pp. 242–252.
[29] T. Theis, The future of Interconnection Technology, *IBM J. Res. and Dev.*, Vol. 44, No. 3, May 2000, pp. 379–390.
[30] J. Walrand and P. Varaiya, *High-Performance Communication Networks,* Morgan Kaufman, San Francisco, 2000.

[31] M. Wolfe, *High-Performance Compilers for Parallel Computing*, Addison-Wesley, Reading, MA, 1995.

[32] S. Winegarden, A bus architecture centric configurable processor system, *IEEE Custom Integrated Circuits Conf.*, May 1999, pp. 627–630.

[33] F. Worm, P. Ienne, P. Thiran, and G. De Micheli, An Adaptive low-power transmission scheme for on-chip networks, *ISSS, Proc. Int. Symp. on System Synthesis*, Kyoto, Japan, October 2002, pp. 92–100.

[34] R. Yoshimura, T. Koat, S. Hatanaka, T. Matsuoka, and K. Taniguchi, DS-CDMA wired bus with simple interconnection topology for parallel processing system LSIs, *IEEE Solid-State Circuits Conf.*, January 2000, pp. 371.

[35] T.T. Ye, L. Benini, and G. De Micheli, Packetized on-chip interconnect communication analysis for MPSoC, *Proc. on Design Automation and Test in Europe*, March 2003, pp. 344–349.

[36] H. Zhang, V. George, and J. Rabaey, Low-swing on-chip signaling techniques: effectiveness and robustness, *IEEE Trans. on VLSI Syst.*, Vol. 8, No. 3, pp. 264–272, June 2000.

[37] H. Zhang, M. Wan, V. George, and J. Rabaey, Interconnect architecture exploration for low-energy configurable single-chip DSPs, *IEEE Computer Society Workshop on VLSI*, April 1999, pp. 2–8.

31

Highly Integrated Ultra-Low Power RF Transceivers for Wireless Sensor Networks

Brian P. Otis
Yuen Hui Chee
Richard Lu
Nathan M. Pletcher
Jan M. Rabaey
University of California—Berkeley

Simone Gambini
Universita di Pisa

31.1 Introduction

31.1.1 Motivation

Technological advances have made it conceivable to build and deploy dense wireless networks of heterogeneous nodes collecting and disseminating wide ranges of environmental data [1]. An inspired reader can easily imagine a multiplicity of scenarios in which these sensor and actuator networks might excel. To mention just a few: environmental control in office buildings, robot control and guidance in automatic manufacturing environments, warehouse inventory, integrated patient monitoring, diagnostics and drug administration in hospitals, interactive toys, the smart home providing security, identification and personalization, and interactive museums. The overwhelming opportunities emerging from this technology

0-8493-1941-2/05/$0.00+$1.50

TABLE 31.1 Power Density of Energy Scavenging Sources

Power Source	Power Density ($\mu W/cm^2$)	Lifetime
Lithium battery	100	1 year
Solar cell	10–15,000 (in $\mu W/cm^2$)	∞
Vibrational converter	300	∞

indeed give rise to new definitions of distributed computing and user interface. Regardless of the specific application, however, they all rely on a network of ubiquitously distributed sensor, compute, and actuation nodes, which are integrated and embedded into the fabrics of our daily living environment. This explains why the name "ambient intelligence" is often attributed to such environments [2].

Widespread deployment of wireless sensor networks requires that some economic and physical realities be met. More precisely, the physical implementation of an individual network node is constrained by three important metrics: power, cost, and size. Of these three, power (or energy, depending on how the node is powered) turns out to be the most fundamental metric. To keep cost down and to allow for a flexible deployment, most nodes must be untethered. Cost considerations also dictate that frequent replacement of the energy source of the node (especially in a ubiquitous deployment scenario) is out of the question. This leads to the general guideline that a network node must be self-sufficient from an energy perspective for the lifetime of the product. This could be multiple years for applications such as smart homes. The energy storage capability of a node is limited by the storage medium (battery or capacitor) and the size constraints [3]. Although a single-time charge could work for applications with life cycles below one year, replenishment of the energy supply using energy scavenging is often a necessity. As a result, the average power dissipation of a node is firmly capped at 1 mW. More realistically, average power dissipation levels around 100 μW are necessary given today's energy generation technologies. Table 31.1 illustrates the finite power density of state-of-the-art energy sources [3].

As listed in the table, the average power consumption of the sensor node must be very low if the energy scavenging volume is limited. From a volume of 1 cm^3, one or a combination of these power sources could supply an average continuous power output of 100 μW. Although this severely restricts the amount of processing that can be done within a node, it also determines the type of wireless connectivity that can be obtained between the nodes.

Ubiquitous deployment of these nodes is only economically feasible if the cost of the individual elements is ignorable, or, in other words, the electronics have become disposable. This translates to price points per node of less than $1. Achieving a node cost this low requires a minimal number of components, a high level of integration, simple and cheap packaging and assembly, and avoidance of any expensive components and/or technologies.

Finally, embedding the components into the daily environment (walls, furniture, clothing, etc.) further requires that the form factor of the entire sensor node must be very small. Typically, sizes smaller than 1cm^3 are necessary. Again, a very high level of integration is mandatory if such small dimensions are to be achieved.

In the design of these sensor nodes, we have experienced that the wireless interface takes up the largest fraction of the power and size budget of the node. Although the demands of the sensing and digital processing components cannot be ignored, their duty cycle is typically very low. Exploitation of advanced sleep and power-down techniques makes it possible to make their average power dissipation virtually ignorable. Thus, in the remainder of this chapter we will focus our discussion on the design of ultra-low power wireless interfaces for wireless sensor networks. Although optical communication approaches offer the potential for very low power and small size, line-of-sight and directivity considerations make them less attractive [4]. Thus, we will limit our discussion to radio-frequency (RF) interfaces.

The previous observations demonstrate that wireless sensor nodes occupy a unique corner of the semiconductor and embedded system design space, and, in a way, push against many traditional design boundaries.

31.1.2 Characteristics of Wireless Sensor Networks

One may wonder if and why an RF interface for a sensor network should be substantially different than the one used, for instance, in a wireless data network (local area network [LAN]). In fact, it turns out that the operation mode of the sensor node is so fundamentally dissimilar that completely different optimization criteria apply. This results mostly from the traffic patterns that affect the power dissipation profile of a node. Data packets in sensor networks tend to be relatively rare and unpredictable events. In most application scenarios, each node in the network sees at most a couple of packets/sec. In addition, the packets are relatively short (typically less than 200 bits/packet), as the payloads normally represent data measurements, which typically require a resolution of less than 24 bit/measurement. Combined, this means that the average data rate of a single node rarely exceeds 1 kbit/sec. These observations are of foremost importance when designing the wireless transceiver, as we will highlight in the following sections.

In the rest of the discussion, we will assume that the sensor networks of interest are dense, which means that the nodes in the network are placed relatively closely (i.e., the average distance between nodes is less than or equal to 10 m).*

31.1.3 Performance Metrics for Sensor Node RF Transceivers

31.1.3.1 Average Power Dissipation

Traditional quality metrics for radios used in wireless LANs are the data throughput (bit/sec), spectral efficiency (bit/sec/Hz), and energy-efficiency (nJ/bit). None of these is truly important for a wireless sensor node because the required average data rates are very low. Because the bits are few and the nodes are closely spaced, the energy/bit is not an important metric either. In fact, the power used for the actual data transmission and reception is only a fraction of the total power dissipated in the front-end. This is best illustrated with a statistical power model of the transceiver [5]. At any point in time, the transceiver is in one of the following states:

- Transmitting state (TX) during transmission of data.
- Receiving state (RX) during reception of data.
- Acquiring state (AQ) while acquiring synchronization at the start of the packet.
- Monitoring state (MN) when the transceiver is monitoring the channel (carrier sense).
- The idle state (IL) when the majority of the transceiver is turned off, and it is considered to be sleeping; it may be assumed that the power dissipation in this state is zero.

The average power dissipation of the transceiver is then expressed as

$$P_{av} = p_{TX}P_{TX} + p_{RX}P_{RX} + p_{AQ}P_{AQ} + p_{MN}P_{MN} + p_{IL}P_{IL}$$

where P_x is the average power dissipation in state x and p_x the probability the transceiver is in that particular state. The power dissipation in the TX state is determined mostly by the dissipation in the power amplifier. The average power dissipation of the RF front end in the three other modes (RX, AQ, and MN) is approximately equal, although P_{AQ} may dominate slightly.

Given the low data rates and duty cycles, the transceiver should be in the idle state for most of the time, given proper sleep disciplines. Among the four other states, the monitoring state (MN) is the most probable, as became apparent from simulations based on this model. Assuming a dense network and sparse traffic, the average power of the transceiver is well approximated as

$$P_{av} = p_{TX}P_{TX} + (p_{RX} + p_{AQ} + p_{MN})P_{RXon} \cong p_{MN}P_{RXon}$$

*In networks with a lower density, TX power rapidly becomes the dominant power factor.

where P_{RXon} is the dissipation when the receiver is on. It is thus fair to state that the average power is dominated by the power of having just the RF receiver turned on (e.g., low-noise amplifier [LNA], down-converter, and synthesizer), independent of the data activity. Minimizing the average power then translates into minimizing the active current draw of the RF front end, and, obviously, the time that the transceiver is turned on. This leads to the important conclusion that simple RF transceivers with a minimal number of active components are the preferred option for use in wireless sensor networks.

If we assume a power budget of 100 μW and the RF module is allotted 20% of this power budget, at a 1% radio duty cycle, this provides the on-state power consumption goal of 2 mW for the entire RF transceiver.

31.1.3.2 Turn-On and Acquisition Time

In an environment where the radio is in idle or off mode most of the time, and where data communications are rare and packets short, it is essential that the radio can start up very quickly. For instance, a 1-Mbps radio with a 500-μs turn-on time would be poorly suited for the transmission of short packets. The on-time to send a 200-bit packet would be only 200 μs. Start-up and acquisition thus represent an overhead that is larger than the actual payload cost, and may very well dominate the power budget (given that channel acquisition is typically the most power-hungry operation).

Thus, fast start-up and acquisition is essential. An agile radio architecture that allows for quick and efficient channel acquisition and synchronization is therefore desirable. Complex wireless transceivers tend to use sophisticated algorithms such as interference cancellation and complex modulation schemes to improve bandwidth efficiency. These techniques translate into complex and lengthy synchronization procedures and may require accurate channel estimations. Packets are spaced almost seconds apart, which is beyond the coherence time of the channel. This means that these procedures have to be repeated for every packet, resulting into major overhead. Simple modulation and communication schemes are thus the desirable solution if agility is a prime requirement.

31.1.3.3 Integration and Cost

In RF circuit design, the term "fully integrated" typically refers to a transceiver that still requires an off-chip quartz crystal and a few assorted passive components. To meet the cost and form-factor requirements of this application, a true fully integrated transceiver is mandatory. In addition to increasing the size, off-chip passives add to the complexity and cost of the board manufacturing and package design.

One method that can be used to achieve a high level of integration is the use of a relatively high carrier frequency. Currently available simple low-power radios, as used in control applications, typically operate at low carrier frequencies between 100 and 800 MHz. A high carrier frequency has the distinct advantage of reducing the required values of the passive components, making integration easier. For example, a 2.53-μH inductance is needed to tune out a 1-pF capacitor in a narrow-band system at 100 MHz, requiring a surface-mount inductor. For a 2-GHz carrier frequency, the inductance needed is only 6.33 nH, which can easily be integrated on-chip using interconnect metallization layers. In addition, the antenna form-factor is very dependent upon carrier frequency. For a given antenna gain, a higher carrier frequency allows for a much smaller antenna. A quarter-wavelength monopole antenna at 100 MHz would be 0.75-m long. At 2 GHz, the size shrinks to 37.5 mm, making board-level integration or use of small chip-antennas possible. The drive to higher carrier frequencies to achieve high integration is in direct conflict with the need for low-power consumption. As the carrier frequency increases, the active devices in the RF signal path must be biased at higher cutoff frequencies, increasing the bias current and decreasing the transconductance-to-current (g_m/I_d) ratio. This results in increased power dissipation at higher carrier frequencies. Thus, an inherent integration/power consumption trade-off must be dealt with through architectural decisions and the use of new technologies.

In conclusion, RF transceivers for wireless sensor networks should be simple, consume a minimum amount of on-current, and operate at higher carrier frequencies. In the rest of the chapter, we explore how these goals can be simultaneously accomplished. The emerging technology of RF microelectromechanical systems (MEMS), which promises the availability of small highly tuned high-frequency

FIGURE 31.1 50-MHz capacitively driven/sensed resonator. (B. Bircumshaw, G. Liu, H. Takeuchi, T.-J. King, R. Howe, O. O'Reilly, and A. Pisano, *Tech. Dig., 12th Int. Conf. on Solid-State Sensors, Actuators, and Microsystems*, Boston, MA, pp. 875–878, June 8–12, 2003. With permission.)

passive components, offers an excellent opportunity of doing so. We commence our discussion with a description of this exciting technology. The rest of the chapter then describes how these components can be used to build power-efficient receivers and transmitters. Next, a number of low-power circuit techniques used in the implementation of these modules are described, followed by a discussion of some system integration techniques and some realized prototypes.

31.2 RF MEMS in Low-Power Radios

The relatively new field of RF MEMS provides unique opportunities to RF transceiver designers. This section provides background on RF MEMS and provides insight into the opportunities presented by these new technologies.

31.2.1 Introduction to RF MEMS

The field of RF MEMS includes the design and utilization of RF filters, resonators, switches, and other passive mechanical structures constructed using integrated circuit fabrication techniques. To date, these devices have been used as discrete board-mounted components, primarily used to enhance the miniaturization of mobile phones [6]; however, RF MEMS components have the potential to be batch fabricated using existing integrated circuit fabrication techniques. New capacitively driven and sensed structures offer the potential of integration on the same substrate as the CMOS circuitry. In addition, because the resonant frequency is set lithographically and not by a deposition layer thickness, it is possible to fabricate devices with many unique resonant frequencies on the same wafer [7]. See Figure 31.1 for an example of this technology.

The structure in Figure 31.1 was constructed of micromachined polysilicon on top of a silicon wafer. The continued improvement in the performance, reliability, and manufacturability of these structures will greatly change the performance and form-factor of RF transceivers. As discussed in this chapter, however, even in their current state, these devices hold the potential to enable new circuit blocks and architectures.

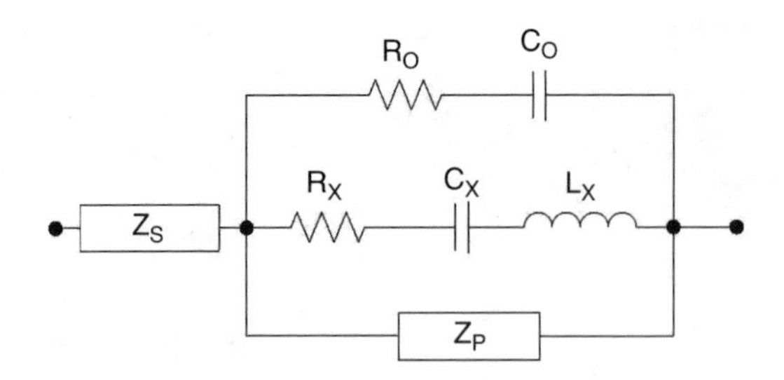

FIGURE 31.2 Simplified circuit equivalent model of MEMS resonator.

31.2.2 Opportunities Offered by RF-MEMS

31.2.2.1 Passives with High Quality Factor

One often-cited benefit of RF MEMS structures is the ability to design resonators with very high-quality factors. As compared with integrated LC tank structures, which typically achieve Q factors of 5 to 10, RF MEMS resonators can achieve Q factors two orders of magnitude higher [8]. See Figure 31.2 for a simplified circuit equivalent model of a MEMS resonator.

In the equivalent schematic, R_x, C_x, and L_x correspond to the motional impedances of the resonator. R_o and C_o represent the finite quality factor of the feed-through capacitance, which affects the Q of the parallel resonance. Finally, Z_s and Z_p represent the impedance load on the resonator due to the CMOS circuitry. It is important to note that these additional impedances have a strong influence on the resonator characteristics, as they affect the loaded resonator quality factor, resonant frequencies, and frequency tolerance.

When used in the design of bandpass filters and duplexers, high Q resonators help to realize the steep skirts necessary to meet cell phone specifications [8]. High Q resonators are further useful in a variety of other transceiver blocks. For example, high Q resonators provide the potential for radio frequency channel select filtering, as their bandwidth is much narrower than what can be obtained from integrated LC filters. This passive channel select filtering can be exploited to simplify the receiver architecture and to reduce the number of active components [9]. In addition, when used in an RF oscillator, RF MEMS resonators provide a vastly improved phase-noise compared with a standard, low Q LC resonator [10]. A design example based on these principles is explored in Section 31.5.

31.2.2.2 Passive Frequency Reference

For all narrowband communication systems, an RF carrier frequency generator is necessary. The absolute frequency reference used is typically a low-frequency quartz crystal oscillator. A frequency synthesizer then multiplies the low-frequency sinusoid up to radio frequencies. This technique has a few disadvantages for low-power radio design. First, even for a fully integrated frequency synthesizer, an off-chip quartz crystal is always necessary, making true full integration impossible. In addition, frequency synthesizers are a huge source of power dissipation in low-power radios [11]. The VCO and frequency dividers tend to dominate the power consumption of frequency synthesizers. Radio frequency MEMS components provide an inherent high-frequency reference without the need for a power hungry frequency synthesizer.

31.2.2.3 MEMS/CMOS Codesign

One of the most exciting aspects of RF micromachined components is the potential for codesigning the MEMS devices with the CMOS circuitry. Until now, passive components included low-quality, on-chip devices (e.g., inductors, capacitors) or high-quality, off-chip components (e.g., inductors, surface acoustic wave (SAW) filters, quartz crystals, duplexers). The on-chip components allow customization to meet the requirements of the circuitry, but their performance is normally poor. High-quality, off-chip passives offer few design degrees of freedom. For example, most filters and duplexers are designed for 50-Ω input and output impedances. This rigid impedance level is very detrimental from a low-power point of view, and has been very troublesome in past receiver implementations [12]. The potential of integrating RF MEMS components and circuitry on the same die or on the same substrate using, for instance, fluidic

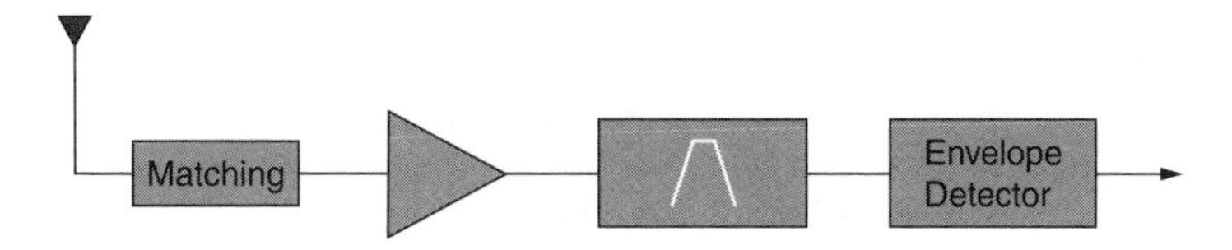

FIGURE 31.3 TRF architecture with envelope detector.

self-assembly (FSA) could allow the circuit designer to size the MEMS components and the circuitry simultaneously [13]. The ability to design these devices alongside the circuitry provides increased system performance and additional designer degrees of freedom.

Overall, the availability of high-quality passive RF components makes it possible to realize transceiver architectures with a minimal number of active components and with a minimum on-current. The following sections evaluate a number of receiver and transmitter architectures that exploit this concept.

31.3 Receivers for Ad Hoc Wireless Sensor Networks

As mentioned earlier, two main considerations in the design of the receiver are ultra-low power consumption and ultra-high integration. The choice of receiver topology has huge implications on the ability to meet these two goals. For example, although aggressive and carefully optimized circuit design can provide low-power consumption, some radio architectures inherently require more active devices biased at higher cutoff frequencies and more off-chip components than other architectures. This section discusses various architectures that can be considered for the implementation of such a radio.

31.3.1 Heterodyne

The omnipresent heterodyne architecture is often the first one to be considered. The process of down-converting the signal allows high gain to be placed at the intermediate frequency, overcoming the noise of the detector and reducing the risk of instabilities. In addition, heterodyning allows channel select filtering to take place at low frequencies, easing the implementation complexity and increasing the potential to integrate these filters. Various flavors of this architecture include high-IF, low-IF, and direct conversion, where the signal is down-converted directly to DC. Regardless of the choice of intermediate frequency, however, an accurate RF frequency reference is needed to drive the mixer in this architecture. Most of the time, this local oscillator (LO) signal is generated from a reference crystal oscillator through a frequency synthesizer. Thus, architectures that eliminate this frequency synthesizer and reduce the number of active devices biased at a high f_t are more appropriate for ultra-low power design.

31.3.2 Tuned Radio Frequency

The tuned radio frequency (TRF) architecture, one of the simplest receiver architectures, eliminates the RF frequency synthesizer and mixers by filtering in the RF domain and directly detecting the RF signal [14]. This architecture relies on sharp RF filters with high-frequency stability — two requirements that are met by RF MEMS components described earlier. The filtered signal can be detected and down-converted in a variety of ways, including envelope detection and subsampling.

31.3.2.1 TRF Envelope Detection

This method of detection, also referred to as diode detection, performs a self-mixing operation on the signal. The RF signal drives a nonlinear element such as a diode or envelope detector, providing a DC component containing the signal spectrum of interest, as depicted in Figure 31.3.

After down-conversion via envelope detection, the signal is simply low-pass filtered to remove the fundamental and higher harmonics, leaving only the baseband signal. One main disadvantage of this approach is that a large RF input signal must be present to induce this self-mixing, as there is no separate

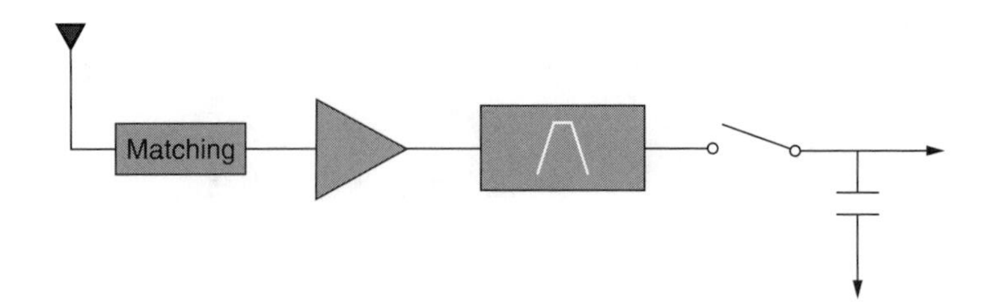

FIGURE 31.4 TRF architecture with subsampling detector.

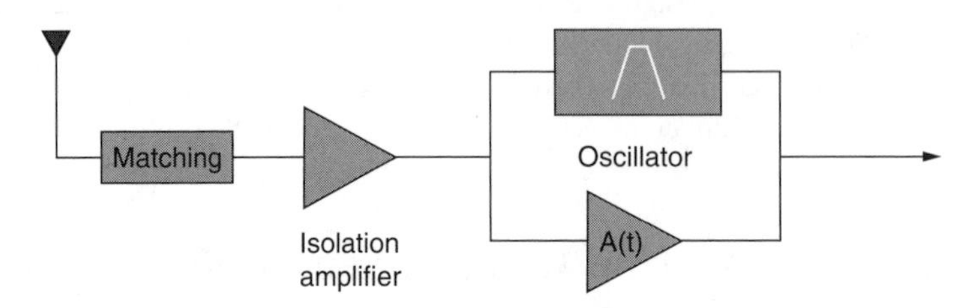

FIGURE 31.5 Super-regenerative detection.

local oscillator driving the system into nonlinearity. Thus, the sensitivity of the detector is poor, and a large RF signal is necessary to achieve detection. The inherently poor noise figure of this detector thus necessitates high RF gain to overcome the noise of the detector.

31.3.2.2 TRF Subsampling Detection

The concept of subsampling overcomes the need for the self-mixing operation needed for envelope detection. As depicted in Figure 31.4, a subsampling architecture samples the signal directly at RF frequencies.

As the name implies, the sampling rate of this detector does not satisfy the Nyquist criterion for sampling the RF signal, and aliasing of the signal occurs. This aliasing effect down-converts the signal to DC, where it is converted to the digital domain by an A/D converter. The sampling rate must only satisfy the Nyquist sampling criterion of the baseband signal. Unfortunately, all noise and sources of interference are also aliased into the baseband bandwidth. Even with very sharp RF filters, the noise of the sampler is problematic. Although the signal is sampled at a low rate, the sampler must track the RF signal, so its bandwidth must be high. This high-bandwidth requirement translates into a relatively small sampling capacitor, which means that the sampler itself is a source of high KT/C noise. Similar to the envelope-detector approach, overcoming the poor noise figure of the subsampler requires high RF gain.

In conclusion, the TRF architecture is promising in its simplicity and reduction of active circuit blocks. They have the advantage that all mixers and synthesizers have been eliminated. The challenge of integrating the steep RF filters is also well suited for RF MEMS technologies. The detection is noisy, however, requiring high gain in the RF amplification stages. This necessitates multiple gain stages biased at high cutoff frequencies (f_t), resulting in high power dissipation in the RF amplifiers.

31.3.3 Super-Regenerative

As discussed in the previous section, the TRF architecture takes advantage of RF MEMS technologies to perform channel selection without a need for mixers or frequency synthesizers. The RF gain needed, however, is very high due to the noisy detection circuitry. A super-regenerative front-end provides extremely high RF amplification and narrowband filtering at low bias-current levels. As depicted in Figure 31.5, the heart of a super-regenerative detector is an RF oscillator with a time-variant loop gain. The isolation amplifier between the antenna and the oscillator performs the following functions: it prevents radiation of the oscillation to the antenna, it provides an input match to the antenna, and it injects the RF input signal current into the oscillator tank without adding significant loading to the oscillator.

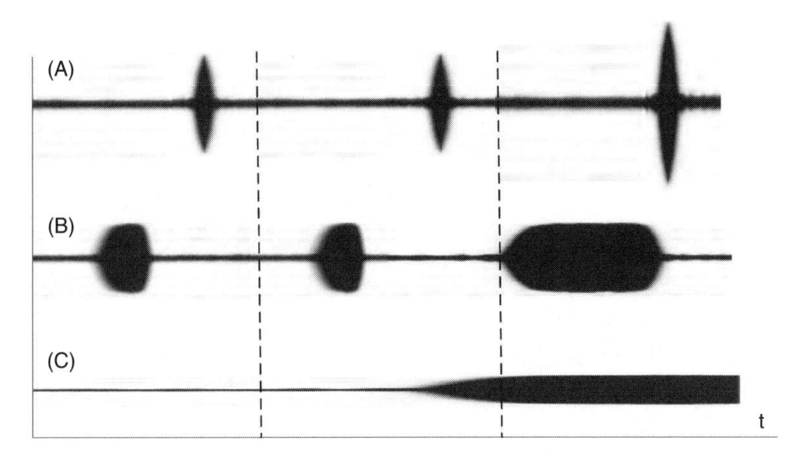

FIGURE 31.6 Super-regenerative detector waveforms.

The time-varying nature of the loop-gain is designed such that the oscillator transconductance periodically exceeds the critical g_m necessary to induce instability. Consequently, the oscillator periodically starts up and shuts off. The start-up time of an oscillator is:

$$t_{rise} = \tau_{rise} \ln\left[\frac{V_{osc}}{V_{initial}}\right]$$

where τ_{rise} is the time constant of the exponentially increasing oscillation envelope, V_{osc} is the zero-peak RF voltage of the saturated oscillator, and $V_{initial}$ is the zero-peak RF signal when the oscillator loop gain is unity (at the onset of oscillation). As this equation demonstrates, the start-up time of the oscillator is exponentially dependent upon the initial voltage in the oscillator tank. This dependency translates into the huge gain attainable by the super-regenerative receiver.

There exists two basic modes of operation that can be utilized: the logarithmic mode or the linear mode [15], as is illustrated in Figure 31.6.

Waveforms (a), (b), and (c) illustrate the detector output in the linear mode, the output in the logarithmic mode, and the RF input signal, respectively. In the linear mode, the level of oscillation is measured before the oscillator reaches saturation, providing a high signal independent gain. As illustrated in waveform (a), the sampled envelope is much larger in the presence of an RF input signal. In the logarithmic mode, detection circuitry senses the area under the oscillation envelope, providing signal dependent gain. Waveform (b) depicts the increased area under the saturated oscillation envelope in the presence of an RF input, resulting from the decreased oscillator start-up time in this condition. Due to the severe fading anticipated in dense indoor sensor networks, a very wide dynamic range is required from the receiver. The logarithmic mode provides an inherent automatic gain control, making its use preferable for this application.

The potential of the super-regenerative receiver to generate large signal gain at very low bias currents makes it the preferred architecture for integrated ultra-low power wireless receivers.

31.4 Transmitters for Ad Hoc Wireless Sensor Networks

As mentioned in Section 31.1, the environment of an ad hoc wireless sensor network differs significantly from those in a conventional wireless network (e.g., GSM, CDMA, Wireless LAN, and Bluetooth). In a typical sensor network, the transmitter sends out sporadic bursts of short data packets to neighboring sensor nodes (< 10 m) [9]. This implies the need for:

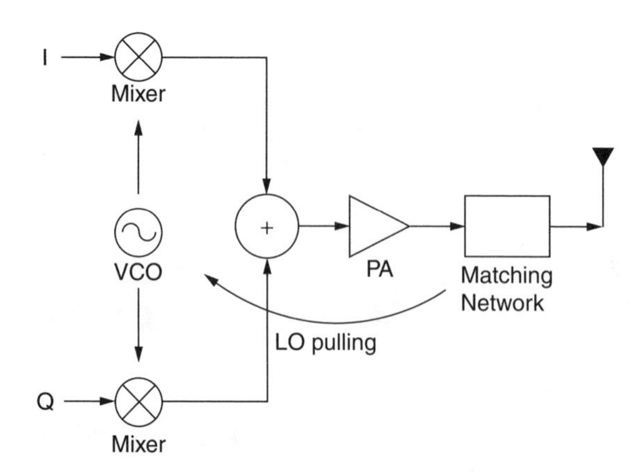

FIGURE 31.7 Direct conversion transmitter architecture.

1. Power shutdown of the transmitter when idle
2. Short turn-on time of the transmitter to minimize overhead power
3. A low transmit power of about 0 dBm (1 mW) for typical receiver sensitivity and indoor multi-path fading conditions

In addition, all the energy dissipated by the transmitter is scavenged from the environment. As scavenged energy is the most precious resource in a sensor node, the efficiency of the transmitter when switched on must be maximized. To put this in into perspective, consider an RF transceiver with an on-state power consumption goal of 2 mW (see Section 31.1.3). An on–off keyed (50% duty cycled) transmitter with 25% efficiency will require 2 mW of power, consuming the entire power budget allocated for the transceiver. A more reasonable division of the power budget between transmitter and receiver is 70% and 30%, respectively. This translates to a required transmitter on-state efficiency of around 36%. Note that this is the required global efficiency of the transmitter, not just the power amplifier.

These requirements have a profound impact on the architecture and implementation of the transmitter for wireless sensor network. In this section, we compare various transmitter architectures and propose a transmitter amenable to low-power ad hoc sensor networks. The low-power design techniques employed in the implementation are discussed in Section 31.5.

31.4.1 Direct-Conversion Transmitter

The direct conversion and the two-step transmitter architectures are most commonly used in conventional radios (e.g., GSM, CDMA, Bluetooth, Wireless LAN) [16]. In the direct conversion transmitter, illustrated in Figure 31.7, the baseband signal is up-converted directly to RF in the modulator, and then efficiently boosted to the required power level by the power amplifier and matching network. The direct conversion transmitter suffers from local oscillator (LO) pulling, in which the output power from the power amplifier leaks into the local oscillator, corrupting the clean LO signal.

31.4.2 Two-Step Transmitter

The effect of LO pulling is alleviated if a two-step conversion as illustrated in Figure 31.8 is employed. The baseband signal is first up-converted to an IF signal, and then converted to the required RF signal in a second mixing step. In this case, the frequency of the oscillator and the RF carrier are different, and LO pulling is reduced.

A careful analysis of the preceding transmitters reveals that they are not ideally suited for sensor applications for two main reasons:

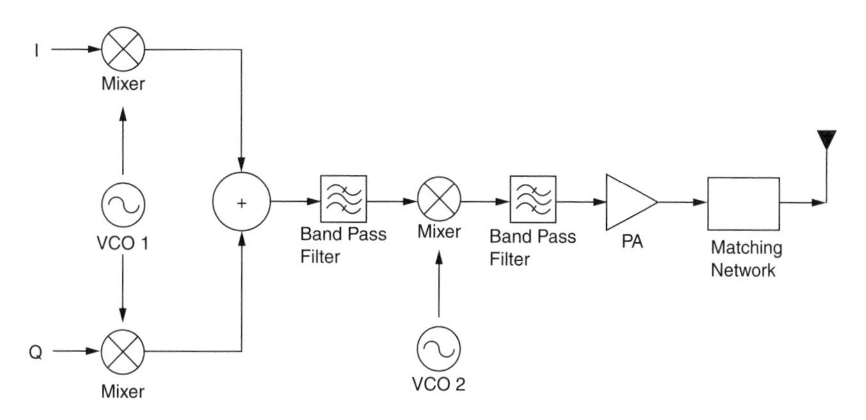

FIGURE 31.8 Two-step transmitter.

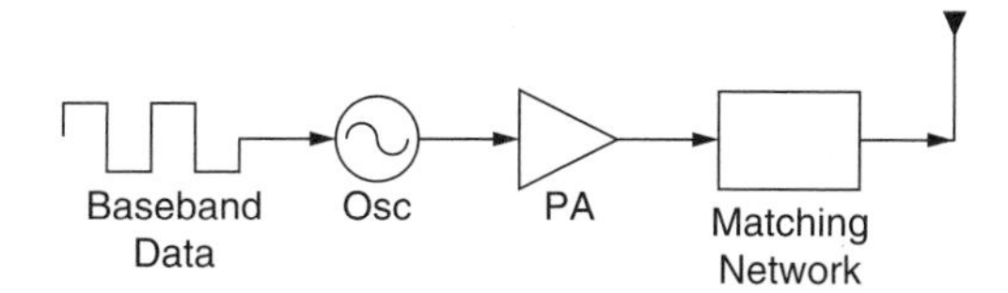

FIGURE 31.9 Direct modulation transmitter.

1. High data rate/bandwidth/spectral efficiency. Conventional radios employ complex modulation schemes (using both AM and PM) to maximize spectral efficiency. Given that the accumulated traffic is approximately 1 kbit/sec/node, maximizing spectral efficiency is not a primary goal. Advanced modulation schemes require the use of a linear power amplifier, which generally have low efficiency. The relaxed requirement on spectral efficiency and linearity means that constant envelope modulation schemes can be employed, which allow for the use of high-efficiency, non-linear power amplifiers.
2. High transmit power. Conventional radios transmit hundreds of mW to 1W, and LO pulling is an important concern; however, the transmitted power from a sensor node is about 1 mW. With an isolation of 20 dB to 30 dB between the power amplifier (PA) and the local oscillator, the leakage power ranges between 10 μW to 1 μW, which is insignificant compared with the LO's output power. Thus, LO pulling is not a significant issue, and the two-step approach is definitely overkill.

31.4.3 Direct-Modulation Transmitter

Taking advantage of the unique characteristics of a wireless sensor network, the direct modulation transmitter in Figure 31.9 is very attractive.

In this architecture, the oscillator is directly modulated by the baseband data (on–off keying) and the nonlinear PA and matching network efficiently boosts the power of the RF signal. The direct-modulation transmitter architecture lends itself to an ultra-low power transmitter for the following reasons:

- Direct modulation of the oscillator eliminates power-hungry mixers. This is enabled by the use of a simple on–off keying modulation scheme. This is deemed acceptable due to the relaxed spectral efficiency requirements of the sensor network.
- On–off keying allows the use of nonlinear high efficiency power amplifiers.
- On–off keying eliminates the use of quadrature channels, thus reducing the number of active components.
- In on–off keying, the transmitter is only turned on when transmitting a one, resulting in a 50% energy savings if the long-term probability of sending a one and a zero is the same.

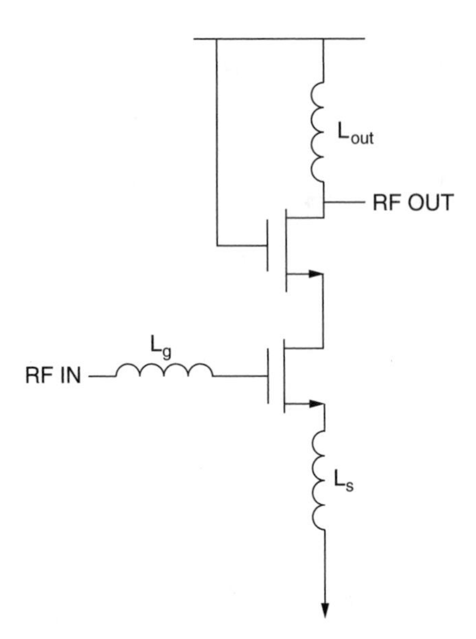

FIGURE 31.10 Inductively degenerated common source amplifier.

31.5 Low-Power Circuit Design Techniques

So far, we have introduced new technologies and promising transmit/receive (Tx/Rx) architectures, which may facilitate the implementation of ultra-low power, high-integration transceivers for sensor networks. Enabling the potential of these approaches requires the availability of the appropriate RF modules, which is the topic of this section. First, we explore the design of low-current front-end amplifiers. Then, an envelope detector implementation is explored, as it is another important component in the receiver. Next, we discuss the modules that compose the transmitter: the RF oscillator and the PA. Finally, we analyze the requirements of low-current proportional to absolute temperature (PTAT) bias circuits.

31.5.1 Low-Current RF Amplification

Whether used to overcome the noise of a subsequent stage, to provide isolation to the antenna, or to provide an input match, an input amplifier is a crucial part of virtually every RF receiver. The considerations for an input amplifier to be used in a sensor-network receiver are quite different from a traditional CMOS LNA. In traditional LNA design, the main goal is to minimize the noise figure while satisfying gain, linearity, and input matching constraints. As became clear in the receiver architecture section, the most important metric for the first stage in the receiver of an ultra-low power, gain-limited receiver is to obtain a large gain at a low current level. In such architectures, the noise figure of the front-end stage is typically swamped by the high noise figure of the detection circuitry. Thus, the goal is to maximize gain for a given power budget, while still maintaining reasonable noise figure, linearity and input matching performances.

An input impedance match is one of the primary architectural considerations for the input amplifier. Many circuit topologies can be used to achieve an input impedance match, including a simple resistive termination circuit. In terms of gain and noise performance, however, one promising topology is the inductively degenerated common source (IDCS) amplifier in Figure 31.10. The advantage of this topology is that it is able to provide high gain and a noiseless real impedance at the resonant frequency.

To explore the difficulty of obtaining sufficient gain at low bias currents in an IDCS amplifier, it is helpful to write down the equation for its overall transconductance G_m:

$$G_m \approx \left(\frac{f_t}{f_o}\right)\frac{1}{2R_S}$$

where R_s is the source impedance, f_t is the transistor cutoff frequency, and f_o is the operating frequency. It can be observed that G_m is not dependent on the g_m of the input device. Instead, it depends on the ratio of the operating frequency to the cutoff frequency and the value of the source resistance. Because the operating frequency and source resistance are set by system constraints, they are not available to the LNA designer as design variables. Consequently, to increase G_m, it is necessary to increase the f_t of the device.

Although it appears attractive that improved process technology (increased f_t) will result in higher gain, system-level integration considerations complicate the design. The problem results from the fact that increases in f_t will result in large values of L_g and very small values of L_s, both of which can pose problems. The small value of L_s is problematic because it is sensitive to parasitic inductances, reducing the accuracy of the input match impedance. The large value of L_g, where the size is determined by the operating frequency, makes it difficult to implement on-chip in modern CMOS processes while still maintaining a high-quality factor. Using an off-chip inductor also introduces a parasitic capacitance at the input, which alters the resonant frequency and decreases the G_m. When biasing low-current RF amplifiers to operate with a maximum f_t, it is essential to include the effect of this capacitance since it may be on the same order of magnitude as C_{gs}. Increasing the bias current allows for larger device widths for the same f_t, which makes the design less sensitive to parasitics and facilitate integration.

To achieve higher gains for a given bias current, it is necessary to deviate from traditional architectures. One example of a current-reuse (stacked) topology is depected in Figure 31.11. The idea is to recycle the

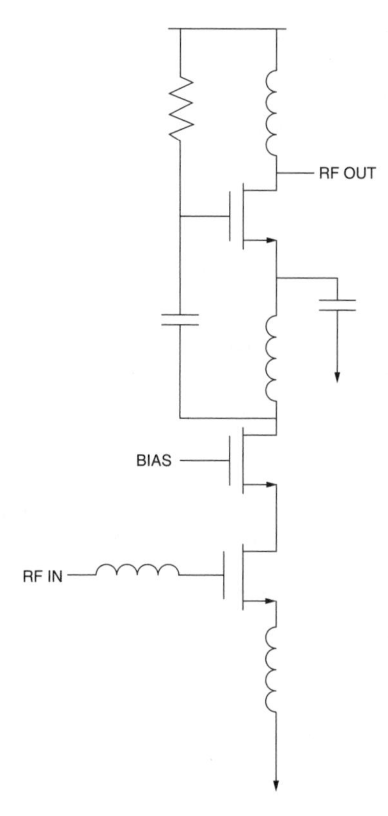

FIGURE 31.11 Current-reuse topology.

bias current so that it can be used by more than one stage. The circuit combines the inductive source degeneration architecture analyzed previously with a common-source amplification stage [17,18]. The gain is increased because the signal is coupled through a capacitor from the drain of the first stage to the gate of the second, providing two stages of amplification. The capacitor at the source of the second stage is made large so that it serves as an AC ground. A potential problem with this topology is the limited headroom and output swing because two stages are stacked on top of each other. In addition, the parasitic bottom plate capacitance of the coupling capacitor could reduce the achievable gain [19]. This topology also consumes more area due to the addition of large passives. Nevertheless, this topology is better suited for this design, and simulation has demonstrated that it is possible to achieve approximately 40 dB of voltage gain at < 1mW of power consumption.

An important function of the receiver input stage is to provide an input impedance match. The input impedance of the IDCS amplifier is:

$$Z_{in}(\omega) = j\omega\left(L_s + L_g\right) + \frac{1}{j\omega C_{gs}} + \frac{g_m L_s}{C_{gs}}$$

In this equation, the non-quasi static gate resistance is ignored [20]. This assumption is acceptable in conventional LNA design, where the g_m of the device is high enough such that $r_{g,NQS}$ is approximately a few ohms, and input matching is not a problem. In low-power designs, however, g_m values are small, and the $r_{g,NQS}$ cannot be ignored making input matching a lot more complicated. The non-quasi static gate resistance term adds directly to the preceding input impedance equation. Input matching is further complicated by the parasitic capacitance of the bonding pad at the input, which alters the frequency at which an input match is achieved, as well as the value of the real impedance. Due to the small device sizes used in low-power design, this capacitance cannot be ignored; thus, accurate modeling of this parasitic is a necessity.

The noise figure of the IDCS amplifier is dominated by the drain noise and the induced gate noise of the input transistor. Because the drain noise is proportional to g_m, and the induced date noise is inversely proportional to g_m, an input transistor size would minimize the noise figure. In modern CMOS processes, the f_t is high enough that acceptable noise figures for low-power sensor networks can be obtained.

31.5.2 Envelope Detector

As discussed in Section 31.3, an envelope detector is a crucial component in a TRF architecture, and is useful to detect the oscillation envelope in amplitude control loops and/or a super-regenerative detector. The following CMOS envelope detector (see Figure 31.12) was found to be quite effective for its use in these applications.

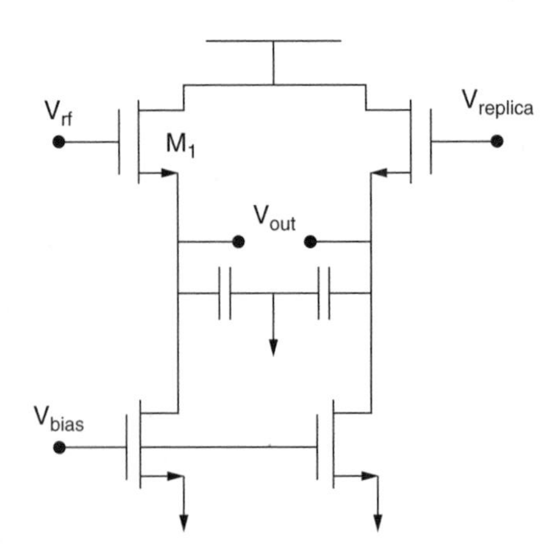

FIGURE 31.12 Simplified envelope detector schematic.

This circuit, which was previously used in bipolar applications, is well suited to CMOS implementations as well [21]. As depicted in the schematic, the single-ended circuit includes a replica half to produce a reference output voltage. The output is pseudo-differential, and the transconductance of M1 and the capacitive loading of the output determine the time constant. Translation of the RF-to-DC spectrum is accomplished though the nonlinearity of M_1. The fundamental and higher harmonics produced by this nonlinearity are filtered, leaving only the DC term. Because the conversion gain of the RF-to-DC spectrum is determined by the nonlinearity of the active devices, the CMOS transistors should be operated in the subthreshold regime, resulting in very low bias currents (< 1 μA). Additionally, the efficiency of the circuit can be increased if transistor M_1 and its replica are designed in a DTMOS configuration (gate connected to body contact). This decreases the subthreshold slope to a value approaching 60 mV/decade, increasing the conversion gain of the detector [22].

31.5.3 RF Oscillator

As mentioned in Section 31.2, recent advancements in RF MEMS offer the potential increase the performance and decrease the power consumption of circuit blocks. The RF frequency synthesizer is an example of a transceiver block that can be greatly influenced by the availability of these new technologies. For example, the use of high-Q RF MEMS resonators can potentially create a very low-power, low-phase-noise VCO for use in a traditional frequency synthesizer. Additionally, although current RF MEMS resonators have poor fabrication tolerance compared with crystal resonators, they can be used to create a RF frequency reference in open-loop mode, without the need for a frequency synthesizer. Although the frequency stability would be worse than a traditional synthesizer, the power consumption and phase noise potentials are very promising. The following is a design example of a 1.9-GHz RF reference using the MEMS/CMOS codesign philosophies discussed in Section 31.2 [10].

The goals of the oscillator design were the following:

- Demonstrate the benefits of RF MEMS/CMOS codesign.
- Provide an ultra-low power, open-loop RF frequency reference without the need for a frequency synthesizer.
- Implement the solution in a very small, reproducible form factor.

The MEMS components used in the design were Agilent thin film bulk acoustic wave (FBAR) resonators [23]. Though typically used in the design of complex filter ladder networks for RF duplexers and transmit filters, they were found to be very well suited for use in RF oscillators. Because it is possible to customize the dimensions and properties of the MEMS resonator, accurate models are necessary to jointly optimize the CMOS and MEMS components. (See Figure 31.2 for a simplified model of the RF MEMS resonator.) A Pierce oscillator topology was used to provide low-phase-noise and low power consumption. Figure 31.13 is a simplified schematic of the oscillator core.

A complete model was developed, including the CMOS transistor models and the MEMS resonator models. The design was optimized for minimum power consumption and a 100-mV output voltage swing. A custom RF MEMS resonator chip and a 0.18-μm CMOS chip were fabricated. The system integration was accomplished using chip-on-board (CoB) technology. See Figure 31.14 for the completed oscillator subsystem.

As discussed in the preceding photograph, the system is easily bonded together with standard CoB technology, resulting in a very small form factor. The oscillator consumes 300 μA from a 1-V supply, and the 100-mV output swing exhibits a phase noise of −120 dBc/Hz at a 100-kHz offset. The phase noise and frequency stability is much better than an integrated LC oscillator, making the oscillator suitable as an RF frequency reference in low-power, low-datarate transceivers. In addition, the start-up time of this frequency reference is approximately 1 μs, providing a very fast turn-on time for the transceiver.

31.5.4 Nonlinear Power Amplifiers

One of the key factors determining the efficiency of the transmitter during transmission is the efficiency of the power amplifier (PA). Power amplifiers can be classified as linear PAs (Class A or AB) or nonlinear

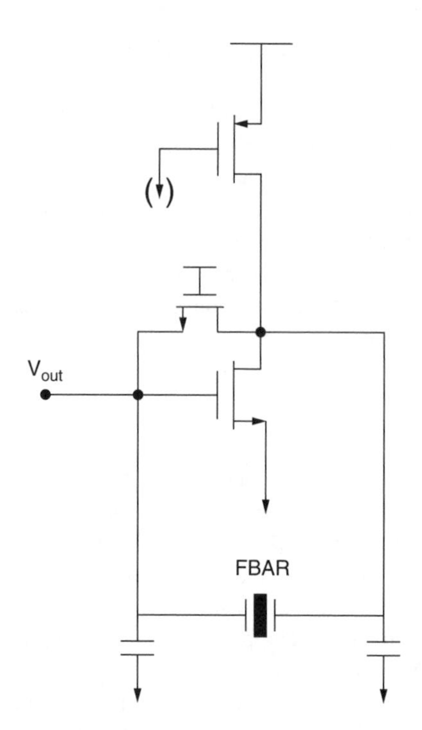

FIGURE 31.13 Simplified oscillator schematic.

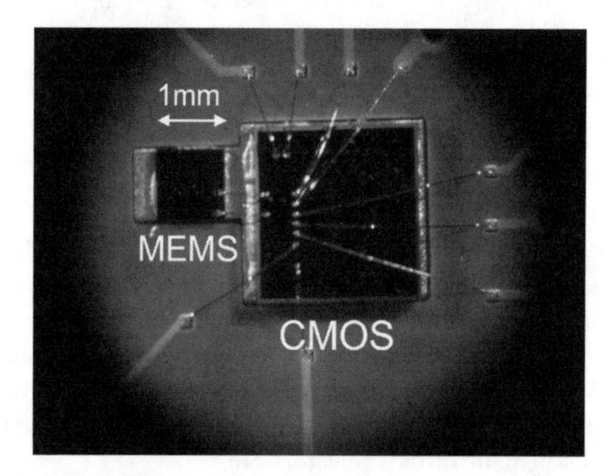

FIGURE 31.14 RF MEMS/CMOS chip-on-board implementation.

PAs (Class C, D, E, or F) [24]. Fully integrated CMOS linear PAs generally suffer from lower efficiency (30 to 40%) as compared with their nonlinear counterparts (40 to 50%) [25]. Thus, linear PAs are generally not suitable for wireless sensor networks. This is evident by considering, as an example, the goal of achieving a global 36% transmitter efficiency (see Section 31.4). If all the transmitter circuits excluding the PA consume 20% of the transmitter power, the efficiency of the PA must be 45%. This exceeds the typical achievable efficiency of fully integrated CMOS linear PAs and is only achievable by nonlinear PAs.

One possible implementation is the Class C PA depicted in Figure 31.15, in which the transistor operates as a current source modulated by the input signal. The high-Q tank filters out the harmonics and ensures that the transmitted signal is sinusoidal. High efficiency is obtained by reducing the conduction angle (< 180°) to minimize the product of the drain current and drain voltage; however, decreasing the conduction angle also reduces the output power for a given input drive amplitude.

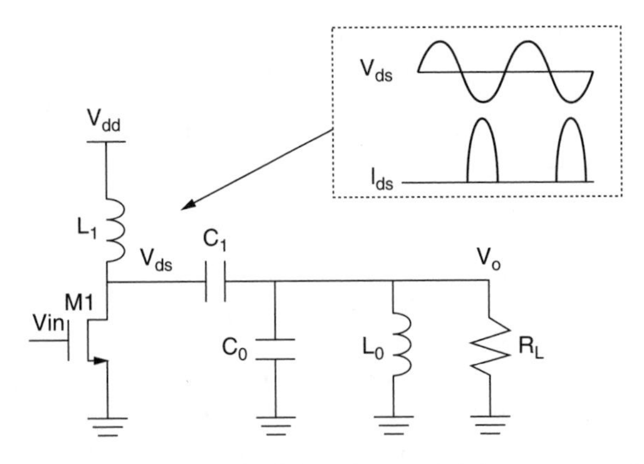

FIGURE 31.15 Class C amplifier. Inset depicts the drain voltage and drain current waveforms.

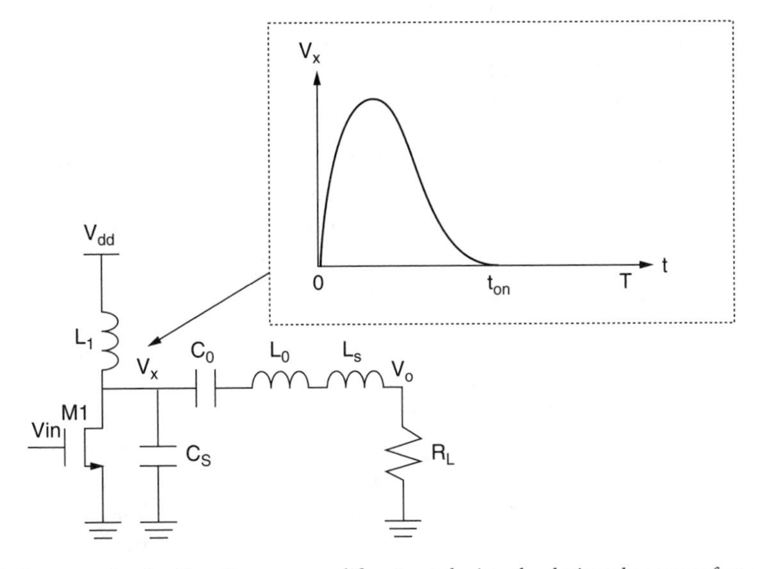

FIGURE 31.16 An example of a Class E power amplifier. Inset depicts the drain voltage waveform.

Using a switching PA, in which the transistor operates as a switch, can alleviate the strong dependence of efficiency on output power. Ideally, the voltage across the switch is zero when the switch is on and the current is zero when the switch is off. This results in zero power dissipation in the switch and thus higher efficiency. An example is the Class E PA depicted in Figure 31.16. The values of the components L_1, L_s, and C_s are chosen such that

$$v_x(t_{on}) = 0 \quad \text{and} \quad \left.\frac{dv_x}{dt}\right|_{t=t_{on}} = 0$$

where t_{on} represents the time at which the switch closes, and V_x is the voltage across the switch. C_0 and L_0 operate as a high Q filter to ensure that the output is sinusoidal.

In addition to the choice of the amplifier topology, some other factors can help to improve the power efficiency of the power amplifier and the transmitter as a whole:

1. Oscillator/PA Driver/PA codesign. The transistor in a nonlinear PA is usually much larger than the one in its linear counterparts. This means that the driver power is significant, especially if the

overall transmitted power is low. For example, if the oscillator output voltage is 160 mV zero to peak and a drive voltage of 800 mV is desired, the driver stage has to provide a gain of 5. For a g_m/I_d of $20V^{-1}$ and an output resistance of 800 Ω, the driver stage will need 313 μW of power with a 1-V supply. To reduce this overhead, it is necessary to design the oscillator, the PA and the PA driver together for overall optimal efficiency. By increasing the drive power of the oscillator, for instance, it is possible to eliminate the driver stage altogether, resulting in lower overall transmitter power consumption.

2. Use of RF MEMS devices. The lack of on-chip high Q passive inductors is detrimental to the efficiency of the transmitter. To reduce losses, high-Q MEMS devices can be employed, for instance, to replace the LC tank in the Class C PA.
3. Power control. The transmitted power of the power amplifier can be reduced when transmitting to nearer sensor nodes. This not only preserves the scavenged energy but also reduces interference with other sensor nodes. The transmitter has to be designed to operate at optimal efficiency at various levels of radiated power.
4. Minimize the overhead turn-on time. The turn-on time of the entire transmitter is limited by the turn-on time of the high-Q local oscillator. During the turn-on time, no data can be transmitted and all energy expended is considered as overhead. To reduce this overhead, the PA can be switched on after a delay to minimize its idle time.

With the incorporation of these low-power design techniques, a nonlinear power amplifier with an efficiency of 50% is achievable. With a 20% overhead power for the oscillator and driver, the entire transmitter can operate with an on-state efficiency of 42%, meeting all the requirements discussed earlier.

31.5.5 On-Chip References and Bias Circuits

This section describes bias generation considerations for low-power transceivers for sensor node applications. This application presents interesting issues, partly due to the sub-threshold operation of most of the transistors present in the transceiver, which is primarily dictated by g_m/I_d efficiency considerations. First, it should be noted that, due to indoor operation and extremely low-power consumption (which eliminates the possibility of circuitry self-heating), the dynamic temperature behavior of the system does not impose aggressive specifications on the design. This allows the circuit design to be carried out at a nominal temperature of approximately 27°C, and successive translation of performance metrics at this temperature into lower bounds for performance metrics over the whole temperature range. The high sensitivity of the aforementioned transceiver architectures to both gain and oscillator startup time result in the need to stabilize device transconductance (g_m) over temperature.

In this scenario, a PTAT current source can be proven to be the optimal solution both for stability and from a power consumption perspective, as it draws from the supply only the amount of current needed to ensure $g_m(t) = g_{m,initial}$ over the entire temperature range. With a nonadaptive biasing technique, the required constant bias current would be determined by the g_m necessary at the worst-case temperature, thus increasing the average power consumption.

From an implementation perspective, a PTAT current source leaves little freedom to the designer: a current mirror, a low thermal coefficient resistor, and a translinear loop, usually implemented with subthreshold devices, are needed. Cascode implementation of the PMOS current mirror (see Figure 31.17) enhances the power supply rejection ratio and suppresses channel length modulation effects, but may be prohibitive if very low supply voltages are used.

Current levels in the mirror should be made as low as possible in order to minimize bias chain current consumption overhead; however, matching considerations limit the practical efficiency for a cascade mirror. For example, assuming an output current of 1 mA and an allowed 3-σ variation of 5% for g_m, matching analysis demonstrates that a mirror current (I_o) of 100 μA is reasonable (~ 85% efficiency), and requires an area of 1000 μm^2 for the bias circuitry. Mirror offset is temperature-dependent, and thus does not simply result in a constant offset for performance metrics, but instead in a distortion of

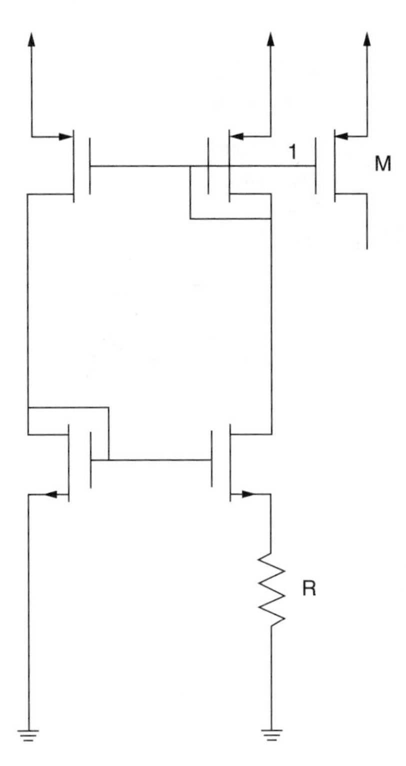

FIGURE 31.17 PTAT current reference schematic.

performance metrics over temperature. If a mirror current (I_o) of 10 μA was used to increase the efficiency (~ 98% efficiency), and the same matching criteria were used, the required die area would increase to 10,000 μm^2, which is not acceptable. An implementation of the mirror based on a simple PMOS pair (see Figure 31.17) would achieve poorer power supply rejection ratio (PSRR) and output impedance, but better performance with respect to matching for given area budget thanks to strong inversion mirror operation. Thus, multiple trade-offs exist in the design of ultra-low power biasing circuits, including area budget, parameter stabilization performance, supply insensitivity, and power efficiency.

31.6 System Integration

The packaging and physical integration strategy of the transceiver is crucial as it dramatically affects the performance, cost, and form-factor of the completed system. The number and size of the components and the assembly technology determine the size of the node. As mentioned previously, careful architectural decisions can help to minimize the part-count. Some components that deserve special attention are the antenna and the energy-supply chain.

As previously stated, using a relatively high carrier frequency can minimize the antenna size. Three main parameters that determine the type of antenna suitable for wireless sensor network are the radiation pattern, size, and bandwidth. As the deployment pattern of the sensor nodes is random, an omni-directional radiation pattern is desired. The form factor of the antenna should be small and allow for easy integration onto a PCB. The bandwidth of the antenna should be large enough to accommodate all channels. Given those considerations, commercially available chip-antennas (using ceramic dielectrics) are good candidates. These antennas have radiation patterns similar to a dipole and measure only about 30 mm by 5 mm. This allows for easy integration with the radio on a printed circuit board (PCB).

The energy-scavenging and the energy-storage (battery or capacitance) devices are the most volume-consuming components of the completed node. As listed in Table 31.1, their respective volumes are

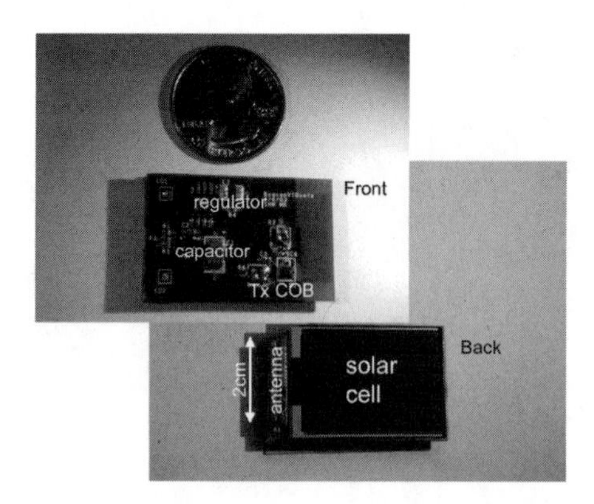

FIGURE 31.18 Photograph of the completed transmit beacon.

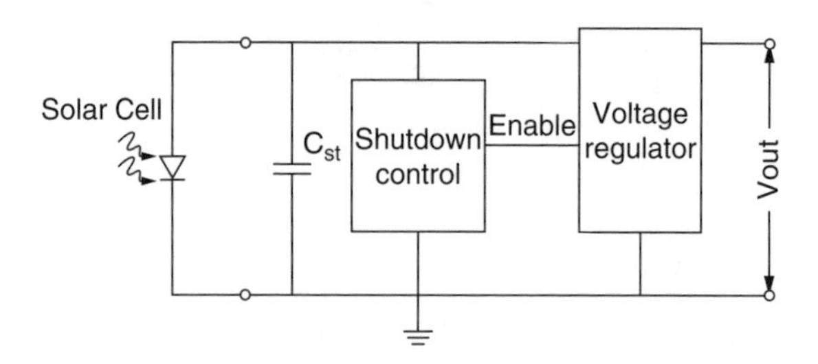

FIGURE 31.19 Transmit beacon powertrain.

directly proportional to the average and peak power dissipation levels of the node. Thus, decreasing the power dissipation of the transceiver — the most power-hungry component of the sensor node ultimately results in a linear decrease in the node size.

These guidelines are best illustrated with the aid of an implementation example, which also serves to demonstrate how the techniques introduced in this chapter effectively lead to ultra-low power and small size. Figure 31.18 is a photograph of a completed 1.9-GHz transmit beacon [26].

This system contains an ultra-low power 1.9-GHz transmitter, a 1.9-GHz chip antenna, a solar panel, an energy storage capacitor, and voltage regulation circuitry. Due to the high carrier frequency used, the chip antenna is very small and takes up little board space. It is mounted on the back of the board. Also on the back of the board is a small solar cell, which is able to provide all the energy necessary for transmitter operation. Figure 31.19 is a conceptual diagram of the transmit beacon power train.

The power train consists of a solar cell, an energy storage capacitor, shutdown control logic, and a linear voltage regulator, which supplies the RF circuitry with a stable 1.2-V supply. The transmitter is fully integrated, requiring only one MEMS resonator. It delivers 0 dBm (1 mW) to the 50-Ω chip antenna. The transmitter is mounted using a chip-on-board technology, and consumes minimal board space. See Figure 31.20 for a photograph of the transmitter circuitry.

As pictured in Figure 31.20, the transmitter requires no crystals or external inductors, and takes up approximately 2 mm × 3 mm of board space. The transmitter can operate from an approximately 1% duty cycle in low light conditions to 100% duty cycle in sunlight. Figure 31.21 depicts pertinent waveforms during the operation of the transmit beacon.

The top waveform is the voltage on the storage capacitor. Once the voltage regulator is enabled (bottom waveform), the RF transmitter turns on (middle waveform). During transmission, the energy in the storage capacitor is dissipated, and the voltage on the storage capacitor decreases. Once the energy on the storage capacitor is depleted, the transmitter is disabled and the charge process starts again. The board will operate indefinitely with no batteries or power supplies.

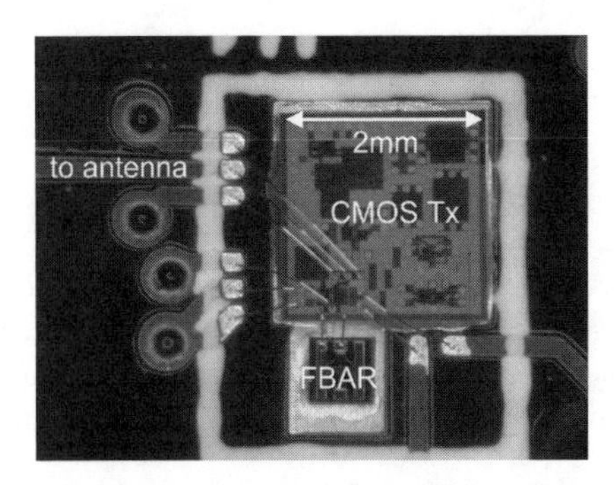

FIGURE 31.20 Transmitter CMOS/MEMS circuitry and antenna feed.

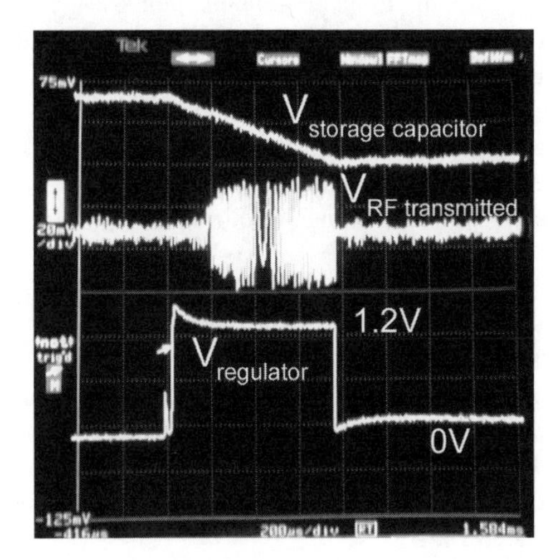

FIGURE 31.21 Transmit beacon ouput waveforms.

31.7 Conclusion

Functional and cost-effective ad hoc sensor networks will only reach ubiquity once the problem of ultra-low-power RF communication is solved. This chapter has discussed the design considerations for achieving ultra-low power and ultra-high integration in transceivers for wireless sensor networks. The concept of CMOS/RF MEMS codesign was introduced as a powerful tool in the successful implementation of these systems. Various transmitter and receiver architectures were analyzed for their applicability to this problem. Low-power design techniques were provided for the implementation of the essential RF modules. Finally, system-level considerations were discussed, including packaging and antennas.

It is the authors' conviction that the field of ultra-low power RF design is only in its infancy, and that exciting new approaches will continue to emerge in the coming years. With this chapter, we hope to have at least provided a glimpse into the myriad of potential solutions and to have projected some of the limiting bounds and constraints.

31.8 Acknowledgments

The authors thank Agilent Technologies for the resonator fabrication and ST Microelectronics for the CMOS fabrication. The generous support of DARPA (under the PACC and the IMT programs) is gratefully acknowledged.

References

[1] J. Rabaey et al., PicoRadio supports ad hoc ultra-low power wireless networking, *IEEE Comput.*, Vol. 33, No. 7, pp. 42–48, July 2000.

[2] F. Boekhorst, Ambient intelligence: the next paradigm for consumer electronics, *Proc. IEEE ISSCC 2002,* San Francisco, CA, February 2002.

[3] S. Roundy, Energy scavenging for wireless sensor nodes with a focus on vibration to electricity conversion, Ph.D. thesis, University of California—Berkeley, May 2003.

[4] L. Zhou, J.M. Kahn, and K.S.J. Pister, Corner-cube retroreflectors based on structure-assisted assembly for free-space optical communication, *IEEE J. Microelectromechanical Syst.*, Vol. 12, pp. 233–242, June 2003.

[5] E.A. Lin, A. Wolitz, and J. Rabaey, Power efficiency analysis of two rendezvous schemes for dense wireless sensor networks, *IEEE Int. Conf. on Commn.*, June 2004.

[6] R. Ruby, P. Bradley, J. Larson III, Y. Oshmyansky, and D. Figueredo, Ultra-miniature high-Q filters and duplexers using FBAR technology, *IEEE ISSCC Dig. Tech. Papers,* pp. 120–121, February 2001.

[7] B. Bircumshaw, G. Liu, H. Takeuchi, T.-J. King, R. Howe, O. O'Reilly, and A. Pisano, The radial bulk annular resonator: towards a 50-Ohm RF MEMS filter, *Tech. Dig., 12th Int. Conf. on Solid-State Sensors, Actuators, and Microsystems,* Boston, MA, pp. 875–878, June 8–12, 2003.

[8] K.M. Lakin, Thin-film resonators and high-frequency filters, http://www.tfrtech.com, retrieved June 2001.

[9] J. Rabaey, J. Ammer, T. Karalar, S. Li, B. Otis, M. Sheets, and T. Tuan, PicoRadios for wireless sensor networks: the next challenge in ultra-low power design, *IEEE ISSCC Dig. Tech. Papers,* pp. 200–201, February 2002.

[10] B. Otis and J. Rabaey. A 300-μW 1.9-GHz CMOS oscillator utilizing micromachined resonators, *IEEE J. Solid-State Circuits,* Vol. 38, pp. 1271–1274, July 2003.

[11] A.-S. Porret, T. Melly, D. Python, C.C. Enz, and E.A. Vittoz, An ultralow-power UHF transceiver integrated in a standard digital CMOS process: architecture and receiver, *IEEE J. Solid-State Circuits,* Vol. 36, pp. 452–466, March 2001.

[12] S. Sheng and R. Brodersen, Low-power CMOS wireless communications, Kluwer, Dordrecht, 1998.

[13] J.S. Smith, High-density, low-parasitic direct integration by fluidic self-assembly (FSA), *Dig. IEEE Int. Electron. Devices Meeting,* pp. 201–204, Piscataway, NJ, 2000.

[14] C. van den Bos and C. Verhoeven, Architecture of a reconfigurable radio receiver front-end using overall feedback, *Proc. ProRISC,* The Netherlands, November 2001.

[15] J.R. Whitehead, *Super-Regenerative Receivers,* Cambridge University Press, U.K., 1950.

[16] B. Razavi, RF transmitter architectures and circuits, custom integrated circuits, 1999. *Proc. IEEE 1999,* May 16–19, 1999, pp. 197–204, San Diego, CA.

[17] A.R. Shahani, D.K. Shaeffer, and T.H. Lee, A 12-mW wide dynamic range CMOS front end for portable GPS receivers, *ISSCC Dig. Tech. Papers,* pp. 368–369, February 1997, San Francisco, CA.

[18] Triquint Semiconductor, TQ9203, low-current RF IC downconverter, Wireless Communication Products, 1995.

[19] B. Razavi, *RF Microelectronics,* Prentice Hall, Upper Saddle River, NJ, pp. 180, 1998.

[20] Y. Tsividis, *The Operation and Modeling of the MOS Transistor,* McGraw-Hill, New York, 1999.

[21] R.G. Meyer, Low-power monolithic RF peak detector analysis, *IEEE J. Solid-State Circuits,* Vol. 30, No. 1, January 1995.

[22] F. Assaderaghi, D. Sinitsky, S. Parke, J. Bokor, P. Ko, and C. Hu, Dynamic threshold-voltage MOSFET (DTMOS) for ultra-low voltage VLSI, *IEEE Trans. on Elec. Devices,* Vol. 44, No. 3, March 1997.

[23] R. Ruby, Micromachined cellular filters, *IEEE MTT-S Int. Microwave Symp. Dig.*, pp. 1149–1152, June 1996.

[24] F.H. Raab, P. Asbeck, S. Cripps, P.B. Kenington, Z.B. Popovic, N. Pothecary, J.F. Sevic, and N.O. Sokal, Power amplifiers and transmitters for RF and microwave, *IEEE Trans. Microwave Theory and Techniques*, Vol. 50, No. 3, pp. 814–826, March 2002.

[25] R. Gupta and D.J. Allstot, Fully monolithic CMOS RF power amplifiers: recent advances, *IEEE Commn. Mag.*, Vol. 37, No. 4, pp. 94–98, April 1999.

[26] S. Roundy, B. Otis, Y.H. Chee, J. Rabaey, and P. Wright, A 1.9-GHz RF transmit beacon using environmentally scavenged energy, *Dig. IEEE Int. Symp. on Low-Power Elec. and Devices*, Seoul, Korea, 2003.

32

Power-Aware On-Demand Routing Protocols for Mobile Ad Hoc Networks

Morteza Maleki
Massoud Pedram
University of Southern California—Los Angeles

32.1 Introduction

Wireless mobile networks may be classified into two general categories:

1. Infrastructure-Based Networks. Wireless networks often extend, instead of replace, wired networks, and are referred to as infrastructure networks. A hierarchy of wide area and local area wired networks (WANs and LANs, respectively) is used as the backbone network. The wired backbone connects to special switching nodes called base stations. They are responsible for coordinating access to one or more transmission channel(s) for mobiles located within their coverage area. The end user nodes communicate via the base station using their respective wireless interfaces. Wireless LANs and WANs are a good example of this type.
2. Mobile Ad Hoc Networks (MANETs). A MANET is composed of a group of mobile wireless nodes that form a network independently of any centralized administration, while forwarding packets to each other in a multi-hop manner. Because the mobile devices are battery-powered, extending the network lifetime has become an important objective. Researchers and practitioners have focused on power-aware design of network protocols for the ad hoc networking environment.

0-8493-1941-2/05/$0.00+$1.50

Because each mobile node in a MANET performs the routing functions for establishing communication among different nodes, the "death" of even a few nodes, due to energy exhaustion, might cause the disruption of service in the entire network. The focus of this chapter is survey and design of power-aware unicast and multicast routing protocols and algorithms for wireless ad hoc networks with special attention to MANETs.

Metrics used by conventional routing protocols for the wired Internet, which is oblivious to an energy budget, typically do not need to consider any energy-related parameters. Thus, routing information protocol (RIP) [1] uses hop count as the sole route quality metric, thereby, selecting minimum-hop paths between the source and destinations. Open shortest path first (OSPF) [2], on the other hand, supports additional link metrics such as available bandwidth and link propagation delay. These algorithms, however, may result in a rapid depletion of the battery energy in the nodes along the most heavily used paths in the network. Routing protocols for wireless ad hoc environments contain special features to reduce the signaling overheads and convergence problems caused by node mobility and potential link failures. While these protocols do not necessarily compute the absolute minimum-cost path, they aim at selecting paths that have lower cost (in terms of metrics, such as hop count or delay). Such protocols must be modified to yield energy-efficient routing solutions.

A large number of researchers have addressed the problem of energy-efficient data transfer in the context of multi-hop wireless networks. Existing protocols may be classified into two distinct categories. One category of protocols is based on minimum-power routing algorithms, which focus on minimizing the power requirements over end-to-end paths. A typical protocol in this category selects a routing path from a source to some destination to minimize the total energy consumption for transmitting a fixed number of packets over that path. Each link cost is set to the energy required for transmitting one packet of data across that link and Dijkstra's shortest path algorithm is used to find the path with the minimum total energy consumption. These protocols traditionally ignore the power dissipated on the receiver side in a node, and therefore, tend to result in routing paths with a large number of short hops. A key disadvantage of these protocols is that they repeatedly select the least-power cost routes between source-destination pairs. As a result, nodes along these least-power cost routes tend to "die" soon by rapidly exhausting their battery energy. This is doubly harmful because the nodes that die early are precisely the ones that are most needed to maintain the network connectivity (and thus increase the useful service life of the network.)

A second category of protocols is based on routing algorithms that attempt to increase the network lifetime by attempting to distribute the forwarding load over multiple different paths. This distribution is performed by either intelligently reducing the set of nodes needed to perform the forwarding duties, thereby, allowing a subset of nodes to sleep over different periods of time, or by using heuristics that consider the residual battery power at different nodes and route around nodes that have a low level of remaining battery energy. In this way, they balance the traffic load inside the MANET to increase the battery lifetime of the nodes and the overall useful life of an ad hoc network. These protocols indeed constitute state-of-the-art power-aware network routing protocols and are the focus of this chapter.

This chapter is organized as follows. Section 32.2 gives a brief classification of the broad domain of ad hoc routing protocols. Section 32.3 gives a brief literature review of research in power-aware ad hoc routing protocols. Section 32.4 describes the rationale and details of the power-aware source routing (PSR) algorithm, and likewise, Section 32.5 describes the rationale and details of the proposed lifetime prediction routing (LPR) algorithm. Section 32.6 contains the experimental results comparing PSR and LPR with other popular ad hoc routing techniques.

32.2 MANET Routing Protocols

Routing protocols in ad hoc networks may be classified into three groups:

1. Proactive (table-driven)
2. Reactive (on-demand)
3. Hybrid

32.2.1 Proactive (Table-Driven) Routing Protocols

These routing protocols are similar to and come as a natural extension of those for the wired networks. In proactive routing, each node has one or more tables that contain the latest information of the routes to any node in the network. Each row has the next hop for reaching to a node/subnet and the cost of this route. Various table-driven protocols differ in the way the information about change in topology is propagated through all nodes in the network.

The two kinds of table updating in proactive protocols are the periodic update and the triggered update [3]. In periodic update, each node periodically broadcasts its table in the network. Each node just arriving in the network receives that table. In triggered update, as soon as a node detects a change in its neighborhood, it broadcasts entries in its routing table that have changed as a result. Examples of this class of ad hoc routing protocols are the destination-sequenced distance-vector (DSDV) [4] and the wireless routing protocol (WRP) [5]. Proactive routing tends to waste bandwidth and power in the network because of the need to broadcast the routing tables/updates. Furthermore, as the number of nodes in the MANET increases, the size of the table will increase; this can become a problem in and of itself.

DSDV, which is known not to be suitable for large dense networks, was described in Perkins [3]. A route table at each node enumerates all available destinations and the corresponding hop-count from the node. Each route table entry is tagged with a sequence number, which is created by a destination node. To maintain consistency of the route tables in a dynamically changing network topology, each node transmits table updates either periodically (periodic update) or when new, significant information is available (triggered update). Routing information is advertised by broadcasting or multicasting. The packets are transmitted periodically and incrementally as topological changes are detected. Topological changes include movement of a node from place to place or the disappearance of the node from the network. Information about the time interval between arrival of the very first routing solution and the arrival of the best routing solution for each particular destination is also maintained. Based on this information, a decision may be made to delay advertising routes that are about to change, thus, reducing fluctuations in the route tables. The advertisement of possible unstable routes is delayed to reduce the number of rebroadcasts of possible route entries that normally arrive with the same sequence number.

32.2.2 Reactive (On-Demand) Protocols

Reactive routing protocols take a lazy approach to routing. They do not maintain or constantly update their route tables with the latest route topology. Instead, when a source node wants to transmit a message, it floods a query into the network to discover the route to the destination. This discovery packet is called the route request *(RREQ)* packet and the mechanism is called route discovery. The destination replies with a route reply *(RREP)* packet. As a result, the source dynamically finds the route to the destination. The discovered route is maintained until the destination node becomes inaccessible or until the route is no longer desired.

The protocols in this class differ in handling cache routes and in the way route discoveries and route replies are handled. Reactive protocols are generally considered efficient when the route discovery is employed rather infrequently in comparison to the data transfer. Although the network topology changes dynamically, the network traffic caused by the route discovery step is low compared to the total communication bandwidth. Examples of reactive routing protocols are the dynamic source routing (DSR) [3,6], the ad hoc on-demand distance vector routing (AODV) [7] and the temporally ordered routing algorithm (TORA) [35]. The proposed power-aware routing algorithms belong to this category of routing algorithms. Because our approach is an enhancement over DSR, a brief description of DSR is warranted.

DSR, which is one of the widely accepted reactive routing protocols, is entirely on demand with no periodic activity of any kind at any level within the network. This pure on-demand behavior allows the number of routing discovery packets for a set of communication patterns to scale to zero when all nodes are approximately stationary. This is because if nodes are not moving about, all the routes employed by the current set of communication patterns will be discovered and will remain unchanged until the communications are completed. As nodes begin to travel or as communication patterns change, the routing packet overhead of the DSR automatically scales only to that which is needed to track the routes currently in use.

In DSR when a node wishes to establish a route, it issues a RREQ to all of its neighbors. Each neighbor broadcasts this RREQ, adding its own address in the header of the packet. When the RREQ is received by the destination or by a node with a route to the destination, a RREP is generated and sent back to the sender along with the addresses accumulated in the RREQ header. The responsibility to assess the status of a route falls to each node in the route. Each node must ensure that packets successfully cross the link to the next node. If the start node does not receive an acknowledgement from the end node of a link on the path, it reports the error back to the source node and leaves it to the source to find and establish a new route. Because this process may consume a lot of bandwidth, DSR provides each node with a route cache to be used aggressively to reduce the number of control messages that must be sent. If a node has a cache entry for the destination when a route request for that destination is received at the node, it will use the cached copy instead of forwarding the request in the network. In addition, it promiscuously listens to other control messages (RREQs and RREPs) for additional routing data to add to its cache. DSR has the advantage in that no routing tables need to be maintained to route a given packet because the entire route is contained in the packet header; however, tables are used to cache routes and enhance performance. The caching of any initiated or overheard routing data can significantly reduce the number of control messages being sent, thus drastically reducing the overhead.

The disadvantages of DSR are twofold. DSR is not scalable to large networks. The *Internet Draft* acknowledges that the protocol assumes the diameter of the network is no greater than 10 hops. Additionally, DSR requires significantly more process resources than most other protocols. To obtain routing information, each node must spend much more time processing any control data it receives, even if that node is not the intended recipient. This is the ability of many network interfaces, to operate the network interface in "promiscuous" receive mode, including most current LAN hardware for broadcast media such as wireless. This mode causes the hardware to deliver every received packet to the network driver software without filtering, based on link-layer destination address. The promiscuous mode increases bandwidth utilization of DSR by reducing the number of control messages being sent out, though the use of promiscuous modes may increase the power consumption of the network interface hardware. Depending on the design of the receiver hardware, and in such cases, DSR can easily be used without the optimizations that depend on the promiscuous receive mode, or can be programmed to only, periodically switch the interface into promiscuous mode. Use of promiscuous receive modes is optional in DSR.

32.2.3 Hybrid Routing Protocols

Both the proactive and reactive protocols work well for networks with a small number of nodes. As the number of nodes increases, hybrid reactive/proactive protocols are used to achieve higher performance. Hybrid protocols attempt to assimilate the advantages of purely proactive and reactive protocols. The key idea is to use a reactive routing procedure at the global network level while employing a proactive routing procedure in a node's local neighborhood.

Zone routing protocol (ZRP) [3] is an example of the hybrid routing protocols. In ZRP, every node has a zone around itself, which includes nodes that are R hops away from that node. R is called the zone radius. ZRP limits the scope of proactive procedure to each node's zone. In this way, ZRP reduces the cost of frequent updates in response to continuously changing network topology by limiting the scope of the updates to the neighborhood of the change. The ZRP route discovery operates as follows. When a source node wants to find a route, it first checks whether the destination is within its zone. If so, the path to the destination is fetched from its table and no further route discovery is required. If the destination is not within the source routing zone, the source broadcasts a route request to its peripheral nodes, which are nodes in the border of the node's zone. The peripheral nodes execute the same algorithm — checking whether the destination is within their zone. If so, a route reply is sent back to the source indicating the route to the destination. If not, peripheral nodes forward the route request to their peripheral nodes, which execute the same procedure.

32.3 Low-Power Routing Protocols

The focus of research on routing protocols in MANETs has been the network performance. A handful of studies on power-aware routing protocols for MANETs have been conducted. Presented next is a review of some of them.

32.3.1 Minimum Power Routing

Singh et al. [8] proposed a routing algorithm based on minimizing the amount of power (or energy per bit) required to get a packet from source to destination. More precisely, the problem is stated as:

$$\underset{\pi}{Min}\left\{\sum_{(i,j)\in\pi} T_{ij}\right\} \tag{32.1}$$

where T_{ij} denotes the power expended for transmitting and receiving between two consecutive nodes i and j (aka cost of link (i,j)) in route π.

This link cost can be defined for two cases:

1. When the transmit power is fixed.
2. When the transmit power is varied dynamically as a function of the distance between the transmitter and intended receiver. Each node chooses the transmission power level for a link so that the signal reaches the receiver node with the same constant received power. To achieve this, clearly, links with larger distances require a higher transmission power than links with smaller distances.

For the first case, all the nodes in the network use a fixed power for all transmissions, which is independent of the link distance. Because the power cost of transmitting and receiving is fixed, then the link cost is fixed and consequently Equation (32.1) results in selecting a path with a minimum number of hops. In fact, assuming lossless links, a path with the minimum number of hops has a minimum number of transmissions and when the transmit power is fixed, then that path will also result in the least total power dissipation [9].

Generally, for a network with 802.11b as media access control (MAC) layer, energy consumption of each operation (i.e., receive, unicast transmit, broadcast, and discard) on a packet is given by Freeney and Nilsson [10]:

$$E(packet) = b \times packet\_size + c \tag{32.2}$$

where b and c are the appropriate coefficients for each operation. Coefficient b denotes the packet size-dependent energy consumption that depends on distance, wireless channel conditions and so on, whereas c is a fixed cost that accounts for acquiring the channel and for MAC layer control negotiation.

The link cost is the sum of all the costs incurred by the source and destination nodes. Traffic is classified as broadcast and unicast (i.e., point-to-point).

For unicast traffic, when receivers are in nonpromiscuous mode operation, the energy cost of the link between sender and receiver may be calculated as follows:

$$T_{SD} = E_{S\_send}(unicast\_packet) + E_{D\_recv}(unicast\_packet) \tag{32.3}$$

where S and D denote the sender and destination of the unicast packet.

In 802.11b, before sending a unicast packet, the source broadcasts a request-to-send (RTS) control message, specifying a destination and data packet size (duration of transmission). The destination responds with a clear-to-send (CTS) message. If the source does not receive the CTS, it may retransmit the RTS message. Upon receiving the CTS, the source sends the DATA and awaits an acknowledge (ACK) from the receiver. For unicast traffic with nonpromiscuous mode operations, the energy cost for all nondestination nodes that can hear the packets is nearly zero because nondestination nodes only consume energy to receive the RTS packet. After this step, they will be discarding packets or even turning off their receivers during the ongoing transaction.

For unicast traffic when receivers are in promiscuous mode operation, the link cost between the sender and destination pair may be calculated as follows:

$$T_{SD} = E_{S\_send}(unicast\_packet) + \sum_{r \in Rs} E_{r\_recv}(unicast\_packet) \tag{32.4}$$

where R_s denotes the set of all nodes that can hear source S, which obviously includes destination D. Notice that $T_{S,D}$ represents an extended link cost. It accounts for the receiver energy cost of the neighboring nodes of the source that can hear the packets sent from the source to the intended destination. According to this link cost function, assuming that all candidate paths have same hop-count, the "best" paths are those that traverse sparse areas of the network where the node density is low.

For broadcast traffic, the sender listens briefly to the channel and sends data if the channel is free. If the channel is busy, the sender waits and retries later. The broadcast cost may be calculated as follows:

$$T_S = E_{S\_send}(broadcast\_packet) + \sum_{r \in R_S} E_{r\_recv}(broadcast\_packet) \tag{32.5}$$

This is not a link cost. Instead, it is a node cost which is assigned to sender(s) of broadcast packets. Broadcast and multicast routing algorithms may make use of this node cost to construct power-aware broadcast or multicast routing trees. These categories of routing algorithms will be explained later in this chapter.

The question of how to make use of the variable transmission power level is more involved. Stojmenovic and Lin [11] propose a local routing algorithm for this case. The authors assume that the power needed for transmission and reception is a linear function of d^α where d is the distance between the two neighboring nodes and α is a parameter that depends on the physical environment. The authors make use of the global positioning system (GPS) information to transmit packets with the minimum required transmit energy. The key requirement of this technique is that nodes in the MANET know the relative positions of themselves as well as all other nodes; however, this information may not be readily available. In addition, the GPS-based routing algorithm has two drawbacks. One is that the GPS cannot provide useful information about the physical environment (blockages and dynamics of wireless channels) to the nodes. The second weakness is that the power dissipation of the GPS is an additional power draw on the battery source of the mobile node.

Heinzelman et al. [12] proposed a minimum transmission energy (MTE) multi-hop routing algorithm for wireless sensor networks. Assuming a first-order radio model for a wireless sensor node and assuming d^n energy loss due to channel transmission where n is between 2 and 4, the article uses the following equations for calculating energy, sending and receiving k bit data over a distance d:

$$\begin{aligned} E_{Tx}(k,d) &= E_{tx\_elec} * k + E_{amp} * k * d^n \\ E_{Rx}(k) &= E_{rx\_elec} * k \end{aligned} \tag{32.6}$$

where E_{tx_elezc} and E_{rx_elezc} are energy dissipated in the transmitter and receiver electronics, and E_{amp} is energy dissipated in the transmit amplifier. If nodes A and B are separated by distance D (as depicted in Figure 32.1), then MTE calculates the optimum number of relaying nodes, K_{opt}, that is required to send data from A to B with minimum transmission energy as follows:

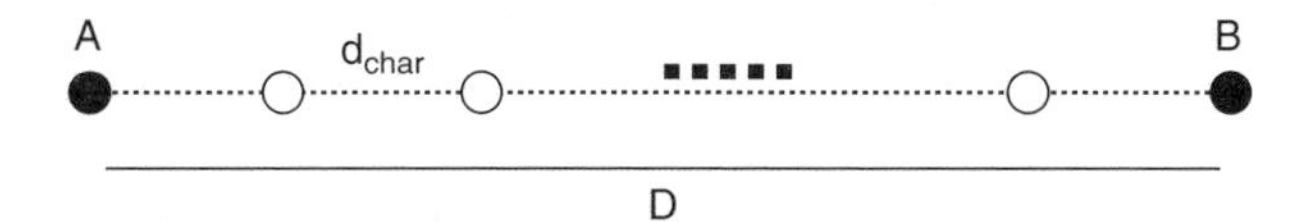

FIGURE 32.1 Relaying nodes are inserted between nodes A and B to reduce the energy of sending a packet from A to B.

$$K_{opt} = \left\lfloor \frac{D}{d_{char}} \right\rfloor \text{ or } \left\lceil \frac{D}{d_{char}} \right\rceil \quad (32.7)$$

where distance d_{char}, called the characteristic distance, is independent of D and is calculated as:

$$d_{char} = \sqrt[n]{\frac{(E_{tx\_elec} + E_{rx\_elec})}{(n-1)E_{amp}}} \quad (32.8)$$

32.3.2 Battery-Cost Lifetime-Aware Routing

The main disadvantage of the problem formulation of Equation (32.1) is that it always selects the least-power cost routes. As a result, nodes along these least-power cost routes tend to "die" soon by rapidly exhausting their battery energy. This is doubly harmful because the nodes that die early are precisely the ones that are most needed to maintain the network connectivity (and thus increase the useful service life of the network.) Therefore, it may be more advantageous to use a higher power cost route if this routing solution avoids using nodes that have low remaining battery energy. This observation has given rise to a number of "battery-cost lifetime-aware routing" algorithms as described next.

The min-sum battery cost routing algorithm [13] minimizes the total cost of the route. More precisely, this algorithm minimizes a summation of the inverse of remaining battery capacities for all nodes on the routing path. One drawback of this algorithm is that it may select a rather short path containing mostly nodes with high remaining battery capacity but also a few nodes with low remaining battery capacity. The cost of such a routing solution may be lower than that of a path with a large number of nodes all having medium level of remaining battery capacity. The former routing solution, however, is generally less desirable from the network longevity point of view because such a path will become disconnected as soon as the very first node on that path dies.

The min-max battery cost routing algorithm is a modification of the minimum battery cost routing to address the previously mentioned weakness. This algorithm attempts to select a route such that has the cost of the most "expensive" link (i.e., one with the minimum remaining battery capacity) on that path is minimum. Thereby, this algorithm results in a more balanced use of the battery capacity of the nodes in the network. One drawback of this algorithm is that because there is no guarantee that paths with the minimum hop-count or with the minimum total power are selected, it can select paths that result in much higher power dissipation to send traffic from a source to destination nodes. This feature does actually lead in shorter network lifetime because in essence the average energy consumption per delivered packet of user data has been increased.

A conditional min-max battery cost routing algorithm was also proposed in Toh [13]. This algorithm, which is a hybrid of the min-sum and the min-max battery cost routing algorithms, chooses the route with minimal total transmission power if there exists at least one feasible routing solution where all nodes in that route have remaining battery capacities higher than some prespecified threshold value. If there is no such routing solution, however, then the min-max routing algorithm is employed to select a route.

Several experiments were reported in Toh [13] to evaluate the effect of different battery cost-aware routing algorithms on the network lifetime. According to the reported results, the min-sum battery cost routing exhibits superior results compared to the min-max battery cost routing in terms of the expiration times of the nodes in the network. Conditional min-max routing demonstrated better or worse results compared with the first two algorithms, depending on how the threshold value was chosen.

Maximum residual packet capacity (MRPC) was proposed in Misra and Banerjee [14]. MRPC is conceptually similar to the conditional min-max battery cost routing; however, MRPC identifies the capacity of a node not only by the residual battery capacity, but also by the expected energy spent in reliably forwarding a packet over a specific link. In fact, the objective function of Equation (32.1) is for a path with lossless links, however, for lossy links, the number of retransmissions in each link increases in proportion to the packet error rate of that link. Misra and Banerjee [14] proposed to rewrite the objective function of Equation (32.1) for reliable minimum total transmission power routing on lossy links and ignoring power expended for receiving packets as follows:

$$\underset{\pi}{Min}\left\{\sum_{(i,j)\in\pi}\frac{\rho_{ij}}{1-e_{ij}}\right\} \tag{32.9}$$

where e_{ij} is the packet error rate of *link(i,j)* (assuming constant packet size) when the transmit power level of the link is ρ_{ij}. Notice that Equation (32.9) is for the case of hop-by-hop retransmission where sender of each individual link provides reliable forwarding to the next hop by using localized packet retransmissions. Hop-by-hop retransmission may be contrasted to end-to-end retransmission where individual links do not provide link-layer retransmissions, and error recovery is achieved only via retransmissions initiated by the source node. For end-to-end retransmission, Equation (32.9) is modified as follows [15]:

$$\underset{\pi}{Min}\left\{\left(\sum_{(i,j)\in\pi}\rho_{ij}\right)\cdot\prod_{(i,j)\in\pi}\frac{1}{1-e_{ij}}\right\} \tag{32.10}$$

Several experiments are reported in Misra and Banerjee [14] to compare the routing method with different battery cost routings and minimum total transmission power routings. According to these results, although the first node dies sooner in the minimum total transmission power routings compared to the battery-cost routing algorithms, the last node dies later in the first case compared to the second case. MRPC, similar to other battery-cost routing algorithms, increases the expiration time of the first node while the death rate of the nodes is as smooth as the minimum total transmission power routing. Performance of MRPC, however, like that of the conditional min-max battery cost routing, depends on a threshold value. This threshold value determines exactly when either the min-max battery cost routing or the reliable minimum total transmission power is applied for route selection.

Chang and Tassiulas [16] describe a multi-path battery-cost routing algorithm to balance the energy consumption of nodes in a static wireless ad hoc sensor network. The routing has been designed for a network of stationary nodes whose task is to detect events inside a monitoring region. Nodes that detect an event (so-called source nodes) send their measurement data to specific destination(s) (so-called gateway) by using multi-hop routing. The article proposes a maximal residual energy path (MREP) routing algorithm, which has a min-max or min-sum objective function for selecting paths where the cost function for each link is as follows:

$$C_{i,j}=(F_i-\lambda.\rho_{ij})^{-1} \tag{32.11}$$

where C_{ij} is the cost of link *(i,j)*, F_i is the full-charge battery capacity of node *i*, ρ_{ij} is transmit energy for sending a bit from node *i* to node *j*, and λ is an augmentation step size. Chang and Tassiulas [16] also propose a flow reduction (FR) algorithm. First, FR finds all possible paths from each source to a single gateway node (single commodity flow) or to several gateway nodes (multi-commodity flow). We define

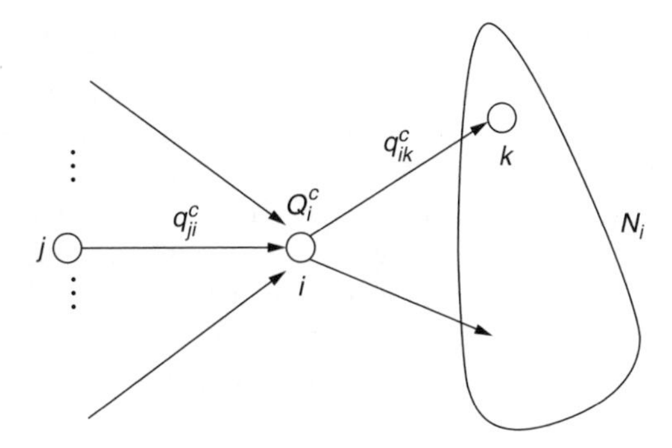

FIGURE 32.2 Node i sends out incoming traffic plus locally generated traffic toward nodes that can be reached from i (nodes in set N_i).

a commodity as data exchanged between a specific source-destination pair. FR defines longevity of a path π, L, as follows:

$$L_\pi(q) = \underset{i\in\pi}{Min}\ \tau_i(q)$$
$$\tau_i(q) = \frac{F_i}{\sum_{j\in N_i}\rho_{ij}\cdot\sum_{c\in C} q^c{}_{ij}} \tag{32.12}$$

where C is set of all commodities (i.e., data communications between various source-destination pairs), $q^c{}_{ij}$ is the rate at which data is sent from node i to node j in commodity c with transmit power $\rho_{i,j}$, N_i is the set of all nodes that can be reached by node i with power level $\rho_{i,j}$ (cf. Figure 32.2), $\tau_i(q)$ is the lifetime of node i, and $L(q)$ is the longevity of path under a given flow $q = \{q_{ij}\}$. The goal of FR is to divide the traffic flow of a source node to a sink (gateway) node on all paths between that pair of source and sink nodes such that all paths have the same lifetimes. FR tries to achieve this goal by redirecting some flow of each commodity from the shortest path (which has minimum longevity) toward longest path (which has maximum longevity) and this is repeated separately for each commodity in several steps until all paths between a source and sink have the same longevity. By defining network lifetime, as the time when the first node dies, Chang and Tassiulas [16] have demonstrated that the problem of finding maximum lifetime of a sensor network may be formulated as a linear programming problem as follows:

$$\begin{aligned}
&Maximize\ \tau\\
&q_{ij}^{(c)} \geq 0, \forall i \in V, \forall j \in N_i, \forall c \in C\\
&\sum_{j\in N_i}\rho_{ij}\sum_{c\in C} q_{ij}^{(c)} \leq \frac{F_i}{\tau}, \forall i \in V\\
&\sum_{j:i\in N_j} q_{ji}^{(c)} + Q_i^{(c)} = \sum_{k\in N_i} q_{ik}^{(c)}, \forall i \in V - S^{(c)}, \forall c \in C
\end{aligned} \tag{32.13}$$

where $Q_i^{(c)}$ denotes the information generation rate at source nodes to be sent to destination nodes (or sink nodes) $S^{(c)}$ for each commodity c. V is set of all nodes, and N_i was defined previously. This linear programming can be solved in polynomial time. The solution to this linear programming problem provides the optimal network lifetime.

32.3.3 Energy-Conserving Techniques for Multi-Hop Ad Hoc Networks

It is known that an idle receiver listening for packets can consume almost as much as power as one doing active reception. More precisely, idle, receive, and transmit energy cost ratios for the transceiver part of a mobile node are 1, 2, 2.5 as per Kasten [17] and 1, 1.2, 1.7 as per Chen et al. [18]. Clearly, energy consumed in idle state of the transceiver cannot be ignored. In addition, Freeney and Nilsson [10] have demonstrated that a major source of extraneous energy consumption is from overhearing (or eavesdropping). Radios have a relatively large broadcast range. All nodes in that range must receive each packet to determine if it is to be received locally or forwarded to some other node in the network. Although most of these packets are immediately discarded, they cause superfluous energy consumption in the mobile node. Because the network interface may often be idle or simply overhearing data, the energy dissipated at these states can be saved by turning the radio off when it is not in use. In practice, however, this approach is not straightforward: a node must arrange to turn its radio on not only to receive packets addressed to it, but also to participate in any higher level routing and control protocols. The need for power-aware routing protocols is particularly acute for multi-hop ad hoc networks.

32.3.3.1 Power-Aware Multiple Access Protocol with Signaling (PAMAS)

PAMAS [19] is a MAC-level protocol that avoids overhearing problem by powering off radios in any of the following cases:

- A node powers off if it is overhearing a transmission and does not have a packet to send.
- If at least one neighbor is transmitting and at least one neighbor is receiving a transmission, a node may power off. This is because, even if the node has a packet to transmit, it cannot do so because of fear of interfering with its neighbor reception.
- If all neighbors of a node are transmitting and the node is not a receiver, it powers itself off.

In PAMAS, nodes attempt to capture the communication channel by exchanging RTS/CTS packets. These packets contain duration of data packet transmission. A node can learn about the times that it can be sleeping (or turn off its radio transceiver) by listening to the RTS/CTS exchange. In PAMAS, this exchange takes place over a separate signaling channel. Thus, this exchange does not interfere with ongoing data transmission. It is possible that a new transmission starts when a node is asleep. In such a case, the node does not know about the duration of data transmission. To solve this problem, nodes probe the signaling channel to find out the length of remaining transmission. Although PAMAS avoids the overhearing problem, it does not address the problem of energy consumption when nodes are idle. Solutions to this latter problem are proposed in GAF and span described next.

32.3.3.2 Geography-Informed Energy Conservation for Ad Hoc Routing

Geographical adaptive fidelity (GAF) [20] employs intelligent node scheduling techniques to conserve the energy. In MANETs, GAF is driven by this observation that when there is significant node redundancy in a MANET, multiple paths will exist between nodes, thus some intermediate nodes can be powered off to conserve energy while still maintaining the network connectivity.

GAF divides the whole area where the nodes are distributed to small virtual grid cells such that every node in each virtual grid cell can communicate with other nodes in that same cell. At any instant of time, exactly one node in each grid is active while all other nodes are in the power saving mode (sleep or discovery). As illustrated in Figure 32.3, nodes make transitions between discovery, sleep, and active states. In the discovery state, which is the initial state, a node identifies all other nodes that are located in the same grid cell by exchanging discovery messages. A node goes to the active state, T_d seconds after it enters the discovery state. A node stays in the active state for T_a seconds after which it goes to the discovery state. A node that is in the discovery or active states enters the sleep state when it finds out that some other node in the same grid is active and will thus handle routing. When transitioning to the sleep state, a node cancels all pending timers and powers down its radio. A node in the sleep state wakes up after an application-dependent sleep time T_s.

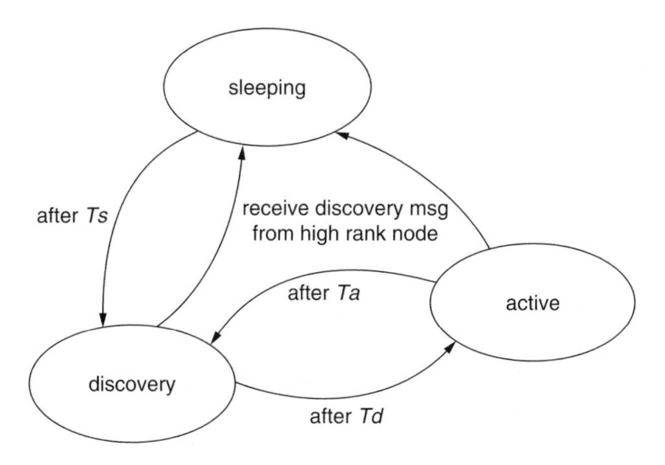

FIGURE 32.3 The state transition graph in GAF.

In GAF, nodes are ranked by several rules. An active node has a higher ranking than a node in the discovery state. For nodes that are in the same state, GAF gives higher ranking to nodes with a higher remaining battery capacity. Thus, a node with higher energy resource has a greater chance to become active. In this way, GAF achieves its load balancing strategy. When nodes have high mobility, it is possible that an active node moves out of its grid cell and leaves the grid cell with no active node in it. This problem significantly increases the packet drop rate. To avoid this problem, each node uses the GPS information to determine its bearing and velocity, thereby estimating the time that it expects to stay in the current grid, and adds this expected time to the discovery message. Any node that enters the sleep state wakes up after a time equal to the minimum of T_s and the expected time for the active node to stay in the grid cell.

Assuming a static network and without accounting for the protocol overhead, the maximum increase in the network lifetime that is achieved by GAF is equal to $(n \cdot R^2 / 5 \cdot A)$ where R is radio range of each node, and n is the total number of nodes distributed in an area A. In order for nodes in each grid cell to be able to communicate with nodes in the neighboring grid cells, the grid side length cannot be greater than $R/\sqrt{5}$. Thus, area A is divided to $A/(R/\sqrt{5})$ virtual grid cells.

32.3.3.3 Topology Maintenance for Energy Efficiency in Ad Hoc Networks (Span)

Span [18] builds on the observation that when there is a region of dense nodes, only a small number of these nodes need to be on at any given time to forward traffic. Span thereby adaptively elects some nodes as coordinators in the network. Coordinators stay awake to maintain connectivity of the network and to route packets in the network. All other nodes go to sleep to save power. These nodes periodically check if they should wake up and become a coordinator. One possible way for some node x to become a coordinator is that two neighbors of x cannot communicate with one another directly or through one or at most two coordinators. In addition, if node x has data to send out, it becomes a coordinator during its data transfer. When a node decides to be a coordinator, it uses a slotting and damping technique to delay its announcement of the fact. The node picks a random slot and delays its announcement until that slot. The random delay helps keep away from contention when several nodes decide to become coordinator at the same time. The delay function is as follows:

$$delay_i = \left((1 - \frac{R_i}{F_i}) + (1 - \frac{P_i}{\binom{N_i}{2}}) + \zeta \right) .N_i.\mu \tag{32.14}$$

where R_i is the remaining energy of the node, F_i is the full-charge battery capacity, P_i denotes the number of pairs of neighbors of i that cannot talk to one another unless through i itself, N_i denotes the number of neighbors of i (i.e., those nodes that can directly be reached from i). Recall that $\binom{N_i}{2}$ gives the number of pairs of neighbors of i. According to this equation, a node is more likely to pick an earlier time slot to become a coordinator if its ratio of remaining energy to full-charge battery capacity is high (close to 1). The node is also more likely to pick an earlier time slot if it can help connect a large number of pairs of its neighbors that would be disconnected without its assistance. μ is a random number that is picked uniformly in the (0,1) range, whereas ζ is the link propagation delay. A node switches from the coordinator to a noncoordinator role if every pair of its neighbors can reach each other directly or through one or two other coordinators. To balance the rate of energy consumption over all nodes, however, a node switches from a coordinator to a noncoordinator after some fixed period. In this way, it allows other nodes to become coordinators.

In span, nodes make all their decisions based on their local information and, unlike GAF, no knowledge of geographical information is required. Nodes find out about their neighbors proactively by broadcasting HELLO messages. These HELLO messages contain the status of the sender (i.e., coordinator or noncoordinator role), a list of its current coordinators, and its current neighbors. The list of the coordinators and neighbors are used by each of the node's neighbors in coordinator election and withdrawal rules that were described previously.

32.3.4 Energy-Aware Multicast Routing Algorithms

The primary goal of the conventional multicast routing protocols and algorithms has been to reduce the route latency because most multicast applications tend to be delay-sensitive audio/video broadcasting. Therefore, most of the multicast routing protocols are designed to construct a multicast tree that minimizes the communication latency. Because the number of hops is a good heuristic metric for capturing this latency, a multicast tree with the minimum number of hops has been favored by most routing protocols [21–23]. We call this tree the minimum hop-count tree (MHT). As has been described, in MANETs, two other criteria that make routing design an even more complex task (i.e., mobility and power efficiency) are used. The issue of mobility has been addressed extensively in the literature. In fact, the performance of multicast routing protocols has been evaluated in regard to their robustness to link failure due to the mobility [21,22,24,25]; however, little work has been accomplished on the development of a wireless multicast routing protocol in which power is key objective or constraint. More precisely, although some studies on the construction of energy-efficient broadcast and multicast tree in ad hoc networks [26,27] have been conducted, most of these works require a global view of the network and cannot be applied in a distributed way where the nodes have only local knowledge.

32.3.4.1 Minimum Energy Broadcasting

The objective of the minimum energy broadcasting is to reach from a specific source to all other nodes in the network by using multi-hop transmission while consuming the minimum total transmission energy and assuming that nodes have variable transmission power. In MANETs, broadcasting takes place by flooding the network from a specific source. Because the main use of flooding is in route discovery, it is important that flooding is done with the minimum total energy. Minimum energy broadcasting has been demonstrated as an NP-hard problem. Several heuristic algorithms for solving this problem have been proposed [26].

32.3.4.2 Energy-Aware Multicast Routing

The goal of energy-efficient multicast routing is to reach a subset of nodes (one-to-many cast) that we will refer to as multicast receivers, from a multicast source, such that we have maximum longevity of the paths between the source and the receivers. The problem of the energy-aware multicast tree is mathematically defined as follows. Consider a network graph $G(V,E)$, when V is set of nodes (or vertices) and

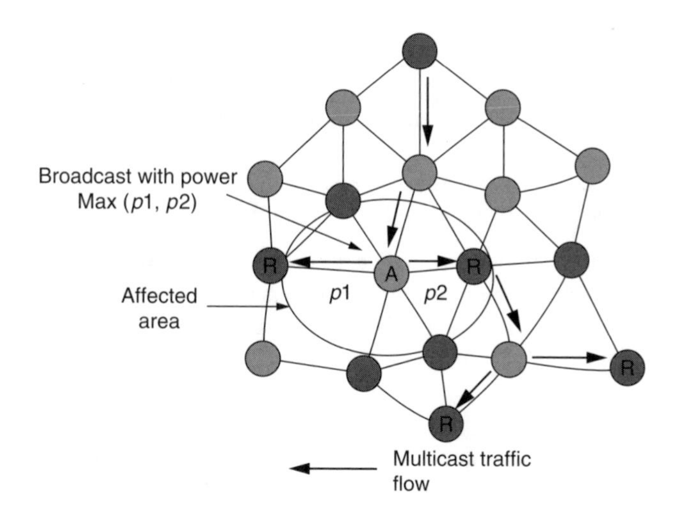

FIGURE 32.4 Neighbor cost effect in wireless networks.

E is set of edges in graph *G*. Let R_S denote the set of muticast receivers, *s* the multicast source, and *c(u,v)* the cost of edge *(u,v)*. The objective function may be stated as follows:

$$Min\ C(M) = \sum_{(u,v)\in M} c(u,v) \qquad (32.15)$$

where *C(M)* is the cost of multicast tree, M, connecting *s* to R_S.

The edge cost function *c(u,v)* may represent the transmit power level needed for sending data from *u* to *v*. In this case, the previously mentioned objective function results in a minimum total transmit power multicast tree. In addition, *c(u,v)* may be a battery-related cost of node *u* if the objective is to extend the lifetime of the network graph G. Figure 32.4 is an example of multicast tree. In general, finding a minimum energy multicast tree is equal to finding a minimum Steiner tree that is known to be an NP-hard problem [28]. Two related works on developing heuristic energy-aware multicast (or broadcast) trees are as follows:

- Least-Cost Shortest Path Tree (LPT). This is a tree obtained by superimposing all the least cost paths (or shortest paths) between the source and each multicast receiver.
- Broadcast Link-Based MST (BLMST). This is a minimum spanning tree where the link cost is set to the transmission energy needed to sustain communication over that link.
- Multicast Incremental Power Tree (MIPT). This tree is obtained from the Broadcast Incremental Power (BIP) tree proposed in Wieselthier et al. [27]. The BIP algorithm consists of the following steps:
  - For all nodes *i* in the tree and all nodes *j* not in the tree, evaluate $\rho'_{ij} = \rho_{ij} - \rho_{i,}$ where ρ_{ij} was defined earlier, ρ_i denotes the power level of node *i*. (Note that ρ'_{ij} provides the incremental cost associated with adding node *j* to the tree.) Initially, the tree includes only the source node (i.e., the broadcast initiator node).
  - A pair *(i,j)* that results in the minimum value of ρ'_{ij} is chosen, and node *j* is added to the tree.
- This procedure is continued until all intended destination nodes are included. The MIPT is generated by pruning the broadcast tree (i.e., by eliminating all sub-paths that are not required to reach the multicast receivers).

32.3.4.3 The Neighbor Cost Effect in Multicast Routing

Assume that a multicast tree from the source to several receivers has been constructed. The packet flow is coming out from the source, and is terminated at the leaves of the tree where the receivers are located. We will refer to those intermediate nodes of the tree that have more than one child in the tree as multi-fanout nodes (e.g., node A in Figure 32.4). In MANETs because the MAC layer does not have the ability of multicasting [10], two distinct methods are used to send out the packets from a multi-fanout node:

- Multiple unicast. The parent node sends unicast packets to every child node in the multicast tree separately.
- Single broadcast. The parent broadcasts the packets to all nodes in its immediate neighborhood (which may include nodes that are not in the multicast tree).

Freeney and Nilsson [10] experimentally studied the power-optimal choice between these two methods. According to its results, the multiple unicast method results in much higher power consumption for the sender (parent node in the multicast tree). The following is empirical energy-cost measurement by Freeney and Nilsson [10] for broadcast and unicast send/receive packets:

	Unicast	Broadcast
Send (μW.sec/byte + μW.sec)	1.9 · packet_size + 454	1.9 · packet_size + 256
Receive (μW.sec/byte + μW.sec)	0.5 · packet_size + 356	0.5 · packet_size +56

These measurements have been completed on Lucent IEEE 802.11 2 MBPS WAVELAN PC card with 2.4-GHZ direct sequence spread spectrum (DSSS).

Based on these results, a single broadcast method in multi-fanout nodes is more energy efficient. When using the single broadcast method, however, all the nodes that are in the radio range of the sender listen to the channel and receive the packet, thereby, unnecessarily consuming power in receiving the packet. As a result, these nodes will find the multiple unicast method to be more beneficial to them from a power dissipation viewpoint. Consequently, one must consider the power consumption cost of all neighbors of nodes that broadcast packets when calculating the cost of a multicast tree, in which multi-fanout nodes use a single broadcast method. This phenomenon, which we will refer to as the neighbor cost effect, makes the problem of finding a multicast tree with optimal cost quite complex. Regarding neighbor cost effect, the general objective function of the multicast tree problem is changed as follows:

$$C(M,t) = \sum_{(u,v)\in M} c(u,v) + (\text{if } \deg(u) \geq 2 \quad \textit{then} \sum_{v \in N_u \wedge j \notin M} c(u,v) \text{ else } 0) \tag{32.16}$$

where *deg(u)* denotes degree of node u in multicast tree M (including incoming and outgoing edges), and N_u refers to the set of nodes that are in the radio range of node u.

Another issue concerning the single broadcast method of multi-fanout nodes is that the farthest child from the parent determines the broadcast transmission power of that transmitting node. For example, in Figure 32.4, the transmission power of node A is $Max(\rho 1, \rho 2)$. Considering the neighbor cost effect in multi-fanout nodes makes the multicast routing problem even more challenging. Recall that finding a minimum energy-cost multicast tree without considering the neighbor cost effect is equivalent to that of finding a minimum Steiner tree, which is NP-hard. As a result, the problem of finding an energy-aware multicast tree with consideration of the neighbor cost effect is also an NP-hard problem.

Many algorithms for finding a tree with near optimal cost are available [29,30]. Although it is possible to modify some of these algorithms to account for the neighbor cost effect at multi-fanout nodes, this approach is ill advised in our context because these algorithms are too complex and require global information about the network connectivity graph to be applied. However, we are interested in finding solutions that can be deployed in an ad hoc network where nodes only have local knowledge about themselves and perhaps their neighboring nodes and must do the route discovery in a distributed, ad hoc manner (no global depository of information exists.) Furthermore, in ad hoc networks, the underlying network topology (connectivity graph) changes dynamically due to the mobility and link failure. Thus, ad hoc routing algorithms should be able to update their routes periodically. The routing update cost should be rather low.

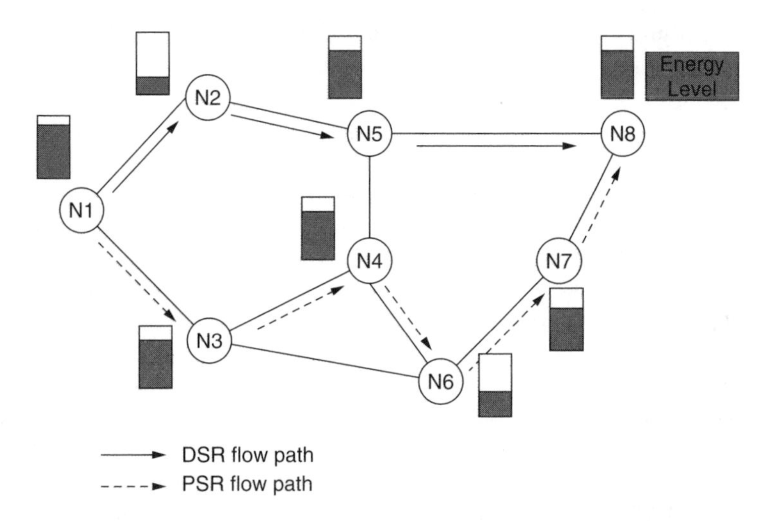

FIGURE 32.5 PSR avoids routes consisting of nodes with low remaining battery capacity.

32.4 Power-Aware Source Routing

32.4.1 Cost Function

The objective of power-aware source routing (PSR) [31] is to extend the useful service life of a MANET. This is highly desirable in the network because node death leads to a possibility of network disconnectedness, rendering other live nodes unreachable. PSR solves the problem of finding a route π at route discovery time t such that the following cost function is minimized:

$$C(\pi,t) = \sum_{i\in\pi} C_i(t)$$
$$C_i(t) = \rho_i \cdot \left(\frac{F_i}{R_i(t)}\right)^{\alpha} \tag{32.17}$$

where ρ_i is transmit power level of node i, F_i, and R_i are full-charge and remaining battery capacities of node i at time t, and α is a positive weighting factor.

PSR uses a graded cost function as explained next. The exponent α is a discrete function of the ratio of the remaining battery capacity over the full-charge battery capacity. As this ratio decreases and successively becomes less than a specified set of threshold values, α increases according to a fixed schedule. In this way, nodes with very low battery capacity contribute a much higher value to the total path cost. In other words, if a path from source to destination has some nodes with a very low residual battery, the cost of the path will be very high, and therefore, PSR will behave similar to the min-max battery cost routing. Figure 32.5 illustrates how PSR avoids routes that include node(s) with low remaining energy. Routing path N1-N2-N5-N8 has the minimum hop-count from N1 to N8, and, therefore, it is selected by DSR; however, this route includes node N2, which has a very low remaining energy capacity. Thus, PSR selects another route: N1-N3-N4-N7-N8.

In DSR, because the route selection is done based on a shortest path finding algorithm (i.e., it selects paths with the minimum number of hops), a selected path may become invalid only due to node movements. In contrast, in PSR, both the node mobility and the node energy depletion may cause a path to become invalid. Because the route discovery and route maintenance processes in PSR are slightly more complicated compared to their counterparts in DSR, these two steps will need to be described in detail. In addition, because PSR is derived from DSR, the PSR description will often be contrasted with that of DSR.

32.5.2 Route Discovery

In DSR, activity begins with the source node flooding the network with RREQ packets when it has data to send. An intermediate node broadcasts the RREQ unless it gets a path to the destination from its cache or it has already broadcast the same RREQ packet. This fact is known from the sequence number of the RREQ and the sender ID. Consequently, intermediate nodes forward only the first received RREQ packet. The destination node only replies to the first arrived RREQ because that packet usually takes the shortest path.

In PSR, all nodes except the destination calculate their link cost (cf. Equation 32.17) and add it to the path cost in the header of the RREQ packet (cf. Equation 32.17). When an intermediate node receives a RREQ packet, it starts a timer (T_r) and keeps the cost in the header of that packet as mincost. If additional RREQs arrive with same destination and sequence number, the cost of the newly arrived RREQ packet is compared with the mincost. If the new packet has a lower cost, mincost is changed to this new value and the new RREQ packet is forwarded. Otherwise, the new RREQ packet is dropped. The destination waits for a threshold (T_r) number of seconds after the first RREQ packet arrives. In that time, the destination examines the cost of the route of every arrived RREQ packet. When the timer T_r expires, the destination node selects the route with minimum cost and replies. Subsequently, it will drop any received RREQ. The reply also contains the cost of the selected path appended to it. Every node that hears this route reply adds this route along with its cost to its route cache table. Although this scheme may somewhat increase the latency of the data transfer, it results in a significant improvement of network lifetime, as discussed later.

32.4.3 Route Maintenance

Route maintenance is needed for two reasons:

1. Mobility. Connections between some nodes on the path are lost due to their movement.
2. Energy Depletion. The energy resources of some nodes on the path may be depleting too quickly.

In the first case, a new RREQ is sent out and the entry in the route cache corresponding to the node that has moved out of range is purged. In the second case, two possible approaches are used:

1. Semi-Global Approach. The source node periodically polls the remaining energy levels of all nodes in the path and purges the corresponding entry in its route cache when the path cost increases by a fixed percentage. Notice that this results in very high overhead because it generates extra traffic.
2. Local Approach. Each intermediate node in the path monitors the decrease in its remaining energy level (thus the increase in its link cost) from the time of route discovery because of forwarding packets along this route. When this link cost increase goes beyond a threshold level, the node sends a route error back to the source as if the route was rendered invalid. This route error message forces the source to initiate route discovery again. This decision is only dependent on the remaining battery capacity of the current node, and thus, is a local decision.

PSR adopts the local approach that minimizes the control traffic. Furthermore, it assumes that all transmit power levels ($\rho_{i,j}$) are constant. This enables PSR to separate the effect of mobility from that of energy depletion during route maintenance. More precisely, for each node i along a path π, we define a "delta cost" function as follows:

$$\Delta C_i(t_a) = C_i(t_a) - C_i(t_d) = \rho_i \cdot \left(\frac{F_i}{R_i(t_a)}\right)^{\alpha} - \rho_i \cdot \left(\frac{F_i}{R_i(t_d)}\right)^{\alpha} \tag{32.18}$$

where t_a denotes the time instance when this route entry is fetched from the cache table of node I; t_d denotes the time instance at which π was added to the cache table of node i; and $C_i(t_a)$ is the fractional cost contributed by node i to total cost of the path π at time t_a, whereas $C_i(t_d)$ is the fractional cost contributed by node i to total cost of the path π at time t_d.

Assuming that α remains unchanged from time t_a to time t_d, then the condition for invalidating route π from the cache table of node i is:

$$\frac{\Delta C_i(t_a)}{C(\pi, t_d)} > \delta \tag{32.19}$$

where δ is a user-specified threshold value.

This condition invalidates a path π in the cache table of node i if the change in the normalized cost of node i exceeds a threshold δ. This metric appears to be a good way of capturing the dynamics of the node usage in MANETs. As the remaining energy of a node decreases, the cost of the node increases. The node will force new routing decisions in the network by invalidating its own cache entries to various destinations. If a path was recently added to the cache table, however, the node will not force a new decision (route finding step) unless the node's remaining energy is depleted by a certain normalized amount, due to messages passing through that path. The effect of δ on the performance of PSR is studied in detail in Section 32.6.

It should be noted that we provision for the reuse of invalidated paths if node i was the source of the message and wanted to continue to send data via this path as follows. When node i has data to send to the destination, it looks up its route cache and chooses a route, if such a route can be found in the cache, irrespective of whether the route was invalidated or not. In this way, we avoid redundant route discoveries in the presence of an existing route. The invalidated cache is purged after a fixed time. The invalid entries are analogous to the victim buffer in the cache structure of general-purpose processors; however, the same does not hold good for relaying data. If a cache entry is invalidated in a node and that node is asked to relay data/reply to the destination of that cache entry, then the node will send a route error back to the source. This reply will invalidate routing entries for all nodes on the trace path back to the source. The PSR function of intermediate nodes is given in Figure 32.6 in pseudo code. The function of the destination node is similar to the intermediate node with the exception that it does not need to check for validation of the path when it refers to its cache because it is the end point for each possible path between that itself and the source.

32.5 Lifetime Prediction Routing

32.5.1 Basic Mechanism

Lifetime prediction routing (LPR) [32] is an on-demand source routing protocol that uses battery lifetime prediction. The objective of this routing protocol is to extend the service life of MANET with dynamic topology. This protocol favors the path with the maximum remaining lifetime. We represent our objective function as follows:

$$\underset{\pi}{Max} \left\{ L_\pi(t) = \underset{i \in \pi}{Min}(\tau_i(t)) \right\} \tag{32.20}$$

where $L_\pi(t)$ is lifetime of path π and $\tau_i(t)$ is the predicted lifetime of node i at time t.

32.5.1.1 Lifetime Prediction

Each node tries to estimate its battery lifetime based on its past activity. This is achieved using a simple moving average (SMA) predictor by keeping track of the last N values of residual energy and the

```
Event: Tx expired
      If reply-route-from-cache
        Add the cost of the path in the cache to the mincost and append
        it to RREP packet and Send it back;
        else
             Drop any coming RREQ packet;
             mincost=0;
Event: RREQ  packet Arrival
If Tx expired
      Drop the packet;
      Exit;
If packet→ cost ≥ mincost
      Drop the packet;
Else
      mincost = packet→cost;
      Add the cost of this node to packet→cost;
If earliest RREQ arrival
        Start Tx timer;
Look up in the Cache;
If any path to the destination exists in the cache
          Check for validation of the fetched path from the cache;
           If path is valid
             Reply back after Tx timer expiration;
            Else Invalid that path and Propagate the RREQ packet;
   Else Propagate the RREQ packet;

Event: RREP or Data packet arrival
Check for the validity of the route used if exists is in the cache
If valid
      Forward it;
 Else Make that path (the cache line) invalid and Send error back
```

FIGURE 32.6 Pseudo code for the key operations performed in the intermediate nodes of a path in the PSR.

corresponding time instances for the last W packets received/relayed by each mobile node. This information is recorded and stored in each node. We have carefully compared the predicted lifetimes based on the SMA approach to the actual lifetimes for different values of W and found $W = 10$ to be a good value.

Our motivation in using lifetime prediction is that mobility introduces different dynamics into the network. In Chang and Tassiulas [16] the lifetime of a node is a function of residual energy in the node and energy to transmit a bit from the node to its neighbors (cf. Equation (32.12)). This metric works well for static networks for which it was proposed; however, it is very difficult to efficiently and reliably compute this metric when we have mobility because the location of the nodes and their neighbors constantly change.

PSR does not use prediction and only uses the remaining battery capacity. LPR is superior to PSR because LPR not only captures the remaining (residual) battery capacity but also accounts for the rate of energy discharge. This makes the cost function of LPR more accurate. This is true in MANETs because mobility can change the traffic patterns through the node, which thereby affects the rate of depletion of its battery. In addition, recent history is a good indicator of the traffic through the node, and thus we chose to employ lifetime prediction.

Our approach is a dynamic distributed load balancing approach that avoids power-congested nodes and chooses paths that are lightly loaded. This helps LPR achieve minimum variance in energy levels of different nodes in the network. As an example, consider the scenario in Figure 32.7. Here, node F has three flows going through it (D → F →, B → F →, and C → F →). Now, if A wants to transmit data to E, the shortest path routing will use A → F → E. LPR will use A → B → C → D → E, however, because E is very power-congested (as a result of relaying multiple flows) and the path passing through F will not be selected by LPR.

Figure 32.8 is an example that depicts how different policies of DSR, PSR, and LPR give different answers with the same scenario. Although PSR avoids choosing a path that goes through node N6, because of low remaining energy, the path selected by LPR (N1-N3-N6-N7-N8) includes N6. The reason is that N6 has a low depletion rate, and its estimated lifetime is high.

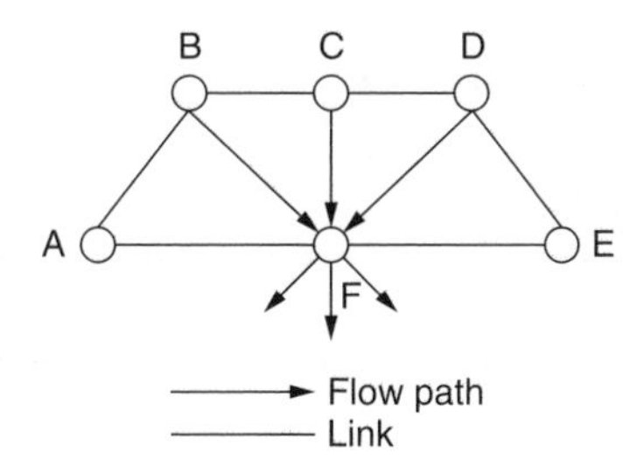

FIGURE 32.7 LPR avoids power-congested paths.

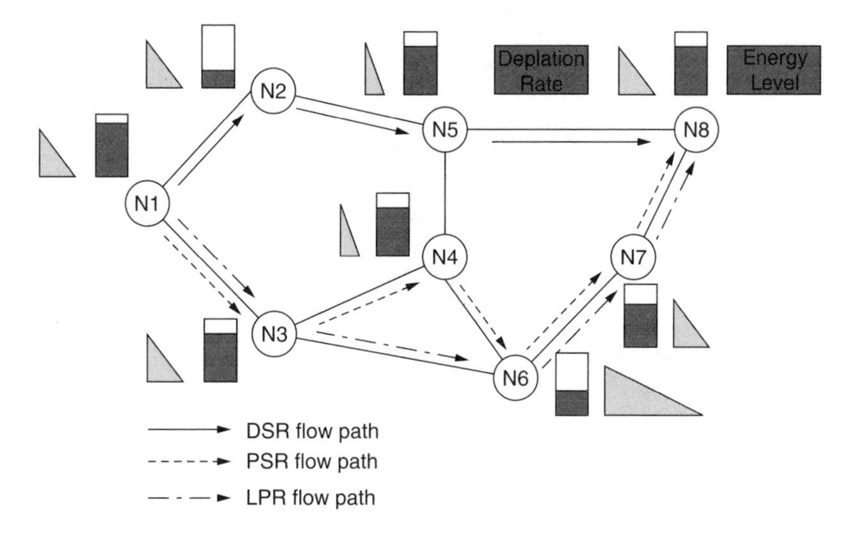

FIGURE 32.8 LPR avoids paths consisting of nodes with high-energy depletion rates.

32.5.2 Route Discovery

Route discovery in LPR is similar to PSR. In LPR, all nodes except the destination calculate their predicted lifetime, τ_i (see Equation (32.21)) and replace the minlifetime in the header with τ_i if τ_i is lower than the existing minlifetime value in the header.

$$\tau_i(t) = \frac{R_{r,i}(t)}{\frac{1}{W-1}\sum_{k=i-W+1}^{i} r_k(t)} \tag{32.21}$$

where $R_{r,i}(t)$ denotes remaining energy at the i_{th} packet is being sent or relayed through the current node, $r_k(t)$ is rate of energy depletion of the current node when the k_{th} packet was sent and is calculated by the ratio of the difference between residual energies of the nodes for packets *k-1* and *k* and the difference between arrival times of these two packets, and *W* is length of the history used for calculating the SMA.

When an intermediate node receives a RREQ packet, it starts a timer (*Tr*) and keeps the min. lifetime in the header of that packet as minlifetime. If additional RREQs arrive with the same destination and sequence number, the cost of the newly arrived RREQ packet is compared with the mincost. If the new packet has a lower cost, mincost is changed to this new value, and the new RREQ packet is forwarded. Otherwise, the new RREQ packet is dropped (see Figure 32.9).

In LPR, the destination waits for a threshold number (*Tr*) of seconds after the first RREQ packet arrives. During that time, the destination examines the cost of the route of every RREQ packet that

```
Predict its lifetime;
If its lifetime < minlifetime
      Replace minlifetime with its lifetime;
If Sequence Number exists
      Compare minlifetime of current RREQ with minlifetime
      of existing one;
      If new minlifetime <= old minlifetime
            Discard new RREQ;
      If new minlifetime > old minlifetime
            Replace old minlifetime with new minlifetime;
            Forward new RREQ;
If Sequence Number does not exist
      Save this minlifetime;
      Forward RREQ;
```

FIGURE 32.9 Pseudo code of functions performed in an intermediate node as it is executing the LPR algorithm.

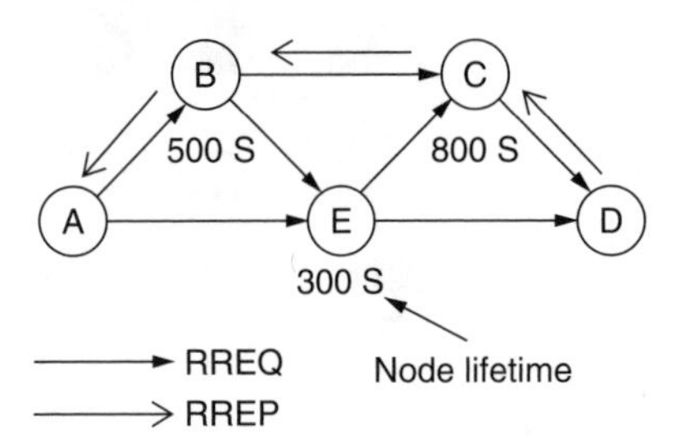

FIGURE 32.10 The route setup process in LPR.

arrived. When the timer (*Tr*) expires, the destination node selects the route with the minimum cost and replies. Subsequently, it will drop any received RREQs. The reply also contains the cost of the selected path appended to it. Every node that hears this route reply adds this route along with its cost to its route cache table. Although this scheme can somewhat increase the latency of the data transfer, it results in a significant power savings, as discussed later. A simple example of this process is illustrated in Figure 32.10. Here, the route A-B-C-D is chosen by LPR over the route A-E-D because the path lifetime of the former is in the 500s, which is greater than the latter.

LPR has a route invalidation timer that invalidates old routes. This helps in removing old routes. This also avoids over usage of particular routes in cases of low mobility.

32.5.3 Route Expiration

Route maintenance is needed for two reasons:

1. Connections between some nodes on the path are lost due to their movement
2. Change in the predicted lifetime

In the first case, a new RREQ is sent out and the entry in the route cache corresponding to the node that moved out of range is purged. The following policy is adopted to tackle the second situation.

Once the route is established, the weakest node in the path (the node with minimum predicted lifetime at path discover time) monitors the decrease in its battery lifetime. When this remaining lifetime decrease goes beyond a threshold level, the node sends a route error back to the destination as if the route was rendered invalid. The destination sends this route error message to the source. This route error message forces the source to initiate route discovery again. This decision is only dependent on the remaining battery capacity of the current node and its discharge rate in the short history, and thus is a local decision. LPR adopts this local approach because this approach minimizes control traffic. Figure 32.11 is an example of the route expiration process.

More precisely, node *i* generates a route error at time t when the following condition is met:

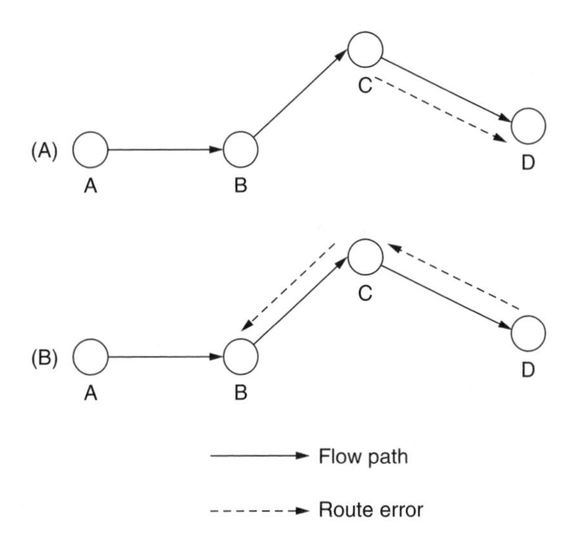

FIGURE 32.11 (a) Node C sends route error to destination node D; (b) Node D sends route error to source A to invalidate the whole path.

$$\tau_i(t_0) - \tau_i(t) \geq \delta \tag{32.22}$$

where t denotes the current time, t_0 is time of the route discovery, and δ is threshold value.

32.6 Quantitative Evaluation of Source Routing Algorithms

32.6.1 Simulation Setup

We used the event driven simulator ns-2 [33] along with the wireless extensions provided by CMU [34]. The simulation consists of a network of 20 nodes confined in a 1000×1000 m^2 area. Random connections were established using CBR traffic (at 4 packets/second) such that each node has chance to connect to every other node. Packet size was 512 bytes and each simulation was executed for 20,000 sec. The initial battery capacity of each node is 100 units. Nodes followed a random waypoint mobility model with a specific max velocity and no pause time. Each packet relayed or transmitted consumes a fixed amount of energy from the battery as given by Equation (32.2); a and b are constants.

The key parameters of study are the network lifetime, node lifetime, and root mean square (RMS) of energy consumption (E_{RMS}) in the network. We vary the speed and radio transmission range and study their effects on these metrics.

32.6.2 Simulation Results

The network lifetime is defined as the time taken for a fixed percentage of the nodes to die due to energy resource exhaustion. Network lifetime of DSR, PSR, and LPR are compared for a given scenario. Here, the speed of each node is 10 m/s and radio transmission range is 125 m. Figure 32.12 plots the time instances at which a certain number of nodes have died when simulating LPR, PSR, and DSR. Note that in Figure 32.12, node death of all 20 nodes is not plotted because some nodes are still alive at the end of the simulation. Some of these nodes, however, are rendered unreachable because many of the nodes have exhausted their energy and thus cannot reach other nodes consistently.

As can be seen, the first node in DSR and PSR dies about 20% earlier than in the case of LPR. Similarly, in DSR 5 nodes die approximately 32% earlier than LPR and 27% earlier than LPR in the case of PSR.

Due to the dynamic nature of the path cost function of PSR (and LPR), a discovered path cannot remain valid for a long time. This is because these connections, if maintained for a long period, may

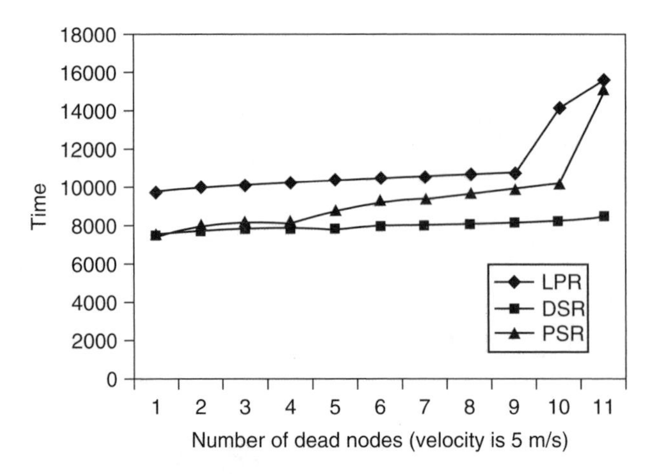

FIGURE 32.12 Number of dead nodes in DSR, PSR, and LPR as a function of the elapsed time.

exhaust the energy of some nodes on that path. Discovered paths are in the cache, however, and can be accessed whenever they are required in DSR, and (as implemented in ns-2) only mobility can invalidate these cache entries. In addition, cache invalidation is very expensive for the network because the route is reconstructed by flooding the network. This is handled in PSR as described in the next paragraph.

When the path is discovered, every node puts its remaining energy and path cost in the cache entry. Intermediate nodes check for validity of this path by computing the cost difference as in Equation (32.19). Here, δ (the threshold) is a metric that decides how often we invalidate the cache. This threshold affects the performance of PSR. If the threshold is very high, we do cache invalidation very rarely, and might end up overexercising some nodes in the path. If it is very low, the cache invalidation rate is very high and may lead to unnecessary flooding in the network. The effect of varying this threshold is plotted in Figure 32.13.

Because LPR outperforms PSR in terms of results in a longer network lifetime, we have selected LPR to compare it with DSR for the rest of simulation.

To increase the lifetime of the network, the variance of the residual energy of the nodes should be minimized. Figure 32.12 is not informative in this regard. A histogram of the snapshots of the energy

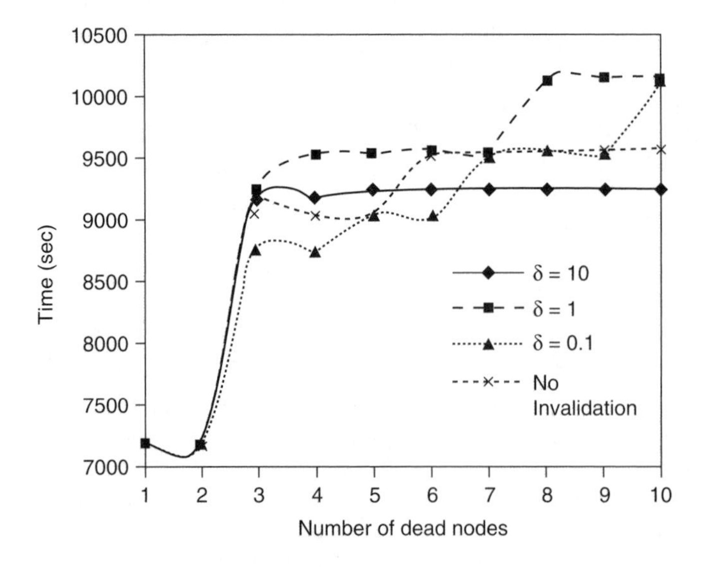

FIGURE 32.13 Effect of the threshold value, δ (for the PSR path invalidation step) on the network lifetime.

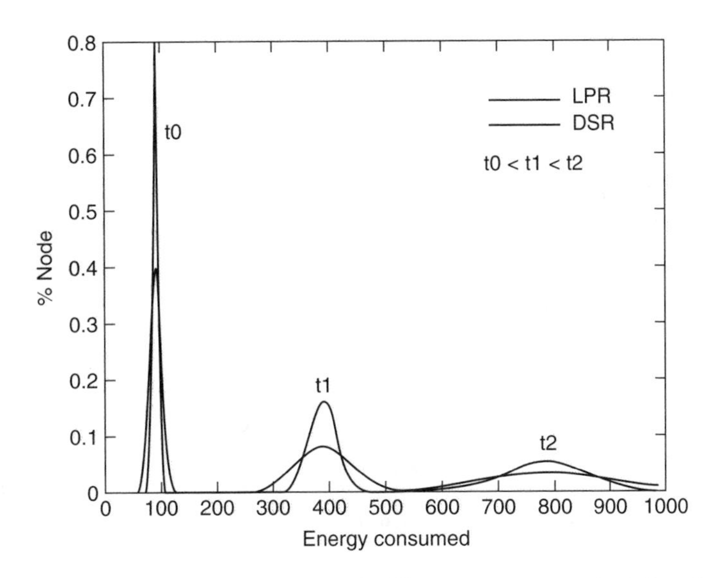

FIGURE 32.14 Distribution of energy consumed at three different time instances for LPR and DSR.

consumption in each of the nodes at different time instances would be more informative. Figure 32.14 depicts this histogram at three time instances. Initially, all nodes have zero energy consumption. As time increases, variance of energy consumption or remaining energy of nodes increases, but the rate of increasing for LPR is more than DSR. One of the ways to compare such histograms would be to look at the RMS of the remaining energy (E_{RMS}) at different time instances. It provides information about the total energy consumed and spread of consumed (residual) energy. Figure 32.15 plots the evaluation of E_{RMS} as a function of time for DSR and LPR before any node dies out. The effect of mobility on E_{RMS} can also be observed in this figure. A linear estimation of E_{RMS} is depicted for ease of comparison. As can be observed, LPR is always better than DSR in terms of E_{RMS} value. This graph is in agreement with our expectations. As the velocity of node movement increases, however, the rate of energy consumption in the network goes up. This is expected because higher velocity of movement implies more route discoveries being performed and consequently higher energy consumption in the network. In addition,

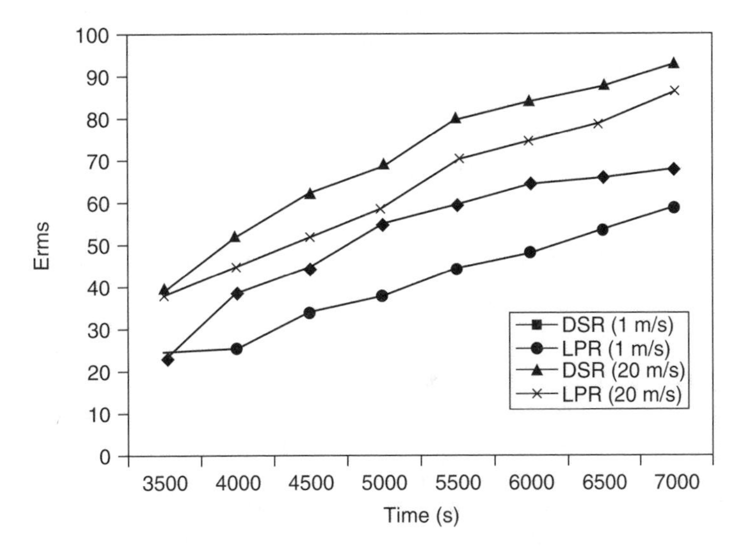

FIGURE 32.15 Evaluation of E_{RMS} for different velocities of node movement.

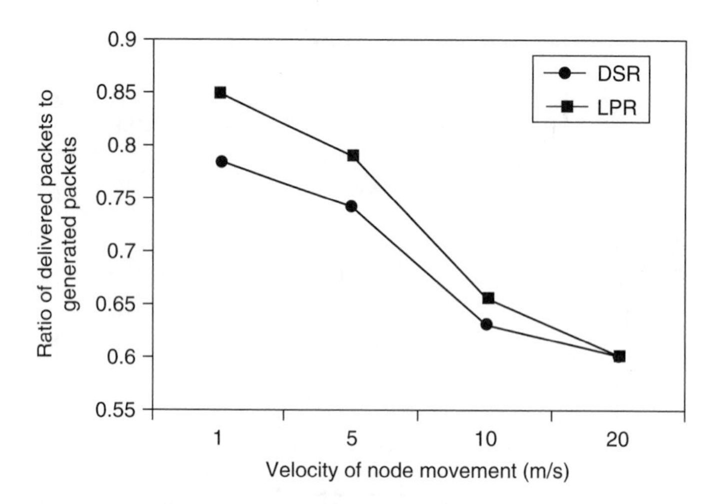

FIGURE 32.16 Packet delivery ratio vs. velocity of node movement.

as the node mobility increases, the difference between DSR and LPR decreases. This could be attributed to two reasons:

1. LPR makes use of the fact that DSR overloads certain nodes and has a big variance between remaining energies of the nodes. As mobility increases, the amount of overhead (control packets for route discovery) increases for both DSR and LPR. Consequently, less room is available for LPR to balance the energy consumption among the nodes in the network and extend its network lifetime.
2. Because more route discoveries occur, no paths are overused even by DSR. Consequently, DSR also achieves load balancing to an extent, decreasing the gain seen by LPR.

Packet delivery ratio is defined as the number of delivered data packets to the number of generated data packets in all nodes. Note that the number of generated packets is the "expected" number of generated packets. We generate as many as 200,000 data packets during the simulation. They are generated between random sources and destination pairs at random times. Many of these might not have reached their intended destination because of the lack of existence of a route between the source and destination for various reasons. In addition, the network lifetime clearly affects this ratio. If the network was alive for longer time, it implies that more data traffic goes through because we establish random connections throughout the time of simulation.

As plotted in Figure 32.16, for lower velocities of node movement, LPR has a greater ratio of delivered packets. As the mobility increases, however, this ratio goes down. The intuition for why LPR does not perform as well in higher velocities was presented earlier.

The transmission range is another parameter that can affect the performance of routing protocols because it changes the connectivity of the network. We changed the transmission range to see the effect of the degree of connectivity on our metric (see Figure 32.17). We assume the same transmission power for all nodes in a simulation. The node transmit range was assigned two different values (125 and 200 m) for the simulations. We make the following observations based on this figure:

- When the transmission range increases, each node covers more nodes. In other words, when a node sends a unicast or broadcast packet, more nodes will receive packets and they consume power in their receiver. Thus, each transmission has a lot of power overhead for the network. As a result, when the range increases, nodes discharge faster.
- The number of hops per route decreases by increasing the transmission range. Thus, nodes have less participation in relaying packets resulting in lower activity for each node and slower discharge of its battery capacity.

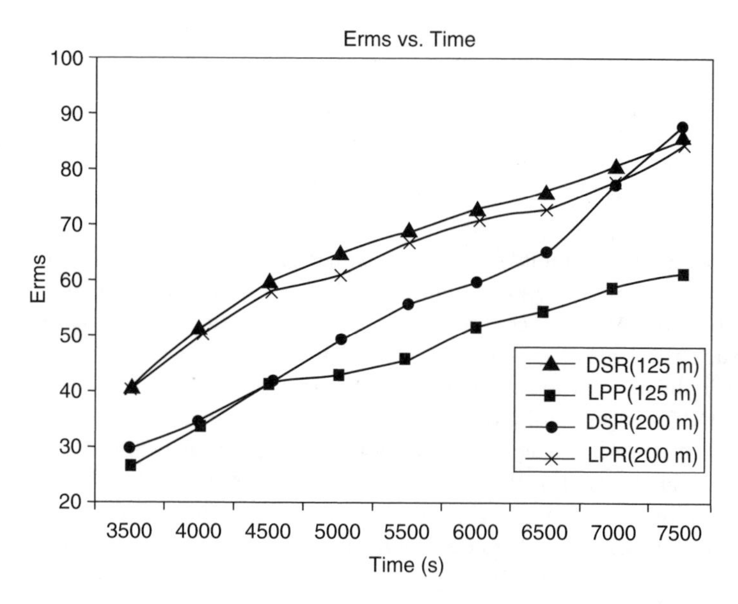

FIGURE 32.17 Effect of transmission range on the E_{RMS} (node velocity is 5 m/s).

When range increases from 125 to 200 m, the dominant effect is the first and the charge rate of the nodes increases drastically. Both of those effects reduce the effect of the LPR scheme and as can be seen, the difference between LPR and DSR decreases, such that when the range is 200 m, the difference is not clear. To reduce the cost of the power due to the second effect, one way is to shut down the nondestined nodes in the range of a transmitting node.

In LPR, route discovery process needs more control packets to be propagated in the network because it needs to compare all possible paths between a source and a sink and selects a path with maximum lifetime. To illustrate the overhead of LPR on the network, we have measured the ratio of the number of control packets to the number of delivered packets in the network. This normalizes the overhead of the routing protocol to the goodput (i.e., number of received packets) in the network. Figure 32.18 plots this ratio for LPR and DSR for different velocities of node movement and for 380 user datagram protocol (UDP) connections. As the velocity of movement increases, routes are valid for a shorter time and more route discoveries are done in the network resulting in more control packets and more difference between

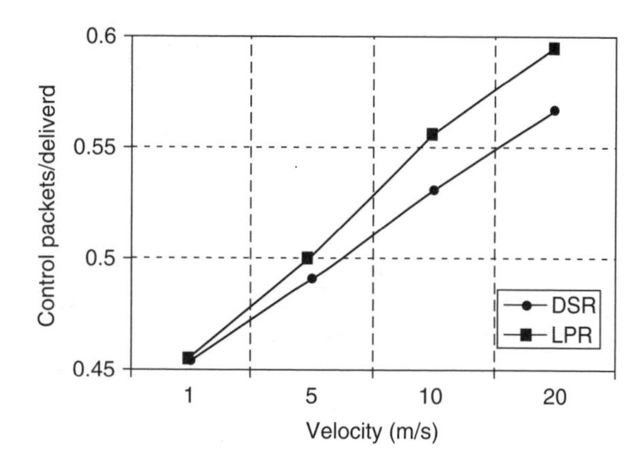

FIGURE 32.18 The ratio of control packets to delivered packets as a function of velocity of node movement for LPR and DSR for 380 UDP connections.

LPR and DSR. LPR increases the ratio of the control packet to transmit a packet less than 4%. The increase in the size of a control packet in DSR to that of the LPR is approximately 1/10, and the overhead in energy for sending such a packet increases by approximately 0.4%. Thus, the additional energy overhead of LPR for route discovery is small.

32.7 Conclusion

One of the main design constraints in MANETs is that they are energy constrained. Thus, network routing algorithms must be developed to consider energy consumption of the nodes in the network as a primary objective. In MANETs, every node has to perform the functions of a router. Therefore, if some nodes die early due to lack of energy so that the network becomes fragmented, then it may not be possible for other nodes in the network to communicate with each other. This chapter presented PSR and LPR protocols for MANETs where the aim is to maximize the network lifetime (which is typically defined as the duration time after which a fixed percentage of the nodes in the network "die" as a result of energy exhaustion). This goal of extending the network lifetime was accomplished by finding routing solutions that tend to minimize the variance of the remaining energies of the nodes in the network. Although these power-aware network routing protocols and algorithms tend to create additional control traffic, simulations reported in this chapter demonstrate that they improve the network lifetime by more than 20% on average.

References

[1] C. Hedrick, Routing information protocol, *RFC 1058,* http://www.faqs.org/rfcs/rfc1058.html, Aug. 2001.

[2] J. Moy, OSPF version 2.0, *RFC 2328,* http://www.faqs.org/rfcs/rfc2328.html, Aug. 2001.

[3] C.E. Perkins, *Ad Hoc Networking,* Addison-Wesley, Reading, MA, 2001.

[4] C. Perkins and P. Bhagwat, Highly dynamic destination-sequenced distance-vector routing (DSDV) for mobile computers, *Proc. of ACM SIGCOMM Conf. on Communications Architectures, Protocols, and Applications,* pp. 234–244, Oct. 1994.

[5] S. Murthy and J.J. Garcia-Luna-Aceves, An efficient routing protocol for wireless networks, *ACM Mobile Networks and Applications J., Special Issue on Routing in Mobile Communication Networks,* vol. 1, no. 2, pp. 183–197, 1996.

[6] D.B. Johnson, D.A. Maltz, Y.-C. Hu, and J.G. Jetcheva, The dynamic source routing for mobile ad hoc wireless networks, *IETF Internet Draft,* http://www.ietf.org/internet-drafts/draft-ietf-manet-dsr-09.txt, Nov. 2001.

[7] C.E. Perkins, E.M. Belding-Royer, and S. Das, Ad hoc on-demand distance vector (AODV) routing, *IETF Internet Draft,* draft-ietf-manet-aodv-12.txt, Nov. 2002.

[8] S. Singh, M. Woo, and C.S. Raghavendra, Power-aware routing in mobile ad hoc networks, *Proc. of Mobile Computing and Networking (Mobicom),* pp. 181–190, 1998.

[9] IEEE Standards Board 802 Part 11: *Wireless LAN Medium Access Control (MAC) and Physical Layer (Phy) Specifications,* Mar. 1999.

[10] L.M. Freeney and M. Nilsson, Investigating the energy consumption of a wireless network interface in an ad hoc networking environment, *Proc. IEEE Infocom,* pp. 1548–1557, Apr. 2001.

[11] Stojmenovic and X. Lin, Power-aware localized routing in wireless networks, *Proc. IEEE Trans. on Parallel and Disrtibuted Systems,* vol. 12, no. 11, pp. 1122–1133, May 2001.

[12] W. Rabiner, W. Heinzelman, A. Chandrakasan, and H. Balakrishnan, Energy-efficient communication protocol for wireless microsensor networks, *Proc. 33rd Annu. Hawaii Int. Conf. on System Sciences,* pp. 3005–3014, Jan. 2000.

[13] C.K. Toh, Maximum battery life routing to support ubiquitous mobile computing in wireless ad hoc networks, *IEEE Communication Mag.,* pp. 138–147, June 2001.

[14] A. Misra and S. Banerjee, MRPC: maximizing network lifetime for reliable routing in wireless environments, *Proc. IEEE Wireless Commn. and Networking Conf.*, pp. 800–806, Aug. 2002.

[15] S. Banerjee and A. Misra, Minimum energy paths for reliable communication in multi-hop wireless networks, *Proc. MobiHoc,* pp. 146–156, June 2002.

[16] J.-H. Chang and L. Tassiulas, Energy-conserving routing in wireless ad hoc networks, *Proc. Infocom,* pp. 22–31, Mar. 2001.

[17] O. Kasten, Energy consumption. ETH-Zurich, Swiss, Federal Institute of Technology. Available at http://www.inf.ethz.ch/~kasten/research/bathtub/energy_consumption.html, Apr. 2001.

[18] B. Chen, K. Jamieson, H. Balakrishnan, and R. Morris, Span: an energy-efficient coordination algorithm for topology maintenance in ad hoc wireless networks, *Proc. Mobile Computing and Networking (Mobicom),* pp. 85–96, July 2001.

[19] S. Singh and C. Raghavendra, PAMAS: power-aware multiple access protocol with signaling for ad hoc networks, *ACM Computer Communication Review,* 28(3):5–26, July 1998.

[20] Y. Xu, J. Heidemann, and D. Estrin, Geography-informed energy conservation for ad hoc routing, *Proc. Mobile Computing and Networking (Mobicom),* pp. 70–84, July 2001.

[21] S.-J. Lee, W. Su, and M. Gerla, On-demand multicast routing protocol (ODMRP) for ad hoc networks, *IETF Internet Draft,* http://www.cs.ucla.edu/NRL/wireless/PAPER/draft-ietf-manet-odmrp-02.txt, Oct. 2002.

[22] J.J. Aceves and E. Madruga, The core-assisted mesh protocol, *IEEE JSAC,* vol. 17, no. 8, pp. 1380–1394, Aug. 1999.

[23] E. Royer and C. Perkins, Multicast ad hoc on-demand distance vector (MAODV) routing, *IETF Internet Draft,* draft-ietf-manet-maodv-00.txt.

[24] S.-J. Lee, W. Su, J. Hsu, M. Gerla, and R. Bagrodia, A performance comparison study of ad hoc wireless multicast protocols, *Proc. IEEE Infocom*, pp. 565–574, Mar. 2000.

[25] M. Gerla, C.-C. Chiang, and L. Zhang, Tree multicast strategies in mobile, multi-hop wireless networks, *ACM/Kluwer Mobile Networks and Applications,* vol. 4, no. 3, http://www.ietf.org/proceedings/00dec/I-D/draft-ietf-manet-maodv-00.txt, Oct. 2002.

[26] M. Cagalj, J. Phubaux, and C. Enz, Minimum-energy broadcast in all-wireless networks: NP-completeness and distribution issues, *Proc. Mobile Computing and Networking (Mobicom),* Sep. 2002.

[27] J.E. Wieselthier, G.D. Nguyen, and A. Ephremides, On the construction of energy-efficient broadcast and multicast trees in wireless networks, *Proc. Infocom,* pp. 585–594, Mar. 2000.

[28] A. Goel and K. Munagala, Extending greedy multicast routing to delay sensitive applications, *J. Algorithmica,* vol. 33, no. 3, pp. 335–352, 2002.

[29] M. Parsa, Q. Zhu, and J.J. Garsia-Luna-Aceves, An iterative algorithm for delay-constrained minimum-cost multicasting, *IEEE/ACM Trans. on Networking,* vol. 6, no. 4, pp. 461–474, Aug. 1998.

[30] J. Cong, A.B. Kahng, G. Robins, M. Sarrafzadeh, and C.K. Wong, Provably good performance-driven global routing, *IEEE Trans. on Computer-Aided Design,* vol. 11, no. 6, pp. 739–752, 1992.

[31] M. Maleki, K. Dantu, and M. Pedram, Power-aware source routing in mobile ad hoc networks, *Proc. Int. Symp. on Low-Power Electronics and Design (ISLPED),* pp. 72–75, Aug. 2002.

[32] M. Maleki, K. Dantu, and M. Pedram, Lifetime prediction routing in mobile ad hoc networks, *Proc. IEEE Wireless Commn. and Networking Conf.,* Mar. 2003.

[33] NS -2 Manual, http://www.isi.edu/nsnam/ns/doc/index.html, Feb. 2002.

[34] CMU Monarch Extensions to ns, http://www.monarch.cs.rice.edu/, Feb. 2002.

[35] V.D. Park and S. Corson, Temporally ordered routing algorithm (TORA) version 1 functional specification, *IETF Internet Draft,* draft-ietf-manet-tora-spec-01.txt, Aug. 1998. Network simulator, Feb. 2002.

33

Modeling Computational, Sensing, and Actuation Surfaces

Phillip Stanley-Marbell
Diana Marculescu
Radu Marculescu
Pradeep K. Khosla
Carnegie Mellon University

33.1 Introduction

Recent years have seen the emergence of many efforts to embed computing resources in everyday environments. These efforts have ranged from the use of wireless sensor networks, to wired ubiquitous computing environments in homes and commercial installations. A promising culmination of these directions is that of general purpose flexible surfaces, with large numbers of computational, sensing, and actuation elements embedded in them. Thin and flexible sheets of general purpose active materials could find use in a variety of commercial and household applications. These materials with embedded computation, sensing, and actuation capabilities, may be deployed cheaply over large surfaces, for both the interiors and exteriors of buildings, automobiles, marine vessels (e.g., as a hull lining), and aerospace applications. Such active or computational surfaces will take advantage of their large contiguous spatial extents, and the ability to actuate these surfaces. Such an active material with embedded actuators might be used in building structures that self-repair, or adapt to weather conditions.

33.1.1 Computational Surfaces

Technologies being developed for wireless sensor networks are targeted at enabling the use of cheap, miniscule, discrete sensing devices for monitoring. Such devices, when either dispersed over large areas

0-8493-1941-2/05/$0.00+$1.50

or embedded into commodity items, enable the gathering of data and monitoring of phenomena. Wireless networked sensors will enable more efficient tracking (e.g., of items in a warehouse) continuous monitoring of inhospitable or remote environments, and will generally enable a significant increase in our abilities to extract useful information from our environments.

Beyond simple sensing tasks which can be encapsulated in discrete networked sensors, many opportunities exist for "intelligent" materials with embedded sensing, computation and more importantly, actuation capabilities. Such a platform is the macro-scale analog of micro-electro-mechanical systems (MEMS): it integrates computational, sensing, and mechanical actuation devices into a general purpose material surface. Surfaces form an integral part of the design of structures. The ability to control and sense phenomena over large surfaces is, however, not easily achieved by straightforward extension of the capabilities of discrete sensors. Although it is possible to employ large numbers of wireless networked sensors over surfaces, the use of individual, discrete sensors does not enable the use of actuation, or take advantage of the possibility of using less power- and cost-intensive wired networking between devices.

The technologies to enable the inclusion of computation, sensing, and actuation arrays in such surfaces, are not a simple extension of any existing technologies. The manner in which the devices will be interconnected will most likely be through the embedding of conductors in the surface, however, it is premature to tell whether this will indeed be the norm. Despite these uncertainties, a common set of issues need to be addressed to make such computational surfaces a reality.

The physical construction of such substrates will vary based on their applications. For example, for the purposes of commercial applications such as "smart" lining materials for applications such as automobiles, aircraft, or marine vessels, a flexible polymer- or carbon-fiber-based substrate might be desirable. On the other hand, for other applications, a woven substrate might be desirable. Regardless of the actual method of fabrication, these substrates will share the following common properties:

- General purpose. It will be desirable to obtain such platforms as a general purpose programmable substrate. Instead of constructing custom systems, they will be obtained in units of area and programmed for application specific purposes. Differentiation between products will be achieved by the area density and types of computational, sensing, and actuation devices, as well as the type of the substrate material (e.g., polymer substrate, carbon fiber, or woven materials).
- Large surface area. The spatial extents of the substrate can be harnessed by applications which benefit from the combination of computational resources, sensing, and actuation over a large area.
- Computation. Large numbers of computational devices embedded into surfaces at low cost. The use of large numbers of devices will be driven not by a requirement for increased performance, but instead for fault-tolerance, as well as the desire to spatially distribute computational resources over the surface. If the platform will be obtained in units of area, a swath of material cut from a larger piece should still be programmable and usable.
- Sensing. A major benefit of the large area systems will be their spatial extent, which makes them particularly useful for sensing applications such as active antenna arrays. Such arrays for many types of signals (e.g., acoustic or RF) can take advantage of such large intelligent surfaces, to augment applications such as speaker location or ultra-wideband radios. For example, a surface being employed as an antenna array might adapt its shape to achieve better signal reception.
- Actuation. It is possible to embed many types of actuation devices in these large area surfaces. Actuating strands such as shape-memory-alloys, can be embedded to enable the surfaces to change shape in response to data obtained from the sensors and driven by the computational elements. Other possible actuators include materials such as those which change color or reflectivity in response to signals, as well as heating and Peltier cooling elements.

Many challenges must be addressed to make such a platform a reality. The materials aspects of these substrates may indeed be the easiest to address. Although significant benefits in integration could be derived from customizing device packaging technologies for these flexible large area computation and sensing arrays, it will nonetheless be possible to employ off-the-shelf components which have been

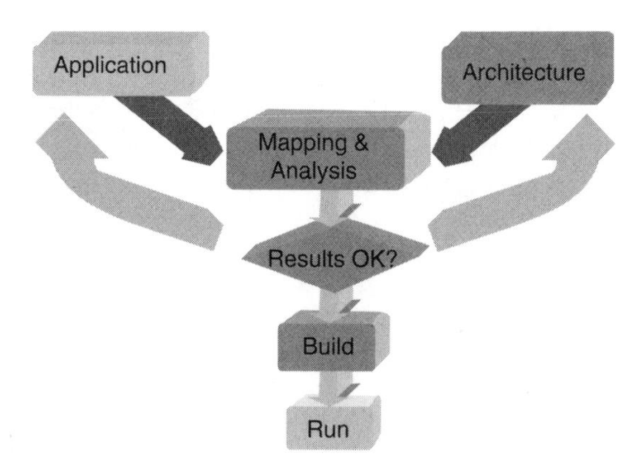

FIGURE 33.1 Classical static design cycle: no remapping occurs after the initial design is built.

packaged and optimized for traditional computing platforms in the first generations of systems. The greater challenges, we believe, lie in ensuring reliability in applications, particularly in the computer-aided design (CAD) methodologies, systems software, and programming language technologies needed to enable the harnessing of the unique capabilities of these systems. To address these issues, modeling frameworks and prototype systems are required to evaluate the use of the platform. To assess the efficacy of proposed designs, appropriate metrics that take into consideration the unique properties (e.g., restricted energy resources, performance constraints, reliability, and battery subsystem nonlinearities) must be employed.

Techniques to program such networks are required, permitting useful applications to be constructed over the defect and fault-prone substrate. In the classical design cycle (Figure 33.1), the application is mapped onto a given platform architecture, under specified constraints (e.g., performance, area, and power consumption). When these constraints are met, the prototype is tested, manufactured, and used for running the application. In the platforms of interest (Figure 33.2), the substrate is comprised of large numbers of interconnected computing elements, with no prescribed functionality. To achieve high yields, as well as high fault-tolerance later in the lifetime cycle, regularity is important. An application must be partitioned to expose concurrency. At system startup, the partitions of the application are mapped to hardware, so as to optimize different metrics of interest (e.g., quality of results, power consumption, operational longevity, and fault-tolerance) and later remapped whenever operating conditions change.

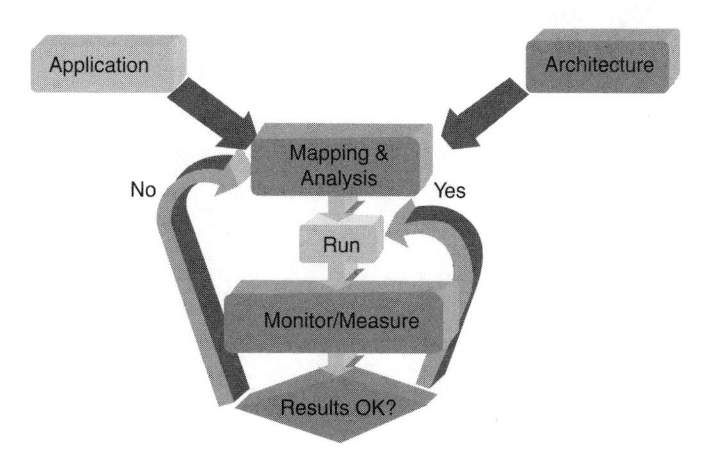

FIGURE 33.2 Dynamic continuously adapting living designs: through continuous online monitoring, the mapping of applications to the hardware substrate may evolve over time.

This chapter presents a conceptual framework and a simulation infrastructure for modeling several aspects of large surface arrays of computational, sensing, and actuation devices. Together, these provide a foundation for possible research directions. The next three sections present the conceptual framework, and Section 33.5 follows with a description of computation, communication, failure, and battery modeling in the simulation framework. The chapter ends with a summary of the presented ideas and possible directions for research in this new field.

33.2 Colloidal Computing

The model of colloidal computing (MC^2) [1] proposes local computation and inexpensive communication among computational elements: simple computation particles are "dispersed" in a communication medium which is inexpensive, possibly unreliable, yet sufficiently fast (Figure 33.3). The concept of colloids is borrowed from physical chemistry [2].* In the case of unstable colloidal suspensions, colloidal particles tend to coalesce or aggregate together due to the Van der Waals and electrostatic forces among them. Coalescing reduces surface area, whereas aggregation keeps all particles together, without merging. Similarly, the resources of a classic system are coalesced together in a compact form, as opposed to the case of colloidal computation where useful work can be spread among many, small, possibly unreliable computational elements that are dynamically aggregated depending on prevailing needs (Figure 33.4). Dynamic or adaptive aggregation is explicitly performed whenever operating conditions change (e.g., failure rate of a device is too high or battery level is too low) — a "stable" configuration is one that achieves the required functionality, within prescribed performance, power consumption, and probability of failure limits. The mapping and reconfiguration process of the application onto the underlying architecture is achieved via explicit mechanisms, as opposed to classic computing systems where mapping and resource management is done via implicit mechanisms.

The MC^2 model [1] was previously proposed to model both the application software and architecture platform. Most of the applications under consideration consist of a number of computational kernels with high spatial locality, but a low degree of communication among them. Such kernels (typically associated with media or signal processing applications) can thus be mapped on separate computational "particles" that communicate infrequently for exchanging results.

Reorganization and remapping requires thin middleware or firmware clients, sufficiently simple to achieve the required goals without prohibitive overhead. In addition, fault and battery modeling and

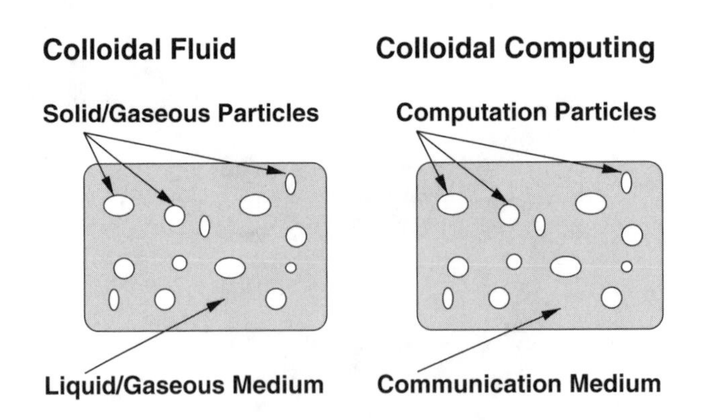

FIGURE 33.3 Colloidal computing model: analogy to colloidal chemistry [1].

*Colloid [käl'oid] = a substance consisting of very tiny particles (between 1 nm and 1000 nm), suspended in a continuous medium, such as liquid, a solid, or a gaseous substance.

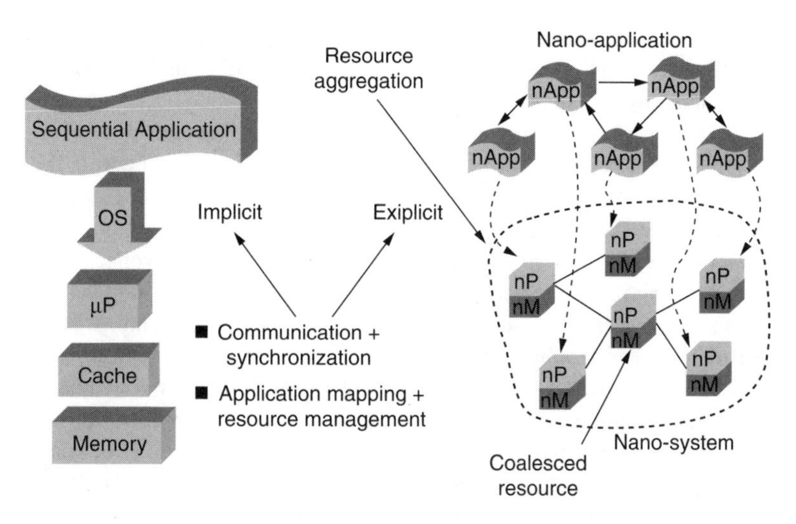

FIGURE 33.4 Coalesced vs. aggregated resources: partitioning applications to expose concurrency [1].

management become intrinsic components for achieving requisite levels of quality of results or operational longevity. We describe in the following sections, some of the issues that are critical to application lifetime, namely, application partitioning, followed by communication and fault management.

33.3 Application Partitioning

An issue of concern in mapping applications to hardware is that of concurrency. Given that the hardware substrate will contain many computational particles distributed on large surfaces, it is of crucial importance to expose the concurrency available in applications. The methods by which such concurrency may be extracted are a very interesting research avenue in their own right.

33.3.1 Driver Application: Beamforming

Beamforming consists of two primary components — source location and signal extraction. It is desired to detect the location of a signal source, and "focus" on this source. In a classic implementation, signals from spatially distributed sensors are sent to a central processor, which processes them to determine the location of the signal source and reconstruct a desired signal. Each received sample is filtered, and this filtering could indeed be performed at the sensor. Figure 33.5 illustrates the organization for a wired network of sensors used to perform beamforming.

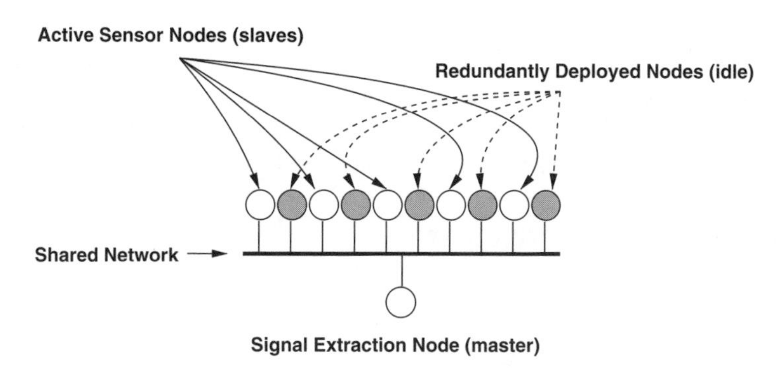

FIGURE 33.5 Beamforming in a wired sensor network.

In contrast to a classic implementation, the beamforming application as depicted in the figure is partitioned for execution over a network of processing devices. The filtering operation on each collected sample can be considered to be independent of other samples, thus it could be performed individually at each sensor node (slave node). The final signal extraction need only be performed at one node (master node). This division of tasks scales well with increasing number of sensors because the complexity of processing at each sensor node remains the same, and the only increase in complexity is in the number of filtered samples collected at the master.

Our example system operates in periods. During each period, all the slaves collect samples, filter them, and send the filtered samples to the master. The duration of the sampling period will vary for different applications of beamforming. In the case of beamforming for speech applications, an overall sampling rate of 8 KHz is sufficient. For geophysical phenomenon, a sampling rate of 1 KHz is enough, whereas for tracking the motions of animals, a sampling rate of 10 Hz is sufficient. In the analysis used throughout the rest of the chapter, a sampling rate of 10 Hz corresponding to a 100 msec sampling period is used.

The communication messages between the master and slave nodes consist of 4-byte data packets containing the digitized sample reading. When the battery level on any of the slave nodes falls below a specified threshold, the slave application attempts to use its remaining energy resources to migrate to one of the redundant nodes. If migration is successful, the slave application resumes execution on the redundant node, and adjusts its behavior for the fact that it is now executing on a different sensor, which is detected when it restarts. The migrated application code and data for the slave application is small (only 14 KB). The application on the processing elements with attached sensors implements a 32-tap FIR filter, and consists of 14 KB of application code and 648 bytes of application state. The application mapped on the sample aggregation (master) node performs a summation over the samples. The sequence of messages that are exchanged between the master and slaves during normal operation, and between the slaves and redundant nodes during migration is illustrated in Figure 33.7.

33.4 Communication Architecture and Fault Management

Achieving reliable computation in the presence of failures has been an active area of research dating back to the early years of computing [3,4]. Unlike large networked systems, in which failure usually occurs only in communication links or in computational nodes and communication links with low correlation, in the platform of interest, nodes and links coexist in close physical proximity and thus witness high correlation of failures.

It is assumed that the application is initially mapped at system startup for given quality of results (QoR), power and fault-tolerance constraints. As operating conditions change (e.g., permanent failures due to wear and tear, or intermittent failures due to battery depletion), the entire application (or portions of it) will have to be remapped, or communication links rerouted (Figure 33.6). Such reconfiguration mechanisms assume that redundancy exists for both nodes and links. In a fixed infrastructure, the logical implementation of redundancy is to replicate resources, with one resource taking over on the failure of the other. Upon such failures, applications must be remapped, for example, by code migration or remote execution. Code migration is generally a difficult problem, as it could, in the worst case, require the movement of the entire state of an executing application; however, migration of running applications can be greatly simplified by restricting the amount of application state that must be preserved.

33.5 Simulation Infrastructure

To investigate system architectures and programming models for the platform, it is necessary to be able to model it at the level of detail of its computation and inter-device communication. Using high-level behavioral models will be insufficient, as such an approach will fail to capture the interplay between computation and communication performance, as well as computation and communication failures, and their effects on performance, power consumption, and system reliability.

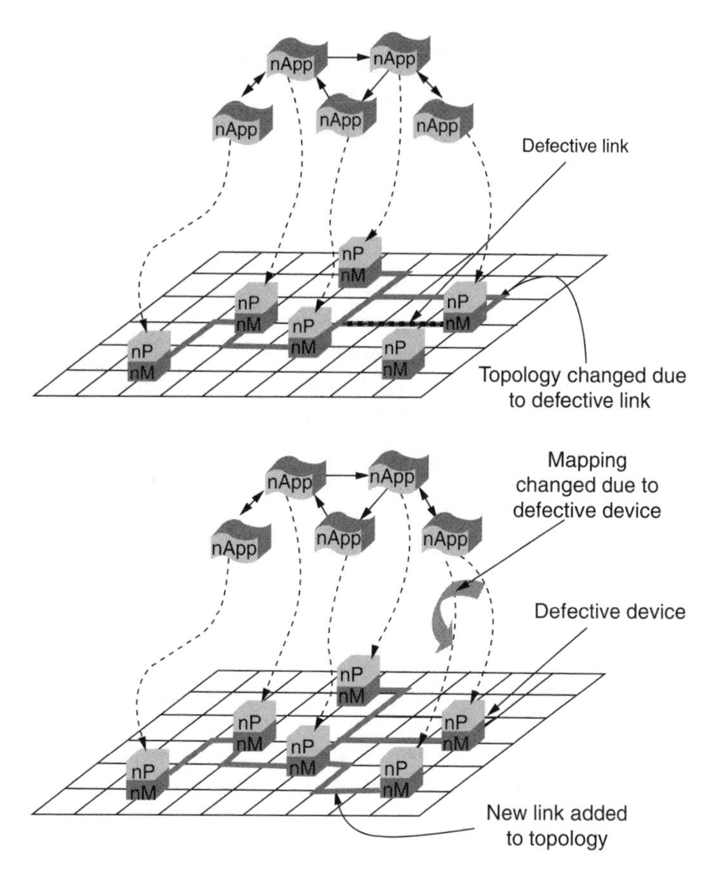

FIGURE 33.6 Dymamic reconfiguration in the presence of failures: in the event of a defective link (top), the network topology mapping is adapted to circumvent the faulty link. Likewise, in the case of a defective node, mapping and topology change to use a redundant node and a different link (bottom) [1].

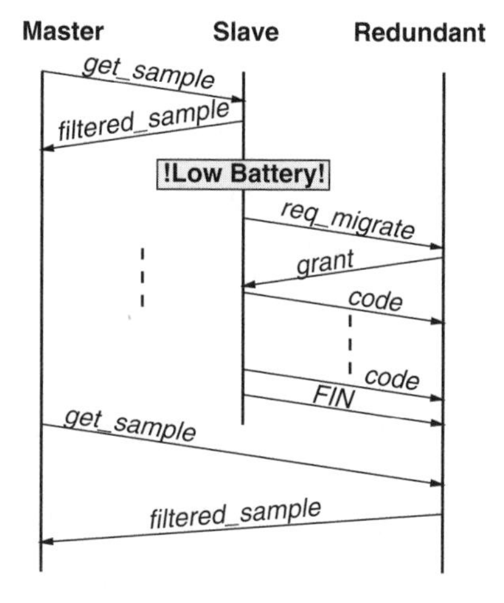

FIGURE 33.7 Messages exchanged in a beamforming application.

Any given platform will consist of multiple processing elements, distributed over a spatial extent in the surface. Such processing elements may either be general purpose programmable devices such as microcontrollers, or might be programmable logic devices. The processing devices will communicate over some communication infrastructure, whose topology might take advantage of the two-dimensional nature of surfaces. Failures in both the devices and in their interconnection networks will be common. These failures might be predictable, as in the case of depletion of energy resources, or they might be unpredictable, such as a defect in a communication or power conductor leading to intermittent communication or device failure.

It is therefore desirable for a simulation infrastructure to support the cycle-accurate simulation of multiple processing elements, for both general purpose processing and programmable logic devices. It must likewise enable the modeling of bit-level communication, in user-defined communication topologies. Failures in both processing devices as well as the communication links that interconnect them should be modeled, along with the relevant power dissipation for both computation and communication, and the effects of the current discharge profiles on power sources such as battery subsystems.

To support the investigation of CAD methodologies, system architectures and programming models for the hardware platforms of interest, we have developed a simulation framework [5] that aims to address the aforementioned modeling issues. The simulator models:

- Processing devices. The simulator permits the instantiation of multiple processing elements. Each instantiated element is modeled at the level of instruction execution. Two different processor architectures are modeled: the Hitachi SH3 architecture and the TI MSP430. Each processing node may have instantiated with it, one or more network interfaces, and each of these can be connected to an interconnection link.
- Interconnection links connecting the processing nodes. Interconnection links may be instantiated as necessary to create networks, and the network interfaces of processing nodes are attached to links. The links may be configured for variable transmission delay (link speed), link frame size, link failure probability, and link failure modes.
- Batteries and DC-DC converters. Each processing node is associated with a source of energy. The first-order effects of discharge rate on the battery cell and DC-DC converter efficiency are modeled by the simulation framework.
- Failures. The failure rate, average failure duration, and failure probability distribution of both processing nodes and interconnection links may be specified. Correlated failures between nodes and links may also be enabled, by specifying appropriate correlation coefficients.

The following describe each of the components of the simulation framework in more detail, motivating the need to perform modeling at the level of abstraction employed.

33.5.1 Processing Devices

At the core of the simulation framework is the modeling of instruction execution. Modeling applications at the level of detail of the simulation of the execution of their compiled code, make it possible to employ the simulation framework as a debugging platform for actual prototypes. It also makes it possible to determine important interactions between the requirements of computation, communication, and reliability, and the effects of these constraints on power consumption.

For example, Figure 33.8 plots the variation in two indicators of performance for the beamforming application implemented over the simulation framework. The beamforming application, as previously described, consists of two phases, which require differing amounts of network bandwidth. For both of these phases, however, increasing network speed does not indefinitely lead to increased performance, and may actually lead to reduced system performance. The "migration" phase of the said application, where the performance is depicted by the "average migration cost" curve in Figure 33.8, can actually not keep up with increasing network performance beyond a link speed of 1.6 Mb/s. This is because the

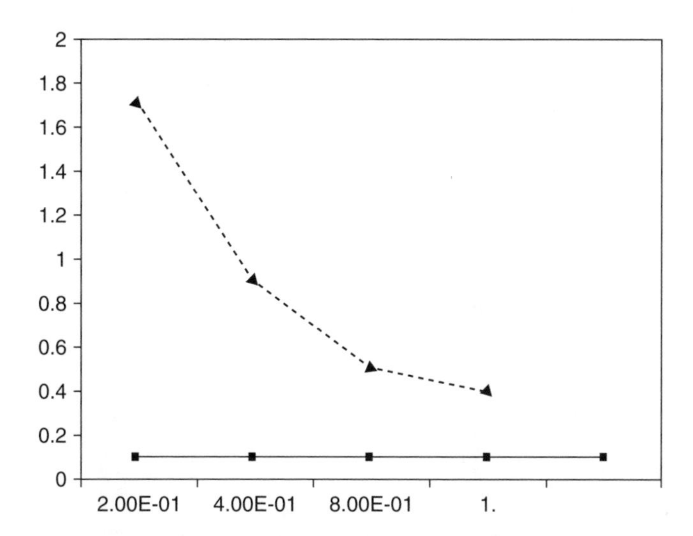

FIGURE 33.8 Variation in performance (average sample inter-arrival time and average migration cost) with communication link speed for an example application.

application attempts to transmit data at the fastest rate possible on the network, however, at this communication rate and for the modeled processing speeds, the receiving node does not have sufficient time (i.e., its computation is not fast enough) between arriving data frames to copy data it has received, before it is overwritten. Such delicate interplay between computational resources and networking resources is easily missed in high-level simulation (e.g., when employing a simple behavioral model of the entire system implemented in a high-level language).

The simulation framework includes two different architectural models, one for the Hitachi SH architecture, based on the Hitachi SH3 SH7708 (Figure 33.9), and the other of the Texas Instruments (TI) MSP430 architecture (Figure 33.10). Support for new architectures is easily added, and requires primarily the addition of code for implementing instruction decode and execution. The modeling of on-chip structures, such as interrupt generation, caches, and memory interfaces, as well as some standard peripherals, such as a network interface, is shared across the different architectures.

The Hitachi SH3 model includes detailed modeling of the CPU core, on-chip cache, and on-chip peripherals such as an RS-232 Universal Asynchronous Receiver/Transmitter (UART). It incorporates two complementary means of estimating the energy cost of application software — an empirical instruction level power model and circuit activity estimation. The instruction level power model functions by assigning to each instruction executed, an energy dissipation based on empirically measured values, scaled if necessary for a given operating voltage and frequency, as the model supports dynamic scaling of both operating voltage and frequency. Employing this simple energy estimation scheme enables fast simulation, which is critical because the framework is often used to simulate such platforms consisting of tens of processing devices. Although simple, the employed instruction level power estimation has been demonstrated to be within 6.5% of measured values for the hardware it models [6]. The instruction level power model can be augmented with a circuit transition activity estimation, which reports, for each simulation cycle, the signal transition activity on the address and data buses, in the register file, the program counter, and pipeline registers. The SH3 core model provides six levels of detailed simulation, enabling a trade-off between power estimation accuracy and simulation speed [6].

The Texas Instruments MSP430 architecture model provides functional simulation of the processor and its peripherals for the MSP430F11 series of microcontrollers. Unlike the SH3 model, it currently provides only functional modeling of the modeled microcontroller to enable applications compiled for a prototype system to be modeled and debugged in the simulation framework.

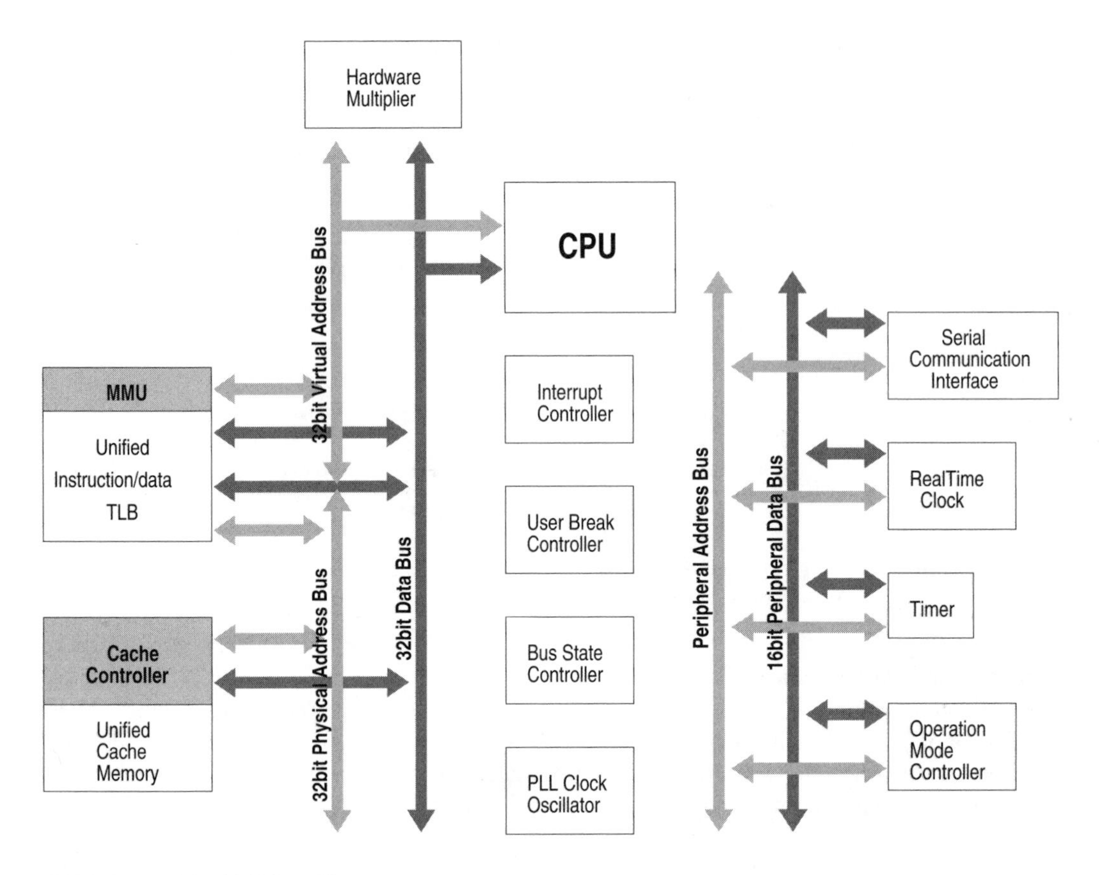

FIGURE 33.9 Hitachi SH3 architecture.

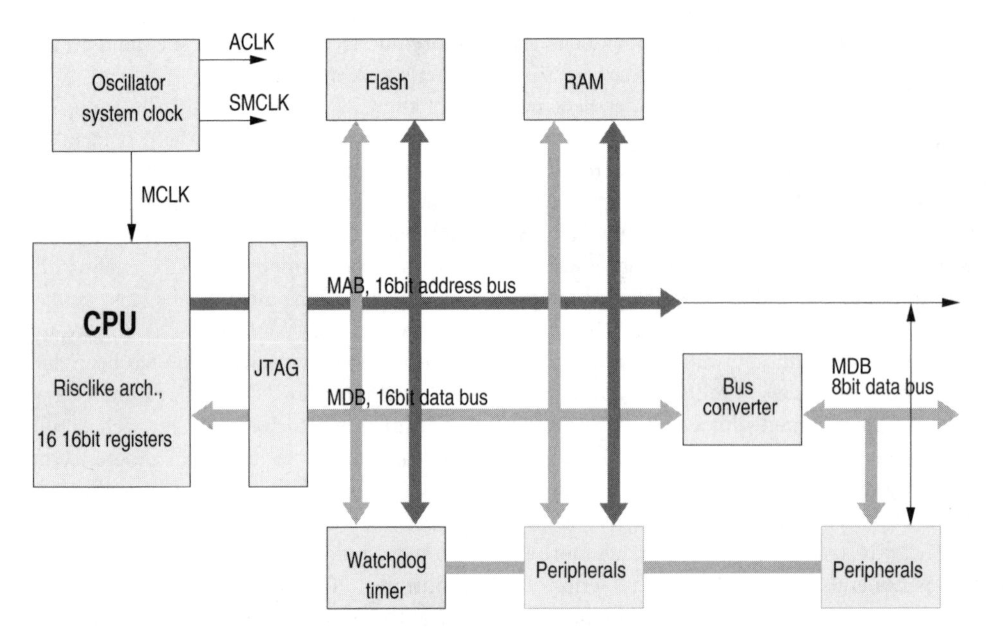

FIGURE 33.10 Texas Instruments MSP430 architecture.

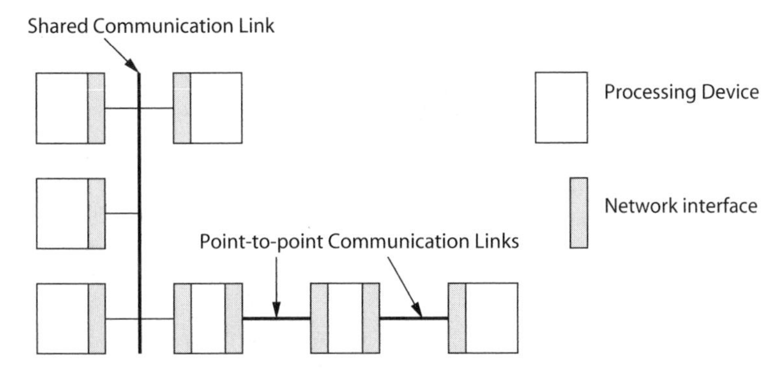

FIGURE 33.11 An example of communication topology.

33.5.2 Communication

The communication modeling architecture in the simulation framework enables the construction of a variety of communication topologies, such as that illustrated in Figure 33.11. Processing elements communicate with each other through their network interfaces, which are connected to communication links. The behavior of the network interface modeling is independent of the processing core model chosen for a given device. The properties of a modeled communication links and interfaces are flexible and parametrizable, enabling them to be configured to model the properties of media ranging from one with properties like RS-232, to one that behaves like Ethernet.

Each communication interface on a device must be associated with a communication link. Each communication link or network segment may be configured for the following specific properties:

- Frame size. Data is transmitted on a communication link in groups of bytes referred to as a "frame."
- Propagation speed. The propagation delay specifies the speed at which a signal travels in the communication medium, over the communication link. When modeling wired communication, this is taken to be the speed of light. Nodes in the simulation can have associated with them a location in three-dimensional space, which will then be used in conjunction with the propagation speed to determine the propagation delay. For most simulation scenarios, however, this parameter can be ignored.
- Transmission speed. The transmission speed specifies the number of bits that are modulated per second, or the bit-rate of the communication medium.
- Maximum simultaneous accesses. Specifying a maximum number of simultaneous accesses permits a medium to be configured to behave, for example, either as a carrier sense multiple access with collision detection (CSMA/CD) medium, or as one that employs frequency division multiplexing (FDM).
- Failure probability and maximum failure duration. These are discussed further in the description of the failure in Section 33.5.4.

To ensure network interfaces are always compatible with the networks to which they are attached, network interfaces inherit the aforementioned properties from a network segment to which they attached. The transmission and receive power consumption of a network interface may, however, be configured independently of the properties of the link with which it is associated. The simulation of data transmission and receipt is kept cycle-accurate with respect to computation. The granularity at which data is transferred from one device to another is determined by the smallest cycle time of all the modeled processing devices.

33.5.3 Battery Subsystem

The simulator includes a detailed discrete-time battery modeling engine based on [7]. In brief, the model takes into account properties of battery cells, such as dependence of battery terminal voltage on the state

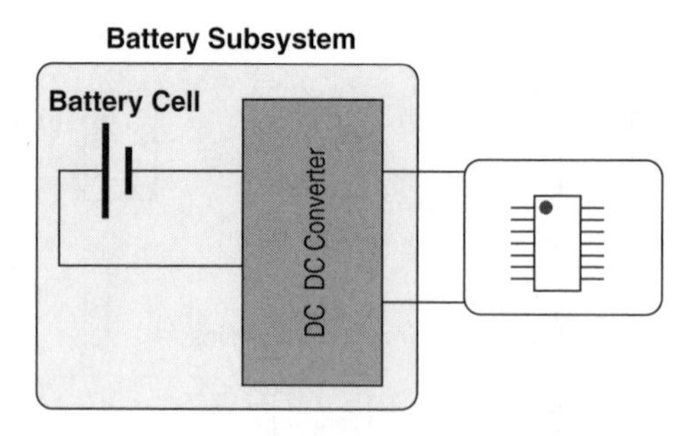

FIGURE 33.12 Organization of battery subsystem: the DC-DC converter is required to obtain a constant voltage to power electronics, due to the dependence of battery cell terminal voltage on battery state of charge.

of charge (SOC) of a battery, dependence of usable capacity on discharge rate, and dependence on the rate of change of current over time. To provide a constant voltage to the powered electronics in the face of variation in battery terminal voltage over time, a DC-DC converter provides voltage stabilization, at the cost of a loss due to inherent inefficiencies in the conversion. A simple organization of a battery-powered system (Figure 33.12) illustrates this further.

To model different types and sizes of batteries and DC-DC converters, the model (and its implementation in the simulator employed in this work) uses lookup tables (LUTs) and additional fixed parameters to store the characteristics of specific batteries. The default battery characteristics employed in our implementation are those for a lithium ion cell from the Panasonic CGR18 family. The DC-DC converter characteristics employed are those for a Dallas semiconductor/Maxim MAX1653 device. User LUTs may be loaded into the simulator to mimic other device's characteristics, for both the battery cell and DC-DC converter. Figure 33.13 plots the dependence of battery terminal voltage with time for a nominal discharge rate of 150 mA. The data in Figure 33.13, although plotting the voltage at the terminals of the battery cell, also includes the effect of DC-DC conversion, and depicts the lumped behavior of the battery cell if the battery subsystem were attached to electronics that had a constant current draw of 150 mA.

The components of the battery properties are illustrated in Figure 33.14. The parameters of interest in this work are V_r, a measure of the rate of discharge, V_{rate}, a time-sluggish (i.e., low-pass filtered) version of V_r, V_{lost}, which models the dependence of battery terminal voltage on the magnitude of V_{rate} for a

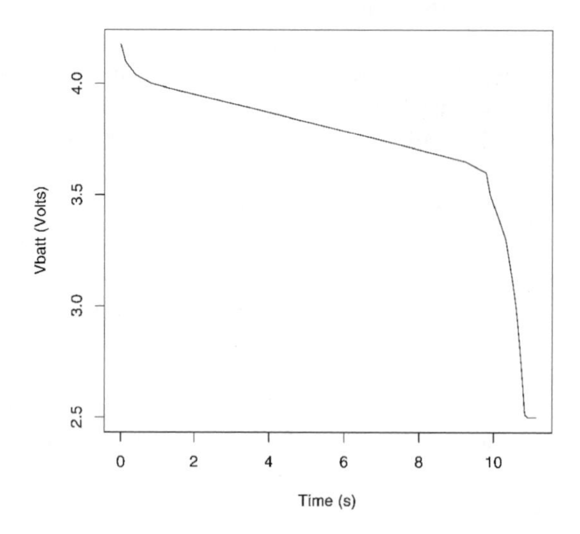

FIGURE 33.13 Variation of battery cell terminal voltage over time for a nominal current draw of 150 mA from outside the battery subsystem.

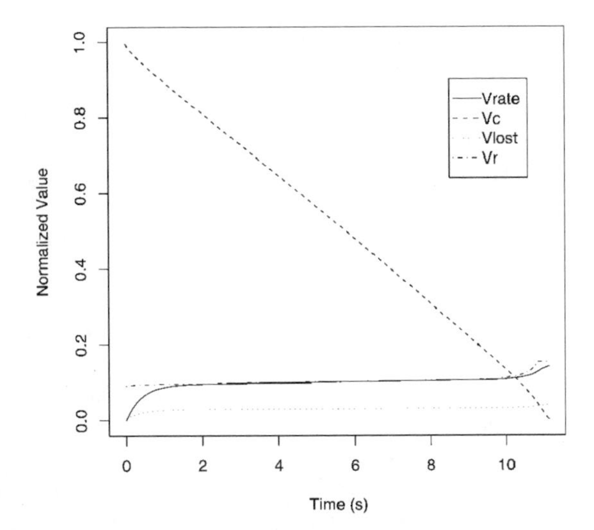

FIGURE 33.14 Variation of components of a battery model with time for a nominal current draw of 150 mA from outside the battery subsystem.

particular battery type (from an LUT). Last, V_C models the instantaneous state of charge, taking into consideration V_{lost}.

33.5.4 Modeling Failures

The simulation framework models failures in both processing devices and communication links. Failures in processing devices manifest as intermittent stalls of the entire processing device for the duration of the failure. Failures in communication links manifest as intermittent loss of carrier for the duration of the failure. Failures, such as bit errors introduced into the communication stream and into device computation, are not currently supported, but are planned. For both failures in devices and communication links, the failure rate and maximum failure duration are configurable. Correlated failures between processing devices and communication links can be modeled by specifying appropriate correlation coefficients for a given node-link pair.

33.6 Conclusion

New technologies often pose new challenges in terms of system architectures, device architectures, and sometimes, models of computation. The technology of interest in this chapter is that of computational, sensing, and actuation surfaces, which are flexible meshes of material containing large number of unreliable, networked computing elements, sensors, and actuators. The challenges addressed herein were those of design methodologies, system architectures, modeling, and fault-tolerance.

Computational, sensing, and actuation surfaces inherently have high defect rates, as well as high fault rates, and thus must by necessity provide mechanisms for extracting useful work out of the unreliable substrate. By employing a detailed simulation infrastructure designed to enable the simulation of the computation, communication, power consumption, and battery discharge characteristics, the dynamic adaptation of applications in the presence of faults can be investigated.

33.7 Acknowledgments

This research was supported in part by the Defense Advanced Research Projects Agency (DARPA) Information Processing Technology Office, under contract F33615-02-1-4004, and the Semiconductor Research Corporation, under grant 2002-RJ-1052G.

References

[1] R. Marculescu and D. Marculescu. Does Q = MC^2? (On the relationship between quality in electronic design and model of colloidal computing). *Proc. IEEE/ACM Intl. Symp. on Quality in Electronic Design (ISQED),* March 2002.

[2] R. Rajagopalan and P.C. Hiemenz. *Principles of Colloid and Surface Chemistry.* Marcel Dekker, New York, 1992.

[3] J. von Neumann. Probabilistic logics and the synthesis of reliable organisms from unreliable components. *Automata Studies,* pp. 43–98, 1956.

[4] M.D. Beaudry. Performance-related reliability measures for computing systems. *IEEE Trans. on Comput.,* c-27(6):540–547, June 1978.

[5] P. Stanley-Marbell. *Myrmigki Simulator Reference Manual.* Technical report, Center for Silicon System Implementation (CSSI), Department of Electrical and Computer Engineering (ECE), Carnegie Mellon University, Pittsburgh, PA, 2003.

[6] P. Stanley-Marbell and M. Hsiao. Fast, flexible, cycle-accurate energy estimation. *Proc. Int. Symp. on Low-Power Electron. and Design (ISLPED),* pp. 141–146, August 2001.

[7] L. Benini, G. Castelli, A. Macii, E. Macii, M. Poncino, and R. Scarsi. A discrete-time battery model for high-level power estimation. *Proc. Conf. on Design, Automation, and Test in Europe (DATE),* pp. 35–39, January 2000.

V

Embedded Software

34

Low-Power Software Techniques

Catherine H. Gebotys
University of Waterloo

34.1 Introduction

Software can have a large impact on the average power, peak power, energy dissipation, and instantaneous power of the embedded processor core. In turn, average power is directly related to battery lifetimes. Peak power constrains the thermal design of the embedded system. In addition, the peak power affects the power supply design and instantaneous power can affect reliability and security. Today, low-power dissipation is critical for wireless communication devices, which demand long battery lifetimes, high reliability, low thermal dissipation, and high security. This chapter discusses the relationship between software and power. First, instruction-level models for predicting the average power and predicting the average energy of applications executing on an embedded processor are reviewed. An example of an instruction-level model combined with statistics is presented for a digital signal processing (DSP) processor. Next, recent research in instruction-level models for predicting instantaneous power of a processor core is discussed. Finally, new emerging applications of instantaneous power design utilizing software, specifically in security, are addressed.

The need for low-power dissipation in many general-purpose processor cores and DSP processor cores has created a need for further understanding of power in system on chip (SoC) devices. Architectural design for low-power and high-level transformations for low-power applications are becoming a well understood area of research [1]. Previous methods of estimating power at a high level using gate-level or architecture-level simulations were either inaccurate or too time consuming. For embedded systems designers, power measurement and optimization for code is very important, however, gate level processor representations are not always available for estimating power. Instruction-level accurate power prediction tools for embedded processor cores are important. The equations for power measurement and prediction are discussed next.

0-8493-1941-2/05/$0.00+$1.50

The energy dissipation of a processor running a program [2], *E,* can be approximated by the product of the time required to execute the program (T), the average current (I), and the supply voltage (V_{dd}) as in Equation (34.1).

$$E = P\ T = I\ V_{dd}\ T = I\ N\ V_{dd} \tag{34.1}$$

This equation ignores the additional power dissipation arising from leakage current and short circuit current [3]. The term T is equal to $N\ \tau$, where N is the number of clock cycles and τ is the clock period.

From Equation (34.1), a reduction in N (equivalently a performance increase) will always provide equivalent or higher reduction in energy as long as the new value of current, I, is equivalent or lower in value. In cases where I increases as N is decreased (i.e., performance improvement could be derived from the use of more parallel instructions which help create higher average current per cycle), the reduction in energy will be less than the reduction in N. In some cases (possibly rare) it may even result in increased energy dissipation if $p_i > p_n/(1 - p_i)$, where p_i is the fraction of I that is increased, and p_n is the fraction of N that is reduced.

For many embedded processor core applications, N is fixed (by the throughput requirements), making energy reduction techniques rely solely on reducing I or reducing power. Thus, I is an important parameter for embedded systems design that needs to be studied and predicted for software energy prediction. The problem becomes how to modify or generate processor code that is energy efficient or meets power constraints. An excellent review of system-level power optimization techniques and tools can be found in Benini and DeMicheli [1]. This chapter provides an in-depth review of experimental setups for measuring current of processors. Additionally, instruction-level average power modeling of embedded processors is reviewed. An example of modeling instantaneous power of a very long instruction word (VLIW) DSP processor at the instruction-level is outlined. Throughout the chapter, current, power, and energy models are discussed. This chapter concludes with emerging areas for instantaneous power prediction in security. The next section discusses predictive models of average power.

34.2 Software Models for Predicting Average Power

Instruction-level models for predicting the power dissipation of processors was investigated in Lee et al. [2], Tiwari et al. [4], Qu et al. [5], and Russell and Jacome [6]. These models were verified by real measurements of average power for both instructions and programs on the target processor board. The first section briefly describes the different experimental setups used in power research, followed by a description of the verified power-prediction models and their accuracy in the next section. A general review of other instruction-level power models is also presented, followed by a detailed example of utilizing a statistical approach to model building. Section 34.3 and Section 34.4 present recent research in building instruction-level instantaneous power models of embedded processors and applications of instantaneous power modeling to security, respectively.

34.2.1 Experimental Setups for Average and Instantaneous Current

Measuring the current drawn by the processor while executing an application has been used by many researchers to verify power models. The equipment setups vary from ammeters to oscilloscopes. The objective of measuring the current drawn by the processor while it is executing an application varies as well. In some cases, current measurements are used to obtain average power readings per instruction or per program, and in other cases the measurement is used to obtain the real execution times of a processor (i.e., including cache misses and memory stalls), which are too difficult to simulate.

For current measurements of a general-purpose processor and a DSP processor in Lee et al. [2] and Tiwari et al. [4], a current meter was used. This setup allowed power measurements for small programs (whose execution times were much less than 100 ms). These programs were repeated several times in a loop until a stable current reading could be obtained. These current readings were taken visually so stability was important. Other researchers [7–9] have used this same type of experimental setup except the current meter (e.g., a Fluke 867B GMM) was more sophisticated, allowing sampled readings of current to be transferred to a workstation. This allowed variations in current readings to be averaged. In addition, the current measurements of longer programs could be supported. The user can set the period over which samples can be averaged and these averaged samples are then transferred to the workstation for further analysis. Thus, the requirement for stability of the readings and limitations of short programs as in Lee et al. [2] and Tiwari et al. [4] was not necessary. This type of setup (using HP34401A digital multimeter) has also interestingly been used to validate a runtime power estimator in Joseph and Martonosi [10]; however, here a shunt resistor was placed between the power supply and the processor's board power terminal. The voltage across the shunt resistor was measured and then divided to obtain current measurements. In Joseph and Martonosi [10], the chipset power was subtracted from the total board power to obtain the CPU power. Sinha and Chandraksan [11] also used a source meter to measure the current drawn by subroutines. This setup was used to analyze leakage current as V_{dd} changed.

An interesting experimental setup described in Chang et al. [12] used capacitors in between the power supply and the processor. This setup allowed the researchers to analyze specific cycle-by-cycle energy [12]. The minimum and maximum voltages on the capacitors were acquired in real time. Each cycle the capacitors were charged up by the current drawn by the processor and in the next cycle the capacitors were fully discharged. This technique allowed measurement of current effects from single cycle activity alone (without the influence of previous cycles). The capacitors were completely discharged before the next clock cycle; thus, only activity in the clock cycle of interest was measured. This setup allowed researchers to isolate cycle-by-cycle influences of instructions in the pipeline on the current draw.

Other researchers have measured power [6,13–16] using an oscilloscope. Oscilloscopes offer higher sampling rates and more accuracy than the digital multimeters. The oscilloscope in Wolf et al. [16] was used to accurately measure the execution time of applications. In Russell and Jacome [6], an oscilloscope was used to obtain the instantaneous power measurement of a single instruction (for an instruction-level power model). The instruction was repeated several times in a loop [6]. In the later case, a resistor was placed in between the power supply and power pin of the processor. Decoupling capacitors were introduced to reduce the voltage noise during current surges. Although instantaneous power was measured in Russell and Jacome [6] using an oscilloscope, the average power was calculated from this waveform over the loop body (which consisted of 100 instances of an instruction). In Wolf et al. [16], a resistor was again used, however, a custom experimental setup using differential amplifiers, an integrator, and an ADC was used to transfer and process the power readings into a logic analyzer. Power measurements per clock cycle were recorded using this setup. Nickolaidis et al. [28] utilized current mirror circuitry to obtain current measurements. This technique avoided the use of a resistor between the supply and the power pin (which typically may cause supply noise problems).

Instantaneous power was also captured in Muresan and Gebotys [13,14] and Muresan [15] by using an inductive probe (instead of a resistor in between the supply and power pin). An oscilloscope and pattern generator were both used to synchronize the program and oscilloscope. Additionally the pattern generator produced the clock signal as well as the trigger signal. This allowed accurate measurements of the instantaneous current over various sections of a program executing on the processor. The purpose of this setup was not to measure average current per instruction, but to measure and model the instantaneous power. Section 34.3 outlines the research resulting from this experimental setup for instantaneous power modeling. Other researchers [17] used a National Instruments data acquisition card to simultaneously read 16 power sources (at rates up to 1 million samples per second) in a portable laptop PC environment. This setup was used to evaluate power management algorithms. Researchers in the emerging area of security have also used oscilloscopes to measure instantaneous power [21,22,26]. Here, instantaneous power is analyzed to ensure no confidential information is leaked from the security application, to be further discussed in Section 34.4. The next section gives a brief overview of instruction-level power models, which have been researched.

34.2.2 Previous Instruction-Level Average-Power Models

Previously researched instruction-level power models for processors are briefly reviewed in this section. Some of these models have been verified with real power measurements or with lower-level power estimation tools. These models are discussed, followed by an illustration of building a power model using statistics and instruction-level power measurements. The power model is then verified with real power measurements of applications executing on the processor.

One of the earlier instruction-level models of power [2,4] was derived from a base power cost per instruction along with an overhead cost related to the next or nearby instruction. They achieved an accuracy of 10% but required characterization not only on a per instruction basis, but also for pairs of instructions and beyond. Some tools were developed that provided performance improvement in code in addition to power improvements. In several cases and where data was available, their results demonstrated that fewer instructions lead to faster code and lower energy. This approach has been utilized by several researchers to build instruction-level power models of various processors.

Russell and Jacome [6] measured instantaneous power of individual instructions across one loop iteration and used this in an instruction-level average-power model. Using statistics, their model concluded that for the two reduced instruction set computer (RISC) 32-bit processors considered, a model utilizing only the average power of all assembly instructions multiplied by the execution time (also determined from the oscilloscope) provided an 8% accurate model for energy with a 99% confidence level. Other researchers have also captured instantaneous power for purposes of building an average energy instruction-level model. In Nikolaidis et al. [28], the oscilloscope captured the instantaneous power of a single instruction embedded by NOPs. The power was integrated over one clock cycle to obtain the average energy measurement for the single instruction. Simunic et al. [18] extended an instruction-level simulator of an ARM processor for power along with energy models for board interconnect and memory. The total energy of this embedded system was modeled overall with a verified accuracy of 5%. Chang et al. [12] studied specific cycle-by-cycle energy, however, a power prediction model was not created. Instead, the research studied what factors influence the power dissipation of each instruction. For example, the analysis indicated that hamming distance between address values, and other switching had a significant influence on the power. Step-power analysis [19] uses power simulators to study effects of clock gating on maximum power consumption. Step power is defined as dp/dt, and it causes reliability concerns. It is studied to identify the causes of high step-power.

34.2.3 Example of Statistically Generated Model for Average Power

This section briefly describes a statistically generated model for power dissipation of a TMS320C5x DSP processor. The DSP processor's datapath has an accumulator register, product register (of the multiplier), and the input register to the multiplier, respectively. The memory addressing can support direct or indirect memory addressing modes. Offset addressing is also supported, but only one offset address register is available. Eight address registers are available, and a three-bit register points to the current one. The processor has address characteristics similar to many popular DSP processors. Similar to the DSP processor used in Lee et al. [2], parallel instructions are used, and some instructions have design features similar to other DSP processors (specifically that of two nonsequential instructions changing the same state). The TMS320C5x DSP processor along with the Fluke 867B digital multimeter was used. All experiments were repeatable. After the board was powered up, a warm-up period was allowed before any experiments were run. Figure 34.1 illustrates the current measurements over time when the board is initially warmed up. A series of NOP (no-operation) instructions were executed before and after each series of programs were run to calibrate any variation in power measurements due to temperature variation. In all cases, standard deviations were lower than 0.03 mA.

The average current for each type of instruction is recorded and used to generate a variable, x_p, which is detailed later in this section. Several DSP benchmark programs are also run with four different types of input data (e.g., voice and pseudorandom), and current measurements are recorded. The power

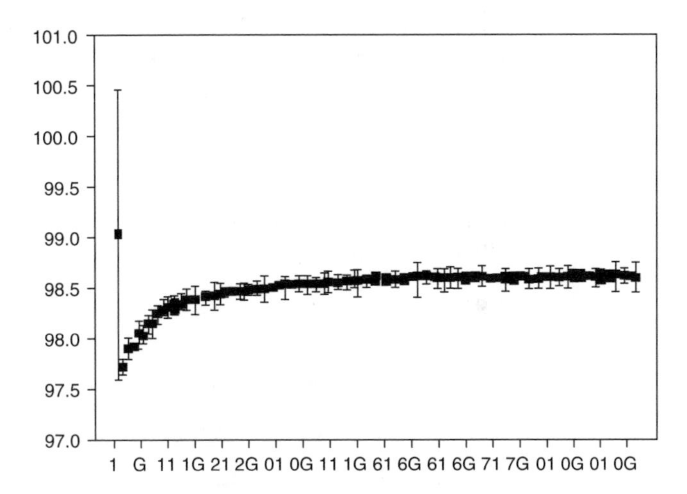

FIGURE 34.1 Warm-up current measurements approaching stability.

predictor model generation is based upon linear regression performed on a number of variables from the benchmark DSP code and x_p. The output is a model (or equation $y = f(x_p, \text{variables})$), which predicts current from only the DSP code itself (or variables extracted from it). The embedded systems designer then generates code for their application and uses it with the power-prediction model to predict current. Code is regenerated (using some technique such as rescheduling or rewriting the application) in an attempt to obtain code that meets the performance constraint and minimizes the predicted power dissipation.

DSP code for embedded types of applications, such as the fast Fourier transform, least means squares, high pass filter, and discrete cosine transform, were generated. The programs ranged from 60 to 150 instructions. Different schedules, addressing arrangements, and coding were used to study power effects. In many cases, different codes for the same filter were created with equivalent performance (e.g., using different schedules and address generation). One set of programs was used to generate the power prediction model, and a different (independent) set of programs were used to verify the model.

For each DSP benchmark program, a straight-line basic block code sequence was repeated several times and then placed within a loop. Each repeat of the program used a different part of the input data. For example if a DSP program used 40 words of speech data as input. The DSP program was repeated 100 times in a loop, performing computations on 4000 words of a continuous speech sample. This study would be repeated with pseudo-random generated data and other types of data. The different types of input data used were:

1. Random data generated from a pseudorandom number generator
2. A second set of pseudorandom numbers
3. Raw voice data from a voice sample
4. A second sample of voice data

Variables obtained directly from analysis of the DSP benchmark code are listed in Table 34.1. For example, IR in Table 34.1 refers to the average switching of data stored in the instruction register (available from the DSP code), whereas DABUS refers to the average switching of the data address bus.

A new variable, x_p, was added to the statistical methodology. The value of this variable, x_p, was created for each DSP program by summing the number of each instruction multiplied by the average current per cycle (measured with this instruction repeated several times in a loop) divided by the total number of instructions in the program. This approach is similar to that used in Lee et al. [2] and Tiwari et al. [4], however we use details of addressing and include this variable, which can be obtained directly from the code, among all other data independent variables to form a model. Furthermore, we do not have to use pairs of instructions and record their currents. We instead model the overhead or state with data-independent variables.

TABLE 34.1 Variables Used to Build Power Prediction Model

Variable	Average switching in the
IR	Instruction register
PC	Program counter
ACC	Accumulator
PREG	P register
TREG	T register
ARi	Address register *i*
ARP	Pointer to address registers
DMEM	Data memory
DBUS	Data bus
DABUS	Data address bus
PMEM	Program memory
PBUS	Program bus
PABUS	Program address bus

A number of linear models were fit using the measured current as the dependent or y variable. Several independent or predictor variables (x) were considered, see Table 34.1, along with x_p. The models were fit using a least squares algorithm to minimize the distance between the observed data and the predicted data under the model. We assume that the y data is some linear function of the x, y = f(x). The least squares equation predicting average power from x is given by the following linear equation $E(y|x) = b_0 + b_1 x_1 + b_2 x_2 + \ldots + b_k x_k$, where the b_i represents the least squares estimates of the population parameters, and $E(y|x)$ is the average or expected value of y given x. A stepwise selection method was used to automatically find the best model for predicting current. The model reported is an excellent model statistically; model adequacy tests have p-values < 0.001, where p-value is the observed level of significance.

The R^2 value is reported indicating the percent of variation in current accounted for by the model. The least squares equation is given predicting average current for the model (all coefficients are highly statistically significant, p-values < 0.001) and standard error of prediction are given for the maximum residual as another measure of accuracy of the model. The normality assumption for statistical tests and confidence intervals was verified using normal probability plots and histograms. The statistical package SPSS [20] was used for all statistical calculations.

Specifically 168 benchmark DSP programs (each repeated several times in a loop) run with different types of input data were executed on the DSP processor and average current was read from the meter. Using the average current measurements (obtained from the single instruction tests) and the variables extracted directly from the DSP programs, the variable x_p was obtained. The variable x_p together with the variables (also obtained directly from the DSP code) from Table 34.1 were then used by the linear regression algorithm (see details of statistical procedure outlined in the experimental section of the *SPSS User's Guide* [20]) to automatically form the equation for predicting current. The automatic power prediction model generation results are presented here.

For 168 cases (DSP benchmark programs), the stepwise selection procedure automatically produced the following model. The x variables automatically chosen due to their significance by the statistical procedure are listed in order of their importance in predicting energy: x_p, IR, DABUS. The value of R^2 for this model is 0.78 or 78% of the variation in current is accounted for by these three variables. The equation for predicting current from these x variables (where x_p, x_4 = IR, x_5 = DABUS) is $y = 27.41 + (0.38)\, x_p + (2.65)\, x_4 + (0.08)\, x_5$.

The standard error of prediction for the largest residual is 0.17. In other words we would be 95% confident that ±0.34 mA of the predicted value of current would contain the average current. The confidence interpretation and tests of significance depend on the assumption of a normal distribution of residuals. A histogram of the residuals was analyzed to verify this assumption. In Figure 34.3, this histogram clearly indicates a normal curve. The worst case residual was -1.7mA, providing a maximum

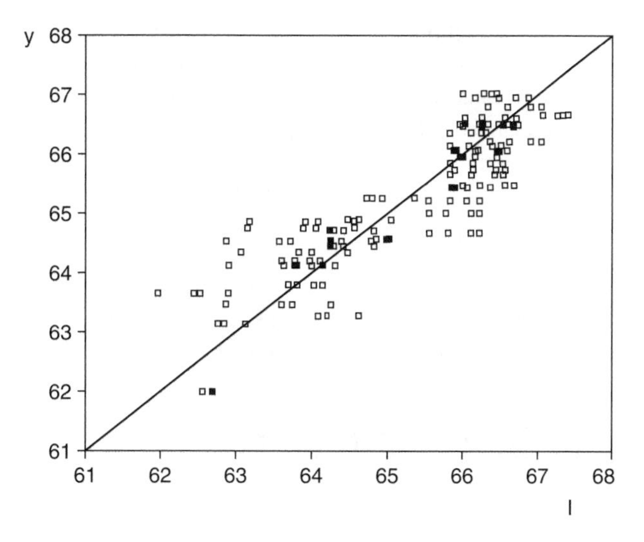

FIGURE 34.2 Measured current (x) vs. predicted current (y) in mA.

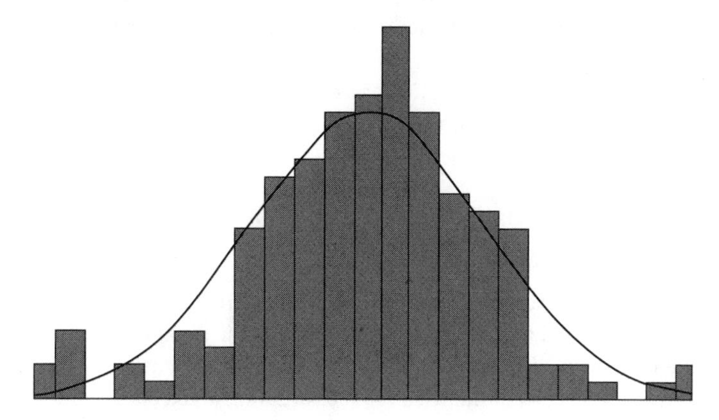

FIGURE 34.3 Histogram of the residuals indicating a normal curve.

worst-case error of 2.7%. The overall fit of the model can be seen in Figure 34.2 where the predicted current, y is plotted against the actual current, I, both measured in mAs.

Table 34.2 compares this statistically derived model to other statistically derived models that have access to higher level algorithmic variables or to more detailed switching details. The prediction ability of a higher-level model (algo for algorithmic) where only the number of each type of operation is used to predict power is very poor (see row 1 of Table 34.2 with R^2 value of only 0.63). For example, the number of additions (and/or subtractions), and the number of multiplications in the application are recorded as variables. A more detailed model (switching level in row 3 of Table 34.2) that records the actual average switching in registers and busses of the DSP processor provides a better R^2 value

TABLE 34.2 Average Power Model Levels, Variables, Maximum Error, and R^2

Model	Example	Maximum Error, R^2
Algorithm level	No. of additions, multiplications	3.7%, 0.63
Instruction level	x_p, IR, DABUS	2.7%, 0.78
Switching level	No. of loads, PABUS, No. of subtracts, IR, DABUS	2.1%, 0.89

[7], however, this requires an instruction-level simulator with switching activity to which embedded systems designers typically do not have access. The instruction-level power model (see row 2 of Table 34.2) compares very well in R^2 value to the more detailed switching level model (of row 3). More important, it is very suitable for embedded systems design because all inputs to the model can be obtained from the generated code itself along a one time only model generation phase using single instruction and DSP benchmark tests together with statistical optimization.

To further independently test out the validity or accuracy of the model, variables from other DSP programs (that were not used to derive the statistical model) were used in the previously presented equation for predicting current, y, and this predicted current was compared with actual current measurements. The predicted power or current value had an error less than 2.7% of the actual measured current for the different types of voice and pseudorandom data input. This approach has been used for various processors including a highly parallel DSP processor [7].

34.3 Instruction-Level Models for Predicting Instantaneous Power

Models of dynamic power at the software level have been researched in Muresan and Gebotys [13,14] and Muresan [15]. The current model is based upon summing instruction-level current models (gamma functions) together and using multiplicative factors to correct for block-to-block current variation at the higher application software level.

A simpler formulation than in Muresan [15], based upon processor clock cycles, is given next. The variable $i_{processor}(c)$ represents the current of the processor at clock cycle c.

$$i_{processor}(c) = i_{base} + \sum_{n=0}^{w} \sum_{instr} (\beta_{instr}) gamma_n x_{instr,c-n}$$

The variable $x_{instr,c-n}$ is a binary variable, which is one of the instructions, and *instr*, is executed at clock cycle *c-n*, otherwise it is zero. The i_{base} is the base current similar to base current in Lee et al. [2] and Tiwari et al. [4]. The parameter β_{instr} represents the amplitude of the current for instruction *instr*. The variable $gamma_n$ represents the current modeled (at 100MHz and targeted for SC140 processor [15] with the parameter 0.0038) as a gamma function

$$gamma_n = (0.0038)^2 (n\tau + \frac{\tau}{2}) e^{-(0.0038)(n\tau + \frac{\tau}{2})}$$

where τ is the clock period, and $gamma_n$ uses the gamma value at the center of the clock period *n* (because this gamma function is actually a function of time but simplified here in clock cycles). For example, $gamma_0$ is the gamma value for a general instruction in clock cycle 0 when it starts executing. Whereas $gamma_4$ is the gamma value for an instruction which started executing four clock cycles ago. Note that more than one instruction can be executed in one clock cycle (thus supporting parallelism) and again the gamma values are summed. Figure 34.4 illustrates the measured power traces and superimposed gamma models for three separate types of instructions: *MOVE.2L (EA), Da:Db, MOVE.L #s32,C4,* and *EOR Da,Dn* in order of highest to lowest amplitudes. The first instruction loads two 32-bit words into two data registers. The second instruction loads an absolute 32-bit value into a control register. The lowest current draw was obtained from the exclusive or on two registers. For plotting purposes, the amplitudes for the exclusive or instruction (*EOR*) were multiplied by an additional factor of two. It is interesting to note that unlike previous research [28] which integrated the single instruction waveform over one clock period, the waveforms of this highly parallel processor extended over many clock cycles [15]. This is most likely due to the larger capacitance of the processor because it contains many more ALU units.

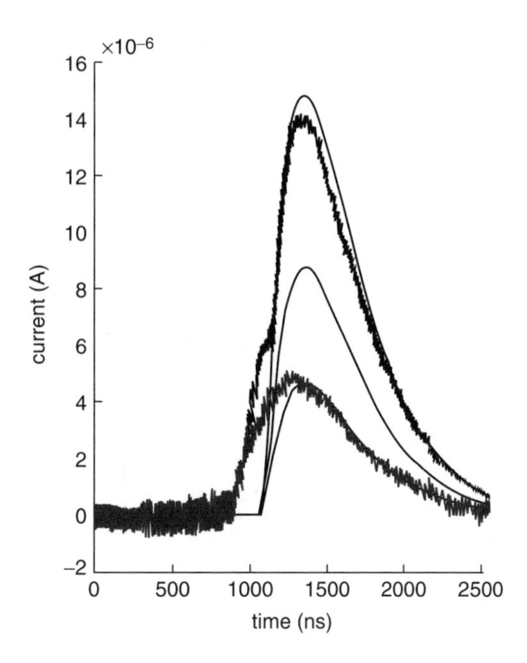

FIGURE 34.4 Gamma functions superimposed with current of *MOVE.2L (EA),Da:Db, MOVE.L #s32,C4,* and *EOR Da,Dn* (illustrated as double the amplitude) from highest to lowest amplitudes, respectively.

The application, where power is being modeled, is divided into blocks where the average instruction parallelism per block and variation of parallelism is used to create multiplicative factors. These factors correct for block-to-block current variation at the higher application software level. The multiplicative factors are derived statistically utilizing a benchmark set of applications. The final multiplicative factors are used for all subsequent current models representing instantaneous current of new application software. Results in Muresan and Gebotys [13,14] and Muresan [15] found that these current models captured over 94% of the real measured current variation. An example of the instantaneous current model is given in Figure 34.5 where the top waveform is the real current measured, and the lower waveform is the instantaneous current model based on gamma functions. The bottom arrow indicates the software execution time. In general, the gamma function could be retargeted to other processors by fitting it to their single instruction instantaneous power waveforms (through modifying 0.0038).

34.4 Emerging Applications of Instantaneous Power Prediction: Security

Security is crucial for today's portable devices including PDAs, cell phones, and other wireless devices. For example, some PDAs or cell phones are Internet-enabled and contain credit card information, others used in the healthcare industry contain confidential health information, and still other portable devices provide access to private corporate networks. In all these cases if the portable device is lost, it must be secure: specifically it must prevent unauthorized users from breaking into the portable device or obtaining any valuable information from the device. Even if the device is not lost, it may still be possible to obtain valuable information from the EM waves being radiated from the device while it is in use. One of the greatest feared attacks on SmartCards arose in the late 1990s [21,22] when it was demonstrated that the secret key could be determined by measuring the power (highly correlated with EM waves) drawn by the SmartCard processor. This is known as a power analysis attack. Since then, much research has concentrated on enhancing SmartCard security. However, portable devices also demand high security, yet are typically more complex than SmartCards. For example, they often have debug modes which can be used by attackers to access data without even knowing the users password or even download hostile code and thus are

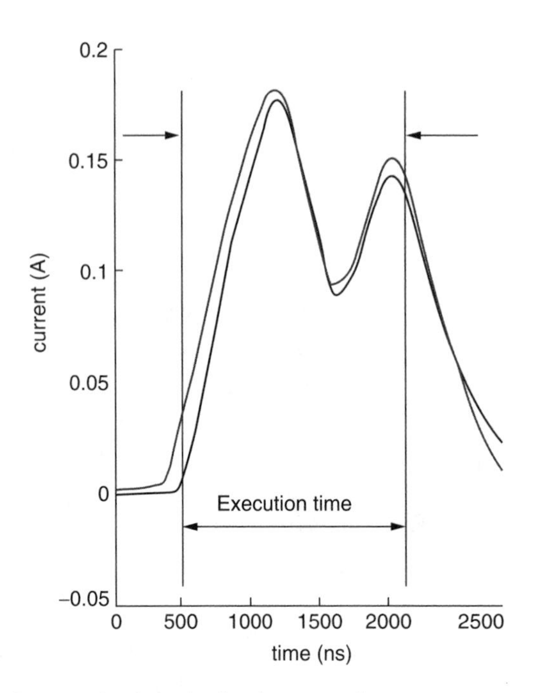

FIGURE 34.5 Real measured current (top) vs. simulated current of a program.

vulnerable to attack. This section introduces the power analysis attacks and gives examples of instantaneous power measurements in this security field.

The measurement of instantaneous power while a processor is executing an application (or a power trace) has been used in power-attacks of cryptographic devices, such as smart cards (typically 8- or 16-bit embedded processors). The equipment setup consists in general of an oscilloscope measuring the voltage over a small resistor placed in series between the processor supply pin (contact point on the smart card) and the supply (external to the smart card). In particular, the analysis of the variation of instantaneous power and statistical computations on a number of power traces can be used to detect data and algorithmic dependencies. This research studied the correlation of power variation with data values being manipulated and instruction sequencing. In the former case, known as differential power analysis attacks (DPA), encryption applications were analyzed [22]. In the latter case, known as a simple power analysis attack (SPA) [21], it was concluded that the correlation was significant and techniques such as random sequencing of instructions have since been researched. Typically, SmartCard applications are not time critical and energy dissipation is not a major concern because power is attained from the card reader (or ATM machine). Power attacks of more sophisticated processors with parallel instruction execution have more recently reported in Gebotys and Gebotys [23].

34.4.1 Simple Power Analysis

As an example of a simple power attack, consider a security algorithm running on a VLIW processor. The security algorithm implements elliptic curve point multiplication for NIST approved elliptic curve $y^2 + xy = x^3 + ax^2 + b$ over 163 bit binary fields ($F_2{}^{163}$) using prime polynomial $x^{163} + x^7 + x^6 + x^3 + 1$ [24,25] (using affine coordinates). Two power traces are plotted in Figure 34.6. The top power trace is a sum routine, and the bottom is a double routine from an elliptic curve point multiplication, widely used in public key cryptography. If an attacker can determine when a sum is being performed and when a double is being performed, the secret key can be easily determined. The detailed differences in the algorithm's power traces can be seen (i.e., sum routine does not have a loop in between the mul and square) from this figure. Thus, it can be seen that instantaneous power models are important for designing security applications. Figure 34.7 illustrates a more secure design where it is now more

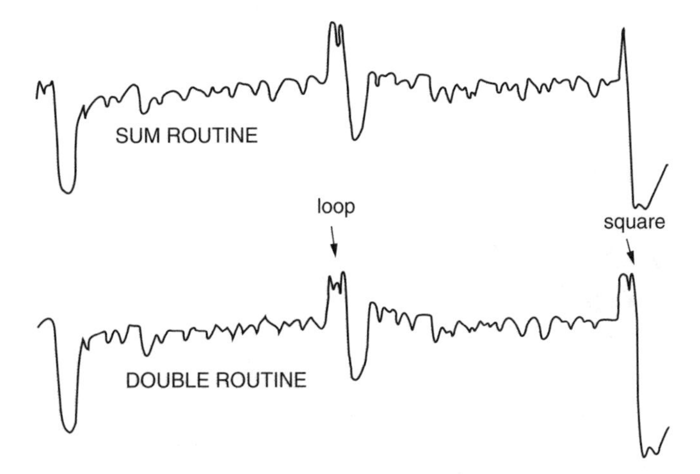

FIGURE 34.6 Current for eliptic sum (top) and double (bottom) routines.

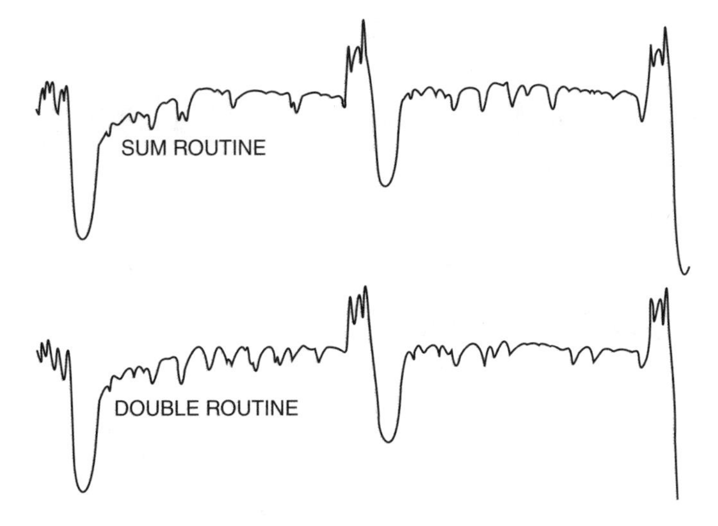

FIGURE 34.7 Secure current/power trace for sum and double eliptic curve routines.

difficult to identify the differences between the sum and double routine from the power traces. Further results can also be found in Gebotys and Gebotys [23] for prime fields.

34.4.2 Differential Power Analysis

Instantaneous power measurement is also crucial for verifying power–analysis security, particularly if an attack is able to acquire a large number of power traces from a SmartCard or some device, it may be possible to again obtain the key or reduce the search space size for enumerating through possible keys. Differential power analysis is an attack based upon the data-dependent switching activity component of power, particularly when data is placed on a processor bus the power reveals information about the data's hamming weight. With a sufficient number of power traces generated with different text inputs, it may again be possible to confirm key bits (and entire key values). Some hamming weights have been measured in Messerges et al. [26] for an 8-bit, 5-V, 4-MHz processor, however, hamming weights for a 32-bit, 2-V, 100-MHz VLIW processor, SC140, as plotted in Figure 34.8 for hamming weights 0, 4, and 6, are more difficult to determine. Nevertheless differential power analysis is still a threat as indicated by the differential signal plotted in Figure 34.9 (second plot from the top), whose differential peaks are greater than two standard deviations [27] (and therefore significant), acquired with 3000 power traces

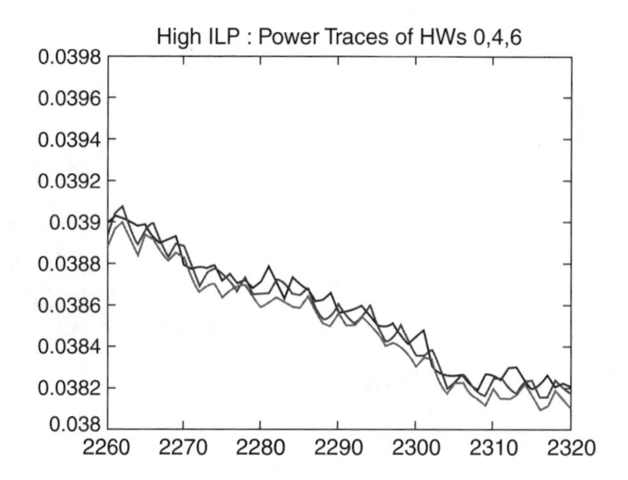

FIGURE 34.8 High, average, and low hamming weights (mA).

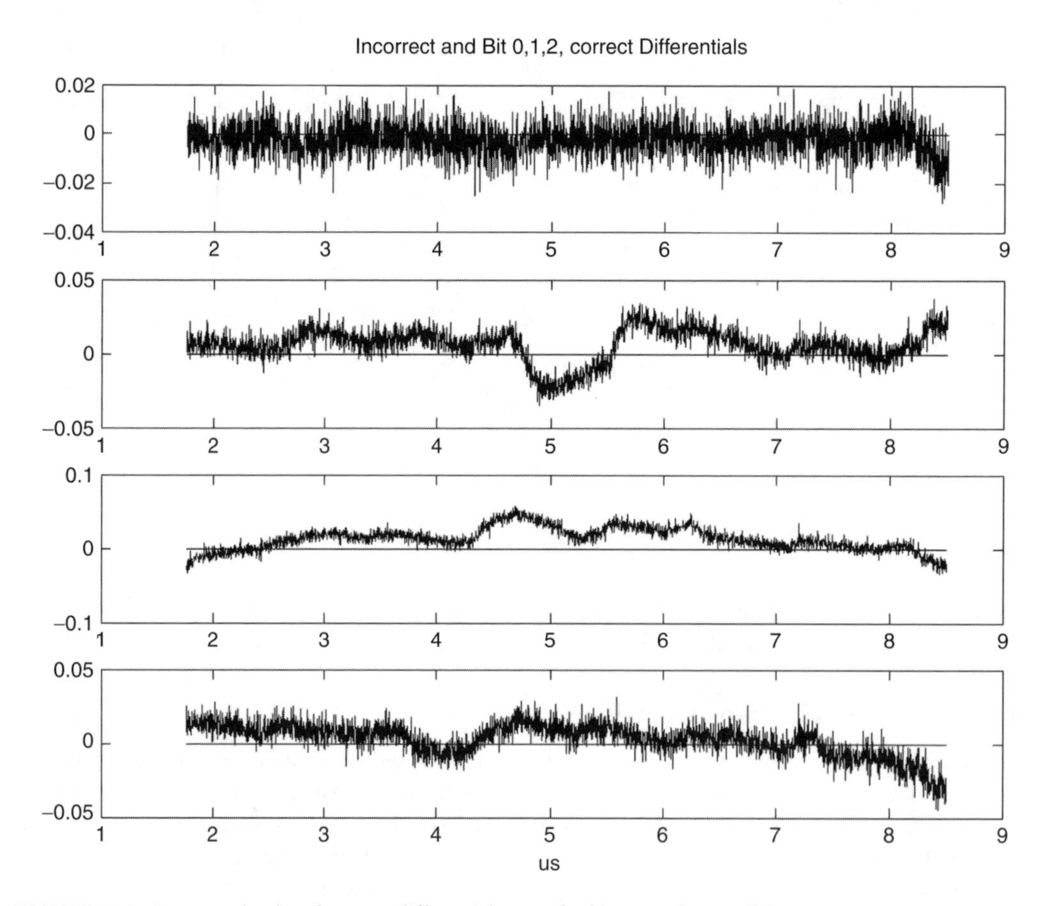

FIGURE 34.9 Incorrect (top) and correct differential traces for bits 0,1,2 for parallel program.

of a parallel program. In this DPA experiment, one group of power traces (group 0) is averaged and subtracted from another group of averaged power traces (group 1). Each power trace is the instantaneous power dissipated by the processor while it runs a specific application. The application involves several memory loads (of 6-bit data) as well as ALU and logical instructions executing in parallel. Group 0 represents the power traces obtained from this program when the memory loads different 6-bit data words, where the LSB (least significant bit) of the data word is always zero. The group 1 traces are acquired with the same application code; however, the six-bit data word has a LSB of one. In theory, the difference of means removes the variation of the power due to the algorithm (or program). The averaging also removes the variation in power due to the higher 5 bits of the data word (assuming sufficient power traces have been obtained to average out the noise and all combinations of 2^5 data words are exercised to average out the power variation due to the upper 5 bits). The difference of means then theoretically represents the power variation solely due to the LSB, which differs from group 0 to group 1. The theory assumes that the hamming weight of the data being placed on the bus has an influence on the instantaneous power. The top differential power signal in Figure 34.9 illustrates an incorrect guess of the bits (where half of the bit i's of the data word are 0 and the other half of bit i's are 1 in each group). The lower 3 differential traces lie off of the zero axis indicating that the power was influenced by bit 0, bit 1, and bit 2 of the data word respectively. The bottom two traces are less clear and indicate it may be more difficult to obtain a DPA on all bits of the bus. This approach has been used in security applications (e.g., to guess the key bit when the data being loaded is the result of the exclusive or operation on the key and the plain text). Here, the user can input a larger number of plain texts and record the power traces for each. The approach has also been used to correctly guess a double or sum in SPA-resistant elliptic curve cryptography (thus also providing key bit information).

The emerging application of high level software models of power, that of security, briefly discussed in this chapter, is crucial for design of many portable devices, such as PDAs and cell phones, that are internet-enabled or support wireless communications. Unlike SmartCard security research, these portable embedded systems must be energy efficient [29] to maintain long battery lifetimes. Thus, the emerging area of low-energy security will be important, and will be built upon the equipment setups for measuring current, power, and energy as well as instruction-level power models developed for many embedded processors.

34.5 Acknowledgment

The author acknowledges the financial support provided by NSERC, CITO, RIM, and Motorola.

References

[1] L. Benini and G. DeMicheli, System-level power optimization: techniques and tools, *International Symposium on Low Power Electronic Design (ISLPED)*, pp. 288–293, 1999.

[2] M. Lee, V. Tiwari, S. Malik, and M.Fujita, Power analysis and minimization techniques for embedded DSP software, *IEEE Trans. on VLSI Design,* March 1997, pp. 123–135.

[3] A. Chandrakasan and R.Brodersen, *Low-Power Digital CMOS Design,* Kluwer Academic Publishers, Dordrecht, 1995.

[4] V. Tiwari, S. Malik, and A. Wolfe, Power analysis of embedded software, *IEEE Trans. on VLSI,* Dec. 1994, pp. 437–445.

[5] G. Qu, N. Kawabe, K. Usami, and M. Potkjonak, Function-level power estimation methodology for microprocessors, *Design Automation Conference (DAC)*, 2000.

[6] J. Russell and M. Jacome, Software power estimation and optimization for high-performance 32-bit embedded processors, *International Conference on Computer Design (ICCD),* 1998.

[7] C. Gebotys and R. Gebotys, Statistically based prediction of power dissipation for complex embedded DSP processors, *Microprocessors and Microsystems,* 23, pp. 135–144, 1999.

[8] C. Gebotys, R. Gebotys, and S. Wiratunga, Power minimization derived from architectural-usage of VLIW processors, *Proc. Design Automation Conf., ACM,* pp. 308–311, June 2000.

[9] C.H. Gebotys and R.J. Gebotys, An empirical comparison of algorithmic, instruction, and architectural power prediction models for high-performance embedded DSP processors, *Proc. IEEE Int. Symp. on Low-Power Electron. Design,* pp. 121–123, August 1998.

[10] R. Joseph and M. Martonosi, Run-time power estimation in high-performance microprocessors, *ISLPED*, pp. 135–140, 2001.

[11] A. Sinha and A. Chandraksan, Energy aware software, *13th Int. Conf. on VLSI Design*, 2000, pp. 50–55.

[12] N. Chang, K. Kim, and H. Lee, Cycle-accurate energy consumption measurement and analysis: case study of ARM7TDMI, *ISLPED*, pp. 185–190, 2000.

[13] R. Muresan and C. Gebotys, Dynamic power simulation model for VLIW DSP processor VLSI cores with secure applications, *Proc. 11th IFIP VLSI-SOC*, December 2001, pp. 67–72.

[14] R. Muresan and C. Gebotys, Current consumption dynamics at instruction and program level for a VLIW DSP processor, *Proc. ACM/IEEE 14th Int. Symp. on Syst. Synthesis (ISSS)*, October 2001, pp. 130–135.

[15] R. Muresan, Measurements, macro-modeling, and applications of current dynamics in complex core processors. Ph.D. thesis, Department of Electrical and Computer Engineering, University of Waterloo, 2003.

[16] F. Wolf, J. Kruse, and R. Ernst, Compact trace generation and power measurement in software emulation, *Proc. SPIE*, Vol. 42, 28, 2000.

[17] Y.H. Lu, L. Benini, and G. DeMicheli, Requester-aware power reduction, *Int. Symp. on System-Level Synthesis*, 2000, pp. 18–23.

[18] T. Simunic, L. Benini, and G. DeMicheli, Source code optimization and profiling of energy consumption in embedded systems, *Int. Symp. on System-Level Synthesis*, 2000, pp. 193–198.

[19] W. El-Essawy, D. Albonesi, and B. Sinharoy, A microarchitectural-level step-power analysis tool, *ISLPED*, pp. 263–266, 2002.

[20] *SPSS User's Guide, Base 8.0 for Windows*, SPSS Inc., 1998.

[21] P. Kocher, Timing attacks on implementations of Diffie–Hellman, RSA, DSS, and other systems, *Lecture Notes in Computer Science (LNCS)*, 1998.

[22] P. Kocher, J. Jaffe, and B. Jun, Differential power analysis, *CRYPTO '99*, pp. 388–397, 1999.

[23] C. Gebotys and R. Gebotys, Designing VLSI cores with secure applications, *Proc. Cryptographic Hardware and Embedded Syst.*, Redwood City, CA, LNCS 2523, August 2002, pp. 114–128.

[24] IEEE Std. 1363-2000, *IEEE Standard Specifications for Public-Key Cryptography*, IEEE Computer Society Press, Washington, DC, 2000.

[25] C. Gebotys and R. Gebotys, A framework for security on NoC technologies, *IEEE Int. Symp. on VLSI*, February 2003.

[26] T. Messerges, E. Dabbish, and R. Sloan, Investigations of power analysis attacks on SmartCards, *USENIX Workshop on SmartCard Technol.*, 1999.

[27] C. Gebotys, Design of secure cryptography against the threat of power-attacks in DSP embedded processors, *ACM Trans. on Embedded Comput. Syst.*, February 2004, pp. 92–113.

[28] S. Nikolaidis, N. Kavvadias, P. Neofotistos, K. Kosmatopoulos, T. Laopoulos, and L. Bisdounis, Instrumentation set-up for instruction level power modeling, *PATMOS 2002*, LNCS 2451, pp. 71–80, 2002.

[29] C. Gebotys and Y. Zhang, Security wrappers and power analysis for SoC technologies, *Int. Symp. on Syst.-Level Synth.-CODES*, 2003.

35

Low-Power/Energy Compiler Optimizations

Ulrich Kremer
Rutgers University

35.1 Introduction

Embedded processors and systems on chip (SoCs) are used in many devices, ranging from pace makers, sensors, phones, and personal digital assistants (PDAs), to general-purpose, handheld computers and laptops. Each of these devices has their own requirements for performance, power dissipation, and energy usage, and typically implements a particular trade-off among these entities. Allowing components of these devices to be controlled by software has opened up opportunities for compilation and operating strategies to reduce power dissipation and energy usage, at the potential cost of performance degradation. Such control includes:

1. Hibernation (i.e., initiating transitions of a component between high-power active states and lower-power hibernating states)
2. Dynamic frequency and voltage scaling, which allows the clock speed and supply voltage to be set explicitly within a range of feasible voltage and frequency combinations
3. Remote task mapping, where power and energy is saved on a mobile device by executing a task remotely on a server

This chapter discusses general issues and challenges related to compilers for power and energy management. A set of compilation strategies are further examined, together with initial results that describe their potential benefits.

35.2 Why Compilers?

Compilers translate a program in a high-level language into a program that can be executed on a target architecture. In other words, compilers support high-level programming models that allow programmers

0-8493-1941-2/05/$0.00+$1.50

to describe the solution to their problem at an abstraction level closer to the particular problem domain. As a result, programs are easier to understand and maintain. Porting a program to another target system requires recompilation on the new system instead of reimplementing the program in the new assembly/machine language; however, these benefits may come at the price of a reduction in overall program performance. Typically, the effectiveness of a compiler and its generated code is measured by comparing it against a code that an "expert" assembly/machine code programmer would have written, or even the best machine code possible. For an optimizing compiler, this difference should not be too large, where the acceptable performance gap depends on the particular application domain. What such a comparison does not capture is the effort needed by an "expert" programmer to come up with such a high-quality code. Modern embedded processors have many features previously found only in high-performance processors, including SIMD instructions, VLIW design, and multiple independent memory banks.

The effort to write efficient or even correct programs may be prohibitively high, particularly for embedded systems with short time-to-market cycles. As a result, high-level languages and their optimizing compilers are becoming a necessary alternative to programming advanced embedded processors in machine and assembly code. Instead of rewriting a set of applications for a new target system, a new compiler has to be provided for that new architecture. Researchers in the embedded systems compiler community have developed and are further investigating new compilation infrastructures that allow the effective retargeting of compilers [9]. Although the issue of retargetability is very important, it is not covered in this chapter.

Optimizing compilers perform program analyses and transformations at different levels of program abstraction, ranging from source code and intermediate code, such as three-address code, to assembly and machine code. Analyses and transformations can have different scopes. They can be performed within a single basic block (local), across basic blocks but within a procedure (global), or across procedure boundaries (interprocedural). Traditionally, optimizing compilers try to reduce overall program execution time or resource usage such as memory. The actual compilation process can be done before program execution (static compilation) or during program execution (dynamic compilation). This large design space is the main challenge for compiler writers. Many trade-offs have to be considered to justify the development and implementation of a particular optimization pass or strategy; however, every compiler optimization needs to address the following three issues:

1. Opportunity. When can the optimization be applied?
2. Safety. Does the optimization preserve program semantics?
3. Profitability. When applied, how much performance improvement can be expected?

Clearly, every program transformation should be safe. Compiler writers would be out of their jobs if safety is ignored. Profitability has to consider any overheads introduced by an optimization, particularly runtime overheads. The combination of opportunity and profitability allows the assessment of the expected overall effectiveness of an optimization.

In principle, hardware- and operating system (OS)-based program improvement strategies face the same challenges as compiler optimizations; however, the trade-off decisions are different based on the acceptable cost of an optimization and the availability of information about dynamic program behavior. Hardware and OS techniques are performed at runtime where more accurate knowledge about control flow and program values may be available. Opportunity, safety, and profitability checks result in execution time overheads, and therefore need to be rather inexpensive. Profitability analyses typically use a limited window of past program behavior to predict future behavior. In contrast, in a static compiler, most of the opportunity, safety, and profitability checks are done at compiler time (i.e., not at program execution time), allowing more aggressive program transformations in terms of affected scope and required analyses. Because the entire program is available to the compiler, future program behavior may be predicted more accurately in the cases where static analysis techniques are effective. Purely static compilers do not perform well in cases where program behavior depends on dynamic values that cannot be determined or approximated at compile time. In many cases, however, the necessary dynamic information can be derived at compile time or code optimization alternatives are limited, allowing the appropriate alternative to be selected at runtime based on compiler-generated tests. The ability of the compiler to reshape program behavior through aggressive whole-program analyses and transformations, which is a key advantage over hardware and OS techniques, exposes optimization opportunities that were not available before. In addition, aggressive whole-program analyses allow optimizations with high runtime overheads that typically require a larger scope to assess their profitability.

The following sections discuss several promising compiler optimization techniques, together with an assessment of their potential benefits. These optimizations include remote task mapping, resource hibernation, and dynamic voltage and frequency scaling.

35.3 Power vs. Energy vs. Performance

Optimizing compilers need underlying performance models and metrics to be able to transform the program code for a specific optimization goal. These models and metrics guide the compiler to make selections among program transformation alternatives. If one optimization goal subsumes another, there is no need to develop separate models and metrics for the subsumed models. This section addresses the question of whether or not power, energy, and performance should be considered separate compiler optimization goals.

35.3.1 Power vs. Energy

Optimizing for minimal power dissipation or minimal energy usage may have different metrics, and therefore result in different optimization strategies. One possible metric for power and energy is that of activity level at any given point during program execution and total amount of activities for a program region, respectively. The more "work" is done at a program point, the more power is dissipated. Given these metrics, is optimizing for power the same as optimizing for energy? The answer depends on the particular definition of "work".

An optimizing compiler may define work as the number of instructions executed at a given point in time. This model assumes that:

1. A fixed amount of power is associated with each executed instruction.
2. The power dissipation of an instruction is independent of its particular operand values or other executing instructions. Figure 35.1 illustrates this case. By reordering or rescheduling instructions, for instance, in a VLIW or superscalar architecture, the initial power profile of a program region as presented on the left of Figure 35.1 may ideally be transformed into the one presented on the right. Although the peak power dissipation is different for both profiles, the energy usage is the same. In other words, activity or work rescheduling can be an effective way to reduce peak power dissipation while having no impact on energy usage. Therefore, peak power reduction may be an optimization objective different from energy reduction.

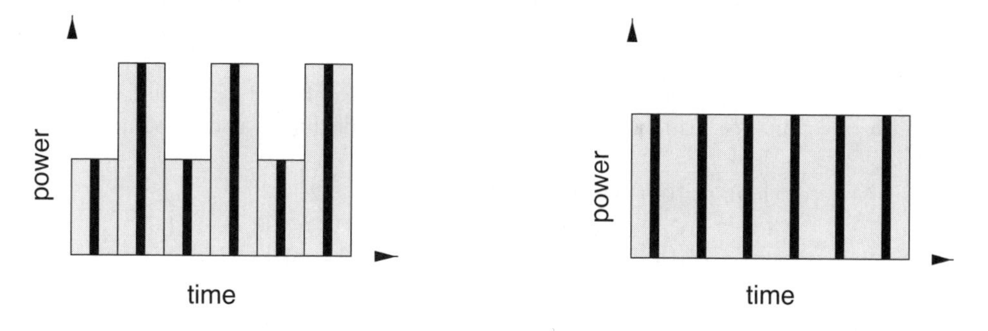

FIGURE 35.1 Optimizing for power vs. energy: two possible power profiles of an example program region.

For power models based on bit-level switching activities as its work notion, rescheduling instructions may also target overall energy usage by grouping instructions based on their particular bit patterns. In addition to instruction scheduling, a careful selection of register names in the code generation phase of a compiler can result in code sequences that have bit patterns with less switching activities, for instance, due to the reuse of "similar" register names [7].

Due to the particular chemical characteristics of some batteries, highly varying discharge rates (i.e., varying power dissipations) may reduce the lifetime of a battery significantly. By "smoothing" the power dissipation profile of an application through instruction scheduling and reordering, the usable energy of a battery can be significantly increased [11].

From now on, we will not distinguish between the optimization objectives of reducing peak power dissipation and overall energy usage unless explicitly stated.

35.3.2 Power/Energy vs. Performance

Early work on optimizing compilers for power and energy management suggested that optimization transformations for performance subsume those for power and energy management. Therefore, power/energy is not an optimization objective in its own right [13]. Traditional optimizations, such as common subexpression elimination, partial redundancy elimination, strength reduction, or dead code elimination increase the performance of a program by reducing the work to be done during program execution [2,12]. Clearly, reducing the workload may also result in power/energy savings. Memory hierarchy optimizations, such as loop tiling and register allocation, try to keep data closer to the processor because such data can be accessed more quickly. Keeping a value in an on-chip cache instead of an off-chip memory, or in a register instead of the cache, also saves power/energy due to reduced switching activities and switching capacitance.

However, a fundamental difference exists between the models and metrics used for performance and those used for power/energy optimizations. Many performance models have the notion of a critical path (i.e., a sequence of instructions or activities that will dominate the overall program execution time). If an optimization introduces activities on the noncritical path, performance is not affected. Therefore, as long as these noncritical activities lead to an overall decrease of the critical path (at least in most cases), the optimization is beneficial. In the context of power/energy optimizations, this is not true. Any activity, whether on or off the critical path, will contribute to the overall power dissipation and energy usage.

Figure 35.2 is an example that illustrates the differences in optimizing for power/energy versus optimizing for performance for a source-level transformation, in this case loop invariant code motion [2,12]. In the example program, the assignment a = b * 2 is assumed to be loop invariant. For a traditional scalar architecture, loop invariant code motion will move the assignment out of the loop, resulting in the code on the right side of Figure 35.2. In a VLIW architecture, the code on the left may be best if empty VLIW instruction slots are available to execute the loop invariant assignment for each iteration of the loop. Although the assignment is done 10 times, it may reduce the overall critical path. Depending on the particular overall compilation strategy used, moving the assignment out of the loop may actually increase the critical path. In the context of power/energy optimization, performing redundant computations should be avoided, and, therefore, moving the invariant assignment out of the loop typically leads to power and energy savings.

Another example where optimizations for power/energy may be different from that for performance is speculative execution. Speculation performs activities "ahead of time" based on some assumptions

```
for (i= 0; i< 10; i++)  {              a = b   2;
    a = b   2;                         for (i= 0; i< 10; i++)  {
    c[i] = d[i] + 2.0;                     c[i] = d[i] + 2.0;
}                                      }
```

FIGURE 35.2 Example code fragment to illustrate power vs. performance optimization strategies.

about the future behavior of the program. If these assumptions turn out to be false, additional work may be necessary to undo the impact of the speculative performed activities. Software prefetching is an example of such a transformation. The compiler may insert prefetch instructions for memory accesses across control branches. Assuming that the target machine allows multiple outstanding loads, this optimization can be very effective. Again, as long as the speculative activity can be hidden on the noncritical execution path, no negative impact on performance will occur. In the context of power/energy optimizations every additional, speculative activity has to be compensated for by the overall power/energy benefit of the optimization to make things not worse. In other words, the window of profitability has to be larger for power/energy optimizations than performance optimizations. This does not mean that speculation cannot be applied for power/energy optimizations, but suggests a less aggressive application of such a transformation by restricting it to the cases where the benefit is likely.

35.3.3 Summary

In recent years, reducing the power dissipation and energy consumption of a program have actually become optimization goals, no longer considered byproducts of traditional performance optimizations that mainly try to reduce program execution times. Power and energy optimizations can be implemented in hardware through circuit design, by the operating system through scheduling techniques that consider the power and energy requirements of active processes, and by the compiler through compile-time analyses, code reshaping, and hints to the operating system. The following issues should be considered during the design of an optimizing compiler for power/energy management:

1. You can run but you cannot hide. All instructions, including instructions on the noncritical path contribute to the overall power dissipation and energy consumption. As a result, power/energy optimizations have a higher threshold for profitability than performance optimization if they require additional instructions to be executed.
2. Keep the overall picture in mind. A power/energy optimization with a slight performance penalty may be profitable for a single system component (e.g., cache, CPU, and memory), it may not be profitable for the overall system due to its impact on the power/energy requirements of other system components. In addition, the power/energy characteristics of other active processes have to be considered in a multi-programming environment.
3. You cannot beat hardware. If an operation is implemented in hardware, and an application can take advantage of this hardware (e.g., floating point unit), a compiler should try to generate code for it. If the hardware dissipates power while idle, the compiler needs to be able to disable it during such idle periods.

35.4 List of Optimizations

The following section discusses three compiler optimizations. These optimizations are just examples, and are presented to illustrate the potential benefits of compile time power/energy management. This list is by no means complete.

35.4.1 Dynamic Voltage and Frequency Scaling

Dynamic voltage scaling (DVS) is recognized as one of the most effective power reduction techniques. It exploits the fact that a major portion of power of CMOS circuitry scales quadratically with the supply voltage [3]. As a result, lowering the supply voltage can significantly reduce power dissipation. For noninteractive applications, such as movie playing, decompression, and encryption, fast processors reduce device idle times, which, in turn, reduce the opportunities for power savings through hibernation strategies. In contrast, DVS techniques are still beneficial in such cases (i.e., DVS reduces power even when these devices are active); however, DVS comes at the cost of performance degradation. An effective

DVS algorithm is one that intelligently determines when to adjust the current frequency-voltage setting (scaling points) and to which frequency-voltage setting (scaling factors), so that considerable savings in energy can be achieved while the required performance is still delivered.

One possible compiler-directed algorithm identifies program regions where the CPU can be slowed down with negligible performance loss [6]. It is implemented as a source-to-source level transformation using the SUIF2 [1] compiler infrastructure. Physical measurements on a laptop with a 600–1200-MHz AMD Athlon 4 processor demonstrate that total system energy savings of up to 23% can be achieved with performance degradation of less than 5% for the SPECfp95 benchmarks. On average, the energy and energy-delay products are reduced by 11% and 9%, respectively, at the cost of the performance slowdown of 2%. It was also discovered that the energy usage of the programs using this DVS algorithm is within 6% from the theoretical lower bound.

35.4.2 Resource Hibernation

A common approach to increase energy efficiency puts idle resources or entire devices in low-power (hibernation) states until they have to be accessed again. The transition to a lower power state usually occurs after a period of inactivity (an inactivity threshold), and the transition back to active state usually occurs on demand. Unfortunately, the transitions to and from the low-power state can consume significant time and energy. Nevertheless, this strategy works well when there is enough idle time to justify incurring such costs.

Source-level transformations can be used to reshape the program behavior such that inactivity thresholds of a device or component are extended, allow hibernation to be more effective. By allowing the compiler to give hints to the operating system about expected idle times of these components and devices, the OS is able to issue deactivation directives earlier and activation directives just in time before the device or component is used again. In addition, the operating system can use these hints to implement the most efficient policy for the set of active processes. The results reported in Heath et al. [5] demonstrate that on a set of streamed and nonstreamed application, the reshaped programs can achieve disk energy reductions ranging from 55% to 89% (70% on average) under a sophisticated energy management policy with only a small performance degradation.

35.4.3 Remote Task Mapping

Mobile devices come in many flavors, including laptop computers, Webphones, pocket computers, PDAs, and intelligent sensors. Many such devices already have wireless communication capabilities, and we expect most future systems to have such capabilities. Two main differences exists between mobile and desk-top computing systems, namely the source of the power supply and the amount of available resources. Mobile systems operate entirely on battery power most or all the time. The resources available on a mobile system can be expected to be at least one order of magnitude less than those of a "wall-powered" desk-top system with similar technology. This fact is mostly due to space, weight, and power limitations placed on mobile platforms. Such resources include the amount and speed of the processor, memory, secondary storage, and I/O. With the development of new and even more power-hungry technology, we expect this gap to widen even more. Remote task mapping is a technique that tries to off-load computation to a remote server, thereby saving power and energy on the mobile devices [8,10].

A possible compilation strategy that generates two versions of the initial application, one to be executed on the mobile device (client), and the other on a machine connected to the mobile device via a wireless network (server) [8]. The client and server codes have to be able to deal with disconnection events. The proposed compilation strategy uses checkpointing techniques to allow the client to monitor program progress on the server, and to request checkpoint data to reduce the performance penalty in case of a possible server and/or network failure.

The reported results have been obtained by actual power measurements of an image processing application (face detection and face recognition) on three client systems:

1. The StrongARM-based, low-power SKIFF system developed at Compaq's Cambridge Research Laboratory.
2. Compaq's commercially available StrongARM-based iPAQ H3600.
3. A Pentium-II-based laptop. Initial experiments demonstrate that energy consumption can be reduced significantly, in some cases, up to one order of magnitude, depending on the selected characteristics of the mobile device, remote host, and wireless network.

35.5 Future Compiler Research for Power/Energy

Compiler research for power and energy management is still in its infancy. Such research requires platforms that expose power and energy management features to higher software levels such as the compiler through standardized interfaces (APIs). Although efforts have been made in some areas (e.g., ACPI [4]), more work needs to be done.

In addition, the lack of a reliable and effective evaluation infrastructures for power and energy optimizations has significantly hampered compiler research. The compiler community relies mostly on physical measurements on existing target systems for a set of representative benchmarks to evaluate the benefits of a given optimization or set of optimizations. Simulation results are accepted as an indication of a potential benefit of an optimization, but are typically not considered sufficient proof that the optimization is worthwhile in practice. What is needed is an evaluation infrastructure for power and energy optimizations that consists of a combination of physical measurements and performance modeling. Physical measurements need to include current and voltage measurements, as well as temperature measurements. Performance models are needed for the CPU, memory subsystems, controllers, communication modules, and I/O devices such as the disk and screen. This technology is crucial to be able to understand and assess the benefits of a proposed optimization for the entire target system, subsets of system components, or single system components.

35.6 Acknowledgment

This work has been partially supported by National Science Foundation (NSF) CAREER Award No. 9985050. Any opinions and conclusions expressed in this chapter are those of the author, and do not necessarily reflect the view of the NSF.

References

[1] National Compiler Infrastructure (NCI) project. Overview available online at http://www-suif.stanford.edu/suif/NCI, Co-funded by NSF/DARPA, 1998.

[2] A.V. Aho, R. Sethi, and J. Ullman. *Compilers: Principles, Techniques, and Tools, 2nd ed.* Addison-Wesley, Reading, MA, 1986.

[3] T. Burd and R. Brodersen. Energy-efficient CMOS microprocessor design. *28th Hawaii Int. Conf. on System Sciences (HICSS-95),* January 1995, pp. 288–297.

[4] Advanced Configuration and Power Interface Specification. Compaq, Intel, Microsoft, Phoenix Technologies, Toshiba, Revision 2.06, October 11, 2002. http://www.ocpi.info.

[5] T. Heath, E. Pinheiro, J. Hom, U. Kremer, and R. Bianchini. Application transformations for energy and performance-aware device management. *Int. Conf. on Parallel Architectures and Compilation Tech. (PACT '02),* Charlottesville, VA, September 2002, pp. 121–130.

[6] C.-H. Hsu and U. Kremer. The design, implementation, and evaluation of a compiler algorithm for CPU energy reduction. *ACM SIGPLAN Conf. on Programming Languages, Design, and Implementation (PLDI '03),* San Diego, CA, June 2003, pp. 38–48.

[7] M. Kandemir, N. Vijaykrishnan, M.J. Irwin, W. Ye, and I. Demirkiran. Register relabeling: a post-compilation technique for energy reduction. *Workshop on Compilers and Operating Syst. for Low Power (COLP '00),* Philadelphia, PA, October 2000.

[8] U. Kremer, J. Hicks, and J. Rehg. A compilation framework for power and energy management on mobile computers. *Int. Workshop on Languages and Compilers for Parallel Computing (LCPC '01)*, Cumberland, KY, August 2001, pp. 115–131.

[9] R. Leupers. Compiler design issues for embedded processors. *IEEE Design Test of Comput.*, 19(4):51–58, July/August 2002.

[10] Z. Li, C. Wang, and R. Xu. Computation offloading to save energy on handheld devices: a partition scheme. *Int. Conf. on Compilers, Architectures, and Synthesis for Embedded Systems (CASES 2001)*, Atlanta, GA, November 2001, pp. 238–246.

[11] T. Martin and D. Siewiorek. The impact of battery capacity and memory bandwidth on CPU speed-setting: a case study. *Int. Symp. on Low-Power Electron. and Design (ISLPED)*, pp. 200–205, San Diego, CA, August 1999.

[12] S.S. Muchnick. *Advanced Compiler Design Implementation.* Morgan Kaufmann Publishers, San Franscisco, CA, 1997.

[13] V. Tiwari, S. Malik, A. Wolfe, and M. Lee. Instruction-level power analysis and optimization of software. *J. VLSI Signal Process.*, 13(2/3):1–18, 1996.

36

Design of Low-Power Processor Cores Using a Retargetable Tool Flow

Gert Goossens
Target Compilers Technologies

Peter Dytrych
Dirk Lanneer
Philips Digital Systems Laboratories

36.1 Introduction

With process geometries shrinking to nanometers, unprecedented levels of silicon integration are now available. This has fuelled the design of complete electronic systems on a single multimillion-transistor chip. To master the design complexity of such systems on chip (SoC), the reuse of processor cores has become an important design paradigm. Different types of predesigned and preverified processor cores can be instantiated and connected as building blocks in a heterogeneous chip architecture; however, power consumption is becoming a major hurdle in the successful design of future SoCs.

This chapter describes a methodology for designing low-power processor cores in SoCs. Its key component is the ability to quickly and adequately customize the instruction-set architecture of the processor core, to match the characteristics of the application. It is demonstrated that this allows for a drastic reduction of the power consumption of the processor, while retaining sufficient design flexibility as offered by a programmable processor. The methodology is supported by a retargetable tool-suite, offering architectural exploration, software development, and verification capabilities. The practical applicability of the methodology and tool-suite is demonstrated by the design of an industrial ultra-low power digital signal processor (DSP) core for audio coding applications, named CoolFlux DSP.

0-8493-1941-2/05/$0.00+$1.50

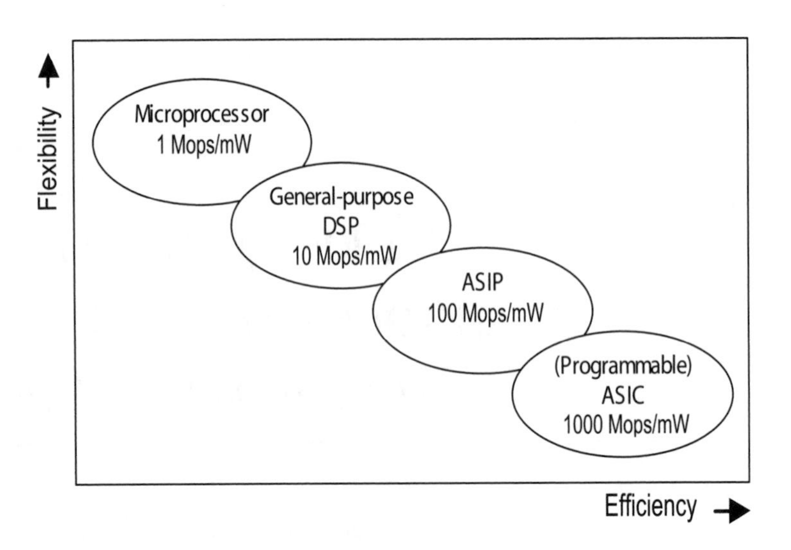

FIGURE 36.1 Classification of processor cores in SoC design.

36.1.1 Processor Cores in SoC Design

Figure 36.1 depicts a classification of processor cores used in SoCs. On the one hand, general-purpose microprocessor and DSP cores, as available from intellectual property (IP) vendors, offer most flexibility. On the other hand, application-specific cores are being designed to implement system functions that are critical in terms of computational throughput and power dissipation. Due to their application-specific nature, specialized semiconductor or system companies often design these cores in-house. Traditionally, these application-specific cores take the form of a fixed-function application-specific integrated circuit (ASIC) block, designed in a hardware description language such as Verilog or VHDL. Such an ASIC core consists of a data-path with special-purpose functional units and interconnections. Control flow is typically restricted, and can be implemented in a small finite-state machine (FSM).

In competitive markets such as telecom and consumer electronics, the flexibility to take into account rapidly changing functional requirements, and the efficiency to cope with high computational throughput and low-power dissipation requirements, are both important. New processor cores must combine the best of both worlds. DSP cores are becoming more application-specific, by extending a general-purpose instruction-set architecture (ISA) with specialized functional units and instructions. This is referred to as application-specific instruction-set processor (ASIP) cores [10]. At the same time, new-generation ASIC cores offer a small layer of software programmability on top of a specialized data-path, to allow for limited functional changes that are crucial to extend the lifetime of these cores. This is referred to as programmable ASIC cores.

Usually, the design of an embedded processor core is significantly influenced by the overall SoC architecture. SoC design addresses the system partitioning, which determines the balance of low power and parallelism, including task parallelism, data parallelism, and instruction-level parallelism (ILP). This defines the broad specification points of the different embedded processor cores in the SoC, which each typically only provide a solution for part of the computational load of the complete system. In this chapter, we assume that the system partitioning and analysis have been performed upfront, although in practice the procedure is rarely this cleanly decoupled and the optimization of a processor core architecture is likely to be done in a complete system context.

As an example, two rather different application scenarios could be a digital hearing instrument and an SoC platform for portable consumer audio. The hearing instrument will have very exacting power and area constraints, but can usually be rather application-specific and have a small application code base (hundreds of assembly lines), for which an ASIP or programmable ASIC (Figure 36.1) is most adequate. The audio platform will tolerate less demanding power constraints and have a very broad code base (hundreds of thousands of assembly lines), requiring a more general-purpose, 24-bit DSP. The audio platform may also be configured as a multiprocessor to allow scalability.

36.1.2 SoC Integration and Low-Power Design

The requirement of low-power dissipation is becoming an important, if not *the* most important, motivation for making application-specific processors. Whereas from an area and gate-count perspective, putting together tens or even hundreds of processor cores on a single chip is not a problem anymore, controlling the heat dissipation and the energy consumption of such a chip becomes a major issue.

To control the power characteristics of the SoC, the overall system architecture, the architectures of the composing processor cores, and the circuit-level implementation are all important [11]. This chapter primarily focuses on the processor core architectural level. The main idea that is explored is that by making the processor architecture application-specific (i.e., by designing an ASIP or a programmable ASIC instead of a general-purpose processor) [Figure 36.1], its power consumption can be reduced drastically.

In the authors' opinion, the following observations are cornerstones of low-power architecture design:

1. Optimizing for minimum cycle count is beneficial for power consumption. The main part of the dynamic power consumption of a circuit is due to the capacitance effect, and is proportional to the average switching frequency (i.e., clock frequency times an activity factor), and to the square of the supply voltage [5,9]:

   $$P = C \times (f_{Clock} \times Act) \times {V_{dd}}^2$$

   Processor architectures that can implement a given software program in a small number of instruction cycles, will generally exhibit better power characteristics. Indeed, a lower cycle count will allow for scaling down f_{Clock}, and especially V_{dd} (known as voltage scaling).

   A low cycle count can be achieved by introducing specialized functional units to accelerate the critical functions of the algorithm, and by providing instruction-level parallelism. These are key features of application-specific processor architectures.
2. Optimizing for reduced memory access is beneficial for power consumption. A substantial portion of the power consumption in a processor is due to memory accesses. This is both related to the switching activity on data and address busses, and to the loading of word lines in the memories [16]. Processor architectures that can implement a given software program using a small number of data and program memory accesses, will generally exhibit better power characteristics.

   Important power savings can be achieved by providing a storage hierarchy. For example, by providing a loop cache, one can avoid excessive program memory fetches for programs containing loop structures. To reduce the required number of data memory accesses, the architecture should be allowed to maintain variables as much as possible in registers local to functional units. Program memory accesses can be reduced among others by designing an application-specific instruction-set with only a small number of instruction bits, as well as by using techniques such as variable-length encoding and instruction compaction. Again, these are key features of application-specific processor architectures.
3. Minimalistic architectures are beneficial for power consumption. An effective architectural design strategy for low power must be minimalistic [6]. By including only those hardware resources that are really needed by the target applications, power consumption can be reduced significantly. Once again, this leads to application-specific processor architectures.
4. Low-power architectural design is holistic. An effective architectural design strategy for low power must be holistic [6]. To effectively reduce the switching activity and capacitance, all aspects of a processor architecture are important, and only the combination of all elements will result in an overall power-efficient architecture.

36.1.3 Architectural Tool Support for Low-Power Processor Design

Instead of attempting to develop an automatic optimization tool for low-power architecture design, an interactive and iterative methodology is proposed that allows architecture designers to explore different architectural trade-offs and obtain rapid feedback about the quality of architectural decisions.

This methodology, which takes into account the holistic nature of low-power architecture design (see Section 36.1.2), is based on a retargetable tool-suite for processor design available from Target Compiler Technologies called CHESS/CHECKERS [2]. Key features of this technology are the following:

- Whereas other architectural design environments are based on a predefined but parameterizable template of a processor architecture [1,3,4,13], one of the key objectives when developing the CHESS/CHECKERS tool-suite was to provide maximum architectural freedom to the designer. In this way, the designer can find an optimal balance between architectural flexibility and specialization, to obtain the best power dissipation characteristics for his or her application.
- THE CHESS/CHECKERS architectural exploration capabilities effectively allow for the discovery of the architectural sweet spots that result in low-power dissipation.

The retargetable tool-suite must be coupled to a complete power-aware very large scale integration (VLSI) design flow. This allows us to simulate instead of to speculate about power consumption. Good architectural candidates can be determined in the retargetable tool flow by using Chess/Checkers retargetable C compiler and getting profiling data from the retargetable instruction-set simulator (ISS). These can then be pushed through the VLSI design flow to ensure that a good, low-power implementation can actually be achieved. Our experience has been that this process is very revealing and really helps to build efficient processor architectures, taking into account that low-power architecture design is holistic.

The Chess/Checkers tool-suite has been applied successfully to design power-efficient, application-specific processor cores for critical applications in wireless and wireline telecommunications, consumer electronics, and medical devices such as hearing aids. The Chess/Checkers tool-suite, and its abilities for low-power architectural exploration, is described in Section 36.2.

Section 36.3 of this chapter surveys a number of important architectural optimizations that make part of a holistic strategy for low-power architectural design. These optimizations are typically explored in the architectural design phase, using Chess/Checkers.

As an illustration of the methodology, the industrial design of an ultra-low power DSP core for audio coding applications is described in Section 36.4. This processor, called CoolFlux DSP, has been designed by Philips Digital Systems Laboratories [6], with the help of the Chess/Checkers tool-suite.

36.2 A Retargetable Tool-Flow for Designing Power-Efficient, Application-Specific Processors

36.2.1 The Chess/Checkers Retargetable Tool-Suite

Chess/Checkers is a retargetable tool-suite that supports the different phases of designing application-specific processor cores, developing application software for these cores, and verifying the correctness of the design. An outline of the Chess/Checkers tool-suite is listed in Figure 36.2. Chess/Checkers consists of the following tools:

- Chess. A retargetable C compiler that translates C source code into machine code for the target processor. Different from conventional compilers such as GCC [12], the Chess compiler uses graph-based modeling and optimization techniques [15], to deliver highly optimized code for specialized architectures exhibiting peculiarities such as complex instruction pipelines, heterogeneous register structures, specialized functional units, and instruction-level parallelism. Chess produces machine code in the Elf object file format, with source-level debug information in the Dwarf 2.0 format.

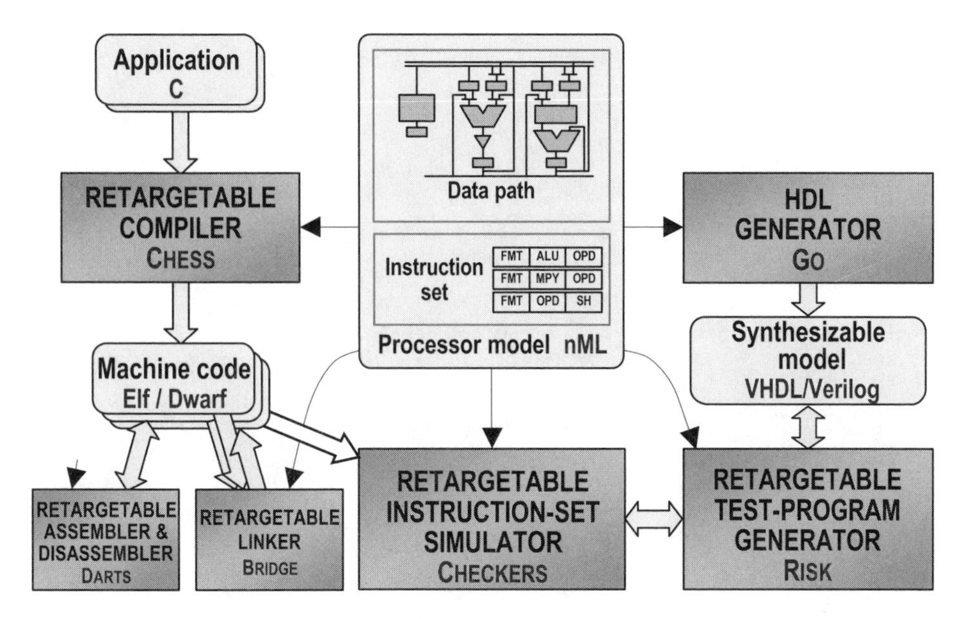

FIGURE 36.2 Outline of the CHESS/CHECKERS tool-suite.

- Bridge. A retargetable linker that builds executable programs from separately compiled Elf/Dwarf object files and libraries.
- Darts. A retargetable assembler and disassembler that translates assembly code into binary Elf/Dwarf object files and back. The assembly language syntax is user-defined.
- Checkers. A retargetable ISS generator that produces a cycle and bit accurate ISS for the target processor. The ISS can be run in a stand-alone mode or be embedded in a co-simulation environment through an application programming interface (API). Checkers comes with a graphical debugger that can connect both to the ISS, as well as to the available processor hardware via a JTAG or debug port for on-chip debugging. Source-level debugging is supported.
- Go. A hardware description language (HDL) generator that produces a synthesizable register-transfer level HDL model of the target processor core. Through APIs, users can plug in their own HDL implementations of functional units and of the memory architecture.
- Risk. A retargetable test-program generator that allows for the quick generation of a large number of assembly-level test-programs for the target processor. These test programs can then be executed both in the ISS and in the HDL model of the processor to check for consistency of both models.

A unique feature of the Chess/Checkers tool-suite is its architectural retargetability, based on the *nML* processor description language. nML is a high-level language that captures a programmer's model of the target processor [7]. This is the abstraction level commonly found in a programmer's manual of a processor. Using nML, an architecture designer can quickly define the ISA of a processor. After reading the nML description, the different Chess/Checkers tools are automatically targeted to the specified architecture.

Figure 36.3 depicts a part of an nML description of a processor. Structural information about the processor is introduced by declaring its storage elements (i.e., memories, registers, and pipeline registers) and its interconnections. The instruction-set is defined using an attributed grammar. The grammar breaks down the instruction set into instruction classes (e.g., `alu_inst`, `mac_inst`, and `shift_inst` in Figure 36.3). The behavior of instructions is specified in action attributes of the grammar rules, using a register-transfer model. In these register-transfer actions, user-defined primitive functions can be called (e.g., `add()`, `sub()`, `and()`, and `or()` in Figure 36.3). To enable instruction-set simulation, the user adds bit-true simulation models for each primitive function. Likewise, to enable hardware generation,

```
// Declaration of storage elements and interconnections:
mem DM[1024]<num,addr>;
reg R[4]<num>;
pipe C<num>;
trn A<num>; trn B<num>;
...

// Definition of instruction set (using attributed grammar):
opn my_core (alu_inst | mac_inst | shift_inst);
...

opn alu_inst (op:opcod, x:c2u, val:c16s, y:c2u) {
  action {
    stage EX1:
      A = R[x];
      B = val;
      switch (op) {
      case add : C = add(A, B) @alu;
      case sub : C = sub(A, B) @alu;
      case and : C = and(A, B) @alu;
      case or : C = or(A, B) @alu;
      }
    stage EX2:
      R[y] = C @alu;
  }
  syntax : op " R" y ", R" x ", " val;
  image : "0"::op::x::y::val;
}
```

FIGURE 36.3 Excerpt of an nML processor description.

the user adds HDL models for each primitive function. Grammar rules also have `syntax` and `image` attributes, defining the assembly language syntax and the binary encoding of the instructions.

36.2.2 Architectural Scope

CHESS/CHECKERS supports a wide range of processor architectures. Retargetability is supported within this range. The following parameters indicate the current architectural scope of the Chess/Checkers tools:

- Data types. Chess/Checkers can support the built-in data-types of the ANSI C language. In addition, users can also introduce any custom data-type. This is useful for application-specific processors, which often contain a variety of specialized data-types. Chess/Checkers allows for the definition of application-specific data types as *C*++ classes. The defined classes can then be used in the nML processor description, to specify the data types of the processor's memories, registers, and interconnections. The same class definitions can also be used in the source program for the Chess compiler.
- Arithmetic functions. Chess/Checkers can cope with standard arithmetic instructions found in general-purpose processors, as required for compiling ANSI C code. Users can, however, also define specialized arithmetic instructions in nML, and specify a mapping from the C source code to these instructions using the concept of intrinsic function calls.
- Memories. Chess/Checkers supports von Neumann and Harvard architectures. The processor may have any number of data memories. Each memory may have one or multiple ports. In case of multiple memories, the user can assign static variables in the C source program to specific memories using a memory qualifier. Several addressing modes are supported for data memories. This includes indexed (or offset), direct, and indirect addressing — optionally with postmodification of address pointers. Special addressing operations, such as modulo and bit-reversed addressing, are supported through intrinsic function calls.

- Instruction format. Chess/Checkers supports a wide range of instruction formats, from orthogonal to highly encoded formats. An orthogonal format consists of fixed control fields that can be set independently from each other. In an encoded format, the interpretation of the instruction bits as control fields is dependent on the instruction. Very long instruction word (VLIW) processors have an orthogonal instruction format. The tools support variable-length instructions, as well as instruction compaction to encode small sequences of instructions in a single instruction word.
- Registers. Chess/Checkers supports a wide variety of register structures, ranging from a homogeneous structure with a single, general-purpose register-file to a heterogeneous structure with special-purpose registers that are dedicated to store operands and results of specific instructions. Chess/Checkers also supports various constraints on the utilization of registers. For example, one may specify that the selection of multiple operand or result registers of an instruction be controlled by a single selection-field in the instruction word. Such register coupling constraints often occur in application-specific processors to save opcode space.
- Instruction pipeline: Chess/Checkers supports instruction pipelines of any depth. Different instructions do not need to have the same number of pipeline stages. Chess/Checkers also supports multi-cycle instructions, multi-word instructions, and instructions with delay slots. The Chess compiler can ensure that pipeline hazards, which are specified in nML, are resolved in the generated code.
- Control flow. Chess/Checkers provides support for subroutines and interrupt service routines. Several mechanisms are available to support the concept of a software stack for storage of automatic variables. Chess/Checkers also supports the concepts of hardware do-loop instructions and of mode bits that determine the behavior of instructions.

36.2.3 Architectural Exploration

As explained in the previous section, Chess/Checkers supports a wide architectural design space. Through architectural exploration, a designer can quickly determine a power-efficient architecture for a specific application domain. As a starting point for exploration, the designer will collect the following inputs:

- Application code for the critical functions that need to run on the processor. A large range of possibilities are available, depending on the nature of the design. At one extreme, only a small number of quite similar algorithms may need to run on an application-specific processor. At the other "general-purpose" extreme, one has to consider a large number of potentially highly diverse applications and attempt to optimize from this "sea of C" to ultimately some fully laid out VLSI design and the corresponding object code for an application. For complete DSP applications, such as an MP3 decoder, the code is usually characterized with a 20/80 rule where 80% of the cycles are spent in 20% of the code. This gives a good spread between unstructured control code and DSP loop kernels.
- Architectural design constraints in terms of area, timing, and power budgets, as well as time scales and other project-related factors. Other more difficult constraints may be present as well, such as backward compatibility to an existing processor architecture with a large legacy code base, scalability to allow coverage of multiple price/performance points, and how application-specific or general-purpose to make the processor architecture. These overall specifications will limit the scope of some architectural choices, such as the number of functional units and the choice of initial ISA.

Based on the preceding inputs, the designer typically makes an initial proposal of an ISA, which is described in nML. Once this starting point is chosen, a more refined architectural exploration can be performed, which will lead to the final design. Note that the initial architecture can be a subset or superset of the finished design, although it is probably more common to start with a subset and add features

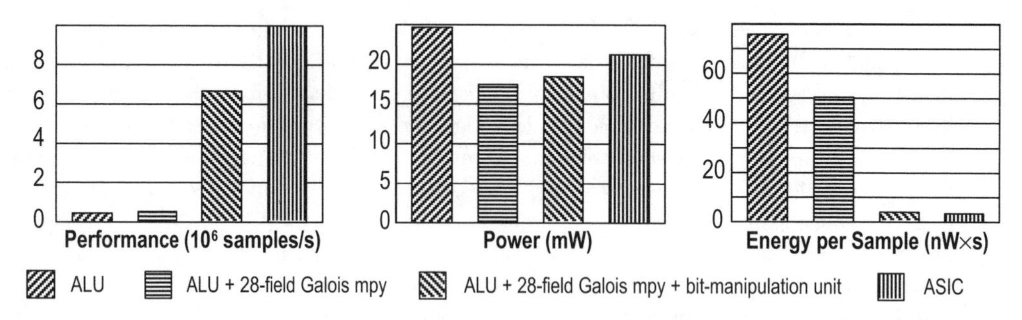

FIGURE 36.4 Performance, power, and energy per sample measurements for different processor architectures for a Reed–Solomon encoding function in an ADSL modem chip.

as required. Section 36.3 discusses a number of important architectural choices for a low-power processor design.

When using Chess/Checkers, a designer can afford to make many iterations. Of each intermediate architecture, the performance can be evaluated by compiling critical C functions with the Chess compiler, and simulating and profiling the resulting machine code with the Checkers ISS. In the first place, the ISS's profiling capabilities allow to evaluate the cycle and instruction count for the application; however; it is also possible to introduce high-level power models in the ISS and to make a comparative power analysis of different architectures (see Section 36.2.4). Using this feedback from the tools, the designer can compare different ISAs and optimize the architecture in nML to obtain a good match between flexibility, throughput, and power characteristics. At some intermediate points, the designer may want to generate synthesizable HDL using the Go HDL generator and enter the VLSI design flow for more accurate measurements.

Figure 36.4 illustrates the architectural exploration capabilities of the Chess/Checkers tool-suite, during the design of an application-specific processor core for Reed-Solomon encoding, for use in an asymmetric digital subscriber line (ADSL) modem SoC. The Reed–Solomon encoding algorithm was described in C source code. As a starting point, a simple microprocessor architecture was used, with a single 32-bit arithmetic and logic unit (ALU). The C code was compiled on the architecture and profiled using the Chess/Checkers tool-suite. The different diagrams depict the computational performance, the power consumption, and the energy that is needed to process one data sample. After profiling the machine code for the single-ALU architecture, it was clear that too many cycles and program memory accesses were spent in the calculation of critical functions such as Galois-field multiplications and the bit-manipulation operations in the Reed-Solomon algorithm. These functions were initially implemented in software on the ALU. In a second design iteration, the designer extended the architecture with a dedicated functional unit capable of computing Galois-field multiplications in a single cycle. This resulted in a moderate increase of the computational performance and an important reduction of the power consumption. In a third iteration, the designer additionally allocated a dedicated functional unit for bit manipulation. This allowed the offloading of the ALU significantly, resulting in a major performance improvement and, likewise, a reduction of the energy needed per data sample.

For comparison, Figure 36.4 also illustrates the characteristics of a hardwired ASIC core for Reed-Solomon encoding, developed in the same process technology. As can be observed, the ASIC core's characteristics are close to the third alternative designed with the Chess/Checkers tools. This comparison illustrates that the Chess/Checkers tool-suite can span a wide range of processor architectures, from general-purpose microprocessors to programmable ASICs. The designer can perform a true architectural exploration and get rapid feedback about the quality of the intermediate results.

36.2.4 Power-Conscious Architectural Design

As explained previously, Chess/Checkers supports an interactive methodology for architecture design. This approach is based on the assumption that automatically generated architectures can never approach the specialization of a human designer. Instead of automating the architecture generation phase within a restricted architectural scope, Chess/Checkers relies on the designer's creativity while supporting a wide scope of processor architectures. This section elaborates on how the Chess/Checkers tool-suite can be used to design power-efficient processor architectures.

When defining an architecture, the designer makes the basic decisions that influence the power efficiency of the architecture. Optimization for power is supported by the tools in the following ways:

- The Chess compiler primarily aims at optimizing the cycle count of the program, with instruction count or code size as the secondary optimization goal. As explained in the introduction, this generally contributes to low-power consumption. With a low cycle count, it is easier to fit a low V_{dd} and f_{Clock}, while a low instruction count reduces the power dissipated in program memory accesses.
- Note that the length of a clock cycle is not known *a priori*, when modeling an architecture in nML. Typically, the designer can make an estimate, but this needs to be verified by running the HDL generator Go and performing logic synthesis on the generated description.
- The architectural scope of the Chess/Checkers tools and of the nML language is wide enough so that the designer can experiment with different architectural techniques for low power. These techniques are described in Section 36.3. In particular, the cycle and instruction count can be reduced by exploiting instruction-level parallelism, by bundling multiple functions in a single instruction, by exploiting special-purpose registers, and by designing highly encoded instruction sets. The Chess compiler contains various optimization phases to make efficient use of these features.
- The tools can give early feedback about cycle count and instruction count. This is mainly obtained through the profiling capability of the ISS. In addition, the designer can check the effective utilization of functional units and registers, and strip those that are not frequently used.
- It is possible to have the ISS automatically calculate an approximate power consumption figure when executing a program. Based on the nML processor model, the Checkers tool generates an ISS in the form of a C++ source program. The generated model is open enough so that the user can integrate instruction-level power models in the ISS.
- Obviously, such power models are library and technology dependent. To construct and tune these models, a basic architecture can be specified in nML and small programs can be run that repeatedly execute specific instructions or instruction sequences, both in the ISS and in the derived HDL model using a tool such as the Synopsys Power Analyzer. The following experimental observations may serve as guidelines when constructing power models for application-specific DSPs and programmable ASICs:
  - In case of an orthogonal instruction format (see Section 36.2.2 on the architectural scope of Chess/Checkers), power models may be constructed per orthogonal subclass of the instruction word. By adding the power consumption of the orthogonal subclasses, a sufficiently accurate figure for the overall power consumption is obtained.
  - Within an instruction class (e.g., `alu_inst` in Figure 36.3), the specific choice of opcodes (e.g., `add`, `sub`, `and`, and `or` in Figure 36.3) has a dominant effect on the relative power consumed by the instruction. In contrast, the choice of operands or results (e.g., `R[0]` vs. `R[1]` as the source or destination register, and the exact bit-pattern of the immediate constant `val` in Figure 36.3) is much less relevant. Therefore, the choice of opcodes is an important parameter in a power model, while the choice of operands or results may be neglected more easily.
  - Power models are best defined for small sequences of instructions, instead of for individual instructions [8]. Experiments have demonstrated that power calculations in the ISS based on power models for pairs of instructions can be within 30% of the actual power of the circuit (as obtained in a gate-level simulations). In case only power models for individual instructions are used, however, the results can differ as much as 80%. To reduce the complexity of the model, the order of the sequence may be neglected (i.e., one may assume that the power consumed by the sequence A|B is the same as for B|A).

- The HDL generator Go contains a number of optimizations that contribute to a power-efficient hardware implementation of the processor core. For example, Go can generate write-enable signals for selected registers, which allows commercially available logic synthesis tools to introduce clock gating to reduce power dissipation. In addition, Go is able to latch the inputs of unused functional units, to prevent toggling of unused logic and thus save power.

36.3 Low-Power Processor Architecture Design

This section addresses some common issues covering the design of low-power processors and introduces how such a design may proceed in practice. The general area of low-power processor architecture design is potentially a very broad subject, and we will limit the discussion to that of embedded, low-power, DSP design and specifically focus on the processor core itself. It should be pointed out that the retargetable tool-suite presented in Section 36.2 is not limited to this domain, and this choice is driven by our actual design experience with a relatively general-purpose audio processor called CoolFlux DSP, which is described in Section 36.4.

36.3.1 General Characteristics

When considering different low-power processor core architectures, it is worth having some sort of power-aware metric with which to compare them. We have used simple metrics that consider some key aspects of a processor core. An example of such a figure of merit for a specific application could be:

$$cost = (m_{app}/m_{max}) \times P \times A$$

where m_{app} is the minimum clock frequency required for the application to achieve real-time operation, m_{max} is the maximum clock frequency of which the processor core is capable, P is the power per MHz, and A is the complete area, including memory. For this particular metric minimizing the cost would be a goal. It is worth noting that this formula contains conflicting factors, so that, for example, increasing parallelism and thus A is likely to decrease m_{app} due to the ability to exploit ILP. Thus, in many respects, an optimized processor core architecture has to find some good compromises among conflicting requirements. This section covers some key aspects regarding the broad architectural choices that have to be made at an early stage.

- Parallelism. Here, we have to consider the type and amount of parallelism a single processor core node will support. A low-power processor core design should try to approach the ideal power/parallelism characteristic as illustrated in Figure 36.5. This is based on minimizing control overhead. The two main types of parallelism exploited here are ILP and data parallelism. ILP is related to the number of operations an instruction can issue to functional units and the pipeline depth of the various functional units. These two factors will define the number of operations in flight at any one time. Given the need for efficient compiler support and the ILP potential of the applications, parallelism needs to also be well balanced in the architectural exploration. We have obtained efficient compilation results for a machine that issued up to eight basic operations per instruction and had four pipeline stages. For many DSP applications data parallelism, as supported in a single-instruction multiple-data (SIMD) machine, is a particularly efficient approach to use as the processor core control overhead is further amortized over several operations on sub-words. Thus, a 32-bit base architecture could also define instructions that perform four 8-bit operations in parallel. This is easily supported by the retargetable tool-suite by the definition of vector data types and operations.
- Pipeline structure. This is a key aspect of low-power processor core design, and many factors need to balance well here. We spent a lot of architectural exploration time in this area when designing our audio DSP (see Section 36.4). We have found that relatively short, simple pipelines (i.e., three to five stages) with limited interlocks and bypassing have given us good

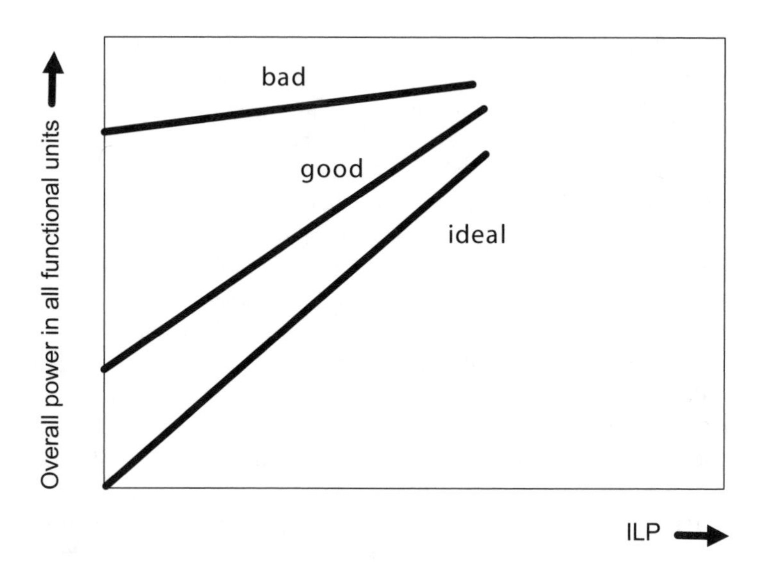

FIGURE 36.5 General relationships between power and ILP used in a processor core (assuming V_{dd} and f_{Clock} are kept constant).

results, with minimal design and verification effort. Instruction issue policy is strictly in order. Generally, pipeline depth has a quickly diminishing improvement on performance. At the same time, cost factors, such as design and verification effort, tend to increase rapidly, as is illustrated in Figure 36.6. As an example of many of the factors concerned, a search for the best pipeline depth would have to consider the following elements: the speedup possible for the system clock (this gives a power advantage from the larger potential voltage scaling with its quadratic power reduction), reduced cycle efficiency due to extra delay slots (or the need for extra bypasses), the potential deglitching effects of pipeline registers, the extra power cost of the clock tree, and the pipeline schedule and length of control transfers. A final comment on pipeline structure is that the CHESS compiler is good at resolving static pipeline scheduling issues. Thus, the onus is on providing a rather exposed pipeline and allowing a minimal hardware solution that is good for low power.

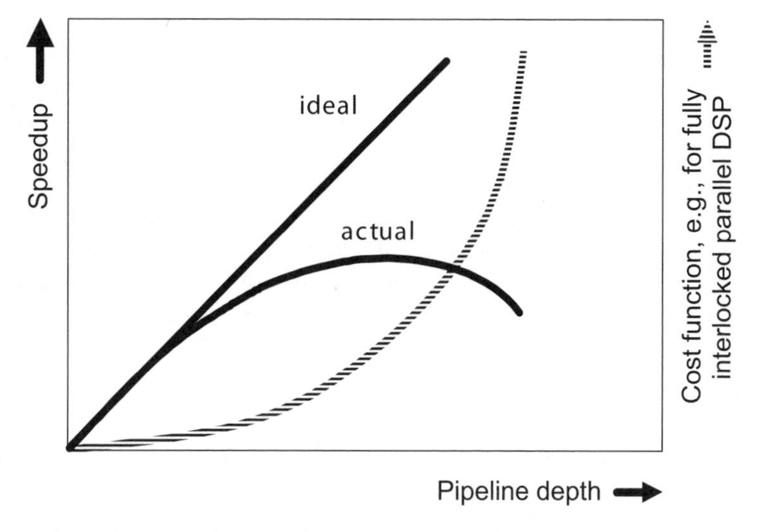

FIGURE 36.6 Speedup and cost as a function of pipeline depth.

- Register file structure. The register file of a processor core can be homogeneous (a central structure) or may be fully heterogeneous (distributed) at the two extremes. For low-power DSP applications, a distributed register file is desirable because this reduces the number of register ports and the sizes of the various files, as well as leveraging locality of storage. The costs associated with a single central register file serving many functional units over many (probably bypassed) pipeline stages are usually prohibitive for a low-power machine. With distributed register files, particularly within the main datapath, however, comes the problem of efficient register allocation and scheduling. Our experience here is that Chess is particularly well positioned to solve this problem and has allowed us to use this low-power distributed register file feature in our designs.
- Memory architecture. Memory access can now be over 50% of the power budget of an embedded DSP, and is likely to rise as geometry shrinks and applications grow. Several issues must be addressed here including aspects of memory hierarchy and separate memory data spaces. The second issue is a standard feature of most DSP designs and will not be elaborated here. Most low-power machines should exploit a data memory hierarchy of some sort, which may span main storage, local tight coupling between memory and register files [16]. Here, intelligent prefetch techniques can have significant advantages over traditional caches, particularly for applications such as video that combine statically known addressing patterns and large data objects.
- For program memory, a small cache, or at least a loop buffer, will reduce overall program memory power consumption for the 20/80 code encountered in DSP applications, as will any techniques used to reduce program memory size. Compact instruction encoding should be sought to reduce the program memory's size. In addition, we have used compression of the program memory contents coupled with the good code density already produced by the Chess compiler to aggressively reduce the program memory footprints.
- Memory addressing. For DSP applications, a well-established set of extended addressing modes are used, usually operating as a post modify on the relevant pointer register. These greatly contribute to reducing the cycle count required for the application, and thus indirectly to low power. Several trade-offs that balance the number of addressing modes against the complexity of the addressing units are possible here. The likely address modes include facilities for cyclic addressing as well as bit reversed addressing for fast Fourier transforms (FFTs) and other butterfly-based computations. To support a C compiler and to maintain efficient data structures that minimize use of data memory, good support of a software stack is also desirable. Thanks to the software stack concept, both the data memory size and the cycle count can be reduced, which contributes to low power. Stack support can be provided in a number of ways. We have typically used a fully indexed stack with a dedicated stack pointer.
- ISA. DSP designs encompass a wide variety of ISA design styles, from orthogonal to highly encoded. For low-power applications, we have favored highly encoded ISA design styles. This approach minimizes program memory size while providing enough parallel instruction classes for the ILP extracting Chess compiler to operate efficiently. Although we provided symmetry across the various parallel views of the machine, parallel operations were only introduced for the common forms of parallelism in DSP applications, such as multiply-accumulate (MAC)-based inner loop kernels. This has also allowed us to have a rather asymmetric datapath where most of the functionality is in the primary ALU, thus allowing a relatively compact design. Some of these issues are further expanded in the Section 36.3.2.

36.3.2 Instruction-Set Architecture

A low-power processor has a carefully optimized ISA, which attempts to strike a fine balance among code size, encoding, instruction decoder complexity, and compiler efficiency in scheduling ILP. This is quite a difficult balance to achieve due to the requirement to maintain enough parallelism in much of the ISA to keep compilation efficient, while maintaining a short instruction word and thus small program memory footprint.

Because DSP code is characterized by the 20/80 rule, there is a need to support rather diverse requirements. The ISA of the processor core can be thought of as having several distinct facets, in the form of instruction classes that implement various styles of computation. For example, in our audio DSP design, there is a relatively non-parallel microcontroller like facet as well as one that codes for maximum parallelism in DSP kernels.

Considering the design of the ISA has repercussions throughout the processor core design, and as far as producing a low-power instruction decoder is concerned, it is important to use regular formats where possible to minimize the amount of field extraction multiplexing that is needed. In addition, to maintain a good pipeline timing balance, certain encoding styles can be used that allow the fast production of time critical control signals and a spread of distributed decoding functions across pipeline stages.

Another issue we addressed aggressively is the potential inefficiencies due to flow control instructions, such as branches. As control flow instructions can occur up to about once every five instructions in general compiled C code, it is very important to maintain high cycle efficiency here. A number of techniques can be used that minimize the power consumption of the processor core. We typically provide a good mixture of flow control instructions with and without exposed delay slots. This allows the Chess compiler to make good code selection choices based on whether delay slots can be scheduled efficiently. We also provide zero overhead hardware looping to maintain high efficiency within inner loop constructs. Again, the compiler handles this automatically and forms software-based loops when the hardware loop stack is fully utilized. Another feature we use is conditional execution of instructions, which eliminates the use of a branch construct and have no exposed delay slots.

Generally, it is best to try to avoid the explicit coding of no-operation (NOP) instructions. This is partly aided by the options that have been provided on flow control instructions, but we have also added a form of NOP compression to some of our designs, which reduces our program memory size by up to 25% for typical compiled applications in our audio DSP. These savings have a significant impact on power and area: we measure an average 25.3 bits/instruction for typical compiled code for our CoolFlux DSP, while the instruction width is 32 bits.

36.3.3 Micro-Architecture

A low-power micro-architecture should attempt to minimize control overhead while keeping the main datapath as efficient as possible, within the bounds of the technology used. This means that pipeline interlocks and bypass networks should be used only when necessary. It will also be useful to run candidate designs right through placement and routing to ensure that cell row utilization during chip layout can be maintained as this will reduce area and thus the capacitance and power associated with many nets while improving timing.

From a clocking perspective, a single edge clocked synchronous design can achieve a better timing balance and clock tree efficiency than designs using both clock edges. We have always tended to carefully limit the number of overall registers in a design in order to keep control of the clock tree size and its significant power consumption (up to 40% of processor core power for semi-custom VLSI design flows). A low-power design will use the standard techniques of micro-clock gating and operand isolation that are now available with many synthesis tools. The Go HDL generator in the Chess/Checkers tool-suite is capable of selectively enabling these techniques in the generated HDL design. These are standard techniques and will not be further discussed here.

The pipeline should already be designed so that good timing balance is achievable and when implementing the micro-architecture this goal must be furthered through the VLSI design flow. Usually, for critical sections, this will mean attention at RTL source, synthesis, and back end of the flow. For the memory subsystems, this is particularly important because many critical timing paths are likely to be present here. A common solution to this problem is to use a write-back buffer, which schedules writes to memory only when free access slots are available. This technique allows a full clock cycle to be allowed for memory access.

The control signals from the instruction pipeline should be held until they are actually needed by a functional unit. From a power perspective it is detrimental to toggle, for example, multiplier input selection lines unless an instruction specifies a valid multiplier operation [14]. If possible, the instruction decoder itself should be designed to minimize internal toggling; this may be achieved by using a distributed design, for example, by instruction class.

Many of the final micro-architectural optimizations are made when the VLSI flow is exercised. This is particularly true for issues such as unnecessarily toggling control logic and optimization of critical timing paths.

36.3.4 Methodology

We have used a design methodology that attempts to provide as much information as possible to the processor core system architect, so that fine design trade-offs can be made using real simulation data. Therefore, we have established a full VLSI design flow as well as having the retargetable compiler tool-suite in place at a very early stage in the project. Thus, RTL design has proceeded in parallel with processor core architectural exploration, and this has allowed very useful insights to be had. These activities have also tended to bond the team together through having access to a common design database.

Most architectural exploration is done within the retargetable tool-flow environment by compiling a suite of applications with Chess and performing profiling with the Checkers ISS. This loop will include changes to the nML processor description, typically things such as ISA, pipeline schedules, and internal computational resources as well as optimization of the application source code are explored here. Good architectural candidates are pushed through synthesis, test insertion, and occasionally full layout to ensure that no problems occur with the realization of the complete processor core. We are particularly interested in maintaining high efficiency through physical design.

This VLSI design flow is power aware, and we use the accurate gate-level power simulator Diesel, which was developed by Philips. This simulator is driven by actual simulation vectors, and typically reaches accuracy within 10% of final silicon for the technologies in use. We had experimented with RTL power estimators before, but have kept with the gate level simulator because we typically needed early area and timing figures anyway. Once the flow is scripted, it is easy to get some accurate figures in an overnight run. This design flow is also complete in that full layouts are produced, and we use fully extracted (Hyperextract 3D) parasitic data when producing final area, timing, and power figures.

The specification of a design has to be carefully managed. We have tended to take a minimalist approach, where additional features are only added if they demonstrate a significant performance/cost benefit. Another factor in this process has been to lock the specification once confidence has been established; any changes beyond this are handled by a strict change request procedure. A factor we have particularly avoided has been "creeping" specifications.

We have been fortunate in being able to build a small effective team of like-minded engineers run by a single system architect who makes any final design related decisions. Of particular significance has been the synergy between engineers at Philips and Target Compiler Technologies and the close cooperation that was achieved. Perhaps a small, tightly knit team is reflected in a compact power optimal processor core design.

36.4 An Ultra-Low Power DSP for Audio Coding Applications

This section illustrates some of the principles outlined before, by considering a few aspects of the actual design of a low-power C programmable audio DSP, the CoolFlux DSP, developed within Philips PDSL.

36.4.1 Background and Goals

The CoolFlux DSP audio core was designed with two main objectives in mind:

1. The need for very low-power consumption
2. The need to be efficiently programmable with the C high-level language

The extra productivity advantages of programming in C outweighed the minor power increase for efficiently providing C language support in the processor core.

The project was actually also a first introduction into the codesign of the ISA and compiler using a retargetable compiler. In many ways, the question of how low-power a general-purpose high-level language programmable DSP could be made was being addressed (e.g., did we have to include application-specific features (in this case for MP3 coding) or programming restrictions to meet our power goals?).

The motivation behind this is that power consumption continues to be a very important issue in applications. This is due to the increase in portable products on the one hand and their higher processor core needs on the other hand. The portable MP3 players that are appearing in the market are a good example. The processor core throughput requirements for these devices increase rapidly. The application algorithms become increasingly computationally demanding and a larger number of functions need to be executed. At the same time, product lifetimes are decreasing, and this puts increasing pressure on product development cycle times, particularly the software component of these projects.

These factors and a large code base of 24/56-bit fixed-point applications were the starting specification points for the CoolFlux DSP. An MP3 decoder program was used as the main driver application because this gave a good mix of unstructured "control-like" code as well as some tight DSP kernels with high ILP potential.

36.4.2 Architecture

The CoolFlux DSP architecture is depicted in Figure 36.7. It is a dual MAC, dual Harvard machine capable of sustaining two MACs, two memory operations, and two pointer updates per instruction making it highly cycle efficient for computationally intensive DSP applications when scheduled by Chess. Due to the unique ISA design, the CoolFlux DSP is also highly efficient at supporting unstructured "control-type" code as well making it very well balanced for complete embedded applications in which 20/80 code mix is typical.

Finally, much attention has been paid to efficiently supporting the operations needed by ANSI C making the CoolFlux DSP an excellent compiler target while maintaining very low-power consumption through the careful realization of the underlying micro-architecture and the entire design flow.

The datapath consists of an X and a Y data processing side (Figure 36.7). The two sides are highly asymmetric. The main arithmetic components that are available are two multipliers, two full ALUs, and two rounding and saturation units.

The X multiplier is coupled to a preadder, thus the third small ALU. This ALU (ALU0) can perform a nonsaturating addition or a subtraction of two datapath registers. The result of this operation is then multiplied by another datapath register. The X multiplier performs all useful combinations of signed and unsigned multiplication. This allows efficient C type support as well as enabling higher precision arithmetic if needed. The Y multiplier is a lot simpler than the one on the X side. There is no preadder function available and only a signed/signed multiplication is possible.

The X ALU is the main ALU in the processor core. It has full 56-bit precision as well as efficient support for C `long` and `int` types. The operations supported are quite extensive and include DSP specific functions such as absolute and maximum. Division support and full four-quadrant division is also efficiently supported from C. The X ALU also primarily generates the condition code flags. The Y ALU is much simpler and has 56-bit precision. It only supports a subset of the X ALU operations. The main datapath has four distributed register banks: the X, Y registers and the A and B accumulators.

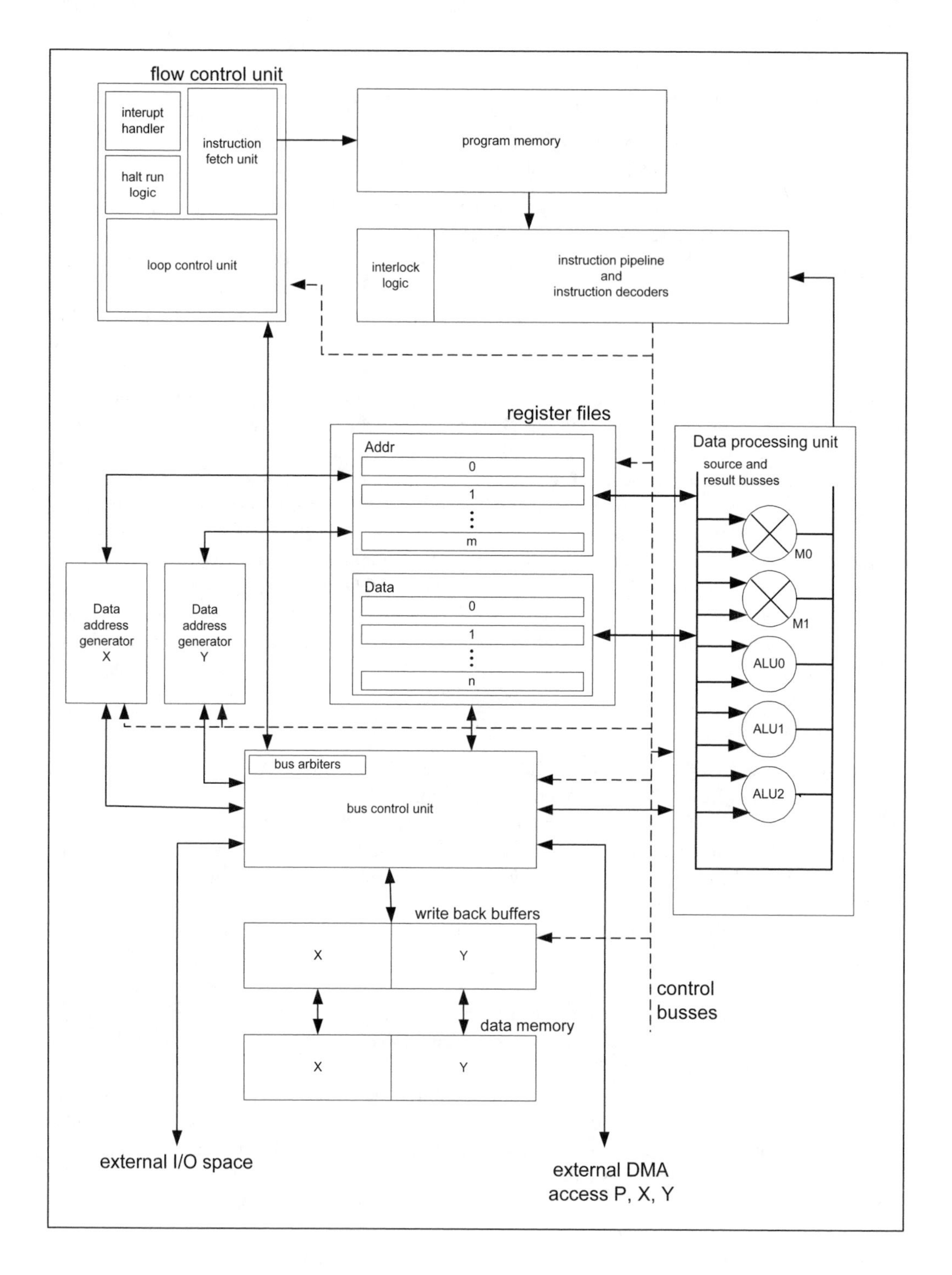

FIGURE 36.7 Block diagram of CoolFlux DSP audio core.

The bus control unit supports moves within the machine. This unit has been designed as a central bus switch used to exchange the contents within and between the X and Y busses as well as supporting move operations on each side. Considerable design effort was spent on this unit to ensure it was power efficient as well as meeting the stringent timing requirements needed.

Besides the main datapath unit, an address generation unit is also used. Two separate units are available for the X and Y memories, respectively. These units contain the address registers, the modulo-protected, and indexing ALUs, as well as the bit reverse addressing logic.

The flow control unit controls the processor core. This fetches the instruction stream, decodes it, and schedules the pipelines by issuing control signals to the entire machine. Interrupts are also processed here, and we have implemented a low latency, vectored interrupt system. This system guarantees interrupt response within a limited number of cycles thus easing the buffering needs of external devices. We support fully interruptible hardware loops, even if they only contain one instruction.

A very important unit is the loop control unit, which provides zero-overhead hardware looping. A maximum of four nested loops is supported, although this can be set as a parameter. This loop control unit is very efficient, both in software overhead and in real hardware parameters such as area and power.

36.4.3 Low-Power Techniques

The processor core uses all of the standard techniques for low-power design, mainly based around micro-clock gating, operand isolation, and general unnecessary toggle reduction techniques. These will not be elaborated here. We also use a technology library, including SRAM, which allows use of aggressive voltage scaling to below 1V.

A pipeline structure was developed that allowed use of a single edge clock while giving memory access a full clock cycle. This will tend to maximize the execution clock rate, thus allowing the most scope for voltage scaling, which gives approximately quadratic power reduction. The pipeline structure is also finely balanced with the rest of the micro-architecture design resulting in minimal and simple interlocks, minimal bypassing, and minimal length control transfers among pipeline segments.

The pipeline also leverages the ISA structure by utilizing a small number of distributed instruction decoders as opposed to one large one, thus reducing unnecessary logic toggling. In addition, some instruction encoding techniques were used to attempt to minimize unneeded logic toggling [17]. Finally, instructions are included that can put the core into various sleep modes, including a deep sleep where the clock can be deactivated.

The processor core uses distributed register files, local to their respective computational resources, to reduce power consumption. The ISA also includes some coupling mechanisms that attempt to reduce the need for copies among register files. Another factor is that the pipeline structure of the processor core was designed so as to minimize the need for bypassing mechanisms. The advantages all add up when compared with using a single multi-ported central register file. For a machine with the parallelism that is available within our processor core, this is a large advantage in both power and maximum clock frequency.

Significant area (cost) and power consumption is now evident in the memory subsystems of processor cores. Several techniques have been used to reduce both the size and power consumption of these memories. Globally, all memory spaces are made up of smaller physical segments such that only one is active in any space at any cycle.

Data memory utilization is optimized by allowing use of efficient data structures that are supported by powerful addressing modes within the processor core. Both cyclic and bit reversed addressing are supported as well as other common DSP addressing modes. Particularly efficient stack support is provided allowing efficient linkage and local storage for functions. These techniques ensure that data memory requirements are minimized as well as execution cycles. On the architectural front, a good balance has been sought between memory subsystem costs and the ability to provide sufficient memory bandwidth for the parallel computational units.

To increase the efficiency of the program memory, innovative techniques for the code size reduction have been included, on top of the very efficient code that is generated by the Chess C compiler anyway (i.e., an efficient method for code compression has been included). A major trade-off in designing the ISA was the width of the instruction word against the amount of encoding used while still leaving enough degrees of freedom for the compiler to perform code selection well. An ISA with key areas that were orthogonal was developed. This ISA was enhanced with a form of NOP compression, which used very little control logic. This has been leveraged by the use of scheduling techniques within the compiler, which favor the generation of instruction sequences, which allow maximal compression to be used. We have measured savings between 20 and 25% in code size for typical applications.

The processor core supports extensive I/O facilities allowing easy, efficient interfacing to other systems. Particularly, interrupts are implemented in a complete and robust way. The interrupt system is characterized by very low latency so that even single instruction hardware loops are fully interruptible. This allows a minimal amount of specific buffering to be implemented, thus keeping system costs low. This flexible I/O system and the ability of the retargetable environment to support intrinsic functions also mean that more application-specific accelerators can be easily added to increase the system power efficiency if needed.

The processor core is implemented in VHDL at RTL level and a power aware semi-custom ASIC VLSI design flow is used. The requirements for low power are carefully considered in all stages of the VLSI design flow as well, including the coding, synthesis, and layout areas.

36.4.4 Results

The CoolFlux DSP design has been initially realized in Philips' 0.18-μ CMOS technology. All the figures given are for simulation data; the timing, area, and power results are using fully extracted parasitic three-dimensional data from the layout database and not wire load estimates.

The power breakdown of the processor core is depicted in Figure 36.8. It demonstrates a low control overhead of approximately 14% in the processor control unit (PCU) logic and power that is dominated by the datapath components (DCU). The overall control overhead with program memory fetches

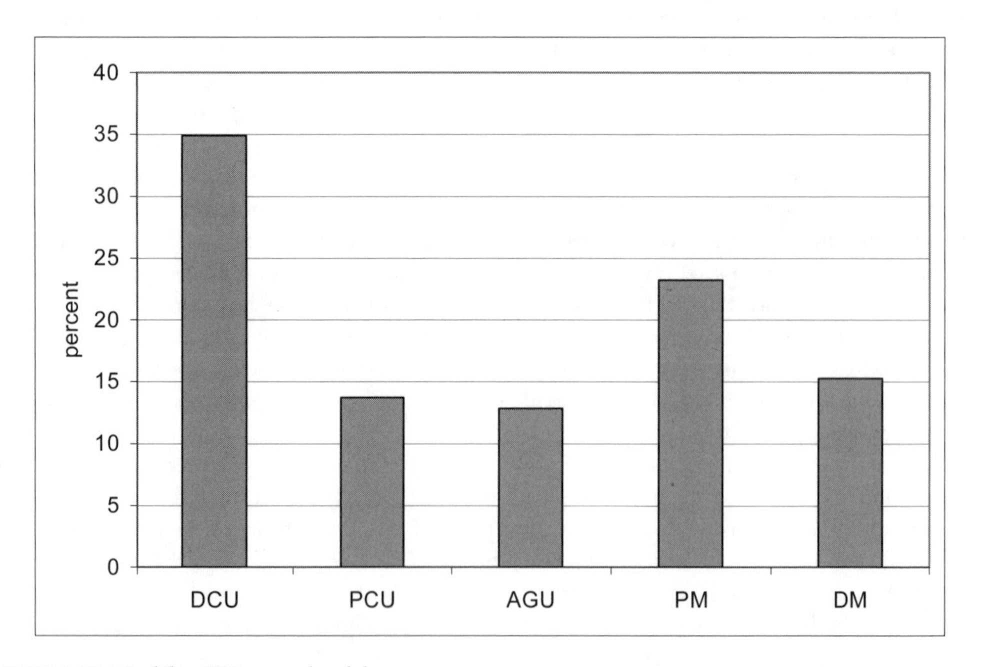

FIGURE 36.8 CoolFlux DSP power breakdown.

TABLE 36.1 CoolFlux DSP Performance Figures for MP3 Decoder

MP3 decode computation load needed	14.9 MIPS (128-Kbps, 44.1-KHz stereo material)
Program memory size	4.6 Kwords × 32 bits
Data memory size	10 Kwords × 24 bits (total)
MP3 core power consumption	1 mW (Philips 0.18 μ standard cell low-leakage CMOS, V_{dd} = 0.9 V, f_{Clock} = 15 MHz)
Maximum clock frequency	135 MHz (worst-case commercial)
Core size	Approximately 45 kGate (NAND2 equivalents)

included is some 37% of the total power consumption. The memories are represented by the program memories and data memories (PM, DM) and the address calculation units are in the AGU module.

Table 36.1 presents some figures obtained for the CoolFlux DSP for an MP3 decoder application and some general core area and timing parameters.

36.5 Conclusions

This chapter described a methodology for the design of low-power programmable processor cores in SoCs. The main idea that is explored is to customize the instruction-set architecture of the core. By making the core application-specific, efficient architectures with minimal overhead, running at lower clock frequencies and exploiting lower supply voltages, can be obtained while retaining much of the flexibility and programmability of a general-purpose processor.

The key infrastructure for making this methodology work in an industrial environment is a retargetable tool-flow that allows for quick and thorough exploration of the architectural design space as well as for efficient software development starting from C source code. In this chapter, the Chess/Checkers tool-suite from Target Compiler Technologies has been introduced for this purpose. In addition, a survey has been given of various architectural techniques that are beneficial for designing low-power processor cores. Most of the presented architectural optimizations are within the architectural solution space of the CHESS/CHECKERS tool-suite.

The effectiveness of the methodology has been demonstrated through the design, by Philips, of an ultra-low power processor core for audio coding applications called COOLFLUX DSP. As an example, this core can run MP3 decoding from compiled C code in less than 15 MIPS, which results in a processor core power consumption of about 1 mW in 0.18-μ standard cell technology.

36.6 Acknowledgment

The authors thank their colleagues at Easics N.V. for sharing some of the data on low-power design experiments.

References

[1] ARCtangent-A4 Core — A Technical Summary, ARC International Inc., http://www.arc.com, 2003.

[2] CHESS/CHECKERS: a retargetable tool-suite for embedded processors, Technical white paper, Target Compiler Technologies, http://www.retarget.com, June 2003.

[3] Xtensa Architecture and Performance, Tensilica Inc., http://www.tensilica.com, Sept. 2002.

[4] R. Camposano and J. Wilberg, Embedded system design, *Design Automation for Embedded Syst.*, Vol. 1, No. 1-2, pp. 5–50, Jan. 1996.

[5] A. Chandrakasan and R. Brodersen, *Low-Power Digital CMOS Design*, Kluwer Academic Publishers, Boston, 1995.

[6] P. Dytrych, M. Adé, J. Coninx, J. David, and P. Vandebroek, The design of a very low-power MP3 decoder accelerator, *DSP Valley Annu. Res. and Technol. Symp.*, Leuven, Belgium, Oct. 2002.

[7] A. Fauth, J. Van Praet, and M. Freericks, Describing instructions set processors using nML, *Proc. European Design and Test Conf.*, pp. 503–507, March 1995.
[8] M.T.-C. Lee, V. Tiwari, S. Malik, and M. Fujita, Power analysis and minimization techniques for embedded DSP software, *IEEE Trans. on VLSI Syst.*, Vol. 5, No. 1, pp. 123–133, March 1997.
[9] J.M. Rabaey and M. Pedram, *Low-Power Design Methodologies*, Kluwer Academic Publishers, Boston, 1995.
[10] J. Sato, M. Imai, T. Hakata, A. Alomary, and N. Hikichi, An integrated design environment for application specific integrated processor, *Proc. Int. Conf. Comput. Design*, pp. 414–417, Oct. 1991.
[11] D. Singh, J. Rabaey, M. Pedram, F. Catthoor, S. Raigopal, N. Seghal, and T. Mozdzen, Power conscious CAD tools and methodologies: a perspective, *Proc. IEEE, Special Issue on Low-Power Electron.*, Vol. 83, No. 4, 1995.
[12] R.M. Stallman, Gnu compiler collection internals, http://gcc.gnu.org, Dec. 2002.
[13] J. Sato, A. Alomary, Y. Honma, T. Nakata, A. Shiomi, N. Hikichi, and M. Imai, PEAS-I: a hardware/software codesign system for ASIP development, *IEICE Trans. on Fundamentals*, Vol. E77-A, No. 3, March 1994.
[14] C. Su, C. Tsui, and A. Despain, Low-power architecture design and compilation techniques for high-performance processors, *Proc. IEEE COMPCON*, Feb. 1994.
[15] J. Van Praet, D. Lanneer, W. Geurts, and G. Goossens, Processor modelling and code selection for retargetable compilation, *ACM Trans. on Design Automation of Electron. Syst.*, Vol. 6, No. 3, pp. 277–307, July 2001.
[16] F. Catthoor, *Custom Memory Management Methodology: Exploration of Memory Organisation for Embedded Multimedia System Design*, Kluwer Academic Publishers, Boston, 1998.
[17] S. Woo, J. Yoon, and J. Kim, Low-power instruction encoding techniques, School of Computer Science and Engineering, Seoul National University, undated.

37

Recent Advances in Low-Power Design and Functional Coverification Automation from the Earliest System-Level Design Stages

Thierry J.-F. Omnès
Philips Semiconductors

Youcef Bouchebaba
University of Nantes

Chidamber Kulkarni
University of California-Berkeley

Fabien Coelho
Ecole des Mines

37.1 Introduction

To meet the cost, power, performance, and programmability constraints of next-generation multimedia devices and platforms in a reasonable design and verification time, introducing a system specification cleaning engine so-called the software washing machine by IMEC's Hugo de Man is key [23]. At this highest level (and first step) in the overall system-level design and verification flow, automation is a very

0-8493-1941-2/05/$0.00+$1.50

difficult problem because system architects typically prefer the expressiveness of C/C++ to the powerful semantics of synchronous languages [6] such as Esterel [16], Lustre [32], and Signal [49], using tools such as Esterel Studio [26], Polychrony [42], and Simulink. They also prefer the expressiveness of C/C++ to the powerful mathematics of geometric data-flow modeling available from languages such as Fortran, Alpha [21], and tools such as paralléliseur interprocedural de programmes scientifiques (PIPS) [37] and MMAlpha [47]. Because data access and transfer have the biggest impact on the cost, power, and performance of embedded systems, bridging the gap between geometric data-flow modeling and C/C++ requires special attention. Section 37.2 tours basic geometric transformations to motivate the introduction of a novel and advanced low-power optimization engine, based on PIPS, with the ability to take restricted but still C code as input. Beyond the productivity gain achieved by automation, we observe superior power savings than typically obtained using the systematic but manual code rewriting techniques that stand for best practice in this field today.

Another well-known source of cost, power, and performance optimization is parallelism. Having performed our novel and advanced transformations from the earliest system-level design stage, we move to task- and data-level parallelism exploitation. On an Intel IXP1200 network processing platform, we partition an IPv4 forwarding application written in C into 4 tasks (so-called micro-engines) and 16 threads for exploiting data parallelism using a YAPI (Y-chart) approach. As illustrated in Section 37.3, this results again in important productivity gains while maintaining near-optimal performance on large packet lengths.

Because our advances rely on sophisticated C/C++ source code transformations and manual partitioning, it is essential to verify their functional correctness. Section 37.4 describes the co-verification methodology currently used within Philips' System and Software Design Environment (SSDE). It is the last but essential step to bridge the gap between advanced low-power techniques and production-quality embedded system design and verification.

37.2 Advanced Loop Transformations for Low Power

Research in the program transformation field has drawn much attention for several years. It consists in finding new techniques that allow the compiler to transform source code to optimize some criteria, such as parallelism, execution time, or data locality, which have a direct effect on the reduction of energy consumption [13,38]. The transformations described here aim at improving data locality to switch costly transfers from the main memory to cheaper cache or register memory. During program optimization, the greatest profit comes from the loop nest optimizations because they use the most time in computation of scientific programs. For many years, several techniques have been proposed to transform these nests. Among these techniques, tiling [36], fusion [39,65], and memory reallocation [24,28] can be cited. Tiling is a good technique for increasing the data locality, but most work of this technique is only dedicated to code with single loop nest. This chapter demonstrates how we combine all these techniques to apply them to sequences of nested loops [7–9]. The codes considered here are signal processing applications, which are sequences of loop nests of equal but arbitrary depth. Each of these nests uses a stencil of data produced in the previous nest, and the references to the same array are equal, up to a shift.

37.2.1 Input Code

As mentioned previously, the input code includes signal-processing applications, which are a sequence of loop nests as depicted in Figure 37.1. Note that the dependencies of this code form a directed acyclic graph, and each of these nests uses a stencil of data produced in the previous nest and represented by a set: $V^k = \{\overrightarrow{v_1^k}, \ldots \overrightarrow{v_{mk}^k}\}$.

Domain D_0 associated with array A_0 is defined by the user. To avoid illegal accesses to the various arrays, the domains D_k $(1 \le k \le p)$ are derived in the following way: $D_k = \{\vec{i} \,/\, \forall\ \vec{v} \in V^k : \vec{i} + \vec{v} \in D_{k-1}\}$.

We suppose that the vectors of the various stencils are lexicographically ordered, so that $\forall\ k : \overrightarrow{v_1^k} \le \cdots \le \overrightarrow{v_{mk}^k}$ ($\le$ is a lexicographic operator).

```
do i⃗ ∈ D_0
   A_1(i⃗) = F_1(A_0)
enddo
⋮
⋮
do i⃗ ∈ D_k
   A_k(i⃗) = F_k(A_{k-1})
enddo
```

FIGURE 37.1 General form of input code.

37.2.2 Loop Fusion

Loop fusion is a transformation technique that combines several loops into one loop. It has several advantages including:

1. Lower cost of loop bound testing [19]
2. Synchronization reduction when loops are distributed among different computation units [12]
3. Data locality increasing [44]

This chapter focuses on this last point, which consists of reducing data transfers among various levels of memory hierarchy, which has a direct effect on energy consumption [13,38].

Generally, the fusion of two loops is valid if and only if they have the same iteration domains and in the merged nest we do not create dependencies from instruction of the second nest to instruction of first nest such as: flow dependence, output dependence or anti-dependence [63]. To merge all nests of code in Figure 37.1, we should make sure that all elements of array A_{k-1} that are necessary for the computation of an element $A_k(\vec{i})$ at iteration $\vec{i}$ in the merged nest have already been computed by previous iterations. To satisfy this condition, we shift [34] the iteration domain of every nest by a delay $\vec{h_k}$. Our fusion will be valid if and only if these various delays satisfy the following condition [7–9]: $\vec{h_k} \geq \overrightarrow{h_{k+1}} - MAX_i(\overrightarrow{v_i^{k+1}})$.

The merged code after shifting the various iteration domains is given in Figure 37.2. S_k is the instruction label and $D_{iter} = \bigcup_{k=1}^{p}(D_k')$, where D_k' is domain D_k shifted by vector $\vec{h_k}$. As instruction S_k might not be executed at each iteration of domain D_{iter}, we guard it by condition: $C_k(\vec{i}) = if(\vec{i} \in D_k')$.

```
do i⃗ ∈ D_iter
   S_1: C_1(i⃗)  A_1(i⃗ − h⃗_1) = F_1(A_0)
   ⋮
   S_1:C_1(i⃗)  A_1(i⃗ − h⃗_1) = F_1(A_{k-1})
   ⋮
   S_0: C_0(i⃗)  A_0(i⃗ − h⃗_0) = F_0(A_{0-1})
enddo
```

FIGURE 37.2 Merged nest.

```
do i∈D_iter
  S_1: C_1(i) B_1(F_1(i))=F_1(A_0)
  ⋮
  S_2: C_2(i) B_2(F_2(i))=F_2(B_2-1)
  ⋮
  S_k: C_k(i) A_k(i-h_k)=F_k(B_k-1)
enddo
```

FIGURE 37.3 Merged nest with buffer allocation.

37.2.3 Fusion with Buffer Allocation

In practice, we only need the first and last arrays (A_o and A_p) because all others arrays only hold temporary data. Thus, we replace arrays A_1 .... A_{p-1} by circular buffers B_1 .... B_{p-1}. Buffer B_i is a one-dimensional array that will contain array A_i live data. This transformation saves memory space and avoids loading the same element several times.

37.2.4 Live Data

An element of array A_{k-1} is said to be live at iteration $\vec{i} \in D'_k$ if and only if:

1. It is produced in the domain D'_{k-1} at iteration $\vec{i}_1 \leq \vec{i}$.
2. It exists in the domain D'_k at iteration $\vec{i}_2 \geq \vec{i}$, which will consume it.

The memory volume $M_k(\vec{i})$ corresponding to an iteration $\vec{i} \in D'_k$ is then given by the number of elements of array A_{k-1} that are live at iteration $\vec{i}$. We can verify easily that $M_k(\vec{i})$ is bounded by a constant, which will be noted Sup_k.

37.2.5 Code Generation

As mentioned it earlier, each array A_k will be replaced by a buffer B_k, which will be managed in a circular and sequential way. The size of buffer B_k is given by Sup_{k+1} defined previously. To store and load the elements of array A_k in the buffer B_k we associate with it an access function: $F_k : D'_k \rightarrow N$ such that:

$$F_k(\vec{O}_k) = 0$$

$$F_k(SUCC(\vec{i}_k)) = (F_k(\vec{i}_k) \bmod Sup_{k+1}$$

The merged code after the replacement of different arrays by buffers will have the form of the code in Figure 37.3.

37.2.6 Tiling

Tiling is one of the most important techniques in the program transformation domain. Generally, it transforms a nest of depth n into another nest of depth $2n$. Much work has been done on tiling [36,62], but most of it is only dedicated to a single loop nest. This chapter presents a simple and effective method that simultaneously applies tiling with fusion to a sequence of loop nest. We are interested only in data that live in the cache memory, thus our tiling is at one level. Our tiling is used as loop transformation and is represented by two matrices: a matrix $A(n, 2n)$ that gives the various coefficients of tiles and a

```
do i⃗
  S1: C1(Ap⁻¹l⃗) A1(Ap⁻¹l⃗−h⃗1)=F1(A0)
  ⋮
  S2: C2(Ap⁻¹l⃗) A2(Ap⁻¹l⃗−h⃗2)=F2(A2-1)
  ⋮
  Sk: Ck(Ap⁻¹l⃗) Ak(Ap⁻¹l⃗−h⃗k)=Fk(Ak-1)
enddo
```

FIGURE 37.4 Tiled code.

permutation matrix $P(2n, 2n)$ that allows for the specification of the organization of tiles and the iterations inside these tiles. As with fusion, the first step before applying tiling with fusion is to shift the iteration domain of every nest by a delay $\overrightarrow{h_k}$. We note by D_k' the shift of domain D_k by vector $\overrightarrow{h_k}$.

37.2.7 Matrix *A*(*n*, 2*n*)

Matrix $A(n, 2n)$ defines the tile size and allows us to transform every point $\vec{i} \in Z^n$ into a point $\vec{i}' \in Z^{2n}$. In the vector, $\vec{i}'$ we have two types of loops:

1. Loops that iterate over tiles
2. Loops that iterate over iterations inside tiles

All the elements of the ith line of this matrix are equal to zero except:

1. $a_{i,2i-1}$, which represents the size of tiles on the ith axis
2. $a_{i,2i}$, which is equal to 1

The relationship between $\vec{i}$ and $\vec{i}'$ is given by $\vec{i} = A\vec{i}'$.

37.2.7 Matrix *P*(2*n*, 2*n*)

Matrix $A(n, 2n)$ has no impact on the execution order of the initial code. Permutation matrix $P(2n, 2n)$ allows for:

1. The placement of all the loops that iterate over tiles before the loops that iterate over iterations inside tiles
2. The specification of the order in which the iterations will be executed

This matrix transforms every point $\vec{i}'$ into a point $\vec{l} \in Z^{2n}$ such as $\vec{l} = P\vec{i}'$. After the application of matrices $A(n, 2n)$ and $P(2n, 2n)$, we obtain the code in Figure 37.4.

37.2.9 Tiling as a Loop Transformation

Contrary to fusion without tiling, the computation of the various delays $\overrightarrow{h_k}$ in the case of fusion with tiling requires more effort. Our fusion with tiling will be represented as transformation:

$$\omega : Z^n_{\vec{i}} \rightarrow Z^{2n}_{\vec{l}}$$

As mentioned in our previous work [7–9], the simultaneous application of tiling with fusion to the input code is valid if and only if:

$$\forall k, \forall \vec{i} \in D_{k+1}, \forall \vec{v} \in V^{k+1} : \omega(\vec{i} + \vec{v} - \overrightarrow{h_{k+1}} + \overrightarrow{h_k}) \leq \omega(\vec{i}).$$

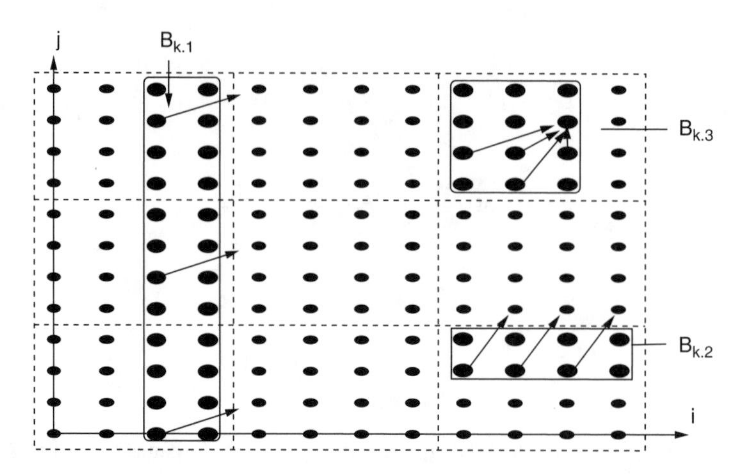

FIGURE 37.5 Multiple buffer allocation.

This condition means that each data-producing iteration must be computed before the iteration that consumes it.

37.2.10 Buffer Allocation

In the case of fusion, we suggest replacing arrays $A_1 \ldots. A_{p-1}$ by circular buffers $B_1 \ldots. B_{p-1}$. A buffer B_i is a one-dimensional array that contains the live data of array A_i. In the case of fusion with tiling, this technique has two drawbacks:

1. Dead data are stored in these buffers to simplify access functions
2. The size of these buffers increases when the tile becomes large

To eliminate these two problems, we replace every array A_k by $2n+1$ buffers. As mentioned earlier, tiling allows transforming a nest of depth n into another of depth $2n$. The n external loops iterate over tiles, while the n internal loops iterate over iterations inside these tiles. For every external loop m, we associate two buffers $B_{k,m}$ and $B'_{k,m}$ k (corresponds to array A_k); for all internal loops, we define single buffer $B_{k,n+1}$, which contains the live data inside the same tile. For example, if the depth of the nests is two, every array A_k will be replaced by five buffers:

- Buffer $B_{k,1}$ contains the data produced by a column of tiles, which will be consumed by the following column.
- Buffer $B_{k,2}$ contains the data produced in a tile, which will be consumed by the following tile.
- Buffer $B_{k,3}$ contains the live elements in the same tile, and it is managed as the circular buffer for the fusion.

In a given tile, we use data that are produced in the previous tile, and we produce other data that will be consumed in the following tile. To avoid destroying data in the buffer $B_{k,2}$, we duplicate it by another buffer $B_{k,2}$. For this same reason, we duplicate the buffer $B_{k,1}$ by another buffer $B_{k,1}$.

37.2.11 Implementation and Tests

As mentioned in the introduction (Section 37.1), our goal is to reduce the energy consumption in signal processing applications, which strongly depends on data transfers between the various levels of the memory hierarchy. The tests presented here are the numbers of cache misses and execution times of various transformations. We have considered only the external cache because it is the only one that generates data transfers between the processor chip and its environment. The two measurements are carried out for Sun Blade 1000, based on a microprocessor UltraSPARCIII with and 8-MB external cache

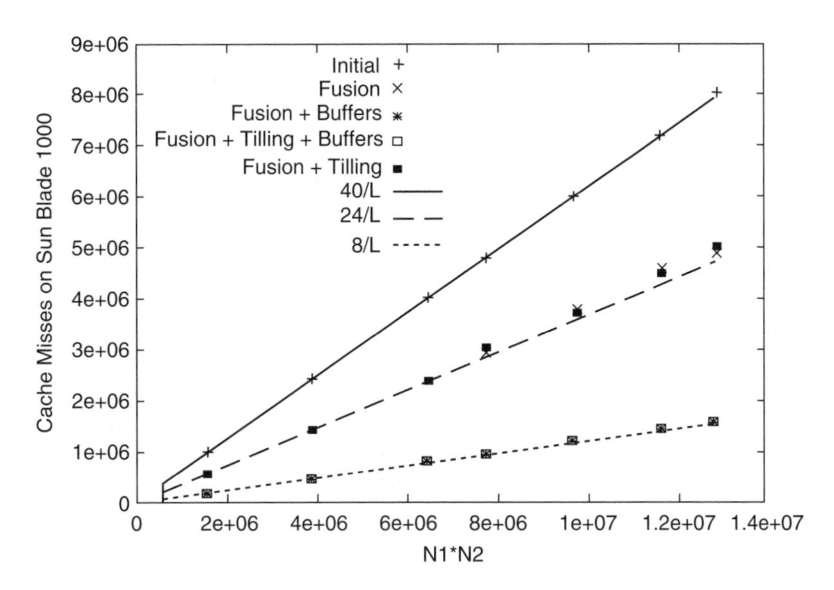

FIGURE 37.6 External cache misses.

and 750 MHZ of clock frequency. The code on which we made our experiments is real application; cavity detection code provided by IMEC, which is a sequence of five nested loops.

37.2.12 Cache Misses

The Figure 37.6 plots cache misses according to array size on Sun Blade 1000. The lines of rate 40/*L*, 24/*L*, and 8/*L* (L is the size of external cache lines) represent theoretical values for cache misses respectively of the initial code, the merged code (tiled and merged code), and the merged code with buffer allocation (merged and tiled code with buffer allocation). As one can observe from Figure 37.6:

- Buffer allocation in fusion and in fusion with tiling decreases considerably the number of external cache misses by almost a factor of 5 when compared with the initial code.
- All tests follow asymptotically well-defined lines that correspond to expected theoretical results.

37.2.13 Execution Time

The objective of our research was to increase data locality. We nevertheless measured the execution times of different transformations of our application. Figure 37.7 gives the execution times according to the array size on Sun-Blade-1000. Notice that the merged code with buffer allocation increases considerably the execution time. This increase is foreseeable because the modulo used in access functions is time-consuming. To improve the execution time of this code we can either use powers of 2 as buffer sizes or eliminate the modulo by unrolling the loops. Figure 37.8 contains the execution times of the various transformations when the buffer sizes are a power of 2.

37.2.14 Conclusion

Section 37.2 discussed the reduction of the energy consumption of signal processing applications executed in embedded systems. The reduction of the energy consumption requires a reduction of memory accesses as demonstrated by IMEC's work [14,30,64]. We have studied several program transformations improving the data locality, and we have extended them.

Much work on loop transformation has been done, but most of it is only dedicated to codes with single loop nests. This section combined loop fusion, tiling, loop shifting, and memory reallocation to apply them to sequence of nested loops. Our method consists in shifting each iteration domain by a delay

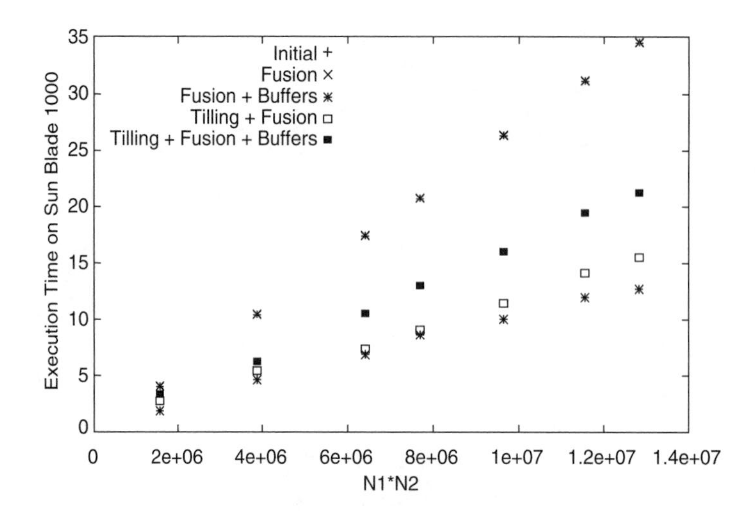

FIGURE 37.7 Execution time of various transformations on Sun-Blade1000.

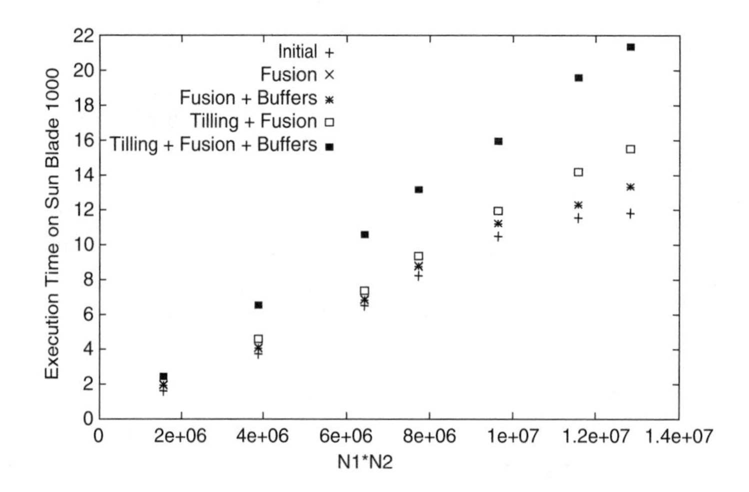

FIGURE 37.8 Reduction of time execution by changing the buffer size (size = power 2).

to ensure the legality of the simultaneous application of fusion with tiling. We then used the concept of live data to replace each array by:

1. Sequential and circular buffers in case of fusion
2. A set of buffers in case of fusion with tiling

All transformations described in this section have been implemented in our source-to-source prototype compiler PIPS, and we have carried out two measurements (cache misses and execution time) on target Sun Blade 1000 machines. These measurements demonstrate that the number of cache misses can be reduced by a factor of up to 5.

37.3 Exploiting Task-Level and Data-Level Parallelism on the Intel IXP1200

Network processors exploit task and packet level parallelism to achieve high throughput. To date, this has resulted in a huge diversity of architectures for similar applications. Driven by practical implementations. This section explores the different trade-offs in network processor design and implementation.

37.3.1 Introduction

Contemporary network processors (NPUs) exhibit a wide range of architectures for performing similar tasks: from simple reduced instruction set computer (RISC) cores with dedicated peripherals, in pipelined or parallel organization, to heterogeneous multiprocessors, based on complex multi-threaded cores with customized instruction sets. Although so diverse, all NPUs exploit task-level concurrency in applications by means of parallel programmable processing elements (PEs) to meet line speed requirements. Thus, the inter-PE communication and the topology of PEs are performance critical aspects of any NPU architecture.

Programming such concurrent systems remains an art. The programmer is not only required to partition and balance the load of the application manually among multiple PEs, it is also necessary to implement each task, often in assembly, before a reliable performance estimation can be obtained. Thus, a robust application mapping strategy for such architectures requires a balance between thread partitioning, scheduling, memory accesses, and input/output (I/O). With current tools, this task becomes time-consuming and error prone, due to trial-and-error methods employed by system implementers based on simulation runs. Therefore, topology, inter-PE communication, and the ease of mapping are likely to be key aspects of the quest for a natural programming model.

For the next generation of network processor based system implementations, we strongly believe that considerable emphasis will be put on performance per cost (e.g., power consumption) aspects and on support of appropriate programming models. Therefore, it is essential to identify and investigate limitations and bottlenecks in system implementation without going all the way down to complete implementations, as is the current practice. Thus, the use of high-level design space exploration and verification tools is required that support a wide range of heterogeneous architectures and enable precise reasoning about different implementation styles and their performance. The main goal of this section is to clearly understand the performance/cost trade-offs for network processor based implementations. In this process, we have implemented two differently mapped versions of our IPv4 benchmark [26] on IXP1200 to gain detailed insight into programmability of existing NPU architectures.

37.3.2 Performance Modeling and Evaluation

Our approach to design space exploration is based on the Y-chart approach. Separate descriptions of the application (workload) and the micro-architecture are bound to each other in an explicit mapping step, describing bindings of tasks and communication onto micro-architecture building blocks. The following evaluation of the system may manually or automatically trigger adaptations of the workload, the allocation of architecture building blocks, or the mapping of the application onto the architecture.

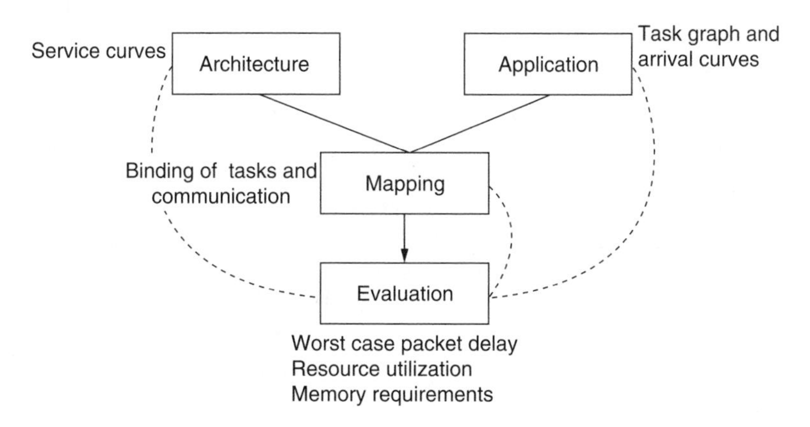

FIGURE 37.9 Design exploration using a Y-chart.

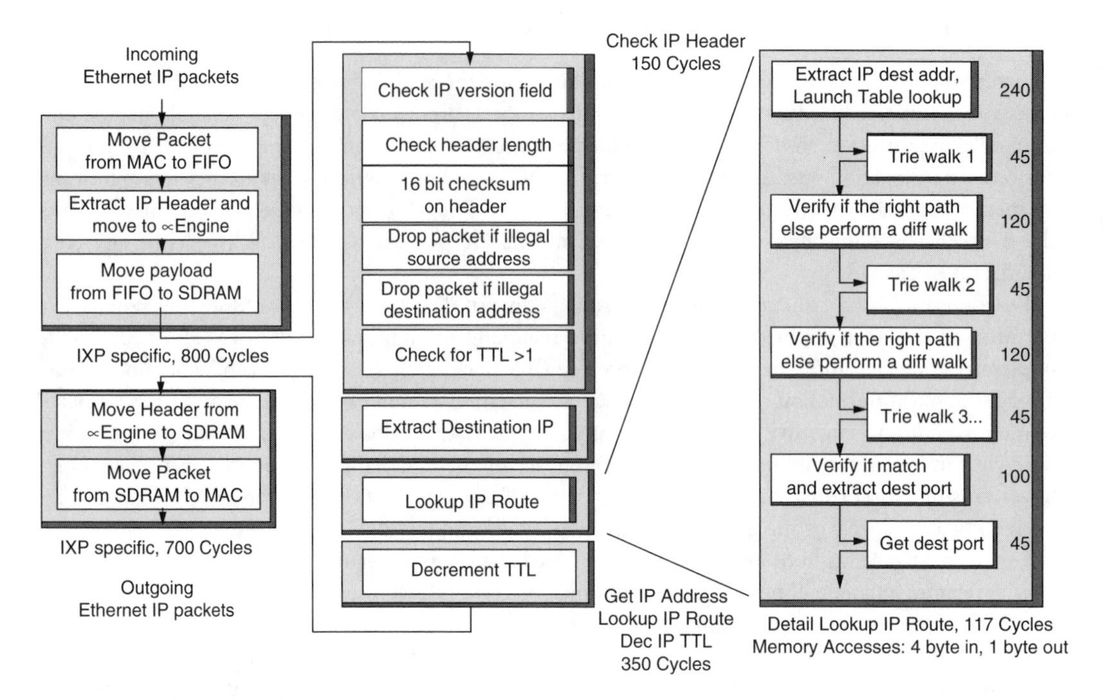

FIGURE 37.10 Instance of annotated IPv4 task graph derived from the application for analysis.

37.3.3 Modeling IPv4 Forwarding Application

A 16 fast Ethernet port IPv4 forwarding switch application is used in this work. Our functional specification of the application is based on RFC 1812 [59]. Figure 37.10 illustrates the main components in the functionality of our benchmark annotated with cycle counts for 64-byte packet size. It is important to note that in addition to the core functionality a number of steps are required to receive the packets from the external media access control (MAC) unit into the IXP1200 and extract the packet header, on which the previously stated operations are performed. Last, the modified packet header and the packet payload need to be written back into the external MAC unit via the IX bus unit. These additional operations, in fact, result in most of the programming effort for our application. For example, 14 detailed tasks are required to perform the core functionality of our benchmark, whereas we need 42 detailed tasks to perform the ingress and egress operations on each packet.

37.3.4 Modeling IXP1200 Architecture

The Intel IXP1200 network processor [48] is targeted for applications performing packet forwarding and classification at layer three and below of the open system interconnection (OSI) model. This section introduces only the main components of the IXP1200 utilized by our application as needed for modeling. The IXP1200 comprises six micro-engines, with four threads on every micro-engine, for computation. There are four unidirectional on-chip buses connecting both the off-chip memories (SRAM and SDRAM) to the micro-engines. External MAC units are connected to the IXP1200 via the IX Bus. The IX bus interface unit has the required logic and memories to receive and transmit packets from/to the external MAC unit. The IX bus unit has a scratchpad memory (SRAM) and two FIFO memories, with each having 16 entries of size 64 bytes. In addition, the SDRAM unit is connected to the IX bus unit via a separate on-chip bus, used to transfer packet payloads directly based on micro-engine commands, and last, an on-chip command bus carries events and signals between micro-engine and the IX bus unit.

This section focuses only on the data plane of the IXP1200 network processor. Thus, aspects related to the StrongARM processor are not modeled. In addition, we have not modeled the PCI bus interface

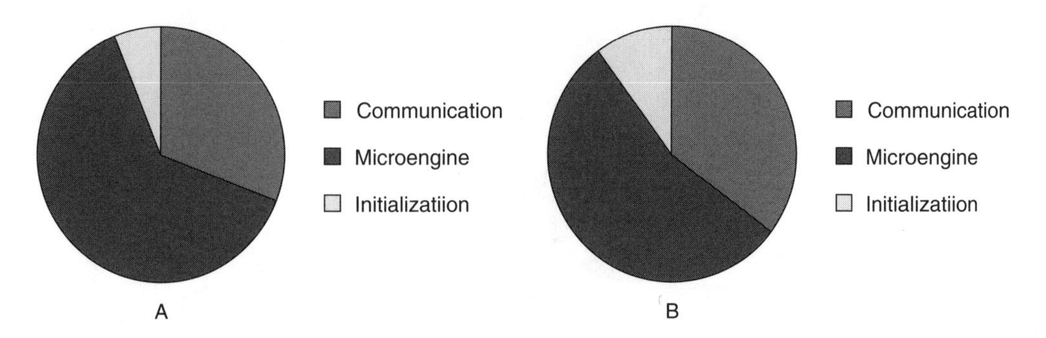

FIGURE 37.11 Assembly and u Engine C code distribution.

as well as the hash engine because we do not utilize these peripherals. A typical packet flow through the IXP1200 follows the steps given in Figure 37.10. Comparing results from analysis and simulation, Section 37.3.5 describes the experimental setup that we used for our implementation.

37.3.5 Experimental Setup

In our case study, we first developed the application in micro-engine C, following the preceding specification based on the Intel reference code. We made a few modifications to the transmit threads to improve the performance and to make the code stable and usable across different packet streams. In addition, we have used the assembly implementation from Intel reference code because it is hand tuned and most performing. The application was partitioned so that 16 threads on 4 micro-engines were assigned 1 port each on the receive (and forwarding) part. The transmit part of the application was assigned eight threads on two micro-engines. This partitioning holds because the end-to-end delay for a packet on the receive part is more than twice that on the transmit part. This implementation was used to derive the per-packet profiling information used to build the task graph for the network calculus-based approach.

Performance on the IXP1200 was measured using version 2.01 of the developer workbench assuming a clock frequency of 200 MHz; the IX bus is 64-bit wide and has a clock frequency of 80 MHz. Two IXF440 external MAC units (with eight duplex fast Ethernet ports each) are connected to the IX bus and Ethernet IP packets are streamed from this unit to the IXP1200 and back. The packets for the application contain destination addresses evenly distributed across the IPv4 32-bit address space. We employ different packet sizes, namely from 40 bytes to 256 bytes. A single packet source for each input port generates an evenly distributed load. In addition, the range of destination addresses and associated next-hop destinations provide an evenly distributed load on every output port.

37.3.6 Results and Analysis

We now present some of our results from the analysis of this implementation.

37.3.6.1 Distribution of Code

Figure 37.11 depicts the total distribution of the IXP1200 assembly and uEngineC code in terms of lines of code devoted to either communication (between different memories and interface units), computation related (includes code for register transfers in micro-engine) and initialization code (related to initialization of ports, etc). The main observation being the percentage distribution of code is not drastically different compared with the assembly, however, the number of lines of code in uEngineC is more compared with the assembly (620 lines compared with 500 lines).

37.3.6.2 Per-Packet Time Distribution

Figure 37.12 depicts the percentage time distribution based on the previous classification for each packet. Observe that the micro-engine spends a significant amount of time in the idle state. In addition, the

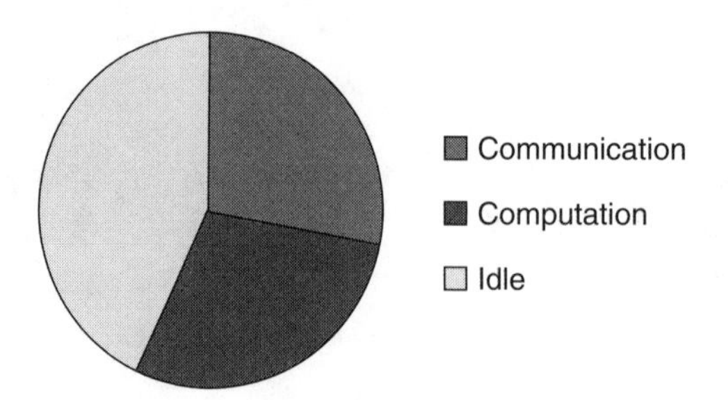

FIGURE 37.12 Per-packet time distribution.

amount of time spent in communication is almost equal to that of computation (Note: Computation here does not indicate arithmetic and logical computations only, but any micro-engine activity including register loads and stores.).

37.3.6.3 Implementation Results

To have a reference set of design points we determined the maximum possible throughput for IPv4 forwarding without packet loss by simulation. We varied the packet length to account for payload storage versus header processing trade-offs. The results are presented in Figure 37.13. Observe that we approach

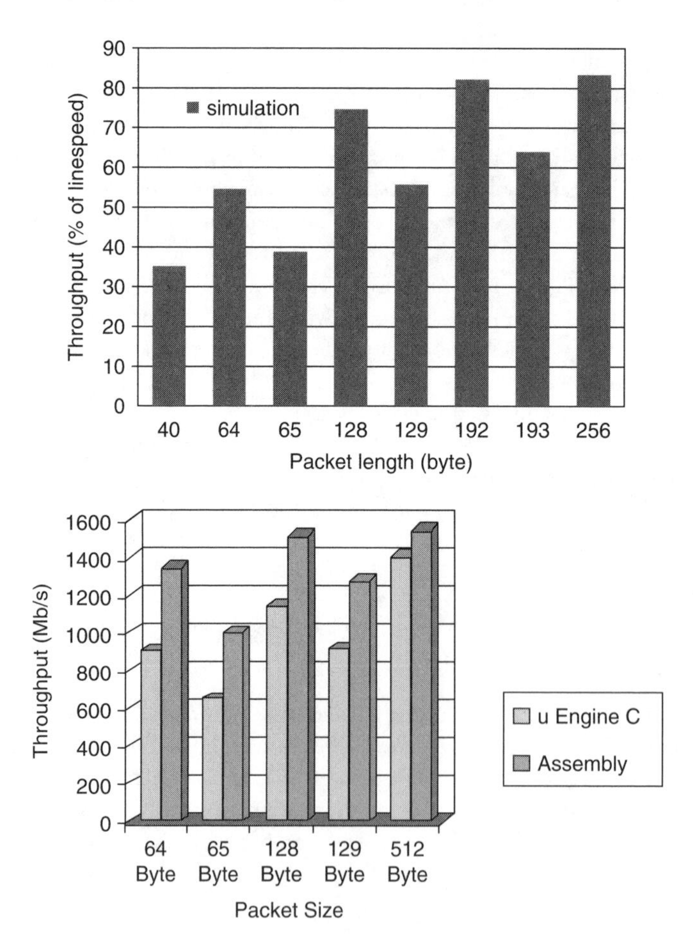

FIGURE 37.13 Throughput for IPv4 forwarding on IXP.

line speed only for larger packet sizes where the micro-engines can keep up with the processing demand of the reduced number of packet arrivals (compared with small packet lengths). We can also recognize the influence of the 64-byte receive and transmit FIFOs in the IX bus unit. As soon as an additional 64-byte segment is needed, the throughput drops, due to the basic unit of data transfer between synchronous dynamic random access memory (SDRAM) and FIFOs being 64 bytes. Thus, for a given delay of two 64-byte transfers we are transferring only 65 bytes (instead of 128 bytes).

Figure 37.13 also presents the comparison between the throughputs obtained by the assembly and uEngineC versions of IXP1200 code. We observe a decrease in performance between uEngineC code and that of assembly code by an average of 25 to 30%. For packets with larger sizes (in our experiments 512 bytes), however, this difference reduces to 5 to 6%. This is because larger packets result in higher throughput for both the cases; however, for the uEngineC code this increase is significant and offsets the difference in performance between the assembly and uEngineC code.

37.3.6.4 Exploring the Implementation Space

Figure 37.14 depicts the variation in throughput with increasing buffer sizes in the external MAC unit (IXF440). We observe that increasing buffer sizes does not contribute to a large increase in the throughput and thus buffer sizes are not the main bottleneck in our set up. Figure 37.14 also presents the variation in the throughput for different IX bus clock frequencies for 64-byte packets. We observe that for both the 80-MHz and 133-MHz, not much change occurs in the throughput; however, by increasing the clock frequency to 200 MHz (the same as the IXP1200 clock frequency), we obtain line rate performance.

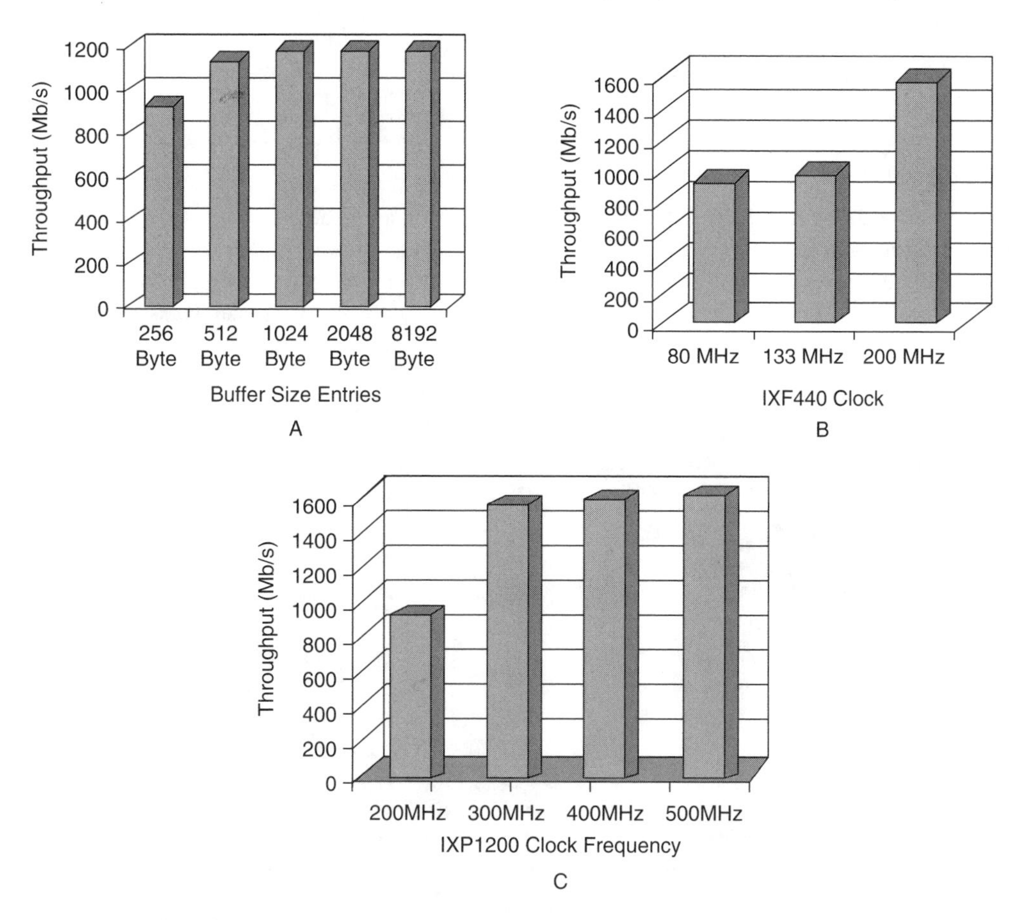

FIGURE 37.14 Throughput for IPv4 forwarding on IXP.

Thus, the IX bus is the bottleneck to attaining line rate performance. Finally, Figure 37.14 depicts the variation in throughput with increasing clock frequency of IXP1200 for IX bus frequency of 80 MHz. We observe that for clock frequencies of 300 MHz and above, we achieve almost a line rate performance. Thus, based on these experiments, we conclude that the two main reasons for performance degradation in IXP1200 are:

1. The speed of IX bus
2. The latency of the IXP1200 itself

Improving either of the two aspects above results in a line rate performance for the given set up.

37.4 Advanced Functional Coverification Using SSDE

Traditionally, software and hardware design activities are clearly separated, as represented by a wall in Figure 37.15. Because of this separation, handing off the hardware IP to the software activity is postponed to the availability of a sufficiently stable and detailed hardware implementation datasheet for the software implementation phase to start. From that point, software IP is developed independent of hardware and hardware engineers can work on finalizing their hardware IP and preparing a rapid prototyping platform like (e.g., an FPGA board). Only after the rapid prototyping platform and the software IP are both available can software and hardware coverification start, usually performed by a separate design integration team.

This traditional approach has two major disadvantages:

1. The software activity remains idle for half of the process, which significantly increases time-to-market.
2. The coverification happens only as a last step in the flow, which does not offer enough room for verifying the system properly.

Moreover, traditional rapid prototyping platforms (e.g., simulation accelerators, FPGA boards, emulators, and early silicon) operate at the Gate-Level (GL), which does not offer the necessary level of visibility for debugging back to the Register-Transfer Level (RTL) and the software programming language level efficiently. As a result, systems are poorly verified; the few bugs found take weeks to fix; several

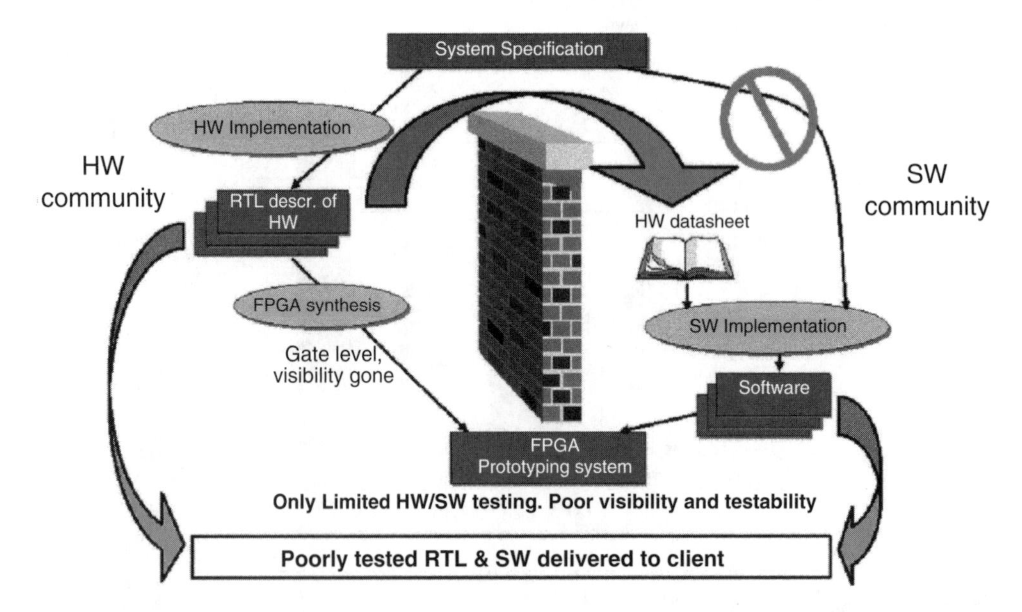

FIGURE 37.15 Traditional wall between the hardware and software communities.

expensive silicon respins are needed before final tape-out; and frustration propagates among the hardware, the software, and the IP integrator communities, not to mention the frustration of the users.

37.4.1 Coverification Using Our System and Software Design Environment (SSDE)

Our proposed SSDE environment alleviates the disadvantages of separating the software and hardware design activities, which was proven on a real USB 2.0 high speed (HS) [29,43], IP9021, business case in cooperation with the Re-Use Technology Group (RTG) Interconnectivity Software Design (ISD) section of the Philips International Technology Center Leuven (ITCL). We illustrate this pilot project in Figure 37.16. Because Seamless from Mentor Graphics, Inc. enables embedded code execution on a simulation model of the hardware, we can use the early RTL-level description of the hardware to start embedded software development at a much earlier stage in the design cycle. From that point, cross-compiling the embedded software into Seamless enables the cycle-accurate software and hardware co-debug with excellent visibility both of the embedded software aspects (i.e., the Seamless XRay * interface provides a conventional software debug interface) and of the hardware signals (i.e., Seamless links to your preferred RTL simulator and waveform viewer). As a result, bugs are found at a much earlier stage and are fixed within a few hours, if not minutes. Because Specman Elite from Verisity, Inc. enables advanced coverage-driven functional verification while offering an e-verification component (eVC) for the USB protocol and protocol checkers for the AHB bus protocol from ARM Ltd., we can verify our full implementation of a USB 2.0HS device in a realistic working environment — including an AHB bus connecting the ARM processor to the USB device and a USB host generating traffic into the USB device — from the early stages of software development. As a result, the system is verified with a much higher confidence.

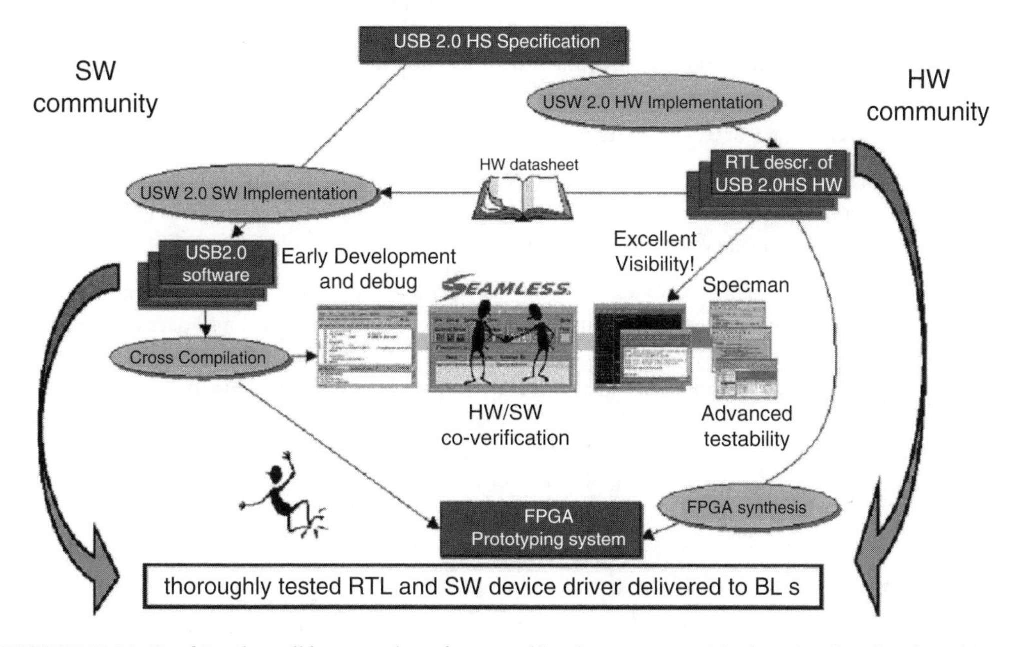

FIGURE 37.16 Breaking the wall between the software and hardware communities by using Seamless from Mentor Graphics, Inc. Boosting the functional verification productivity by using Specman Elite from Verisity, Inc.

*Note that XRay is only one of the software debuggers that may be supported by a processor support protocol (PSP). Others include the ARM debugger and gdb (mainly used for MIPS PSPs). However, only Xray currently supports the interface with Specman via Seamless CVE.

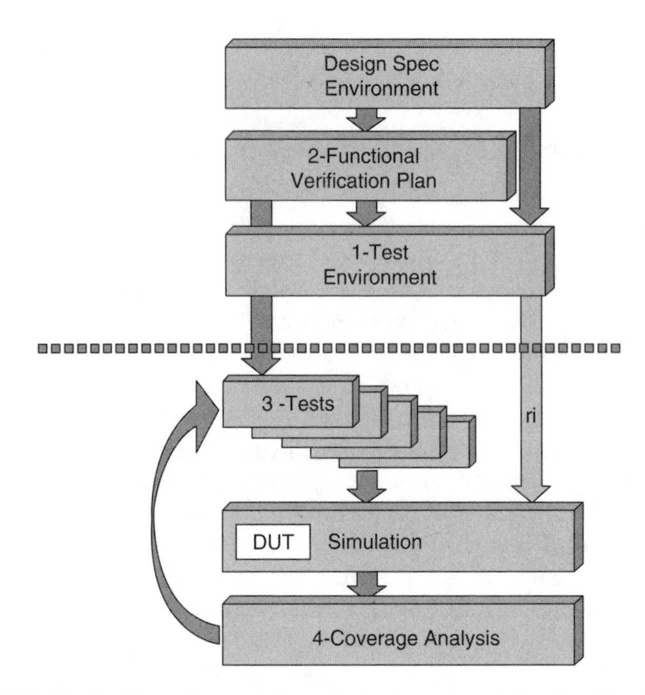

FIGURE 37.17 SSDE Verification Flow: four steps into advanced coverage-driven functional verification.

Because this Seamless/Specman rapid system-level prototyping environment cannot reach very high-speed or even real-time execution, it is still useful to rely on traditional prototyping platforms such as FPGA boards, simulation accelerators, and emulators after the functional specification is verified.

37.4.2 Should I Consider Using SSDE?

Because functional coverification is a broad and weakly defined topic, we need to separate concerns before proceeding further with the presentation of our SSDE methodology. As illustrated in Figure 37.17, we are addressing four steps in the coverification flow:

1. SSDE Test Environment (or Database): the delivery of a proven software and hardware test environment, currently based on Seamless from Mentor Graphics, Inc. and Specman Elite from Verisity, Inc.
2. Functional Test Plan: the tools and methodology support for capturing a functional coverification plan
3. Tests: the tools and methodology support for generating appropriate system tests
4. Coverage Analysis: the tools and methodology support for measuring the coverage quality of the generated tests

37.4.3 Our Generic SSDE Setup

Using our generic SSDE setup in full, Seamless from Mentor Graphics, Inc. reads in the software and fires the cosimulation session. Specman Elite from Verisity, Inc. reads in the e-language test-bench; automatically generates random signal-level tests; performs "on the fly" data and temporal checks; and measures functional coverage.

37.4.4 Overview of Seamless

Starting from a programmable platform design under test (DUT), which may, for example, consist of a CPU, a memory, a direct memory access (DMA) controller, and an I/O block, we use Seamless from

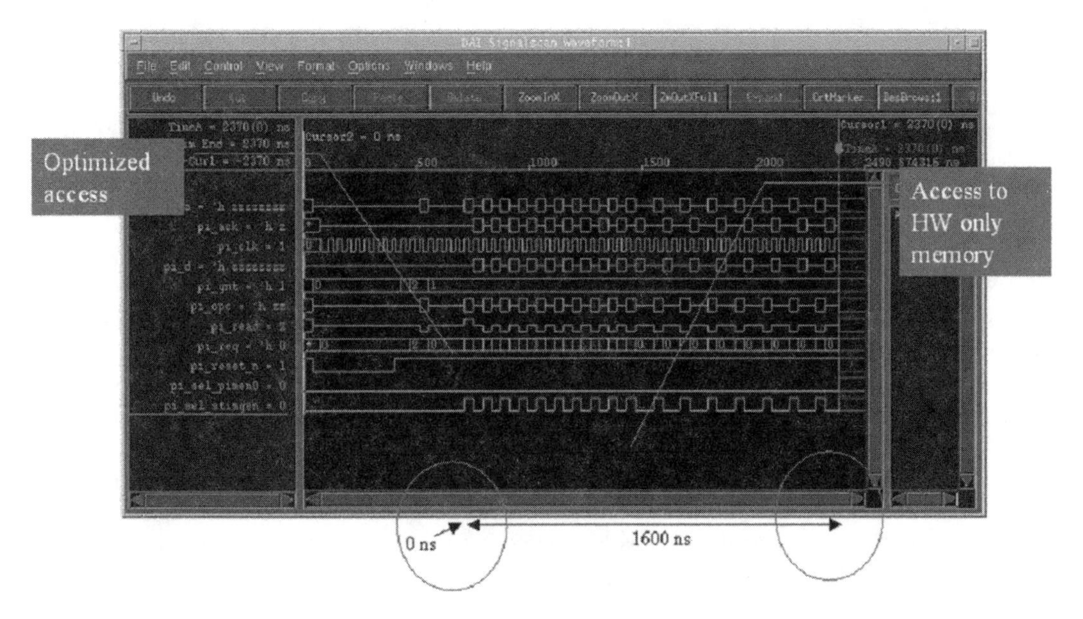

FIGURE 37.18 Optimized software and hardware co-verification using Seamless from Mentor Graphics, Inc.

Mentor Graphics, Inc. to abstract away the CPU and the memory from the DUT. In practice, this requires purchase of a so-called Processor Support Package (PSP) for the processor being used (e.g., ARM or MIPS) along with the basic Seamless license. Please note that Seamless does not support multi-core debugging, which will be addressed by future SSDE releases only. From that point, software can be directly cross-compiled into Seamless that will automatically take care of applying tests to the DUT while allowing cycle-accurate co-debug with excellent visibility of the software and the hardware altogether. As a result, tests can be abstracted away from the detailed signal-level implementation of the test-bench, enabling an easier test-suite portability across, for example, various RTL simulators, various FPGA prototyping boards, emulators, and final silicon [10] while leveraging engineering productivity in the tedious process of test-bench creation. Currently, no standard Transaction Simulation Language (TSL) has been defined to capture the abstracted test-suite, but discussions are under way within the Philips Semiconductors Advanced Functional Verification Workgroup (AFV-Wg) and the SystemC Verification Working Group [54]. Within Philips Semiconductors, many business lines (BLs) already have experience in developing their own standard language and test-suite reuse infrastructure, which should offer an ideal transition path to SSDE.

Figure 37.18 illustrates a typical waveform produced by a Seamless-based cosimulation. Within such a Seamless run, we distinguish three types of events:

1. Access to optimizable memory: These are accesses to memories that support high-level modeling within the Seamless environment.
2. Access to unoptimizable memory: These are accesses to memories that do not support high-level modeling within the Seamless environment, which is often the case for memories embedded inside the hardware part.
3. Standard event: This is any other type of event.

As illustrated in Figure 37.18, when switching Seamless to the high-level modeling mode, accesses to optimizable memories are removed from the cycle-accurate cosimulation. As an effect, optimized Seamless cosimulations can easily be 10× faster than a conventional cosimulation, in the range of 100 kcycles/sec.

37.4.5 Overview of Specman Elite

Starting either from a Seamless setup or from the RTL description of a hardware block alone, Specman Elite from Verisity, Inc. offers all the necessary support for introducing an advanced coverage-driven verification methodology [60]. The idea is to iteratively constrain the test-suite to match the needs captured in the functional test plan with 100% accuracy, so-called coverage. This process is in detail. Using the fully integrated Seamless/Specman environment, Specman has visibility over the full software and hardware design. It thus becomes possible to verify any piece of functionality that was intentionally implemented partly in software and partly in hardware.

37.4.6 Functional Verification Plan

37.4.6.1 Specification-Based Verification

Because bugs are typically hidden in design corner cases that humans and sophisticated design tools do not capture effectively, extensive functional verification is about checking (or proving) that a system behaves correctly for all possible corner cases. To achieve this task effectively, best practice recommends starting with a definition of a functional test plan from the actual system functional specification [57] instead of a detailed paper or RTL implementation, which makes the task of extracting corner cases even harder. In the case of a standardized protocol, such as USB, this information is publicly and freely accessible in documents such as Ganssle [29] and Greef et al. [43]. For more specific developments, it is very important that this information is clearly defined in the system specification document presented in the SSDE1.0 flow from Figure 37.17. Compared with a design-centric flow, an advanced functional verification strategy requires that much more attention is spent on defining (or just collecting) a clean and Golden System Specification from the early stages of the design cycle, which Verisity, Inc. calls "specification-based verification" [57]. An important task (and second step in our SSDE1.0 flow presented in Figure 37.17) is to derive the functional test plan from this system specification.

37.4.6.2 e-Based Executable Test Plan

Many approaches are used for capturing a functional test plan. The traditional approach consists in making an explicit list of items to be tested. This approach clearly does not scale as the number of items grows exponentially for complex systems. Therefore, engineers need to compromise either for an incomplete list (i.e., a low coverage list) or for an ambiguous list (i.e., a list that does not really say what should be done). Both compromises are of course not recommendable. The research approach consists of automatically generating tests from a formal and, therefore, unambiguous specification [3,5]. This approach scales much better than the latter, but it still cannot address the complexity of today's SOC designs. It can be used for specific needs only. Between these two approaches is the e-language from Verisity, Inc., which allows for the capture of a higher-level (and therefore incomplete) list of items to be tested from a verification language, which is nicely complemented by automatic random test generation [61]. Provided the system under consideration is well suited for random test generation [1], scalability and coverage can be very high, which cannot be achieved by any other approach. Examples of a good fit are core datapath verification, where computed data can take any random value, and telecommunication application verification, where transmitted packets can carry any random information. Examples of a bad fit are complex protocols requiring high order coverage [2], which cannot be reached by chance.

As illustrated in Figure 37.19, engineers typically think about electronic and software design in two orthogonal dimensions. The first dimension relates to pieces of information, such as a packet, a channel, an Ethernet frame, and a memory buffer, that are ideally modeled using object-programming techniques [27,45]. The second dimension relates to pieces of functionality such as basic definitions, interfaces to the DUT, assertion checkers, coverage computations, and specific corner case tests. Such design aspects typically span over several objects and are therefore not captured effectively by standard object programming. By offering a clean separation of objects and aspects [33], the e-language enables mixed object-aspect programming. To extend an aspect definition, you do not need to edit several base objects and vice versa. This is extremely useful for capturing many corner cases in the least amount of development effort and time.

More precisely, let us study some mixed object-aspect programming examples. Imagine that you want to functionally verify a communication agent that sends transactions over a network. These transactions

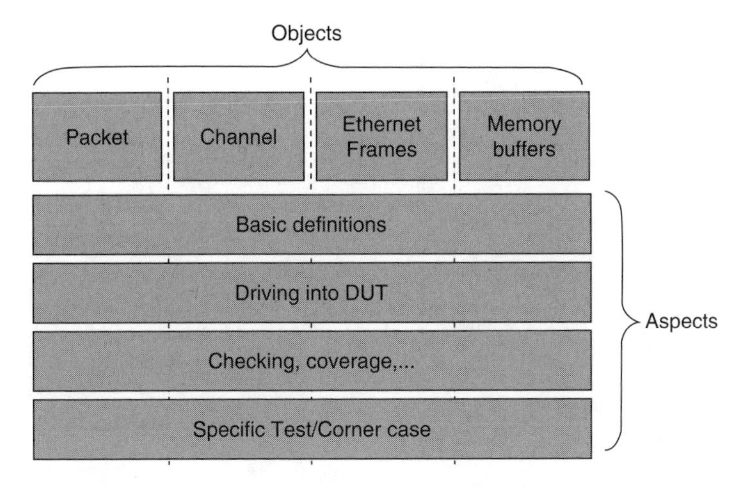

FIGURE 37.19 Objects and aspects are orthogonal programming paradigms.

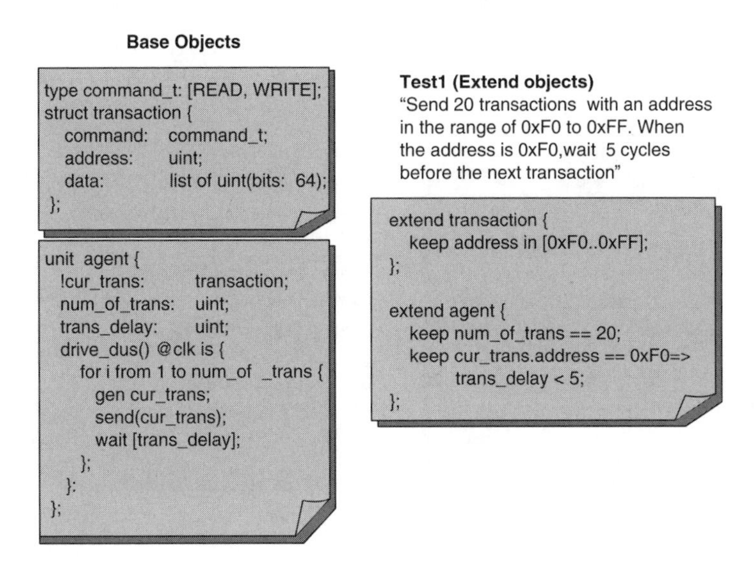

FIGURE 37.20 Using object-aspect programming, file editing is minimized for refining the functional test plan with corner conditions.

consist of a command, a destination address, and some data, which can be easily captured from an object description, as illustrated in Figure 37.20. As a first incomplete approximation of your functional test plan, you wish to generate a large number of such transactions and drive them through the bus on the rising edge of a clock. You also want to wait a number of cycles between two transactions. A first refinement of this functional test plan is to constrain the number of transactions, the address space, and the delay. This is ideally captured by an aspect extension using the extend construct in the e-language.

Imagine that you now want to add a debug feature by capturing the time and content of each transaction sent. As presented in Figure 37.21, this is made possible by extending the transaction object with a start time — the agent object with a list of transactions and the send method from the agent object with a time stamp and history addition.

Finally, imagine that a new derivative of the DUT supports burst mode accesses for transactions of size 4, 8, 16, 32, and 64. As presented in Figure 37.22, this new feature can be tested by constraining the command type to include a BURST mode and the data instance variable from the transaction object to match the required size for burst execution. In conclusion, using mixed object-aspect programming for

Base Objects

```
type command_t: [READ, WRITE]
struct transaction {
   command:    command_t;
   address:    uint;
   data:       list of uint(bits: 64);
};
```

```
unit agent {
  !cur_trans:       transaction;
  num_of_trans:   uint;
  trans_delay:     uint;
  drive_dus() @clk is {
     for i from 1 to num_of _trans {
        gen cur_trans;
        send(cur_trans);
        wait [trans_delay];
     };
  }:
};
```

New Debug Feature

"Create a transaction history list with transaction timestamps for each test."

```
extend transaction {
   start_time:       time;
};

extend agent {
   !history_list:  of transation;
   send()  is first {
      cur_trans. start_time = sys.time;
      history_list.add(cur_trans);
   };
};
```

FIGURE 37.21 Using object-aspect programming, debugging features are easier to introduce and manipulate.

Base Objects

```
type command_t: [READ, WRITE];
struct transaction {
   command:    command_t;
   address:    uint;
   data:       list of uint(bits: 64);
};
```

```
unit agent {
  !cur_trans:       transaction;
  num_of_trans:   uint;
  trans_delay:     uint;
  drive_dus() @clk is {
     for i from 1 to num_of _trans {
        gen cur_trans;
        send(cur_trans);
        wait [trans_delay];
     };
  }:
};
```

Version 2 DUT

"The new version of the DUT supports burst types of commands."

```
extend command_t: [BURST];
extend transaction {
   when BURST transaction {
      burst_size: uint;
      keep burst_size in [4,8,16,32,64];
      keep data,size() == burst_size;
   };
};
```

FIGURE 37.22 Using object-aspect programming, file editing is minimized for testing a derivative design.

capturing the functional test plan is superior not only to conventional verification techniques for productivity, but is also superior for maintainability and debugability purposes.

37.4.7 Random Test Generation

As soon as the functional test plan is available, random test generation can start [61]. When captured using the object-aspect programming capabilities of the e-language from Verisity, Inc., this process is completely automatic. Section 37.4.8 describes the advantages and disadvantages of this technology.

37.4.8 Manual Tests Development

As already mentioned, the traditional approach of making an explicit list of items to be tested and implementing them does not scale. This verification effort is proportional to the square of capacity in the best case, and, therefore, at least doubles every 6 to 9 months according to Moore's Law. Moreover,

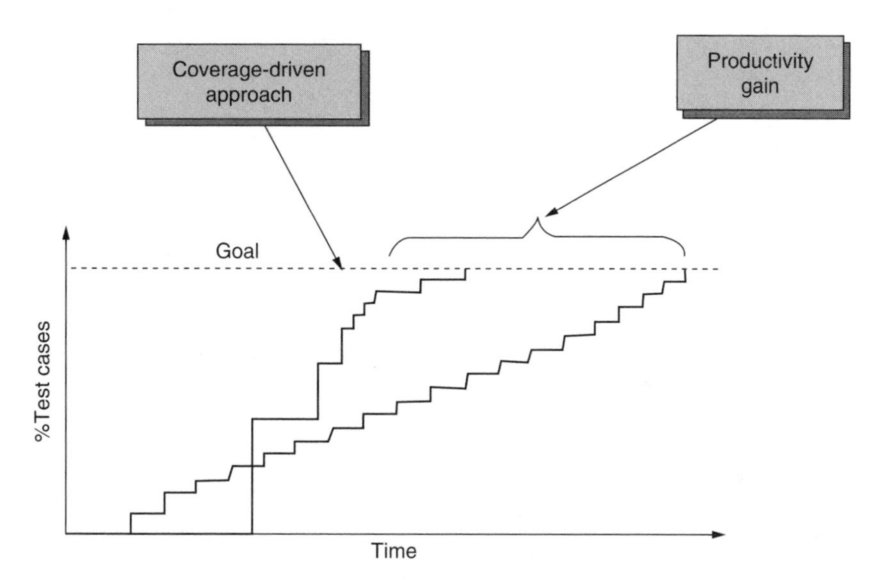

FIGURE 37.23 Coverage-driven verification reuse process.

when performing this task manually in an ad-hoc way, verification reuse across similar verification needs and portability across verification platforms and environments is hard, if not impossible, further extending the waste in non-reusable engineering (NRE) cost. Figure 37.23 illustrates the process of developing tests manually to achieve a goal specified in the functional test plan. Whenever this goal falls beyond the human perception, the only possible behavior for the designer is to write several small tests hoping that the global verification objective will be solved. In the best case, this process is tedious and slow. In the worst case, the designer quickly gets confused and introduces highly redundant tests that may never converge to the wanted goal. This redundancy typically accounts for a large part of the NRE overhead for complex systems. Moreover, by artificially growing the verification complexity, it may also account for the need to rely on very expensive high-capacity verification infrastructure such as emulators.

37.4.9 Automatic Test Pattern Generation

Instead of writing tests by hand, it is possible to rely on automatic test pattern generation (ATPG). Several kinds of ATPG exist, but we focus here on the random type [1,61], using a simplified three-step version of the methodology evolution concept from Verisity, Inc. [59]. The first step consists in relying on fully automatic random test generation from the functional test plan captured in the e-language. Because tests are generated from an intelligent engine, their statistical repartition is homogeneous, which tends to achieve less redundancy than conventional tests written by hand. Because design corner cases are by definition unfit for human perception, randomly generated tests tend to reach them more easily. As a result, automatic random test generation is likely to converge to verification objectives in less time and effort than the manual approach. This productivity gain is illustrated in Figure 37.23.

The second step consists of introducing coverage metrics to drive the quality of tests [60]. Because tests are steered by a measured objective, most of the redundancy can be removed and productivity is largely improved especially toward the end of the verification project, as illustrated in Figure 37.23.

The third step consists in structuring the verification IP in a reusable form. Whenever this IP is reused, a significant portion of the verification effort is saved. This results in an earlier time-to-first-test and a higher productivity win-win situation, as illustrated in Figure 37.23.

In conclusion, by properly exploiting the capabilities of random test generation, it is possible to improve both the availability of first results and the overall project productivity. Whenever coverage-driven reuse applies, the initial tool support and training investment can be recovered from the first day compared with traditional manual and RTL-level practice. In this situation, there should be no hesitation in adopting our SSDE methodology.

References

[1] V. Agrawal and R. Mercer. Deterministic versus random testing. *Int. Test Conf.*, Los Angeles, CA, September 1986, pp. 718–718.

[2] G. Apostol (Brecis Communications, Inc.). Network processor designer tackles verification nightmare. *EE Design*, November 5, 2001.

[3] L. Arditi, H. Boufaied, A. Cavanie, and V. Stehle. Coverage directed generation of system-level test cases for the validation of a DSP system. *Int. Symp. on FME 2001: Formal Methods for Increasing Software Productivity*, LNCS, Vol. 1, 2001.

[4] F. Baker. *Requirements for IP Version 4 Routers. RFC1812*, Internet Engineering Task Force (IETF), June 1995.

[5] M. Benjamin, D. Geist, A. Hartman, G. Mas, R. Smeets, and Y. Wolfsthal. A feasibility study in formal coverage driven test generation. Technical Report, IBM Haifa Laboratories, Haifa, Israel, June 1999. http://www.haifa.il.ibm.com/projects/verification/gtcb/.

[6] G. Berry. *The Foundations of Esterel.* G. Plotkin, C. Stirling, and M. Tofte, Eds., MIT Press, 2000.

[7] Y. Bouchebaba. Optimisation des transferts de données pour le traitement du signal. Ph.D. thesis, Ecole Nationale Supérieure des Mines de Paris, 2002.

[8] Y. Bouchebaba and F. Coelho. Pavage pour une séquence de nids de boucles. *J. Technique et science informatiques. Parallélisme et systèmes distribués* 21, numéro 5, 2002, pp. 579–603.

[9] Y. Bouchebaba and F. Coelho. Tiling and memory reuse for sequences of nested loops. *Euro-Par 2002, Parallel Process., 8th Int. Euro-Par Conf. Proc. Lecture Notes in Computer Science 2400*, 2002, pp. 255–264.

[10] R. Brackebusch, S. Muller, G.S.-Y. Sokomak, F. Grassert, and D. Timmermann. A new synthesizable architecture approach for verification environments applying transaction-based methodology. *Proc. 40th Design Automation Conf. (DAC '03)*, Anaheim, CA, June 2003.

[11] A. Bruce and J. Goodenough (ARM, Ltd.). Re-usable hard-ware /software co-verification of IP blocks. Verisity Design, Inc. Club Verification, June 2002.

[12] D. Callahan. A global approach to detection of parallelism. Ph.D. thesis, Rice University, Houston, TX, 1987.

[13] F. Catthoor et al. *Custom Memory Management Methodology — Exploration of Memory Organization for Embedded Multimedia System Design*, Kluwer Academic Publishers, Dordrecht, 1988.

[14] F. Catthoor, F. Franssen, S. Wuytack, L. Nachtergaele, and H. DeMan. Global communication and memory optimizing transformations for low power signal processing systems. *IEEE Workshop on VLSI Signal Process.*, 1994, pp. 178–187.

[15] H. Chang, L. Cooke, M. Hunt, G. Martin (Cadence Design Systems, Inc.), A. McNelly and L. Todd (Simutech, Inc.). *Surviving the SOC Revolution: A Guide to Platform-Based Design.* Kluwer Academic Publishers, Dordrecht, 1999.

[16] A. Chatelain, Y. Mathys, G. Placido (Motorola, Inc.), A. La Rosa and Luciano Lavagno (Politecnico di Torino). High-level architectural co-simulation using Esterel and C. *CODES '01*, pp. 189–194.

[17] Intel Corporation, *Intel IXP1200 Network Processor Family: Hardware Reference Manual, Revision 8*, Intel Corporation, Santa Clara, CA, pp. 225–228, 2001.

[18] P. Crowley, M. Fiuczynski, J. Baer, and B. Bershad. Characterizing processor architectures for programmable network interfaces, *Proc. 2000 Int. Conf. on Supercomputing*, Santa Fe, NM, May 2000.

[19] A. Darte. On the complexity of loop fusion, *Parallel Computing*, Vol. 26, No. 9, 2000, pp. 1175–1193.

[20] D. Dempster and M. Stuart (TransEDA, Ltd.). *Verification Methodology Manual: Techniques for Verifying HDL Designs.* Kluwer Academic Publishers, Dordrecht, 2002.

[21] F. de Dinechin, P. Quinton, and T. Risset. Structuration of the Alpha language, in *Massively Parallel Programming Models*, IEEE Computer Society Press, Berlin, Germany, 1995, pp. 18–24..

[22] E.A. de Kock, W.J.M. Smits, P. van der Wolf, J.-Y. Brunel, W.M. Kruijtzer, P. Lieverse, K.A. Vissers, and G. Essink. YAPI: application modeling for signal processing systems. *Proc. 37th Conf. on Design Automation (DAC-00),* NY, June 5–9 2000, pp. 402–405.

[23] H. de Man. Washing machine: the key to low-power. *EE Times,* March 6, 2002, http://www.electronicstimes.com/.

[24] C. Eisenbeis, W. Jalby, D. Windheiser, and F. Bodin. A strategy for array management in local memory, *J. Mathematical Programming: Series A,* Vol. 63, No. 3, 1994, pp. 331–370.

[25] C. Eisner and D. Fisman. Sugar 2.0: an introduction. Technical Report, IBM Haifa Research Laboratory, Haifa, Israel, 2002.

[26] Esterel Technologies, S.A. Esterel technologies develops top-level validation methodology for STMicroelectronics: Speeds functional verification of chips with multiple design blocks. Available at http://www.esterel-technologies.com, June 2002.

[27] E. Gamma, R. Helm, R. Johnson, and J. Vlissides. *Elements of Reusable Object-Oriented Software, Professional Computing Series,* Addison-Wesley, Reading, MA, 1994.

[28] D. Gannon, W. Jalby, and K. Gallivan. Strategies for cache and local memory management by global program transformation. *J. Parallel and Distributed Computing, Vol.* 5, No. 10, 1988, pp. 587–616.

[29] J.G. Ganssle. An introduction to USB development. Embedded Systems Programming, 2002. Available at http://www.embedded.com.

[30] E. de Greef, F. Catthoor, and H. de Man. Reducing storage size for static control programs mapped onto parallel architectures, presented at *Dagstuhl Seminar on Loop Parallelisation,* Schloss Dagstuhl, Germany, April 1996.

[31] M. Gries, C. Kulkarni, C. Sauer, and K. Keutzer. Comparing analytical modeling with simulation for network processors: a case study, *Design Automation and Test in Europe (DATE),* Munich, Germany, March 2003.

[32] N. Halbwachs, P. Caspi, P. Raymond, and D. Pilaud. The synchronous data-flow programming language LUSTRE. *Proc. IEEE,* Vol. 79, No. 9, pp. 1305–1320, September 1991.

[33] Y. Hollander, M. Morley, and A. Noy. The e-language: a fresh separation of concerns. *Proc. Technology of Object-Oriented Languages and Syst. (TOOLS) Europe,* Zurich, Switzerland, March 1999, pp. 41–50.

[34] G. Huard. Algorithmique du décalage d'instructions. Ph.D. thesis, École Normale Supérieure de Lyon, 2001.

[35] International Technology Roadmap for Semiconductors (ITRS). Design chapter of the 2001 edition. Technical Report, EECA, JEITA, KSIA, TSIA, SIA, and International SEMATECH, 2001. Available at http://public.itrs.net.

[36] F. Irigoin and R. Triolet. Supernode partitioning. *Proc. 15th Annu. ACM Symp. on Principles of Programming Languages,* San Diego, CA, 1988, pp. 319–329.

[37] F. Irigoin, P. Jouvelot, and R. Triolet. Overview of the PIPS project. *Proc. Int. Workshop on Compilers for Parallel Computers,* Paris, France, November 1990.

[38] M. Kandemir, N. Vijaykrishnan, M.J. Irwin, and H.S Kim. Experimental evaluation of energy behavior of iteration space tiling, *LCPC 2000,* Yorktown Heights, NY, 2000, pp. 142–157.

[39] K. Kennedy. Fast greedy weighted fusion. *Int. J. Parallel Programming,* Vol. 29, No. 5, 2001, pp. 463–491.

[40] B. Kienhuis, E. Deprettere, K.A. Vissers and P. Van Der Wolf. An approach for quantitative analysis of application-specific dataflow architectures. *Proc. Int. Conf. on Application-Specific Syst., Architectures and Processors (ASAP '97),* pp. 338–349, 1997.

[41] E. Kohler, R. Morris, B. Chen, J. Jannotti, and M. Kaashoek. The Click Modular Router. *ACM Trans. on Computer Syst.,* Vol. 18, No. 3, pp. 263–297, August 2000.

[42] P. Le Guernic, J.-P. Talpin, and J.-C. Le Lann. Polychrony for system design. Technical Report RR-4715, IRISA, Environnement de spécification de programmes réactifs synchrones (ESPRESSO), June 2003.

[43] C.-W. Leong. Understanding the universal serial bus (USB). USB developer, 2002. Available at http://www.USBDeveloper.com.

[44] K. McKinley and K. Kennedy. Maximizing loop parallelism and improving data locality via loop fusion and distribution, languages and compilers for parallel computing, *6th Int. Workshop*, Portland, Oregon, 1993, pp. 301–320.

[45] Bertrand Meyer. Object oriented software construction. In C.A.R. Hoare, Ed., *Series in Computer Science*. Prentice Hall International, Inc., Englewood Cliffs, NJ, 1988.

[46] A. Mihal, C. Kulkarni, M. Moskewicz, M. Tsai, N. Shah, S. Weber, Y. Jin, K. Keutzer, C. Sauer, K. Vissers, and S. Malik. Developing architectural platforms: a disciplined approach, *IEEE Design and Test of Computers*, Vol. 19, No. 6, pp. 6–16, November/December 2002.

[47] A. Mozipo, D. Massicote, P. Quinton, and T. Risset. Automatic synthesis of a parallel architecture for Kalman filtering using MMAlpha. *IEEE Canadian Conf. on Electrical and Comput. Eng.*, Edmonton, Canada, May 1999.

[48] J. Nickolls, L.J. Madar III, S. Johnson, V. Rustagi, K. Unger, and M. Choudhury, Broadcom Calisto: a multi-channel multi-service communication platform, *Hot-Chips Symp.*, 2002.

[49] T. Pascalin Amagbegnon, P. Le Guernic, H. Marchand, and E. Rutten. Signal. *Lecture Notes in Computer Science*, Vol. 891, pp. 113–, 1995.

[50] Philips Semiconductors, B.V. Nexperia pnx8500: Home entertainment engine. Functional Overview, 2000. Available at http://www.semiconductors.philips.com/nexperia.

[51] S.K. Roy (Synplicity, Inc.), S. Ramesh, S. Chakraborty (IITBombay), T. Nakata, and S.P. Rajan (Futjitsu Laboratories). Functional verification of systems on chip (SOCs) — practices, issues and challenges. *ASP-DAC/VLSI Design 2002*, January 7–11, 2002, Bangalore, India, pp. 11–.

[52] M. Scott, J. Dickerson, and B. Payne. Panel probes SOC problems, solutions. *EE Design*, February 2002. http://www.eedesign.com/news/OEG20000202S0044.

[53] N. Shah, Understanding network processors. Master's thesis, Department of Electrical Engineering and Computer Sciences, University of California-Berkeley, September 2001.

[54] SystemC Verification Working Group. SystemC verification standard specification. Technical Report, Open SystemC Initiative (OSCI), November 2002. Available at http://www.systemc.org.

[55] Teja Technologies, IPv4 forwarding application performance, White Paper, July 2002. Available at http://www.teja.com/library/ip4_whitepaper.html.

[56] M. Tsai, C. Kulkarni, C. Sauer, N. Shah, and K. Keutzer, A benchmarking methodology for network processors, *First Workshop on Network Processors at the 8th Int. Symp. on High Performance Computer Architecture (HPCA8)*, Cambridge, MA, February 2002.

[57] Verisity Design, Inc. Spec-based verification. White Paper, 1999. Available at http://www.verisity.com/resources/whitepaper /.

[58] Verisity Design, Inc. e-reuse methodology (eRM) developer manual: maximizing verification productivity. Technical Report, Verisity Design, Inc., 2001.

[59] Verisity Design, Inc. The evolution of verification methodology. Technical Report, Verisity Design, Inc., 2001.

[60] Verisity Design, Inc. Coverage-driven functional verification: using coverage to speed verification and ensure completeness. White Paper, September 2001. Available at www.verisity.com/resources/whitepaper /.

[61] J.A. Waicukauski, E. Lindbloom, E.B. Eichelberger, and O.P. Forlenza. A method for generating weighted random test patterns. *IBM J. Res. and Dev.*, Vol. 33, No. 2, pp. 149–161, March 1989.

[62] M.E. Wolf. Improving locality and parallelism in nested loops. Ph.D. thesis, Stanford University, Stanford, CA, 1992.

[63] M. Wolfe. High-*Performance Compilers for Parallel Computing*, Addison-Wesley, Reading, MA, 1996.

[64] S. Wuytack, J.P. Diguet, F. Catthoor, and H. De Man. Formalized methodology for data reuse exploration for low-power hierarchical memory mappings, *IEEE Trans. on VLSI Syst., Special Issue ISLPED '97*, Vol. 4, No. 6, pp. 529—537, Dec. 1998.

[65] H.P. Zima and B.M. Chapman. *Supercompilers for Parallel and Vector Computers*, Addison-Wesley, Reading, MA, 1990.

VI

CAD Tools for Low Power

38

High-Level Power Estimation and Analysis

Wolfgang Nebel
Oldenburg University

Domenik Helms
OFFIS

38.1 Introduction

"Big things always start small." This wisdom also applies to microelectronic design; and it is at the beginning, when the complexity is still small and can well be understood under different aspects, that the important decisions are made, which will lead to success or failure. Once a design has been developed to a large structure of logic and wires, it is difficult to cure problems, which, in many cases, also started small and eventually became large, hard to solve, and without major design respins, these problems may cost months of design time, major engineering resources, and can be responsible for missed marketing opportunities.

This chapter covers the area of early system-level power analysis and algorithmic-level power estimation. The techniques presented here shall enable the reader to understand the underlying concepts as well as the chances and limitations of tools, which shall guide the designers in optimizing the global system architecture for low power and help them selecting and further optimizing the algorithms to be implemented at lower levels. The figure of merit in reducing the power consumption by making the right decisions during this early phase covers several orders of magnitude. Just to illustrate the potential: there exist dozens of known and well-understood sorting algorithms. They all perform exactly the same task: take a set of objects and put them in an order according to the chosen sorting criterion. Despite the exactly same functional behavior, however, they all perform differently with respect to the computation time, memory usage, and the power consumption. Similarly, different algorithms with equivalent functionality are known for Fourier transform, compression, and many other functions, which are copiously used in mobile multimedia applications. Selecting the most power efficient one can be a product-differentiating factor.

0-8493-1941-2/05/$0.00+$1.50

TABLE 38.1 Breakdown of Power Consumption

Technology	Switched Cap. Power	%	Short-Circuit Power	%	Leakage Power	%
150 nm	439 nW	71.1	173 nW	28.0	5.6 nW	0.9
130 nm	317 nW	71.8	118 nW	26.7	6.7 nW	1.5
100 nm	236 nW	73.5	75 nW	23.4	10 nW	3.1
90 nm	183 nW	70.1	67 nW	25.7	11 nW	4.2
70 nm	139 nW	56.3	55 nW	22.3	53 nW	21.4
45 nm	74 nW	36.8	30 nW	14.9	97 nW	48.3
32 nm	51 nW	23.3	28 nW	12.8	140 nW	63.9
22 nm	20 nW	13.9	14 nW	9.7	110 nW	76.4

38.1.1 Analysis vs. Estimation

Although the terms estimation and analysis are frequently used in the low-power community without careful distinction, we would like to clarify the terminology here. Analysis is based on an existing design at any level (i.e., the structure is given, typically in terms of a netlist of components). These modules are predesigned and for each one a power model exists. These power models can be evaluated based on the activation of the modules. Thus, power analysis is the task of evaluating the power consumption of an existing design at any level. It is used to verify that a design meets its power and reliability constraints (e.g., no electromigration occurs, no hot spots will burn the device, and no voltage drops will cause spurious timing violations). Power analysis finally helps to select the most cost efficient chip package.

In contrast, estimation builds on incomplete information about the structure of the design or part of the design under consideration. The design does not yet exist and can only be generated based on assumptions about the later physical implementation of the design, its modules, its interconnect structure, and physical layout. In summary, estimation requires design prediction followed by analysis; for instance, if the floorplan of a design is not yet available, interconnect power estimation first requires a floorplan prediction. Power estimation is applied to assess the impact of design decisions and compare different design alternatives on incomplete design data. It allows to efficiently exploring the design space without the need for a detailed implementation of all different design options.

38.1.2 Sources of Power Consumption

This section briefly revisits the physical basics of power consumption, which is also the basis for high-level power analysis. Table 38.1 lists a breakdown of the estimated power consumption of a single transistor for high performance logic. The data are our own calculations based on the 2002 update of the International Technology Roadmap for Semiconductors [1]. Our assumptions include a 1% expected switching activity compared with the maximum transistor operation frequency, which is typical for processing components.

These data clearly demonstrate that today, for high performance applications, the switched capacitance power consumption is still dominating. Considering the fact that the short-circuit power is typically captured as part of the power models for dynamic power, and further, that the data in Table 38.1 do not include dynamic power related to interconnect capacitances, we can safely assume that the power consumption of computation intensive devices at 70 nm and larger technology nodes will be dominated by the dynamic power consumption. For mobile applications, language power is an important source of power consumption already at 90 nm.

38.1.2.1 Switched Capacity Power

Equation (38.1) allows for the calculation of the power consumption of a switched capacitor. At the transistor-level C_{load} includes the parasitic gate overlap and fringing capacitances as well as the Miller capacity. α models the switching probability of the transistor during a cycle of the clock toggling at frequency f. V_{dd} is the supply voltage. We will demonstrate later that applying this formula at higher levels of abstraction will require a revised interpretation of some of these parameters.

$$P_{swcap} = \frac{1}{2} C_{load} \cdot \alpha \cdot V_{dd}^2 \cdot f \quad (38.1)$$

Equation (38.1) holds for unnecessary transitions (glitches), while it needs refinement for modeling a sequence of n incomplete transitions within a period of T and with a voltage swing of ΔV_n (Equation (38.2)).

$$P_{incompleteswcap} = \frac{1}{2T} \cdot C_{load} \cdot V_{dd} \cdot \sum_{i=1}^{n} \Delta V_n \quad (38.2)$$

38.1.2.2 Short-Circuit Power

Short-circuit power is the second part of the dynamic power consumption. It occurs when during a short period both the pull-up and the pull-down networks of static CMOS-gates are conducting. Equation (38.3) gives a simple model of the short-circuit power with β modeling the transistors' conductivity per voltage factoring the linear region, T is the inputs' rise/fall time, and τ is the gate delay.

$$P_{shortcircuit} = \frac{\beta}{12} \left(V_{dd} - 2V_{th}\right)^3 \frac{\tau}{T} \quad (38.3)$$

Equation (38.3) is an overestimation by up to a factor of three. For an accurate analysis, transistor level models and transient analyses are needed [2]; however, $P_{shortcircuit}$ within modules can be captured as part of the dynamic power models of the modules.

38.1.2.3 Leakage

The leakage power consumption is mostly due to leakage currents flowing through the channel in weak inversion even when the gate-source voltage is below threshold and due to carriers tunneling through the gate oxide. The leakage power depends on the state of the circuit. Analysis, modeling, and optimization of the leakage power are currently subject of intensive research. It is covered in depth in Chapter 3 of this book.

38.2 Generic Design Flow for Low-Power Applications

This section introduces the common principles of power analysis at any level before we present the details of system-level power analysis and algorithmic-level power estimation in the following sections.

38.2.1 Generic Power Estimation and Analysis Flow

Generally, any analysis tool for the dynamic power consumption needs to evaluate Equation (38.1) and Equation (38.2). This can be done at different degrees of abstraction. For instance, for high-level power analysis, the entire dynamic power consumption of a module will be described by a single power model instead of by all individual capacitances inside the module. Similarly, the switching probability of all capacitances is lumped together into an activity model for the module. Figure 38.1 depicts a generic power estimation and analysis flow that can be applied at any level of abstraction.

The upper part of Figure 38.1 needs to be applied when a preimplementation power estimate is needed. In this case, the architecture of the design is unknown yet. The set of components to be allocated to implement the device is still to be determined. Consequently, neither their interconnection and communication structure nor their activation patterns are defined. Thus, an evaluation of Equation (38.1) and Equation (38.2) is not possible yet, even if higher-level capacitance and activation models were applied.

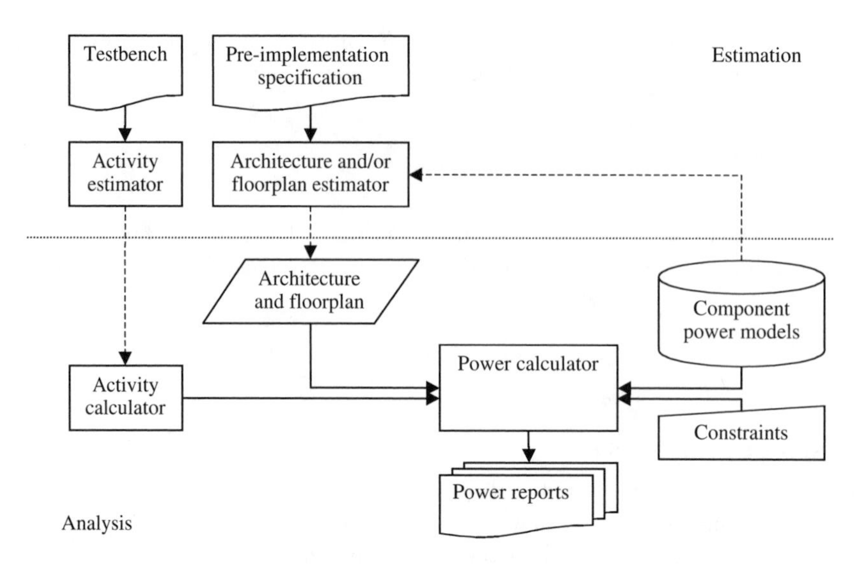

FIGURE 38.1 Generic power estimation and analysis tool flow.

It is the task of the architecture and the floorplan estimators to predict the component allocations, their physical floorplan, which is needed for the estimation of the interconnect and clock tree, and the scheduling of the operations, which again is needed to predict the activation of the components. These estimation techniques will be discussed in more detail in Section 38.4.

Once this is done, the kind of information is adequate for a power analysis of this predicted architecture. The relevance of the results of the following power analysis step, however, strongly depends on the quality of the predicted architecture.

Let us discuss two application scenarios for high-level power estimation. First, we are interested at the first possible instance to get an estimate of the to-be-expected final power consumption of a system (e.g., to validate the feasibility of a certain package). In this case, a reasonable absolute accuracy is needed. A reliable power estimate for this case requires that the predicted architecture is very similar to the final architecture once the system has been fully designed and optimized through all levels of abstraction. An architecture estimator for this application has to take into account the specific design styles, circuit technologies, design skills, and the tool flow applied to generate a sufficiently accurate architecture prediction.

In the second use case, we are interested in a fast comparison of different design options. We want to know which out of several paths through the design space to follow. Thus, we are looking for relative power figures for each of the options we have in mind. In this case, it is important, that the solution, which had been estimated to be the most power efficient one, really proves to be the least power consuming one. For this scenario, the predicted architectures for each of the options should be similar in their power efficiency, even if they are not exactly identical to the final implementation. The predicted power figure of the different solutions may differ to some extent from the final power figure after implementation as long as the order between them is maintained.

The lower part of Figure 38.1 depicts the generic power analysis flow. The power calculator collects the parameters of Equation (38.1) at the respective level of abstraction. The floorplan and the architecture of modules, each having a power model attached, determine the physical and structural architecture and represent the load capacitance C_{load} of Equation (38.1) — maybe at an abstract level. The activity calculator produces an activation profile for each of the components modeling the switching probability α of Equation (38.1). Finally, the supply voltage V_{dd} and the clock frequency f are part of the constraints provided by the designer.

As we can easily see, power analysis at any level requires three main input models: The architecture model and its component models as well as an activation model.

38.2.2 Low-Power Design Flow

This section exemplifies the generic power estimation and analysis flow of the previous section for each of the highest levels of abstraction. Figure 38.2 is a generic power conscious design flow, which, however,

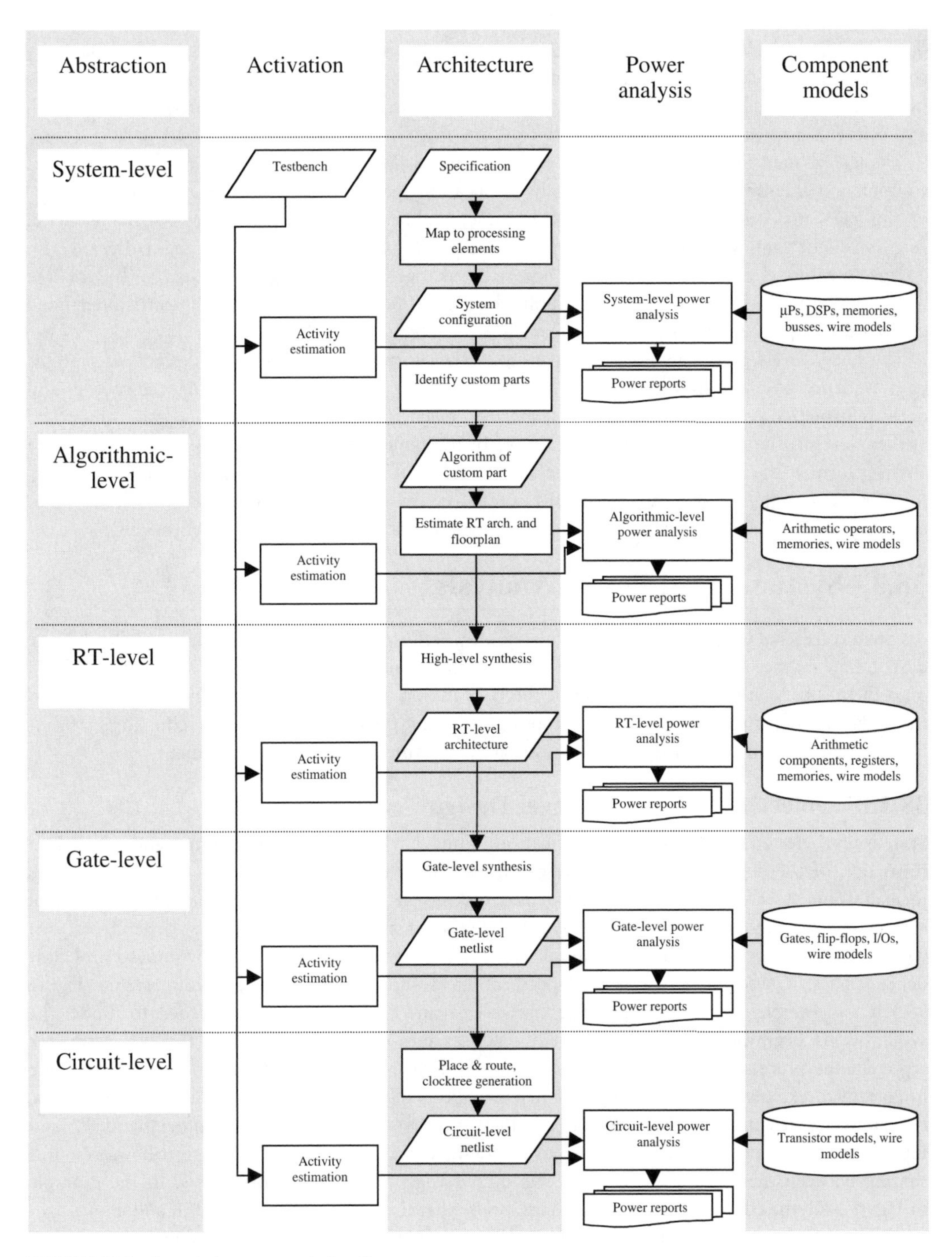

FIGURE 38.2 Generic low-power design flow.

is also applicable to platform based design and incremental designs. In the sequel, we will walk through this design flow in a top-down manner.

At the system-level, the design objective is to map a given, possibly informal, specification onto a target architecture. In many cases, this architecture will be constrained to a platform consisting of fixed architecture elements such as processors, micro-controllers, digital signal processors (DSPs), and a bus standard. The design objective is to find an optimized mapping, which meets functional and performance constraints at least cost and power consumption. A system-level power management function can be included at this level and needs to be considered during power analysis. The most frequently used tool at this level is a spreadsheet program.

The largest reduction in power consumption can most likely be gained by implementing the most computation intensive parts of the system by application specific logic. Due to the custom character of this part, no predefined module exists. Its functionality is best defined by an executable model (e.g., an algorithm written in a programming or a hardware description language). Because no architecture model exists yet, this has to be predicted together with an interconnect and clock tree model. Based on this estimated design combined with real power models for the allocated predefined components, a power analysis can be performed.

The lower levels of design follow by consecutively generating more detailed design descriptions by a sequence of synthesis steps, refining the test-bench by including bit width, data encoding, and delay information, and by providing the respective lower-level power models. The design objectives at these levels include further local optimizations of the same cost function already applied in a more abstract form at the higher levels of abstraction. A variety of commercial tools are available including, for instance, at the algorithmic-level: ORINOCO by ChipVision [3]; and at the RT-level: PowerChecker by BullDAST [4], PowerTheater by Sequence [5], and PrimePower by Synopsys [6].

38.3 System-Level Power Analysis

A system consists of a set of components, which jointly perform a common task. Definitions like this one describe the essence of system-level design: allocation of components, partitioning of the system's task onto these subsystems, and organization of the cooperation of the components. This section presents methods and tools that can be applied to exploit the largest possible gain in power reduction by partitioning the system in a power-optimal way as well as by introducing power management.

38.3.1 Objectives of System-Level Design

System-level design starts from a specification, some environmental constraints, and possibly a restriction of the design space. The specification can be given informally and, in this case, requires formalization. A well-established formalism is a task graph [7]. A task graph is a representation of a task depicting the subtasks (processes) and their data as well as control flow dependencies. It consists of vertices representing the subtasks and edges representing the data flow and control flow dependencies. A task graph is a system specification exhibiting parallelism and concurrency. Figure 38.3 is an example of a task graph. The start-vertex and the end-vertex are needed to model the synchronous beginning and termination of each execution loop of the task, the other vertices represent the processes P1 to P6 of the task. The solid edges represent data dependencies, which are important to exploit resource sharing. For instance, P6 is data-dependent on the results of P2. Thus, P6 cannot be executed in parallel to P2 (i.e., they can share resources). On the other hand, P3 and P2 are concurrent processes (i.e., it is up to the designer's choice whether he or she allows resource sharing between these processes or not). The dotted edges model the control flow. In the example of Figure 38.3, the edge between end and start specify that this task will be executed in a loop.

The environmental constraints typically include minimum performance requirements, maximum cost, and power constraints, as well as some form factors and I/O loads.

Finally, the third input can be a restriction of the design space (e.g., by requesting to use a given set of processors, DSPs, memories, available custom area, and bus structures). These elements,

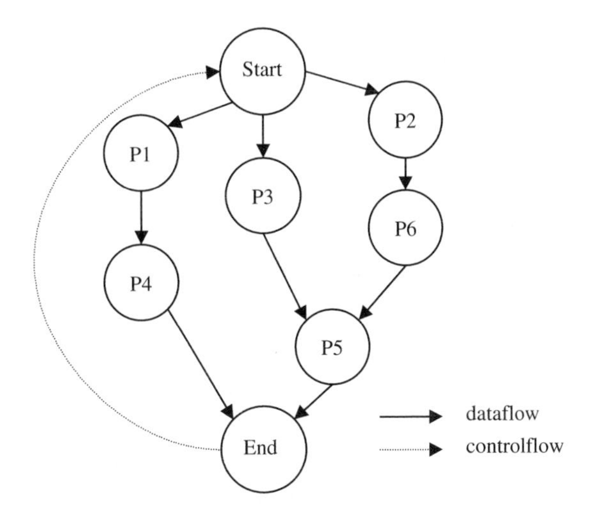

FIGURE 38.3 Task graph.

together with the respective software infrastructure, make a platform. Figure 38.4 is a generic system-level architecture, which could be a platform or an existing design. It is the objective of the system-level designer, the so-called system-architect, to allocate a set of components, map each process of the task graph onto exactly one component (frequently called the processing element), and define the necessary control and communication structure to implement the task graph. The optimization criterion is to achieve an architecture within the given design space, which meets all performance constraints and is an optimal trade-off between cost and power consumption. Note that at this level, the basic structures of a power management policy need to be defined.

A straightforward way for power and performance optimization is to identify so-called computational kernels [8]. These are the inner loops of computation intensive processes. Implementing them by application specific hardware will not only increase the performance and allow using a less expensive processor, but also reduce the power consumption because of the optimized datapath and hard-wired control. Although such a decision may be obvious, however, it requires a detailed understanding of the implications. For example, the communication between the processor and the application specific hardware needs to be considered. The memory architecture may become more

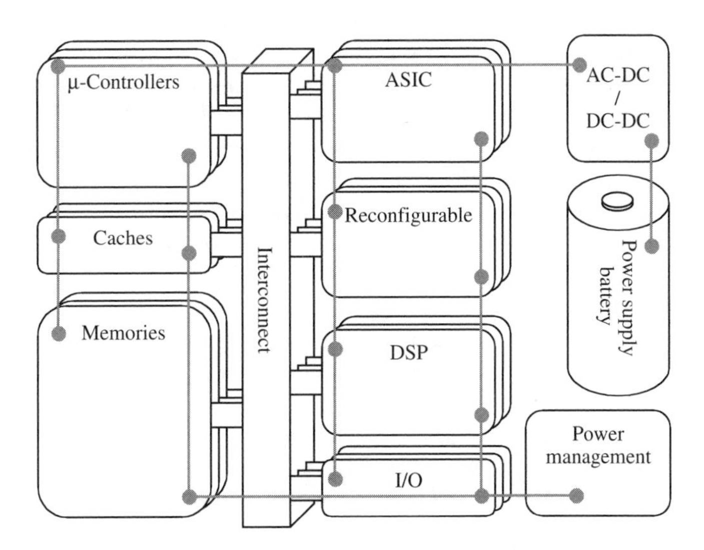

FIGURE 38.4 System-level architecture.

complicated and require multi-port memories. Thus, an adequate tool support is needed to enable sound decisions. An excellent overview of the optimization techniques at this level can be found in Benini and de Micheli [9].

38.3.2 Analysis of an Implementation Model

A frequent design scenario is a platform based design flow. In this case, the designers have to use a given hardware (HW)-/software (SW)-platform to implement the specification. In many cases, an executable model of the system does not yet exist; however, experience with previous systems on the same platform is available. The power estimation can only be based on the general application know-how of the design team, existing power data of the architecture and its components, assumptions about application specific logic, and an intended mapping of the processes of the task onto the various processing elements of the platform, such as processors, DSPs, application specific logic, or memories.

The frequently used straightforward spreadsheet approach of power estimation is based on collecting power related information from data sheets, semiconductor vendors, the application, and experience in a spreadsheet, which implements a power model of the entire system. Often, the experience or data are not yet available for the intended target semiconductor technology of the new design, but only for previous technologies. In these cases, technology-scaling models need to be applied to estimate the power consumption of a module to be processed in a new technology from data of a recent one.

The sources of the power figures for the spreadsheet depend on the kind of the module under consideration. For processors and DSPs, power figures are available from the data sheets of the processor vendors. At this level, power data given by the vendors are typically not detailed to the instruction level or even data dependent, but a single figure in terms of power per megahertz for the processor working at a given supply voltage. Similarly, power data for embedded memories need to be collected from the memory provider or semiconductor vendor.

The interfaces of the design are typically well defined at this level. The system specification should exactly name the I/Os of the system and their required load. The data sheets of the semiconductor vendor offer detailed power figures for these cells.

So far, the power model is an analysis one. Even with rough power models of the components, there was no need to make assumptions about the system architecture. Estimation starts with the consideration of application specific logic. This part may only consume a small fraction of the entire power of the system, but it is the key to reduce the processor power and possibly downgrade the processor to a less expensive and less power-consuming version. At this stage, little is known about the application specific logic. At best, an algorithmic description of this part is available, which can be used for algorithmic-level power estimation and will be covered in Section 38.4.2. If such an algorithm does not yet exist, power estimation is based on the experience of the designer who can predict the number of gates and registers most likely needed to implement the required function within the given performance constraints. Application knowledge can help to estimate the expected activity of such a module. ASIC vendors can provide average power figures for logic in a given technology based on their experience and characterizations. This figure will come in terms of milliwatts per megahertz per kilogate and needs to be weighted with an activity ratio expected for the application. The number of registers can be a useful input to an estimate of the clock tree power.

Table 38.2 presents an example of such a spreadsheet. A refined approach to this principle has been developed as the Web-based tool PowerPlay [10]. Another concept at his level is the power state machine [11], which includes a dynamic model of the activation and deactivation of the various system components. It is well suited for evaluating and optimizing power management policies.

Given the rough granularity of the power models used to create a spreadsheet system-level power model, this can be useful to support a first check whether power constraints of the system will be met or not. It can be used to analyze the impact of moving the design to a new technology node or replacing a processor by another one; however, even these conclusions from the model have to be

TABLE 38.2 Example Spreadsheet for System-Level Power Estimation

	Design				Reference Technology				Scaled Technology			
Example	No. Inst.	Complexity (k gates)	Registers	Activity	Frequency (MHz)	V_{dd} (V)	Power (μW/ MHz/k gate)	Module Power (mW)	Frequency (MHz)	V_{dd}	Scaling Factor	Module Power [mW]
Processors												
Proc.1	1				200	1,8		160	250	1,2		90,0
DSP	1				200	1,8		110	250	1,2		33,0
Memories												
SRAM1	1				200	1,8		70,0	250	1,2		21,0
ROM1	1				200	1,8		55,0	250	1,2		16,5
ROM2	1				200	1,8		55,0	250	1,2		16,5
ASIC logic												
Mod. 1	1	25	500	0,3	200	1,8	4,0	1,5	250	1,2	0,72	0,6
Mod. 2	3	60	260	0,4	200	1,8	4,0	14,4	250	1,2	0,72	5,8
...												
Mod. N	2	18	200	0,1	200	1,8	4,0	0,7	250	1,2	0,72	0,3
I/O												
Inputs	38			0,2	200	1,8		0,2	250	1,2		0,1
Outputs	18			0,5	200	1,8		70,2	250	1,2		46,8
Total								537,0				230,5

drawn carefully because many important parameters could not yet be captured. These include the communication power consumed by the data transfers between processors, memories, application specific logic, and I/Os. The clock network, which may consume a considerable part of the total power, is not yet designed. Issues like cross coupling and the second order effects of the scaling theory are out of the scope of such a model. Finally, neither the impact of the software structure nor of the data has been considered.

38.3.3 Analysis of an Execution Model

A more accurate power analysis at the system-level is possible once executable models of the system processes to be implemented exist. An executable model can have the form of a program written in a programming language, such as C, a hardware description language, such as VHDL or verilog, or a system-level language such as SystemC. Alternatively, a heterogeneous model, combining several languages and models of computation into a single framework, can model the system. In processor design, the system model also can be an executable performance model. By executing any of these models, an understanding of the dynamic behavior of the system can be achieved. This allows a more detailed power analysis under consideration of the real activity in the system. Still, power models of the various components of the system must be at hand, which can be combined with the system architecture and the component activation patterns to a power analysis as given in Figure 38.1 and Figure 38.2. The components of a system-level design include: software, memories, and other existing or yet to be designed modules.

Software-implementing algorithms and running on predefined and power-characterized cores are covered in Section 38.4, Algorithmic-Level Power Estimation and Analysis. If the actual processor is still being developed and optimized, a more detailed bus functional model of the processor execution is needed. Besides functional models, this includes activation models of the processor components (e.g., the issue queue, the branch prediction unit, the execution units, the cache, and the register file). During a simulation of this model, the various components are activated and this activity information is captured. It can be used to evaluate the power models of the components and provide a power analysis of the intended processor architecture [12]. In the case of multipurpose modules controlled by control signals, different power models are required associate to each of the operation modes. Their runtime percentages can be used to calculate the total module power consumption [13].

Memories may consume a considerable percentage of the total power consumption. Consequently, they offer a large power reduction potential. Research has proposed a number of methodologies for memory power optimization. The probably most holistic methodology, called DTSE, has been developed by Catthoor [14]. Arguing that memory power is the dominating part of the system power consumption in signal processing applications, Catthoor advocates that a memory power optimization should be performed before any other power optimizations. The key idea is to apply a sequence of optimizations on the specifications, which, partly automated, perform global loop and control-flow transformations, a data-reuse analysis, a storage cycle distribution, memory allocation and assignment, and, finally, an in-place optimization. The objective is to increase data locality, avoid memory access, and design an optimized memory hierarchy. The result is a system and memory description that can be synthesized.

To assess the power consumption regardless of the optimization methodology requires power models of the various memory types. Power models for memories are difficult to create. They shall be flexible and parameterized (at least with respect to their size), they shall be accurate, and they shall be generated efficiently. The task of characterization of a parameterized memory power model requires a simulation of different instances of the memory. Consequently, simulation models need to be available. Due to the flat structural hierarchy of memories, typically only transistor level models are available. They cause a prohibitive simulation time when executed for memories of practical size. Thus, abstractions have to be used in memory power modeling. These abstractions can be used to model parts of the memory cell array, particularly its capacitive load; however, this abstraction requires access to the internal data of the memory, which is sensitive proprietary information of the memory vendor. To overcome the confidentiality issue, the power model itself should not disclose any information about the internal structure of the memory. Thus, functional power models are adequate, which can be generated using regression techniques. An approach by Schmidt et al. [15] includes nonlinear terms in the regression, which are needed to accurately model, for instance, the address decoder, which is a logarithmic structure.

For other (nonsoftware) components, the activation has to be captured from the executable system model and mapped to a state dependent power macro-model. If application-specific logic (e.g., a block of standard or reconfigurable cells), is to be part of the architecture, this is not yet power characterized. The executable model in this case is a pure functional model (e.g., an algorithm), and the problem of power estimation is the same as discussed with the spreadsheet approach: it requires architecture estimation (see Section 38.4).

Due to long wires and heavily loaded system busses, interconnect power can exhibit a significant percentage of the system power. Analyzing the power consumption of interconnect requires input of physical layout and material properties. This can be partly available for a platform based on measurements or simulations. Off-chip interconnect capacitive loads, which can easily be several orders of magnitude larger than on-chip loads, can be derived from the system specification. The power analysis becomes more difficult for on-chip interconnect and in case of complex bus encoding schemes. The interconnect prediction problem for on-chip wires are discussed in more detail in Section 38.4.

Complex bus encoding schemes have been proven to allow a significant reduction of the switched capacitance of busses (e.g., Fornaciari et al. [16] report a 48% reduction in address busses with Gray Code. Because bus encoding is defined at this level, the impact on the power consumption needs to be taken into account including the overhead for the encoders and decoders.

Similarly, power management needs to be included in system-level power estimation. Its optimization is part of the system-level design. It requires models for the power management policies under consideration as well as for the shutdown and wake-up power penalty. The power management policies can be integrated into the execution model of the system or they can be modeled by a power state-machine [11]. Similarly, dynamic power management techniques, which are typically implemented in the software or the real time operating system (RTOS), require respective models of the policy [9].

38.4 Algorithmic-Level Power Estimation and Analysis

The design tasks at the algorithmic-level of abstraction include optimizing algorithms, which are to be implemented either by software, application specific hardware, or by a combination of both. The objectives include performance, cost, and power optimizations. Means of improvement include selection of the most suitable algorithm performing the requested function, optimizing this algorithm, and partitioning the algorithm into parts, which will finally be implemented in software, and others, which will be realized by application specific hardware.

Selecting the most power-efficient algorithm out of a repertoire of available and functionally equivalent ones requires an estimate of the expected power consumption of an implementation of the different algorithms. Of course, the comparison must be based on power-efficient realizations of these algorithms without the need to really implement them.

Once an algorithm has been chosen, it can be optimized for low power. First, the control flow can be optimized to reduce the number of control statements (e.g., by different kinds of loop unrolling strategies). Additionally, these transformations extend the scope of local statement reordering and pave the way to local memory access optimizations. An example of a sequence of such optimizations is presented in Sarker et al. [17]. The data of the algorithms is typically specified in terms of floating-point variables and arrays. For a hardware implementation, a more efficient data representation is possible (e.g., fixed-point data types of adequate precision for the intended application). Algorithmic-level power estimation is applied to evaluate the impact of the algorithmic transformations and design decisions mentioned in Stammermann et al. [18].

These optimizations, however, have to be made, while considering the target hardware. Moving the computational kernels of the algorithms to power optimized application specific hardware is the most

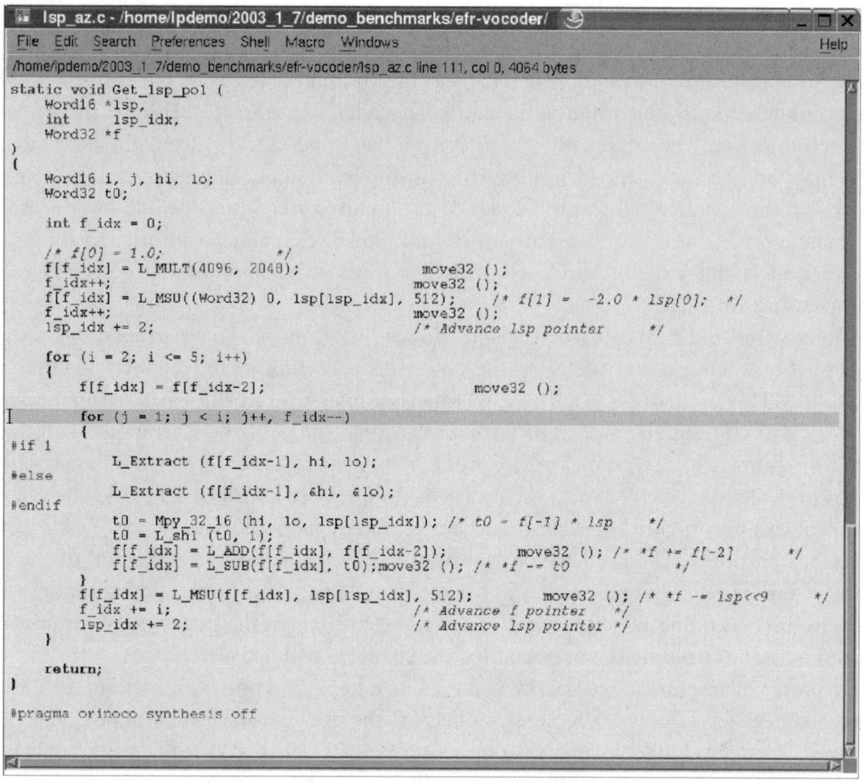

```
lsp_az.c - /home/lpdemo/2003_1_7/demo_benchmarks/efr-vocoder/
File  Edit  Search  Preferences  Shell  Macro  Windows                              Help
/home/lpdemo/2003_1_7/demo_benchmarks/efr-vocoder/lsp_az.c line 111, col 0, 4064 bytes
static void Get_lsp_pol (
    Word16 *lsp,
    int    lsp_idx,
    Word32 *f
)
{
    Word16 i, j, hi, lo;
    Word32 t0;

    int f_idx = 0;

    /* f[0] = 1.0;              */
    f[f_idx] = L_MULT(4096, 2048);                    move32 ();
    f_idx++;                                         move32 ();
    f[f_idx] = L_MSU((Word32) 0, lsp[lsp_idx], 512);    /* f[1] =  -2.0 * lsp[0];  */
    f_idx++;                                         move32 ();
    lsp_idx += 2;                                    /* Advance lsp pointer     */

    for (i = 2; i <= 5; i++)
    {
        f[f_idx] = f[f_idx-2];                                move32 ();

        for (j = 1; j < i; j++, f_idx--)
        {
#if 1
            L_Extract (f[f_idx-1], hi, lo);
#else
            L_Extract (f[f_idx-1], &hi, &lo);
#endif
            t0 = Mpy_32_16 (hi, lo, lsp[lsp_idx]); /* t0 = f[-1] * lsp    */
            t0 = L_shl (t0, 1);
            f[f_idx] = L_ADD(f[f_idx], f[f_idx-2]);        move32 (); /* *f += f[-2]      */
            f[f_idx] = L_SUB(f[f_idx], t0);move32 (); /* *f -= t0             */
        }
        f[f_idx] = L_MSU(f[f_idx], lsp[lsp_idx], 512);        move32 (); /* *f -= lsp<<9     */
        f_idx += i;                                  /* Advance f pointer   */
        lsp_idx += 2;                                /* Advance lsp pointer */
    }

    return;
}

#pragma orinoco synthesis off
```

FIGURE 38.5 Algorithmic specification (C-source-code).

promising path to the largest gain in power consumption. The reasons are simple: the application specific hardware has a hard-wired controller and no need for consecutive control steps to perform a single instruction. No memory access is needed to find out what to do next. The datapath just contains the minimum amount of hardware to perform the operation, and, finally yet importantly, concurrency can be exploited to a much larger degree than this is possible on a processor core. All this avoids wasting energy [19]. Thus, HW/SW partitioning is another important design step, which requires algorithmic-level power estimation to support a trade-off analysis between application specific hardware implementations of parts of the design vs. software implementations. Due to the different nature of software and hardware, dedicated tools are needed for software power analysis and algorithmic-level power estimation for hardware implementations. Both are covered later in this section.

Design input at the algorithmic-level is an algorithmic description, typically executable, or a functional model describing the I/O relation, and a set of constraints. It is important to note that the algorithm is not yet meant as an implementation, but just as a prototype, which needs optimization and implementation. Figure 38.5 is an example of parts of a vocoder design [20]. The function `Get_lsp_pol` is invoked as part of the entire design. It shall serve as an example of a process, which shall be implemented by application specific hardware. It consists of two nested for-loops with some arithmetic operations in the inner loop.

The algorithm can formally be represented by a control and data flow graph (CDFG) [21,22]. The vertices of the CDFG represent either the arithmetic or logic statements of the algorithm, or the control statements. The edges model the data and control flow dependencies. A CDFG implies a partial order on the execution of the statements as required by the data and control dependencies of the algorithm. Figure 38.6 presents the CDFG of the function pictured in Figure 38.5. Because the function contains a nested loop, a hierarchical CDFG is useful, which allows to partly unfolding

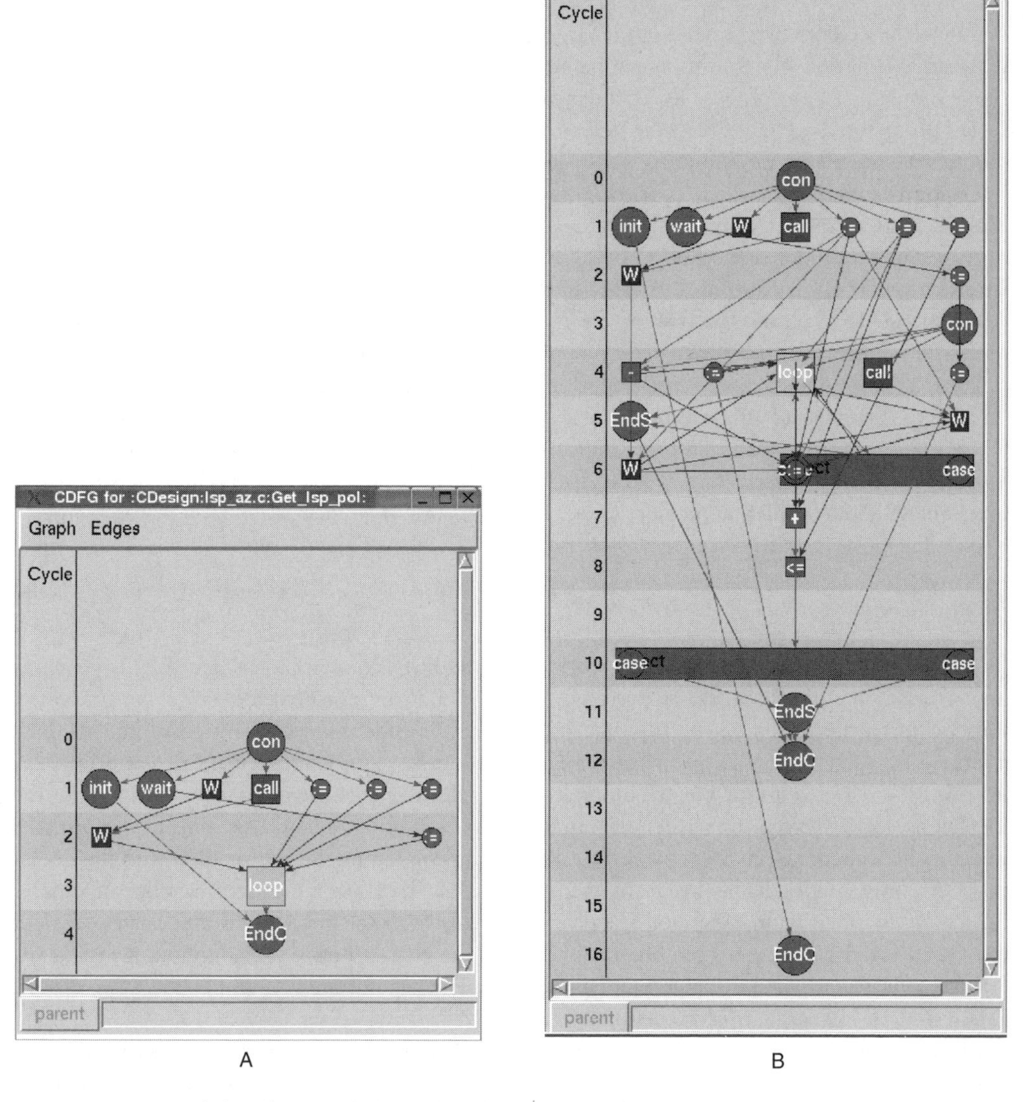

FIGURE 38.6 Control data flow graph (CDFG): (A) CDFG with folded outer loop, (B) CDFG with unfolded outer loop.

the loops as demonstrated in Figure 38.6(B). Comparing Figure 38.6(A) and Figure 38.6(B) demonstrate that the CDFG does not yet imply a schedule but only a partial order.

The output of the algorithmic-level design phase is an algorithm that can be compiled to the target architecture by a software compiler in case the target is software implementation, or an architectural synthesis tool in case of a custom hardware.

38.4.1 Software Power Analysis

Software power analysis is applied to processes to be implemented as software on embedded processors, which can be μ-controllers or DSPs. Software power analysis can be performed at three different levels of granularity:

1. The source-code-level
2. The instruction-level
3. Functional-bus-model-level

They are all based on power models of the target processor and an execution of the software to capture the dynamic behavior. The execution can be performed on the target processor, on another processor, or on a simulator. At any of these levels, the power analysis can be used to compare different programs, to select processors, and to optimize the software. The different levels offer a trade-off in accuracy versus effort to generate the power reports.

The source-code-level, which is the highest abstraction level, provides the fastest turn around times for software power estimation because it avoids the generation of the machine code for the target machine. Brandolese et al. [23] have demonstrated that the execution time of a program on a given processor can be used as a measure of its energy dissipation. Following this idea, the problem of source-code-level power analysis can be reduced to the estimation of the number of execution cycles. This number can be estimated by mapping the source code on instruction classes, which have been empirically characterized with respect to the instructions per cycle of each class. The total energy $E_{program}$, needed for the execution of a program, can thus be estimated using Equation (38.4) with $T_{execution}$ being the total execution time of the program and E_{proc} the energy per MHz clock frequency of the processor. The accuracy of the approach for the average power of single issue processors without considering memory power is within 20% of an instruction level power analysis.

$$E_{program} = T_{execution} \cdot E_{proc} \cdot f \qquad (38.4)$$

A higher accuracy can be achieved by working on the instruction set for which code is generated. Power estimation at this level was pioneered by Tiwari et al. who measured the power consumption of individual instructions and the effects of inter-instruction dependencies [24]. The measurements can be performed by running long loops of the same instruction or sequence of instructions and physically measuring the power consumed by the processor. Through these measurements, a power figure for each pair of consecutively executed instructions can be obtained. A power analysis can thus be performed by capturing the sequence of instructions being executed and combining this information with the instruction-level power model. It has been observed, that an abstraction of the large number of different instructions and their addressing modes is possible by clustering the instructions into classes of similar power behavior. It has further been observed [23] that the relative power consumption of instructions of different classes is similar for different processors. This significantly simplifies the task of power characterization of a large set of processors.

Software power analysis at the levels mentioned so far is limited in accuracy because many aspects of the program execution cannot be considered. These aspects become an increasingly important with the deployment of more complex embedded μ-controllers and hierarchical memory architectures for systems on chip (e.g., pipelined RISC processors, multi-threaded CPUs, out-of-order execution, and embedded caches). Ideally, the processor power models should include the complex relationships between issue queue, execution unit, multiple threads, speculative execution, data dependencies, and cache hit- and miss-rates. Accurately analyzing the power consumption of such architectures requires a profiling of the software on an instruction set simulator with access to the system bus. For instance, such a model has been developed for the ARM processor [25] with an accuracy of 5%. Consideration of these architectural aspects during power analysis allows optimizing either the program for a given processor or the processor configuration for a given program. For example, Simunic et al. [25] could achieve a power reduction of more than 50% by replacing an L2 cache by a burst synchronous dynamic random access memory (SDRAM), while even improving the throughput of a signal processing system. The disadvantage of working at this low level is the long execution time of the simulation. Generating synthetic programs, demonstrating the same performance, and power consumption as the original program but with fewer executed instructions, can speed up the analysis by several orders of magnitude [26].

38.4.2 Algorithmic-Level Power Estimation for Hardware Implementations

As we have discussed, implementing an algorithm or part of it in application specific hardware can significantly reduce the power consumption and relieve the processor from computation intensive task. This may allow downgrading the processor to a cheaper and less power-consuming type. The problem of power estimation at this level, however, is different from the software power analysis problem. The main difference is that the target hardware is not designed yet. The building blocks of that hardware are not yet allocated; the control and data communication between these components is yet to be defined. Thus, before being able to predict the power consumption, it is necessary to estimate the target architecture (see Figure 38.1). The problem of algorithmic-level power estimation for hardware implementations is thus: given the CDFG of an algorithm (Figure 38.6), predict the power expected to be consumed by a power optimized custom hardware implementation of this algorithm. That is, predictions of the target architecture and the activation of the components of the architecture are needed as well as the prediction of the communication and storage. At the algorithmic-level, Equation (38.1) can be replaced by Equation (38.5) [27]:

$$P_{dynamic} = N_a \cdot C_{avg.} \cdot V^2 \cdot f_{comp} \tag{38.5}$$

N_a is the number of activations of the respective module per computation iteration (per sample), $C_{avg.}$ is the average switched capacitance of the module per activation, V the supply voltage of the component, and f_{comp} is the iteration (sampling) frequency of the algorithm. The number of modules and their activation strongly depend on the scheduling, allocation, and binding, which have not yet been performed at the algorithmic-level. To evaluate Equation (38.5), assumptions about the scheduling, the allocation, and binding, as well as the interconnect and storage architecture have to be made.

Additionally, power models for the components must be available. In the case of standard components, these models can be generated by simulation and power characterization based on lower-level power analysis tools [4] and appropriate power models [28,29]. Thus, algorithmic-level power analysis includes the following steps: architecture estimation (i.e., scheduling, allocation, binding of operations and memory accesses, and communication architecture estimation, including wire length prediction), activation estimation, and power model evaluation.

The main challenge of algorithmic-level power estimation for hardware implementations is the difficulty to predict the structural and physical properties of a yet to be designed power optimized circuit. Existing approaches to solving this problem rely on a power-optimizing architectural synthesis of the design before power analysis. The accuracy of the power analysis depends on how well the assumed architecture matches the final architecture. This final architecture is subject to many parameters (e.g., the design style specific architecture templates), which are the main differentiating factors in times of fabless semiconductor vendors, or the tool chain applied at the later phases of the design process (e.g., RT-level synthesis, floorplanning, routing, and clock tree generation). Thus, an architecture estimator should either consider the design flow and style applied to the real design, or generate an architecture of such high quality that it can be implemented without further changes.

38.4.2.1 Target Architecture

Architectural synthesis maps a CDFG onto an architecture template. Figure 38.7 depicts such a generic target architecture for the hardware implementation of a CDFG. It consists of three parts:

1. The datapath, which implements the dataflow of the CDFG
2. The controller, which organizes the dataflow and the control flow
3. The clock tree

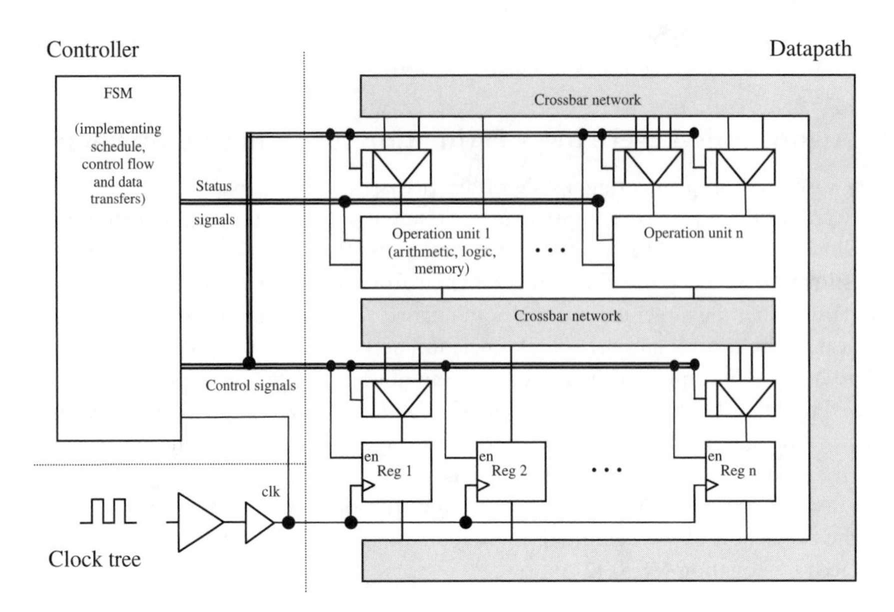

FIGURE 38.7 Generic target architecture.

It is the task of the architecture synthesis to schedule the operations under timing and resource constraints, as well as to allocate the required resources in terms of operation units. The operation units can be arithmetic or logic modules as well as memories. One result of the architecture synthesis is the set of operation units and registers allocated as well as the steering logic, which implements the data transfer connections between the operation units and the registers. The second output is the controller, which is a state machine generating the necessary control signals to steer the multiplexers, operation units, and enable signals of the registers. To do so, it needs to implement the control flow and the schedule based on the status signals of comparator operation units in the datapath. Early work on architectural synthesis for low power has analyzed the impact of binding and allocation during high-level synthesis on the power consumption and integrated power optimizations into high-level synthesis tools [22,30].

38.4.2.2 Scheduling

The schedule of a datapath defines at which control step each of the operations is performed. It has an impact on the power consumption. It defines the level of parallelism in the datapath and thus the number of required resources. The schedule determines the usage of pipelining and chaining. Although pipelining can be a means to reduce power by isolating the propagation of unnecessary signal transitions even within one operation unit, chaining causes the propagation of such glitches through several operation units in one clock cycle and thus increase the power consumption. Musoll and Cortadella [31] have proposed an approach to utilize operations of the CDFG with multiple fan-outs to reduce the power consumption by binding the successor nodes of the CDFG to the same resource in consecutive control steps if they are operationally compatible. This reduces the input activity of these operation units.

Figure 38.8 is the scheduled CDFG of the vocoder example introduced earlier. Comparing the scheduled CDFG with Figure 38.6(b) reveals that some operations have been moved to other control steps and that the additions and subtractions have been bound to specific instances of operation units.

38.4.2.3 Resource Allocation Binding and Sharing

The allocation of resources defines which and how many resources are to be used to implement the CDFG. The binding assigns exactly one operation unit to each of the operations of the CDFG. Several operations can be assigned to the same operation unit if they are scheduled into disjoint control steps and the operation belongs to a subset of the operations that can be implemented by the same unit. These

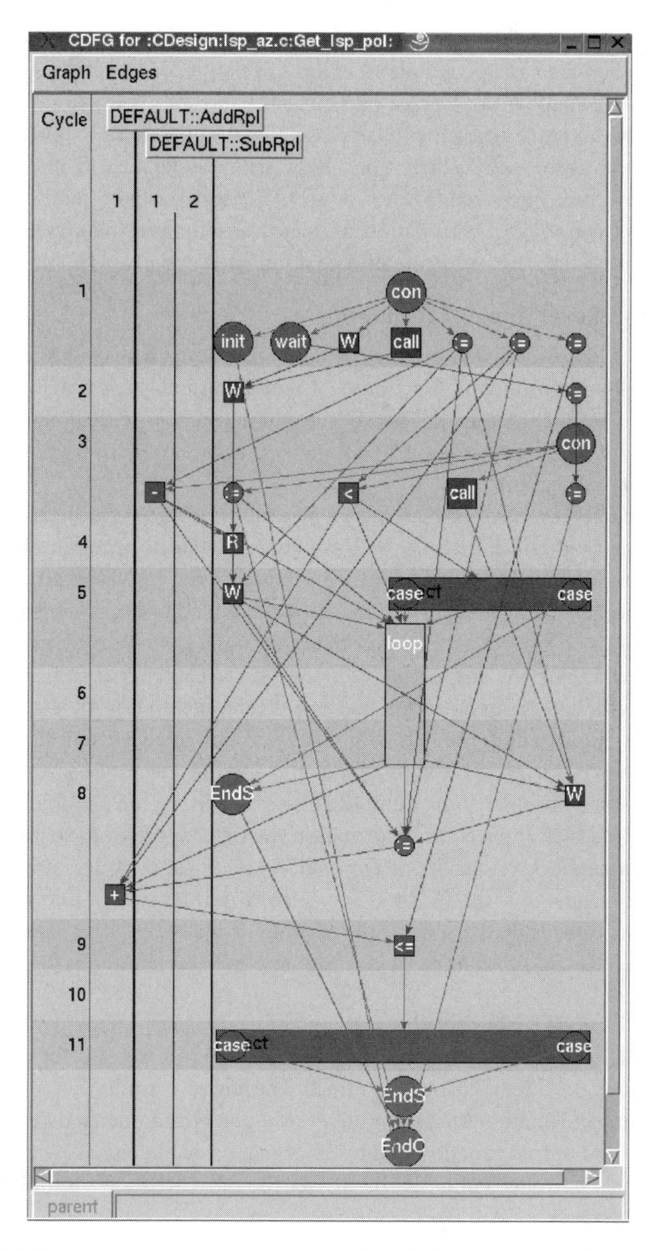

FIGURE 38.8 Scheduled CDFG.

operation units are predesigned and power-characterized modules, such as multipliers, memories, adders, ALUs, comparators, and subtractors.

The valid set of target units of the resource binding depends on the set of operations these units can perform. This opens further possibilities for power optimization because more than one type of operation unit can be chosen as target unit, influencing the resulting power consumption. For example, an addition can be bound to a carry-look-ahead adder, a carry-save adder or an arithmetic logic unit (ALU). Similarly, variables and arrays can be mapped to registers or memories. Typically, arrays are mapped to memories, while single variables are mapped to registers.

The resource allocation and binding affects the power consumption of the datapath due to several effects. The power consumption of each operation unit strongly depends on the switching activity of its inputs. In a routine processing real input data, the internal data applied to the operation units will usually not be independent, but highly correlated in a similar way over a wide range of input data. Applying consecutive input data of high correlation to an operation unit reduces its power consumption. An established measure for the input switching activity is the average hamming distance of a sequence of input patterns [28]. Analyzing the input streams of the operations allows assigning the operations to operation units in a power optimized way by exploiting these data correlations. Because this assignment is an NP-complete problem, different heuristics have been proposed. Khouri et al. [32] use an activity matrix to capture this data dependency and include control flow information and state transition probabilities into the power analysis, while Kruse et al. focus on the iterative nature of data dominated designs [33].

38.4.2.4 Behavioral-Level Power Estimation

In addition to the operations discussed in the previous subsection, an algorithmic specification and its CDFG may contain calls to nonstandard functions (e.g., combinational logic functions, which are defined by their I/O behavior). Because these are not part of the power-characterized library, they require a special treatment during algorithmic-level power estimation. Two main approaches are possible in principle: synthesis or complexity estimation.

Surely, the most accurate results could be achieved by a fully optimized synthesis of the function under observation. This large synthesis effort may be prohibitive for quick turnaround times desired when exploring the algorithmic-level design space. A quick synthesis can be a workaround if it is combined with a calibration procedure, which reliably estimates possible further improvements from the outcome of the quick synthesis. It is obvious that this approach lacks accuracy, while delivering relatively fast results.

The second approach builds on complexity estimates. Müller-Glaser et al. [34] integrated an area and power estimator into a design planning and management tool. Its input is the expected number of gate equivalences, which is empirically calibrated with respect to design styles, tool flow, and technology to produce area and power estimates. This approach is also used in the spreadsheets presented earlier in this chapter. If the required estimate of the number of gates needed is not available because, for instance, no experience exists for a new application, information theoretic approaches step in.

Their input is the functional I/O behavior of a module. A key indicator of the computational complexity and thus of the energy required, is the entropy. The entropy is a measure of uncertainty. The larger the uncertainty of the function's result, the larger is the effort to compute this value. The entropy of a module output can thus be used as an indicator of its computing power consumption [35–37].

38.4.2.5 Controller Power Estimation

Scheduling, resource allocation, and binding have defined the requirements for the controller. Yet, its structure and implementation are still to be determined. The power consumption of the controller (Figure 38.7) depends on its implementation (i.e., the number of registers and their activity, the implementation of the state-transition and output functions, and their signal probabilities). As with the behavioral-level power estimation, a full controller synthesis will deliver the most accurate controller model, which can be used for power analysis.

To reduce the power estimation time, empirical power models [27] have been proposed, which use regression techniques to generate power models for controllers.

The input parameters for the regression include: the number of states, inputs, outputs, and the state coding, as well as the input signal probabilities, which can be extracted from the schedule, the status, and control signals.

38.4.2.6 Interconnect Power Estimation

So far, we have discussed the algorithmic-level power estimation of software and the various hardware components of an embedded system. These components can be separately analyzed or estimated once they have been allocated and their input activity was captured. The power consumption of the communication between these components and their synchronization by the clock, however, requires physical information of the placement of these components and their interconnect as well as their clock tree. As we will see, it is important to consider the effect of the interconnect on the total power consumption during the different steps of the architecture definition. A power aware interconnect design can significantly reduce the total power consumption of the system. For example, Zhong and Jha [38] and Stammermann [39] report power reductions of more than 20% by an interconnect aware high-level synthesis for low power.

This interconnect aware power optimization requires an estimation technique for the interconnect and its power consumption. The interconnect power models applied so far are based on the switched capacitance of the wires, as formulated in Equation (38.1). For a global power estimate, it is sufficient to estimate the total switched capacitance. Empirical wire models like Rent's Rule [40] can be applied to predict the number and average length of wires; however, because a power estimate of an optimized floorplan is needed, this average figure is too pessimistic. Such a power optimal floorplan will locate components, which are communicating at a high data rate as close together as possible and thus save power. Thus, to steer interconnect power optimization, the capacitance and switching activity of individual wires must be known.

The problem of interconnect power estimation is to estimate the capacitance and activity of each wire of an RT-level architecture. Because the activity can be derived from the activity data of the modules, which have been discussed previously, the remaining problem is to estimate the wire capacitance, which is primarily determined by the wire's length, the physical layers used to implement the wire, and the number of vias. The wire length depends on the location of the modules of the design on the floorplan and the routing structure between the modules. The capacitance of a wire, including the effects of vias and multiple connection layers, typically correlates with the wire's length in a nonlinear way [41]. Thus, the main problem remaining is to calculate the expected length of each wire in a power-optimized floorplan. This requires including floorplanning and routing into the estimation. Because of the large impact of the floorplan on the total power consumption, a separated floorplanning phase, once the architecture is fixed, will create suboptimal solutions. Existing approaches, which consider interconnect power during power analysis and optimization at the system-level and algorithmic-level, attack the problem by integrating floorplanning and routing into the architecture optimization discussed previously.

Traditionally, high-level synthesis consists of the phases: allocation, scheduling, and binding, which are typically performed in a sequential manner. High-level synthesis for low power adds a further step: interconnect optimization. Each of these steps is a NP-complete problem, thus the entire problem is NP -complete (i.e., a guaranteed optimal solution cannot be found in reasonable computation time). An optimal design can further not be achieved by applying these steps sequentially because the optimizations are not independent. Consequently, because power analysis at this level requires a detailed understanding of the target architecture, heuristics are needed to synthesize such architecture in a power-optimized way under simultaneous consideration of allocation, scheduling, binding, and floorplanning.

First approaches to combine several of these tasks of high-level synthesis into one optimization loop have been proposed [38,41,42]. The common feature of these optimization flows is to apply a set of moves on a preliminary design, to evaluate the impact of these moves, and to follow an optimizing heuristic such as simulated annealing, thus applying further moves until a stopping criterion is fulfilled.

Prabhakaran et al. [42] apply moves changing the schedule and the binding. Before evaluating the cost function, they perform a floorplanning step during each iteration. Zhong and Jha [38] use allocation and binding moves followed by a floorplanning step for cost estimation. Stammermann et al. [41] include allocation, binding, and floorplanning moves into their optimization heuristics (see Figure 38.9). The upper part of the figure presents the outer loop of the optimization, during which binding and allocation moves are performed. If, based on a preliminary power estimate, a binding/allocation move is promising, then the floorplan is updated and optimized by several floorplan moves in an inner loop, as presented in the lower part of Figure 38.9.

The floorplan and the allocated registers are also the basis for the generation of a clock tree model, which can be used for clock power prediction.

The result is a power-optimized architecture automatically generated from an algorithmic-level description. The expected power consumption of this architecture is analyzed during the optimization loops. This power figure will have a high relative accuracy and can serve as an estimate of the power consumption of the input algorithm. It can be taken as a guide for optimizing the input algorithm for low power.

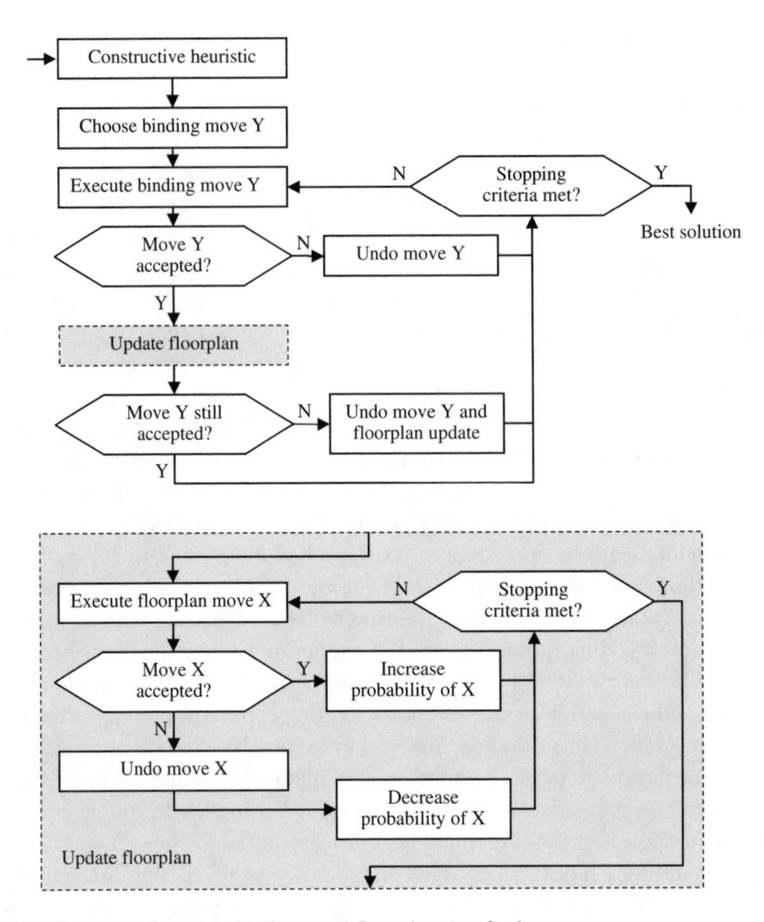

FIGURE 38.9 Simultaneous allocation, binding, and floorplanning for low power.

If, however, the implementation style assumed during the optimization fits the implementation style of the final SoC design, and the architectural parameters (e.g., allocation, scheduling, binding, and floorplanning) of the generated architecture are applied to the final implementation, then the power estimate is a good prediction of the absolute power consumption to be expected for the design.

38.5 ORINOCO: A Tool for Algorithmic-Level Power Estimation

The algorithmic-level power estimation approaches presented so far are results of academic research. Besides these, the ORINOCO tool [3] is commercially available. It is partly based on the research results presented here [15,18,28,29,33,39,41].

Figure 38.10 illustrates the ORINOCO workflow. ORINOCO accepts algorithmic design specifications in the C or SystemC languages. The input description is analyzed and automatically instrumented. The analysis generates the CDFG of the algorithm, which is needed for optimization. The code instrumentation inserts protocol statements, which capture the activity of the algorithm during execution.

In the analysis and optimization phase, presented on the right-hand side of Figure 38.10, the optimization steps described in the previous subsections are applied to generate a power optimized architecture, which is the base of power calculation. The power models are automatically generated by characterization tools, which are part of the ORINOCO tool suite.

Figure 38.11 presents a window of the ORINOCO design browser after power estimation of the vocoder design. The left column lists the various processes of the vocoder; the other columns present parameters of the design and the results of the power analysis. The graphical power reports enable the designer to

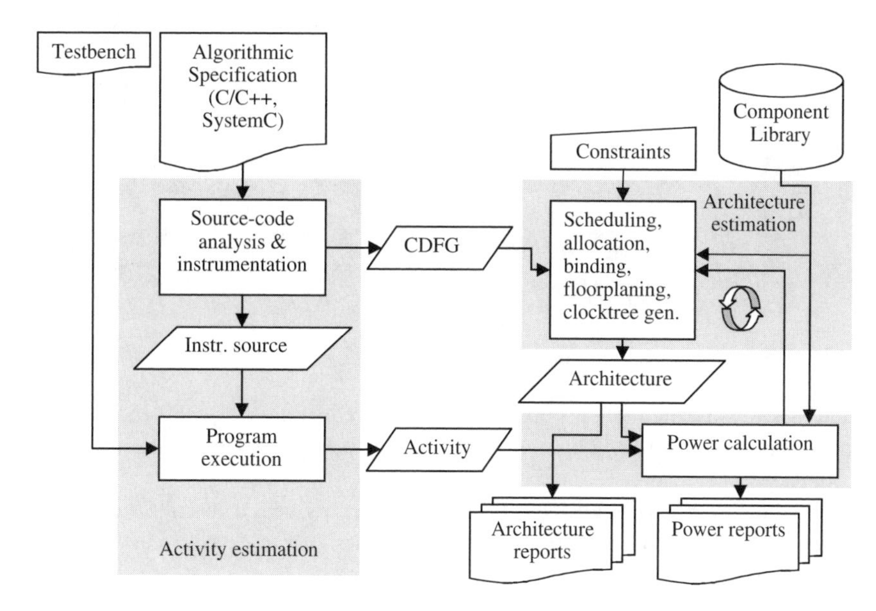

FIGURE 38.10 Workflow ORINOCO Algorithmic-level power estimation.

ORINOCO: efr_vocoder

File Design Settings Display Macro Help

Setup | Read Activity | Estimate | Change constraints and restart the estimation or change source and restart the whole process.

Process	Energy [Ws]	Code Coverage [%]	Voltage [V]	Frequency [MHz]	Critical Path Length/ Constraint [cycles]
CDesign	12.00 μWs	75.68	1.20	100.00	auto
dtx.c	11.04 pWs	82.11	1.20	100.00	auto
vad.c	66.27 nWs	89.38	1.20	100.00	auto
lsp_az.c	60.12 nWs	99.17	1.20	100.00	auto
Lsp_Az	10.05 nWs	100.00	1.20	100.00	8/auto
Get_lsp_pol	50.07 nWs	98.33	1.20	100.00	11/auto
oper_32b.c	21.70 nWs	100.00	1.20	100.00	auto
q_plsf_5.c	6.617 nWs	93.47	1.20	100.00	auto
int_lpc.c	3.658 nWs	100.00	1.20	100.00	auto
syn_filt.c	743.4 nWs	98.63	1.20	100.00	auto
lag_wind.c	5.027 nWs	100.00	1.20	100.00	auto
levinson.c	99.67 nWs	97.55	1.20	100.00	auto
weight_a.c	89.97 nWs	95.24	1.20	100.00	auto
residu.c	2.670 μWs	92.11	1.20	100.00	auto
pitch_ol.c	837.7 nWs	93.85	1.20	100.00	auto
basicop2.c	73.13 nWs	18.18	1.20	100.00	auto
negate_instr	NA	0.00	1.20	100.00	4/auto
L_msu_instr	72.86 nWs	100.00	1.20	100.00	2/auto
L_macNs_instr	NA	0.00	1.20	100.00	2/auto
L_msuNs_instr	NA	0.00	1.20	100.00	2/auto
L_add_c_instr	NA	0.00	1.20	100.00	3/auto
L_sub_c_instr	NA	0.00	1.20	100.00	3/auto
L_negate_instr	NA	0.00	1.20	100.00	4/auto
mult_r_instr	273.6 pWs	100.00	1.20	100.00	2/auto
mac_r	NA	0.00	1.20	100.00	2/auto
msu_r	NA	0.00	1.20	100.00	2/auto
div_s_instr	NA	0.00	1.20	100.00	9/auto
convolve.c	4.134 μWs	94.59	1.20	100.00	auto
autocorr.c	3.636 μWs	97.04	1.20	100.00	auto
az_lsp.c	48.99 nWs	98.92	1.20	100.00	auto
cod_12k2.c	4.560 pWs	100.00	1.20	100.00	auto
I/O Pads	2.715 nWs				
Memories	7.745 μWs				

```
Power estimation successful.

Power estimation took 27.05 seconds.
Power values stored.
report_power -file result.txt
```

Prompt | ready

FIGURE 38.11 ORINOCO browser.

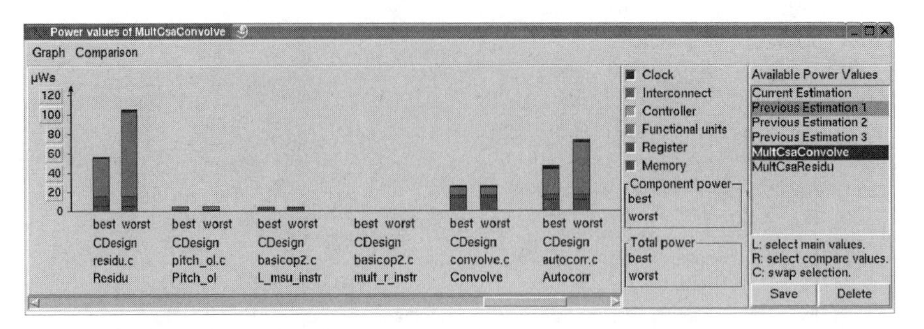

FIGURE 38.12 ORINOCO process power report.

efficiently detect hot spots, the highest potential for power reduction of the design, and the sources of the power consumption. Figure 38.12 is a typical output: the power breakdown of some processes of the vocoder example.

In addition to the power reports, ORINOCO generates outputs, which describe the parameters of the architecture and the floorplan. These outputs can guide the designer to develop a low-power implementation of the current algorithm.

38.6 Conclusion

We described methodologies and techniques used to analyze the power consumption of systems on chip (SoCs) at the earliest possible phase of the design. It is this phase that offers the largest impact on the design, and thus the best chances to optimize the design for performance, power, and cost. Second, performing an early analysis of the expected design properties will help to avoid design iterations from later design phases.

Estimating power at the system-level or the algorithmic-level requires anticipating all the design steps the design will have to go through during the later stages of the design process. It has to predict the impact of the process technology as well as of the design style. It has to consider the application-specific properties of the design, and the performance and cost constraints imposed.

A complete high-level, low-power design flow has been explained with the necessary tools and data to perform power analysis and estimation at the system-level and at the algorithmic-level. This flow is based on the collection, and, if needed, generation as well as the combination of the necessary amount of design data to predict the power consumption of a later implementation in a reliable and sufficiently accurate way.

Nowadays, such a prediction is possible by combining the available techniques into a design and tool flow. The first commercial tools supporting this flow are on the market.

References

[1] 2002 Update of International Technology Roadmap for Semiconductors. Available at http://public.itrs.net/Files/2002Update/Home.pdf.

[2] Hedenstierna, N. and Jeppson, K., CMOS circuit speed and buffer optimization, *IEEE Transactions on CAD*, Vol. 6, No. 3, 1987.

[3] http://www.chipvision.com/dokumente/ORINOCO_WhitePaper.Ext. ChipVision Design Systems AG, *White Paper ORINOCO.*

[4] http://www.bulldast.com/Pee.pdf BullDAST s.r.l., *Power Checker.* Sequence Design, *Power Theater.*

[5] http://www.sequencedesign.com/2_solutions/2b_power_theater.html.

[6] http://www.synopsys.com/products/solutions/galaxy/power/power.html Synopsys, *Galaxy Power Management.*

[7] Rammamoorthy, C.V. and Gonzalez, M.J., Recognition and representation of parallel processable streams in computer programs—II, *Proc. ACM/CSR-ER, Proc. 1969 24th National Conf.*, 1969.

[8] Henkel, J., A low-power hardware/software partitioning approach for core-based embedded systems, *Proc. Design Automation Conf.*, New Orleans, LA, June 1999.

[9] Benini, L. and de Micheli, G., System-level power optimization: techniques and tools, *ACM Trans. on Design Automation of Electronic Syst.*, Vol. 5, No. 2, 115–192, 2000.

[10] Lidsky, D. and Rabaey, J.M., Early power exploration — a World Wide Web application, *Proc. Design Automation Conf.*, Las Vegas, NV, June 1996.

[11] Benini, L., Hodgson, R., and Siegel, P., System-level power estimation and optimization, *Proc. Int. Symp. on Low-Power Electron- and Design*, Monterey, CA, August 1998.

[12] Brooks, D., Tiwar, V., and Martonosi, M., Wattch: a framework for architectural-level power analysis, *Proc. Int. Symp. on Computer Architecture*, Vancouver, Canada, 2000.

[13] Liu, X. and Papefthymiou, M.C., HyPE: hybrid power estimation for IP-based programmable systems, *Proc. Design Automation Conf.*, Anaheim, CA, June 2003.

[14] Catthoor, F., *Custom Memory Management Methodology: Exploration of Memory Organisation for Embedded Multimedia System Design*, Boston, MA: Kluwer Academic Publishers, 1998.

[15] Schmidt, E., von Cölln, G., Kruse, L., Theeuwen, F., and Nebel, W., Memory power models for multilevel power estimation and optimization, *IEEE Transactions on Very Large-Scale Integration Syst.*, Vol. 10, No. 2, 2002.

[16] Fornaciari, W., Sciuto, D., and Silvano, C., Power estimation for architectural exploration of HW/SW communication on system-level buses, *Proc. 7th Int. Workshop on Hardware/Software Codesign (CODES)*, Rome, Italy, 1999.

[17] Sarker, B., Nebel, W., and Schulte, M., Low-power optimization techniques in overlap add algorithmus, *Proc. Int. Conf. on Computer, Commn. and Control Technologies: CCCT '03*, Orlando, FL, July/August, 2003.

[18] Stammermann, A., Kruse, L., Nebel, W., Pratsch, A., Schmidt, E., Schulte, M., and Schulz, A., System-level optimization and design space exploration for low power, *Proc. Int. Symp. on System Synthesis*, Montreal, Canada, September, 2001.

[19] Henkel, J. and Li, Y., Energy-conscious HW/SW-partitioning of embedded systems: a case study on an MPEG-2 encoder, *Proc. 6th Int. Workshop on Hardware/Software Codesign (CODES)*, Seattle, WA, 1998.

[20] European Telecommunications Standards Institute. Available at http://www.etsi.org/.

[21] Girczyc, E.F. and Knight, J.P., An ADA to standard cell hardware compiler based on graph grammers and scheduling, *Proc. IEEE Int. Conf. on Computer Design*, October, 1984.

[22] Raghunathan, A. and Jha, N.K., Behavioral synthesis for low power, *Proc. IEEE Int. Conf. on Computer Design*, October, 1994.

[23] Brandolese, C., Fornaciari, W., Pomante, L., Salice, F., and Sciuto, D., A multi-level strategy for software power estimation, *Proc. Int. Symp. on System Synthesis*, Madrid, Spain, 2000.

[24] Tiwari, V., Malik, S., and Wolfe, A., Power analysis of embedded software: a first step towards software power minimization, *Proc. Int. Conf. on Computer-Aided Design*, San Jose, CA, November 1994.

[25] Simunic, T., Benini, L., and de Micheli, G., Cycle-accurate simulation of energy consumption in embedded systems, *Proc. Design Automation Conf.*, New Orleans, LA, June 1999.

[26] Hsieh, C.-T., Pedram, M., Mehta, G., and Rastgar, F., Profile-driven program synthesis for evaluation of system power dissipation, *Proc. Design Automation Conf.*, Anaheim, CA, June, 1997.

[27] Mehra, R. and Rabaey, J., Behavioral-level power estimation and exploration, *Proc. 1st Int. Workshop on Low-Power Design*, Napa Valley, CA, April, 1994.

[28] Von Cölln, G., Kruse, L., Schmidt, E., Stammermann, A., and Nebel, W., Power macro-modelling for firm-macros, *Proc. PATMOS*, Göttingen, Germany, September, 2000.

[29] Schmidt, E., von Cölln, G., Kruse, L., Theeuwen, F., and Nebel, W., Automatic nonlinear memory power modelling, *Proc. Design, Automation, and Test in Europe (DATE),* Munich, Germany, March, 2001.
[30] Martin, R.S. and Knight, J.P., Power-profiler: optimizing ASICs power consumption at the behavioral level, *Proc. Design Automation Conf.,* San Francisco, CA, June, 1995.
[31] Musoll, E. and Cortadella, J., Scheduling and resource binding for low power, *Proc. Int. Symp. on System Synthesis,* Cannes, France, September, 1995.
[32] Khouri, K.S., Lakshminarayana, G., and Jha, N.K., Fast high-level power estimation for control-flow intensive designs, *Proc. Int. Symp. on Low-Power Electron. and Design,* Monterey, CA, August, 1998.
[33] Kruse, L., Schmidt, E., Jochens, G., Stammermann, A., Schulz, A., Macii, E., and Nebel, W., Estimation of lower and upper bounds on the power consumption from scheduled data flow graphs, *IEEE Trans. on Very Large-Scale Integration (VLSI) Syst.,* Vol. 9, No. 1, February, 2001.
[34] Müller-Glaser, K.D., Kirsch, K., and Neusinger, K., Estimating essential design characteristics to support project planning for ASIC design management, *Proc. Int. Conf. on Computer-Aided Design,* San Jose, CA, November 1991.
[35] Marculescu, D., Marculescu, R., and Pedram, M., Information theoretic measures of energy consumption at register transfer level, *Proc. Int. Symp. on Low-Power Electron. and Design,* Dana Point, CA, April, 1995.
[36] Nemani, M. and Najm, F.N., High-level area and power estimation for VLSI circuits, *Proc. Int. Conf. on Computer-Aided Design,* San Jose, CA, November, 1997.
[37] Ferrandi, F., Fummi, F., Macii, E., Poncino, M., and Sciuto, D., Power estimation of behavioral descriptions, *Proc. Design, Automation, and Test in Europe (DATE),* Paris, France, March, 1998.
[38] Zhong, L. and Jha, N.K., Interconnect-aware high-level synthesis for low power, *Proc. Conf. on Computer-Aided Design,* San Jose, CA, November, 2002.
[39] Stammermann, A., Helms, D., Schulte, M., and Nebel, W., Interconnect-driven low-power high-level synthesis, *Proc. PATMOS,* Torino, Italy, September, 2003.
[40] Christie, P. and Stroobandt, D., The interpretation and application of Rent's Rule, *IEEE Trans. on VLSI Syst.,* Vol. 8, No. 6, December, 2000.
[41] Stammermann, A., Helms, D., Schulte, M., Schulz, A., and Nebel, W., Binding, allocation and floorplanning in low-power high-level synthesis, *Proc. Int. Conf. on Computer-Aided Design,* San Jose, CA, November 2003.
[42] Prabhakaran, P, Banerjee, P., Crenshaw, J., and Sarrafzadeh, M., Simultaneous scheduling, binding, and floorplanning for interconnect power optimization, *Proc. VLSI Design,* Goa, India, January, 1999.

39

Power Macro-Models for High-Level Power Estimation

Enrico Macii
Politecnico di Torino

Massimo Poncino
Università di Verona

39.1 Introduction

As clearly stated at the beginning of Chapter 38, the addition of the power dimension to the already large area/speed design space dramatically expands the number of available design alternatives. Therefore, in a power-conscious design flow, the ability of estimating the impact of the various choices made by the designers or the effects of automatic optimizations on the final power budget is of utmost importance.

Most of the research on power estimation has initially focused on gate and transistor levels; where, due to the available information on the structure and the macroscopic parameters of the devices, accurate power estimates are expected and satisfactory methods are available.

More recently, techniques for high-level (i.e, register transfer level (RTL) and algorithmic-level) power estimation have been proposed to enable designers to cope with increased design complexity and time-to-market requirements. These approaches are usually based on the construction and the evaluation of abstract power models. This information is supposed to guide the designer in exploring the impact of his or her choices on the quality of the final design.

RTL power estimation is at the transition point between research and industrial applications. Power estimators at the RTL are available as commercial tools [1–3], but they have not yet gained widespread acceptance in the design practice, due to two key technical reasons for this. First, the accuracy gap between gate-level and RTL power estimation has not been fully quantified in an industrial setting. Second, RTL estimators are based on macro-modeling, which requires a preliminary characterization step, where a power macro-model is created for the basic functional components in RTL libraries (e.g., adders and multipliers).

0-8493-1941-2/05/$0.00+$1.50

Automated macro-model characterization is a fundamental requirement for the acceptance of RTL power estimation flows in the industrial practice, and this step has been initially overlooked by EDA developers. Fortunately, effective automatic macro-model characterization approaches do exist and are now implemented in the newest commercial tools [3].

As already discussed in Chapter 38, the availability of accurate RTL power models goes beyond the usage in RTL estimation tools; such models are, in fact, also at the basis of the success of power estimators operating at higher levels of abstraction (i.e., algorithmic). This chapter digs deep inside RTL power modeling, and provides a detailed insight to the different facets of the problem of building accurate, yet efficient and easy to characterize RTL power models. Section 39.2 begins by investigating the modeling problem from the theoretical stand-point and by analyzing what are the main dimensions that need to be explored during model definition. In particular, the issues related to the choice of the physical quantities (i.e., the parameters) upon which the models depend are discussed. The semantics associated with the models are also discussed. We conclude by looking at the problem of model representation and storage.

Section 39.3 continues with a discussion of the most relevant steps of a typical macro-modeling flow. Advantages and drawbacks of the macro-modeling technology are outlined, and we demonstrate how such a technology can be exploited for the definition of an RTL power estimation methodology, which is at the foundation of modern EDA tools for low-power design.

Section 39.4 addresses the issues related to the integration of the macro-modeling based approach to RTL power estimation into industry-strength design flows. We discuss how we can deal with RTL descriptions specified through hardware description language (HDLs) and the implications that HDLs may have on the required power modeling capabilities. In addition, we illustrate how models can be enhanced to account for the effects that synthesis and technology may have on the power consumed by the final implementation of the design. Finally, Section 39.5 closes the chapter with some concluding remarks.

39.2 RTL Power Modeling

The problem of power estimation at the RTL amounts to building a power model that relates the power consumption of the target design to suitable quantities. In formula, $P = P(X_1,\ldots,X_n)$, where $X_i, i = 1,\ldots,n$ are the n model parameters.

The construction of a model P implies addressing the following issues:

- The granularity of the model (i.e., to what types of components the power model is referred); this issue implies the definition of the reference architectural model for RTL power estimation.
- The choice of the model parameters (i.e., what and how many parameters upon which the model should depend).
- The semantics of the model (i.e., what is the interpretation of the values returned by the model).
- The way the model is built (i.e., how parameters are put in relation with power); this issue also involves how the model is represented and stored.

The rest of this section discusses the preceding list of four modeling space dimensions, which result in various modeling alternatives.

39.2.1 Model Granularity

In principle, building an RTL power model for an RTL design could be done by considering the design as a monolithic entity. In this case, the model should relate power consumption to properties of the description that can be observed from its RTL I/O behavior.

The choice of a single, monolithic model has several drawbacks:

- Its construction will be extremely time-consuming for RTL designs of realistic size.

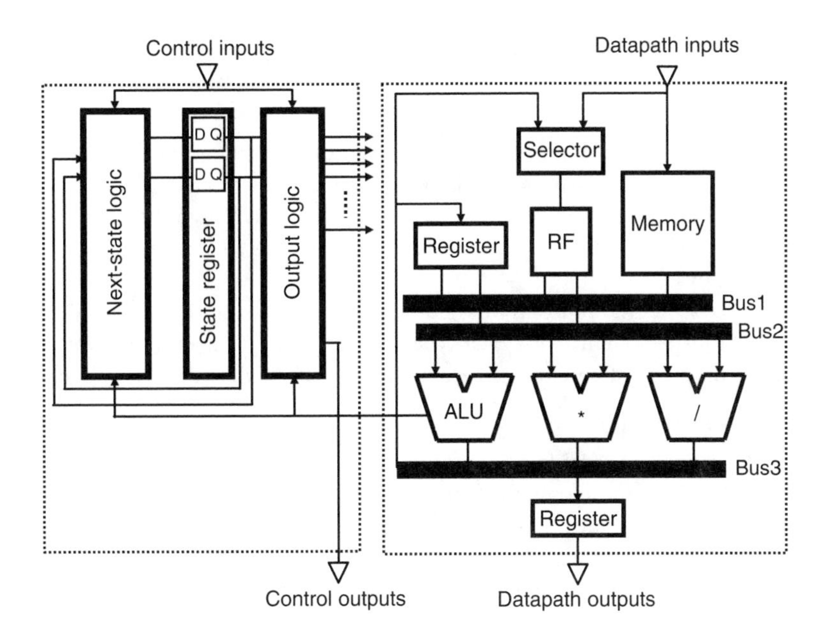

FIGURE 39.1 Architectural template for RTL power estimation.

- Its accuracy will be rather low because we would be trying to express a complex behavior with a single model.
- Its reusability will be quite limited, given its highly customized nature.

These shortcomings prompt for a smaller granularity of the power models. This can be achieved by defining an architectural template and by assuming that all RTL descriptions will map onto that template. The advantage of such a template is that of defining some finer grain objects, for which specific power models can be built. The previous limitations are thus removed by construction: target blocks are smaller, so model construction is simpler and more accurate. Furthermore, because all designs will map onto the same template, reuse will be maximum.

Most of the scientific contributions on RTL power modeling assume (sometimes with minor variants) an architectural template that views an RTL design as the interaction of a datapath and a controller, which fits well the so-called finite state machine with datapath (FSMD) model [4]. Variants to the base FSMD model concern the structure of the controller (e.g., sparse logic implementation vs. wired logic), the structure of the interconnect (e.g., number, type, and size of the buses), or the supported arithmetic operations (e.g., number and type of available functional units). Figure 39.1 depicts an example of a conventional FSMD organization, in which the basic building blocks are explicitly exposed: the controller (shown on the left) and the datapath (shown on the right), with the latter consisting of a register file (and possibly some sparse registers), a memory, various interconnection buses, and some functional units (integer or floating-point).

Fitting an RTL description to this template allows us to restrict the scope and the granularity of the power models to those of the following five main building components: the controller, the registers, the memory, the buses, and the functional units. This approach is followed by most RTL power estimation approaches proposed in the literature [5–12].

39.2.2 Model Parameters

The choice of what parameters should be included in the model is constrained by the fact that they must be quantities that are observable at the RTL. Given the architectural assumptions described in the previous subsection, the traditional model for switching power translates into the following high-level expression:

$$P_{total} = k \cdot \sum_{\forall \, module \, i} A_i C_i \quad (39.1)$$

where A_i and C_i denote the switching activity and the physical capacitance of the generic component i, respectively. k represents the $f \cdot V_{dd}^2$ term, which can be considered as a scaling factor at the RTL because V_{dd} is a technological variable, and the clock frequency is specified up front in the RTL description. Notice that the subscript *total* refers to the fact that power is computed over all components; P_{total} is actually the value of the average power.

Equation (39.1) decouples the problem of building a power model into that of building a model for activity and a model for capacitance, for each type of component. Generally, activity and capacitance models will depend on different parameters because they are affected by different physical quantities.

39.2.2.1 Activity Parameters

Activity is generally easier to model because, even at the RTL, it is a well-defined quantity; however, because RTL simulation is able to monitor activity only at the granularity of the clock cycle, RTL activity is an approximation of the actual one. Therefore, intra-cycle activity, such as that caused by glitches, cannot be extracted from RTL simulation.

Activity models rely on activity parameters, including:

- Bit-wise static and transition probabilities (i.e., quantities referred to specific input or output signals of a component)
- Word-wise static and transition probabilities (i.e., quantities referred to input or output values of a component)

Most activity models proposed in the literature use these probability measures as parameters; the choice between bit- and word-wise quantities allows us to trade off model accuracy for model complexity (i.e., the number of parameters). A n-input, m-output component will require $n + m$ bit-wise parameters, but only two word-wise parameters.

These basic models may be enhanced by using additional activity parameters such as:

- Transition density [13], defined as an average (over time) switching rate. At the RTL, where time is intrinsically discretized to clock edges, density can be regarded as a quantity that incorporates switching and clock frequency information. Density is usually more useful at the gate level, because it allows to capture activity also for nonperiodic (i.e., nonclocked) signals.
- Correlation measures, which take into account spatial correlation between individual inputs or outputs of a component. Spatial correlation, defined as the joint probability of two signals being one, is roughly equivalent to computing the correlation coefficient between these signals. As for probabilities, spatial correlation can be computed bit-wise (one value for each signal pair), or word-wise, by averaging all bit-wise values over the number of signals, to get a single quantity [14]. Notice that transition probabilities already account for temporal correlation (between signals or words).
- Entropy [15,16], which can be used in place of transition probability. In fact, it can be demonstrated that the entropy of a digital signal vs. its static probability p is given by the formula: $p\log_2(1/p)+(1-p)\log_2(1/1-p)$, which closely mimics the behavior of transition probability, with an expression of $2p(1 - p)$. In this sense, entropy does not provide particular advantages over transition probability as a parameter, although it can be used also as a measure of complexity [17].

As a general comment, all activity parameters can be used to represent the switching activity of any of the components of the FSMD model discussed in the previous subsection (i.e., registers, datapath components, memories, buses, and a controller). We can say then that activity parameters are independent of the architecture of the components, and a generic activity model template could be used for all types of components.

39.2.2.2 Complexity Parameters

Modeling physical capacitance is more difficult than modeling switching activity. As a matter of fact, the term "physical" suggests that it is unlikely that we are able to link capacitance to quantities observable at the RTL. For example, the relation between a generic datapath component (e.g., an adder) and its physical capacitance may not be so intuitive. We thus expect capacitance models to be generally less accurate than activity models.

Despite this, RTL capacitance models can be derived with a reasonable degree of accuracy. They all rely on the intuitive observation that capacitance will be roughly related to the number of "objects" (i.e., gates, transistors, or similar lower-level primitives) of the target component. In other words, physical capacitance at the RTL is approximated by complexity, and we thus speak of complexity models, based on complexity parameters.

At the RTL, only a few complexity parameters are available. Those that map onto the basic building blocks of our architectural template are:

- The width of a component, meant as its number of inputs and outputs. This parameter applies to any type of component.
- The number of states. This parameter applies only to the controller, for which the notion of state is explicit.

As mentioned in the previous section, some works have also used entropy as an approximation of complexity [15,16,18], although the relation between the two quantities is quite weak and mostly valid for obsolete circuit technologies.

Any complexity parameter different from the two listed above would require some additional information derived from back-annotation of physical information of previous implementations.

Given the limited choice of parameters, it is quite common that capacitance models exploit information about the "architecture" or the implementation style of a given component to customize the form of the model, and impose some predefined mathematical dependencies. For instance, the model for a *N*-bit ripple-carry adder might be something like $C_{rpc\_adder} = k_1 \cdot N$, recognizing its "linear" complexity in the number of input bits. Using the same line of reasoning, a $N \times N$ array multiplier might have a model of the type: $C_{arr\_mult} = k_2 \cdot N^2$ [19,20].

This implies that capacitance models for the various components will not fit to a single generic template, but they will have different shapes (e.g., equations) for each type of component.

39.2.3 Model Semantics

So far, the discussion has focused on models for average power, which is normally used as a metric to track battery lifetime or average heat dissipation. In this case, the semantics of the model is that of having a single figure to represent the consumption of the target description. Average power models are called cumulative power models [21].

However, the notion of cycle intrinsic of RTL simulation allows us to obtain a power model with a richer semantics by simply changing the way we collect statistics. The first step in this direction consists of modifying the model of Equation (39.1) as follows:

$$P_{total} = k \sum_{\forall\, cycle\ j} P_j = k \sum_{\forall\, cycle\ j} \sum_{\forall\, module\ i} A_{ij} C_{ij} \tag{39.2}$$

where P_j denotes the power consumption at cycle *j*, which can be obtained by summing the power consumption for each component (as in Equation (39.1)), this time using activities and capacitances of component *i* at each cycle *j*.

The semantics of the model of Equation (39.2)is cycle-accurate, because it allows to track cycle-by-cycle (total) power. Equation (39.2) can be thought of as a particular case of a more general model in which power is computed over a sliding window of size W; in this case, $W = 1$ corresponds to the cycle-accurate model, whereas when W equals the length of the simulation stream we fall into the cumulative model of Equation (39.1) [22,23].

The use of a cycle-accurate model clearly affects the choice of the model parameters. For example, transition or static probabilities are not suitable quantities anymore because, as statistical measures, they are intrinsically "average." Conversely, cycle-accurate models should use cycle-based activity measures, such as the number of bit toggles between consecutive patterns (i.e., the hamming distance) [23–25], or the values of consecutive input patterns [21,22]. In the former case, one single parameter is able to capture the information, while in the latter case, the parameter space consists of all possible word pairs, and may thus become quite large.

A cycle-accurate model provides several advantages over a cumulative one. First, it goes beyond the bare evaluation of average power and can be used to perform sophisticated analysis of power consumption over time, which may be required in some application, such as reliability, noise, or IR drop analysis [23]. In addition, a cycle-accurate model is more accurate than a cumulative one, not just because it provides a series of power values as opposed to a single one. In fact, the relation between input statistics and power is nonlinear: average consumption is usually different from the consumption associated to average input statistics, especially when power consumption varies significantly over time. Therefore, even when average power is the objective, averaging the series of cycle-by-cycle values will yield a more accurate estimate than a model of average power. On the negative side, cycle-accurate models require significant larger storage space than cumulative ones.

39.2.4 Model Construction and Storage

The last modeling issue to be considered concerns the way models are built and stored. In our context, the two dimensions are relatively independent, so that we can analyze them separately.

39.2.4.1 Model Construction

To options are used to build an RTL power model: a top-down (or analytical) approach, and a bottom-up (or empirical) one [5]. All power models proposed in the literature fall into one of these two categories.

Top-down approaches relate the power consumption (but also the activity and the physical capacitance) of an RTL component to the model parameters through a closed formula. The term "top-down" refers to the fact that the model is derived directly from the RTL description, and it is not based on lower-level information. For this reason, such a formula normally has a physical interpretation. Analytical models are particularly useful in two cases:

1. When dealing with a newly designed circuit, for which no information of previous implementations is available
2. When the implementation of the circuit, even if not available, follows some predictable template, which can be exploited to force some specific relation between the model parameters

Memories are good examples of blocks for which analytical models are particularly suitable. Their internal organization is well-known and relatively fixed (e.g., cell array, bit-lines, word-lines, decoders, MUXs, and sense amps), thus allowing accurate modeling based on various "internal" parameters [26,27].

If we exclude these special cases, however, top-down models are not very accurate because their link to the implementation (e.g., technology, or synthesis constraints) are quite weak. For instance, the analytical models based on entropy [15,16] are totally insensitive to technology and timing information, and are mostly useful for architectural exploration rather than actual estimation.

Bottom-up approaches, conversely, are based on estimating the power consumption of existing implementations, from which the actual power model is derived. Typically, the template of the power model (i.e., the parameters and a set of coefficient used to weigh the parameters) is defined up front; statistical techniques are then used to fit the model template to the measure of power values. This approach is known as macro-modeling, and has proved to be a very accurate and robust methodology for RTL power

Table 39.1 Model Construction Space

	Equation	Lookup Table
Top-Down	√	Not Used
Bottom-Up	√	√

estimation, and can be considered the state-of-the-art solution. Section 39.3 is devoted entirely to the detailed description of the macro-modeling flow and of its role in the definition of an RTL power estimation methodology.

39.2.4.2 Model Storage

The issue of model storage is concerned with the form of the model. Because models express a mathematical relation between power and a set of parameters, the problem amounts to that of representing such a relation. The two options are:

1. Equation-based models
2. Table-based models

The classification is self-explanatory, and it corresponds to the choice of representing a relation as a continuous function (equation-based models), or a discrete-function approximated by points (table-based models).

These two types of models differ in their storage requirements and robustness; the latter is a measure of model sensitivity to the conditions (i.e., the experiments) used for the construction. In that sense, robustness is an issue only for empirical models. Section 39.3 provides further insight on model robustness.

Concerning storage requirements, equation-based models are clearly much more compact than table-based ones. Generally, an equation will only require the storage of the coefficients of the model, as opposed to a full table. In addition, the accuracy of a table-based model is directly proportional to its size (the denser the table, the higher the accuracy), whereas the accuracy of an equation-based model is independent of the model size.

To summarize, the model construction approach and the model shape are substantially independent characteristics, and, in principle, all combinations of options are feasible. As presented in Table 39.1, however, top-down approaches do not resort to table-based models because they naturally try to construct a formula. In this case, approximating the function by points would not bring any particular advantage.

39.2.5 Accuracy Issues

One important point to be addressed concerns the estimation accuracy that RTL power models can guarantee. In the literature, a 20% estimation error with respect to gate- or transistor-level estimates appears to be accepted as the mark that defines "accurate" models; however, power estimation accuracy is affected by so many factors that expressing it as a single figure can have a very poor meaning. Moreover, sometimes even the definition of accuracy itself is not well understood. This section analyzes what affects the evaluation of the estimation accuracy, and provides some hints on how to critically interpret the results available in the literature.

Having accuracy as a target, we refer to empirical models, which are intrinsically more accurate than analytical ones.

39.2.5.1 Accuracy Metrics

In principle, accuracy can be defined as the (absolute or relative) error of the estimate with respect to the "measured" quantity, that is,

$$E = \frac{|P_e - P|}{\max(P_e, P)},$$

where P_e is the power obtained from model evaluation and P the power obtained from gate- or transistor-level simulation. P_e and P refer to a single evaluation of the model (i.e., for one assignment of the parameters) and to a single low-level simulation run, respectively. Therefore, E is a good indicator for a specific execution of the estimation flow.

However, a correct assessment of the accuracy of the model requires that the dependence of power consumption on the input statistics is taken into account. We call robustness the capability of a power model to provide accurate power estimates over a wide range of parameter values (i.e., statistics).

Assuming that S estimation runs have been performed (using different input values), robustness can be computed by averaging the error E over S experiments, that is:

$$E_{avg} = \frac{\sum_{i=1,\ldots,S} E_i}{S}$$

where subscript i denotes the generic experiment.

Sensitivity to input conditions can be assessed by way of the standard deviation of the relative error, that is:

$$SD = \sqrt{\frac{1}{S} \sum_{i=1,\ldots,S} (E_i - E_{avg})}$$

In the case of cycle-accurate models, the root-mean-square (RMS) of the error can be used instead of the average error to track the robustness.

39.2.5.2 Choice of Experiments

Another important element to be considered when evaluating the accuracy of a model is the choice of the set of experiments used to construct the model. This issue is critical in the case of empirical models, where the model is built based on a set of measured points. A common mistake in this case is to evaluate model accuracy on the same set of experiments (usually, the input/output (I/O) statistics) used to build the model. Such error represents the intrinsic error of the model (typically, very low), but it is not a significant measure of the quality of the model under generic conditions, which can be very different from those used for model training.

This is again a matter of model robustness; in fact, we should distinguish between the in-sample accuracy (the intrinsic error of the model) and the out-of-sample accuracy (the error under all other conditions) [28].

The problem of ensuring robustness has been partially solved by resorting to table-based models, where the parameter space is discretized into equivalence classes, and the dependence of the model on the specific values used for model training is weaker. However, a thorough evaluation of even the most robust and (intrinsically) accurate models has demonstrated that estimation errors higher than 100% can be obtained for specific corner cases (namely, input conditions where the activity parameters have very low values [23]).

A realistic evaluation of model accuracy should be carried out by measuring the suitable metrics (e.g., average error or RMS error, and the relative standard deviation) over a very large set of parameters configuration, including pathological cases. Under these assumptions, an average accuracy of 20% is quite hard to achieve, and figures around 30 to 40% are by far more realistic.

39.3 RTL Power Macro-Modeling and Estimation

In statistical analysis, the term macro-model defines models with a "coarse" level of details, which are thus used for overview purposes (as opposed to micro-models, which incorporate finer levels of details).

The definition of macro-model fits well to RTL power models because the latter are employed to relate quantities pertaining to different abstraction levels, such as RTL parameters to the actual power.

In RTL power estimation, the term macro-model has a more restricted meaning because it is used to identify empirical models, without further distinction between other characteristics.

A number of articles dealing with power macro-models have appeared in the literature. Besides those referenced in Section 39.2, other approaches have been proposed recently [29–36], thus leading to a large variety of macro-models spanning many points of the power modeling space. In fact, cycle accurate and cumulative, activity-based and complexity-based, and equation-based and table-based models have been successfully demonstrated.

Despite the differences that do exist between the various macro-models, the design of a macro-model goes through a well-defined sequence of steps; the following describes the macro-modeling flow in detail.

39.3.1 Macro-Modeling Flow

The construction of an accurate macro-model consists of the following four major steps:

1. Choice of model parameters. Although this step applies also to nonempirical models, it is particularly important for macro-models because it defines the parameter space and it affects the complexity of the next macro-modeling steps. Generally, the goal of this phase consists of choosing what and how many parameters X_i will be part of the model.
2. Design of the training set. The training set is a representative subset of the set of all possible pairs of input vectors that will be used to construct the model. The decisions to be made during this phase concern the size of the training set (i.e., the total number of pairs of input vectors) and the statistical distribution of the pattern pairs in the training set. Although the former issue has to do with the total simulation time, the second is more critical; a bad statistical distribution of the training set may easily offset the advantage of a large number of vector pairs.

   What defines a "good" distribution depends on what are the parameters chosen for the model. A general requirement for the training set is that it should span the domain of all the model parameters as much as possible. When one or more domains are not sufficiently covered by the training set, we say that the model is insufficiently trained.

   For instance, if the parameter of the model is the switching activity, the choice of random patterns as a training set would not be a good one because only a very small portion of the activity domain (i.e., the one around a switching activity of 0.5) would be exercised.

   Although the statistical distribution of the training set is important, it is not the only criterion to be used for choosing the training set. In fact, it is also important to consider how the training set is representative of the actual conditions under which the target component for which power is being modeled will be used. For instance, if we take switching activity as a parameter, we should consider that input vectors with low switching activity will be more frequent than those with high switching activity, in normal operating conditions of the component. In this case, a larger number of low-to-medium activity vectors should be included in the training set.

   In the case of table-based models, the generation of the training set is subject to additional constraints. In fact, models stored as lookup tables are defined only for a set of discrete points, corresponding to specific values of the parameters. This implies that the training set is constrained to vector pairs with values of the parameters corresponding to such discrete points only.
3. Characterization. This step entails the usage of the training set to generate a set of points in the power-parameter space. For each element in the training set (i.e., a vector pair), a corresponding value of power is obtained by means of an accurate, low-level power simulator (i.e., a gate-level or a circuit-level simulator).

   More sophisticated schemes introduce an additional averaging of the samples, by grouping a set of vector pairs and associating a single power value to the set (instead of to each vector pair). This solution is often preferred because simulating a set of patterns instead of a single one increases the confidence of the resulting power value.

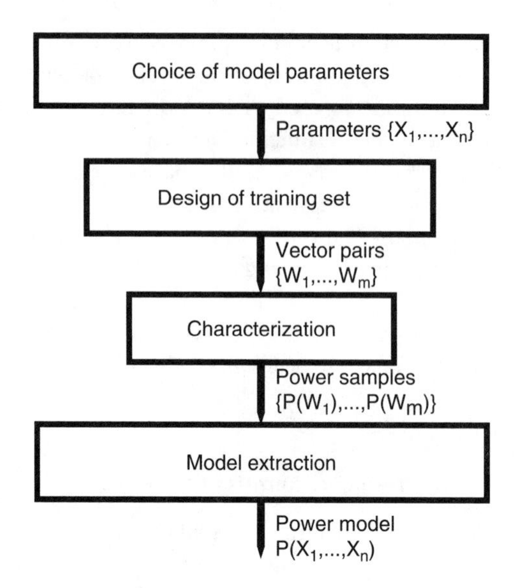

FIGURE 39.2 Summary of macro-modeling flow.

4. Model extraction. This phase consists of deriving the model from the set of power data obtained in the previous step. The actual calculation depends on how the model is stored. For equation-based models, a least-mean-square (LMS) regression engine is applied to the sample generated during characterization. Depending on the options offered by the regression engine, different equation families can be built (e.g., linear, polynomial, logarithmic/exponential).

   For table-based models, the extraction of the model consists of collecting the power values for each of the discrete points of the parameter space. After characterization, each point (i.e., table entry) may contain many power values; the decision on whether to store a single value (e.g., the average of the values) or the complete list of values in each table entry depends on how much room is available for model storage and on how the model will be used for power estimation.

Figure 39.2 summarizes the four steps of the power macro-modeling flow, emphasizing the inputs and the outputs of each phase.

39.3.2 Macro-Modeling Example

Let us apply the macro-modeling flow to the case of a 16-bit ripple-carry adder. To emphasize the difference between equation-based and table-based models, we deal with the two cases separately.

1. Choice of model parameters. For the sake of illustration, we consider a power macro-model containing only one parameter, namely the average switching activity of the inputs S_{in} to the adder. This is a real number between 0 and 1, and it is computed as the number of input transitions between input pattern pairs, divided by the total number of inputs (here, 32). This step is common to both equation-based and table-based models.
2. Design of the training set.
   - Equation-based model. We choose as training set a set $(w_1,\ldots,w_m)$ of $m = 3000$ pattern pairs, with a uniform distribution of S_{in} between 0 and 1.
   - Table-based model. Because the lookup table will have a finite number of entries, we have to choose a proper discretization of the parameter. We use intervals of S_{in} of 0.1; the (one-dimensional) table representing the model will thus have 10 entries. Then, we select 300 pattern pairs for each discrete value of Sin $(0.1,0.2,\ldots,1.0)$, for a total of 3000 vector pairs. Notice that, for table-based models, the goodness of the statistical distribution of the training set is

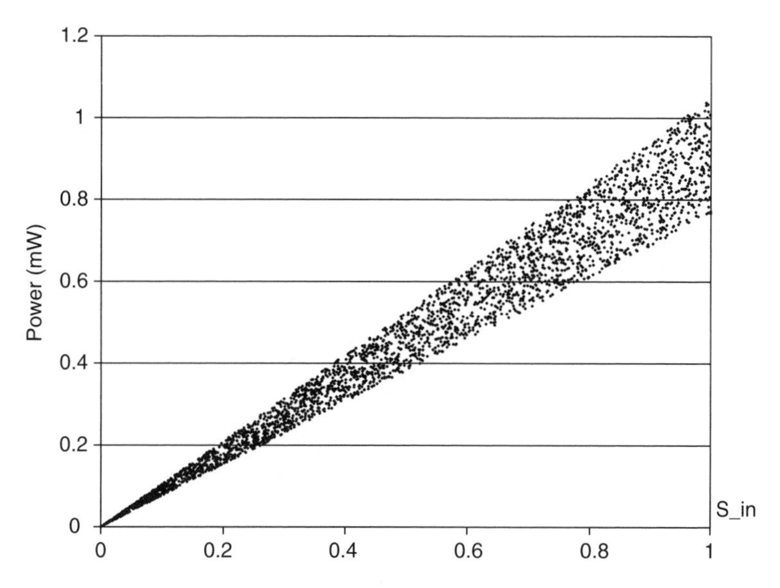

FIGURE 39.3 Scatter plot for the equation-based model.

not an issue. The discretization forces the training set to be designed based on specific values of the parameters.

3. Characterization.
   - Equation-based model. For each vector pair *wi* the corresponding power *P*(*wi*) is determined by means of a low-level simulator. Each simulated value is plotted in the (*P*,*Sin*) space, as depicted in Figure 39.3. Note how the cloud of points illustrates the intuitive trend of an increased power consumption for higher values of the input switching activity.
   - Table-based model. As for the equation-based model, for each vector pair *wi* the corresponding power *P*(*wi*) is computed. Figure 39.4 is a pictorial representation of the set of power data points, which emphasizes the discrete nature of the model.
4. Model extraction.
   - Equation-based model. Given the scatter plot of Figure 39.3, we can derive an equation from it by simply running LMS regression on the set of raw data. Assuming that linear regression is used, and that some corrections are applied to force the intercept with the *y*-axis to be 0, we obtain the following model: $P(S_{in}) = 0.9079 \cdot S_{in} [mW]$.
   - Table-based model. One possible option to store the model is to build a table with one row for each value of S^i_{in}, and a single entry, calculated as the average of the corresponding values $P(S^i_{in})$. With this choice, our table-based model *P* will be as follows:

0.0	0.000
0.1	0.091
0.2	0.181
0.3	0.271
0.4	0.361
0.5	0.451
0.6	0.541
0.7	0.631
0.8	0.721
0.9	0.821
1.0	0.911

   - Using the average as a single representative is based on the assumption of a uniform distribution of the values. More sophisticated solutions analyze the actual distribution of the values and

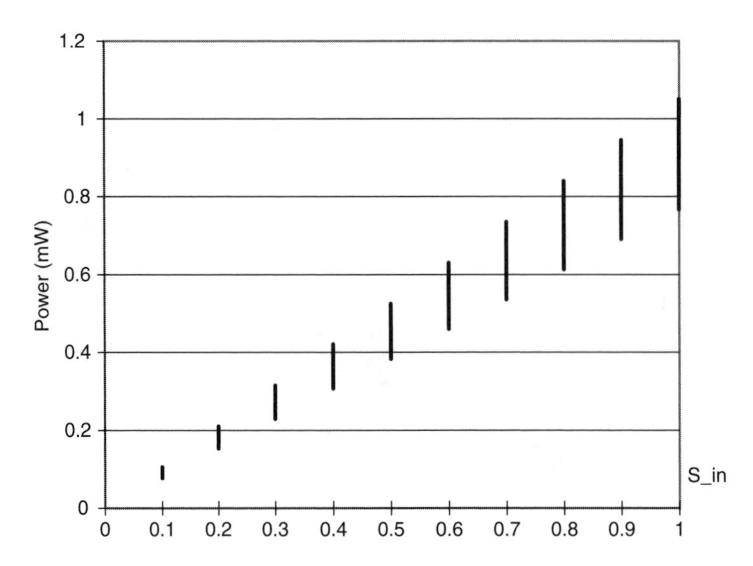

FIGURE 39.4 Plot of power data points for the table-based model.

determine an average value. Some approaches even apply linear regression locally, thus storing an equation instead of a single value [37].

39.3.3 RTL Power Estimation Based on Macro-Modeling

The macro-modeling technology discussed in the previous subsections can be successfully used to enhance state-of-the-art RTL-to-physical design flows with power estimation capabilities.

Assuming that the design to be estimated is described by an FSMD, the estimation procedure consists of the following basic steps [10]:

1. Identification of the individual components in the FSMD. This implies identifying and separating the datapath components among each others and from the finite state machine that represents the control. This is needed to enable the power estimator to generate the power macro-models for each component in the FSMD.
2. Simulation of the FSMD. An RTL simulator is used to trace all the internal signals that define the boundaries between the various components in the FSMD. This is of fundamental importance for the evaluation of the macro-models, which is necessary to complete the estimation procedure.
3. Power estimation. The design hierarchy is traversed at the top level, from the inputs toward the outputs. For each component, the following operations are carried out:
   - Model construction. For each component, the proper model is built using the macro-modeling flow illustrated earlier in this section. A caching strategy can be used to limit the number of times that macro-models are built. More specifically, after the macro-model for a given component is built, it is stored into a cache, so that it can be reused in subsequent runs of the power estimator, should the technology and design constraints not change.
   - Model evaluation. The proper parameter values obtained during the RTL simulation phase are plugged into each model to get the actual values of power consumption.

The total power information for the design is then obtained by summing up the contribution of the model of each component, so that both a total power budget and a power breakdown can be reported to the user. Estimation of the individual components present in the hierarchical description can also be provided. This is possible because components represent the finest level of granularity in the RTL description.

39.4 RTL Power Estimation in Real-Life Settings

Section 39.3 has defined an effective and robust methodology for RTL power estimation based on the macro-modeling paradigm. This section revisits the problem of RTL power estimation from a designer's perspective; and in particular, we try to answer the question of how the macro-modeling approach can be integrated into a standard design flow.

As mentioned in Section 39.2, RTL power macro-modeling relies on a view of an RTL description that fits into the FSMD model. We call this abstraction level structural RTL to distinguish it from the typical designer's view of RTL, usually that of an HDL description containing clocked processes communicating through signals. We call the latter abstraction level cycle-accurate RTL. The gap between these two views of RTL amounts to the presence or absence of structure in the specification, respectively.

In a traditional RTL synthesis flow, the cycle-accurate description undergoes a compilation process that transforms it into an internal format in which structure is made explicit. In VHDL terminology, the compilation process includes the steps of analysis and elaboration; the latter takes care of the delicate task of inferring RTL components from HDL operators.

Therefore, the HDL compilation appears able to fill the gap between cycle-accurate and structural RTL, and, therefore, consistency exists between the two styles. Unfortunately, this consistency is only apparent: HDL compilation builds a structure based on automatic inferencing rules that have limited semantic power. Therefore, the view of structural RTL as an FSMD is ideal. In a realistic design flow, HDL compilation results into the instantiation of four basic types of primitives: RTL components, logic gates, multiplexers, and flip-flops.

At first glance, the difference with the FSMD model is limited to how the controller is specified. Because RTL components are the equivalent of RTL operators, gates, selectors, and memory elements should represent the building blocks that constitute the controller. In practice, however, the degree of inferencing is quite limited, and only the basic HDL operators are inferred ("+", "–", "*",...). For example, registers of the datapath are not instantiated as a single component, and are translated to a set of memory elements. Figure 39.5 depicts how a fragment of an RTL VHDL description is transformed by analysis and elaboration.

The RTL code on the left of Figure 39.5 describes the computation of $z = |x - y|$, which, after HDL compilation, is transformed into the internal database presented on the right of Figure 39.5 (the notation is not bound to any specific synthesis tool). We notice that the "$\geq$" and the "–" operations are translated into the corresponding synthetic operators: One `GTE (component001)` and two `SUB (component002 and component003)`. There is no clear notion of a controller, however, and the rest of the database consists of six gates, one multiplexer (MUX), and four individual flip-flops. Even in this simple example, where the control is limited, a nonnegligible amount of sparse logic exists.

This limited degree of inferencing is due to the fact that the compiled HDL database is meant for being a starting point for synthesis, which will transform primitives into actual library cells. Thus, the instantiation of larger blocks is not so useful to the synthesis tool.

A more detailed analysis on a set of industrial designs confirms the results of the previous example [39]. Although the distribution of the various block types varies significantly across different benchmarks, the relative importance of other primitives with respect to RTL components remains high.

All this discussion demonstrates that, in a realistic design flow, the granularity of a structural RTL design is usually very small, and only a limited number of RTL components is exposed in the RTL internal database. As a consequence, accurate RTL power models for the non-RTL primitives are as important as the macro-models for the RTL components to achieve satisfactory power estimates.

Solutions such as those offered by Synopsys DesignWare [41] represent a step toward a higher degree of inferencing: a library of RTL components is linked to the design, and each RTL component (e.g., a floating-point divider) comes with a VHDL/verilog API that allows designers to replace standard HDL operators (e.g., "/") with a function call (e.g., `DWF_DIVF()`). Designers tend to be uncomfortable with this paradigm, however, because it complicates the reuse of existing designs.

```
ENTITY example IS
GENERIC (W : integer := 4);
PORT(
  clk : IN bit;   -- Global clock
  xin : IN std_logic_vector(W-1 downto 0);
  yin : IN std_logic_vector(W-1 downto 0);
  oup : OUT std_logic_vector(W-1 downto 0));

END example;

ARCHITECTURE rtl OF example IS
BEGIN
  main: PROCESS
  VARIABLE   x   : integer;
  VARIABLE   y   : integer;
  VARIABLE   z   : integer;
  BEGIN
    WAIT UNTIL clk = '1';
    x <= conv_std_logic_vector(xin,W);
    y <= conv_std_logic_vector(yin,W);

    IF (x >= y) THEN
       z := x - y;
    ELSE
       z := y - x;
    END IF;
    oup <= conv_std_logic_vector(z,W);
  END PROCESS main;
END rtl;
```

```
BUF     gate001;
ANDNOT  gate002;
ONE     gate003;
BUF     gate004;
NOT     gate005;
AND2    gate006;
--------------------
MUX     mux001;
--------------------
DFF     ff001_0;
DFF     ff001_1;
DFF     ff001_2;
DFF     ff001_3;
--------------------
GTE     component001;
SUB     component002;
SUB     component003;
```

FIGURE 39.5 Initial RTL VHDL code and internal database after compilation.

The preceding discussion highlights the two main requirements of RTL power estimation in a realistic, HDL-based design flow:

- Power macro-models for the basic RTL components (e.g., "+", "−", "*", "/", "=", " ≠ ", " ≥ ", " ≤ ").
- Power models for the other types of primitives, namely gates, multiplexers, and flip-flops at the RTL. Hereafter, we group these primitives under the name of non-synthetic operators.

Macro-models for RTL components have been extensively discussed in Section 39.2 and Section 39.3. In the sequel, we address the problem of constructing RTL power models for non-synthetic operators.

39.4.1 Power Models of Non-Synthetic Operators

Power estimation of non-synthetic operators at the RTL is different from the case of RTL components. In principle, their power models are quite well understood (e.g., the power model of a NAND gate), and the difficulty arises from two main facts:

- The RTL netlist is different from the actual netlist that will be produced by synthesis, which will optimize the design, possibly under some design constraints. Optimization will reduce the total number of primitives as well as their distribution.
- The RTL netlist is expressed in terms of technology-independent primitives; synthesis, on the contrary, will map primitives onto instances of library cells.

These two facts complicate the problem because our underlying assumption is that RTL power estimation must not rely on RTL synthesis. We usually talk of RTL power estimation as pre-synthesis power estimation.

This leads us to the main issue behind power estimation at the RTL for these types of primitives: estimating the impact of RTL synthesis on power consumption.

Solutions to this problem are not addressed in the literature because the topic is deemed as a practical issue, thus of limited interest from the research point of view.

One exception is the approach followed in [38], which in fact targets an industrial design flow. In that work, the RTL description is decomposed into a set of fine-grain power primitives by a process that combines elaboration and low-effort synthesis. Power primitives are non-synthetic operators for which straightforward power models can be used. The technology independence issue is solved by resorting to fast synthesis.

The rest of this section describes one possible approach that solves both the synthesis estimation and the technology independence issues in an integrated way [39]. This solution, which has been implemented into BullDAST PowerChecker, has provided an estimation accuracy of about 20% with respect to post-synthesis estimates.

One important point to understand is that to achieve acceptable estimation accuracy, we cannot be completely independent of the synthesis tool used in the actual design flow. In the following, without loss of generality, we refer to the Synopsys DesignCompiler [40] flow; similar considerations may apply to any other synthesis flow.

39.4.1.1 Estimating the Effects of Synthesis

Estimating the effects of synthesis consists of relating the pre-synthesis netlist to the postsynthesis one by means of an empirical macro-model, parameterized with respect to complexity and activity parameters.

In this context, complexity parameters are N_R, the number of flip-flops (FF), N_M, the number of MUXs, and N_G, the number of gates in the structural RTL description. The activity parameters extracted from RTL simulation are A_R, the average toggle rate of the FFs, A_M, the average toggle rate of the MUXs, and A_G, the average toggle rate of the gates.

The characterization phase is based on the application of the synthesis flow to a set of RTL benchmarks used as a sample. Regression analysis is applied to the points corresponding to the gate-level power estimation of the synthesized benchmarks, to generate an equation $P = P(N_R, N_M, N_G, A_R, A_M, A_G)$.

The exploration described in Bruno et al. [39] yields a second-order model that is able to provide an average accuracy of about 10%.

39.4.1.2 Achieving Technology Independence

The model described in the previous subsection is technology-dependent because the characterization process refers to a given technology library. The synthesis onto a different technology library (e.g., with different cell types and complexities, or a different feature size) would result in different synthesized implementations, and thus different models.

One way of achieving a model independent of the technology library is that of defining a model scaling mechanism [42]. Starting from a reference technology library L_{ref} (i.e., the one used to build the original power model), a technology scaling factor K is determined for a new technology, such that if P_{ref} is the reference power model, the model for the new library L_{new} will be obtained as $P_{new} = K \cdot P_{ref}$.

The determination of K can be done by defining a sort of "golden" RTL design G to be used for the calibration of the model. G is first synthesized onto L_{ref} and its power consumption P^G_{ref} is evaluated; then, G is synthesized onto L_{new} to determine its power consumption P^G_{new}. K is thus simply obtained as $K = P^G_{new} / P^G_{ref}$.

Concerning the choice of G, a relatively simple description is preferred because it maps onto a small number of non-synthetic operators. For instance, an AND gate is used in Bruno et al. [39], achieving an estimation accuracy of about 20%.

39.5 Conclusions

The ability of accurately characterizing power consumption of complex digital components is at the basis of the setup of power estimation capabilities usable at high levels of abstraction.

This chapter has addressed the problem of building power models for RTL components, and we have discussed how such models can be used for RTL power estimation, a key feature for the enhancement of state-of-the-art RTL-to-physical design flows.

We have illustrated in detail the basic principles of RTL power macro-modeling, which represents today's most advanced technology for RTL power estimation. Both theoretical and practical issues have been considered, thus providing a comprehensive overview of the problem and a review of the solutions that are currently available.

39.6 Acknowledgments

The authors thank BullDAST s.r.l. for the provision of the valuable experimental data included in Section 39.3 and Section 39.4. In particular, the help and support offered by Fabrizio Pro and Maurizio Bruno (BullDAST s.r.l. R&D Division) is acknowledged.

References

[1] Synopsys PowerCompiler, available at http://www.synopsys.com, April 2004.

[2] Sequence PowerTheater, available at http://www.sequencedesign.com, April 2004.

[3] BullDAST PowerChecker, available at http://www.bulldast.com, April 2004.

[4] D.D. Gajski, N.D. Dutt, A.C.-H. Wu, and S.Y.-L. Lin, *High-Level Synthesis: Introduction to Chip and System Design*, Kluwer Academic Publishers, Boston, 1992.

[5] P. Landman, High-level power estimation, *ISLPED-96: ACM/IEEE Int. Symp. on Low-Power Electron. and Design*, pp. 29–35, Monterey, CA, August 1996.

[6] P. Landman and J. Rabaey, Activity-sensitive architectural power analysis, *IEEE Trans. Computer-Aided Design*, Vol. 15, No. 6, pp. 571–587, June 1996.

[7] P. Landman, R. Mehra, and J. Rabaey, An integrated CAD environment for low-power design, *IEEE Design Test of Computers*, Vol. 13, No. 2, pp. 72–82, Summer 1996.

[8] A. Raghunathan, S. Dey, and N. Jha, Register-transfer level estimation techniques for switching activity and power consumption, *ICCAD-96: IEEE/ACM Int. Conf. in Computer-Aided Design*, pp. 158–165, San Jose, CA, November 1996.

[9] S. Katkoori and R. Vemuri, Architectural power estimation based on behavioral profiling, *J. VLSI Design*, Vol. 7, No. 3, pp. 255–270, 1998.

[10] A. Bogliolo, I. Colonescu, R. Corgnati, E. Macii, and M. Poncino, An RTL power estimation tool with on-line model building capabilities, *PATMOS-01: Int. Workshop on Power and Timing Modeling, Optimization and Simulation*, pp. 2.3.1–2.3.10, Yverdon-les-Bains, Switzerland, September 2001.

[11] S. Ravi, A. Raghunathan, and S. Chakradhar, Efficient RTL power estimation for large designs, *IEEE Int. Conf. on VLSI Design*, pp. 431–439, New Delhi, India, January 2003.

[12] D. Helms, E. Schmidt, A. Schulz, A. Stammermann, and W. Nebel, An improved power macro-model for arithmetic datapath components, *PATMOS-02: Int. Workshop on Power and Timing Modeling, Optimization, and Simulation*, pp. 16–24, Sevilla, Spain, September 2002.

[13] F. Najm, Transition density: a new measure of activity in digital circuits, *IEEE Trans. on Computer-Aided Design*, Vol. 12, No. 4, pp. 310–323, April 1993.

[14] S. Gupta and F. Najm, Power macromodeling for high-level power estimation, *DAC-34: ACM/IEEE Design Automation Conf.*, pp. 365–370, Anaheim, CA, June 1997.

[15] M. Nemani and F. Najm, Towards a high-level power estimation capability, *IEEE Trans. on Computer-Aided Design*, Vol. 15, No. 6, pp. 588–598, June 1996.

[16] D. Marculescu, R. Marculescu, M. Pedram, Information theoretic measures for power analysis, *IEEE Trans. on Computer-Aided Design*, Vol. 15, No. 6, pp. 599–609, June 1996.

[17] K.-T. Cheng and V.D. Agrawal, An entropy measure for the complexity of multi-output boolean functions, *DAC-27: ACM/IEEE Design Automation Conf.*, pp. 302–305, Orlando, FL, June 1990.

[18] M. Nemani and F. Najm, High-level area and power estimation for VLSI circuits, *IEEE Trans. on Computer-Aided Design,* Vol. 18, No. 6, pp. 697–713, June 1999.

[19] P. Landman and J. Rabaey, Black-box capacitance models for architectural power analysis, *IWLPD-94: ACM/IEEE Int. Workshop on Low-Power Design,* pp. 165–170, Napa Valley, CA, April 1994.

[20] P. Landman and J. Rabaey, Architectural power analysis: the dual-bit-type model, *IEEE Trans. on VLSI Syst.,* Vol. 3, No. 1, pp. 173–187, March 1995.

[21] Q. Qiu, Q. Wu, C.-S. Ding, and M. Pedram, Cycle-accurate macro-models for RT-level power analysis, *IEEE Trans. on VLSI Syst.,* Vol. 6, No. 4, pp. 520–528, December 1998.

[22] L. Benini, A. Bogliolo, M. Favalli, and G. De Micheli, Regression models for behavioral power estimation, *PATMOS-96: Int. Workshop on Power and Timing Modeling, Optimization, and Simulation,* pp. 125–130, Bologna, Italy, October 1996.

[23] C. Anton, A. Bogliolo, P. Civera, I. Colonescu, E. Macii, and M. Poncino, RTL macromodels for non-stationary workloads, *PATMOS-99: Int. Workshop on Power and Timing Modeling, Optimization, and Simulation,* pp. 313–322, Kos, Greece, October 1999.

[24] H. Mehta, R.M. Owens, and M.J. Irwin, Energy characterization based on clustering, *DAC-33: ACM/IEEE Design Automation Conf.,* pp. 702–707, Las Vegas, NV, June 1996.

[25] S. Gupta and F. Najm, Energy-per-cycle estimation at RTL, *ISLPED-99: ACM/IEEE Int. Symp. on Low-Power Electron. and Design,* pp. 16–17, Monterey, CA, August 1999.

[26] D. Liu and C. Svensson, Power consumption estimation in CMOS VLSI chips, *IEEE J. Solid-State Circuits,* Vol. 29, No. 6, pp. 663–671, June 1994.

[27] E. Schmidt, G. Jochens, L. Kruse, F. Theeuwen, and W. Nebel, Memory power models for multilevel power estimation and optimization, *IEEE Trans. on VLSI Syst.,* Vol. 10, No. 2, pp. 106–109, April 2002.

[28] A. Bogliolo and L. Benini, Robust RTL power macromodels, *IEEE Trans. on VLSI Syst.,* Vol. 6, No. 4, pp. 578–581, December 1998.

[29] A. Bogliolo, L. Benini, and G. De Micheli, Adaptive least mean square behavioral power modeling, *EDTC-97: IEEE European Design and Test Conf.,* pp. 404–410, Paris, France, March 1997.

[30] Z. Chen and K. Roy, Estimation of power dissipation using a novel power macromodeling technique, *IEEE Trans. on Computer-Aided Design,* Vol. 19, No. 11, pp. 1363–1369, November 2000.

[31] M. Barocci, L. Benini, A. Bogliolo, B. Riccò, and G. De Micheli, Lookup table power macro-models for behavioral library components, *IEEE Alessandro Volta Memorial Workshop on Low-Power Design,* pp. 173–181, Como, Italy, March 1999.

[32] A. Bogliolo, E. Macii, V. Mihailovici, M. Poncino, Combinational characterization-based power macro-models for sequential macros, *PATMOS-99: Int. Workshop on Power and Timing Modeling, Optimization, and Simulation,* pp. 293–302, Kos, Greece, October 1999.

[33] G. Jochens, L. Kruse, E. Schmidt, and W. Nebel, A new parameterizable power macro-model for datapath components, *DATE-99: IEEE Design Automation and Test in Europe,* pp. 29–36, Munich, Germany, March 1999.

[34] G. Bernacchia and M.C. Papaefthymiou, Analytical macromodeling for high-level power estimation, *ICCAD-99: IEEE/ACM Int. Conf. on Computer-Aided Design,* pp. 280–283, San Jose, CA, November 1999

[35] A. Bogliolo, L. Benini, and G. De Micheli, Regression-based RTL power modeling, *ACM Trans. on Design Automation of Electronic Syst.,* Vol. 5, No. 3, pp. 337–372, July 2000.

[36] L. Benini, A. Bogliolo, E. Macii, M. Poncino, and M. Surmei, Regression-based RTL power models for controllers, *GLSVLSI-00: ACM/IEEE Great Lakes Symp. on VLSI,* pp. 147–152, Evanston, IL, March 2000.

[37] M. Anton, I. Colonescu, E. Macii, and M. Poncino, Fast characterization of RTL power macromodels, *ICECS-01: IEEE Int. Conf. on Electron., Circuits and Syst.,* pp. 1591–1594, La Valletta, Malta, September 2001.

[38] R. Peset Llopis and K. Goossens, The Petrol approach to high-level power estimation, *ISLPED-98: ACM/IEEE Int. Symp. on Low-Power Electron. and Design,* pp. 130–132, Monterey, CA, August 1998.

[39] M. Bruno, A. Macii, and M. Poncino, A statistical power model for non-synthetic RTL operators, *PATMOS-03: Int. Workshop on Power and Timing Modeling, Optimization, and Simulation*, pp. 208–218, Torino, Italy, September 2003.
[40] Synopsys DesignCompiler, available at http://www.synopsys.com, April 2004.
[41] Synopsys DesignWare Library, available at http://www.synopsys.com, April 2004.
[42] A. Bogliolo, R. Corgnati, E. Macii, and M. Poncino, Parameterized RTL power models for combinational soft macros, *IEEE Trans. on VLSI Syst.*, pp. 880–887, Vol. 9, No. 6, December 2001.

40

Synopsys Low-Power Design Flow

Renu Mehra
Barry Pangrle
Synopsys, Inc.

40.1 Introduction

Design automation tools for analysis and optimization are key enablers for low-power design. With million-gate designs becoming commonplace and design sizes reaching the tens of millions mark, it is impossible to get a power-efficient implementation without appropriate automation. Automated optimization techniques are the fastest way to low-power design and, in fact, sometimes the only way for the highly complex chips of today. For optimization, a comprehensive set of register transfer level (RTL) and gate-level techniques are needed. These include clock gating, operand isolation, and many logic optimizations. In addition, several capabilities that allow both multi-threshold design and multi-voltage design are becoming increasingly important. A good analysis capability provides a basis for understanding the power needs of the design, identifying bottlenecks, and aiding in making correct decisions to reduce power. The power analysis tools need to provide a detailed, time-based power analysis capability for full chips.

Figure 40.1 presents some sources of power dissipation using an example inverter. Dynamic power is consumed by internal or short-circuit switching current that occurs when both transistors are on as when the input is in transition. Also contributing to dynamic power is the switching of the output that causes

0-8493-1941-2/05/$0.00+$1.50

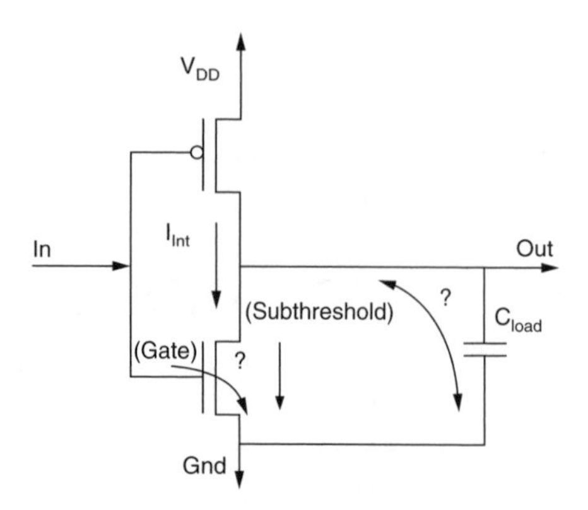

FIGURE 40.1 Example sources of power dissipation.

the charging and discharging of the output capacitive load. Static or leakage power is consumed largely by subthreshold leakage that is becoming more prevalent as threshold voltages are dropping and by gate leakage currents that are on the rise with thinner gate oxides being used.

The rest of this chapter gives an overview of the capabilities needed in today's power optimization and analysis tools. Various automated optimization techniques are discussed in Section 40.2 through Section 40.7. Section 40.8 and Section 40.9 cover the basics of power analysis in computer-aided design (CAD) tools in terms of modeling of the various components and provide an idea of the analysis flow from the designer's point of view.

40.2 Clock Gating

Clock gating involves dynamically shutting off the clock to portions of a design that are idle or are not performing useful computation. This technique is one of the most successful and widely used techniques for power reduction [1–4]. Figure 40.2(b) depicts the concept of clock gating using an AND gate. The basic idea is to AND the clock with an enable signal, so that the register receives a clock signal only when the enable is high.

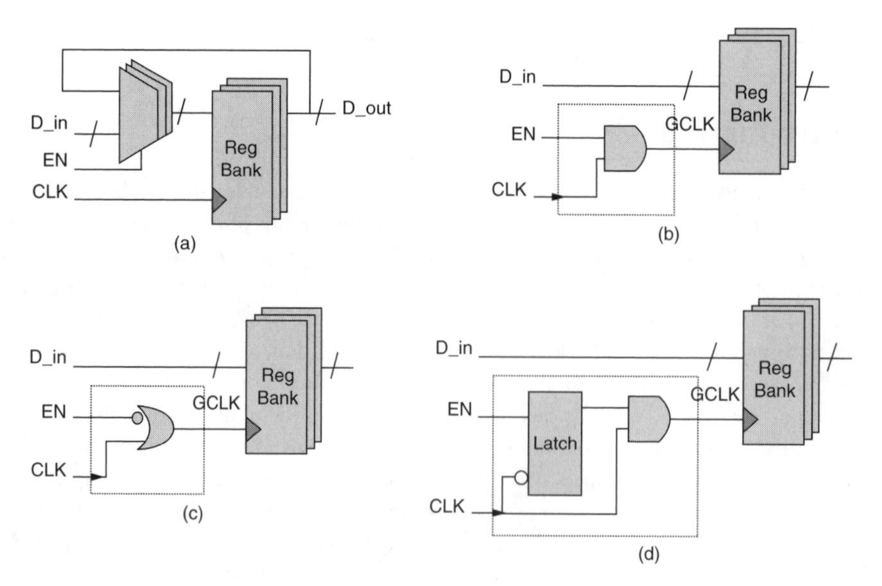

FIGURE 40.2 Clock gating: (a) traditional load-enabled register bank implementation, (b) clock-gated implementation using AND gate only, (c) clock gating using an OR gate, (d) clock gating using a latch and AND gate.

The granularity of the circuit block at which clock gating is applied greatly affects the power savings that can be achieved because gating larger blocks results in higher power savings in the "off" clock cycle, but allows fewer number of "off" clock cycles. Three levels of granularity can be distinguished and are discussed next.

40.2.1 Module-Level Clock Gating

This involves shutting off an entire block or module in the design. Usually, this decision is taken by the system or RTL designer. This method saves large amounts of power when the entire block is not functioning. Perfect cases for this are when a block is used only for a specific mode of operation (e.g., the receiver and transmitter parts in a transceiver may not be active at the same time), and the receiver can be shut off during transmit stages or vice versa. The opportunities for this kind of clock gating are limited and must be identified by the designer and incorporated into the RTL code.

40.2.2 Register-Level Clock Gating

In this method of clock gating, the clock to a single register or set of registers is gated. Synchronous load-enabled registers are usually implemented using a clocked D flip-flop and a recirculating multiplexer as depicted in Figure 40.2(a), with the D flip-flop being clocked every cycle. Clock-gated versions of the same register are depicted in Figure 40.2(b), Figure 40.2(c), and Figure 40.2(d). In the clock-gated versions, the register does not get the clock signal in the cycles when no new data is loaded, thereby saving power. Eliminating the multiplexer also saves power. Gating a single bit register, however, has the associated penalty of power consumption in the clock-gating logic. The key, therefore, is to amortize this penalty over a large number of registers, saving the flip-flop clocking power and the multiplexer power of all of them using a single clock-gating circuit.

Although power saving per clock-gate is much less with register-level clock gating than that obtained with module-level clock gating, this method detects many more opportunities to shut off clocks than would be possible with module-level clock gating. In addition, it lends itself well to automated insertion and can result in very large number of clock-gating cells in the design or massively clock-gated designs. Massively clock-gated designs cause several issues with automated flows most of which are discussed in Section 40.3.

40.2.3 Cell-Level Clock Gating

The cell designer usually introduces cell-level clock gating. For example, a register-bank can be designed such that the registers in the bank receive the clock only when the register is loading new data. Similarly, a memory block may be clocked only during active access cycles. Although this is an easier method of implementing clock gating with no flow issues, it may not be the most efficient from the area and power point of view. It has an area overhead and limits the amount of power savings because all the registers in the design would need to be predesigned with clock gating. In addition, it does not allow the sharing of clock-gating logic across many registers. The previous two methods have additional power savings due to reduced switching in the capacitance of the clock lines from the clock gate to the register, which are not realized with this implementation.

40.3 Automated Clock Gating at the Register Level

In a traditional automated ASIC design flow, all registers are assumed to synchronously read data. This paradigm is violated with register-level clock gating that introduces several challenges in an ASIC design flow. This section presents the challenges of using massively gated clocks in a practical design flow and discuss some of the solutions. For ease of explanation, we assume that the registers being clock gated are positive-edged; however, all concepts will apply, with inverted clock sense, for negative-edged registers.

40.3.1 Practical Gating Circuits

In the simple clock-gating configuration depicted in Figure 40.2(b), glitches on the enable signal that occur when the clock is high are propagated to the clock pin of the register. Although most glitches can be prevented by applying appropriate setup and hold constraints on the enable signal of the AND gate, any spurious changes during runtime (due to coupling with other signals, etc.) can cause wrong values to be latched into the gated register. A slightly safer method is to use an OR gate as depicted in Figure 40.2(c), which holds the output values at logic "1" when the enable signal, *en*, is high. Spurious runtime glitches in this configuration can cause wrong values to be loaded into the flip-flop but the final value loaded at the "valid" rising edge of the clock will be correct.

A better way to avoid this potential problem is to add a level-sensitive, active-low latch on the enable path as depicted in Figure 40.2(d). This freezes the latch output at the rising edge of the clock, and ensures that the new enable signal, *en1*, at the AND gate is stable when the clock is high. In addition, the enable signal can time-borrow from the latch, so that it essentially has the entire clock period to propagate to the latch.

Alternatively, a falling-edge flip-flop can be used instead of the latch to ensure a clean signal to the AND gate. This requires that the enable signal be stable before the falling edge of the clock, however, resulting in a stricter timing constraint on the enable path.

40.3.2 Clock Latency

To maximize the power savings, a single clock gate may be used to gate several flip-flops, if the enable signal is common to all of them; but the clock gate may not have the drive strength required to drive all these registers, requiring a clock tree at its output. If a clock tree is introduced between the clock gate and the registers it gates, the clock signal at the gating logic arrives much before the clock signal at the registers and the enable signal must be ready before the clock arrives at the gating logic. This applies strict timing requirements on the enable signal, which must be addressed during synthesis. Otherwise, this can result in large timing violations after clock-tree synthesis.

One way to address this issue is to specify the clock latency at the clock gate to be smaller than that at the registers themselves during synthesis. The difference in the latencies is the delay of the clock-gating circuit and the clock tree between the clock gate and the registers. This forces the synthesis tool to ensure that the enable signal arrives on time. The difficulty with this approach is that the designer should be able to estimate the latency difference at the two points far in advance of the actual clock-tree synthesis step. The designer can either use a conservative (worst-case) estimate of the clock-tree synthesis delay, or "force" a specific delay by limiting the fanout of each clock-gating cell.

40.3.3 Effect of Clock Skew

Another problem in the latch-based architecture comes from the fact that clock skew between the latch and the AND gate can result in glitches at the gated-clock output. This is explained in Figure 40.3. Figure 40.3(a) illustrates the case when the clock arrives much earlier at the AND gate than at the latch. Here, the clock-skew between the latch and the AND gate should be less than the clock-to-output delay of the latch for the circuit to function properly. Figure 40.3(b) illustrates the case when the clock arrives earlier at the latch. Here, the clock-skew between the AND gate and latch should be less than the sum of the setup time of the latch and the input-to-output delay of the latch to function properly. Therefore, the clock-skew between the latch and AND gate, C_s, should be carefully controlled according to the following equation:

$$-(s + d_{in}) < C_s < d_{clk}$$

where s is the setup time of the latch, d_{in} is the input-to-output delay of the latch, C_s is the difference in clock arrival time between the latch and the AND gate (the clock arrival time at the AND gate minus the clock arrival time at the latch), and d_{clk} is the clock-to-output delay of the latch.

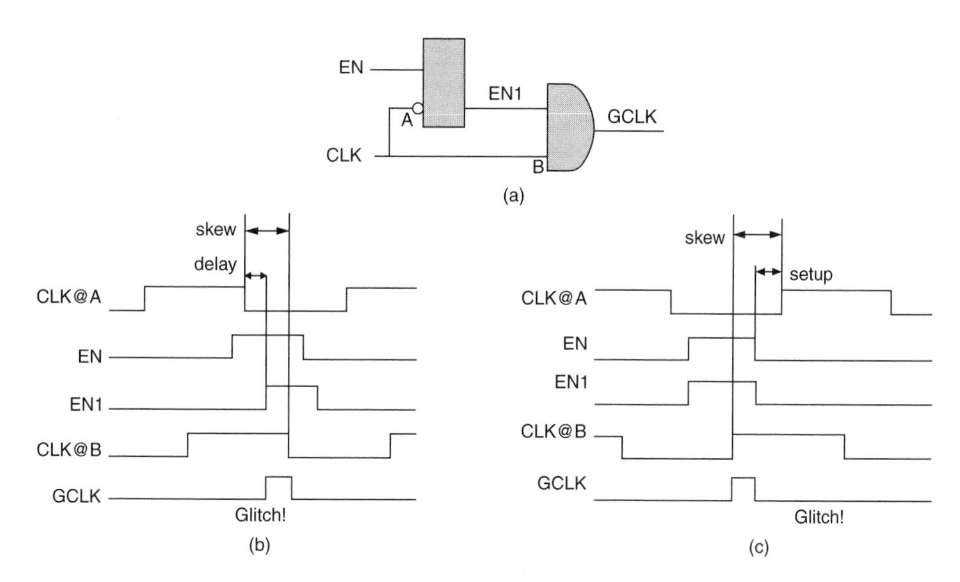

FIGURE 40.3 Clock skew within clock-gating logic: (a) clock-gating logic, (b) glitches due to positive clock skew between the latch and AND gate, (c) glitches due to negative clock skew between the latch and AND gate.

Depending on the relative placement of the latch and the AND gate, these requirements may pose very stringent constraints on the clock-tree synthesis tool.

The best way to control the relative timing of the two clock signals is to keep the entire structure in a single cell, called the integrated clock gating (ICG) cell. The cell should be designed specifically for clock gating, with the explicit requirements discussed previously. Because this cell cannot be modeled either as a combinational or sequential cell, the new "state-table" model in Liberty library format [24] is used for this.

Another way to address this issue is to ensure that the latch and AND gate are close to each other during the placement phase of the design, by placing hard constraints on the distance between them. This makes it simpler for clock-tree synthesis tools to reduce the clock skew between them during clock-routing phase.

40.3.4 Clock-Tree Synthesis

In an automated ASIC design environment, the clock signal typically remains untouched during synthesis, and clock-tree synthesis is done as one of the last steps in the design flow after placement and routing. Because manual (module-level) clock gating introduces only few clock gates in the design, clock-tree synthesis tools can work with these with some manual intervention. In the presence of massively gated clocks, however, clock-tree synthesis tools must automatically address the presence of clock gates on the clock line to be a viable solution. The requirements from the clock tree synthesis are:

- Optimization of the clock-tree in the presence of logic.
- Support for the integrated clock-gating cell on the clock tree. This is a sequential cell, but is not an end point on the clock-tree and, therefore, must be handled in a special way.
- Support for different relative latency requirements at different points in the clock tree.
- Stringent control of clock skew between the latch and the AND gate if the integrated clock-gating cell is not used.

40.3.5 Physical Clock Gating

Physical clock gating simultaneously takes into account the factors mentioned in the three previous subsections, namely latency and clock-tree synthesis issues.

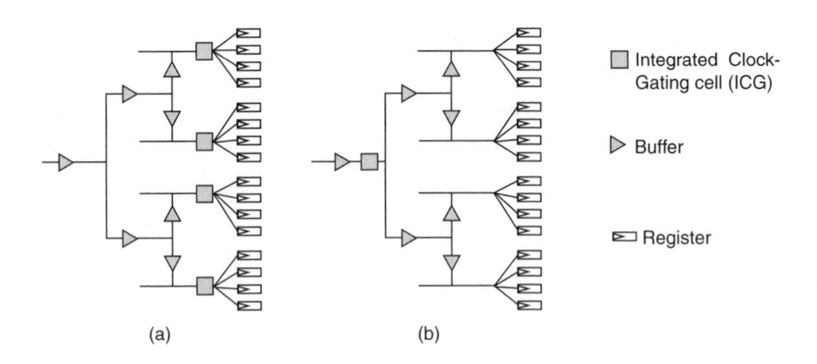

FIGURE 40.4 Clock-gating cell placement: (a) clock gating close to registers, (b) clock gating with post-gate buffering.

There exists a spectrum of clock-gating approaches with regard to the placement of clock-gating cell in the clock tree. Designers often opt to place the clock-gating cells as close as possible to the final placement of their corresponding registers as depicted in Figure 40.4(a). This placement can be enforced during physical synthesis by specifying a bound for the proximity of the clock-gating cells to the registers.

Some advantages of this approach are that it makes it easier to estimate the latency from the clock-gating cell, and it increases the amount of available slack for the arrival of the enable signal. The impact on the clock-tree is minimal because the clock-gating cells are placed close to the registers and can eliminate the need for post clock-gating cell buffer insertion.

A disadvantage to this approach is that it leaves the majority of the clock tree switching even when branches are leading to registers that will have the clock blocked by a clock gate. To save as much power as possible, it is desirable to gate as many buffers on the clock tree as possible. This is difficult for the designer to do without the knowledge of the actual physically induced timing constraints.

In a physically aware clock-gating system, clock-tree synthesis works in conjunction with placement and clock gating to determine an optimized placement and insertion of the clock-gating cells into the clock-tree. This information is used to balance the delay on the enable signal with the amount of potential power saved by placing the clock-gating cell closer to the root of the clock tree as depicted in 40.4(b).

By creating a system that has access to the physical timing information while inserting clock gating and synthesizing the clock-tree, selective clock skewing can also be used to improve timing and peak power characteristics of the design. In general, a tighter clock skew constraint forces more activity into a narrower window of time and that forces peak power higher (see Figure 40.5(a)). If these events are dispersed over a longer period, the peak can be flattened out (see Figure 40.5(b)), thus lowering the impact of peak power and IR drop.

40.3.6 Testability Concerns

Clock gating reduces test-coverage of the circuit because clock-gated registers are not clocked unless the enable signal is high. During test or scan modes, test-vectors need to be loaded into the registers, thus they must be clocked irrespective of the value of the enable signal. One way to address this is to include a control point or control-gate at the enable signal, as illustrated in Figure 40.6(a). This allows the clock-gating signal, *en*, to be overridden during the scanning in or out of vectors by the test mode signal. In this way, during the test clock cycles, the clock signal is not gated by the enable signal, *en*, and the register can be tested to see if it holds the correct state.

Further, the test mode signal is held at logic "1" during test-mode, making any stuck-at faults on the enable signal unobservable. If full observability is required, this signal must be explicitly made observable by tapping it into an observability XOR tree, as illustrated in Figure 40.6(b).

A growing concern around scan-based testing is the power consumed during the scanning in and out of test vectors. [5,6] The changing register values during scanning can create activity levels that are

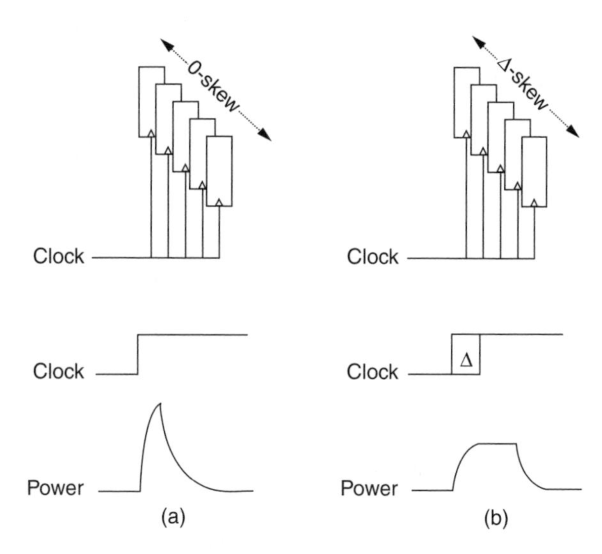

FIGURE 40.5 Selective clock skew to reduce peak power: (a) 0-skew, (b) Δ-skew.

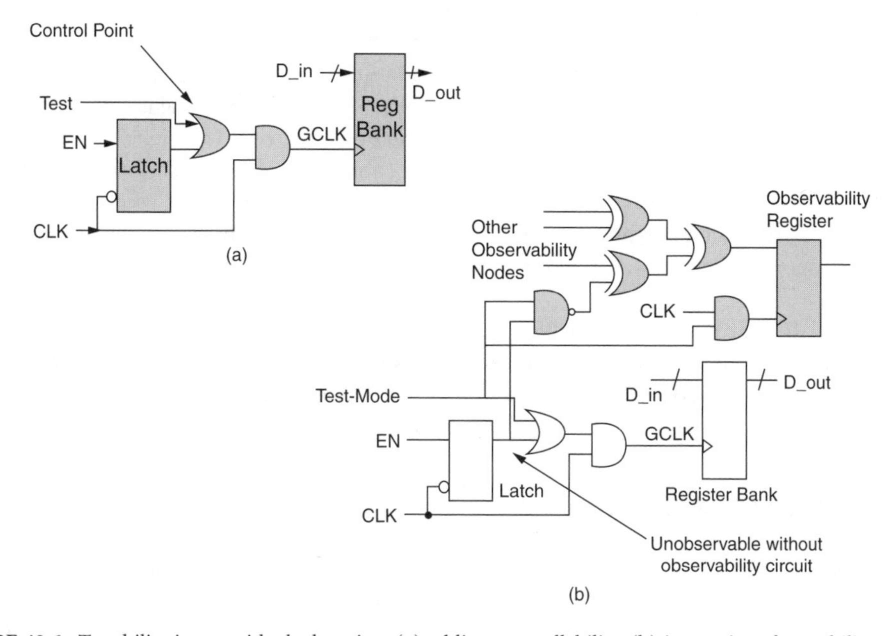

FIGURE 40.6 Testability issues with clock gating: (a) adding controllability, (b) improving observability.

much higher than those experienced during "normal" operation, and can lead to "good" chips failing during testing.

40.4 Operand Isolation

Although clock gating saves power dissipation in the clocked or sequential parts of the design, power savings in the combinational portions are untapped. At the RT level, the most popular technique for power reduction in the combinational parts is operand isolation [7]. Similar to clock gating, the basic concept here is to "shut-off" logic blocks in clock cycles when they do not perform any useful computation.

"Shutting-off" a combinational block involves preventing the activity in the block by not allowing the inputs to toggle in clock cycles in which the block output is not used. The basic concept is depicted in

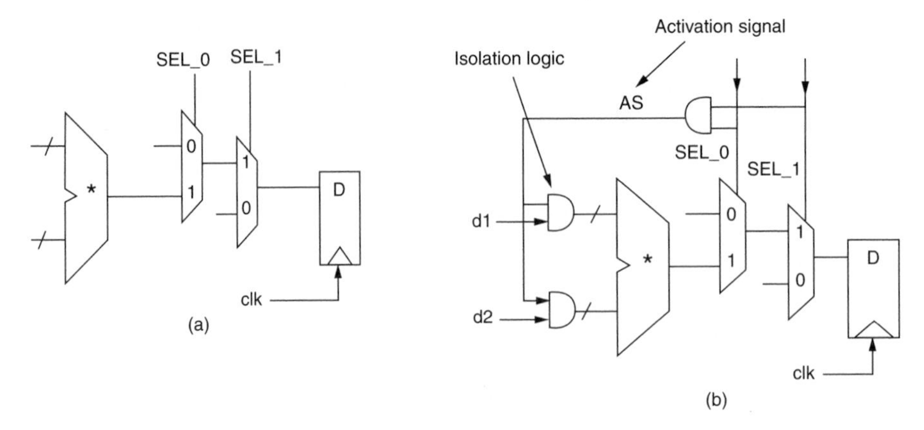

FIGURE 40.7 Operand isolation: (a) original circuit, (b) after operand isolation.

Figure 40.7. In Figure 40.7(a), we notice that the output of the multiplier is only used when the control signals to the multiplexers, *SEL_0* and *SEL_1*, are both high. In cycles when either of the control signals is low, if the multiplier inputs change, the multiplier performs computation, but its result is not used. The wasted power may be substantial if these idle cycles occur for long periods.

Figure 40.7(b) illustrates the operand isolation applied to this multiplier circuit. First, the activation signal, *AS*, is created to detect the idle cycles of the multiplier. The activation signal is high in the "active" clock cycles when the multiplier output is being used, and low otherwise. This signal is used to isolate the multiplier by freezing its inputs during idle cycles using a set of gates, called isolation logic. In Figure 40.7(b), AND gates are used as isolation logic but OR gates or latches may also be used. Using AND/OR gates avoids the introduction of new sequential elements and reduces the impact on the rest of the flow. In addition, in our experiments, we found that AND/OR gates are cheaper and give better power savings overall.

Operand isolation saves power by reducing switching in the operator being isolated, but it also introduces timing, area, and power overhead from the additional circuitry for the activation signal and the isolation logic. This overhead must be carefully evaluated against the power savings obtained to ensure a net power saving without too much delay or area penalty.

40.5 Logic Optimization

Logic-level optimizations reduce the power of a given circuit by transforming them into a different but functionally equivalent implementation. These transformations include RTL and gate-level techniques. Due to the wide acceptance of logic-synthesis tools in the design market today, these techniques are good candidates for automatic optimizations.

These techniques usually aim at reducing either the dynamic or short-circuit power. Because both of these components occur during switching, the techniques rely heavily on the availability of activity statistics at the input pins of all the cells of the circuit. To provide this information to the optimization tool, the user can either simulate the circuit to obtain the switching statistics at input pins of all cells or simply provide activity statistics at the primary inputs. In the latter case, the optimization tool must propagate the activity from the primary inputs to the internal cell inputs using either binary decision diagram (BDD)-based probabilistic propagation techniques [8] or some internal simulation.

These techniques also require estimates of capacitance values of the nets and input/output pins of cells in the circuit — values that are provided through library models of cells, wires, etc.

The next few subsections discuss some of the most popularly used logic transformations for power reduction.

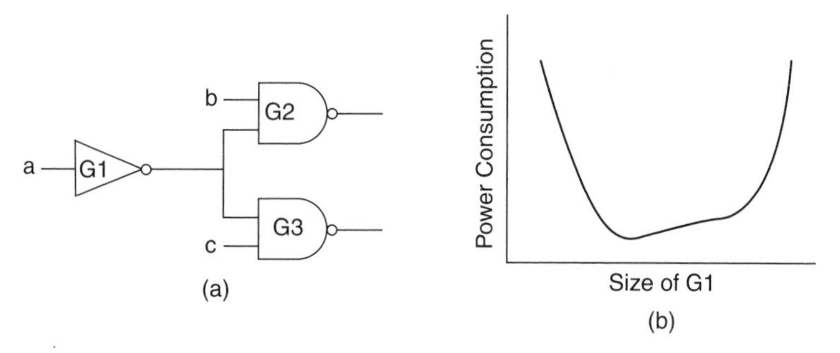

FIGURE 40.8 Gate sizing and power: (a) original circuit, (b) impact of sizing on power dissipation.

40.5.1 Sizing and Buffering

Gate sizes affect the power consumption of the design in the following ways:

1. A larger cell contributes higher capacitance that increases the overall dynamic power dissipation.
2. For a given input transition time, a larger gate has higher short-circuit currents.
3. The output of a gate with high drive has a sharper slope compared to one with lower drive strength, directly decreasing the short-circuit dynamic currents in the fanout gates.

Buffering has a similar effect as sizing because the addition of a buffer contributes both extra capacitance and short-circuit currents but improves the slope of the output signals.

These concepts can be more clearly illustrated using an example. Although we demonstrate the ideas using a sizing example, the points made here also apply to buffering. Consider Figure 40.8 where gate G_1 that fans out to two other gates G_2 and G_3. Let us assume that the input transition times for G_1, G_2, and G_3 are t_i, t_o, and t_o; their gate sizes are W_1, W_2, and W_3; and their output switching frequencies are f_1, f_2, and f_3, respectively. In addition, G_1 has an input switching frequency of f_i and input and output capacitances of C_i and C_o, respectively. Using the equations for the switching and short-circuit power of the gates, the power consumption of the circuit is given by:

$$P = C_o V_{dd}^2 f_1 + C_i V_{dd}^2 f_i + k(W_1 t_i f_1 + W_2 t_o f_2 + W_3 t_o f_3)$$

Let us consider each of the terms. The first term represents the power consumed in switching the capacitance C_o. If we assume that the source/drain capacitance of the gate is a small component of C_o, this term is almost constant and increases slightly with the sizing up of the gate, G_1. The second term represents the switching power at the input of the gate G_1 and increases linearly with the size of G_1. The last three terms represent the short-circuit power of each of the three gates. As the size of G_1 is increased, this short-circuit current of G_1 increases due to an increase in W_1, while the short-circuit current of G_2 and G_3 decrease due to a reduction in the transition time of their input signals. Therefore, as the size of the gate G_1 is increased, the power decreases first and then starts rising, showing a clear minima and a potential for optimization [9].

Sizing and buffering are also used for "path balancing." Different arrival times of the different inputs of a gate cause spurious switching or glitches at the output causing excess power dissipation. These techniques can be used to equalize the arrival times at the inputs of a given gate, thereby reducing glitches.

40.5.2 Technology Mapping

Technology mapping involves optimal mapping of a block of logic using cells from a given library. Technology mapping for low power is driven by three observations:

1. Internal nodes of library cells usually have lower capacitance than external nodes.
2. Gate size has a large impact on power consumption, as discussed in the previous subsection.

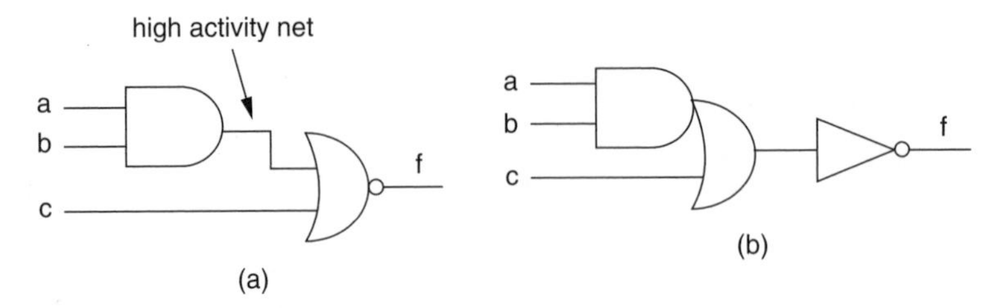

FIGURE 40.9 Technology mapping for low power: (a) original circuit, (b) remapped circuit.

3. The lowest-power mapping does not coincide with either the minimum delay or area mapping [10,11].

The first observation indicates that it is better to map nodes with high switching activity in a design to internal nodes of a cell. This reduces power by reducing the capacitance on the highly active nodes. The second observation indicates that the cell-size selection process must explore the trade-off between the driving capacity and the power consumption of the cells. The third observation indicates that power must explicitly be part of the cost function to achieve a power-sensitive technology mapping solution.

Two implementations of a simple and-or-invert (AOI) are shown in Figure 40.9. We are given that node $x = a \cdot b$ is a high activity node. The circuit in Figure 40.9(b) implements the high activity node, x, as an internal node and therefore dissipates less power.

40.5.3 Phase Assignment

Phase assignment inverts the inputs to an operation and, at the same time, also inverting the output. This transformation reduces power in the following ways. First, because this transformation adds inverters on nets that previously did not have inverters, it creates opportunity for several other transformations: Two inverters next to each other can be merged and removed, and an inverter at the output of a gate may be absorbed into the gate using a composite gate from the library. Second, it can be used to remove inverters from high-activity nets and move them to lower-activity nets.

40.5.4 Algebraic Transformations

Algebraic transformations use algebraic properties to derive equivalent implementations of a given circuit to reduce power. The most popularly used properties for power reduction include commutativity, associativity, and distributivity.

Commutativity is used in a transformation called pin swapping. At the gate level, many Boolean operations, such as AND, OR, NAND, NOR, and XOR, are commutative (i.e., their inputs can be interchanged without affecting the functionality). Based on the capacitance and toggle-rates on pins, the input pins of a gate can be swapped, connecting the lower input-capacitance pin to the net with the higher toggle rate, thus reducing power consumption. The same technique can be used at the RT level with larger commutative blocks such as adders and multipliers. Besides directly reducing power consumption, this technique generates more opportunities for the other techniques and helps to pull the overall algorithm out of local minima.

The associative and distributive properties of gate-level operations, such as AND, OR, and XOR, as well as RTL operators, such as adders and multipliers, are used in a transformation for low power called factoring. Factoring is based on the idea that the power dissipated by an operation and the activity at its outputs depends on the activity at its inputs. Therefore, the power dissipation caused by a certain net depends on the number of operations to which its activity is propagated. The goal of factoring is to reduce the logic depth connected to high-activity nets. This is illustrated in the Figure 40.10, which presents two implementations of the function $ab + bc + cd$. Of the inputs, input b has the highest activity. The thick lines in the figure identify high-activity nets. In this case, the implementation in Figure 40.10(b) dissipates less power because b is propagated only through one gate.

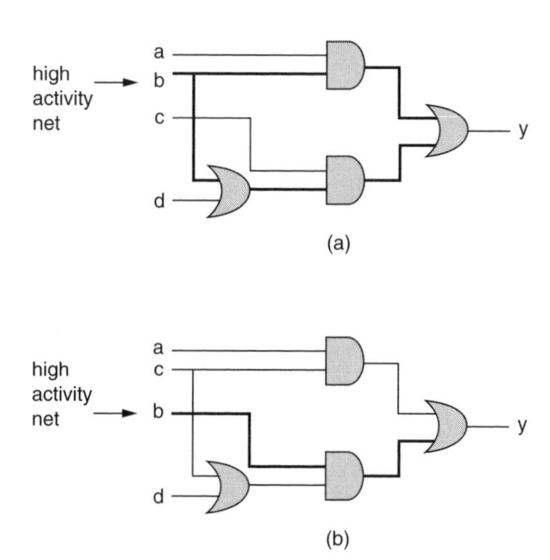

FIGURE 40.10 Factoring for low power: (b) is the lower power implementation.

40.6 Leakage Control — Managing Thresholds

Since the .5-µm technology node, core voltage levels have been scaling at approximately 1 V/0.1 µm. As technologies scale to 100 nm and below, the reduced operating voltages are forcing threshold voltages down to .25 V and below. This has had a major impact on the leakage current of the transistors built in these technologies. For approximately every 65 mV to 85 mV decrease in threshold voltage (depending on temperature), there is an order of magnitude increase in subthreshold leakage current.

As the next equation demonstrates, the subthreshold leakage current grows exponentially as the threshold voltage decreases.

$$I_{sub} = I_0(e^{[-Vth/S]}\ [1 - e^{-qVds/kT}])\ (\text{at } V_{gs} = 0)$$

40.6.1 Multi-Threshold Design

Silicon foundries have started to offer multiple threshold devices at the same process node to address the need to control leakage current and enabling designers to trade off leakage and performance [12]. From the standard V_{th}, a low- and high-V_{th} transistor may be offered. It is not uncommon for the low-V_{th} device to have an order of magnitude higher leakage than the standard V_{th} device and the high-V_{th} device to have leakage characteristics an order of magnitude below the standard. For special applications, a special low-leakage device may exist that will reduce the leakage further by another order of magnitude. This reduction in leakage is not free, however, and it comes at the expense of the speed of the device. There could be a 20% to more than 2× delay penalty between the standard and the high-V_{th} devices and an increase in the cost of the fabrication process.

The challenge for EDA tools is to use the available characteristics of the cells in the design library to create an implementation that will meet the timing constraints, while reducing the leakage current as much as possible.

A simplistic approach is to use the percentage of high- and low-threshold cells in the design as a quality measure. It is important to keep in mind that the goal is to reduce the total leakage. A design with a higher percentage of high-V_{th} cells may appear to be better at first glance, but it could also be inferior to a design with a lower percentage of high-V_{th} cells that uses fewer cells overall. It is for this reason that these trade-offs should be considered early in the synthesis process.

A better approach for synthesis using multi V_{th} cells is to create a post-logic-synthesis routine that replaces low-V_{th} cells with high-V_{th} cells as long as the timing constraints and other design rules and constraints are not violated. This can be a useful "clean-up" step for an existing gate-level netlist design.

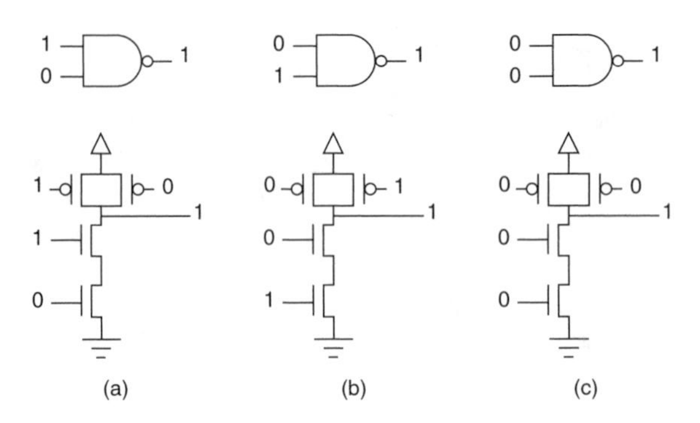

FIGURE 40.11 Leakage varies depending on input values.

The cells may be created to have the same "footprint" to facilitate this type of optimization in a post-placement phase.

A more sophisticated approach used during synthesis incorporates the leakage power as a component of the optimization's objective function. Here, it is important to have libraries that have been properly characterized to perform this type of optimization. In particular, state-dependent leakage information can be used to reduce the leakage in the design. It is understood that the amount of current that a cell will leak is dependent on the state of its inputs. In particular, transistors in series will have varying amounts of leakage current depending on the values placed on their gates. This is depicted in Figure 40.11, where the leakage current decreases from (a) to (c) even though no change occurs in the output value of the gate. In this case, as in dynamic power reduction, pin-swapping techniques may also be used to reduce the average leakage component or the leakage for a given state. To perform this optimization, it is necessary for the libraries to include state-dependent leakage information for the cells. Another necessary component is the state probabilities on the inputs. The switching activity interchange format (SAIF) [13] can be used to annotate this information onto a design, which can then be internally propagated and updated for use in optimizing pin selection for leakage reduction. Some details on the SAIF format are presented in Section 40.8.

Another relevant area for multiple V_{th} designs is the impact on noise sensitivity. Higher-V_{th} devices, by their nature, will have a lower sensitivity to noise. Optimization tools that simultaneously consider the signal integrity impact can make use of these cells to improve signal integrity properties as well as leakage.

40.6.2 Variable Threshold Biasing

The threshold voltage and therefore the leakage current in a CMOS transistor are controllable by varying the back biasing. The change in the threshold voltage is roughly proportional to the square root of the back bias voltage. As threshold voltages drop below .25 V, variable back biasing may gain more appeal.

A distinct advantage of this approach is that during periods when heavy processing is needed, the threshold voltage can be reduced, thus speeding up the cells. When the cells are in a slower drowsy or idle mode, the threshold voltage can be raised, thus lowering the leakage.

One significant impact of using variable back biasing is that two new terminals for each cell need to be routed. A common ASIC design practice is to create cells that tie the N-well regions to V_{dd} and the P-well regions to ground. In the physical implementation, these are simply predefined contacts designed into the cell, which are connected as part of the power and ground routes. To enable back-biasing, new voltage lines are routed to control the bias. These can be to individual cells or, more likely, to regions that contain multiple cells sharing the same WELL and a common tie-cell to control the WELL bias [14,15]. Figure 40.12 illustrates the concept of variable body biasing. Figure 40.12(a) is a traditional implementation with wells tied to V_{dd} and V_{ss}, and Figure 40.12(b) presents wells biased for leakage reduction.

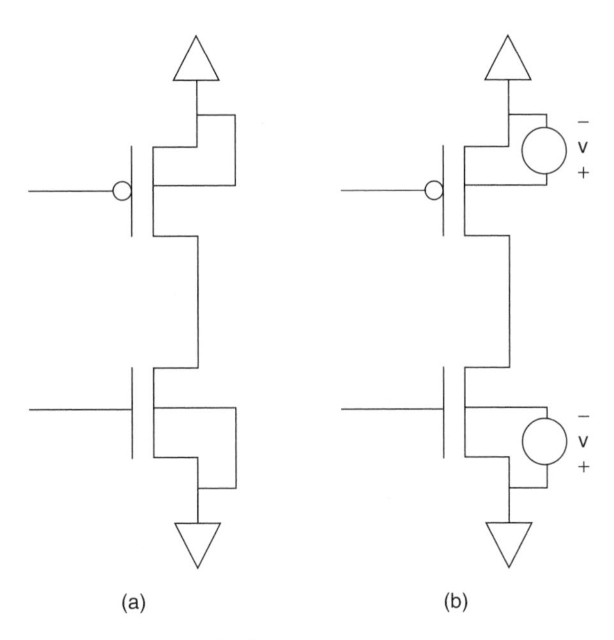

FIGURE 40.12 Using bias to control threshold voltages.

40.7 Voltage Scaling

The switching power consumed in a design is proportional to the square of supply voltage. This relationship of power to voltage makes voltage scaling a prime candidate for reducing the power in CMOS designs.

One approach is to create separate voltage islands that operate at levels that best suit the power and performance levels for each block of logic [16–20]. This can range from having each island run at one voltage and selectively turning blocks of logic completely off, to dynamically varying the voltages supplied to those blocks. These design techniques create interesting dynamics on the chip. Turning the voltage ON or OFF to a block can cause large transients on the power grid, affecting many other blocks on the chip. Further, it is necessary to provide isolation on the outputs of the block that is shut down. It is also possible to use register structures that have a second voltage rail that provides power to retention logic that can be used to save and then restore the state to a block that has had its power shutdown [21,22]. An example of this type of structure is given in Figure 40.13.

A major impact on the design flow when multiple voltages are used is the need to treat the supply line as another variable. For most previous mainstream designs, logical netlists only specified the input and output connections between gates. V_{dd} and V_{ss} were constants, and the V_{dd} pins for all the cells (as well as the V_{ss} pins) were attached to the same net. Figure 40.14 is a simple diagram indicating the need for cells to be able to handle new voltage terminals. A level shifter, for instance, needs to handle two V_{dd} levels as well as separate potential well biases.

The tools have to manage cells that have more than one supply rail and circuitry that can vary or completely shut down the supply voltage to a block. Communication between blocks operating at different voltages requires the insertion of voltage level shifters to transform signals to the appropriate levels. Clock-tree generators need to account for buffers that operate at different voltages to provide clock signals to each block and the router needs to account for buffer placement in the context of different voltage regions on the chip. Routing a feed-through signal via a region may now require the insertion of level shifters to adequately drive the signal. Analysis tools need to understand these different situations — tracking all these new voltage based modes — and provide useful feedback to the designer.

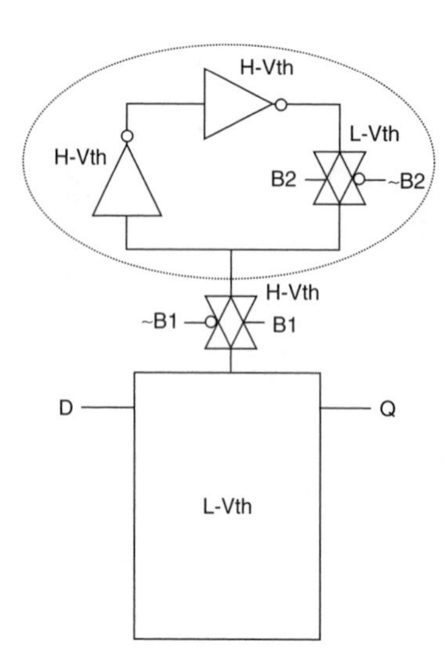

FIGURE 40.13 Low-power state-saving register.

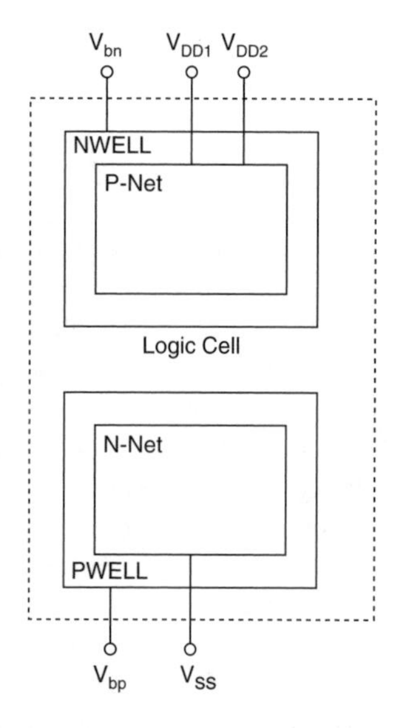

FIGURE 40.14 Handling multiple supply voltages.

Another design implication that optimization and analysis tools must account for is the impact of driving some lines at higher voltages than others. The higher-voltage lines can cause larger spikes in neighboring low-voltage lines than other lower-voltage aggressors, which impacts timing analysis, power, and the routing of lines on the chip. This is presented in Figure 40.15.

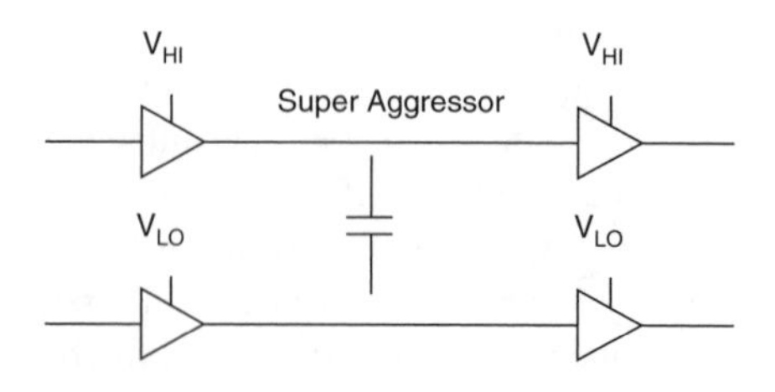

FIGURE 40.15 Signal integrity impact with multiple voltages.

Implementation tools need to account for the effects of the different operating conditions. Instead of designing for the typical *Min* and *Max* conditions, the circuitry now has to deal with a much broader range of conditions. Complicating matters, as the voltages have been decreasing, the current on chip has also been increasing, thus raising the sensitivity to IR-drop and *L di/dt* effects.

40.8 Modeling Basics

The previous sections presented a set of automated power-optimization solutions. To be effective and powerful, these optimization solutions should be supported by robust and accurate power modeling and analysis techniques. Current EDA tools provide comprehensive support for modeling power consumption on logic blocks and for analyzing the power consumption of designs. This section looks at some of the basics of modeling power for gates [24]. Section 40.9 presents typical power analysis flows.

40.8.1 Switching Power

Switching power consumption is computed using the famous CV^2f formula. To compute switching power accurately, the library must provide voltage and capacitance data. Capacitance is computed from the pin capacitances, which are specified on the library cells, and wire capacitances, which may either be back annotated from physical tools or calculated from wire-load models.

40.8.2 Internal Power

The internal power of a cell includes the power consumed due to the switching activity on the internal nodes of the cell and the short-circuit power consumed by the cell, as depicted in Figure 40.1. Internal power of a cell depends on the input rise or fall times and the output capacitances. It is modeled in the library as a lookup table, also called the nonlinear power model (NLPM), which is indexed by two variables: the input rise/fall time and the output capacitance.

A lookup table is specified by a table template that specifies the values of the indices to the table for the different entries. For specifying power values, a specific table template is used, and only the power values at the different points in the table are specified. An example of a one-dimensional table template, *Power_1D*, and a two-dimensional one, *Power_2D*, is presented next. These are used in power models presented later in this section.

```
power_lut_template(power_1D) {
variable_1 : input_transition_time;
index_1 ("1000, 1004, 1005, 1006");
}
power_lut_template(power_2D) {
variable_1 : input_transition_time;
variable_2 : total_output_net_capacitance;
index_1 ("1000, 1001, 1002");
index_2 ("1000, 1001, 1002, 1003, 1004, 1005, 1006");
}
```

A more advanced modeling method that allows multiple voltages and more variables to be modeled is called the scalable polynomial power model (SPPM). This format is explained later in this section.

Internal power can be state-dependent or path-dependent, or both. State dependency captures the fact that the internal power dissipation of a cell can be different based on the states of the inputs or outputs of the cell that are not switching. For example, the power dissipated when the clock pin of a RAM memory block switches depends on whether the RAM is reading, writing, or idle. This kind of effect can be captured by the state-dependency construct "when." The state-dependent power model for a RAM cell given next demonstrates that the power is separately specified for the three states: read, write, and idle.

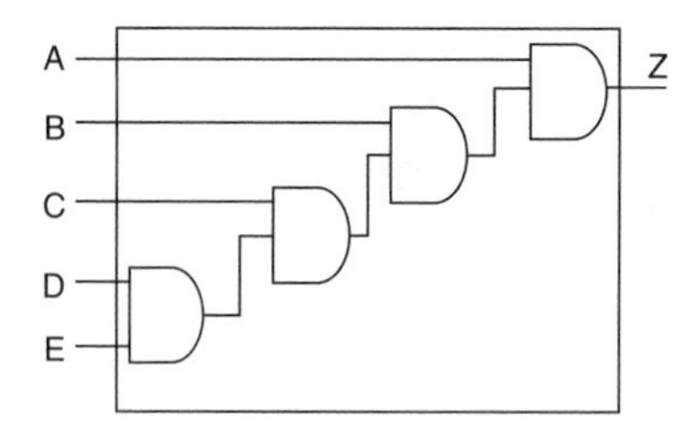

FIGURE 40.16 Path dependency: changes in output caused by a toggle at input E dissipate more power than those caused by changes in input A.

```
cell (RAM)
.........
pin (clk)
internal power {
when (!RD & WR & CS) {/* write state */
rise_power(power_1D)
      values ("1.5 2.5 5.6 7.7") }
fall_power (power_1D)
      values ("1.7 2.4 5.3 7.2") }
when (RD & !WR & CS) {/* read state */
rise_power(power_1D)
      values ("2.5 3.5 6.6 8.7") }
fall_power (power_1D)
      values ("2.7 3.4 6.3 8.2") }
when (!CS) {          /* idle state */
power (power_1D)
values ("0.6 2.0 1.6 4.7") }
}
```

Path dependency captures the fact that the internal power dissipation of a cell is different depending on which input change caused the switching behavior. In the example depicted in Figure 40.16, the power consumption is more if the output change is caused by a change in input D, instead of a change in input A. This effect can be captured by the path-dependency construct "related pin" for internal power presented next.

```
cell (AND5)
pin (Z)
internal power {
related_pin A {
power (power_2D)
values ("0.5 1.5 3.6," "0.6 1.7 4.0,"Ö.)}
ÖÖÖÖ
related_pin D {
power (power_2D)
values ("1.5 2.5 5.6," "1.6 2.7. 5.7,"Ö.) }
Ö......Ö
}
```

40.8.3 Leakage or Static Power Modeling

As discussed in previous sections, leakage may happen in a circuit from several different sources: sub-threshold leakage, drain induced barrier lowering, gate induced drain leakage, etc. [23]. Regardless of

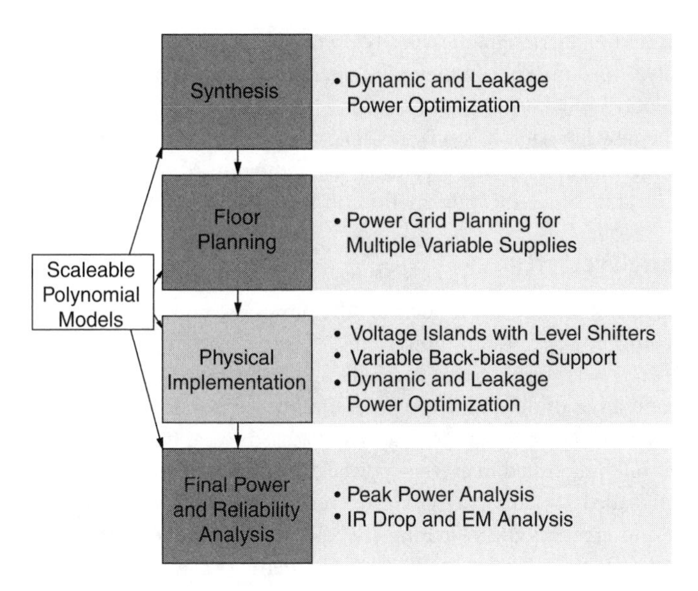

Figure 40.17 A flow for energy-efficient design.

the physical reasons for leakage power, library developers can annotate a cell with the total leakage power dissipated by it. The leakage model can be state dependent and can specify different leakage values for different input states of the cell. Here is an example of the leakage model on a cell:

```
cell my_cell() {
leakage_power () {
when: "A & B"
value: 5.5
}
cell_leakage_power: 4.0
}
```

This specifies that the cell `my_cell` consumes 5.5 units of leakage when both A and B inputs are high and consumes 4.0 units otherwise.

40.8.4 Scalable Polynomial Power Models (SPPMs)

It was once possible to create tables based on a few characterization points and use scaling factors, commonly called k-factors, to extrapolate to uncharacterized points in between. With multiple voltages, multiple thresholds, and back biasing, however, that methodology is now failing to provide the accuracy required. More sophisticated techniques are now needed to enable energy-efficient design. In addition, as chip sizes become larger and transistor sizes shrink, intra-chip temperature variations are becoming significant and must be appropriately modeled.

Scalable polynomial models provide an efficient and faster alternative to nonlinear lookup tables. These models capture library characterization information in an accurate equation-based format. Each of the voltages used for supply and back biasing become a variable in an equation, which is stored with the cell's information in the library. This equation-based format allows the tools to obtain precise data on the timing and power characteristics of a cell across a broad operating range of conditions. The cells are characterized for timing using scalable polynomial delay models (SPDMs), for power using SPPMs, and for leakage using scalable polynomial leakage models (SPLMs). Due to their compact representation, they can be used to model much higher degrees of freedom. For example, they allow us to also model the impact of supply voltage and temperature variations without blowing up in size as would happen with lookup table models.

The SPPM syntax allows the designer to specify up to seven variables. For very large data with abrupt changes, a single polynomial may not fit the entire operating range of interest. In this case, the piecewise or adaptive domain polynomial syntax can be used.

Equipped with the new libraries, it is now possible to ensure that the chip functions correctly across the expected operating ranges of process, voltage, and temperature. For this, design optimization and analysis tools must be able to use this information on a per instance basis.

40.8.5 Modeling Activity

The previous subsections discussed the models for the physical components that are required for power consumption, switching power, internal power, and leakage. The other component that needs to be appropriately modeled is switching activity. Based on the analysis flow that is used, switching activity can be modeled in different levels of detail. If a complete time-based power profile view of the chip is desired, the value change dump (VCD) or VCD+ formats can be used to capture detailed switching activity.

If, however, one is only interested in average power dissipation, a much more compact representation for switching activity called switching activity interchange format (SAIF) [13] can be used. SAIF is an open ASCII format and captures the switching statistics for each node in the design in terms of static and dynamic attributes that can be state and path dependent. The attributes captured are listed next:

40.8.5.1 Static Attributes

- T0: time spent in 0 state
- T1: time spent in 1 state
- TX: time spent in unknown X state
- TZ: time spent in floating Z state
- TB: time spent in bus-contention state (two or more drivers simultaneously driving same object)

40.8.5.2 Dynamic Attributes

- TC: number of 0→1 or 1→0 transitions. This can be split into rise and fall transitions.
- TG: number of transport glitches. These are glitches on the output where the output pulse width is more than gate delay. These consume the same power as a full transition.
- IG: number of inertial glitches. These are glitches where the output pulse width is less than gate delay. These do not consume same power as full transition, and a derating factor is used for power dissipation calculation for these.

Besides these basic attributes, the SAIF language provides state and path dependency constructs to capture more specific information about the switching statistics on a particular node or cell. Both the static and dynamic attributes can be state-dependent, thus capturing the switching statistics separately for the different states of a cell. State dependent static attributes are useful for computing state dependent leakage power and for computing dynamic power.

The dynamic attributes can also be path dependent, capturing separate switching statistics based on which input path caused the transition on the pin.

The modeling techniques for the various power components discussed in this section provide the underlying fabric for analysis and optimization tools discussed in this chapter.

40.9 Analysis Flows

Power optimization within power compiler [25] provides average power and uses the SAIF compact representation for switching activity. You can use SAIF from RTL simulation or gate-level simulation for your power analysis. Although not as accurate as gate-level simulation based analysis, RTL-simulation based analysis has the huge advantage in that it avoids the long time-consuming gate-level simulations.

The RTL-simulation based power analysis flow within power compiler is depicted in Figure 40.18. Switching activity is captured via RTL simulation at the synthesis invariant points in the design. These

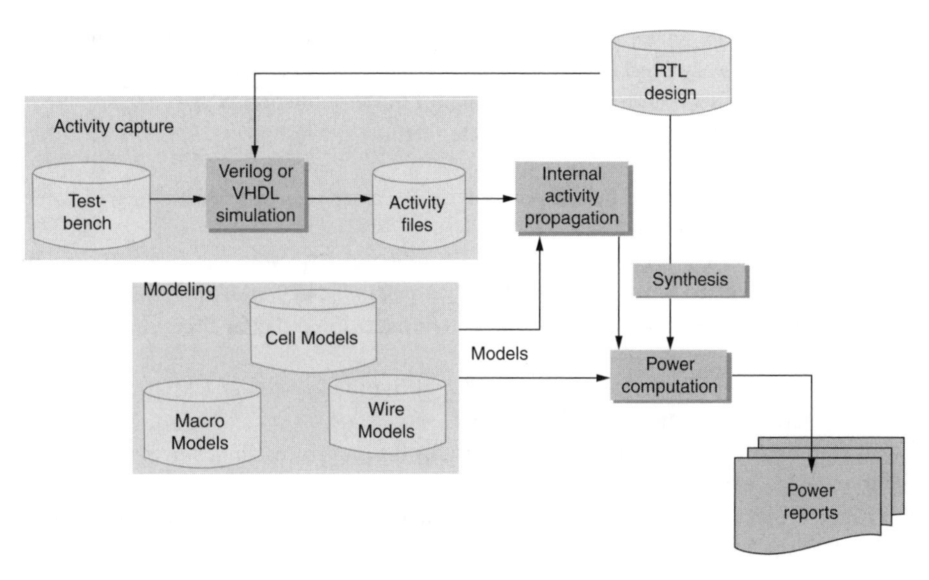

FIGURE 40.18 RTL simulation-based power analysis flow.

include the hierarchy boundaries and sequential elements. Capacitance and power models for wires and gates are taken from the library. The RTL design is synthesized to gate level along with all the constraints. Activity information that is captured at the synthesis invariant points is propagated to all the input pins of all cells in the gate-level design. This information is passed to the power computation engine, which reports the power for the entire design.

The gate-level simulation based flow is similar, except that no internal activity propagation is required because activity is captured at the input pins of all the cells in the gate-level netlist via gate-level simulation. Because this activity is captured in full detail, it is possible to use the state and path dependent information in the library models and in SAIF to perform a more accurate power analysis. In a post-placement or a post-routing netlist, the wire capacitances can be back annotated is for more accuracy.

PrimePower [26] provides a detailed analysis of the power dissipation in a design and relies on the more complete VCD switching activity format. It works on a gate-level netlist with gate-level simulation data and is targeted for full-chip capacity. Along with the average power numbers, it also gives the time-based waveforms of power consumption in different parts of the design. This allows the designer to do more detailed debugging of hot spots in the design.

The analysis tools and flows presented in this section enable the designer to understand and optimize the power of his or her design, and provide the basis for the optimization capabilities discussed earlier in this chapter.

40.10 Conclusion

This chapter described some of the power optimization technology currently available or in development in commercial CAD solutions. Both synthesis and analysis solutions were discussed. Power models that form the basis of such a solution were also presented. The authors thank the Power Compiler and PrimePower teams for their support.

References

[1] M. Gowan, L.L. Biro, and D.B. Jackson, Power considerations in the design of the Alpha 21264 microprocessor, *Proc. Design Automation Conf.*, 1998, pp. 726–731.

[2] A. Correale, Overview of the power minimization techniques employed in IBM PowerPC 4xx embedded controllers, *Int. Symp. on Low-power Design*, 1995, pp. 75–80.

[3] V. Tiwari, D. Singh, S. Rajgopal, G. Mehta, R. Patel, and F. Baez, Reducing power in high-performance microprocessors, *Proc. Design Automation Conf.,* 1998, pp. 726–731.
[4] Z. Khan and G. Mehta, Automatic clock gating for power reduction, *SNUG '99.*
[5] B. Pouya and A. Crouch, Optimization for vector volume and test power, *Proc. Int. Test Conf.,* 2000, pp. 873–881.
[6] J. Saxena, K. Butler, and L. Whetsel, An analysis of power reduction techniques in scan testing, *Proc. Int. Test Conf.,* 2001, pp. 670–677.
[7] M. Muench, B. Wurth, R. Mehra, and J. Sproch, Automating RT-level operand isolation to minimize power consumption in datapaths, *Proc. Design Automation and Test in Europe,* 2000, pp. 624–631.
[8] F. Najm, Transition density, a stochastic measure of activity in digital circuits, *Proc. Design Automation Conf.,* 1991, pp. 644–649.
[9] M. Borah, R. Owens, and M. Irwin, Transistor sizing for low-power CMOS circuits, *Trans. on Computer-Aided Design,* June 1996, pp. 665–671.
[10] O. Coudert and R. Haddad, Integrated resynthesis for low power, *Proc. Int. Symp. on Low-Power Electron. and Design,* 1996, pp. 169–174.
[11] V. Tiwari, P. Ashar, and S. Malik, Technology mapping for low power, *Proc. Design Automation Conf.,* 1993, pp. 74–79.
[12] S. Svilan, J.B. Burr, and G.L. Tyler, Effects of elevated temperature on tunable near-zero threshold CMOS, *Proc. Int. Symp. on Low-Power Electron. and Design,* 2001, pp. 255–258.
[13] Switching Activity Interchange Format (SAIF), http://www.synopsys.com/partners/tapin/saif.html.
[14] K. Flautner, D. Flynn, and M. Rives, A combined hardware-software approach for low-power SoCs: applying adaptive voltage scaling and the vertigo performance-setting algorithms, *Proc. Design Conf.,* 2003.
[15] S. Martin, K. Flautner, T. Mudge, and D. Blaauw, Combined dynamic voltage scaling and adaptive body biasing for lower power microprocessors under dynamic workloads, *Proc. Int. Conf. on Computer-Aided Design,* 2002, pp. 721–725.
[16] D. Tamura, B. Pangrle, and R. Maheshwary, Techniques for energy-efficient SoC design, http://www.eedesign.com/features/exclusive/OEG20030724S0044.
[17] D.E. Lackey, S. Gould, T.R. Bednar, J. Cohn, and P.S. Zuchowski, Managing power and performance for system-on-chip designs using voltage islands, *Proc. Int. Conf. on Computer-Aided Design,* 2002, pp. 195–202.
[18] K. Usami, M. Igarashi, F. Minami, T. Ishikawa, M. Kawakawa, M. Ichida, and K. Nogami, Automated low-power technique exploiting multiple supply voltages applied to media processor, *IEEE J. Solid-State Circuits,* Vol. 33, No. 3, 1998, pp. 463–472.
[19] L. Wei, K. Roy, and V. De, Low-power, low-voltage CMOS design techniques for deep submicron ICs, *Proc. Int. Conf. on VLSI Design,* 2000, pp. 24–29.
[20] F. Ishihara, F. Sheikh, and B. Nikolic, Level conversion for dual supply systems, *Proc. Int. Symp. on Low-Power Electron. and Design,* 2003, pp. 164–167.
[21] S. Shigematsu, S. Mutoh, Y. Matsuya, Y. Tanabe, and J. Yamada, A 1-V high-speed MTCMOS circuit scheme for power-down application circuits, *IEEE J. Solid-State Circuits,* Vol. 32, June 1997, pp. 861–869.
[22] V. Zyuban and S. Kosonocky, Low-power integrated scan-retention mechanism, *Proc. Int. Symp. on Low-power Electron. and Design,* 2002, pp. 98–102.
[23] J. Rabaey, A. Chandrakasan, and B. Nikolic, *Digital Integrated Circuits, 2nd ed.,* 2003, Prentice Hall/Pearson.
[24] *Library Compiler User Guide: Modeling Timing and Power Technology Libraries,* Synopsys.
[25] *Power Compiler Reference Manual,* Synopsys.
[26] *PrimePower Reference Manual,* Synopsys.

41

Magma Low-Power Flow

Ed Huijbregts
Lars Kruse
Eric Seelen
Magma Design Automation

41.1 Introduction

In the case of today's increasingly large and complex digital integrated circuit (IC) and system on chip (SoC) designs, design power closure and circuit power integrity are becoming one of the main drains on engineering resources, thereby impacting the device's total time-to-market.

The shear amount of power consumed by some devices can cause significant design problems. For example, a recently announced CPU consumes 100 A at 1.3 V, which equates to 130 W. This class of device requires expensive packaging and heat sinks, the heat gradient across the chip can cause mechanical stress leading to early breakdown, and the act of physically delivering all this power into the chip is nontrivial. Thus, even in the case of devices intended for use in nonportable equipment where ample power is readily available, power-aware designs can offer competitive advantages with respect to such considerations as the size and cost of the power supply and cooling systems.

The majority of power considerations are exacerbated in the case of low-power designs. The increasing use of battery-powered portable (often wireless) electronic systems is driving the demand for IC and SoC devices that consume the smallest possible amounts of power.

Whenever the industry moves from one technology (i.e., feature size) to another, existing power constraints are tightened and new constraints emerge. Power-related constraints are now being imposed throughout the entire design flow to maximize the performance and reliability of devices. In the case of today's extremely large and complex designs, implementing a reliable power network and minimizing power dissipation have become major challenges for design teams.

Creating optimal low-power designs involves making trade-offs such as timing vs. power and area vs. power at different stages of the design flow. To enable designers to perform these trade-offs accurately

0-8493-1941-2/05/$0.00+$1.50

and efficiently, it is necessary for low-power optimization techniques to be integrated with — and applied throughout — the entire RTL-to-GDSII flow.

41.1.1 Integrated Tool Suite

A number of very sophisticated power analysis tools are available to designers; however, these tools are typically provided as third-party point-solutions that are not tightly integrated into the main design environment. Either these tools require the use of multiple databases or they combine disparate data models into one database. This means that design environments based on these tools have to perform internal or external data translations and file transfers, making data management cumbersome, time-consuming, and error-prone.

Correlating results from different point-tools can be difficult, which means that problems may be discovered late in the design cycle or may never be detected at all. Perhaps the most significant problem with existing design environments, however, is that power, timing, and signal integrity effects are strongly interrelated in the nanometer domain, but conventional point-solution design tools do not have the capability to consider all of these effects and their interrelationships concurrently.

The lack of integration between power analysis tools and the rest of the environment can result in a tremendous amount of false errors, such as minor voltage drops in portions of the design that will not affect the performance or functionality of the device. Engineers often overcompensate for these false errors and modify the power grid unnecessarily. In turn, this can cause these portions of the design to fail to meet their area constraints and to become congested, and compensating for this can cause ripple effects throughout the rest of the design.

Even worse, the lack of integration between power analysis tools and the rest of the environment — coupled with extremely limited (if any) repair capabilities — means that, when the results from the power analysis are used to locate and isolate timing or signal integrity problems, the act of fixing these problems may introduce new problems into the power network. This can result in numerous, time-consuming design iterations.

Ultimately, using point-solution power analysis tools can result in nonconvergent solutions that prevent designs from achieving their time-to-market windows (or from being realized at all). Thus, a true low-power design environment should have all of the power analysis tools operating concurrently with the implementation tools, including synthesis, place-and-route, clock-tree, extraction, timing, and signal integrity analysis. Furthermore, all the tools in the environment should employ a single, unified database.

The rest of this chapter describes capabilities of Magma's Blast Fusion and Blast Rail* tool. Built on top of Magma's unique data model, they offer an integrated analysis and optimization engine, combining synthesis, place and route engines together with extraction, timing, power, and rail analysis capabilities.

41.2 Power Dissipation

This section discusses only complementary metal oxide semiconductor (CMOS) devices only because this is currently the most prevalent digital IC implementation technology.

41.2.1 Dynamic Power

Dynamic power dissipation occurs in logic gates that are in the process of switching from one state to another. During the act of switching, any internal capacitance associated with the gate's transistors has to be charged, thereby consuming power. Of more significance, the gate also has to charge any external (load) capacitances, which are comprised of parasitic wire capacitances and the input capacitances associated with any downstream logic gates.

*Blast Fusion, Blast Rail, GlassBox are (registered) trademarks of Magma Design Automation, Incorporated.

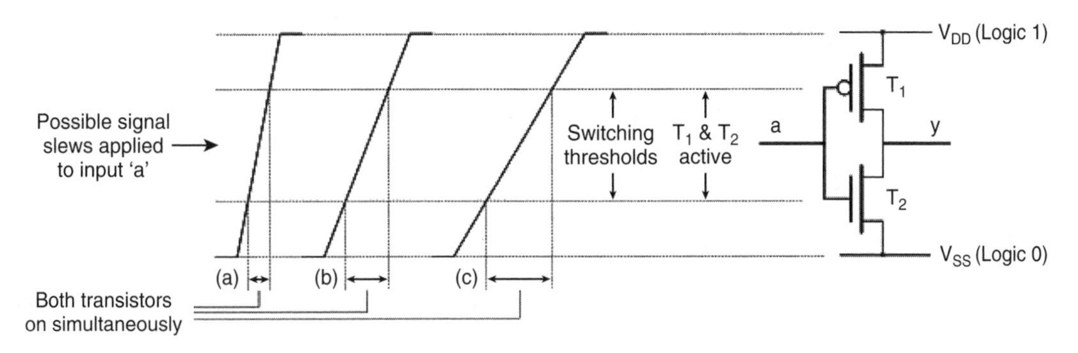

FIGURE 41.1 While the gate is switching, both transistors may be active simultaneously.

Refer to Figure 41.1. Consider a simple inverter gate, in which only one of transistors T_1 and T_2 is usually on at any particular time. When the gate is in the process of switching from one state to another, however, both T_1 and T_2 will actually be on simultaneously for a fraction of a second. This causes a momentary short circuit between the V_{DD} (logic 1, power) and V_{SS} (logic 0, ground) rails, and the ensuing crowbar current results in a transitory power surge.

The amount of time the two transistors are simultaneously active is a function of their input switching thresholds and the slew (slope) of the input signal driving the gate.

For the purposes of this chapter, the amount of dynamic power dissipation may be represented as:

$$Dynamic\ Power \sim af \times C \times V^2 \tag{41.1}$$

where *af* is the amount of activity as a function of the clock frequency *f*, *C* is the amount of capacitance being driven/switched, and *V* is the supply voltage.

This equation demonstrates that minimizing the circuit activity, reducing the capacitance being driven, or reducing the supply voltage may reduce the dynamic power dissipation.

41.2.2 Static Power

Static power dissipation is associated with logic gates when they are inactive (static); that is, not currently switching from one state to another. In this case, these gates should theoretically not be consuming any power at all. In reality, however, there is always some amount of leakage current passing through the transistors, which means they do consume a certain amount of power.

Even though the static power consumption associated with an individual logic gate is extremely small, the total effect becomes significant when we come to consider today's ICs, which can contain tens of millions of gates. Furthermore, as transistors shrink in size when the industry moves from one technology to another, the level of doping has to be increased, thereby causing leakage currents to become relatively larger. The result is that, even if a large portion of the device is completely inactive, it may still be consuming a significant amount of power. In fact, static power dissipation is expected to exceed dynamic power dissipation for many devices in the near future.

Two key equations need to be considered when it comes to addressing static power dissipation. The first describes the leakage associated with the transistors as:

$$Leakage \sim exp\ (-qV_t/kT) \tag{41.2}$$

where q is the elementary charge, V_t is the transistor's threshold voltage, k is Boltzmann's constant, and T is the temperature.

One important point about this equation is that it shows that static power dissipation has an exponential dependence on temperature. This means that as the chip heats up, its static power dissipation increases exponentially. Furthermore, we see that static power dissipation has an inverse exponential dependence on the switching threshold of the transistors.

The second equation describes the delay (switching time) associated with a transistor affected by its threshold voltage and the supply voltage V_{DD} as:

$$Delay \sim V_{DD} \times (V_{DD} - V_t)^{-\alpha}, \text{ with } 1 < \alpha < 2 \tag{41.3}$$

From this, we see that delay goes up if the threshold increases.

41.3 Power Analysis

Power analysis consists of calculating leakage power, internal power, and switched capacitance power (which accounts for wire and input capacitance) for each cell in the design. Reporting is then done in textual reports of various types and drawing colored power maps in the layout graphical user interface (GUI).

For accurate power analysis, a number of inputs are required:

- Accurate switching activity information for all signals in the design
- Correct capacitance values for interconnect and cell inputs
- Library information for internal and leakage power at the desired operating conditions

41.3.1 Activity

The activity in the circuit has direct impact on the power dissipated by the circuit, as is shown by Equation (41.1). Therefore, it is very important to use correct activity numbers when analyzing and optimizing for power.

The activity of a signal is defined by a pair (*pr, tr*), where *pr* denotes the probability that a signal is a logic one and *tr* denotes the toggle rate. Activity can be specified in a number of ways:

- User annotation, where the user specifies the pair for a signal net, a signal pin, or an entire model
- From timing constraints, which specify clock frequencies, and user-specified activity ratios, which relate clock frequencies to data activities for each specific clock domain
- Reading activity obtained via simulation using formats such as value change dump (VCD), global activity format (GAF), and switching activity interchange format (SAIF)
- Automatic activity propagation from primary inputs and flip-flop outputs

41.3.2 Interconnect Modeling

For modern technologies, interconnect capacitance dominates the input capacitance of standard cells and therefore directly affects accuracy of power analysis.

In a synthesis, place and route can flow, and information that is more detailed is added along the way. Mapping, sizing, cell placement, global routing, detailed routing, and metal fill all impact the type of cells and their locations and gradually refine the interconnect estimates up until the point where exact wires and aggressor wires are known. It is only then that detailed extraction can compute wire capacitances with accuracy of within a few percent. At any stage before this point, interconnect capacitance estimation has to be used, exploiting the available information. We therefore distinguish six different interconnect models: constant, wire load, Manhattan, global routing, track routing (i.e., refined global routing), and detailed routing model. Model selection is typically done automatically and different blocks in a design can use different interconnect models depending on where they are in the flow.

41.3.3 Multiple Corner Analysis

A library describes the leakage power and the internal power for all event arcs that are possibly interesting: combinations of any switching input, any switching output, and possibly a set of values for the remaining pins. Depending on the arc type, the internal power is typically described as function of input slew and output load.

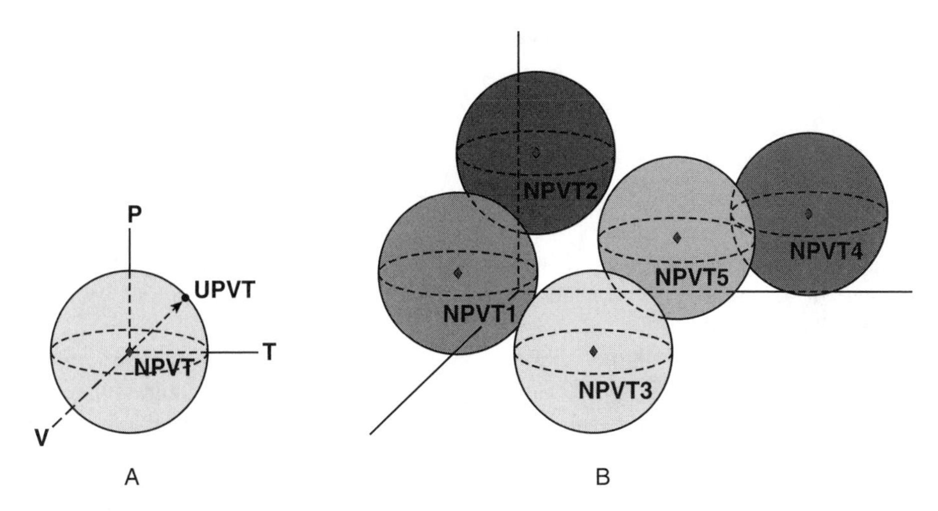

FIGURE 41.2 (a) Region around nominal PVT (NPVT) for which derating to the user defined PVT (UPVT) can be done accurately; (b) increased accuracy by using multiple characterization points.

A typical problem during analysis and optimization for different operating conditions (and, in general, for any scenario where process, voltage, and temperature can change) is the correct selection and derating of the available cell characterization information.

For instance, by changing the voltage of a supply net, the timing and power behavior of all cells supplied by that net changes. If the voltage differs only slightly from the characterization point, then derating might suffice (refer to Figure 41.2(a)). Typically, k-factor derating is then used, although obtaining correct k-factors turns out to be a problem in practice. Better derating models, such as the scalable polynomial delay model (SPDM) and similar models for power (SPPM) and leakage (SPLM), might perform better here.

To improve accuracy or to avoid derating altogether, an approach that characterizes the cell library at many different operating conditions is frequently used (refer to Figure 41.2(b)). All this data is read into the tool. The user or the tool then has to select that characterization data that best fits the desired operating condition.

Therefore, either the user selects the operating condition for which he or she wants to analyze/optimize, or automatic characterization selection is available. The latter takes full advantage of the amount of libraries because it automatically selects the right characterization point (or nearest set of surrounding points) to determine delay and energy values for the individual cells as soon as any process, voltage, or temperature changes. Derating can still be used, but now from the nearest operating condition available.

41.4 Power Optimization

The majority of today's design environments concentrate on analyzing and addressing power considerations toward the back end of the physical portion of the design process. This makes it almost impossible to fix any problems caused by poor decisions made during the early stages of the design.

A key requirement for a true low-power design environment is to provide an early analysis of the effects, such as voltage drop, using whatever data is available at the time, and to successively refine the analysis as more accurate data becomes available. This allows potential problems to be identified and resolved as soon as possible.

Creating optimal low-power designs involves making trade-offs such as timing vs. power and area vs. power at different stages of the design flow. To enable designers to perform these trade-offs accurately and efficiently, it is necessary for low-power optimization techniques to be integrated with, and applied throughout, the entire RTL-to-GDSII flow.

A wide variety of power-aware design optimization techniques can be brought into play. During the early (presynthesis) stages of the design, the RTL can be modified to employ architectural optimizations, such as replacing a single instantiation of a high-powered logic function with multiple instantiations of low-powered equivalents. The design may also be partitioned for implementation in multiple voltage domains (aka voltage islands), and power-aware clock gating techniques can be automatically applied.

In the following paragraphs, all low-power techniques minimize one or more of the factors in Equation (41.1).

During synthesis, power-aware mapping techniques may be used to optimize the netlist. These techniques include mapping highly active nodes into specific cells and mapping highly active input signals onto low-capacitance input pins.

Lowering the supply voltage dramatically reduces a logic gate's power consumption, but this also significantly reduces the switching speed of the gate. One solution is to use multiple voltage domains, allowing different areas of the chip running at different voltages (aka voltage islands). In this case, any performance-critical functions would be located in a higher voltage domain, while noncritical functions would be allocated to a lower voltage domain.

Advanced techniques also enable optimization for power during floorplanning and placement. To correctly implement multiple voltage domains, it is necessary to separate the different power meshes for each domain. The results from early voltage drop analysis can be used to determine better locations for any buffers that are to be inserted. Advanced clustering techniques can also be applied to clock-trees to reduce power consumption.

One way to reduce the amount of switching activity is to reduce the frequency of the system clock. Obviously, this will have a corresponding impact on the performance of the device. Another technique is to employ clock gating, which restricts the distribution of the clock to only those portions of the device that are actually performing useful tasks at that time. It is also possible to minimize local data activity (glitches and hazards) by applying appropriate delay balancing.

The amount of capacitance may be reduced in a number of ways. One approach is to downsize the gates driving overdriven wires, thereby lowering the capacitances associated with these gates. Another technique is to use power-aware cell placement, based on weighing nets according to their activity. The idea is to minimize the total weighed net length to minimize the switched capacitance thereby minimizing dynamic power consumption. Yet another alternative is to exploit technology options such as using low-k dielectric (insulating) materials and low-resistance/capacitance copper (Cu) tracks.

Interesting trade-offs can also be made between functional parallelism and frequency or voltage during the algorithmic and architectural stages of the design flow. For example, replacing one block of logic running at frequency f and voltage V with two copies of that block, each of which performs half of the task, and each of which is running at a lower frequency or a lower voltage. In this case, the total power consumption of this function may be reduced while maintaining performance at the expense of using more silicon real estate.

The following subsections highlight the three techniques that were mentioned previously.

41.4.1 Power Management

One of the frequently used techniques in power management solutions is partitioning the design into different blocks, each operating on block-specific voltages.

Having block specific power supply allows for switching off the supply entirely, to minimize leakage power when the block does not have to perform any logic function. Second, it allows for selecting the block supply voltage(s) that gives just the desired performance, thus minimizing dynamic power. This selection can de done once, at design time, or can be done over time using dynamic voltage scaling (DVS) schemes. Obviously, supply switching techniques and voltage-scaling techniques are orthogonal, and can be combined as such.

This section does not discuss the typical system and architectural issues that need to be addressed, such as the data retention problem for switched blocks and creating the partition of the design into

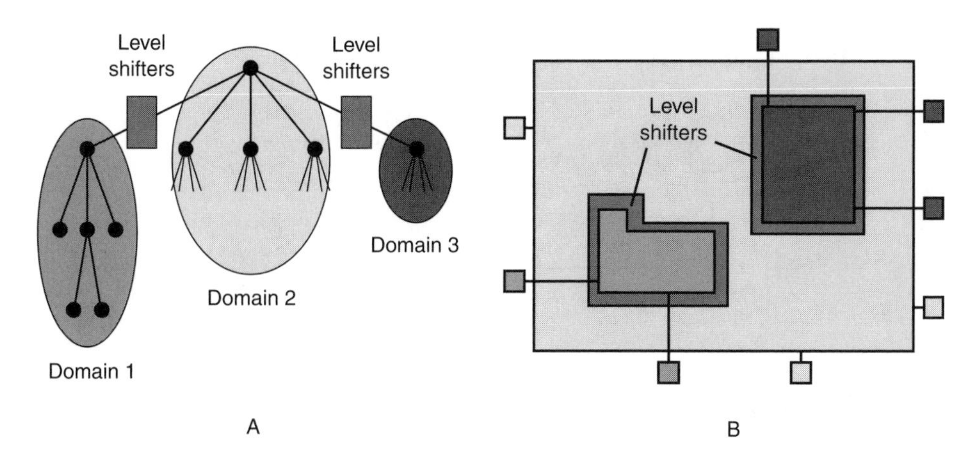

FIGURE 41.3 (a) Design partitioned into electrical domains; (b) design partitioned geometrically into physical floorplans.

separately supplied blocks. Addressed instead are the needs for the synthesis and place-and-route (P&R) flow to correctly handle such a design, such that the electrical behavior of the design is guaranteed.

We describe three types of design partitioning: functional partitioning using logical hierarchy, electrical partitioning in so-called domains, and physical partitioning in so-called (sub)floorplans (refer to Figure 41.3). Domains and floorplans can be introduced early in the flow, at or before RTL input.

A logical hierarchy starts with the top model. Each model contains model pins, cells, and nets. Nets connect to model pins and cell pins. A cell is an instantiation of a model, thus allowing for hierarchical design descriptions.

Clearly, a model represents a group of cells. The same holds for domains and floorplans.

A domain defines the operating conditions for a group of cells and the supply nets together with the recipe how to hook up each library cell to these supply nets. Domain membership is enough to fully define a cell's operating conditions and pin voltages. This information will be used during any analysis (e.g., delay, timing, power, and rail analysis) and optimization (e.g., mapping, sizing, cloning, and buffering). For instance, at the beginning of the design cycle, many cells that will be in the final design still need to be inserted. Knowing to what domain they belong is enough to do analysis correctly.

A floorplan describes a rectilinear area of cell rows in which all cells associated with the floorplan need to be placed. This information is used to guide P&R. Furthermore, each floorplan is required to be associated with one domain, from which power routing deduces what supply nets need to be routed as the rails in the cell rows, and what supply connections needs to be made by point-to-point routing.

Domains can specify more than one power and one ground net. This is required to connect complex cells, such as level shifters, which oftentimes have two power and one ground connection, cells with supply as well as bias lines, and macros. Per net, a voltage level is specified (actually, per analysis case as well) as well as the supply type. The supply type describes the behavior of the net's voltage level over time. The four supply types are:

- Constant. The voltage level is constant over time.
- Switched constant. Constant, but it may be become floating (undefined).
- Variable. The voltage level varies over time.
- Switched variable. Variable, and it may become floating (undefined).

Apart from voltage levels, a domain can also describe the process and temperature for its cells.

Assuming a cell is supplied by a specific power and ground net, its output will carry the voltage of the power (ground) net when driven to a logical one (zero). Likewise, each of its inputs will assume the power (ground) supply voltage to drive it to a logical one (zero). We extend this to multi-supplied cells, where we associate a supply line with each input and output.

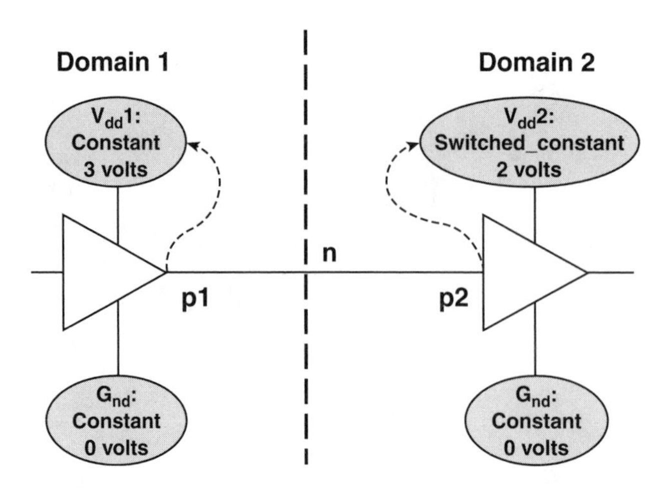

FIGURE 41.4 Net n needs level shifter with isolation capabilities.

Now, suppose we have a net that is connected to a driver cell (source) and a number of cell input pins (sinks). With each of the connected cell pins, we can associate a supply net. These supply nets have their voltage and supply-type defined on the domains. Using this information we can determine what nets need level-shifters and or isolation cells. Level–shifters are used to up-shift (and sometimes explicitly down-shift) a signal's voltage level to overcome the difference in source and sink swing. Isolation cells are cells that keep their output at a predefined logical level when the input becomes floating, and are typically required on signals that travel from switched to unswitched blocks and vice versa. Refer to Figure 41.4 for an example.

Any net crossing domain boundaries needs to be inspected for special situations described previously. In addition, however, cells within one domain can be connected to different supply nets and, therefore, should be checked as well. This, however, depends on the setup of your domains.

41.4.2 Gate Sizing

Refer to Figure 41.1. One of the factors controlling the slew of the signal being presented to the inverter's input is the size of the transistors forming the logic gate driving this signal. These need to be sufficiently large such that the signal transitions fast enough to keep the amount of time the inverter's transistors are both active to a reasonable level (Figure 41.1(b)).

Now consider what happens if the driving gate's transistors are too large and the driving gate is overpowered. In this case, the power savings achieved by minimizing the time where the inverter's transistors are both on (Figure 41.1(a)) will be negated by the driving gate having to charge the increased capacitance associated with its oversized transistors, thereby consuming excessive amounts of power. Furthermore, the high speed of the signal's transitions will also cause signal integrity problems in the form of noise, overshoot, undershoot, and cross talk.

By comparison, if the driving gate's transistors are too small and the driving gate is underpowered, the inverter's transistors will both be on for a significant amount of time (Figure 41.1(c)), thereby causing the inverter to consume unwarranted amounts of power (the under-driven input signal will also be susceptible to noise and cross talk coupling effects from other signals).

The gain-based scenario that we use selects minimum gate sizes, subject to maximum slew limits and maximum load limits. Minimizing gate size is good for cell area and cell power dissipation, since it implies minimization of cell parasitic capacitances. In doing so, all paths are automatically made as slow as possible, thereby balancing the delay of the paths as much as possible. This is good for glitch and hazard power as well.

41.4.3 Multiple Thresholds

To address low-power designs, IC foundries offer multiple V_t libraries, in which each type of logic gate is available in two (or more) flavors:

1. With low-threshold transistors that switch quickly but have higher leakage and consume more power
2. With high-threshold transistors that have lower leakage and consume less power but switch more slowly

One option is to use low-V_t transistors only on timing-critical paths and to use high V_t transistors on noncritical paths, the so-called dual V_t approach. Another solution is to use multiple voltage domains to implement MTCMOS (i.e., to selectively power-down leaking blocks using nonleaking transistors whenever those portions of the device are not required), for example, when those portions are placed in a standby mode. These two solutions may of course be used in conjunction.

Given the possibility to use multiple thresholds and (dynamic) voltage selection for blocks makes that the engineers have to perform a complicated balancing act. For instance, lowering the supply voltage reduces the amount of heat being generated, which, in turn, lowers the static power dissipation; however, lowering the supply voltage also increases gate delays. By comparison, lowering the transistors' switching thresholds speeds them up, but this exponentially increases their leakage and therefore their static power dissipation. In addition, switching entire blocks on and off can cause dramatic current surges and thus generate *Ldi/dt* voltage drops, which may require the use of additional circuitry to provide a soft (staged) power on/off for these blocks.

41.5 Rail Analysis

Deep submicron (DSM) and ultra-deep submicron (UDSM) devices are prone to voltage drop effects, which are caused by the resistance associated with the network of wires used to distribute power and ground from the external pins to the internal circuitry.

Purely for the purposes of providing a simple example, consider a chain of inverter gates connected to the same power and ground tracks. Refer to Figure 41.5. Every power and ground track segment has a small amount of resistance associated with it. This means that the logic gate closest to the IC's primary power or ground pins (gate G1 in this example) is presented with the optimal supply. The next gate in the chain (G2 in this example) will be presented with a slightly degraded supply, and so on down the chain.

The problem is exacerbated in the case of transient or alternating current (AC) voltage drop effects. These occur when gates are switching from one value to another or, even worse, when entire blocks are switched on and off. This causes transitory power surges, which momentarily reduce the voltage supply to gates farther down the power supply chain. The simple example circuit shown in Figure 41.5 consists

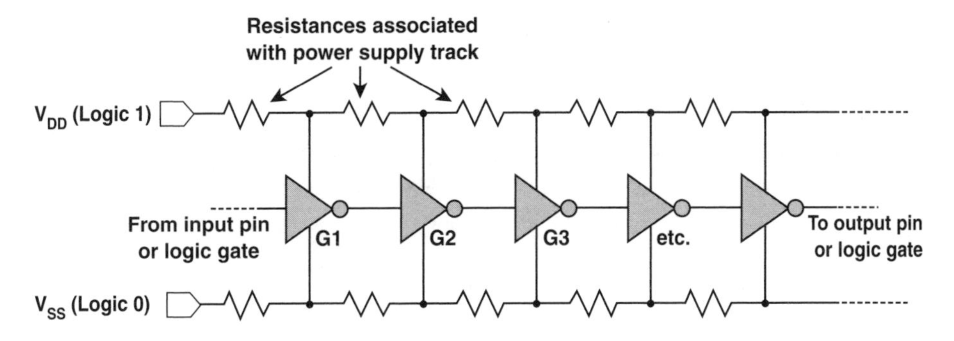

FIGURE 41.5 A chain of invertors connected to the same power and ground tracks.

only of inverter gates, but a real design typically contains tens of thousands of register (storage) elements triggered by a clock signal. The clock can cause large numbers of register elements to switch simultaneously, resulting in significant glitches in the power supply. To analyze and address these effects, it is necessary to consider resistive, inductive, and capacitive effects.

Voltage drop caused by resistive effects is often referred to as IR drop and, erroneously, associated with DC currents only. Compare this with the inductive voltage drop effects caused by transient currents through inductors. Of course, IR drop also occurs in this last case.

The reason voltage drop effects are so important is that the input-to-output delays across a logic gate increase as the voltage supplied to that gate is reduced, which can cause the gate to miss its timing specifications. An increase in the interconnect delays associated with wires driven by underpowered gates also occurs. Furthermore, a gate's input switching thresholds are modified when its supply is reduced, which causes that gate to become more susceptible to noise.

41.5.1 Analysis Flow

Blast Rail is a full-chip voltage drop analysis and repair tool completely integrated into the RTL-to-GDSII design environment. The tight integration makes error-prone data transfer between tools superfluous. A voltage drop analysis based on power estimation (refer to Figure 41.6) consists of six steps:

1. Layout extraction of power and ground nets
2. Determination of current sources from power analysis
3. Creation of an electrical network suitable for voltage drop computation
4. Matrix solving
5. Reporting
6. Update cell timing information using computed cell voltages

For static DC voltage drop analysis, the electrical network model of the power and ground nets consist of electrical nodes, resistances that connect nodes, and current as well as voltage sources. Resistance values are derived from the geometries of the extracted segments and the layer specific sheet resistance, which might be width dependent. Voltage sources are due to connected supply pads or bump cells in flip chip designs. All other cells, such as standard cells, macros, or I/O pads, result in current sources in the electrical network. Their values are derived from a power analysis. The position of the current sources within the network depends on when the analysis is performed within the design flow. After detail placement, the position is easily derived from the placement location of the corresponding cells and macros. A preplaced design is analyzed by assuming a uniform distribution of the current sources while placement blockages are honored.

A transient analysis additionally requires the extraction or specification of inductances and capacitances. On-chip wire inductances are negligible for current technologies; however, bonding wires have inductances that must be taken into account. Capacitive effects are mainly due to decoupling capacitors. Decoupling capacitors are placed close to high switching cells to act as a charge supply. The result is a low-pass filtering of current spikes.

41.5.2 Voltage-Drop-Induced Analysis

The flow described previously assumes that the current sources have known values. These values are derived from a power consumption analysis of the entire design; however, the initial power and timing analysis is based on a fixed voltage supply for all cells and macros (i.e., voltage drop is not considered at this stage). This results in optimistic slews and delays but a pessimistic power consumption behavior of the cells and macros. Blast Rail offers the feature to feed back the voltage drop values for timing and power analysis. The user can then iterate the process of timing, power, and voltage drop analysis. It turns out that the analysis results converge very quickly after only a few iterations (two to three iterations). Since Blast Rail is integrated into Magma's unified data model, the updated timing information can also be used for timing driven optimizations such as buffer sizing and placement.

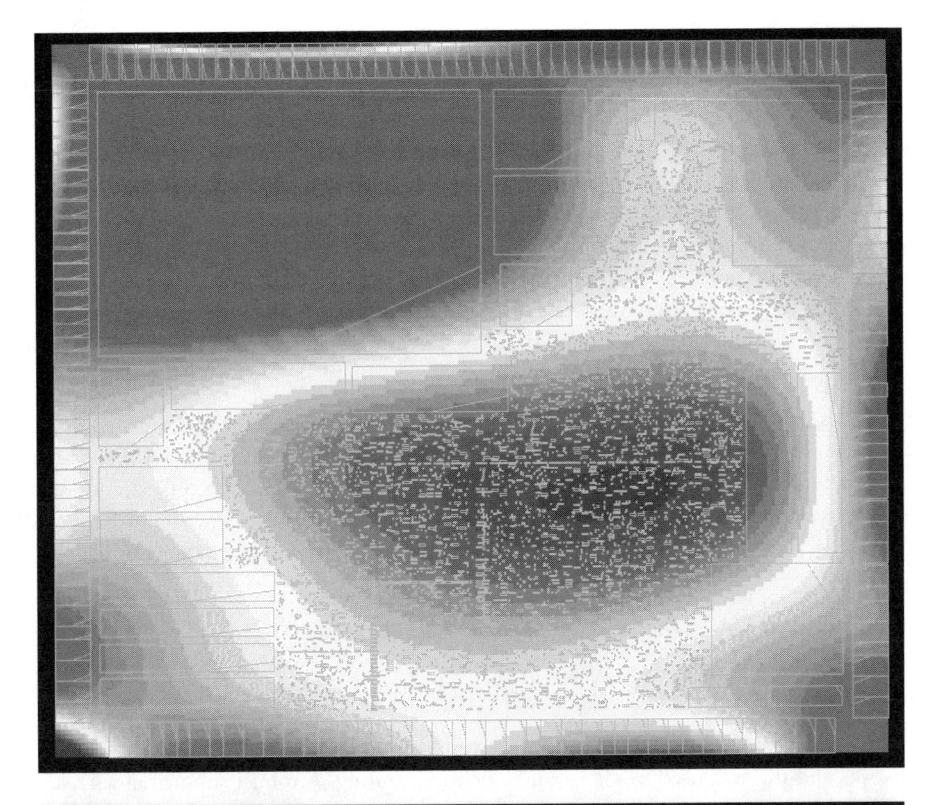

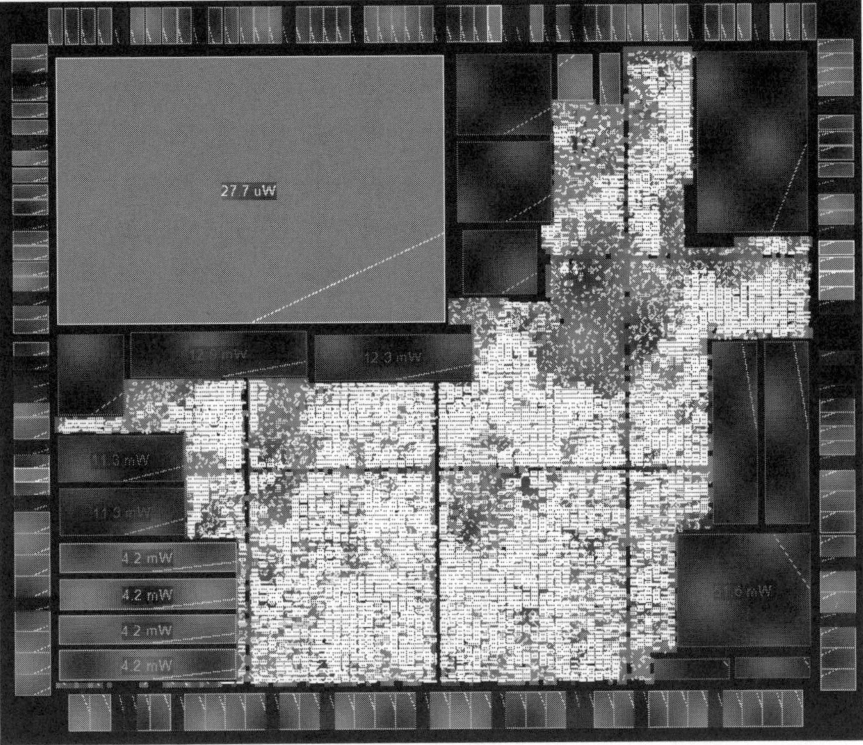

FIGURE 41.6 (Top) Power map showing the power dissipation for each cell of a design after power analysis and (Bottom) the corresponding voltage-drop oil map after rail analysis.

41.5.3 Abstraction

To handle large SoCs (such as 10 million gates or beyond), we offer our GlassBox abstraction technique for hierarchical chip design. The basic idea is to abstract away from a block as much data as possible. Only boundary information is kept such that for instance accurate top-level timing analysis and optimization is still possible. Blast Rail also supports the GlassBox abstraction technique. The power and ground layout information is deleted, while a reduced electrical network is kept. The reduced network mimics the same electrical behavior of the block at the top-level as the original block. Typical reduction of up to 95% is obtained with accuracy loss of less than 1%.

After reduction, the voltage drop inside the block cannot be observed while doing a top-level analysis. An excessive drop inside the block due to a poor supply network might remain undetected. Three different solutions can be applied to circumvent this drawback:

1. Assume pessimistic voltages at the boundaries while doing block level analysis.
2. The user can specify regions (aka view ports) inside the block, which are not reduced and thus stay observable. This combines the best of both, namely observability in combination with high-reduction factors.
3. The electrical network of the GlassBox is not reduced at all (but all other abstractions are applied).

These hierarchical analysis techniques offer the designer to control the trade-off between memory reduction and observability of voltages and currents within a block.

41.5.4 What-If Analysis

Designing the power and ground distribution networks amounts to trading off the chip area used by these networks and the area used by signal and clock-net routing. A robust supply network ideally has a dense mesh with as many via connections to rails and rings as possible. It should consist of wide wires and a lot of connected power pads or bumps; however, such a network results in highly congested routing layers, making it harder or impossible to route signal- and clock-nets.

What-if analyses support the designer to deal with this trade-off. It allows for analyzing different power network configurations without generating the corresponding layout. The designer can add or delete resistances, capacitances, inductances, current sources, and voltage sources directly in the electrical network. An analysis of the modified network can then be done without performing an extraction. Adding a resistance means that a connection is made in the geometric domain, such as dropping a via between the mesh and a rail. Replacing an extracted resistance by a smaller resistance corresponds to widening a wire or creating a larger via array. Introducing or removing capacitances and inductances are used to model different decoupling alternatives or chip package models respectively. Adding current sources allows for testing the robustness of the network. For example, the designer can distribute a certain amount of current over a region of the chip where a macro will be placed later on in the design flow. Adding or removing voltage sources allows for instance to play with a different number of supply pads or different pad locations.

41.5.5 Partial Grids

Manually adding resistances to an extracted electrical network becomes a tedious and error prone task if entire power routing steps are to be simulated. Blast Rail offers the feature of partial grid analysis, which automatically creates resistances to mimic the routing steps of via dropping and pin tapping. Pin tapping is the task to create the proper connections of macro pins to rings and meshes during power routing.

The automatic resistance creation reduces design as well as extraction time. It is typically used when designing the power rings and meshes, when the designer is focusing on the global aspects of the power grid and does not want to be disturbed by detailed connectivity issues such as pin tapping and via dropping.

In this phase of the flow, positioning and connectivity of current sources also play a role. During preplacement phases, such as floorplanning and power grid planning, unplaced cells are modeled as current sources that are uniformly distributed over the unused standard cell area as defined by the floorplan.

41.5.6 Electromigration

Electromigration occurs when the current density (current per cross-sectional area) in tracks is too high. In the case of power and ground tracks, electromigration effects are DC-based. The so-called electron wind induced by the current flowing through a track causes metal ions in the track to migrate. This migration creates voids in the upwind direction, while metal ions can accumulate downwind to form features called hillocks and whiskers. The increased track resistance associated with a void can result in a corresponding voltage drop and thus might cause timing problems or even functional errors due to undersupplied logic. Major functional errors can also occur when the voids eventually lead to open circuits or when the hillocks and whiskers may cause short circuits to neighboring wires.

Electromigration rules are defined as a maximum current per width of a wire (and not per area) because the height of each wire is constant and a parameter of the applied process technology (neglecting intra-die variation effects on layer thickness). Modern process technologies require the definition of width dependent electromigration rules for each layer. These rules are usually staircase functions where wider wires have a higher current density limit. Vias are treated differently compared with routing wires. Via electromigration rules define a maximum current per via cut.

41.6 Power Grid Synthesis

To accommodate variations in operating temperature and supply voltage, designers have traditionally been obliged to pad device characteristics and design margins; however, creating a device's power network using excessively conservative design practices consumes valuable silicon real estate and results in performance that is significantly below the silicon's full potential. This is simply not an option in today's highly competitive marketplace.

Voltage drop effects are becoming increasingly significant, because the resistance of the power and ground tracks rises as a function of decreasing feature sizes (e.g., track widths). Increasing the width of power and ground tracks can minimize these effects, but can cause routing congestion problems. To solve these problems, the logic functions have to be spaced farther apart, which increases delays (and power consumption) due to longer signal tracks. Thus, implementing an optimal power network requires the balancing of many diverse factors.

41.6.1 Grid Synthesis

The process of designing the power distribution network should be based on the results of early rail analysis performed when the power grids are still incomplete (refer to Section 41.5.5, "Partial Grids"). Correct distribution of dissipating elements across the chip can avoid hot spots and local voltage drop problems, and special wire-widening algorithms can be used to address voltage drop and electromigration issues.

Voltage drop problems can be fixed by wire widening, via insertion, and by adapting mesh frequencies. Support for automatic determination of the mesh parameters, namely mesh frequency and wire width, is available as part of automatic power grid synthesis. Mesh optimization is done for all meshes simultaneously and can be guided by the user.

Appropriate on-chip decoupling capacitors should be added to minimize the inductive voltage drop effects caused by off-chip current variations over time. The transient effects should be kept to a minimum (i.e., the charge gets to the cells in time). Thus, the voltage-drop reduction problem is made as *DC* as possible. To lower the current-per-pad and bond-wire inductance, many pads are allocated for power and ground, thereby making the analysis of pad placement a nontrivial task. Flip-chip packaging technologies can be used to increase the number of pads connected to the power and ground supplies, thereby lowering the current-per-pad and lowering the inductance.

Electromigration violations in power and ground tracks are fixed by widening the wires, which cause the violations. Electromigration rules for vias are defined as current per via cut. This means that fixing an electromigration problem in a via translates to increasing the number of via cuts. If the via already consists of the maximum number of via cuts for the overlapping top and bottom routing wires, then the wires have to be widened as well. The electromigration report generated by Blast Rail prints for each violated wire the required width and the number of required via cuts per violated via, and may apply these suggestions.

41.6.2 Packaging Considerations

Power consumption — both static and dynamic — increases a device's operating temperature. In turn, this may require engineers to employ expensive device packaging and external cooling technology.

Yet, another consideration is that the on-chip temperature gradient (i.e., the difference in temperatures at different portions of the device caused by unbalanced power consumption) can produce mechanical stress, which may degrade the device's reliability.

When it comes to power distribution, the first problem is to get the power from the outside world, through the device's package, to the silicon chip itself. Typically, the amount of current per pin is limited requiring many pins for supply. The wires used to distribute power throughout the chip have resistances associated with them — the longer the wires, the larger the resistance, and the larger the resistance, the greater the associated voltage drops. This means that traditional packaging technologies based on peripheral power pads are no longer an acceptable option in the case of today's extremely large and complex designs.

One solution is to use a flip-chip packaging technology, in which pads located across the face of the die are used to deliver power from the external power supply directly to the internal areas of the chip. In addition to being able to support many more power and ground pads, this minimizes the distance the power has to travel to reach the internal logic. Furthermore, the inductance of the solder bumps used in flip-chip packages is significantly lower than that of the bonding wires used with traditional packaging techniques.

41.7 Conclusion

Addressing the problems associated with DSM and UDSM devices requires power design and analysis tools that work throughout the entire RTL-to-GDSII design flow. Identifying and resolving power problems late in the flow may result in expensive, time-consuming iteration cycles. What is required is to identify and resolve these problems throughout the flow and to be able to "forget" issues once they have been addressed and made "safe."

To handle complex interrelationships between diverse effects, it is necessary for all of the power tools to be fully integrated with each other, and also with other analysis engines in the flow, including synthesis, place-and-route, timing, and signal integrity analysis. This requires that all design and analysis tools have concurrent access to a single, unified design database, and that any changes made by one tool are immediately tested and validated by the others. This results in a convergent algorithm that quickly determines optimal solutions with a minimum of time-consuming iterations.

42

Sequence Design Flow for Power-Sensitive Design

Jerry Frenkil
Sequence Design

42.1 Introduction

Integrated circuit (IC) power consumption has become a significant issue for most applications. For wireless and battery-powered applications, the key issue is that of maximizing battery life. This issue is exacerbated by the continuously increasing amounts of computing required for advanced functionality, such as color displays and full-motion video. For tethered applications, where batteries do not limit the power budgets, power is also an issue because it directly affects critical manufacturing parameters, such as die size, packaging choices, and unit cost.

These issues and concerns are not new. Various forms of low-power design have been practiced for decades, but what is new is the criticality of effectively addressing the various facets of power consumption. Simply put, power threatens to derail the constant progress of Moore's law [1]. In many ways, the advances delivered by process engineers in the form of smaller line widths and thinner oxides are now creating as many problems as they solve. Moreover, some of the problems, such as transistor leakage, appear unlikely to be solved in the near future by the processing advances, so it is the design community that will have to step up with solutions to the power problem.

The earliest forms of low-power design involved the basic practice of reducing the power supply voltage, either in the entire design or in certain parts. The attractiveness of this approach was, and still is, the quadratic relationship between the supply voltage and the resulting power consumption. An additional method involved the use of more advanced semiconductor processes with narrower transistors and shorter wires; all else being equal, the reduction in parasitic capacitances due to the smaller geometries resulted in less dynamic power.

0-8493-1941-2/05/$0.00+$1.50

Today, these approaches are no longer enough and, in some cases, such as subthreshold leakage, they are counterproductive. Thus having employed the most basic approaches, designers have turned to more design-oriented techniques in the architecture, logic, and physical design spaces. In doing so, designers have found it necessary to employ various forms of design automation tools. In some cases, the design tools enable the designer to intelligently choose between various design alternatives based on power characteristics. In other cases, the tools evaluate the choices and make the decisions automatically.

42.2 Design Flow Overview

The vast majority of digital ICs designed today are built in complementary metal-oxide semiconductor (CMOS) technology. Once viewed as a low-power technology, CMOS chips can consume as little as a few microwatts or as much as 100 W. Where a chip's power consumption characteristics fall on this continuum depend on a large number of variables, not the least of which is the amount of attention paid to power consumption during the design process.

42.2.1 CMOS Power Consumption

Generally, power-efficient CMOS design involves the minimization of one or more of the terms in the basic power consumption equation

$$P = C_L V_{dd}^2 f + V_{dd} I_{dd} \tag{42.1}$$

or, in its more detailed version

$$P = C_L V_{dd} V_{swing} f + V_{dd} Q_{sc} f + V_{dd} I_{lkg} + V_{dd} I_{through} \tag{42.2}$$

where P represents the total power consumed, V_{dd} represents the supply voltage, I_{dd} represents the static current drawn from the supply, C_L represents the load capacitance, and f represents the switching frequency. The VI term represents the static, or DC power consumption, while the CV^2f term represents the dynamic power consumption. In the more detailed version, V_{swing} represents the signal voltage swing (which for CMOS is usually equal to V_{dd}), Q_{sc} represents the charge consumed due to the short-circuit momentary current (also known as crowbar current) drawn from the supply during switching events, I_{lkg} represents the parasitic leakage current, and $I_{through}$ represents the (by design) quiescent static current. The first two terms of this equation represent the dynamic power consumption, while the latter two represent the static power consumption.

Until relatively recently, the design and design automation communities viewed low-power design as being primarily focused on the CV^2f component; however, with chips inexorably becoming bigger, faster, and more power-hungry, it has become clear that the problems, as well as their solutions, are much more complicated.

42.2.2 Power-Sensitive Design Challenges

Today, what is commonly known as low-power design actually comprises two different but related design activities. The first is power minimization — the reduction of the power consumption characteristics of the design. The second is power integrity management — managing the delivery of power to the various portions of the design as well as the effects of a nonideal power source on the design's timing and functionality. Together, these two design activities are sometimes referred to as power-aware design or power-sensitive design.

Power minimization seeks to reduce power consumption, be it average power or instantaneous power or both. It may be directed at all modes of operation, or only a particular power mode, such as standby or sleep mode. It may focus on only dynamic power, only on leakage power, or on the total. By comparison, power integrity management seeks to illuminate and minimize the effects of power on the design. These effects include timing, noise, reliability, and cost.

The determination and optimization of these various effects becomes more complex as designers go to greater lengths to control the effects. For example, the reduction of power supply voltages over the last several years, from the long standing standard of 5 V to around 1 V, exacerbates already challenging issues such as large on-chip supply currents and miniscule noise margins. The use of multiple voltages to obtain higher performance or to interface to devices running at higher voltages creates additional issues in physical design.

The implications of these challenges are that a variety of design tools are needed to address the various power issues — the power problem is sufficiently critical and complex that a "one size fits all" approach is inadequate. Fortunately, a number of design automation solutions are available now. For example, power can be calculated early in the design process by utilizing tools that estimate power from a high-level register-transfer level (RTL) description, while later in the design process, detailed analyses of dynamic supply currents flowing through on-chip power distribution networks can be obtained via the use of power rail analyzers. In between, power can be calculated and optimized at the gate level, after synthesis and before physical design.

42.2.3 Feed Forward Design Flow

Given an appropriate variety of tools, effective use is often dependent upon a well-structured design flow. For example, for power optimization as for other parameters, such as performance and cost, it is critical to architect the system properly at the beginning and successively refine it as the project proceeds. Such a multilevel approach increases the likelihood of meeting design goals by providing both early visibilities into critical issues as well as multiple opportunities for mitigation.

Much of digital design is performed today utilizing a top-down or modified top-down design flow. Here top refers to the higher levels of design abstraction, such as the system, behavior, and register-transfer (RT) levels, and time flows downward toward the lower levels of design abstraction, such as the gate and transistor levels. In this case, flow refers to the sequence of tasks; however, the flow of detailed design information is somewhat less clear.

In conventional practice, detailed design information tends to follow a feedback design flow, wherein information about particular power characteristics does not become available until the design has progressed to the lower abstraction levels. A feedback design flow features a relatively lengthy feedback loop from the analysis results obtained at the gate or transistor level back up to the design tasks at the RT-level and above. Thus, information about the design's power characteristics is not obtained until quite late in the design process. Once this information is available, it is fed back to the higher abstraction levels to be used in determining how to deal with the power issues of concern. The farther the lower-level power analysis results exceed the target specification, the higher the abstraction level in which the design must be changed.

By comparison, a feed forward approach, illustrated in Figure 42.1, replaces these lengthy, cross-abstraction feedback loops with more efficient abstraction specific loops. Thus, the design that is fed forward to the lower abstraction levels is much less likely to be fed back for reworking, and the analysis performed at the lower levels becomes essentially a verification task. The key concept is to identify, as early as possible, the design parameters and trade-offs that are required to meet the project's power specs. This helps to ensure that the design being fed forward is fundamentally capable of achieving the power targets. Later in the design flow, optimizations at the lower levels can be used to further minimize the power as desired.

The feed forward flow is enabled by a high-level analysis tool, such as PowerTheater [2], which can accurately predict power characteristics. These early, high-level analysis capabilities are employed to make informed trade-offs, such as which algorithms and architectures to employ, without having to resort to detailed design efforts or low-level implementations to assess performance against the target power

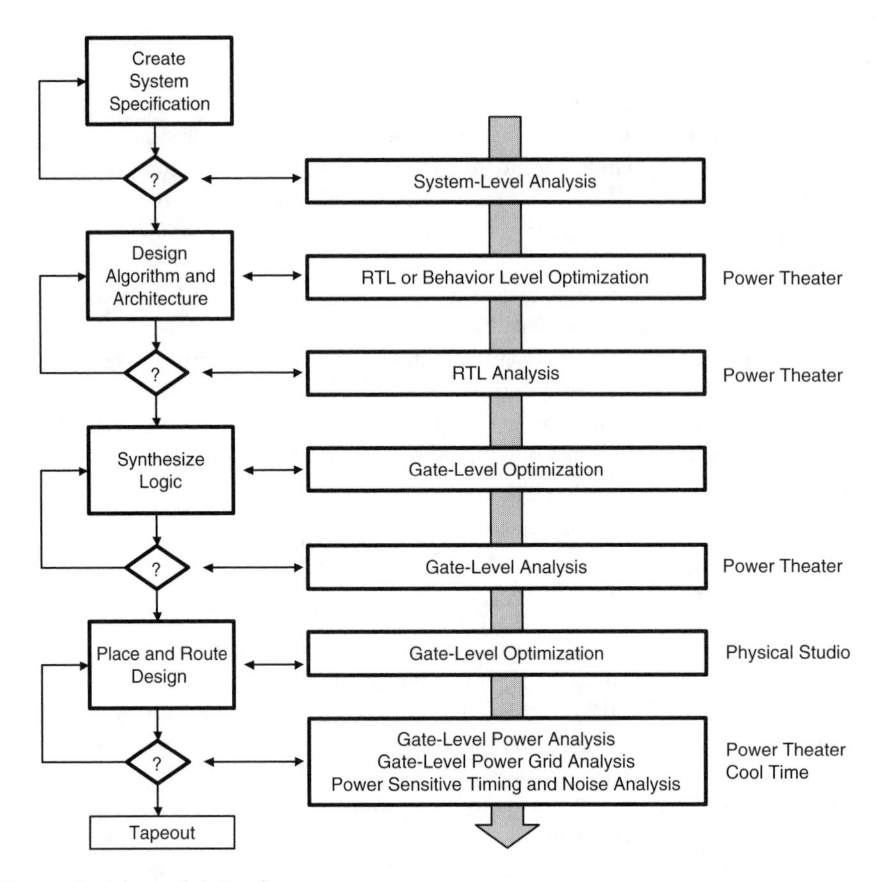

FIGURE 42.1 Feed forward design flow.

specification. Compared with the traditional top-down methods, the key difference and advantage is added by the early prediction technology.

Proceeding in parallel with, or sometimes ahead of, the architecture development is the design of the library macro functions and custom elements, such as datapath cells. These are used in the subsequent implementation phase in which the RTL design is converted into a gate-level netlist. At this point, appropriate optimizations are performed again, and power is reestimated with more detailed information, such as floorplanned wiring capacitances. The power grid is planned and laid out using this power data.

Once the design has been synthesized into a technology mapped gate-level netlist, lower level power optimizations can be employed, using a tool such as PhysicalStudio [3], to further reduce dynamic or leakage power consumption. Specific goals or issues, such as battery life or noise margin repair, will determine the particular optimizations employed.

These optimizations can be performed either before (using estimated wiring parasitics) or after routing (using extracted wiring parasitics). In either case, after the design has been routed and optimized, a final tape-out verification and electrical verification check is performed with an electrical sign-off tool such as CoolTime [4]. In this step, power is calculated and used to compute and validate key design parameters, such as total power consumption in active and standby modes, junction temperatures, power supply droop, noise margins, and signal delays.

Thus, power is analyzed and optimized multiple times, at each abstraction layer following the feed forward approach. Each analysis is successively refined from the previous analysis by using information fed forward from prior design decisions along with new details produced by the most recent design activities. Each optimization, at the various abstraction layers, results in more efficient logic structures to feed forward to the downstream design tasks, thereby successively squeezing out the wasted power.

This approach encourages design efforts to be spent up front, at the higher abstraction levels, where design efforts are most effective in terms of minimizing and controlling power [5]. In addition, because power-sensitive issues are tracked from the beginning to the end, the likelihood of a late surprise issue is minimized.

42.3 Sequence Tools for Power-Sensitive Design

Sequence design provides several different tools for use in a comprehensive power-sensitive design flow. These tools are described next, in the order that they would be utilized in a feed forward methodology.

42.3.1 PowerTheater

PowerTheater is a mixed-level power analysis tool that analyzes RTL designs, gate-level designs, and mixed- — RTL and gate — level designs. It reads in the design description, technology libraries, and environmental data, such as power supply values and external loadings, and activity information. It produces a detailed report listing the amount of power consumed by the various portions of the design along with additional power debug information.

PowerTheater's architecture is illustrated in Figure 42.2. Its front end includes a language parser and inference engine that reads designs described in Verilog, VHDL, or mixed Verilog/VHDL, translates the design into an internal representation, and loads the internal database. The power calculation engine reads the internal database, and, using the specified simulation activities, environmental data, and technology libraries, calculates the power for each portion of the design [2].

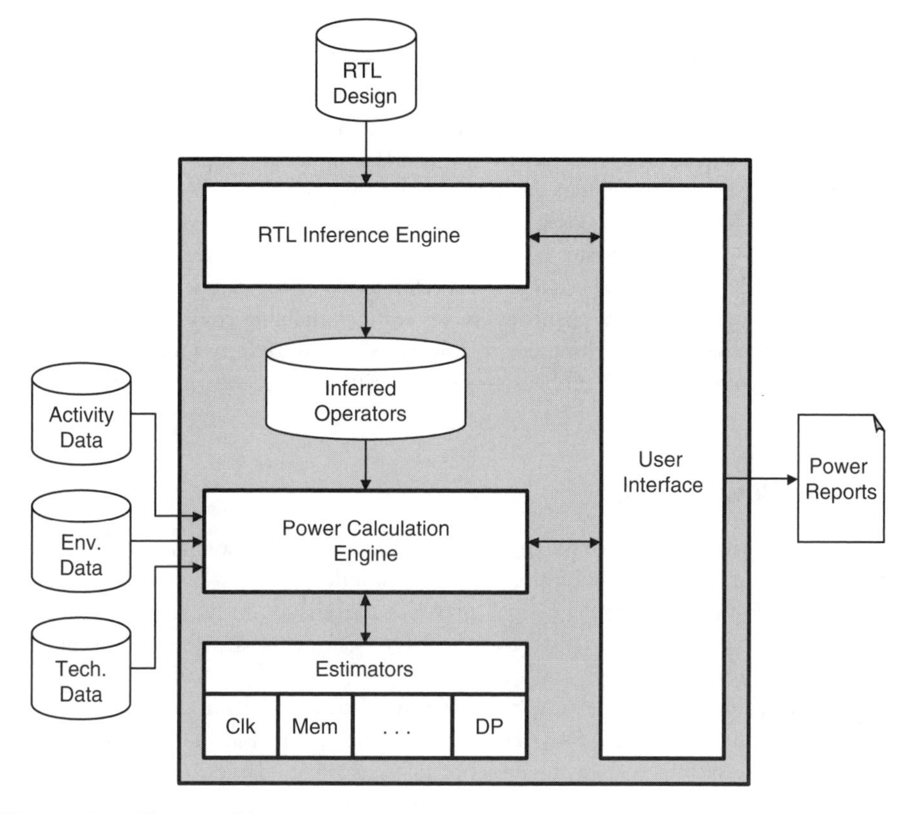

FIGURE 42.2 PowerTheater architecture.

The technology library describes the electrical characteristics of the circuit primitives to be employed in the design. Data, such as input capacitances, state dependent static and dynamic power consumption, logical function, and physical size, are all used during various computations leading up the final power calculation. The data can be read in any of several industry standard formats, such as Liberty, advanced library format (ALF), or open library architecture (OLA) [6–8].

Upon encountering RTL code, PowerTheater infers the hardware that would be produced by a synthesizer for that code; however, unlike a full logic synthesizer, PowerTheater does not produce a gate-level netlist. Instead, it produces a micro-architectural netlist that contains technology unmapped, inferred operators, such as multi-bit registers, arithmetic units, multiplexers, and decoders. Although similar to the first steps executed by a conventional gate-level synthesizer, this approach results in faster execution time, a smaller internal database, and an ability to easily cross-reference the original RTL code with the inferred operators — the latter, especially, being a key advantage to those writing the RTL code. By contrast, when encountering instantiated objects, such as compiled memories or gate-level primitives, PowerTheater passes these objects directly to the internal database.

Once the inferencing step is completed and the internal database is loaded, the design is ready for analysis. The calculation engine loads the database, along with activity data (which can come from a simulation trace, in the form of a value change dump (VCD) file, or from a vectorless activity specification), technology data, and environmental data. For gate-level power analyses, the calculations are relatively straightforward: for each instance, PowerTheater determines the particular stimulus from the activity data, looks up the power characteristics for that stimulus in the technology library, and computes the instance's power using the specified environmental data.

For RTL analyses, the operation is similar but with a key difference: instead of processing gate-level instances, the engine calculates power for each inferred instance. This is accomplished by elaborating a parameterized model for each inferred instance. The elaboration process involves evaluating a built-in parameterized power equation for each instance utilizing power information from the technology library along with the activity and environmental data. Each inferred operator has its own unique power equation, thus enabling PowerTheater to separately calculate power for the different design structures and objects that are found in the RTL. For example, datapath operators are inferred and evaluated separately from control and clock operators. This divide and conquer approach enables faster and more accurate calculations (more accurate than the "one size fits all" algorithm common to gate-level tools) along with a reporting format that is closely linked to the RTL source code. For example, clock power, input/output buffer (I/O) power, register power, and random logic power are all reported separately. Power consumed in driving wiring capacitances, and cell internal power can be reported separately.

The various reporting styles and mechanisms are key features enabling PowerTheater to be used as a design tool. Given the objective of writing power-efficient RTL code, effective analysis, and debugging tools, capable of pinpointing power problem areas, are critical for identifying power minimization opportunities.

42.3.2 Using PowerTheater

PowerTheater is primarily used for two different purposes: power minimization and power verification. In the former case, the objective is to minimize power consumption by writing power-efficient RTL code and by providing early visibility into the design's power characteristics. In the latter case, the objective is to check or verify that the design, whether at the RT or gate level, is within an acceptable power consumption limit.

Producing power-efficient RTL code requires the ability to identify the amount and source of power consumption along with its underlying causes. For example, considering Equation (42.2) above, isolating the largest contributing factors enables the designer to focus on the largest opportunities for power reduction. Although many of these factors, such as V_{dd} or C_L, are fixed by either technology or environmental dictates, others, such as nodal switching frequency, can be affected by coding styles.

Conventional Code: 64x32 Register File using Flip-Flops	**Low Power Code: 64x32 Register File using Latches**

```
input web, oe,clk;
input [31:0] di;
input [5:0] aadr, badr;
output [31:0] do;

// define storage array
reg[31:0] array [63:0];
reg[31:0] do;

// Write Cycle: edge trigger
//              implies flops
always @ (posedge clk) begin
    if (web == 0) array[aadr] = di;
end

// Read Cycle - a sync read
always @ (badr or oe) begin
    if (oe == 1) do = array[badr];
end
```

```
input web, oe,clk;  // clk not needed
input [31:0] di;
input [5:0] aadr, badr;
output [31:0] do;

// define storage array
reg[31:0] array [63:0];
reg[31:0] do;

// Write Cycle: level trigger
//              implies latches
always @ (aadr or web or di) begin
    if (web == 0) array[aadr] = di;
end

// Read Cycle - a sync read
always @ (badr or oe) begin
    if (oe == 1) do = array[badr];
end
```

FIGURE 42.3 Register file RTL code.

In a power minimization methodology, PowerTheater is used to estimate power as the RTL code is written, thus enabling the designer to quickly understand the impact of design decisions and to optimize the code during the creation process, which is the essence of low-power design. If a simulation test bench is available, then the design is simulated to collect activities for the subsequent power calculation. If a simulation test bench is not available, then PowerTheater's vectorless activity function is used to generate activity and state information for use in the power calculations. Although the accuracy of the resulting calculations is much better with simulated data, vectorless activity specification can be used to determine which of several coding alternatives will consume the least power when simulation data is not yet available.

A comparative example of different RTL codes for a given function is listed in Figure 42.3. Consider a register array organized as 64 words by 32 bits. Such a storage function can be coded in several different ways, two of which are presented here. Although the amount of code is the same, the power difference is substantial: the latch-based array consumes only about half as much power as the conventionally coded register array.

In this case, the power reduction principle at work is the use of latches instead of flip-flops — latches being more power-efficient than flip-flops. This targets the $V_{dd}Q_{sc}f$ term in Equation (42.2) — latches, implemented with fewer transistors than flip-flops, consume less internal current; however, a more common method of power reduction is the minimization of effective switching frequencies, targeting the two frequency dependent terms in Equation (42.2), $C_L V_{dd} V_{swing} f$ and $V_{dd}Q_{sc}f$. Here, the concept is to reduce unwanted or unnecessary toggles, which in turn reduces the dynamic power consumption.

The classic example of reducing effective switching frequency is the use of gated clocks to inhibit the clocking of storage elements when it is known that stale data would be latched. Thus, inhibiting the clock edge saves power without changing overall functionality. A subtler example involves operator isolation. Consider the case of a multiplexer that must select between an operand and a multiplied version of that operand as presented in the code in Figure 42.4. In the original version, the multiplier multiplies all the time, whether the multiplexer is set to select the multiplier's result or not — the multiplier multiplies any time a change occurs on either of its inputs. In the modified code, however, the multiplier only multiplies when its output results will be used, thus the effective switching output frequency is reduced because data is not allowed to flow into the multiplier unless its output will be selected.

PowerTheater provides two different methods of finding such opportunities, among others, for power reduction. The first method involves the use of the graphical user interface (GUI) for interactive power

Conventional Code: Selecting Operand or Multiplied Operand	Low Power Code: Selecting Operand or Multiplied Operand

`assign muxout = sel ? A : A*B;`	`assign isoA = sel ? 0 : A;`
	`assign isoB = sel ? 0 : B;`
	`assign muxout = sel ? A : isoA*isoB;`

FIGURE 42.4 Operator isolation RTL code.

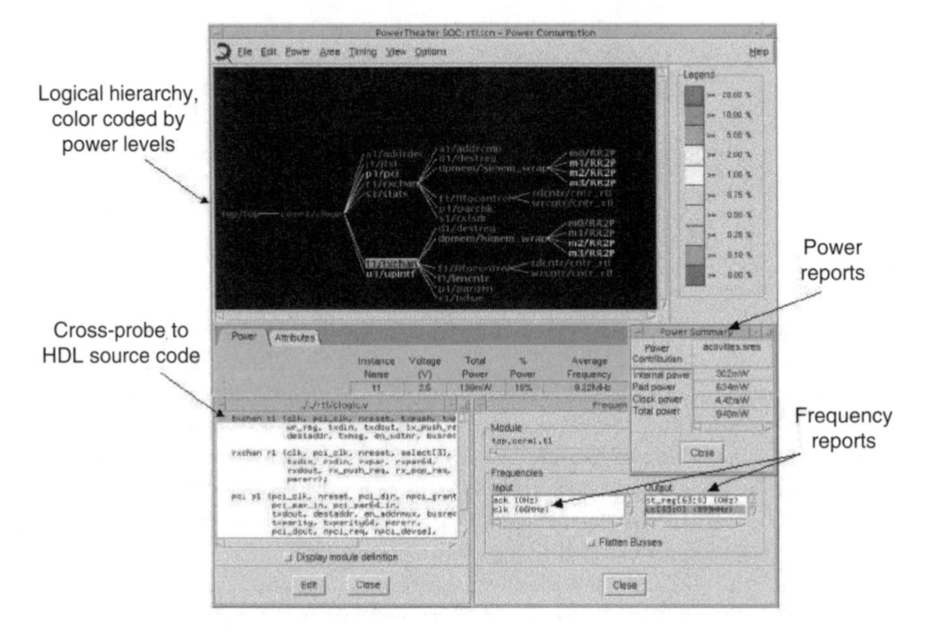

FIGURE 42.5 PowerTheater GUI.

debugging. As presented in Figure 42.5, the PowerTheater GUI displays a hierarchical representation of the design's inferred operators. The display is colorized to facilitate discovery of the design's power-hungry portions via a quick visual inspection. Once a particular "hot spot" is identified, debugging proceeds by clicking on the inferred operator of interest, which will bring up a display of the particular lines of RTL code along with the power consumed by those lines of code. The root cause of that consumption can be investigated by displaying the nodal frequencies of the operator under consideration.

PowerTheater's second method for finding power waste in a design is fully automatic. This method utilizes design search modules known as "WattBots" that walk the inferred netlist looking for various types of power inefficient structures. In addition to finding several different types of clock gating opportunities, the WattBots will also identify operator isolation opportunities, inefficient memory structures, bus conflicts and floating busses, and glitchy nets and control signals. The WattBots will also estimate the amount of power wasted in each instance so that the designer can make informed decisions as to which changes will produce the largest power savings. This is especially important in the case of clock gating because gating too many clocks presents significant problems later during physical design with respect to clock skew management. Thus, it becomes critical to identify which clocks are worth gating (the ones that result in the largest power savings) as well as the ones that should not be gated (the ones that result in the least power savings). The WattBot reports support this type of decision making by presenting the potential power savings for each identified opportunity. A sample WattBot report is pictured in Figure 42.6.

PowerTheater is also used at the RT level for power verification. For many designs, gate-level simulation is impractical due the number of gates and the slowness of simulation speed relative to the operational

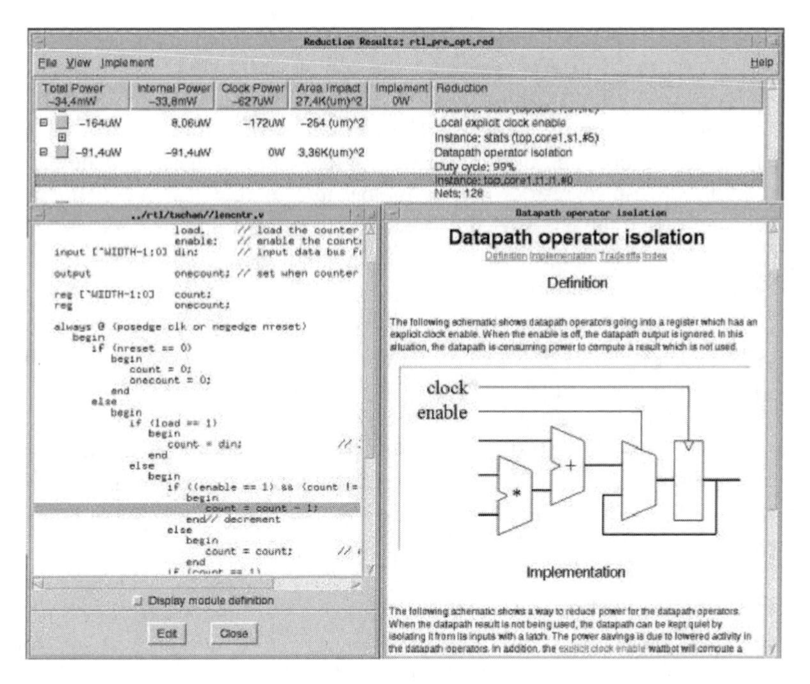

FIGURE 42.6 WattBot report.

speed — significant numbers of operational cycles cannot be run in a reasonable period of time. For these designs, power analysis of the RTL code is the only practical method of obtaining accurate full-chip power estimates because roughly an order of magnitude more cycles can be simulated in a given period using an RTL representation as opposed to a gate-level representation. Another motivation for RTL power verification is the desire to check a design's conformance to its power specification as early as possible. This, in particular, is a key methodology step in a feed forward design flow.

Nevertheless, some designs are verified at the gate level, and PowerTheater is used for this task as well. The use model for gate-level verification is identical to that for RTL verification: the (gate-level) design is simulated, and the resulting activities and the design itself are loaded into PowerTheater for analysis. The analysis can be performed using either estimated or extracted wiring capacitances. If an RT level analysis was previously run, then the same setup and command files can be used to run the gate-level analysis.

To aid in choosing the most appropriate simulation activities for power analysis, PowerTheater provides a simulation activity viewer, which enables the user to view aggregate activity over time. This viewer is used to find, for example, the point in the simulation in which the activities reach a sustained maximum — the most appropriate period for calculating worst-case average power. The viewer can also be used for power debugging, such as searching for those modules that should be inactive but instead are toggling unexpectedly.

42.3.4 PhysicalStudio

PhysicalStudio is a cell-level physical design closure tool for analyzing and optimizing power, timing, and signal integrity issues concurrently in both pre- and post-route designs. It includes a static timing analysis (STA) engine, a delay calculator, and a signal integrity (SI) analyzer for both coupling-delay effects and glitching, along with placement aware optimizations. The combination of the various analysis capabilities, operating off a single database, enables concurrent optimizations in which individual power optimizations are implemented only if they do not break either timing or noise margin limits [3].

PhysicalStudio produces as output an optimized physical design in LEF/DEF format [9]. The optimizations may be performed on a preroute database, in which case the output is placement optimized DEF and corresponding Verilog netlist, or the optimizations may be performed post-route, in which case the output is placement optimized DEF and corresponding Verilog netlist along with a routing ECO file.

PhysicalStudio loads design information and technology libraries to build an internal database from which analyses and optimizations are launched. The required design information includes the design netlist, the placement description, and timing constraints; for routed designs, the required design information includes extracted wiring parasitics. The technology information includes cell libraries for logical, timing, power, signal integrity, and layout views. PhysicalStudio reads the same libraries as PowerTheater, in either Liberty or ALF formats, but it also requires layout definitions in the LEF format.

Central to PhysicalStudio's optimizations are the delay calculator and STA engine. Delays are computed using full three-dimensional coupling capacitances, including aggressor and victim analyses. Similarly, glitch injection and propagation are modeled to evaluate effective noise margins for all instances. Subsequently, during the optimization phase, any potential transform is evaluated against how it affects timing and noise margins. The transform is committed to the database only if it does not violate any of the other constraints; otherwise, the transform is discarded and subsequent optimizations are considered.

PhysicalStudio employs two different types of power optimizations, one aimed primarily at reducing dynamic power consumption and the other aimed at leakage power reduction. Both of the optimizations employ the same fundamental trade-off: speed is traded for power on those paths that have positive slack timings.

This trade-off is implemented as follows. All timing paths are analyzed and slack timings (timing margins) are computed. For each path with positive slack timing, PhysicalStudio replaces individual cells along the path with lower power equivalents as long as the slack timing remains positive. If a potential cell replacement causes the slack timing to become negative, the replacement is not implemented. The overall post-optimization result is that fewer paths exhibit substantial positive slack and a commensurately larger number exhibit less slack, thus indicating that those paths have become slower. Nevertheless, no path is allowed to exceed the clock period timing constraint so that the circuit will still function as desired, but with reduced power consumption.

The specific power reductions that PhysicalStudio employs to trade-off slack timing for power are cell resizing and dual-Vt cell swapping. In the former case, the $C_L V_{dd} V_{swing} f + V_{dd} Q_{sc} f$ terms in Equation (42.2) are targeted, and cell drive strengths (or sizes) are reduced as far as possible without introducing timing or noise problems. The reduction in size reduces several parameters: occupied area, cell crowbar current, and most importantly capacitive loading for the fan-in logic. The dual-Vt optimization targets the $V_{dd} I_{lkg}$ term in Equation (42.2) and cells employing a high-Vt threshold implant, resulting in reduced leakage albeit with lengthier delays, are substituted for low-Vt cells wherever timing slack permits [10].

42.3.4 Using PhysicalStudio

PhysicalStudio is used in two different optimization modes: one before routing and one after routing. The preroute mode is used to prepare the design to avoid or prevent timing, noise, or power problems from arising after routing. The post-route mode is used to fix any problems that persist after the physical design and routing have been completed.

In preroute mode, PhysicalStudio reads the placement and estimates route lengths from the placement. Wire parasitics are estimated from these route lengths and the parasitics are, in turn, used to calculate delays. PhysicalStudio then performs a static timing analysis and a noise analysis to determine available timing slack and noise margins after glitching effects have been considered. Timing constraints are used to guide the static timing analysis. If either timing violations or noise violations are detected, PhysicalStudio will repair the violations using timing and noise optimizations, respectively. Once the design is timing and noise violation free, physical power optimizations are applied.

Dynamic power is optimized by utilizing the resizing command in the optimization script, along with the target minimum amount of timing slack. PhysicalStudio will then examine the sizings of all instances, and decrease the instances' sizes wherever possible, so long as the minimum timing slack parameter is not violated.

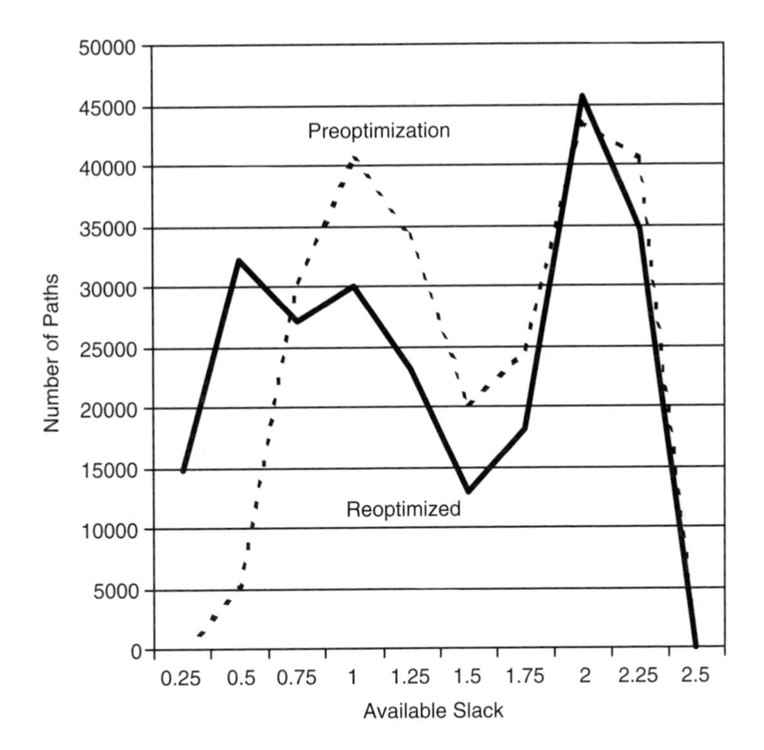

FIGURE 42.7 Power optimization and delay redistribution.

PhysicalStudio supports two different leakage power reduction flows. The first flow works similarly to dynamic power optimization with one significant difference — a second library is required. This second library is functionally and physically equivalent to the first, but each cell has different timing and leakage characteristics due to the use of a different transistor threshold than that in the first library. PhysicalStudio loads both libraries — the primary low-Vt as well as the second high-Vt version to be used for replacement — and automatically establishes functional equivalencies between them. After this point, the flow and use model are identical to that of resizing. PhysicalStudio replaces a low-Vt cell in the original netlist with a high-Vt cell wherever possible as long as the replacement does not violate the minimum timing slack parameters. Leakage power is reduced because the high-Vt cells exhibit much less leakage than the low-Vt cells. An example of this type of optimization is plotted in Figure 42.7. In this particular case, the design consisted of 255,000 cells, and the dual-Vt optimization resulted in 84% of the low-Vt cells being swapped for high-Vt cells. Leakage power was reduced by 43% without changing any of the critical timing.

The second leakage power reduction flow can be thought of as a timing closure flow that starts with a design utilizing all high-Vt cells, but later selectively substitutes low-Vt cells to repair paths with negative timing slack [11]. In this flow, the design is initially synthesized and placed using the slower, high-Vt library as the primary target, and the timing is optimized as much as possible with this single library. For those paths that cannot meet timing using high-Vt cells alone, PhysicalStudio uses low-Vt cells to replace as many of the high-Vt cells as necessary to meet the specified timing constraints.

The second flow generally results in less leakage power consumption, but may require more area as the synthesizer will attempt to close timing with the single, slower library by adding additional logic or buffering for heavily loaded nets.

PhysicalStudio supports both of the leakage reduction flows as well as the dynamic power reduction capability in the post-route mode in addition to the preroute mode described previously. The only difference between the two modes is that in post-route mode the wiring parasitics are known exactly instead of being estimated. This knowledge enables PhysicalStudio to reduce both dynamic and leakage power even further because no uncertainty exists regarding the timing slack (with the exception of on-chip process variation). Both dynamic power and leakage power can be optimized together, although priority must be given to one over the other.

42.3.5 CoolTime

CoolTime is a cell-based electrical integrity analysis tool for analyzing the effects of power on timing, noise, and reliability. Similar to PhysicalStudio, CoolTime includes a STA engine, a delay calculator, and a signal integrity (SI) analyzer for both coupling-delay effects and glitching. It additionally includes a power rail parasitic extractor, a power calculator for both average and instantaneous power, a power rail voltage solver, and several power rail display capabilities [4]. The CoolTime architecture is shown in Figure 42.8.

CoolTime also employs a cell electrical modeler, ElMo, to characterize each cell for its timing and glitching characteristics under various voltage conditions. ElMo reads Liberty and SPICE models, and then runs numerous SPICE simulations to create a set of glitch models and voltage derating factors for each characterized cell. These models and derating factors are subsequently used by CoolTime to compute the response of each individual instance to the calculated voltage variations.

CoolTime produces reports, both textural and graphical, of the power rail voltage variation across the design. This variation data is used to calculate the effects of the power rail voltages and currents upon timing and noise.

CoolTime loads the same design and technology information as PhysicalStudio along with additional information, such as extraction rule definitions, decoupling capacitor definitions, and package parasitics. Once the power rails are extracted, CoolTime uses a combination of internal estimation algorithms, along

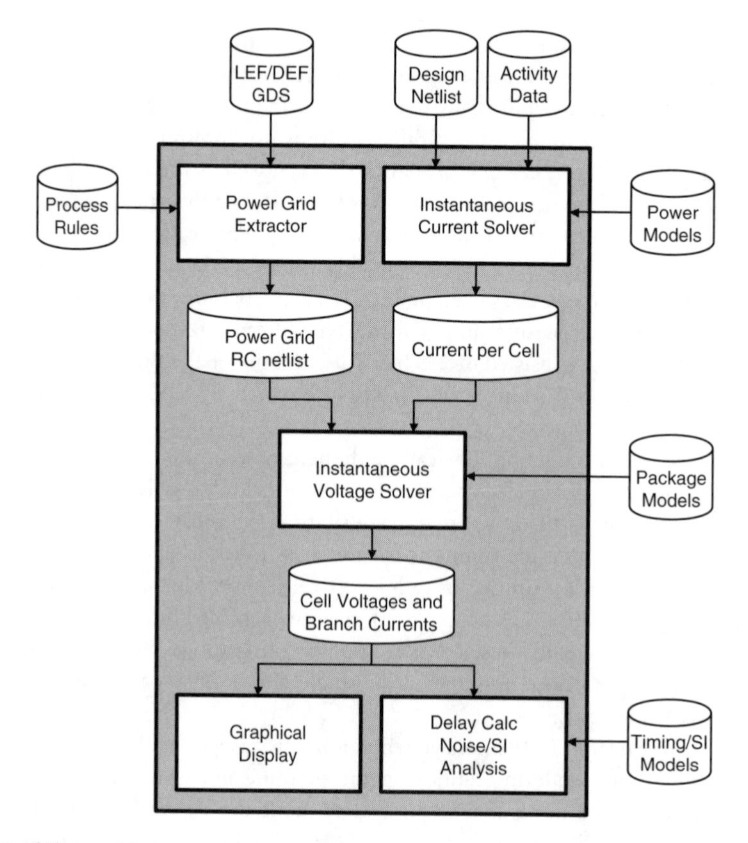

FIGURE 42.8 CoolTime architecture.

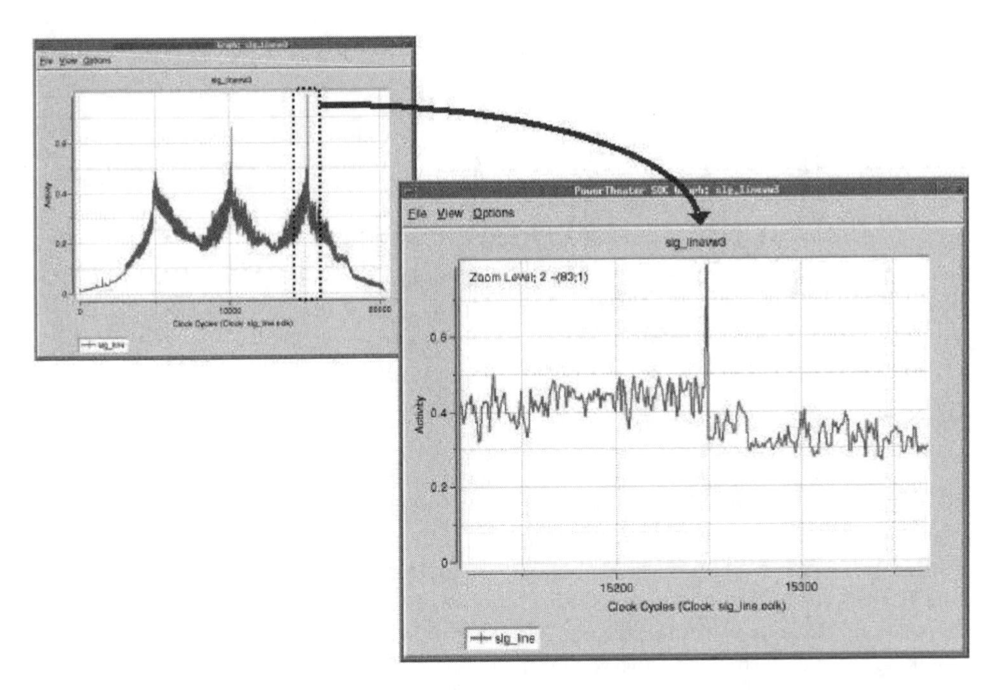

FIGURE 42.9 Finding a simulation cycle to use in CoolTime.

with timing constraints to compute current waveforms for each node in the power rail network. These current waveforms are then fed, along with the extracted power rail network and package parasitics, to a matrix solver to solve for nodal voltages. The resulting output is a set of time varying voltage waveforms for each node in the power rail network. These voltages are subsequently used to recalculate delays and noise margins for each instance.

CoolTime utilizes two different approaches for computing power. The first approach is a vectorless approach in which no external stimulus is needed. The second is a simulation-based approach that relies upon a PowerTheater analysis of the simulation's activities to find the simulation cycle with the most activity as pictured in Figure 42.9. Once found, the state points in that selected cycle are fed to CoolTime as a seed vector from which CoolTime determines which nodes switch and when. By contrast, the vectorless approach requires no simulation data at all but instead makes nodal switching activity determinations based upon the input design constraints and a topological analysis of the netlist. The advantages of the vectorless approach are that no logic simulations are required and that the resulting current and power estimates will be conservative. The advantage of the simulation-based approach is that the results will represent the actual conditions for a particular cycle of interest, although the simulation time required to reach that cycle may be excessive, and as with all simulation results, there can be no assurance that the results represent a worst-case condition.

Either of these two power calculation approaches can be used to compute both average and instantaneous power. For average power, CoolTime computes an average current for one cycle, I_{avg}, which is then used to compute a time-averaged voltage drop:

$$V_{avg} = I_{avg} R \tag{42.3}$$

where I_{avg} is the sum of all currents consumed during the period of interest divided by the length of time of that period.

CoolTime also computes the time varying, instantaneous voltages, $V(t)$, according to the following equation:

$$V(t) = (I(t) + Cdv/dt)R + Ldi/dt \tag{42.4}$$

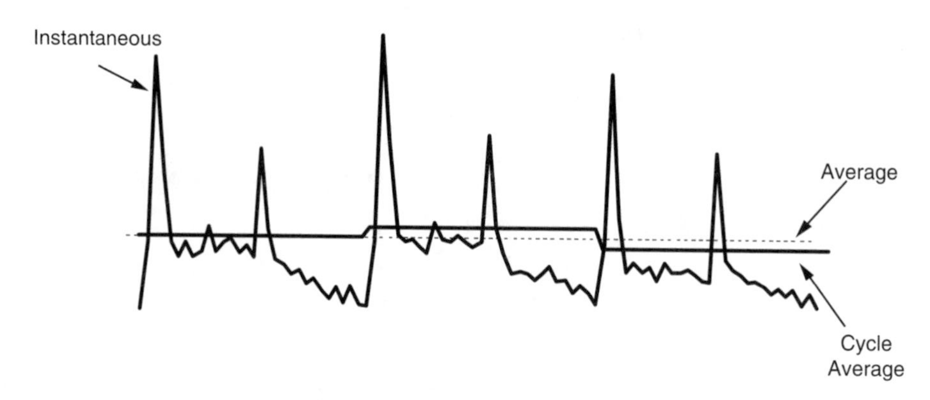

FIGURE 42.10 Current waveform comparison.

where C represents the sum of various parasitic capacitances such as power rail capacitances, well capacitances, and explicitly inserted decoupling capacitances, and L represents the sum of various parasitic inductances such as power rail inductance, bond-wire inductance, and package pin inductance.

Figure 42.10 illustrates the significance of using time-averaged or instantaneous currents for the voltage drop calculations. Not only are the magnitudes materially different, but also the rapid magnitude changes due to spiking can lead to large inductive voltage fluctuations. In effect, the use of time-averaged currents negates the $(Cdv/dt)R$ and Ldi/dt terms in Equation (42.4), and is thus inappropriate for detailed analyses of the power rail network.

The time varying currents and voltages can be viewed using CoolTime's voltage and current recorder (VCR) as depicted in Figure 42.11. This visualization capability enables the user to single step through time, either forward or backward, to see where the "hot spots" occur in the layout along with when they occur. The CoolTime VCR contains four panes, one each for V_{dd} voltage and current as well as V_{ss} voltage and current, providing the ability to view animations of the dynamic current and voltage variations on both V_{dd} and V_{ss}.

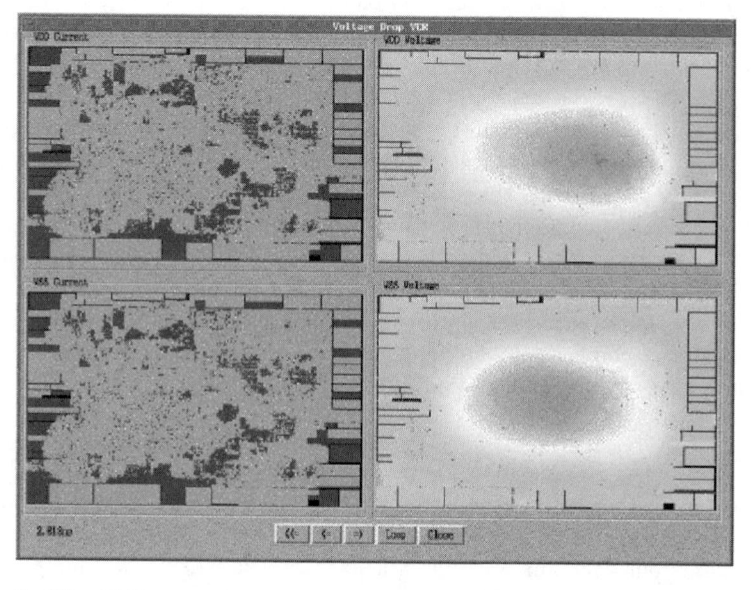

FIGURE 42.11 CoolTime VCR.

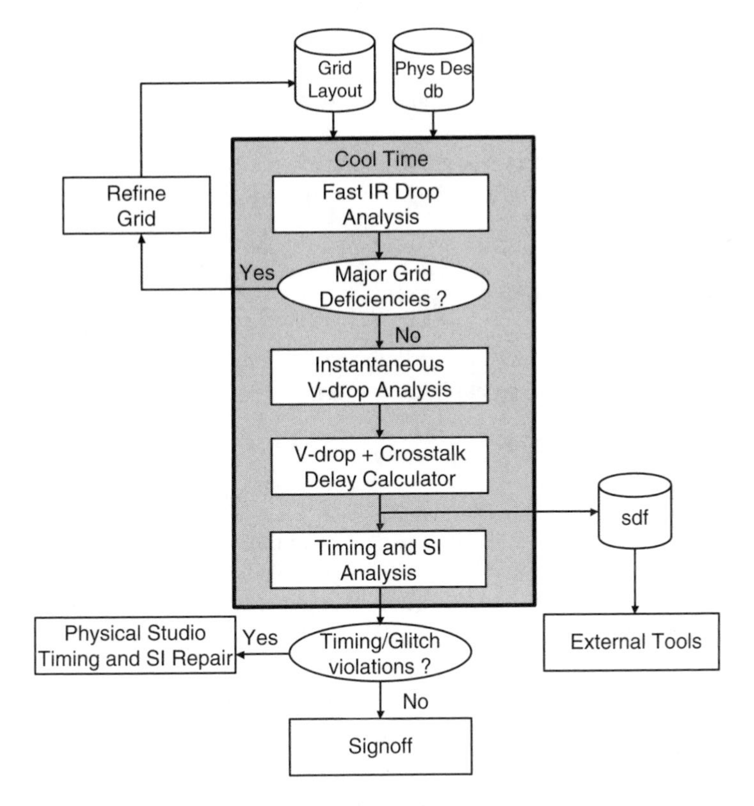

FIGURE 42.12 CoolTime methodology for power grid integrity.

42.3.6 Using CoolTime

CoolTime is primarily intended to be used as an electrical verification and sign-off tool, however it also serves a role in the design and implementation process. A guide to CoolTime's usage in the overall IC development flow is depicted in Figure 42.12.

During initial power rail design and layout, CoolTime is used to check basic power rail sizes, using its fast average IR drop calculator. The intent at this point is to identify any basic power rail deficiencies before proceeding to more detailed physical design. To perform this analysis, the power rails are extracted and the power is calculated, either from a simulation-based analysis or from a vectorless analysis. In either case, the average power computation method is selected because the intent is to isolate any resistive issues in the rails' design. If excessive IR drop is found at any point in the power rail network, the rail sizes or topologies are adjusted to reduce the IR drop to acceptable levels.

Once the physical design has been refined and nears completion, CoolTime is used to analyze instantaneous current and voltage drop effects to verify timing and reliability for sign-off. The power grid and signal parasitics are extracted. Pad voltages and package parasitic models are specified along with decoupling capacitor definitions. As before, power and current consumption are calculated; however, in this case, instantaneous calculations are employed to incorporate the effects of the various capacitances and inductances, which are needed to verify timing and reliability with the highest confidence. The results are a set of instantaneous voltage waveforms for all nodes in the power rail network. These instance-specific, time-varying voltages are fed to the delay calculator, along with fully coupled signal parasitics, to compute voltage derated delays, where the voltage derating incorporates the effects of driver and receiver voltages being different than the library characterization conditions represented in the cell library. These voltages are also used by the signal integrity analyzer to adjust the noise margins according to the nonideal V_{dd} and V_{ss} voltages that power each cell.

CoolTime also evaluates electromigration sensitivity at this point. Using the extracted power rail network, along with the computed branch currents, CoolTime will calculate the current density in the various power rail network branches. The results can be displayed by the GUI for a visual inspection; violations will be highlighted as well as written to a violation report file.

The conclusion of all these analyses is a set of timing, power, and reliability reports that include the effects of power on each parameter. If no violations are found, then the design is ready for electrical sign-off.

42.4 A Design Example

The usage of the sequence tools in a feed forward power-sensitive design flow is perhaps best illustrated by example. Consider the design of an application-specific digital signal processor slated for use in a wireless device.

42.4.1 High-Level Design

Following a feed forward design flow, the first steps involve the system specification and design which includes setting the power supply voltage(s) as low as possible within the constraints of performance, battery availability, and the voltage requirements of other chips in the system. A power budget is set for the entire design and apportioned for the various modules or design components. Once the system specification solidifies, RTL design begins.

As the RTL coding proceeds, PowerTheater is used to evaluate the RTL code's power characteristics. Whenever available, activities derived from RTL simulations, such as those of the inner loops of the most frequently used DSP algorithms, are used as the input stimulus. Power debugging proceeds by using PowerTheater to identify the design constructs that consume the largest portions of the power budget, as well as those that are unexpectedly high. Debugging continues by using PowerTheater's reporting mechanisms, especially the frequency reports, to isolate the causes behind any excessive power consumption. Once the root causes have been identified and rectified by code modifications, the code-simulate-analyze local loop is repeated until the power consumption target is met. PowerTheater's automatic power linting utilities are also employed at this point to highlight any overlooked opportunities for code optimization.

During this process, particular attention is paid to clocking, datapaths, and memories. Clock power can be reduced by incorporating clock gating, although large numbers of gated clocks often pose problems in clock skew management by downstream physical tools. PowerTheater's WattBots are used to identify the most effective clock gating opportunities so that the total number of gated clocks can be minimized, thus avoiding later issues in clock tree synthesis and layout.

Inspecting the switching frequencies of intermediate nodes identifies power waste in datapaths. Code with quiescent outputs and active inputs represent opportunities for improvement by moving gating logic as far upstream as possible Conversely, code with glitchy control inputs is rearranged so that the glitches are prevented or blocked. If this is not possible, then the function should be coded such that the glitchy inputs are fed as deeply into the datapath logic as possible to limit the amount of logic through which the glitches could propagate.

Memories and data storage structures warrant particular attention as they often consume the largest portion of the total power. Because PowerTheater computes memories' read-and-write access rates, inspection of PowerTheater's frequency reports can uncover inadvertent memory accesses, which waste power. More aggressive power consumption goals may dictate revising the entire data storage architecture to minimize the total number of accesses; in this case, PowerTheater provides the mechanism by which to evaluate which architecture will be the most power efficient for that particular application. One example of this situation is the consideration of asynchronous vs. synchronous memories — which type of memory will be more power efficient depends not only on its internal structure, but also on the details of how it is employed within the particular target system.

42.4.2 Physical Design

Once the RTL has been demonstrated to meet the target power specification, the design is synthesized or otherwise converted into a cell level netlist suitable for placement. After placement and timing closure, PhysicalStudio is used to minimize the dynamic power consumption through resizing and, if desired, minimize leakage power through dual-Vt cell swapping. The output of these operations is a timing-closed, power-minimized physical design ready for initial power rail sizing. CoolTime can now be used to evaluate the basic integrity of the power rail sizing by analyzing the design's IR drop. If this analysis indicates a sufficiently small amount of IR drop, then the design is fed to the router. On the other hand, if the IR drop is judged as excessive then mitigating steps are undertaken to rectify the issues. The rails may be resized or even redesigned altogether, after which CoolTime is used again to verify the IR drop. Once the IR drop is within the target spec the design is fully routed.

After routing PhysicalStudio is employed again to further squeeze both dynamic and leakage power using similar transformations to those employed preroute. Additional reductions are usually possible at this point because extracted parasitics are used in lieu of estimated wire lengths and the timing slacks are known with much greater certainty. Any changes other than simple cell swaps will necessitate another route, but once the design has been routed after the last optimization, the design is ready for electrical sign-off verification using CoolTime.

42.4.3 Electrical Sign-Off

At this point, the completed physical design is reextracted (if the layout changed at all, otherwise the original extracted data is used) to produce both signal and power rail parasitics, which are loaded into CoolTime along with the package parasitics definition. CoolTime then computes the amount of supply droop and bounce along with their effects on timing and noise margin degradation. This data is, in turn, used by the static timing analysis and noise analysis engines to verify electrical performance and noise immunity in the presence of power rail voltage variations. Electromigration limits are also calculated in all power rail branches. If none of the limits for timing, noise, or reliability are violated, then the design is ready for tape-out.

42.5 Conclusion

Power-sensitive design has become an essential focus in this age of wireless and multimedia computing, but it is no longer directed at simply reducing the amount of power consumption. Power-related issues now directly affect many facets of design and these issues are sufficiently complex as to require significant amounts of design automation.

Sequence design develops advanced tools to address these critical issues and views power-sensitive design as a multilevel endeavor. To that end, PowerTheater supports RTL design and analysis early in the design process, PhysicalStudio optimizes power, timing, and noise characteristics at the physical level, and CoolTime verifies cell-based electrical characteristics for verification and sign-off. As presented in this chapter, these tools are used in a comprehensive methodology for developing robust, power-efficient designs.

References

[1] Wilson, R. and Lammers, D., Grove calls leakage chip designers' top problem, *EE Times*, December 13, 2002.

[2] Sequence Design, *PowerTheater Reference Manual*, Sequence Design Inc., Santa Clara, CA, 2003.

[3] Sequence Design, *PhysicalStudio Reference Manual*, Sequence Design Inc., Santa Clara, CA, 2003.

[4] Sequence Design, *CoolTime User Manual*, Sequence Design Inc., Santa Clara, CA, 2003.

[5] Landman, P. et al., An integrated CAD environment for low-power design, *IEEE Design and Test of Computers*, vol. 13, Summer 1996, pp. 72–82.

[6] Synopsys, *Liberty User Guide, Version 2001.08,* Synopsys, Inc., Mountain View, CA, 2001.

[7] IEEE, *P1603/D9, A Draft Standard for Advanced Library Format (ALF),* IEEE, New York, 2003.

[8] IEEE, *1481, Standard for Delay & Power Calculation Language Reference Manual,* IEEE, New York, 1999.

[9] Cadence Design Systems, *LEF/DEF Language Reference,* Cadence Design Systems, Inc., San Jose, CA, January 2003.

[10] Lee, W. et al., A 1V DSP for Wireless Communications, *Proc. ISSCC,* February 1997, pp. 92–93.

[11] Wang, Q. and Vrudhula, S.B.K., Algorithms for minimizing standby power in deep submicrometer, dual-Vt CMOS circuits, *IEEE Trans. on Computer-Aided Design of Integrated Circuits and Syst.,* vol. 21, no. 3, March 2002, pp. 306–318.

VII

Battery Cells, Sources of Energy, and Chip Cooling

43

Battery Lifetime Optimization for Energy-Aware Circuits

Davide Bertozzi
Luca Benini
University of Bologna

43.1 Introduction

During the development of a low-power digital system, the attention of designers is focused on the minimization of the power dissipated by the circuits and interfaces that perform computations, storage, and data transfer/communication. Accurate and efficient power models for digital circuits at various levels of abstraction have been developed to support design space exploration [1]. Unfortunately, much less attention has been dedicated to power supply models. In many cases, it is implicitly assumed that the power supply provides a constant voltage and delivers a fixed amount of energy. This assumption is not valid in the case of battery-operated devices.

A battery is not an ideal finite-charge power supply. The energy stored in a fully charged battery cannot be supplied to the digital circuitry to its full extent, and the usable energy cannot be supplied at a constant rate. This happens because the amount of energy a battery can provide depends on the current drawn from the battery itself. The higher the discharge current, the higher the energy waste of the battery. An overview of battery non-idealities is reported in Section 43.2.

Battery simulation models have been developed to help designers estimating the discharge characteristics of common batteries, much before such characteristics can be measured by connecting the actual battery to a system prototype. In this way, the risk of violating lifetime/weight specifications is minimized,

0-8493-1941-2/05/$0.00+$1.50

avoiding the need for an expensive redesign step. This chapter provides a detailed report of battery models developed so far, and goes into the details of a discrete-time battery and DC-DC converter model that helps in bridging the efficiency gap between electrical-level and high-level simulation. The model is therefore suitable for the purpose of battery lifetime estimation of systems described at a very high level of abstraction.

The availability of accurate and computationally tractable battery models is key to assessing the effectiveness of battery-driven system design approaches. For instance, they make it possible to analyze the discharge behavior of a battery under different design choices, such as system architectures and power management policies.

This chapter addresses battery-efficient system design both for single- as well as multi-battery systems. In both cases, designing a system with careful consideration of the battery and its characteristics promises to provide further improvements in battery life beyond what can be achieved by conventional low-power design techniques. The main techniques for battery lifetime optimization will be presented, from frequency and voltage scaling to task scheduling and power management. In the context of multi-battery systems, where batteries are usually sequentially discharged, the effectiveness of alternative approaches is demonstrated, consisting of the design of efficient power supply subsystems that optimally exploit the energy obtainable from existing battery packs.

The chapter finally provides a brief insight on emerging industry standards for implementing smart battery systems. In fact, the availability of accurate information about battery state by the system is key to the implementation of effective power management policies and to the consequent battery lifetime optimization. The smart battery system specification goes in this direction, beyond promoting interoperability between products of different vendors.

43.2 Non-Idealities of Real-Life Battery Cells

In most of the work on low-power design, batteries are implicitly viewed as ideal charge reservoirs, containing a fixed amount of charge, and providing a fixed output voltage until the charge is fully depleted. In reality, batteries are nowhere close to being ideal charge storage units. Main non-idealities of real-life battery cells include:

1. Battery output voltage depends nonlinearly on its state of charge (SOC). Voltage drops progressively as the cell discharges and it plummets very rapidly when the charge is exhausted. Cells are rated with a fully discharged voltage, which is usually fixed at the onset of the rapidly decreasing region, and it is significantly lower than nominal, fully charged voltage. Because of this fact, batteries cannot be directly connected to electronic circuits, but their output voltage must be shifted and stabilized by feedback-based DC-DC conversion circuitry.
2. Battery capacity depends on the current load. At high current, the effective capacity (i.e., the total amount of charge that can be extracted from a battery) decreases. Thus, it is important not to assume that charge can be extracted from a cell at an arbitrary high rate. Most batteries are in fact rated for maximum discharge current, but, at this load level, capacity is significantly degraded.
3. The "frequency" of the discharge current affects the amount of charge the battery can deliver. The battery does not react instantaneously to load changes, but it shows considerable inertia, caused by the large time constants that characterize electrochemical phenomena.
4. Batteries have some (limited) recovery capacity when they are discharged at high current loads. If a battery is discharged at a high current for a short period of time, and it is then allowed to rest for some time at low (or zero) load, its output voltage increases, and it can be used again, even if its output voltage had dropped to full-discharge level at the end of the high-current burst.
5. Nominally equal battery cells may exhibit significant differences in internal resistance, output voltage, and voltage vs. current characteristics. These differences are due to a number of causes, such as aging, different number of charge-discharge cycles (i.e., long-term memory effects), and

different manufacturers. For these reasons, connecting batteries in parallel is not considered a safe design practice, and it is usually avoided.

43.3 Battery Modelling

Battery simulation models have been developed to help designers estimating the real discharge characteristics of common batteries, much before such characteristics can be measured by connecting the actual battery to a system prototype.

The most accurate battery models represent the fine-grained, electro-chemical phenomena determining battery cells discharge by means of partial differential equations (PDE). These electro-chemical models [2–5] are significantly more detailed than others and are able to capture most nonideal effects, and this involves that they are most of times strictly tied to specific batteries. Furthermore, they are extremely computationally intensive, and cannot be used, for instance, for design space exploration, which requires a higher level of abstraction.

Higher-abstraction battery models can be formulated at the circuit level [6–9]. PSPICE equivalent circuits fall into this category. Unfortunately, a continuous-time circuit-level battery model requires a load model at the same level of abstraction. Obviously, modelling the entire system loading the battery at the circuit level is a challenging task. Furthermore, circuit-level simulation of a system over the typical lifetime of a battery would require an enormous amount of time. Moreover, such models are usually capable of capturing rate-capacity and thermal effects but not the recovery effect.

The recovery effect has been efficiently modelled so far by means of stochastic models, wherein the discharge behavior of the battery is modelled using discrete-time transient stochastic processes [10,11]. Although they do not take into account thermal effects, they represent a good trade-off between accuracy and computation requirements and can be employed in system level simulations.

Finally, analytical models often represent a tractable representation of the battery behavior [12,13]. Actual battery capacity and lifetime can be therefore analytically derived from discharge current values, operating environment characteristics and physical properties of the batteries. All of these models capture rate-capacity effects, some capture thermal effects but none addresses recovery effects during idle periods. Advantages of analytical models are flexibility (easy configurability for specific batteries) and computational efficiency, requiring evaluation of simple analytical expressions.

Some relevant battery models are described in the following subsections.

43.3.1 Peukert Equation

The simplest (analytical) high-level battery model accounts for the nonlinear relation between capacity and current for a typical battery and was empirically determined by Peukert in the late 1800s. Peukert's formula for the energy capacity C of a battery is

$$C = K / I^{\alpha} \tag{43.1}$$

where K is a constant determined by the chemical family and physical design of a battery, and I is the discharge current. For an ideal battery, $\alpha = 0$ (i.e., the capacity is constant), but for real batteries α ranges up to 0.7 for most loads, and, similar to K, is determined by the chemical family and physical design of the battery [14].

Peukert's formula does not capture the recovery effect, but it is useful as a first-order approximation of battery life for most loads and is easy to manipulate mathematically. However, the main drawback of this model is that it is limited to constant loads within the range of loads for which α is empirically determined. This is not the case of the intermittent and variable loads induced by power management techniques, currently employed to preserve battery lifetime in portable devices.

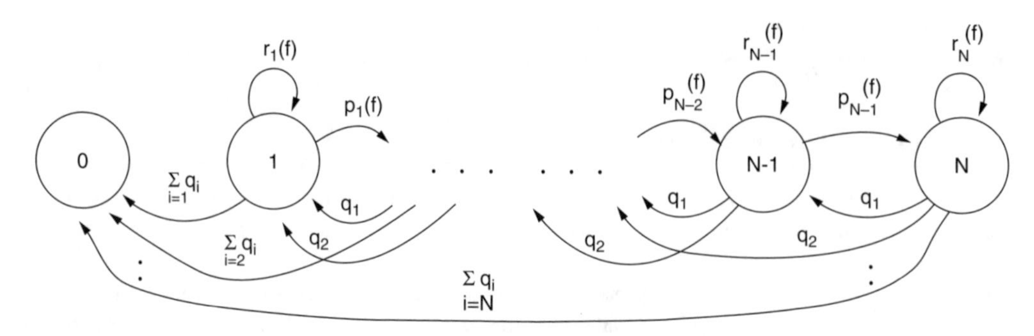

FIGURE 43.1 Battery cell stochastic model that tracks the state of charge.

43.3.2 Markov Chain-Based Model

The distinctive feature of current load in presence of power management is a discharge profile consisting of short activity periods followed by idle periods or relaxation times during which the battery can partially recover the capacity lost in previous discharges. This recovery effect is the core of a battery stochastic model wherein battery charge states are modelled as a Markov chain [11]: forward transitions correspond to discharge, while backward ones represent recovery effects. The load is expressed as a stochastic demand on charge units (the smallest amount of capacity that can be discharged). If in a given time slot a certain amount of charge units are "discharged," then an appropriate forward transition takes place. On the contrary, if no charge units are demanded in a time slot, an appropriate backward transition is carried out. In this way, both charge delivery nonlinearity and charge recovery effects can be taken into account.

The cell behavior is modelled as a discrete time transient stochastic process, which tracks the cell state of charge, as depicted in Figure 43.1. At each time unit, the state of charge decreases from state i to state $i - n$ if n charge units are demanded from the battery. Alternatively, if no charge units are demanded, the battery may recover from its current state of charge to a higher state. The stochastic process starts from the state of full charge denoted by N (nominal capacity) and terminates when the absorbing state 0 is reached, or the maximum available capacity T is exhausted. By allowing idle periods in between discharges, the battery can partially recover its charge during the idle times, and thus we can drain a number of charge units greater than N before reaching state 0.

The probability that in one time slot i charge units are demanded is denoted by q_i . Thus, starting from N, at each time slot i charge units are lost with probability q_i and the cell state moves from state z to $z - i$. On the other hand, with probability q_0 an idle slot occurs and the cell may recover one charge unit (i.e., the cell state changes from state z to $z + 1$) or remain in the same state.

The recovery effect is represented as a decreasing exponential function of the state of charge of the battery, with the exponential decay coefficient that takes different values as a function of the discharged capacity.

This model has been validated by means of a comparison with a PDE model of a dual lithium ion insertion cell, and exhibits a maximum error equal to 4% and an average error equal to 1%.

43.3.3 Efficiency Factor

The stochastic battery model approximates a continuous behavior through a set of discretizations (such as number of charge states and minimum absorbed charge), and it is therefore extremely hard to characterize. A simpler approach is proposed in [12], wherein a battery efficiency factor is introduced, that accounts for charge delivery nonlinearity. In other words, some amount of battery energy is considered to be wasted when the battery delivers the energy required by the circuit.

In analytical form, given a fixed battery output voltage, if the circuit current requirement for the battery is I, the actual current that is taken out of the battery is:

$$I_{act} = \frac{I}{\mu}, \qquad 0 \le \mu \le 1 \tag{43.2}$$

where μ is called the battery efficiency (or utilization) factor. I_{act} is always larger than or equal to I. Let us now define CAP_0 as the amount of energy that is stored in a new (or fully charged) battery and CAP_{act} as the actual energy that can be used by the circuit. It follows from Equation (43.2) that

$$CAP_{act} = CAP_0 \cdot \mu \tag{43.3}$$

The efficiency factor μ is a function of the discharge current I:

$$\mu = f(I) \tag{43.4}$$

where f is a monotonic decreasing function [15]. Only the low-frequency part of the current is relevant to changing the battery efficiency [7]. Therefore, I must be the average output current of the battery over some period of time, which may be as large as a few seconds [7]. The actual capacity of the battery decreases when the discharge current increases.

Function f can be approximated in two ways:

$$\mu = 1 - \beta \cdot I \tag{43.5}$$

$$\mu = 1 - \gamma \cdot I^2 \tag{43.6}$$

where β and γ are positive constant numbers. Both analytical forms have been demonstrated to provide good modelling for the capacity-current relation of lithium batteries as long as the appropriate values of β and γ are chosen.

43.3.4 Chemical-Kinetics-Derived Model

Another approach to analytical battery models is to derive the analytical relations from the fundamental physical laws, and to use statistical techniques to estimate the parameters. As an example, the variable load model proposed in Rakhmatov and Vrudhula [16] assumes the simple case of one-dimensional diffusion across the electrolyte. Through partial differential equations and setting proper boundary conditions, an analytical form for the concentration of species at the electrode surface is derived, and the time at which it drops below the cut-off level (so that no reaction can take place) can be determined. The time when the reaction can no longer take place at the electrode surface is the time-to-failure.

Overall, this procedure provides a general expression relating the load and the time-to-failure, with only two unknown parameters that can be estimated based on experimental data. The model allows accurate predictions of battery lifetime, and the average error with respect to simulation results is within 3% for constant discharge, interrupted discharge, and general variable discharge conditions.

43.3.5 Discrete Time Model for System-Level Design

Previously described approaches have two major drawbacks. First, they relate battery lifetime to the average current absorbed by active circuits. However, lifetime of actual batteries does not depend only on average current, but also on the profile of the time-domain current waveform. Second, they neglect the presence of voltage converters that can be responsible for a significant fraction of the total power. These inaccuracies may prevent accurate design-space exploration, especially in the case of power-managed systems, which exhibit highly nonstationary current waveforms.

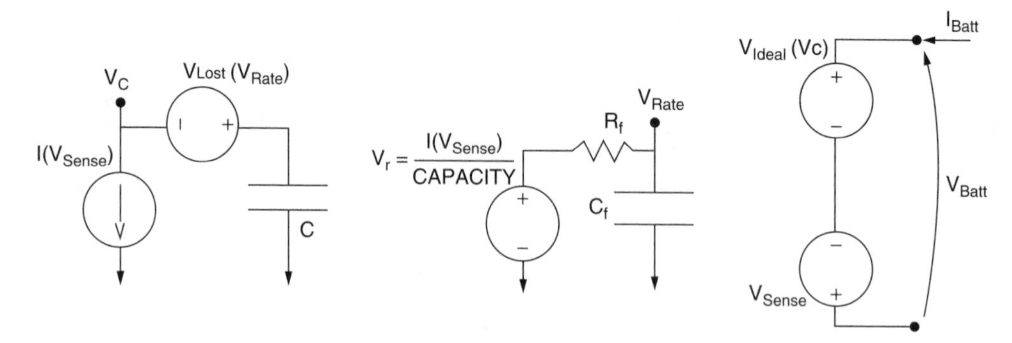

FIGURE 43.2 First-order continuous time battery model.

This subsection introduces discrete-time battery and DC-DC converter models that help in bridging the efficiency gap between electrical-level and high-level simulation, without incurring the accuracy losses that are normally imposed by simplified battery-conscious power metrics (such as those defined in Martin and Sewiorek [15] and Pedram and Wu [36]). The model takes into account first-order effects, such as dependence of battery voltage on its state of charge, discharge rate, and discharge frequency. Second-order effects, such as battery output resistance and dependencies on the temperature, are also considered.

The complete discrete-time model can be implemented within any system level simulation environment, while still accurately approximating the continuous-time behavior. Therefore, it can be used for the purpose of battery lifetime estimation for systems described at a very high level of abstraction.

Some key features of continuous-time battery models need to be reproduced in a discrete-time setting to achieve accurate lifetime estimation, and are hereafter discussed.

43.3.5.1 Battery

A first-order continuous time battery model is reported in Figure 43.2. Dependency of battery voltage on its SOC ($V_{ideal}(V_C)$) is realized by storing several points of the curve into a lookup table (LUT) addressed by the value of the state of charge V_C. The model is accurate up to a minimum *cut-off voltage*, after which the battery is considered fully discharged.

Dependency between the actual capacity of a battery and the magnitude of the discharge current is modelled with a voltage source V_{lost} in series with the charge storage capacitor. Voltage V_{lost} reduces the apparent charge of the battery, which controls battery voltage V_{Batt}. The value of V_{lost} is a nonlinear function of the discharge rate, which can be modelled by another LUT. Dependency on the discharge frequency, and the time-domain transient behavior of the battery are modelled by averaging the instantaneous discharge rate used to control V_{lost} through a low-pass filter $\left(R_f,C_f\right)$. The low-pass filter models the relative insensitivity of batteries to high-frequency changes in discharge current.

Notice that V_{sense} is a zero-valued voltage source added in series with the output voltage functions as the discharge-current (I_{Batt}) sensor. *CAPACITY* is the total capacity of the battery. According to Hageman [6] and Gold [7], this model fits measured data fairly well (within 15%). This accuracy is acceptable because the actual capacity of any group of cells may vary as much as 20% between identical units, when we take into account manufacturing variances [6].

Among the various secondary phenomena that affect battery voltage [6], the most sizable one is due to the offset in the output voltage caused by the heat released by the cell. This effect is particularly evident for high discharge currents. The effect of temperature can be modelled as a voltage loop similar to that of V_{Rate} in Figure 43.2, as illustrated in Figure 43.3 [7]. The state variable in the thermal loop on the left V_{Cell_Temp} causes an offset V_{Temp} in the cell output voltage. V_{Cell_Temp} is obtained as the sum of the equivalent voltage V_e of the environmental temperature and the voltage source V_{Rise}, proportional to the temperature rise [6]. In the model, R_{TH} lumps the effects of both thermal resistance and thermal capacitance, while R_{Int} represents the internal resistance.

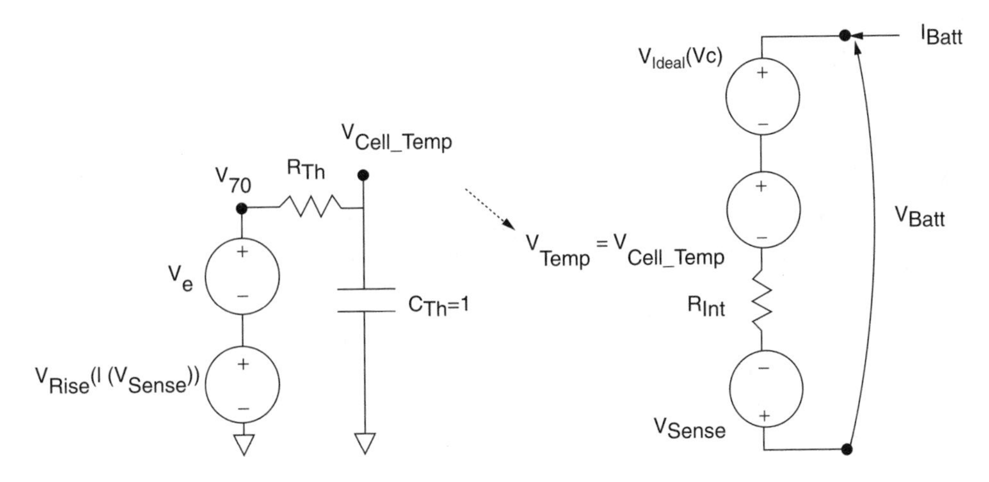

FIGURE 43.3 Second-order continuous time battery model.

43.3.5.2 DC-DC Converter

The output voltage of a battery depends on its chemistry and its state of charge. During operation, battery voltage is not well controlled. Thus, the battery cell cannot be connected directly to active circuits, but it requires the presence of a DC-DC converter for shifting and stabilizing the voltage supply. The most common DC-DC converter circuits for battery-operated devices are switching converters [17]. A basic switching down-converter known as buck converter is depicted in Figure 43.4.

A single-pole, double-throw switch is alternatively connected to the DC input voltage and to ground. The switch output is connected to a LC low-pass filter. If the switch position is changed periodically, at a frequency $f >> 1/2\pi\sqrt{LC}$ and with duty-cycle $D \le 1$, the output voltage of the converter is nominally $V_{Out} = DV_{In}$; thus, the buck converter performs voltage down conversion.

Buck converter is only one of the many switching converters described in the literature, and is adopted here for the sake of explanation, although the high-level model can be used for generic converters. All real-life DC-DC converters have sizable losses, usually collapsed in a single figure of merit called efficiency:

$$\eta = \frac{P_{Out}}{P_{In}} \tag{43.7}$$

Typical efficiencies are within the range (0.8,0.9). For a given fixed output voltage value V_{Out}, we can represent η as a nonlinear function $\eta(V_{In}, I_{Out})$. Efficiency curves are usually plotted in the data sheets of commercial DC-DC converters, and are used by system designers to choose among different converters and to set the operating point. They can also be used to create behavioral black-box models for fast simulation of DC-DC converters: such models do not contain any information on the internal structure but just mimic I/O characteristics, thus providing simulation efficiency.

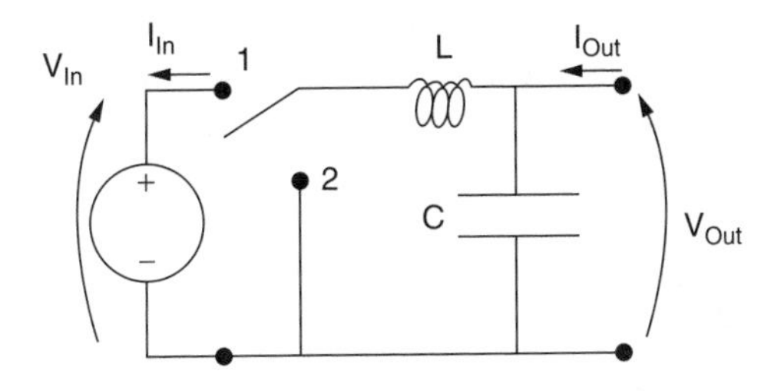

FIGURE 43.4 Simple DC-DC buck converter scheme.

```
entity battery is
  port(I_Batt : in amps; update : in std_logic; V_Batt : out real);
end battery;

architecture behavior of battery is
begin
  V_Batt <= PWL(V_C) + V_Cell_Temp - R_Int * I_Batt;
  -- V_Cell_Temp is 0.0 if no second order effects are considered
  Compute_V_C : process (I_Batt, update, V_Lost)
  begin
    cap_act := (cap_act - I_BattOld * (NOW - chgt));
(*) V_C <= (cap_act/ cap_i - V_Lost);
    I_BattOld = I_Batt;
    chgt = NOW;
  end process;
  Compute_V_Lost : process (I_Batt, update, Compute)
  begin
    V_r := I_BattOld/ CAPACITY;
(**) V_Rate := (V_RateOld - V_r) * exp(-(NOW - chgt)/(R_f * C_f)) + V_r;
    V_Lost:= PWL(V_Rate);
    if (I_Batt'event) then
      V_RateOld := V_Rate;
      Compute <= '1' after (τ / 5.0),
                 '0' after (τ / 5.0 * 2.0),
                 '1' after (τ / 5.0 * 3.0),
                 '0' after (τ / 5.0 * 4.0),
                 '1' after (τ / 5.0 * 5.0),
                 '0' after (τ / 5.0 * 6.0);
    end if;
  end process;
end behavior;
```

FIGURE 43.5 High-level VHDL code of the battery model.

43.3.5 Discrete Time Model for System-Level Design

Based on previous considerations, a discrete-time power supply model can be easily implemented within any system-level design environment. For the sake of concreteness, a VHDL implementation of just the battery model is reported in Figure 43.5. Similar considerations hold for the DC-DC converter.

The battery can be defined as a VHDL entity with two inputs: I_{Batt}, representing the current absorbed by the DC-DC converter, and update, a periodical signal used to update the values in the model. The output of the battery entity is V_{Batt}, which represents the voltage supplied by the cell to the DC-DC converter (see Figure 43.2).

The internal structure is based on the circuit-level model of Figure 43.2 and consists of two concurrent, communicating processes. The first one (*Compute_VC*) computes the value of node V_C in Figure 43.2, the instantaneous state of charge of the battery (taking into account losses due to high discharge rate). The second process (*Compute_VLost*) computes the value of V_{Lost} (i.e., it implements the low-pass filter depicted in Figure 43.2).

The changes in V_{Rate} (and V_{Lost}) in response to a variation in I_{Batt} are not instantaneous, but follow a transient with the time constant of the battery's low-pass filter $\tau = R_f C_f$ (for real-life batteries, this interval is in the order of 1 second). Thus, a $\Delta t = \tau/5$ is sufficient to model the transient behavior of node V_{Rate} in response to changes of I_{Batt}.

Second-order effects can be taken into account through a separate process. In summary, the models for battery and DC-DC converter can have limited complexity, and are well suited to work with system-level descriptions without sizable simulation overhead.

Validation of this model against experimental data provides an average error of only 0.52%, and it can be easily adapted to different types of batteries such as lithium-ion, nickel-cadmium, alkaline batteries, or lead-acid [9].

43.4 Battery-Driven System Design

Research has demonstrated that the amount of energy that can be supplied by a given battery varies significantly depending on how the energy is drawn. Consequently, new battery-driven approaches to system design have been developed, which deliver battery life improvements over and beyond what can be achieved through conventional low-power design techniques.

This chapter provides an exploration of techniques for battery-driven system design, and follows the grouping criteria of Lahiri et al. [18].

Battery lifetime optimization requires an introductory work to characterize battery properties and to incorporate battery considerations into the system design process. To this purpose, the definition of battery aware metrics is key to capturing battery discharge characteristics. One relevant result given by these metrics is that battery life is better predicted by peak power instead of average power [19]. This means that reducing active power is more effective than reducing idle power from the battery lifetime viewpoint. Moreover, reduction of both active power and active time leads to a decrease of average power, but this latter results in a longer battery lifetime.

Another derived concept is that the actual capacity of batteries depends strongly on the mean value and the profile (distribution) of the current discharged from the battery. More precisely, a higher portion of the battery capacity is wasted at a higher discharge current (see Equation 43.2 through Equation 43.6). High-rate (current) discharge can indeed cause dramatic waste of the initial capacity of the battery. Furthermore, even for the same mean value of discharge current, the battery efficiency may change by as much as 25% as a result of the discharge current profile.

Considerations derived from battery aware metrics are at the basis of techniques for battery efficient system design. In particular, a first class of techniques based on system architecture optimization will be presented hereafter, including battery-driven policies for power management, task scheduling, voltage, and frequency scaling. Finally, a second class of techniques addressing multi-battery systems will be considered.

43.4.1 Battery-Driven Dynamic Power Management

Dynamic power management (DPM) techniques for the reduction of system level average power have been extensively investigated [20,21]. They consist of selectively shutting down system components during inactivity periods. According to the system workload, a power manager dictates how and when the power state transition has to be carried out.

Previous subsections unfortunately show that average power reduction and battery lifetime extension may be numerically far apart. This implies that optimizations for minimum average power may not be equally effective in extending battery lifetime, and vice versa.

Several DPM policies specifically tailored to battery lifetime maximization are described hereafter. In particular, a class of closed-loop policies is introduced, where the decision rule used to control the state of operation of the system is based on the observation of battery's output voltage, which is nonlinearly related with the SOC. This is in contrast with open-loop (i.e., workload-driven) solutions, which take decisions about component shutdown independently of battery voltage measurement.

Open-loop policies are normally simpler, but less effective, than closed-loop ones; therefore, they are the only viable option when cost constraints prevent the use of a voltage sensor on battery terminals. On the other hand, the distinguishing feature of closed-loop policies is that they control system operation based on the observation of both system workload and battery output voltage. Consequently, they can dynamically adapt the component shut down scheme to the actual SOC of the battery.

To illustrate battery-driven DPM, the system-level description of an MPEG 2 Layer 3 (MP3) digital audio player is considered [23]. System components can be power managed through signals issued by a DPM unit in accordance with the selected DPM policy.

The system can operate in five different states:

1. Off. The system is completely turned off and consumes no power.
2. Sleep. The system is in sleep state and absorbs 33 mA.
3. Idle. The system is idle and absorbs 38 mA.
4. RawMusic. The system plays low-quality music and dissipates 46 mA.
5. FineMusic. The system plays high-quality music and dissipates 57 mA.

When the system moves from one of the quiescent states (i.e., Off, Sleep, and Idle) to one of the active states (i.e., RawMusic and FineMusic), it absorbs some additional current, as summarized in the following table:

	RawMusic	FineMusic
OFF	23 mA	28 mA
SLEEP	14 mA	17 mA
IDLE	10 mA	11 mA

43.4.1.1 Open-Loop Time-Out Policy

Let us consider a simple open-loop time-out policy first. When the system stops playing, it enters immediately the Idle state; it waits there for a first time-out, T1, then it transitions to the Sleep state. After a second time-out, T2, if the system is still quiescent, it is forced to the Off state. Clearly, this policy aims at increasing battery lifetime by reducing the current absorbed by the system while it is not playing any music (Sleep and Off states are less current demanding than the Idle state), but also by reducing the overhead due to transitions from states Sleep and Off to FineMusic (These states are not entered until time-outs have expired). The time-out policy is workload-driven, and it does not take into account battery characteristics.

43.4.4.2 Closed-Loop Policy

The simplest closed-loop policy is threshold-based. It aims at maximizing battery lifetime by playing low-quality music when the battery is almost discharged. If the battery is fully charged, the system is kept in the FineMusic state. When the battery's output voltage falls below a threshold V_{th}, the system is forced into the RawMusic state until the battery is fully discharged. The rationale for this policy is to provide graceful degradation of system performance as the battery discharges. Clearly, the choice of V_{th} is critical for trading off music quality with battery lifetime.

Let us now introduce the quality factor Q as a quality metric. Q is defined as the ratio between the time the system is in the FineMusic state T_{Fine} and the total time of operation $T_{Fine} + T_{Raw}$. In symbols:

$$Q = \frac{T_{Fine}}{T_{Fine} + T_{Raw}} \tag{43.8}$$

The trade-off between lifetime and music quality is captured by the product between (normalized) battery lifetime *NLT* and quality factor:

$$P = NLT x Q \tag{43.9}$$

The optimal value V_{th}^{*} that maximizes P depends on both system and battery characteristics. It is important to notice that the time-out and the voltage threshold policy are not mutually exclusive, and

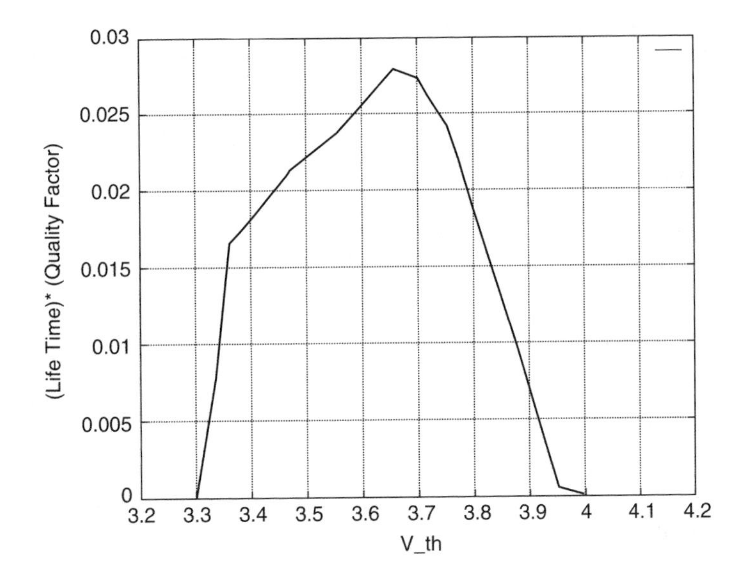

FIGURE 43.6 Lifetime quality product as a function of the threshold.

they should be applied together for best results. The hybrid policy exploits quiescent intervals in the workload, but it also trades off quality for battery lifetime.

43.4.4.3 Experimental DPM Policies Characterization

For the typical usage of the MP3 player over a period of about 1 hour, the open-loop time-out policy turns out to extend battery lifetime by approximately 121%. Application of the closed-loop policy first requires the identification of the threshold voltage V_{th}, which discriminates between system operation in FineMusic and RawMusic.

Figure 43.6 plots product *P* as a function of V_{th}. Because *NLT* increases monotonically as V_{th} increases, while *Q* decreases (still monotonically but with a different shape and slope), the product curve exhibits a maximum value for $V_{th} = V_{th}^* = 3.65$ V. This value of V_{th} is used in the implementation of the battery-driven, closed-loop policy.

When the policy is applied in isolation (i.e., with the time-out policy disabled), a lifetime extension of 119% is obtained. Moreover, if the two policies are combined together, lifetime extension becomes 132% higher than the non-managed mode.

43.4.2 Battery-Aware Task Scheduling

Many real-time scheduling techniques have been proposed so far, and many of them address power issues with the objective of minimizing power usage. More recently, instead of reducing power consumption alone, researchers have begun to study the battery behavior and the effect of the battery discharge pattern on the battery capacity as well. Generally, battery-aware scheduling algorithms include both task sequencing algorithms (supply voltage or operating frequency of the processing elements is not changed) and variable-voltage/frequency scheduling schemes. The former policies aim at exploiting battery charge recovery effects or at minimizing the peak power, while the latter ones try to decrease the average discharge current level.

Task sequencing has a strong impact on battery lifetime as it determines the load profile and thus the battery discharge profile. This subsection will address sequencing approaches both for single-processor load scheduling and for heterogeneous multi-processor systems. Battery-aware voltage and frequency scaling will be examined in next subsection.

An aperiodic task sequencing algorithm for single processor systems has been proposed by Rakhmatov et al. [25]. An analytical cost function is defined, that intuitively expresses the charge that the battery has lost by a given time T. T represents the length of a predefined load profile, which is described by three sets: the set of currents S_I drawn by each task, the set of their durations S_Δ, and the set of the start times S_t.

The input for the battery-aware task sequencing program consists of sets S_I and S_Δ, a directed acyclic task graph G, the delay budget B, and two battery parameters. The objective is to minimize the cost function subject to the following three constraints:

1. Dependency constraint. Task dependencies are preserved.
2. Delay constraint. The latency does not exceed the delay budget.
3. Endurance constraint. At any time within a schedule, the battery is alive.

The sequencing problem is tackled in three steps:

1. Greedy sequencing
2. Incremental recovering
3. Local compressing

In the first greedy sequencing step, no idling is allowed and the profile is determined only based on task dependencies. In this case, the delay constraint is already satisfied, under the assumption that B is never less than the sum of all task durations. However, the endurance constraint may be violated, and this explains why a second step is performed, called incremental recovering. It consists of the insertion of idle periods so that the load profile no longer fails. As much battery recovery as necessary is exercised, and the load ordering is not changed. Consequently, the endurance constraint and dependencies are not violated, although the delay budget may be exceeded due to recovery delay penalty. In the third step, local compressing, an attempt is made to place light loads inside idle periods, subject to dependencies, so that load relaxation is still present but recovery delay penalty is reduced. Thus, the purpose of the third step is to reduce the profile length T to the budget B.

The synthesis of battery efficient load profiles has been addressed also for high performance battery powered distributed embedded systems, which are generally composed of a network of heterogeneous processing elements (PEs). The input specification of such systems is typically in the form of task graphs. A task graph is a directed acyclic graph in which each node is associated with a task and each edge is associated with the amount of data that must be transferred between the two connected tasks. The period associated with a task graph indicates the time interval after which it executes again. A hard deadline (i.e., the time by which the task associated with the node must complete its execution) exists for every sink node and some intermediate nodes. All the hard deadlines must be met. The goal of real-time scheduling algorithms is to guarantee the deadlines of periodic task graphs, while honoring the precedence relationship among tasks. Due to the importance of energy in battery-powered systems, the scheduling scheme should be energy-aware and battery-efficient.

For those multiprocessor systems that can be statically scheduled, Luo and Jha propose a battery-aware static scheduling policy [26], where the goal is to extend the battery life span, while meeting the hard real-time constraints and precedence relationships among tasks.

The scheduling algorithm is able to manage the power profile of the whole system in order to achieve improved battery efficiency. The optimization of the discharge current profile is achieved through a series of schedule transformations starting from an initially valid schedule.

In practice, the algorithm starts from the highest power consumption time point to the lowest point, and tries to interchange adjacent events or shift forward or shift backward events around that time point, with the goal to reduce the average power consumed by tasks in every schedule slot. No local schedule transformation is performed if it violates the precedence relationship or hard timing constraints or it is useless in terms of the actual power drawn by the battery.

This transformation scheme is greedy, and strongly depends on a good initial solution. Such a good solution is obtained by means of a battery-aware global shifting stage, which tries to shift the schedule slots in a global manner with the goal of reducing the peak power consumption and increasing the flexibility of the schedule. The process starts from an initial schedule where every scheduled event is

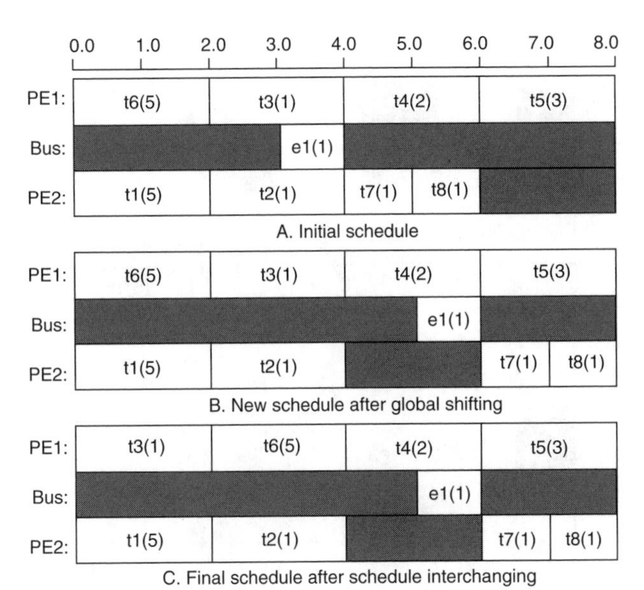

FIGURE 43.7 Example of two-step battery aware optimization.

shifted backward to its as early as possible position. Then tasks and communication events are shifted in the processing queue as late as possible, so that in the new position the overall average power consumption for the schedule slot does not exceed some given threshold value, while the negative effects resulting from the change in grouping of the idle periods are less than some threshold value.

A simple example of this scheduling algorithm is reported in Figure 43.7. Eight processing events and one communication event are mapped on two processing elements and a bus. Average power consumptions of the tasks are reported in brackets, while their execution times are reported on the x-axis. The average power consumption is taken as the power threshold (4.675), while the side effect one is set to zero. Forward global shifting has been applied in b). For instance, *t8* is shifted to the as late as possible slot. The average power for the new time period (7,8) of *t8* is 4, thus the power threshold is not exceeded. Finally, a simple local transformation is carried out in c) (*t3* exchanged with *t6*) to get the final schedule that minimizes the actual power drawn out of the battery, according to the metric defined in Pedram and Wu [12].

Another power-aware scheduling under timing constraints for heterogeneous embedded systems has been proposed in Liu et al. [24]. The authors used a NASA/JPL Mars Pathfinder rover as a motivating application with two power sources: a nonrechargeable battery and a solar panel. The objective was to utilize the solar panel (the "free" energy source) as much as possible and minimize the energy drawn from the battery. However, the scheduler was only aware of the presence of an alternative energy source, not the actual battery behavior.

Domain-specific knowledge about power sources, battery models, and other operating conditions must be expressible in terms of supported types of constraints on both timing and power. They are represented by min and max timing constraints on task, as well as min and max power constraints on the system. Min/max timing constraints subsume deadlines and precedence dependencies and can express dependencies across subsystems. The max power constraint tracks the budget imposed by the power sources and constrains the system-level power curve under a budgeted level. The min power constraint is counterintuitive in that it forces the power manager to maintain a certain level of activity. The primary motivation is that energy from free sources that cannot be stored should be utilized greedily. The power-aware scheduler is defined as the one that satisfies the rigorous min/max timing constraints and the max power budget, while making the best effort to meet the min power goal.

A schedule assigns a start time to each task and is called time-valid if all the start time assignments do not violate any timing constraints and tasks that share the same resource are serialized. The schedule

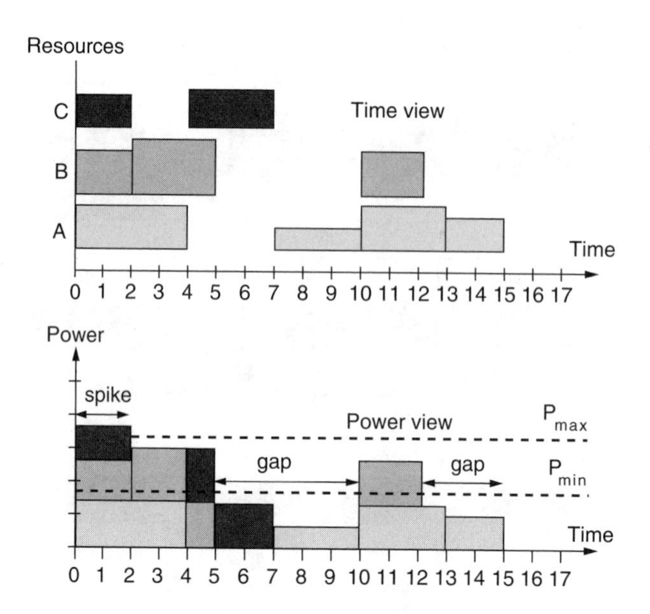

FIGURE 43.8 Power-aware gantt chart of a time-valid schedule.

has a power profile representing the instantaneous power consumption during the execution of the schedule. The power profile is constrained by two parameters: max power, which specifies the maximum budget of supply power that can be provided to support task execution, and min power, which specifies the level of power consumption to maintain a preferred level of activity. The max power constraint being a hard constraint, a schedule is called power valid if it is time-valid and its power profile does not exceed the max power constraint. Otherwise, the time interval wherein the max power constraint is violated is called a power spike. Figure 43.8 reports a Gantt chart that provides a visual representation of the problem. In the time view, tasks are displayed as bins placed on several rows that represent resources executing in parallel. The power view depicts the power profile of the schedule with min and max power constraints.

In Liu et al. [24], three steps are required to come up with a time-valid and power-valid scheduling with minimum power gaps. First, based on the constraint graph of the problem, a time-valid schedule is set. Second, the max power constraint is applied to the time-valid schedule to remove power spikes using heuristics. Finally, the min power constraint is applied, and tasks are reordered within their slacks to reduce power gaps.

43.4.3 Supply Voltage Scaling

Several voltage and frequency setting policies for variable voltage and frequency systems have been developed, which take into account battery non-idealities. The main rationale of these works is that battery lifetime maximization is a different optimization objective than energy minimization: intuitively, there is an extra premium in scaling down voltage supply because the reduced current draw also increases the battery's effective capacity.

Supply voltage scaling is critical in shaping the system load profile appropriately. In Pedram and Wu [12], an analytical technique is presented to optimally select the supply voltage V_{dd} to find the best trade-off between battery capacity and performance. The objective is to find a value of V_{dd} that minimizes the battery discharge-delay product, which is given by the product of the actual charge drained from the battery and the delay of the circuit for a given task. To this purpose, the battery discharge-delay product is mathematically expressed in terms of V_{dd} and other known (or measurable) parameters. The analysis uses an analytical model of the battery, the distribution of the discharge current profile and a model for CMOS circuit delay in terms of V_{dd} and V_{th}. Note that this approach does not attempt to modify the shape of the current discharge profile, either statically or dynamically. Instead, if the current discharge profile (or distribution) can be statically determined, this technique can be used to select a constant supply voltage to jointly optimize circuit delay and battery life.

The approach presented in [26] applies to statically scheduled real-time systems, and is based on reallocating slack times for tasks to enable supply voltage scaling. In this technique, an initial static schedule, which satisfies all performance constraints, is subjected to a set of global and local transformations to shape the current discharge profile with knowledge of the battery rate capacity characteristics, and facilitate reallocation of the amount of slack assigned to each task. Slack available for various tasks is then exploited by voltage scaling, with the objective of flattening the discharge profile, while meeting all real-time constraints. A similar approach was proposed by Rakhmatov et al. [25]

Though the approaches of Pedram and Wu [12] and Luo and Jha [26] are both static approaches, the important difference between the two is that the latter results in an explicit modification to the shape of the discharge current profile because the voltages of different tasks may be scaled differently.

43.4.4 Frequency Scaling

For battery-operated devices where voltage scaling is not possible, frequency scaling can be used to reduce peak power. Frequency scaling approaches use information from a battery model to vary the clock frequency of system components dynamically at run time. Because they also use workload characteristics (run- and idle-time percentages), and models of system power and performance, these approaches can be used to ensure efficient use of the battery without significantly compromising system performance.

A commonly used history-based policy can be employed for battery-driven CPU frequency scaling [27]. It dynamically calculates the CPU frequency for the next time interval based on run- and idle-time percentages of the previous time interval, as well as work left over, in case the CPU frequency was too slow in the previous time interval. A larger percentage of idle-time in the previous time interval results in decreasing the CPU clock frequency by a small constant for the next time interval, while not going below a lower bound. A larger percentage of run-time results in increasing CPU frequency by a small constant, while not going beyond the CPU's maximum clock frequency.

It was demonstrated that by using the preceding history-based, frequency-scaling policy, important factors such as nonideal battery behavior are neglected. Thus, the lower bound on CPU clock frequency was redefined to maximize a metric combining battery capacity, performance, and power [19].

Note that if in addition to frequency scaling, the supply voltage is scaled to match the frequency, the energy drawn from the battery is decreased as well, and battery lifetime can be significantly enhanced.

43.5 Multi-Battery Systems

Many wearable and portable devices, such as personal communicators, cellular phones, and laptops, are equipped with two or more battery packs to increase user flexibility in selecting the optimal form-factor/weight vs. required lifetime trade-off. For instance, the Compaq IPAQ PDA [28] is equipped with an add-on module that contains PCMCIA expansion and an auxiliary battery pack. Similarly, the HP OmniBook 500 notebook [29], when latched to its docking station, is powered by one 11.1 V, 3100 mAH primary battery, and two 14.8 V, 3400 mAH secondary batteries that can be plugged in and out independently.

In this context, one degree of freedom is the policy to be used for discharging the available batteries. The basic discharge policy adopted in existing products consists of fixing, for the last time during system design, the order in which batteries have to be discharged. A battery is not disconnected from the current load until it is exhausted.

For example, in the HP OmniBook 500, secondary battery no. 2 is discharged first, followed by secondary battery no. 1, and, last, by the primary battery. The rationale for this solution stands on the assumption that batteries well approximate ideal charge storage. Unfortunately, the behavior of a real battery is different from the ideal case [9,16,30,31].

In particular, the load-dependent capacity of batteries has profound implications for multi-battery systems. Primarily, the commonly accepted sequential discharge schedule is a very inefficient policy from a battery lifetime viewpoint, as observed in Benini et al. [22] and Wu et al. [32]. More efficient discharge techniques are required.

Wu et al. [32] move from the observation that batteries with different chemistries can be exploited for different discharge currents. They propose a current-controlled battery selection scheme that selects which battery to connect to the load based on runtime measurement of current draw, in an effort to match a load current to the battery that better responds to it. The main limitation of this approach is that it does not attempt to change the current profile to improve battery utilization. In contrast, Chiasserini and Rao [31] and Benini et al. [34] propose current shaping policies to extend battery lifetime. Chiasserini and Rao exploit charge recovery effects for pulsed current loads. The main idea is to steer a current absorption pulse toward the battery that has had more time to recover. Unfortunately, this policy, although helpful, does not address the main source of nonideal behavior in batteries, namely, the reduced effective capacity caused by high discharge currents, because every time the entire current load is absorbed by a single battery.

Benini et al. [33,34] address this limitation, observing that, if multiple batteries could be connected in parallel to the load, each one of them would perceive only a fraction of the total current, thereby achieving higher effective capacity. However, batteries cannot be connected in parallel. Thus, the authors propose a policy that alternatively connects the load to one battery at a time, but it switches between batteries at a very fast rate.

This is an extension of the pulse width modulation scheme (PWM) utilized by most DC-DC converters: by switching very rapidly between full load and no load, the battery perceives an effective averaged discharge current that can be obtained by multiplying the total current by the duty cycle of the switch control waveform. In other words, if the battery is connected to the load current I for $F < 1$ fraction of the switching period, it will perceive a load current $I \times F$. For instance, if the system has two identical batteries, we can switch between them with duty cycle $F = 0.5$ and each battery will perceive an $I/2$ current load. This is formally equivalent to connecting the two batteries in parallel, but it avoids the mutual discharge of the battery packs. For this reason, we can call this policy virtual parallel.

Although the idea of the virtual parallel policy may be relatively simple, its implementation requires some support by the hardware, in particular to realize the alternating connection between a battery and the load. An example is provided in Figure 43.9. The key elements of the circuit are the power MOS transistors M1 and M2. The power MOS transistors are periodically switched on and off by control signals Ch1 and Ch2. The Schmitt triggers (components A1a, A1b) and the common-emitter, level-shifting amplifiers with emitter degeneration (Q1, R7, R3 and Q2, R2, R4) drive the gates of the power transistors. The power switch stage feeds an external DC-DC converter that is used to stabilize the output voltage to 3.3 V, over a variable load.

43.5.1 The Virtual Parallel vs. Serial Policy

A comparison between sequential and virtual parallel policies is summarized in Figure 43.10, based on experiments with real life batteries [35]. Performance of sequential vs. virtual parallel policies is measured in relative terms over sequential discharge for the case of two battery packs. Under the virtual parallel policy, batteries are switched with a duty cycle of 50% at different frequencies, ranging from 120 Hz to 12 kHz.

Two effects are visible. First, notice how setting a low switching frequency leads to inferior results because batteries are exposed for a longer time to the full current load. The degradation is more marked at higher current loads. Thus, it is important to set the switching frequency in the 10-kHz range to get the full benefits of this technique.

Second, for the highest current load, lifetime extensions are extremely high. This indicates that the virtual parallel policy extends the current load range for the battery system. In other words, at a 3.0-A current load, the single battery pack is well outside the operational range, but the two virtual parallel packs can still provide acceptable lifetime.

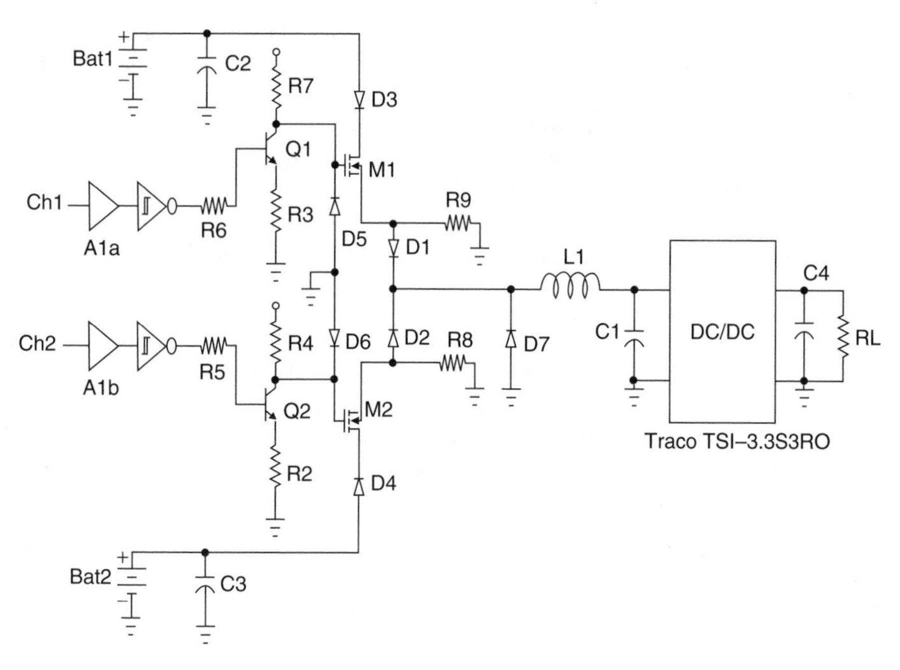

FIGURE 43.9 DC-DC converter for virtual parallel policy.

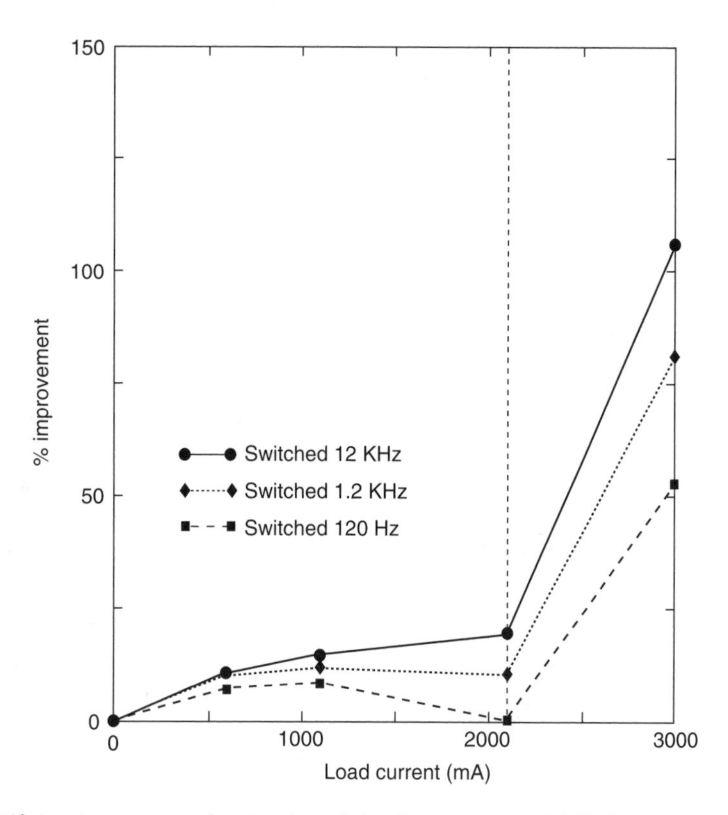

FIGURE 43.10 Lifetime improvement for virtual parallel policy over sequential discharge.

Finally, notice that, as current load goes to zero, the sequential and virtual parallel policies converge to the same lifetime. This is expected because, at low load, batteries tend to behave as ideal charge storage units.

The virtual parallel policy considered so far assumed that the current is equally split between the two batteries as they have the same capacity. Clearly, the current-steering policy can be utilized even when battery packs are not identical, but, intuitively, we observe that it may be convenient to tailor the current loading of every pack to its discharge characteristics. The new policy will be referred to as proportional current steering, while the policy that equally splits the current to all batteries as uniform current steering.

43.5.2 Proportional Current Steering

The proportional current steering problem is a load optimization problem, whose formulation is based on the assumption that all the N batteries are discharged concurrently and the load current is partitioned, in general not equally, among them. This is equivalent to assuming a round-robin switching policy that connects each battery to the load for a time proportional to the fraction of the load current we wish to absorb from that battery. The cycle frequency of the round-robin schedule is fast enough to:

1. Let the batteries perceive only its time-averaged effect (i.e., a constant current equal to a fraction of the load current).
2. Allow us to neglect charge recovery effects in the batteries because of the very short times of disconnection from the load.

The load optimization problem can be formulated as a nonlinear optimization problem that, despite the simplicity of the objective function, is far from having a trivial solution because of some nonlinear constraints [33]. A local minimum can be found using standard continuous nonlinear optimizers [37] (e.g., quasi-Newton, gradient). Notice that the size of the problems is not a major concern because the number of load levels and the number of battery cells in the system is unlikely to be very large.

A simple instance of the problem demonstrates that the current steering is superior to a sequential discharge scheme. Consider a system with $N = 2$ batteries and a constant current load of 2 A. For simplicity, let us assume that the two batteries have the following linear capacity equations (in some unit of charge):

$$C_1(I) = 10 - I$$

$$C_2(I) = 15 - 2 \cdot I$$

With the preceding methodology and given the characteristics of the two batteries, it can be stated that the best choice is to steer 41.95% of the current to the first battery, and 58.05% to the second battery. The conventional scheme consisting of an equal distribution of the current over the two batteries yields a lifetime that is approximately 2% shorter. Finally, the sequential discharge of the batteries leads to an overall lifetime, which is 13% shorter than the optimal value.

43.5.3 Simultaneous Discharge

The previous formulation allows each battery to discharge independently of the others. However, the early discharge of some of the batteries intuitively seems to be in contradiction with the idea that inspires this method, that is, sharing the load among multiple batteries to achieve higher battery efficiency. The maximum possible degree of sharing should be achieved when all the batteries discharge concurrently in such a way that they all discharge at the same time T (i.e., the lifetime of the overall battery pack).

Under this assumption, any differences in time-to-total discharge among batteries can be seen as inefficiency in the current steering policy and should be eliminated by construction. From the point of view of the formulation, this is a simplifying assumption that leads to much faster execution times.

Under this assumption, experimental characterization of the virtual parallel policy with load proportional current steering yields lifetime extensions over uniform current steering as high as 12%. Performance is maximized in presence of heavy workloads, meaning that the higher the currents, the higher the margin for current steering.

Moreover, by artificially generating workloads, it can be stated that proportional current steering outperforms uniform current steering, particularly for those workloads in which the asymmetry of the batteries can be fully exploited (i.e., higher current levels and larger variance in the levels).

Finally, with respect to a sequential discharge policy, performance of the proportional steering technique can be as high as 160%.

43.6 Smart Battery Systems

The development of techniques for battery lifetime optimization makes it necessary to have accurate sensing of battery data by the system. In order to meet this need, an industry consortium called the smart battery system (SBS) implementers forum is currently developing the specification of an emerging industry standard relating smart batteries and the systems that deploy them [38].

Initially developed by Intel and Duracell, smart battery systems have the distinctive feature of an increased awareness of battery conditions by means of a better communication with the battery subsystem, and they, therefore, generally exhibit higher battery efficiency. Another objective of the SBS forum is to promote interoperability between products from battery, software, semiconductor, and system vendors. In this way, systems will be able to use batteries of any chemistry type and new technologies can be directly utilized by existing SBS compatible systems.

The SBS specifications include the following components [18]:

- System management bus (SMBus), which defines the protocols for the battery to communicate with other system components. Because the SMBus can also be used as a control bus for other low-speed system communications, the SMBus specifications are defined by an independent forum [39].
- Battery data set, which defines the information that is provided by a smart battery to the system host. Its accuracy is covered by the specifications.
- Smart battery charger, which enables the charging characteristics to be controlled by the actual batteries, in contrast to conventional chargers that have fixed charging characteristics hardwired for specific battery chemistries and configurations.
- Smart battery system selector, which is used in multi-battery systems to select the battery that will actually supply power to the load system. It is also responsible for reporting any changes in the selector state to the system's power management software.
- Smart battery system manager, which manages the usage of all the smart batteries in a system. The SBS manager is an alternative to the smart battery selector that provides for added functionality, including the possibility for multiple batteries to simultaneously power the system. It also schedules and controls the charging of multiple batteries, and reports the characteristics of batteries powering the system to the management software.

Smart battery systems result in improvements in system safety, usable energy, and charging time, and, therefore, push the development of battery friendly system architectures.

43.7 Conclusions

This chapter has addressed the problem of battery lifetime optimization in the context of battery-operated portable electronic devices. After providing insights about the main effects that make real-life batteries differ from ideal charge storage units, the main battery models proposed in the literature are examined. Modelling actual battery behavior is key to battery-efficient system design, and the effectiveness of techniques such as frequency and voltage scaling as well as task scheduling and power management can be assessed.

Approaches for battery lifetime optimization in the context of multi-battery systems were also described. Finally, emerging industry standards for implementing smart battery systems were briefly presented.

References

[1] E. Macii, M. Pedram, and F. Somenzi, High-level power modeling, estimation, and optimization, *IEEE Trans. CAD,* vol. 17, pp. 1061–1079, Nov. 1998.
[2] S. Li and J. W. Evans, Electrochemical-thermal model of lithium polymer batteries, *J. Electrochem. Soc.,* vol. 147, pp. 2086–2095, June 2000.
[3] M. Doyle, T. F. Fuller, and J. S. Newman, Modeling of galvanostatic charge and discharge of lithium/polymer/insertion cell, *J. Electrochem. Soc.,* vol. 140, pp. 1526–1533, June 1993.
[4] T. F. Fuller, M. Doyle, and J. S. Newman, Relaxation phenomena in lithium-ion insertion cells, *J. Electrochem. Soc.,* vol. 141, pp. 982–990, April 1994.
[5] W. B. Gu and C. Y. Wang, Thermal-electrochemical modeling of battery systems, *J. Electrochem. Soc.,* vol. 147, no. 8, pp. 2910–2922, Aug. 2000.
[6] S. Hageman, Simple PSPICE models let you simulate common battery types, *Electronic Design News,* vol. 38, pp. 117–132, Oct. 1993.
[7] S. Gold, A PSPICE macromodel for lithium-ion batteries, *Proc. Annu. Battery Conf. on Applications and Advances,* pp. 9–15, Jan. 1997.
[8] M. Glass, Battery electrochemical non-linear dynamic SPICE model, *Energy Conversion Eng. Conf.,* pp. 292–297, Aug. 1996.
[9] L. Benini, G. Castelli, A. Macii, E. Macii, M. Poncino, and R. Scarsi, Discrete time battery models for system-level low-power design, *Trans. on VLSI Syst.,* vol. 9, no. 5, pp. 630–639, Oct. 2001.
[10] C. F. Chiasserini and R. R. Rao, Pulsed battery discharge in communication devices, *Proc. MOBICOMM,* pp. 88–95, Aug. 1999.
[11] D. Panigrahi, C. F. Chiasserini, S. Dey, R. R. Rao, A. Raghunathan, and K. Lahiri, Battery life estimation for mobile embedded systems, *Proc. Int. Conf. on VLSI Design,* pp. 55–63, Jan. 2001.
[12] M. Pedram and Q. Wu, Design considerations for battery-powered electronics, *DAC 1999,* pp. 861–866, June 1999.
[13] D. Rakhmatov and S. B. K. Vrudhula, Time to failure estimation for batteries in portable electronic systems, *Proc. Int. Symp. Low-Power Electronics and Design,* pp. 88–91, Aug. 2001.
[14] D. Linden and T. Reddy, *Handbook of Batteries,* McGraw-Hill, New York, Aug. 2001.
[15] T. Martin and D. Sewiorek, A power metric for mobile systems, *ISLPED 1996,* pp. 37–42, Aug. 1996.
[16] D. Rakhmatov and S. Vrudhula, An analytical high-level battery model for use in energy management of portable electronic systems, *Proc. ICCAD,* pp. 488–493, Nov. 2001.
[17] R. Erickson, *Fundamentals of Power Electronics,* Chapman Hall, New York, May 1997.
[18] K. Lahiri, A. Raghunathan, S. Dey, and D. Panigrahi, Battery-driven system design: a new frontier in low-power design, *Proc. Int. Conf. on VLSI Design, VLSID '02,* pp. 261–267, Jan. 2002.
[19] T. Martin, Balancing batteries, power, and performance: system issues in CPU speed-setting for mobile computing, Ph.D. dissertation, Carnegie Mellon University, Aug. 1999.
[20] L. Benini and G. De Micheli, *Dynamic Power Management: Design Techniques and CAD Tools,* Kluwer Academic Publishers, Norwell, MA, Nov. 1997.
[21] L. Benini, A. Bogliolo, and G. De Micheli, A survey of design techniques for system level dynamic power management, *IEEE Trans. on VLSI Syst.,* vol. 8, pp. 299–316, June 2000.
[22] L. Benini, G. Castelli, A. Macii, and R. Scarsi, Battery-driven dynamic power management, *IEEE Design and Test of Comput.,* vol. 18, pp. 53–60, April 2000.
[23] Cirrus Logic, EP7209 Ultra-Low-Power Audio Decoder SoC, 2002, http://www.cirrus.com/en/.
[24] J. Liu, P. H. Chou, N. Bagherzadeh, and F. Kurdahi, Power-aware scheduling under timing constraints for mission-critical embedded systems, *Proc. DAC 2001,* pp. 840–845, June 2001.
[25] D. Rakhmatov, S. Vrudhula, and C. Chakrabarti, Battery-conscious task sequencing for portable devices including voltage/clock scaling, *Proc. DAC 2002,* pp. 189–194, June 2002.

[26] J. Luo and N. Jha, Battery-aware static scheduling for distributed real-time embedded systems, *Proc. DAC 2001,* pp. 444–449, June 2001.
[27] W. Weiser, B. Welch, A. Demers, and S. Shenker, Scheduling for reduced CPU energy, *Proc. USENIX Symp. on Operating Syst. Design,* pp. 13–23, Nov. 1994.
[28] Compaq IPAQ PDA Overview and Characteristics, 2003, http://www.compaq.com.
[29] HP OmniBook 500, 2003, http://www.hp.com.
[30] T. Martin and D. Sewiorek, Non-ideal battery and main memory effects on CPU speed-setting for low power, *IEEE Trans. VLSI Syst.,* vol. 9, no. 1, pp. 29–34, Feb. 2001.
[31] C. Chiasserini and R. Rao, Energy efficient battery management, *IEEE J. Selected Areas in Commn.,* vol. 19, no. 7, pp. 1235–1245, July 2001.
[32] Q. Wu, Q. Qiu, and M. Pedram, An interleaved dual-battery power supply for battery-operated electronics, *Proc. IEEE Asia and South Pacific DAC,* pp. 387–390, Jan. 2000.
[33] L. Benini, D. Bruni, A. Macii, E. Macii, and M. Poncino, Discharge current steering for battery lifetime optimization, *IEEE Trans. on Comput.,* vol. 52, no. 8, Aug. 2003.
[34] L. Benini, G. Castelli, A. Macii, E. Macii, M. Poncino, and R. Scarsi, Extending lifetime of portable systems by battery scheduling, *Proc. DATE-01,* pp. 197–201, March 2001.
[35] 1HR-AAAU Sanyo Battery Datasheet, http://sanyo.wslogic.com.
[36] M. Pedram and Q. Wu, Battery-powered digital CMOS design, *Proc. DATE-99,* pp. 72–76, March 1999.
[37] K. Schittowski, NLQPL: A FORTRAN-subroutine solving constrained nonlinear programming problems, *Annals Operations Res.,* vol. 5, pp. 485–500, 1985.
[38] Smart Battery System Implementers Forum, 2003, http://www.sbs-forum.org.
[39] System Management Bus, 2003, http://www.smbus.org.

44

Miniature Fuel Cells for Portable Applications

Didier Bloch
CEA-LETI-DIHS

44.1 The Market of Power Sources for Portable Applications

44.1.1 Portable Energy Sources: Market Figures and Applications

According to Avicenne Développement Consulting Group [1], the market of secondary batteries for portable equipment in 2002 represented 16% (roughly $4.5 billion) of the total $28 billion turnover of the primary and secondary batteries.

After a maximum of $5.4 billion in fiscal year 2000, followed by the turmoil due to the telecom crisis, the 3% value increase in fiscal year 2002 compared with 2001 (Figure 44.1) may indicate a stabilization of the market. (Lithium-ion and lithium polymer may be considered as two variations of the same technology.)

Three main types of battery technologies, namely nickel-cadmium (Ni-Cd), nickel-hydride (Ni-MH), and lithium batteries (Li-Ion or Li-Polymer), are used in countless portable equipment, such as cellular phones, portable PCs, personal digital assistants (PDAs), camcorders, digital cameras, cordless tools, cordless phones, household devices, e-bikes, security lighting, and medical devices.

44.1.2 Energy, Power, and Service Requirements of Next-Generation Portable Equipment

Despite significant recent improvements of the battery technologies described next, users of energy-demanding portable devices, such as mobile phones, notebook computers, or PDAs, do not appear to be satisfied with their batteries:

0-8493-1941-2/05/$0.00+$1.50

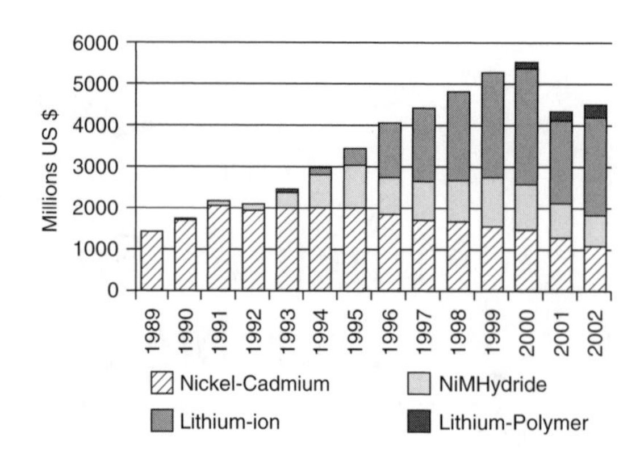

FIGURE 44.1 Worldwide rechargeable battery sales, 1993–2002. (From Avicenne Développement. With permission.)

- They complain constantly about running out of batteries at critical moments or about forgetting the battery charger when they are away from home.
- They desire freedom from recharging of batteries.
- Flying abroad, they become upset with AC plugs that do not fit with their charger connection.

Moreover, the increasing functionalities of next-generation portable devices makes users ever more energy- and power-hungry (e.g., large color displays, data processing, multimedia, and software). All these new high power drain features require additional embedded energy, whereas the required volume in the device is not necessary available.

In summary:

- A strong demand is emerging for improved "ubiquitous" energy sources offering longer operating time, an instant recharging process, and available whenever needed.
- The "plateauing" performances (in terms of energy density) of the commercial lithium battery technologies may not be compatible with the ever-growing energy requirements of new portables devices.

If they better fit to these requirements than secondary batteries, miniature fuel cells may be promised to a nice future. Kun Soo Lee, senior analyst at WestLB Securities Pacific, forecasted on March 2003 [2] that about 17% of portable devices, including mobile phones and notebook PCs, will use fuel cells in 2010. He estimated that this would roughly amount to a market of 850 billion Japanese yen ($6.1 billion). Compare this with the expected $4 billion for the global market of portable batteries in 2003.

44.2 Commercial Technologies

Miniature fuel cells will have to compete with existing energy sources. Their acceptance by the consumer market will depend on their better performances compared with conventional technologies, which have to be briefly introduced for that purpose.

44.2.1 Secondary Batteries

A more detailed examination of Figure 44.1 illustrates differences between market share evolutions of the different battery technologies: one observes a constant decrease of the nickel-cadmium (Ni-Cd) and of the nickel metal hydride (Ni-MH) technologies (5% turnover decrease between 2001 and 2002), whereas the lithium batteries achieve a 32% turnover progression in the same period of time.

This trend can be explained by many reasons:

- The gravimetric Watt.hour per kilogram (Wh/kg) and volumetric Watt.hour per liter (Wh/l) energy densities of the lithium technologies have constantly increased (Figure 44.2) during the

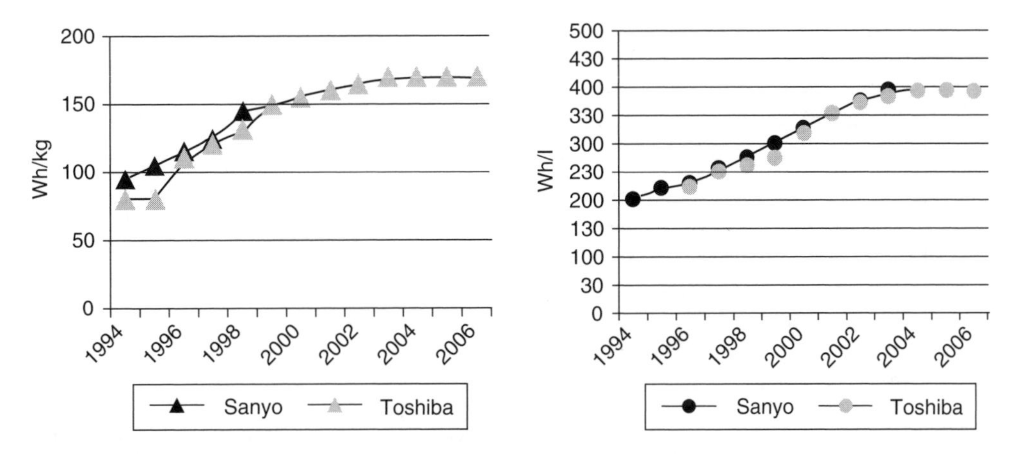

FIGURE 44.2 Gravimetric and volumetric energy density of lithium-ion technology. (From Avicenne Développement. With permission.)

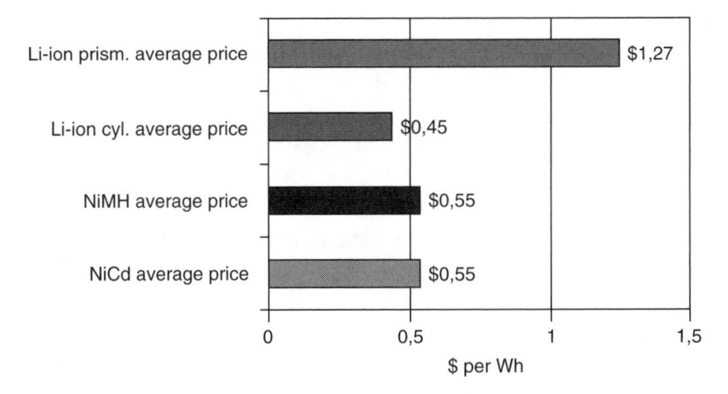

FIGURE 44.3 Average price per Wh for each technology in 2002. (From Avicenne Développement. With permission.)

last ten years (8 to 10% annual pace). From initial values not far from 90 Wh/kg and 200 Wh/l in 1993, the performances achieved 10 years later (≈ 170 Wh/kg and ≈ 350 Wh/l) make them far more interesting for energy demanding applications than the conventional Ni-Cd (50 Wh/kg) or Ni-MH (75 Wh/kg) technologies.

- The cost per Wh of the lithium technology dramatically decreased during the last 2 years: the average cost per Wh of cylindrical lithium-ion cells (≈ 0.5 $/Wh) is now lower than the average Ni-Cd or Ni-MH price (≈ 0.55 $/Wh) (Figure 44.3). This trend will certainly be confirmed in the next few years (Figure 44.4).

Moreover, the main drawbacks of the lithium technology, namely its discharge rate capability and its limited cycle life will most likely be overcome in the next few years. As a matter of fact, the safety concerns that have been legitimately put in evidence at the beginning of the commercial development do not prevent millions of lithium cells from traveling every day in planes, embedded in every laptop and cellular phone, without any major incident. The safety issue will instead be a concern for power units that will be developed for hybrid car or electric vehicles (EVs) applications.

Finally, one can expect a continuous, although not unlimited, improvement of the performances of the lithium battery technology. Thanks to the development of new electrodes materials, cells offering performances as high as 200 Wh/kg and 400 Wh/l will probably be commercialized before 2005.

With such performances, it becomes reasonable to believe that the use of lithium secondary batteries will soon not only be limited to cellular phones, laptops, or PDAs, but will also spread to numerous

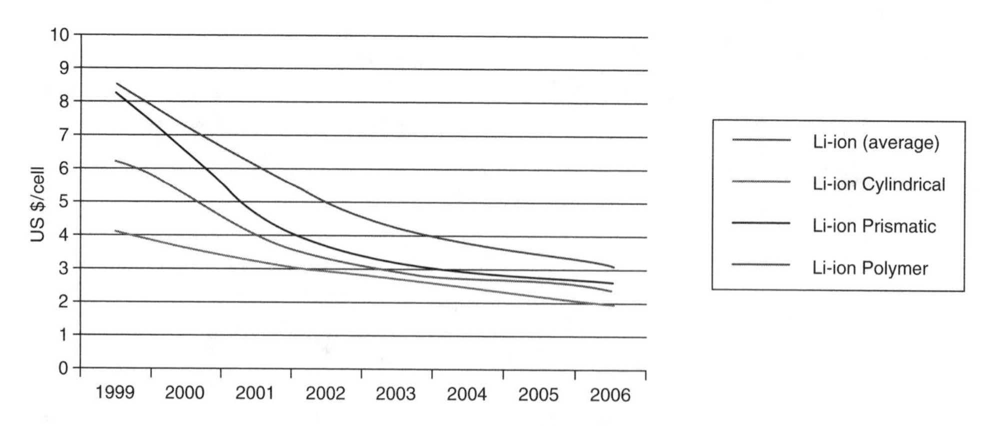

FIGURE 44.4 Lithium ion price decreases 1999–2006: cylindrical/prismatic. (From Avicenne Développement. With permission.)

applications up to now "reserved" to mature, low-cost technologies, such as home appliances, cordless tools, e-bikes, or sport utility vehicles (SUVs).

If miniature fuel cells pretend to play a role on the portable equipment market, they should offer at least an equivalent performance/cost ratio, with additional characteristics that will be considered of prime importance by the final customer.

44.2.2 Primary Batteries

Primary systems have to be mentioned here as an elegant option to comply with many of these characteristics. As a matter of fact, primary batteries:

- Are available all over the world in every retail store
- Do not need any long recharging process or additional charger

Their operating time is only limited by the number of available additional "recharging units" (primary batteries here) available.

These are, incidentally, exactly the advantages that fuel cell promoters are claiming to substantiate their developments.

However, despite performances (≈ 170 Wh/kg and ≈ 450 Wh/l for an AA standard type), comparable or even slightly higher than lithium battery performances, alkaline primary batteries are most often not preferred to lithium secondary batteries for applications requiring standby operation and frequent use, because:

- They do not allow any decisive additional operating time compared with lithium secondary batteries, which results in a frequent replacement of the "energy cartridge" (primary batteries cannot be recharged on the AC network).
- The user has to cope with a permanent reserve of bulky and heavy primary batteries that cannot be easily carried in pockets.
- The customer is soon confronted to a considerable increase of the exploitation cost, which becomes rapidly unacceptable (and unaffordable in the case lithium primary batteries would be used instead of alkaline batteries).
- The actual form factor of most of the primary batteries (cylindrical cells) is not considered as favorable for a customer acceptance: cartridges for portable applications should ideally be easily carried in wallets, such as credit cards or prepaid phone cards.

In this context, one can, however, notice the recent remarkable initiative coming from Duracell, which introduced in March 2003 [3] the first-ever alkaline prismatic battery: with its flat profile (6.1-mm thick), the so-called "LP1" battery may offer a form factor that may provide a back up solution for devices with interchangeable (i.e., primary and secondary) power sources.

Therefore, what can other systems do to offer more? The decisive benefit that can be expected with miniature fuel cells compared with primary batteries may be related to the fact that the energy conversion device — the fuel cell core and its peripherals — is embedded once for its life in the final equipment. Once empty the fuel unit only has to be replaced or refilled, exactly as an internal combustion engine coupled with a gasoline tank. Thus, an expected objective gain of three or four times operating time for the same weight and volume as a primary battery.

44.2.3 Metal-Air Systems

As fuel cells, metal/air systems present the considerable advantage, compared with other conventional batteries because the oxidant that feeds the positive electrode (air) may be extracted from ambient air instead of stored in the battery. This leads to an increase in the specific energy density.

In an alkaline solution, air coming from ambient air may be used as an oxidant to oxidize metallic zinc thanks to the following theoretical reactions:

- Negative electrode:

$$Zn + 4\ OH^- \rightarrow Zn(OH)_4^{2-} + 2\ e^- \qquad E_0 = -1.266\ V\ vs.\ NHE$$

- Positive electrode:

$$1/2\ O_2 + H_2O + 2e^- \rightarrow 2OH^- \qquad E_0 = 0.401\ V\ vs.\ NHE$$

- Overall reaction:

$$Zn + H_2O + 1/2\ O_2 \rightarrow ZnO \qquad E_0 = 1.667\ V$$

Due to its high theoretical energy density, the transformable free energy (ΔG) of the oxidation reaction of zinc with oxygen is equal to 1370 Wh/kg. This kind of system focused the interest of developers as early as 1960 [4].

The reaction starts in the presence of oxygen. Once the package has been opened by the user to establish the contact with air, it continues until the complete consumption of zinc.

This explains why these systems are appropriate for continuous use, and have been widely developed for devices as hearing aids (button cells) or for fence electrification for the cattle.

Electric Fuel Limited (EFL) commercializes zinc-air primary systems that can be used instead of batteries for cellular phones. They target frequent travelers, who are ready to pay for longer capacity and want to be free from losing contact at a critical moment. The effective specific energy density is in the 200 Wh/kg range. Other companies, such as AER Energy Resources Inc., however, failed to find a market in the same domain, for apparently the same reasons that make alkaline primary batteries not adapted to the market of energy-demanding portable devices (see the previous paragraph).

Many attempts have been made to develop a reversible system, either through an electrical recharging process or through the mechanical replacement, after use, of the active material (the zinc-based cassette).

The commercial development of rechargeable systems has been up to now hindered by many bottlenecks: electrolyte carbonation with CO_2 coming from the air, electrode mechanical damage through repetitive charging, and components corrosion.

Practical performances appear to remain lower than other electrochemical systems. The energy density of rechargeable systems is in the range 70 to 100 Wh/kg and reversibility remains quite poor: only 200 to 450 cycles have been reported [5].

However, air-zinc systems are of interest for many applications. They should at the end benefit of their intrinsic advantages: low cost, simple to use, and environmentally safe (i.e., they contain no heavy or noble metal or hazardous compound).

44.2.4 Other Systems: Energy Recovery Systems

Much research activity is currently dedicated to the investigation of energy recovery systems. For example, results on miniature thermoelectric devices were disclosed in April 2003 by Nippon Telegraph and Telecom (NTT) [5] Microsystems Integration (Kanagawa, Japan). Thermoelectric converters generate 1 V from a 5°C temperature difference. However, these systems are limited to very low-power densities and do not appear to be able, in the long run, to fit with energy requirements of portable devices.

In summary, the sizing of any new energy source for portable equipment storing a total energy amount in the range 1 to 50 Wh should consider the lithium technology as a reference.

This can easily be described in some figures. Assuming a miniature fuel cell should release 10 Wh of electrical energy, which is the typical energy amount required for a lithium battery embedded in a PDA, the requirements of the complete system would be:

- A cost lower than 5€
- A weight much lower than 50 g
- A volume much lower than 25 cm^3
- A lifetime longer than the lifetime of the device
- A very high level of safety
- A very low cost for a simple recharging process
- An operating temperature range between −10°C and +50°C

These simple elements illustrate the challenge that miniature fuel cells have to face.

44.3 Fuel Cells

44.3.1 Principle

Following the invention by Alessandro Volta in 1800 of the first galvanic cell, the decomposition of water in oxygen and hydrogen through the use of a galvanic cell was observed as early as 1802 by Sir Humprey Davy. Thirty-seven years later, in 1839, William Grove, based on the work of Christian Schönbein, confirmed the reversibility of this process when he presented the first fuel cell ever manufactured.

Almost 200 years later, fuel cells are embedded in space shuttles and are used for backup power supplies in niche markets, whereas a great number of research and demonstration programs validate the technical feasibility of fuel cells powering cars, buses, portable, and stationary applications.

Many technological options are available for fuel cells, which mainly depend on cell temperature operation. SOFC (solid oxide fuel cells), for example, use ceramic-based electrolyte materials, which become ionic conductors at elevated temperatures, namely 800 to 850°C.

Experts believe that the miniature fuel cell technology will most likely be derived from the so-called "PEM" (proton exchange membrane or polymer exchange membrane) technology. Its principle is described in Figure 44.5. The core part of the fuel cell unit cell consists of two electrodes separated by an ion-conducting polymeric membrane (the electrolyte). Fuel (hydrogen in the example) is transformed on catalytic sites at the negative electrode and form protons (H^+), on one hand, which cross the membrane, and electrons, on the other hand, which produce a current running outside the cell. Electrical power is produced by the fuel cell when electrons recombine at the positive electrode with protons (H^+) coming from the negative electrode and oxygen coming from the air. With electricity, the byproducts of the reaction are water and heat, according to the following reactions:

Negative electrode:

$$H_2 + H_2O \rightarrow 2H_3O^+ + 2e^-$$

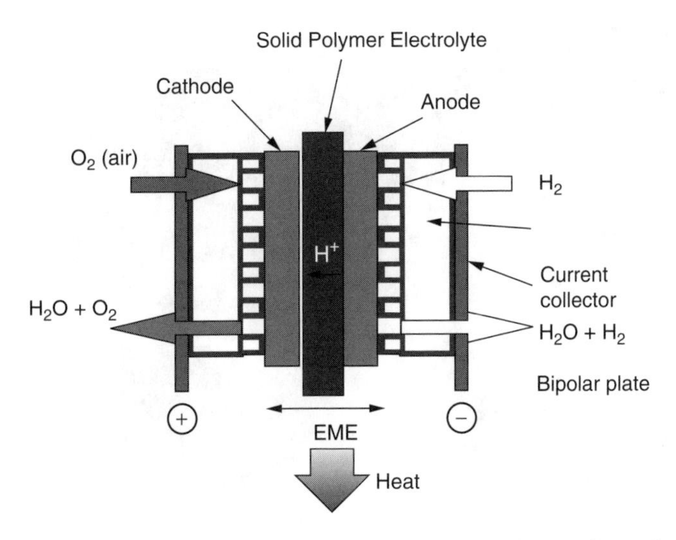

FIGURE 44.5 Proton exchange membrane (PEM) fuel cell principle. (From the French atomic energy commission, Commissariat à l'Energie Atomique (CEA). With permission.)

Positive electrode:

$$1/2\ O_2 + 2H_3O^+ + 2e^- \rightarrow 3H_2O$$

Overall reaction:

$$H_2 + 1/2\ O_2 \rightarrow H_2O \quad E_0 = 1.229\ V$$

During operation, the fuel cell core (membrane and electrodes) is partly flooded with water, which is required for proper ionic conduction.

The operating voltage and the global efficiency of such a unit cell depends on parameters such as: electrical load, electrical resistance losses, electrodes overpotential, electrocatalytic efficiency, and temperature. It is usually in the range of 0.3 to 0.5 V at near-ambient temperature, and in the range of 0.6 to 0.8 V at 80 to 90°C under nominal current load.

Conventional fuel cell stacks are made by assemblies of unit cells in series. Figure 44.6 illustrates a 500-W fuel cell stack. Bulky, electrically conductive, graphite-based or metallic bipolar plates are feeding the electrodes with fuel and oxygen, and collect the current. Heavy end plates allow a proper assembly of seals usually integrated in self-supported electrode-membrane-electrode (EME), and prevent gas or liquid leakage. These components are designed to operate under severe conditions: elevated electrode power densities — 700 mW/cm^2 — temperature as high as 60 to 80°C, pressurized gases — typically 2 to 4 bar — and large thermal losses to evacuate. End plates and bipolar plates represent usually about 80% of the weight of the stack.

A complete fuel cell system (Figure 44.7) includes the fuel system (fuel + container), the fuel cell core (the stack), and its associated electrical, thermal, and fluidic management systems.

44.3.2 Applications

Fuel cells are estimated to be more efficient at converting chemical energy into work (through electrical and mechanical energy conversion) than internal combustion engines. Because their only byproduct is water, they are considered as promising alternatives for clean energy in automotive or stationary equipment. Provided that defined operating conditions are gathered, a global efficiency of 50% is claimed to be achievable, including peripherals: water management and recovery system, heat management system, and fuel storage and management.

FIGURE 44.6 Conventional 500 Watt PEM fuel cell stack. (From CEA. With permission.)

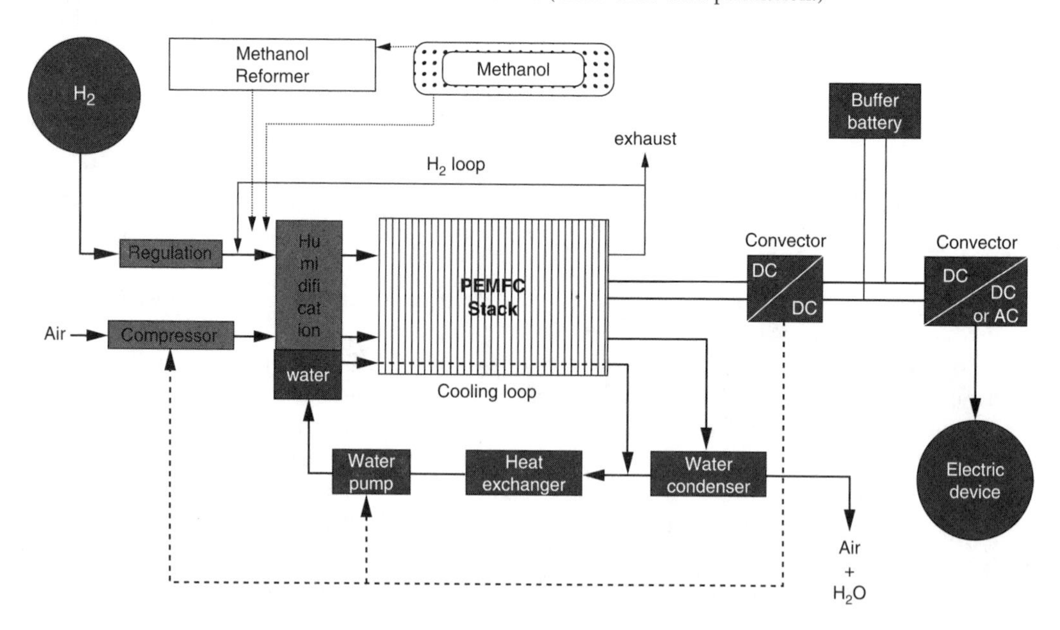

FIGURE 44.7 Simplified overview of a PEM fuel cell system. (From CEA. With permission.)

This would mean that a fuel cell system fed with only 10 g of hydrogen (hydrogen lower heating value: 33.33 Wh/g) every minute (600 g/h) would release approximately $10 \times 33.33 \times 60/2 = 10{,}000$ Wh of electrical energy every hour (10 kW fuel cell).

Despite significant technology investment and substantial improvements in materials and components (e.g., membrane, catalysts, and bipolar plates), however, progress in cost reduction, reliability, and system efficiency remain necessary to envisage a real market entry. Time span for real industrial development depends on further advance in all these fields. The selection of the fuel that will be used is also of critical interest, and greatly depends on political decisions concerning fuel infrastructure development.

Developers of large power fuel cells are confronted with technical, legislative, and policy limitations that will probably hinder significant mass-market penetration until 2020.

Quite surprisingly, first commercialization may soon come from where it was not expected only a few years ago. Prototypes of miniature fuel cells are now developed by a broad set of big electronics manufacturers: Toshiba, Sony, Samsung, Motorola, and Intel are among these major players, which develop in-house proprietary technologies. Venture capital is also funding countless start-ups in the field.

This situation is not only due to an evolution of the mindset of fuel cell manufacturers, seeking for new profitable markets. Thanks to new miniaturization approaches, it also appears that fuel cells could eventually be integrated in small devices.

These new ideas emerged at the end of the 1990s. At that time, some research groups began to think about the possibility to make fuel cells work in conditions drastically different than large power fuel cells used to up to now (i.e., in ambient temperature and air breathing conditions), with limited place for current collection, heat dissipation, or ancillary subsystems.

44.3.3 Miniature Fuel Cells

Portable equipment concerns a large variety of applications, from cellular phones to e-bikes. Large differences in power output requirements, from some hundreds of mW to hundreds of W, separate these applications. Two main approaches, presented in the next paragraph, cope with these different requirements.

44.3.3.1 Architecture

Two approaches are are used:

1. The first approach investigates the miniaturization possibilities of the conventional so-called "bipolar" technology. It is commonly estimated that this strategy will reveal pertinent for fuel cells of intermediate power, in the 25- to 500-W power range.
2. A second approach is privileged by other research groups, targeting the lower 0.1- to 25-W power range. In this case, it is of common belief that starting from the existing concepts makes the challenge very difficult, if not impossible, to achieve. These groups decided to avoid traditional fuel cell thinking and created new concepts from scratch. This option has been selected by various teams such as Medis, Integrated Fuel Technology (IFTC), Neah Power, Case Western University, Sony, or CEA together with Centre Suisse d'Electro-Mécanique (CSEM).

44.3.3.1.1 Conventional "Bipolar" Technology

To some extent, it is possible to design miniaturized, thin bipolar plates, end plates, and electrode-membrane-electrode (EME).

Heinzel and Hebling [6] refer, for example, to the prototypes developed by companies or research administrations, such as Smart Fuel Cell GmbH (Germany) (25- and 40-W complete systems including methanol-based fuel cartridge; see Figure 44.8 and Figure 44.9), the Fraunhofer Institute for Solar Energy Systems (Germany, 10-W energy supply for a camcorder), Novars (Germany, e-bike), or Manhattan Scientifics (U.S.). Some of these prototypes operate without any auxiliary device. Numerous prestigious other groups are also investigating this field, such as the Jet Propulsion Laboratory (U.S.) together with Giner Electrochemical Systems Inc., and the Los Alamos National Laboratory (U.S.) together with Ball Aerospace. In Japan, Toshiba, Casio, and NEC Corporation (NEC) appear to explore the same concept (only very limited information is available from these studies).

The same article [6] clearly details why miniature fuel cells deriving from conventional bipolar technology should prove their advantages for systems above ≈ 25 to 30 Wh. At lower power output, the power density and the cost of the conventional stack assembly appears to lack the ability to compete with existing lithium batteries technologies.

44.3.3.1.2 Fuel Cells Developed According to Microfabrication Techniques

EME self-supported films used with bipolar architecture are usually constructed using continuous casting techniques. They are assembled with bipolar plates to build the fuel cell stack. They do not require any substrate for deposition.

The process is quite different for fuel cells developed according to microfabrication techniques. In this case, the elimination of bipolar and end plates with their screws and clamps is considered essential to

FIGURE 44.8 Modules from SFC (Smart Fuel Cell, Germany). (Modified with permission of SFC, http://www.smartfuelcell.de.)

FIGURE 44.9 Modules from SFC (Smart Fuel Cell, Germany). (Modified with permission of SFC, http://www.smartfuelcell.de.)

successfully achieve the drastic weight and volume reduction requirements. Components that have to deal with current collection, thermal dissipation, or gas management have to be manufactured because of radically different micro-manufacturing techniques. The use of very thin (with a few micrometers width) electrodes and membrane films that cannot be self-supported appears mandatory to achieve the performances targets at low temperatures.

The latter requirement is opportunely feasible because of the less stringent operating conditions that prevail for portable applications, namely low current densities and atmospheric pressure.

On the other hand, a thin substrate is designed to:

- Process and deliver fuel and air to the electrodes
- Be used as a substrate for the thin films that constitute the fuel cell core
- Properly extract the excess of water generated during the electrochemical reaction
- Dissipate excess heat

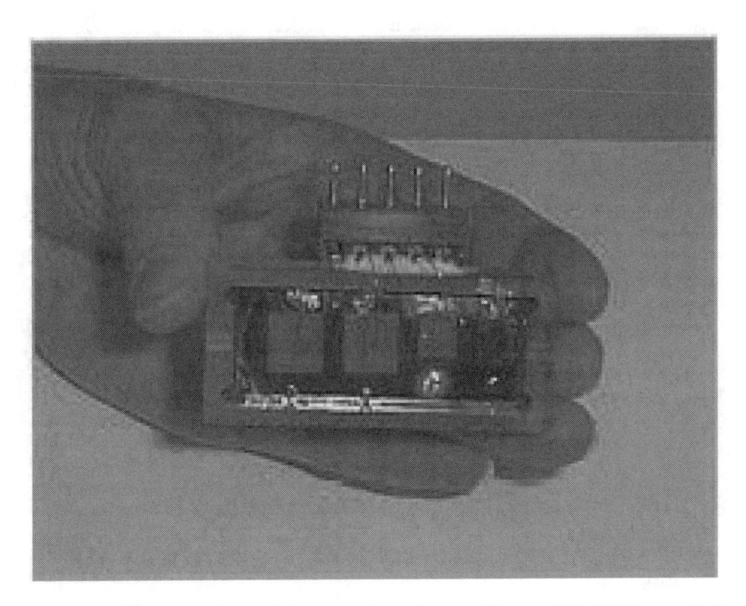

FIGURE 44.10 Miniature fuel cell core assembly on a planar layout. (From CEA. With permission.)

The usual architecture selected is usually a monopolar and planar configuration.

As many cells as required are connected in series by electrical wiring on a planar layout (Figure 44.10). This simplifies the design of the system compared with a standard stack because an adequate sizing allows for passive thermal and water management.

Firms and research centers, such as Neah Power, Integrated Fuel Cell Technologies, the French Atomic Energy Commission, and the Case Western University, introduced the "silicon" approach (Figure 44.11 and Figure 44.12). Thin silicon wafers are specially grown and treated with lithography techniques to develop specific architectures, which offer enhanced electrode surface area compared with conventional approaches.

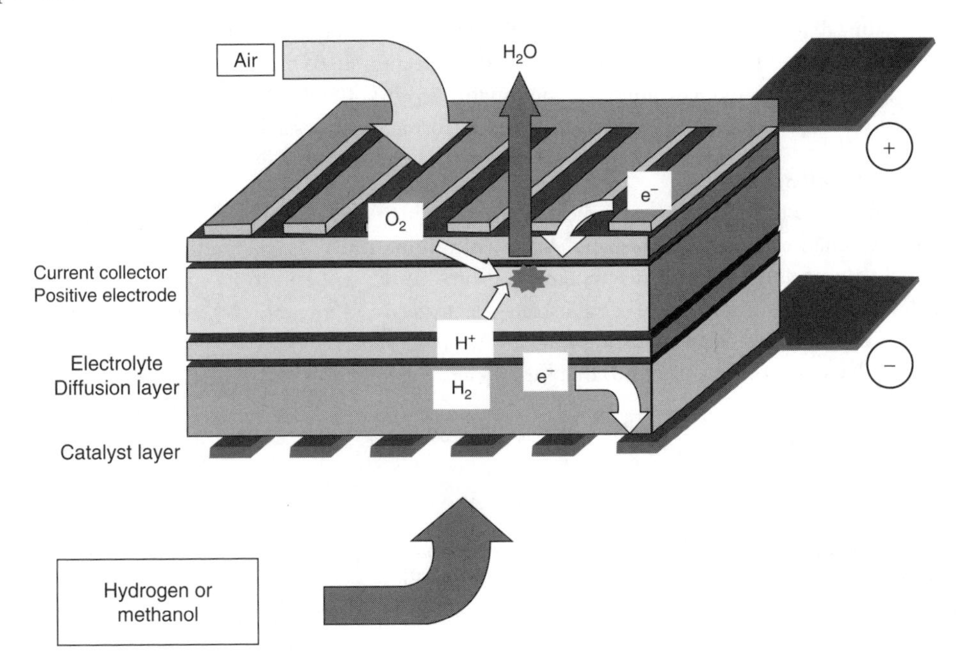

FIGURE 44.11 Fuel cell core architecture on silicon. (From CEA. With permission.)

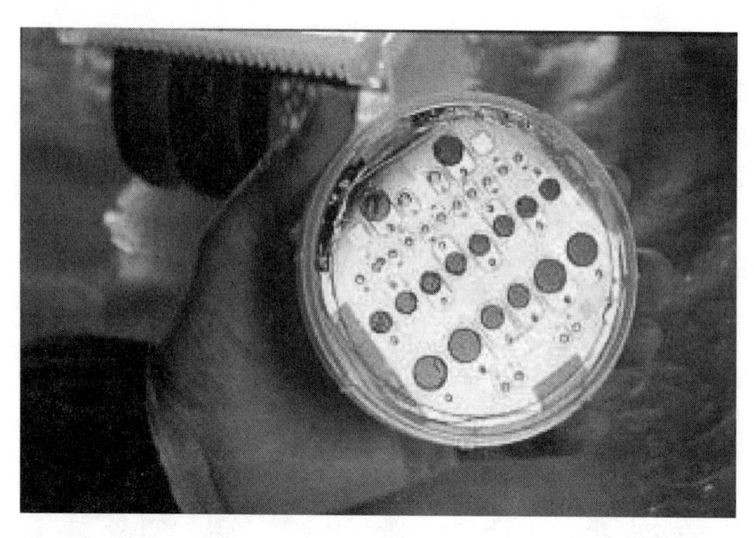

FIGURE 44.12 Fuel cell core architecture on silicon wafer. Each dark dot on the planar layout is a single fuel cell. (From CEA. With permission.)

With the same technological approach, scientists at Motorola developed an integrated multilayer ceramic technology.

44.3.3.2 Basic Issues: General Specifications for Portable Applications

The following subsections focus our interest on low-power portable equipment. Some kind of wireless handheld device can be imagined, which may look like a next-generation mobile phone or PDA. For intermediate 25- to 500-W power levels, the reader is invited to refer to the article mentioned previously [6].

The power requirement of such a small portable device is not known precisely today, but it can be fairly estimated to be in the range of 0.3 to 1 W (continuous output), with power pulses in the range 4 to 10 W.

Let us assume that the energy amount that will be stored in this equipment will be approximately 10 Wh.

As discussed in Section 44.2, the equivalent next-generation 10-Wh lithium secondary battery will weigh 50 g (200 Wh/kg), and its volume will be roughly 25 cm^3 (400 Wh/l). A miniature fuel cell should thus at least offer the same performances, or better performance if possible.

44.3.3.2.1 Overall Sizing

It may here be useful to consider some indicative figures, which may not be precise but pertinent enough for a rough estimation of the system sizing. A fair assumption leads to estimate that approximately half of the total volume and weight of the system is devoted to the fuel and to its storage system, whereas 30% is devoted to the fuel cell core (i.e., substrate, electrodes, and membranes), and 20% to the system peripherals (i.e., electronic management, pumps, valves, and fans).

For a 10-Wh energy source, this means that the 50 g and the 25 cm^3 are distributed in the following ways:

$\approx$ 25 g and $\approx$ 12.5 cm^3 available for the fuel and its storage system
$\approx$ 15 g and $\approx$ 7.5 cm^3 for the fuel cell core
$\approx$ 10 g and $\approx$ 5 cm^3 for peripherals

Another assumption has to be made concerning the fuel efficiency. At this stage, let us estimate that the overall efficiency of the electrochemical reaction will be $\approx$ 33%. This value is quite low compared with fuel cells for cars or stationary equipment, which target overall efficiencies of 50% or higher, but we must not forget that fuel cells for consumer electronics will operate at lower temperature, which makes the catalytic reaction less efficient than at 70 to 90°C.

TABLE 44.1 Lower Heating Values of Some Available Fuels

Fuel	Nature	Volumetric Energy Density (Wh/1)	Gravimetric Energy Density (Wh/kg)
Hydrogen	Gas	3	33,330
Hydrogen	Liquid	2,359	33,330
Methanol	Liquid	4,440	5,470
Methane	Liquid	9,970	13,900
Diesel	Liquid	10,000	11,900

This means that to produce ≈ 10 Wh of electrical energy (10 Wh_e), the total amount of stored energy will be ≈ 30 Wh. This also means that ≈ 20 Wh of thermal energy (20 Wh_{th}) will be released during the operation of the fuel cell.

With these figures in mind, it is now possible to proceed with the evaluation of the low-power miniature fuel cell system.

44.3.3.2.2 Technical Hurdles

Some authors believe that the use of ion-exchange membrane fuel cell technology in battery-replacement type applications such as cellular phones, laptops, and handheld devices is a market ripe for a product that can free users from the aggravation of repetitive charging and limited usage cycles [7]. Despite the great attractiveness of this idea, it appears that most people underestimate the complexity of the system. The following paragraphs describe some of the key hurdles that must be overcome to make miniature fuel cells a reality.

44.3.3.2.3 The Fuel Unit

Figures in Table 44.1 compare the lower heating values of some available fuels. They indicate that the energy content of 1 kg of hydrogen is equivalent to 2.75 kg of gasoline. Unfortunately, it also reveals that 1 Nm^3 (1000 liters under atmospheric pressure) of gaseous hydrogen is equivalent to only 0.34 liters of gasoline, whereas liquefied at −253°C (20 K), the energy content of 1 liter of liquid hydrogen is only equivalent to 0.27 liters of gasoline.

This clearly demonstrates the very interesting performances of the hydrogen gravimetric energy density and its very poor volumetric energy density performances under standard conditions.

44.3.3.2.3.1 Hydrogen — As a chemical reactant, gaseous hydrogen remains the best possible choice as a fuel: the electrochemical reaction, described in the beginning of this section, cannot be simpler. At the same time, no byproduct is created, no intermediate species is formed, and there is no concern about electrode poisoning.

Hydrogen, however, is the lightest of all gases. Although its gravimetric energy density is very high, namely 33 kWh/kg, its volumetric energy density is very low, namely 3 Wh/l under 1 atmosphere (i.e., 3 kWh/Nm^3).

The 30 Wh that are needed to produce 10 Wh_e represented here by 0.9 grams of hydrogen that have to be stored in an appropriate device.

Are solutions available to store or to produce gaseous hydrogen that fulfill the ≈ 25 g and ≈ 12.5 cm^3 requirements of the fuel and its storage system? The following subsections evaluate the possible answers. The case of the storage of liquid hydrogen in a small, super-isolated container will not be discussed because storage can only be envisaged in large, isolated containers with small thermal losses.

44.3.3.2.3.2 Pressurized Hydrogen — Hydrogen may be stored in pressurized vessels. Considering the specifications given previously, for the fuel and its storage system, this means that 0.9 grams of hydrogen have to be stored in 12.5 cm^3. Although this is technically feasible, this would mean that billions of very expensive, specific composite tiny tanks inflated to approximately 700 bar would have to be commercialized through a distribution network, used for consumer applications by adults and children, recovered, and refilled for a new use. This points out, of course, safety concerns related to the handling of pressurized vessels, including mechanical failure or abuse conditions hazards, which may reveal very difficult to manage for consumer electronics.

As a matter of fact, pressurized hydrogen, which appears to be a very pertinent solution for automotive applications, may simply not be acceptable for portable equipment.

44.3.3.2.3.3 Metallic Hydrides — Extensive experience has been acquired on metallic hydrides over the last 30 years. These specific "AB_2" or "AB_5" ("A" being, most likely, lanthanum and "B" being nickel) alloys are able to absorb and release gaseous hydrogen under temperatures close to ambient and intermediate pressure (a few bar up to 10 bar) conditions.

Many issues related to the long-term reversibility of the charging/discharging processes have been overcome during the last decade.

The potential interest of using hydrides is relying on:

- An interesting gravimetric (450 Wh/kg) and volumetric energy density (1050 Wh/l at 1.3 wt% typical storage density) of the material itself (without the container)
- A low operating pressure (safety)
- An ability to release pure hydrogen
- A good reversibility
- A proven technology

Moreover, metallic hydrides are not harmful to the environment.

Assuming a gravimetric energy density of ≈ 1.5% wt (i.e., 1.5 g of hydrogen can be stored in 100 g of material) for an MH operating at ambient temperature and atmospheric pressure, as much as 60 g of MH metallic powder would be necessary to store 0.9 g hydrogen, not counting the actual MH container, which is most likely heavier than the actual active material itself. This figure still has to be compared with the 25 g available.

Moreover, MHs have severe drawbacks, related to their high cost or to the necessity of a refueling infrastructure (as for pressurized containers).

This explains why metallic hydrides have been, until recently, considered ineffective storage means for very small power applications.

The near future may, however, pave the way to some surprises in this field, thanks to the development of new promising compounds. This option has to be carefully explored.

44.3.3.2.3.4 Carbon Nanotubes — Richard E. Smalley discovered fullerenes (soccer-ball-like carbon molecules) in 1985, and was awarded the 1996 Nobel Prize in chemistry for this discovery. In 1991, Sumio Iijima, working at NEC Corporation, synthesized the first carbon nanotube. Since that date, the imagination of researchers has played with these tiny carbon nanotubes structures, which may also look like horns, cones, or cylinders, reveal to be single or double walled, and which may be 100 times stronger than steel. Among the numerous revolutionary ideas that have been suggested, the rumor spread in 1997 that a laboratory in South Eastern Asia claimed to have demonstrated the capability of carbon nanotubes structures to store large quantities of hydrogen. Unfortunately, it appears that no definitive confirmation of these preliminary promising results has been achieved up to now, and carbon nanotubes presently available appear, most likely, to be able to store hydrogen with efficiencies around 1 wt%, which makes them for the moment not more interesting than metallic hydrides.

44.3.3.2.3.5 Steam Reforming — The direct use of methanol in fuel cells may be confronted to severe bottlenecks. This explains why scientists from Motorola, Toshiba, Casio, or Battelle are investigating the field of *in situ* conversion of methanol into hydrogen and CO_2 thanks to a small-scale methanol steam reformer. In a presence of a catalyst, methanol may be converted at 250 to 300°C in a rich hydrogen and CO_2 mixture.

The dimensions of the 200 mW reforming capacity Battelle prototype, developed in cooperation with Case Western University, which includes two vaporizers, a heat exchanger, a combustor, and a steam reformer, are quite stunning — less than 10 mm^3.

As early as March 2002, Casio [8] claimed that it had succeeded in realizing prototypes of the PEM fuel cell with integrated methanol reformers. This would have meant that they succeeded in overcoming hurdles related to the removal of carbon monoxide (CO) traces generated during the reforming reaction (CO has to be removed because it causes poisoning of catalyst located in PEM electrodes.) and also the issues related to the thermal management of a subsystem operating at 300°C integrated in a handheld device.

44.3.3.2.3.6 Chemical Hydrogen Storage Devices: $NaBH_4$ Solutions — In contact with a catalyst (usually a ruthenium-based material), chemical compounds as borohydride $NaBH_4$ diluted in aqueous alkaline solutions are known to produce hydrogen according to the chemical catalytic hydrolysis reactions:

$$NaBH_4 + 2\ H_2O \rightarrow 4\ H_2 + NaBO_2$$

The heat released during the catalytic reaction may be used to vaporize a fraction of the water, so that hydrogen is mixed with moisture, which may be used with profit for the humidification the fuel cell membrane.

$NaBO_2$ is soluble in water, is environmentally benign, and can be recycled for the generation of new sodium borohydride.

The amount of hydrogen released by the reaction depends on the $NaBH_4$ concentration in water. A theoretical efficiency of 10.84 wt% is calculated at 51.2 wt% $NaBH_4$ in water. The gravimetric energy density of the fuel mixture then becomes ≈3600 Wh/kg, not considering the packaging and the fuel supply devices.

The great interest of the process may be pointed out here: if this theoretical figure can be achieved practically, only 8.30 g of fuel mixture would be needed to store the required 0.9 g of hydrogen calculated previously, whereas the overall requirement for the fuel system, including the fuel container, is 25 g.

Millenium Cell, founded in 1998, investigates this promising field and targets automotive applications as well as consumer electronics [9].

44.3.3.2.3.7 Hydrogen from Pyrotechnic Materials — Some recent patents and papers [10,11], suggest that it may be possible to release pure hydrogen from a pyrotechnic reaction. The concept is very similar to the air-bag technology: once initiated by a small electric current, the solid-solid combustion of pyrotechnic materials releases a gas. The interesting point appears to be the weight efficiency, which may be in the range of 8 to 10 wt%. As for borohydride, this invention could reveal an interesting breakthrough because it would offer an efficient way for the storage of small to intermediate quantities of hydrogen.

44.3.3.2.3.8 Methanol — Methanol appears to be a very sympathetic solution. As a liquid, it is easy to handle and to distribute, and disposable recharging cartridges can be imagined as ink cartridges for pens. Its volumetric energy density is far higher than hydrogen (4440 Wh/l).

Considering the practical concerns for the handling and storage of gaseous hydrogen, the vast majority of the miniature fuel cells developers have decided to work with methanol. The so-called "direct methanol fuel cell" (DMFC) technology is currently the "standard" model for miniature fuel cell investigation. This option allows using a small container of liquid methanol as a recharging cartridge. A specific platinum and ruthenium-based catalyst located in the negative electrode is supposed to convert directly methanol in hydrogen, and finally produce electricity, according to the reactions:

- Negative electrode:

$$CH_3OH + H_2O \rightarrow C\ O_2 + 6H^+ + 6e^- \tag{44.1}$$

- Positive electrode:

$$6H^+ + 6e^- + 1.5\ O_2 \rightarrow 3\ H_2O \tag{44.2}$$

- Overall reaction:

$$CH_3OH + 1.5\ O_2 \rightarrow CO_2 + 2H_2O \tag{44.3}$$

The preceding equations indicate that the complexity of the system must not be underestimated, and many additional technical bottlenecks remain to be overcome.

The first issue is related to the fact that methanol behaves like water. During operation, a part of the methanol mixed with water (Equation (44.1)) does not react as expected at the negative electrode, but instead crosses the electrolyte and joins the positive electrode. There, the direct reaction between the fuel (methanol) and the oxygen is no longer an electrochemical reaction, but instead a chemical reaction. This phenomenon, called the "crossover" effect, causes a loss of efficiency of the fuel cell, increases thermal losses, and in some cases partly causes the degradation of the membrane. To limit this effect, pure methanol is not used, but it is mixed with more water than the simple equation (Equation 44.1) indicates, which means that water produced from the reaction at the positive electrode has to be partly recycled back and mixed with the methanol that comes to the negative electrode. At the early stages of development, DMFC developers were obliged to limit the fuel concentration to very low levels as 1M (≈ 5 wt% in water), which limited the efficiency of the methanol option. Thanks to the development of specific selective membranes with reduced crossover compared with conventional Nafion membranes, it now appears possible to use more concentrated solutions — up to 25 wt% or even more. Firms, such as Polyfuel and Giner Electrochemical Systems, adopted this approach [12].

A second issue is related to the formation of carbon dioxide on the negative electrode (Equation (44.1)). This gas has to be regularly drained off to avoid electrode damping, which requires liquid-gas separation components.

However, not indicated by the preceding electrochemical reactions, under circumstances depending on temperature or reaction speed, intermediate compounds may be formed, which poisons the Pt-Ru-based catalysts. This has for consequence a shorter lifetime than expected, due to electrode progressive degradation.

In any case, the methanol option suffers from severe drawbacks: many power-consuming peripherals, such as pumps, sensors, gas-liquid separators, heat exchangers, and valves, are required to monitor continuously methanol concentration, fluid (i.e, water and methanol) levels and rate of flow, to release the carbon dioxide formed at the negative electrode and the excess heat, and to prevent, if possible, the conditions of formation of intermediate compounds. Overcoming all these bottlenecks will not be an easy task, but it appears that joint efforts of major developers, such as Toshiba, NEC, or Samsung, are on the verge of success.

Finally, methanol is listed as a toxic fuel. It may take a long time to get codes and standards from administrations for airplane transportation or agreements for consumer electronics domestic applications. These elements will certainly play a key-role in the development of DMFCs.

44.3.3.2.4 Air Feeding of the Positive Electrode

Miniature fuel cells cannot afford, as space shuttle fuel cells do, to use pressurized pure oxygen. Most of the technologies developed rely on air breathing positive electrodes; however, nitrogen represents 70% of the gases contained in air and this causes performances limitations of the fuel cell. This explains why low-consumption micro fans and other smart devices are often used to accelerate the oxygen diffusion at this site.

44.3.2.3.5 Fuel Cell Core

Thin film technologies should permit the realization of substrate-supported EME with an overall width of 500 μm. Thus, the sizing of the fuel cell core should not cause much concern: with 7.5 cm^3 available, ≈ 20 cmΣ of electrode surface could be integrated in a flat packaging. This means that electrode current densities should reach 50 mW/cmΣ to release 1 W of electrical power. This level of performance is without any doubt achievable with pure hydrogen at ambient temperature. With DMFC, the challenge should be achievable on a short term: electrodes power densities as high as 67 mW/cmΣ at 55 to 60°C with over-stoichiometric air-flow, and 35 mW/cmΣ at 30°C, operating on ambient pressure airflow, have been reported [12].

Beyond concerns related to electrode efficiency and low-cost fabrication techniques, a major issue is related to the way a fuel cell has to be dimensioned versus the application requirements: as an internal combustion engine, the fuel cell efficiency is higher if it is correctly sized, with electrode surface corresponding to the nominal output power. If this was not the case, either the fuel cell would be undersized compared with the nominal output power (and the voltage drop causes a strong decrease of the efficiency because of electrode resistance losses), or the fuel cell core would be oversized compared with the real need, thus penalizing the fuel cell in terms of volume and weight. In other words, a better efficiency can be achieved with an adequate sizing of a fuel cell operating at a constant nominal output power. This situation has consequences, which are explained further (Section 4.4) because portable devices do not work on a constant mode, but instead on mixed standby modes followed by peak power pulses.

The following short paragraphs point out the main issues related to the core components of fuel cells. Because they concern strategic parts of the system, only a few studies have been published.

44.3.2.3.6 Electrodes

Catalytic activity control is of key importance with miniature fuel cell systems because, at ambient temperature, the catalytic performance is expected to be low. Two extreme options are possible, which, of course, have to be mixed to target an adequate trade-off:

1. In the first option, a specific architecture of the fuel cell system allows the fuel cell core (as a tungsten wire in a lamp bulb) to operate locally at intermediate temperature (namely 60 to 70°C). This trick allows in reality the fuel cell to work in operating conditions similar to those of power fuel cells. For such small equipment as a portable device, however, the compromise is not easy to find between heat management (which requires isolating materials), current collection, and other issues.
2. The second option explored relies on electrode surface development. The largest the electrode surface area, the highest the power delivered for the same planar surface. Thus, companies, such as Sony and NEC, are developing specific 100-nanometer carbon nanohorns clumps that are covered by nanoparticles of catalyst and used to manufacture highly efficient electrodes. So far, no experimental result has been published, but this approach could reveal very promising.

44.3.2.3.7 Electrolyte

The choice of the electrolyte mainly depends on the nature of the fuel that has been selected to feed the fuel cell (Section 44.4). Nafion-based polymers seem to be widely used with hydrogen as a fuel, whereas specific membranes are developed with methanol.

When the nature of the membrane has been selected, the second step concerns the process that is used for its thin-film deposition. In this field, all possibilities remain open including combinations between solvent casting methods and physical or chemical vapor deposition techniques.

44.3.2.3.8 Peripherals

Peripherals are compulsory for proper operation: fuel feeding and electronic management. At the same time, they penalize the energy density, and they have an influence on the cost of the equipment. All actors explore passive systems that use capillarity effects observed in micro-channels and natural fluid convection.

44.3.4 Safety Issues

Safety concerns are always of great importance for consumer electronics. The portable devices will have to be transported in every airplane or car by adults and children, withstanding high temperatures on a car passenger seat during summer.

It will surely take a longer time than expected to get codes and standards from administrations. Issues will, of course, arise related to possible methanol evaporation in a confined car atmosphere with children inside or to the use of gaseous hydrogen. Considering the energy amount and the quantities that will be concerned, however, supporters of miniature fuel cells will also have strong arguments. Nothing is settled, but smart safety devices should eventually, most likely, allow proper and safe use.

44.3.5 Energy Source System Management: Hybrid Systems

Materials issues have been introduced, but additional issues may be raised:

- How do we manage the peak currents?
- How can we be sure that the device will start immediately when required?
- How does the system operate in cold weather, when the temperature drops below the freezing point and the water in the fuel cell may freeze?
- How do we save fuel when the "open to air" device is suddenly turned off, with fuel inside that may evaporate or volatilize?
- How does the system stay "on line" when the recharging cartridge has to be replaced?
- How do we use the equipment when there is no recharging cartridge immediately available?

All these elements give additional complexity to the system, but militate for the hybridization of a lithium battery with the miniature fuel cell. Lithium miniature power batteries will be soon able to withstand high peak current without major capacity loss. Consequently, the preceding questions will be answered favorably, provided a plug is available on the wire in the surroundings.

44.3.6 The Players

Although few actors in the domain existed a few years ago, it would be now impossible to list all players on the miniature fuel cell field.

Major players are, of course, integrated Japanese and Korean companies, such as Toshiba, Casio, Hitachi, NEC, Sony, and Samsung. U.S. firms also appear to be on track, including Motorola, Intel, Hewlett Packard, Duracell, and Bic, followed by numerous startups supported by venture capital: MTI, Neah Power, Manhattan Scientifics, Polyfuel, Medis, and IFTC.

In Germany, credible outsiders, such as Smart Fuel Cell GmbH or Novars, may eventually play a role.

National research centers all over the world, such as LANL, Case Western University, Pacific Northwest National Lab, Battelle, Fraunhofer ISE-IZM, KIST, and CEA, together with CSEM, etc. are developing new concepts that will allow breakthroughs and facilitate a rapid development of this new technology.

The first company that will succeed in entering the market and distribute a recharging cartridge together with a fuel cell-based energy source will have the opportunity to create a new standard. Although the technology is not yet mature, the race is already open for commercialization.

44.4 Prospective

44.4.1 Commercialization Issues

In March 2003, Toshiba introduced to the visitors of the CeBIT 2003 in Hanover the world's first prototype of a 12-W nominal output DMFC system able to run a notebook computer [13]. Similar to Mechanical Technology Institute (MTI) [14], Toshiba publicly targeted a market entry date of late 2004 for first products. NEC introduced nearly the same achievements a few weeks earlier.

A day does not pass without new announcements that miniature fuel cells will soon be integrated in various products, such as wireless handsets, notebook computers, cameras, camcorders, PDAs, and mobile phones, which will require power supplies offering ever growing energy density performances.

Neah Power Systems is more cautious and does not expect sales to mainstream notebook PC markets until 2006, after it has built up an infrastructure for its retail methanol cartridges [15]. Considering technical and regulations hurdles that remain to be overcome, Neah Power may be closer to the reality.

44.4.2 Expected Costs

44.4.2.1 Cartridges

Polyfuel is talking with industrial partners about creating a retail market for $2 to $3 disposable methanol cartridges that would power fuel cells [15].

44.4.2.2 Fuel Cell System

One of the key technical challenges of fuel cells is to make them at an affordable cost. Today, the recharging process on the plug costs nearly nothing. The "cost of freedom" offered by a new "wire-free" power source must not be outrageous for the end-user. Low-cost, embedded products, manufactured according to large-scale manufacturing processes should probably be privileged. The cost target should in any case be as low as 0.5 $/Wh, which is a very ambitious goal.

44.5 Conclusion

Miniature fuel cells are a hot topic. Research consulting firms, such as Allied Business Intelligence (Oyster Bay, NY), estimate that 200 to 500 million miniature fuel cells could be sold by 2011, for an annual revenue that could be as much as $5 billion [16]. Original equipment manufacturers, such as Toshiba, Sony, Sanyo, NEC, Hitachi, Casio, Matsuchita, Motorola, or Samsung, investigate basic research and development as well as industrialization issues. They may clearly benefit from the fact that they control many market entry parameters, but technological breakthroughs may also permit some pioneer outsiders and start-ups to play an essential role.

Miniature fuel cell technologies and architectures will probably differ according to the requirements of the targeted applications:

- For intermediate 25- to 500-W power range, conventional "bipolar" technology may be envisaged, using methanol or reformed methanol as a fuel. The operating conditions could be quite similar to the conditions prevailing for large power fuel cells. Companies, such as SFC, claim to be ready for commercialization [17].
- For low-power applications in the range of 0.1 to 25 W, micro-fabrication techniques seem quite an interesting solution to permit operation at ambient temperature. The fuel would ideally be hydrogen, provided a solution is confirmed for an efficient storage of this gas, but other types of liquid alcohol-based fuels may also be envisaged.

In both cases, key technical and commercial decisions have to be taken by developers for fuel selection, fuel cell design optimization, power source integration, and distribution channels. Technical and regulation barriers will have to be overcome as well as for liquid-based and gaseous operating fuel cells.

The latest versions of sophisticated prototypes are introduced at every new conference.

All miniature fuel cell developments must cope with decreasing cost and increasing performances over lithium batteries. Marriage between both systems will be most likely compulsory because they are complementary: Battery offers its ability to withstand peak currents and supply energy whenever required. The fuel cell allows rid of the wire and to virtually unlimited operating time, provided the user gets enough recharging cartridges in his pockets. Fuel cells offer freedom and safety, which are advantages that address the sensitivity of users who may be ready to pay for them.

References

[1] C. Pillot (*Avicenne Développement)*, The worldwide rechargeable battery market, *Batteries 2003, 5th Batteries Int. Symp.*, April 16–18, 2003, Paris.

[2] K. Ishibashi, Interview: WestLB bets fuel cells Japan's next megahit, *Dow Jones Newswire*, March 18, 2003.

[3] The Gillette Company, Duracell introduces first-ever alkaline prismatic battery for portable digital audio devices, Press Release, The Gillette Company, March 17, 2002.

[4] O. Haas, F. Holzer, K. Müller, and S. Müller, Metal/air batteries: the zinc/air case, in *Handbook of Fuel Cells, Volume 1, Fundamentals and Survey of Systems*, John Wiley & Sons, New York, 2003.

[5] R. Wilson, Researchers disclose work on inventive materials, Press Release, *EE Times*, February 14, 2003.

[6] A. Heinzel and C. Hebling, Portable PEM systems, in *Handbook of Fuel Cells, Volume 4, Fuel Cell Technology and Applications,* John Wiley & Sons, New York, 2003.

[7] C. Stone and A.E. Morrison, From curiosity to power to change the world, *Solid-State Ionics,* 152–153 (2002) 1–13.

[8] Casio, Inc., Success in R&D of optimal, small scale, high performance fuel cells for portable devices, News Release, http://www.casio.com, March 13, 2002.

[9] http://www.milleniumcell.com.

[10] D. Bloch, G. Delapierre, J.Y. Laurent, T. Priem, and D. Marsacq, Pile à combustible pour l'alimentation d'appareils électroniques, notamment portables, French Patent No. Publication 2,818,808, 2000.

[11] QinetiQ, *The Engineer,* May 16, 2003.

[12] S.R Narayanan and T.I Valdez, Portable direct methanol fuel cells systems, in *Handbook of Fuel Cells, Volume 4, Fuel Cell Technology and Applications,* John Wiley & Sons, New York.

[13] Toshiba announces world's first small form factor DMFC fort portable PC's, Press Release, *Business Wire,* March 4, 2003.

[14] S. Gottesfeld, Development of direct methanol fuel cells for consumer electronics applications at MTI microfuel cells, *Small Fuel Cells Conf.,* April 21–23, 2002, Washington, D.C.

[15] R. Merritt, Two start-ups take different paths to fuel cells for notebooks, Press Release, *EE Times,* May 7, 2003.

[16] B.J. Feder, For far smaller fuel cells, a far shorter wait, *The New York Times,* March 16, 2003.

[17] M. Stefener, *Last developments in miniature fuel cells,* Excellence in fuel cells seminar, Hanover Fair, April 2003.

45

Human-Generated Power for Mobile Electronics

Thad E. Starner
Georgia Institute of Technology

Joseph A. Paradiso
Massachusetts Institute of Technology

45.1 Introduction

Since the 1990s, mobile computing has transformed its penetration from niche markets and early prototypes to ubiquity. Personal digital assistants (PDAs) evolved from GRiD's PalmPad and Apple's Newton in 1993 to the Palm, Handspring, and Microsoft-based models that support the multibillion dollar industry today. Although BellSouth/IBM's Simon may have been the only mobile phone to offer e-mail connectivity in 1994, almost every modern mobile phone provides data services today. Compact digital music players have replaced cassette and CD-based systems, and these "MP3 players" are evolving into portable repositories for music videos, movies, photos, and personal information such as e-mail. Laptops, which were massive and inconvenient briefcase devices in the late 1980s, now outsell desktops in some markets. Yet, all these devices still have a common, difficult problem to overcome: power.

This chapter reviews trends in mobile computing over the past decade and describe how batteries affect design tradeoffs for mobile device manufacturers. This analysis leads to an interesting question: is

0-8493-1941-2/05/$0.00+$1.50

there an alternative to batteries? Although the answer has many components that range from power management through energy storage [142], the bulk of this chapter discusses the history and state of the art in harvesting power from the user to support body-worn mobile electronics.

45.2 Technology Trends in Mobile Computing

Mobile phone companies often sell more batteries than phones to consumers. The phones sold to users include a rechargeable battery so that the device is immediately useful, but a certain number of consumers are expected to own more than one battery during the life of their phone. The same can probably be said for laptops and camcorders. Yet, there is little incentive for consumers to buy new batteries except for when they fail or when the consumer feels the need for a larger battery. Unlike other areas of mobile computing that benefit from exponential improvements in performance, battery energy density (as measured by joules per kilogram or joules per cubic centimeter) changes slowly so that consumers are under little pressure to upgrade.

45.2.1 Battery Energy Density as a Lagging Trend

As Figure 45.1 indicates, battery energy is one of the most lagging trends in mobile computing. Figure 45.1 depicts the progression of technology in the last 13 years for laptop computers, a technology now mostly mature. In general, the laptop technology represented in the graph would, if repackaged in a body-worn device, weigh 7 pounds or less and could be used while standing on a street corner in a major U.S. city. Although some mobile computers existed before 1990, most weighed over 10 pounds, or did not include hard drives. In addition, commercial wireless data networks in the U.S. were not openly available before 1990 or required amateur radio licenses to operate.

The graph depicts increases in performance as multiples of the state of the technology from 1990 (e.g., the amount of RAM available in a laptop increased by 256× from 1990 to 2003). Due to the exponential nature of the improvements, the y-axis in Figure 45.1 is on a logarithmic scale.

The laptop specifications depicted were determined by examining advertisements in the December issues of popular computing magazines (e.g., *Byte, PC Computing,* etc.) for each year. The numbers used reflect a composite from the highest-end machines available at that time. An example of a high-end machine from 1990 (the base value of 1 in the graph) would be a 16-MHz 80386 with 8 megabytes (MB) of RAM and 40 MB of hard drive space using a nickel-cadmium battery and communicating at 4800 baud over the Advanced Radio Information Service (ARDIS) network. Processor performance is compared in terms of Intel's iCOMP® index, as derived from http://www.cpuscorecard.com; RAM and disk storage are compared by size; wireless networks are compared by maximum bits per second of data transfer; and battery energy density is determined by the type of technology used (i.e., nickel cadmium, nickel metal hydride, or lithium ion) and the progression these technologies made in increasing the joules stored per kilogram (J/kg). The wireless connectivity graph represents the first author's pursuit of the commercial citywide networks available in the U.S. (cellular standards; not emerging 802.11 "hotspots").

Although disk storage density has increased over 4000× since 1990, the lowly battery has only increased a factor of three in energy density. New materials, along with nano and micro-fabrication technologies, have recently enabled "micro fuel cells" [145] aimed at recharging handhelds, such as cell phones with power plants the size of a small candy bar [177], and eventually powering wireless sensor nodes with fuel cells on a chip [102,120,163]. Although the technology is rapidly advancing [60], laptop-sized plants (e.g., 30 to 50 Wh) have tended to be in an awkward place for fuel cells — too big to directly power with micro cells, but small enough that the overhead in mass needed to handle the standard fuel cell chemistry is significant (not to mention safety factors associated with the fuel and the high expense of the platinum membrane). Nonetheless, several companies have announced prototypes designed for laptops [32], which should make it to market over the next couple of years and gradually improve.

More exotic emerging power technologies tend to have characteristics that force them into niche applications (e.g., radioactive batteries [81] can last for decades, but provide very little current, while

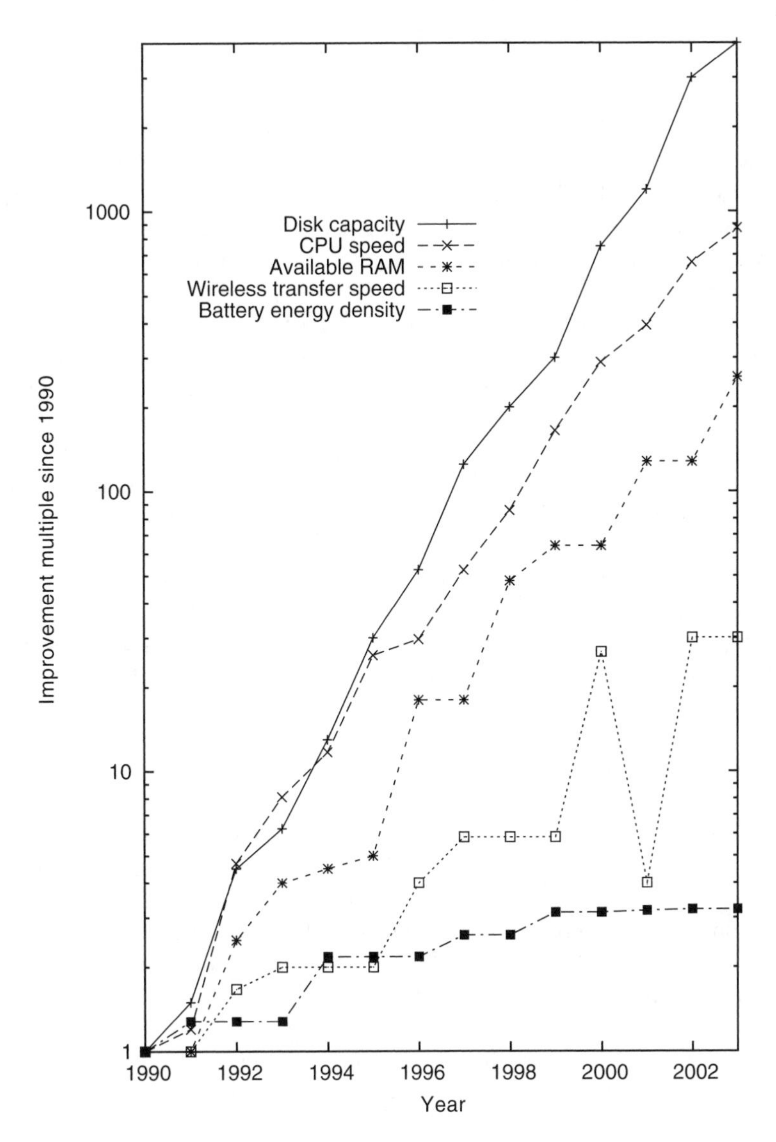

FIGURE 45.1 Improvements in mobile computing technology from 1990–2003. Note that the wireless connectivity curve considers only cellular standards, not short-range 802.11 "hotspots."

devices that actually burn fuel [3], such as microturbines [63] and microengines [72] have potential issues with safety and byproducts, such as exhaust, heat, noise, and thrust).

The lesson to mobile device designers is clear: specify the battery or power source first, then design the mobile device's electronics around it. Battery technology is the least likely element to change in the 12-month development cycle and may be the most limiting factor in the design with respect to size, weight, and cost.

45.2.2 Trading Storage and Processing for Wireless Connectivity

Wireless connectivity is also a conundrum for mobile designers. Whereas the designer can control the CPU, RAM, disk, and battery in his or her device, another party often provides wireless connectivity. In the extreme case, a wireless provider may go out of business and significantly affect the quality of service that can be expected. Such a situation is reflected in Figure 45.1 where the removal of the Metricom network reduced the maximum available throughput from 128,000 bits per second to 19,200 in several major U.S. markets.

Even on a minute-by-minute basis, a wireless connection may or may not be available at any given moment. The device designer must either cache information for the user or refuse service when the network is not available [107]. Thus, many devices, such as wireless PDAs, have nonvolatile RAM or disks so that the user can work "off-line." Using mass storage strategically can save significantly on battery consumption because both receiving and transmitting data from cellular and 802.11 networks require substantial power [174]. More specifically, the power needed for transmitting is proportional to the distance to the fourth power [47]. Given the exponential trends in disk density discussed previously, it may soon be a viable power-saving strategy to cache a good fraction of static Internet content for a mobile web surfer instead of connecting over power-hungry and potentially expensive wireless networks! One can imagine a system that examines the user's e-mail, Web history, and downloads and, based on this data, continuously updates the user's mobile cache while the device has wired (or low-power) connectivity.

An interesting illustration of this point is to compare the power required to retrieve information from modern flash memory with the power required to transmit a request of that information from a remote source. Suppose that we have the option of storing information on a cellular phone in the form of a flash disk or sending a wireless request for the information to the network. Reading a bit from modern flash memory requires approximately 10 pJ or 1×10^{-11} J/bit [13]; however, transmitting a single bit at 0.6 W from a mobile phone at an aggressive 1 Mbps rate would require 6×10^{-7} J. Thus, for every bit transmitted in the wireless request for information, the same amount of energy could be used to read 60,000 bits from a flash drive. This calculation is conservative because it ignores the inefficiencies in the radio, the overhead generally associated with transmission error checking, and the amount of power that would be required to receive, process, and store the response from the network. Thus, a mobile device designer should always consider how much information could be stored or cached on the actual device as opposed to depending on wireless services.

The sensor network community is very concerned with a similar trade-off (i.e., how much data to process locally at a sensor node vs. how much data to wirelessly transmit). Because it takes between 100 and 10,000 times more power to transmit one bit across even a short range than to execute a single processor instruction (depending on the implementation) [153], it is often advantageous to analyze or compress the node's data before broadcasting [151]. To reduce the node's power requirements down to the point where ambient energy harvesting is practical, researchers are pursuing joint optimization of the processor hardware, radio circuitry, and network protocols [132]. Although we do not explicitly consider the amount of power required to receive the information, this is often not negligible, especially in short-range networks, where it can take more power to receive and decode a bit than to transmit one [153].

45.3 Power from Incident Radiation

45.3.1 Catching the Ambience

With so many RF transmitters of various sorts distributed throughout today's urban environments, one might consider background RF as a potential power reservoir for mobile devices. Electronic systems that harvest energy from ambient radiation sources tend to be extremely power-limited, however, and generally require a large collection area or need to be located very close to the radiating source. A classic example can be found in old-fashioned crystal radio kits [106] that draw their power directly from AM radio stations, which play audibly through high-impedance headphones without needing a local source of energy. The size of the required antenna, however, can be prohibitive for wearable applications unless the bearer is very close to a transmitter, and access to a good ground is usually required. Even so, the received power is very limited in a standard crystal radio, where set builders typically see received powers on the order of 10's of μW, approaching a mW for proximate stations. An interesting adaptation of a crystal radio set is described in U.S. Patent No. 2,813,242 [56], where a resonant tank circuit tuned to a strong, nearby station provides enough power to run a single-transistor radio with a small loudspeaker that can be tuned to other stations. An analysis of RF power scavenging at higher frequencies by Yeatman [194] crudely approximates the power density produced by a receiving antenna as E^2/Z_0, where Z_0 is the radiation resistance of free space (377 Ohms). An electric field (E) of 10 V/m thus yields 26 μW/cm^2 at the antenna. Field strengths of even a few volts per meter are rare in habitated environments, however, except when very close to a powerful transmitter [123]. In a related note, power can also be extracted from the earth, across a large ground loop, tapping the AC potential difference between grounds at different locations. A harvest of 1.4 mW has been reported using a pair of grounds separated by 50 feet [169].

An example of ambient radio frequency (RF) power harvesting in the mobile sphere at higher frequency is found in aftermarket modules that flash LEDs when your cell phone rings. Several of these designs are batteryless, but need to be extremely close (or right against) the cell phone's antenna to work, as they draw their energy through near-field capacitive or inductive coupling. Perhaps another mobility example, much further afield, comes from the strange, scattered and usually anecdotal reports of people receiving strong, nearby radio broadcasts from spontaneous detectors formed by loose fillings in their teeth [86,40], and the passive implantable receiver design that this has inspired [7,152].

Higher up in the electromagnetic spectrum, it is not uncommon to see very low-power consumer items, such as simple calculators, run off photovoltaics with ambient illumination. The energy conversion efficiency of easily available and relatively inexpensive crystalline silicon solar cell modules (without going to integrated circuit (IC)-grade silicon or stacked junction structures) is generally below 20%, and closer to 10% for flexible amorphous silicon panels [26]. Accordingly, mobile applications, which generally imply limited surface area, tend to be constrained, especially in scenarios without strong and consistent sunlight (standard solar cells produce roughly 100 mW/cm^2 in bright sun and 100 μW/cm^2 in a typically illuminated office). Nonetheless, products such as solar battery chargers for cell phones that purport to produce 2 W of power [117] and PDAs that run off a panel of solar cells lining their case [164] currently exist, and researchers continually strive to refine solar cell materials [26,92] and technologies [27] to increase efficiency [78], as well as explore unusual form factors, such as flexible photovoltaic fibers [97], which promise to be more amenable to wearable implementations.

45.3.2 Get on the Beam

Instead of relying on the limited energy that can be scavenged from ambient radiation, other approaches actively beam power from a transmitter to remote devices. The wireless transfer of power originates with Heinrich Hertz who, ushering the dawn of radio in the late 1800s, induced sympathetic sparks across a gap interrupting a resonantly tuned ring placed several yards away from a transmitting antenna that was directed with a parabolic reflector [178]. The dream of wirelessly broadcasting power to an urban area dates back to the turn of the 20th century and Nicola Tesla [180], who experimented with grandiose concepts of global resonance and gigantic step-up coils that radiated strong, 150-kHz electromagnetic fields able to illuminate gas-filled light bulbs attached to a local antenna and ground at large distances [50]. Wireless power research continued with the work of H.V. Noble [38], who in the early 1930s at the Westinghouse Laboratory, demonstrated the transfer of several hundred watts between 100-kHz antennas separated by 25 feet, leading to public demonstrations of this technology at the Chicago World's Fair in 1933. The development of radar [39], and thus powerful microwave transmitters, enabled further work in directed energy transmission, a highlight of which was the wireless powering of a small helicopter by William C. Brown in 1964 [38]. Microwave-to-DC converters, termed "rectannas" can be extremely efficient; efficiencies of over 90% have been produced in laboratory experiments and 30 kW have been transferred across more than a mile at 84% efficiency [38]. This has led to proposals for beaming massive amounts of power to earth from solar collectors in space [75] and remotely beaming propulsion to interstellar probes from an earth-orbiting 10-GW transmitter [69].

Closer to home, FCC and safety regulations (e.g., IEEE/ANSI C95.1) along with public perception [70] have restricted the beaming of any significant amount of power in the proximity of people. Nonetheless, researchers have experimented with microwave transmission of power in domestic environments, transferring several mW across meters to sensors for ubiquitous and wearable computing applications

[19]. At much lower power levels, short-range wireless power transmission is now commonplace in passive radio frequency identification (RFID) systems [65], which derive their energy inductively, capacitively, or radiatively from the tag reader. Because most RFID chips talk back to the reader by dynamically changing their impedance or reflection coefficient, they require minimal power, generally between 1 and 100 μW, depending on their implementation and operating frequency (lower-frequency, magnetically coupled tags consume less power). Today, people commonly carry RFID transponders, most often for keyless entry systems. Simple resonant RF tags that change their tuned frequency or Q as a function of a local or environmental parameter have been used as passive sensors in several applications [99,155]. Examples include LC (inductive-capacitive) tags for wireless displacement and pressure sensors in human-computer interfaces [143], measuring tire inflation with pressure-varying backscatter from crystal bulk resonators [24], tracking tire strain with surface-acoustic wave (SAW) devices [149], and proposed studies for using such SAW sensors as implantable blood pressure monitors [129].

The reverse, where people carry the reader to interrogate tags in the environment, is not as feasible because the readers tend to be power-hungry and large (e.g., several orders of magnitude more massive than the tags). Researchers, however, in wearable and ubiquitous computing have adapted reader circuits to identify tagged objects when handled with reader-integrated gloves [146,165] or put into coil-lined pockets [94], and small, single-chip readers are now becoming available by companies like EM Microelectronic and Innovision Research & Technology for very short-range, lower-power applications [6].

45.4 Power from the People

Potentially, there is a way around the limitations of batteries and the very restricted amount of energy available to siphon off common ambient environments: scavenge power from the user [175]. The human body is a tremendous storehouse of energy. Just one gram of fat stores nine dietary calories, which is equivalent to 9000 calories or

$$\left(\frac{9{,}000\text{ calories}}{1\text{g}_{\text{fat}}}\right)\left(\frac{4.19\text{J}}{\text{calorie}}\right) = 37{,}700\text{ J per gram of fat}$$

An average person of 68 kg (150 lbs) with 15% body fat stores energy approximately equivalent to

$$0.15(68\text{ kg})\left(\frac{1000\text{ g}}{1\text{ kg}}\right)\left(\frac{37{,}700\text{ J}}{1\text{ g}_{\text{fat}}}\right) = 384\text{ MJ}$$

Thus, if even a small fraction of this stored energy could be scavenged, a mobile device would have a large and renewable resource to draw upon. That said, the devil is in the details. Although researchers are working to develop in-vivo fuel cells [105] that oxidize blood glucose to provide a very small trickle of energy (of order 1 mW) to power low duty-cycle implants (e.g., a valve to aid incontinence, efficient biomedical sensors, or low-power transmitters for tracking animals) [29,157], tapping directly into the biological processes that turn fat into energy is beyond currently available technology.

On the other hand, power might be scavenged indirectly from the user's everyday actions or might be intentionally generated by the user. Indeed, products (e.g., flashlights, radios, and watches) have been on the market for years that operate in this mode and researchers are driven to leverage other devices into this niche, while finding alternative ways to tap excess energy from human activity [54,104]. Table 45.1 provides a perspective on the amount of power used by the human body during various activities. Everyday human activity consumes power at a rate of 81 to 1630 W, a factor of 20 in energy use. Bearing in mind that any technique that parasitically harvests background energy from unrelated human activity must be totally unobtrusive to be commonly adopted, perhaps a couple of watts might be scavenged somewhere for a mobile phone or on-body computer without putting an onerous load on the user.

TABLE 45.1 Human Energy Expenditures for Selected Activities

Activity	Kilocal/hr	Watts
Sleeping	70	81
Lying quietly	80	93
Sitting	100	116
Standing at ease	110	128
Conversation	110	128
Eating a meal	110	128
Strolling	140	163
Driving a car	140	163
Playing the violin or piano	140	163
Housekeeping	150	175
Carpentry	230	268
Hiking, 4 mph	350	407
Swimming	500	582
Mountain climbing	600	698
Long-distance run	900	1048
Sprinting	1400	1630

Source: Derived from D. Morton. *Human Locomotion and Body Form.* Williams & Wilkens, Baltimore, MD, 1952.

The following sections examine this possibility with respect to power recovery from body heat, breathing, blood pressure, typing, arm motion, pedaling, and walking. A summary of the potentially scavengable power and the total power from various body-centered actions is provided in Figure 45.2. Note, however, that energy harvested from the user may require considerable conditioning (e.g., storage, voltage/current, or impedance conversion) before it can be used for an application. Although we touch on a few important

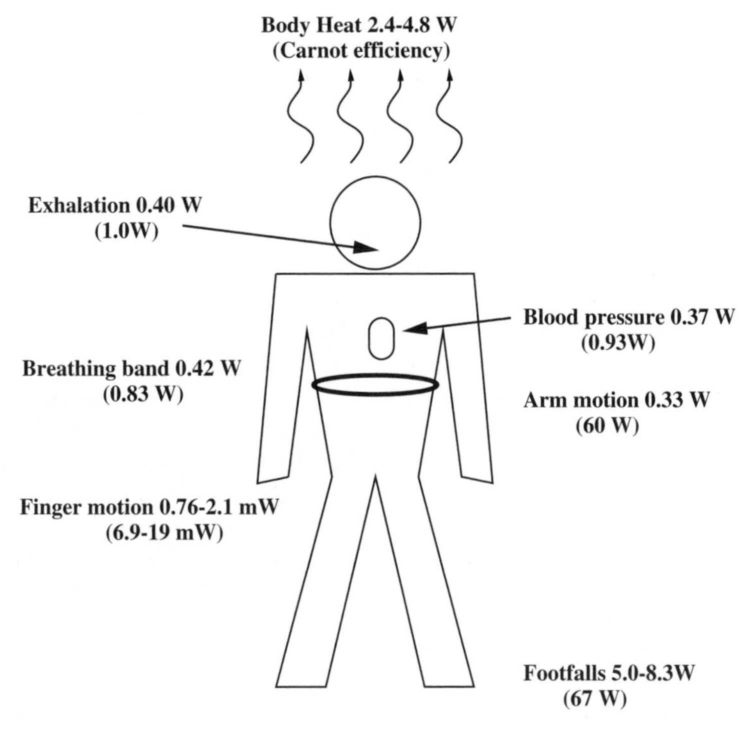

FIGURE 45.2 Possible power recovery from body-centered sources. Total power for each action is included in parentheses.

issues in conditioning power for piezoelectric generators, Chapter 7 of Edgar Callaway's *Wireless Sensor Networks* [44] provides an introduction to this topic specific to energy harvesting systems, and several other references on power conditioning are provided for the reader's convenience [46,140].

45.5 Hot Bodies: Power from Body Heat

Because the human body emits energy as heat, it follows naturally to try to harness this energy; however, Carnot efficiency puts an upper limit on how well this waste heat can be recovered. Assuming normal body temperature and a relatively low room temperature (20°C), the Carnot efficiency is

$$\frac{T_{body} - T_{ambient}}{T_{body}} = \frac{(310K - 293K)}{310K} = 5.5\%$$

In a warmer environment (27°C) the Carnot efficiency drops to

$$\frac{T_{body} - T_{ambient}}{T_{body}} = \frac{(310K - 300K)}{310K} = 3.2\%$$

This calculation provides an ideal value. Today's thermoelectric generators that might harness this energy do not approach Carnot efficiency in energy conversion. Although work on new materials [68,186] and new approaches to thermoelectrics [58,82] promise to somewhat improve conversion efficiencies, today's standard thermopiles are 0.2% to 0.8% efficient for temperature differences of 5 to 20°C [176], as expected for a wearable system in temperate environments. For the sake of discussion, the theoretical Carnot limit will be used in the analysis below, hence the numbers are optimistic.

Table 45.1 indicates that a total of 116 W of power is available while sitting. Using a Carnot engine to model the recoverable energy yields 3.7–6.4 W of power. In more extreme temperature differences, higher efficiencies may be achieved, but robbing the user of heat in adverse environmental temperatures is not practical.

Evaporative heat loss from humans accounts for 25% of their total heat dissipation (basal, non-sweating) even under the best of conditions. This "insensible perspiration" consists of water diffusing through the skin, sweat glands keeping the skin of the palms and soles pliable, and the expulsion of water-saturated air from the lungs [73]. Thus, the maximum power available without trying to reclaim heat expended by the latent heat of vaporization drops to 2.8–4.8 W.

The above efficiencies assume that all of the heat radiated by the body is captured and perfectly transformed into power. However, such a system would encapsulate the user in something similar to a wetsuit. The reduced temperature at the location of the heat exchanger would cause the body to restrict blood flow to that area [73]. When the skin surface encounters cold air, a rapid constriction of the blood vessels in the skin allows the skin temperature to approach the temperature of the interface so that heat exchange is reduced. This self-regulation causes the location of the heat pump to become the coolest part of the body, further diminishing the returns of the Carnot engine unless a wetsuit is employed as part of the design.

Although a full wetsuit or even a torso body suit is unsuitable for many applications, the neck offers a good location for a tight seal, access to major centers of blood flow, and easy removal by the user. The neck is approximately 1/15 of the surface area of the "core" region (those parts that the body tries to keep warm at all times). As a rough estimate, assuming even heat dissipation over the body, a maximum of 0.20 to 0.32 W could be recovered conveniently by such a neck brace. The head may also be a convenient heat source for some applications where protective hoods are already in place — the head is also a very convenient spot for coupling sensory input to the user. The surface area of the head is approximately three times that of the neck and could provide 0.60 to 0.96 W of power, given optimal conversion. Even so, the practicality, comfort, and efficacy of such a system are relatively limited.

Even given all the limitations mentioned previously, practical body-worn, thermally powered systems have been created. The Seiko Thermic® wristwatch uses 10 thermoelectric modules to generate sufficient μW to run its mechanical clock movement from the small thermal gradient provided by body heat over ambient temperature [100]. Although this commercial example is recent, the idea was articulated in 1978 in U.S. Patent No. 4106279 [126] with variations reflected in later patents [66,67]. Given the commercial production of such thermogenerative systems, one can imagine small on-body sensor networks working on the same principle. These systems can store power during periods of higher ΔT in order to continue to run during periods of warmer ambient temperatures. In addition, such storage may be useful for communicating sensor readings, perhaps from a medical device, in bursts to some central server. Indeed, such a product has been recently announced by Applied Digital Solutions as the "Thermo Life Generator," a half-square-centimeter thermoelectric device claimed to be capable of generating 10 μW or more at 3 V when in contact with the body [145].

45.6 Heavy Breathing: Power From Respiration

An average person of 68 kg has an approximate air intake rate of 30 liters per minute [137]; however, available breath pressure is only 2% above atmospheric pressure [147,192]. Studies indicate that the power consumed by pulmonary ventilation (breathing) is between 0.1 and 40 W [53]. Increasing the effort required for intake of breath may have adverse physiological effects [147] so only exhalation will be considered for generation of energy. Thus, the maximum available power is

$$W = p\Delta V =$$

$$0.02\left(\frac{1.013\times10^{5}\ \text{kg}}{\text{m-sec}^{2}}\right)\left(\frac{30\ \text{l}}{1\ \text{min}}\right)\left(\frac{1\ \text{min}}{60\ \text{sec}}\right)\left(\frac{1\ \text{m}^{3}}{1000\ \text{l}}\right) = 1.0\ \text{W}$$

During sleep, the breathing rate, and therefore the available power, may drop in half, while increased activity increases the breathing rate. Forcing an elevated breath pressure with an aircraft-style pressure mask can increase the available power by a factor of 2.5, but it causes significant stress on the user [73].

For some professionals, such as military aircraft pilots, astronauts, or handlers of hazardous materials, such masks are already in place. The efficiency of a turbine and generator combination is only about 40% [84], however, and any attempt to tap this energy source would provide additional load on the user. Thus, the benefit of the estimated 0.40 W of recoverable power has to be weighed against the other, more convenient methods discussed in the following sections.

Another way to generate power from breathing is to fasten a tight band around the chest of the user. From empirical measurements, there is a 2.5-cm change in chest circumference when breathing normally and up to a 5-cm change when breathing deeply. A large amount of force can be maintained over this interval. Assuming a respiration rate of 10 breaths per minute and an ambitious 100-N force applied over the maximal 0.05 m distance, the total power that can be generated is

$$(100\ \text{N})(0.05\ \text{m})\left(\frac{10\ \text{breaths}}{1\ \text{min}}\right)\left(\frac{1\ \text{min}}{60\ \text{sec}}\right) = 0.83\ \text{W}$$

A ratchet and flywheel or a stretchable dielectric elastomer generator (see Section 45.10 and Section 45.12.3) attached to an elastic band around the chest might be used to recover this energy; however, friction due to the small size of the parts may cause some energy loss. With careful design, a significant fraction of this power might be recovered, but the resulting 0.42 W is a relatively small amount of power for the inconvenience.

Although such a chest band may, at first, appear inappropriate, some popular breath and heart rate monitors sold as exercise equipment use similar chest bands for their sensors. Interestingly, the idea of using a chest band for recovering power from the user is quite old. Chapuis reports a similar mechanism for winding watches in the historical record from the 1600s [48].

Researchers have explored tapping the energy of breathing for powering implantable electronics; *in vivo* animal tests of a piezoelectric foil laminate that's bonded to a pair of ribs that stretch the foil during breathing have generated 17 μW in a dog. With improvements the researchers claim to be able to attain 1 mW [87].

45.7 And the Beat Goes on: Power from Blood Pressure

Although powering electronics with blood pressure may appear impractical, the numbers are actually quite surprising. Assuming an average blood pressure of 100 mm of Hg (normal desired blood pressure is 120/80 above atmospheric pressure), a resting heart rate of 60 beats per minute, and a heart stroke volume of 70 ml passing through the aorta per beat [31], then the power generated is

$$(100\ \text{mmHg})\left(\frac{1.013\times10^{5}\ \text{kg/m-sec}^{2}}{760\ \text{mmHg}}\right)\left(\frac{60\ \text{beats}}{1\ \text{min}}\right)\left(\frac{1\ \text{min}}{60\ \text{sec}}\right)\left(\frac{0.07\ \text{l}}{\text{beat}}\right)\left(\frac{1\ \text{m}^{3}}{1000\ \text{l}}\right)=0.93\ \text{W}$$

Although this energy rate can easily double when running, harnessing this power is difficult. Adding a turbine to the system would increase the load on the heart, perhaps dangerously so; however, even if 2% of this power is harnessed, low-power microprocessors and sensors could run. Thus, self-powering medical sensors and prostheses could be created. Ramsay and Clark [154] performed a design study on a variant of this idea using blood pressure to drive a piezoelectric generator (more details on piezoelectric materials are provided in Section 45.12.2). Their results indicate that a generator using a square centimeter of piezoelectric material should be capable of providing power on the level of μW continuously and mW intermittently.

45.8 Shaking It Up: Power from Inertial Microsystems

Capturing enough energy from vibrations to power sensors and telemetry has a long history in vehicles, where considerable mechnical excitation is available. In the patent literature, one can find techniques ranging from linear motor generators with bouncing spring-mounted magnet arrays for use in trucks and trains [181] to piezoelectric generators embedded in tires and wheels for monitoring air pressure and tire conditions [109,172,184]. Similarly, pure mechnical devices exist that scavenge energy from dynamics — for example, a self-powered hour meter that integrates time when it is exposed to vibration from sources such as operating machine tools [16].

In the world of mobility, pocket watches and wristwatches, in some senses the precursors to wearable computers, addressed the issue of power scavenging with the advent of the self-winding watch. These watches use the motion of the user's body during walking ("pedometer" watches) or the motion of the user's arm during everyday actions to wind their mechanisms. The first known self-winding pedometer watch was created circa 1770 by Abraham-Louis Perrelet, though there are indications that earlier watches may have been made in the 1600s [48]; however, widespread adoption of these systems did not occur until after the 1930s when watch cases could be hermetically sealed to protect the mechanism from dust.

Taking apart a modern self-winding wristwatch reveals a 2-gram "proof" mass mounted off-center on a spindle. As the user moves during the day, the mass rotates on the spindle and winds the mechanism. A simple variant would use the same off center mass design except that the mass would be a magnet. As the magnet spins past coils of wire mounted in the sides of the watch, it induces an electrical current that can be used to run low-power electronics.

An electrical version of this concept has proven successful in the form of the ETA autoquartz self-winding electric watch (see Figure 45.3) [74]. The proof mass winds a spring which, when enough mechanical energy is stored, drives a micro generator at its optimal rate of 15,000 rotations per minute

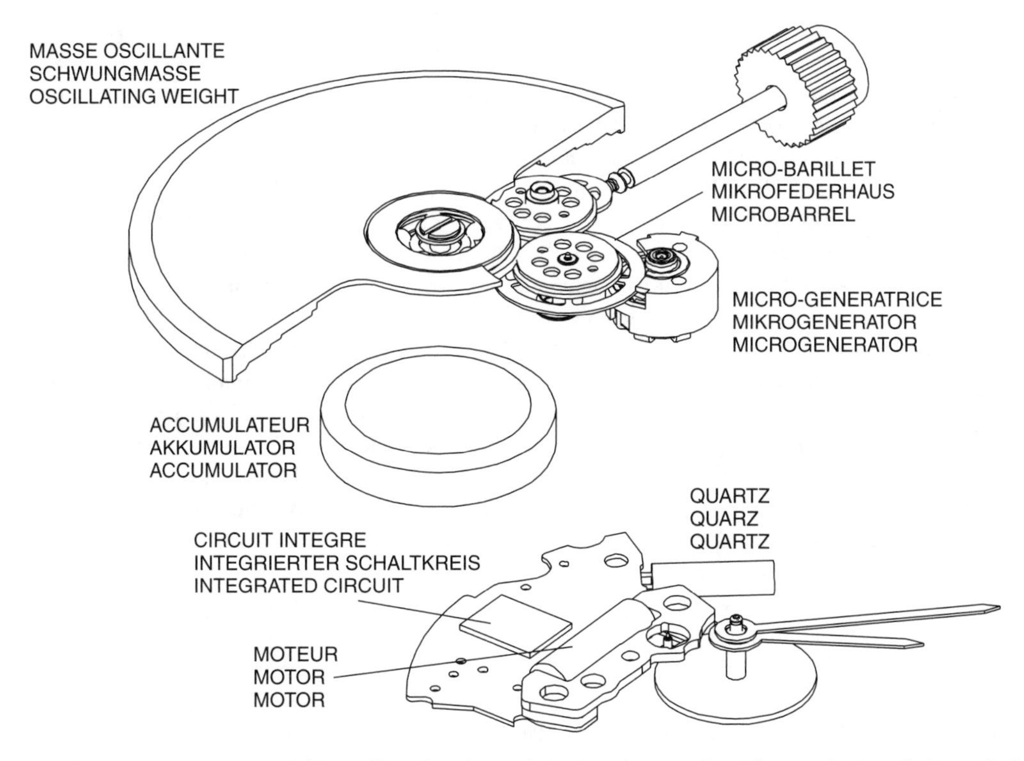

FIGURE 45.3 ETA Autoquartz Self-Winding Electric Watch. Image courtesy of B. Gilomen, the Swatch Group [74].

(RPM). The generator is pulsed for 50 ms at a time yielding 6 mA at greater than 16V. The generated power is stored in a capacitor for later use.

Another watch-sized electrical inertial generation system is the Seiko AGS [127]. This system provides a more direct connection from the proof mass to the generator and creates 5 μW on average when the watch is worn and 1 mW when the watch is forcibly shaken. The Seiko AGS system has been scaled up to provide a power source for sensors mounted on marine mammals. The idea is that some of the motion of the marine mammal will be recovered and converted to electrical power for the on-body system. This system, because it can be larger, generates 5 to 10 mW of power in an approximately 5 × 5 cm package. In a similar vein, light-emitting diode (LED) flashlights are on the market that power themselves through active shaking [185], which causes a proof-mass magnet to oscillate through a solenoidal coil, bouncing efficiently against rubber bumpers placed at each end. We have tested the generating mechanism of one of these products: weighing 150 grams, it was capable of generating 200 mW with a steady shake at its mechanical resonance (roughly 200 cycles/minute).

Inertial power generators do not have to be limited to systems that are handheld or mounted on the wrist. At the International Symposium on Wearable Computers (ISWC) in 2003, von Büren, Lukowicz, and Tröster at ETH in Zurich theorized a similar approach for passive excitation using a spring-mounted one gram mass [41]. Their experiments demonstrated that up to 200 μW of power could be parasitically generated due to the vibrations of the mass while the user walks.

Meanwhile, several compact vibration-based microgenerators are appearing in the sensors and actuators literature; Mitcheson and colleagues give an excellent review in [134]. Several of these are compact magnetic generators based on small, sub-cm structures with moving magnets or coils, similar in concept to a phonograph cartridge. Accordingly, Ching et al. have manufactured 1 cm^3 micro-spring Faraday generators that generate 830 μW at a constant vibration of 60 to 100 Hz at an amplitude of appoximately 200 μm [51], while James et al. [98] have built a similar, but somewhat larger device that generates over a milliwatt at these excitation frequencies. El-Hami et al. [62] have built a 240 mm^3 magnetic generator that surpasses a milliwatt at 320 Hz. Such devices have begun entering the commercial realm with products such as the Energy Harvester from Ferro Solutions in Cambridge, MA, which claims an energy density of 120 μW per cubic centimeter from 100 mG vibrations at 21 Hz (power output scales exponentially with acceleration — a 20 mG vibration at 21 Hz generates 0.4 mW from their 75-cm^3 generator) [5,156]. Taking another tack, Roundy and collaborators have developed a compact piezoelectric generator from a tip-loaded, cantelevered beam made from a pair of laminated PZT strips to form a bimorph that produces nearly 100 μW when shaken at resonance [145,159,161]. In an application closer to wearables, a small piezoelectric tip-loaded cantilever was proposed for powering bioelectric implants back in 1967; a prototype device was claimed to have produce 150 μW when mechanically coupled to 80-Hz heartbeats [108]. Piezoelectric-based, vibration-driven generators are now reaching the commercial market through products such as the Harvester package from Continuum Control [156].

Other projects proceed to the MEMS scale with variable-capacitance electrostatic generators [131,133,160]. Unlike magnetic and piezoelectric generators, electrostatic generators need to be "bootstrapped" with an external power source (e.g., a battery) that applies an initial voltage across the device's capacitance before it begins producing power. Because these devices tend to be quite small, they are designed to be driven at frequencies ranging from hundreds of Hz to several kHz and, depending on their excitation and power conditioning, typically yield on the order of 10 μW. Thus they are intended to support extremely low-power applications, perhaps sited on the same chip as the generator. Although these excitation frequencies cannot be commonly expected when mounted on the human body, some energy is produced in these regions from shocks or rapid motion (perhaps frequency-translated by introduced mechanical nonlinearity).

Some researchers have exploited larger electrostatic generators with highly resonant mechanical coupling to work at lower frequencies. Miyazaki [136] and collaborators have used a 45-Hz electrostatic generator with a Q of 30 to extract hundreds of nanowatts from micron-level wall vibrations, while Tashiro and collaborators [179] derived 58 μW from such a device resonating at 4.76 Hz with the aim of parasitically exploiting body motion to power biomedical implants. Görge, Kirstein, and Erbel, who describe initial experiments of using inertial electrical generators for powering pacemakers, have also pursued this goal. Their systems generated only 1 to 10% of the necessary power during normal office work, but they indicate that their systems have not yet been optimized in weight, orientation, and efficiency [77]. Because environmental vibration, especially from human motion, can occur over a range of generally lower frequencies, Mitcheson and collaborators have developed an electric generator using a nonresonant snap-action restoring forces on the proof mass instead of the communications spring of a standard tuned system. Laboratory prototypes have yielded 0.3 μJ of energy per mechanical cycle [135].

Vibration-driven microgenerators would allow for small, wireless, self-powered sensors that could be distributed on the body. By simply reporting the amount of vibration of the mass, these devices can act as a crude accelerometer. Applications could include systems that monitor the tremors of Parkinson's patients for better diagnosis and adjustment of medical dosage [189], gesture recognition systems, sports devices such as pedometers, and devices that monitor activities of daily living for older adults with Alzheimer's or with a high risk of stroke or heart disease. Signals from small cantilevered piezoelectric sensors can be large enough to selectively activate electronics in deep sleep when stimulated by typical human motion, enabling highly efficient, hybrid battery-powered systems that quiescently take essentially no current, but passively "wake up" when subject to an impulse above a certain level [64].

45.9 Power Typing

Keyboards will continue to be a major interface for computers into the next decade [173]. As such, typing may provide a useful source of energy. On a one-handed chording keyboard (HandyKey's Twiddler®), it is necessary to apply 130 grams of pressure to depress a key the required 1 mm for it to register. Thus,

$$\left(\frac{0.13\ \text{kg}}{\text{key stroke}}\right)\left(\frac{9.8\ \text{m}}{\text{sec}^2}\right)(0.001\ \text{m}) = 1.3\ \text{mJ per key stroke}$$

is necessary to type. Assuming a moderately skilled typist (40 wpm [122]), and taking into account multiple keystroke combinations, an average of

$$\left(\frac{1.3 \text{ mJ}}{\text{key stroke}}\right)\left(\frac{5.3 \text{ key strokes}}{\text{sec}}\right) = 6.9 \text{ mW}$$

of power is generated. A fast QWERTY typist (90 wpm) depresses 7.5 keys per second. A typical keyboard requires 40 to 50 grams of pressure to depress a key the 0.5 cm necessary to register a keystroke (measured on a DEC PC 433 DX LP). Thus, a QWERTY typist may generate

$$\left(\frac{0.05 \text{ kg}}{\text{key stroke}}\right)\left(\frac{9.8 \text{ m}}{\text{sec}^2}\right)(0.005 \text{ m})\left(\frac{7.5 \text{ key strokes}}{\text{sec}}\right) = 19 \text{ mW}$$

of power. Unfortunately, neither method provides enough continuous power to sustain a portable computer, especially because the user would not be continuously typing on the keyboard; however, there may be enough energy in each keystroke for each key to "announce" its character to a nearby receiver.

Self-powered buttons are not a new idea. Zenith televisions in the 1950s featured a self-powered remote control where a button, when pressed, would strike one of several tuned aluminum rods that resonated at ultrasonic frequencies [1]. This sound pulse was decoded at the TV, which changed channels appropriately. Paradiso and Feldmeier took this theme further by using a piezoelectric element with resonantly matched transformer and conditioning electronics that, when struck by a button, generates approximately 0.5 mJ at 3V per 15N push, enough power to run a digital encoder and a radio that can transmit up to 50 feet [141]. This innovation enables compact digital controllers (for example, a light switch) to be placed freely, without needing any wiring or batteries and their associated maintenance. A recent working prototype of this device is depicted in Figure 45.4. Another version of a self-powered piezoelectric radio button has recently been marketed in Germany by a company called EnOcean, an affiliate of Siemens [148,195], which uses a bistable piezoelectric cantilever that snaps when pressed and released, conditioned by a switching regulator. We have seen this device produce about 100 μJ per 8N push at 3.3V.

Another option, potentially feasible for keyboards, is to make a keyboard with permanent magnets in its base. Each key would then have an embedded coil that would generate a current when the key was pressed. This concept was presented by the authors in 1996 [175] and appears in U.S. Patent No. 5,911,529 [55].

One can imagine other on-body input devices communicating wirelessly using power scavenged from the user's actions. For example, a finger or wrist-mounted trackball could be "self-powered." Moving the

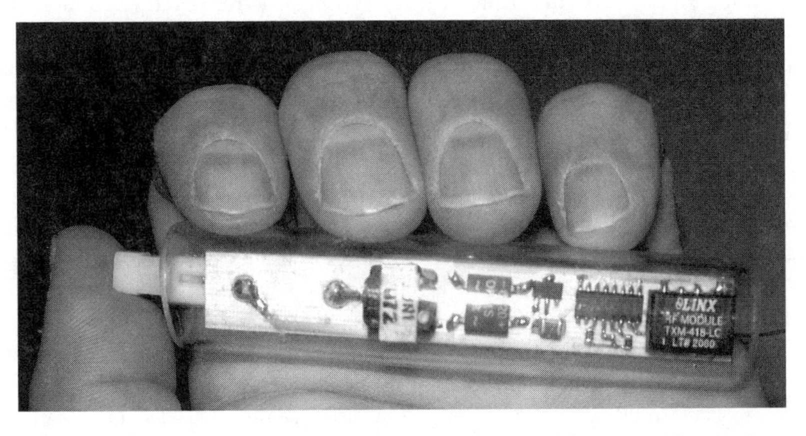

FIGURE 45.4 The MIT self-powered wireless button — power generated from piezoelectric element transmits a digital RF code after a single push.

trackball would turn the wheel encoders inside the device, both registering the movement and powering the device.

45.10 Hand Waving: Power from Arm Motion

Whereas finger motion might allow for powering buttons or keyboards, intentional arm motion might generate enough power for notebook computing. The comparison of the activities listed in Table 45.1 indicates that violin playing and housekeeping use up to 30 kcal/hr, or

$$\frac{30\text{ kcal}}{1\text{ hr}}\left(\frac{4.19\text{ J}}{1\text{ calorie}}\right)\left(\frac{1\text{ hr}}{3600\text{ sec}}\right)=35\text{ W}$$

more power than standing. Moving the upper limbs generates most of this power. Empirical studies done at the turn of the century demonstrate that for a particular 58.7-kg man, the lower arm plus hand masses 1.4 kg, the upper arm 1.8 kg, and the whole arm 3.2 kg [30]. The distance through which the center of mass of the lower arm moves for a full bicep curl is 0.335 m, while raising the arm fully over the head moves the center of mass of the whole arm 0.725 m. Empirically, bicep curls can be performed at a maximum rate of 2 curls/sec and lifting the arms above the head at 1.3 lifts/sec. Thus, the maximum power generated by bicep curls is

$$(1.8\text{ kg})\left(\frac{9.8\text{ m}}{\text{sec}^2}\right)(0.335\text{ m})\left(\frac{2\text{ curls}}{\text{sec}}\right)(2\text{ arms})=24\text{ W}$$

while the maximum power consumed by arm lifts is

$$(3.2\text{ kg})\left(\frac{9.8\text{ m}}{\text{sec}^2}\right)(0.725\text{ m})\left(\frac{1.3\text{ lifts}}{\text{sec}}\right)(2\text{ arms})=60\text{ W}$$

Obviously, housekeeping and violin playing do not involve as much strenuous activity as these experiments; however, these calculations do show that there is plenty of energy to be recovered from an active user. Consequently, the task at hand is to recover a useful amount of energy without burdening the user. A much more reasonable number, even for a user in an enthusiastic gestural conversation, is attained by dividing the bicep curl power by a factor of eight. Thus, the user might make one arm gesture every 2 seconds. This activity, then, generates a total of 3 W of power. By doubling the normal load on the user's arms and mounting a pulley system on the belt, 1.5 W might be recovered (assuming 50% efficiency from loss due to friction and the small parts involved), but the system would be extremely inconvenient.

Driving the generators via an arm-fitting exoskeleton may be a slightly better approach, although still very bulky and uncomfortable for common use. A somewhat less encumbering variant might involve mounted pulley systems in the elbows of a jacket. The take-up reel of the pulley system could be spring-loaded to counterbalance the weight of the user's arm. Thus, the system would generate power from the change in potential energy of the arm on the downstroke and not require additional energy by the user on the upstroke. The energy generation system, the CPU, and the interface devices could be incorporated into the jacket. Thus, the user would simply don his jacket to use his computer; however, any pulley or piston generation system would involve many inconvenient moving parts and the addition of significant mass to the user.

A more innovative solution would be to use electroactive materials at the joints, which would generate current when pushed or pulled via the movement of the user. Thus, no moving parts per se would be involved, and the jacket would not be significantly heavier than a normal jacket. One might naively think that piezoelectric polymers (e.g., PVDF) would be a candidate material. As outlined in the last section, piezoelectric foils need to be pulled along their most sensitive axis to generate charge, and as the maximum effective strain supported by such a foil is approximately 1 to 2%, this would lead to a rather stiff arm. Piezoelectrics bonded to stiffer structural members in bridges, buildings, or aircraft, however, are able to achieve sufficient strain to generate some energy — researchers at Sandia National Laboratories [4] and Microstrain Corporation [52,156], for example, have developed such strain-based harvesters for powering logging monitors and wireless sensor nodes mounted on large structures subject to flex and vibration. Another option would be to squeeze dielectric elastomers [20,144]: these are soft, rubbery, compliant materials, capable of supporting 50 to 100% area strain, which are sandwiched between the plates of a capacitor. As a charged dielectric elastomer is compressed and released, the voltage across the capacitor changes in proportion to the capacitance shift, producing power. As the power scales with the square of the voltage across the capacitor plates, several thousand volts (typically 1 to 6 kV) are applied to these devices. In response, a full compression/expansion cycle can produce well over a Joule of energy. Dialectric elastomers (sometimes called electroactive polymers) are discussed further in the section on extracting energy from heel strikes (Section 45.12.3), to which they are well suited.

A more practical solution of having the user deliberately impart energy that is stored in the device can be seen in consumer items. Wind-up magnetic generators housed with flashlights have been around since the beginning of the 20th century [57,121]; their descendant, the shake-driven flashlight, was introduced earlier, in Section 45.8. More recently, radios designed by South African inventor Trevor Baylis and sold by Freeplay and Radio Shack allow the user to wind them up and thereby store enough power for 30 to 60 minutes of operation [96,104]. In a typical windup radio, 60 turns (1 min of cranking) stores 500 joules of energy in a spring, which drives a magnetic generator that is 40% efficient [95], metering out enough power for up to an hour of play (magnetic generators with efficiencies better than 90% are available, but are much too expensive for such a mass-market product). Using a wind-up system, the mobile user can spend some directed effort at generating and storing power for a mobile computer or phone followed by a period of use of the device, either eliminating the battery (by storing energy in a spring) or charging it [21]. Indeed, windup cell phone chargers have become standard items — a collaboration between Motorola and Freeplay resulted in the "Freecharge," a cigarette-pack sized generator that, after cranking for 45 seconds, allows a 4- to 5-minute phone call [12,54], while Innovative Technologies produces the even smaller "SideWinder." Weighing under 80 grams and featuring a side-mounted crank, the SideWinder provides over 6 minutes of talk time after 2 minutes of cranking [15]. Japan-based Nissho Engineering has produced innovative hand-cranked generators for many years under the brand name "Aladdin" [14]. Nissho's "AladdinPower" is a handheld electromagnetic generator with a lever that one cranks by squeezing; it produces 1.6 W of power when the handle is squeezed at 90 times per minute, and was built for general applications that include charging cell phones or running flashlights. Nissho's Tug Power series of generators operate in a different mode: it is an 80-gram device in a similar cigarette-pack form factor, but here, one grabs a finger ring and repeatedly pulls a spring-return cord extending from the bottom of the unit. As energy is stored in a flywheel that drives the generator, and the arm motion involved in pulling is less strenuous than hand cranking. This device produces more power; the manufacturer rates it at 2.5 W. Saul Griffith of the MIT. Media Lab has designed another kind of strung-up power source. Hailing from Australia, Griffith took inspiration from the "bull roarer," an indigenous musical wind instrument attached to a length of rope that a player whips around their head. Saul's device, termed a "Bettery" [79], employs a small cusioned ball tethered to a hand-held generator. By revolving this 100- to 200-gram proof mass around on a .3 to .5 meter long string, the generator turns at a 1- to 2-Hz rotation rate, and 3 to 5 W are produced [80]. Most test subjects felt comfortable swinging the Bettery for up to 2 to 4 minutes. The user needs to be surrounded by a clear region of 1- to 2-meters, however, to avoid hitting objects and other people with the revolving proof mass. Taking a similar direction, the "ReGen" design study explores embedding a wireless MP3 player and magnetic generator into a yo-yo, claiming that a dozen vigorous tosses can provide up to an hour of continuous music play [23].

Although most laptops would require over an hour of cranking to achieve a reasonable charge, the concept of wind-up portable computers has been around for a while [101]. Carnegie-Mellon University

(CMU) demonstrated this concept by making a wind-up generator for their 1-W StrongArm-based wearable computer called the Metronaut [170].

45.11 Pedal Power

Manually cranked electronics were common in the 1940s, when shortwave radios were taken into the Australian outback. Soldiers and adventurers needed a way of communicating with the rest of the world without the support of an electrical grid. Accordingly, companies began making miniature bicycle pedal arrangements, similar to those sold in today's gadget magazines for under-the-office-desk exercising, to generate power for the user's two-way shortwave radio. Up to 60 W can be obtained in this manner. Because the legs tend to be stronger and more enduring than the arms, and because legs naturally project the force of the body's weight, generators pumped by leg motion are an ideal way to obtain more power through deliberate action. Today, the electronics in some fitness club exercise bicycles are powered by the user's actions, and it is not unusual to see personal computers powered by stationary bicycles in developing regions like rural India and Laos [18]. In fact, some Indian schools combine physical education with computer class; one half of the students bicycle to provide the power for the other half's computers!

Other foot-driven generators are not based on bicycles, but instead use a small, stationary pedal coupled to an embedded magnetic generator. A perfect example is the "Stepcharger" [14], also manufactured by Nissho Engineering, which can generate up to 6 W when the pedal is vigorously pumped. Concluding this section with a more whimsical example, an inventor named Henderson is reported to have developed hydraulically driven generators powered by bladders that are placed onto roadways to scavenge power when run over by passing cars, on farms to glean power when stepped on by passing cattle, and on sidewalks to harness energy from the footsteps of pedestrians [162].

45.12 March of Dynes

45.12.1 Power from Walking

An obvious extension from pedal systems is to design a power recovery system for walking. Using the legs is one of the most energy consuming activities the human body performs. In fact, a 68-kg man walking at 3.5 mph, or 2 steps per second, uses 280 kcal/hr or 324 W of power [137]. Comparing this with standing or a strolling rate implies that up to half this power is being used for moving the legs. While walking, the traveler puts up to 30% more force on the balls of his feet than that provided by his body weight (Figure 45.5). However, calculating the power that can be generated by simply using the fall of the heel through 5 cm (the approximate vertical distance that a heel travels in the human gait [30]) reveals that

$$(68\ \text{kg})\left(\frac{9.8\ \text{m}}{\text{sec}^2}\right)(0.05\ \text{m})\left(\frac{2\ \text{steps}}{\text{sec}}\right) = 67\ \text{W}$$

of power is available. Even though walking is not continuous like breathing, some of the power could be stored, providing a constant power supply even when the user is not walking. The following sections outline the feasibility of harnessing this power via piezoelectric, electrostatic, and rotary generators.

The 67-W result given previously is indeed a maximum number in that utilizing the full 5-cm stroke would result in considerable additional load on the user (while creating a significant trip hazard); it is like continuously ascending a circa 5° grade and would produce the feeling of "walking in sand." Indeed, the body continually optimizes the gait to minimize energy expenditure in various types of ambulation [76] — any significant energy load must be limited and carefully extracted to avoid ambulatory fatigue or even podiatric injury after significant use. A 1-cm stroke, roughly the amount of deflection that one can see in a padded running shoe [168], can be considered as an upper practical limit that can be tolerated

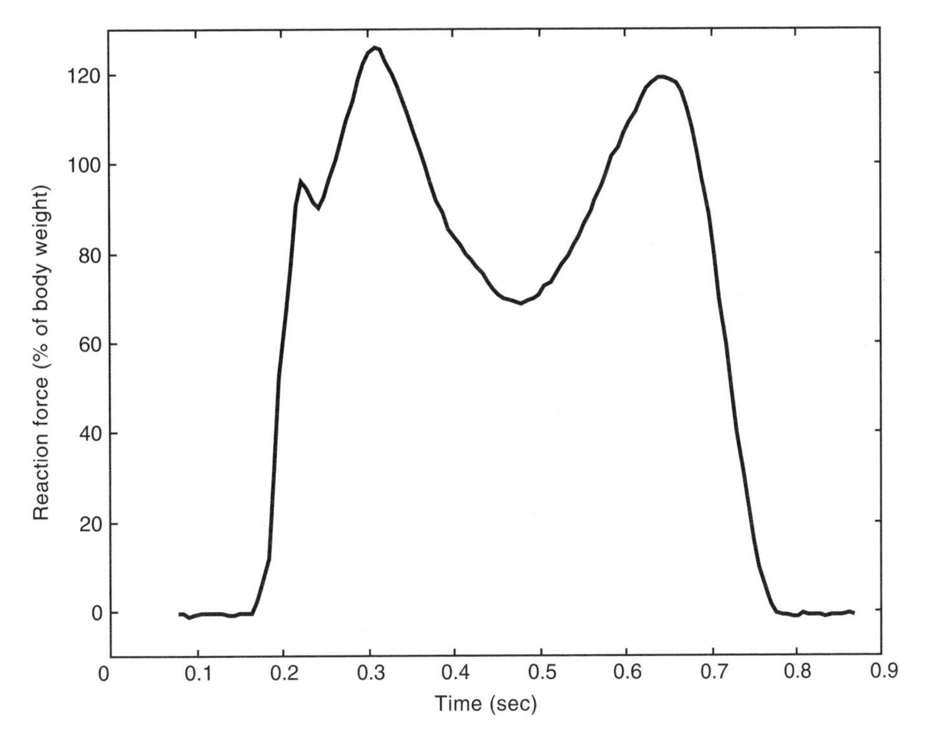

FIGURE 45.5 Empirical data taken for a healthy 52 kg woman, showing the time dependence of the vertical reaction force over a single footstep in a standard walk. The narrow spike at left is the "heel strike transient," caused by rapid flattening of the calcaneal fat pad. It is followed by the major force peaks occurring at heel down and just before toes off, which can exceed 120% of body weight. This curve is typical for human gait. (Data courtesy of the Biomotion Laboratory at the Massachusetts General Hospital.)

by general users [125], resulting in a maximum of 13 W available with full body weight. Even so, we continue to use a 5-cm stroke in the calculations that follow to determine a theoretically maximum power that might be obtained.

45.12.2 Piezoelectric Materials

Piezoelectric materials create electrical charge when mechanically stressed. Among the natural materials with this property are quartz, human skin, and human bone, though the latter two have very low coupling efficiencies [25]. Table 45.2 lists properties of common industrial piezoelectric materials: polyvinylidene fluoride (PVDF) and lead zirconate titanate (PZT). For convenience, references for data sheets and several advanced treatments of piezoelectricity are included at the end of this chapter [8,11,43,61,71,93,158].

The coupling constant in Table 45.2 is the efficiency with which a material converts mechanical energy to electrical. The subscripts on some of the constants indicate the direction or mode of the mechanical and electrical interactions (see Figure 45.6). The surfaces normal to axis 3 are typically metallized (e.g., with sputtered aluminum, silver epoxy, or, in cases of extreme strain, carbon) to facilitate electrical connection. "31 mode" indicates that strain is caused to axis 1 by electrical charge applied to axis 3. Conversely, strain on axis 1 will produce an electrical charge along axis 3, hence (since the device in Figure 45.6 forms a capacitor), pulling the piezo material along the 1 axis develops a voltage across the 3 axis. Bending elements, termed unimorphs (piezo material on one side of the element) and bimorphs (piezo material on both sides), exhibiting an expanding upper layer and a contracting bottom layer, are commonly exploited in industry. In practice, such bending elements have an effective coupling constant of 75% of the theoretical due to storage of mechanical energy in the mount and shim center layer.

TABLE 45.2 Piezoelectric Characteristics of PVDF and PZT

Property	Units	PVDF	PZT
Density	$\frac{g}{cm^3}$	1.78	7.6
Relative permitivity	$\frac{\varepsilon}{\varepsilon_0}$	12	1700
Elastic modulus	$\frac{10^{10}N}{m}$	0.3	4.9
Piezoelectric constant	$\frac{10^{-12}C}{N}$	$d31 = 20$ $d33 = 30$	$d31 = 180$ $d33 = 360$
Coupling constant	$\frac{CV}{Nm}$	0.11	$k31 = 0.35$ $k33 = 0.69$

Source: Adapted from [8,9,71].

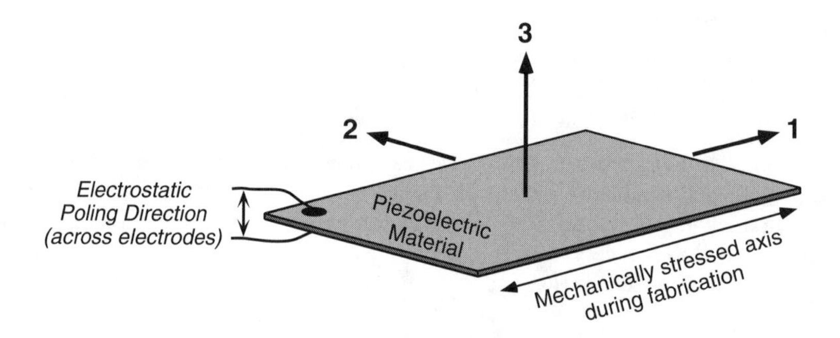

FIGURE 45.6 Definition of axes for piezoelectric materials. Note that the electrodes are mounted across the 3-axis.

The most efficient energy conversion, as indicated by the coupling constants in Table 45.2, comes from compressing PZT (d_{33}). Even so, the amount of effective power that could be transferred this way is minimal because compression follows the formula

$$\Delta H = \frac{FH}{AY}$$

where F is force, H is the unloaded height, A is the area over which the force is applied, and Y is the elastic modulus. The elastic modulus for PZT is 4.9×10^{10} N/m^2. Thus, it would take an incredible force to compress the material a small amount. Because energy is defined as force through distance, the effective energy generated through direct human-powered compression of PZT would be vanishingly small, even with perfect conversion.

On the other hand, bending a piece of piezoelectric material to take advantage of its 31 mode is much easier. Because it is hard and brittle, unprocessed PZT does not have much range of motion in this direction. Maximum surface strain for this material is 5×10^{-4}. Surface strain can be defined as

$$S = \frac{xt}{L_c^{\ 2}}$$

where x is the deflection, t is the thickness of the beam, and L_c is the cantilever length. Thus, the maximum deflection or bending for a beam (20 cm) of a piezoceramic thin sheet (0.002 cm) before failure is

$$x = \frac{(S)(L_c^{\ 2})}{t} = \frac{(5x10^{-4})(0.2m)^2}{0.00002m} = 1cm$$

Thus, unprocessed PZT is unsuitable for jacket design or applications where flexibility is necessary, although a piezoceramic-laden composite is available [89] that offers a limited amount of flexibility.

PVDF, on the other hand, is very flexible. In addition, it is easy to handle and shape, exhibits good stability over time, and does not depolarize when subjected to very high alternating fields. The drawback, however, is that PVDF's coupling constant is significantly lower than PZT's. In addition, shaping PVDF can reduce the effective coupling of mechanical and electrical energies due to edge effects. Furthermore, the material's efficiency degrades, depending on the operating climate and the number of plies used. In addition, because PVDF is compliant, one cannot as easily exploit power transfer through a mechanical resonance, although the frequencies involved in walking are far below the natural frequencies of any reasonably sized piece of PZT. Fortunately, from an industry representative [85], we know a 116-cm^2 40-ply triangular plate with a center metal shim deflected 5 cm by 68 kg three times every 5 seconds results in the generation of 1.5 W of power, as developed for an application to harness energy from ocean waves [45]. This result is a perfect starting point for the calculations in the next section. For the newcomer to piezoelectric film, several basic articles on power generation with PVDF by Brown [33–37] are available from the website of Measurement Specialties (MSI), the prime manufacturer of piezoelectric PVDF.

45.12.3 Piezoelectric Shoe Inserts and Elastomer Heels

Consider using PVDF shoe inserts for recovering some of the power in the process of walking. There are many advantages to this tactic. First, a 40-ply pile would be only (28 μm)(40) = 1.1-mm thick (without electrodes). In addition, the natural flexing of the shoe when walking provides the necessary deflection for generating power from the piezoelectric pile. PVDF is easy to cut into an appropriate shape and is very durable [8,71]. In fact, PVDF might be used as a direct replacement for normal shoe stiffeners. Thus, the inserts could be easily put into shoes without moving parts or seriously redesigning the shoe.

A small women's shoe has a footprint of approximately 116 cm^2. Knowing that the maximum effective force applied at the end of a user's walking step increases the apparent mass up to 30%, the user needs only 52 kg (115 lbs) of mass to deflect the PVDF plate a full 5 cm. Although the numbers given in the last section were for a 15.2 × 15.2 cm triangular 40 ply pile, these values can be used to approximate the amount of power an appropriately shaped piezoelectric insert could produce. Thus, scaling the previous 1.5 W at 0.6 deflections per second to 2 steps per second, these numbers indicate that

$$(1.5\text{W})\left(\frac{2 \text{ steps / sec}}{0.6 \text{ steps / sec}}\right) = 5 \text{ W}$$

of electrical power could be generated by a 52-kg user at a brisk walking pace.

Such a considerable predicted power harvest is encouraging, but it was derived by ad-hoc normalization of results from the ocean-wave generator element, which potentially misses some important details of the insole implementation. A more accurate relation for such an insole that accounts for the displacement current delivered to a load from the strain of a bending piezoelectric element (or "stave"), has been derived by Toda [183] and adapted by Kendall [103]:

$$P_{peak} = \frac{(e_{31} A S_1 \omega)^2 R}{1 + \omega^2 C^2 R^2}$$

where:

P_{peak} = Peak power produced in Watts

e_{31} = Piezo Stress Constant = $d_{31}Y$, where d_{31} is the Piezo Strain Constant, and Y is Young's Modulus

A = Total area of the piezoelectric material (the area of the stave scaled by the number or piezoelectric layers)

ω = Dominant angular frequency of excitation

R = Load resistance

C = Total capacitance of all piezoelectric layers (note that all layers are connected in parallel to minimize the stave impedance, and layers on opposite sides of the center need to be electrically reversed to account for the opposite strain, thus a change in polarity)

S_1 = Net strain along axis 1:

$$S_1 = \frac{h\Delta y}{\left(\frac{1}{2}L\right)^2}$$

where

h = Thickness of the stave

Δy = Maximum bending deflection of the stave

L = Length of stave along the bending direction (axis 1)

Plugging in the previous numbers, together with an expected 3 μF stave capacitance and a dominant frequency of roughly 5 Hz (as seen in waveforms taken with people walking on similar piezoelectric insoles [113]) and applying a matching load resistance that delivers the most power at this excitation frequency (note that this occurs when the denominator of the relation for P_{peak} is 2), we obtain P_{peak} = 2.5 W. Note that the average power is about a factor of 5 lower than this because a single-cycle, roughly sinusoidal, 5-Hz pulse is produced once per step per shoe in a standard 1-Hz per foot (2-Hz aggregate) gait, hence we expect a mean power of $\langle P \rangle$ = 250 mW for one such shoe.

This number, however, is based on several unwarranted assumptions. A 40-ply PVDF stave risks suffering differential slippage between layers, thus lowering the amount of actual strain. Also, the 5 cm of deflection (Δy) is not realistic because the bulk of the bending in a shoe sole occurs in a limited area under the metatarsals, thus the strain is not distributed evenly along the "1" axis, as assumed in the relations presented previously.

Accordingly, a MIT Media Lab team led by Paradiso developed a somewhat more modest piezoelectric insole. The stave was a 10-cm long, truncated diamond-tapered (to maximize the strain distribution), 16-layer bimorph, with eight 28 μm PVDF sheets laminated on either side of a 1 mm neutral plastic insert, that fit into a men's U.S. size 11 1/2 shoe (see Figure 45.7). The actual device was seen to produce peak powers into a matched resistive load (empirically found to be 250 $K\Omega$) of roughly 15 mW at heel up and slightly less at toe off; over the course of a 1 step/second per foot standard walk, the average power was 1.3 mW [113]. If we plug the parameters that define this stave into the previous relation, but assume a maximum bending deflection of 7 mm (appropriate when we consider the limited area over which strain is applied), we obtain a peak power prediction of P_{peak} = 16 mW with the 250 $K\Omega$ load used in the actual tests, which matches the experimental results. Note that the preceding equation predicts an optimum match at 100 $K\Omega$, however, with 50% more power produced. In the experiment, the power was seen to decrease with loads under 200 $K\Omega$, indicating the relevance of other features (e.g., excitation dynamics, leakage, and strain distribution effects) that are not properly captured in this model or our assumptions.

This team produced another piezoelectric insert that tapped into heel strike dynamics. This generator was a unimorph, with a flexible piezoceramic composite laminated on a curved piece of spring steel (manufactured as the Thunder™ by Face International) [89] that pressed flat when the heel came down,

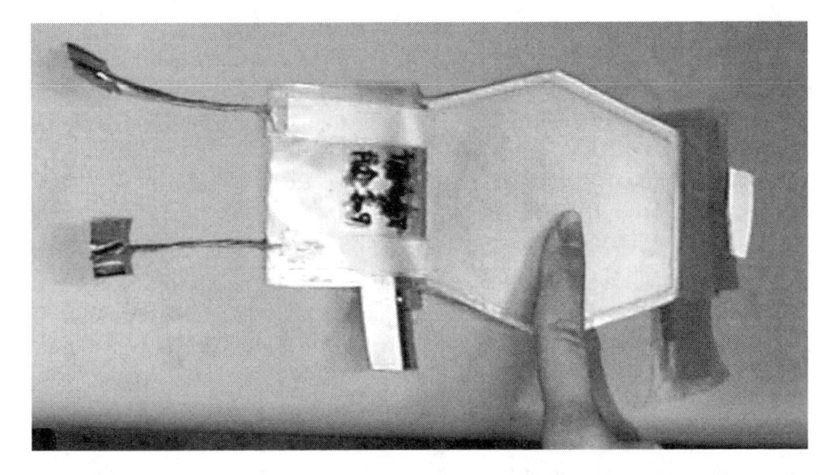

FIGURE 45.7 Prototype PVDF bimorph generator for a shoe insole. Power is generated through the mechnical bending of the sole.

accordingly straining the piezoelectric material and producing considerably more peak power (60 mW). Due to the brief, impulsive nature of the impact, however, only 1.8 mW of average power was obtained [113]. Shenck and Paradiso [167] subsequently made a clamshell out of two paralleled Thunder™ unimorphs placed back-to-back. Again sited under the heel (Figure 45.8), this generator was excited by 5 mm of displacement and produced similar peak power, but since its voltage pulses were wider, the

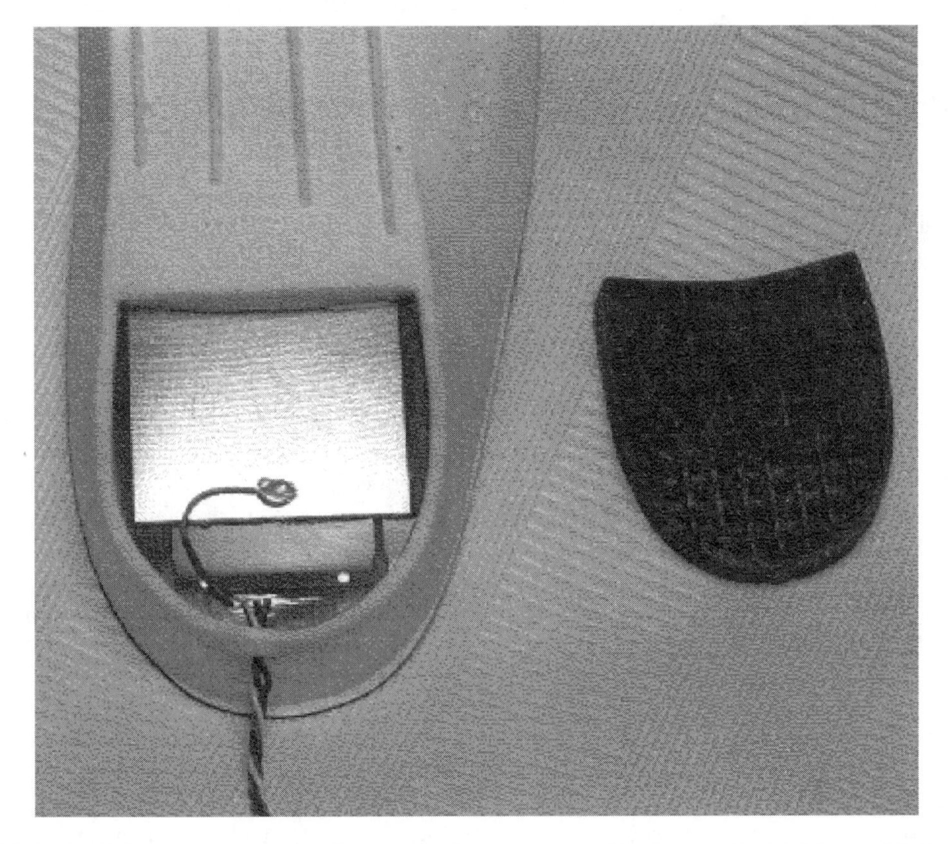

FIGURE 45.8 PZT generator embedded in an insole. Power is generated by the user's heel striking (and flattening) the PZT clamshell.

average power increased to 8.4 mW. Due to the indirect strain applied from bending the somewhat stiff neutral layer on these laminates, these systems had very limited mechanical-to-electrical conversion efficiency, of order 1% or less. On the other hand, the generators were worn inside standard jogging sneakers under the insoles, involved essentially no modification to the shoes, and were unnoticeable to the users.

The results given previously were for power delivered to a resistor empirically matched to the magnitude of the piezoelectric generator's capacitive impedance for the dynamics of a standard gait. Because piezoelectrics are high-impedance devices, the voltages tend to be very large (e.g., up to hundreds of volts) and currents very small (e.g., hundreds of μA). To power useful electronics, these values must be efficiently transformed to 3 to 5 V with currents of milliamperes or more (e.g., the impedance must be lowered). The original study of Paradiso and collaborators [113] used a simple system that full-wave-rectified the generator's signal, then directly applied it directly to a tank capacitor (several orders of magnitude larger than the piezoelectric's capacitance) to accumulate charge. When enough voltage (at least 12 V) appeared across this capacitor, a 5-V series regulator was activated, powering a load until the tank capacitor's voltage dropped below 5 V. Although this did not make an efficient match to the piezoelectric source, it produced enough power after four to five steps to operate a digital encoder and short-range RF transmitter. Each shoe would then broadcast several cycles of a 12-bit ID code to the vicinity, enabling the wearer to be wirelessly tracked as they moved about; an application originally served by battery-powered IR-transmitting badges at the dawn of ubiquitous computing in the early 1990s [187].

Other power conversion techniques can be significantly more efficient. Because stored power is linearly proportional to capacitance, but proportional to the square of voltage, there is a large gain in not loading the piezoelectric source until it attains its maximum potential. One approach [171] involves switching the tank capacitor into the piezoelectric source only at its peak voltage, avoiding such inefficiency from continuous capacitive loading. Another approach involves introducing a series inductor to match the piezoelectric's capacitance, producing a LC resonance at the frequency of excitation. For the Hz-level frequencies produced by walking, however, the required inductor would be impractically massive. Exploitation of a synchronous technique [42], which switches the inductor across the piezoelectric element at the extremes of its voltage swing, can gate a higher-frequency LC resonance to be synchronized to a much lower-frequency stimulation, such as from walking, and make the power conditioning electronics considerably more efficient.

Switching regulators can provide an efficient coupling to the piezoelectric source, and are much more efficient than linear regulators for large potential drops. One must optimize their design, however, for the very low currents and high voltages involved. A forward-switching converter designed by Shenck [166,167] to condition the power from the piezoelectric clamshell mounted in the heel of a Navy boot achieved a conversion efficiency of 17.6%, better than twice the efficiency of the bucket capacitor and linear regulator originally used in the original Media Lab study [113]. Researchers are further evolving this approach by exploring adaptive switching regulators for piezoelectric power harvesting that dynamically adjust the switcher's duty cycle to maximize the output current [140].

Other researchers and inventors have embedded rigid piezoelectrics in shoes for power scavenging from the heel strike. As mentioned earlier, materials like PZT have much higher piezoelectric coefficients and can be driven efficiently at resonance. They are brittle, however, and because the resonant frequencies of manageable pieces are quite high (e.g., many KHz), the low frequencies of walking must be translated up to this level by a nonlinear mechanism (e.g., modulated hydraulics or mechanical impacts). In 2000, Trevor Baylis and collaborators at the Electric Shoe Company in the UK claimed to have generated 100 to 150 mW of power from heel inserts embedded with a piezoelectric crystal; Baylis demostrated the system by using it to partially charge a cellular phone battery after a 5-day trek through the Namibian Desert [59]. Better-documented shoe generators that drive PZT elements via active and passive hydraulics are described in more detail in Section 45.12.5.

As part of a Defense Advanced Research Projects Agency (DARPA) initiative on energy harvesting [139], Pelrine, Kornbluh, and collaborators at SRI International have developed electrostatic generators based around materials called electroactive polymers or dielectric elastomers [20,144], which were intro-

FIGURE 45.9 An electrostatic generator based on compression of a charged dielectric elastomer during heel strike. Prototype implementation in a boot (left) and close up of the generator (right) showing bellows on bottom and retaining frame on top. Photos courtesy SRI, International.

duced previously in Section 45.10. Dielectric elastomers, made from components such as silicone rubber or soft acrylics, are extremely compliant — a displacement of 2 to 6 mm can easily drive these materials to 50 to 100% area strain, depending on the generator's configuration — thus, they are ideal substitutes for the rubbery heel of a running shoe [168], for example. They can also be highly efficient, with a practical device achieving energy densities of 0.2 J/g and calculations indicating a possibility of approaching 1.5 J/g.

The SRI team has built an elastomer generator into the heel of a boot, as presented in Figure 45.9. The generator's structure can be gleaned from the right photo — an elastomer membrane is mounted between a bellows filled with a fluid or gel and a rigid frame riddled with holes (the wires at either side connect to the electrodes on each face of the elastomer). As indicated in the left photo, the generator is mounted in a hole cut out of the center of a heel made from compliant foam (the foam only supports the prototype generator, which could technically make up the entire heel). Accordingly, when the heel presses down, the bellows compress, applying pressure to the elastomer membrane, which balloons into the holes in the frame, producing strain. Thus, when voltage is applied across the electrodes, it produces power. They have achieved an energy output of 0.8 J per step with this boot [112] with a heel compression of only 3 mm (limited by Army footwear specifications), yielding 800 mW of power per shoe at a 2-steps/sec pace. Benchtop testing has indicated that the material will last for at least 100,000 cycles, but they believe that improved packaging and design can increase the lifetime to beyond 1 million cycles — enough to meet the required lifetime of commercial footwear. More compression (up to 5 to 9 mm) is feasible in a commercial shoe, therefore, they anticipate being able to extract 1 W of power, allowing for a 50% voltage conversion (from several kV applied across the elastomer to 3–5 V that can power standard electronics) and storage efficiency [111].

Finally, to round out this discussion, we consider another type of footwear. Instead of extracting the power generated by a piezoelectric element embedded in or bonded to a structure, it can be applied via simple filtering and conditioning electronics to another piezoelectric element in the same structure [28], or just dissipated into a passive load, in order to damp vibrations and artificially "stiffen" the structure [2]. The first commercial application of such "smart structure" research to hit the mass market has been the K2 ski [119], designed by the MIT spinoff company known as Active Control eXperts (ACX). The K2 ski uses a piece of piezoceramic inserted between the skiboot attachment and the ski, coupled to passive electronics that damp vibrations. Sufficient power is generated to flash an LED when the ski flexes, yielding a visual indication of the device's operation [118]. (Note that, contrary to frequent assumption, the famous flashing sneakers made by companies such LA Gear drive the LEDs in their soles by an embedded battery connected to an inertially triggered tamper switch [190], not via parasitically extracted power.)

45.12.4 Rotary Generator Conversion

Using a cam and piston or ratchet and flywheel mechanism, the motion of the heel might be converted to electrical energy through more traditional rotary generators. The efficiency for industrial electrical generators can be very good; however, the added mechanical friction of the stroke-to-rotary converter reduces this efficiency. A normal car engine, which contains all of these mechanisms and suffers from inefficient fuel combustion, attains 25% efficiency. Thus, for the purposes of this section, 50% conversion efficiency will be assumed for this method, which suggests that, conservatively, 17 to 34 W might be recovered from a "mechanical" generator.

How can this energy be recovered without creating a disagreeable load on the user? A possibility is to improve the energy return efficiency of the shoe and tap some of this recovered energy to generate power. Specifically, a spring system, mounted in the heel, would be compressed as a matter of course in the human gait. The energy stored in this compressed spring can then be returned later in the gait to the user. Normally this energy is lost to friction, noise, vibration, and the inelasticity of the runner's muscles and tendons (humans, unlike kangaroos, become less efficient the faster they run [137]). Spring systems have approximately 95% energy return efficiency, while typical running shoes range from 40 to 60% efficiency [10,90,168]. Indeed, shoe soles with embedded heel springs have been developed to augment human gait capacity [91]. Volumetric oxygen studies have demonstrated a 2 to 3% improvement in running economy using such spring systems over typical running shoes [90]. Similarly suggestive are the "tuned" running track experiments of McMahon [130]. The stiffness of the surface of the indoor track was adjusted to decrease foot contact time and increase step length. The result was a 2 to 3% decrease in running times and seven new world records in the first two seasons of the track. Additionally, a reduction in injuries and increase of comfort was observed. Thus, if a similar spring mechanism could be designed for the gait of normal walking, and a ratchet and flywheel system is coupled to the upstroke of the spring, it may be possible to generate energy, while still giving the user an improved sense of comfort. In fact, active control of the loading of the generation system may be used to adapt energy recovery based on the type of gait at any given time.

Although constant-force springs are available and used in products such as clocks, the simplest mechanical springs do not provide constant force over the fall of the heel but instead a linear increase, hence, only about half of the calculated energy (for the ideal spring) would be stored on the downstep. An open question is what fraction of the spring's return energy can be sapped on the upstep, while still providing the user with the sense of an improved "spring in the step" gait. Initial mock–ups have not addressed this issue directly, but a modern running shoe returns approximately 50% of the 10J it receives during each compression cycle [10,168] (such "air cushion" designs were considered a revolutionary step forward over the hard leather standard several decades ago). Given a similar energy return over the longer distance of the spring system, the energy storage of the spring, and the conversion efficiency of the generator, 12.5% of the initial 67 W is harnessed for 8.4 W of available power.

The idea of embedding a spring and rotary generator into the heel of a boot or shoe has not escaped the attention of various inventors. Patents of this ilk date back to the 1920s and seem to reappear periodically in different incarnations [22,49,114]. Paradiso and his team evaluated these ideas by building a prototype shoe attachment with a simple spring, flywheel, and generator system that produced peak powers near 1 W (average power roughly 250 mW), while exploiting a 3-cm deflection in the fall of the heel during a normal walk (Figure 45.10). This provided enough power to play loud music from a radio speaker as the user walked about [113]. The mechanical system proved quite obtrusive, however, which brings into doubt how much deflection might be utilized without annoying the user. Hayashida created an improved model with the generator integrated into the sole (Figure 45.11), yet mechanical wear

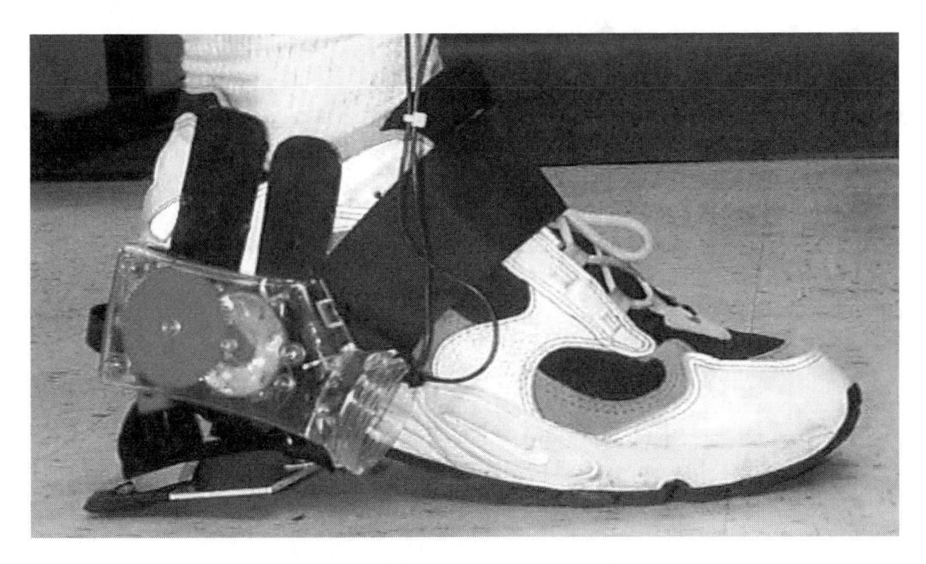

FIGURE 45.10 Rotary generator fit to a shoe for proof-of-concept studies at MIT.

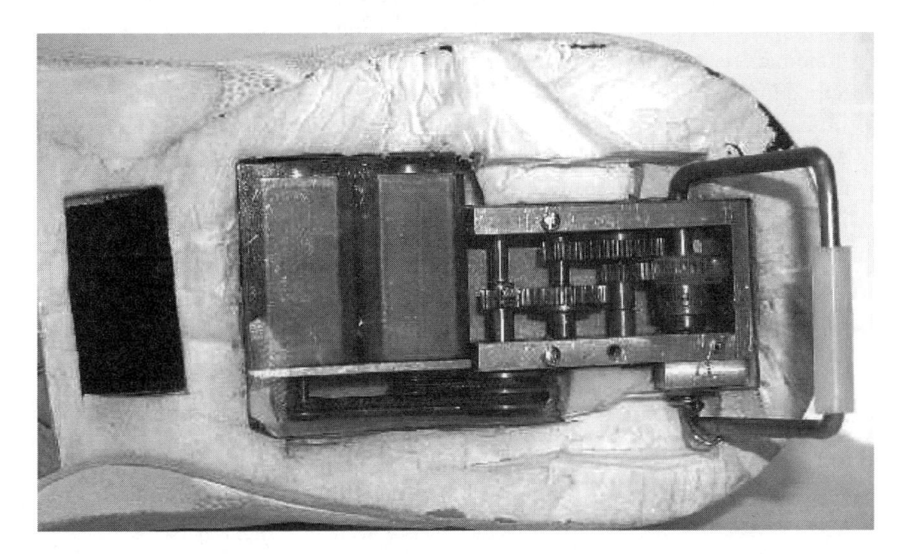

FIGURE 45.11 Improved rotary shoe generator, with gearbox and two magnetic generators built into the sole of a sneaker (bottom view).

remained an issue [88]. Because rotary generators need to spin rapidly to achieve efficiency, these systems all involve significant gear ratios, which introduce considerable mechanical complexity and fairly high torque, leading to a high probability of breakage. Indeed, a heel-mounted dynamo was developed by Jim Gilbert from the University of Hull for Trevor Baylis' walk through the Namibian Desert; although the piezoelectric insole appeared to last for several days, the mechanical shoe's crankshaft broke after only a few hours of use [59]. A linear motor may provide a simpler mechanical interface to footfalls, and thus has been proposed for adaptation to shoe generators [110].

U.S. Patent No. 4,845,338 gives an interesting variant of the mechanical generator idea [115]. The patent describes an "inflatable boot liner with electrical generator and heater." Footfalls not only generate electricity through a flywheel system to power an electrical resistance heater but also pump air through the boot to distribute the heat. This idea of using footfalls to create fluidic pressure leads to yet another approach to generating power from walking, which we address in the next section.

45.12.5 Hydraulic and Pneumatic Systems

In 1971, McLeish and Marsh tested a hydraulic pump system in the heel of the user's shoe for powering the user's bionic arm [128]. This system had a relatively small, 0.375-inch throw, which the user reported did not hinder his normal gait; however, this system recovered, on average, 5 W of power while the user was walking. Although quite impressive in power recovery, one can only imagine that the hydraulic line running up the pants leg would chafe the wearer. Surprisingly, Marsden and Montgomery also reported a separate but similar system in the same volume [125]. Note that such hydraulic systems can also drive a rotary magnetic generator via a turbine [116] or impeller and also provide a means of moving the bulky generator mechanism away from the highly constrained and hazard-prone neighborhood of the shoe sole [182] without cumbersome mechanical linkages.

Antaki and collaborators presented a shoe-mounted piezoelectric generator in 1995 that was developed for the purpose of powering artificial organs [17]. Their device, which looked like a large platform shoe, incorporated two cylindrical tubes in the insole, each of which housed a PZT stack stimulated by a passive hydraulic pulser-amplifier that converted low-frequency footfall energy into an intense series of high-frequency impulses that drove the PZT at its mechanical resonance. The hydraulic reservoir was differentially compressed during heel strike and toe off, thus power was extracted across the entire gait. Although the prototype was somewhat bulky and heavy, the entire generator was embedded in the shoe, and average powers of 250 to 700 mW were extracted from walking (depending on the type of gait and weight of the user) and over 2 W could be gleaned from a simulated jog.

A more ergonomic version of the concept, also developed under DARPA's Energy Harvesting Program [139], has been explored in Nesbit Hagood's lab at MIT [83,193]. The heel strike compresses a hydraulic bladder by 8 mm. This pressure is then routed through an active valve that chops the fluid flow in order to hammer a PZT stack at its resonant frequency of 20 kHz. As tested in components, the system is 40% efficient and produces 3 W of power with three one cubic centimeter PZT units per shoe (these generator chambers are not anticipated to be mounted in the insole, keeping a compact shoe with attached power nodule).

45.12.6 Getting Off Your Feet

Although these systems are still laboratory protoypes, walking seems a fruitful area to exploit for power generation in the future. A significant concern about the overall market penetration of shoe-generator systems, however, is the potential inconvenience of transporting power from the shoes to devices at other parts of the body. Although some researchers have explored sending high frequency AC current through the body to power wearable electronics [138,191], safety considerations and coupling limitations can severely restrict the available power. A more practical possibility, as exploited by Baylis' shoes, is to charge a battery mounted on the shoe itself, which is then moved into the device once charging is complete [59]. This competes with the utility of charging your batteries at home, however, which is certainly simpler unless you are far off the power grid (e.g., backpackers or military deployments). Another possibility is to exploit a network of wires integrated into clothing — certainly an area of interest for researchers in wearable computing [94,124,150], but not yet on the fashion racks. A third strategy is to just use the power in a local application sited at the foot. Although inventors seem to love generator-powered boot heaters [49,114,115], and several *in situ* footwear applications have been demonstrated with lower-power systems (e.g., the self-powered wireless tracking system [113] and the damped ski [119]), new developments, such as the SRI elastomer generators [144], hint at an ergonomic package that produces enough power to support a considerable amount of computation. One can perhaps envision central, wearable "personal servers" [188] embedded into the footwear, where they glean all the power that they need from their host's ambulation. These servers need no hardwired connection to other devices — they would communicate wirelessly with peripherals scattered around the body (powered by locally scavenged energy where appropriate) and the external environment.

45.13 Conclusion

In the design of mobile electronics, power is one of the most difficult restrictions to overcome, and current trends indicate this will continue to be an issue in the future. Designers must weigh wireless connectivity, CPU speed, and other functionality versus battery life in the creation of any mobile device. Power generation from the user may alleviate such design restrictions and may enable new products such as batteryless on-body sensors. Power may be recovered passively from body heat, arm motion, typing, and walking or actively through user actions, such as winding or pedaling. In cases where the devices are not actively driven, only limited power can generally be scavenged (with the possible exception of tapping into heel strike energy) without inconveniencing or annoying the user. That said, as detailed elsewhere in this volume, clever power management techniques combined with new fabrication and device technologies are steadily decreasing the energy needed for electronics to perform useful functions, providing an increasingly relevant niche for power harvesting in mobile systems.

References

[1] R. Adler, P. Desmares, and J. Spracklen. Ultrasonic remote control for home receivers. *IEEE Transactions on Consumer Electronics*, CE-28(1):123–128, February 1982.

[2] J.B. Aldrich, N.W. Hagood, A. von Flotow, and D.W. Vos. Design of passive piezoelectric damping for space structures. *Proceedings of the SPIE - Smart Structures and Materials 1993*, 1917(2):692–705, 1993.

[3] A. Ananthaswamy. Give it some gas. *New Scientist*, 182(2452):26–29, June 19, 2004.

[4] Annonymous. Sandia designs vibe-powered sensor that transforms shakes into electricity. Technical report, Sandia National Laboratory, http://www.sandia.gov/media/NewsRel/NR2002/vibepoweredsensor.htm, 2002.

[5] Annonymous. Energy harvester converts low-level vibrations into usable electricity. Technical report, Ferro Solutions, Inc., http://www.ferrosi.com/files/FS product sheet wint04.pdf, 2004.

[6] Annonymous. Near Field Communication - Interface and Protocol (NFCIP-1). Technical Report ISO/IEC 18092 and ECMA-340, International Organization for Standardization, Geneva, Switzerland, 2004.

[7] Anonymous. Miniature radio can be hidden in tooth. *Dental Survey*, 51:45, September 1975.

[8] Anonymous. Piezo film sensors technical manual. Technical Report 65751, Measurement Specialties (MSI), Inc. (formerly AMP Inc.), Valley Forge, PA, 1995.

[9] Anonymous. *Piezoelectric Motor/Actuator Kit, Introduction to Piezoelectricity*. Piezo System, Inc., Cambridge, MA, 1995. See: http://www.piezo.com.

[10] Anonymous. Running training section. *Endurance Training Journal*, 1995.

[11] Anonymous. *Piezoelectric Ceramics: Principles and Applications*. American Piezo Ceramics (APC International), Mackeyville, PA, 2001.

[12] Anonymous. Turn the handle and talk. *The Economist*, pages 10, Technology Quarterly Section, December 8, 2001.

[13] Anonymous. Power calculator for flash memory. *http://www.micron.com*, 2003.

[14] Anonymous. Nissho power products. *http://www.nseg.co.jp/english/products/line up.htm*, 2004.

[15] Anonymous. Sidewinder specifications. *http://www.windupradio.com/sidewinder/index.htm*, 2004.

[16] Anonymous. Vibration-powered hour meters. *McMaster-Carr Catalog*, 110:1423, 2004.

[17] J.F. Antaki, F.E. Bertocci, E.C. Green, A. Nadeem, T. Rintoul, R.L. Kormos, and B.P. Griffith. A gait powered autologous battery charging system for artificial organs. *ASAIO Journal: Proc. 1995 American Society of Artificial Internal Organs Conf.*, 41(3):M588–M595, July-September 1995.

[18] A. Applewhite. IT takes a village. *IEEE Spectrum*, 40(9):40–45, September 2003.

[19] J.U. Martinez Arazia. Wireless transmission of power for sensors in context aware spaces. Master's thesis, MIT Media Laboratory, June 2002.

[20] S. Ashley. Artificial muscles. *Scientific American*, 289(4):52–59, October 2003.

[21] C.J. Bader and R.P. Branco. 3,657,113: System for the generation of electrical power having a spring powered prime mover responsive to output voltage. *US Patent*, November 24, 1972.

[22] J. Barbieri. 1,506,282: Electric shoe. *US Patent*, August 26, 1924.

[23] T. Bartczak and C. Aimone. Around-The-World Music. *Popular Science*, page 28, January 2004.

[24] O. Bartels. 6,378,360: Apparatus for wire-free transmission from moving parts. *US Patent*, April 30 2002.

[25] C.A.L. Bassett. Biologic significance of piezoelectricity. *Calcified Tissue Research*, 1:252–272, 1968.

[26] J.P. Benner and L. Kazmerski. Photovoltaics gaining greater visibility. *IEEE Spectrum*, 26(9):34–42, September 1999.

[27] B. Berland. Photovoltaic technologies beyond the horizon: Optical rectenna solar cell - Final report, 1 August 2001 – 30 September 2002. Technical Report NREL/SR-520-33263, National Renewable Energy Laboratory (NREL), Golden, Colorado, February 2003.

[28] A.S. Bicos. 5,315,203: Apparatus for passive damping of a structure. *US Patent*, May 24, 1994.

[29] G. Binyamin, T. Chen, Y.-C. Zhang, and A. Heller. Design and stability of implanted biofuel cell anodes. In R. Nowak, editor, *Proc. of the DARPA Energy Harvesting Program Review*, Washington, D.C., April 13-14, 2000. DARPA.

[30] W. Braune and O. Fischer. *The Human Gait*. Springer-Verlag, Berlin, 1895–1904; reprinted 1987.

[31] E. Braunwald, editor. *Heart Disease: A Textbook of Cardiovascular Medicine*. W. B. Saunders Company, Philadelphia, 1980.

[32] H. Bray. Your next battery: a fuel cell? *The Boston Globe*, pages F1,F5, November 25, 2003.

[33] R.H. Brown. The LED report. Technical report, MSI, Piezo Sensors Division, Valley Forge, PA, http://www.msiusa.com/download/pdf/english/piezo/RB EG 05.pdf , September 2, 1987.

[34] R.H. Brown. Energy generation using piezo film (I). Technical report, MSI, Piezo Sensors Division, Valley Forge, PA, http://www.msiusa.com/download/pdf/english/piezo/RB EG 01.pdf, 1991.

[35] R.H. Brown. Energy generation using piezo film (II). Technical report, MSI, Piezo Sensors Division, Valley Forge, PA, http://www.msiusa.com/download/pdf/english/piezo/RB EG 02.pdf, 1991.

[36] R.H. Brown. Power generation using PVDF on a credit-card. Technical report, MSI, Piezo Sensors Division, Valley Forge, PA, http://www.msiusa.com/download/pdf/english/piezo/RB EG 04.pdf, July 27, 1998.

[37] R.H. Brown. Energy generation using piezo film (III) - thermal destruction of piezo film. Technical report, MSI, Piezo Sensors Division, Valley Forge, PA, http://www.msiusa.com/download/pdf/english/piezo/RB EG 03.pdf , September 14, 1999.

[38] W.C. Brown. The history of power transmission by radio waves. *IEEE Trans. Microwave Theory Tech., Special Centennial Historical Issue*, MTT-32(9):1230–1242, September 1984.

[39] R. Buderi. *The Invention That Changed the World: How a Small Group of Radar Pioneers Won the SecondWorld War and Launched a Technical Revolution*. Simon & Schuster - Touchstone Books, New York, 1997.

[40] A. Van Buren. Dear Abby - Yet another reason to floss regularly. *Chicago Tribune Magazine*, page 36, October 18, 1988.

[41] T. Büren, P. Lukowicz, and G. Tröster. Kinetic energy powered computing - an experimental feasibility study. In *ISWC*, pages 22–24, 2003.

[42] J.R. Burns, P. Smalser, G.W. Taylor, and T.R. Welsh. 6,528,928: Switched resonant power conversion electronics. *US Patent*, March 4, 2003.

[43] W. Cady. *Piezoelectricity*. Dover Publishers, Inc., New York, 1964.

[44] Edgar Callaway. *Wireless Sensor Networks: Architectures and Protocols*. CRC Press LLC, Boca Raton, FL, 2004.

[45] C.B. Carroll. 5,814,921: Frequency multiplying piezoelectric generators. *US Patent*, September 29, 1998.

[46] A. Chandrakasan, R. Amirtharajah, S. Cho, J. Goodman, G. Konduri, J. Kulik, W. Rabiner, and A. Wang. Design considerations for distributed microsensor systems. In *IEEE 1999 Custom Integrated Circuits Conference*, pages 279–286. IEEE, May 1999.

[47] J. Chang and L. Tassiulas. Energy conserving routing in wireless ad-hoc networks. In *INFOCOM (1)*, pages 22–31, 2000.

[48] A. Chapuis and E. Jaquet. *The History of the Self-Winding Watch*. Roto-Sadag S.A., Geneva, 1956.

[49] S-H. Chen. 5,495,682: Dynamoelectric shoes. *US Patent*, March 5, 1996.

[50] M. Cheney. *Tesla: Man Out of Time*. A Laurel Book, Dell Publishing Co., New York, 1981.

[51] N. Ching, H. Wong, W. Li, P. Leong, and Z. Wen. A laser-micromachined multi-modal resonating power transducer for wireless sensing systems. In *Sensors and Actuators, A: Physical*, volume 97–98, pages 685–690, April 2002.

[52] D.L. Churchill, M.J. Hamel, C.P. Townsend, and S.W. Arms. Strain energy harvesting for wireless sensor networks. *Proceedings of the SPIE - Smart Structures and Materials 2003*, 5055:319–327, March 2003.

[53] J.H. Comroe. *Physiology of Respiration, Second Edition*. Year Book Medical Publishers, Chicago, 1979.

[54] M. Cooper. Batteries not included. *New Scientist*, 171(2307):24, September 24, 2001.

[55] A. Crisan. 5,911,529: Typing power. *United States Patent*, June 15, 1999.

[56] L.R. Crump. 2,813,242: Powering electrical devices with energy abstracted from the atmosphere. *US Patent*, November 12, 1957.

[57] H.R. Van Deventer. 1,184056: Self-contained generating and lighting unit. *US Patent*, May 23, 1916.

[58] R.S. DiMatteo, P. Greiff, S.L. Finberg, K.A. Young-Waithe, H.K.H. Choy, M.M. Masaki, and C.G. Fonstad. Microngap thermophotovoltaics (MTPV). In *Proc. of the Fifth Conference on Thermophotovoltaic Generation of Electricity (Rome, Italy)*, pages 232–240, Melville, NY, September 16-19 2002. AIP Conference Proceedings (No. 653).

[59] J. Drake. The greatest shoe on earth. *Wired*, 9(2):90–100, February 2001.

[60] C.K. Dyer. Fuel cells for portable applications. *Journal of Power Sources*, 106(1-2):31–34, 2002.

[61] T. Eggborn. Analytical models to predict power harvesting with piezoelectric materials. Master's thesis, Virginia Polytechnic Institute and State University, Blacksburg, VA, May 2003.

[62] M. El-hami, P. Glynne-Jones, N. White, M. Hill, S. Beeby, E. James, A. Brown, and J. Ross. Design and fabrication of a new vibration-based electromechanical power generator. *Sensors and Actuators, A: Physical*, 92(1-3):335–342, August 2001.

[63] A.H. Epstein. Millimeter-scale, MEMS gas turbine engines. In *American Society of Mechanical Engineers, International Gas Turbine Institute, Turbo Expo (Publication) IGTI*, pages 669–696, Atlanta, Georgia, June 16-19, 2003. ASME.

[64] M. Feldmeier and J.A. Paradiso. Giveaway wireless sensors for large-group interaction. In *Human Factors in Computing Systems (CHI 2004 Proceedings) - Extended Abstracts*, pages 1291–1292, Vienna, Austria, April 27-29, 2004. ACM Press.

[65] K. Finkenzeller. *RFID Handbook: Fundamentals and applications in contactless smart cards and identification*. John Wiley & Sons, New York, 2003.

[66] J.-P. Fleurial, T. Olson, A. Borschevsky, T. Caillat, E. Kolawa, M. Ryan, and W. Philips. 6,288,321: Electronic device featuring thermoelectric power generation. *United States Patent*, September 11, 2001.

[67] J.-P. Fleurial, M. Ryan, A. Borschevsky, W. Phillips, E. Kolawa, J. Snyder, T. Caillat, T. Kascich, and P. Mueller. 6,388,185: Microfabricated thermoelectric power-generation devices. *United States Patent*, May 14, 2002.

[68] J.-P. Fleurial, G.J. Snyder, J.A. Herman, M. Smart, P. Shakkottai, P.H. Giauque, and M.A. Nicolet. Miniaturized thermoelectric power sources. In *Proc. of the 34th Intersociety Energy Conversion Engineering Conference*, Paper 1999–01–2569, Vancouver, BC, Canada, August 2-5, 1999. Society of Automotive Engineers.

[69] R.L. Forward. Starwisp: An ultralight interstellar probe. *Journal of Spacecraft and Rockets*, 22(3):345–350, 1985.

[70] K.R. Foster, L.S. Erdreich, and J.E. Moulder. Weak electromagnetic fields and cancer in the context of risk assessment. *Proceedings of the IEEE*, 85(5):733–746, May 1997.

[71] J. Fraden. *Handbook of Modern Sensors, Third Edition*. Springer Verlag, Berlin, 2004.

[72] K. Fu, A.J. Knobloch, F.C. Martinez, D.C. Walther, C. Fernandez-Pello, A.P. Pisano, and D. Liepmann. Design and fabrication of a silicon-based MEMS rotary engine. *ASME, Advanced Energy Systems Division (Publication) AES*, 41:303–308, 2001.

[73] J. Gillies, editor. *A Textbook of Aviation Physiology*. Pergamon Press, Oxford, 1965.

[74] B. Gilomen and P. Schmidli. Mouvement á quartz dame dont l'énergie est fournie par une génératrice, calibre ETA 204.911. In *Congrès Européen de Chronométrie*, Geneva, September 2000.

[75] P. Glaser. Power from the sun; its future. *Science*, 162(22):857–861, November 1968.

[76] E.G. Gonzalez and P.J. Corcoran. Energy expenditure during ambulation. In J.A. Downey, S.J. Myers, E.G. Gonzalez, and J.S. Lieberman, editors, *The Physiological Basis of Rehabilitation Medicine*, pages 413–446. Butterworth-Heinemann, Boston, MA, 1994.

[77] G. Görge, M. Kirstein, and R. Erbel. Microgenerators for energy autarkic pacemakers and defibrillators: Fact or fiction? *Herz*, 26(1):64–68, 2001.

[78] M.A. Green, K. Emery, D.L. King, and W. Warta S. Igari. Solar cell efficiency tables (Version 22). *Progress in Photovoltaics: Research and Applications*, 11:347–352, 2003.

[79] S. Griffith. Bettery homepage. *http://web.media.mit.edu/˜saul/bettery/index.htm*, 2004.

[80] S. Griffith. E-mail correspondence. *MIT Media Laboratory*, February 22 2004.

[81] H. Guo and A. Lal. Nanopower betavoltaic microbatteries. In *Proceedings of Transducers '03*, volume 1, pages 36–39, Boston, MA, June 8-12, 2003. IEEE Press.

[82] P.L. Hagelstein and Y. Kucherov. Enhanced figure of merit in thermal to electrical energy conversion using diode structures. *Physics Letters*, 81(3):559–561, July 15, 2002.

[83] N. Hagood, D. Roberts, L. Saggere, M. Schmidt, M. Spearing, K. Breuer, R. Mlcak, J. Carretero, F. Ganji, H. Chen, Y. Su, and S. Pulitzer. Development of micro-hydraulic transducer technology. In *Intl. Conf. Adaptive Structures and Technologies*, pages 91–102, Lancaster, PA, 1999. Technomic Publishing.

[84] D. Halliday, R. Resnick, and K. Krane. *Physics, vol. 1 & 2 Extended, 4th Edition*. Wiley & Sons Inc., New York, 1992.

[85] D. Halvorsen. Private correspondence. *AMP Inc.*, May 1995.

[86] W.G. Harris. *Lucy & Desi: The Legendary Love Story of Television's Most Famous Couple*, page 113. Simon & Schuster, New York, hardcover edition, 1991.

[87] E. Häusler, L. Stein, and G. Harbauer. Implantable physiological power supply with PVDF film. *Ferroelectrics*, 60:277–282, 1984.

[88] J. Hayashida. Unobtrusive integration of magnetic generator systems into common footwear. *Bachelor's Thesis, MIT Department of Mechanical Engineering*, June 2000.

[89] R.F. Hellbaum, R.G. Bryant, and R.L.Fox. 5,632,841: Thin layer composite unimorph ferroelectric driver and sensor. *US Patent*, May 27, 1997.

[90] H. Herr. Private correspondence. *MIT Leg Laboratory*, April 1996.

[91] H.M. Herr and R.I. Gamow. 6,029,374: Shoe and foot prosthesis with bending beam spring structures. *US Patent*, February 29, 2000.

[92] J. Hogan. Now we can soak up the rainbow. *New Scientist*, 176(2372):24, December 7, 2002.

[93] R. Holland and E. EerNisse. *Design of Resonant Piezoelectric Devices (Research Monograph no. 56)*. MIT Press, Cambridge, MA, 1969.

[94] A.P.J. Hum. Fabric area network: a new wireless communications infrastructure to enable ubiquitous networking and sensing on intelligent clothing. *Computer Networks*, 35(4):391–399, 2001.

[95] J. Hutchinson. History and status of personal power devices for the commercial market. In M. Rose, editor, *Prospector IX: Human-Powered Systems Technologies*, pages 201–210, Auburn, AL, November 1997. Space Power Institute, Auburn Univ.

[96] J.E. Hutchinson and P. Becker. 6,472,846: Power source. *US Patent*, October 29, 2002.

[97] W.U. Huynh, J.J. Dittmer, and A.P. Alivisatos. Hybrid nanorod-polymer solar cells. *Science*, 295:2425–2427, March 29, 2002.

[98] E. James, M. Tudor, S. Beeby, N. Harris, P. Glynne-Jones, J. Ross, and N. White. A wireless self powered microsystem for condition monitoring. In *Eurosensors XVI*, Prague, September 2002. Carolina - Eurosensors XVI.

[99] W.B. Spillman Jr., S. Durkee, and W.W. Kuhns. Remotely interrogated sensor electronics (RISE) for smart structures applications. In *Proc. of the SPIE Second European Conference on Smart Structures and Materials (Glasgow UK)*, volume 2361, pages 282–284. SPIE, October 1994.

[100] T. Kanesaka. Development of a thermal energy watch. In *Societé Suisse De Chronométrie*, September 1999.

[101] N. Karaki and O. Miyazawa. 5,630,155: Portable computer system with mechanism for accumulating mechanical energy for powering the system. *US Patent*, May 13, 1997.

[102] S.C. Kelley, G.A. Deluga, and W.H. Smyrl. Miniature fuel cells fabricated on silicon substrates. *AIChE Journal*, 48(5):1071–1082, 2002.

[103] J.C. Kendall. Parasitic power collection in shoe mounted devices. *Bachelor's Thesis, MIT Physics Department*, June 1998.

[104] C. Kenneally. Power from the people breaks the hold of batteries and plugs. *New York Times*, page G9, August 3, 2000.

[105] H.-H. Kim, N. Mano, Y. Zhang, and A. Heller. A miniature membrane-less biofuel cell operating under physiological conditions at 0.5 V. *Journal of The Electrochemical Society*, 150(2):A209–A213, 2003.

[106] P. Kinzie. *Crystal Radio: History, Fundamentals and Design*. Xtal Set Society, Lawrence, Kansas, 1996.

[107] J. Kistler and M. Satyanarayanan. Disconnected operation in the Coda File System. *ACMTrans. on Computer Systems*, 10(1), February 1992.

[108] W.H. Ko. 3,456,134: Piezoelectric energy converter for electronic implants. *US Patent*, July 15, 1969.

[109] W.H. Ko. 6,438,193: Self-powered tire revolution counter. *US Patent*, August 20, 2002.

[110] J.A. Konotchick. 5,818,132: Linear motion electric power generator. *US Patent*, October 6, 1998.

[111] R. Kornbluh. E-mail correspondence. *SRI International*, January 6, 2004.

[112] R.D. Kornbluh, R.E. Pelrine, Q. Pei, R. Heydt, S.E. Stanford, S. Oh, and J. Eckerle. Electroelastomers: Applications of dielectric elastomer transducers for actuation, generation and smart structures. In A. McGowan, editor, *Proceedings of the SPIE - Smart Structures and Materials 2002: Industrial and Commercial Applications of Smart Structures Technologies*, volume 4698, pages 254–270. SPIE Press, 2002.

[113] J. Kymissis, C. Kendall, J. Paradiso, and N. Gershenfeld. Parasitic power harvesting in shoes. In *IEEE Intl. Symp. On Wearable Computers*, pages 132–139, IEEE Computer Society Press, October 1998.

[114] N. Lakic. 4,674,199: Shoe with internal foot warmer. *US Patent*, June 23, 1987.

[115] N. Lakic. 4,845,338: Inflatable boot liner with electrical generator and heater. *United States Patent*, July 4, 1989.

[116] N. Landry. 6,201,314: Shoe sole with liquid-powered electrical generator. *United States Patent*, March 13, 2001.

[117] P.K. Lau and S.M. Peress. 6,650,085: Modular solar battery charger. *US Patent*, November 18, 2003.

[118] K.B. Lazarus and J.W. Moore. D404,100: Ski damper. *US Patent*, January 12, 1999.

[119] K.B. Lazarus, J.W. Moore, R.N. Jacques, F.M. Russo, and R. Spangler. 6,102,426: Adaptive sports implement with tuned damping. *US Patent*, August 15, 2000.

[120] S.J. Lee, A. Chang-Chien, S.W. Cha, R. O'Hayre, Y.I. Park, Y. Saito, and F.B. Prinz. Design and fabrication of a micro fuel cell array with 'flip-flop' interconnection. *Journal of Power Sources*, 112:410–418, 2002.

[121] A. Luzy. 1,472,335: Magneto flash light. *US Patent*, October 30, 1923.

[122] K. Lyons, T. Starner, D. Plaisted, J. Fusia, A. Lyons, A. Drew, and E. Looney. Twiddler typing: One-handed chording text entry for mobile phones. In *Human Factors in Computing Systems (CHI 2004 Proceedings)*, pages 671 – 678, Vienna, Austria, April 27-29, 2004. ACM Press.

[123] E.D. Mantiply, K.R. Pohl, S.W. Poppell, and J.A. Murphy. Summary of measured radiofrequency electric and magnetic fields (10 kHz to 30 GHz) in the general and work environment. *Bioelectromagnetics*, 18(8):563–577, 1997.

[124] D. Marculescu, R. Marculescu, S. Park, and S. Jayaraman. Ready to ware. *IEEE Spectrum*, 40(10):29–32, October 2003.

[125] J. Marsden and S. Montgomery. *Human Locomotor Engineering*, chapter Plantar Power for Arm Prosthesis Using BodyWeight Transfer, pages 277–282. Inst. of Mechanical Engineers Press, London, 1971.

[126] J. Martin and C. Piguet. 4,106,279: Wrist watch incorporating a thermoelectric generator. *United States Patent*, August 15, 1978.

[127] K. Matsuzawa and M. Saka. Seiko human powered quartz watch. In M. Rose, editor, *Prospector IX: Human-Powered Systems Technologies*, pages 359–384, Auburn, AL, November 1997. Space Power Institute, Auburn Univ.

[128] R. McLeish and J. Marsh. *Human Locomotor Engineering*, chapter Hydraulic Power from the Heel, pages 126–132. Inst. of Mechanical Engineers Press, London, 1971.

[129] C. McLeod, R. Dickinson, A. Sabkha, and C. Tormazou. Applications for implantable SAW pressure sensors. In G-Z. Yang, editor, *Proc. of the International Workshop on Wearable and Implantable Body Sensor Networks*, pages 22–23, London, April 6-7, 2004. Imperial College.

[130] T. McMahon. *Muscles, Reflexes, and Locomotion*. Princeton University Press, Princeton, NJ, 1984.

[131] S. Meninger, J.O. Mur-Miranda, R. Amirtharajah, A.P. Chandrakasan, and J.H. Lang. Vibration-to-electric energy conversion. *IEEE Transactions On Very Large Scale Integration (VLSI) Systems*, 9(1):64–76, February 2001.

[132] R. Min, M. Bhardwaj, S.-W. Cho, N. Ickes, E. Shih, A. Sinha, A.Wang, and A. Chandrakasan. Energy-centric enabling technologies for wireless sensor networks. *IEEE Wireless Communications*, 9(4):28–39, August 2002.

[133] J.O. Mur Miranda. *Electrostatic Vibration-to-Electric Energy Conversion*. PhD thesis, MIT EECS Department, February 2004.

[134] P.D. Mitcheson, T.C. Green, E.M. Yeatman, and A.S. Holmes. Architectures for vibration-driven micro-power generators. *IEEE/ASME Journal of Microelectromechanical Systems*, 13(3):429- 440, June 2004.

[135] P.D. Mitcheson, P. Miao, B.H. Stark, A.S. Holmes, and T.C. Green. MEMS electrostatic micro-power generator for low frequency operation. *Sensors and Actuators, Part A*, to appear.

[136] M. Miyazaki, H. Tanaka, G. Ono, T. Nagano, N. Ohkubo, T. Kawahara, and K. Yano. Electric-energy generation using variable-capacitive resonator for power-free LSI: Efficiency analysis and fundamental experiment. In *Proceedings of the 2003 international symposium on Low power electronics and design (ISLPED '03)*, pages 193–198. ACM Press, August 25-27, 2003.

[137] D. Morton. *Human Locomotion and Body Form*. The Williams & Wilkins Co., Baltimore, MD, 1952.

[138] B. Nivi. Passive wearable electrostatic tags. Master's thesis, MIT Department of Electrical Engineering and Computer Science, September 1997.

[139] R. Nowak, editor. *DARPA Energy Harvesting Program Review*, http://www.darpa.mil/dso/trans/energy/briefing.html, Washington, D.C., April 13-14, 2000. Defense Advanced Research Projects Agency.

[140] G.K. Ottman, H.F. Hofmann, A.C. Bhatt, and G.A Lesieutre. Adaptive piezoelectric energy harvesting circuit for wireless, remote power supply. *IEEE Transactions on Power Electronics*, 7(5):669–676, September 2002.

[141] J. Paradiso and M. Feldmeier. A compact, wireless, self-powered pushbutton controller. In G. Abowd, B. Brumitt, and S. Shafer, editors, *Ubicomp 2001: Ubiquitous Computing*, pages 299–304, Atlanta, GA, September 2001. Springer-Verlag.

[142] J.A. Paradiso. Renewable energy sources for the future of mobile and embedded computing. In *Invited talk given at the Intel Computing Continuum Conference, San Francisco, California*, Carnegie Mellon University, March 16, 2000. Posted at The Universal Library - http://www.ulib.org.

[143] J.A. Paradiso, L.S. Pardue, K-Y. Hsiao, and A.Y. Benbasat. Electromagnetic tagging for electronic music interfaces. *Journal of New Music Research*, 32(4):395–409, December 2003.

[144] R. Pelrine, R. Kornbluh, J. Eckerle, P. Jeuck, S. Oh, Q. Pei, and S. Stanford. Dielectric elastomers: Generator mode fundamentals and applications. In Y. Bar-Cohen, editor, *SPIE Electroactive Polymer Actuators and Devices*, volume 4329, pages 148–156, Newport Beach, CA, March 2001.

[145] D. Pescovitz. The power of small tech. *SmallTimes*, 2(1):21–31,51, 2002.

[146] M. Philipose, K.P. Fishkin, M. Perkowitz, D. Patterson, and D. Haehnel. The probabalistic activity toolkit: Towards enabling activity-aware computer interfaces. Technical Report IRS-TR-03-013, Intel Research, Seattle,WA, December 2003.

[147] P. Picot. Private correspondence. *Medical Doctor, Robarts Research Institute*, May 1995.

[148] K. Pistor and F. Schmidt. WO 10/91315 A2: Energy self-sufficient high frequency transmitter. *German Patent*, November 29, 2001.

[149] A. Pohl, R. Steindl, and L. Reindl. The 'intelligent tire': Utilizing passive SAW sensors-measurement of tire friction. *IEEE Transactions on Instrumentation and Measurement*, 48(6):1041–1046, 1999.

[150] E. Post and M. Orth. Smart fabric, or wearable clothing. In *IEEE Intl. Symp. on Wearable Computers*, pages 167–168, Cambridge, MA, 1997.

[151] S.S. Pradhan, J. Kusuma, and K. Ramchandran. Distributed compression in a dense microsensor network. *IEEE Signal Processing Magazine*, 19(2):51–60, March 2002.

[152] H.K. Puharich. 2,995,633: Means for aiding hearing. *US Patent*, August 8, 1961.

[153] V. Raghunathan, C. Schurgers, S. Park, and M.B. Srivastava. Energy-aware wireless microsensor networks. *IEEE Signal Processing Magazine*, 19(2):40–50, March 2002.

[154] M. Ramsay andW. Clark. Piezoelectric energy harvesting for bio MEMs applications. In *Proc. of SPIE*, volume 4332, pages 429–439, 2001.

[155] L. M. Reindl, A. Pohl, G. Scholl, and R.Weigel. SAW-based radio sensor systems. *IEEE Sensors Journal*, 1(1):69–78, June 2001.

[156] T. Riedel. Power considerations for wireless sensor networks. *Sensors*, 21(3):38–41, March 2004.

[157] S. Ritter. Biofuel cells get smaller. *Chemical & Engineering News*, 79(36):10, 2001.

[158] N. Rogacheva. *The Theory of Piezoelectric Shells and Plates*. CRC Press, Boca Raton, 1994.

[159] S. Roundy. *Energy Scavenging with a Focus on Vibration-to-Electricity Conversion for Low Power Wireless Devices*. Ph.D. thesis, University of California, Berkeley - Mechanical Engineering Dept., May 2003.

[160] S. Roundy, P. Wright, and K. Pister. Micro-electrostatic vibration-to-electricity converters. In *International Mechanical Engineering Congress and Exposition*, pages 487–496, New Orleans, November 17-22, 2002. ASME, MEMS Division Publication.

[161] J. Rabaey S. Roundy, P.K. Wright. A study of low level vibrations as a power source for wireless sensor nodes. *Computer Communications*, 26(11):1131–1144, July 2003.

[162] I. Sample. Juice on the loose. *New Scientist*, 175(2354):36–38, August 3, 2002.

[163] R.F. Savinell, J.S.Wainright, L. Dudik, and C.C. Liu. Recent advances in microfabricated fuel cells. In *New Materials for Electrochemical Systems IV. Extended Abstracts of the Fourth International Symposium on New Materials for Electrochemical Systems*, pages 371–372, Montreal, Que., Canada, July 9-13, 2001. Ecole Polytechnique de Montreal.

[164] H. Schmidhuber and C. Hebling. First experiences and measurements with a solar powered personal digital assistant (PDA). In *Proceedings of the 17'th European Photovoltaic Solar Energy Conference*, pages 658–662, Munich, Germany, October 22-26 2001. WIP.

[165] A. Schmidt, H.W. Gellersen, and C. Merz. Enabling implicit human computer interaction: A wearable RFID-tag reader. *Proc. of the 2000 IEEE Intl. Symp. on Wearable Computers (ISWC)*, pages 193–194, 2000.

[166] N.S. Shenck. A demonstration of useful electric energy generation from piezoceramics in a shoe. Master's thesis, MIT EECS Department, Cambridge, MA, May 1999.

[167] N.S. Shenck and J.A. Paradiso. Energy scavenging with shoe-mounted piezoelectrics. *IEEE Micro*, 21(3):30–42, May 2001.

[168] M.R. Shorten. Energetics of running and running shoes. *Journal of Biomechanics*, 26(Suppl. 1):41–51, 1993.

[169] B. Simes. A ground-noise powered receiver. In P.N. Anderson, editor, *Crystal Sets: The Xtal Set Society Newsletter*, volume 5, pages 66–69, Lawrence, Kansas, 1996. The Xtal Set Society.

[170] A. Smailagic and R. Martin. Metronaut: A wearable computer with sensing and global communication capabilities. In *IEEE Intl. Symp. on Wearable Computers*. IEEE Computer Society Press, 1997.

[171] P. Smalser. 5,703,474: Power transfer of piezoelectric generated energy. *US Patent*, December 30, 1997.

[172] D.S. Snyder. 4,510,484: Piezoelectric reed power supply for use in abnormal tire condition warning systems. *US Patent*, April 9, 1985.

[173] T. Starner. The cyborgs are coming. Technical Report 318, Perceptual Computing, MIT Media Laboratory, January 1994.

[174] T. Starner. Thick clients for personal wireless devices. *IEEE Computer*, 35(1):133–135, January 2002.

[175] T. Starner, S. Mann, B. Rhodes, J. Levine, J. Healey, D. Kirsch, R. Picard, and A. Pentland. Augmented reality through wearable computing. Technical Report 397, MIT Media Lab, Perceptual Computing Group, October 1996.

[176] J. Stevens. Optimized thermal design of small thermoelectric generators. In *Proc. of the 34th Intersociety Energy Conversion Engineering Conference*, pages Paper 1999–01–2564, Vancouver, BC, Canada, August 2-5, 1999. Society of Automotive Engineers.

[177] C. Stuart. Powerful year coming in mobile fuel cells. *SmallTimes*, 2(2):60, 2002.

[178] C. Susskind. *Heinrich Hertz: A Short Life*. San Francisco Press, San Francisco, 1995.

[179] R. Tashiro, N. Kabei, K. Katayama, Y. Ishizuka, F. Tuboi, and K. Tsuchiya. Development of an electrostatic generator that harnesses the motion of a living body. *Journal of the Japan Society of Mechanical Engineers, Series C*, 43(4):916–922, December 2000.

[180] N. Tesla. The transmission of electric energy without wires. *Electrical World and Engineer*, XLIII:429–431, March 5, 1904.

[181] J.J. Tiemann. 5,578,877: Apparatus for converting vibratory motion to electrical energy. *US Patent*, November 26, 1996.

[182] E. Tkaczyk. Technology summary. In M. Rose, editor, *Prospector IX: Human-Powered Systems Technologies*, pages 38–43, Auburn, AL, November 1997. Space Power Institute, Auburn Univ.

[183] M. Toda. Shoe generator: Power generation mechanism. Technical report, AMP Sensors, now the Piezo Sensors Division of MSI, Valley Forge, PA, August 1, 1997.

[184] C.G. Triplett. 4,504,761: Vehicular mounted piezoelectric generator. *US Patent*, March 12, 1985.

[185] S.R. Vetorino, J.V. Platt, and D.A. Springer. 6,220,719: Renewable energy flashlight. *US Patent*, April 24, 2001.

[186] C.B. Vining. Semiconductors are cool. *Nature*, 413:577–578, October 11, 2001.

[187] R. Want, A. Hopper, V. Falcao, and J. Gibbons. The active badge location system. *ACM Transactions on Information Systems*, 10(1):91–102, January 1992.

[188] R. Want, T. Pering, G. Danneels, and M. Kumar. The Personal Server - changing the way we think about ubiquitous computing. In G. Borriello and L.E. Holmquist, editors, *UbiComp 2002: 4th International Conference on Ubiquitous Computing*, pages 194–209, Berlin and Heidelberg, October 2002. Springer-Verlag.

[189] J.A. Weaver. A wearable health monitor to aid Parkinson Disease treatment. Master's thesis, MIT Media Lab, June 2003.

[190] M.C. Tsu W.I. Hwang. 5,396,720: Fixing structure for lightening circuit of 2-stage switch on lightening shoe. *US Patent*, March 14, 1995.

[191] L. Williams, W. Vablais, and S.N. Bathiche. 6,754,472: Method and apparatus for transmitting power and data using the human body. *US Patent*, June 22, 2004.

[192] D. Wren. Medical Doctor. Private correspondence, May 1995.

[193] O. Yaglioglu. Modeling and design considerations for a micro-hydraulic piezoelectric power generator. Master's thesis, MIT EECS Department, Cambridge, MA, February 2002.

[194] E.M. Yeatman. Advances in power sources for wireless sensor nodes. In G-Z. Yang, editor, *Proc. of the International Workshop on Wearable and Implantable Body Sensor Networks*, pages 20–21, London, April 6-7, 2004. Imperial College.

[195] U. Zechbauer. Neue funkschalter funktionieren ohne strom. *Handelsblatt*, (98):18, May 24-25, 2002.

46

Chip Cooling: Why – How

Yervant Zorian
VirageLogic

Dimitris Gizopoulos
University of Piraeus

46.1 Introduction and Technology Situation

The electronic products industry has been able to provide end users with a variety of devices featuring amazing functionality, which can be purchased at a decreasing cost per device. This is the outcome of advances in manufacturing technologies, which enabled the integration of enormous numbers of transistors, gates, and storage elements in very small areas of silicon. Reduction of the circuit elements' dimensions was the result of manufacturing industry efforts and investments over the last three decades and is usually called feature scaling or technology scaling. The International Technology Roadmap for Semiconductors (ITRS) [1] summarizes the main implications that feature scaling has in integrated circuits (ICs) development. Feature scaling, as presented in Table 46.1, drives important improvements in IC characteristics.

It has been demonstrated that Moore's law remains perfectly valid (the number of circuit elements – transistors – doubles every 1.5 years) and multimillion gates ICs are being manufactured, while at the same time, their operating speeds keep increasing into the GHz range of frequencies. Due to the increased importance of portable devices, IC power supply voltages are reduced to enable the usage of smaller and lighter power storage devices. Other factors that make supply voltage reduction a necessity are the reduction of system power dissipation, the reduced transistor channel length, and increase in reliability of gate dielectrics. Table 46.2 lists power-supply voltage trends [1].

Although supply voltage decreases and is expected to continue decreasing, as presented by Table 46.2, the faster increase in the number of devices per chip as well as the increase in operating frequencies dominate and determine the overall electrical power that the ICs consume. Depending on the nature of an electronic application, power consumption and management of the thermal behavior of the IC due to its power consumption are critical and very serious design considerations. Thermal management criticality is becoming more important over the years because increased functionality and frequencies of the chips lead to increased power consumption (although supply voltage reduction as we mentioned

0-8493-1941-2/05/$0.00+$1.50

TABLE 46.1 Improvement Trends for ICs Enabled by Feature Scaling [1]

Allowable Maximum Power (W)	2002	2003	2004	2005	2006	2007
High-performance (W)	140	150	160	170	180	190
Cost-performance (W)	75	81	85	92	98	104
Hand-held (W)	2.6	2.8	3.2	3.2	3.5	3.5

TABLE 46.2 Power Supply Voltage (V_{dd}) Trends [1]

Chip Category	Chip Cooling
Low cost	Free convection
Hand held	Free convection + heat spreaders
Cost-performance	Forced convection + heat sinks
High-performance	Forced convection + heat sinks + heat pipes + liquid
Harsh environment	All together

before), and thus the heat produced by complex chips of our days gets very high. This tremendous heat increase puts manufacturability of the chips in danger if appropriate heat removal and transfer mechanisms are not employed.

The existence of a large variety of sophisticated chip cooling techniques applied at the package level or externally to the device is always a necessary chip infrastructure that allows the IC to successfully operate below its thermal limits without damaging its package or affecting its fault-free operation. Depending on the type of the electronic system, its application domain, and its overall financial framework, chip cooling cost is among the major factors that determine the chip maximum power consumption limits and applicable cooling mechanisms. In many application domains, the excessive chip cooling mechanism's cost is the reason for the cut-down in system's functionality because power consumption and heat generation cannot be handled by less-expensive cooling mechanisms.

Electronic products are classified into five major categories based on the characteristics and demands of their application domain:

1. Low-cost
2. Handheld (portable)
3. Cost-performance
4. High-performance
5. Harsh environment

ITRS provides some figures for the maximum allowable power for three of the preceding types of electronic products (today's value and forecast at the near-term). Table 46.3 presents this information.

It is apparent that for three types of systems where the preceding chip classes are used, there are completely different cost calculations for the chip cooling environment and mechanisms. For example, in the first case of high-performance class, state-of-the-art heat sinks at the chip packaging can be used

TABLE 46.3 Allowable Maximum Power for Three Electronic Products Categories [1]

Improvement Trend	Example
Integration level	Components per chip increase, Moore's law still valid
Cost	Cost per function reduces
Speed	Microprocessor clock rate increases, GHz level frequencies
Power	Laptop or cell phone battery life increases
Compactness	Small and light-weight products are getting more important
Functionality	Nonvolatile memory, imager

for thermal management, while in the cost-performance case, the desirable solution is the one that achieves the best heat removal performance under specific economical restrictions. In the third case, of battery-operated, portable systems, the use of special chip cooling mechanisms is usually out of the question, and this is the reason for the extremely low limit of power consumption in such applications.

Chip cooling techniques always face the challenge to follow advances in functionality and integration level increase. Chips are getting larger and more sophisticated but they also dissipate much more power and produce much more thermal energy. If not appropriately cooled, below the levels at which they malfunction, systems based on such chips are not reliable and lead to errors in operation with tremendous impact on costs, environmental, and human resources, depending on the application domain of each system.

In the following sections, we briefly describe power dissipation sources in ICs, the effects that temperature increase may have to the manufactured chip, an overview of chip cooling mechanisms and future research directions in the domain.

46.2 Sources of Chip Power Consumption and Effects of Heat Production

The main source of chip power dissipation at the active circuits devices is the dynamic power dissipation due to toggles at circuit node logic values. In this case, the power that the circuit consumed is proportional to the node capacitance, the power supply voltage, and the operating frequency of the chip (frequency at which the nodes change their values).

A secondary source of power dissipation is the leakage current's flow at the silicon substrates. In previous technologies, this factor of power dissipation was considered limited and in many cases was not analyzed at all. In modern feature sizes brought to us by technology scaling, leaking current power consumption is getting more important than ever and should be seriously taken into consideration when building power consumption and heat dissipation models for ICs.

Whereas in previous technologies, circuit capacitance was determined by the transistor's capacitances, in modern deep submicron technologies, interconnections have the dominating role in the capacitance of the circuit and this is factor to be seriously considered in power and thermal models because power is consumed and heat is generated mostly when capacitances are charged and discharged in digital ICs. Interconnection capacitances used to participate in thermal and power model moderately, which should not be the case today.

A last factor not to be neglected in the self-heating of interconnects which is due to their Joule heating because of current flowing through them. Joule heating of interconnects depends on the current that flows through them and their electrical resistance. Interconnects heating due to current flow is a parameter that recently entered thermal calculations in ICs. Temperature increase in interconnects may lead to significant problems of operation due to electromigration effects.

It has been demonstrated that technology scaling has a serious impact on interconnections thermal behavior because in lower circuit feature sizes, the chip interconnects reach higher temperatures at much lower power consumption levels, both for classical dielectric materials and for low-k dielectric materials.

A major effect of excessive heat dissipation in ICs is that due to electromigration effects, the circuits deteriorate faster than when operating with smaller heat dissipation (in lower temperatures). This fact directly affects the IC reliability and significantly reduced the expected time to failure of the device. Chip cooling is therefore essential for temperature reduction that will lead to smoother circuit operation and will prolong its lifecycle making it much more cost-effective than without chip cooling. We must not forget that the application domain and criticality of the chip determines the cost and type of the cooling mechanism.

Even if we see the effect of temperature increase from just the performance point of view of a specific circuit, we can still see how seriously the chip is affected when operating at excessive temperatures. Due to the previously mentioned self-heating scenario of interconnects, significant delays are added to signal propagations via the circuit wires. Although, high-frequency chips are developed to operate with high performance, if not appropriately cooled, their operation will not be as expected.

With the increase in the complexity of modern ICs and the emergence of the system on chip (SoC) design paradigm, many different types of circuit types, such as processor and memory, are integrated in a single substrate. During circuit operation, different parts of the circuit present different activity, and thus heat distribution is not uniform throughout the circuit. This may lead to a circuit malfunction although the average heat dissipation throughout the chip never goes beyond the expected limits. Therefore, power consumption should be uniformly distributed in the entire chip area. Even if this is not possible (a usual case because not all components of a chip can be given uniform activity at any time), however, appropriate heat distribution mechanisms must be provided to spread heat produced at specific areas to the rest of the chip, so that correct operation and material characteristics are minimally affected.

46.3 Chip Cooling Strategies

Depending on the category that each circuit belongs (i.e., low-cost, handheld, cost-performance, high-performance, or harsh environment) different chip cooling strategies can be applied. Several different heat transfer-related mechanisms and thermal management building blocks could be used [2–4]:

- Free convection
- Heat spreaders
- Interface materials
- Air-cooled heat sinks
- Liquid-cooled cold plates
- Direct "immersion" cooling
- Vapor compression
- Solid-state refrigeration

We briefly describe several heat transfer "vehicles" giving an idea of the way they achieve heat transfer and thus reduction of temperature in the chip. Through the brief descriptions given in the following subsections, the basic definitions related to chip cooling and heat transfer are provided to the reader. For detailed information of techniques and methodologies, the reader may refer to [3–5] among many important works available in the open literature.

46.3.1 Free Convection

Heat transfer due to simple free convection is described by Newton's law of cooling, where heat transfer rate is proportional to the device surface area and the temperature difference between the device and the free-stream fluid out of it. Heat is transferred due to the relative motion between fluid and surface due to buoyancy within the fluid (i.e., temperature or mass gradient). No extra mechanism is used to enhance convection.

46.3.2 Forced Convection

Forced convection is still governed by Newton's law of cooling but in this case, the relative motion between fluid and surface is maintained by external means (such as a fan or pump). This is a relatively inexpensive chip cooling mechanism applied in low-cost applications. Several options exist for the implementation of external means of maintaining the relative motion.

46.3.3 Heat Spreaders

Heat spreaders are used to externally cover the packaged chips and transfer the heat produced by the chip to the outside air. They are very popular in memory devices. Heat spreaders are inexpensive devices with high thermal conductivity. Through careful optimization, the thermal and thermodynamic design of heat spreaders is improved, for example, using new types of materials or heat spreaders mounting techniques.

46.3.4 Interface Materials

Special materials, such as nanoparticle/nanotubed-filled thermal pastes, epoxies, and elastomers, are used for heat transfer when used in conjunction with classical chip manufacturing materials. Interface materials improve the overall thermal conductivity of the manufactured chip. They differ from heat spreaders because they are not external to the chip, but instead are part of the chip manufacturing process.

46.3.5 Heat Sink

A device attached to a bare (unpackaged) chip to keep it from overheating by absorbing its heat and dissipating it into the air. Heat sinks are manufactured as an integral part of the chip, and their design can be based on different materials. The most effective heat sinks are manufactured by metals, due to their very good thermal conductivity. Furthermore, composite materials are developed with even better thermal properties for advanced applications with higher heat transfer requirements. The term heat sink is a generic one that characterizes several on-chip heat transfer mechanisms. The heat sinks are usually categorized in five classes [6]:

1. Passive heat sinks, where no specific supply of airflow is provided for heat transfer.
2. Semi-active heat sinks, where heat transfer is based on leveraging existing fans in the system.
3. Active heat sinks, where specific fans are used for heat transfer and this is done by involving mechanical movement, thus its success and effectiveness depends on the quality of the mechanical parts design.
4. Liquid cooled cold plates, where heat transfer is conducted by tubes-in-block or milled paths for the use of pumped water or other liquids such as oil.
5. Phase change recirculating system, where heat transfer is performed by a combination of a boiler and condenser in a passive, self-driven mechanism; it also includes solid-to-liquid systems for transient temperature gradients instead of heat dissipation.

46.3.6 Heat Pipes

The concept of very small micro heat pipes incorporated into semiconductor devices has been proposed exactly for the application of electronic cooling. Heat pipes are flexible, high-flux, low-cost means for heat removal in the actual chip.

46.3.7 Direct "Immersion" Cooling

In this advanced chip cooling method, no physical wall separates the electronic chips and the surface of the substrate from the liquid coolant. Direct "immersion" cooling removes heat directly from the chip with no intervening thermal conduction resistance, other than that between the device heat source and chip surface in contact with the liquid.

46.3.8 MEMS for Chip Cooling

Micro-electro-mechanical systems (MEMS) have gained importance in demanding chip cooling situations where quick and massive heat transfer must take place. MEMS can be combined with other heat transfer mechanisms to optimize convective and ebullient heat transfer.

46.3.9 Chip Cooling for Different Chip Categories

From the brief description of the different cooling options it is obvious that there are different cooling strategies that can be followed depending on the complexity of the system and its heat transfer requirements. An outline of the suggested chip cooling mechanisms is given in The 2002 NEMI Packaging

TABLE 46.4 Recommended Chip Cooling Strategies [2]

Power Supply Voltage (V)	2002	2003	2004	2005	2006	2007
Vdd (high performance)	1.0	1.0	1.0	0.9	0.9	0.7
Vdd (low operating power, high V_{dd} transistors)	1.2	1.1	1.1	1.0	1.0	0.9
Vdd (low standby power, high V_{dd} transistors)	1.2	1.2	1.2	1.2	1.2	1.1

Roadmap [2]. The recommended strategies for each of the different types of chip classification are given in Table 46.4 [2].

Several research directions for the future are very important in the chip cooling technical domain [3]. Sophisticated heat transfer, thermofluid, and thermomechanical research is needed to define new opportunities and means for heat removal and to improve the predictability and reliability of the manufactured chips during design, verification, and production [3]. Special emphasis is put on the use of MEMS for heat removal because it is believed that the limits of classical approaches for chip cooling have been reached.

46.4 Impact on Design for Manufacturability

Appropriate chip-cooling mechanisms used in today's microelectronics products are considered as a necessary infrastructure so that correct and operational manufacturing of the device is guaranteed. Design automation tools for digital and analog systems have evolved quickly during the last decades and they have led to products designed quickly and correctly. Importance of post-design factors such as chip cooling are becoming increasingly important during the stages of circuit and system implementation and manufacturing. Toward the target of successful integration of chip cooling mechanisms design into the chip design process, thus guaranteeing chip manufacturability, two issues are important.

There should be detailed models of the deep submicron technology chip, describing its power and thermal behavior under the actual operating conditions where it is going to be used. These models will guide the chip design process, so that the same functionality is obtained from the device with less power consumed or with less heat produced. After the best optimization of these factors is achieved, the next issue is important.

There should be available appropriate heat transfer mechanism models to describe and analyze the behavior of the chip when a specific chip cooling mechanism is employed. We must not overlook the fact that a detailed cost analysis of the cooling system will have a direct impact of the final chip cost, system cost, and the final ability of the manufacturer to produce the chip. A trade-off analysis should always take place before an actual design is released to the electronics market. The design should provide users the desired functionality and performance under several constraints most of which come from cost analysis issues. The packaging and chip cooling cost have proven to be determining factors in the overall system cost.

A loosely analyzed chip cooling strategy of a complex design can potentially lead to a chip that cannot be manufactured or, even worse, although manufactured does not operate correctly due to malfunctions caused by insufficient heat removal and chip overheating.

46.5 Conclusions

VLSI devices implemented in earlier manufacturing technologies were usually not faced with chip cooling problems that could significantly affect their manufacturability and later their correct operation in the field. In modern complex SoC architectures, tremendous numbers of transistors and gates are integrated in single substrates. Power consumption of these devices may lead to excessive heat production that can seriously affect chip operation as well as destroy the package of the device and disconnect it from the mounting points. Avoidance of such situations can be guaranteed only if appropriate on-chip and off-chip cooling mechanisms are employed.

We have outlined the problems related to chip overheating and cooling needs along with forecast based on ITRS roadmap predictions. The most important chip cooling alternative mechanisms have been also outlined, along with impact on design for manufacturability.

Chip cooling modeling (i.e., packages, materials, and processes), design automation, and integration with the chip manufacturing process are already considered very critical problems that must be resolved to ensure the correct and easy realization of future complex microelectronics products. Such products are expected to have more complex functionality, consume more electrical power, and, therefore, produce more heat to be transferred away from the chip and system.

References

[1] The International Technology Roadmap for Semiconductors (ITRS), 2002 Update, http://public.itrs.net. Retrieval date: August 2003.

[2] The 2002 NEMI Packaging Roadmap (Thermal Management Roadmap), http://www.nemi.org. Retrieval date: August 2003.

[3] S.V. Garimella, Y.K. Joshi, A. Bar-Cohen, R. Mahajan, K.C. Toh, V.P. Carey, M. Baelmans, J. Lohan, B. Sammakia, and F. Andros, Thermal challenges in next-generation electronic systems — summary of panel presentations and discussions, *IEEE Trans. on Components and Packaging Technologies,* vol. 25, no. 4, pp. 569–575, December 2002.

[4] R.J. Goldstein, E.R.G. Eckert, W.E. Ibele, S.V. Patankar, T.W. Simon, T.H. Kuehn, P.J. Strykowski, K.K. Tamma, A. Bar-Cohen, J.V.R. Heberlein, J.H. Davidson, J. Bischof, F.A. Kulacki, U. Kortshagen, and S. Garrick, Heat transfer — a review of 2000 literature, *Int. J. Heat and Mass Transfer,* vol. 45, pp. 2853–2957, 2002.

[5] A. Bar-Cohen, M. Iyengar, Design and optimization of air-cooled heat sinks for sustainable development, *IEEE Trans. on Components and Packaging Technologies,* vol. 25, no. 4, pp. 584–591, December 2002.

[6] S. Lee, Optimum design and selection of heat sinks, *IEEE Trans. on Components, Packaging, and Manufacturing Technol. – Part A,* vol. 18, no. 4, pp. 812–817, December 1995.

Index

A

B

C

D

E

F

G

H

I

J

K

L

M

N

O

P

Q

R

S

T

V

W

X

Z